T0180555

Advances in Intelligent Systems and Computing

Volume 895

The series "Advances in Intelligent Systems and Computing" contains publications on theory, applications, and design methods of Intelligent Systems and Intelligent Computing. Virtually all disciplines such as engineering, natural sciences, computer and information science, ICT, economics, business, e-commerce, environment, healthcare, life science are covered. The list of topics spans all the areas of modern intelligent systems and computing such as: computational intelligence, soft computing including neural networks, fuzzy systems, evolutionary computing and the fusion of these paradigms, social intelligence, ambient intelligence, computational neuroscience, artificial life, virtual worlds and society, cognitive science and systems, Perception and Vision, DNA and immune based systems, self-organizing and adaptive systems, e-Learning and teaching, human-centered and human-centric computing, recommender systems, intelligent control, robotics and mechatronics including human-machine teaming, knowledge-based paradigms, learning paradigms, machine ethics, intelligent data analysis, knowledge management, intelligent agents, intelligent decision making and support, intelligent network security, trust management, interactive entertainment, Web intelligence and multimedia.

The publications within "Advances in Intelligent Systems and Computing" are primarily proceedings of important conferences, symposia and congresses. They cover significant recent developments in the field, both of a foundational and applicable character. An important characteristic feature of the series is the short publication time and world-wide distribution. This permits a rapid and broad dissemination of research results.

**** Indexing: The books of this series are submitted to ISI Proceedings, EI-Compendex, DBLP, SCOPUS, Google Scholar and Springerlink ****

More information about this series at http://www.springer.com/series/11156

Ching-Nung Yang · Sheng-Lung Peng ·
Lakhmi C. Jain
Editors

Security with Intelligent Computing and Big-data Services

Proceedings of the Second International Conference on Security with Intelligent Computing and Big Data Services (SICBS-2018)

Editors
Ching-Nung Yang
Department of Computer Science
and Information Engineering
National Dong Hwa University
Hualien, Taiwan

Sheng-Lung Peng
Department of Computer Science
and Information Engineering
National Dong Hwa University
Hualien, Taiwan

Lakhmi C. Jain
University of Canberra
Canberra, ACT, Australia

University of Technology Sydney
Broadway, Australia

Liverpool Hope University
Liverpool, UK

KES International
Shoreham-by-sea, UK

ISSN 2194-5357 ISSN 2194-5365 (electronic)
Advances in Intelligent Systems and Computing
ISBN 978-3-030-16945-9 ISBN 978-3-030-16946-6 (eBook)
https://doi.org/10.1007/978-3-030-16946-6

This Springer imprint is published by the registered company Springer Nature Switzerland AG
The registered company address is: Gewerbestrasse 11, 6330 Cham, Switzerland

Preface

With the proliferation of security with intelligent computing and big-data services, the issues of information security, big data, intelligent computing, blockchain technology, and network security have attracted a growing number of researchers. The purpose of 2018 International Conference on Security with Intelligent Computing and Big-data Services (SICBS 2018 for short) is to provide a platform for researchers, engineers, academicians, and industrial professionals from all over the world to present their research results in security-related areas.

SICBS 2018 was held at Guilin, China, from December 14 to 16, 2018. The conference brought together researchers from all regions around the world working on a variety of fields and provided a stimulating forum for them to exchange ideas and report on their researches. The proceedings of SICBS 2018 consists of 75 selected papers which were submitted to the conferences and peer-reviewed by conference committee members and international reviewers. All these papers cover the following topics: Blockchain Technology and Application, Multimedia Security, Information Processing, Security of Network, Cloud and IoT, Cryptography and Cryptosystem, Learning and Intelligent Computing, and Information Hiding.

On behalf of sponsors and conference committees, we would like to express our gratitude to all of the authors, the reviewers, and the attendees for their contributions and participation for SICBS 2018. Their high competence and expertise knowledge enable us to prepare this high-quality program and make the conferences successful. Finally, we would like to wish you have good success in your presentations and social networking. Your strong supports are critical to the success of this conference. In addition, we are grateful to Springer for publishing the conference proceeding.

Finally, we would like to wish you have good success in your presentations and social networking. Your strong supports are critical to the success of this conference. We hope that the participants will not only enjoy the technical program in conference but also discover many beautiful places in Guilin, which has China's most beautiful karst landscapes Li River and Yangshuo. Wishing you a fruitful and enjoyable SICBS 2018.

Ching-Nung Yang
Sheng-Lung Peng
Lakhmi C. Jain

Organization

2018 International Conference on Security with Intelligent Computing and Big-data Services (SICBS 2018)

Honorary Chairs

Han-Chieh Chao	National Dong Hwa University, Taiwan
Lakhmi C. Jain	University of Canberra, Australia, University of Technology Sydney, Australia, Liverpool Hope University, UK, and KES International, UK

General Chairs

Ching-Nung Yang	National Dong Hwa University, Taiwan
Mingzhe Liu	Chengdu University of Technology, China

Organizing Chairs

Yi-Ning Liu	Guilin University of Electronic Technology, China
Rushi Lan	Guilin University of Electronic Technology, China

Program Chairs

Sheng-Lung Peng	National Dong Hwa University, Taiwan
Shiuh-Jeng Wang	Central Police University, Taiwan

Technical Program Committee

Jinsuk Baek	Winston-Salem State University, USA
Zhifeng Bao	RMIT University, Australia
Jalel Ben-Othman	Université Paris 13, France
Nadia Bennani	University of Lyon, France
Tao-Ku Chang	National Dong Hwa University, Taiwan
Chao Chen	Xidian University, China
Chia-Mei Chen	National Sun Yat-sen University, Taiwan
Jiageng Chen	Central China Normal University, China
Tzung-Her Chen	National Chiayi University, Taiwan
Yu-Chi Chen	Yuan Ze University, Taiwan
Bo-Chao Cheng	National Chung Cheng University, Taiwan
Hung-Yu Chien	National Chi Nan University, Taiwan
Isao Echizen	National Institute of Informatics, Japan
Atilla Elci	Aksaray University, Turkey
Chun-I Fan	National Sun Yat-sen University, Taiwan
Zhangjie Fu	Nanjing University of Information Science and Technology, China
Clemente Galdi	University of Napoli, Italy
Andy Guo	University of Chinese Academy of Sciences, China
Lein Harn	University of Missouri—Kansas City, USA
Ji-Han Jiang	National Formosa University, Taiwan
Cheonshik Kim	Sejong University, Korea
Wen-Chung Kuo	National Yunlin University of Science and Technology, Taiwan
Narn-Yih Lee	Southern Taiwan University of Science and Technology, Taiwan
Zne-Jung Lee	Huafan University, Taiwan
Min Lei	Beijing University of Posts and Telecommunications, China
Li Li	Hangzhou Dianzi University, China
Peng Li	North China Electric Power University, China
Changlu Lin	Fujian Normal University, China
Feng Liu	University of Chinese Academy of Sciences, China
Qi Liu	Nanjing University of Information Science and Technology, China
Quansheng Liu	Université Bretagne Sud, France
Wen-Jie Liu	Nanjing University of Information Science and Technology, China
Yanxiao Liu	Xi'an University of Technology, China
Der-Chyuan Lou	Chang Gung University, Taiwan
Wei Lu	Sun Yat-sen University, China
Jiafa Mao	Zhejiang University of Technology, China

Contents

Cryptography and Cryptosystem

Learning and Intelligent Computing

Information Hiding

Blockchain Technology and Application

Fast Adaptive Blockchain's Consensus Algorithm via Wlan Mesh Network

Xin Jiang[1], Mingzhe Liu[1(✉)], Feixiang Zhao[1], Qin Zhou[1], and Ruili Wang[2]

[1] State Key Laboratory of Geohazard Prevention and Geoenvironment Protection, Chengdu University of Technology, Chengdu, Sichuan, China
liumz@cdut.edu.cn

[2] Institute of Natural and Mathematical Sciences, Massey University, Auckland, New Zealand

Abstract. This paper presents a decentralised and fast adaptive block chain's consensus algorithm with maximum voter privacy using wlan mesh network. The algorithm is suitable for consortium blockchain and private blockchain, and is written as a smart contract for Hyperledger Fabric. Unlike previously proposed blockchain's consensus protocols, this is the first implementation that does not rely on any trusted authority to compute the tally or to protect the voter's privacy. Instead, the algorithm is a fast adaptive protocol, and each voter is in control of the privacy of their own vote such that it can only be breached by a full collusion involving all other voters. The execution of the protocol is enforced using the consensus mechanism that also secures the Fabric blockchain. This paper tests the implementation on Fabric's official test network to demonstrate its feasibility. Also, this paper provides a computational breakdown of its execution cost.

Keywords: Consensus algorithm · Blockchain · Wlan mesh · Hyperledger Fabric

1 Introduction

Hyperledger Fabric is the most popular cryptocurrency of blockchain as of 2018. It relies on the same innovation behind Bitcoin [1]: namely, the blockchain which is an append-only ledger. The blockchain is maintained by a decentralised and open-membership peer-to-peer network. The purpose of the blockchain was to remove the centralised role of banks for maintaining a financial ledger. Today, researchers are trying to re-use the Blockchain to solve further open problems such as coordinating the Internet of Things [2], carbon dating [3], and healthcare [4]. However, How to reach consensus fastly has become a bottleneck in the development of blockchain in above fields.

Supported by organization X.

C.-N. Yang et al. (Eds.): SICBS 2018, AISC 895, pp. 3–16, 2020.
https://doi.org/10.1007/978-3-030-16946-6_1

In this paper, we focus on decentralised internet consensus of blockchain using wlan mesh network. This algorithm is designed as a voting protocol. E-voting protoools that support verifiability normally assume the existence of a public bulletin board that provides a consistent view to all voters. In practice, an example of implementing the public bulletin board can be seen in the yearly elections of the International Association of Cryptologic Research (IACR) [5]. They use the Helios voting system [6] whose bulletin board is implemented as a single web server. This server is trusted to provide a consistent view to all voters. Instead of such a trust assumption, we explore the feasibility of using the Blockchain as a public bulletin board. Furthermore, we consider a decentralised election setting in which the voters are responsible for coordinating the communication amongst themselves. Thus, we also examine the suitability.

2 Related Works

In recent years, the unique advantages of blockchain technology have been paid attention to by academic circles. There are some research methods of blockchain in consensus algorithm. BBLAST system is recent one of them, it solved the problem of low trust centralized ledger held by a single third party, high trust decentralized forms held by different entities, or in other words, verification nodes [7].

Gramoli [8] discussed the mainstream blockchain consistency algorithm and the classic Byzantine consensus to re-examine the blockchain context. To against bitcoin and the Ethum consensus algorithm, the document [9] worked to prove consensus algorithm of mining dilemma in the process of analysising PoW (Proof of the work) strategy choice for the existence of Nash equilibrium. In general, this solution came from the perspective of game theory, analyzed the PoW consensus algorithm, and provided new ideas and methods for the further design of the consensus algorithm based on game theory.

To overcome performance problem, document [10] proposed a public supply chain system based on double-chain structure. That results of test showed that the two-chain structure of supply chain based on agricultural product supply chain could consider the openness and security of transaction information as well as the privacy of enterprise information, and could achieve rent-seeking and matching adaptively. Ref. [11] proposed a new blockchain consistency algorithm, introduced the two-stage delivery and quorum vote process, and solved the legitimacy verification problem in decentralized environment by using the distributed ledger feature of the blockchain protocol. Compared with the traditional Byzantine consensus, the algorithm reduced the number of message passing, improved system fault tolerance. Artical [12] instantiated a provably secure OKSA solution at transport stage, reduced a round interaction and constant communication cost. The analysis and evaluation showed that the search chain could remain reasonably cost effective without loss of retrieval privacy. The literature [13] proposed a security certification scheme that it could manage human-centric solutions (SAMS) to verify resource information in participating mobile devices and processing in MRM resource pools.

In order to verify the SAMS of MRM, the data of FalsifICA was tested. The test results showed that data tampering was impossible. To achieve the fault tolerance of the XFT consensus algorithm, the solution proposed by the document [14] presented a Byzantine consensus algorithm based on the Gossip protocol, which could enable the system to tolerate less than half of the nodes as byzantines. At the same time, the system has better scalability, and it is helpful to identify malicious nodes for correct nodes in the blockchain system since the unified data structure is adopted. In order to improve operation efficiency of the consensus algorithm in blockchain system, the document [15] firstly introduced various potential of blocks in the chain of consensus algorithm optimization scheme, then optimized dBFT consensus process by combining with aggregate signature technology and bilinear mapping technology. The optimized aggregate dBFT consensus algorithm can effectively reduce the space complexity of signature in blockchain system. The document [16] also explored how different network conditions can change the results of consistency between nodes. Besides, Bach's analysis [17] focused on the algorithm steps: the scalability algorithm adopted by each consistent algorithm, and the time and security risks of the algorithm reward validator verification block exist in the algorithm.

As above of all, generalizability is a basic problem in distributed computing, although the question is known that existing agreements have been designed over the past three years to address consensus. Under various assumptions with the advent of chains, various consensuses have come to high performance. However, the Byzantine fault tolerance (BFT) protocol is not suitable. For a high number of participants, it must be accommodated. Regarding to performance and security, one can use novel network techniques and trusted computing [18,19].

3 Method

3.1 Consensus Algorithm

The consensus algorithm is a decentralized two-round protocol designed for supporting small-scale boardroom voting. In the first round, all voters register their intention to vote in the election, and in the second round, all voters cast their vote. The systems assumes an authenticated broadcast channel is available to all voters. The self-tallying property allows anyone (including non-voters) to compute the tally after observing messages from the other voters. In this paper, we only consider an election with two options, e.g., yes/no. Extending to multiple voting options, and a security proof of the protocol can be found in [20].

A description of the Open Vote Network is as follows. First, all n voters agree on (G, g) where G denotes a finite cyclic group of prime order q n which the Decisional Diffie-Hellman (DDH) problem is intractable, and g is a generator in G. A list of eligible voters $(P_1, P_2, ..., P_n)$ is established and each eligible voter P_i selects a random value as their private voting key.

Round 1. Every voter P_i broadcasts their voting key g^{x_i} and a (non-interactive) zero knowledge proof $ZMP(x_i)$ to prove knowledge of the exponent

x_i on the public bulletin board. $ZMP(x_i)$ is implemented as a Schnorr proof [21] made non-interactive using the Fiat Shamir heuristic [22].

At the end, all voters check the validity of all zero knowledge proofs before computing a list of reconstructed keys:

$$Y_i = \prod_{j=1}^{i-1} g^{x_j} / \prod_{j=i+1}^{n} g^{x_j} \tag{1}$$

Implicitly setting $Y_i = g^{y_i}$, the above calculation ensures that $\sum_i x_i y_i = 0$.

Round 2. Every voter broadcasts $g^{x_i y_i} g^{v_i}$ and a (non-interactive) zero knowledge proof to prove that v_i is either no or yes (with respect to 0 or 1) vote. This one-out-of-two zero knowledge proof is implemented using the Cramer, Damgård and Schoenmakers (CDS) technique.

All zero knowledge proofs must be verified before computing the tally to ensure the encrypted votes are well-formed. Once the final vote has been cast, then anyone (including non-voters) can compute $\prod_i g^{x_i y_i} g^{v_i}$ and calculate $g^{\sum_i v_i}$ since $\prod_i g^{x_i y_i} = 1$. The discrete logarithm of $g^{\sum_i v_i}$ is bounded by the number of voters and is a relatively small value. Hence the tally of yes votes can be calculated subsequently by exhaustive search.

Note that for the election tally to be computable, all the voters who have broadcast their voting key in Round 1 must broadcast their encrypted vote in Round 2. Also note that in Round 2, the last voter to publish their encrypted vote has the ability to compute the tally before broadcasting their encrypted vote (by simulating that he would send a no-vote). Depending on the computed tally, he may change his vote choice. In our implementation, we address this issue by requiring all voters to commit to their votes before revealing them, which adds another round of commitment to the protocol.

The decentralised nature of the consensus algorithm makes it suitable to implement over wlan mesh network. Wlan mesh could be used as the private blockchain to store the voting data for the consensus algorithm.

3.2 Structure of Implementation

There are two smart contracts that are both written in Fabric's Solidity language. The first contract is called the voting contract. It implements the voting protocol, controls the election process and verifies the two types of zero knowledge proofs we have in the Open Vote Network. The second contract is called the cryptography contract. It distributes the code for creating the two types of zero knowledge proofs 3. This provides all voters with the same cryptography code that can be used locally without interacting with the Ethereum network. We have also provided three HTML5/JavaScript pages for the users, as shown in the Fig. 1. We assume that voters and the election administrator have their own Fabric accounts. The Web3 framework is provided by the Ethereum Foundation to faciltiate communication between a user's web browser and their Fabric client. The user can unlock their Fabric account (decrypt their Fabric private key using a password) and authorise transactions directly from the web browser. There is no need for the user to interact with an Ethereum wallet, and the Fabric client can run in the background as a daemon.

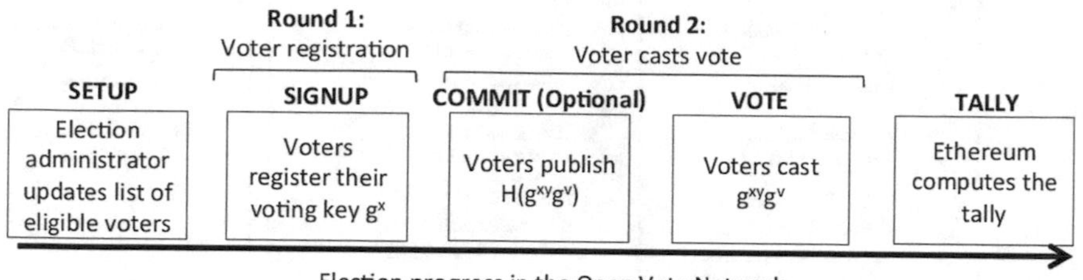

Fig. 1. There are five stages to the election.

Election Administrator. This includes establishing the list of eligible voters, setting the election question, and activating a list of timers to ensure the election progresses in a timely manner. The latter includes notifying Ethereum to begin registration, to close registration and begin the election, and to close voting and compute the tally.

Voter. Voter can register for an election, and once registered must cast their vote.

Observer. Observer can watch the election's progress consisting of the election administrator starting and closing each stage and voters registering and casting votes. The running tally is not computable.

3.3 Setup Flow

We defined 2 typical user setup scenarios: (1) New nodes – How to privatize your nodes first time. (2) Privatized nodes – How to re-privatize your nodes. When first open one node never been privatized, our nodes will be in public mesh state, user can use phone APP to privatize the devices and setup AP configurations. Figure 2 is an example showing the new devices setup sequence: Fig. 3 is an example showing a new device how to join a privatized group:

Definitions, Acronyms and Abbreviations:

WMD: WIFI manager daemon.
Master: The device which is in private mesh network and connected to AP.
Slave: The device which is in private mesh network.
Public Mesh: Mesh network with all known mesh ID.
Private Mesh: Mesh network with privatized mesh ID.

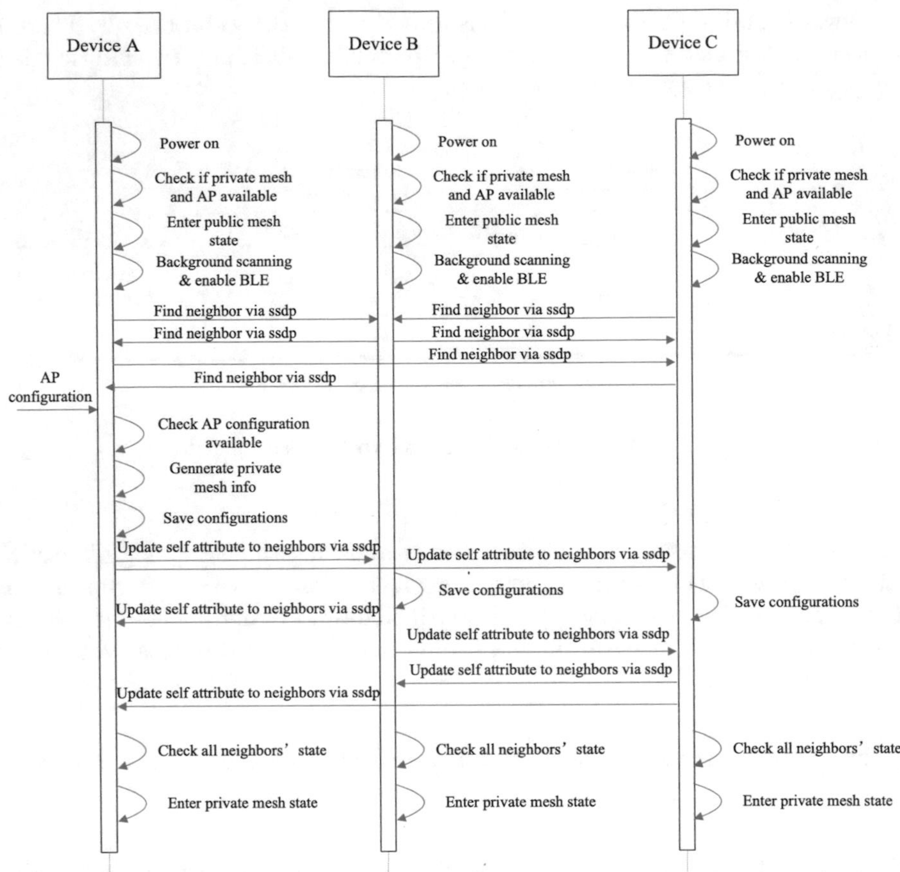

Fig. 2. Flow of new nodes.

If there is a privatized device want to join another privatized group, what we need to do is press reset button to clear privatized information and reboot. Figure 4 is an example showing a privatized device how to join another privatized group.

3.4 Node Discovery Protocol

We use SSDP protocol to discovery node, the structure of SSDP is designed as Fig. 5. According to the protocol, when a control point (client) accesses the network, it can send "ssdp: discover" messages to a SSDP port with a specific multicast address using the M-SEARCH method. When the device listens for messages sent by the control point on the reserved multicast address, it analyzes the service requested by the control point. If it provides the service requested by the control point itself, the device will directly respond to the request by unicast, see Fig. 6.

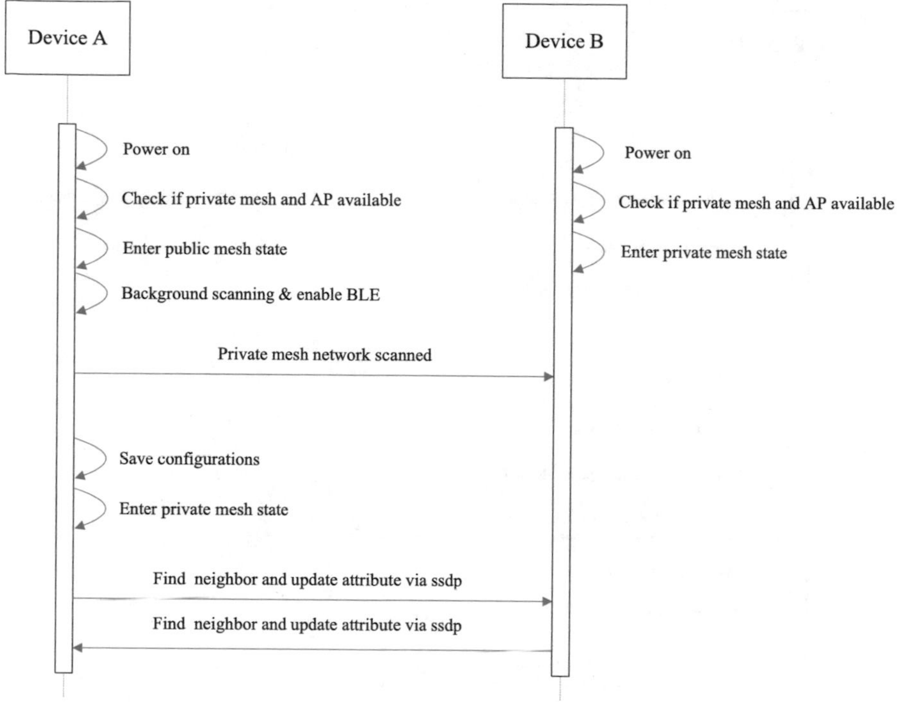

Fig. 3. Flow of resetting privatized nodes.

3.5 Master Vote Strategy

This paper first define Parameters as:

Role: define the role of the node.
AP & RSSI: define the rssi of the AP found (if there are multi AP, there should be multi RSSI for different AP).
Master: every node in the group should save the master's UUID and IP.
Self: every node should save their own UUID and IP in the service.
Master vote strategy is presented as Fig. 7.

State: No Role, Role is None; Send broadcast "request master information without role"; Receive "request master information without role" broadcast, compare with self IP. If self is smaller, then keep state. If self is bigger, then transfer to "Pre-Role" state. Receive "request master information with slave role" broadcast, compare with self IP. If self is smaller, then keep state. If self is bigger, then transfer to "Pre-Role" state. Receive "Definder" broadcast, change to "Pre-Role" state. Master or slave responses the broadcast to set the master information, then save master information and change to "Slave" state. After 8 s timeout, change to "Definer" state.

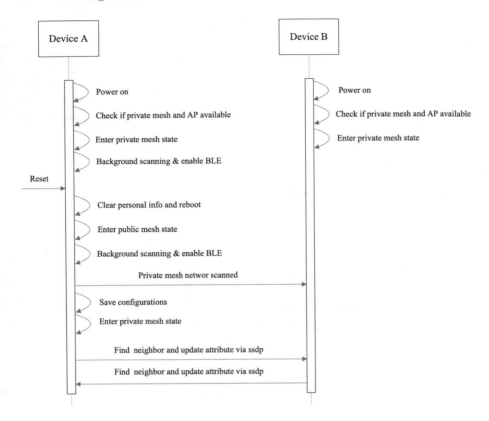

Fig. 4. Flow of finding privatized nodes.

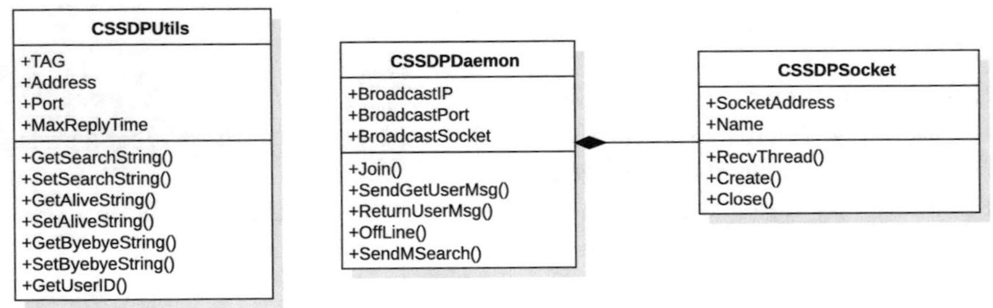

Fig. 5. Internal structure of SSDP.

State: Pre-Role, Role is "slave without master information" (could be slave role but without master parameters values.); Send broadcast "request master information with slave role"; Definer responses the broadcast to get the AP RSSI value; Definer responses the broadcast to set it to master role, then connect the AP and change to "Master" state. If connect AP failed, it will transfer to "Definer" state. Master or slave responses the broadcast to set the master

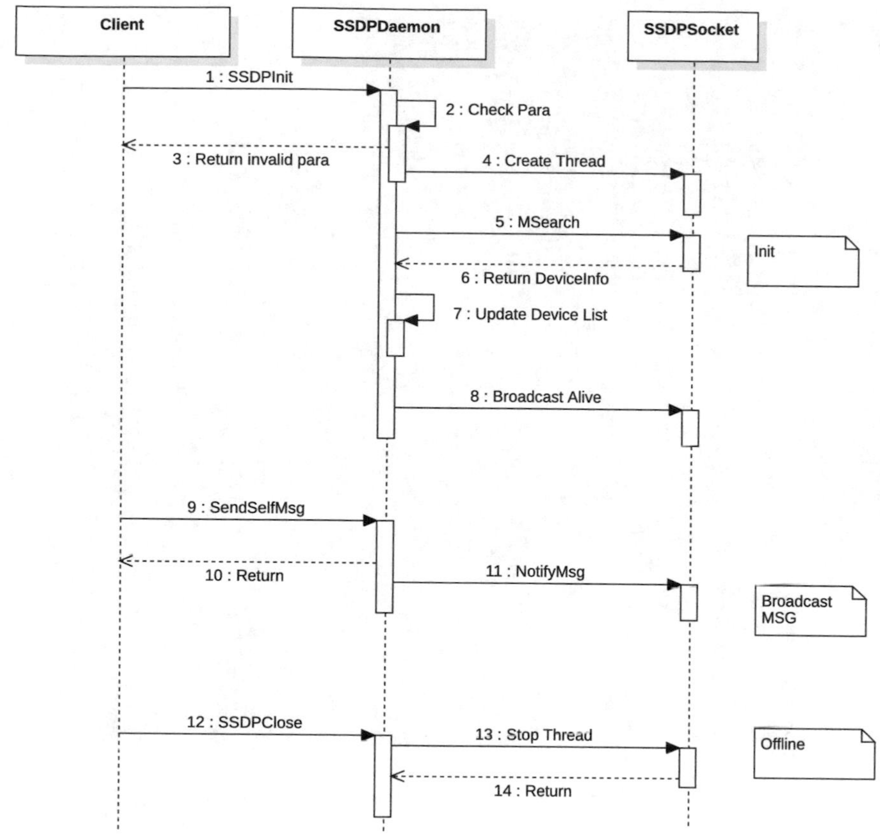

Fig. 6. Setup flow of SSDP.

information, then save master information and change to "Slave" state. If there is no change on the state after 8 s timeout, will transfer back to "No Role" state.

State: Definer, Role is definer (or slave but send "definer" broadcast); Send broadcast "definer"; Receive "request master information with slave role" to get the AP RSSI value from the broadcaster; Receive "Definer" broadcast, check IP (or UUID), if small than self's, then change to "Pre-Role" state; After 2 s timeout, set the highest RSSI slave to the master and change to "Pre-Role" state (or change to "Slave" state), or change self to the "Master" state if self is the highest RSSI one or no slave found; Master or slave responses the broadcast to set the master information, then save master information and change to "Slave" state.

State: Slave. Role is slave with master information saved; Receive "request master information without role/with slave role" and "definer" broadcast, set the master information to the broadcaster; Receive master's heartbeat packet.

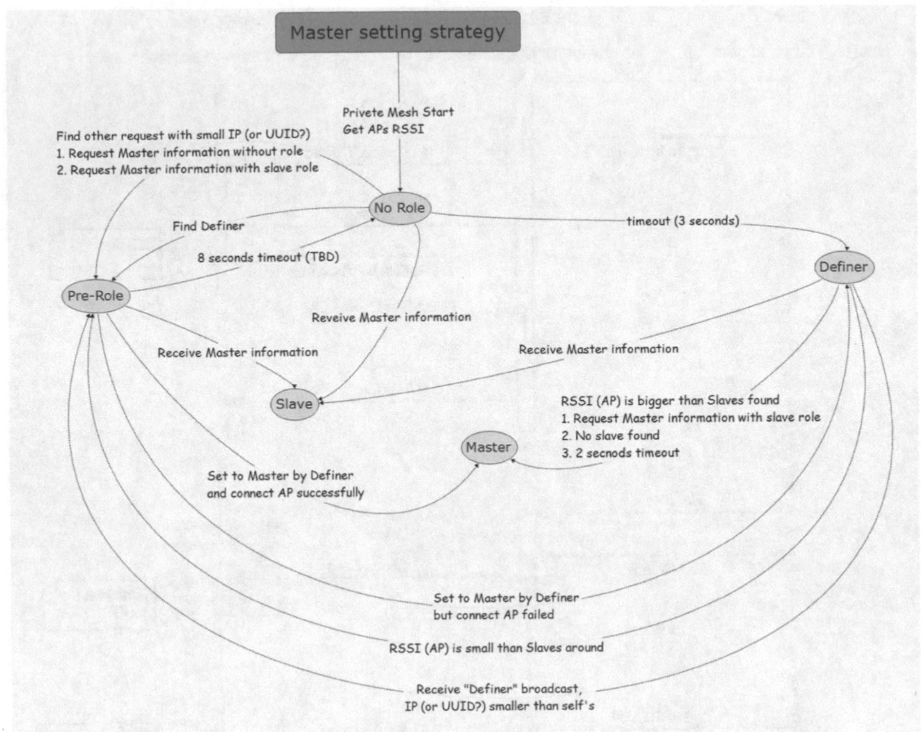

Fig. 7. Master vote strategy.

State: Master. Role is master with master information saved; Receive "request master information without role/with slave role" and "definer" broadcast, set the master information to the broadcaster; Send out heartbeat packet.

Special case: When master leave actively, master set new master and send to neighbors' the change; When master leave suddenly, slave has timeout and change to "No Role" state. Only neighbor slaves monitor the master's heartbeat packets. If master leaves suddenly, only neighbor slaves change to "No Role" state. Next jump slaves will be set to "Pre-Role" by neighbor slaves or next jump slaves.

4 Experiments and Evaluation

4.1 Experiments

Our implementation was deployed on Fabric's official test network that can mimic the production network. We sent 126 transactions to simulate forty voters participating in the protocol. Each transaction's computational cost is outlined

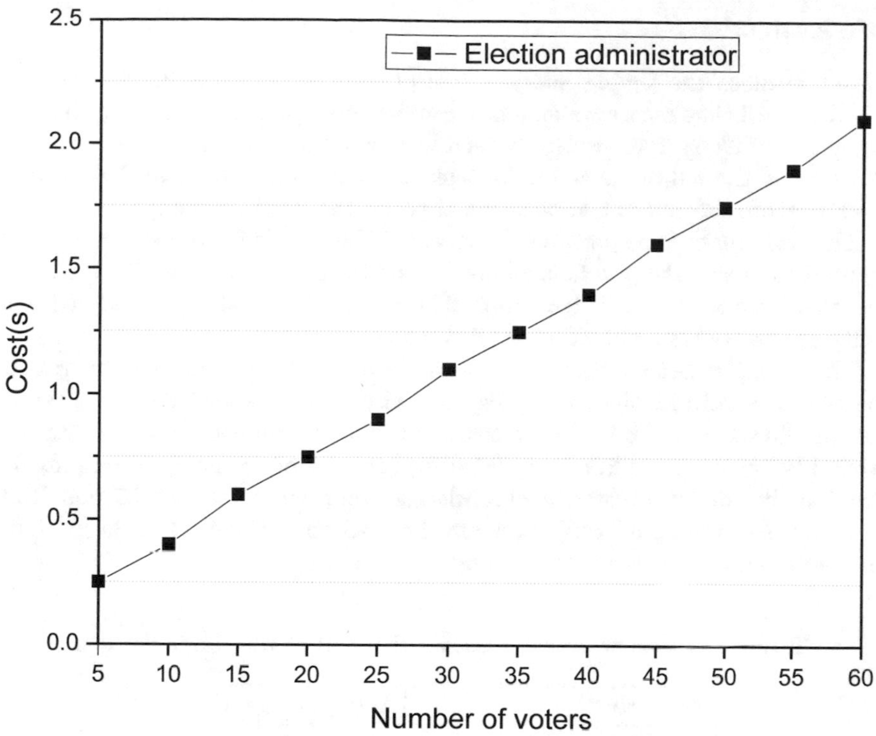

Fig. 8. The time cost for the election administrator based on the number of voters participating in the election.

in Table 1. Each transaction by the election administrator is broadcast only once, and each transaction by a voter is broadcast once per voter, i.e., a total of 40 times.

Figure 8 demonstrates the results of our experiment and highlights the breakdown of the election administrator's time consumption. Except for opening registration, the time cost for each task increases linearly with the number of voters. The time limit for a block was set at 4.7 s by the miners before the recent DoS attacks. This means that the smart contract reaches the computation and storage limit if it is computing the voter's reconstructed keys for around sixty registered voters. This limit exists as all keys are computed in a single transaction and the time used must be less than the block's time limit. To avoid reaching this block limit, we currently recommend a safe upper limit of around 50 voters. However, the contract can be modified to perform the processing in batches and allow multiple transactions to complete the task.

4.2 Evaluation

Table 1 outlines the timing analysis measurements for tasks in the consensus algorithm. All time measurements are rounded up to the next whole millisecond. We use the Web3 framework to facilitate communication between the web browser and the Fabric daemon. All tasks are executed using .call() that allows us to measure the code's computation time on the local daemon.

The wlan mesh is responsible for creating the proofs for the voter. The time required to create the proofs is 81 ms for the Schnorr proof and 461 ms for the one-out-of-two zero knowledge proof. These actions are always executed using .call() as this contract should never receive transactions.

The voting smart contract is responsible for enforcing the election process. Registering a vote involves verifying the Schnorr zero knowledge proof and in total requires 142 ms. To begin the election requires computing the reconstructed public keys which takes 277 ms in total for forty voters. Casting a vote involves verifying the one-out-of-two zero knowledge proof which requires 573 ms. Tallying involves summing all cast votes and brute-forcing the discrete logarithm of the result and on average takes around 132 ms.

Table 1. A time analysis for actions that run on the Fabric daemon.

Action	Avg. Time (ms)
Create ZKP(x)	81
Register voting key	142
Begin election	277
Create 1-out-of-2 ZKP	461
Cast vote	573
Tally	132

5 Conclusion

In this paper, we have presented a fast adaptive blockchain consensus algorithm implementation by using aximum voter privacy that runs on Fabric. Our implementation was tested on the official Fabric test network with forty simulated voters. We have shown that our implementation can be readily used with minimal setup for elections at a little cost per voter. The cost can be considered reasonable as this voting protocol provides maximum voter privacy and is publicly verifiable. This is the first implementation of a decentralised internet consensus protocol running on wlan mesh network. It uses the Fabric blockchain not just as a public bulletin board, but more importantly, as a platform for consensus computing that enforces the correct execution of the voting protocol.

In future work, we will investigate the feasibility of running a national-scale election over the blockchain. Based on the knowledge gained from this paper, we believe that if such a perspective is ever considered possible, its implementation will almost certainly require a dedicated blockchain. For example, this can be an Fabric-like blockchain that only stores the e-voting smart contract. This new blockchain can have a larger block size to store more transactions on-chain and may be maintained in a centralised manner similar to RSCoin [9].

References

1. Nakamoto, S.: Bitcoin: a peer-to-peer electronic cash system (2008). https://bitcoin.org/bitcoin.pdf
2. Higgins, S.: IBM invests $ 200 million in blockchain-powered IoT (2016). http://www.coindesk.com/ibm-blockchain-iot-office/
3. Clark, J., Essex, A.: CommitCoin: carbon dating commitments with bitcoin. In: Financial Cryptography and Data Security. Springer, Heidelberg (2012)
4. Ekblaw, A., Azaria, A., Halamka, J. D., Lippman, A.: A case study for blockchain in healthcare:medrec prototype for electronic health records and medical research data (2016). http://dci.mit.edu/assets/papers/eckblaw.pdf
5. International Association for Cryptologic Research: About the helios system (2016). http://www.iacr.org/elections/eVoting/about-helios.html
6. Adida, B.: Helios: Web-based open-audit voting. In: USENIX Security Symposium, vol. 17, pp. 335–348 (2008)
7. Nguyen, G.T., Kim, K.: A survey about consensus algorithms used in blockchain. J. Inf. Process. Syst. **14**(1), 101–128 (2018)
8. Gramoli, V.: From blockchain consensus back to Byzantine consensus. Futur. Gener. Comput. Syst. (2017)
9. Tang, C., Yang, Z., Zheng, Z.L.: Game dilemma analysis and optimization of PoW consensus algorithm. Acta Autom. Sin. **43**(9), 1520–1531 (2017)
10. Leng, K., Bi, Y., Jing, L., Fu, H.C.: Research on agricultural supply chain system with double chain architecture based on blockchain technology. Futur. Gener. Comput. Syst. **86**, 641–649 (2018)
11. Wu, T., Huang, X., Zhou, L.L.: Research on blockchain consistency algorithm with state legality verification. Comput. Eng. **44**(1), 160–164 (2018)
12. Jiang, P., Guo, F., Liang, K., Lai, J., Wen, Q.: Searchain: blockchain-based private keyword search in decentralized storage. Futur. Gener. Comput. Syst. (2017)
13. Kim, H.W., Jeong, Y.S.: Secure authentication-management human-centric scheme for trusting personal resource information on mobile cloud computing with blockchain. Hum. Centric Comput. Inf. Sci. **8**(1), 11 (2018)
14. Zhang, S.J., Cai, J., Chen, Z.H.: Byzantine consensus algorithm based on Gossip protocol. Comput. Sci. **45**(2), 20–24 (2018)
15. Yuan, C., Xu, M.X., Si, X.M.: Optimization scheme of consensus algorithm based on aggregation signature. Comput. Sci. **45**(2) (2018)
16. Porat, A., Pratap, A., Shah, P.: Blockchain consensus: an analysis of proof-of-work and its applications (2018). http://www.scs.stanford.edu
17. Bach, L., Mihaljević, B., Žagar, M.: Comparative analysis of blockchain consensus algorithms. In: 41st International Convention for Information and Communication Technology, Electronics and Microelectronics (2018)
18. Feng, L., Zhang, H., Tsai, W.T., Sun, S.: System architecture for high-performance permissioned blockchains. Front. Comput. Sci., 1–15 (2018)

19. David, S., Noah, Y., Arthur, B.: The Ripple Protocol Consensus Algorithm. Ripple Labs Inc., San Francisco (2014)
20. Hao, F., Ryan, P.Y., Zielinski, P.: Anonymous voting by two-round public discussion. IET Inf. Secur. **4**(2), 62–67 (2010)
21. Schnorr, C.P.: Efficient signature generation by smart cards. J. Cryptol. **4**(3), 161–174 (1991)
22. Fiat, A., Shamir, A.: How to prove yourself: practical solutions to identification and signature problems. In: Odlyzko, A.M. (eds.) Crypto 1986. LNCS, vol. 263, pp. 186–194 (1987)

Toward a Blockchain Based Image Network Copyright Transaction Protection Approach

Chengqiang Zhao, Mingzhe Liu(✉), Yanhan Yang, Feixiang Zhao, and Shijie Chen

State Key Laboratory of Geohazard Prevention and Geoenvironment Protection, Chengdu University of Technology, Chengdu 610059, China
liumz@cdut.edu.cn

Abstract. In existing image network copyright transaction, the original attribute of the transaction image content cannot be identified. In this paper, we propose an image network copyright transaction protection approach based on blockchain technology. In this approach the entire copyright transaction process is protected and the attribute identification of image content is identified. The advantages of blockchain applied to the copyright transaction of an image network are discussed. On this basis, a realization structure of copyright market transaction based on blockchain is described, and a specific business scheme of image copyright transaction is presented. Comparing and analyzing with the existing implementation schemes, the results show that the proposed scheme has great advantages in terms of reliability and practicality.

Keywords: Blockchain · Image copyright protection · Copyright transaction

1 Introduction

In network copyright transaction of images, generally the copyright owner uploads the image through a third-party platform, the demander purchases through the third-party platform, and the third-party platform extracts a certain proportion of service fee. However, in the entire process, there are several drawbacks such as the infringement of trading works, imbalance of interest distribution, frequent occurrence of copyright disputes, and difficulty of network copyright protection. Many of the problems exposed by the "centralization" model involved in the third-party platform cannot be avoided. With the rapid development of Internet technology, a new peer-to-peer copyright trading model is presented as an alternative mean [1].

Blockchain technology is one of the new technologies to subvert traditional copyright transactions in recent years. The asymmetric encryption and timestamp techniques of blockchain can accurately trace the copyright attribution of picture and information content of transaction. Although blockchain technology has been widely used in financial transactions and asset management, it has a great potential in the application of digital copyright protection [2, 3]. The research on related aspects is in the discussion, especially the application of the blockchain to the copyright transaction of images [4–7].

C.-N. Yang et al. (Eds.): SICBS 2018, AISC 895, pp. 17–28, 2020.
https://doi.org/10.1007/978-3-030-16946-6_2

Huminski et al. [6, 7] discussed the core features of decentralization of blockchain, hash algorithm, digital signature, distributed consensus mechanism, timestamp, smart contracts, and other technical advantages. There is a bright future in digital copyright protection. Savelyev et al. [8] discussed the related legal issues involved in the application of blockchain technology in the field of digital copyright and proposed that users should use the copyrighted works recorded through the blockchain for formal purposes as well as transactions and intelligence on cryptocurrencies, legalization of contracts and other issues. Refs. [4, 9] discussed that although blockchain can guarantee the security of information such as confirmation copyrights and transaction links during digital copyright protection, it is difficult to guarantee the originality of copyright content information, i.e., it cannot guarantee that the content information is original or modified. On account of the image content can only be detected by the relevant image content antitamper detection technology. To solve this problem, traditional digital watermarking technology can be used to encrypt the image [10–13], but new problems occur: the technical cost is high, and it is difficult for the work encryption technology modified by the physical processing to exert its effect. Therefore, an integrated approach to the original identification of copyright transactions and image content information is needed.

In this regard, an image network copyright transaction protection scheme considering image content is proposed based on blockchain technology. On one hand, it provides a strong guarantee for copyright attribution, copyright protection, and copyright infringement in image network transactions. On the other hand, through image processing and detection-related technology, the original content is tamper-proof detection of images. To a certain extent, the image copyright is clearly recognized, and the image content is identified. First, this paper discusses the feasibility of blockchain technology in solving the security problem of image network copyright transaction and further designs a technical architecture scheme that can protect the image copyright transaction and image attribute identification. The key point is to add the image content tampering detection and performing attribute identification, i.e., distinguishing the original work and the original creative work, and realizing all-round protection of all links in the image network copyright transaction. The scheme is analyzed to obtain the protection principle, and the reliability and practical advantages of the scheme is evaluated by comparing with the existing implementation schemes.

2 Technical Framework and Protection Mechanism

As shown in the characteristics of blockchain, it is a systematic project completed by multidisciplinary crossover technology [14]. An excellent architecture design should be able to improve and develop a system function with strong flexibility, and the system's own security, ease of operation, stability, other performance should be persistent. Therefore, system design will pay attention to the characteristics of entire technology architecture, including scalability, maintainability, security, and flexibility.

Unlike the popular blockchain system architecture, the copyright transaction object targeted in this paper is the image, and an image has its own content information characteristics in the network copyright transaction, especially in the copyright confirmation link and the transaction copyright changes. The image content tampering detection is added to ensure that the image creation properties are clearer. The technical realization structure scheme design is shown in Fig. 1.

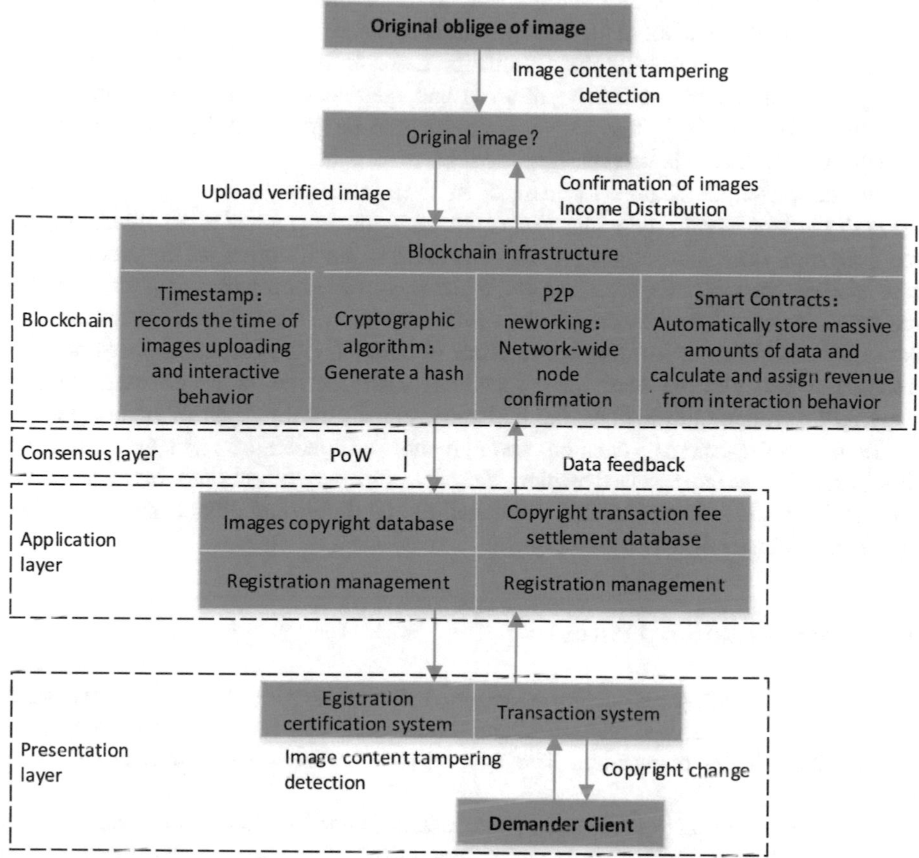

Fig. 1. Image network copyright transaction protection architecture scheme of blockchain technology

In the image content tampering detection, the concern of the main detection is the part of image content which is manipulated and tampered. The deep learning neural network is designed by studying the RGB, content and noise characteristics of images. By learning a large data set composed of clipped images and recognizing content tampering operation part, the image processing detection can be realized, and the design effectively solves the originality identify of image content in the image network copyright transaction. In this attribute identification, an image in which the image content pixel has not changed is called as an original work, and a creative work that falsifies the image content but does not involve any copyright problem is called as an original creative work [15].

In the image copyright confirmation, the system encrypts and uploads an image that has been verified for image content attributes. This is verified by the blockchain network and its owner's unique ID, uploading time of work, and other information of work. The unique ID is used to realize the authentication and confirmation of work. The initial upload time by timestamp is used to prove who the initial owner is. Through the consensus mechanism [16], during the entire stateless verification of transactions process, each node records all the information and confirms it and performs distributed storage, which can be arbitrarily retrieved and queried, and the cost of information modification is relatively high, so it is not easy to be modified. The copyright confirmation of an image is protected.

In the copyright change of images, in the entire image copyright transaction, through the blockchain, the realization of image works is efficient and convenient, and the copyright use of all levels enjoys convenience. Each copyright demander is not restricted by time domain or region and realizes convenient transactions. This satisfies the needs of entire trading market. In addition, each participating node, through the consensus mechanism, introduces the Proof of Work (PoW) mechanism, rewards the interaction behavior of participating nodes, cultivates the netizens' awareness of copyright payment, and realizes the maximum benefit of copyright of image [17].

In the infringement protection, through the verification of right authentication, timestamp, image tampering detection, etc., the efficiency of enumerating evidence to protect rights and interests are improved, and more reliable and authoritative copyright evidence is obtained.

3 Implementation Process

The network copyright transaction of an image can be roughly divided into two parts. One is the copyright confirmation of the work itself, and the other is the copyright change through the transaction. The approximate realization structure is shown in Fig. 2.

The underlying platform based on blockchain is used to record and store relevant data. Before the transaction, the owner of image works first performs the authentication of right, and the owner of work confirms the copyright to the primary trading market or issues the copyright. The primary trading market includes copyright certification organizations and copyright distribution organizations, while the copyright demander and secondary trading markets obtain the copyright to use the works through purchase. The secondary trading market can develop the multilevel market after being authorized. The transaction interaction behavior is carried out in a blockchain-based system. Each nodes of the system submits transaction information to the blockchain system. Second, it also checks the transaction status. Each nodes participant is subject to the network contract. Each node confirms the transactions independent verification, records, and stores it in a distributed manner in the database. Each copyright transaction in the database has the corresponding smart contracts, and all relevant information is written into the system blockchain with the contract [16].

If the secondary market again issues copyright transaction through the blockchain system and completes the transaction, the transaction is fully confirmed, then the

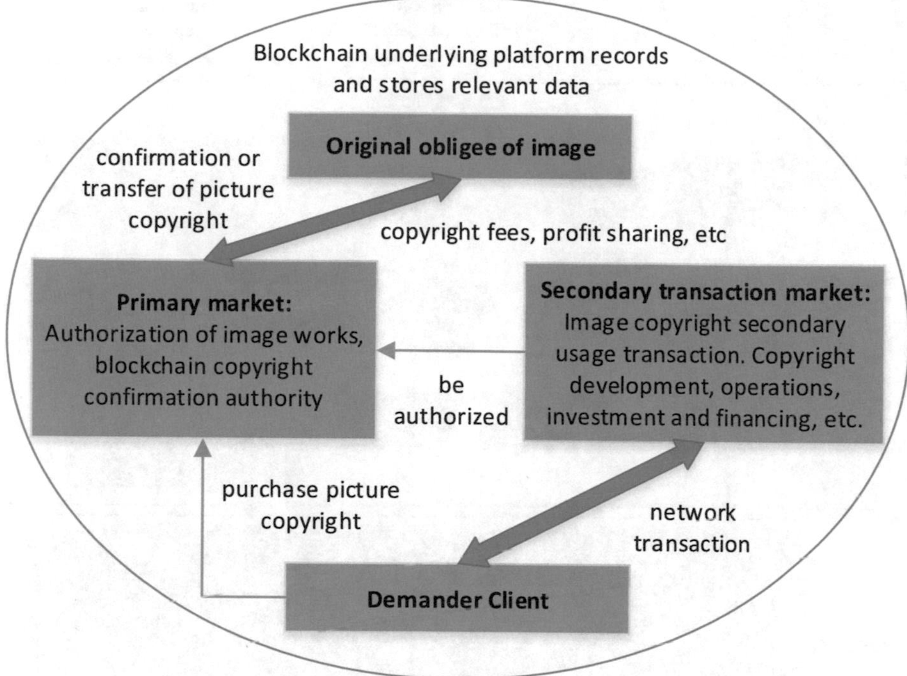

Fig. 2. Blockchain-based image copyright market transaction realization structure

transaction amount will be written into the blockchain, and the relevant income information will also be recorded and stored. Regardless of whether the transaction occurs in the primary or secondary market, all transaction process execution and transaction income distribution are carried out through the consensus mechanism under the blockchain system, which can simplify the transaction process and optimize the transaction income distribution. This improves the efficiency of copyright transaction operation to some extent. The specific trading business scheme is shown in Fig. 3.

Combined with Figs. 1, 2, and 3, the protection scheme designed in this paper relies on the consensus algorithm for the maintenance system. The application layer uses the P2P network as the support, and the digital currency uses the common Bitcoin. In the process of image copyright transaction, the image content attribute can be identified when performing image copyright authentication, and the transaction related information must be registered and authenticated at the certification center. The implementation process of the two major links in the transaction is carried out as follows:

(1) The copyright owner initiates copyright authentication.
First, the original owner of the image is registered as a member through network certification, becomes a peer node in the network, generates a digital currency address for the original owner of image, uploads the image through the network, and executes the attribute identification of the image content in Fig. 1 before uploading.

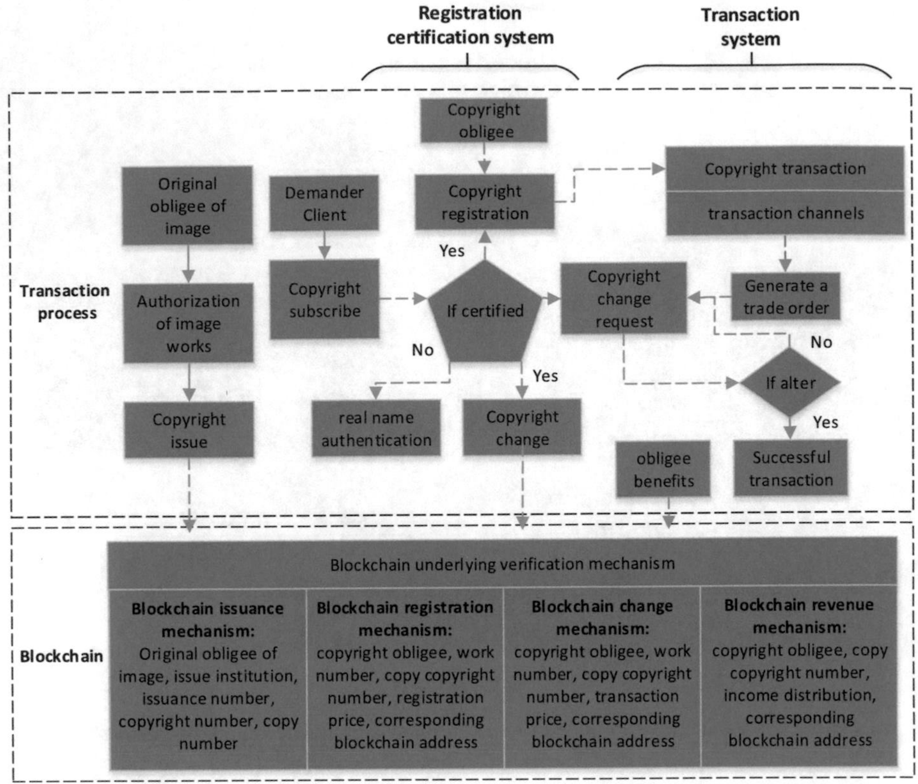

Fig. 3. Specific business implementation model

After the content attribute of the image is identified, and the original attribute related information and other information of the image are stored together in the system. After the image is uploaded successfully, the server calculates the hash of image file through hash algorithm and uses it as a unique identifier of the image [18]. The network node queries the generated hash through the copyright database system of image. If the hash already exists, it indicates that the image has been authenticated. If it does not exist, the network system initiates a transaction to the original owner of image, and the original owner of image pays the corresponding Bitcoin to the Bitcoin address of network [19]. The amount of payment is determined by the consensus mechanism workload revenue mechanism.

Upon the completion of above transaction, the network node writes the hash string to the Bitcoin transaction in the above process and initiates a transaction to the Bitcoin network. The network node confirms and records the transaction information and stamps the timestamp and records the transaction information. For example, the Bitcoin transaction ID, timestamp, and image hash and other transaction details are stored in a distributed manner in the website database [20]. The hash of image file and transaction timestamp can be used as the declaration of copyright of image, and the authentication of image is confirmed. If the original owner of image wants to conduct a copyright transaction, it can be authorized and issued a copyright in the network system.

(2) The demander client initiates the transaction
The copyright demander client registers as a member through P2P network authentication, i.e. becomes a network node, performs copyright subscription authentication, and then indexes the target image information according to its own needs. The network uses the hash algorithm to output the target image hash and passes the system authentication. The network node finds the record information of target image in the network, generates an order, initiates a copyright change application to the system, and initiates a transaction to image demander, prompting to pay a certain amount of Bitcoin to the Bitcoin address of target image right holder. At this time, the target image of transaction is registered by copyright according to the blockchain-based registration mechanism and accepts the copyright transaction, and the system obtains the transaction information through various transaction channels. After the image demander pays, the function of obtaining the image can directly initiate the transaction to the Bitcoin network based on the blockchain technology and add the image hash of copyright change to the transaction as the additional information, and the website saves the Bitcoin transaction ID, the time stamp, the image hash of the copyright change, and original image hash are saved in the website database.

On the other hand, the original image can be modified, the network transaction can be initiated, and the new image hash can be associated with the original hash, so that the creation process can be recorded. The entire process achieves image copyright changes through a blockchain-based change mechanism. The right holders who obtain the copyright change can further carry out the transaction with the support of system contract, and the income distribution in all the transaction is distributed by the proof of workload under the consensus mechanism [21].

4 Evaluation and Discussion of Transaction Protection Scheme

4.1 Comparative Analysis of Transaction Protection Scheme

The scheme designed in this paper was evaluated by comparative analysis. Several expression schemes were selected for comparative analysis. A comparison of other solution functions is shown in Table 1.

The selected image copyright protection representative schemes are all single or multi-faceted protections on the content of the image and the copyright transaction process. Among them, Liu et al. [22, 23] mainly encrypted the image by image watermark encryption. Meng et al. [20, 24–26] mainly protected the image rights using techniques such as blockchain, mainly applied to copyright management system. Ebookchain [27] is a blockchain-based commercial copyright trading system, but it does not test the image content attributes of trading works. The scheme designed in this

Table 1. Comparison of several image copyright protection schemes

Program	Contrast content			
	If blockchain-based	If for image design	If to operate on image content	If for network transaction protection
Liu et al. [22]	No	Yes	Yes	No
Sae-Tang et al. [23]	No	Yes	Yes	Yes
Meng et al. [20, 24]	Yes	Yes	No	No
Kishigami et al. [25, 26]	Yes	No	No	No
Ebookchain [27]	Yes	Yes	No	Yes
This paper	Yes	Yes	Yes	Yes

paper not only provides protection for the copyright transaction of image, but also provides attribute identification for the image content of the image being traded.

In Table 1, the functions and application scenarios of each scheme can be compared and analyzed, but the technical advantages and performance characteristics of each scheme cannot be specifically compared. Table 2 further analyzes the characteristic performance of traditional copyright protection method and the scheme designed in this study.

4.2 Discussion

As an emerging technology, blockchain technology is not without any problems in its application to image network copyright transaction. Although it can solve many existing limitations, it also faces some challenges for its perfect use.

(1) Blockchain is a system engineering based on multidisciplinary crossover. Except for the actual successful operation in Bitcoin, it has only recently used in other applications. Many technical specifications and standards have not yet been established. The application of some technologies is still in the stage of experimental development and verification. Some technical risks have not been discovered yet, and there have also been risk accidents such as "The DAO Attack" [28, 29]. From the perspective of blockchain application industry, only the financial transaction and asset management industries have been successfully applied, and other industries are developing.

(2) Although copyright protection based on blockchain has technical advantages, the birth of copyright system is a mechanism for regulating rights and interests. This should adapt to many factors such as law, policy, and humanities, etc. It is impossible to develop further by relying on the advantages of a single technology, so perfect compatibility with existing legal systems is the key to development.

Table 2. Comparison of traditional copyright protection methods and blockchain-based protection schemes

	Traditional technical means	The method designed in this paper
Method of implementation	1. Public cognition confirmation 2. Network registration and password verification 3. Electronic filing and copyright registration law	Toward a Blockchain Based Image Network Copyright Transaction Protection Approach
Characteristics of analysis	1. Deficiency of authority and limited credibility; 2. Relying on third-party platforms; it is vulnerable to attack; it has strong maneuverability and no strict and reliable traceability basis; 3. Copyright registration, high cost, long cycle, and many links; 4. The cost of piracy is at the bottom, and it is difficult to obtain evidence.	1. Decentralized, secure and trustworthy: The system does not rely on the hardware or management system of any centralized third-party platform, and all the data recording, storage, transmission, verification, and other aspects are based on cryptography technology. The asymmetric encryption algorithm guarantees that the transaction data cannot be easily falsified. Through the intelligent contract to execute the entire transaction process of copyright, the condition triggers automatic completion, reduces the intermediate links, reduces the circulation cost, and improves the efficiency. 2. Open, transparent, and reliable: The data blocks in the system are jointly maintained by multiple nodes of get the accounting rights in the network. Each time the accounting records are jointly confirmed, the query can be arbitrarily retrieved, and the modification is difficult. The database uses distributed storage, plus time dimensions for high reliability. 3. Traceable, immutability: The timestamp feature guarantees that all the trading behaviors can be traced back, facilitating the proof of rights.

The current legal trust in blockchain are still to be improved, so the development of blockchain in the copyright protection industry still suffers from many limitations [9, 30, 31].

(3) The shared and free business model is also a challenge based on the copyright protection of blockchain. Many current business models attract users' attention through massive amounts of free information and profit from the number of users, clicks, data traffic, etc. Therefore, this poses a certain challenge to paid viewing or digital copyright protection.

(4) Image copyright transaction protection and value are embodied under specific application. The image is different from other digital media information, and it displays value by providing its own image content information. In the blockchain system, when the image file uploading, only the encrypted number of file is stored, but the ontology of work is not stored, even though the content of work is also modified to be tamper-detected, and the content of work is guaranteed to a certain extent. However, this storage method hinders the public's direct contact with the image works. This is not conducive to the spread of its network value. If we only consider the technical advantages of blockchain technology in copyright confirmation and do not consider the issue of copyright circulation and display value of image works, how to maximize the value of traded goods through the market trading of blockchain system is a question to be considered. This is because especially for trades like images, it can be easily obtained in various ways.

5 Conclusion

Digital copyright protection highly depends on various information security technologies. Therefore, digital copyright protection based on blockchain has attracted much attention. Because of image own characteristic content information expression form and coupled with market transaction circulation, the research on blockchain technology in the protection of image network copyright transaction helps to promote the development of copyright protection.

A scheme is designed for the protection of network copyright transaction of image, and the implementation principle of this scheme is introduced. On this basis, the realization structure of copyright market transaction based on blockchain is introduced, and a specific business model of image copyright transaction is described. The advantages of the copyright transaction protection scheme designed in this paper are compared and analyzed, and the challenges faced by blockchain in the image network copyright transaction are discussed. Comparing and analyzing with the existing implementation schemes, the overall results show that the proposed scheme has great advantages in terms of reliability and practicality.

Acknowledgments. The authors would like to acknowledge the support provided by the 2018 Sichuan Science and Technology Innovation Seedling Project Funding Project (No.: 2018133) and 2018 National Students' project for innovation and entrepreneurship training program (No.: 201810616031x). The authors would like to thank the editor and the anonymous reviewers for their careful review.

References

1. Zhao, F., Zhou, W.: Analysis of digital copyright protection based on block chain technology. Sci. Technol. Law (2017)
2. Zheng, Z., Xie, S., Dai, H., Chen, X., Wang, H.: Blockchain challenges and opportunities: a survey. Int. J. Web and Grid Serv. **14**, 352–375 (2017)
3. Li, X., Jiang, P., Chen, T., Luo, X., Wen, Q.: A survey on the security of blockchain systems. Future Gener. Comput. Syst. (2017)
4. Jia, Y.: Research on network copyright transaction based on blockchain technology. Technol. Publ. **07**, 90–98 (2018)
5. Swan, M.: Blockchain: Blueprint for a New Economy. O'Reilly Media, Sebastopol (2015)
6. Huminski, P.: The technology behind bitcoin could revolutionize these 8 industries in the next few years. https://www.businessinsider.com/author/peter-huminski
7. Garzik, J., Donnelly, J.: Chapter 8 - Blockchain 101: An Introduction to the Future. Handbook of Blockchain, Digital Finance, and Inclusion, vol. 2, pp. 179–186. Elsevier (2018)
8. Savelyev, A.: Copyright in the blockchain era: promises and challenges. SSRN Electron. J. (2017)
9. Herian, R.: Blockchain and the (re)imagining of trusts jurisprudence. Strateg. Change **26**, 453–460 (2017)
10. Lee, G. J., Yoon, E. J., Yoo, K. Y.: A novel multiple digital watermarking scheme for the copyright protection of image. In: Fourth International Conference on Innovative Computing, Information and Control, pp. 756–759. IEEE Computer Society (2009). https://doi.org/10.1109/icicic.2009.38
11. Rawat, S., Raman, B.: A publicly verifiable lossless watermarking scheme for copyright protection and ownership assertion. AEU Int. J. Electron. Commun. **66**(11), 955–962 (2012). https://doi.org/10.1016/j.aeue.2012.04.004
12. Rani, A., Raman, B.: An image copyright protection scheme by encrypting secret data with the host image. Multimedia Tools Appl. **75**(2), 1–16 (2016). https://doi.org/10.1007/s11042-014-2344-0
13. Gorbachev, V., Kaynarova, E., Makarov, A., Yakovleva, E.: Digital image watermarking using DWT basis matrices. In: 2017 21st Conference of Open Innovations Association, pp. 127–133. IEEE (2017)
14. Pilkington, M.: Blockchain Technology: Principles and Applications. Social Science Electronic Publishing, Rochester (2015)
15. Zhou, P., Han, X., Morariu, V., Davis, L.: Learning rich features for image manipulation detection. In: The IEEE Conference on Computer Vision and Pattern Recognition, pp. 1053–1061. IEEE (2018)
16. Gramoli, V.: From blockchain consensus back to Byzantine consensus. Future Gener. Comput. Syst. (2017)
17. Nakamoto, S.: Bitcoin: A peer-to-peer electronic cash system (2008). https://bitcoin.org/bitcoin.pdf. Accessed 6 August 2018
18. Zhao, W., Luo, H., Peng, J., Fan, J.: Spatial pyramid deep hashing for large-scale image retrieval. Neurocomputing **243**(C), 166–173 (2017)
19. Decker, C., Wattenhofer, R.: A fast and scalable payment network with bitcoin duplex micropayment channels. In: 17th International Symposium on Stabilization, Safety, and Security of Distributed Systems, pp. 3–18. Springer (2015)

20. Meng, Z., Morizumi, T., Miyata, S., Kinoshita, H.: Design scheme of copyright management system based on digital watermarking and blockchain. In: IEEE 42nd Annual Computer Software and Applications Conference, pp. 359–364. IEEE (2018)
21. Stathakopoulou, C.: A Faster Bitcoin Network (2015)
22. Liu, H.: Based on wavelet transform image copyright protection system design. Comput. Knowl. Technol. **8**, 2347–2351 (2012)
23. Wannida, S., Fujiyoshi, M., Kiya, H.: Encryption-then-compression-based copyright- and privacy-protected image trading system. In: International Conference on Advances in Image Processing, pp. 66–71. ACM (2017). https://doi.org/10.1145/3133264.3133281
24. Xu, R., Zhang, L., Zhao, H., Peng, Y.: Design of network media's digital rights management scheme based on blockchain technology. In: IEEE International Symposium on Autonomous Decentralized System, pp. 128–133. IEEE (2017). https://doi.org/10.1109/isads.2017.21
25. Fujimura, S., Watanabe, H., Nakadaira, A., Yamada, T., Akutsu, A., Kishigami, J.: BRIGHT: A concept for a decentralized rights management system based on blockchain. In: IEEE International Conference on Consumer Electronics, pp. 345–346. IEEE (2016). https://doi.org/10.1109/icce-berlin.2015.7391275
26. Kishigami, J., Fujimura, S., Watanabe, H., Nakadaira, A., Akutsu, A.: The blockchain-based digital content distribution system. In: IEEE Fifth International Conference on Big Data and Cloud Computing, pp. 187–190. IEEE (2015). https://doi.org/10.1109/bdcloud.2015.60
27. ebookchain.pdf. http://www.ebookchain.org/ebookchain.pdf
28. Siegel, D.: Understanding the DAO attack. http://www.coindesk.com/understanding-dao-hack-journalists/
29. Li, X., et al.: A Survey on the security of blockchain systems. Future Gener. Comput. Syst. (2018). https://doi.org/10.1016/j.future.2017.08.020
30. Schrepel, T.: Is Blockchain the death of antitrust law? The blockchain antitrust paradox. SSRN Electron. J. (2018). https://doi.org/10.2139/ssrn.3193576
31. Kevin, W.: Trust, But Verify: Why the Blockchain Needs the Law. Social Science Electronic Publishing, Rochester (2016). https://doi.org/10.2139/ssrn.2844409

A Blockchain-Based Scheme for Secure Sharing of X-Ray Medical Images

Bingqi Liu[1], Mingzhe Liu[1(✉)], Xin Jiang[1], Feixiang Zhao[1], and Ruili Wang[2]

[1] State Key Laboratory of Geohazard Prevention and Geoenvironment Protection, Chengdu University of Technology, Chengdu, Sichuan, China
liumz@cdut.edu.cn

[2] Institute of Natural and Mathematical Sciences, Massey University, Auckland, New Zealand

Abstract. This paper proposes a secure sharing and trading scheme of X-Ray medical image data based on block chain, as it can be used for further scientific research processing. The specific representation includes the following steps: i. The original X-Ray data of medical instruments are transmitted to the cloud platform through the MQTT protocol; ii. The patient information (name, medical card number, etc.) of the image data is encrypted by hashing algorithm to protect privacy; iii. Applying a watermark to the image data; iv. Generating blocks through the consensus mechanism of block chain. The scheme proposed in this paper overcomes the security challenges faced by the traditional cloud-based image data management solution: user privacy disclosure, illegal tampering, and the risk of data being stolen and sold, and realizes a secure transaction system connecting users with data needs, which makes the huge image data on clinical medicine have higher scientific research value.

Keywords: Cloud platform · Blockchain · X-Ray medical image · Public ledger

1 Introduction

X-Ray medical images are used for radiotherapy, as well as have high scientific research and market value. How to obtain these valuable data has become a bottleneck in the development of artificial intelligence in the medical field. The current centralized, cloud-based image data management solution cannot be expanded, nor can it solve the security challenges faced by large hospitals.

Although large hospitals already have relevant big data platforms in this field, most artificial intelligence products have too little training data and require a large amount of manpower and material resources to label training data. The hospital's big data platform is dedicated to off-line data analysis inside hospitals and basically does not have a real data sharing system and related products

Supported by organization x.

C.-N. Yang et al. (Eds.): SICBS 2018, AISC 895, pp. 29–42, 2020.
https://doi.org/10.1007/978-3-030-16946-6_3

for enterprises outside hospitals, especially the development and development of a safe data sharing platform that can guarantee patients' privacy, prevent data tampering and illegal circulation. In view of the huge demand for safe data sharing platforms in the intelligent medical industry and other artificial intelligence research and markets, and the fact that the research and product development in actual depth study have to rely heavily on labeled data, it is of great research significance to successfully develop a safe image data platform.

2 Related Works

2.1 Blockchain Technology

Up to now, block chain technology has been used in electronic medical record cases in the medical field, and future applications in the medical field may also include image data, health insurance, biomedical research, drug supply and procurement processes and medical education [1]. Zhou et al. [2] propose a threshold based on block chain, this system has gained some special advantages, such as dispersion, tamper resistance and recording nodes to help users verify verifiable public information. Patel [3] uses block chain as a distributed data storage to establish a ledger of radiology research and patient-defined access rights, eliminates third-party access to protected health information, meets many standards of interoperable health systems, and is easily extended to areas other than medical imaging, but the complexity and security of the framework's privacy and security model are difficult to be guaranteed.

To overcome security problem, the solution proposed by the document [4] outlines the framework of the cloud platform, internal work and protocols for processing heterogeneous medical data. Dagher et al. [5] proposed a high-level decentralized block chain system named Ancile, while acknowledging that some nodes should have higher permissions. This study shows that it is not possible to completely hide all information and maintain an accessible and interoperable system, but by using smart contracts to separate information, Ancile still provides significant privacy protection and data Integrity. Li et al. [6] proposed DPS, the user can permanently save important data, The original data can be Verified. The literature [7] uses the MEDREC of the shared chain to record genomic sequencing data, allowing individuals to sell access to their entire genome. The solution proposed by the document [8] based on the block chain of geographical space uses a cryptographic spatial coordinate system to add an immutable spatial context so that these geographical spatial chains can record a specific time of an entry, also require verification by both parties.

The document [9] creates a sensor system that uses the licensing mechanism of block chains to collect intelligent contracts for evaluating patient information, which trigger alarms to apply to patients and medical service providers. The document [10] designs a block chain network called "MedBlock" for electronic medical records, which combines customized access control protocol and symmetric cipher technology to show high information security. The document [11] proposes a practical group optimization algorithm using bionic methods to improve the security of medical data management, and examines the characteristics of access control with a machine learning prediction method to enhance detection of unknown root.

As above of all, the security problem of block chain technology have been paid attention to by academic circles. There are some research methods of block chain in electronic medical record storage and application examples of distributed database. However, the access control protocol is mainly used in security of block chain for medical images, and useful image data information has not been used for security reporting.

2.2 Digital Watermarking

The remote transmission of digital medical images is an important part of telemedicine. During the network transmission, medical images may encounter some unexpected situations such as illegal tampering and illegal copying. In order to prevent and detect this situation in time, some researchers proposed to introduce digital watermarking technology into the remote transmission of digital medical images.

Generally speaking, the digital medical image watermarking algorithm is divided into three parts: watermark generation, watermark embedding and watermark extraction [12,13,18]. These algorithms are divided into two categories, the first category is mainly to process images in pixel domain and the second category is to process images in various transform domains. In the field of digital watermarking of medical images, quite a number of algorithms have been produced in recent years [15]. The digital watermarking algorithm in pixel domain usually uses LSB substitution and various chaotic sequences [14,16,17,19]. The digital watermarking algorithms in the transform domain generally use wavelet transform, discrete cosine transform and Fourier transform [20–23].

Medical images are different from ordinary natural images in that the former has a strong regional value, that is, only a part of the medical images are valuable, and the rest are often meaningless. Meaningful regions are called regions of interest (ROI) and meaningless regions are called regions of non-interest (RONI). In general, the watermark should be added completely away from the region of interest in order to keep information beneficial to diagnosis as much as possible. Guo et al. [24] proposes a medical image watermarking scheme that requires a diagnostic doctor to specify ROI region.

Duan et al. [25] proposed an energy conduction model ECM (Energy Conduction Model) for accurate segmentation of ROI and RONI of medical images, but the algorithm is inefficient. In addition, some medical image watermarking algorithms [26–28] which partition ROI and Roni have achieved good results, but these algorithms often lack protection for the most important parts of medical images and cannot restore the original medical images.Therefore, some researchers have proposed some digital watermarking schemes [29–32] that can completely restore the original medical image, but these methods often need to transmit some information about watermark extraction and original medical image restoration to the receiver on another completely secure channel.

All of the above watermarking algorithms will change the original medical image to a certain extent, so some researchers put forward the concept of 'zero watermarking', that is, to generate a watermark from the image without modifying the original medical image and upload it to the database, only a unique watermark can be generated from the same medical image, and copyright protection can be achieved by this method [33–35]. However, this approach requires the construction of a complex and completely fair authentication system [36].

3 Blockchain-Based System Design

In this paper, we realize the guidance of all historical data, helps the hospital to construct the image data and realize the full-position analysis and response of the image data, which can be used not only in the scientific research of tumor treatment, but also in the clinical response of tumor treatment, and can also guide the improvement and upgrading of tumor treatment drugs.

3.1 System Structure

The main functions include digital watermarking of image data, privacy protection of image data, platform authentication access, internet of things access and block chain sharing scheme (see Fig. 1).

(i) X-Ray image encryption module
Based on packet key management, the data transmission network with strong anti-interference ability, high transmission speed and self-adaptive matching of keys is controlled through the authenticator, key server and authentication server.

(ii) Data transmission module
Using the algorithm of digital signature, the abstract of the image data to be signed is generated by using hash function, and the ciphertext obtained by encrypting the abstract with public key is the digital signature.

(iii) Digital watermarking module
The watermark application function of different image data is realized, and the security of image data is guaranteed and will not be tampered with.

(iv) Block chain privacy protection module
Using block chain isolation verification technology can not only ensure the legitimate use of the data demander, but also protect the user's privacy to the maximum extent. For example, the patient's name, medical card number and other information are encrypted by hash algorithm, and the authenticity of the data is guaranteed by isolation verification.

(v) Block Chain Consensus Module
Using the open source Fabric smart contract, data traceability can confirm data ownership and circulation channels. If data is illegally re - disseminated by users, it can provide proof materials for infringement complaint stage and provide a more credible big data trading environment.

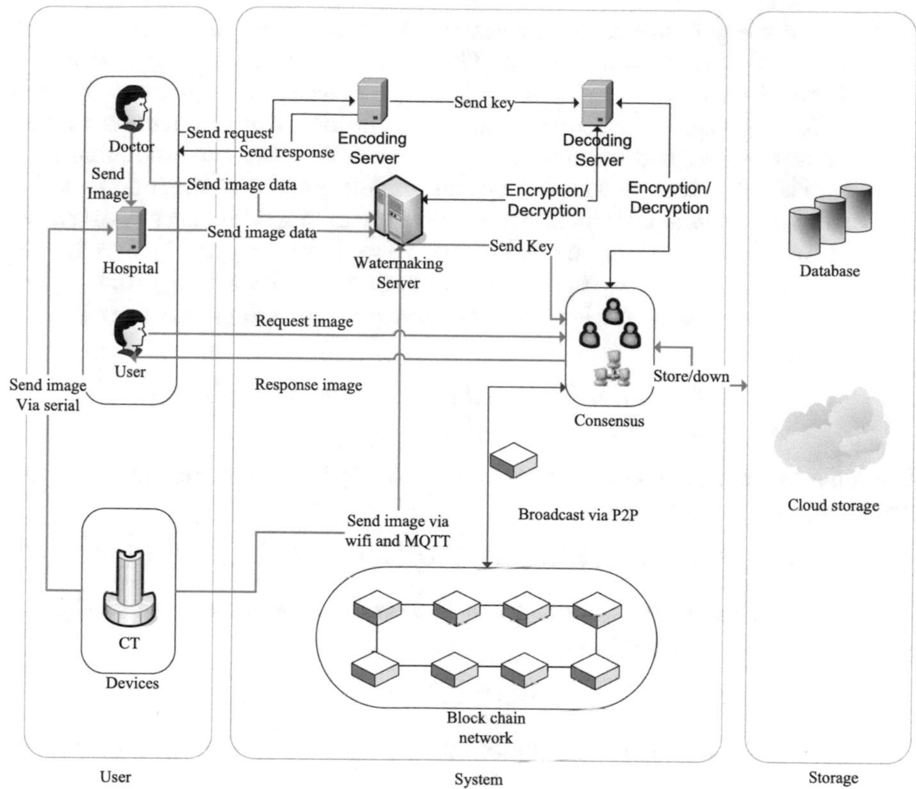

Fig. 1. System module.

3.2 X-Ray Image Encryption

For the sparse representation of an X-Ray image to be useful within the current trend of medical technology developments it should be suitable to be encapsulated in a small file. Accordingly, the coefficients of the atomic decompositions need to be converted into integer numbers. This operation is known as quantization. We adopt a simple and commonly used uniform quantization technique. For $q = 1, ..., Q$ the absolute value coefficients $c_q|(n)|, n = 1, ..., k_q$ are converted to integers as follows:

$$c_q^{\Delta}(n) = \begin{cases} \lceil \frac{|c_q(n)|-\theta}{\Delta} \rceil, \; if \; |c_q(n)| \geq \theta \\ 0 \; otherwise, \end{cases} \tag{1}$$

where $\lceil x \rceil$ indicates the smallest integer number greater than or equal to x, Δ is the quantization parameter, and θ the threshold to disregard coefficients of small magnitude. The signs of the coefficient are encoded separately, as a vector s_q using a binary alphabet.

In order to store the information about the particular atoms presented in the approximation of each block, we proceed as follows: firstly each pair of indices

$(\ell_n^{x,q}, \ell_n^{y,q})$ corresponding to the atoms in the decompositions of the block I_q is mapped into a single index $o_q(n)$. Then the set $o_q(1), ..., o_q(k_q)$ is sorted in ascending order $o_q(n) \to \tilde{o}_q(n), n = 1, ..., k_q$. This guarantees that, for each q-value, $\tilde{o}_q(i) < \tilde{o}_q(i+1), i = 1, ..., k_q - 1$. The order of the indices induces an order in the unsigned coefficients, $c_q^{\Delta} \to \tilde{c}_q^{\Delta}$ and in the corresponding signs $s_q \to \tilde{s}_q$. The advantage introduced by the ascending order of the indices is that they can be stored as smaller positive numbers, by taking differences between two consecutive values. Certainly by defining $\delta_q(n) = \tilde{o}_q(n) - \tilde{o}_q(n-1), n = 2, ..., k_q$ the string $\tilde{o}_q(1), \delta_q(2), ..., \delta_q(k_q)$ stores the indices for the block q with unique recovery. The number 0 is then used to separate the strings corresponding to different blocks.

$$\begin{aligned} st_{ind} = [&\tilde{o}_1(1), \delta_1(2), ..., \delta_1(k_1), 0, \tilde{o}_2(1), \delta_2(2), ..., \delta_2(k_2), \\ &0, ..., \tilde{o}_{k_Q}(1), \delta_{k_Q}(2), ..., \delta_{k_Q}(k_Q)] \end{aligned} \tag{2}$$

The quantized magnitude of the re-ordered coefficients are concatenated in the strings st_{cf} as follows:

$$st_{cf} = [\tilde{c}_1^{\Delta}(1), ..., \tilde{c}_1^{\Delta}(k_1), \tilde{c}_2^{\Delta}(1), ..., \tilde{c}_2^{\Delta}(k_2), ..., \tilde{c}_{k_Q}^{\Delta}(1), ..., \tilde{c}_{k_Q}^{\Delta}(k_Q)] \tag{3}$$

Using 0 if the sign is positive and 1 if it is negative, the signs of the coefficients are placed in the string, st_{sg} as

$$st_{ind} = [\tilde{s}_1(1), ..., \tilde{s}_1(k_1), \tilde{s}_2(1), ..., \tilde{s}_2(k_2), ..., \tilde{s}_{k_Q}(1), ..., \tilde{s}_{k_Q}(k_Q)] \tag{4}$$

The next encoding/decoding scheme summarizes the above described procedure.

Encoding. Given an image partition $I_q \in \mathbb{R}^{N_b \times N_b}, q = 1, ..., Q$, approximate each element of the partition by the atomic decomposition:

$$I_q^{k_q} = \sum_{n=1}^{k_q} c_q(n) d^x_{\ell_n^{x,q}} (d^y_{\ell_n^{y,q}})^T \tag{5}$$

The approximation is carried out on each block, independently of the others, until the stopping criterion is reached.

For each q quantize as in Eq. (1) the magnitude of the coefficients in the decomposition Eq. (5) to obtain $c_q^{\Delta}(n), n = 1, ..., k_q$. Store the signs of the non-zero coefficient as components of a vector s_q. For each q map the pair of indices $(\ell_n^{x,q}, \ell_n^{y,q}), n = 1, ..., k_q$ in Eq. (5) into a single index $o_q(n), n = 1, ..., k_q$ and sort these numbers in ascending order to have the re-ordered sets:

$\tilde{o}_q(1)..., \tilde{o}_q(k_q)$; $\tilde{c}_q^{\Delta}(1)..., \tilde{c}_q^{\Delta}(k_q)$ and $\tilde{s}_q(1)..., \tilde{s}_q(k_q)$ to create the strings: st_{ind}, as in Eq. (2), and st_{cf}, and st_{sg} as in Eqs. (3) and (4) respectively. Let's recall the content of the file encoded by the above steps:

st_{ind} contains the difference of indices corresponding to the atoms in the approximation of each of the blocks in the image partition.

st_{cf} contains the magnitude of the corresponding coefficients (quantized to integer numbers).

st_{sg} contains the signs of the coefficients in binary format. The quantization parameter Δ also needs to be stored in the file. We fix $\theta = 1.3\Delta$ or all the images.

Decoding. Recover the indices from their difference. This operation also gives the information about the number of coefficients in each block. Read the quantized unsigned coefficients from the string st_{cf} and transform them into real numbers as $|\tilde{c}_q^r(n)| = \Delta \tilde{c}_q^{\Delta}(n) + (\theta - \Delta/2)$. Read the corresponding signs from the string st_{sg}. Recover the approximated partition, for each block, through the liner combination

$$I_q^{r,k_q} = \sum_{n=1}^{k_q} \tilde{s}_q(n)|\tilde{c}_q^r(n)| d^x_{\tilde{\ell}_n^{x,q}} (d^y_{\tilde{\ell}_n^{y,q}})^T \tag{6}$$

Assemble the recovered image as

$$I^{r,K} = \hat{J}_{q=1}^{Q} I_q^{r,k_q}, \tag{7}$$

where the $\hat{J}$ indicates the operation for joining the blocks to restore the image.

3.3 Digital Watermarking for Multimodality Image

Due to the defect of the implementation principle of position encryption, although the watermark image after position encryption has already been scrambled in visual effect, it cannot change the histogram of the original image. It is easy to leak the statistical information of the watermark image. Therefore, in order to enhance security, this paper intends to encrypt the watermark image by combining position encryption with gray value encryption (two - factor encryption). This encryption method (see Fig. 2) can reduce the probability that the watermark will be decoded when the watermark image is extracted and ensure the security and confidentiality of the watermark. From the point of view of watermark encryption, the two-factor encryption mechanism generates two chaotic matrices when encrypting digital watermarks. The marked spatial coordinate information and volume size information are used for position encryption, and the information such as terminal number, geographical location and shooting time are used for gray-scale encryption. By combining the two, chaotic sequences are unpredictable. Therefore, even if the watermark sequences are known, they cannot be cracked without the chaotic encryption key matrix. In order to meet the requirements of watermark security, it has played a better role in protecting copyright.

3.4 Blockchain Implementation

The block chain network of the data sharing platform provides a variety of block link ports. Its transaction interface is first for message queue data transmission, and this submission is real-time submission. The core block chain module receives the authentication request and notifies the consensus node through TCP for verification. After verification is completed, the consensus node will send out TCP asynchronous messages, and the block chain module will receive the verification result message. If a consensus is reached, the transaction will first be written into the local MySQL account book, and then the block information of

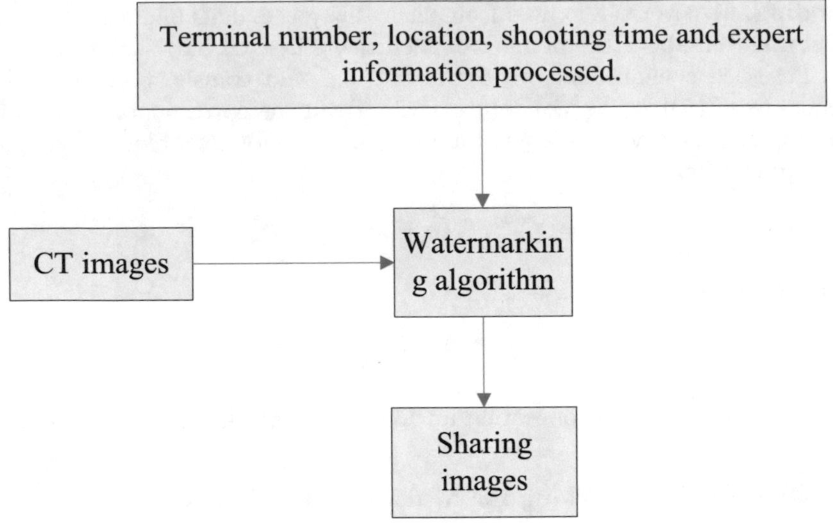

Fig. 2. Digital watermark.

the block chain will be broadcast to the block chain network through P2P. The result is a phased notification to each node. As shown in the Fig. 3, MySQL technology is used to store the public ledger information of the block chain locally.

4 Experiments and Evaluation

4.1 Experiments

We illustrate the effectiveness of the proposed sharing scheme by storing the outputs of high quality sparse approximation of two data sets: the Lung Nodule Analysis 2016 dataset (abbreviated as LUNA) and the training set of Data Science Bowl 2017 (abbreviated as DSB). The LUNA dataset includes 1186 nodule labels in 888 patients annotated by radiologists, while the DSB dataset only includes the per-subject binary labels indicating whether this subject was diagnosed with lung cancer in the year after the scanning. The DSB dataset includes 1397, 198, 506 persons (cases) in its training, validation, and test set respectively.

The quality of the approximation (the classical Peak Signal-to-Noise Ratio) is calculated as:

$$PSNR = 10\log_{10}(\frac{(2^8-1)^2}{MSE}) \tag{8}$$

MSE indicates as:

$$MSE = \frac{\|I - I^{r,K}\|_F^2}{N_x N_y} \tag{9}$$

This quality guarantees that the approximation is indistinguishable from the image, in the original size. For the comparison with standard formats all the PSNRs are fixed as values for which JPEG and JPEG2 produce the required

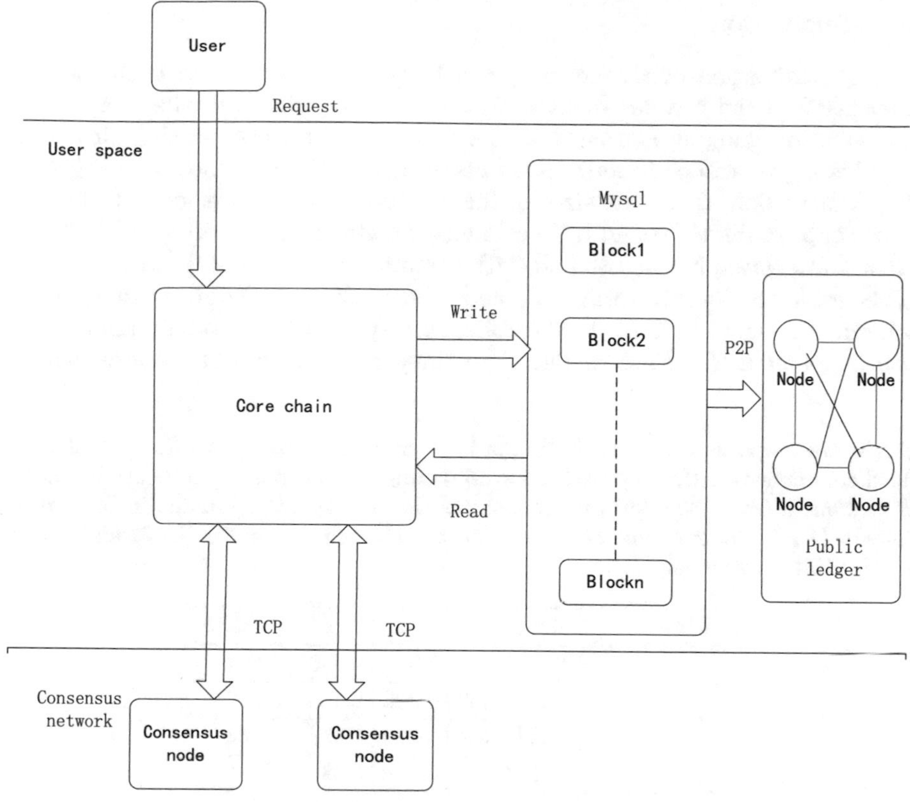

Fig. 3. Internal structure of block chain.

MSSIM. For producing a requested PSNR with the sparse representation approach we proceed as follows: the approximation routine is set to yield a slightly larger value of PSNR and the required one is then obtained by tuning the quantization parameter Δ.

As a measure of sparsity we use the Sparsity Ratio, which is defined as:

$$SR = \frac{Number\ of\ pixels\ in\ the\ image}{Number\ of\ coefficients\ in\ the\ representation} \tag{10}$$

Accordingly, the sparsity of a representation is manifested as a high value of SR.

In addition to the SR, which is a global measure of sparsity, a meaningful description of the variation of the image content throughout the partition is rendered by the local sparsity ratio, which is given as

$$sr(q) = \frac{N_b^2}{k_q}, q = 1, ..., Q, \tag{11}$$

where k_q is the number of coefficients in the decomposition of the q-block and N_b^2 is the number of pixels in the block.

4.2 Evaluation

The approximation of all the images in DSB are performed in both the pixel intensity and the wavelet domain. The size of the blocks in the image partition is fixed taking into account previously reported results, which indicate that 16–16 is a good trade-off between the resulting sparsity and the processing time. The information about the sizes of the corresponding files is given in bits per pixel (bpp) in the third and forth columns of Table 1. The results for JPEG and JPEG2 are placed in the fifth and sixth columns, respectively. The last two rows of the table are the mean value of standard deviation (std) of the corresponding columns. As shown in Table 1, all the files corresponding to approximations in the wavelet domain (S_{wd}) are smaller than those corresponding to approximations

Table 1. Comparison of size rate (in bpp) for the DSB, listed in the first column. The third column shows the bpp values corresponding to the sparse representation in the pixel domain (S_{pd}). The forth column shows the corresponding results in the wavelet domain (S_{wd}). The fifth and sixth columns are the bpp values for the formats JPEG and JPEG2, respectively.

Image	dB	S_{pd}	S_{wd}	JPEG	JPEG2
1	48.1	0.443	0.286	0.436	0.234
2	48.6	0.462	0.306	0.449	0.247
3	47.4	0.441	0.320	0.419	0.244
4	48.0	0.488	0.316	0.485	0.286
5	48.1	0.624	0.391	0.566	0.335
6	47.1	0.676	0.416	0.575	0.334
7	48.8	0.612	0.424	0.586	0.371
8	46.4	0.599	0.452	0.546	0.346
9	49.1	0.612	0.419	0.573	0.364
10	45.8	0.697	0.453	0.594	0.358
11	44.3	0.519	0.465	0.605	0.393
12	44.3	0.691	0.629	0.832	0.521
13	44.1	0.827	0.686	0.874	0.629
14	43.4	0.816	0.693	0.924	0.619
15	48.9	1.000	0.867	1.152	0.759
16	49.2	1.384	1.240	1.584	1.056
17	44.3	1.596	1.418	1.828	1.248
18	44.4	1.606	1.435	1.827	1.310
19	47.0	2.131	1.922	2.463	1.630
20	47.4	2.395	2.298	2.764	1.902
Mean value	46.7	0.938	0.727	1.004	0.659
S_{td}	1.9	0.573	0.521	0.707	0.500

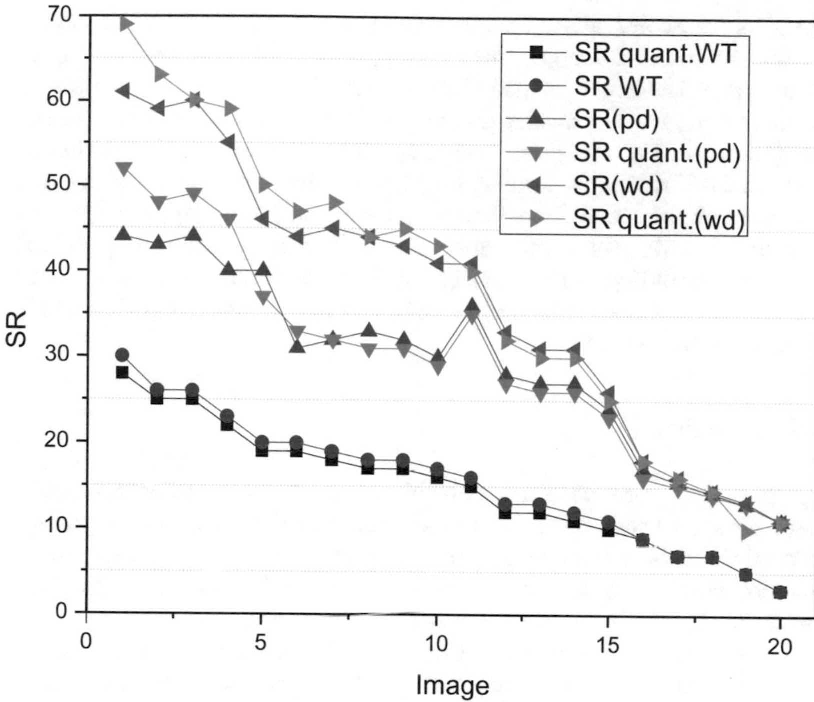

Fig. 4. Comparison of the SRs, before and after quantization, corresponding to the dictionary approaching both the pixel intensity and the wavelet domain and to the wavelet approximation by nonlinear thresholding.

in the pixel intensity domain (S_{pd}), and also smaller than the JPEG ones. On average the files with the sparse representation in the wavelet domain are 22% smaller than those with the representation in the pixel domain, and 27% smaller than the JPEG files. In the present form JPEG2 produces the smallest files (on average 9% smaller than the files with the representation in the wavelet domain). However, both JPEG and JPEG2 formats involve an entropy coding step, which is not included in our scheme. Instead, the outputs of our algorithm are stored in HDF5 format.

A very interesting feature of the numerical results is that the quantization process, intrinsic to the economic store of the coefficients in the image approximation does not reduce the sparsity. For the sake of comparison in addition to calculating the SR_s obtained with the dictionary approach in both domains, before and after quantization, we have also calculated the corresponding SR_s produced by nonlinear thresholding of the wavelet coefficients. The results are shown in Fig. 4. As already discussed, since the quantization of coefficients degrades quality, to achieve the required PSNR the approximation of the image has to be carry out up to a higher PSNR value. Nevertheless, because the quantization process maps some coefficients to zero, for the corpus of 20 images in this study, quantization does not affect sparsity. On the contrary, as can be observed in Fig. 4, for the sparsest images (first 5 images in Table 1) sparsity actually benefits from quantization. It is also clear that for the images in the upper part of the table the SR in the wavelet domain is significantly larger than in the pixel intensity domain. However, the level of sparsity achieved by the dictionary approach is, in both domains, significantly higher than that achieved by nonlinear thresholding of the wavelet coefficients. If the dictionary approach operates in the wavelet domain, then after quantization the mean value gain in SR with respect to thresholding of the wavelet coefficients is 163%, with standard deviation of 17%. In the pixel intensity domain the corresponding gain is 113%, with standard deviation of 30%.

5 Conclusion

In this paper, we proposed a blockchain-based mechanism to mediate secure problem between users and a pool of shared (sensitive) data. Compared to the traditional cloud-based image data management network, we constructed a scalable (redesigned to allow speedy transactions) and lightweight blockchain to demonstrate the efficiency of our design which permits data sharing in a secure manner and protects the privacy of data. In the proposed system, communication, authentication protocols and consensus algorithms between entities were not fully investigated. It would be interesting to extend this work by fully exploring these in future studies. We state that the architecture described in this paper is a top layer of the blockchain-based sharing control system that is under implementation and testing. In our future work, an experimental study will be conducted to improve the system's efficiency and to obtain empirical data for further studies.

References

1. Radanović, R., I., Likić, R.: MIStore: opportunities for use of blockchain technology in medicine. Appl. Health Econ. Health Policy, 1–8 (2018)
2. Zhou, L., Wang, L., Sun, Y.: MIStore: a blockchain-based medical insurance storage system. J. Med. Syst. **42**(8), 149 (2018)
3. Patel, V.: A framework for secure and decentralized sharing of medical imaging data via blockchain consensus. Health Inform. J. (2018)
4. Kaur, H., Alam, M.A., Jameel, R., Mourya, A.K., Chang, V.: A proposed solution and future direction for blockchain-based heterogeneous medicare data in cloud environment. J. Med. Syst. **42**(8), 156 (2018)
5. Dagher, G.G., Mohler, J., Milojkovic, M., Marella, P.B.: Ancile: privacy-preserving framework for access control and interoperability of electronic health records using blockchain technology. Sustain. Cities Soc. **39**, 283–297 (2018)
6. Li, H., Zhu, L., Shen, M., Gao, F., Tao, X., Liu, S.: Blockchain-based data preservation system for medical data. J. Med. Syst. **42**(8), 141 (2018)
7. Gammon, K.: Experimenting with blockchain: can one technology boost both data integrity and patients' pocketbooks? Nat. Med. **24**(4), 378–381 (2018)

8. Kamel, M.B., Wilson, J.T., Clauson, K.A.: Geospatial blockchain: promises, challenges, and scenarios in health and healthcare. Int. J. Health Geogr. **17**(1), 25 (2018)
9. Griggs, K., Ossipova, O., Kohlios, C.: Healthcare blockchain system using smart contracts for secure automated remote patient monitoring. J. Med. Syst. **42**(7), 130 (2018)
10. Fan, K., Wang, S., Ren, Y., Li, H., Yang, Y.: Medblock: efficient and secure medical data sharing via blockchain. J. Med. Syst. **42**(8), 136 (2018)
11. Firdaus, A., Anuar, N.B., Razak, M.F.A., Hashem, I.A.T., Bachok, S., Sangaiah, A.K.: Root exploit detection and features optimization: mobile device and blockchain based medical data management. J. Med. Syst. **42**(6), 112 (2018)
12. Hartung, F., Kutter, M.: Multimedia watermarking techniques. Proc. IEEE **87**(7), 1079–1107 (1999)
13. Nyeem, H., Boles, W., Boyd, C.: A review of medical image watermarking requirements for teleradiology. J. Digit. Imaging **26**(2), 326–343 (2013)
14. Ping, N.L., Ee, K.B., Wei, G.C.: A study of digital watermarking on medical image. In: World Congress on Medical Physics and Biomedical Engineering (2007)
15. Navas, K.A., Sasikumar, M.: Survey of medical image watermarking algorithms. Clin. Immunol. Immunopathol. **73**(3), 338–343 (2007)
16. Boucherkha, S., Benmohamed, M.: A lossless watermarking based authentication system for medical images. In: Proceedings of International Conference on Computational Intelligence, ICCI 2004, Istanbul, Turkey, pp. 240–243 (2008)
17. Memon, N.: Watermarking of medical images for content authentication and copyright protection. Ph.D. thesis, Pakistan: Faculty of Computer Science and Engineering, GIK Institute of Engineering Sciences and Technology (2010)
18. Bouslimi, D., Coatrieux, G., Cozic, M.: A joint encryption/watermarking system for verifying the reliability of medical images. IEEE Trans. Inf. Technol. Biomed. **16**(5), 891–899 (2012)
19. Abd-Eldayem, M.M.: A proposed security technique based on watermarking and encryption for digital imaging and communications in medicine. Egypt. Inform. J. **14**(1), 1–13 (2013)
20. Memon, N.A., Gilani, S.A.M.: Adaptive data hiding scheme for medical images using integer wavelet transform. In: International Conference on Emerging Technologies, pp. 221–224. IEEE (2009)
21. Ahmed, F., Moskowitz, I.S.: A semi-reversible watermark for medical image authentication. In: Transdisciplinary Conference on Distributed Diagnosis and Home Healthcare, pp. 59–62. IEEE (2006)
22. Mehto, A., Mehra, N.: Adaptive Lossless Medical Image Watermarking Algorithm Based on DCT & DWT. Elsevier Science Publishers B. V., Amsterdam (2016)
23. Hanki, R., Borra, S., Dwivedi, V.: An efficient medical image watermarking scheme based on FDCuT–DCT. Eng. Sci. Technol. Int. J. **20**, 1366–1379 (2017)
24. Guo, X., Zhuang, T.G.: A region-based lossless watermarking scheme for enhancing security of medical data. J. Digit. Imaging **22**(1), 53–64 (2009)
25. Duan, C.J., Jingfeng, M.A., Zhang, Y.B.: Energy conduction model and its application in medical image segmentation: energy conduction model and its application in medical image segmentation. J. Softw. **20**(5), 1106–1115 (2009)
26. Rahimi, F., Rabbani, H.: A dual adaptive watermarking scheme in contourlet domain for DICOM images. Biomed. Eng. Online **10**(1), 53 (2011)
27. Al-Qershi, O.M., Khoo, B.E.: A dual adaptive watermarking scheme in contourlet domain for DICOM images. J. Digit. Imaging **24**(1), 114–125 (2011)

28. Tan, C.K., Ng, J.C., Xu, X., Poh, C.L., Guan, Y.L., Sheah, K.: Security protection of dicom medical images using dual-layer reversible watermarking with tamper detection capability. J. Digit. Imaging **24**(3), 528–540 (2011)
29. Pan, W., Coatrieux, G., Montagner, J., Cuppens, N.: Comparison of some reversible watermarking methods in application to medical images. In: International Conference of the IEEE Engineering in Medicine & Biology Society, p. 2172 (2009)
30. Tsai, H.H., Tseng, H.C., Lai, Y.S.: Robust lossless image watermarking based on α-trimmed mean algorithm and support vector machine. J. Syst. Softw. **83**(6), 1015–1028 (2011)
31. Rahmani, H., Mortezaei, R., Moghaddam, M.E.: A new lossless watermarking scheme based on DCT coefficients. In: International Conference on Digital Content, Multimedia Technology and ITS Applications, pp. 28–33. IEEE (2010)
32. Kallel, I.F., Bouhlel, M.S., Lapayre, J.C.: Improved Tian's method for medical image reversible watermarking. GVIP J. **7**, 1–5 (2007)
33. Fu, R.D., Jin, W.: A wavelet-based method of zero-watermark utilizing visual cryptography. In: International Conference on Multimedia Technology, pp. 1–4. IEEE (2010)
34. Wei, J., Jinxiang, L.I., Yin, C.Q.: An image zero-watermarking scheme based on visual cryptography utilizing contour-wavelet. J. Optoelectron. Laser **20**(5), 653–656 (2009)
35. Qu, C., Yang, X., Yuan, D.: Zero-watermarking visual cryptography algorithm in the wavelet domain. J. Image Graph. **19**, 365–372 (2014)
36. Aleš, R., Karel, S., Otto, D., Michal, J.: A new approach to fully-reversible watermarking in medical imaging with breakthrough visibility parameters. Biomed. Signal Process. Control. **29**, 44–52 (2016)

Blockchain Based Data Trust Sharing Mechanism in the Supply Chain

Luya Wang(✉) and Shaoyong Guo

Beijing University of Posts and Telecommunications, Beijing 100876, China
wangluyadei@163.com

Abstract. Currently, the trust and privacy problem is an obstacle during data sharing process in supply chain. To solve it, this paper proposes a blockchain based data trust sharing mechanism in the supply chain. Firstly, we design the ar-chitecture to introduce the system framework, service process and data model for data trust sharing. Secondly, we implement the blockchain-based supply chain platform, consisting of account management module and data request processing module with open data index name extension (ODINE). At last, we state an use case to analyze this platform.

Keywords: Blockchain · Supply chain · Data storage and access

1 Introduction

Supply chains are networks with big data attributes that connect providers, manufacturers, operators, retailers and consumers. In the supply chain, information cooperation sharing plays an important role in improving the resource utilization of upstream and downstream enterprises. However, at present, the problem of low efficiency in information access and value mining and privacy leakage limit the sustainable development of the supply chain system [1]. Blockchain is a distributed computing paradigm with features of centralization, trust, collective maintenance, reliable storage, and automatic operation. From this perspective, blockchain is a natural remedy for supply chain system, which can achieve data transparency, form a good transaction order and commercial ecology under the condition of trust, and provide high-level data security protection.

Since Nakamoto took the lead in proposing a set of Internet governance mechanisms [2] based on cryptography and trust in "Bitcoin: A Peer-to-Peer Electronic Cash System", many scholars began to study blockchain technology in the supply chain applications in the field. The rapid development of B2B (Business-to-Business) trading theory and integration platform establishes an efficient and secure supply chain blockchain transaction prototype [3,6], but the current low level of system interoperability lead to high investment costs, which limit the realization of potential benefits [7,8]. Skinner et al. [9] proposed to improve the digital activities in the internet supply chain finance to achieve commercial operation. Korpela [10] and others developed a digital supply chain (DSC)

C.-N. Yang et al. (Eds.): SICBS 2018, AISC 895, pp. 43–53, 2020.
https://doi.org/10.1007/978-3-030-16946-6_4

conceptual model, but it has limited scalability. Abeyratne et al. [11] discuss a blockchain-based manufacturing supply chain system to improve data transparency, but this technology requires some IT infrastructure for all participants. For data governance issue, Azaria [12], Liu [13] and others based on blockchain technology to give a model for recording and accessing medical big data, but two further exploration is needed to maintain the auditability of the chain. For the problem of data security storage and sharing mechanism [14], Zyskind et al. [15] proposed to store information and parse it by using bitcoin, but it is very expensive. Kosba et al. [16] introduced a third party database, and Zhang et al. [17] introduced an audit center for system data storage and extended access, but the two models for special cases lacked other models. Aizhan et al. [18] used multi-signature technology to protect information security in distributed energy supply chain systems through, but it did not involve data processing integration and industry analysis.

For those problems, we combined encryption technology and named query technology to data sharing processing in supply chain for designing our architecture. And then we implement the blockchain-based supply chain platform, consisting of account management module and data request processing module. In data request processing module, we construct a new open data index name extension for the data index. At last, we state an use case to analyze this mechanism.

2 Data Trust Sharing Architecture

2.1 Blockchain Based Transaction Processing Model

The architecture consists of three main parts: Data collection, Data procession and Data storage and access. In the first part, we use intelligent contracts to support data sharing processing between the upstream and downstream of the supply chain to improve the level of data trust. In the second part, we analyze and extract the value of the integrated data. Finally, we use the ODINE name index standard to access data safely and effectively (Fig. 1).

In the supply chain system, data is mainly derived from peripheral device acquisition, transaction processing, and personal information input. The process is roughly as follows:

Step 1: Summarizing data and extracting the transactions and information recorded in the blockchain. Using intelligent contracts for loading, noise reduction and for-mat conversion.
Integrating various industry data and tapping the value.
Classifying the identification objects according to the ODINE naming rules.
Storing the acquired industry information sets and user information sets through the blockchain API interface according to the naming rules.
After verifying the identity and permissions, the visitor can send a data request and find the data location through the ODINE parsing library.

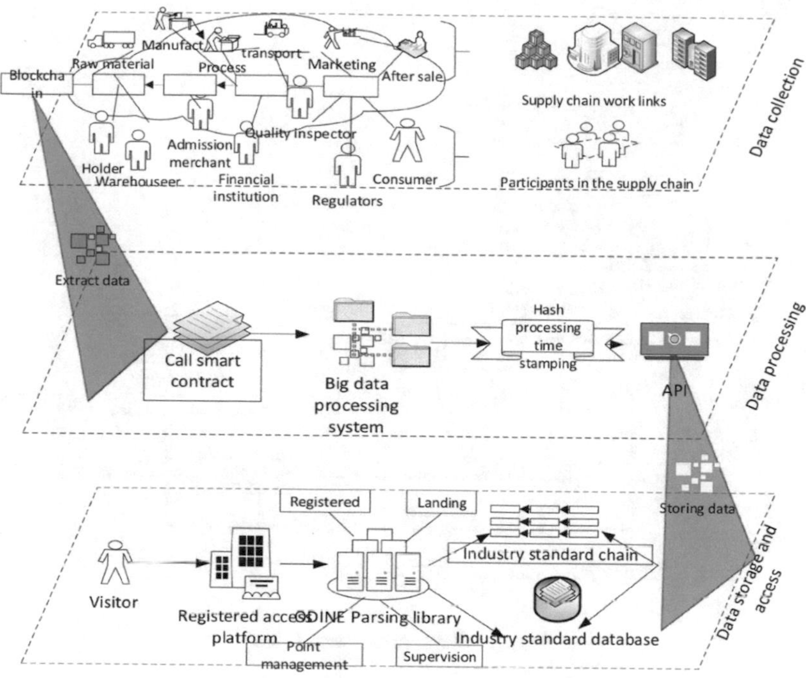

Fig. 1. Overview of the process flow

Among them, the intelligent contract can be automatically executed when the access condition is satisfied. The timestamp of the tag verifies the authenticity of the original data.

2.2 Blockchain Based Data Processing

After processing, the data can generate industry value. In the data processing part, the flow of the data is (Fig. 2):

Step 1: The transaction information and industry data are stored in the blockchain through the P2P transaction module.
Step 2: The node calls the intelligent contract to clean the data for inspection, fault diagnosis, and information specification.
Step 3: The data is classified, integrated, and reorganized by the big data processing module, and named according to certain rules (In Sect. 4.3).
Step 4: The data is restored to the blockchain by calling the API interface of the blockchain system.
When a visitor accesses data information he can obtain data by a certain authority audit (In Sect. 4.1) and access rules (In Sect. 4.2).

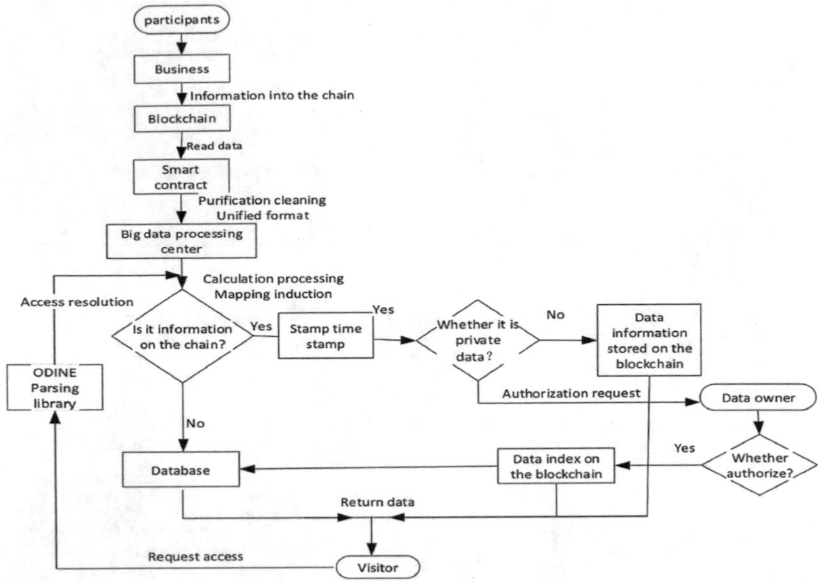

Fig. 2. Data processing flow

2.3 Inter-chain Data Correlation Model of Supply Chain

Integrating and mining the potential value of data is one of the keys to this paper, so data correlation analysis is required.

The value of the data is gradually tapped and aggregated. When business partners conduct business activities directly or indirectly between supply chains, they generate a large amount of industry data that can be used as a data source for big data processing after purification. By arranging messy data into industry data sets or personal information sets, we can further analyze the intrinsic connection of abstract data for market analysis. In order to achieve secure storage and privacy protection of the integrated data, the data is stored in different modes. The key data is stored in the block-chain, the common data is put into the database, and the private data is set access rights (Table 1).

After registering, the users can drive the value realization. They can implement business transactions, access information and decide how to respond to other people's access requests.

Table 1. Meaning of model parameters

Function	Definition	Function	Definition
TxHash .Hash	The root hash of the transaction tree	Extra []byte	Block additional data
ReceiptHash .Hash	The root hash of the receipt tree	MixDigest .Hash	Hash value, combined with Nonce for work-load calculation
Difficulty *Big.Int	The difficulty of this block	Nonce BlockNonce	Random value when the block is generated
Number *Big.Int	Block number of this block	Datai.Int	Transaction data corresponding to each initial hash
Time *big.Int	Timestamp	Datacategory.Int	Data category
[SCHEME]	Data ODINE identification string	Databelong.Int	Data attribution account
Account-need*Big.Int	Account demand information	Datatype.Int	Data transaction
Accounttrade*Big.Int	Account transaction information	Datachange.Int	Data rate of change
Authority.Int	Access permission	MerkleRoot .Hash	Block root hash value

3 Blockchain-Based Supply Chain Platform Implementation

3.1 Account Management Module

To ensure secure and efficient access to information, an account management module has been established. When an account is being registered, some corresponding in-formation must be submitted. When a user logs in, the client can authenticate. The registration process includes:

Step 1: The applicant initiates a registration request and sends his personal ID.
Step 2: If the initialization ID is legally registered, the calculated parameters will be issued and sent to the application account.
Step 3: Registrant sends his personal information and public key.
The platform generates the user's private key according to the user's public key and parameters.
Step 4: The user inputs personal information encrypted by the private key into the registration platform, and the platform executes the corresponding access authority setting flag according to the identity information.

When data is requested, the visitor sends the information encrypted by private key, and the authentication platform matches the information to identify the visitor's per-missions. Finally, the access platform sends the user request and user account address to the ODINE parsing library for data request processing (Figs. 3 and 4).

The registration process is as follows (Tables 2 and 3):

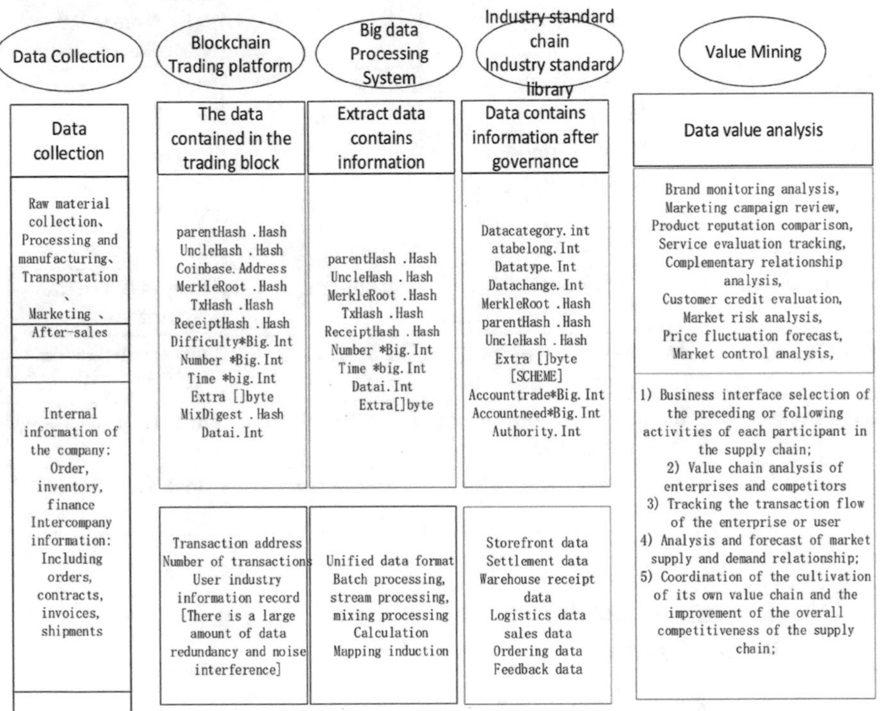

Fig. 3. Inter-chain data correlation model

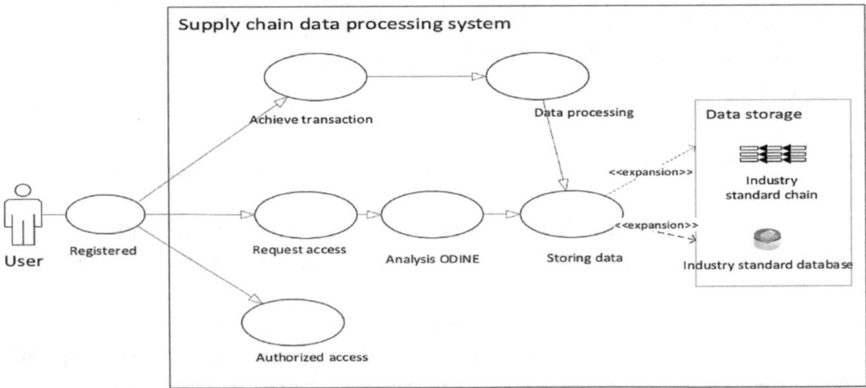

Fig. 4. Value extraction

Table 2. Participants, Run functions, and Parameter definitions

Participating object	Abbreviation	Participating object	Abbreviation
User (participants)	U	Registration certification platform	(R)
Database	D	Certification platform	(C)
Named resolution library operating	ODINE meaning	Integration platform	(I)
U wants to access its visitor record in D	Data1	Modify information in the industry standard library chain	Data3
Access information in the industry standard library chain	Data2	Delete information in the industry standard library chain	Data4
Have access	Token1	Have the right to delete	Token3
Have the right to draft	Token2		

Table 3. Function definition

Function	Definition	Function	Definition
Permission type tag	Authority( )	Blockchain address (identity)	Address
Generating a public key	Generate( )	Non-blockchain address (identity)	ID
Generate access parameters	Calculate( )	Direct insertion transaction	Input( )
Match verification	Match( )	Create index	Create( )
Query by index	Find( )	Identity legality verification	Legal( )
Insert key-value pairs	Insert( )	Pre-inserted key value	Preinsert( )

3.2 Data Request Processing Module

When making data access, certain naming rules are required to improve efficiency. In this paper, a new big data information access naming rule (ODINE Open Data Index Naming Extension) is applied to the supply chain system by combining the URI (Uniform Resource Identifier) specification and the ODIN [19] naming convention. It is an index name and identification of the data resource, which will become part of the data resource. It is independent of the actual address and makes the data unique and traceable. The specific naming rules are:

[1] Data in the block:
[WHETHER_VIP][IF_PRI][BLOCK_SN].[TRANS_INDEX][AUTHOR ITY]RESOURCE_ID#[DATA_SN].[CHUNK_INDEX]

[2] Access restricted data in the block:
[WHETHER_VIP][IF_PRI][BLOCK_SN].[TRANS_INDEX][AUTHOR ITY]RESOURCE_ID.OWNER_ID#[DATA_SN].[CHUNK_INDEX]
[3] Subject information in the database:
[WHETHER_VIP][IF_PRI] [AUTHORITY]RESOURCE_ID[FORM]#[KEY]

[WHETHER_VIP] is used to distinguish whether the data is important industry data stored in the block; [I_PRI] is used to identify whether the data needs special authentication by the ODINE owner; SN is the serial number [AUTHORITY] is access rights identification; RESOURCE_ID is the resource identifier; OWNER_ID is the point resolution of the ODINS owner; [DATA_SN] is in the industry standard chain The block number; [CHUNK_INDEX] is the index of the subdata in the block; [FORM] is the database header; [KEY] is the primary key value of the data accessed in the data base. The ODINE, metadata, and URL information of the open data re-source can be stored in the ODINE parsing library or the ODINE owner's database in the form of Json encoding (Fig. 5).

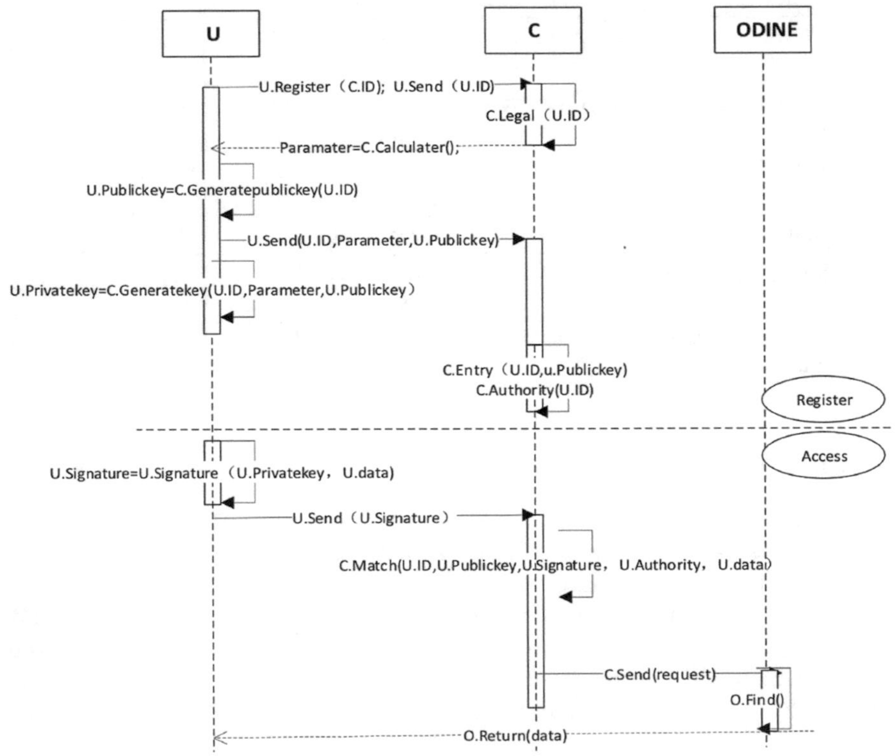

Fig. 5. User registration and access sequence diagram

When a visitor sends a data access request, the ODINE parsing library resolves the location and finds the data location. The corresponding processing of the above three cases is as follows:

[1] The analysis result of the data maps the blockchain to acquire data.
[2] The parsing library sends the access request to the ODINE owner for information confirmation, and accesses the industry data after obtaining the permission.
[3] Acquire the requested data according to the URL of the metadata obtained by the ODINE string and the actual data address (Fig. 6).

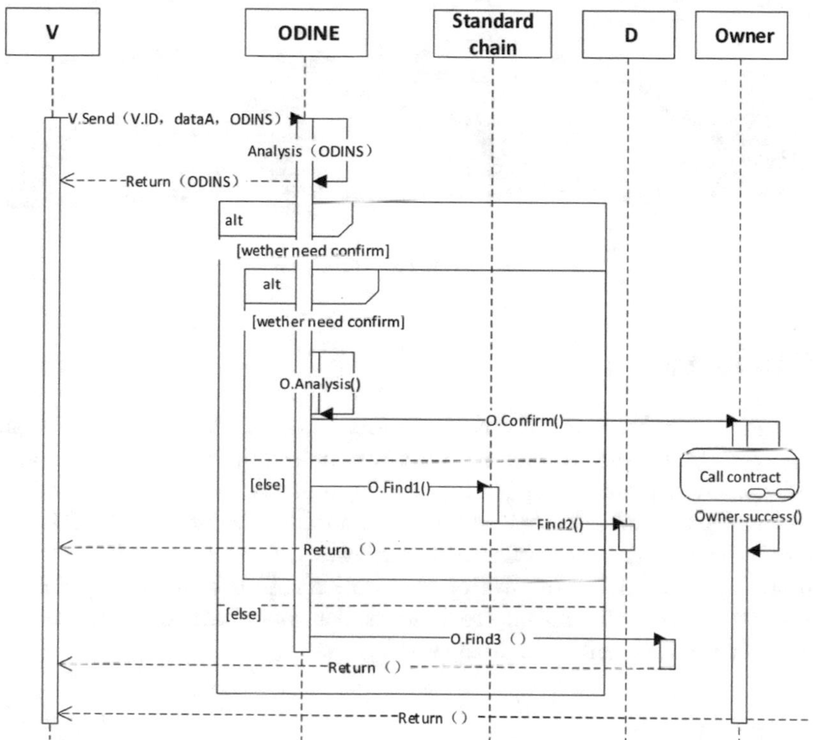

Fig. 6. Data request processing flow

This access rule not only enables efficient storage and access to supply chain integration information, but also protects data security and respects personal privacy data.

3.3 Use Case Analysis

We take the energy-saving and environmental protection enterprise as an example to realize the application analysis of this architecture. Based on this mechanism, enterprises can register and access the API interface of the system to obtain effective industry reference for product design, raw material procurement, sales and other aspects. Sewage treatment and recycling can be recorded in the blockchain for national supervision. When the core data of the supply chain is obtained, the competitiveness of the enterprises will be greatly improved.

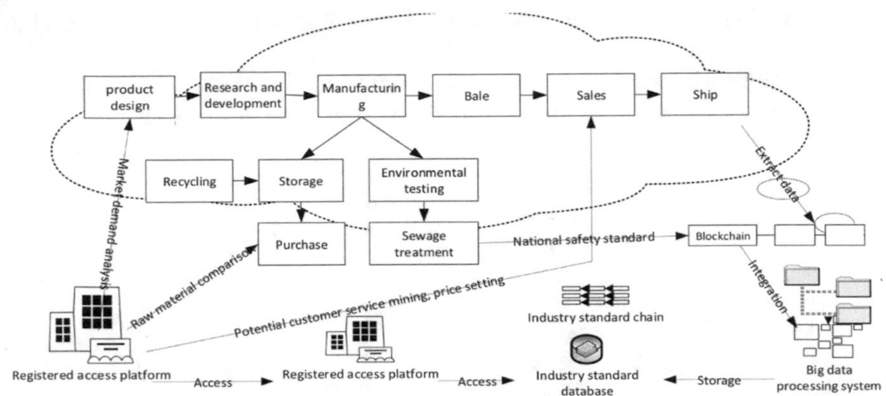

Fig. 7. Use case analysis

4 Summary

Aiming at the problem of information processing and trust sharing in supply chain system, this paper proposes a data processing model based on blockchain technology. We explain the operating principle of the mechanism by giving the overall framework and interpreting the main model. We design the ODINE naming rules for accessing to data effectively and safely. Some algorithms in the mechanism still require further innovation and optimization to increase efficiency. In the next study, we will continue to research the potential relationship analysis of data between different industries (Fig. 7).

References

1. Beth, S., Burt, D.N., Copacino, W., et al.: Supply chain challenges. Building relationships. Harv. Bus. Rev. **81**(7), 64–73, 117 (2003)
2. Nakamoto, S.: Bitcoin: A Peer-to-Peer Electronic Cash System [EB/OL]. https://bitcoin.org/bitcoin.pdf
3. Hazen, B.T., Byrd, T.A.: Toward creating competitive advantage with logistics information technology. Int. J. Phys. Distrib. Logist. Manag. **42**(1), 8–35 (2012)
4. Jeffers, P.I., Muhanna, W.A., Nault, B.R.: Information technology and process performance: an empirical investigation of the interaction between IT and non-IT resources. Decis. Sci. **39**(4), 703–735 (2008)
5. Ordanini, A., Rubera, G.: Strategic capabilities and internet resources in procurement: a resource-based view of B-to-B buying. Int. J. Oper. Prod. Manag. **28**(1), 27–52 (2008)

6. Croom, S.R.: The impact of e-business on supply chain management: an empirical study of key developments. Int. J. Oper. Prod. Manag. **25**, 55–73 (2005)
7. Evangelista, P., Kilpala, H.: The perception on ICT use among small logistics service providers: a comparison between Northern and Southern Europe (2007)
8. Murphy, P.R., Daley, J.M.: EDI benefits and barriers: comparing international freight forwarders and their customers. Int. J. Phys. Distrib. Logist. Manag. **29**(3), 207–217 (1999)
9. Skinner, C.: Blockchain is Fintech's real game-changer. Americanbanker **55**, 1 (2016)
10. Korpela, K., Hallikas, J., Dahlberg, T.: Digital supply chain transformation toward blockchain integration. In: Proceedings of the 50th Hawaii International Conference on System Sciences (2017)
11. Abeyratne, S.A., Monfared, R.P.: Blockchain ready manufacturing supply chain using distributed ledger (2016)
12. Azaria, A., Ekblaw, A., Vieira, T., et al.: MedRec: using blockchain for medical data access and permission management. In: International Conference on Open and Big Data (OBD), pp. 25–30. IEEE (2016)
13. Liu, P.T.S.: Medical record system using blockchain, big data and tokenization. In: International Conference on Information and Communications Security, pp. 254–261. Springer, Cham (2016)
14. Rouibah, K., Ould-Ali, S.: Dynamic data sharing and security in a collaborative product definition management system. Robot. Comput. Integr. Manuf. **23**(2), 217–233 (2007)
15. Zyskind, G., Nayhan, O., Pentland, A.: Decentralizing privacy: using blockchain to protect personal data. In: IEEE Security and Privacy Workshops, pp. 180–184. IEEE (2015)
16. Kosba, A., Miller, A., Shi, E., et al.: Hawk: the blockchain model of cryptography and privacy-preserving smart contracts. In: Security and Privacy, pp. 839–858. IEEE (2016)
17. Zhang, N., Zhong, S.: Personal privacy protection mechanism based on block-chain. J. Comput. Appl. **37**(10), 2787–2793 (2017)
18. Aitzhan, N.Z., Svetinovic, D.: Security and privacy in decentralized energy trading through multi-signatures, blockchain and anonymous messaging streams. IEEE Trans. Dependable Secur. Comput. **15**(5), 840–852 (2018)
19. Wang, J., Gao, L., Dong, A., et al.: Research on data security sharing network system based on blockchain. J. Comput. Res. Dev. **54**(4), 742–749 (2017). https://doi.org/10.7544/issn1000-1239.2017.20160991

Designing Smart-Contract Based Auctions

Chiara Braghin[1(✉)], Stelvio Cimato[1], Ernesto Damiani[1,2], and Michael Baronchelli[1]

[1] Dipartimento di Informatica, Università degli Studi di Milano, Milano, Italy
{chiara.braghin,stelvio.cimato,ernesto.damiani}@unimi.it, michael.baronchelli@studenti.unimi.it

[2] Centre on Cyber-Physical Systems, Khalifa University, Abu Dhabi, UAE
ernesto.damiani@kustar.ac.ae

Abstract. In this paper, we developed an online auction system based on Ethereum smart contracts. A smart contract is executable code that runs on top of the blockchain to facilitate, execute and enforce an agreement between untrusted parties without the involvement of a trusted third party. A decentralised auction guarantees greater transparency and avoids cheating auctioneers. Since in Ethereum computation is expensive as transactions are executed and verified by all the nodes on Ethereum network, we analysed our implementation in terms of cost and time efficiency, obtaining promising results.

Keywords: Auction · Blockchain · Smart contract

1 Introduction

Auctions are platforms for selling goods in a public forum through open and competitive bidding. Commonly, the auction winner is the bidder who submitted the highest price, however, there are a variety of other rules to determine the winner. The auctioneer leads the auction according to the rules and always charge a fee to the vendor for his services, usually a percentage of the gross selling price of the good.

In electronic auctions (*e-auctions*), the interaction among auctioneers, suppliers and buyers takes place by means of a centralised intermediary providing a platform for posting products, checking bids validity, and committing the winner. Such a third party, that could be impersonated by the auctioneer himself, may ask a commission as well, and has to be trusted.

In this paper, we design and implement the most common types of auctions by means of Ethereum smart contracts. Ethereum is the second most popular blockchain, with a built-in Turing complete programming language that allows running smart contracts in a global virtual machine environment known as Ethereum Virtual Machine (EVM), without depending on any third-party.

C.-N. Yang et al. (Eds.): SICBS 2018, AISC 895, pp. 54–64, 2020.
https://doi.org/10.1007/978-3-030-16946-6_5

The rules required by the different types of auctions and the actions of the different actors can be coded directly in a smart contract. Thus, by inheriting some properties from the blockchain, in our proposed solution:

1. every transaction executed in the smart contract is visible and verifiable to the entire network: bidders and the auctioneer cannot cheat and the winning bidder can use the blockchain as a proof;
2. a trusted third party is not required, promising low transactions fees compared to traditional systems;
3. neither the code, nor the data stored in the blockchain can be modified;
4. the identity of the actors is always verified before any action is taken.

However, in Ethereum, computation is expensive as transactions are executed and verified by all the nodes on Ethereum network. Therefore, Ethereum defines a *gas* metric to measure the computation efforts and storage cost associated with transactions. That is, each transaction has a fee (i.e., consumed gas) that is paid by the transaction's sender in *Ether* (Ethereum currency). There is also a block gas limit that defines the maximum amount of gas that can be consumed by all transactions combined in a single block. Therefore, smart contracts cannot include very expensive computations that exceed the block gas limit.

In this paper, our main goal was to evaluate the cost of a prototype implementation of the four classical types of auction (English auction, Dutch auction, First-price sealed-bid auction, and Second-price sealed-bid auction). Thanks to the promising experimental results, we plan to formally verify the correctness of the smart contracts, and to further investigate the security and privacy issues on the blockchain.

The structure of this paper is as follows. Section 2 discusses background information about auctions, blockchain and smart contracts technologies. Section 3 describes how smart contracts for auctions were designed, developed, and tested, and discusses the experimental results we obtained, in terms of gas costs and time. Section 4 concludes the paper.

2 Background and Related Work

This section presents some background information on auctions, blockchain and smart contracts technologies. It also describes the few current works presenting auction systems based on blockchain technologies.

2.1 Auction

Auctions can be described as games of incomplete information that involve a certain number of actors: a *vendor* V, willing to sell a good, an *auctioneer* A, leading the auction according to some predefined rules and asking a commission for his services, and a set of *bidders* B, willing to buy the good at the best price.

There exist traditionally four types of auctions [5] for the allocation of a single item, differing on the rules to determine the winner, or the final buying price:

- *English auction*, also called *open ascending price auction*, where participants bid openly against one another, with each subsequent bid required to be higher than the previous bid. There are many variants depending on how bidders signal their will to attend and to go on with attending, or on the way the subsequents bids are given. For example, an auctioneer may announce prices raising them with small increments as long as there are at least two interested bidders, or bidders may call out their bids themselves (or have a proxy call out a bid on their behalf). The auction ends when a single bidder remains signaling his interest, at which point the highest bidder pays his bid. The seller may also define a *reserve price*, that is the minimum price he will accept as the winning bid in the auction.
- *Dutch auction*, also called *open descending price auction*, used for perishable commodities such as flowers, fish and tobacco. In the traditional Dutch auction, the auctioneer begins with a high price and then continously lowers it until a bidder is willing to accept the auctioneer's price, or until the seller's reserve price is met.
- *First-price sealed-bid auction* (FPSB), also called *blind auction*, where all bidders submit bids in a sealed envelop to the auctioneer, so that no bidder knows the bid of any other participant. Later, the auctioneer opens the envelope to determine the winner who submitted the higher bid.
- *Vickrey auction*, also known as *Second-price sealed-bid auction* (SPSB), which is identical to the first-price sealed-bid auction, with the exception that the winning bidder pays the second-highest bid rather than his own.

2.2 Blockchain and Smart Contract

A blockchain [6] is a distributed peer-to-peer network where non-trusting members can interact with each other without a trusted intermediary, in a verifiable manner. Smart contracts are scripts that reside on the blockchain that allow for the automation of multi-step processes. In this section, we give the details of how they work in order to be able to understand the implementation details that will be given in the next section.

Blockchain Technology. A blockchain is a *distributed ledger*, typically managed by a peer-to-peer network collectively adhering to a protocol for inter-node communication and for validating new blocks. It is distributed in the sense that there is no central repository of information, no database on a file server which can be hacked: it is a distributed data structure that is replicated and shared among all the members of a network. It is a ledger in the sense that it records and stores all transactions that have ever occurred in the blockchain network in a permanent and verifiable way.

In a blockchain, each block is identified by its cryptographic hash. Each block contains the cryptographic hash of the previous block, resulting in a chain of blocks (see Fig. 1), a timestamp, and a set of transactions (generally represented also as a Merkle tree root hash in the block header). Thus, by design, a blockchain is resistant to modification of the transaction data since, once recorded, the data in any given block cannot be altered retroactively without alteration of all subsequent blocks.

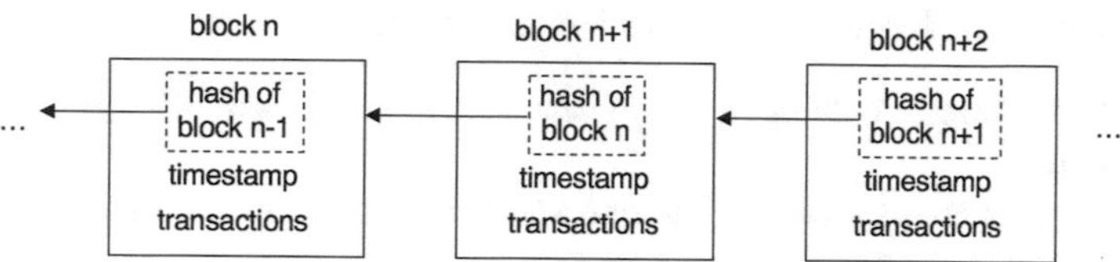

Fig. 1. Structure of a blockchain.

A blockchain network consists of a set of nodes that operate on the same blockchain via the copy each one holds. Users interact with the blockchain via a pair of private/public keys: they use their private key to sign their own transactions, and they are addressable on the network via their public key.

Transactions that occurred in the network are verified by special nodes (called *miners*). Verifying a transaction means checking the sender and the content of the transaction. Miners generate a new block of transactions after solving a computationally expensive task (called *Proof of Work*) and then propagate that block to the network. Other nodes in the network can validate the correctness of the generated block.

Regardless of specific implementations, a blockchain gives us the following benefits out of the box:

- a blockchain is resistant to modification of data;
- a blockchain is tolerant of node failures;
- a blockchain is a distributed peer-to-peer system able to reach consensus without the need of a third trusted party;
- all the information in the blockchain is available to all the nodes of the network;
- a blockchain is a method for tagging different pieces of information as belonging to different participants, and enforcing this form of data ownership without a central authority.

Smart Contracts. The term *smart contract* was introduced by Nick Szabo in 1994 as "a computerized transaction protocol that executes the terms of a contract" [7]. The idea was to translate contractual clauses (collateral, bonding, etc.) into code, and embed them into property (hardware, or software) that could self-enforce them, in order to minimize the need for trusted intermediaries between transacting parties, and the occurrence of malicious or accidental exceptions.

Within the blockchain context [8], a smart contract is executable code stored and running on the blockchain to facilitate, execute and enforce the terms of an agreement. We trigger a smart contract by addressing a transaction to it. It then executes independently and automatically in a prescribed manner on every node in the network, according to the data that was included in the triggering transaction (this implies that every node in a smart contract enabled blockchain is running a virtual machine (VM), and that the blockchain network acts as a distributed VM).

Thus, the main features of a smart contract are:

1. a smart contract is deterministic: the same input will always produce the same output;
2. a smart contract is stored on the blockchain, thus it can be inspected by every network participant;
3. since all the interactions with a contract occur via signed messages on the blockchain, all network participants get a cryptographically verifiable trace of the contract's operations.

2.3 Blockchain-Based Bidding Systems

As far as we know, this is the first attempt to implement the four most common types of auctions, and to analyse the implementations in terms of cost and time efficiency. Previous works focused on blind auctions and on security issues, since anonymity and confidentiality seem to collide against the idea of distributed ledger and the inherent transparency and lack of privacy of blockchains.

In [4], the authors implement on top of the Ethereum blockchain a protocol to achieve bids privacy in sealed-bid auctions. In [1], a protocol guaranteeing bid confidentiality against fully-malicious parties is described.

In [3], the authors discuss with little details a possible implementation with Ethereum smart contracts of sealed-bid auctions.

3 Design and Implementation of a Smart-Contract Based Bidding System

We implemented the four classical types of auction (English auction, Dutch auction, First-price sealed-bid auction, and Second-price sealed-bid auction) by means of four Ethereum smart contracts. In the rest of this Section, we will describe how Ethereum works, then we will show some code fragments implementing the major operations for two of the auctions. Finally, we will evaluate the implementations in terms of cost and time efficiency.

3.1 Ethereum

Smart contracts can be developed and deployed in different blockchain platforms (e.g., Ethereum, Bitcoin and NXT). At the moment, Ethereum [2] is the most popular public blockchain platform for developing smart contracts, since it provides a built-in Turing-complete language called Solidity, that can be used to write any smart contract and decentralised application. The code written in Solidity can be compiled into Ethereum Virtual Machine (EVM) bytecodes to be uploaded (and possibly executed) on the blockchain. Solidity is a contract-oriented, high-level language influenced by C++, Python and JavaScript. It is statically typed, and it supports inheritance, libraries and complex user-defined types among other features.

Ethereum Accounts. Accounts (also called *state objects* or *entities*) play a central role in Ethereum: the state of the Ethereum network consists of the state of all its accounts. There are two types of accounts: *externally owned accounts* (EOAs), and *contract accounts*, which represent users interacting via transactions with the blockchain, or smart contracts, respectively. An external account has an Ether balance, can send transactions (either to transfer Ether, or to trigger contract code), is controlled by private keys, has no associated code. A smart contract has an account balance, a private storage and an associated executable code. Once it is deployed into the blockchain, the contract code cannot be changed. Every time a contract account receives a transaction fired by a user, or a message (a sort of function call) sent by another contract, its code is executed by the EVM on each node. The contract may, based on the transaction it receives, read/write to its private storage, store money into its account balance, send/receive money from users or other contracts, or send messages to other contracts, triggering their execution. The contract's state will then be updated accordingly.

Ether and Gas. *Ether* (ETH) is the name of the currency used within Ethereum. It is used to pay for computation within the blockchain and the EVM, although transactions' execution fee is computed in terms of *gas*: one unit of gas corresponds to the execution of one atomic instruction, i.e., a computational step. Gas and ether are decoupled deliberately since units of gas align with computation units having a natural cost, while the price of ether generally fluctuates as a result of market forces.

Transaction and Messages. The term *transaction* is used in Ethereum to refer to the signed data package that stores a message to be sent from an externally owned account to another account on the blockchain. Contracts have also the ability to send *messages* to other contracts: messages are virtual objects that exist only in the Ethereum execution environment, they can be conceived of as *function calls*. Essentially, a message is like a transaction, except it is produced by a contract and not by an external actor.

They both contain the recipient, a value field indicating the amount of wei (1 *ether* is 1e18 *wei*) to transfer from the sender to the recipient, an optional data field that represents the input data to the contract, a gasLimit field representing the maximum number of computational steps the transaction or code execution is allowed to take to be used to compute the cost of the computation. Miners can refuse to process a transaction with a gas price lower than their minimum limit. A transaction contains also a signature identifying the user and a gasPrice field, representing the fee the sender is willing to pay for gas.

When a contract is executed as a result of being triggered by a message or transaction, every instruction is executed on every node of the network. This has a cost: for every executed operation there is a specified cost, expressed in a number of gas units. The maximum total ether cost of a transaction is then equal to gasLimit $*$ gasPrice.

3.2 Tools and Environment

For the implementation and deployment of smart contracts, we used Remix to write Solidity smart contracts; Ropsten, one of Ethereum public testing networks, to deploy and run the smart contracts; MetaMask to manage the accounts of the bidders, and Etherscan, an online service allowing to explore and search the Ethereum blockchain and its public testnets, to verify the correctness of execution of the smart contracts.

3.3 Implementation Details

We have implemented a prototype consisting of four different smart contracts, one for each type of auction. In the rest of the Section, we describe in detail some of the functions in the smart contracts of the English auction and of the First-price sealed-bid auction, since the others only differ slightly.

English Auction. In an English auction, participants bid openly against one another, with each subsequent bid required to be higher than the previous bid, starting from a *reserve price* chosen by the seller. At the end, the highest bid wins.

The key aspects that need to be set up by the constructor are (see Fig. 2 for a Solidity fragment): (*i*) the beneficiary, an external account representing the seller of the good that starts the auction by creating the contract; (*ii*) endTime, the time after which no further bids are accepted; (*iii*) deliveryLimit, the time lapse by which the winner has to receive the good; (*iv*) reservePrice, the minimum price the seller will accept as bid in the auction; and (*v*) minIncrement, the minimum difference between two bids.

The bidding function bid() is reported in Fig. 3. The first four lines of code express the main rules of the auction: a bid must be done before the auctions ends, it must be higher than the previous bid and of the reserve price, and it must come from a valid bidder and not from the seller. If the bid satisfies all the constraints, it will be accepted and recorded as the highest bid.

```
constructor(
    uint _biddingTime,
    uint _deliveryTime,
    uint _reservePrice,
    uint _minIncrement
) public {
    beneficiary = msg.sender;
    endTime = now + _biddingTime;
    deliveryLimit = endTime + _deliveryTime;
    reservePrice = _reservePrice;
    minIncrement = _minIncrement;
}
```

Fig. 2. English auction: the constructor.

```
function bid() public payable {
    require(now <= endTime);
    require(msg.value >= highestBid+minIncrement);
    require(msg.value >= reservePrice);
    require(msg.sender != beneficiary);

    if (highestBid != 0) {
        highestBidder.transfer(highestBid);
    }
    highestBidder = msg.sender;
    highestBid = msg.value;
    emit NewHighestBid(msg.sender, msg.value);
}
```

Fig. 3. English auction: Bidding function.

First-Price Sealed-Bid Auction. In a First-price sealed-bid auction (FPSB), all bidders submit bids in a sealed envelop to the auctioneer, so that no bidder knows the bid of any other participant. Later, the auctioneer opens the envelope to determine the winner who submitted the higher bid. In the constructor, the difference with the English auction (see Fig. 4) is given by the revealEnd, i.e., the time after the end of the auction when it is possible to reveal the bid.

In this auction, the bid must be sealed. A bidder (see Fig. 5) sends his own blinded bid, which is the hash of the concatenation of the actual value of the bid and a secret, used as a salt. In this case the value of the bid cannot be read also by the auctioneer/smart contract, since at this point he does not know the secret used to compute the hash.

```
 constructor(
     uint _biddingTime,
     uint _revealTime,
     uint _deliveryTime,
     uint _reservePrice
) public {
     beneficiary = msg.sender;
     biddingEnd = now + _biddingTime;
     revealEnd = biddingEnd + _revealTime;
     deliveryLimit = revealEnd + _deliveryTime;
     reservePrice = _reservePrice;
}
```

Fig. 4. First-price sealed-bid auction: the constructor.

```
function bid(bytes32 _blindedBid)
    public
    payable
    onlyBefore(biddingEnd)
{
    require(msg.sender != beneficiary);
    bids[msg.sender].push(Bid({
        blindedBid: _blindedBid,
        deposit: msg.value
    }));
}
```

Fig. 5. First-price sealed-bid auction: Bidding function.

3.4 Evaluation of the System

Cost Analysis. As described in Sect. 3.1, the computation within the EVM has a cost. In this Section, we show the *gas* costs and the corresponding prices in Euro for the deployment and the major operations of the smart contracts implementing the four different types of auctions. At the time of carrying out the experiments, October 2018, the *ether* exchange rate was 1 ETH = 192.80 €, and the median gasPrice was approximately 0.0000002 ETH (20 Gwei) (Tables 1, 2, 3 and 4).

As expected, the smart contract deployment cost is more expensive in the two blind actions, since it involves a larger number of data and of set-up operations. As for the other operations, the cost depends on the complexity of the operation: for example, a sum costs 1 *gas*, whereas the execution of a SHA3 hash costs 20 *gas*.

Table 1. Costs of an *English auction.*

Operation	Gas used	Cost (ETH)	Cost (Euro)
Contract deployment	2345	0.000469	0.09
Bidding	320	0.000064	0.01
End the auction	270	0.000054	0.01
Payment	275	0.000055	0.01

Table 2. Costs of an *Dutch auction.*

Operation	Gas used	Cost (ETH)	Cost (Euro)
Contract deployment	2535	0.000507	0.10
Price decreasing	140	0.000028	0.01
Bidding	410	0.000084	0.02
End the auction	195	0.000039	0.01
Payment	200	0.00004	0.01

Table 3. Costs of an *First-Price Sealed-Bid auction.*

Operation	Gas used	Cost (ETH)	Cost (Euro)
Contract deployment	3635	0.000727	0.14
Bidding	425	0.000085	0.02
Bid opening	370	0.000074	0.01
End the auction	270	0.000054	0.01
Payment	275	0.000055	0.01

Table 4. Costs of an *Second-Price Sealed-Bid auction.*

Operation	Gas used	Cost (ETH)	Cost (Euro)
Contract deployment	3805	0.000761	0.15
Bidding	425	0.000085	0.02
Bid opening	470	0.000094	0.02
End the auction	270	0.000054	0.01
Payment	275	0.000055	0.01

Time Analysis. In Ethereum, a new block is validated every 15 s and the transaction rate is of around 20 s. During the test in the Ropsten network, the effective average transaction time (i.e., transaction processing, validation and inclusion in a new block) was of 10 s. The effective time is the result of the propagation time within all the nodes in the network, and of the time lapse between the creation of two new block. This value may be a problem when bidding just before the end of the auction.

Privacy Considerations. At the moment, in our prototype implementation, users do not provide any personal data, since their identity is given by their account address, which is public since Ethereum is a public blockchain. For this reason, our implementation can be considered pseudo-anonymous: a user will be linked to a public Ethereum address, but no one will get to know the actual name or address. However, by observing the blockchain evolution, it is possible to collect all the past transactions by a single user, that can be seen as lack transactional privacy. However, this is not a main issue for some of the auctions we implemented.

4 Conclusion

In this paper, we showed how it is possible to build different types of auctions on top of Ethereum blockchain. We analysed the implementation in terms of cost and time efficiency. The advantages of a smart-contract based auction with respect to a classical eAuction are:

- *transparency:* the bidding history is public and verifiable by all the users;
- *non-repudiation:* a bidder cannot negate his bid;
- *integrity:* a bidder cannot modify his own bid, or other bidders' bid;
- *no trusted third party.*

For future works, we intend to apply formal verification techniques on our code to detect potential flaws and vulnerabilities. Moreover, we want to further investigate the security and privacy issues.

References

1. Blass, E., Kerschbaum, F.: Strain: a secure auction for blockchains. In: Computer Security - 23rd European Symposium on Research in Computer Security, ESORICS 2018, Part I, Proceedings, Barcelona, Spain, 3–7 September 2018, pp. 87–110 (2018)
2. Buterin, V.: Ethereum white paper: a next-generation smart contract and decentralized application platform (2013)
3. Chen, Y., Chen, S., Lin, I.: Blockchain based smart contract for bidding system. In: 2018 IEEE International Conference on Applied System Invention (ICASI), pp. 208–211, April 2018. https://doi.org/10.1109/ICASI.2018.8394569
4. Galal, H.S., Youssef, A.M.: Succinctly verifiable sealed-bid auction smart contract. In: Garcia-Alfaro, J., Herrera-Joancomartí, J., Livraga, G., Rios, R. (eds.) Data Privacy Management, Cryptocurrencies and Blockchain Technology. LNCS, pp. 3–19. Springer, Cham (2018)
5. Krishna, V.: Auction Theory, 2nd edn. Academic Press, New York (2010)
6. Nakamoto, S.: Bitcoin: a peer-to-peer electronic cash system (2008)
7. Szabo, N.: Smart contracts (1994)
8. Wood, D.G.: Ethereum: a secure decentralised generalised transaction ledger (2014)

Blockchain-Based Autonomous Peer to Peer Information Interaction System in Financial Audit

Yu Zhuo[1(✉)], Dong Aiqiang[1], and Yan Yong[2]

[1] Beijing China-Power Information Technology Co., Ltd., Beijing 100192, China
yuzhuo@sgitg.sgcc.com.cn
[2] Electric Power Research Institute of State Grid Zhejiang Electric Power Company, Hangzhou 310009, China

Abstract. The current distributed energy Internet financial audit has problems in information sharing and lacks an efficient peer-to-peer security transmission strategy. Facing the problems of financial auditing in distributed energy internet, we construct an autonomous peer-to-peer information interaction system combining with blockchain and named data network, and research on data acquisition and secure transmission of this system. For data acquisition, a data identification parsing module based on blockchain is designed, and the corresponding data is obtained by parsing the data identifier. For secure transmission, this paper designs a blockchain-based encryption and Decryption module to ensure the secure transmission of data in the system environment and the controllability of data access by encrypting and decrypting data assets. This paper builds an experimental environment based on the campus network, and performs functional testing and verification on the two modules.

Keywords: Energy internet · Blockchain · Information interaction · Data acquisition · Secure transmission

1 Introduction

With the progress of the energy revolution, the distributed energy network system has become a necessary way to develop and utilize energy that can be generated in a large scale. The energy Internet should provide a fast, credible and automatic energy trading model between the supply and demand sides. Help both the supply and demand sides to efficiently establish and complete transactions. A large number of energy transactions pose challenges for energy Internet financial auditing, lacking an efficient peer-to-peer security interaction strategy to achieve efficient and secure sharing of audit data. Facing the problems of distributed energy internet, this paper combines blockchain and named data network to construct an autonomous peer-to-peer data interaction system, which is of great significance for the efficient and secure transmission of distributed energy Internet audit information. As a decentralized trusted network with decentralized distributed storage, the blockchain realizes the reliability and non-tamperability of transaction data storage through computer technologies such as distributed data storage,

C.-N. Yang et al. (Eds.): SICBS 2018, AISC 895, pp. 65–75, 2020.
https://doi.org/10.1007/978-3-030-16946-6_6

peer-to-peer transmission, consensus mechanism and encryption algorithm. The core advantage is decentralization, which can realize peer-to-peer transaction-based peer-to-peer transaction, coordination and collaboration in a distributed system where nodes do not need to trust each other by means of data encryption, time stamping, distributed consensus and economic incentives [1–7] to provide solutions to the problems of high cost, low efficiency and insecure data storage that are common in centralized organizations.

In this paper, the blockchain is combined with the Named Data Network (NDN) [8–11] to study the problems of the current distributed energy Internet financial audit, and to study the autonomous peer-to-peer data interaction system that can meet the needs of the energy Internet. In order to ensure the correct and efficient information interaction of the system, this paper will design the data identification and data encryption and decryption module based on blockchain in detail, and provide a new data transmission and data request mode for information communication, realizing data in the network. Safe and reliable transmission.

2 System Architecture Model

Building a blockchain-based financial auditing autonomous peer-to-peer information interaction system architecture as shown in Fig. 1, the system includes three entity objects: data publisher, data requester, and relay node. For a sensor or data providing point device of an access system under a certain local area network, a uniquely determined “name” needs to be assigned to the device. When the sensor data needs to be acquired, only the request for the data identifier needs to be returned. Encrypted and signed corresponding data to achieve data access.

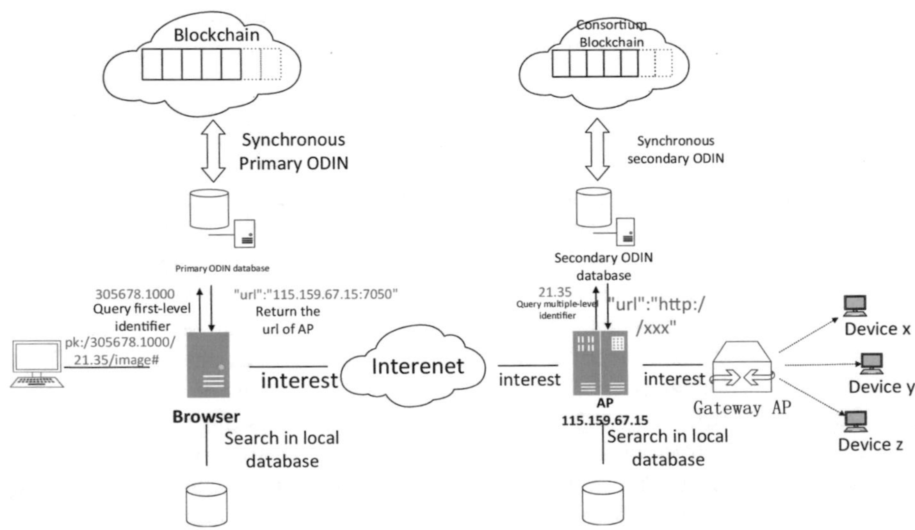

Fig. 1. Blockchain-based autonomous peer-to-peer information interaction system

3 System Function Module

The system combines blockchain technology with a content center network to achieve point-to-point reliable transmission of energy Internet audit data. This section mainly introduces the system function requirements, briefly describes the functions that each functional module needs to implement and the corresponding implementation methods. The system is divided into three sub-modules according to functional entities, 8and the sub-modules are subdivided into multiple modules in function. The overall functional module is divided as shown in Fig. 2. The data identification analysis and the design of the encryption and decryption module will be studied in detail.

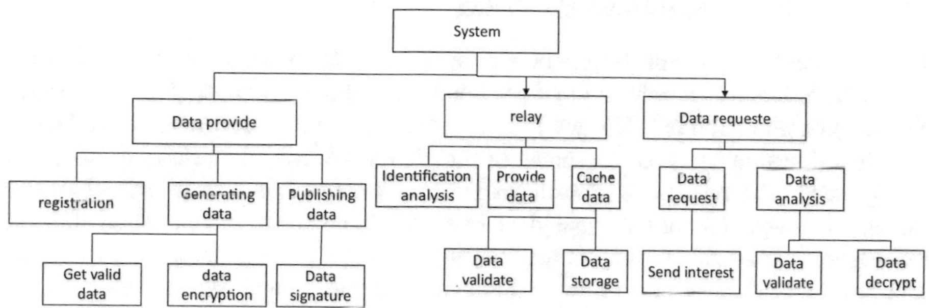

Fig. 2. System overall function block diagram

3.1 Design of Data Identification Based on Blockchain

The design and implementation of the data identification parsing module depends on the design of the data identifier. This section will briefly introduce the autonomous and secure open data index naming identifier [12] based on the blockchain.

The open data index naming identifier is divided into three parts: naming system, prefix and suffix, separated by slashes. The prefix is divided into two parts by decimal point, where [BTC_BLOCK_SN] is the registration of the open data index naming identifier on the bitcoin blockchain. The digital serial number of the blockchain in which the record is located, [BTC_TRANS_INDEX] is the Arabic numeral number of the specific storage location in the block where the open data index naming the registration record on the blockchain, and the blockchain is not modified. The non-deformable modification of the ODIN logo is guaranteed. The suffix [DSS] is optional and arbitrarily given by the open data index naming identifier, and the uniqueness is ensured by itself, thereby determining the specific data content identifier corresponding to the open data index naming identifier. The naming rules for open data index naming identifiers are as follows:

Standard structure of primary ODIN:

ppk: [BTC_BLOCK_SN]. [BTC_TRANS_INDEX] */DSS*

Multi-level extended ODIN standard structure:

ppk: [PARENT_ODIN_PREFIX] /[SUB_BLOCK_SN]. [SUB_TRANS_INDEX] /DSS

The [PARENT_ODIN_PREFIX] field of the multi-level extension ODIN corresponds to the upper-level open data index naming identifier prefix. In the design of the data identifier parsing module, the data identifier prefix registered on the blockchain corresponding to the field is similar to the prefix part in the first-level identifier, [SUB_BLOCK_SN]. The [SUB_TRANS_INDEX] field corresponds to the data identifier prefix registered on the federated chain.

3.2 Data Identification Analysis Module

The data identification analysis module combines the characteristics of the NDN network and the data hierarchical structure naming method. The data identification resolution process is divided into two types: first-level identification analysis and multi-level identification analysis. As shown in Fig. 3, the primary identification function is mainly used. The module is blockchain information synchronization, interest packet parsing, message forwarding, and data backhaul. The main functional modules of multi-level identification analysis are interest message parsing, message forwarding, data backhaul, alliance chain information synchronization, and dynamic discovery of local area network data access points.

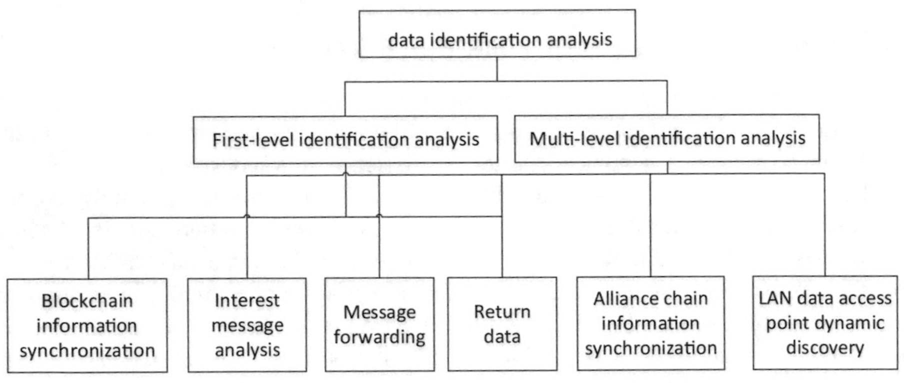

Fig. 3. Data structure analysis module function structure diagram

The data identifier parsing module is mainly used to implement parsing of the data identifier based on the blockchain, and depends on the identifier registration module. The identifier that the parsing face may be the identifier of a certain data content, or may be the identifier of a certain sensor, and the identifier is unique and not repeated. The main service scenarios implemented by the parsing module include first-level identifier resolution and multi-level identifier resolution, as shown in Fig. 4(a) and (b). The user inputs the data identifier on the client, and the client packages the identifier into a packet of the specified format to the local relay node, and the local relay node receives the interest packet to perform an interest packet parsing to obtain the data identifier, and if the local data has no corresponding data. Go to the first-level ODIN ID resolution server to obtain the forwarding destination address and forward the interest packet to the corresponding server relay. The server relay determines whether it needs to perform secondary resolution, and performs corresponding actions, repeats parsing and forwarding until the requested content package is found, and returns according to the interest packet path.

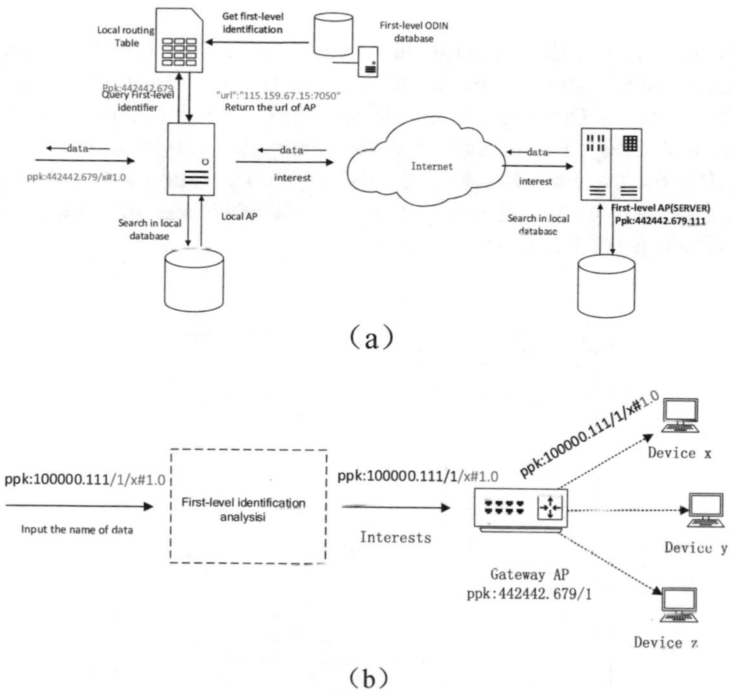

Fig. 4. ID resolution

3.3 Data Encryption and Decryption Module

The data encryption and decryption module uses the blockchain encryption and decryption technology [11], combines the blockchain to generate public and private keys, encrypts and signs the data, and aims to realize the secure, credible and non-tamperable transmission of data in the naming center network. The function structure of the data encryption and decryption module is shown in Fig. 5. The module is divided into two sub-modules: data signature, encryption sub-module, and verification and decryption sub-module. The data signature and encryption sub-module need to implement data generation and data release, and the data verification and decryption sub-module realizes data verification and decryption.

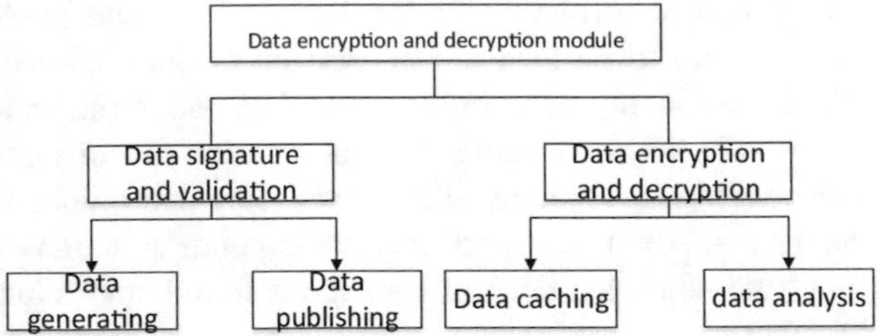

Fig. 5. Data encryption and decryption module function structure diagram

The encryption and decryption of information is the key link of the blockchain. Many excellent encryption algorithms are used in the blockchain to ensure the reliability of the system. Drawing on and using blockchain, this paper designs a data encryption and decryption module that can encrypt, decrypt and verify the data transmitted in the system network. The module works mainly at the data providing point and the data requester. The relay node module will There are some calls, and the overall business process is shown in Fig. 6.

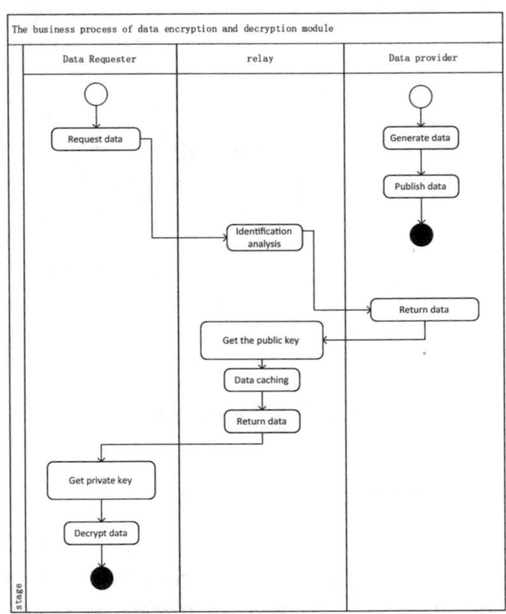

Fig. 6. Data encryption and decryption module business flow chart

When the data providing point needs to publish data, the data is encrypted by using the public key in a pair of asymmetric encryption keys generated by the publisher when registering, the plaintext is changed into ciphertext, and then the publishing data module is called, and the data encryptor is used. The module constructs a content packet that meets the transmission requirements for the ciphertext and signs the content packet with another private key. The configuration information for the public key applied to the verification is stored as a data identifier on the blockchain. After the publishing data module completes the construction of the data packet, the data is placed in the cache waiting for the request of the data requester. After the data requester sends the interest packet to perform the data request, the queried data packet is returned according to the interest packet path. During the return process, each relay node that passes through will call the cached data after receiving the data packet. Module, the module calls the data verification sub-module, the data verification sub-module obtains the public key corresponding to the data verification from the blockchain, and verifies the data, and the data is buffered and the data packet is returned after the verification is passed. Module, complete the return of the message. When the message is sent back to the requesting client of the data requester, the client obtains the data ciphertext and queries its own decrypted private key pool. If there is a corresponding decrypted private key, the ciphertext is decrypted. If not, the privilege is not accessed. Clear text data.

4 System Construction and Experimental Effect Analysis

As shown in Fig. 7, the network test platform is constructed and divided into multiple network segments. NET1 includes one or more data request points, and the data request point is composed of a browser and a local relay. Currently, a relay node configured with a server identity is called a server. The server has a unique data identifier registered on the blockchain, which can provide data or secondary resolution. NET2, NET3, …, NETn are LAN subnets, and each network segment has a locally unique secondary identifier registered on the superbook. NETi (i >= 2) The gateway relay node connects

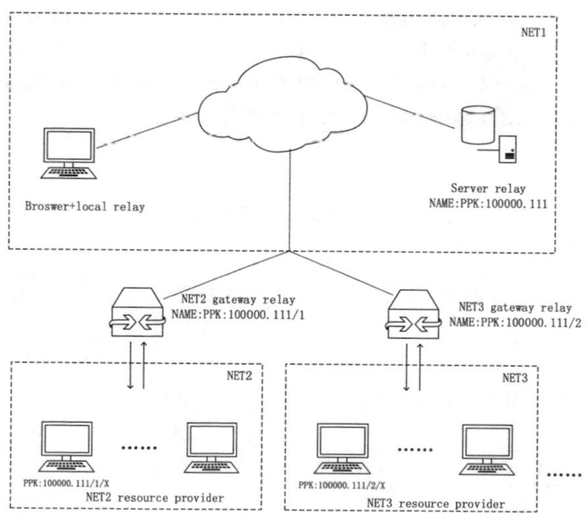

Fig. 7. Network test platform

to the public network and the local area network to provide forwarding of various packets. There are resource providers in the NETi (i >= 2) network to implement data distribution.

As shown in the figure, each server device of the NET1 network segment has a unique first-level identifier, and the data providers of each subnet of NETi (i >= 2) have uniquely determined secondary identifiers. Before the networking is completed, you need to use the registration tool to register the blockchain and the super-book identifier. After completing the registration, start each relay node and start the module test.

The specific network environment configuration is shown in Table 1. The NET1 local relay mainly implements the verification of the received content package, and does not need to allocate an open data index naming identifier, and forms a data request client in combination with a simple browser.

Table 1. Network environment configuration.

Device	Data identification	Is there a data resource
NET1 local relay	no	no
Server relay relay	ppk:100000.111	Yes
NET2 gateway relay relay	ppk:100000.111/1	no
NET3 gateway relay relay	ppk:100000.111/2	no
……	……	……
NET $_n$ gateway relay relay	ppk:100000.111/n	no
NET2 resource provider x	ppk:100000.111/1/x	Yes
NET2 resource provider y	ppk:100000.111/1/y	Yes

The audit data resource is generated on the device capable of generating resources. When the data is generated, the data is encrypted and the data packet is signed, and the generated data content is cached in the CS table, waiting for the interest packet request. Configure network resources on the device that can implement data resource caching. The specific resource configuration is shown in Table 2:

Table 2. Network resource data

Device type	Data identification
ODIN resolution server	ppk:100000.111/Note#1.0
	ppk:100000.111/Note#2.0
	ppk:100000.111/Packet#2.0
NET-2 ODIN resource provider	ppk:100000.111/2/a#1.0
	ppk:100000.111/2/a#3.0
NET-3 ODIN resource provider	ppk:100000.111/3/b#2.0

4.1 Request Primary Identification

The data with the resource name ppk:100000.111/Note#1.0, ppk:100000.111/Note#2.0, ppk:100000.111/Packet#2.0 is cached on the server relay node, where ppk:100000.111/Note#1.0, ppk:100000.111/Note #2.0 is a different release of the same name data, of which 2.0 is the latest version. The request is made for the latest version data of ppk:100000.111/Note#1.0, ppk:100000.111/Note#2.0, ppk:100000.111/Note, and ppk:100000.111/Packet#2.0, and the corresponding data result is obtained:

Data request is made to ppk:100000.111/Note#1.0. As shown in Fig. 8(a), the data resources on the server relay relay can be successfully obtained, and the data parsing function in the data encryption/decryption module can be realized normally. Data request for ppk:100000.111/Note#2.0, as shown in Fig. 8(b), can successfully obtain the data resources on the server relay relay. Data request is made to ppk:100000.111/Note#, and the latest version data of the ppk:100000.111/Note is obtained, as shown in Fig. 8(c), the data resources on the server relay relay can be successfully obtained. Data request is made to ppk:100000.111/Packet#2.0, as shown in Fig. 8(d), the data resources on the server relay relay can be successfully obtained.

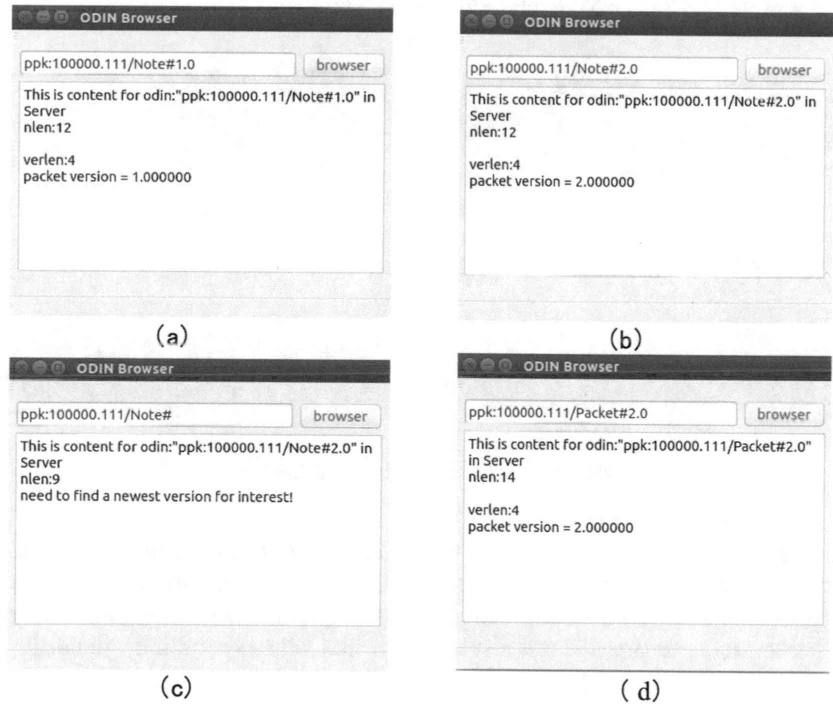

Fig. 8. Request primary Identification

4.2 Request Secondary Identification

The resource provider in the NET2 and NET3 networks has the data resources corresponding to the secondary identifier, and the resource names are ppk:100000.111/2/a#1.0, ppk:100000.111/2/a#3.0, ppk:100000.111/3/b. #2.0. Ppk:100000.111/2/a#1.0, ppk:100000.111/2/a#3.0 are different versions of the data. Data requests are made to ppk:100000.111/2/a#3.0, ppk:100000.111/3/b#2.0, respectively, and the corresponding results are obtained.

Data request for ppk:100000.111/3/b#2.0, as shown in Fig. 9(a), can correctly obtain the data resources in the NET2 subnet environment, indicating that the secondary ID resolution module can correctly implement the secondary identification. Parsing, and the request is transmitted through the server relay relay, indicating that the cache data module in the data encryption and decryption module is correctly designed. Data request for ppk:100000.111/2/a#3.0, as shown in Fig. 9(b), because the NET1 client does not have data access rights to the NET2 data, that is, the NET1 client does not have a corresponding data decryption private key, and the client can receive The encrypted data cannot be decrypted by the validated data message, so the data plaintext cannot be obtained. It shows that data encryption can ensure that data is not obtained by clients that do not have access, and the data is "hidden".

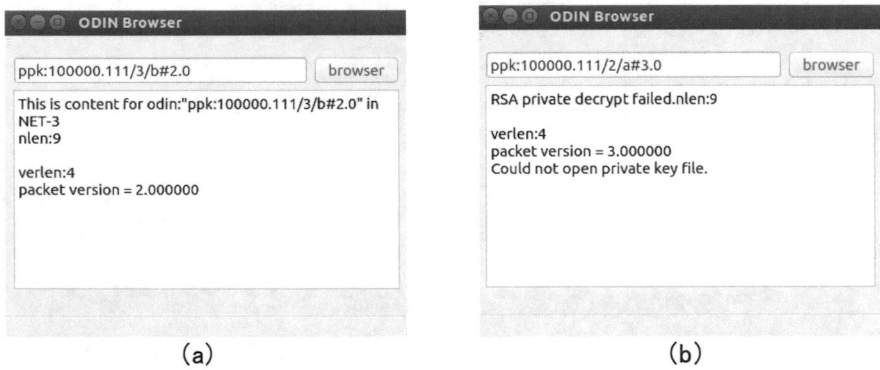

Fig. 9. Request secondary identification

It can be seen from the test result of the data identification request that the data identification parsing module can correctly process the request for data, and the data encryption and decryption module can provide security and reliability guarantee for the data packet. In summary, the test results are in line with expectations, indicating that the module design is correct.

5 Conclusion

This paper proposes an autonomous peer-to-peer data interaction system for energy Internet audit data transmission needs. The system combines the blockchain with the content center network to achieve secure and efficient sharing of energy Internet audit data. Secondly, a detailed research and design is carried out for the data identification and parsing module in the relay module and the data encryption and decryption module involving multiple sub-modules. Building a modular test platform, using the test platform to complete the module test, analyzing different scenarios, and obtaining test results, indicating the correct feasibility of the design. In the process of implementation, this paper simplifies the construction of the network environment. If it needs to be applied to the real network environment, the next step needs to consider more practical network requirements, and the multi-signature technology and blind signature proposed by the current blockchain. Research on technology to further ensure data security and reliability.

Acknowledgement. This work is funded by the Science and Technology Project of State Grid Corporation (52110417000G).

References

1. Ping, Z., Yu, D., Bin, L., et al.: White Paper on China's Blockchain Technology and Application Development, October 2016
2. Nakamoto, S.: Bitcoin: a peer-to-peer electronic cash system. Consulted (2008). http://www.bitcoin.org
3. Yong, Y., Feiyue, W.: Current status and prospects of blockchain technology development. Chin. J. Autom. **42**(4), 4:481–4:491 (2016)
4. Christidis, K., Devetsikiotis, M.: Blockchains and smart contracts for the internet of things. IEEE Access **4**, 2292–2303 (2016)
5. Jia, C., Feng, H.: Blockchain: From Digital Currency to Credit Society. CITIC Publishing House, Beijing (2016)
6. Narayanan, A., et al.: Blockchain: Technology Driven Finance. CITIC Publishing House, Beijing (2016)
7. Yan, Z., et al.: Security research in key technologies of blockchain. Inf. Secur. Res. **2**(12), 1090–1097 (2016)
8. Kai, L.: Information Center Network and Named Data Network. Peking University Press, Beijing (2015)
9. Youfeng, W.: Research and implementation of browser based on CCN network in Windows environment. Thesis Beijing University of Posts and Telecommunications, Beijing University of Posts and Telecommunications (2014)
10. Zhen, C., Junwei, C., Hao, Y.: Content Center Network Architecture. Tsinghua University Press, Beijing (2014)
11. NDN Project Team.: NDN Packet Format Specification. http://named-data.net/doc/ndn-tlv/
12. PPk Open Group.: ODIN (Open Data Index Name) - Open Data Index Naming Identification Technical Specification. http://www.ppkpub.org/ppk_odin_spec_cn.html

Multimedia Security

Relations Between Secret Sharing and Secret Image Sharing

Xuehu Yan[1(✉)], Jinming Li[2], Yuliang Lu[1], Lintao Liu[1], Guozheng Yang[1], and Huixian Chen[3]

[1] National University of Defense Technology, Hefei 230037, China
publictiger@126.com
[2] 32126 Institute, Shenyang 110000, China
[3] Artillery Air Defense Academy, Hefei 230037, China

Abstract. Secret sharing (SS) for (k, n) threshold generates secret data into n shadows, where any k or more shadows can reconstruct the secret while any $k-1$ or less shadows reconstruct nothing of the secret. SS is useful for cloud computing security, block chain security and so on. Since nowadays image covers more information, secret image sharing (SIS) is studied widely. Although most SIS principles are derived from SS and SIS belongs to SS, SIS has its specific features comparing to SS due to image characteristics, such as, pixel value range, region relationship and so on. In this paper, first we discuss the relations between SIS and SS, where differences between them are mainly considered. Then, some typical sharing principles are employed to further indicate the differences and analyze the possible ways to deal with the issues when directly applying SS to SIS. Finally, we perform experiments to verify our analyses.

Keywords: Information hiding · Secret sharing · Secret image sharing · Polynomial · Chinese remainder theorem

1 Introduction

Shamir [1] and Blakley [2] introduced secret sharing (SS), respectively. In SS for (k, n) threshold, the secret data is generated into n noise-like shadows (also called shares), which are then distributed among multiple owners to achieve access control. Any k or more shadows can reconstruct the secret while any $k-1$ or less shadows reconstruct nothing of the secret. SS is useful for cloud computing security, block chain security, information hiding, authentication, watermarking, access control, transmitting passwords, distributed storage and computing, and so on [3–5].

Polynomial-based scheme [1], Chinese remainder theorem (CRT) [6], and visual secret sharing (VSS) [7,8] also called visual cryptography scheme (VCS), are the main principles in SS.

Since nowadays image covers more information, secret image sharing (SIS) is studied widely, where most principles of SIS are derived from SS. Although SIS

C.-N. Yang et al. (Eds.): SICBS 2018, AISC 895, pp. 79–93, 2020.
https://doi.org/10.1007/978-3-030-16946-6_7

belongs to SS, SIS has its specific features comparing to SS due to image characteristics, such as, pixel value range, region relationship and so on. Therefore, there are some issues when directly applying SS principles to SIS, such as, lossy recovery, information leakage, and so on. We will take polynomial-based scheme as an example to indicate these issues.

In original polynomial-based SS [1] for (k, n) threshold, the secret data is generated into the constant coefficient of a random $(k-1)$-degree polynomial to output n shadows, which are then also distributed to n associated owners. The secret data can be reconstructed losslessly modulo 251 based on Lagrange's interpolation when collecting any k or more shadows. Following original scheme and employing all the coefficients of the polynomial for embedding the secret image pixels, Thien and Lin [9] applied original polynomial-based SS to image. Their algorithm reduced shadow size $1/k$ times to the original secret image. However, it suffers from lossy reconstruction and auxiliary encryption [5], where lossy reconstruction is due to modular 251 when pixel value is larger than 250, and auxiliary encryption is performed on the secret image before sharing to avoid secret information leakage. To deal with these two issues and to obtain more features, some following polynomial-based schemes [10–12] were therefore proposed. Unfortunately, these two issues are still not well addressed. Lossy reconstruction is caused by that grayscale pixel value range is $[0, 255]$ and modular 251 can not cover it; Adjacent pixel values are in general close to each other and there will be secret information leakage when employing all the coefficients of the polynomial for embedding secret image pixels, so that auxiliary encryption is employed to avoid secret information leakage.

In VSS [13–16] for (k, n) threshold, the generated n shadows are first printed onto transparencies and then assigned to n associated owners. The strength of VSS lies in, the secret image can be reconstructed by stacking any k or more shadows and human naked eyes with no cryptographic computation. Less than k shadows will generally give no clue about the secret even if infinite computation power is available. After original work, the associated VSS problems and its physical properties are further researched, such as threshold [17], multiple secrets [11], contrast [18,19], pixel expansion [13,20,21], meaningful shadows [15,22–25], and so on [26,27]. It is noted that since VSS is originally designed for image other than data, it doesn't face the issues when applying to SIS.

In a word, although most SIS principles are derived from SS and SIS belongs to SS, SIS has its specific features comparing to SS due to image characteristics, such as, pixel value range, region relationship and so on. Thus, we need to discuss the relations between SIS and SS as well as the ways to deal with the issues when directly applying SS to SIS.

In this paper, first we discuss the relations between SIS and SS, where differences between them are mainly considered. Then, some typical sharing principles are employed to further indicate the differences and analyze the possible ways to deal with the issues when directly applying SS to SIS. Finally, we perform experiments to verify our analyses.

The rest of the paper is organized as follows. Section 2 discusses some typical secure technique relations. In Sect. 3, some typical principles are further analyzed as examples. Section 4 is devoted to experimental results. Finally, Sect. 5 concludes this paper.

2 Typical Secure Technique Relations

Encryption, information hiding, SS and SIS are significant techniques to protect privacy. They have relations while differences.

2.1 Relations Between Encryption, Information Hiding, SS and SIS

Category relation between encryption, information hiding, SS and SIS is exploited in Fig. 1.

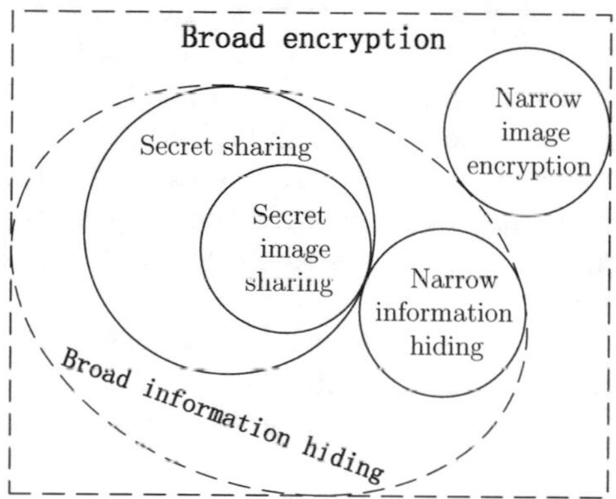

Fig. 1. Category relation between encryption, information hiding, SS and SIS

1. In a broad sense, narrow image encryption, information hiding, SS and SIS all belongs to encryption.
2. In a broad sense, SS and SIS all belongs to broad information hiding.
3. In general, narrow image encryption and narrow information hiding encode or embed the secret into single stego-cover, e.g., stego-image, while SS generates it into multiple shadows. Narrow information hiding usually outputs comprehensible stego-image, while narrow image encryption, or SS generates uncomprehensible stego-image or shadows.
4. Any two or more of narrow image encryption, narrow information hiding, and SS can be combined in some scheme to achieve specific feature at the cost of high algorithmic complexity or other drawbacks, so that the scheme belongs to which technique is hard to define. As an example, we can easily utilize SS to generate n noise-like shadows, which are then embedded into n natural

cover images to obtain n comprehensible stego-shadows. In such a way, SS with meaningful shadow is achieved at the cost of high algorithmic complexity and large pixel expansion. In a research, we had better focus on improving single technique other than combining skills.
5. To define the secure levels is a serious issue for all the techniques.

2.2 The Feature Analysis of Image

Digital image belongs to data, while it is also different from pure secret data.

1. An image is composed of pixels, and there is a certain correlation between pixels, such as texture, structure, edge and other related information. In a local region of an image, the pixel value of one pixel is close to its adjacent pixels values. The features are related to security and thus should be considered in algorithm design.
2. An image contains many pixels and thus a big amount of data, therefore we should consider the efficiency in algorithmic design.
3. An image has its specific storage file structure. Taking grayscale image and SIS design as an example, its pixel value range is $[0, 255]$, which should be considered in the algorithmic design, such as, the input value, the output value, and other relative parameters, should be in the range.
4. As well known, an image is a special form of the data, in which each grayscale pixel is represented by one byte and each binary pixel is by one bit of a byte, thus image protecting technique is easily to be generalized for secret data protecting technique.
5. In general, secret data can be used when it is losslessly reconstructed. However, an image can be reconstructed with some errors due to the low pass filter feature of human eyes so that progressiveness makes sense to image protecting technique.

2.3 SS and SIS

As SIS belongs to SS and most SIS principles are derived from SS, why we need to further research SIS.

Based on the above feature analyses of image, we mainly discuss as follows.

1. Since an image is composed of pixels, and there is a certain correlation between pixels, the security of an image protecting technique should include two aspects, i.e., corresponding single pixel security and region security. While a data protecting technique in general only considers data block.
2. Since an image contains a big amount of data, therefore we should consider the generating and reconstructing algorithmic efficiency in SIS design for practice and saving energy.
3. Since an image has its specific storage file structure, in the SIS design we should consider more aspects, such as, the secret pixel value range, the shadow pixel value range, and other parameters. While a data protecting technique in general only considers that the output shadows can be stored even with high expansion and more big blocks.

4. Since an image is a special form of the data, in which each grayscale pixel is represented by one byte and each binary pixel is by one bit of a byte, thus SIS is easily to be generalized for SS.
5. In general, SS has the property of "all-or-nothing", thus lossy data makes none of the business. While progressiveness makes sense to SIS, which is applicable to art-work image vending, multi-level representation, Pay- TV/Music, degraded encryption and so on.
6. In general, SS rarely generates the secret data into comprehensible shadows due to each shadow is only data block. While comprehensible shadow is important to SIS, due to which increases management efficiency as well as decreases encryption suspiciousness.

3 Some Typical Principles Analyses

Moreover, as examples, some typical SS principles and their derived SIS schemes will be employed to further indicate the differences and analyze the ways to deal with the issues when directly applying SS to SIS.

3.1 Polynomial-Based SS and SIS Schemes

Well-known polynomial-based SS and SIS will be presented first.

In order to generate secret data $s = a_0$ into n shadows, original polynomial-based SS generates a $k-1$ degree polynomial as Eq. (1), in which a_i is random in $[0, P]$, for $i = 1, 2, \cdots k-1$, and P is any prime greater than a_0.

$$f(x) = (a_0 + a_1 x + \cdots + a_{k-1} x^{k-1}) \bmod P \tag{1}$$

For any given x, $f(x)$ will be obtained according to Eq. (1) and these n pairs $\{(i, sc_i)\}_{i=1}^{n}$ are distributed to n owners. In the reconstructed stage of original polynomial-based SS, given any k pairs of these n pairs $\{(i, sc_i)\}_{i=1}^{n}$, where i serves as an identifying index or an order label corresponding to the ith owner, we can obtain the coefficients of $f(i)$ by the Lagrange's interpolation, and then evaluate $s = f(0)$. And the secret data a_0 cannot be overall reconstructed with less than k shadows.

Following original polynomial-based SS scheme and utilizing all coefficients of the polynomial for embedding secret, Thien and Lin [9] reduced shadow size $1/k$ times to the secret image as follows. Thien and Lin's contributions are applying polynomial-based SS to SIS with smaller shadow size.

The differences between original polynomial-based SS and polynomial-based SIS with smaller shadow size (namely follow-up polynomial-based SIS) lie in:

1. Only the first coefficient a_0 in Eq. (1) covers the secret in original polynomial-based SS, while all the k coefficients $a_0, a_1, \cdots a_{k-1}$ cover the secret pixels in follow-up polynomial-based SIS. Thus, shadow size of follow-up polynomial-based SIS is reduced $1/k$ times to that of original polynomial-based SIS.

Thien and Lin's polynomial-based SIS (follow-up polynomial-based SIS).
Input: A $M \times N$ grayscale secret image S, the threshold parameters (k, n) and permutation *key*.
Output: n shadows $SC_1, SC_2, \cdots SC_n$.
Step 1: A prime number $P = 251$ is selected. For each position $(h, w) \in \{(h, w)\|1 \leq h \leq M, 1 \leq w \leq N\}$, if $s = S(h, w) > P - 1$ we set $s = P - 1$. **Step 2:** Encrypt the secret image to obtain S_1 using permutation method with the input *key*. **Step 3:** Every k not-shared-yet pixels (a block) of S_1 are sequentially picked up to obtain a block, denoted as $B_0, B_1, \cdots B_{k-1}$. **Step 4:** Generate a $k-1$ degree polynomial $$f(x) = (a_0 + a_1x + \cdots + a_{k-1}x^{k-1}) \bmod P \quad (2)$$ in which $a_i = B_i$, $i = 0, 1, 2, \cdots k-1$. **Step 5:** Compute $$sc_1 = f(1), \cdots, sc_i = f(i), \cdots, sc_n = f(n). \quad (3)$$ **Step 6:** Arrange $sc_1, sc_2, \cdots sc_n$ to the corresponding pixel values of $SC_1, SC_2, \cdots SC_n$. **Step 7:** Repeat Steps 3-6 until all the pixels of S_1 are processed. **Step 8:** Output the n shadows $SC_1, SC_2, \cdots SC_n$

2. Before the really sharing processing, the secret image S is generated to obtain secret image S_1 in follow-up polynomial-based SIS. Then the encrypted secret image S_1 is the secret image for sharing.
3. P is any prime number larger than secret a_0 in original polynomial-based SS; $P = 251$ in follow-up polynomial-based SIS.

Therefore, we need to deal with following issues when original polynomial-based SS is applied to SIS.

1. Since an image contains a big amount of data, therefore following follow-up polynomial-based SIS schemes intend to share more secret pixels once to improve efficiency. However, since the correlation between pixels of an image, sometimes it is not secure without auxiliary encryption. All the k coefficients $a_0, a_1, \cdots a_{k-1}$ cover the secret pixels in follow-up polynomial-based SIS, and the pixel value of one pixel is close to its adjacent pixels value, thus we have $a_0 \approx a_1, \cdots \approx a_{k-1}$. Then, from $f(1) = (a_0 + a_1 + \cdots + a_{k-2} + a_{k-1}) \bmod P$, we know $s_1' = \frac{f(1)}{k} \approx a_0$. Hence, $s_1' \approx f(s_1)$. s_1' has a relation with the encrypted secret result. s_1' will give clue about the secret s without encryption technique; on the other hand, if auxiliary permutation encryption is performed, the security is relied on encryption other than SIS and permutation *key* transmission will lead to another issue.
2. P cannot be any prime greater than a_0 due to $0 \leq sc_i \leq 255$, for $i = 0, 1, 2, \cdots k-1$. The primes close to 255 are 257 or 251. If 257 is selected, we cannot store the i-th shadow pixel when $sc_i = 256$; otherwise 251 is selected, we cannot reconstruct the secret pixel a_0 when $a_0 \geq 251$.

We further analyze the ways to deal with the above issues when directly applying SS to SIS.

1. The statistical feature of the secret image can be used as permutation *key* to partially avoid the permutation *key* transmission issue [28] when auxiliary encryption has to be used.
2. If $P = 257$ is selected and only a_0 is utilized to cover the secret pixel, we can screen the random numbers, i.e., $a_1, a_2, \cdots, a_{k-1}$, to avoid shadow pixel values larger than 255, thus to store shadow pixels [28] and to achieve lossless reconstruction.
3. If only a_0 is utilized to cover the secret pixels, we can utilize primitive polynomial for $GF(2^8)$ to achieve lossless reconstruction and to avoid shadow pixel values larger than 255 at the cost of larger computation.

3.2 CRT-Based SS and SIS Schemes

In Chinese southern and northern dynasties, CRT was formally introduced, which can solve a set of linear congruence equations.

First an integers set, denoted as $m_i (i = 1, 2, \cdots, k)$, is chosen satisfying $\gcd(m_i, m_j) = 1, i \neq j$, then according to CRT we have a unique solution $y \equiv \left(a_1 M_1 M_1^{-1} + a_2 M_2 M_2^{-1} + \cdots + a_k M_k M_k^{-1}\right) \pmod M$, $(y \in [0, M-1])$ to Eq. (4).

$$
\begin{aligned}
y &\equiv a_1 \pmod{m_1} \\
y &\equiv a_2 \pmod{m_2} \\
&\cdots \\
y &\equiv a_{k-1} \pmod{m_{k-1}} \\
y &\equiv a_k \pmod{m_k}
\end{aligned}
\tag{4}
$$

where $M = \prod_{i=1}^{k} m_i$, $M_i = M/m_i$ and $M_i M_i^{-1} \equiv 1 \pmod{m_i}$.

Obviously, we can set $s = y$ to achieve SS for (k, k) threshold, which can be applied to encryption and secret fusion [29]. The significant work is to achieve SS for (k, n) threshold. Asmuth and Bloom [6] introduced one CRT-based SS for (k, n) threshold, as follows.

When applying Asmuth and Bloom's CRT-based SS for (k, n) threshold to SIS, we need to consider the following issues.

1. In Step 1 of Asmuth and Bloom's CRT-based SS, if s covers the grayscale secret image pixel and sc_i is served as the i-th shadow pixel value for $i = 1, 2, \cdots, n$, since $0 \le s \le 255$ and $s < p$, we have $255 \le p$; since $0 \le a_i \le 255$ and $a_i \le m_i - 1$, we have $m_i \le 256$; $255 \le p$ and $m_i \le 256$ are in contradiction with $\{p < m_1 < m_2 \cdots < m_n\}$.

Asmuth and Bloom's CRT-based SS for (k, n) threshold.
Input: Secret s and threshold (k, n). **Output**: n shadows $SC_1, SC_2, \cdots SC_n$.
Step 1: Choose a set of integers $\{s < p < m_1 < m_2 \cdots < m_n\}$ satisfying 1. $\gcd(m_i, m_j) = 1, i \neq j$. 2. $\gcd(m_i, p) = 1$ for $i = 1, 2, \cdots, n$. 3. $M > pN$ where $M = \prod_{i=1}^{k} m_i$, $N = \prod_{i=1}^{k-1} m_{n-i+1}$ and p is public among all the owners. **Step 2:** Let $x = s$, choose a random integer A in $\left[\left\lceil \frac{N}{p} \right\rceil, \left\lfloor \frac{M}{p} - 1 \right\rfloor\right]$ and let $y = x + Ap$. **Step 3:** Calculate $a_i \equiv y \,(\bmod\, m_i)$ and let $sc_i = a_i$ for $i = 1, 2, \cdots, n$. **Step 4:** Output n shadows $sc_1, sc_2, \cdots sc_n$ and their corresponding privacy modular integers $m_1, m_2, \cdots m_n$.

2. We can divide the grayscale secret image pixel value range into two parts, such as, $0 \leq s < 128$ and $128 \leq s \leq 255$, or $0 \leq s < 131$ and $131 \leq s \leq 255$, i.e., $p = 128/131$, thus we can satisfy $\{s < p < m_1 < m_2 \cdots < m_n \leq 255\}$ and achieve SIS when $0 \leq s < p$.
3. When $p \leq s \leq 255$, we can Calculate $1 \leq s - p < p$. We need to design an identity to separate $0 \leq s < p$ and $p \leq s \leq 255$, such as, the interval division of A in Step 2 of Asmuth and Bloom's CRT-based SS.
4. We suggest that p is as small as possible for security as well as m_i is as large as possible so that the pixel values in shadows can randomly lie in large range.

According to the above analyses, we can derive CRT-based SIS for (k, n) threshold [30] and its reconstructing algorithm as follows.

Based on CRT-based SIS for (k, n) threshold, we can extend more schemes with admirable features, such as, reduced shadow size [31], multiple decryptions [32], information hiding, and so on [33].

Since polynomial-based SIS and CRT-based SIS are studied, respectively. We discuss their differences as follows.

1. The number of owners is not limited in polynomial-based SIS, while that in CRT-based SIS is in general small, such as, $n \leq 6$, since as n increases the available values of m_i decrease, which will affect the distribution of shadow pixels values and thus further lead to secure issue.
2. The principle of CRT-based SIS is complex, which is hard to be understood.
3. The shadow size of polynomial-based SIS is easy to be reduced.
4. CRT-based SIS can achieve lossless reconstruction, while most polynomial-based SIS schemes are lossy.
5. The reconstruction operation is Lagrange's interpolation ($O(k \log^2 k)$ in polynomial-based SIS, while that in CRT-based SIS is modular operation ($O(k)$) [6], thus CRT-based SIS needs lower computation cost than polynomial-based SIS to reconstruct the secret image.

CRT-based SIS for (k,n) threshold.
Input: The original secret image S with size of $H \times W$ and threshold (k,n). **Output**: n shadows $SC_1, SC_2, \cdots SC_n$.
Step 1: Choose a set of integers $\{128 \le p < m_1 < m_2 \cdots < m_n \le 255\}$ satisfying 1. $\gcd(m_i, m_j) = 1, i \ne j$. 2. $\gcd(m_i, p) = 1$ for $i = 1, 2, \cdots, n$. 3. $M > pN$ where $M = \prod_{i=1}^{k} m_i$, $N = \prod_{i=1}^{k-1} m_{n-i+1}$ and p is public among all the owners. **Step 2:** Calculate $T = \left\lfloor \frac{\left\lfloor \frac{M}{p} - 1 \right\rfloor - \left\lceil \frac{N}{p} \right\rceil}{2} + \left\lceil \frac{N}{p} \right\rceil \right\rfloor$ and T is public among all the owners as well. For each position $(h,w) \in \{(h,w) \mid 1 \le h \le H, 1 \le w \le W\}$, repeat Steps 3-4 . **Step 3:** Let $x = S(h,w)$. If $0 \le x < p$, choose a random integer A in $\left[T+1, \left\lfloor \frac{M}{p} - 1 \right\rfloor\right]$ and let $y = x + Ap$. Else choose a random integer A in $\left[\left\lceil \frac{N}{p} \right\rceil, T\right)$ and let $y = x - p + Ap$. **Step 4:** Calculate $a_i \equiv y \,(\text{mod } m_i)$ and let $SC_i(h,w) = a_i$ for $i = 1, 2, \cdots, n$. **Step 5:** Output n shadows $SC_1, SC_2, \cdots SC_n$ and their corresponding privacy modular integers $m_1, m_2, \cdots m_n$.

Secret image reconstructed method of CRT-based SIS for (k,n) threshold.
Input:k shadows $SC_{i_1}, SC_{i_2}, \cdots SC_{i_k}$, their corresponding privacy modular integers $m_{i_1}, m_{i_2}, \cdots m_{i_k}$, p and T. **Output**: A $H \times W$ reconstructed secret image S'.
Step 1: For each position $(h,w) \in \{(h,w) \mid 1 \le h \le H, 1 \le w \le W\}$, repeat Steps 2-3. **Step 2:** Let $a_{i_j} = SC_{i_j}(h,w)$ for $j = 1, 2, \cdots, k$. Solve the following linear equations by CRT. $$\begin{aligned} y &\equiv a_{i_1} \,(\text{mod } m_{i_1}) \\ y &\equiv a_{i_2} \,(\text{mod } m_{i_2}) \\ &\cdots \\ y &\equiv a_{i_{k-1}} \,(\text{mod } m_{i_{k-1}}) \\ y &= a_{i_k} \,(\text{mod } m_{i_k}) \end{aligned} \quad (5)$$ **Step 3:** Calculate $T^* = \left\lfloor \frac{y}{p} \right\rfloor$. If $T^* \ge T$, let $x \equiv y \,(\text{mod } p)$. Else let $x = y \,(\text{mod } p) + p$. Set $S'(h,w) = x$. **Step 4:** Output the reconstructed secret image S'

3.3 VSS

In RG-based VSS [34], '0' denotes a white pixel and '1' denotes a black pixel. The generation and reconstruction stages of one typical original $(2,2)$ RG-based VSS are as follows.

Generation step 1: Randomly generate 1 RG SC_1 according to coin flipping function.

Generation step 2: Calculate SC_2 as in Eq. (6).

Reconstruction stage: $S' = SC_1 \otimes SC_2$ as Eq. (7). If a secret pixel $s = S(i, j)$ of S is 1, the reconstructed bit $sc_1 \otimes sc_2 = 1$ is always black. If the secret pixel is 0, the reconstructed bit $sc_1 \otimes sc_2 = SC_1(i, j) \otimes SC_1(i, j)$ has half chance to be black or white since sc_1 is generated randomly.

$$SC_2(i,j) = \begin{cases} SC_1(i,j) & if\ S(i,j) = 0 \\ \overline{SC_1(i,j)} & if\ S(i,j) = 1 \end{cases} \tag{6}$$

$$\begin{aligned} S'(i,j) =& SC_1(i,j) \otimes SC_2(i,j) \\ =& \begin{cases} SC_1(i,j) \otimes SC_1(i,j) & if\ S(i,j) = 0 \\ SC_1(i,j) \otimes \overline{SC_1(i,j)} = 1 & if\ S(i,j) = 1 \end{cases} \end{aligned} \tag{7}$$

The above method can be extended to (k, n) threshold scheme by repeatedly applying the above process to the first k bits and generating the last $n - k$ bits to be equal to subset of the first k bits or even to 0. For some (k, n) RG-based VSS examples, please refer to [19,35].

Recall that, VSS can utilize naked human eyes to reveal the original secret image and image responds to human eyes, which is thus originally designed for image other than data, in such as way VSS doesn't face the issues when applying to SIS.

4 Experimental Results and Analyses

In this section, we perform experiments to verify our discussions on relations between SS and SIS.

4.1 Polynomial-Based SIS Experiments

First, we will share a secret image by original polynomial-based SIS for $(3, 4)$ threshold, $P = 251$ and gray secret image in Fig. 2(a). As Fig. 2(b), the first shadow SC_1 of the generated four shadows is noise-like. The secret image reconstructed by the first two or more shadows are exhibited in Fig. 2(c)–(e), where $S'_{t=2}$ denotes the reconstructed secret image by the first two shadows. The reconstructed secret image is lossy due to modular 251 when pixel value is larger than 250.

Second, Fig. 3 indicates the experimental result of follow-up polynomial-based SIS without secret image permutation for $(3, 4)$ threshold. Figure 3(a) denotes the grayscale secret image whose four generated shadows SC_1, SC_2, SC_3 and SC_4 are demonstrated in Fig. 3(b)–(e). The reconstructed results by different number of shadows are in Fig. 3(f)–(h). Although reduced shadow size is achieved, follow-up polynomial-based SIS without secret image permutation suffers from lossy reconstruction and information leakage.

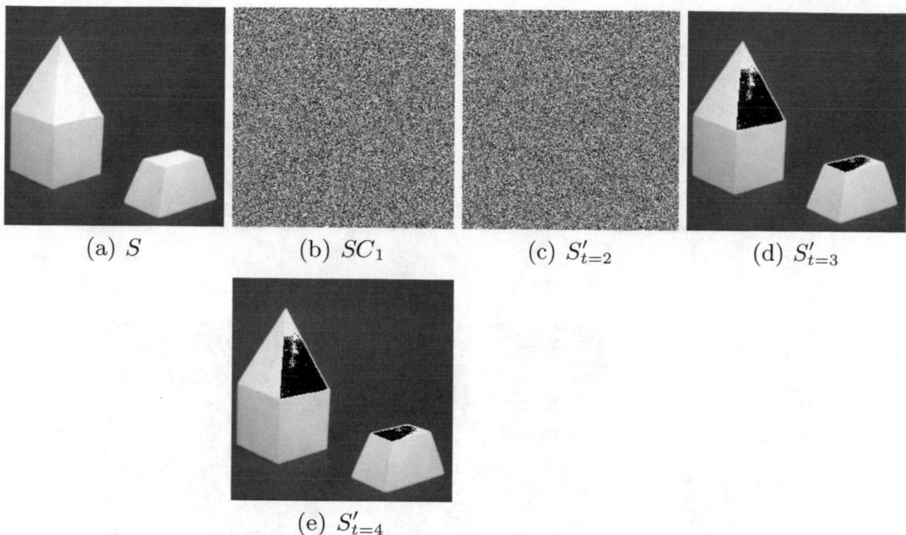

(a) S (b) SC_1 (c) $S'_{t=2}$ (d) $S'_{t=3}$

(e) $S'_{t=4}$

Fig. 2. Experimental results of original polynomial-based SIS method modulo 251, where $k = 3, n = 4$. (a) The gray secret image; (b) first shadow SC_1; (c)–(e) reconstructed results by different number of shadows.

Finally, Fig. 4 shows the results of polynomial-based SIS modulo 257 based on screening for (3, 4) threshold. Figure 4(a) presents the original grayscale secret image. The four generated shadows SC_1, SC_2, SC_3 and SC_4 are illustrated in Fig. 4(b)–(e), which are noise-like. The reconstructed results by different number of shadows are in Fig. 4(f)–(h). The reconstructed secret image is lossless without information leakage.

4.2 CRT-Based SIS Experiments

Figure 5 shows the results of CRT-based SIS for (k, n) threshold, where $k = 3, n = 4$, $p = 131, m_1 = 251, m_2 = 253, m_3 = 254$ and $m_4 = 255$. Figure 5(a) illustrates the original grayscale secret image. The four generated shadows SC_1, SC_2, SC_3 and SC_4 are illustrated in Fig. 5(b)–(e), which are noise-like. The reconstructed results by different number of shadows are in Fig. 5(f)–(h). The reconstructed secret image is lossless without information leakage by only modular operation.

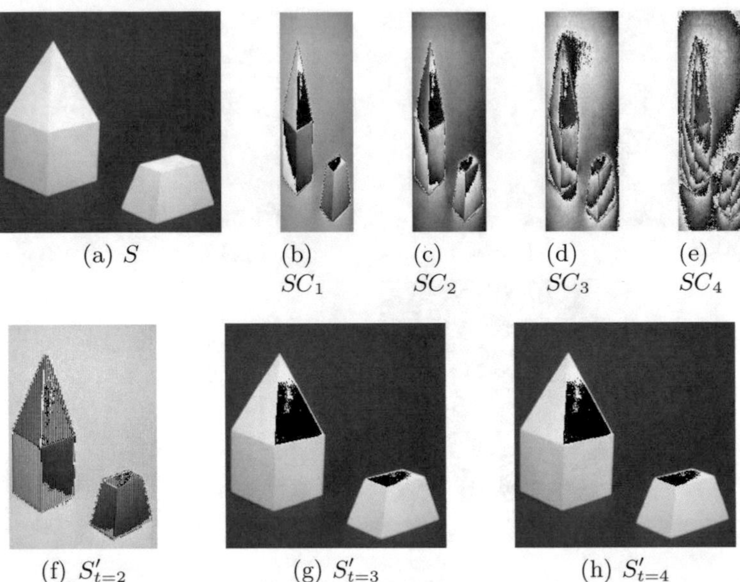

Fig. 3. Experimental results of follow-up polynomial-based SIS modulo 251 without secret image encryption, where $k = 3, n = 4$. (a) The secret image; (b)–(e) four shadows SC_1, SC_2, SC_3 and SC_4; (f)–(h) reconstructed results by different number of shadows

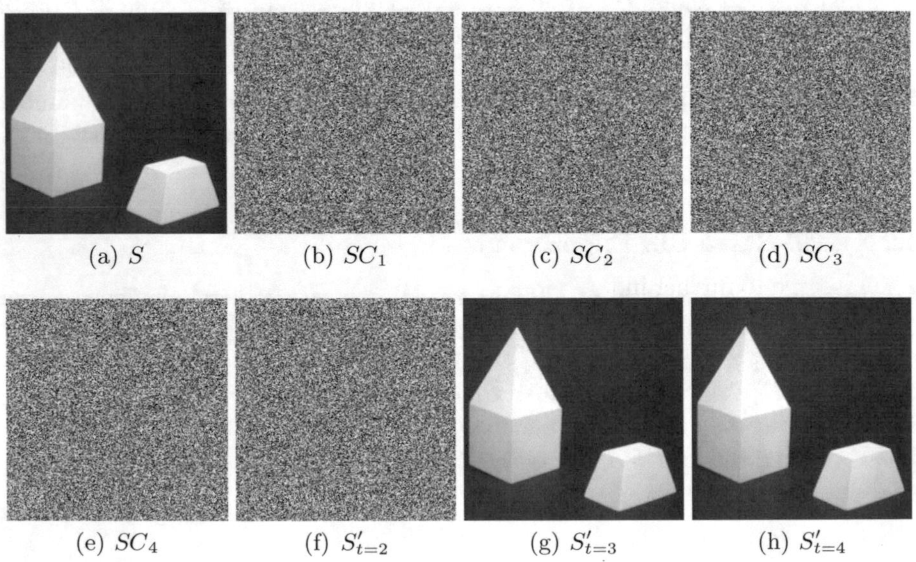

Fig. 4. Experimental results of polynomial-based SIS modulo 257 based on screening, where $k = 3, n = 4$. (a) The secret image; (b)–(e) four shadows SC_1, SC_2, SC_3 and SC_4; (f)–(h) reconstructed results by different number of shadows.

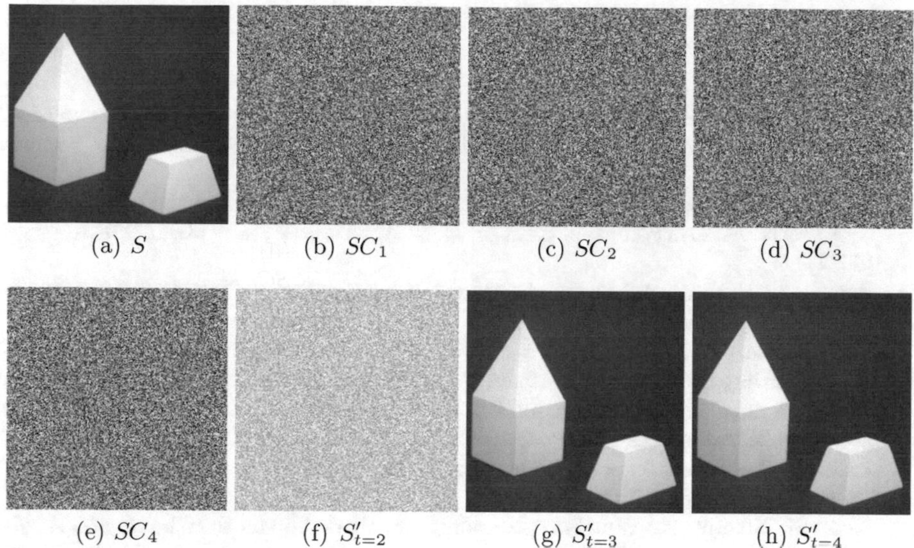

Fig. 5. Experimental results of CRT-based SIS for (k, n) threshold, where $k = 3, n = 4$, $p = 131, m_1 = 251, m_2 = 253, m_3 = 254$ and $m_4 = 255$. (a) The secret image; (b)–(e) four shadows SC_1, SC_2, SC_3 and SC_4; (f)–(h) reconstructed results by different number of shadows.

Based on the above results we can verify our analyses.

5 Conclusion

In this paper, first we discuss the relations between encryption, information hiding, SS and SIS, especially the relations between SIS and SS, where differences between SIS and SS are mainly considered. Then, some typical sharing principles are employed to further indicate the differences and analyze the ways to deal with the issues when directly applying SS to SIS, such as, lossy reconstruction and auxiliary encryption, where some possible ways are provided as well. Finally, we perform experiments to verify our analyses. Analyzing and evaluating the security level of SIS will be our future work.

Acknowledgment. The authors would like to thank the anonymous reviewers for their valuable comments. This work is supported by the National Natural Science Foundation of China (Grant Number: 61602491) and Key Program of National University of Defense Technology.

References

1. Shamir, A.: How to share a secret. Commun. ACM **22**(11), 612–613 (1979)
2. Blakley, G.R.: Safeguarding cryptographic keys. In: Proceedings of the National Computer Conference, New York, USA, pp. 313–317. IEEE Computer Society. IEEE (1979)
3. Yan, X., Lu, Y., Liu, L., Wan, S., Ding, W., Liu, H.: Exploiting the homomorphic property of visual cryptography. Int. J. Digit. Crime Forensics (IJDCF) **9**(2), 45–56 (2017)
4. Belazi, A., El-Latif, A.A.A.: A simple yet efficient s-box method based on chaotic sine map. Optik Int. J. Light Electron Opt. **130**, 1438–1444 (2017)
5. Yan, X., Lu, Y., Liu, L., Wan, S., Ding, W., Liu, H.: Security analysis of secret image sharing. In: Proceedings of the Data Science: Third International Conference of Pioneering Computer Scientists, Engineers and Educators, ICPCSEE 2017, Part I, Changsha, China, 22–24 September 2017, pp. 305–316 (2017)
6. Asmuth, C., Bloom, J.: A modular approach to key safeguarding. IEEE Trans. Inf. Theory **29**(2), 208–210 (1983)
7. Naor, M., Shamir, A.: Visual cryptography. In: Workshop on the Theory and Application of Cryptographic Techniques on Advances in Cryptology-EUROCRYPT 1994. Lecture Notes in Computer Science, Perugia, Italy, 9–12 May 1994, pp. 1–12. Springer (1995)
8. Wang, G., Liu, F., Yan, W.Q.: Basic visual cryptography using Braille. Int. J. Digit. Crime Forensics **8**(3), 85–93 (2016)
9. Thien, C.C., Lin, J.C.: Secret image sharing. Comput. Graph. **26**(5), 765–770 (2002)
10. Yang, C.N., Ciou, C.B.: Image secret sharing method with two-decoding-options: lossless recovery and previewing capability. Image Vis. Comput. **28**(12), 1600–1610 (2010)
11. Li, P., Ma, P.J., Su, X.H., Yang, C.N.: Improvements of a two-in-one image secret sharing scheme based on gray mixing model. J. Vis. Commun. Image Represent. **23**(3), 441–453 (2012)
12. Li, P., Yang, C.N., Kong, Q.: A novel two-in-one image secret sharing scheme based on perfect black visual cryptography. J. Real-Time Image Proc. **14**(1), 41–50 (2018)
13. Cimato, S., De Prisco, R., De Santis, A.: Probabilistic visual cryptography schemes. Comput. J. **49**(1), 97–107 (2006)
14. Wang, D., Zhang, L., Ma, N., Li, X.: Two secret sharing schemes based on boolean operations. Pattern Recognit. **40**(10), 2776–2785 (2007)
15. Wang, Z., Arce, G.R., Di Crescenzo, G.: Halftone visual cryptography via error diffusion. IEEE Trans. Inf. Forensics Secur. **4**(3), 383–396 (2009)
16. Weir, J., Yan, W.: A comprehensive study of visual cryptography. In: Shi, Y.Q. (ed.) Transactions on Data Hiding and Multimedia Security V. LNCS, vol. 6010, pp. 70–105. Springer, Heidelberg (2010)
17. Yan, X., Wang, S., Niu, X.: Threshold construction from specific cases in visual cryptography without the pixel expansion. Sig. Process. **105**, 389–398 (2014)
18. Wu, X., Sun, W.: Improving the visual quality of random grid-based visual secret sharing. Sig. Process. **93**(5), 977–995 (2013)
19. Yan, X., Liu, X., Yang, C.N.: An enhanced threshold visual secret sharing based on random grids. J. Real-Time Image Proc. **14**(1), 61–73 (2018)
20. Guo, T., Liu, F., Wu, C.: Threshold visual secret sharing by random grids with improved contrast. J. Syst. Softw. **86**(8), 2094–2109 (2013)

21. Fu, Z.x., Yu, B.: Visual cryptography and random grids schemes. In: Shi, Y., Kim, H.J., Pérez-González, F. (eds.) Digital-Forensics and Watermarking, pp. 109–122. Springer, Heidelberg (2014)
22. Yan, X., Wang, S., Niu, X., Yang, C.N.: Generalized random grids-based threshold visual cryptography with meaningful shares. Sig. Process. **109**, 317–333 (2015)
23. Yan, X., Wang, S., Li, L., El-Latif, A.A.A., Wei, Z., Niu, X.: A new assessment measure of shadow image quality based on error diffusion techniques. J. Inf. Hiding Multimedia Signal Process. (JIHMSP) **4**(2), 118–126 (2013)
24. Liu, F., Wu, C.: Embedded extended visual cryptography schemes. IEEE Trans. Inf. Forensics Secur. **6**(2), 307–322 (2011)
25. Yan, X., Wang, S., Niu, X., Yang, C.N.: Halftone visual cryptography with minimum auxiliary black pixels and uniform image quality. Digit. Sig. Process. **38**, 53–65 (2015)
26. Ateniese, G., Blundo, C., De Santis, A., Stinson, D.R.: Visual cryptography for general access structures. Inf. Comput. **129**(2), 86–106 (1996)
27. Yan, X., Lu, Y.: Progressive visual secret sharing for general access structure with multiple decryptions. Multimedia Tools Appl. **77**(2), 2653–2672 (2018)
28. Zhou, X., Lu, Y., Yan, X., Wang, Y., Liu, L.: Lossless and efficient polynomial-based secret image sharing with reduced shadow size. Symmetry **10**(7), 249 (2018)
29. Yan, X., Lu, Y., Liu, L., Liu, J., Yang, G.: Secret data fusion based on Chinese remainder theorem. In: 2018 3rd IEEE International Conference on Image, Vision and Computing, pp. 380–385. IEEE (2018)
30. Yan, X., Lu, Y., Liu, L., Wan, S., Ding, W., Liu, H.: Chinese remainder theorem-based secret image sharing for (k, n) threshold. In: Cloud Computing and Security: Third International Conference, ICCCS 2017, Part II. Revised Selected Papers, Nanjing, China, 16–18 June 2017, pp. 433–440 (2017)
31. Chen, J., Liu, K., Yan, X., Liu, L., Zhou, X., Tan, L.: Chinese remainder theorem-based secret image sharing with small-sized shadow images. Symmetry **10**(8), 340 (2018)
32. Yan, X., Lu, Y., Liu, L., Liu, J., Yang, G.: Chinese remainder theorem-based two-in-one image secret sharing with three decoding options. Digit. Sig. Process. **82**, 80–90 (2018)
33. Yan, X., Lu, Y., Liu, L.: General meaningful shadow construction in secret image sharing. IEEE Access **6**(1), 45246–45255 (2018)
34. Kafri, O., Keren, E.: Encryption of pictures and shapes by random grids. Opt. Lett. **12**(6), 377–379 (1987)
35. Yan, X., Lu, Y., Liu, L., Wan, S.: Random grids-based threshold visual secret sharing with improved visual quality. In: Digital Forensics and Watermarking: 15th International Workshop, IWDW 2017. Revised Selected Papers, Beijing, China, 17–19 September 2016, pp. 209–222 (2016)

Enhanced Secret Image Sharing Using POB Number System

Yan-Xiao Liu[1(✉)] and Ching-Nung Yang[2]

[1] Xi'an University of Technology, Xi'an, China
liuyanxiao@xaut.edu.cn

[2] National Dong Hwa University, Hualien County, Taiwan
cnyang@gms.ndhu.edu.tw

Abstract. In 2017, Singh et al. constructed a secure cloud-based $(2, 2)$ secret image sharing with the capability of tampering detection using (r, n) Permutation Ordered Binary (POB) number system. This scheme can solve the problem of secure storage of image-shares at cloud servers, and is also capable of authenticating image-shares at pixel level. However, Singh et al.'s scheme assumes that participants already know the parameters r in (r, n) POB, on which they can recover each image pixel efficiently. In fact, the distribution of parameters r is not that easy as described in their scheme, it needs secure channels and the storage of r is almost same as the share. In Singh et al.'s scheme, the authors did not show how to share the information of r among participants. In this paper, we proposed a new secret image sharing using (r, n) POB number system which is extended from Singh et al.'s framework. In our scheme, the parameters r for all image pixels are identical and are publicly published. Thus, the problem of Singh et al.'s scheme is solved. In addition, our scheme can also authenticate tampered shares at pixel level.

Keywords: Image secret sharing · Permutation Ordered Binary (POB) · Authentication

1 Introduction

Secret image sharing is an important issue in the field of information security that has been researched for decades. In (k, n) secret image sharing, an secret image with important information is encrypted into n shadows in such way that only k or more shadows can decode secret image, less than k shadows get no information on the image. Using secret image sharing, n participants can collaborate to safely guarding a secret image. There were many different approaches of secret image sharing, the schemes [1–3] were based on interpolation polynomial, [4–6] were based on visual cryptography, the schemes [7,8] were using Boolean operation at pixel level, and the scheme [9] was based on the data hiding scheme GEMD, and the schemes [10,11] was based on POB number system. Each approach has its own advantage in shadow size or quality of recovered image respectively.

C.-N. Yang et al. (Eds.): SICBS 2018, AISC 895, pp. 94–102, 2020.
https://doi.org/10.1007/978-3-030-16946-6_8

In 2017, Singh et al. [12] proposed a POB based (2, 2) secret image sharing in background of cloud computing. In their scheme, a secret image is encrypted into 2 shadows using (r, n) POB number system, and each shadow is distributed at a cloud data center respectively. The reconstruction of secret image can be only accomplished with both the two shadows, and any tamper on the shadow can be also authenticated. The secure problem [13, 14] is an important issue in cloud computing, the data which is stored in cloud center could be tampered by intruders, thus the integrity authentication is necessary when retrieving data from cloud center. There were schemes [15–17] that considered the secure problem in cloud computing. Comparing to those schemes, the POB based secret image sharing has higher efficiency in integrity authentication, it is capable of detecting tampered shadows at the pixel level.

In this work, we proposed an enhanced POB based secret image sharing which is extended from the Singh et al.'s work. In Singh et al.'s scheme, each pixel of secret image is encrypted into two shares using (r, n) POB respectively, with the acknowledgement on r and two shares, the original pixel can be recovered. It is assumed in [12] that the different parameters r for all pixels are already known by all users. In fact, the distribution of parameters r is complicated, it not only increases the cost the transmission, but also the storage of shares. Since storage for each r is as same as one pixel. In proposed scheme, we modify Singh et al.'s scheme by using a fixed parameter r for all pixels, and the parameter r is published to all users. In this way, users do not need to store the parameter r, it saves the cost of communication and storage from Singh et al.'s scheme. In addition, our scheme can also authenticate shadows in pixel level efficiently when retrieving it from cloud center.

The rest of this paper is organized as follows. In Sect. 2, we describe the definition of POB number system and Singh et al.'s POB based (2, 2) secret image sharing scheme respectively. In Sect. 3, we proposed our POB based secret image sharing scheme which is extended from Singh et al.'s scheme and capable of the problem of distribution on parameter r. In Sect. 4, experimental results show the reconstructed image using our scheme and the theoretical analysis on tampering detection is also described in this section. The conclusion is made in Sect. 5.

2 Preliminaries

In this section, we describe POB number system and the POB based secret image sharing scheme in [12] respectively.

2.1 POB Number System

POB number system was first introduced in the work [10]. It is combined with two parameters $r, n, (r \leq n)$, and is denoted as $POB(r, n)$. In this number system, each decimal integer in the set $\{0, 1, ..., C_n^r - 1\}$ can be presented as a binary string $B = b_{n-1}b_{n-2}...b_1b_0$, where this n-length string contains exactly r 1s and $n-r$ 0s. The value of POB number $B = b_{n-1}b_{n-2}...b_1b_0$ is:

$$V(B) = \sum_{j=0}^{n-1} b_j \times C_{p_j}^{j}, where\ p_j = \sum_{i=0}^{j} b_i \quad (1)$$

For instance, the value of $(4,9)$ POB number 001101010_P is 29. It is also proved that each decimal integer in $\{0,1,...,C_n^r - 1\}$ has a unique (r,n) POB number. The algorithm for finding a (r,n) POB number of an integer in $\{0,1,...,C_n^r - 1\}$ is described in following Algorithm 1.

Algorithm 1. Finding POB number for an integer

Input: r, n, an integer $W \in [0, C_n^r - 1]$; Output: (r,n) POB number $B = b_{n-1}b_{n-2}...b_1b_0$
Set $j = n, temp = W$
For $(k = r$ to $1)$
  Repeat $\{j = j-1;\ p = C_j^k;$
    if $(temp \geq p)$
      $\{\ temp = temp - p;\ b_j = 1;\}$
    else $b_j = 0$
      $\}$ (until $b_j == 1$)
if $(j > 0)$
For $(k = j-1$ to $0)$
  $b_k = 0;$

2.2 POB Based (2, 2) Secret Image Sharing

In [12], Singh et al. constructed a POB based (2, 2) secret image sharing scheme which can be adopted in cloud computing. In their scheme, a secret image is divided into 2 shadows using POB number system, which are stored in different cloud data centers. The image can be reconstructed with both the two shadows, and any tamper of shadows can be authenticated efficiently at the pixel level. The scheme is described as follows.

SCHEME 1: *POB based* $(2,2)$ *secret image sharing*
Shadow Generation phase:
Input secret image O with $u \times v$ pixels $O = \{s_{i,j} | 1 \leq i \leq u, 1 \leq j \leq v\}$; Output two shadows S_1, S_2.

1 For each pixel $s_{i,j}, i \in [1,u], j \in [1,v]$, divide $s_{i,j}$ into two shares $s_{i,j}^1, s_{i,j}^2$ that satisfy $s_{i,j} = s_{i,j}^1 \oplus s_{i,j}^2$.
2 Generating two extra bits $Auth_1, Auth_2$ for authentication of each share, and then all shares are extended to binary strings with length 10, $B_{i,j}^t = s_{i,j}^t || Auth_1 || Auth_2, t \in \{1,2\}$. The methods of generating $Auth_1, Auth_2$ from a share are described in following Algorithms 2 and 3 respectively.

3 Each binary string $B^t_{i,j}$ is treated as a $(r, 10)$ POB number, and compute the value $V(B^t_{i,j})$ of the $(r, 10)$ POB number $B^t_{i,j}$. r is the number of 1s in the string $B^t_{i,j}$.
4 The two shadows S_1, S_2 are

$$S_t = \{V(B^t_{i,j})|i \in [1, u], j \in [1, v]\}t = 1, 2 \tag{2}$$

Image Reconstruction phase:
Input: two shadows S_1, S_2; Output: secret image O.

1 Extracting each $V(B^t_{i,j})$ from $S_t, t = 1, 2$, transfer $V(B^t_{i,j})$ to its $(r, 10)$ POB number $B^t_{i,j}$.
2 Each $B^t_{i,j}$ is a 10 bits length binary string $B^t_{i,j} = s^t_{i,j}||Auth_1Auth_2$, where $s^t_{i,j}$ is a byte and $Auth_1, Auth_2$ are two bits for authentication. Then authenticating each byte $s^t_{i,j}$ using $Auth_1, Auth_2$ according to Algorithms 2 and 3.
   i if all $s^t_{i,j}$ pass authentication, then the secret image O can be recovered by: $O = \{s_{i,j}|i \in [1, u], j \in [1, v]\}$, where $s_{i,j} = s^1_{i,j} \oplus s^2_{i,j}$.
   ii else, output the tampered share $s^t_{i,j}$.

Algorithm 2. Generation of $Auth_1$

Input: a share $s^t_{i,j} = b_7b_6...b_1b_0$ of 8-length binary string; Output: a bit $Auth_1$.

- Transform i, j to corresponding binary strings $\mathcal{B}(i), \mathcal{B}(j)$
- Compute $\delta = s^t_{i,j} \oplus \mathcal{B}(i) \oplus \mathcal{B}(j)$.
- Suppose there are l of 1s in δ, and the positions of these 1s are $q_1, q_2, ..., q_l$, then the bit $Auth_1$ is computed as: $Auth_1 = (\sum_{i=1}^{l} q_i) mod2$

Algorithm 3. Generation on $Auth_2$ of share $s^t_{i,j}$

Input: a share $s^t_{i,j}$ and its neighbor share $s^t_{i,j+1}$; Output: a bit $Auth_2$.

- Compute $\lambda = s^t_{i,j} \oplus s^t_{i,j+1}$.
- Suppose there are l of 1s in λ, and the positions of these 1s are $q_1, q_2, ..., q_l$, then the bit $Auth_2$ is computed as: $Auth_2 = (\sum_{i=1}^{l} q_i) mod2$

---

3 Proposed Scheme

As described previously, the POB based secret image sharing scheme [12] encrypts each pixel of secret image into two shares using a $(r, 10)$ POB number system. The secret image cannot be reconstructed without acknowledgement of r. In [12], it is assumed that all users have obtained the information of all r before image reconstruction, however, the distribution of all r to users is complicated. First, there are totally uv different r (one for each pixel), the transmission would cause secure problem and increase communication cost; second, each r takes about $log10 = 4$ bits for storage, which is half of a share. Thus the distribution of r would greatly increase storage cost for each user. In this part, we propose a new POB based secret image sharing which is extended from the scheme [12], that can solve the problem of distribution of r. In our scheme, all pixels in secret image is encrypted into two shares using $(r, 10)$ POB number system, where the parameter r for all different pixels is fixed, i.e. $r = 5$. The value of r is published to each user, thus the step of distribution of r can be omitted. On the other hand, our scheme is also capable of authenticating any tampered shadows in pixel level. Our scheme is shown in following Scheme 2.

Scheme 2: *Proposed POB based* $(2, 2)$ *secret image sharing*
Shadow Generation phase:
Input secret image O with $u \times v$ pixels $O = \{s_{i,j} | 1 \leq i \leq u, 1 \leq j \leq v\}$; Output two shadows S_1, S_2.

1 For each pixel $s_{i,j}, i \in [1, u], j \in [1, v]$, divide $s_{i,j}$ into two shares $s^1_{i,j}, s^2_{i,j}$ according to following Algorithm 4.
2 For each share $s^t_{i,j}$, generating authentication bits $Auth_1$ using Algorithm 2. The other authentication bit $Auth_2$ is computed according to the value of $Auth_1$:
If $Auth_1 = 0$, then $Auth_2 = 0$; else $Auth_2 = 1$.
Then each share $s^t_{i,j}$ is extended to binary string $B^t_{i,j}$ with length 10, $B^t_{i,j} = s^t_{i,j} || Auth_1 || Auth_2, t \in \{1, 2\}$.
3 Each binary string $B^t_{i,j}$ is treated as a $(r, 10)$ POB number, and compute the value $V(B^t_{i,j})$ of the $(r, 10)$ POB number $B^t_{i,j}$, where r is a fixed parameter $r = 5$, and is published to each user.
4 The two shadows S_1, S_2 are

$$S_t = \{V(B^t_{i,j}) | i \in [1, u], j \in [1, v]\}, t = 1, 2 \tag{3}$$

Image Reconstruction phase:
Input: two shadows S_1, S_2; Output: secret image O.

1 Extracting each $V(B^t_{i,j})$ from $S_t, t = 1, 2$, transfer $V(B^t_{i,j})$ to its $(5, 10)$ POB number $B^t_{i,j}$.
2 Each $B^t_{i,j}$ is a 10 bits length binary string $B^t_{i,j} = s^t_{i,j} || Auth_1 Auth_2$, where $s^t_{i,j}$ is a byte and $Auth_1, Auth_2$ are two bits for authentication. Then authenticating each byte $s^t_{i,j}$ using $Auth_1, Auth_2$ according to Algorithm 4.
   i if all $s^t_{i,j}$ pass authentication, then the secret image O can be recovered by: $O = \{s_{i,j} | i \in [1, u], j \in [1, v]\}$, where $s_{i,j} = s^1_{i,j} \oplus s^2_{i,j}$.
   ii else, output the tampered share $s^t_{i,j}$.

In fact, the method of finding $s^*_{i,j}$ with even Hamming Weight closest to $s_{i,j}$ can be simply achieved by changing LSB of $s_{i,j}$. For instance, if $s_{i,j} = 44 = (00101100)$, then $s^*_{i,j}$ can be obtained by changing LSB of $s_{i,j}$, thus $s^*_{i,j} = (00101101) = 45$. It is easy to verify that $s^*_{i,j} = 45$ is closest to $s_{i,j} = 44$ with even Hamming Weight.

Algorithm 4. Generation of two shares from a pixel

Input: one pixel $s_{i,j}$; Output: two shares $s^1_{i,j}, s^2_{i,j}$.

- If the Hamming Weight of pixel $s_{i,j}$ is even, $s_{i,j}$ is divided into two shares $s^1_{i,j}, s^2_{i,j}$, where both $s^1_{i,j}$ and $s^2_{i,j}$ have Hamming Weight 4, and satisfy $s_{i,j} = s^1_{i,j} \oplus s^2_{i,j}$.
- Else, transform $s_{i,j}$ into $s^*_{i,j}$, where $s^*_{i,j}$ is the closest pixel to $s_{i,j}$ with even Hamming Weight. Then $s^*_{i,j}$ is divided into two shares $s^1_{i,j}, s^2_{i,j}$, where both $s^1_{i,j}$ and $s^2_{i,j}$ have Hamming Weight 4, and satisfy $s^*_{i,j} = s^1_{i,j} \oplus s^2_{i,j}$.
- Output: $s^1_{i,j}, s^2_{i,j}$.

In Algorithm 4, each pixel of even Hamming Weight is divided into two shares with Hamming Weight 4. In fact this is executable according to the following Lemma 1.

Lemma 1. Each pixel s with even Hamming Weight can be divided into two shares s_1, s_2, where s_1, s_2 have Hamming weight 4, and satisfy $s = s_1 \oplus s_2$.

Proof. Suppose s has Hamming Weight $2k, k \in \{0, 1, 2, 3, 4\}$, and the positions of these 1s are $w_1, w_2, ..., w_{2k}$. Then these $2k$ positions are divided into two non-overlapping groups W_1, W_2 where each group includes k positions. Each s_1 and s_2 has k 1s on the positions in W_1 and W_2, and other $4 - k$ 1s on the same positions other than $w_1, w_2, ..., w_{2k}$. In this way, any pixel s with even Hamming Weight can be divided into two shares s_1, s_2, where s_1, s_2 have Hamming weight 4, and satisfy $s = s_1 \oplus s_2$.

For instance, let $s = (01101100)$, the Hamming Weight is $2k = 4$, and the positions of 1s are $\{2, 3, 5, 6\}$. We divide the set $\{2, 3, 5, 6\}$ into two groups $\{2, 5\}$ and $\{3, 6\}$, then s_1 has two 1s on positions $2, 5$ and s_2 has two 1s on positions $3, 6$. On the other hand, s_1 and s_2 have other two 1s on the same positions other than $\{2, 3, 5, 6\}$ respectively. Thus, s_1 and s_2 can be simply obtained using this approach: $s_1 = (10100110)$, $s_2 = (11001010)$.

4 Experimental Results

In this part, we use experimental results to show our scheme. Firstly we analyze the PSNR value of reconstructed image theoretically using our scheme. As we know, The quality of reconstructed image is commonly evaluated in terms of the peak-signal-to-noise ratio (PSNR). Higher PSNR reflects better image quality. Normally, a reconstructed image can be visually distinguished from the original image when the PSNR is lower than 30 dB. The PSNR and mean square error (MSE) are presented as follows.

$$
\begin{aligned}
MSE &= \frac{1}{M \times N} \sum_{x=1}^{M} \sum_{y=1}^{N} (I(x, y) - I^{'}(x, y))^2 \\
PSNR &= 10 \times log_{10} \frac{255^2}{MSE}
\end{aligned}
\tag{4}
$$

(a) Cameraman (b) Lena (c) Jetplane

Fig. 1. Three original images

where M and N represent the length and width of the image. In our scheme, each pixel with even Hamming Weight of original image can be correctly reconstructed, and each pixel with odd Hamming Weight can be reconstructed with distortion ± 1. Therefore the theoretical MSE value of reconstructed image is:

$$MES = \frac{1 + 1 + ... + 1}{256} = \frac{128}{256} = 0.5 \tag{5}$$

The theoretical $PSNR$ value of reconstructed image is our scheme is:

$$PSNR = 10 \times log_{10} \frac{255^2}{0.5} = 51.14 \tag{6}$$

Since 51.14 is much larger than 30, no difference can be distinguished from reconstructed image from the original image. The following Fig. 1 lists three original images (cameraman, lena, jetplane) that are tested using our POB based $(2, 2)$ secret image sharing, and Fig. 2 lists the shares and reconstructed image for each original image in Fig. 1. The PSNR value of each reconstructed image is listed in Table 1, which are close to its theoretical value in Eq. (6).

Table 1. PSNR value of recovered images

Recovered image	Cameraman	Lena	Jetplane
PSNR	51.16	51.13	51.19

Our scheme is also capable of authenticate tampered shares, since there are two bits $Auth_1, Auth_2$ for authentication in POB number for each share. According to our scheme, there are exact one 1 and one 0 for $Auth_1, Auth_2$, the probability for successful detecting tampered shares is 50%. On the other hand, the authentication bits in scheme [12] can be either $(00, 01, 10, 11)$, the probability for successful detecting tampered shares is 75%. It means that the capability of authentication in scheme [12] is stronger than our scheme. But our scheme can solve the problem of distribution r between the cloud center to each user, which greatly saves cost of communication and storage.

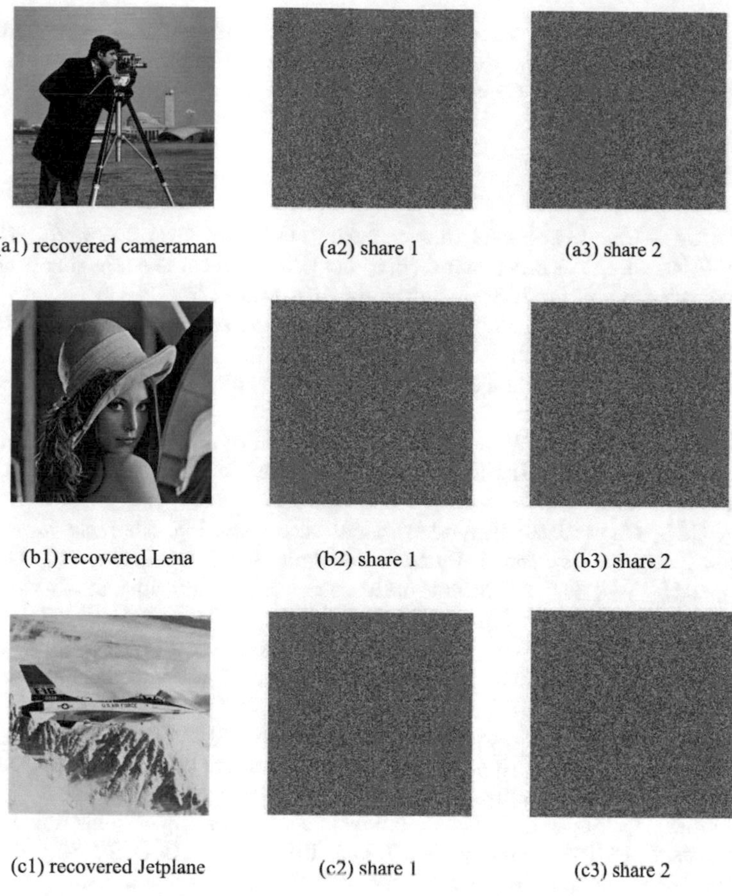

Fig. 2. Recovered images and shares

5 Conclusion

In this paper, we propose a new $(r, 10)$ POB based $(2, 2)$ threshold secret image sharing scheme which is expended from Singh et al.'s work. Their scheme can solve the problem of secure storage of image-shares at cloud servers, and is also capable of authenticating image-shares at pixel level. The image reconstruction can be accomplished with acknowledgement on the two shadows and the parameter r for each secret pixel. In Singh et al.'s scheme, they assume that the parameter r for each pixel have already been distributed to each user. However, the distribution of r is not that easy, the secure transport and storage on r would spend huge cost. In this work, we construct a new $(r, 10)$ POB based $(2, 2)$ threshold secret image sharing scheme where the parameter $r = 5$ is fixed for each secret pixel. Using this approach, the cost on distribution r can be saved and our scheme is also capable of authenticating tampered shadows at pixel level.

Acknowledgement. This work was supported by National Natural Science Foundation of China under Grant No. 61502384, and in part by Ministry of Science and Technology (MOST), under Grant 107-2221-E-259-007.

References

1. Thien, C.C., Lin, J.C.: Secret image sharing. Comput. Graph. **26**, 765–770 (2002)
2. Yang, C.N., Chu, Y.Y.: A general (k, n) scalable secret image sharing scheme with the smooth scalability. J. Syst. Softw. **84**(10), 1726–1733 (2011)
3. Liu, Y.X., Yang, C.N.: Scalable secret image sharing scheme with essential shadows. Sig. Process. Image Commun. **58**, 49–55 (2017)
4. Wang, R.Z.: Region incrementing visual cryptography. IEEE Signal Process. Lett. **16**(8), 659–662 (2009)
5. Yang, C.N., Shih, H.W., Wu, C.C., Harn, L.: k out of n region incrementing scheme in visual cryptography. IEEE Trans. Circuits Syst. Video Technol. **22**(5), 799–809 (2012)
6. Yang, C.N., Chen, T.S.: Extended visual secret sharing schemes: improving the shadow image quality. Int. J. Pattern Recognit. Artif. Intell. **21**, 879–898 (2007)
7. Chen, T.H., Wu, C.S.: Efficient multi-secret image sharing based on Boolean-operations. Sig. Process. **91**(1), 90–97 (2011)
8. Liu, Y.X., Yang, C.N., Wu, S.Y., Chou, Y.S.: Progressive (k, n) secret image sharing schemes based on Boolean operations and covering codes. Sig. Process. Image Commun. **66**, 77–86 (2018)
9. Liu, Y.X., Yang, C.N., Chou, Y.S., Wu, S.Y., Sun, Q.D.: Progressive (k, n) secret image sharing scheme with meaningful shadow images by GEMD and RGEMD. J. Vis. Commun. Image Represent. **55**, 766–777 (2018)
10. Sreekumar, A., Sundar, S.B.: An efficient secret sharing scheme for n out of n scheme using POB-number system. Int. J. Inf. Process. **3**(4), 77–83 (2009)
11. Deepika, M.P., Sreekumar, A.: A novel secret sharing scheme using POB number system and CRT. Int. J. Appl. Eng. Res. **11**(3), 2049–2054 (2016)
12. Singh, P., Raman, B., Agarwal, N., Atrey, P.K.: Secure cloud-based image tampering detection and localization using POB number system. ACM Trans. Multimedia Comput. Commun. Appl. **13**(3), 1–23 (2017)
13. Zissis, D., Lekkas, D.: Addressing cloud computing security issues. Future Gener. Comput. Syst. **28**(3), 583–592 (2012)
14. Kuyoro, S.O., Ibikunle, F., Awodele, O.: Cloud computing security issues and challenges. Int. J. Comput. Netw. **3**(5), 445–454 (2011)
15. Mohamed, H., Kianoosh, M.: Authentication schemes for multimedia streams: quantitative analysis and comparison. ACM Trans. Multimedia Comput. Commun. Appl. **6**(1), 1–24 (2010)
16. Sanjay, R., Balasubramanian, R.: A chaotic system based fragile watermarking scheme for image tamper detection. AEU Int. J. Electron. Commun. **65**(10), 840–847 (2011)
17. Bhatnagar, G., Wu, Q.M.J., Atrey, P.K.: Secure randomized image watermarking based on singular value decomposition. ACM Trans. Multimedia Comput. Commun. Appl. **10**(1), 1–21 (2013)

Study on Security Enhancing of Generalized Exploiting Modification Directions in Data Hiding

Wen-Chung Kuo[1(✉)], Ren-Jun Xiao[1], Chun-Cheng Wang[2(✉)], and Yu-Chih Huang[3(✉)]

[1] Department of Computer Science and Information Engineering, National Yunlin University of Science & Technology, Yunlin, Taiwan, R. O. C.
simonkuo@yuntech.edu.tw, sun420504@gmail.com

[2] National Center for High-Performance Computing, National Applied Research Laboratories, Tainan 744, Taiwan, R.O.C.
1703158@narlabs.org.tw

[3] Department of Information Management, Tainan University of Technology, Tainan City 71002, Taiwan, R.O.C.
t00232@mai.tutl.edu.tw

Abstract. In 2017, Liu et al. proposed the high capacity GEMD scheme which increases the capacity of GEMD by divides n adjacent pixels into multiple groups. The dividing modes among the image are the same in their scheme. As a result, it is easy to be attacked by an attacker, In order to improve this shortcoming, we change the original embedding procedure, which makes secret message more secure after embedding. The concept is combining Liu's scheme with the incremental sequence. We change each size of embedding secret message to increase security and maintain stable image quality.

Keywords: Data hiding · EMD · GEMD · Incremental sequence

1 Introduction

With the rapid development of technology, the public's dependence on the Internet is increasing, people can quickly get information from the Internet, and the Internet is an open environment, so the issue of information security is worth exploring. In the transmission of information, whether through text or image transmission, it is a common occurrence to be intercepted or tampered by people who are interested in it. Therefore, it is more important to protect the security of information transmission. In order to solve this problem, it is available. Cryptography and data hiding methods are used to complete. After cryptography encrypts important messages into ciphertext, the recipient can only retrieve the secret message through the key. However, the ciphertext is meaningless garbled and is easily intercepted by the attacker. Data hiding is the characteristic of embedding the secret information into the original image. After the embedded image is very similar to the original image, it is not easy to be noticed by the people who are interested. The unscrupulous person will not find secret information

C.-N. Yang et al. (Eds.): SICBS 2018, AISC 895, pp. 103–115, 2020.
https://doi.org/10.1007/978-3-030-16946-6_9

embedding, thereby achieving information security. This paper is a data hiding method proposed by GEMD, which improves the original method to achieve the effect of increasing safety.

The structure of this paper: The second chapter uses the relevant methods for this thesis, mainly introduces the related data embedding methods of the EMD (Exploiting Modification Directions) and then the improved GEMD (Generalized Exploiting Modification Directions) then the GEMD that increases the amount of capacity through the concept of grouping; the third chapter proposes the way of incremental sequence establishment in this article, and explains how to convert the length of the embedded message into an incremental sequence mode and through the length of the embedding, Pixel group grouping and other behavior; the fourth chapter is the experimental environment and results, compared with other methods, and finally the safety analysis of this paper; the fifth chapter is the conclusion.

2 Related Works

2.1 EMD (Exploiting Modification Direction) [10]

In 2006, Zhang and Wang proposed the EMD data hiding method. The method mainly divides n pixels into one pixel group, and then performs the modulo operation using Eq. (1), and hides the secret message by adjusting one pixel in the group. As shown in formula (1):

$$(x_1, x_2, \cdots, x_n) = \left[\sum_{i=1}^{n} x_i \times i\right] mod(2n+1) \tag{1}$$

n represents the number of pixels in the group, x_i is the pixel value of the i-th pixel, and i represents the weight value.

The following are the embedding steps of EMD:

Step 1. Calculate f_e using Eq. (1).
Step 2. Convert the secret message S to $(2n + 1)$.
Step 3. Use the secret message S to subtract the f_e value to find the difference $d = (S - f_e) \bmod (2n + 1)$.
Step 4. Use the calculated difference to adjust the pixel value.
 Step 4-1. When $d = 0$, the pixel value is not adjusted.
 Step 4-2. When $d \leq n$, an action of +1 is performed on the dth pixel value.
 Step 4-3. When $d > n$, an action of −1 is performed at the $(2n + 1 - d)$th pixel value.
 Step 4-4. Repeat steps one through four until all secret messages are hidden.

The following are the extracting steps of EMD:

Step 1. Calculate the obtained function value f_e by using the obtained n embedded pixels and using Eq. (1).
Step 2. Convert f_e to binary.

Step 3. Repeat steps one and two until all the secret messages are extracted and concatenated to get the secret message.

2.2 GEMD (General Exploiting Modification Direction) [5]

The EMD data hiding method will increase rapidly as the number of pixels increases. In order to improve this problem, in 2013, Kuo and Wang proposed the GEMD data hiding method. For modifying the modulus value and weight, the modulus is changed from $2n$ +1 to 2^{n+1}, and the weight value is also changed from i to $2^i - 1$, by adjusting a number of pixel values, the effect of increasing the EMD capacity is achieved. The extraction formula is as shown in Eq. (2):

$$f_b(x_1, x_2, \cdots, x_n) = \left[\sum_{i=1}^{n} x_i \times \left(2^i - 1\right)\right] mod2^{n+1} \quad (2)$$

n represents the number of pixels in the group, x_i is the pixel value of the i-th pixel, and $(2^i - 1)$ represents the weight value of the i-th pixel.

The steps for embedding proposed by Kuo and Wang are as follows:

Step 1. Calculate the f_b value by substituting Eq. (2) according to the number of pixels and the pixel value.
Step 2. Select n + 1 bit as the secret message S and convert the secret message into 2^{n+1}.
Step 3. Subtract the secret message from f_b to find the difference d, that is, d = (S − f_b)mod 2^{n+1}.
Step 4. Make the following adjustments to the pixel values based on the difference d.
Step 4-1. If d = 0, the pixel values are the same before and after hiding.
Step 4-2. If d = 2^n, the pixel value is x_1 +1, and the pixel value is x_n +1.
Step 4-3. If d < 2^n, convert the d value to $(d_n d_{n-1} \cdots d_1 d_0)_2$ when $d_i = 0$ and $d_{i-1} = 1$. Then, the pixel value $x_i + 1$, if $d_i = 1$ and $d_{i-1} = 0$, the pixel value $x_i - 1$.
Step 4-4. If d > 2^n, then d = 2^{n+1} − d), and then convert the d value $(d_n d_{n-1} \cdots d_1 d_0)_2$. When d_i = 0 and d_{i-1} = 1, the pixel value x_i− 1,if d_i = 1 and d_{i-1} = 0, the pixel value $x_i + 1$.
Step 5. Repeat steps one through four until all secret messages are embedded.
The following are the extracting steps of GEMD:
Step 1. Bring the obtained n stego pixels back to formula (2).
Step 2. The f_b value is converted to binary.
Step 3. Repeat steps one and two until all secret messages are extracted and concatenated to get a secret message.

2.3 Enhance Embedding Capacity of Generalized Exploiting Modification Directions in Data Hiding [7]

In 2017, Liu et al. improved based on GEMD's method, which only used a group of n pixel groups and then divided into k pixel groups, so that the amount of n + 1 bits that could be embed in the original was increased to the n + k bit is used to increase the GEMD capacity. This method not only increases the amount of capacity, but also maintains a fairly good image quality. The method for hiding Liu et al. is described as follows:

The following are embedded steps:

Step 1. Convert the binary secret message S into decimal s_{10}.
Step 2. Convert the secret message into the formula mode, and satisfy r < 2^{n_2+1} and c < 2^{n_1+1}.

$$s_{10} = 2^{n_1+1} \times \mathrm{c} + \mathrm{r} \quad (3)$$

Where n_1 represents the number of groups of pixel groups in the first group, n_2 represents the number of groups of pixel groups in the second group, c is the value embedded in n_2, and r is the value embedded in n_1.

Step 3. After finding c and r, use GEMD to hide it in the segmented pixel group.
Step 4. Repeat steps one through three until all messages are embedded.
The following are extracted steps:
Step 1. The divided pixel groups is obtained by using GEMD.
Step 2. After finding c and r, use Eq. (3) to find the secret message.
Step 3. Turn the s_{10} message back to the n + 2 bit message.
Step 4. Repeat steps one through three until all secret messages are removed.

3 Proposed Method

This research method is based on the idea of high-capacity GEMD (Generalized Exploiting Modification Directions) which divides the original group of pixel groups into multiple groups, However, this method uses the same group cutting method for the whole picture, so it is easy for the attacker to crack. In order to improve this problem, we propose this method. By establishing a incremental sequence, a combination of secret message lengths is set up. The receiver also has in the case of the sequence of the incremental, there is a way to solve the secret message length, and in the case of pixel grouping, the grouping of the selected incremental sequence is used to achieve the effect of protecting the message. The following Sect. 3.1 describes how to establish the incremental sequence, while the 3.2 section describes how to select the combination of secret message lengths through the incremental sequence and convert it into a incremental sequence mode (Sect. 3.3). We will show how to divide the message length into multiple pixel groups, Sects. 3.4 and 3.5. It is an introduction to the algorithm for embedding and extracting out this paper, and Sect. 3.6 is an example of this method.

3.1 Incremental Sequence Creation

This section is explaining the algorithm created by incremental creation.

Algorithm1:
Input : random seed r
Output : incremental sequence $\{a_1, a_2 \ldots, a_n\}$
Step 1. Generate random sequence $\{r_1, r_2, \ldots\ldots, r_n\}$,by random seed r, where $r_i \in \{2,3\}, \forall i \in \{1, 2, \cdots, n\}$.
Step 2. $a_1 = r_1 + 2$;
Step 3. $S_1 = a_1$;
Step 4. For $i = 2$ to n {
$a_i = S_{i-1} + r_i$;
$S_i = S_{i-1} + a_i$;}

Example 3.1.1
Assume that the resulting incremental sequence is (4, 6, 12), and its maximum representation is 22 bits ($4 \times 1 + 6 \times 1 + 12 \times 1$).

3.2 Convert the Length of the Message Through an Incremental Sequence

We use the incremental sequence to convert the length of the secret message to be represented, thereby achieving a security effect.

Step 1: Establish an incremental sequence.
Step 2: Select the number of secret message length bits, find the representable combination, and convert it into binary.

Example 3.2.1
Assuming an increasing sequence (4, 6, 12, 24) and the number of embed secret message bits is 16 bits, it can be obtained using the concept of the knapsack problem, $16 = 4 \times 1 + 6 \times 0 + 12 \times 1 + 24 \times 0$ The weight combination of is $(1010)_2$, and then $(1010)_2$ is turned into $(10)_{10}$.

After the incremental sequence is combined with the secret message, the GEMD method is used to embed the length of the message in the image.

Since the sequence cannot represent all the embedded combinations, only the length that can be represented can be selected.

3.3 Pixel Grouping

When clustering, a new random seed is picked, and algorithm 1 is used to generate a new sequence. If the secret message length bit cannot satisfy the combination of this incremental sequence, the following three conditions may occur when the group is divide:

Step 1. The secret message length bit coincides with the combination of this increment sequence.
Step 2. When the remaining secret message length is 1 to 2 bits: Add the remaining number of bits to the first bit group.
Step 3. When the length of the remaining secret message is greater than 2 bits: then the remaining bits are grouped by themselves.

Example 3.3.1 (The secret message length bit is just finished)
Assume that the selected sequence is (3, 5, 11). And if the number of embedded bits is 14 bits, then $14 = 3 \times 1 + 5 \times 0 + 11 \times 1$, so the pixel group is divided into 2 pixels and 10 pixels, and two groups of pixels are embedded.

Example 3.3.2 (the length of the remaining secret message is equal to 1–2 bits)
Assume that the selected sequence is (3, 5, 11). And if the number of hidden bits is 15 bits, then after using this (3, 5, 11) sequence, there is still 1 bit that cannot be represented. At this time, he is merged into the first group of bits, and becomes (4 + 11). So the pixel group is divided into 3 and 10, and the two groups of pixels are embedded.

Example 3.3.3 (remaining secret message is longer than 2 bits)
Assume that (3, 5, 11) is selected. And if the number of embedded bits is 22 bits, then after using this (3, 5, 11) sequence, there are still 3 bits left, and the remaining 3 bits are divided into a pixel group (3 + 5 + 11 + 3), the pixels are divided into 2, 4, 10, 2, and four groups of pixels to embed.

3.4 A High Security Hiding Method

This method converts the length of the secret message through an incremental sequence. Then use GEMD to embed in the image first. If the receiver does not have the same sequence, the correct secret message length cannot be obtained. We use an incremental sequence to split the secret message length into multiple pixel groups. The following is the embedding algorithm of the method.

Algorithm 2:
Input: Cover image and Secret message.
Output: Stego image.
Step 1. Incremental sequence with random numbers r_1.
Step 2. Determine the length of the secret message, and select the length that can be represented by the incremental sequence.
Step 3. Select the number of secret message length bits, use the concept of the backpack problem, find the representable combination, and convert it into binary, then use GEMD to embed.
Step 4. Using the random seed r_2, another set of incremental sequences is established, and the pixel groups is divided according to the length of the secret message.
Step 5. The divided pixel groups use GEMD to embed c and r respectively.
Step 6. Repeat the above steps until the secret message is embedded.

3.5 Secret Message Extraction Method

Using the GEMD method, after the length of the secret message is taken out, we then perform secret message extraction on multiple pixel groups through the incremental sequence. The following is the extracted algorithm.

Algorithm 3:
Input: Stego image.
Output: Secret message.
Step 1. Use the random number seed r_1 to establish an incremental sequence.
Step 2. Use the GEMD to capture the embedded length after the conversion by the stego image.
Step 3. By incrementing the sequence, switch back to the correct embedding length.
Step 4. After knowing the length, use the random number seed r_2 to create a sequence, and use this sequence to split the pixel group.
Step 5. Calculate the GEMD value of each pixel group to obtain c and r.
Step 6. Find the secret message by using equation (3) with c and r.

3.6 This Method Example

Assuming that the secret message is $(10101000)_2$ and the pixel group is (20, 15, 10, 11, 7, 5, 10, 9), the stego image can be obtained by this method (21, 15, 11, 11 7, 5, 10, 9).

Embedding:

Step 1. Create a incremental sequence (5, 8, 15).
Step 2. The length of the embedding is 8 bits, and the length of the secret message is represented by the sequence of the incremental $(010)_2 = 2$.
Step 3. Use 2 pixels (20, 15) to hide 2 in GEMD mode. $f_b(20, 15) = (20 \times 1 + 15 \times 3)$ mod8 = 1, d = (2 − 1)mod8 = 1. The embedded pixels (21, 15) are obtained.
Step 4. Select the (3, 5, 11) sequence, divide the 8-bit into two groups (5, 3), use 4 pixels, and 2 pixels, a total of 6 pixels for grouping.
Step 5. Divide n = 6, pixel groups (10, 11, 7, 5, 10, 9) into $I_1 = (10, 11, 7, 5)$, $I_2 = (10, 9)$ two sets of pixel groups.
Step 6. Convert the secret message s = $(10101000)_2$ to $168 = 2^{n_1+1} \times 5 + 8$, c = 5, r = 8, 8 = (1 × 10 + 3 × 11 + 7 × 7 + 15 × 5) mod32, 5 = (1 × 10 + 3 × 9) mod8, will be embedded. The following pixel group $I_1 = (11, 11, 7, 5)$, $I_2 = (10, 9)$.

Extraction:

Step 1. Generate the same incremental sequence (5, 8, 15).
Step 2. Calculate using stego image (21, 15), $f_b(21, 15) = (21 \times 1 + 15 \times 3)$ mod8 = 2, convert 2 to binary $(010)_2$, and use the incremental sequence to switch back to 8.
Step 3. Know the embedding length of 8 bits.

Step 4. Use the same sequence (3, 5, 11), divided into I_1 = (11, 11, 7, 5), I_2 = (10, 9), two Pixel group.
Step 5. Take c = 5 by GEMD, r = 8, 8 = (1 × 11 + 3×11 + 7 × 7 + 5 × 15) mod32, 5 = (1 × 10 + 3 × 9) mod8, using Eq. (5) to find $2^{4+1} \times 5 + 8 = 128 = (10101000)_2$.

Assuming that the secret message is $(1001001001010)_2$ and the pixel group is (20, 15, 4, 3, 5, 3, 2, 2, 5, 6, 4, 2, 2), the stego image can be obtained by this method. For (21, 15, 4, 4, 5, 4, 3, 2, 6, 5, 5, 3, 2).

Embedding:

Step 1. Create a incremental sequence (4, 7, 13).
Step 2. The length of the embedding is 13 bits, and the length of the secret message is represented by a sequence of incremental $(001)_2$ = 1. Step 3: Use 1 pixel (20, 15) to embed 1 in GEMD mode. $f_b(20, 15) = (20 \times 1 + 15 \times 3)\text{mod}8 = 1$, d = hx2009;(1 − 1)mod8 = 0. Obtained embedded pixels (20, 15).
Step 3. Select the (4, 7, 14) sequence. Since the (4, 7) difference is 2 bits, the 2 bits are added to the first group. The 13 bits are divided into two groups (7, 6), using 6 pixels. Grouping with 5 pixels and a total of 11 pixels.
Step 4. Divide n = 11, pixel groups (4, 3, 5, 3, 2, 2, 5, 6, 4, 2, 2) into I_1 = (4, 3, 5, 3, 2, 2), I_2 = (5, 6, 4, 2, 2), two groups of pixels.
Step 5. Convert the secret message s = $(1001001001010)_2$ to $2^{n_1+1} \times 36 + 74$, c = 36, r = 74, 74 = (1 × 4 + 3 × 3 + 7 × 5 + 15 × 3 + 31 × 2 + 63 × 2) mod128, 36 = (1 × 5 + 3 × 6 + 7 × 4 + 15 × 2 + 31 × 2) mod64, the pixel group after embedding will be obtained.I_1 = (4, 4, 5, 4, 3, 2), I_2 = (6, 5, 5, 3, 2).

Extraction:

Step 1. Produce the same incremental sequence (4, 7, 13).
Step 2. Calculate $f_b(20, 15) = (20 \times 1 + 15 \times 3)\text{mo}8 = 1$ using GEMD, convert 1 to binary $(001)_2$, turn back to 13 with the incremental sequence.
Step 3. Know the 13-bit length of the embedding.
Step 4. Use the same sequence (4, 7, 14), divided into I_1 = (4, 4, 5, 4, 3, 2), I_2 = (6, 5, 5, 3, 2), two pixel group.
Step 5. Take c = 36 by GEMD, r = 74, 74 = (1 × 4 + 3 × 4 + 7 × 5 + 15 × 4 + 31 × 3 + 63 × 2) mod128, 36 = (1 × 6 + 3 × 5 + 7 × 5 + 15 × 3 + 31 2) mod64, using the formula (3) to find $2^{6+1} \times 36 + 74 = 4682 = (1001001001010)_2$.

4 Experimental

Data hiding technology, the main standards of measurement, security, image quality, storage and other conditions, in addition to improving safety, this paper can maintain good image quality.

4.1 Experimental Environment

This thesis is implemented using Matlab, and the experiment is carried out using 512×512 grayscale images such as Lena, Statue, Baboon, etc. (Fig. 1).

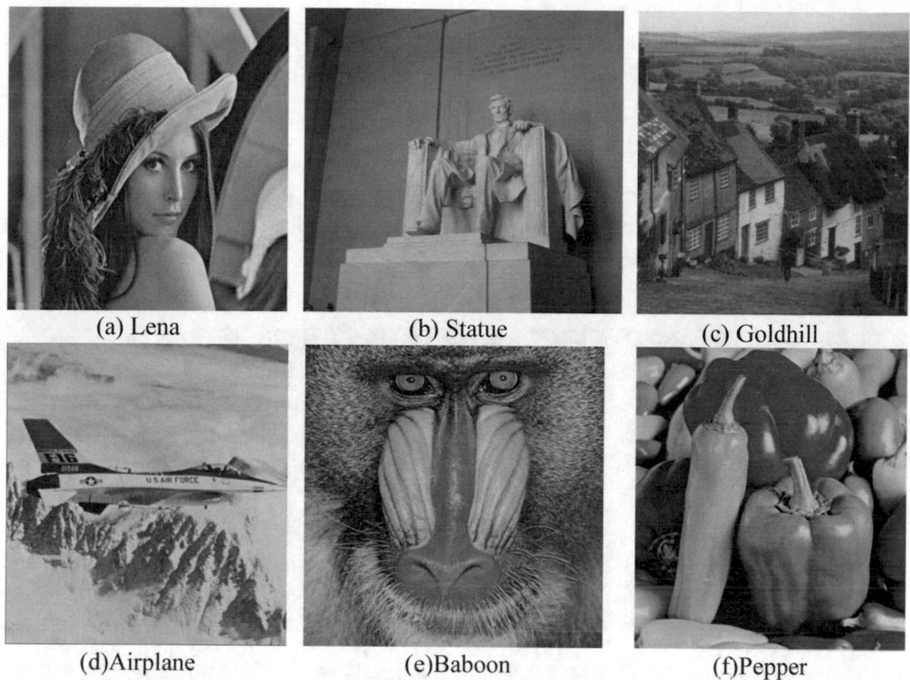

Fig. 1. Cover image.

4.2 Image Quality Assessment

After the secret information is embedded in the cover image, the stego image is obtained and in the process of transmitting the stego image, it is quite important not to be perceived by the illegal person, and the difference between the cover image and the stego image is evaluated, and the commonly used evaluation standard, it is the Peak Signal to Noise Ratio (PSNR). If the PSNR value is higher, the smaller the degree of image distortion is, the more similar it is to the original image. If the human eye is above 30 dB, it will be impossible to judge the difference. The PSNR formula (4) is as follows:

$$\mathrm{PSNR} = 10 \times \log_{10} \frac{255^2}{\mathrm{MSE}} \tag{4}$$

Where MSE is Mean Square Error, and formula (5) is as follows:

$$\mathrm{MSE} = \frac{1}{\mathrm{M} \times \mathrm{N}} \sum_{i=1}^{\mathrm{M}} \sum_{j=1}^{\mathrm{N}} (\mathrm{I_c}(i,j) - \mathrm{I_s}(i,j))^2 \tag{5}$$

Where M and N are the size of the image, $\mathrm{I_c}(i,j)$ is the pixel value before hiding, and $\mathrm{I_s}(i,j)$ is the pixel value after embedding.

4.3 Experimental Results

Through the images before and after embedding, we can get the image quality. From the experimental results, we can see the PSNR value, which is about 50.90 dB, which can maintain quite good image quality (Fig. 2).

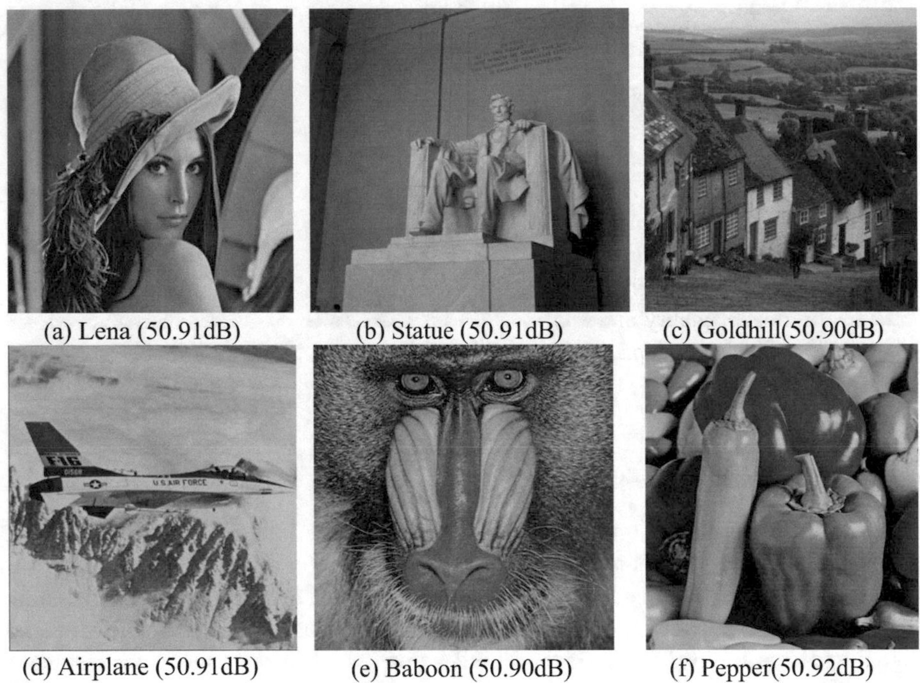

(a) Lena (50.91dB) (b) Statue (50.91dB) (c) Goldhill(50.90dB)

(d) Airplane (50.91dB) (e) Baboon (50.90dB) (f) Pepper(50.92dB)

Fig. 2. Stego image and PSNR.

4.4 Comparison of Related Research

This section compares the image quality of this method with other methods [5, 7, 10] and the amount of capacity, as shown in Tables 1 and 2, where n is the number of pixels used n = 2, 3,…, 8.

Table 1. Related studies comparing PSNR (dB).

Method (dB)	EMD [10]	GEMD [5]	Enhanced GEMD [7]	Propose method
n = 2	52.14	50.82	47.32	50.66–50.93
n = 3	53.51	50.92	47.70	
n = 4	54.54	51.03	47.97	
n = 5	55.32	51.10	48.08	
n = 6	56.18	51.12	48.15	
n = 7	56.57	51.13	48.24	
n = 8	57.34	51.16	48.31	

Table 2. Related research comparison capacity (bpp).

Method (bpp)	EMD [10]	GEMD [5]	Enhanced GEMD [7]	Propose method
n = 2	1.16	1.50	2.00	0.95–1.03
n = 3	0.93	1.33	1.67	
n = 4	0.79	1.25	1.50	
n = 5	0.69	1.20	1.40	
n = 6	0.61	1.16	1.33	
n = 7	0.55	1.14	1.29	
n = 8	0.51	1.12	1.25	

It can be seen from Fig. 3 that while increasing the security, we also take into account the image quality and storage. The image quality can be maintained close to 51 dB, and the storage capacity is maintained above 1 bpp.

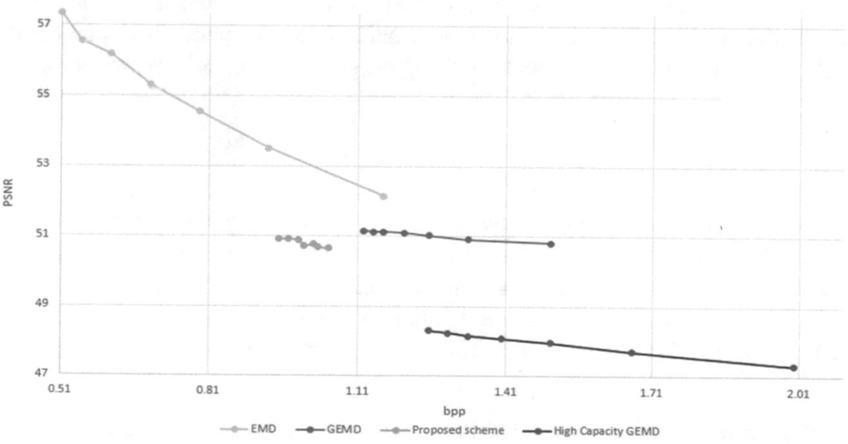

Fig. 3. The results of Tables 1 and 2 can be used to obtain the relationship between PSNR and bpp of various methods.

4.5 Characteristic Analysis

As can be seen from Table 3, in addition to the detection by RS, the method also has a storage capacity of approximately 1, and the number of selected pixels can be adjusted each time, and the security of the image is increased.

Table 3. Characteristic analysis.

	EMD [10]	GEMD [5]	High capacity GEMD [7]	Proposed scheme
Adjustable pixel group	No	No	No	Yes
Bpp	$\frac{\log_2(2n+1)}{n}$	$\frac{(n+1)}{n}$	$\frac{(n+k)}{n}$	≈ 1
PSNR (n = 8)	57.34	51.16	48.31	50.93
RS	pass	pass	pass	pass
Security	NO	NO	NO	YES

4.6 Security Analysis

The Liu et al. method is based on the improvement of GEMD to increase the amount of storage. The pixel group that originally had only one group is divided into groups of pixel groups. After the segmentation, the whole picture uses the same segmentation and embedding method. The main purpose of this method is to make the whole image use different cutting methods. First, we will use a part of the image to hide the length of the secret message, and convert the length of the secret message into a way that the sequence of the incremental can be represented. The length of the post message is embedded in the image, and if the receiver does not have the same sequence, the correct message length cannot be taken out, and according to the length of the secret message, different pixel group segmentation methods can be performed, and the paper The map uses an unfixed split pixel group approach to achieve increased security.

5 Conclusions

In this paper, based on the high-capacity GEMD data hiding method, it is improved. The main purpose of this method is to make the whole image use different dividing methods. We use the incremental sequence to convert the length of the secret message, and use the sequence of the secret message length to segment the multiple pixel groups, thereby increasing the security effect. The following are some of the advantages of this method.

1. Better security: anyone can not extract secret data without random seed r.
2. PSNR: more than 50 dB.
3. Capacity: almost be 1 bpp.

Acknowledgement. This work was supported in part by the Ministry of Science and Technology of the Republic of China under Contract No. MOST 107-2218-E-110-014- and MOST 107-2221-E-224-008-MY2.

References

1. Fridrich, J., Goljan, M., Du, R.: Detecting LSB steganography in color and grayscale images. IEEE Multimedia **8**(4), 22–28 (2001)
2. Gu, S.S., Hao, T.: A pointer network based deep learning algorithm for 0–1 knapsack problem. In: 2018 Tenth International Conference on Advanced Computational Intelligence (ICACI), pp. 473–477 (2018)
3. Jung, K.H., Yoo, K.Y.: Improved exploiting modification direction method by modulus operation. Int. J. Signal Process. Image Process. Pattern **2**(1), 79–88 (2009)
4. Kuo, W.C., Kuo, S.H., Chang, H.Y., Wuu, L.C.: Authenticated secret sharing scheme based on GMEMD. In: International Workshop on Digital-Forensics and Watermarking 2015, 07–10 October 2015
5. Kuo, W.C., Wang, C.C.: Data hiding based on generalized exploiting modification direction method. Imaging Sci. J. **61**(6), 484–490 (2013)
6. Kuo, W.C., Wuu, L.C., Kuo, S.H.: The high embedding steganographic method based on general Multi-EMD. In: 2012 International Conference on Information Security and Intelligent Control (ISIC 2012), Yunlin, Taiwan, pp. 286–289, 14–16 August 2012
7. Liu, Y.X., Yang, C.G., Sun, Q.D.: Enhance embedding capacity of generalized exploiting modification directions in data hiding. IEEE Access, pp. 1–5 (2017)
8. Petitcolas, F.A.P., Anderson, R.J., Kuhn, M.G.: Information hiding – a survey. In: The proceedings of the IEEE, special issue on protection of multimedia content, pp. 1062–1078 (1999)
9. Wang, C.C., Kuo, W.C., Huang, Y.C., Wuu, L.C.: A high capacity data hiding scheme based on re-adjusted GEMD. Multimedia Tools Appl. pp. 1–15 (2017)
10. Zhang, X., Wang, S.: Efficient steganographic embedding by exploiting modification direction. IEEE Commun. Lett. **10**(11), 781–783 (2006)

High-Fidelity Reversible Data Hiding in JPEG Images Based on Two-Dimensional Histogram

Tengfei Dong[1], Zhigao Hong[1], and Zhaoxia Yin[1,2(✉)]

[1] Key Laboratory of Intelligent Computing & Signal Processing, Ministry of Education, Anhui University, Hefei 230601, People's Republic of China
yinzhaoxia@ahu.edu.cn

[2] Department of Computer Science, Anhui University, Hefei 230601, People's Republic of China

Abstract. Recently, a series of excellent reversible data hiding (RDH) algorithms based on two-dimensional histogram modification have been proposed. However, the coefficient-pairs of these algorithms are obtained via fixed combination method, which limits the further improvement of algorithm performance. Based on this consideration, a new combination method with pairwise coefficients is proposed in this paper. Firstly, the coefficients are divided into 32 positions according to Zigzag scanning sequence. Then, we pair the coefficients in the same position, and combination mode of the coefficients is determined by the number of expandable pairs. Unlike the fixed combination method, the coefficients of proposed method are adaptively combined into pairs according to image features. Moreover, we propose a coefficient-pairs selection strategy. The coefficient-pairs which have little effect on image quality are selected for payload embedding, thereby distortion of the mark image is limited. The obtained algorithm has better peak signal to noise ratio (PSNR) and embedding capacity over some state-of-the-art algorithms, showing the superiority of our method.

Keywords: Reversible data hiding · JPEG · Two-dimensional histogram · Pairwise coefficients combination

1 Introduction

Reversible data hiding (RDH), an important branch of information hiding, can realize the lossless recovery of the original medium and embedded data from the covered medium [1]. Therefore, RDH has been widely used in various fields, including military, medicine, finance, commerce, etc.

In general, RDH algorithms in spatial domain can be classified into three categories: lossless compression [2], difference expansion (DE) [3], and histogram shift (HS) [4]. Among them, histogram shift algorithm has been the hotspot of research on RDH. In 2013, Li et al. firstly employed two-dimensional histogram to design RDH [5]. By modifying the pixel differences jointly, the image redundancy can be exploited

C.-N. Yang et al. (Eds.): SICBS 2018, AISC 895, pp. 116–128, 2020.
https://doi.org/10.1007/978-3-030-16946-6_10

better to achieve an improved performance. Then, Ou et al. [6] optimize Li's method by exploiting every two adjacent prediction-errors jointly to generate two-dimensional histogram.

JPEG format is widely accepted as the most popular image format on the web, which makes the design of RDH for JPEG images valuable. However, RDH for JPEG images is more challenging than that for uncompressed images due to the less redundancy, increased file size and sensitive modification.

Currently, most of RDH algorithms for JPEG images has been developed based on modifying quantified DCT coefficients. In 2016, Huang et al. [7] proposed a new HS-based RDH method for JPEG image. In Huang's scheme, coefficients with value '0' hold the line, whereas coefficients with values '1' and '−1' are used for data embedding. Moreover, a block selection strategy was proposed, which means that the DCT blocks with more zero coefficients are selected to carry messages. Recently, Cheng et al. [8] have presented a new RDH scheme for JPEG images based on the two-dimensional coefficient histogram. Compared with Huang's scheme, Cheng's scheme has advantage in increased file size. Furthermore, Cheng designs a selection strategy based on the optimal DCT frequency band to make the quality of the marked image well maintained.

Herein, the two-dimensional coefficient histogram modification proposed by Cheng was employed. Notably, the pairwise coefficient combination method was firstly proposed, which further greatly improve the embedding capacity. In addition, the image visual quality can be improved because of proposed coefficient-pairs selection strategy. The rest of this paper is organized as follows. Section 2 briefly introduces JPEG compression and Cheng et al.'s coefficient-pair-mapping [8]. Section 3 presents the proposed scheme in detail. Experimental results and related analysis are given in Sect. 4. Finally, Sect. 5 concludes this paper.

2 Related Works

JPEG compression and Cheng's CPM have been elaborated in this section for a better understanding of our proposed method.

2.1 JPEG Compression

The process of JPEG compression coding is depicted in Fig. 1. First, the original image is divided into multiple non-overlapping 8×8 blocks, and then DCT transform is applied to each block. After that, original image data is changed from spatial domain to frequency domain. Then, these DCT coefficients are quantified by a quantization table. The quantified DCT coefficients are arranged in the Zigzag scanning order as shown in Fig. 2. Differential pulse code modulation (DPCM) is applied to the DC coefficients, and AC coefficients are encoded by the run length encoding (RLE). Finally, the symbol string is converted to a bitstream by Huffman coding, and the final JPEG image can be obtained after pre-pending the header.

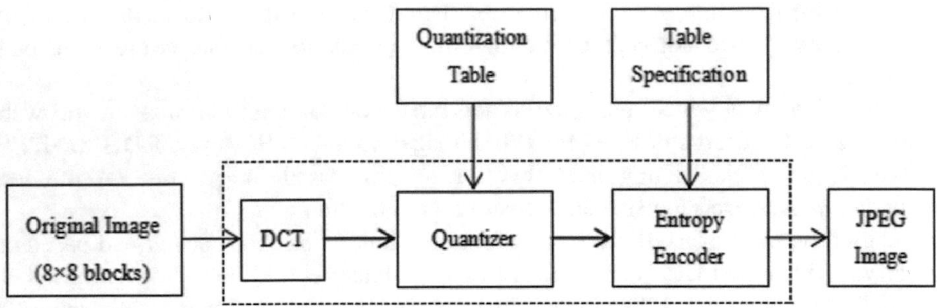

Fig. 1. JPEG compression process

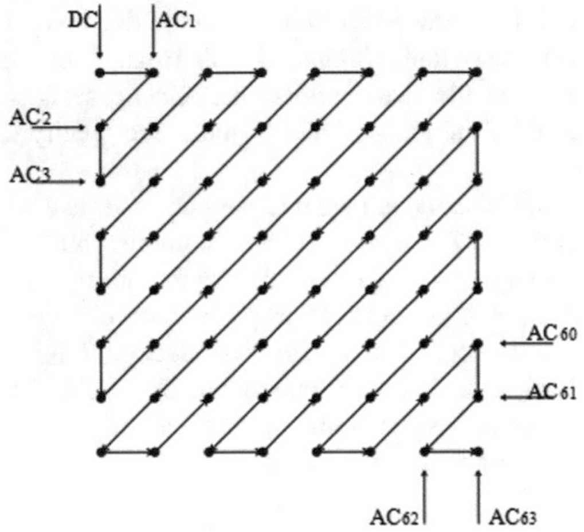

Fig. 2. Zigzag sequence

2.2 Coefficient-Pair-Mapping (CPM) [8]

This coefficient-pair-mapping is put forward by Cheng, which has very excellent performance.

Quantized DCT coefficients are combined into pairs by specific combination method, and the two-dimensional coefficient-histogram is constructed by the DCT coefficient-pairs. Cheng's coefficient-pair-mapping (CPM) on the coefficient-pairs is shown in Fig. 3. The points with different colors represent the different types of coefficient-pairs. All coefficient-pairs are classified into six types, namely the Type A-F. The coefficient-pairs of Type A-C are expandable pairs, which can carry the message bits. The coefficient-pairs of Type D-E are shiftable pairs, which can be shifted and don't carry message bits. The coefficient-pairs of Type F are unchangeable pairs, which remain unchanged among the embedding process.

- Type A: $(c_1, c_2) = \{(1, 1), (-1, 1), (-1, -1), (1, -1)\}$. Red points represent coefficient-pairs of this type, and red arrows indicate the modification directions.

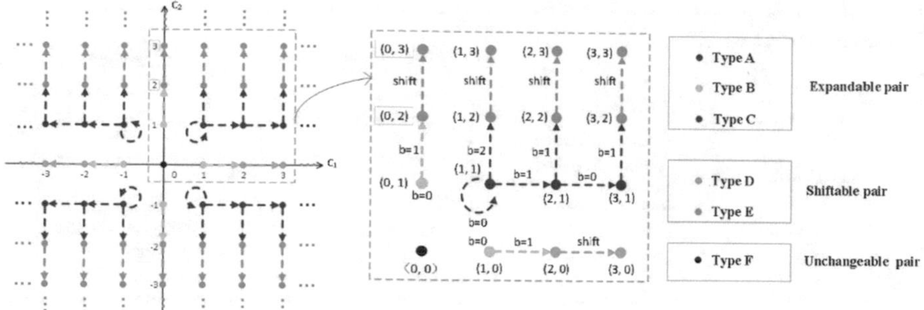

Fig. 3. Coefficient-pair-mapping (CPM)

Each point in type A has three candidate modification directions for data embedding, and it can be embedded with $\log_2 3$ bits.

- Type B: $(c_1, c_2) = \{(1, 0), (0, 1), (-1, 0), (0, -1)\}$. Yellow points represent coefficient-pairs of this type. If the to-be-embedded data bit $b = 0$, the point of this type keeps unchanged. And when $b = 1$, the modification direction is oriented by a yellow arrow.
- Type C: $(c_1, c_2) = \{c_1 > 1 \text{ or } c_1 < -1, c_2 = \pm 1\}$. Purple points represent coefficient-pairs of this type, and purple arrows indicate the modification directions. If $b = 0$, the modification direction is 'horizontal'. If $b = 1$, the modification direction is 'vertical'.
- Type D: $(c_1, c_2) = \{c_1 > 1 \text{ or } c_1 < -1, c_2 = 0\}$. Green points represent coefficient-pairs of this type. Each point of this type is shifted to the horizontally adjacent point, which is oriented by a green arrow.
- Type E: $(c_1, c_2) = \{c_2 > 1 \text{ or } c_2 < -1\}$. Blue points represent coefficient-pairs of this type. Each point of this type is shifted to the vertically adjacent point, which is indicated by a blue arrow.
- Type F: $(c_1, c_2) = (0, 0)$. Black points represent coefficient-pairs of this type. Each point of this type remains unchanged.

3 Proposed Method

In general, a complete RDH algorithm based on two-dimensional histogram contains pairwise coefficient combination, 2D histogram generation and 2D mapping design [6]. The 2D mapping in Cheng's work is utilized in this work due to its excellent performance (Fig. 3). We design a new pairwise coefficient combination to endow corresponding algorithm with a larger embedding capacity, which is particularly introduced in Sect. 3.1. Moreover, we propose a coefficient-pairs selection strategy, which can reduce the distortion of the marked image. This selection strategy is introduced in Sect. 3.2. The framework of the proposed method is depicted in Fig. 4.

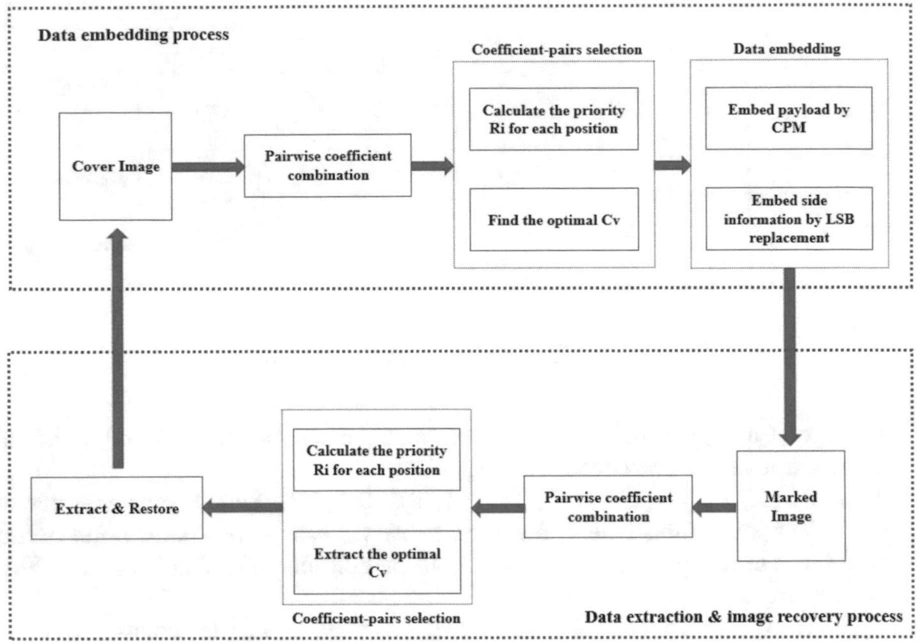

Fig. 4. The framework of proposed method

3.1 Pairwise Coefficient Combination Based on Internal Exchange

Cheng's pairwise coefficient combination: for each 8×8 block, the quantized DCT coefficients are divided into 32 pairs in Zigzag scanning order.

In this way, the DC coefficient of each block will be paired with AC_1, and the AC_2 coefficient of each block will be paired with AC_3…

Two definitions have been made for a better understanding:

(1) Define DC as AC_0;
(2) Define AC_{2i-2}, AC_{2i-1} ($1 \leq i \leq 32$) at position i (in Fig. 5).

	1	2	3	4	5	…	4096
position1 →	AC_0 AC_1	AC_0 AC_1	AC_0 AC_1	AC_0 AC_1	AC_0 AC_1	…	AC_0 AC_1
position2 →	AC_2 AC_3	AC_2 AC_3	AC_2 AC_3	AC_2 AC_3	AC_2 AC_3	…	AC_2 AC_3
position3 →	AC_4 AC_5	AC_4 AC_5	AC_4 AC_5	AC_4 AC_5	AC_4 AC_5	…	AC_4 AC_5
position4 →	AC_6 AC_7	AC_6 AC_7	AC_6 AC_7	AC_6 AC_7	AC_6 AC_7	…	AC_6 AC_7
position5 →	AC_8 AC_9	AC_8 AC_9	AC_8 AC_9	AC_8 AC_9	AC_8 AC_9	…	AC_8 AC_9
	⋮	⋮	⋮	⋮	⋮	⋮	⋮
position32 →	AC62 AC63	AC62 AC63	AC62 AC63	AC62 AC63	AC62 AC63	…	AC62 AC63

Fig. 5. Position of coefficient

For Cheng's coefficient combination method, AC_{2i-2} will pair with AC_{2i-1} ($1 \leq i \leq 32$), and they are combined into pair in the form of (AC_{2i-2}, AC_{2i-1}). The problem is that pairing in the form of (AC_{2i-1}, AC_{2i-2}) in some cases may produce better results. Here, in our approach, AC_{2i-2} will still pair with AC_{2i-1}. The difference is, it chooses to pair in the form of (AC_{2i-2}, AC_{2i-1}) or (AC_{2i-1}, AC_{2i-2}) depending on the image characteristics.

Pairwise coefficient combination we designed (sender): For the coefficients of position i, the coefficients of this position are firstly all paired in the form of (AC_{2i-2}, AC_{2i-1}), and the number of Type C coefficient-pairs is counted as N_1; After that, the coefficients are paired in the form of (AC_{2i-1}, AC_{2i-2}), and the number of Type C coefficient-pairs is counted as N_2. If $N_1 \geq N_2$, then the coefficients of position i are all paired in the form of (AC_{2i-2}, AC_{2i-1}); On the contrary, if $N_1 < N_2$, the coefficients of position i are all paired in the form of (AC_{2i-1}, AC_{2i-2}).

In this way, more coefficient-pairs of Type C could be obtained compared to Cheng's coefficient combination method. Thus, more expandable pairs could be got (i.e. Type A, Type B, Type C), which has been supported by the experimental results (Tables 1 and 2).

Table 1. Comparison for the number of expandable pairs between the proposed method and the method of Cheng et al. (QF = 80)

	Number of expandable pairs	
JPEG image (QF = 80)	Proposed method	Cheng's method [8]
Baboon	37789	37043
Boat	23489	23248
Lake	25137	24838
Lax	40268	39834
Peppers	21988	21820
Splash	18205	17979

Table 2. Comparison for the number of expandable pairs between the proposed method and the method of Cheng et al. (QF = 90)

	Number of expandable pairs	
JPEG image (QF = 90)	Proposed method	Cheng's method [8]
Baboon	45934	44811
Boat	34998	34417
Lake	39120	38517
Lax	52008	51232
Peppers	34377	34013
Splash	25463	25087

After embedding the data, the number of Type C coefficient-pairs has been changed. To ensure reversibility, we set up a 32-length exchange table, which records the coefficients combination mode for each position. For the position i ($1 \leq i \leq 32$), if the coefficients combination mode of this position is (AC_{2i-2}, AC_{2i-1}), the i-th element of the exchange table is set to 0; Conversely, if the coefficients combination mode of this position is (AC_{2i-1}, AC_{2i-2}), the i-th element of the exchange table is set to 1.

The pairwise coefficient combination (receiver): For the coefficients of the position i, if the i-th element of the exchange table is 0, the coefficients of the position i are all paired by (AC_{2i-2}, AC_{2i-1}); Conversely, if the i-th element of the exchange table is 1, the coefficients of the position i are all paired in the form of (AC_{2i-1}, AC_{2i-2}).

3.2 Coefficient-Pairs Selection Strategy

Before introducing our coefficient-pairs selection strategy, the block selection strategy we use must be given. That is, the zero coefficients (not include AC_0, AC_1) in each 8×8 block are calculated. The embedding sequence of block depends on the number of zero coefficients in the block. To be clear, a block with more zero coefficients is preferential for payload embedding.

Our coefficient-pairs selection strategy is as follows:

First, calculate the priority Ri ($1 \leq i \leq 32$) for each position by the following formula.

$$R_i = \frac{2}{Q_{2i-2} + Q_{2i-1}} \tag{1}$$

Where Q_{2i-2} represents the quantization table value of AC_{2i-2}, Q_{2i-1} represents the quantization table value of AC_{2i-1} (The standard JPEG quantization table is shown in Fig. 6). For example, the priority R_1 of the position 1 is calculated, $R_1 = 2/(Q_0 + Q_1) = 2/(16 + 11) = 0.074$. This formula is chosen here because the modification of the coefficients with small quantization value has little impact on the quality of the image.

16	11	10	16	24	40	51	61
12	12	14	19	26	58	60	55
14	13	16	24	40	57	69	56
14	17	22	29	51	87	80	62
18	22	37	56	68	109	103	77
24	35	55	64	81	104	113	92
49	64	78	87	103	121	120	101
72	92	95	98	112	100	103	99

Fig. 6. Standard JPEG quantification table

The higher the priority of the position, the more preferential the coefficient-pairs at this position are used for payload embedding.

Second, cutoff value C_v ($1 \leq C_v \leq 32$) is set to record the positions selected for payload hiding. For example, if $C_v = 3$, the positions of top three priority are selected and the coefficient-pairs at these positions are used for payload embedding. It must be noted that, we default to the position 1 not selected for payload embedding, and this position is used for the side information embedding (explanations will be given in Sect. 3.3). The way to determine the value of C_v: exhaustion method, the C_v corresponding to the maximum PSNR is what we need.

To sum up, the essence of our coefficient-pairs selection strategy is to select the positions of the top C_v priority, and the coefficient-pairs at these positions are used for payload embedding.

3.3 Side Information

In order to ensure the reversibility of the algorithm, the side information {C_v, exchange table, information length L} is embedded into AC_1. Considering value of Cv is in the range [1, 32], 6 bits are used to represent it. The exchange table can be represented in the binary vector of size 32 bits. In general, the length of the message does not exceed 2^18 bits, so information length L can be represented by 18 bits. Therefore, the total side information is 56 bits, which is embedded in the least significant bits (LSB) of the first 56 AC_1. In order to recover the LSB of the original AC values perfectly, they (i.e. LSB of the first 56 AC_1) are appended as part of the payload before embedding.

3.4 Data Embedding and Extraction

The data embedding process is shown as follows:

(1) The least significant bits of the first 56AC_1 are extracted and appended as part of the payload.
(2) Obtain the exchange table according to the contents of 3.1, and complete the pairwise coefficient combination.
(3) Calculate the priority Ri ($1 \leq i \leq 32$) for each position.
(4) Find the optimal C_v according to the content of 3.2.
(5) Embed the payload by the CPM shown in Fig. 3.
(6) Embed the side information {Cv, exchange table, information length L} into the least significant bit of first AC_1.

The data extraction & image recovery process is shown as follows:

(1) Read the least significant bits of the first 56 AC_1 to extract C_v, exchange table, and information length L.
(2) According to the exchange table, we perform pairwise coefficient combination according to the contents of 3.1.

(3) Calculate the priority R_i ($1 \leq i \leq 32$) for each position.
(4) According to C_v and L, the information is extracted by inverse mapping of CPM, and the coefficients are restored at the same time.
(5) Replace the LSBs of the first 56 AC_1 coefficients with the original values (which were extracted along with the payload).

4 Experimental Results and Analysis

In this section, several experiments are conducted on six grayscale images as shown in Fig. 7. Two image quality factors (i.e., QF = 80 and 90) are tested. We evaluate our proposed approach by comparing the visual quality and the embedding capacity with Cheng's method [8]. Here, the embedding capacity is calculated by this formula:

$$C = \text{round}(log_2 3 \times NA + NB + NC - Side) \tag{2}$$

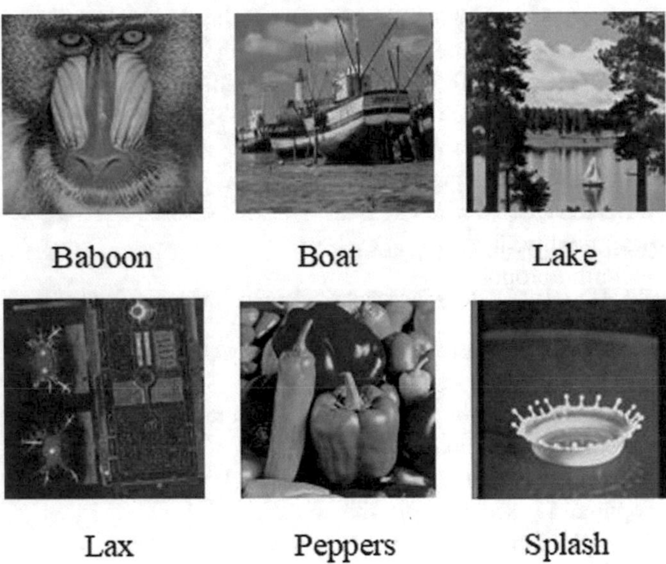

Fig. 7. Test images

Where NA represents the number of Type A coefficient-pairs, NB represents the number of Type B coefficient-pairs, NC represents the number of Type C coefficient-pairs, and Side is the length of the side information. Side's value is 30 in Cheng's method, whereas Side's value is 56 in our proposed method. And round is a rounding function. The experimental results are shown in Tables 3, 4, Figs. 8 and 9.

Table 3. Embedding capacity for six test images (QF = 80)

	Embedding capacity (bit)	
JPEG image (QF = 80)	Proposed method	Cheng's method [8]
Baboon	42669	41949
Boat	26309	26094
Lake	28257	27984
Lax	45583	45175
Peppers	24769	24627
Splash	20478	20278

Table 4. Embedding capacity for six test images (QF = 90)

	Embedding capacity (bit)	
JPEG image (QF = 90)	Proposed method	Cheng's method [8]
Baboon	52215	51118
Boat	39220	38665
Lake	43960	43383
Lax	59435	58685
Peppers	38608	38270
Splash	28766	28416

For the visual quality, it can be observed that our method is better than Cheng's. The reason lies in the proposed coefficient-pairs selection strategy. We not only consider the embedding order of the blocks, but also consider the embedding order of the coefficient-pairs, which Cheng did not expect. For the embedding capacity, it can be seen that our approach is completely better than Cheng's. The reason lies in our proposed pairwise coefficient combination. In fact, Cheng's coefficient combination method is just a special case of our combination method. Our method can obtain more expandable coefficient-pairs, so it is inevitable that the embedding capacity is higher than Cheng's.

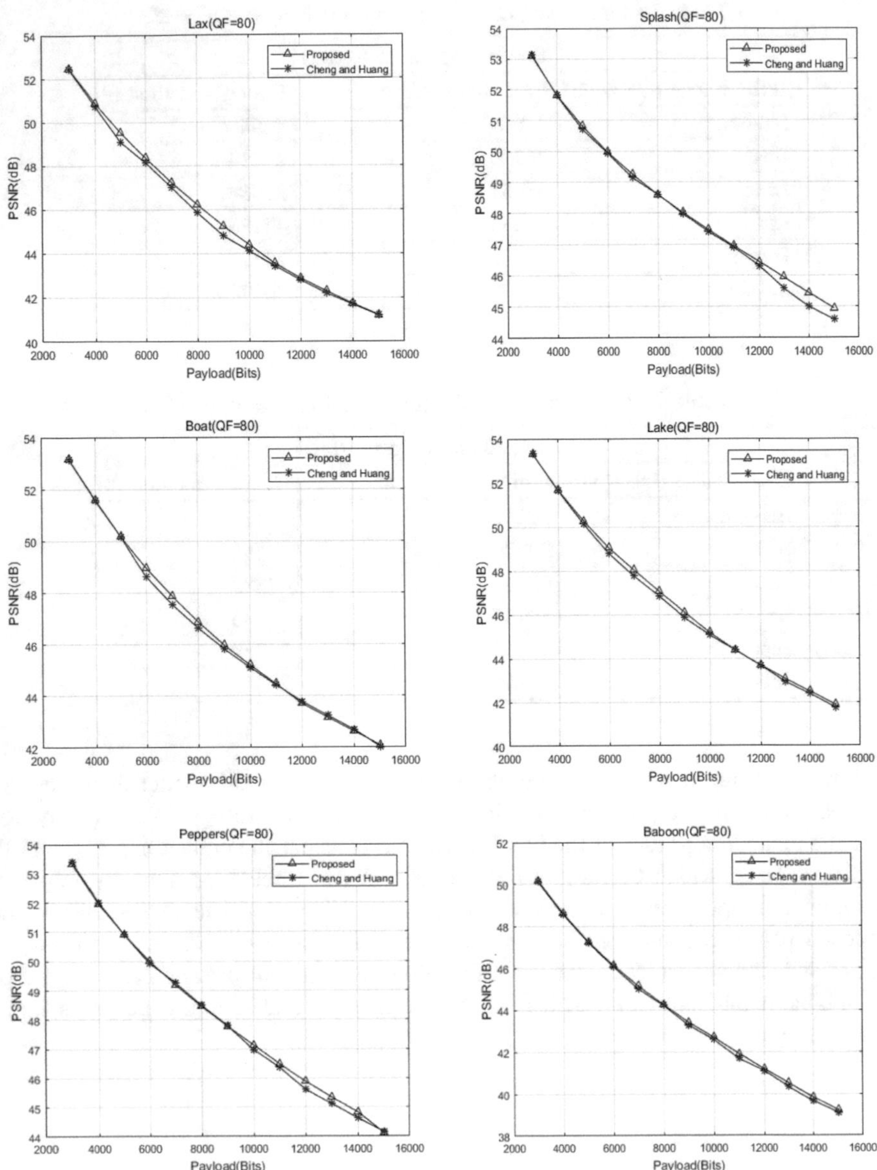

Fig. 8. Visual qualities for six test images (QF = 80)

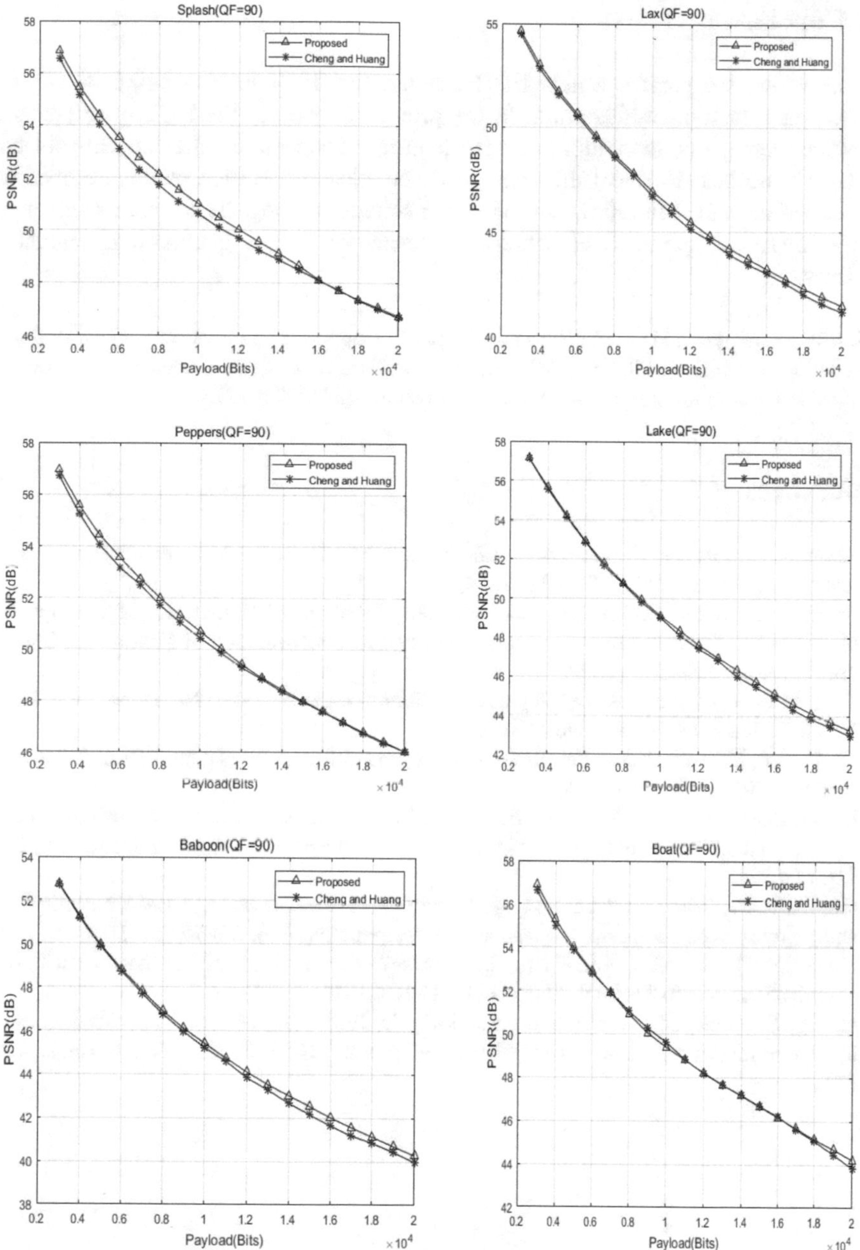

Fig. 9. Visual qualities for six test images (QF = 90)

5 Conclusion

In this paper, we present a new RDH scheme for JPEG images based on the two-dimensional coefficient-histogram. In the proposed method, the DCT coefficients are combined into pairs according to image features. Then the coefficient-pairs selection strategy is utilized to select the appropriate coefficient-pairs for payload embedding. The experimental results show our method is superior among the current excellent two-dimensional histogram RDH scheme in terms of image quality and embedding capacity.

Acknowledgments. This research work is partly supported by National Natural Science Foundation of China (61502009, 61872003), and National College Students' innovation and entrepreneurship training program of Anhui University (201810357075).

References

1. Khan, A., Siddiqa, A., Munib, S., Malik, S.A.: A recent survey of reversible watermarking techniques. Inf. Sci. **279**, 251–272 (2014)
2. Fridrich, J., Goljan, M.: Lossless data embedding for all image formats. In: SPIE Proceedings of Photonics West, Electronic Imaging, Security and Watermarking of Multimedia Contents, vol. 4675, pp. 572–583 (2002)
3. Tian, J.: Reversible data embedding using a difference expansion. IEEE Trans. Circuits Syst. Video Technol. **13**(8), 890–896 (2003)
4. Ni, Z., Shi, Y., Ansari, N., Wei, S.: Reversible data hiding. IEEE Trans. Circuits Syst. Video Technol. **16**(3), 354–362 (2006)
5. Li, X., Zhang, W., Gui, X., Yang, B.: A novel reversible data hiding scheme based on two-dimensional difference-histogram modification. IEEE Trans. Inf. Forensics Secur. **8**(7), 1091–1100 (2013)
6. Ou, B., Li, X., Wang, J., Peng, F.: High-fidelity reversible data hiding based on geodesic path and pairwise prediction-error expansion. Neurocomputing **226**, 23–34 (2017)
7. Huang, F., Qu, X., Kim, H.J., Huang, J.: Reversible data hiding in JPEG images. IEEE Trans. Circuits Syst. Video Technol. **26**(9), 1610–1621 (2016)
8. Cheng, S., Huang, F.: Reversible data hiding in JPEG images based on two-dimensional histogram modification. In: ICCCS 2018, LNCS, vol. 11066, pp. 392–403. Springer, Cham (2018)

A Lossless Polynomial-Based Secret Image Sharing Scheme Utilizing the Filtering Operation

Lintao Liu(✉), Yuliang Lu, Xuehu Yan, Wanmeng Ding, and Qitian Xuan

National University of Defense Technology, Hefei, Anhui, China
liuta1989@163.com

Abstract. As a basic method, Shamir's polynomial-based secret sharing (PSS) is utilized to achieve polynomial-based secret image sharing schemes (PSISSs) with other better performances, such as meaningful shares, two-in-one property and et al. Since the modulus of the sharing polynomial must be a prime, e.g., 251 for PSISSs, Shamir's PSS cannot deal with pixel values more than 250 well. On account of the importance of precise recovery in the image encryption field, there exist several solutions to achieve precise recovery for PSISSs, which have natural drawbacks, such as large computational costs, random shape changes and extra division for sharing. In this paper, we utilize a simple filtering operation to achieve the proposed lossless PSISS, namely LPSISS, without any side effects. We choose 257 rather than 251 as the sharing modulus, and filter sets of shared values to abandon sets with pixel value more than 255. This solution to lossless recovery is simple but effective, which is valuable as a basic technique to achieve better performances. Experimental results are conducted to prove the effectiveness of the proposed scheme.

Keywords: Secret image sharing · Polynomial-based secret sharing · Filtering · Precise recovery

1 Introduction

A secret sharing scheme (SSS) scheme divides the secret into several shares without any secret information revealed, and the secret can be reconstructed with sufficient shares in a certain privileged coalitions. In comparison with other cryptographic techniques, such as symmetric cryptography, asymmetric encryption and information hiding, SSSs have an unique property, namely loss-tolerance, which means the secret information can be still recovered even though parts of shares are lost or destroyed. Therefore, it is beneficial in certain application scenarios, such as access control and communications in unreliable public channels [3]. In recent years, SSs are also utilized in cloud storage [1,5,16] and anonymous e-voting [13].

C.-N. Yang et al. (Eds.): SICBS 2018, AISC 895, pp. 129–139, 2020.
https://doi.org/10.1007/978-3-030-16946-6_11

In 1979, Shamir firstly proposed a (k, n) threshold polynomial-based secret sharing scheme (PSS) on number domain [15], where the secret value can be recovered if and only if sufficient shared values participate in recovery. Based on Shamir's PSSS, Thien and Lin firstly introduced a polynomial-based secret image sharing scheme (PSISS) [17]. From then on, plenty of PSISSs have emerged to realize more interesting performances, such as meaningful shares [6,11], two-in-one recovery [9,20] and shares with different priorities [4,10].

Since the modulus of the sharing polynomial in Shamir's PSSS must be a prime, the original PSISSs based on Thien and Lin's adopt 251 (the maximum prime less than 256) as the modulus as shown in Eq. (1). Considering the value range of a pixel is from 0 to 255, Eq. (1) cannot deal with well the pixel values from 251 to 255. As a result, PSISSs directly based on Shamir's PSSS cannot realize precise recovery of the secret image, and this drawback would greatly limit the practical application of PSISSs in the image encryption field.

$$f(x) = (a_0 + a_1x + a_2x^2 + \cdots + a_{k-1}x^{k-1})\ mod\ 251 \tag{1}$$

In order to achieve precise recovery for PSISSs, researchers have provided several solutions in their works [2,12,14,17–19]. Three primary lossless solutions are discussed in Sect. 2. Thien and Lin realized the important of precise recovery of secret images in the first paper of PSISSs [17]. In their lossless PSISS, they divided pixel values more than 250 into two parts, and then shared two parts with two sharing phases respectively. Similarly to Thien-and-Lin's idea, in Ding et al.'s scheme [2], pixel values more than 250 also need to divide, but then both parts were embedded into two different coefficients of the sharing polynomial during one single sharing phase. However, their solutions also bring in some negative effects, such as extra division and recombination, random shape changes or large computational complexity. In 1989, Rabin proposed an effective method to realize lossless PSSS in the range of $[0, 255]$: he utilized operations on Galois field GF (2^8) instead of integer computations in the finite field. Afterwards, Yang et al. [19] introduced this technique to achieve lossless PSISS. Rabin's technique does not need extra divisions and combinations, but it has a larger time cost than the classic.

In this paper, we proposed a novel solution to precise recovery of the classic PSISS with a simple filtering operation. In the proposed Lossless PSISS (LPSISS) scheme, 257 (the minimum prime larger than 256) is utilized as the modulus in the sharing polynomial, so the range of both secret value and shared value is from 0 to 256. Since the range of the pixel value is from 0 to 255, a filtering operation is added to prevent the set of shared values including 256 being chosen as the final shared combination. Since only one integer (256) needs to be removed by the filtering operation, the time cost of sharing increase quite little while the time cost of recovery remains unchanged. Importantly, its security is also guaranteed, which is described in Sect. 4. Besides, the proposed scheme can be directly utilized to achieve PSISSs with other performances instead of classic PSISSs.

The rest of this paper is organized as follows. In Sect. 2, we introduce Shamir's PSSS and three solutions to lossless recovery of PSISSs in detail. The proposed LPSISS is proposed in Sect. 3. Furthermore, theoretical analyses of its security are given in Sect. 4, and afterwards we provide experimental results in Sect. 5. Finally, we conclude our contributions and envision the future work in Sect. 6.

2 Preliminaries

2.1 Polynomial-Based Secret Image Sharing

In 1979, Shamir proposed (k, n) threshold PSS, which is based on $k-1$ degree polynomial as shown in Eq. (2). In Eq. (2), the modulus p must be a prime to guarantee the recoverability. Furthermore, the coefficient a_0 is utilized to embed the secret value, while the other $k-1$ coefficients including $a_1, a_2, \cdots, a_{k-1}$ are randomly assigned during one sharing phase. Therefore, the function value $f(x)$ is unrelated to a_0, which is the shared value in correspnding to one certain serial x. With n different serials, n shared values $f(x_1), \cdots, f(x_n)$ are generated for distribution. When obtaining k shared values, the secret value a_0 will be precisely decrypted by the Lagrange interpolation.

$$f(x) = (a_0 + a_1x + a_2x^2 + \cdots + a_{k-1}x^{k-1})\ mod\ p \tag{2}$$

Shamir's proposed PSS has the "all or nothing" property, which can be widely used for the encryption in the number domain. Furthermore, Shamir's proposed PSS scheme can be directly utilized for the encryption of images, where the prime p is 251. Experimental results of (3, 4) threshold PSIS based on Shamir's proposed PSS are given in Fig. 1. One out of four shadow images SC_1 as Fig. 1(b) reveals nothing secret as well as the recovered image $S'_{t=2}$ with insufficient shares. The similar image to the secret one can be recovered with 3 or more shares.

However, there exist some errors in the recovered images, e.g. the top right surface of the left indor and the top surface of the right indor should have been recovered to white as the original secret image in Fig. 1(a), but they are wrongly restored into black. Since $p = 251$, all the values in Eq. (2), such as $x, f(x), a_0, \cdots, a_{k-1}$, are limited in the range $[0, 250]$. However, the grayscale image has 256 gray levels from 0 to 255. As a result, some pixel values more than 250 can not be processed, so Shamir's proposed PSIS is treated as SIS with lossy recovery. Currently, many researchers ingored this kind of error in PSIS by truncate values more than 250 to 250. Although the recovered images by this technique look similar to the secret image, they can not satisfy the requirement of lossless recovery in certain application scenarios.

2.2 Solutions to PSISS with Lossless Recovery

Currently, there are mainly three solutions to PSIS with lossless recovery, while some integrated schemes [7, 18] with lossless recovery are not mentioned due to much larger time costs.

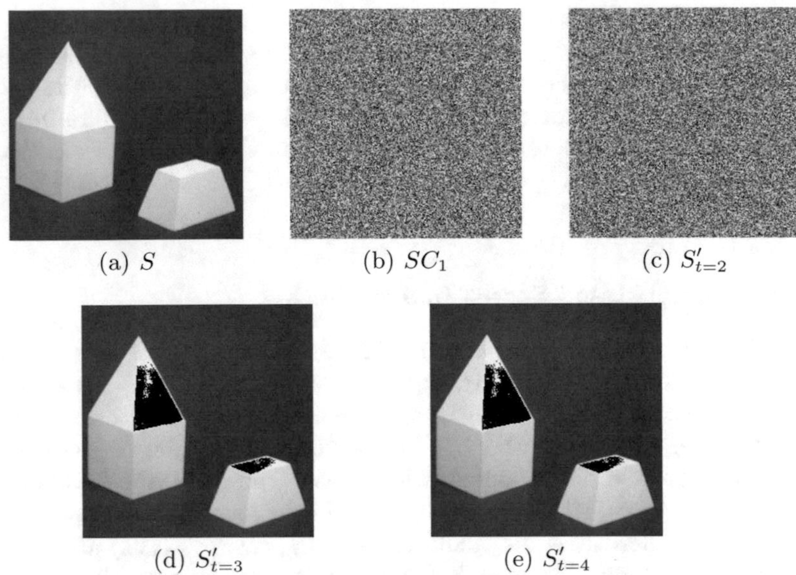

(a) S (b) SC_1 (c) $S'_{t=2}$

(d) $S'_{t=3}$ (e) $S'_{t=4}$

Fig. 1. Shamir's proposed $(3, 4)$ threshold polynomial-based secret image sharing.

In Thien-and-Lin's design with lossless recovery [17], secret values equal to and more than 250 are divided into two parts, including 250 and the remainder modulo 250. Then, two parts are shared with two sharing phases apart. During recovery, if the first recovered value s'_1 is 250, the second value s'_2 also needs to be recovered. The original secret value s' is equal to $s'_1 + s'_2$. By this technique, lossless recovery is achieved, but there exist two obvious drawbacks: (a) extra division and recombination phases are needed, so it costs more computations; (b) it results in random pixel expansion of shares due to the random number of secret pixel values in $[250, 255]$, so shares should be treated as data rather than images.

Similarly, Ding et al. introduced a solution to lossless recovery [2]. Integers from 250 to 255 are also divided into two parts, but both parts are shared with different coefficients during one sharing phase. For example, in $(2, 2)$ threshold scheme, a_0 and a_1 are utilized to embed 250 and the remainder, respectively. In order to guarantee the security, it needs a technique to increase the randomness: a_1 should be random integer multiples of the remainder r, which is also no more than 250. No matter whether the secret pixel is more than 250, it is always recovered with k shared values. In comparison with Thien-and-Lin's solution, the size of shadow images is equal to that of the secret image, as shown in Fig. 2. However, it also needs extra division and recombination phase.

In 1989, Rabin [14] firstly proposed to use other fields instead of Z_p. He also gave a example: for 8-bit bytes we can directly use the Galois field $E = GF(2^8)$ of characteristic 2 and having 256 elements, and all we need is an irreducible polynomial $p(x) \in Z_2[x]$ of degree 8 to allow us to effectively compute in E.

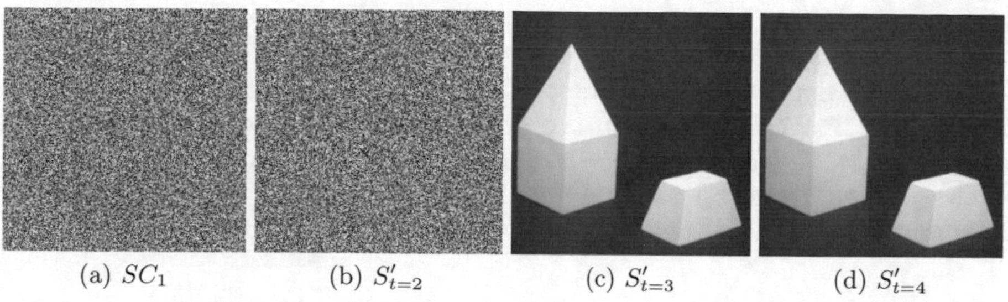

(a) SC_1 (b) $S'_{t=2}$ (c) $S'_{t=3}$ (d) $S'_{t=4}$

Fig. 2. Ding's proposed (3, 4) threshold polynomial-based secret image sharing with lossless recovery.

In 2007, Yang et al. [19] utilized the same idea as a solution to realize a lossless PSISS, where all integer operations are replaced with polynomial operations in GF(2^8). Afterwards, several researches [8,9] referred Yang's proposed PSIS with lossless recovery to build schemes with other properties. The solution based on computations in Galois field can handle with the problem quite well. However, the sharing and recovery phases based on Galois Field have much larger costs than classic PSIS schemes.

3 Lossless Polynomial-Based Secret Image Sharing

3.1 Design Concept

In the classic PSIS, 251, the maximum prime no more than 255, is specified as the prime p, so all generated shared values are limited in [0, 250], which can not be represented by 256 gray levels.

Differently, 257, the minimum prime more than 255, is selected as p, so all secret values in [250, 255] can be processed by the new sharing polynomial as Eq. (3). Furthermore, a filtering operation to shared values is added to guarantee that all shared values must be in [0, 255]. Therefore, the only integer, 256, which can not be represented by 8 bits, is deleted, and then shared values can be stored in pixels. As a result, the secret value from 0 to 255 can be recovered precisely by the Lagrange interpolation.

$$f(x) = (a_0 + a_1x + a_2x^2 + \cdots + a_{k-1}x^{k-1})\ mod\ 257 \tag{3}$$

Remark 1. In fact, this solution to precise recovery of PSISS can be easily expanded into lossless encryption for arbitrary integer infinite ranges from 0 to q. The minimum prime larger than q, written as $p_{min(p \geq q)}$, is used as the modulus of the sharing polynomial. By filtering, only shared combinations with values in the range of $[0, q]$ are reserved as final shared combinations.

3.2 Proposed Scheme

The sharing algorithm of the proposed LPSISS is given in Algorithm 1. The equation in line 7 is the key to the proposed scheme: (*a*) $p = 257$; (*b*) $k - 1$ coefficients $a_1, \cdots, a_{k-1}$ except a_0 is randomly assigned in line 4 to guarantee the security. The filtering operation occurs in line 8–11 and 13–16, which continues to generate shared values until none of them is larger than 255.

Algorithm 1. The algorithm of the proposed (k, n) threshold LPSISS with lossless recovery

Require: A grayscale secret image with size of $M \times N$, S; Threshold, k; Number of shares, n; The list of ID numbers, $serials[n]$.
Ensure: n grayscale shadow images, $SC_1, SC_2, \cdots, SC_n$.
1: **for** $i = 0$ to $M - 1$ **do**
2:     **for** $j = 0$ to $N - 1$ **do**
3:         **while** 1 **do**
4:             Generate a random integer a_{k-1};
5:             $Cond = True$;
6:             **for** $t = 1$ to n **do**
7:                 $g_t = (S(i,j) + a_1 \times serials[t] + \cdots + a_{k-2} \times serials[t]^{k-2} + a_{k-1} \times serial[t]^{k-1})\ mod\ 257$;
8:                 **if** $g_t > 255$ **then**
9:                     $Cond = False$;
10:                     Break;
11:                 **end if**
12:             **end for**
13:             **if** $Cond == True$ **then**
14:                 $[SC_1(i,j), \cdots, SC_n(i,j)] = [g_1, \cdots, g_n]$;
15:                 Break;
16:             **end if**
17:         **end while**
18:     **end for**
19: **end for**

In comparison with Thien-and-Lin's lossless scheme, there exist three differences in the recovery phase of the proposed LPSISS, including: (*a*) $p = 257$; (*b*) a filtering operation in the sharing phase; (*c*) no need for extra decryption.

4 Performance Analysis

4.1 Discussion on Security

In comparison with classic PSISSs, the filtering operation is added into the sharing phase, after the prime 257 is specified as the prime p. The filtering operation deletes certain sets of n shared values, which contain one or more shared values equal to 256. In the other hand, the security and effectiveness of

PSIS are closely related to the number of sets of shared values corresponding to a single secret value. For example, in $(2, 2)$ threshold Shamir's classic PSIS, there are 251 sets of shared values corresponding to a single secret value. In fact, the sharing phase of PSIS is equivalent to the procedure where a set of shared values is randomly selected from all sets corresponding to the secret value. As a result, the security of PSIS increases monotonously with more candidate sets corresponding to a single secret value.

The filtering operation will delete parts of candidate shared combinations. Averagely, the probability, that each of shared values is equal to 256, is $\frac{1}{257}$. Furthermore, the probability of unsatisfied sets for (k, n) threshold scheme is approximately equal to $\frac{n}{257}$. As the result, the rate between unsatisfied sets and all sets is approximately $\frac{n}{257}$. Considering that there exist 257^k sets of shared values, there would be $\frac{n^k}{257}$ unsatisfied sets, the number of which is quite few. Therefore, the number of candidate combinations decreases quite few, and the security of the proposed lossless scheme also decreases very little. When the malicious attacker obtain insufficient shared values, e.g. $k-1$ shared values, they can only eliminate approximately $\frac{n-k+1}{257}$ candidate secret values out of all candidate secret values from 0 to 255.

In summary, the proposed LPSISS is not unconditional security as the same as the classic PSSS, but it is still computational security so that the secret value cannot be deduced with insufficient shares.

4.2 Discussion on Time Costs

As mentioned in Sect. 4.1, the rate between unsatisfied sets and all sets is approximately $\frac{n}{257}$. Furthermore, this conclusion means that it is $\frac{n}{257}$ probability to randomly re-generate a new set of shared values. Therefore, its time costs also increases quite little in comparison with the classic PSSS, and statistical comparisons between them are provided in Sect. 5.

For recovery, the proposed LPSISS has the same time cost with the classic PSSS, since both can utilize the Lagrange interpolation to decrypt.

5 Experiments and Comparisons

In this section, image illustrations are firstly provided to show performances of the proposed scheme, including precise recovery and security. Then, we give statistics about comparisons of time costs between the proposed LPSISS and classic PSISSs, in order to demonstrate its high efficiency.

5.1 Image Illustration

Firstly, experimental results of $(2, 2)$ threshold LPSISS are given in Fig. 3 to illustrate performances of the proposed scheme, with ID numbers $(1, 2)$. Figure 3(a) is the first one out of four shadow images SC_1 without any secret information revealing, while its histogram, as shown in Fig. 3(b), follows the uniform distribution in [0, 255] which provides an effective proof of its security. Figures 3(c) and (d) about the second shadow image provide the same results. When 2 shares participate in recovery, the secret image is reconstructed precisely, as shown in Figs. 3(e) and (f).

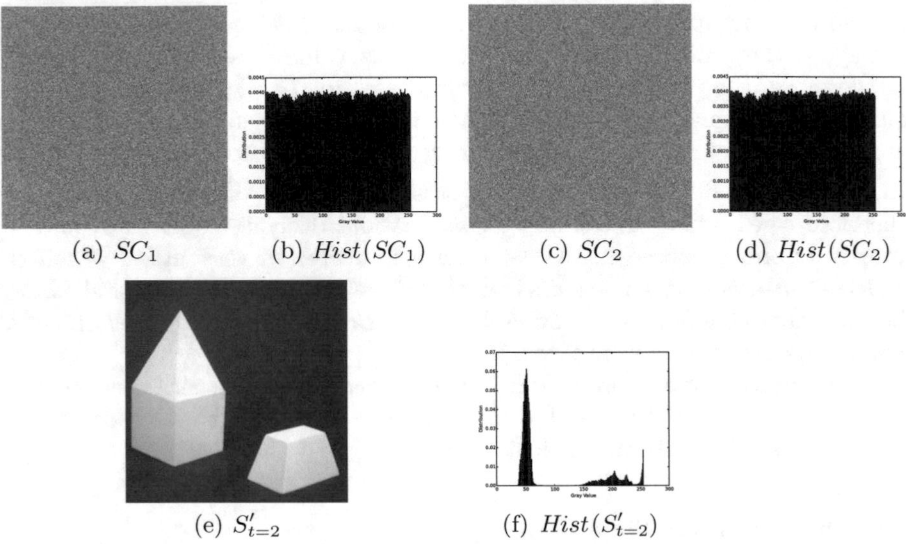

(a) SC_1 (b) $Hist(SC_1)$ (c) SC_2 (d) $Hist(SC_2)$

(e) $S'_{t=2}$ (f) $Hist(S'_{t=2})$

Fig. 3. (2, 2) threshold polynomial-based secret image sharing with lossless recovery, in which $(x_1, x_2) = (1, 2)$.

Then, experimental results of (3, 4) threshold LPSISS are provided in Fig. 4. Similarly, the shadow images SC_1 in Fig. 4(a) is noise-like, in which pixel values follow the uniform distribution in [0, 255] as shown in Fig. 4(b). With 2 shadow images participating in recovery, as shown in Fig. 4(c), the reconstructed image $S'_{t=2}$ is also noise-like. Except 255, pixel values in $S'_{t=2}$ also follow the uniform distribution in [0, 254] as shown in Fig. 4(d). Since 256 is probably recovered with insufficient shares and is truncated into 255 for image storage, pixel values in $S'_{t=2}$ follow the uniform distribution from 0 to 256. with more than 3 shares, the secret image can be precisely recovered as shown in Fig. 4(e)–(h).

From experimental results above, the properties of the proposed PSIS are concluded as follows:

- **Security:** There is no leakage of secret information from shadow images and recovered images with less than k shares.
- **Precise Recovery:** The secret image can be reconstructed precisely with k or more shares.
- **No Pixel Expansion:** Shadow images have the same size as the secret image, this is, one secret pixel are corresponding to one shared pixel in each shares.

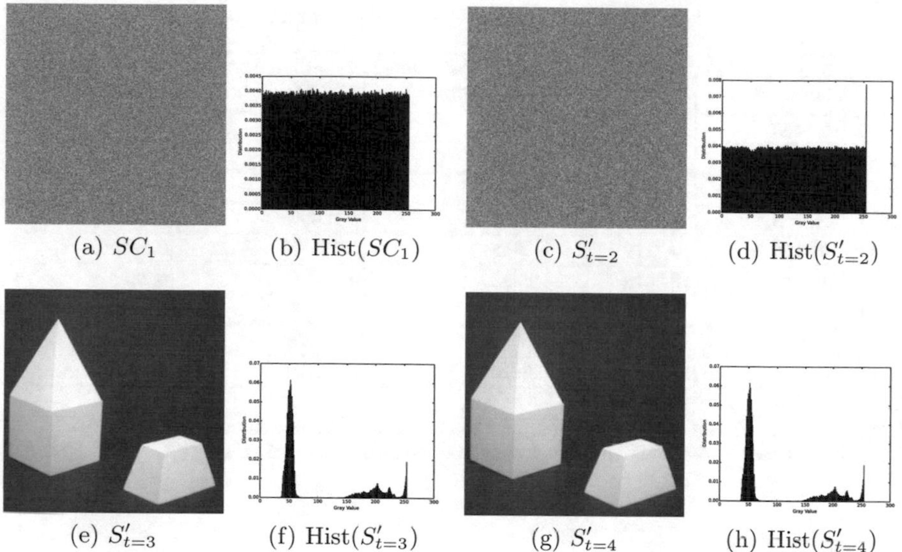

(a) SC_1 (b) Hist(SC_1) (c) $S'_{t=2}$ (d) Hist($S'_{t=2}$)

(e) $S'_{t=3}$ (f) Hist($S'_{t=3}$) (g) $S'_{t=4}$ (h) Hist($S'_{t=4}$)

Fig. 4. (3, 4) threshold polynomial-based secret image sharing with lossless recovery, in which $(x_1, x_2, x_3, x_4) = (1, 2, 3, 4)$.

5.2 Statistics on Time Costs

For statistics on time costs, We carried out experiments on Macbook Pro, in which CPU is 2.5 GHz Intel Core i7. The proposed LPSISS and classic PSISS are written in Python, and executed in a single-threaded manner. The same secret image *indor* with size of 1024 × 1024 is encrypted 10 times for each of (k, n) threshold schemes. The final statistics are the average of corresponding 10 experimental results, including time costs of the sharing phase and the number of filtered sets.

As Table 1 shown, time costs of LPSISS is a little more than that of classic PSISS for all (k, n) threshold schemes, since the filtering operation will filter unsatisfied sets including 256 until generating satisfied shared sets. The number of filtered sets is recorded to prove our conclusion in Sect. 4: the rate between unsatisfied sets and all sets is approximately $\frac{n}{257}$. As shown in columns 5 and 6, the experimental rate is quite similar to the theoretical rate, so the conclusion is proved. Furthermore, it is concluded that the time cost of LPSISS is $\frac{n}{257}$ times more than that of Shamir's PSISS. Therefore, the proposed LPSISS with the filtering operation is still high-efficiency.

As a result, advantages of the proposed LPSISS in efficiency are concluded as follows:

- **High-efficiency:** The time cost of the proposed LPSISS is just $\frac{n}{257}$ more than that of classic PSISS.
- **Simple recovery:** The Lagrange interpolation with prime $p = 257$ is directly utilized to decrypt for the proposed LPSISS, similarly to classic PSISS.

Table 1. Statistics on time costs.

(k, n)	Time cost (\s) classic PSISS	Time cost (\s) LPSISS	Number of filtered sets - Num_f	Experimental rate[a] $(\frac{Num_f}{Sum})_e$	Theoretic rate[b] $(\frac{Num_f}{Sum})_t$
$(2, 2)$	5.7420	5.9049	8333	0.7884%	0.7782%
$(2, 3)$	6.3977	6.5537	12398	0.1168%	0.1167%
$(3, 3)$	8.6397	9.0122	12299	0.1159%	0.1167%
$(2, 4)$	7.1073	7.2878	16688	0.1566%	0.1556%
$(3, 4)$	9.3125	9.7172	16543	0.1553%	0.1556%
$(4, 4)$	11.8576	12.1873	16523	0.1551%	0.1556%
$(2, 5)$	7.9230	8.2907	20815	0.1946%	0.1946%
$(3, 5)$	10.3739	10.7847	20641	0.1930%	0.1946%
$(4, 5)$	13.0451	13.6748	20676	0.1934%	0.1946%
$(5, 5)$	15.7033	16.2453	20551	0.1922%	0.1946%

[a] Experimental rate between the number of filtered sets and all generated sets.
[b] Theoretical rate between the number filtered sets and all generated sets - $\frac{n}{257}$.

6 Conclusion

Based on Shamir's classic PSISS scheme, a (k, n) threshold lossless PSISS, namely LPSISS, is proposed in this paper. For lossless recovery, the significant change of the prime p in the sharing polynomial is from 251 to 257, and then the additional filtering operation can ensure each of shared values in the range $[0, 255]$. In comparison with other solutions to lossless recovery, the proposed scheme is achieved with the least modification on classic PSIS, and has no side effects, such as large computational costs, random shape changes and extra division for sharing. By theoretical analyses and experiments, its security and effectiveness are proved. Our future work is to utilize the proposed scheme to achieve PSISSs with other interesting properties.

Acknowledgment. This work is supported by the National Natural Science Foundation of China (Grant Number: 61602491).

References

1. Attasena, V., Darmont, J., Harbi, N.: Secret sharing for cloud data security: a survey. VLDB J. **2**, 1–25 (2017)
2. Ding, W., Liu, K., Yan, X., Liu, L.: Polynomial-based secret image sharing scheme with fully lossless recovery. Int. J. Digital Crime Forensics **10**, 120–136 (2018)
3. Dingledine, R., Freedman, M.J., Molnar, D.: The free haven project: distributed anonymous storage service. In: International Workshop on Designing Privacy Enhancing Technologies: Design Issues in Anonymity and Unobservability, pp. 67–95 (2001)
4. Guo, C., Chang, C.C., Qin, C.: A hierarchical threshold secret image sharing. Pattern Recogn. Lett. **33**(1), 83–91 (2012)
5. Hadavi, M.A., Jalili, R., Damiani, E., Cimato, S.: Security and searchability in secret sharing-based data outsourcing. Int. J. Inf. Secur. **14**(6), 1–17 (2015)

6. He, J., Lan, W., Tang, S.: A secure image sharing scheme with high quality stego-images based on steganography. Multimedia Tools Appl. **76**(6), 7677–7698 (2017)
7. Jin, D., Yan, W.Q., Kankanhalli, M.S.: Progressive color visual cryptography. J. Electron. Imaging **14**(3), 033019 (2005)
8. Li, P., Ma, P.J., Su, X.H., Yang, C.N.: Improvements of a two-in-one image secret sharing scheme based on gray mixing model. J. Vis. Commun. Image Represent. **23**(3), 441–453 (2012)
9. Li, P., Yang, C.N., Kong, Q., Ma, Y., Liu, Z.: Sharing more information in gray visual cryptography scheme. J. Vis. Commun. Image Represent. **24**(8), 1380–1393 (2013)
10. Li, P., Yang, C.N., Wu, C.C., Kong, Q., Ma, Y.: Essential secret image sharing scheme with different importance of shadows. J. Vis. Commun. Image Represent. **24**(7), 1106–1114 (2013)
11. Lin, P.Y., Chan, C.S.: Invertible secret image sharing with steganography. Pattern Recogn. Lett. **31**(13), 1–7 (2010). https://doi.org/10.1016/j.patrec.2010.01.019
12. Lin, S.J., Lin, J.C.: VCPSS: a two-in-one two-decoding-options image sharing method combining visual cryptography (VC) and polynomial-style sharing (PSS) approaches. Pattern Recogn. **40**(12), 3652–3666 (2007)
13. Liu, Y., Zhao, Q.: E-voting scheme using secret sharing and k-anonymity. World Wide Web Internet Web Inf. Syst. pp. 1–11 (2018)
14. Rabin, M.O.: Efficient dispersal of information for security, load balancing, and fault tolerance. J. ACM (JACM) **36**(2), 335–348 (1989)
15. Shamir, A.: How to share a secret. Commun. ACM **22**(11), 612–613 (1979)
16. Singh, P., Agarwal, N., Raman, B.: Secure data deduplication using secretsharing schemes over cloud. Future Gener. Comput. Syst. **88**, 156–167 (2018)
17. Thien, C.C., Lin, J.C.: Secret image sharing. Comput. Graph. **26**(5), 765–770 (2002)
18. Ulutas, G., Nabiyev, V.V., Ulutas, M.: Polynomial approach in a secret image sharing using quadratic residue. In: International Symposium on Computer and Information Sciences, pp. 586–591 (2009)
19. Yang, C.N., Chen, T.S., Yu, K.H., Wang, C.C.: Improvements of image sharing with steganography and authentication. J. Syst. Softw. **80**(7), 1070–1076 (2007)
20. Yang, C.N., Ciou, C.B.: Image secret sharing method with two-decoding-options: lossless recovery and previewing capability. Image Vis. Comput. **28**(12), 1600–1610 (2010)

CT Image Secret Sharing Based on Chaotic Map and Singular Value Decomposition

Feixiang Zhao, Mingzhe Liu(✉), Xianghe Liu, Xin Jiang, and Zhirong Tang

State Key Laboratory of Geohazard Prevention and Geoenvironment Protection, Chengdu University of Technology, Sichuan, China
liumz@cdut.edu.cn

Abstract. In order to prevent CT images from being stolen and tampered in telemedicine, a secret sharing method of (*n*, *n*) structure is designed and combined with the classical chaotic digital image encryption method. The method consists of three parts: In the first part, singular value decomposition is used to generate the key of the Henon chaotic map and CT subgraphs; in the second part, the Henon map is used to generate chaotic sequences of scramble CT subgraphs. Then a valueless image selected by the user is used as an impurity image, the diffusing operation is completed by a XOR operation between the impurity image and the image obtained by the scrambling operation. Finally, the image obtained by the diffusing operation is divided into multiple parts by the secret sharing method designed in this paper and embedded into the cover images selected by the user to generate several shadow images. The third part is the decryption part, and the decryptor completes the decryption when all the shadow images, impurity image and keys are collected. Several experiments have shown that this method has excellent performance in the test of ciphertext statistical characteristics, sensitivity, PSNR and other typical indicators for measuring the performance of image encryption methods.

Keywords: Singular value decomposition · Chaotic map · Secret sharing · CT image

1 Introduction

With the rapid development of telemedicine technology, more and more patients benefit from it. The remote transmission of CT images is an important part of telemedicine, and the widespread use of telemedicine is accompanied by the transmission of a large number of CT images on the Internet. CT images have the characteristics of high acquisition cost and great scientific value. At the same time, some CT images of important figures in the political, military and economic fields have high confidentiality. Therefore, it is of great practical value to encrypt CT images before transmitting them.

As a digital image with a large amount of data and strong correlation between pixels, ordinary text encryption systems such as DES (Data Encryption Standard) and AES (Advanced Encryption Standard) are not suitable for encrypting CT images.

C.-N. Yang et al. (Eds.): SICBS 2018, AISC 895, pp. 140–155, 2020.
https://doi.org/10.1007/978-3-030-16946-6_12

The urgent need of the market is prompting researchers in related fields to constantly propose new digital image encryption algorithms. At present, the number of existing algorithms in this field is huge. Generally speaking, it can be divided into three types. The first one is to encrypt images in the spatial domain [1–11], and the second is to encrypt images in the frequency domain [12–14, 21, 22], the third is to operate in both the spatial domain and the transform domain [15, 23]. The spatial domain image encryption algorithms mainly uses the chaotic maps [1–5, 39–41], Latin square [6], cellular automata [7, 8], and their mixed use, for example, the mixture of Latin squares and chaotic maps [9, 11] and the mixture of cellular automata and chaotic maps [10]. A common drawback of spatial domain algorithms is that they cannot resist selective plaintext attacks. In recent years, a quite number of algorithms have been proved to be insecure or directly cracked [30–35]. The time-frequency transform tools commonly used in image encryption schemes include Discrete Cosine Transformation (DCT) [21], Discrete Sine Transform (DST) [13] and Discrete Wavelet Transform (DWT) [22]. In general, the frequency domain encryption algorithms is more secure than the spatial domain encryption algorithms, but the former cannot meet the requirements of non-destruction, and the decrypted image obtained by the inverse transform will be different from the original image. Therefore, it is not suitable for encrypting CT images.

All the algorithms above require the decryptor to hold the key, however, some researchers believe that using a key to decrypt in some application scenarios is not a wise decision. In 1979, Shamir [16] first proposed a data secret sharing scheme with (t, n) threshold structure based on the interpolation polynomial on the prime domain $F_p = \{0, 1, 2, \ldots p-1\}$. Specifically, the data is divided into n shares ($n > t$), as long as the receiver collects any t shares, the decryption can be completed without a key, but less than t shares cannot obtain any useful information. Drawing on the theory of secret sharing, Naor and Shamir [17] proposed visual cryptography (VC) in 1994, and applied secret sharing methods to the field of image encryption for the first time. VC splits the image to be hidden into n shadow images, and the decryptor can see the secret image only when the shadow images of not less than t are combined in a specific order.

Inspired by the work of Naor and Shamir, some researchers have come up with some algorithms with better performance [18–20]. Although the operation of these visual cryptography-based algorithms is simple, the visual effect of the restored image is somewhat distorted compared to the original secret image, and the integrity of the pixel data cannot be guaranteed. In addition to visual cryptography, another tool widely used in image secret sharing algorithms is Lagrange interpolation [24]. In 2002, Thien and Lin [25] innovatively introduced Lagrange interpolation into the field of image secret sharing. They constructed a t-1 order polynomial in the finite field and regarded the t pixel values of the image as coefficients, resulting in several shadow images, each containing partial information of the original secret image. When the number of collected images is greater than or equal to t, the original secret image can be completely restored by the Lagrange interpolation formula of finite field. Since then, some researchers have produced numerous research results based on the improvement of the algorithm [26–29]. In general, secret sharing is highly resistant to selective plaintext attacks.

Based on the research status and the high security requirement of CT image encryption, this paper designs a new method of image secret sharing and combines it with singular value decomposition and chaotic map to form a novel CT image secret sharing algorithm. Considering the strict confidentiality requirement of CT images, unlike the (*t*, *n*) threshold structure of the traditional image secret sharing algorithm, the secret sharing scheme designed in this paper is (*n*, *n*) structure. At the same time, inspired by Liu et al. [28, 29], we introduced Least Significant Bit Substitution (LSBR) in the proposed algorithm, so that the shadow images finally produced in the encryption phase will not cause suspicion to the thieves in appearance. The measure further enhances the security of the algorithm.

The rest of the article is structured as follows. Theoretical representation and detailed steps of the algorithm are presented in second section. The experimental results, security and performance analysis of the algorithm are presented in the third section. The fourth section is the conclusion.

2 The Proposed Scheme

The CT image secret sharing method proposed in this paper includes three modules. The first module is responsible for the generation of CT subgraphs and keys. The function of the second module is to produce several shadow images. The third module is used for the recovery of secret CT images. The block diagram of the operation of this method is shown in Fig. 1.

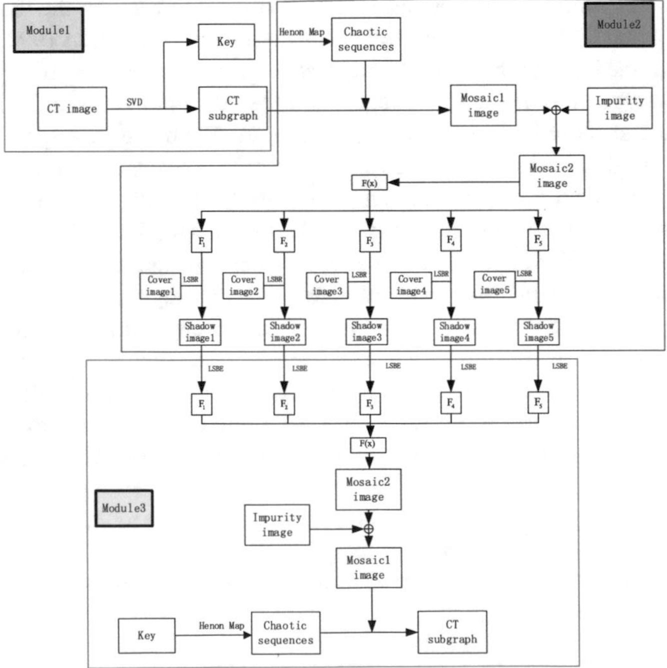

Fig. 1. Diagram of the encryption and decryption process

2.1 Generation of CT Subgraphs

Singular value decomposition (SVD) is a typical matrix decomposition method. Since the singular values of the pixel value matrix of digital image has good robustness, singular value decomposition has been widely used in the field of image processing. Let the pixel value matrix of a CT image be $X, X \in R^{m \times n}$. Its SVD is shown in formula (1).

$$X = UDV^T \tag{1}$$

Where U is the unitary matrix in $R^{m \times m}$, called the left singular matrix. V is the unitary matrix in $R^{n \times n}$, called the right singular matrix. Each row of UD is called a principal component of X, and the corresponding column of V is called a load. D is a diagonal matrix, and the elements on the diagonal are singular values of the matrix X arranged from largest to smallest. Assuming $m > n$, A can be rendered by the outer product expansion, as shown in Eq. (2).

$$A = u_1\sigma_1 v_1^T + u_2\sigma_2 v_2^T + \cdots + u_n\sigma_n v_n^T \tag{2}$$

Where $u_i, i = 1, 2, \ldots, n$ is the i-th column vector of the matrix U, $v_i, i = 1, 2, \ldots, n$ is the i-th row vector of the matrix V^T, $\sigma_i, i = 1, 2, \ldots, n$ is the i-th singular value arranged in descending order of matrix X. The first k sub-forms ($k < n$) in (2) constitute A_k, and the larger the value of k, the closer the obtained image is to the original image. When $k = n$, the original image is obtained, that is, $A = X$. In this way, CT subgraphs of different definitions can be obtained. According to the privilege level of the receiver, the sender can select a CT subgraph of different resolutions as a secret image and send it to the receiver.

2.2 Production of Several Shadow Images

2.2.1 Permutation of CT Subgraphs

Henon map is a two-dimensional discrete chaotic system proposed by Henon in 1976 [36], as shown in formula (3).

$$\begin{cases} x_{n+1} = 1 - ax_n^2 + y_n \\ y_{n+1} = bx_n \end{cases} \tag{3}$$

When $1.07 \leq a \leq 1.40$, $0.2 \leq b \leq 0.314$, the system is in a chaotic state. Assuming that the size of the CT image to be encrypted is $M \times N$, the two chaotic sequences generated by the Henon map, $x_m \in \{1, 2, \ldots, M\}$, $m = 1, 2, \ldots, M$ and $y_n \in \{1, 2, \ldots, N\}, n = 1, 2, \ldots, N$, are respectively used as the row and column coordinates to scrambling the CT image, as shown in formula (4).

$$Mosaic1\_X(m, n) = X(x_m, y_n), m = 1, 2, \ldots, M, n = 1, 2, \ldots, N \tag{4}$$

The pixel value matrix corresponding to the scrambled image Mosaic1 image is *Mosaic1*_X, and X represents the original CT image matrix.

The first k singular values $\sigma_i, i = 1, 2, \ldots, k$ in the formula (2) are scaled to the interval [1.07, 1.40] and a value is randomly selected as a. The first k column vector $v_i, i = 1, 2, \ldots, k$ in the formula (2) are scaled to the interval [0.2, 0.314] and a value is randomly selected as b. The values of x_0 and y_0 are derived from the arbitrary selection of elements in the fractional matrix obtained by rounding down the elements in matrix A.

2.2.2 Diffusion of CT Subgraphs

In 1998, Fridrich [1] first proposed an image encryption algorithm based on permutation of pixel positions and diffusion of pixel values. At present, the permutation-diffusion structure has become a basic encryption structure widely used in image encryption algorithms.

In general, the change of the pixel position does not change the pixel value, and the algorithms containing only the pixel position scrambling function has poor statistical characteristics, and the ability of anti-decipher is not strong [38]. In order to overcome this shortcoming, the diffusion operation can be used to change the individual pixel values of the image, thereby changing the statistical characteristics of the image and enhancing the ability of anti-decipher. In the method proposed herein, the diffusion operation of the CT image is completed by an XOR operation between the Mosaic1 image and the Impurity image selected by the sender. Specifically, assuming that the size of the pixel value matrix *Ip_X* of the Impurity image and *Mosaic1_X* is $M \times N$, the process of generating a diffusion image of the same size is shown in formula (5).

$$\begin{cases} Mosaic2\_X(1,n) = Mosaic1\_X(1,n) \oplus Ip\_X(1,n); n = 1,2,\ldots,N \\ Mosaic2\_X(i,n) = Mosaic1\_X(i,n) \oplus Mosaic1\_X(i-1,n) \oplus Ip\_X(i,n); i = 2,3,\ldots,M, n = 1,2,\ldots,N \end{cases} \tag{5}$$

It can be clearly seen from formula (5) that the diffusion operation in this method is plaintext-related, and this feature improves the ability of anti-decipher of the algorithm.

2.2.3 Production of Several Shadow Images

This section mainly consists of two tasks, the first one is the cutting of Mosaic2 image generated by the diffusion operation; The second is to convert each pixel of the cover images into a binary sequence and embed each sub-block obtained by the cutting into the least significant bits of the binary sequence.

A cubic polynomial as shown in Eq. (6) is defined herein. Each pixel of the Mosaic2 image corresponds to one such cubic polynomial. Where $a_i, i = 0, 1, 2, 3$ are coefficients, and the value ranges from 0 to 3. The coefficients of the cubic polynomial are derived from the recombination of the bits of the pixel value in the Mosaic2 image, as shown in Fig. 2.

$$F(x) = a_0 \oplus a_1 x \oplus (a_2 x^2 mod 16) \oplus (a_3 x^3 mod 16) \tag{6}$$

Bringing $x_1 = 1,\ x_2 = 2,\ x_3 = 3,\ x_4 = 4,\ x_5 = 5$ to $F(x)$, respectively, and obtaining $F(x_1)$, $F(x_2)$, $F(x_3)$, $F(x_4)$ and $F(x_5)$. Since there are four values for the four coefficients, $F(x_1)$, $F(x_2)$, $F(x_3)$, $F(x_4)$ and $F(x_5)$ have 4^4 = 256 combinations. All combinations and their corresponding coefficient values are listed in the Decryption

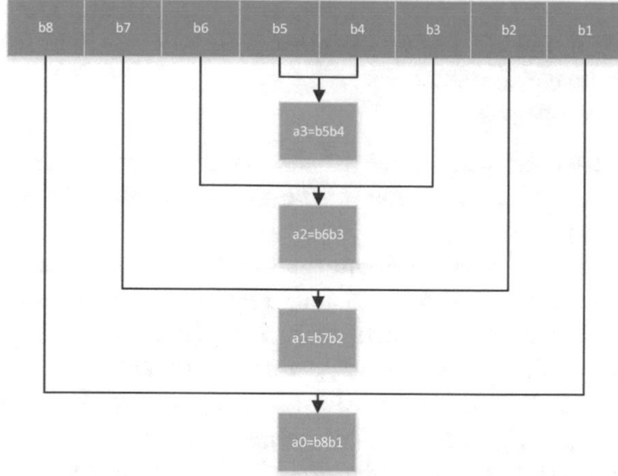

Fig. 2. Recombination of binary pixel values into polynomial coefficients

Table. The five function values of all the pixels constitute five matrices: F_1, F_2, F_3, F_4 and F_5. The dimensions of the five matrices are identical to the original CT image. Since the mod 16 operation is performed, the values of all the elements of F_1, F_2, F_3, F_4 and F_5 are at [0, 15]. At the same time, since the elements in F_1 are obtained by the XOR operation of four 2b binary numbers, the value of the elements in F_1 is in [0, 3].

The next step is to select five worthless images C_1, C_2, C_3, C_4 and C_5 as cover images, and replace the lowest two bits with the above five matrices to generate shadow images S_1, S_2, S_3, S_4 and S_5 respectively. Since the values of the elements of F_2, F_3, F_4, and F_5 are converted from decimal to binary, there are up to 4 bits, which requires two pixels to accommodate. Therefore, the cover images C_2, C_3, C_4, and C_5 need to be twice as large as the original CT image. If the size of the original CT image is $M \times N$, the cover images C_2, C_3, C_4, and C_5 need to be $M \times 2N$. C_1 only needs to be the same size as the original CT image.

The specific process for the lowest two significant bits replacement (LSBR) of C_1 is as follows:

Inputs: Cover image C_1, matrix F_1;
Output: Shadow image S_1.

Step 1: Convert each pixel value of the cover image C_1 into an 8-bit binary sequence, for example, pixel value of the m-th row and n-th column $C_1(m, n) = c_8c_7c_6c_5c_4c_3c_2c_1$;
Step 2: Convert each element of F_1 into an 8-bit binary sequence, for example, pixel value of the m-th row and n-th column $F_1(m, n) = 000000f_2f_1$;
Step 3: Replace the lowest two bits of $C_1(m, n)$ with the lowest two bits of $F_1(m, n)$, then get $S_1(m, n) = c_8c_7c_6c_5c_4c_3f_2f_1$;
Step 4: Repeat step 3 until all pixel values in C_1 have been replaced.

Given $i \in \{2,3,4,5\}, m \in \{1,2,\ldots,M\}, n \in \{1,2,\ldots,N\}$, The specific process for the lowest two significant bits replacement (LSBR) of C_i is as follows:

Inputs: Cover image C_i, matrix F_i;

Output: Shadow image S_i.

Step 1: Convert each pixel value of the cover image C_i into an 8-bit binary sequence, for example, pixel value of the m-th row and n-th column $C_i(m,n) = c_8c_7c_6c_5c_4c_3c_2c_1$;

Step 2: Convert each element of F_i into an 8-bit binary sequence, for example, pixel value of the m-th row and n-th column $F_i(m,n) = 0000f_4f_3f_2f_1$. Then take the lowest three and four bits of $F_i(m,n)$ to form $F_{i,1}(m,n) = 000000f_4f_3$, and take the lowest two bits of $F_i(m,n)$ to form $F_{i,2}(m,n) = 000000f_2f_1$;

Step 3: Replace the lowest two bits of $C_i(m, 2 \times n - 1)$ with the lowest two bits of $F_{i,1}(m,n)$, then get $S_i(m, 2 \times n - 1) = c_8c_7c_6c_5c_4c_3f_4f_3$;

Step 4: Replace the lowest two bits of $C_i(m, 2 \times n)$ with the lowest two bits of $F_{i,2}(m,n)$, then get $S_i(m, 2 \times n) = c_8c_7c_6c_5c_4c_3f_2f_1$;

Step 5: Repeat step 3 and step 4 until all pixel values in C_i have been replaced.

2.3 Recovery of Secret CT Image

The recovery of the CT subgraph requires the participation of several recipients. In general, there are several situations:

Situation 1: If only n ($n < 5$) receivers that hold the Shadow images participate in the decryption, no information about the secret CT subgraph can be obtained;

Situation 2: If there are 5 recipients who save different shadow images to participate in the decryption, Mosaic2 image can be restored;

Situation 3: If there are 5 recipients who save different shadow images and the recipient who saves Impurity image participates in the decryption, Mosaic1 image can be restored;

Situation 4: If there are 5 recipients who store different shadow images, the recipient who saves the Impurity image and the recipient who saves the key participate in the decryption at the same time, the secret CT subgraph can be completely restored.

The decryption steps in Situations 1, 2, and 3 are only part of the decryption steps of Situation 4. The decryption steps of Situation 4 are specifically explained below.

Inputs: Five shadow images, Impurity image, key;

Output: CT subgraph.

Step 1: Confirm S_1 according to the size of the image, convert each element of S_1 into an 8-bit binary sequence, for example, element of the m-th row and n-th column $S_1(m,n) = s_8s_7s_6s_5s_4s_3s_2s_1$. Then extract the lowest two bits of all elements, forming the elements of the corresponding position of the matrix F_1, for example $F_1(m,n) = 000000s_2s_1$;

Step 2: Name the remaining four shadow images as S_a, S_b, S_c, $S_d, a, b, c, d \in \{2,3,4,5\}$, then covert each element of S_a, S_b, S_c, S_d into an 8-bit binary sequence, for example, for S_b, element of the m-th row and n-th column

$S_b(m,n) = s_8s_7s_6s_5s_4s_3s_2s_1$. Finally, *according* to formula (7), generate F_a, F_b, F_c, $F_d, a, b, c, d \in \{2,3,4,5\}$;

$$F_i(m,n) = (S_i(m, 2\times n-1) \wedge 00001100) \vee (S_i(m, 2\times n) \wedge 00000011),\ i = a,b,c,d \tag{7}$$

Step 3: From $F_1(1,1)$, $F_i(1,1), i = a,b,c,d$ to $F_1(M,N)$, $F_i(M,N), i = a,b,c,d$, the values of the four coefficients corresponding to *F*(m, n) are sequentially found by the look-up table method, and are recombined into pixel of corresponding position in the Mosaic2 image. Finally, get the Mosaic2 image;
Step 4: Combine Mosaic2 image with Impurity image and get Mosaic1 image according to formula (8);

$$\begin{cases} Mosaic1\_X(1,n) = Mosaic2\_X(1,n) \oplus Ip\_X(1,n); n = 1,2,\ldots,N \\ Mosaic1\_X(i,n) = Mosaic2\_X(i,n) \oplus Mosaic1\_X(i-1,n) \oplus Ip\_X(i,n); i = 2,3,\ldots,M, n = 1,2,\ldots,N \end{cases} \tag{8}$$

Step 5: Bringing the keys into the Henon map then yields two chaotic sequences, through which the pixels in the Mosaic1 image are transferred to the correct position, and then the secret CT subgraph is restored.

3 Test Results and Performance Analysis

In this section, some typical performance metrics for image encryption algorithms were chosen to measure the quality of the proposed algorithm. Including: key space analysis, statistical analysis, ciphertext sensitivity analysis and PSNR. The program included in this algorithm and all test programs are run using MATLAB-R2016b and Dev-C++. The main configuration of the computer used is as follows: Windows 7, 64-bit; Core-i7 5820K, 3.3 GHz; 24G memory. This section uses a total of three CT images from the same hospital: *pelvis*, *abdomina*, and *spine*, all of which have obtained legal use rights. The above three pictures are all 8b grayscale images. When k = 350, the original image, permutation image, diffusion image, and cover image and shadow image of the three images are shown in Figs. 3, 4 and 5, respectively. Impurity image shown in

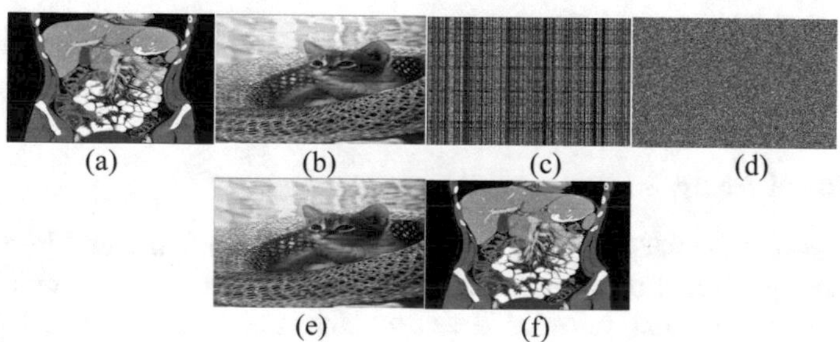

Fig. 3. *Abdomina* (397 × 388), (a) original image, (b) cover image, (c) permutation image, (d) diffusion image, (e) shadow image, (f) revealed image

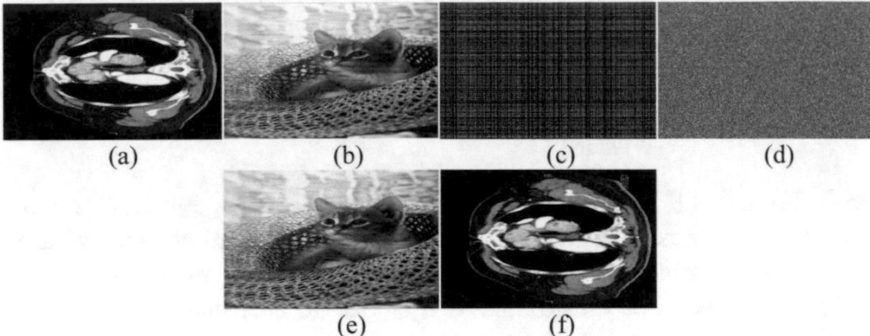

Fig. 4. *Pelvis* (752 × 489), (a) original image, (b) cover image, (c) permutation image, (d) diffusion image, (e) shadow image, (f) revealed image

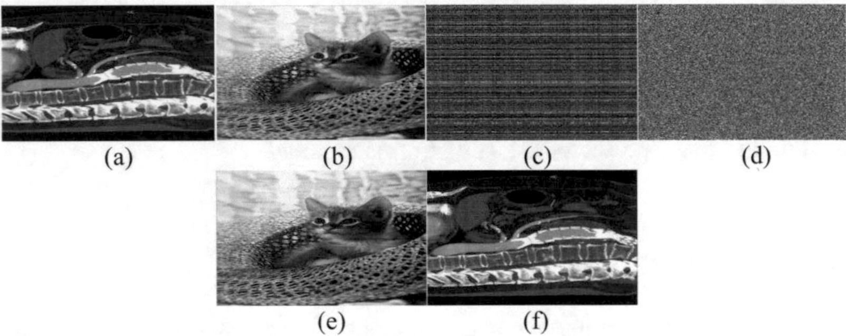

Fig. 5. *Spine* (476 × 388), (a) original image, (b) cover image, (c) permutation image, (d) diffusion image, (e) shadow image, (f) revealed image

Fig. 6. Impurity image

Fig. 6. One point that needs to be specifically stated is that this article only presents one of the five shadow images, because the five shadow images are identical in appearance.

3.1 Statistical Analysis

In order to prevent the thief from using statistical means to mine useful information, an esoteric image generated by an excellent image encryption system should satisfy the irregularity. In this section, in order to test whether the thief can extract information from the low-significant bits (LSBE) of a certain cover image, the low-significant bits of the cover image, namely the matrices F_1, F_2, F_3, F_4 and F_5, are statistically analyzed by the chi-square statistical test.

Given a set of observed frequency distribution $o_i, i = 0, 1, \ldots, n$, assume that its theoretical frequency distribution is $t_i, i = 0, 1, \ldots, n$. When the assumption is true, formula (9) is called Pearson Chi-square statistic, it obeys Chi square distribution with a degree of freedom n.

$$\chi^2 = \sum\nolimits_{i=0}^{n} \frac{(o_i - t_i)^2}{t_i} \tag{9}$$

For a grayscale image with a gray scale of n and a size of $M \times N$, it is assumed that the pixel frequency $o_i, i = 0, 1, \ldots, n$ of each gray value in the histogram obeys a uniform distribution, i.e., $t_i = t = M \times N/n$. Equation (10) obeys Chi square distribution with a degree of freedom n.

$$\chi^2 = \sum\nolimits_{i=0}^{n} \frac{(o_i - t)^2}{t} \tag{10}$$

In order to verify the reliability of the hypothesis, a small significance level $\alpha = 0.01$ is given. Therefore, the formula (11) is obtained.

$$P\{\chi^2 \geq \chi^2_\alpha(n)\} = \alpha \tag{11}$$

That is, when $\chi^2 < \chi^2_\alpha(n)$, o_i has a great probability to obey the uniform distribution.

The range of pixel values in F_1 is [0, 3], and the range of pixel values in F_2, F_3, F_4 and F_5 is [0, 15]. The calculated Pearson Chi-square statistic of F_1, F_2, F_3, F_4 and F_5 corresponding to the three CT subgraph is listed in Table 1.

Table 1. Calculated Pearson Chi-square statistic

	Abdomina	*Pelvis*	*Spine*
F_1	0.0808	0.1641	0.0733
F_2	4.7117	6.2610	5.2009
F_3	4.5367	5.8430	5.0752
F_4	4.6612	5.9222	5.4015
F_5	4.7060	6.2097	5.2044

$\chi^2_{0.01}(3) = 11.3450$, $\chi^2_{0.01}(15) = 30.5780$ was obtained by Chi square distribution table. Comparing with the data in Table 1, it can be found that the Pearson Chi-square statistic of F_1 is significantly smaller than $\chi^2_{0.01}(3) = 11.3450$, and the Pearson Chi-square statistic of F_2, F_3, F_4 and F_5 is significantly smaller than $\chi^2_{0.01}(15) = 30.5780$. Therefore, it can be considered that all the block diagrams are subject to uniform distribution. In other words, it can be considered that any of F_1, F_2, F_3, F_4 and F_5 is noise.

3.2 Key Space Analysis

An excellent encryption algorithm should be extremely sensitive to keys, and its key space should be large enough to effectively combat exhaustive attacks. Key space analysis includes key quantity analysis and key sensitivity analysis.

3.2.1 Key Quantity Analysis

For the proposed algorithm, the keys is the initial values of the chaotic Henon system, and the initial values is related to the order k of the subgraph obtained by the singular value decomposition, so, $K = \{a_0, b_0, x_0, y_0, k\}$. The first four parameters are of type double and k is integer. So the size of the key space is about $S = (2^{64})^4 \times k = 2^{256} \times k$. The larger the size of the image, the more sub-forms obtained by singular value decomposition, the larger k, and therefore the larger the key space. According to [37], The algorithm proposed in this paper is capable of resisting exhaustive attacks.

3.2.2 Key Sensitivity Analysis

The purpose of key sensitivity analysis is to measure the validity of the key. When the key changes slightly, if the two ciphertext images obtained by encrypting the same plaintext image are significantly different, the encryption scheme has strong key sensitivity.

Take *abdomina* as an example. Take the key pair $\{a_0, b_0, x_0, y_0\}$ of the diffused image in 0. Fig. 3(c) as the initial values, and use 10^{-13} as the change amount to arbitrarily change one of the keys to get four new key pairs. Encrypting *abdomina* with the new four key pairs, get the new four diffusion images and their difference images from the original diffusion image, they are listed together in Fig. 7.

It can be seen from Fig. 7 that the four difference images all show noise form, which indicates that the small change of the key value causes significant difference of the ciphertext image, so the algorithm proposed in this paper has high key sensitivity.

3.3 Peak Signal to Noise Ratio

Peak Signal to Noise Ratio (PSNR), this indicator is often used to test whether a digital image encryption algorithm can restore an encrypted picture with high quality. The *PSNR* is defined as shown in Eq. (12).

$$PSNR = 10 \times log_{10}(\frac{255^2}{MSE})d \tag{12}$$

Where *MSE* is Mean Square Error. Let the plaintext image size be $M \times N$, the pixel value of the i-th row and the j-th column of the plaintext image is $p_{i,j}$, and the pixel value of the i-th row and the j-th column of the plaintext image recovered from the ciphertext image is $p'_{i,j}$. MSE is defined as (13).

$$MSE = \frac{1}{M \times N} \sum\nolimits_{i=1}^{M} \sum\nolimits_{j=1}^{N} (p_{i,j} - p'_{i,j})^2 \tag{13}$$

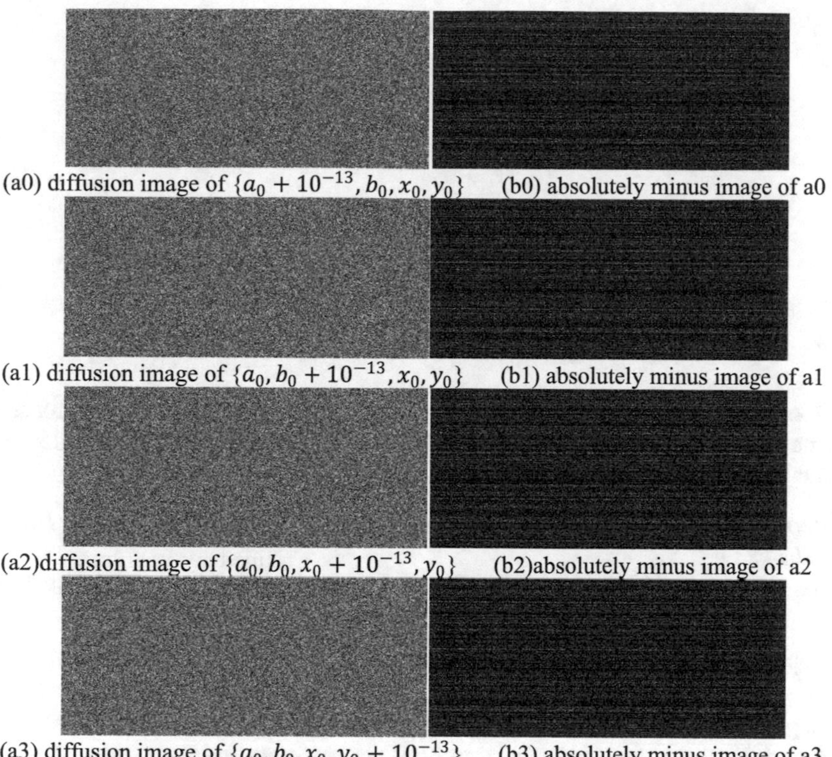

(a0) diffusion image of $\{a_0 + 10^{-13}, b_0, x_0, y_0\}$
(b0) absolutely minus image of a0
(a1) diffusion image of $\{a_0, b_0 + 10^{-13}, x_0, y_0\}$
(b1) absolutely minus image of a1
(a2)diffusion image of $\{a_0, b_0, x_0 + 10^{-13}, y_0\}$
(b2)absolutely minus image of a2
(a3) diffusion image of $\{a_0, b_0, x_0, y_0 + 10^{-13}\}$
(b3) absolutely minus image of a3

Fig. 7. The diffusion images obtained by the new key pairs and the corresponding absolutely minus images

Table 2. The PSNR of the three CT images

	Abdomina	*Pelvis*	*Spine*
PSNR/dB	83.10	86.88	83.89

The PSNR of the three CT images is listed in Table 2. It can be seen that the CT image recovered by the algorithm proposed in this paper has extremely high quality.

3.4 Ciphertext Sensitivity Analysis

The purpose of ciphertext sensitivity analysis is to analyze the difference between the plaintext image recovered by the decryption system and the original plaintext image after a small change in the ciphertext image. For an image encryption system with excellent performance, the difference between the former and the latter is obvious. Commonly used comparison indicators include number of pixels change rate (NPCR) and unified average changing intensity (UACI). Ideally, the NPCR for a random image and a given image is approximately 99.6904% and the UACI is approximately

33.4635%. Given two images P and P' whose size is M × N, NPCR and UACI are defined in Eqs. (14) and (15).

$$\mathrm{NPCR}(P,P') = \frac{1}{M \times N}\sum\nolimits_{i=1}^{M}\sum\nolimits_{j=1}^{N} D(P_{i,j} - P'_{i,j}) \times 100\% \tag{14}$$

$$\text{Where } D(x) = \begin{cases} 0, x = 0 \\ 1, x \neq 0 \end{cases}.$$

$$\mathrm{UACI} = \frac{1}{M \times N}\sum\nolimits_{i=1}^{M}\sum\nolimits_{j=1}^{N} \frac{\left|P_{i,j} - P'_{i,j}\right|}{255} \times 100\% \tag{15}$$

In order to verify the validity of the (*n, n*) structure of the CT image secret sharing scheme proposed in this paper, this section selects the CT image *pelvis* and sets it as *P*. The following experimental steps are designed.

Step 1: Select a set of keys to encrypt *pelvis* and get five block images F_1, F_2, F_3, F_4 and F_5;
Step 2: Choose a block image $F_i, i = 1, 2, \ldots, 5$ arbitrarily, +1 all of its pixel values to get F_i';
Step 3: Obtain a new CT image P_i by performing decryption operation with F_i' and the remaining four block images;
Step 4: The difference between P_i and P is compared by NPCR and UACI;
Step 5: Repeat step 3 and step 4 until all comparisons are completed.

Table 3. NPCR and UACI of P_i, P

	NPCR/%	UACI/%
P_1	97.5418	33.1720
P_2	97.0020	33.0397
P_3	98.0262	32.9510
P_4	97.1725	33.0059
P_5	97.6773	33.1044

The comparison results are listed in Table 3, and P_i is shown in Fig. 8.

From Fig. 8, we can see that as soon as a small change occurs in one of the five ciphertext images, the recovered plaintext image is noise. At the same time, through Table 3, it can be concluded that five newly recovered CT images are approximated as random images. This shows that the (*n, n*) structure proposed in this paper is completely effective, which can effectively prevent the thief from recovering the original CT image by selective ciphertext attack or known ciphertext attack when stealing some shadow images and Impurity image.

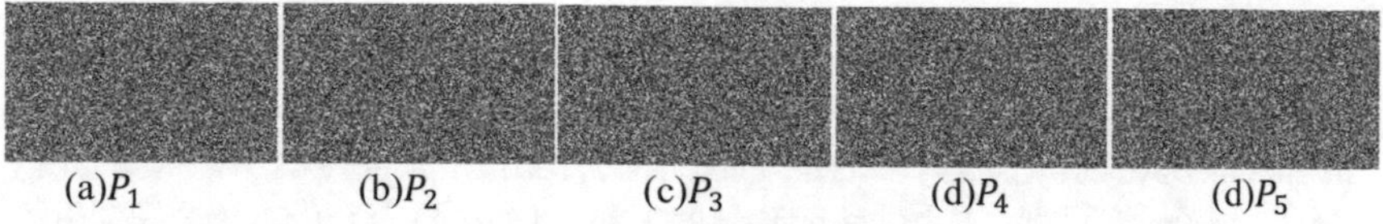

Fig. 8. New recovered image P_i

4 Conclusion

In this paper, a secret sharing method of (n, n) structure is designed and combined with the classical permutation-diffusion structure image encryption method. Based on this, a novel secret sharing method for CT images is proposed. Combined with the idea of secret sharing, this method overcomes the shortcoming of traditional image encryption scheme which can't resist selective plaintext attack and known plaintext attack. Simultaneously, in view of the high confidentiality requirements of CT images, this paper improves the (t, n) structure in traditional secret sharing to (n, n) structure, which further improves security. A series of experiments verify the excellent performance of the algorithm proposed in this paper, which is of sufficient application value in the remote secure transmission of CT images.

References

1. Fridrich, J.: Symmetric ciphers based on two-dimensional chaotic maps. Int. J. Bifurcat. Chaos **08**, 1259–1284 (1998)
2. Wang, X., Yang, L., Liu, R., Kadir, A.: A chaotic image encryption algorithm based on perceptron model. Nonlinear Dyn. **62**, 615–621 (2010)
3. Kanso, A., Ghebleh, M.: A novel image encryption algorithm based on a 3D chaotic map. Commun. Nonlinear Sci. Numer. Simul. **17**, 2943–2959 (2012)
4. Chen, G., Mao, Y., Chui, C.: A symmetric image encryption scheme based on 3D chaotic cat maps. Chaos Solitons Fractals **21**, 749–761 (2004)
5. Liu, L., Zhang, Q., Wei, X., Zhou, C.: Image encryption algorithm based on chaotic modulation of Arnold dual scrambling and DNA computing. Adv. Sci. Lett. **4**, 3537–3542 (2011)
6. Wu, Y., Zhou, Y., Noonan, J., Agaian, S.: Design of image cipher using latin squares. Inf. Sci. **264**, 317–339 (2014)
7. Lafe, O.: Data compression and encryption using cellular automata transforms. Eng. Appl. Artif. Intell. **10**, 581–591 (1997)
8. Chen, R., Lai, J.: Image security system using recursive cellular automata substitution. Pattern Recogn. **40**, 1621–1631 (2007)
9. Machkour, M., Saaidi, A., Benmaati, M.: A novel image encryption algorithm based on the two-dimensional logistic map and the latin square image cipher. 3D Res. **6**, 18 (2015)
10. Bakhshandeh, A., Eslami, Z.: An authenticated image encryption scheme based on chaotic maps and memory cellular automata. Opt. Lasers Eng. **51**, 665–673 (2013)
11. Panduranga, H., Naveen Kumar, S., Kiran: Image encryption based on permutation-substitution using chaotic map and Latin Square Image Cipher. Eur Phys. J. Spec. Top. **223**, 1663–1677 (2014)

12. Liao, X., Lai, S., Zhou, Q.: A novel image encryption algorithm based on self-adaptive wave transmission. Signal Process. **90**, 2714–2722 (2010)
13. Kekre, H., Sarode, T., Halarnkar, P.: Partial image scrambling using walsh sequency in sinusoidal wavelet transform domain. Intell. Syst. Technol. Appl. **384**, 471–484 (2016)
14. Lai, S., Liao, X., Zhou, Q.: Novel image encryption algorithm based on wave transmission. J. Comput. Appl. **29**, 2210–2212 (2009)
15. Ye, G.: A block image encryption algorithm based on wave transmission and chaotic systems. Nonlinear Dyn. **75**, 417–427 (2013)
16. Shamir, A.: How to share a secret. Commun. ACM **22**, 612–613 (1979)
17. Naor, M., Shamir, A.: Visual cryptography. In: Lecture Notes in Computer Science, vol. 950, pp. 1–12. Springer, Berlin (1994)
18. Wang, D., Zhang, L., Ma, N., Li, X.: Two secret sharing schemes based on Boolean operations. Pattern Recogn. **40**, 2776–2785 (2007)
19. Wu, H., Wang, H., Yu, R.: Color visual cryptography scheme using meaningful shares. In: Eighth International Conference on Intelligent Systems Design and Applications. pp. 173–178. IEEE (2008)
20. Yang, C., Yang, Y.: New extended visual cryptography schemes with clearer shadow images. Inf. Sci. **271**, 246–263 (2014)
21. Zhou, Y., Agaian, S., Joyner, V., Panetta, K.: Two Fibonacci P-code based image scrambling algorithms. In: Astola, J., Egiazarian, K., Dougherty, E. (eds.) Image Processing: Algorithms and Systems VI. SPIE (2008)
22. Podoba, T., Giesl, J., Vlcek, K.: Image encryption in wavelet domain based on chaotic maps. In: 2009 2nd International Congress on Image and Signal Processing, pp. 1–5. IEEE (2009)
23. Gu, G.S., Han, G.Q.: The application of chaos and DWT in image scrambling. In: 5th International Conference on Machine Learning and Cybernetics, pp. 3729–3733. IEEE (2006)
24. Yang, C., Huang, S.: Constructions and properties of k out of n scalable secret image sharing. Opt. Commun. **283**, 1750–1762 (2010)
25. Thien, C.C., Lin, J.C.: Secret image sharing. Comput. Graph. **26**, 765–770 (2002)
26. Wang, R., Su, C.: Secret image sharing with smaller shadow images. Pattern Recogn. Lett. **27**, 551–555 (2006)
27. Wu, K.: A secret image sharing scheme for light images. EURASIP J. Adv. Signal Process. **2013**, 49 (2013)
28. Liu, Y., Zhong, Q., Shen, J., Chang, C.: A novel image protection scheme using bit-plane compression and secret sharing. J. Chin. Inst. Eng. **40**, 161–169 (2017)
29. Liu, Y., Wu, Z.: An improved threshold multi-level image recovery scheme. J. Inf. Secur. Appl. **40**, 166–172 (2018)
30. Ahmad, M., Ahmad, F.: Cryptanalysis of image encryption based on permutation-substitution using chaotic map and latin square image cipher. In: 3rd International Conference on Frontiers of Intelligent Computing: Theory and Applications, pp. 481–488. Springer, Berlin (2014)
31. Rhouma, R., Belghith, S.: Cryptanalysis of a new image encryption algorithm based on hyper-chaos. Phys. Lett. A **372**, 5973–5978 (2008)
32. Çokal, C., Solak, E.: Cryptanalysis of a chaos-based image encryption algorithm. Phys. Lett. A **373**, 1357–1360 (2009)
33. Zhang, Y.: Cryptanalysis of an image encryption algorithm based on chaotic modulation of Arnold dual scrambling and DNA computing. Adv. Sci. Focus **2**, 67–82 (2014)
34. Xie, T., Liu, Y., Tang, J.: Breaking a novel image fusion encryption algorithm based on DNA sequence operation and hyper-chaotic system. Optik Int. J. Light Electron Optics **125**, 7166–7169 (2014)

35. Wu, J., Liao, X., Yang, B.: Cryptanalysis and enhancements of image encryption based on three-dimensional bit matrix permutation. Signal Process. **142**, 292–300 (2018)
36. Grassberger, P., Kantz, H., Moenig, U.: On the symbolic dynamics of the henon map. J. Phys. A: Gen. Phys. **22**(22), 5217 (1989)
37. Alvarez, G., Li, S.: Some basic cryptographic requirements for chaos-based cryptosystems. Int. J. Bifurcat. Chaos **16**, 2129–2151 (2006)
38. Li, C., Lo, K.: Optimal quantitative cryptanalysis of permutation-only multimedia ciphers against plaintext attacks. Signal Process. **91**, 949–954 (2011)
39. Lan, R., He, J., Wang, S., Liu, Y., Luo, X.: A parameter-selection-based chaotic system. IEEE Trans. Circuits Syst. II Express Briefs **PP**, 1 (2018)
40. Lan, R., He, J., Wang, S., Gu, T., Luo, X.: Integrated chaotic systems for image encryption. Signal Process. **147**, 133–145 (2018)
41. Hua, Z., Jin, F., Xu, B., Huang, H.: 2D Logistic-Sine-coupling map for image encryption. Signal Process. **149**, 148–161 (2018)

Information Processing

Robust Speaker Recognition Using Improved GFCC and Adaptive Feature Selection

Xingyu Zhang, Xia Zou, Meng Sun(✉), and Penglong Wu

Army Engineering University, Nanjing, China
zxynbnb@126.com, sunmengccjs@163.com

Abstract. Speaker recognition systems have shown good performance in noise-free environments, but the performance will severely deteriorate in the presence of noises. At the front end of the systems, Mel-Frequency Cepstral Coefficient (MFCC), or a relatively noise-robust feature Gammatone Frequency Cepstral Coefficients (GFCC), is commonly used as time-frequency feature. To further improve the noise-robustness of GFCC, signal processing techniques, such as DC removal, pre-emphasis and Cepstral Mean Variance Normalization (CMVN), are investigated in the extraction of GFCC. Being aware the advantages and disadvantages of MFCC and GFCC, an adaptive strategy was proposed to make feature selection based on the quality of speech. Experiments were conducted on TIMIT dataset to evaluate our approach. Compared with ordinary GFCC and MFCC features, our method significantly reduced the EER in speech data with miscellaneous SNRs.

Keywords: Gammatone Frequency Cepstrum Coefficients (GFCC) · i-vector · Robust speaker recognition · Mel-Frequency Cepstrum Coefficient (MFCC) · Adaptive feature selection

1 Introduction

Speaker recognition systems have obtained good results on clean speech sets. However, in real-world scenarios, when voice is corrupted by noises, the performance of recognizer will drastically deteriorate. Besides the corruption of noises, channel distortion and variants of speaking styles can also make negative contributions to the

This work is supported by the Natural Science Foundation of Jiangsu Province for Excellent Young Scholars (BK20180080).

X. Zhang is working for a master degree at the Lab of Intelligent Information Processing of PLA Army Engineering University. His research topic is speaker recognition.

X. Zou is now an associate professor at the Lab of Intelligent Information Processing of PLA Army Engineering University. His research interest is speech signal processing.

M. Sun is now a researcher at the Lab of Intelligent Information Processing of PLA Army Engineering University. His research interests are speech processing, unsupervised/semi-supervised machine learning and sequential pattern recognition.

P. Wu is working for a master degree at the Lab of Intelligent Information Processing of PLA Army Engineering University. His research topic is speech signal processing.

C.-N. Yang et al. (Eds.): SICBS 2018, AISC 895, pp. 159–169, 2020.
https://doi.org/10.1007/978-3-030-16946-6_13

accuracy of speaker recognition. Therefore, improving the robustness of a speaker recognition system is a key stage before putting it into practical usage.

State-of-the-art speaker recognition systems use techniques such as joint factor analysis (JFA) and probabilistic linear discriminant analysis (PLDA) based on i-vectors [1, 2] The algorithm uses a low-dimensional subspace to embed speech from different speakers into different subspaces where each utterance is represented by a fixed low-dimensional vector (named by i-vector). Therefore, in this paper, we also choose i-vector to eliminate the influence of channel variations.

In real-world scenarios, speech is always corrupted by noises. In [3, 4], the authors proposed a new feature, Gammatone Frequency Cepstrum Coefficient (GFCC) in speaker recognition tasks. The feature performs better than the commonly used Mel Frequency Cepstrum Coefficient (MFCC). By combining GFCC and JFA, the speaker verification system proposed in [5] yielded better results than the GMM-UBM based one. In [6], Shi et al. used a modified GFCC for speaker recognition where they found that GFCC improved MFCC significantly under noisy environments, especially when the SNR is below 10 dB. However, GFCC did not outperform MFCC in the case of high SNR. Therefore, to obtain robust recognition, we have two choices: the first one is to enhance the noise robustness of GFCC by imposing some signal processing operations, the second one is to treat clean and noisy speech separately. A similar work with respect to our idea was reported in [7], where GFCC + i-vectors was adopted to tackle noise environments and session variability simultaneously. But their experimental results on NIST-2003 seemed not convincing.

This article has two major contributions. The first is to improve the conventional GFCC extraction approach by imposing techniques like removing of DC components, pre-emphasis and cepstrum mean variance normalization, etc. An analytical study on why GFCC is able to outperform MFCC in noisy conditions is also reported. The second is to propose an adaptive feature selection algorithm to improve the recognition performance for both clean and noisy cases. Our experiments were conducted on the TIMIT dataset. The results showed that the proposed method significantly improved the performance with respect to the baselines for all noisy levels.

2 Adaptive Feature Selection as the Front End

2.1 Improved GFCC

GFCC Feature Extraction

DC component of the speech signal is first removed to reduce the possible bias on the algorithm. A low-pass filter is imposed subsequently to remove high-frequency noises. The signal is then pre-emphasized to enhance the high-frequency part of the signal and improve the SNR of the signal. FFT of the signal is then computed and passed through a 64-channel gammatone filter bank. At each channel, fully rectify the filter response (i.e. take absolute value) and decimate it to 100 Hz as a way of time windowing. Then taking absolute value afterwards. This creates a time frequency (T-F) representation that is a variant of cochleagram [8]. The amplitude of the signal is compressed using the cubic root as the nonlinear transformation function. This is different from the MFCC extraction. In MFCC extraction we choose log as nonlinear rectification function. The cubic root is more robust than log showed in [5]. DCT transform is then performed, and at the end, a Cepstral Mean Variance Normalization (CMVN) operation is performed (Fig. 1).

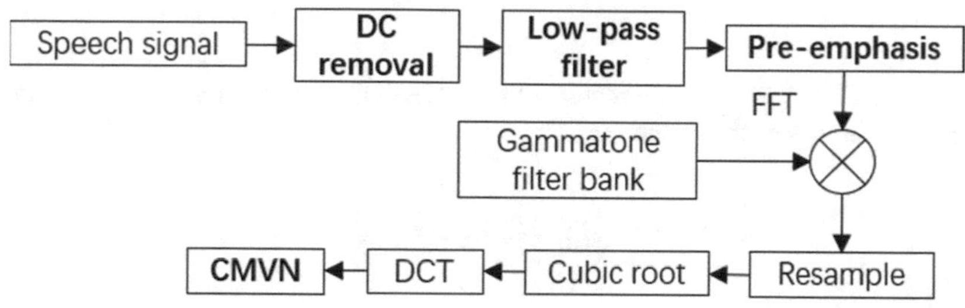

Fig. 1. Improved GFCC extraction process

Gammatone Filter [5]
The input signal passes through a 64-channel gammatone filter bank. In the time domain, the filter response is given by:

$$h(t) = gt^{n-1}\exp(-2\pi dt)\cos(2\pi f_C t + \varphi) \tag{1}$$

In the above equation, g is the output gain, d is the attenuation factor, n is the filter order, f_C is the filter center frequency, and φ is the phase. Where d determines the bandwidth of the response filter, and the relationship between d and the center frequency f_C is:

$$\mathrm{d} = 1.019(24.74.37 \times \frac{f_c}{1000} + 1)\mathrm{Hz} \tag{2}$$

DC Removal
The speech signal usually contains a DC component, which will deviate the signal amplitude from zero. In this paper, the input speech signal s is first filtered to obtain a DC component estimate v. The input speech signal is then subtracted from the DC component estimate to obtain the speech signal y with the DC component removed.

Low-Pass Filter
In our previous experiences, we found that the performance of GFCC will be degraded in the presence of high-frequency noise, so low-pass filter is imposed here in order to preserve the speakers' voice identity information by filtering out the high-frequency noises.

Pre-emphasis
We pre-emphasized the high-frequency part of the remaining speech to remove the effects of lip radiation and increase the high-frequency resolution of the speech. It is

worth noting that the pre-emphasis is operated after the above low-pass filter where high frequency noises had been removed. Thus, high frequency noises existing in the original signals will not be pre-emphasized any more. The high-pass filter used in this paper is:

$$H(z) = 1 - 0.95z^{-1} \tag{3}$$

DCT

DCT is used to derive cepstrum features to reduce their dimensions and disassociate the features. Finally, we obtain the GFCC:

$$C_i[m] = \sqrt{\frac{2}{\mathrm{N}}} \Sigma_{C=0}^{N-1} G_c[m] \cos\left(\frac{\mathrm{i}\pi}{2N}(2c+1)\right)$$

$$i = 0, \cdots\cdots\cdots\cdots, N-1 \tag{4}$$

2.2 Adaptive Feature Selection

In our experiments, we found that GFCC has better recognition performance than MFCC when SNR is low. However, the effect is not as good as MFCC when the SNR is high. In order to achieve a good recognition effect in both clean and noisy conditions, the adaptive extraction algorithm is proposed (Fig. 2).

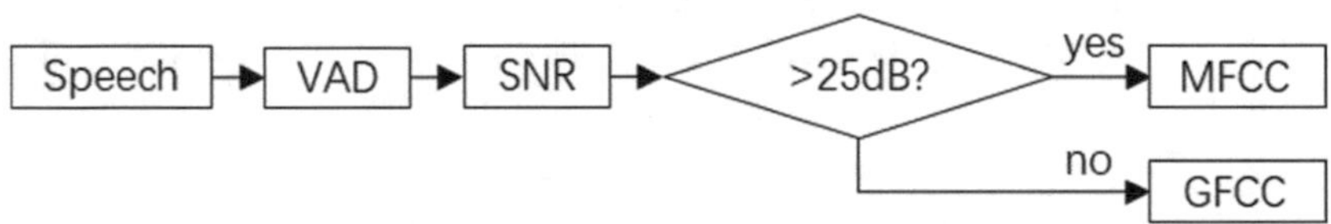

Fig. 2. Adaptive feature selection algorithm

Voice Activity Detection (VAD) is used to get background noise information. The square of the modulus of the value of FFT using the first frame signal is taken as the noise initial energy and a small initial value is added to avoid infinity SNR. After that, the SNR of each frame is calculated by using the initial value, and finally the SNR of all frames is averaged to obtain the SNR of the speech.

According to our previous experience, we choose 25 dB as the threshold. When the speech signal SNR is higher than 25 dB, the noise in speech has little effect on the speaker's voice. So MFCC is extracted. When SNR is lower than 25 dB, the noise in the speech has a greater impact on the speaker's voice. So GFCC is extracted.

This adaptive mechanism enables the speaker recognition system to perform well for both high and low SNRs. By combining the robustness of GFCC in noisy speech and the robustness of MFCC in clean speech, we have improved the recognition effect of the system adaptively in different environments.

3 I-vector-PLDA System

3.1 The Flowcharts for I-vector-PLDA Model

See Fig. 3.

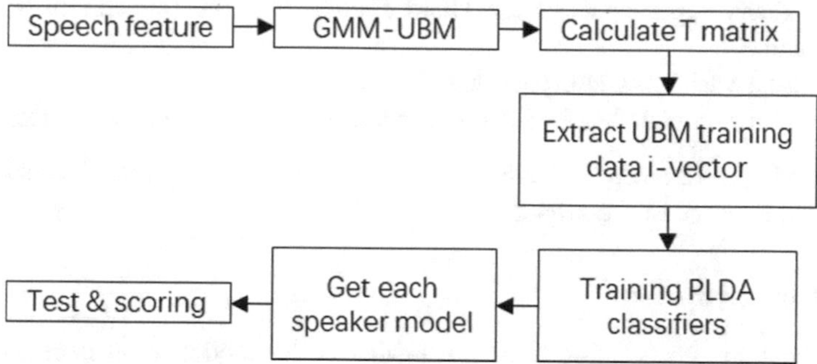

Fig. 3. I-vector-PLDA system

3.2 The Theory of I-vector Speaker Recognition System

The speaker's voiceprint vector is modeled by the likelihood probability density function of the Gaussian mixture model as follows:

$$p\left(x|\Delta^{(s)}\right) = \sum\nolimits_{j=1}^{M} \lambda_j^{(s)} p(x|\mu_j^{(s)}, \sum\nolimits_j^{(s)}) \tag{5}$$

Gaussian mixture model parameters are expressed as:

$$\Delta^{(s)} = \{\lambda_j^{(s)}, \mu_j^{(s)}, \sum\nolimits_j^{(s)}\}_{j=1}^{M} \tag{6}$$

The Universal Background Model (UBM) is a special Gaussian mixture model trained by many speakers' data:

$$p\left(x|\Delta^{(ubm)}\right) = \sum\nolimits_{j=1}^{M} \lambda_j^{(ubm)} p(x|\mu_j^{(ubm)}, \sum\nolimits_j^{(ubm)}) \tag{7}$$

In the same way, the parameters of the UBM model are:

$$\Delta^{(ubm)} = \{\lambda_j^{(ubm)}, \mu_j^{(ubm)}, \sum\nolimits_j^{(ubm)}\}_{j=1}^{M} \tag{8}$$

And then we adjusted to the various speaker models through the maximum posterior probability (MAP) algorithm [9].

The mean value of the speaker's GMM model obtained by UBM is taken as the supervector. Assuming that UBM has C Gaussian components and the feature dimension is F, the dimension of the resulting supervector is $C{\cdot}F$. The following is the formula to get i-vector.

$$s = m + T\omega \tag{9}$$

s: Speaker mean supervector. It correlates with speakers and channels.
m: The mean supervector of the UBM model. It is independent of speaker and channel.
T: The total-variability subspace matrix.
ω: A vector associated with the speaker and the channel called i-vector.

The s and m in formula (8) have been obtained before. EM algorithm is used to get T. The value of ω can be gotten by formula (8).

3.3 I-vector Extraction

Before extracting the i-vector from the GMM-UBM model, it is necessary to first calculate the statistics of the UBM model's supervector. If the feature of a speech segment s is represented by $x_{s,t}$, the Gaussian mixture model's component coefficient is c. The zeroth, first, and second order Baum-Welch statistics are given by:

$$N_{c,s} = \sum\nolimits_t \gamma_{c,s,t} \tag{10}$$

$$F_{c,s} = \sum\nolimits_t \gamma_{c,s,t}\left(x_{s,t} - \mu_c\right) \tag{11}$$

$$S_{c,s} = diag\{\sum\nolimits_t \gamma_{c,s,t}(x_{s,t} - \mu_c)(x_{s,t} - \mu_c)^T\} \tag{12}$$

Where $\gamma_{r,s,t}$ represents the posterior probability of the *c-th* Gaussian component and μ_c represents the mean supervector of the *c-th* Gaussian component.

After obtaining the statistics, EM algorithm can be used to estimate the total-variability subspace matrix T. The steps are as follows [10]:

(1) Initialization
Calculate Baum-Welch statistic.
(2) E step
For each speech segment, calculate their expectations:

$$L_s = I + T^T \Sigma^{-1} N_s T \tag{13}$$

$$\mathrm{E}[\omega_s] = L_s^{-1} T^T \Sigma^{-1} F_s \tag{14}$$

$$\mathrm{E}\left[\omega_s \omega_s^T\right] = \mathrm{E}[\omega_s]\mathrm{E}\left[\omega_s^T\right] + L_s^{-1} \tag{15}$$

(3) M step
Update the T matrix after solving the equation:

$$\Sigma_s N_s TE\left[\omega_s \omega_s^T\right] = \Sigma_s FE[\omega_s] \tag{16}$$

(4) Iteration
Iterate a fixed number of times or until the objective function converges.

4 Experiments

4.1 Experiment Setup

Experiments were conducted using the National Institute of Standards and Technology TIMIT Speech Corpus (NIST TIMIT) [11] and the NOISEUS standard noise database. *White* noise, *babble* noise, *f16* noise, and *cicadas* noise are used here.

The sampling frequency of speech in TIMIT dataset is 16 kHz. TIMIT contains a total of 6300 sentences. Each of the 630 individuals from the eight major dialect regions of the United States spoke 10 sentences, and 70% of the speakers are male.

In this paper, 5300 sentences of 530 individuals were selected to train the UBM model and the PLDA model. The remaining data from 100 speakers was registered in 9 sentences per person to obtain a model containing 100 speakers.

In order to make fair comparison between GFCC and MFCC, we used 39-dimensional MFCC (13 static + 13 delta + 13 delta delta) and 39-diensional GFCC (13 static + 13 delta + 13 delta delta) in all the experiments. The number of Gaussian components of the UBM is 256, and the dimension of i-vector is 400.

We first use the adaptive feature selection algorithm to extract features according to the process described in the Sect. 2, and then use the i-vector speaker recognition model to perform the recognition.

4.2 Experiment Results

Influence of Noise Frequency on the Noise Robustness
In the experiments, we found that the noise robustness of GFCC will decrease with the increase of the main frequency of noises. In this section, we made a group of toy experiments to illustrate how the noise's frequency can influence the performance. Considering that the frequency range of the filter bank used in feature extraction is 100 Hz–5 kHz, we added low frequency (0.5 kHz), intermediate frequency (2 kHz), and high frequency (4 kHz) sinusoidal noises to the clean speech. The following figure shows their performance.

DET Graph
In Fig. 4 we directly used the GFCC extraction method as proposed in [7]. The EER was 1.23. By using the improved GFCC extraction method, the EER was reduced to 0.94. It is obvious that the improved method proposed in this paper can further enhance the noise robustness of GFCC (Fig. 5).

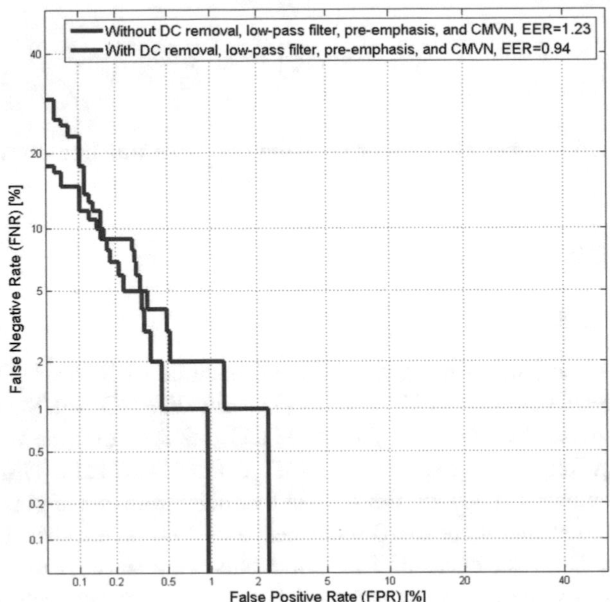

Fig. 4. DET graph. Noise type: babble, SNR = 10 dB.

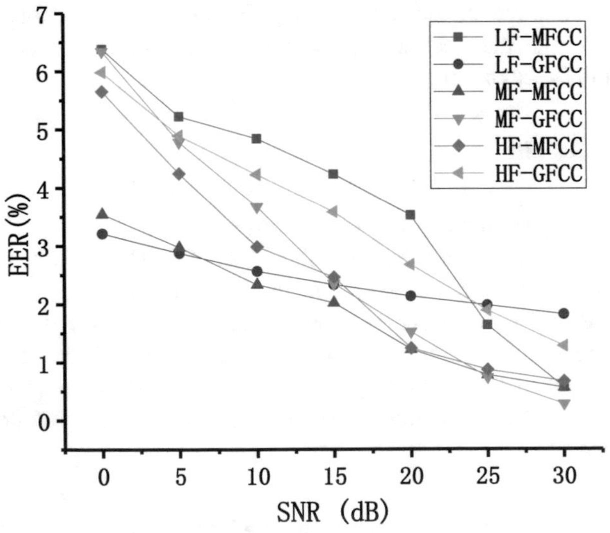

Fig. 5. Performance of MFCC and GFCC when adding high frequency (4 kHz), medium frequency (2 kHz) and low frequency (0.5 kHz) noise

It can be found that GFCC can achieve better performance when adding lower frequency noise. When the noise frequency is high, GFCC performance is inferior to MFCC. The frequencies of human speech are usually in the range of 85–1100 Hz, while human ears can only perceive sounds with frequency between 20 and 20 kHz. Because GFCC simulates the frequency response of the human auditory system, high and medium frequency noises have significant impacts on the recognition results. That is why a low-pass filter is added to filter out high-frequency noise as explained in Sect. 2.

System Performance
Table 1 shows the EER values for using MFCC, GFCC and Improved GFCC in details. The best performance is denoted in bold fonts. It can be found that the EERs of the Improved GFCC + i-vector system under these four noise conditions are lower than those of the other two. Moreover, by observing Improved GFCC, it can be found that as the SNR decreases, the EERs of the Improved GFCC + i-vector system slowly increases.

Table 2 shows the performance of adaptive feature selection on recognition. When the SNR is lower than 25 dB, the Improved GFCC feature is used for identification. When the SNR is higher than 25 dB, the MFCC is used for identification. It can be seen that the system can better compensate for the poor performance of the Improved GFCC under cleaner speech.

Table 1. Equal Error Rate (EER) of MFCC, GFCC and improved GFCC features combined with i-vector

Noise	SNR (dB)	EER (%)		
		MFCC	GFCC	Improved GFCC
White	15	4.13	2.00	**1.28**
	10	4.00	3.00	**2.58**
	5	4.62	3.58	**3.02**
	0	5.00	4.13	**3.53**
Babble	15	1.13	0.66	**0.37**
	10	1.34	1.23	**0.94**
	5	1.91	1.98	**0.86**
	0	2.16	2.35	**0.95**
F16	15	0.89	0.23	**0.04**
	10	**1.00**	1.08	**1.00**
	5	1.23	1.54	**0.95**
	0	2.00	1.67	**1.00**
Cicadas	15	1.32	0.83	**0.55**
	10	1.56	1.28	**1.00**
	5	1.88	1.29	**1.12**
	0	2.23	1.87	**1.42**

Table 2. Adaptive feature selection system vs. improved GFCC

Noise	SNR (dB)	EER (%)	
		Improved GFCC	Adaptive feature selection system
Clean	-	3.00	**0.59**
White	30	2.78	**0.88**
	20	**2.34**	**2.34**
	10	**2.58**	**2.58**
	5	**3.02**	**3.02**
Babble	30	1.56	**0.67**
	20	**1.02**	**1.02**
	10	**1.23**	**1.23**
	5	**1.98**	**1.98**
F16	30	1.23	**0.56**
	20	**1.00**	**1.00**
	10	**1.08**	**1.08**
	5	**1.54**	**1.54**
Cicadas	30	1.33	**0.67**
	20	**1.18**	**1.18**
	10	**1.28**	**1.28**
	5	**1.29**	**1.29**

5 Conclusion

GFCC uses a gammatone filter bank and a nonlinear cubic-root compression, which has stronger noise robustness than MFCC. This paper combines i-vector system and Improved GFCC features to obtain a speaker recognition system with stronger noise robustness. The experimental results show that the noise robustness of this system is better than the commonly used MFCC + i-vector system. GFCC can reduce the influence of additive noise and suppress the identification instability caused by additive noise. But GFCC is not as effective as MFCC for the recognition of clean speech. Therefore, an adaptive feature selection algorithm is proposed to improve the performance of the system for all noise levels. The noise robustness of GFCC is also analyzed by designing intuitive experiments. The scheme and analysis will help us to propose more noise robust features for speaker recognition task in the future.

References

1. Dehak, N., Kenny, P.J., Dehak, R., Dumouchel, P., Ouellet, P.: Front-end factor analysis for speaker verification. IEEE Trans. Audio Speech Lang. Process. **19**(4), 788–798 (2011)
2. Burget, L., Plchot, O., Cumani, S., Glembek, O., Matějka, P., Brümmer, N.: Discriminatively trained Probabilistic Linear Discriminant Analysis for speaker verification. In: IEEE International Conference on Acoustics, Speech and Signal Processing, vol. 125, pp. 4832–4835. IEEE (2011)

3. Zhao, X., Shao, Y., Wang, D.L.: Casa-based robust speaker identification. IEEE Trans. Audio Speech Lang. Process. **20**(5), 1608–1616 (2012)
4. Shao, Y., Srinivasan, S., Wang, D.L.: Incorporating auditory feature uncertainties in robust speaker identification. In: IEEE International Conference on Acoustics, Speech and Signal Processing, vol. 4, pp. IV-277–IV-280. IEEE (2007)
5. Das, P., Bhattacharjee, U.: Robust speaker verification using GFCC and joint factor analysis. In: International Conference on Computing, Communication and Networking Technologies, pp. 1–4. IEEE (2014)
6. Shi, X., Yang, H., Zhou, P.: Robust speaker recognition based on improved GFCC. In: IEEE International Conference on Computer and Communications, pp. 1927–1931. IEEE (2017)
7. Jeevan, M., Dhingra, A., Hanmandlu, M., Panigrahi, B.K.: Robust speaker verification using GFCC based i-vectors. In: Proceedings of the International Conference on Signal, Networks, Computing, and Systems, pp. 85–91. Springer India (2017)
8. Zhao, X., Wang, D.: Analyzing noise robustness of MFCC and GFCC features in speaker identification. In: IEEE International Conference on Acoustics, Speech and Signal Processing, pp. 7204–7208. IEEE (2013)
9. Reynolds, D.A., Quatieri, T.F., Dunn, R.B.: Speaker verification using adapted gaussian mixture models. Dig. Signal Process. **10**, (1–3), 19–41 (2000)
10. Zhiyi, L.I., Liang, H.E., Zhang, W., Liu, J.: Speaker recognition based on discriminant i-vector local distance preserving projcction. J. Tsinghua Univ. (Sci. Technol.) **52**(5), 598–601 (2012)
11. Lamel, L.: Speech database development: design and analysis of the acoustic-phonetic corpus. In: Proceedings of DARPA Speech Recognition Workshop (1986)

Outlier Detection Based on Local Density of Vector Dot Product in Data Stream

Zhaoyu Shou[1(✉)], Fengbo Zou[1(✉)], Hao Tian[1], and Simin Li[2]

[1] School of Information and Communication Engineering, Guilin University of Electronic Technology, Guilin 541004, China
guilinshou@guet.edu.cn, zfbb.cool@163.com, marcus_tian@163.com
[2] Key Laboratory of Cognitive Radio and Information Processing, Ministry of Education, Guilin 541004, China
siminl@guet.edu.cn

Abstract. Outlier detection in data stream is an increasingly important research in many fields. To deal with the data stream with the properties of high dimension, rapid arrival in order, high cost of storing all data in memory and so on, an outlier detection algorithm based on local density of vector dot product in data stream (LDVP-OD) is proposed. LDVP-OD uses the model based on sliding window and multiple validations to decrease the false alarm rate, which divides the data stream into uniform-sized blocks. Local density of vector dot product (LDVP) is described in order to precisely evaluate the outlierness of data in data stream. Furthermore, an outlier judgment criterion based on supreme slope is introduced, which can determine the exact outliers without requiring the number of outliers or other parameters beforehand. Comparison experiments with existing algorithms on synthetic and real datasets prove the high detection rate, good stability, strong adaptability of LDVP-OD.

Keywords: Outlier detection · Data stream · Local density of vector dot product · Multiple validations

1 Introduction

Outlier detection is to quickly detect abnormal objects that do not meet the expected behavior from the complex data environment, providing deep analysis and understanding for users [22]. With the rapid development of network technology and growing popularity of society informatization, the amount of information keeps on increasing explosively. Many fields are generating high-speed, infinite and dynamic data streams. Outlier detection in data stream has been applied to many domains such as network security [19], video surveillance [25], credit card fraud detection [5] and social networking platform [18]. The purpose is to discover the abnormal situation of applications in time, providing some valuable information and calling attention to take actions promptly. As data stream evolves during the time, traditional methods for static dataset cannot

C.-N. Yang et al. (Eds.): SICBS 2018, AISC 895, pp. 170–184, 2020.
https://doi.org/10.1007/978-3-030-16946-6_14

perform well on them. An outlier detection algorithm that applies to dynamic data stream well becomes necessary.

Most existing outlier detection algorithms have been proposed in data stream. It can be roughly divided into distance-based [2,4,15,20], density-based [9,13, 21], clustering-based [23] and angle-based [24].

The distance-based algorithms was widely used for the effectiveness and simplification. Distance-based outliers have been first introduced by Knorr and Ng in a static dataset [14]. Paper [15] by integrating the safe inlier concept of STORM [2] into an event queue, so that it can efficiently schedule the necessary checks that have to be made when points expire. On this basis, LEAP [4] approach is proposed, it exploits the concept of minimal probing and lifespan-aware prioritization. This approach avoids the full range query searches, so the CPU efficiency is improved and the memory consumption is reduced. Distance-based algorithms are simple and straightforward, but they are unable to deal with complex datasets. What's worse, when the dimension increases, distance-based algorithms will lose effectiveness.

Density-based algorithms use density to evaluate the outlierness degree of each object, and update the outlier factor of each data dynamically. LOF (local outlier factor) [3] is a popular density-based algorithm used in static dataset. IncLOF (incremental LOF) [21] applies LOF iteratively after insertion of each new data. N-IncLOF [9] and I-IncLOF (improved IncLOF) [13] introduce sliding window to cut down the consumption of memory resource. Density-based algorithms can handle dataset with complex distribution.

In clustering-based algorithms, outliers are those objects which do not belong to any cluster or deviate far away from the most objects in their clusters. Many traditional clustering algorithms such as DBSCAN (density-based spatial clustering of applications with noise) [8] work well on static datasets. Clustering is the first and important step of outlier detection in clustering-based outlier detection algorithms, and it directly affects the result of outlier detection. In [23], an unsupervised outlier detection algorithm based on weighted clustering (denoted as Algorithm Y in shorthand in following parts) is proposed, which divides the data stream into blocks. Algorithm Y clusters each block and detects outliers in each block. In clustering part, it combines DBSCAN and W-K-Mean (weighted-K-Mean clustering) [12], and updates the parameters needed. In outlier detection part, it treats small clusters as outlier groups and determines the scattered outliers based on distance. Algorithm Y is accurate but tedious. Clustering-based algorithm can get a high detection rate, but the process is complicated. And in many clustering-based algorithms, distance is used to evaluate the outlierness, which lower the detection rate in high-dimensional space because of the obstacle of dimension disaster.

To deal with the problem of dimension disaster, angled-based algorithms introduce ABOF (angle-based outlier factor) to measure the deviation degree of each data more precisely in high-dimensional dataset. Kriegel et al. propose ABOD (angle-based outlier detection) [17] to detect outliers in static dataset. Based on ABOD, DSABOD (data stream angle-based outlier detection) algorithm [24] is presented to detect outliers in high-dimensional data stream. DSABOD updates ABOF of each data and declares those data with high ABOF value as outliers. Angle is more stable than many other measurements in high-dimensional space, and the angle-based algorithm provides a new perspective to estimate the outlierness degree of objects.

In this paper, we propose a new way named local density of vector dot product (LDVP) to evaluate the outlierness of data. This method can measure the deviation degree of each data point accurately in complicated space and puts forward more concise expression. Meanwhile, we also introduce a new outlier judgment criterion to determine outliers in a more accurate way. Different from the traditional outlier judgment criterions, the outlier judgment criterion based on supreme slope does not need artificial parameters and has high adaptivity in evolving data stream.

The rest of this paper is organized as follows: Sect. 2 introduces the detail of LDVP-OD. In Sect. 3, various experiment results are comparatively analyzed to evaluate the performance of the proposed approach. Finally, Sect. 4 presents conclusion and future work.

2 LDVP-OD: Outlier Detection Based on Local Density of Vector Dot Product in Data Stream

In this section, an outlier detection based on local density of vector dot product in data stream is introduced. There are two stages in the process of LDVP-OD: a model based on sliding window and multiple validations and outlier detection. In the first stage, the stream data is divided into uniform-sized blocks and ε blocks form a sliding window. Meanwhile, Candidate outliers in the former sliding window will be remained to the next sliding window, and the dataset in the sliding

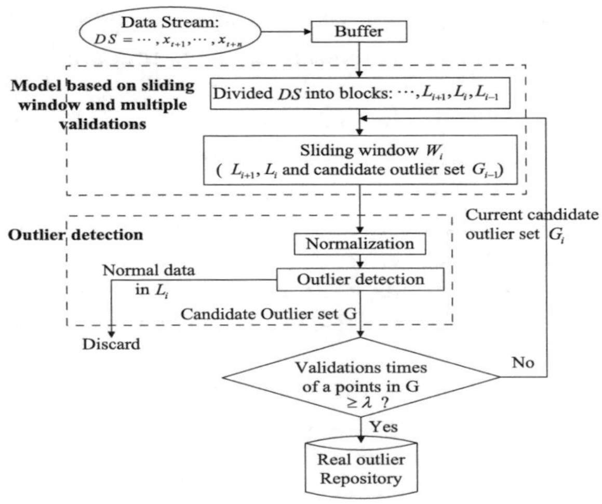

Fig. 1. The block diagram of LDVP-OD.

window is denoted as S^m. In the second stage, the dataset S^m is normalized, and the outlier detection algorithm (Algorithm 1) is executed by the local density of vector dot product (LDVP) and a new outlier judgment criterion, where the candidate outliers are determined when outlier detection is finished. Only when candidate outliers are identified to be outliers during the whole multiple validations can they be declared as real outliers, where the times of multiple validations is denoted as λ. The block diagram of LDVP-OD is shown in Fig. 1.

2.1 Model Based on Sliding Window and Multiple Validations

A data stream consists of a series of data, which is infinite, dynamic and arriving continuously. Due to the high cost of storing all data in memory, an efficient data stream processing model based on sliding window and multiple validations is described in this subsection.

Different from the traditional sliding window [9,10,13] which moves from point to point and keeps a constant width. In this paper, the coming data stream is divided into uniform-sized blocks, and several blocks form a sliding window. Meanwhile, the sliding window constructed moves from data block to data block, and its width may change a little depending on conditions. Data stream is being loaded into memory with new block joining in and historical block moving out, and the sliding window only reserves the valuable information. Due to the dynamic nature of data stream, data behavior may change during the time. As is shown in Fig. 2(a) and (b), at the time of t_1, p' shows up like an outlier. While as the sliding window moves and new data block loaded in, p' belongs to a new dense cluster at t_2. So judging an object for outlierness when it arrives may lead to wrong decisions. To solve the problem that wrong decisions, the multiple validations are employed in this framework. Declare those new coming data which deviate far away from the most other data as candidate outliers, and reserve them in the sliding window and examine their outlierness when following blocks move in. If candidate outliers remain anomalous after specified times (denoted as λ) of multiple validations, declare them as real outliers, otherwise remove them from memory as normal data.

Figure 3 provides an insight about the way of how the model based on the sliding window and multiple validations works. $L_0, L_1, L_2, \cdots$ are the data blocks

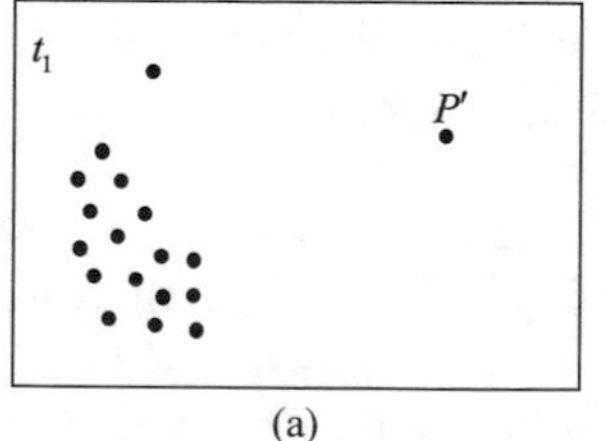

(a)

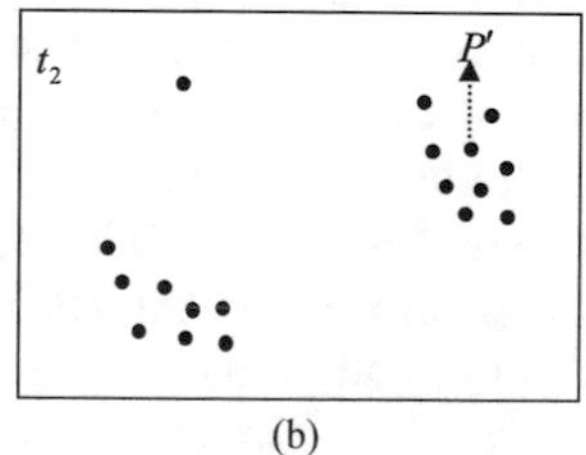

(b)

Fig. 2. Data distribution in sliding window.

divided, ε ($\varepsilon = 2$ in Fig. 3) blocks form a sliding window. As the sliding window W_i at the time of T_i moves to sliding window W_{i+1} at the time of T_{i+1}, block L_{i+1} joins in and the historical block L_{i-1} moves out. At the same time, the candidate outliers in W_i validated at the time of T_i are kept in the sliding window for the next validation.

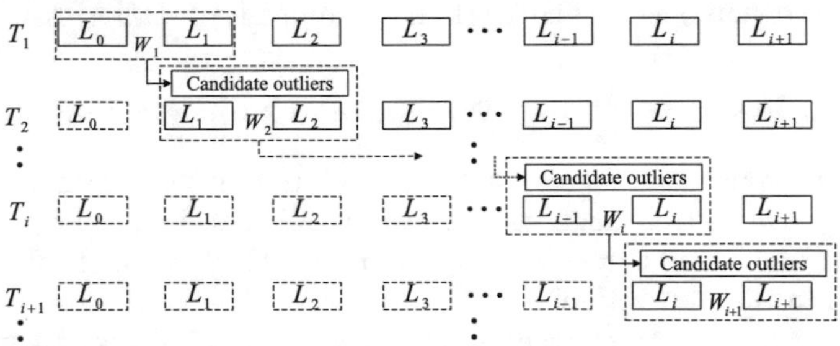

Fig. 3. Model based on sliding window and multiple validations.

2.2 Outlier Detection

Related Definitions of LDVP. Local density of vector dot product (LDVP) is a new concept to evaluate the outlierness of object. In this subsection, the related definitions will be introduced.

Definition 1 *(r-neighborhood). Given a m-dimensional dataset S^m, for point p in S^m, the r-neighborhood of point p is the neighbor dataset of p, where points lay within a distance of r to p. The r-neighborhood of point p can be denoted as $N_r(p)$:*

$$N_r(p) = \{q \in S^m | dist(p, q) \leq r\} \tag{1}$$

Way to choose the best r: assuming that k_{dist} is the distance between a point and its $k-th$ nearest point. Sort the k_{dist}s of all points in descending order and map each point to its corresponding k_{dist}, the graph is called sorted k_{dist} graph. Then the best r can be set to the value of k_{dist} of the point in the first "valley" (see Fig. 4), and usually k is set to 4 [8].

A way to find the first "valley" is shown in Fig. 4. Firstly, connect the highest point to the lowest point with a line in sorted $k_{dist}(k = 4)$ graph, the line is a base line. Then, calculate the distance of each point to the base line. Finally, the first "valley" is where the point with the longest distance to the base line. And best r is the value of k_{dist} of the point in the first "valley". Figure 4 shows the sorted k_{dist} graph of a dataset (40 points). From Fig. 4 it can be known that point 10 has the longest distance to base line ($L_2 > L_3 > L_1$), so point 10 is the first "valley" and the k_{dist} of point 10 can be set as the best r. In evolving data stream, this method to choose the best r is flexible and self-adaptive to real-time data.

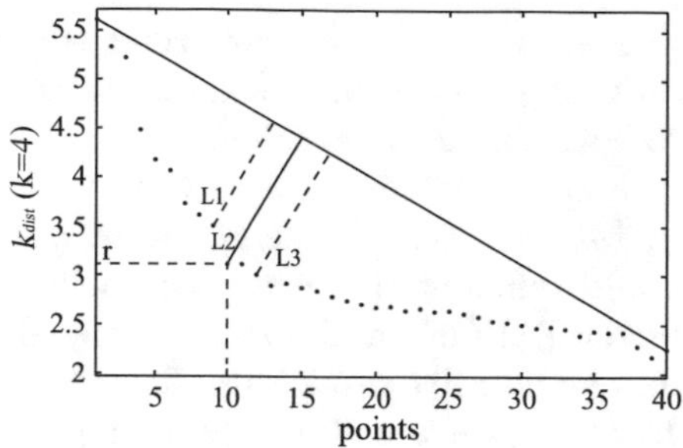

Fig. 4. Sort $k_{dist}(k = 4)$ graph.

Definition 2 *(Mean of vector dot product(MVP)). Assuming that there are n points in S^m, and given a point **A**. For any two points **B** and **C**, $\overline{AB}$ and $\overline{AC}$ respectively denote the difference vectors $\boldsymbol{B} - \boldsymbol{A}$ and $\boldsymbol{C} - \boldsymbol{A}$. The $\angle BAC$ is the angle formed by $\overline{AB}$ and $\overline{AC}$. The mean of vector dot product of point **A** can be defined as $MVP(\boldsymbol{A})$:*

$$\begin{aligned} MVP(\boldsymbol{A}) &= \frac{1}{\frac{1}{2}(n-1)(n-2)} \sum_{B,C \in S^m/\{A\}} (\cos(\angle BAC) \cdot \|\overline{AB}\| \, \|\overline{AC}\|) \\ &= \frac{1}{\frac{1}{2}(n-1)(n-2)} \sum_{B,C \in S^m/\{A\}} (\overline{AB} \cdot \overline{AC}) \end{aligned} \tag{2}$$

where $\frac{1}{2}(n-1)(n-2)$ is the total number of angles formed by $\boldsymbol{A}$ and other points in S^m, and each angle is counted only once. The $\|\cdot\|$ is a norm. As we can see the MVP can be written as a form of dot product of vectors by simplifying the formula.

MVP is formed by the product of two parts: the cosine of angle and the length of vector. The further a point deviates from most normal data, the smaller the angles formed by the point and other points are. And the cosine is getting greater, because the cosine is monotone decreasing when the angle varies in $[0, \pi]$. At the same time, the length of the vector is getting longer. Therefore, the mean of vector dot product (MVP) can denote the distribution of data in some way.

To eliminate the dominance of length, it is necessary to carry out normalization. Linear normalization is conducted in case of changing the relative distribution of the dataset. The normalization can avoid the dominance and make sure that the local density of vector dot product can get its most effectiveness.

Definition 3 *(Local density of vector dot product (LDVP)). The local density of vector dot product of point p can be denoted as $LDVP(p)$:*

$$LDVP(p) = \sum_{q \in N_r(p)} e^{-MVP(q)} \tag{3}$$

LDVP can measure the deviation degree of each data point accurately in complicated space. The further the point deviate from the normal cluster, the smaller the LDVP of the point is, and the more abnormal the point is. LDVP does not only combine the advantages of angle-based measurement and density-based measurement, but also have simple calculation.

A synthetic dataset is chosen to testify the accuracy of LDVP, as is shown in Fig. 5. The synthetic dataset consists of four scattered outliers and two clusters with different shapes and different densities. The local outlier factor (LOF), angle-based outlier factor (ABOF) and local density of vector dot product (LDVP) are tested. The comparative results are shown in Fig. 5(a)–(c). Numbers in figures represent the outlier degree of corresponding points. The smaller the number is, the greater outlierness the point shows (for example, point marked with number "1" is the most outlier point). From Fig. 5(a), it can be seen that LOF has high accuracy in measuring the scattered outliers. However the measured outlierness of points on the edge of left cluster are greater than those of points on the edge of right cluster, which does not match the fact. So, it can be known that LOF is oversensitive to points on the edge of cluster with irregular shape and insensitive to points on the edge of sparse cluster with regular shape (such as circle and sphere). In Fig. 5(b), ABOF has good performance in measuring the outlierness of points in clusters with different densities, especially points on the edge. However, the point between clusters is marked with number "4", which does not coincide with the fact (the fact is that the point between clusters is the second most abnormal point in the dataset). So the outlierness of points between clusters cannot be estimated correctly by ABOF. In Fig. 5(c), it is obvious that LDVP can not only measure the outlierness of scattered outliers correctly but also has high accuracy in estimating the outlierness of points on the edge of clusters and points between clusters.

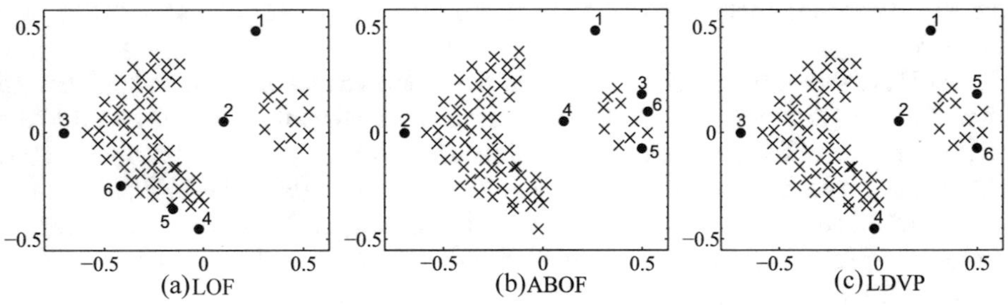

Fig. 5. The most 6 outlier points in (a) LOF, (b) ABOF and (c) LDVP.

Outlier Judgment Criterion Based on Supreme Slope and Outlier Detection. The criterions of outlier judgment in traditional algorithms can be classified into three categories: the first category determines outliers according to the feature of outlier factor constructed such as I-INCLOF, N-INCLOF and TEDA [1] algorithms. The second category is TOP N [6,11,22]. And the third category determines outliers according to the rules in statistics [16]. In this paper, the proposed outlier judgment criterion based on supreme slope belongs to the third category, which determines outliers based on statistics. Considering the fact that outliers stand out from the others and the amount of outliers is usually less than 5% of the dataset [3,13], outlier judgment criterion based on supreme slope can pick out outliers according to the distribution of outlier factors of all the data in dataset without requiring the number of outliers or any other parameters. To verify the feasibility of the proposed outlier judgment criterion based on supreme slope, a small dataset with three scattered outliers and a normal cluster is created. Figure 6(a) shows the dataset, and Fig. 6(b) shows the distribution of LDVP of all data in descending order. The numbers in Fig. 6(b) stand for the order of outlier degree of the corresponding point with the same number in Fig. 6(a). In Fig. 6(b), the highest point and the lowest 8 points (20% of the dataset) are connected with lines. It is obvious that the solid line has the supreme absolute slope, so the point 3 connected with it is a critical point (threshold). And 20% of total amount of data in the dataset are enough to choose a proper critical point. Points with no greater LDVP than that of the critical point are candidate outliers. The outlier judgment criterion based on supreme slope supplies a novelty way to determine outliers accurately. What's more, the criterion can adapt to evolving data stream.

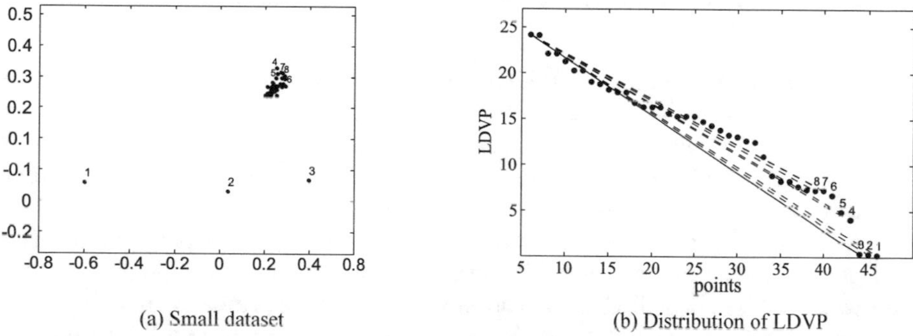

(a) Small dataset

(b) Distribution of LDVP

Fig. 6. The verification of outlier judgment criterion based on supreme slope.

The details of outlier detection are described in Algorithm 1. Only when candidate outliers are identified to be outliers during the whole multiple validations can they be declared as real outliers.

2.3 Determining the Parameters ε and λ

The parameters needed in LDVP-OD are the number of data blocks contained in a sliding window ε and the times of multiple validations λ.

The number of data blocks contained in a sliding window ε. As soon as the sliding window updates a data block, LDVP-OD needs to be performed once. So ε has influence on the efficiency of the process. For a sliding window with a fixed size, the greater ε is, the more times the algorithm needs to be performed,

Algorithm 1. Outlier detection

Input: Dataset in sliding window at a certain time S^m
Output: Current candidate outlier set G

Normalize the dataset S^m, and get the normalized dataseet S
According to the $k_{dist}(k = 4)$ graph to find the best radius of neighborhood r
for each data point x_i in S **do**
  Calculated the r-neighborhood $N_r(x_i)$ of x_i according to the best radius of neighborhood r.
  Calculated the mean of vector dot product $MVP(x_i)$
  Calculated the local density of vector dot product $LDVP(x_i)$
end for
Sort all LDVPs in descending order
According to the outlier judgment criterion based on supreme slope, the best threshold $LDVP(x_c)$ is determined
for each x_i in S **do**
  if $LDVP(x_i) \leq LDVP(x_c)$ **then**
    x_i is identified as an candidate outlier
  end if
end for
Save all candidate outliers into the candidate outlier set G
return The current candidate outlier set G

so more time it will takes. On the contrary, if ε is set to be smaller, the times LDVP-OD needs to be run are reduced and time consuming is less. However this will omit the relationship between blocks in front and behind, resulting in one-sided result and bad performance. Usually ε can be set to 2–5. In this paper, ε is set to 2.

The times of multiple validations λ. The purpose of multiple validations is to decrease the false alarm rate. The greater λ is, the lower false alarm rate the algorithm can get. But if the λ is too large, the outliers can not be determined in time. What's worse, too large λ will result in too many candidate outliers remaining in a sliding window, which will disturb the judgment of outliers. In most cases, λ can be set to 2 or 3. In this paper, λ is set to 3.

3 Experimental Result and Analysis

In order to verify the effectiveness of LDVP-OD, contrast experiments with Algorithm Y, I-IncLOF and DSABOD are performed over several synthetic and real datasets. The information about the datasets is shown in Table 1. The 6 real datasets represent different kinds of datasets with different number of dimensions, different sizes, different distributions. All experiments are conducted in Matlab R2014a on Intel Core i5-3230M, 2.6 GHz with 8 GB memory running on Windows 7x64.

The 6 real datasets are selected from UCI machine learning repository [7]. They are originally used for the evaluation of classification methods. However, the proportion of the outliers is very small in the whole dataset (no more than

Table 1. Information of the real datasets.

Dataset name	Normal class	Outlier class	Numeric dims	samples	Outliers
Yeast	1	5, 6, 7, 8, 9	8	496	33 (6.65%)
Abalone	5–16	1–3, 19–29	8	4020	111 (2.76%)
Breast cancer	Benign, Malignant	User-defined	10	733	34 (4.64%)
Mushroom	Edible	Poisonous	22	4347	139 (3.20%)
Ionosphere	Good	Bad	34	237	12 (5.06%)
KDD1999	normal	Probe, U2R, R2L	41	1015061	42280 (4.17%)

10% of the total number of points). For the purpose of outlier detection, some of the large classes are considered as the normal and some of the small classes are consider as outliers. In Table 1, Both of classes in dataset BREAST CANCER are taken as the normal data, and 34 outlier objects are planted into the dataset based on the statistical characteristics of attributes for better performance analysis. KDD1999 contains the records of 7 weeks of network traffic. There are 972781 instances of normal data, whereas the number of attack records is too high to be considered as outliers (3925650). In order to make the dataset more realistic, the rare attack types (Probe, U2R and R2L) are selected as outliers so that outliers become a small ratio of normal instances.

3.1 Experiment on Synthetic Datasets

In order to prove the robustness of LDVP-OD, a group of synthetic 2-dimensional datasets are created. There are 1500 normal objects in the raw synthetic dataset, including three clusters which satisfy Gauss distribution. The original dataset is shown in Fig. 7(a). Scattered outliers are added in the raw dataset in varying proportions 2%, 4%, 6%,... , 20%, which forms 10 datasets with different proportions of outliers. A ROC curve (false positive rate vs. true positive rate) is depicted for the LDVP values in all experiments such that each point on the curve is related to a threshold (deriving from the specific proportion of lowest points described in Sect. 2.2) of the resulting LDVP values. AUC (Area Under ROC cure) is used to evaluate the performance just like the detection accuracy. Figure 7(b) shows the results of LDVP-OD in comparison with Algorithm Y, I-IncLOF and DSABOD.

It can be seen in Fig. 7(b) that as the proportion of outliers increases, the AUCs of all algorithms decrease to different degrees. Algorithm Y combines the advantages of DBSCAN and W-K-Mean, so Algorithm Y shows good performance on low-dimensional dataset and the AUC of it remains roughly stable shown as the curve of Algorithm Y in Fig. 7(b). I-IncLOF uses LOF to evaluate the outlierness of data point. According to the analysis of effectiveness of LOF in Fig. 5(a) in Sect. 2.2, when the proportion of outliers exceeds 12%, outliers will form sparse clusters. Therefore the performance of I-IncLOF declines sharply.

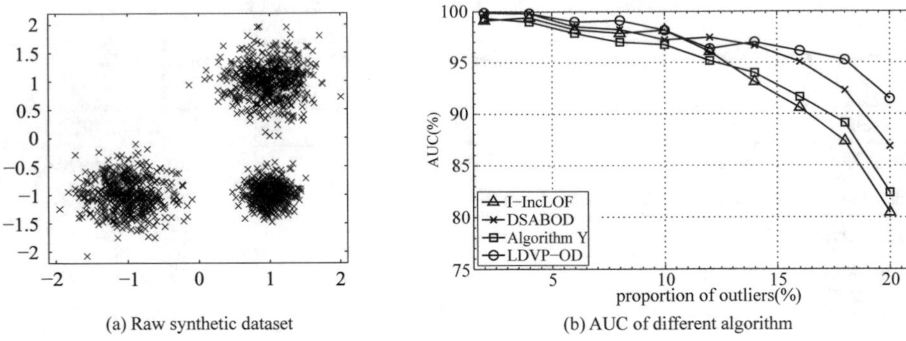

Fig. 7. The robustness of LDVP-OD in comparison with Algorithm Y, I-IncLOF and DSABOD.

From the curve of DSABOD in Fig. 7(b), it can be seen that when the proportion of outliers is less than 12%, DSABOD is less effective than the other algorithms. It's because DSABOD utilizes ABOF to measure the outlierness of data point, which is not accurate enough for outliers between clusters. However ABOF is sensitive to outliers in sparse cluster, when the proportion of outliers is more than 12% which form sparse clusters, DSABOD has better performance than I-IncLOF. Due to the fact that LDVP not only is sensitive to scattered outliers and edge points, but also have high accuracy in evaluating the outlierness of points in sparse cluster, LDVP-OD has better performance in experiment generally. From the groups of experiments on synthetic datasets with different proportions of outliers in comparison with Algorithm Y, I-IncLOF and DSABOD, it can be concluded that LDVP-OD has good performance and robustness.

3.2 Experiments on Real Datasets

Experiments on real datasets are conducted in comparison with Algorithm Y, I-IncLOF and DSABOD. The real datasets are Yeast, Abalone, Breast Cancer, Mushroom, Ionosphere and KDD1999, selected from the UCI machine learning repository. Outlier detection rate and false alarm rate are used to evaluate the performance of algorithms. Tables 2 and 3 show detail comparative experiment results of LDVP-OD vs. Algorithm Y, I-IncLOF and DSABOD on the 6 real datasets.

Table 2 shows the outlier detection rate and Table 3 presents the false alarm rate. It can be observed from Tables 2 and 3 that generally LDVP-OD outperforms Algorithm Y, I-IncLOF and DSABOD with higher outlier detection rate and lower false alarm rate. Algorithm Y does well in some low-dimensional datasets, such as Yeast, Abalone and Breast Cancer. In these datasets, Algorithm Y has lower false alarm rate than LDVP-OD shown in Table 3, but the outlier detection rate is lower than LDVP-OD to shown in Table 2. In higher dimensional space, the performance of Algorithm Y is worse due to its distance-based and density-based nature. I-IncLOF has few advantages in high-dimensional datasets,

Table 2. Outlier detection rate of LDVP-OD vs. Algorithm Y, I-IncLOF and DSA-BOD.

Dataset	Algorithm Y/%	I-IncLOF/%	DSABOD/%	LDVP-OD/%
Yeast	95.13	93.94	75.76	**96.74**
Abalone	70.12	69.23	65.38	**87.31**
Breast cancer	91.17	82.35	91.18	**97.16**
Mushroom	92.28	97.73	97.15	**99.00**
Ionosphere	72.73	72.73	**91.67**	**91.67**
KDD1999	91.09	86.14	57.43	**92.08**

Table 3. False alarm rate of LDVP-OD vs. Algorithm Y, I-IncLOF and DSABOD.

Dataset	Algorithm Y/%	I-IncLOF/%	DSABOD/%	LDVP-OD/%
Yeast	**2.39**	4.35	20.83	2.51
Abalone	**2.35**	5.93	11.76	2.41
Breast cancer	**0.43**	7.22	18.92	0.86
Mushroom	12.38	2.24	24.81	**0.26**
Ionosphere	9.09	54.55	31.25	**3.56**
KDD1999	14.02	32.56	31.58	**6.56**

so I-IncLOF does not have outstanding results in Table 2 or Table 3. ABOF used in DSABOD is effective in high-dimensional datasets to some extent, But when there are outlier clusters in dataset (such as Yeast, Abalone) and the distribution is unbalanced (such as KDD1999), DSABOD loses effectiveness. LDVP-OD utilizes more accurate way to estimate outlierness, more effective criterion to determine outliers and multiple validations to decrease false alarm rate. Although the dimension increases and characteristics vary, generally LDVP-OD performs well and outperforms Algorithm Y, I-IncLOF and DSABOD.

3.3 Time Complexity Evaluation

The complexity of both r value and the mean of vector dot product are $\frac{1}{2}O(n^2)$, thus the time complexity of LDVP-OD is $O(n^2)$. The time complexity of I-IncLOF and Algorithm Y is $O(n{\cdot}log(n))$. And for DSABOD, the outlier factors of all the history data records need to be updated, so the complexity of DSABOD is $O(n^3)$. We can know that the complexity of LDVP-OD is better than DSABOD, but worse than Algorithm Y and I-IncLOF.

It is big challenge to find real world data streams for the evaluation of our proposed outlier detection approach. In order to observe the actual complexity of each algorithm in data stream, we generated synthetic data streams with 15 dimensions and 1000, 2000, 3000, 4000 and 5000 objects. We use basically the same method as in [24] to generate the synthetic data streams for fairness. The

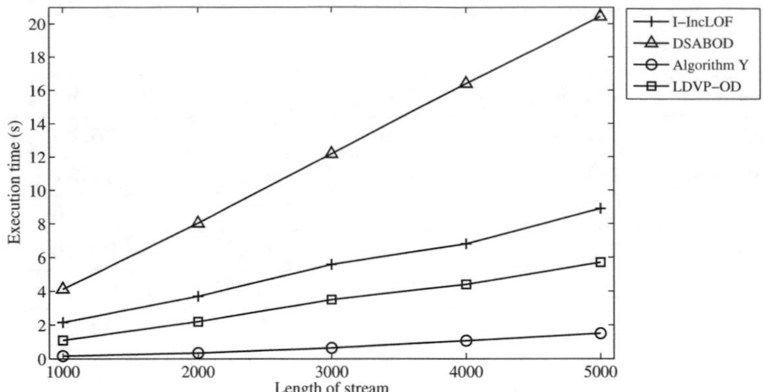

Fig. 8. Execution times of I-IncLOF, Algorithm Y, DSABOD and LDVP-OD with 15 dimensions.

execution times of I-IncLOF, DSABOD, Algorithm Y and LDVP-OD with the number of streams varied from 1000 to 5000 are shown in Fig. 8. In Fig. 8, our proposed algorithm is second only to Algorithm Y and the execution time is less than I-IncLOF. Because of the LDVP-OD divided the data stream into block but the I-IncLOF not. Generally, the experimental results above prove that the proposed algorithm LDVP-OD is efficient when it offers a good performance improvement over Algorithm Y, I-IncLOF and DSABOD in terms of outlier detection rate and false alarm rate.

4 Conclusion and Future Work

In this paper, an outlier detection algorithm based on local density of vector dot product in data stream (LDVP-OD) is proposed. LDVP-OD presents the model based on sliding window and multiple validations to divide the data stream into blocks and decrease the false alarm rate. Local density of vector dot product (LDVP) is proposed, which is a more accurate way to evaluate the deviation degree of data. And outlier judgment criterion based on supreme slope is described to determine candidate outliers without requiring the number of outliers or other parameters beforehand. Furthermore, LDVP-OD reserves and removes data selectively, which makes sure that the limited memory is consumed properly. Experiments on several synthetic and real datasets demonstrate that LDVP-OD has good robustness and outperforms some existing algorithms with higher outlier detection rate and lower false alarm rate. For future work, more efforts will be put into the investigation of automatically adjusting the size of sliding window to improve the self-adaptivity of algorithm.

Acknowledgments. This work is supported by the following foundations: the national Natural Science Foundation of China (61662013, 61362021, U1501252); Natural Science Foundation of Guangxi province (2016GXNSFAA380149); Guangxi

Innovation-Driven Development Project (Science and Technology Major Project) (AA17202024); the Key Laboratory of Cognitive Radio and Information Processing Ministry of Education (2011KF11); Innovation Project of GUET Graduate Education (2017YJCX34, 2018YJCX37).

References

1. Angelov, P.: Outside the box: an alternative data analytics framework. J. Autom. Mob. Rob. Intell. Syst. **8**(2), 29–35 (2014). https://doi.org/10.14313/JAMRIS_2-2014/16
2. Angiulli, F., Fassetti, F.: Distance-based outlier queries in data streams: the novel task and algorithms. Data Min. Knowl. Disc. **20**(2), 290–324 (2010). https://doi.org/10.1007/s10618-009-0159-9
3. Breunig, M.M., Kriegel, H.P., Ng, R.T., Sander, J.: LOF: identifying density-based local outliers. In: ACM Sigmod Record. vol. 29, pp. 93–104. ACM (2000). https://doi.org/10.1145/335191.335388
4. Cao, L., Yang, D., Wang, Q., Yu, Y., Wang, J., Rundensteiner, E.A.: Scalable distance-based outlier detection over high-volume data streams. In: 2014 IEEE 30th International Conference on Data Engineering (ICDE), pp. 76–87. IEEE (2014). https://doi.org/10.1109/ICDE.2014.6816641
5. Carcillo, F., Dal Pozzolo, A., Le Borgne, Y.A., Caelen, O., Mazzer, Y., Bontempi, G.: SCARFF: a scalable framework for streaming credit card fraud detection with spark. Inf. Fusion **41**, 182–194 (2018). https://doi.org/10.1016/j.inffus.2017.09.005
6. Cremonesi, P., Koren, Y., Turrin, R.: Performance of recommender algorithms on top-n recommendation tasks. In: Proceedings of the Fourth ACM Conference on Recommender Systems. pp. 39–46. ACM (2010). https://doi.org/10.1145/1864708.1864721
7. Dheeru, D., Karra Taniskidou, E.: UCI machine learning repository (2017). http://archive.ics.uci.edu/ml
8. Ester, M., Kriegel, H.P., Sander, J., Xu, X., et al.: A density-based algorithm for discovering clusters in large spatial databases with noise. In: KDD, vol. 96, pp. 226–231 (1996). http://dl.acm.org/citation.cfm?id=3001460.3001507
9. Gao, K., Shao, F.J., Sun, R.C.: n-INCLOF: a dynamic local outlier detection algorithm for data streams. In: 2010 2nd International Conference on Signal Processing Systems (ICSPS), vol. 2, p. V2–179. IEEE (2010). https://doi.org/10.1109/ICSPS.2010.5555276
10. Golab, L., Özsu, M.T.: Issues in data stream management. ACM Sigmod Rec. **32**(2), 5–14 (2003). https://doi.org/10.1145/776985.776986
11. Ha, J., Seok, S., Lee, J.S.: Robust outlier detection using the instability factor. Knowl. Based Syst. **63**, 15–23 (2014). https://doi.org/10.1016/j.knosys.2014.03.001
12. Huang, J.Z., Ng, M.K., Rong, H., Li, Z.: Automated variable weighting in k-means type clustering. IEEE Trans. Patt. Anal. Mach. Intell. **27**(5), 657–668 (2005). https://doi.org/10.1109/TPAMI.2005.95
13. Karimian, S.H., Kelarestaghi, M., Hashemi, S.: I-incLOF: improved incremental local outlier detection for data streams. In: 2012 16th CSI International Symposium on Artificial Intelligence and Signal Processing (AISP), pp. 023–028. IEEE (2012). https://doi.org/10.1109/AISP.2012.6313711

14. Knorr, E.M., Ng, R.T.: Algorithms for mining distance-based outliers in large datasets. In: Proceedings of the 24th International Conference on Very Large Data Bases, VLDB 1998. pp. 392–403. Morgan Kaufmann Publishers Inc., San Francisco (1998). http://dl.acm.org/citation.cfm?id=645924.671334
15. Kontaki, M., Gounaris, A., Papadopoulos, A.N., Tsichlas, K., Manolopoulos, Y.: Continuous monitoring of distance-based outliers over data streams. In: 2011 IEEE 27th International Conference on Data Engineering (ICDE), pp. 135–146. IEEE (2011). https://doi.org/10.1109/ICDE.2011.5767923
16. Kriegel, H.P., Kröger, P., Schubert, E., Zimek, A.: Loop: local outlier probabilities. In: Proceedings of the 18th ACM Conference on Information and Knowledge Management. pp. 1649–1652. ACM (2009). https://doi.org/10.1145/1645953.1646195
17. Kriegel, H.P., Zimek, A., et al.: Angle-based outlier detection in high-dimensional data. In: Proceedings of the 14th ACM SIGKDD International Conference on Knowledge Discovery and Data Mining, pp. 444–452. ACM (2008). https://doi.org/10.1145/1401890.1401946
18. Miller, Z., Dickinson, B., Deitrick, W., Hu, W., Wang, A.H.: Twitter spammer detection using data stream clustering. Inf. Sci. **260**, 64–73 (2014). https://doi.org/10.1016/j.ins.2013.11.016
19. Neeraj, K.: Anomaly-based network intrusion detection: an outlier detection techniques. In: Proceedings of the Eighth International Conference on Soft Computing and Pattern Recognition. vol. 614. Springer, Cham (2018). https://doi.org/10.1007/978-3-319-60618-7_26
20. Pang, G., Cao, L., Chen, L., Liu, H.: Learning representations of ultrahigh-dimensional data for random distance-based outlier detection. arXiv preprint arXiv:1806.04808 (2018)
21. Pokrajac, D., Lazarevic, A., Latecki, L.J.: Incremental local outlier detection for data streams. In: IEEE Symposium on Computational Intelligence and Data Mining, CIDM 2007, pp. 504–515. IEEE (2007). https://doi.org/10.1109/CIDM.2007.368917
22. Shou, Z.Y., Li, M.Y., Li, S.M.: Outlier detection based on multi-dimensional clustering and local density. J. Cent. S. Univ. **24**(6), 1299–1306 (2017). https://doi.org/10.1007/s11771-017-3535-4
23. Thakran, Y., Toshniwal, D.: Unsupervised outlier detection in streaming data using weighted clustering. In: 2012 12th International Conference on Intelligent Systems Design and Applications (ISDA), pp. 947–952. IEEE (2012). https://doi.org/10.1109/ISDA.2012.6416666
24. Ye, H., Kitagawa, H., Xiao, J.: Continuous angle-based outlier detection on high-dimensional data streams. In: Proceedings of the 19th International Database Engineering & Applications Symposium, pp. 162–167. ACM (2015). https://doi.org/10.1145/2790755.2790775
25. Zhou, J., Kwan, C.: Anomaly detection in low quality traffic monitoring videos using optical flow. In: Pattern Recognition and Tracking XXIX, vol. 10649, p. 106490F. International Society for Optics and Photonics (2018). https://doi.org/10.1117/12.2303651

Multi-level Competitive Swarm Optimizer for Large Scale Optimization

Li Zhang, Yu Zhu, Si Zhong(✉), Rushi Lan, and Xiaonan Luo

Guangxi Key Laboratory of Intelligent Processing of Computer Images and Graphics, Guilin University of Electronic Technology, Guilin 541004, China
421938207@qq.com

Abstract. In this paper, a new multi-level competitive swarm optimizer (MLCSO) is proposed for large scale optimization. As a variant of particle swarm optimization (PSO), MLCSO first divides the particles of original swarm into two groups randomly and then compares the particles according to their fitness values. The loser with worse fitness value will be put into the first level. The winner with better fitness becomes a new little swarm. New little swarm continues to be divided and compared until the new swarm has only one particle. This process forms a multi-level mechanism. The loser will be updated by the winner. It not only shows a great balance between exploration and exploitation but also enhances the diversity. 20 different kinds of test functions are selected for the experiments. Despite MLCSO algorithm is simple, the experimental results on high-dimension by comparing it with five state-of-the-art algorithms demonstrated its effectiveness.

Keywords: Large scale optimizer · Multi-level · Competitive · Exploration and Exploitation

1 Introduction

Stochastic algorithms such as evolutionary algorithms (EAs) have been shown to be effective optimization techniques [1–3]. However, most of these optimization algorithms will suffer from poor performance due to the dimensionality of the search space increases. Therefore, it is important to search new ways of simplifying a given large scale optimization problem [6,8]. There are several algorithms have been proposed to solve large scale problems [4,9].

One such method is to decompose the original large scale problem into a group of smaller and simpler subproblems. General speaking, it implies that it is easy to manage and solve. Cooperative coevolution (CC) is a framework

This work was supported in part by the National Natural Science Foundation of China (Nos. 61762028, 61772149, U1701267 and 61320106008), and by Guangxi Colleges and Universities Key Laboratory of Intelligent Processing of Computer Images and Graphics (No. GIIP201703).

C.-N. Yang et al. (Eds.): SICBS 2018, AISC 895, pp. 185–197, 2020.
https://doi.org/10.1007/978-3-030-16946-6_15

implementing the divide-and-conquer technique introduced by Potter [20]. Van den Bergh and Engelbrecht applied the cooperative particle swarm optimizer (CPSO), employing cooperative behavior to significantly improve the performance of the original algorithm [7]. Then, a new cooperative coevolving particle swarm optimization (CCPSO2) employed an effective variable grouping technique random grouping was introduced [10]. In 2014, an decomposition strategy called differential grouping with CC (CCDG) allowed an automatic near-optimal decomposition of decision variables, which achieves high accuracy and efficiency in large scale optimization problems [14]. There are two popular DG variants, global differential grouping (GDG) [5] and extended differential grouping (XDG) [9]. CCEAs [29] achieve good performance for large scale optimization, but they suffer from two problems. First, the decomposition strategy of CCEAs is benefit to optimize the large scale problems. However, there is often insufficient knowledge about the structure of a given problem to be able to manually devise a suitable decomposition strategy [14]. Second, numerous function evaluations are required to the optimization process, which increases the complexity of the optimization.

From another perspective, various new updating strategies for classical EAs which can cope with large scale optimization problems are devoted by researchers. The PSO, first introduced by Kennedy and Eberhart [21], is one of the most important swarm intelligence paradigm [1]. The classical PSO algorithm can easily get trapped in the local optima. Thus, two most important goals are accelerating convergence speed and avoiding the local optima. In 2006, comprehensive learning particle swarm optimizer (CLPSO) used all other particles' historical best information to update a particle's velocity [11]. In 2009, an adaptive particle swarm optimization (APSO) that features better search efficiency than classical particle swarm optimization (PSO) is presented [22]. The authors in [12] introduced social learning mechanisms into particle swarm optimization to develop a social learning PSO (SLPSO), in which each particle can learn from any better particles instead of historical information (global best position or personal best position) [12]. In 2015, a competitive swarm optimizer for large scale optimization (CSO) was developed [13]. In 2017, a level-based learning swarm optimizer for large scale optimization (LLSO) [17] that adds a dynamic selection strategy for the number of levels.

Although these PSO variants show better performance, most of them encounter the problem of premature convergence. To address this problem, this paper proposes a new variant, called multi-level competitive Swarm optimizer (MLCSO), for large scale optimization. We introduce multi-level mechanism and competitive learning into PSO. In the proposed MLCSO, using competitive learning, the global best solution and personal best solutions will not be stored. In addition, multi-level mechanism enhances the diversity of particles, which is good for the optimization of the large scale problems.

The rest of this paper is organized as follows. Section 2 presents some preliminaries as background knowledges, followed by the whole framework of MLCSO in Sect. 3. In Sect. 4, some experimental results prove the good scalability and efficiency of MLCSO. Finally, conclusions are given in Sect. 5.

2 Preliminaries

2.1 Problem Statement

Without loss of generality, in this paper, we consider the minimization problems defined as follows:

$$\min f(X),\ X = [x^1, x^2, \cdots, x^D] \tag{1}$$

where D is the number of variables to be optimized. With D increasing, it is very difficult to optimize [14]. On the one hand, multimodal problems easily trapped into local optimal values. On the other hand, the search space is explosively increased. This is referred to as the curse of dimensionality. Evolutionary algorithms [14] are effective optimization methods and have been extensively used for solving a wide range of optimization.

2.2 PSO and Its Developments

PSO. PSO is an abstract framework based on the intelligence of social animal groups. In particle swarm optimization algorithm, the particle of a group represents a potential solution, and it flies in the problem search space to find the optimal solution. These particles broadcast their current location to neighboring particles. The previously determined good position is then used by the bee colony as a starting point for further searches where individual particles adjust their current position and velocity [10].

In a classical PSO, a swarm of particles have a n-dimensional search space. The particles learning can be represented as a current velocity and a current position. The velocity $V_i(t)$ and position $X_i(t)$ of the ith particle are updated as follows:

$$\begin{aligned} V_i(t+1) = \omega V_i(t) + c_1 R_1(t)(pbest_i(t) - X_i(t)) \\ + c_2 R_2(t)(gbest(t) - X_i(t)) \end{aligned} \tag{2}$$

$$X_i(t+1) = X_i(t) + V_i(t+1) \tag{3}$$

where t is the generation (iteration) number, ω is termed inertia weight [23], c_1 and c_2 [24] are two acceleration coefficients, and r_1 as well as r_2 is uniformly randomized within [0,1], $V_i(t)$ and $X_i(t)$ are the velocity vector of the i-th particle. $pbest_i(t)$ and $gbest_i(t)$ is personal best position and global best position of i-th particle respectively. Kennedy [21] considered the second part of Eq. (2) as the cognitive component. And the third part is the social component, respectively.

Some Variants of PSO. Since the introduction of the PSO algorithm, the PSO has become a popular optimizer because of its simple concept and effectiveness. Therefore, numerous PSO variants have been proposed to further improve the efficacy of PSO. To name a few, research on algorithm theory hopes that the search ability and convergence ability of the algorithm will be proved; inspired by the parameter research to improve the performance of the algorithm. In

Eq. (2), it shows that ω, $c1$ and $c2$ are three control parameters. ω is the inertia weight. It was proposed by Shi and Eberhart [25] to balance between the local optimization and the global search. Some PSO variants adjust the value of ω. Ratnaweera introduced time-varying acceleration coefficients [13]. What is more, a time-varying PSO were introduced in [26]; inspired by different topology [26,27] that the global search and the local search can be balanced. For example, a fully informed PSO (FIPS) [16] updated each particle by the positions of several neighbors; inspired from differential evolution to improve the diversity of algorithm.

Specially, there are two state-of-art topological algorithms, SL-PSO and CSO. SL-PSO is composed of swarm sorting and behavior learning. The swarm is first sorted according to the particles' fitness values [12]. Then, every particle will learn from its corresponding demonstrators. CSO uses a competition mechanism which shows good performance on high-dimensional problems. A competition is made between the two particles in each couple. What is more, the particle loses a competition will learn from the winner particle [13]. The loser particles will update their velocity using the following learning strategy:

$$\begin{aligned} V_{l,k}(t+1) = R_1(k,t)V_{l,k}(t) + R_2(k,t)\big(X_{w,k}(t) - X_{l,k}(t)\big) \\ + \varphi R_3(k,t)\big(\overline{X}_k(t) - X_{l,k}(t)\big) \end{aligned} \tag{4}$$

Therefore, the position of losers' in the competition are updated as follow:

$$X_{l,k}(t+1) = X_{l,k}(t) + V_{l,k}(t+1) \tag{5}$$

In Eq. (4), the second part is selected from better particles, the third part is also the mean position of the whole swarm as CSO and SL-PSO. Neither the personal best position of each particle nor the global best position (or neighborhood best positions) is involved in updating the particles [13]. Although these variants PSO have enhance performance, most of them get into the local optima.

3 Multi-level Competitive Swarm Optimizer

In the following, we present the multi-level competitive mechanism for the proposed MLCSO in detail.

3.1 Motivation

With the increasing of dimension size, it becomes more and more difficult to the solve optimization problems. It takes an optimizer a larger number of fitness evaluations to the locate optima due to the increasing search space and the higher computational complexity of the problem. In addition, it may easy get into local optima or premature convergence. Thus, in this paper, we use a multi-level and competitive learning mechanism to improve performance. The proposed method does not use the historical information, neither *gbest* nor *pbest*. Multi-level strategy treats particles differently that enhances the diversity of the particle swarm. The inspiration comes from the human society. Particularly, in the World Cup, different national football teams will first be divided into different groups to compete. The countries that lose the matches can be put into same level, and the countries that won the competition will be remain to be regrouped for the next competition. Following the idea in [13], the particles of each level will learn from other levels which win the competition, the loser particles will be updated velocity and position from the winner particles. Thereby the proposed multi-level competitive swarm optimizer (MLCSO) has promising performance on high-dimensional problems.

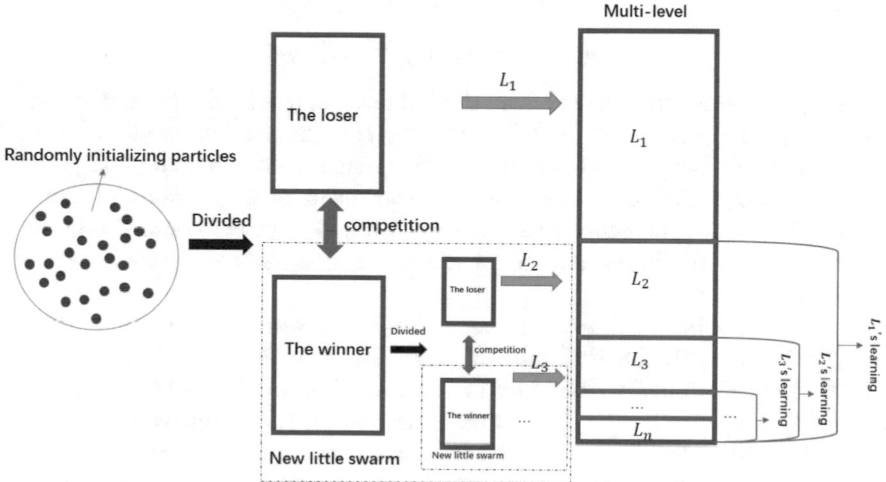

Fig. 1. Multi-level strategy and competitive learning in MLCSO. First, the particles are divided into two groups randomly. Two groups of particles are pairwise for competitions. The particles whose fitness values are better will be put in the winner. Others will be put in the loser. Thus, the loser forms the first level L_1. The winner becomes a new little swarm which will repeat the process. Then, particles in L_i learn from the L_{i+1}, L_{i+2}, ..., L_n.

3.2 MLCSO

It can be seen from Fig. 1 that, multi-level strategy and competitive learning are the most important components in MLCSO. First, we partition particles into different levels. Particles are divided into SL levels, each level denotes by L_i. A swarm contains m particles is randomly initialized, where m is the swarm size. Every particle has its own fitness value. When we divide particles into two same number of groups, particles from different groups will compare their fitness values. The particles have a better fitness called the winner. The particles that loses the competition, the loser, will be put into L_i waiting to be updated.

Then, the winner will become a new little swarm, it is regrouped and compared into winner and loser. This process continues until the new swarm is left with one particle.

In competitive learning, the winner levels and loser levels are generated by compete the fitness values. After learning from the winner, the loser will be passed to next generation.

Combining the above two strategies, the update rule for each particle is rewritten as follows:

$$\begin{aligned} V_{i,j}(t+1) = R_1(t)V_{i,j}(t) + R_2(t)\big(X_{L_w l,j}(t) - X_{L_i,j}(t)\big) \\ + \varphi R_3(t)\big(\overline{X}_j(t) - X_{b,j}(t)\big) \end{aligned} \tag{6}$$

$$X_{i,j}(t+1) = X_{i,j}(t) + V_{i,j}(t+1) \tag{7}$$

where $X_{i,j}(t)$ is the jth particle from the ith level L_i's behavior vector in generation $t(i \in \{1, 2, \cdots, m\}, j \in \{1, 2, \cdots n\})$. $V_{i,j}(t)$ is the particle's speed. $X_{L_w l,j}(t)$ is the particle's behavior vector from the particle that wins the competition. $X_{L_i,j}(t)$ is from the loser level. $\overline{X}_j(t)$ is the mean position value of the relevant particles. φ is the control parameter within [0,1] in charge of the influence of the $\overline{X}_j(t)$. $R_1(t), R_2(t)$, and $R_3(t)$ are three randomly generated vectors in generation t.

To provide a better understanding of MLCSO, we further analyze Eqs. (6) and (7). In Eq. (6), the first part $R_1(t)V_{i,j}(t)$ ensures the stability of search process that is similar to the inertia term in the classical PSO. The second part $R_2(t)\big(X_{L_i+1,j}(t) - X_{L_i,j}(t)\big)$ always represents the cognitive learning. The particle that loses the competition in ith level learns from the particle that wins the competition. Its personal best position will not be used. The third part $R_3(t)\big(\overline{X}_j(t) - X_{b,j}(t)\big)$ is the social learning. The neighborhood control, $\overline{X}_j(t)$, can help enhance PSOs performance on multimodal function by maintaining a higher degree of swarm diversity [13], which potentially increases swarm diversity and improves the search performance of MLCSO.

We randomly select particles from swarm without sorting the fitness values. Then, particles form different levels after the competition. It enhances the diversity of the swarm. The particles of different levels have different abilities in exploration and exploitation. Therefore, learning from particles in different levels and a neighborhood control $\overline{X}_j(t)$, particles have a potential compromise between exploration and exploitation. These improvements of MLCSO can play an important role in large scale optimization.

The pseudo code of MLCSO is shown in Algorithm 1. Because of the maintenance of canonical PSO framework, it is easy to implement MLCSO. It can be seen that lines 4 to 17 are dividing different particles to different levels and updating particles to next generation.

Algorithm 1. The pseudocode of MLCSO.

Input: control parameter φ, maximum number of fitness evaluations $maxfe$, the original swarm size is denoted by m, gs is the group size, ns denotes new swarm size, t is the generation number, the new swarm is divided into two group randomly, $group_1$ and $group_2$

Output: the best fitness of swarm

```
Initialize the particles of swarm randomly and calculate the fitness value
  of each particle;
t = 0, i = 1, ns = m;
while terminal condition FES < maxfe is not satisfied do
  while ns>=1
      divide swarm into two groups randomly;
      X_1(t) in group_1, X_2(t) in group_2
        if gs<=0
          if f(X_1(t)) <= f(X_2(t))
          X_{L_wl,j}(t) = X_1(t), X_{L_i,j}(t) = X_2(t);
        else
            X_{L_wl,j}(t) = X_2(t), X_{L_i,j}(t) = X_1(t);
        end if
         all of X_{L_i,j}(t) will be put into ith level;
          i = i + 1, ns = ns/2;
         Then the loser will learn from the winner using (6) and (7);
         Add the updated particles into P(t);
         swarm = the winner;
     end while
     t = t + 1;
     FES+ = m
end while
```

4 Experiments

4.1 Experimental Setup

In order to evaluate the feasibility and efficiency of the MLCSO, a set of 20 benchmark functions were used. We conduct a series of experiments on a broadly used: the CEC'10 special session on large scale global optimization [10]. The CEC'08 included only seven functions which were either separable or fully-nonseparable. However, investigating the behavior of algorithms on high-dimensional problems in different scenarios, the test suit of CEC'10 provides an improved platform by incorporating the partially-separable functions. The CEC'10 benchmark functions are able to divide into five groups in Table 1.

In addition, the proposed MLCSO will be tested on high-dimension (1000-D) problems to proof its scalability, where D is the dimension size of the test functions. The maximum number of fitness evaluations is set to 3000D. What is more, all the experimental results are achieved by averaging over 25 independent runs.

Table 1. The details of the functions

Types	Function numberName
f_1 to f_3	Separable functions
f_4 to f_8	Single-group m-nonseparable functions
f_9 to f_{13}	$D/2m$-group m-nonseparable functions
f_{14} to f_{18}	D/m-group m-nonseparable functions
f_{19} to f_{20}	Nonseparable functions

Table 2. Parameter settings for the compared PSO variants.

φ	0	0.1	0.15
f_1	2.87E+00	1.65E−17	2.89−17
f_4	4.89E+15	4.89E+15	4.89E+15
f_9	3.02E+08	2.15E+07	1.36E+07
f_{14}	1.54E+09	7.46E+07	3.97E+07
f_{19}	1.74E+07	9.56E+06	8.67E+06

4.2 Parameter Settings

Control Parameter. In Eq. (6), the third part is social learning of particles. The control parameter φ is the influence of the social component. From the statistical results summarized in Table 2, φ is varied from 0 to 0.15 on the five functions. When φ is changes from 0 to 0.1, the fitness value of separable function f_1 has changed a lot. 0.1 or 0.15 is more suitable for separable functions. The fitness values of f_4 do not change. However, f_9, f_{14} and f_{19}, the values of them are gradually getting better.

Swarm Size. The swarm size is an indispensable parameter in the most swarm optimization algorithms [13]. On the one hand, if the swarm size is too large, it is impractical for calculating expensive problems when the FES becomes a large number during every generation. On the other hand, if the swarm size is small, the search space is not well explored when the particles tend to converge very fast. This is one of the reasons for the premature convergence. Generally speaking, the swarm size is based on the experience. In DMS-PSO, a larger swarm size 450 was adopted for the optimization of 500-D functions [13]. In CSO, the swarm size is 500. In this paper, the swarm size m is set to 500 for the dimension of 1000-D.

4.3 Comparisons with State-of-the-Art Methods

In the experiments, we compare MLCSO with some state-of-the-art algorithms on large scale optimization. There are three PSO variants, include a

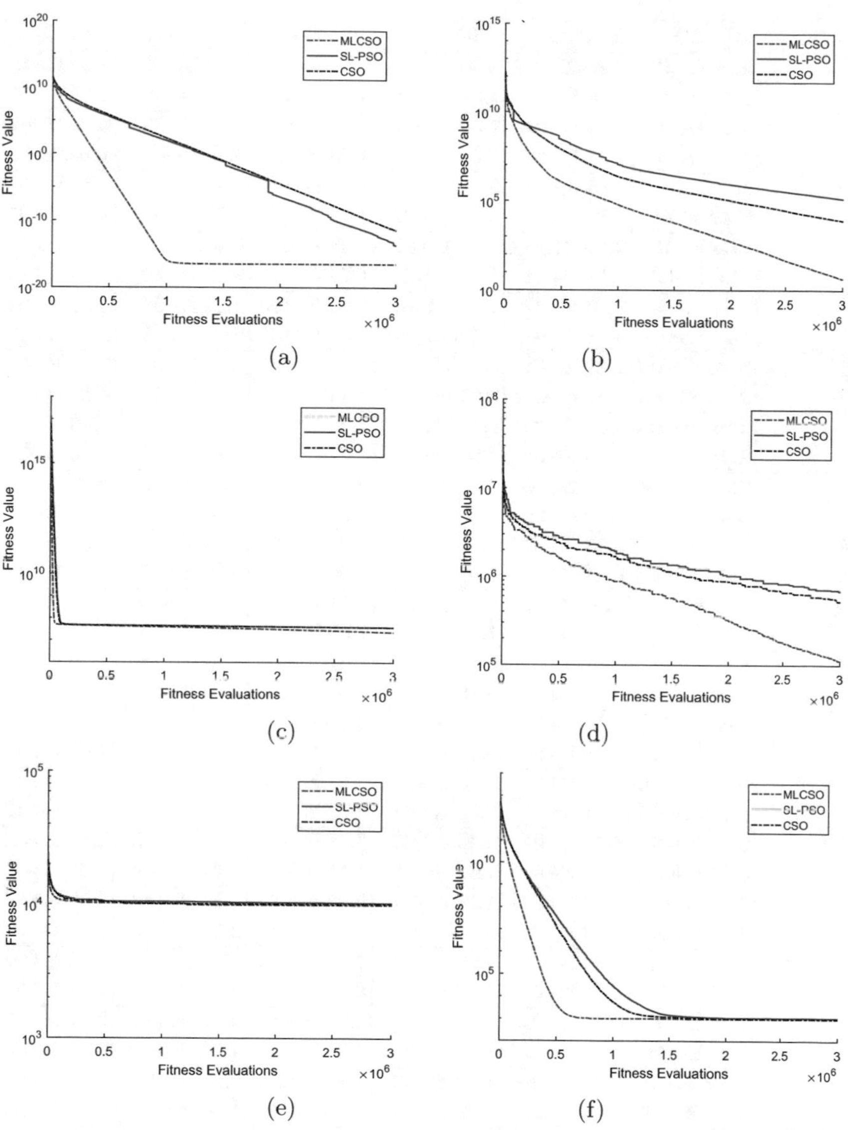

Fig. 2. Comparison of MLCSO with CSO and SL-PSO by testing representative functions. (a) to (f) correspond to the following functions: f_1, f_3, f_8, f_{12}, f_{15}, and f_{20}.

Table 3. Optimization errors on CEC'10 basic functions (1000-D).

Functions	MLCSO	SL-PSO	CSO	DECC-DG	DMS-L-PSO	MLCC
f_1	2.85 E−17	**7.89 E−18**	4.36 E−12	1.86 E+04	1.62 E+07	8.59 E−13
	1.10 E−18	8.56 E−18	4.85 E−13	4.67 E+04	1.38 E+06	2.86 E−12
f_2	6.23 E+03	1.96 E+03	7.39 E+03	4.52 E+03	5.48 E+03	**2.90 E+00**
	4.05 E+00	1.21 E+02	2.85 E+02	1.76 E+02	5.38 E+02	**1.51 E+00**
f_3	**1.25 E−12**	1.85 E+00	2.61 E−09	1.66 E+01	1.55 E+01	2.11 E−07
	2.37 E−14	3.29 E+00	2.60 E−10	3.08 E−01	1.09 E−01	1.18 E−06
f_4	4.89 E+15	**2.98 E+11**	7.31 E+11	5.05 E+12	4.46 E+11	1.79 E+13
	0.00 E+00	7.18 E+11	1.24 E+11	1.76 E+12	8.12 E+10	1.76 E+13
f_5	5.84 E+06	3.30 E+07	**2.89 E+06**	1.56 E+08	9.12 E+07	5.01 E+08
	1.95 E+06	6.28 E+06	**1.78 E+06**	2.22 E+07	9.01 E+06	1.12 E+08
f_6	**5.96 E−07**	2.18 E+01	8.23 E−07	1.65 E+01	3.79 E+01	1.69 E+07
	2.11 E−08	2.54 E+00	2.64 E−08	3.46 E−01	1.20 E+01	4.55 E+06
f_7	**4.07 E−04**	6.51 E+04	2.12 E+04	1.39 E+04	3.47 E+06	1.56 E+08
	2.24 E−04	5.63 E+04	3.68 E+03	1.31 E+04	1.16 E+05	1.46 E+08
f_8	**7.33 E+06**	7.88 E+06	3.58 E+07	2.76 E+07	2.31 E+07	6.68 E+07
	2.45 E+05	1.48 E+06	6.12 E+04	2.27 E+07	1.76 E+06	3.36 E+07
f_9	**1.85 E+07**	3.30 E+07	7.38 E+07	2.69 E+07	4.22 E+07	6.59 E+07
	1.46 E+06	4.46 E+06	5.96 E+06	2.56 E+07	3.56 E+07	3.47 E+07
f_{10}	1.06 E+04	**2.63 E+03**	9.54 E+03	4.51 E+03	5.38 E+03	4.21 E+03
	5.98 E+01	2.23 E+02	7.65 E+01	1.35 E+02	4.32 E+02	1.56 E+03
f_{11}	**1.03 E−11**	2.31 E+01	4.23 E−08	1.01 E+01	1.65 E+02	1.98 E+02
	1.40 E−13	2.11 E+00	5.16 E−09	8.65 E−01	1.69 E+00	1.12 E+00
f_{12}	3.00 E+04	1.69 E+04	4.23 E+05	2.86 E+03	**2.41 E+01**	1.14 E+05
	3.29 E+03	9.23 E+03	6.28 E+04	1.10 E+03	**9.87 E+00**	1.65 E+04
f_{13}	**5.26 E+02**	9.59 E+02	6.35 E+02	6.27 E+03	1.25 E+05	4.31 E+03
	9.268 E+02	3.74 E+02	2.65 E+02	3.68 E+03	6.23 E+04	4.68 E+03
f_{14}	6.75 E+07	8.98 E+07	2.49 E+08	3.48 E+08	**1.18 E+07**	5.89 E+08
	3.02 E+06	6.37 E+06	1.62 E+07	2.56 E+07	**1.56 E+06**	5.51 E+07
f_{15}	9.99 E+03	1.23 E+04	1.11 E+04	5.86 E+03	**5.56 E+03**	8.67 E+03
	4.07 E+02	8.36 E+01	5.36 E+01	1.05 E+02	**3.26 E+02**	2.11 E+03
f_{16}	1.83 E−11	2.56 E+01	5.69 E−08	**7.59 E−13**	3.28 E+02	3.58 E+02
	2.50 E−13	1.21 E+01	5.62 E−09	**6.25 E−14**	2.15 E+00	5.78 E+01
f_{17}	1.02 E+06	9.03 E+04	2.31 E+06	4.12 E+04	**4.25 E+01**	3.47 E+05
	6.94 E+04	1.46 E+04	1.54 E+05	2.30 E+03	**1.15 E+01**	3.21 E+04
f_{18}	**1.17 E+03**	2.77 E+03	1.78 E+03	1.38 E+10	2.56 E+04	1.79 E+04
	1.85 E+02	8.33 E+02	5.22 E+02	2.16 E+09	1.14 E+04	9.58 E+03
f_{19}	8.78 E+06	5.23 E+06	1.03 E+07	1.85 E+06	2.18 E+06	**2.04 E+06**
	4.13 E+05	7.12 E+05	5.64 E+05	1.23 E+05	1.63 E+05	**1.38 E+05**
f_{20}	9.70 E+03	1.89 E+03	1.26 E+03	6.45 E+10	**9.87 E+02**	2.34 E+03
	1.91 E+01	2.60 E+02	1.51 E+02	6.98 E+09	**1.40 E+01**	2.31 E+02
$w/t/l$		12/4/4	15/2/3	12/2/6	13/2/5	13/3/4

social learning particle swarm optimization algorithm for scalable optimization (SL-PSO) [12], competitive swarm optimizer for large scale optimization (CSO) [13] and dynamic multi-swarm particle swarm optimizer with local search (DMS-L-PSO). In addition, two algorithms are also selected that are based on the cooperative coevolution (CC) framework, namely multilevel cooperative coevolution for large scale optimization (MLCC) [28] and DECC-DG [14]. The CC is an efficient method dealing with high-dimensional problems. The optimization errors on 1000-D functions are showed in Table 3. The mean values are listed in the first line, and the standard deviations are listed in the second line. In addition, the $w/t/l$ in the last row shows the number of HSSO wins on w functions, ties on t functions, and losses on l functions.

In Table 3, the proposed MLCSO shows its better performance on more than 12 functions. f_1 is a separable and unimodal function. For unimodal functions, when exploitation is put a little more emphasis, it will achieve the fast convergence. It can be seen that, MLCSO converges much faster than SL-PSO, DECC-DG and DMS-L-PSO. The f_6 to f_8 are single-group m-nonseparable functions, and it shows that HSSO demonstrates good performance. In comparison with three PSO variants (such as DMS-L-PSO, CSO and SL-PSO), MLCSO defeats them down on f_9 and f_{11} functions respectively. f_{11} and f_{18} are multimodal functions. For multimodal functions, if exploration can be dynamically and properly biased without serious loss of exploitation, it may avoid premature convergence and stagnation [17]. Thus, MLCSO still achieves better performance than MLCC and DECC-DG.

From Fig. 2, MLCSO uses different kinds of functions to compare with SL-PSO and CSO. In Fig. 2(a), it is easy to observe that MLCSO has better best fitness values than CSO and SL-PSO. In Fig. 2(b) and (d), the exploration performance should be emphasized. Enhancing the exploration should improve the diversity of swarm. It can help particles to escape from local areas. f_3 and f_8 are nonseparable and multimodal functions respectively. It is clear that MLCSO has better ability on exploration than SL-PSO and CSO according to the fitness values dropping rapidly. In Fig. 2(c) and (e), it shows that MLCSO has not only the fast convergence but also better fitness values. Therefore, MLCSO has good performance in different kinds of functions: unimodal or multimodal, single-group or $D/2m$-group, separable or nonseparable.

5 Conclusion

This paper presented a new MLCSO for tackling large scale optimization problems. In MLCSO, we demonstrated that the multi-level mechanism is an efficiency approach to improve the ability of the classical PSO, and then the losers in different levels learn from the winners instead of using the personal best value. It can enhance the diversity of the swarm. MLCSO was compared with five state-of-the-art algorithms on various high-dimension optimization benchmark functions. The results indicated that MLCSO achieved satisfactory performance dealing with large scale problems.

References

1. Kennedy, J., Eberhart, R.C., Shi, Y.: Swarm Intelligence, pp. 187–219. Morgan Kaufmann Publishers Inc., San Francisco (2001)
2. Yang, Q., Chen, W.N., Yu, Z., Gu, T., Li, Y., Zhang, H., Zhang, J.: Adaptive multimodal continuous ant colony optimization. IEEE Trans. Evol. Comput. **21**(2), 191–205 (2017)
3. Jia, Y.H., Chen, W.N., Gu, T., Zhang, H., Yuan, H., Lin, Y., Zhang, J.: A dynamic logistic dispatching system with set-based particle swarm optimization. IEEE Trans. Syst. Man Cybern. Syst. **48**(9), 1607–1621 (2018)
4. Yang, Z., Tang, K., Yao, X.: Large scale evolutionary optimization using cooperative coevolution. Comput. Sci. Intell. Syst. Appl. Int. J. **178**(15), 2985–2999 (2008)
5. Mei, Y., Omidvar, M.N., Li, X., Yao, X.: A competitive divide-and-conquer algorithm for unconstrained large-scale black-box optimization. ACM Trans. Math. Softw. **42**(2), 13 (2016)
6. Jia, Y.H., Chen, W.N., Gu, T., Zhang, H., Yuan, H., Kwong, S., Zhang, J.: A dynamic logistic dispatching system with set-based particle swarm optimization. IEEE Trans. Syst. Man Cybern. Syst. **48**(9), 1607–1621 (2018)
7. Frans, V.D.B., Engelbrecht, A.P.: A cooperative approach to particle swarm optimization. IEEE Trans. Evol. Comput. **8**(3), 225–239 (2004)
8. Yang, Q., Chen, W.N., Gu, T.: Segment-based predominant learning swarm optimizer for large-scale optimization. IEEE Trans. Cyber. **47**(9), 2896–2910 (2017)
9. Sun, Y., Kirley, M., Halgamuge, S.K.: Extended differential grouping for large scale global optimization with direct and indirect variable interactions. In: Conference on Genetic and Evolutionary Computation, vol. 50 (Anno 26), pp. 313–320. ACM (2015)
10. Li, X., Yao, X.: Cooperatively coevolving particle swarms for large scale optimization. IEEE Trans. Evol. Comput. **16**(2), 210–224 (2012)
11. Liang, J.J., Qin, A.K., Suganthan, P.N., Baskar, S.: Comprehensive learning particle swarm optimizer for global optimization of multimodal functions. IEEE Trans. Evol. Comput. **10**(3), 281–295 (2006)
12. Cheng, R., Jin, Y.: A social learning particle swarm optimization algorithm for scalable optimization. Inf. Sci. **291**, 43–60 (2015)
13. Cheng, R., Jin, Y.: A competitive swarm optimizer for large scale optimization. IEEE Trans. Cyber. **45**(2), 191–204 (2015)
14. Omidvar, M.N., Li, X., Mei, Y., Yao, X.: Cooperative co-evolution with differential grouping for large scale optimization. IEEE Trans. Evol. Comput. **18**(3), 378–393 (2014)
15. Ratnaweera, A., Halgamuge, S.K., Watson, H.C.: Self-organizing hierarchical particle swarm optimizer with time-varying acceleration coefficients. IEEE Trans. Evol. Comput. **8**(3), 240–255 (2004)
16. Mendes, R., Kennedy, J., Neves, J.: The fully informed particle swarm: simpler, maybe better. IEEE Trans. Evol. Comput. **8**, 204–210 (2004)
17. Yang, Q., Chen, W.N., Da Deng, J., Li, Y., Gu, T., Zhang, J.: A level-based learning swarm optimizer for large-scale optimization. IEEE Trans. Evol. Comput. **22**(4), 578–594 (2018)
18. Fogel, D.B.: Evolutionary optimization. In: 1992 Conference Record of the Twenty-Sixth Asilomar Conference on Signals, Systems and Computers, vol. 1, vol. 48, pp. 409–414. IEEE (1992)

19. Back, T., Fogel, D.B., Michalewicz, Z.: Handbook of Evolutionary Computation. IOP Publishing Ltd., Bristol (1997)
20. Potter, M.A.: The design and analysis of a computational model of cooperative coevolution. George Mason University, Fairfax (1997)
21. Kennedy, J., Eberhart, R.: Particle swarm optimization. In: Proceedings of IEEE International Conference on Neural Networks IV, vol. 1000 (1995)
22. Yasuda, K., Ide, A., Iwasaki, N.: Adaptive particle swarm optimization. In: IEEE International Conference on Systems, Man and Cybernetics. vol. 2, pp. 1554–1559 (2003)
23. Shi, Y., Eberhart, R.: A modified particle swarm optimizer. In: The 1998 IEEE International Conference on Evolutionary Computation, pp. 69–73 (1998)
24. Eberhart, R., Kennedy, J. : A new optimizer using particle swarm theory. In: Proceedings of the Sixth International Symposium on Micro Machine and Human Science, pp. 39–43. IEEE (1995)
25. Shi, Y., Eberhart, R: Parameter selection in particle swarm optimization. In: Evolutionary Programming VII, pp. 591–600. Springer, Berlin (1998)
26. Suganthan, P.N.: Particle swarm optimiser with neighbourhood operator. In: Proceedings of IEEE Congress on Evolutionary Computation, vol. 3, pp. 1958–1962 (1999)
27. Kennedy, J.: Small worlds and mega-minds: effects of neighborhood topology on particle swarm performance. In: Proceedings of IEEE Congress on Evolutionary Computation, vol. 3, pp. 1931–1938. IEEE (1999)
28. Yang, Z., Tang, K., Yao, X.: Multilevel cooperative coevolution for large scale optimization. In: IEEE World Congress on Computational Intelligence, pp. 1663–1670 (2008)
29. Chen, W.N., Jia, Y.H., Zhao, F., Luo, X.N., Jia, X.D., Zhang, J.: A cooperative coevolutionary approach to large-scale multisource water distribution network optimization. IEEE Trans. Evol. Comput. (2018)

Experimental Comparison of Free IP Geolocation Services

Wei Xu[1(✉)], Yaodong Tao[2,3], and Xin Guan[4]

[1] University of Science and Technology of China, Hefei, China
xu5ei@mail.ustc.edu.cn

[2] National Joint Engineering Lab for ICS Security of 360ESG, Shenyang, China

[3] Shenyang Institute of Computing Technology of CAS, Shenyang, China

[4] Shanghai Ocean University, Shanghai, China

Abstract. IP geolocation services is used to locate the geographical location of an IP address, which has an important means for mapping cyberspace to the physical world. There are many location methods, such as delay-based method, database-driven method and topology-driven method. Currently, the mainstream IP geolocation services generally use a combination of different methods. We have collected some free IP geolocation services and studied their usage. Online geolocation services include: ip-api.com, ip.taobao.com, Baidu Map API, and Sina IPLOOKUP. Local services include: GeoLite2, ChunZhen database, IPPLUS360 and ipip.net. In the project cooperation, we obtained the IP address data of a city in China. Using the IP address data of the city as a data set, we conducted an experimental comparison of these IP geolocation services. The analysis and comparison of these IP geolocation services were carried out from three angles of data richness, query accuracy and query speed. It is found that the accuracy of online geolocation services is generally higher than that of local services, but the query speed is generally lower than that of local services. Specifically, different query service accuracy and query speed are different, and the difference is obvious, and few geolocation services can provide county-level information. We also found that the accuracy of some geolocation services needs to be improved. These results can be used as references for researchers with similar needs.

Keywords: IP geolocation · Comparative study · Experimental test · Accuracy · Speed

1 Introduction

The resources on the Internet can be located by IP addresses, and the physical world is usually located through such geographic information as country, region, city, district, street and longitude and latitude. The mapping of internet resources to physical locations, a process known as geolocation, has attracted intense interest from researchers.

C.-N. Yang et al. (Eds.): SICBS 2018, AISC 895, pp. 198–208, 2020.
https://doi.org/10.1007/978-3-030-16946-6_16

According to information sources, geolocation methods can be generally divided into three categories: delay-based methods, database-driven methods, and topology-driven methods [1]. Delay-based methods usually collect delay data from a set of known geographic landmarks, and use that knowledge to triangulate a target IP address [2,3]. Database-driven methods collect or aggregate static mapping information and form databases through specific methods [4]. Topology-driven methods deduce geolocations by assuming that topologically close addresses are physically close to each other [5]. Using these methods comprehensively, various third-party IP address geolocation services have been developed. Most geolocation services provide free services as well as business services with different quality of service [6,7].

In order to determine the accuracy of different IP address geolocation services, the researchers compared and analyzed different geolocation services. Among them, the most authoritative is CAIDA's geolocation tools comparison report [8]. CAIDA is a famous Internet measurement organization [9]. In 2011, under the support of the US Department of Homeland Security, they have compared and analyzed various IP address geolocation services.

Since 2011, many new IP address geolocation services have been generated, and the accuracy of the original service may also change. We want to choose an IP address geolocation service because of the project needs. We choose some free IP address location services which is popular in China and compared them. Using the IP address of a city in China that has obtained from an authoritative organization as a test set, the IP address geolocation services are compared and analyzed from three perspectives: data richness, query accuracy and query speed.

2 Backgroud

We collected information from multiple geolocation service providers on the network and found that these geolocation services can be divided into local and online. The online service can be updated in time, occupying less server resources, but often have access restrictions, the speed is also affected by a variety of factors, relatively slow. The speed of the local query services is fast, and the database is updated by downloading data packets, but the local storage resources are required.

2.1 Local IP Geolocation Services

There are three free local IP geolocation services we tested: MaxMind GeoLite2, ChunZhen database, IPPLUS360 and ipip.net offline database.

MaxMind GeoLite2. MaxMind is a Massachusetts-based digital mapping company that provides location data for IP addresses all over the world [10]. Its IP address geolocation service enjoys a worldwide reputation and is the most widely used geolocation service, many papers use it, such as [11]. It provides two versions of the IP address geolocation services Geoip and Geoip2, and provides open source data GeoLite2 [12], which is a free version of GeoIP2, with a slightly lower accuracy than the paid version.

The GeoLite2 database consists of several parts: GeoLite2 country database, GeoLite2 city database and Geolite2 ASN database. They implement different functions. The GeoLite2 country database can only query the country and continent where the IP address is located; the GeoLite2 city database can query the country, region, city, latitude, longitude, and zip code where the IP address is located, and the Geolite2 ASN database is used to query the Autonomous Domain (AS) of Internet Service Provider (ISP) of the IP address.

GeoLite2 offline database is updated once a month, which can download compressed files of MaxMind DB format through official website. MaxMind provides API support for 7 programming languages or software, including C#, C, Java, Perl, PHP, Python, Apache (mod_maxminddb), and there are many third-party APIs for languages and platforms that they do not provide support for. GeoLite2 supports multiple languages including English, Chinese, Russian, Japanese, Spanish and so on.

ChunZhen Database. ChunZhen database is the most famous open IP address location database in China [13]. It collected latest accurate IP address data of ISPs including China Telecom, China Unicom, China Mobile, Great Wall Broadband and Juyou Broadband. Its database is updated every 5 days. You can download a MicroSoft Windows program installation file from the official website. After installation, you can input the IP address to the software to find out their geographic location.

The query result can be divided into two fields: IP address and geolocation. The geolocation field is divided into the country and the detailed address. The detailed address is no division of multiple fields according to the administrative district. We can see that the geolocation field contains information of continent, province, city, ISP and so on, but does not divide multiple fields step by step according to administrative regions. The query results only support Chinese.

IPPLUS360. The IPPLUS360 is a big data technology software developer specializing in geolocation service [14]. It provides a series of IP address geolocation products and services. Registered users can download the offline package of their county-level IP geolocation database for free, and the company also provides higher-accuracy street-level IP address geolocation services to paid users.

The downloaded offline query file can be imported into the MySQL database and queried through database query statements. The results of the search include several fields such as continent, country, province, city, county, latitude, longitude, and ISP. The query results only support Chinese.

ipip.net. Ipip.net is an IP geolocation service that has gradually become popular in recent years. This service is based on the ISPs and network services BGP/ASN datas analytical processing. They have more than 300 network monitoring points around the world [15].

Users can download offline data packages free of charge from ipip.net. The offline data packages for free services contain fewer data fields than those for charging services. The website provides open source query code in multiple languages, including PHP, java, python, golang, C, C++, Ruby, etc.

2.2 Online IP Geolocation Services

There are four online IP geolocation services we tested: ip-api.com, ip.taobao.com, Sina IPLOOKUP, and Baidu Map API.

Ip-api.com. The ip-api.com provides free usage of its Geo IP API through multiple response formats [16]. Support IPv4 and IPv6. It provides a variety of return formats, including XML, JSON, CSV, Newline Separated, and Serialized PHP. To receive the response in JSON format, send a GET request to http://ip-api.com/json/x.x.x.x (x.x.x.x represents an IP address).

The response JSON contains multiple fields, and the 'Status' field is used to determine whether the query is successful. It also contains fields such as 'country', 'region', 'city', 'ISP' and so on. It supports multiple languages including English, Chinese, Russian, Japanese, Spanish and so on.

The ip-api.com has explicit access restrictions. Their system will automatically ban any IP addresses doing over 150 requests per minute.

Ip.taobao.com. The ip.taobao.com is the IP positioning service provided by Alibaba, a famous e-commerce company [17]. According to the IP address provided by the user, the service can quickly find out the location and other related information of the IP address, including the country, the province, the city and the ISP. It is said that the province's accuracy is over 99.8%, and the city's accuracy is over 96.8%.

It can be queried by REpresentational State Transfer (REST) API, Request interface (GET): /service/getIpInfo.php? Ip= [IP address string].

The response information is JSON format, which contains information of the country, province, city (county), ISP, etc. This query has no language options and only supports Chinese.

This service has access restrictions. In order to ensure the normal operation of the service, each user's access frequency needs to be less than 10 qps.

Sina IPLOOKUP. Sina IP positioning is also invoked using the REST API, support js format, json format, csv format and other formats. The json format call method is as follows: http://int.dpool.sina.com.cn/iplookup/iplookup.php?format=json&ip=[IP address].

In the response, the 'ret' field indicates whether the query is successful, '1' indicates success, and '0' indicates failure. There are also countries, provinces, cities, districts, ISP and other fields. No official document was found, so it is not certain that the site has access restrictions.

Baidu Map API. Baidu Map is China's leading online map service provider. Its IP geolocation service claims that it can get the approximate location of the IP address. Including latitude and longitude, province, city, and other address information. They mentioned that the latitude and longitude is only a general location, usually the center of the city.

This service supports the REST API. The format of the request URL is "http://api.map.baidu.com/location/ip?ip=x.x.x.x&ak=Your AK&coor=bd09ll". In addition to IP, the parameters include the developer key "AK" and "SN", and the "coor" parameter is used to specify the type of latitude and longitude coordinates, which is divided into national coordinates system and Baidu coordinates system. Baidu coordinate system can be used in conjunction with Baidu map.

This service has explicit access restrictions. The access restrictions of registered developers are: 100,000 times per day, 6,000 times per minute. If the personal authentication is completed, the access limit will be increased to 300,000 times per day, 12,000 times per minute, and the access restriction will be further relaxed after becoming a paid business user.

3 Comparative Experiment

3.1 Test Dataset

In our project cooperation, the partner provided the IP address of a city in China, and the specific city name is not disclosed in this paper. The IP addresses data include more than 1.2 million IP addresses of different ISPs. The distribution of these IP addresses among ISPs is shown in Fig. 1. Therefore, we can make sure the country, province, city, and ISP where these IP addresses are located, but we cannot ensure the district and county where they are located. Our project

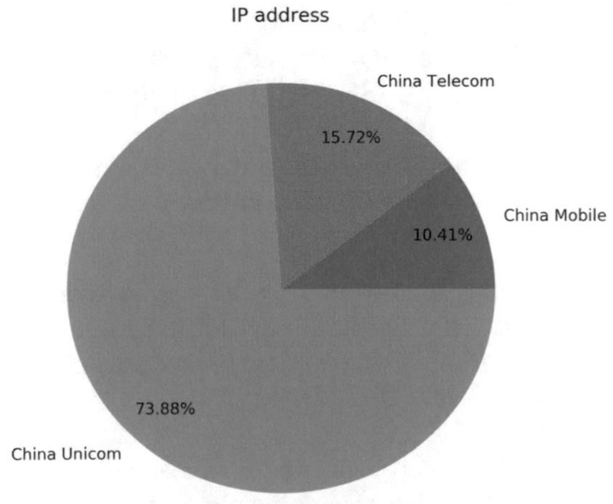

Fig. 1. IP address distribution chart.

needs to use the IP address to locate the device, so, we need to use a free IP location service to complete this task.

3.2 Experimental Method

In order to compare and analyze the free IP geolocation services, their respective characteristics, and the accuracy of positioning, we first collected several free IP geolocation services at home and abroad, study their usage, then, we use the city's IP address as a test set to test and compare these services performances.

We originally planned to use the different IP geolocation services to query all the IP addresses in the data set, and then compare the performance of different IP geolocation services through statistics. However, it was later found that the online IP geolocation service has a query speed limit, and the time for querying all IP addresses is particularly long. Finally, for the local IP geolocation services, we performed a query on all 1.2 million IP addresses and used the query results to calculate the query accuracy, and for the online IP geolocation services, we randomly extract ten thousand IP addresses from these IP addresses, and these ten thousand IP addresses are randomly selected from different ISPs in proportion.

We store the query results of different IP geolocation services in the database. In addition, we have performed multiple queries and recorded the time required for each IP geolocation service. We did data cleaning and normalization for these data. Finally, through data analysis, we compare different IP geolocation services from three aspects: data richness, query accuracy and query speed. The comparison results will be introduced in the following sections.

3.3 Data Richness Comparison

From Table 1, we can see that different services have different geographic information and different data field. The black dot indicates that this field is included and contains this information, the hollow dot indicates that it does not contain

Table 1. Data richness comparison.

Service name	Continent	Country	Province	City	District (county)	Coordinate	ISP
ip-api.com		●	●	●		●	●
ip.taobao.com		●	●	●		●	●
Baidu Map API		○	●	●	●	●	○
Sina IPLOOKUP		●	●	●	●		●
GeoLite2	●	●	●	●	○	●	●
ChunZhen database		○	○	○	○		●
IPPLUS360	●	●	●	●	●	●	●
ipip.net		●	●	●			

separate fields, but the information has included in other fields, the blank indicates that this information is not included.

As can be seen from Table 1, the main difference is whether it includes the three information: continent, district (county) and coordinate. There are two services containing continent information: GeoLite2 and IPPLUS360. There are 5 services that contain district and county information: Sina IPLOOKUP, Baidu Map API, IPPLUS360, ChunZhen database and GeoLite2.

In general, the IPPLUS360 and GeoLite2 service contain the most information.

All of these services provide Chinese query results, but only GeoLite2 and ip-api.com with a higher degree of internationalization provide query results in English and some other languages in addition to Chinese. GeoLite2 and ip-api.com are originally English services, supporting Chinese, but there are some problems. Our test found that some city's names are mixed in English and Chinese, query Chinese city names, some of the return information is in English, and some city names are wrong. There are some names that are not city names, sometimes use districts and streets name as city names, or use ancient names of city names.

3.4 Comparison of Query Accuracy

As we can see from Table 2 and Fig. 2, the query accuracy of different services is not the same, some information is not very different and some information is obviously different.

Table 2. Comparison of query accuracy.

Service name	ISP	Province	City	District (county)*	Number of samples
ip-api.com	98.59%	94.29%	53.67%	1.70%	10,000
ip.taobao.com	100%	100%	99.52%	0.00%	10,000
Baidu Map API	100%	99.84%	96.34%	0.00%	10,000
Sina IPLOOKUP	0.00%	100%	99.52%	0.00%	10,000
GeoLite2	85.69%	97.90%	14.84%	0.00%	1,209,835
ChunZhen database	27.47%	99.89%	75.09%	8.13%	1,209,835
IPPLUS360	98.20%	100%	90.52%	90.19%	1,209,835
ipip.net	-	100%	99.47%	-	1,209,835

The ISP query accuracy of each service is distinctly different. The ip.taobao.com and Baidu Map APIs have the highest ISP query accuracy, all reach 100%, and the ISP query accuracy rates of ip-api.com and IPPLUS360 are closely followed by 98.59% and 98.20%, respectively. The ISP query accuracy rates of GeoLite2 is 85.69%, and that of ChunZhen database is 27.24%. Although Sina IPLOOKUP supports ISP queries, its query results are all empty

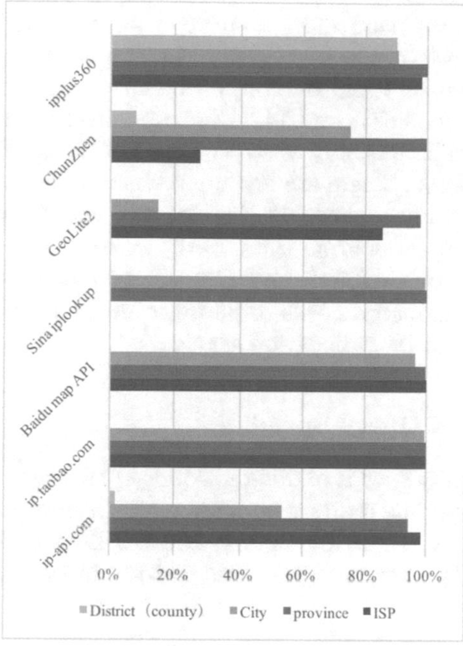

Fig. 2. Comparison of query accuracy.

so its ISP query accuracy rate is zero. The ipip.net free service does not provide ISP information, and may be available in the paid version.

There is little difference in the provincial query accuracy rates of different services. The provincial query accuracy rates of ip.taobao.com, Sina IPLOOKUP, IPPLUS360 and ipip.net has reached 100%. The provincial query accuracy rates of ChunZhen database and Baidu Map API is also close to 100%, 99.89% and 99.84% respectively, and the provincial query accuracy rates of GeoLite2 and ip-api.com are also 97.90% and 94.29% respectively.

The accuracy of cities in different services has already shown great differences. The accuracy of the cities of ip.taobao.com, Sina, ipip.net and Baidu Map API is high, reaching 99.52%, 99.52%, 99.47% and 96.34% respectively. The accuracy of the cities of IPPLUS360 is also 90.52%. The accuracy rate of the city of ChunZhen database and ip-api.com is 75.09% and 53.67% respectively, and the city's accuracy of GeoLite2 is only 14.84%.

Our test dataset does not contain district and county information, so we cannot confirm the accuracy of district queries. We only examine the coverage of district and county information, that is, how many of the results include district and county information when the city information is correct. The results show that most of the services cannot query district and county information, only IPPLUS360's district and county coverage reached 90.19%, the ChunZhen database address field also contains some district and county information, the coverage rate is 8.13%, and the ip-api.com also contains a small number of district and county information, the coverage rate is 1.70%.

In summary, the provincial query accuracy rates for each query service is very high and the difference is not significant, but the accuracy rates of cities and ISP for different services is very different. The city accuracy rates of ip-api.com and GeoLite2 are only 53.67% and 14.84%, respectively, while the ISP accuracy of the ChunZhen database is only 27.47%, Sina IPLOOKUP and ipip.net didn't find any ISP information. There are few query services that provide district and county information. Only ipplus360's district and county information coverage rate reaches more than 90%. From these data, we can also see that the accuracy of some information of GeoLite2 and ChunZhen database, which are widely used in research and industry, needs to be improved, and the developers need to reconsider the choice of IP geolocation service.

3.5 Comparison of Query Speed

We also tested the query speeds of different query platforms and compared them. We conducted an experimental test on a server with Intel(R) Core(TM) i7-6700 3.40 GHz processor, 8 GB DDR3 memory, 100 MB fiber network, and CentOS 7 operating system. The query program was written in the Python language.

Table 3. Comparison of query speed.

Service name	Number of samples	Average query time (s)	Query speed (ip/s)	Speed limit (ip/s)
ip-api.com	10,000	5938	2	2.5
ip.taobao.com	10,000	1958	5	10
Baidu Map API	10,000	1031	10	100
Sina IPLOOKUP	10,000	793	13	-
GeoLite2	1,209,835	512	2364	-
ChunZhen database	1,209,835	31	39012	-
IPPLUS360	1,209,835	135	8958	-
ipip.net	1,209,835	268	4512	-

As shown in Table 3, the query speed of local services can generally reach thousands or tens of thousands of IP addresses per second, the fastest is ChunZhen database, which queries 39,012 IP addresses per second, followed by IPPLUS360 with 8,958 IP addresses per second, followed by ipip.net with 4,512 IP addresses per second, GeoLite2 can also query 2,364 IP addresses per second.

The query speed of online services is relatively slow and generally has query restrictions. We convert these restrictions to the upper limit of query speed. Because of the impact of the implementation method and the network environment, the query speed of actual testing has a certain gap with the upper limit of speed.

Sina IPLOOKUP has the fastest query speed in online services, reaching 13 ip/s and we haven't found any relevant documents about the query limit; the Baidu Map API limits the querying of 100 IP addresses per second, the test speed is only 10 ip/s; the speed limit of ip.taobao.com is 10 ip/s, the actual test speed is 5 ip/s; ip-api.com limits 2.5 ip addresses per second, and the actual test speed is 2 ip/s.

Through experimental tests, we can see that the speed of local query is obviously higher than that of online query.

4 Conclusion

We use the IP address data of a city as a test set to conduct comparative tests on several mainstream IP address geolocation services. Online geolocation services include: ip-api.com, ip.taobao.com, Baidu Map API, and Sina IPLOOKUP. Local services include: GeoLite2, ChunZhen database, IPPLUS360 and ipip.net. We conducted an experimental comparison of these ip location services, we first compare the data richness of these query services, and then compare the data accuracy and query speed through experimental tests. It is found that the accuracy of online geolocation services is generally higher than that of local services, but the query speed is generally lower than that of local services. Specifically, different geolocation service accuracy and query speed are different, and the difference is obvious, and few geolocation services can provide county-level information. We also found that the accuracy of some information of GeoLite2 and ChunZhen database, which are widely used in research and industry, needs to be improved, and IPPLUS360 is outstanding in terms of data richness and query speed, Ipip.net is more accurate. These results can be used as references for researchers with similar needs.

References

1. Huffaker, B., Fomenkov, M., Claffy, K.: Geocompare: a comparison of public and commercial geolocation databases. In: Proceedings of NMMC, pp. 1–12 (2011)
2. Arif, M.J., Karunasekera, S., Kulkarni, S., Gunatilaka, A., Ristic, B.: Internet host geolocation using maximum likelihood estimation technique. In: 2010 24th IEEE International Conference on Advanced Information Networking and Applications (AINA), pp. 422–429. IEEE (2010)
3. Muir, J.A., Oorschot, P.C.V.: Internet geolocation: evasion and counterevasion. ACM Comput. Surv. (CSUR) **42**, 4 (2009)
4. Guo, C., Liu, Y., Shen, W., Wang, H.J., Yu, Q., Zhang, Y.: Mining the web and the internet for accurate IP address geolocations. In: INFOCOM 2009, IEEE, pp. 2841–2845. IEEE (2009)
5. Laki, S., Mátray, P., Hága, P., Csabai, I., Vattay, G.: A model based approach for improving router geolocation. Comput. Netw. **54**, 1490–1501 (2010)
6. Zhao, F., Luo, X., Liu, F.: Research on cyberspace surveying and mapping technology. Chin. J. Netw. Inf. Secur. **2**(9), 1–11 (2016)

7. Zhou, Y., Xu, Q., Luo, X., Liu, F., Zhang, L., Hu, X.: Research on definition and technological system of cyberspace surveying and mapping. Comput. Sci. **2018**(5), 1–7 (2018)
8. CAIDA. http://www.caida.org/projects/cybersecurity/geolocation/. Accessed 8 Nov 2018
9. CAIDA: Center for Applied Internet Data Analysis. http://www.caida.org/home/. Accessed 8 Nov 2018
10. MaxMind. https://www.maxmind.com/en/company. Accessed 8 Nov 2018
11. Xu, W., Tao, Y., Guan, X.: The landscape of industrial control systems (ICS) devices on the internet. In: 2018 International Conference on Cyber Situational Awareness, Data Analytics and Assessment (Cyber SA), pp. 1–8. IEEE (2018)
12. GeoLite2 Free Downloadable Databases MaxMind Developer Site. https://dev.maxmind.com/geoip/geoip2/geolite2/. Accessed 8 Nov 2018
13. ChunZhen. http://www.cz88.net/. Accessed 8 Nov 2018
14. IPPLUS360. http://www.IPPLUS360.com/. Accessed 8 Nov 2018
15. IPIP.NET. https://www.ipip.net/. Accessed 8 Nov 2018
16. IP Geolocation API. http://ip-api.com/docs/#. Accessed 8 Nov 2018
17. ip.taobao.com. http://ip.taobao.com/. Accessed 8 Nov 2018

Integrating Multiple Feature Descriptors for Computed Tomography Image Retrieval

Xiaoqin Wang, Huadeng Wang(✉), Rushi Lan, and Xiaonan Luo

Guangxi Key Laboratory of Intelligent Processing of Computer Images and Graphics, Guilin University of Electronic Technology, Guilin 541004, China
4469206@qq.com

Abstract. Integrating multiple feature descriptors has recently shown to give excellent results for image retrieval. In this paper, we integrate multiple feature descriptors for computed tomography (CT) image retrieval, whose descriptors include the principal components descriptor, scale invariant feature transform descriptor and roberts gradient descriptor. First, we describe the retrieving image based on principal components descriptor, which is a technology of reducing the dimensions and extracting principal component. Second, we extract the scale invariant feature transform descriptor based on scale invariant feature transform algorithm. Third, the roberts gradient descriptor is obtained by roberts operator. Finally, we integrate principal components descriptor, scale invariant feature transform descriptor and roberts gradient descriptor into a retrieval vector to represent the CT image. Experimental results based on a subset of EXACT09-CT, named CASE23 and TCIA-CT show that our approach significantly outperforms the methods of the related works.

Keywords: CT image retrieval · Multiple feature descriptors · PCA · SIFT · Roberts operator

1 Introduction

With the development of medical imaging technology and the spread of medical information networks, hospital produces a large number of medical images containing physiological, pathological and anatomical information of patients every day. These images are very important for doctors to conduct clinical diagnosis, disease tracking, surgical planning, prognosis research, and differential diagnosis. The traditional medical image retrieval methods are based on keywords, but its

This work was supported in part by the National Natural Science Foundation of China (Nos. 61762028, 61772149, U1701267, and 61320106008), and by Guangxi Colleges and Universities Key Laboratory of Intelligent Processing of Computer Images and Graphics (No. GIIP201703).

C.-N. Yang et al. (Eds.): SICBS 2018, AISC 895, pp. 209–220, 2020.
https://doi.org/10.1007/978-3-030-16946-6_17

manual annotation has strong subjectivity. It is difficult for text to fully express the rich semantic information contained in the image.

In recent years, content-based image retrieval (CBIR) [1] has been one of the most vivid research topic in the field of image processing. Many CBIR methods have been achieved accurate and efficient image retrieval. The CBIR technology also have been applied in the field of medical image processing, which can be named as content-based medical image retrieval (CBMIR) [16]. CBMIR extracts grayscale shape, texture, topology, and high-level semantic features to form the feature vector, which can describe the medical image content. And the objective basis for indexing and matching criteria to retrieve the desired image is based on the feature vector. Then CBMIR technology has become a very chippy research direction in the field of biomedical engineering.

Many algorithms based on CBMIR have been proposed. Local binary pattern (LBP) [22] is fundamental properties of local image texture and their occurrence histograms are proven to be a very powerful texture feature. For an medical image, LBP extracts local texture feature by calculating the differences between the center pixel and its surrounding ones, and then converts the intensity of center pixel to another decimal number by encoding the sign of these differences. Some representative methods of LBP model have been proposed. For example, local ternary co-occurrence patterns (LTCoP) [17] encodes the co-occurrence of similar ternary edges which are calculated based on the gray values of center pixel and its surrounding neighbors. Local mesh patterns (LMeP) [18] encodes the relationship among the surrounding neighbors for a given referenced pixel in an image. Spherical symmetric 3D local ternary patterns (SS-3D-LTP) [19] and local wavelet pattern (LWP) [3] are also based on the framework of LBP. Another model of CBMIR is based on transform. The Gabor [24] and wavelet transform [7] have been widely applied to medical image retrieval. And the scattering transform, which is an invariant of convolutional neural networks [14] and provides translation invariant representations of an image, has been applied to derive features of medical images. And the histogram of compressed scattering coefficients (HCSC) [11] has been achieved satisfactory performance in medical image retrieval. However, the above CBMIR methods only extract a single feature descriptor of the medical image and cannot fully describe the semantic information of the image.

In this paper, we integrate multiple feature descriptors for computed tomography (CT) image retrieval. Inspired by the feature fusion [6], which combines the texture, localisation and color features to represent an image, and the local descriptor of quaternionic weber local descriptor (QWLD) [12] and quaternion-Michelson descriptor (QMD) [10], our proposed approach first divides an CT image into non-overlapping sub-regions based on region-based segmentation method [21], and then extracts the principal components descriptor (PCD) based on principal components analysis (PCA) [15] algorithm. Second, we will use the scale invariant feature transform (SIFT) [4] to extract scale invariant feature transform descriptor (SIFTD). Third, the roberts gradient descriptor (RGD) based on roberts operator is obtained. Finally, we will integrate the three feature descriptors (PCD, SIFTD and RGD) to describe a CT image and to be retrieval vector. Compared with the existing feature descriptors, our proposed approach takes account of more semantic characteristics of CT images. Experimental comparison results have demonstrated that our proposed approach obtains better performance than other related work.

The rest of this paper is organized as follows: Sect. 2 introduces our computed tomography image feature descriptors in detail, and Sect. 3 shows the experimental results. Section 4 draws the conclusion.

2 Computed Tomography Image Descriptors

In this section, we will introduce multiple feature descriptors to represent the CT image, which are used in our proposed approach, namely principal components descriptor (PCD), scale invariant feature transform descriptor (SIFTD) and roberts gradient descriptor (RGD). And Fig. 1 shows the main stages of our approach that includes the processes of extracting the feature descriptors and integrating the multiple feature descriptors.

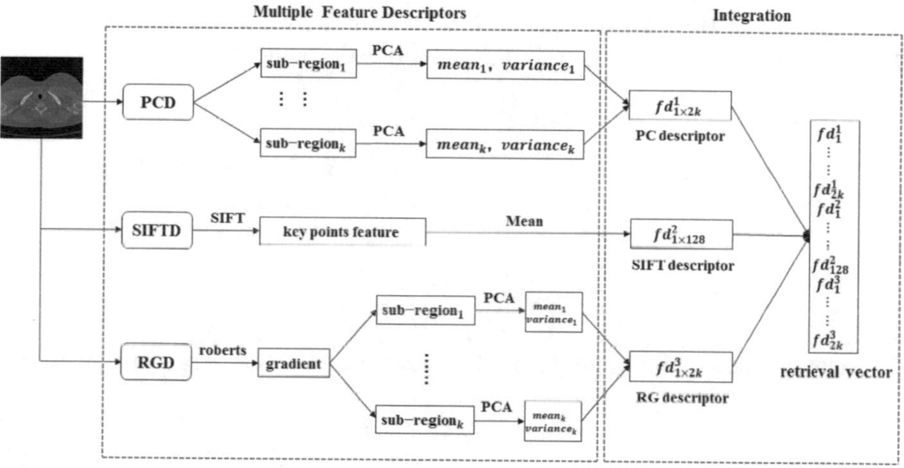

Fig. 1. Integrating multiple feature descriptors for an CT image. The CT image can be describe by PCD, SIFTD and RGD, respectively, and then the PC descriptor, SIFT descriptor and RG descriptor are integrated into a retrieval vector.

2.1 Principal Components Descriptor (PCD)

Principal components analysis (PCA) algorithm [20] is a multivariate statistical analysis method that selects fewer important variables from multiple variables by linear transformation. When dealing with signals of multivariate and high-dimensional vectors, it is often desirable to obtain a larger amount of information with fewer low-dimensional vectors. Under normal circumstances, the signals contained multivariate and high-dimensional vectors also has a certain correlation between the two signals. And the information reflected by the two vectors is overlapping. The task of PCA is to represent the original vector with a small number of new vectors that do not overlap in each other. And the information contained in them will be as much as the original information. PCA algorithm have widely used in various applications, especially in the field of computer vision, such as image processing and pattern recognition.

In this work, we will first use the PCA algorithm to obtain the CT image feature descriptor, namely principal components descriptor (PCD). PCA algorithm is based on the $K-L$ transform, and the specific steps of extracting the PCD are as follows.

For an CT image, we will first segment the image into some non-overlapping sub-regions based on the region-based segmentation method. Let I denotes an CT image. After the segmentation, the I can be divided into k sub-regions, which are non-overlapping, namely $I = \{x^1, x^2, \ldots, x^k\}$. After that, we will extract PCD for each sub-region, and the steps of PCD can be introduced as follow.

We will select some CT images in each category in our databases, and then divide them into many sub-regions to compose the training set. Convert each sub-region x^i of the training sample library into a column vector, which is the n-dimensional vector and the n is the pixel of sub-image, by:

$$x^i = [x^i_1, x^i_2, \ldots, x^i_n]^T \tag{1}$$

The training set is $X = \{x^1, x^2, \ldots, x^p\} \in R^{n \times p}$, where p is the number of training sample. Average all the columns of X, that can get:

$$\mu = \frac{1}{p}\sum_{i=1}^{p} x^i, \overline{x}^i = x^i - \mu \tag{2}$$

That centralize all training samples. That all of $\overline{x}^i$ can compose an matrix $\overline{X}_{n \times p}$.

$$\overline{X} = [\overline{x}^1, \overline{x}^2, \ldots, \overline{x}^p] \tag{3}$$

Multiply the device matrix of matrix $\overline{X}$ by itself can obtain a covariance matrix Ω:

$$\Omega = \overline{X}^T \overline{X} \tag{4}$$

Decompose covariance matrix Ω and obtain K non-zero eigenvalues. Then the all of eigenvalues will be composed diagonal matrix $\Lambda = diag[\lambda_1, \lambda_2, \ldots, \lambda_d]$ and $\lambda_1 \geq \lambda_2 \geq \ldots, \lambda_d$ corresponding to the eigenvector matrix $Q = [q_1, q_2, \cdots, q_d]$. After that, we can obtain the feature space by Eq. 5:

$$U = XQ\Lambda^{-1/2} \tag{5}$$

Then calculate the projection of a vector x^i on the feature subspace:

$$\widehat{x}^i = U^T \overline{x}^i \tag{6}$$

Where $\overline{x}^i$ is the vector after centralizing. And for testing sample, we first centralize as follow:

$$\overline{y}^i = y^i - \mu, \mu = \frac{1}{p}\sum_{i=1}^{p} x^i \tag{7}$$

Then, calculate its projection on the feature face subspace:

$$\widehat{y}^i = U^T \overline{y}^i \tag{8}$$

Finally, we can obtain the vector $\widehat{y}^i$, which reduce the dimension and can express primary information of each sub-region. Then we will calculate the mean and variance of $\widehat{y}^i$, and combine the two value to represent the ith sub-region. Then we will combine all of an mean and variance of all sub-regions to represent feature descriptor (fd^1) of an image: $fd^1_{1\times(2k)}$. $fd^1_{1\times(2k)}$ (k is the number of sub-region of a CT image) is can be called as the principal components descriptor (PCD), which is an global descriptor for a CT image.

2.2 Scale Invariant Feature Transform Descriptor (SIFTD)

Scale invariant feature transform (SIFT) [23] is a computer vision algorithm. It is used to detect and describe local features of images. It finds extreme points in spatial scale and extracts its position, scale, and rotation invariants. SIFT is local feature descriptor and has been widely used in many applications, such as object recognition, robot map perception and navigation, image stitching, 3D model creation, gesture recognition and so on.

In this work, we use the SIFT algorithm to describe the CT image feature descriptor, namely scale invariant feature transform descriptor (SIFTD). The whole algorithm is divided into the following parts. The first step is to construct DOG scale space [8]. The second step is to detect DoG scale space extreme points and remove bad feature points. The third step is that assign a 128-dimensional direction parameter to the feature point. Each key point contains three pieces of information: location, scale, and direction. Finally, we can obtain the SIFT feature matrix, which can be defined as: $sf_{N\times128}$, where the N is the num of keypoints and the 128 is the dimension of each keypoint. More details on implementing the SIFT algorithm can be found in [9]. To achieve our feature descriptor, namely SIFTD, we finally calculate the mean of all N keypoints to obtain the SIFTD, namely the feature descriptor vector $fd^2_{1\times128}$ of an CT image.

SIFTD integrates the feature information of all keypoints, which extract by SIFT algorithm. Although SIFT is a local descriptor of the image, SIFTD is a global descriptor of the image. SIFTD also save some characteristics of SIFT. For example, SIFTD feature has large amount of feature information and can be easily combined with other forms of feature vectors.

2.3 Roberts Gradient Descriptor (RGD)

The importance of edge in image processing is self-evident. At present, the most advanced technology of Artificial Intelligence is deep learning, and the features required for deep learning model in image are starting from the edge and constantly forming a higher level feature description. roberts operator [5] is a classic edge detection operator.

In this work, we will use the roberts gradient to describe each CT image. The roberts operator uses two 2×2 operator templates to calculate the gradient. The two 2×2 operator templates are showed in Eq. 9. Let I represent a CT image, and the roberts gradient can be introduced as follow:

$$R_x = \begin{bmatrix} 1 & 0 \\ 0 & -1 \end{bmatrix}, R_y = \begin{bmatrix} 0 & 1 \\ -1 & 0 \end{bmatrix} \tag{9}$$

$$G_x = R_x * I; G_y = R_y * I \tag{10}$$

where $*$ denotes the 2-D convolution operation, and G_x and G_y denote the horizontal gradients and vertical gradients, respectively. Then the strength of an edge in gray level within the 2×2 neighborhood pixels can be denoted as:

$$G = |G_x| + |G_y| \tag{11}$$

Where G is the gradient vector of roberts.

To obtain the roberts gradient descriptor (RGD), we first use the roberts operator to calculate the roberts gradient vector G. Then the CT image is divide into k sub-regions. For each sub-region, we will extract key feature information by the PCA algorithm. Finally, all the sub-region features extracted by the above steps are compressed into the mean and variance, and all the means and variances for an image are combined into a feature descriptor vector, namely $fd^3_{1\times(2k)}$, where the k is the num of sub-regions for each CT image.

After the extraction of the above three feature descriptors, we will integrate the vectors of $fd^1_{1\times(2k)}$, $fd^2_{1\times128}$ and $fd^3_{1\times(2k)}$ into the retrieval feature rv. And our proposed approach that integrates multiple feature descriptors for CT image retrieval can be achieved as follow:

3 Experimental Results

In this section, we will first introduce the databases and evaluation criteria to verify our proposed approach. After that, the computed tomography image retrieval results are presented.

3.1 Database and Evaluation Criteria

We evaluate our CT image retrieval approach based on EXACT09-CT database [3] and TCIA-CT [2]. EXACT09-CT database ranges from clinical dose to ultra low dose scans, healthy volunteers to patients with severe lung disease and full inspiration to full expiration. All images are stored as anonymized DICOM slices. This database contains 675 chest CT images, and each image is a size of 512×512. All those images can be grouped into 19 categories. Figure 2 illustrates the example image of this database with one CT image of each category.

Algorithm 1. Integrating multiple feature descriptors of our proposed method

Input: Training set of CT images, PCA, SIFT and roberts operator
Output: Each CT image retrieval vector rv.
Procedure:

1. : Extract the PCD based on PCA algorithm and region-based segmentation method;
2. : Extract SIFTD based on the SIFT algorithm;
3. : Extract RGD based on roberts gradient and PCA algorithm;
4. : Integrate the PCD, SIFTD and RGD to form the retrieval vector rv of an CT image.

TCIA-CT database is a large-scale public database containing CT images of common tumors (lung cancer, prostate cancer, etc.) and corresponding clinical information (treatment details, genes, pathology, etc.). The images modalities include MRI, CT, etc. TCIA-CT database is composed by 604 CT images. These images are grouped into 8 categories and the numbers of each category are 75, 50, 58, 140, 70, 92, 78 and 41. The Fig. 3 is showed the example images of this database with one image of each category.

In order to form a comparative experiment with the previous algorithm, we use three evaluation criteria [13] to evaluate the performance of our proposed approach, namely average retrieval precision (P) and average retrieval rate (R), and the F_{score}, which can be obtained by:

$$F_{score} = \frac{2 \times P \times R}{P + R} \tag{12}$$

Where F_{score} is a combination of both P and R. In all experiments, each CT image is chosen as the query image and matched with the rest images. We will select the top s largest similar value FI, which can be defined as Eq. 13:

$$FI(F_1, F_2) = \sum_{i=1} min(F_1(i), F_2(i)) \tag{13}$$

Where the larger $FI(F_1, F_2)$ indicates more similarity between F_1 and F_2, and F_1 and F_2 are two CT image feature descriptors obtained by our proposed approach.

3.2 Computed Tomography Image Retrieval Results

In all experiments we follow the same feature extraction processes. First, we will extract the $fd^1_{1\times 2k}$, where the $k = 64$. Then the PCD of an image is the size of 1×128 (also the same as SIFTD and RGD). The $fd^1_{1\times 2k}$ computational process can be referred part the second section. Second, the $fd^2_{1\times 128}$ is extracted by SIFT and calculating the mean and variance of all its keypoints. Third, we extract $fd^3_{1\times 2k}$ by roberts gradient algorithm and PCA algorithm, and the $k = 64$ (each CT image is divided into 8×8 sub-regions). Finally, we will integrate the

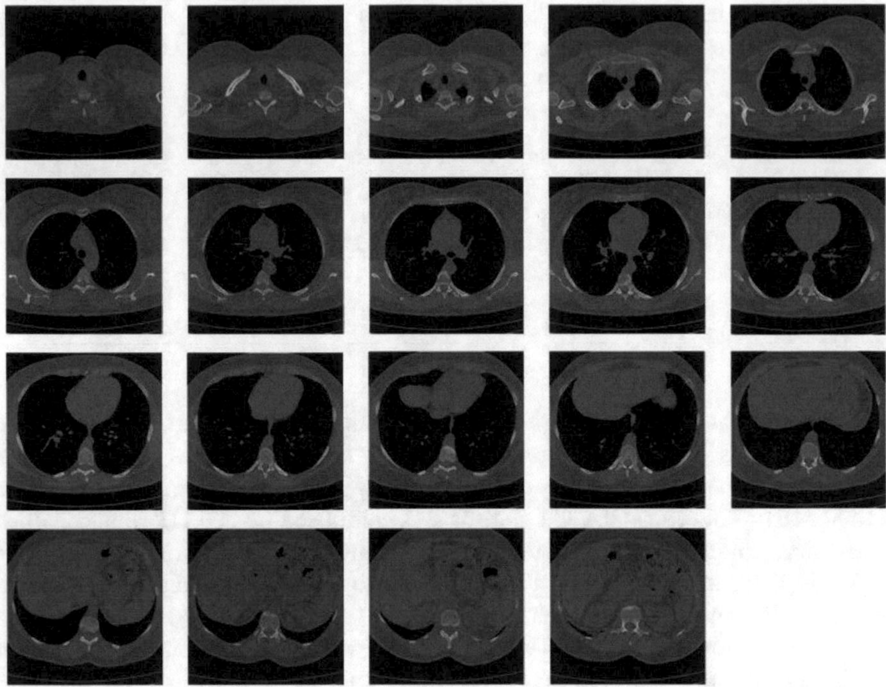

Fig. 2. Some CT images of EXACT09-CT database: each one is selected from the corresponding category.

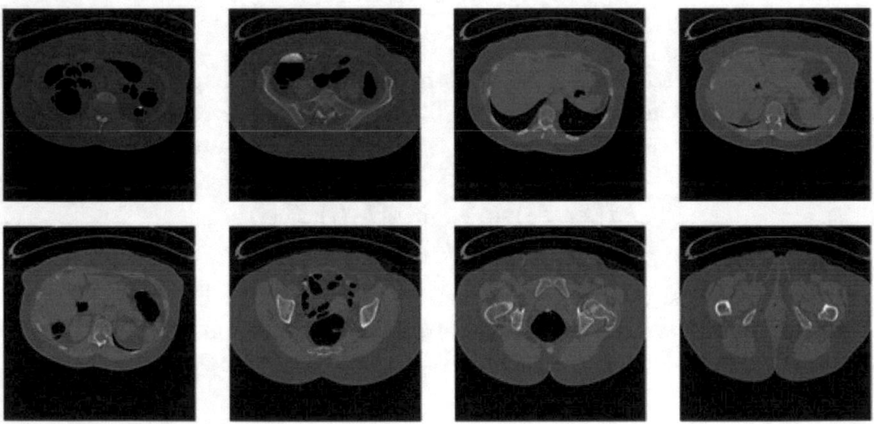

Fig. 3. Some images of TCIA-CT database: each one represents a corresponding category.

descriptors PCD, SIFTD and RGD of an image into a retrieval vector, which can be named as $rv = [fd^1_{1\times 2k}, fd^1_{1\times 128}, fd^3_{1\times 2k}]$ and $rv \in R^{1\times 384}$, where 384 is the sum of $2 \times k$, 128 and $2 \times k$.

Some experimental parameters are introduced as follow. We select the energy, which the cumulative contribution rates are 84% and 86% in the database EXACT09-CT and TCIA-CT, respectively. When we extract descriptors of PCD, SIFTD and RGD, two images of each category in the database are randomly selected. And then we segment each image into 8×8 sub-regions to compose the training set. In the CT image retrieval step, we choose the top $s = 10$ largest similar value $FI(F_1, F_2)$ to be the retrieval results and calculate evaluation criteria P, R and F_{score}.

Through the above introduction, we assess our proposed approach based on the EXACT09-CT database and TCIA-CT database. The two experiments of computed tomography image retrieval results are showed in the Tables 1 and 2, respectively.

Table 1. Comparison results of different methods based on the EXACT09-CT.

Methods	P	R	F_{score}
LBP	0.6503	0.1951	0.3002
LTP	0.6209	0.1854	0.2855
LDP	0.5440	0.1619	0.2495
LTrP	0.5782	0.1729	0.2662
LTCoP	0.7348	0.2216	0.3405
LMeP	0.6323	0.1891	0.2911
SS-3D-LTP	0.6700	0.2009	0.3091
LWP	0.8300	0.2487	0.3827
HCSC	0.9150	0.2883	0.4384
Our method	**0.9269**	**0.2917**	**0.4438**

The retrieval performance of the compared algorithms in this paper are presented in the Table 1. Table 1 shows that retrieval results obtained by our approach yield highest values of P, R and F_{score}. We observe that our proposed approach obtains better results than HCSC method, which bases on scattering transform. And our proposed method has a certain improvement in performance over HCSC, i.e., $\{1.3\%, 1.18\%, 1.22\%\}$, corresponding criteria of P, R and F_{score}.

Table 2 presents the retrieval results of P, R and F_{score} for the compared methods based on TCIA-CT database. And we can visually see that the results of LBP results are the worst than previous method. Our proposed approach improve the $\{43.1\%, 49.59\%, 48.76\%\}$ than LBP method in P, R and F_{score}, respectively.

According to the above experimental results, we can verify that our proposed approach can obtain better performance, using our evaluation criteria. That is the reasons: First, we describe CT image by PCD method, which can extract the feature after projection transformation. The PCD extracts the most of information of image and can lower the dimension of image feature. Second, we describe

Table 2. Comparison results of different methods based on the TCIA-CT.

Methods	P	R	F_{score}
LBP	0.6691	0.0974	0.1700
LTP	0.7183	0.1033	0.1806
LDP	0.6906	0.1005	0.1755
LTrP	0.7469	0.1095	0.1910
LTCoP	0.7440	0.1092	0.1904
LMeP	0.7371	0.1077	0.1879
SS-3D-LTP	0.8054	0.1171	0.2045
LWP	0.8840	0.1309	0.2280
HCSC	0.9512	0.1452	0.2520
Our method	**0.9572**	**0.1457**	**0.2529**

CT image using the SIFTD, which is based on SIFT algorithm. Although the SIFT is local descriptor, the SIFTD is the mean of all key point. So it can be considered as a global descriptor. Third, we describe image by RGD, which is based on roberts gradient and PCA algorithm. It describes the edge information of an image. Finally, in order to fully describe image feature, we integrate the above descriptors, namely PCD, SIFTD and RGD, into a retrieval vector to represent CT image feature for image retrieval. Our experimental results fully verify the feasibility of our proposed approach.

4 Conclusion

This paper proposes the method of integrating multiple feature descriptors for computed tomography image retrieval, which is main based on PCA, SIFT and roberts operator. To comprehensively explore the descriptors, our proposed approach first extracts principal components descriptor (PCD) via PCA algorithm and region-based segmentation method. Second, we extract SIFTD based on SIFT algorithm. Third, the RGD is produced by roberts gradient and PCA algorithm. Finally, we integrate all the descriptors that are obtained by our approach into a retrieval vector. Experimental results obtained by experimenting on the EXACT09-CT and TCIA-CT database show that our proposed approach can achieve better performance than HCSC method and other contrast algorithms.

References

1. Akgül, C.B., Rubin, D.L., Napel, S., Beaulieu, C.F., Greenspan, H., Acar, B.: Content-based image retrieval in radiology: current status and future directions. J. Digit. Imaging **24**(2), 208–222 (2011)
2. Clark, K., Vendt, B., Smith, K., Freymann, J., Kirby, J., Koppel, P., Moore, S., Phillips, S., Maffitt, D., Pringle, M., et al.: The cancer imaging archive (tcia): maintaining and operating a public information repository. J. Digit. Imaging **26**(6), 1045–1057 (2013)
3. Dubey, S.R., Singh, S.K., Singh, R.K.: Local wavelet pattern: a new feature descriptor for image retrieval in medical CT databases. IEEE Trans. Image Process. **24**(12), 5892–5903 (2015)
4. Giveki, D., Soltanshahi, M.A., Montazer, G.A.: A new image feature descriptor for content based image retrieval using scale invariant feature transform and local derivative pattern. Opt. Int. J. Light Electron. Opt. **131**, 242–254 (2017)
5. Günen, M.A., Atasever, Ü.H., Beşdok, E.: A novel edge detection approach based on backtracking search optimization algorithm (BSA) clustering. In: 2017 8th International Conference on Information Technology (ICIT), pp. 116–122. IEEE (2017)
6. Howarth, P., Yavlinsky, A., Heesch, D., Rüger, S.: Medical image retrieval using texture, locality and colour. In: Workshop of the Cross-Language Evaluation Forum for European Languages, pp. 740–749. Springer (2004)
7. Khatami, A., Khosravi, A., Nguyen, T., Lim, C.P., Nahavandi, S.: Medical image analysis using wavelet transform and deep belief networks. Expert Syst. Appl. **86**, 190–198 (2017)
8. Kitagawa, M., Shimizu, I., Sara, R.: High accuracy local stereo matching using dog scale map. In: 2017 Fifteenth IAPR International Conference on Machine Vision Applications (MVA), pp. 258–261. IEEE (2017)
9. Lakshmi, K.D., Vaithiyanathan, V.: Image registration techniques based on the scale invariant feature transform. IETE Tech. Rev. **34**(1), 22–29 (2017)
10. Lan, R., Zhou, Y.: Quaternion-michelson descriptor for color image classification. IEEE Trans. Image Process. **25**(11), 5281–5292 (2016)
11. Lan, R., Zhou, Y.: Medical image retrieval via histogram of compressed scattering coefficients. IEEE J. Biomed. Health Inform. **21**(5), 1338–1346 (2017)
12. Lan, R., Zhou, Y., Tang, Y.Y.: Quaternionic weber local descriptor of color images. IEEE Trans. Circuits Syst. Video Technol. **27**(2), 261–274 (2017)
13. Li, Z., Zhang, X., Müller, H., Zhang, S.: Large-scale retrieval for medical image analytics: a comprehensive review. Med. Image Anal. **43**, 66–84 (2018)
14. Lu, H., Li, B., Zhu, J., Li, Y., Li, Y., Xu, X., He, L., Li, X., Li, J., Serikawa, S.: Wound intensity correction and segmentation with convolutional neural networks. Concurrency Comput. Pract. Experience **29**(6), e3927 (2017)
15. Memon, M.H., Li, J.-P., Memon, I., Arain, Q.A.: Geo matching regions: multiple regions of interests using content based image retrieval based on relative locations. Multimedia Tools Appl. **76**(14), 15 377–15 411 (2017)
16. Müller, H., Michoux, N., Bandon, D., Geissbuhler, A.: A review of content-based image retrieval systems in medical applicationsclinical benefits and future directions. Int. J. Med. Inform. **73**(1), 1–23 (2004)
17. Murala, S., Wu, Q.J.: Local ternary co-occurrence patterns: a new feature descriptor for MRI and CT image retrieval. Neurocomputing **119**, 399–412 (2013)
18. Murala, S., Wu, Q.J.: Local mesh patterns versus local binary patterns: biomedical image indexing and retrieval. IEEE J. Biomed. Health Inform. **18**(3), 929–938 (2014)

19. Murala, S., Wu, Q.J.: Spherical symmetric 3D local ternary patterns for natural, texture and biomedical image indexing and retrieval. Neurocomputing **149**, 1502–1514 (2015)
20. Naidu, V., Raol, J.R.: Pixel-level image fusion using wavelets and principal component analysis. Defence Sci. J. **58**(3), 338 (2008)
21. Niu, S., Chen, Q., De Sisternes, L., Ji, Z., Zhou, Z., Rubin, D.L.: Robust noise region-based active contour model via local similarity factor for image segmentation. Pattern Recogn. **61**, 104–119 (2017)
22. Ojala, T., Pietikainen, M., Maenpaa, T.: Multiresolution gray-scale and rotation invariant texture classification with local binary patterns. IEEE Trans. Pattern Anal. Mach. Intell. **24**(7), 971–987 (2002)
23. Velmurugan, K.: A survey of content-based image retrieval systems using scale-invariant feature transform (sift). In: International Journal of Advanced Re-search in Computer Science and Software Engineering, vol. 4 (2014)
24. Zhang, G., Ma, Z.-M.: Texture feature extraction and description using gabor wavelet in content-based medical image retrieval. In: International Conference on Wavelet Analysis and Pattern Recognition. ICWAPR 2007, vol. 1, pp. 169–173. IEEE (2007)

Automatic Forgery Localization via Artifacts Analysis of JPEG and Resampling

Hongbin Wei[1], Heng Yao[1,2(✉)], Chuan Qin[1], and Zhenjun Tang[2]

[1] School of Optical-Electrical and Computer Engineering, University of Shanghai for Science and Technology, Shanghai 200093, China
hyao@usst.edu.cn

[2] Guangxi Key Lab of Multi-source Information Mining and Security, Guangxi Normal University, Guilin 541004, China

Abstract. With the availability of highly sophisticated editing tools, the authenticity of digital images has now become questionable. The level of image tampering is getting higher and higher, and the tampering procedures become more and more complicated. To recognize the tampering area of the original image, the tampered image is usually executed a series of post-processing. This behavior has greatly increased the difficulty of forgery detection. In this paper, a blind JPEG image forgery detection and localization technique based on JPEG and resampling artifacts analysis is proposed. The process of tampering is to first tamper with JPEG images by bitmaps. Then original JPEG image and tampered area are manipulated by a series of operations, that is, the image is enlarged and then saved as JPEG. A novel tampering localization method is presented based on resampling and JPEG blockness artifacts. Theoretical analysis and experimental results show that the proposed method can effectively identify and locate the tampered region of a spliced image with a JPEG-resampling-JPEG operation chain.

Keywords: Digital forensics · JPEG compression · Resampling effect · Operation chain

1 Introduction

The widespread use of digital imaging devices provide convenience for the acquisition of digital images and it is easier for the users to modify the content of those images due to the rapid development of image editing software. Some people disseminate some elaborately forged images with various malicious intents. These images bring misleading information and even dangerous information to the whole society. Therefore, it is very important to protect and identify the authenticity and integrity of the digital images.

Image forensics can be generally divided into two types: active [1, 2] and passive [3, 4] approaches. Active forensic technology generally includes digital watermarking [5, 6] and digital signature [7, 8], which can achieve the functions of anti-counterfeiting, tamper-proof and copyright protection. However, the main limitation of active technique is that the authentication information deserve to be embedded into

C.-N. Yang et al. (Eds.): SICBS 2018, AISC 895, pp. 221–234, 2020.
https://doi.org/10.1007/978-3-030-16946-6_18

the host image in advance. The passive forensic methods are put forward in order to overcome this limitation. It recognizes whether the image has been tampered or not merely according to the inconsistency of extracted features due to the tampering operation. Passive forensics are commonly used to detect image source devices [9, 10], resampling artifacts [11–13], color filter array (CFA) artifacts [14, 15], noise inconsistencies [16, 17], perspective inconsistency [18], and JPEG compression [19–21], etc.

JPEG compression is widely used in the field of image processing due to its good compression efficiency and reconstruction quality. Therefore, image saved as JPEG format are often considered in many image forensic methods. One of the most significant problem in digital image forensics is how many times the image is compressed. To recover the history of JPEG compression, first, it is necessary to check whether a bitmap image has experienced at least one JPEG compression [22, 23]. In addition, for a JPEG image, although its compression history can be obtained by reading its JPEG header file, whether this image has undergone two or even multiple compression is unknown. So far, many effective methods, e.g., [24–27], have been proposed to identify the history of JPEG compression.

Note that detection of pictures which have been JPEG compressed still does not sufficiently prove that the pictures have experienced malicious tampering. In general, if some parts of an image have been compressed from different source in JPEG, the image can be considered as a spliced image. Therefore, location detection in tampered area in JPEG is also particularly important. In [28], Farid proposed a method to detect tampered areas of double compressed images whose primary compression quality is lower than that of the rest of the image. However, this method requires the use of each possible main quality factor to generate differential images. In [29], Wang et al. detected local tampering area of tampered images based on probability distribution of different DCT coefficients in tampered and invariant regions. In [30], Iakovidou et al. proposed a new image forgery detection method by locating grid alignment anomalies. The method can evaluate multiple grid positions with respect to a fitting function, and areas of lower contribution are identified as grid discontinuities and possibly tampered areas.

Now consider a new problem, when people tamper with images, they often conduct more than one tampering operations, such as the combination of JPEG and resampling technology. Considering the cases where multiple operations may be involved, Chen et al. [31] proposed a method for JPEG-resampling-JPEG operation chain detection. This method analyzes pixels in the operation chain and transform block artifacts (TBAG) in the discrete cosine transform (DCT) domain to identify the resampled JPEG images as well as the JPEG images undergone resampling and then JPEG recompression.

Driven by the complexity of tampering operations, in this paper, we devote to extending tampering detection tools by proposing a new approach. In this paper, we first analyze the interaction between JPEG compression and resampling. Then we solve the confusion problem of resampling factor estimation in JPEG image by using the 8×8 grid which is unique to JPEG image. Finally, we optimize the detection results of the proposed method by a comprehensive operation chain analysis. The experimental results demonstrate the robustness of the method in different tampering scenarios.

The rest of this paper is organized as follows. In Sect. 2, the artifacts of JPEG compression and resampling are presented. The framework for detecting tampered regions in complex situations is proposed in Sect. 3. In Sect. 4, we present the performance results of our method for detecting tampered regions. Section 5 concludes this paper.

2 Analysis of JPEG and Resampling Artifacts

JPEG recompression and Resampling are both common operations during the image editing process. In this section, we briefly introduce the artifacts of JPEG compression and resampling.

2.1 JPEG Compression Model

2.1.1 JPEG Compression Process

JPEG compression is a block-based compression method. The image is first divided into 8 × 8 non-overlapping pixel blocks. As shown in Fig. 1, the image block B is first transformed from the spatial domain to the frequency domain to obtain its corresponding DCT coefficient block D^B. Next the D^B is uniformly quantized to generate quantized coefficients C_q by the 8 × 8 quantization matrix Q. The quantization matrix Q is independently defined by an integer quality factor q (q = 1, 2, …, 99, 100). Each element in the quantization matrix Q is called quantization step Q_q.

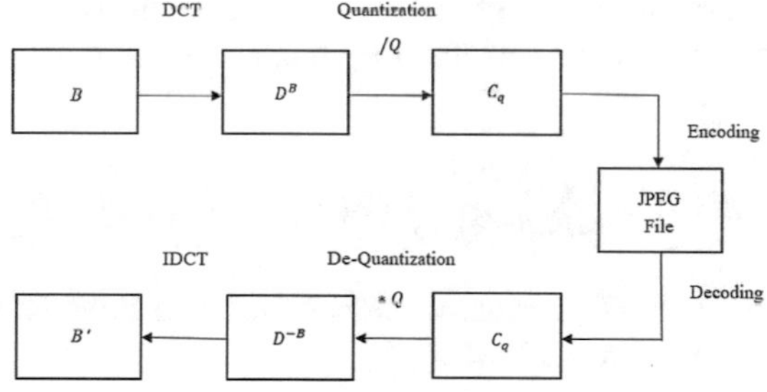

Fig. 1. JPEG compression and decompression

$$C_q = round\left(\frac{D^B}{Q}\right) \tag{1}$$

Note that the decompression process is not an inverse function of the compression process because of the lossy compression caused by the rounding function $round(\cdot)$. The dequantized DCT coefficient block D^{-B} is recovered by multiplying the quantized coefficients C_q with the corresponding quantization matrix Q in Eq. (2).

$$D^{-B} = C_q \times Q \tag{2}$$

2.1.2 JPEG-GHOST Phenomenon

When a JPEG image with factor q is recompressed with factor q'. The quantization matrix corresponding to quantization factor q' is Q'. The recompressed DCT coefficient block $D^{-B'}$ becomes

$$D^{-B'} = round(round\left(\frac{D^B}{Q}\right) \times \frac{Q}{Q'}) \times Q' \tag{3}$$

When $Q' = Q$, Eq. (3) can be rewritten as

$$D^{-B'} = round\left(\frac{D^B}{Q}\right) \times Q \tag{4}$$

Therefore, the de-quantized DCT coefficient block $D^{-B'}$ of the second compression is equal to the de-quantized DCT coefficient block D^{-B} of the first compression.

In [28], Farid explained the principle of JPEG-GHOST. Suppose an image F has undergone two JPEG compression. The two JPEG compression factors are $q1$, $q2$, respectively, and $q1 < q2$. Let E represents the error matrix of the image before and after compression. This image is compressed by JPEG factor $q3$ $(1 \leq q3 \leq 100)$, respectively. In case of $q3 = q2$, that is to say $E(i,j) = 0, \forall(i,j)$. And in case of $q3 = q1$, second minimum values of matrix E will appear. This second minimum is called a JPEG ghost since it reveals that the image was previously compressed with lower quality.

$$E(i,j) = \frac{1}{3}\sum\nolimits_{l \in \{R,G,B\}} [(F(x,y,l) - F_q(x,y,l)]^2 \tag{5}$$

where $F(x,y,l)$, $l \in \{R,G,B\}$, represents each of original three RGB color channels, and $F_{\text{q}}(\cdot)$ is the compression result of $F(\cdot)$.

2.1.3 Blocking Effect Model

If the image is JPEG compressed, the differences across blocks should be different due to block artifacts. To explain the block effect of JPEG images, differences within a block and spanning across a block boundary need to be calculated.

To accurately describe the block effect of 8 × 8 blocks in JPEG images, we propose a new block effect model. Suppose the JPEG image F size is m × n. The pixel value in the image is represented by $F(i,j)(1 \leq i \leq m, 1 \leq j \leq n)$. The pixel difference between adjacent pixels is represented by $Z(i,j)$. For the image $F(i,j)$, we compute

$$Z(i,j) = |F(i,j) + F(i+1,j+1) - F(i,j+1) - F(i+1,j)| \tag{6}$$

Then $L(i,j)$ is calculated as

$$L(i,j) = Z(i,j) - Z(i+4,j+4) \tag{7}$$

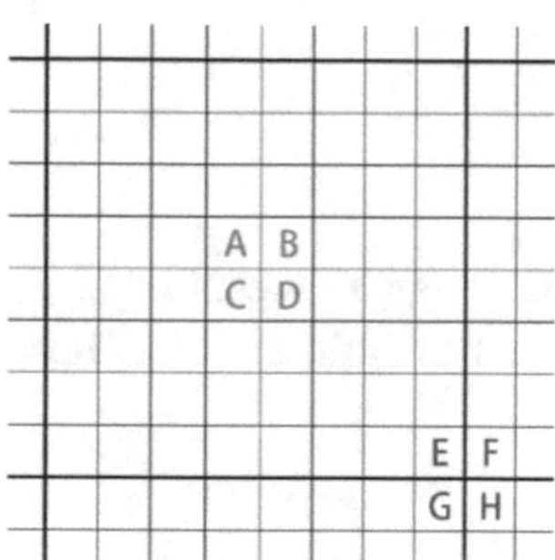

Fig. 2. Blocking artifact

As shown in Fig. 2, when (i,j) is located at the location A, the value of $L(i,j)$ is significantly smaller than the surrounding location E.

$$Z(i,j) = |\mathrm{A} + \mathrm{D} - \mathrm{B} - \mathrm{C}|, Z(i+4,j+4) = |\mathrm{E} + \mathrm{H} - \mathrm{F} - \mathrm{G}| \tag{8}$$

where letters A–H are the values of the pixels in the position. For example, for a JPEG image with quality factor 80, the mean value $L_{(x,y)}(1 \leq x,y \leq 8)$ of $L(i,j)$ in 8 × 8 blocks are calculated. Figure 3 shows $L_{(x,y)}$ is the smallest when (x, y) is (4, 4) and $L_{(x,y)}$ is the biggest when (x, y) is (8, 8).

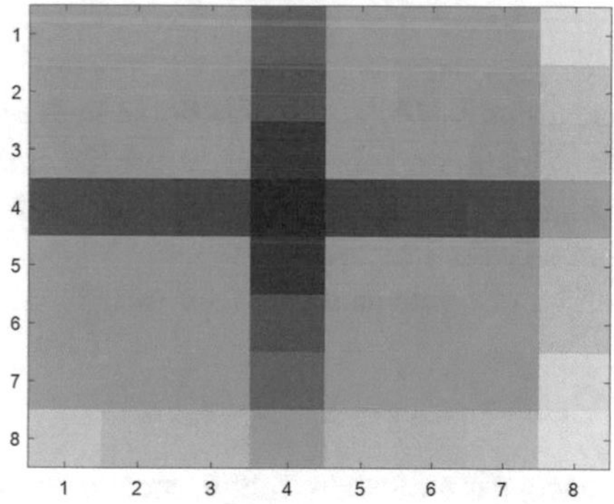

Fig. 3. $L_{(x,y)}$ in 8 × 8 block

We can use Eq. (9) to describe this characteristic of JPEG images.

$$\frac{\sum_{y=1}^{8} \boldsymbol{L}_{(4,y)} + \sum_{x=1}^{8} \boldsymbol{L}_{(x,4)}}{16} < \frac{\sum_{x=1}^{3}\sum_{y=1}^{3} \boldsymbol{L}_{(x,y)} + \sum_{x=1}^{3}\sum_{y=5}^{7} \boldsymbol{L}_{(x,y)} + \sum_{x=5}^{7}\sum_{y=1}^{3} \boldsymbol{L}_{(x,y)} + \sum_{x=5}^{7}\sum_{y=5}^{7} \boldsymbol{L}_{(x,y)}}{36} < \frac{\sum_{y=1}^{8} \boldsymbol{L}_{(8,y)} + \sum_{x=1}^{8} \boldsymbol{L}_{(x,8)}}{16} \tag{9}$$

If the image is not a JPEG image, the resulting $L_{(x,y)}$ will not conform to Eq. (9). Therefore, we can judge whether the grid of JPEG image is correct.

2.2 Analysis of Periodicity of Interpolation

Interpolation is a way to redistribute the pixels of the original image to change the number of pixels. Suppose a signal is $y(x)$, $x \in Z.I(\cdot)$ is the interpolation function. The resampled signal $f(x)$, $x \in R$ can be reconstructed as

$$f(x) = \sum_{k=-\infty}^{\infty} y(k) I\left(\frac{x}{\Delta x} - k\right) \tag{10}$$

where $\Delta \mathrm{x} \in R^{+}$ denote the sampling step.

Let $D^{(n)}f(\mathrm{x})$ be n-th derivative of $f(x)$. It is defined as $D^{(n)}f(x) = f(x)$ for $n = 0$ and $D^{(n)}f(x) = \partial^n f(x)/\partial x^n$ for $n \in N^{+}$. For discrete signals derivative is typically approximated by computing the finite difference between adjacent samples. Assume f is a signal with a constant variance σ^2.

According to [32], variance of $D^{(\mathrm{n})}f(\mathrm{x})$ as a function of x is expressed as,

$$V\{D^{(n)}f(x + \mu\Delta x)\} = V\{D^{(n)}f(x)\}, \mu \in Z \tag{11}$$

where V is the variance.

From Eq. (11) it is clear that $\mathrm{V}\left(D^{(\mathrm{n})}f(\mathrm{x})\right)$ is periodic with respect to x with a period Δx as the sampling rate. This periodicity depends on the interpolation kernel used. Nearest neighbor interpolation, bilinear interpolation and cubic convolution interpolation are the most commonly used methods.

The above discussion can be extended to two-dimensional image interpolation as,

$$r(i,j) = \sum_{m=-\infty}^{\infty} \sum_{n=-\infty}^{\infty} o(m,n) \Phi(i\Delta x - m, j\Delta y - n) \tag{12}$$

where o is the original image, r is a rescaled image, Δx and Δy are the sampling steps in the horizontal and vertical directions respectively, and Φ is the interpolation kernel.

For linear interpolation kernel and cubic interpolation kernel, the first difference between a pair of the original samples is equal. Therefore, the second difference will generate a zero in that position. Characteristics of the resampling signal are periodicity, which can be used to detect the existence of interpolation operations. The zero-crossings of the second difference of a resampled sequence exhibits a periodicity that is absent in an original sequence.

After that, second differences are constructed and the zero crossing of the obtained sequences is calculated. A binary sequence is constructed using the equation below,

$$\mathrm{p}[k] = \begin{cases} 1, \text{if} x''[k] \times x''[k+1] \leq 0 \\ 0, \text{otherwise} \end{cases} \quad (13)$$

By calculating the DFT of the binary sequence of the interpolation signal, the spectral representation is obtained. And the frequency f_n of the factor correlation peak can be found as

$$f_n = \begin{cases} 1 - \frac{1}{R_f}, \text{if } 1 < R_f \leq 2 \\ \frac{1}{R_f}, \text{if } R_f > 2 \end{cases} \quad (14)$$

where R_f is the rescaling factor. When the interpolation factor is larger than 1, the interpolation rate can be estimated as

$$R_f = \frac{1}{f_n} or\, R_f = \frac{1}{1 - f_n} \quad (15)$$

3 Proposed Method

First, we describe the process of the generation of spliced images as shown in Fig. 4. A part of the original JPEG image with factor q_1 is spliced with another image. Here, assume that the spliced image is an uncompressed bitmap image. Then the entire composite image is resampled by a resampling factor R_f to conceal original splicing trace. Finally, the image is recompressed with higher factor q_2. Here we think the factor is larger than 90.

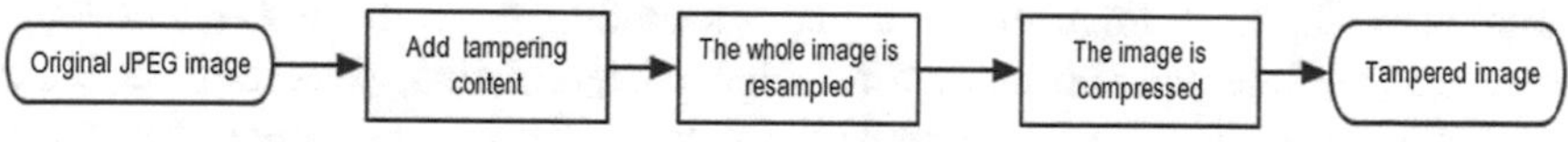

Fig. 4. Flow chart of spliced image generation

Based on the analysis of Sect. 2, we propose a detection method to reveal the forgery shown in Fig. 4. Suppose the suspect image F is with the size of $M \times N$. The proposed method consists of 15 steps:

(1) The color JPEG image F is converted to the gray scale image F' through

$$F' = 0.2989 \times F_R + 0.5870 \times F_G + 0.1140 \times F_B \quad (16)$$

where F_R, F_G, and F_B are R, G, B components of F, respectively.

(2) Compute the second difference of F' through

$$S(i,j) = 2F'(i,j) - F'(i,j+1) - F'(i,j-1) \tag{17}$$

(3) Get a vector to calculate the average of the $S(i,j)$ along the column direction.
(4) Construct a binary sequence according to Eq. (13).
(5) Get a vector by calculating the auto covariance of the binary sequence.
(6) Get a DFT magnitude plot by computing vector's discrete Fourier transformation (DFT).
(7) Ignore 1/8, 1/4 and 3/8 peaks and search for the peak in the frequency spectrum of the range of (0–0.5).
(8) Seek two candidate resampling factors R_{f1} and R_{f2} according to Eq. (15) in the DFT magnitude plot.
(9) Determine the actual resampling factor R_f from R_{f1} and R_{f2} according to a blockness grid analysis.
(10) Resample F to R with the factor of R_f.
(11) Estimate the quantization factor of the first JPEG compression in R.
(12) Compress the image R with the optimum quality factor q^*, producing image R_q.
(13) Generate the difference image D between R and R_q.
(14) Enhance D by a 5×5 sized median filtering operation.
(15) Obtain the final binary image S by a binarization processing and a block merge operation. Specifically, after dividing the image into un-overlapped 8×8 blocks, in each block, if the total number of value "0" pixels is greater than 25, the entire block is regarded as normal block and set all the values to "0" in this block.

We first estimate the resampling factor of F. Here, we design a method capable of resampling factor estimation using properties of the zero-crossings of the second difference. It is worth noting that for Step (7), since the amplitude of DFT is symmetric with respect to 0.5, we merely look for peak value in the frequency of 0–0.5. Once the peaks are located at 1/8, 1/4 and 3/8, they have to be ignored due to the effect of JPEG compression. Hence, for our method, it is impossible to detect the interpolation operations with the factors of 8, 4, 8/3, 8/5, 4/3, and 8/7.

For Step (8), to distinguish the actual resampling factor from R_{f1} and R_{f2}, we assume that the interpolation method is bilinear interpolation, since bilinear interpolation is the most widely used interpolation method. Then we resample the image F once again with the factor of the reciprocal of R_{f1} and R_{f2}, respectively, to generate the resampled images R_1 and R_2, respectively. Then the blockness grid of R_1 and R_2 can be detected according to Eq. (9). The actual resampling factor R_f can be determined by selecting the factor whose corresponding grid is closer to an 8×8 sized grid.

For Step (9), we compress the image R at varying quality factor q_i, where $q_i \in \{40, 41, \cdots, 84, 85\}$. Here we do not consider the case of factor below 40, since the JPEG images with compression factor less than 40 rarely exist in our practical occasions. In addition, due to the limitation of JPEG-GHOST method, we set up the traversal high line to 85. For each JPEG compressed image of the image R, the error matrix E_q is computed according to Eq. (5). Next, the number of elements equaling to 0 in each matrix E_q is recorded with different quality factor q_i and denoted as N_{qi}. The original JPEG factor q_1 can be determined by seeking the maximum number of changes in N_{qi}.

4 Experimental Results

The proposed technique was simulated in MATLAB 2017a. To evaluate the performance of the proposed technique, we used 1338 uncompressed color images (TIFF format) with size 512 × 384 or 384 × 512 from the UCID image database [33].

Tampered region localization can be considered as a binary classification problem. We use specificity and sensitivity to describe the accuracy of the method. We denote the pixels in the tampered region as positive samples (P) and those in the areas that have not been tampered with as negative samples (N). let TP, TN, FP and FN are true positive, true negative, false positive and false negative samples, respectively. Then, specificity and sensitivity are defined as

$$\text{specificity} = \frac{|\text{TN}|}{|\text{TN} + \text{FP}|}, \text{sensitivity} = \frac{|\text{TP}|}{|\text{TP} + \text{FN}|} \tag{18}$$

where $|\cdot|$ takes the number of elements in a set.

To demonstrate the efficacy of the proposed method, Tables 1, 2, 3 and 4 show the detection results with different sizes of manipulated regions, ranging from 32 × 32 to 256 × 256, and all results in the tables are presented as specificity/sensitivity pairs. In addition, the default-resampling factor used in Tables 1, 2, 3 and 4 between the two stages of JPEG compression is set at 1.7.

Table 1. Detection specificity and sensitivity of tampered area size of 32 × 32

q_1	q_2					
	90	92	94	96	98	100
50	89.6/83.1	90.7/82.3	91.2/83.4	91.5/83.0	92.5/80.7	92.9/80.6
60	87.8/81.6	88.0/79.5	88.0/82.0	89.2/81.1	89.4/80.3	90.1/78.7
70	85.2/80.6	86.5/77.3	86.8/79.4	87.9/79.5	88.9/80.7	89.6/77.6
80	82.8/82.7	83.4/78.8	83.5/83.7	84.2/83.4	84.7/82.4	86.2/80.6

Observed from Tables 1, 2, 3 and 4, it can be seen that with the increase of δ, where $\delta = q_2 - q_1$, the specificity and sensitivity of detection technology will be improved. With the increase of the size of the tampering area, the specificity of detection technology will be reduced. When the size of the tampering area is small or too large, the sensitivity of the detection technology will be reduced.

Next, an example of the tampering detection and localization process is presented in Fig. 5. The original image is a JPEG image with the factor of 80, and the resampling factor is 1.8. Finally, it is saved as JPEG with the factor of 95. Figure 5(a) is a tampered image. Figure 5(b) and (c) show the blockness artifact analysis results during the Step

Table 2. Detection specificity and sensitivity of tampered area size of 64 × 64

q_1	q_2					
	90	92	94	96	98	100
50	89.4/98.0	90.5/97.4	90.9/97.1	91.2/96.5	92.2/96.5	92.4/96.5
60	87.3/97.9	87.7/97.7	88.1/97.0	88.9/96.8	89.2/96.4	90.0/96.6
70	84.6/97.4	85.9/97.2	86.5/96.3	87.6/96.5	88.7/96.5	89.7/96.3
80	81.7/97.5	82.9/97.3	82.7/96.6	83.5/96.5	84.4/96.9	85.7/91.1

Table 3. Detection specificity and sensitivity of tampered area size of 128 × 128

q_1	q_2					
	90	92	94	96	98	100
50	88.0/94.9	89.4/94.2	90.2/94.4	90.8/94.1	91.7/94.5	91.9/94.5
60	85.7/94.3	86.6/93.7	87.1/93.0	87.9/92.6	88.8/92.7	89.5/92.0
70	83.8/93.4	85.0/92.8	85.6/91.9	87.0/91.8	88.4/92.1	88.9/91.4
80	79.2/93.3	80.7/93.2	80.9/92.9	80.9/92.6	82.5/92.2	83.0/91.7

Table 4. Detection specificity and sensitivity of tampered area size of 256 × 256

q_1	q_2					
	90	92	94	96	98	100
50	85.8/85.5	87.0/86.0	88.4/86.7	88.7/86.8	89.4/86.6	89.6/86.8
60	83.6/84.4	84.4/84.5	85.6/84.8	85.7/84.6	86.8/84.8	87.4/84.5
70	81.5/83.6	82.5/83.8	84.2/84.1	84.4/83.9	86.2/83.9	87.0/83.2
80	75.1/82.5	76.2/82.3	77.4/82.5	78.0/82.2	79.1/82.6	79.9/82.0

(9) of our proposed method, where (b) corresponds to the incorrect resampling factor, while (c) corresponds to the correct one. Figure 5(d) and (e) are the differential images before and after median filtering, respectively. Figure 5(f) is the final detection result of the proposed method where white regions indicate the splicing regions.

Figure 6 shows several examples of the proposed method. In Fig. 6, the original images are JPEG images with the factors of 75, 65, and 55, respectively. All of the images are then compressed with the factor of 95, after resampled with the factors of 2.3, 1.7, and 1.8, respectively. Figure 6(a), (c), and (e) show the three tampered images, respectively, and Fig. 6 (b), (d), (f) are the corresponding detection results (a), (c), and (e), respectively. It can be seen in Fig. 6, most the forgery can be precisely detected via using our proposed method.

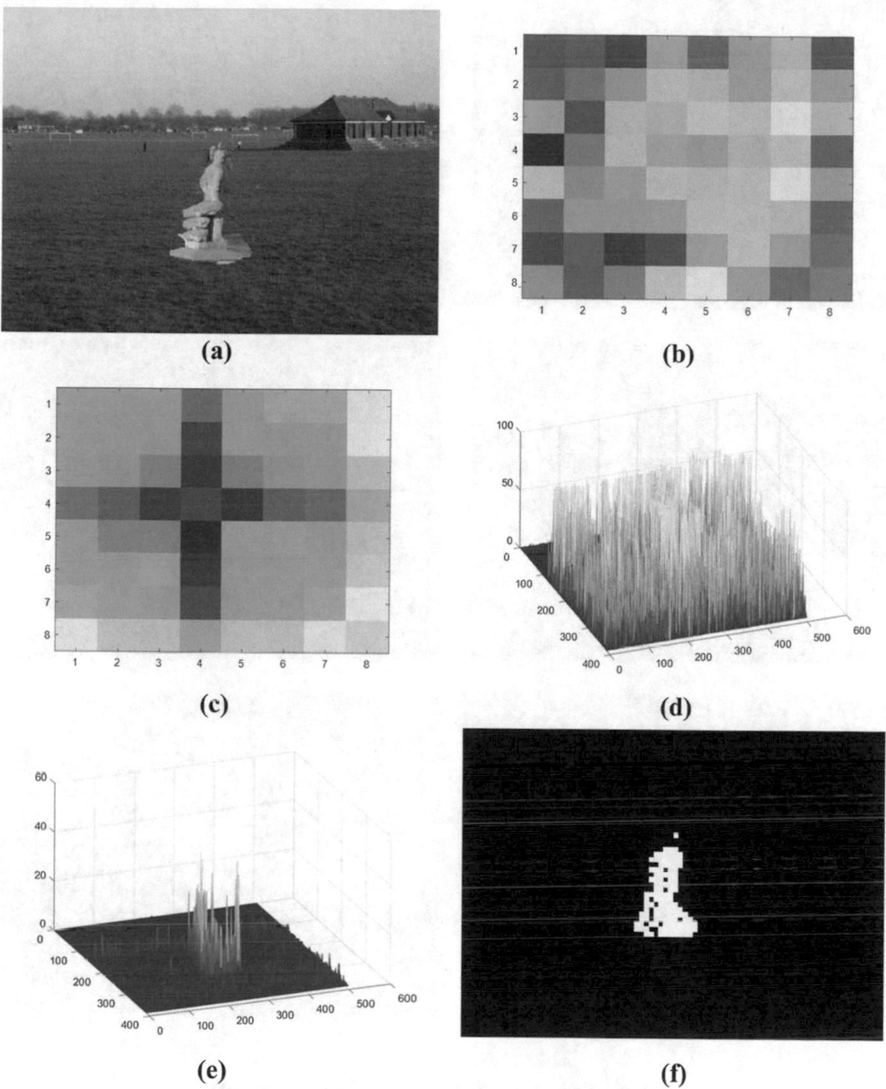

Fig. 5. Tamper localization experiments

Fig. 6. Examples of tampering detection results

5 Conclusion

In this paper, we presented a novel tampering localization method based on a comprehensive analysis of resampling and JPEG blockness artifacts. Forgery detection and localization process can be executed automatically without any human intervention. The most significant advantage of our method is the efficacy that the spliced image can be well detected even the spliced image is resampled globally before saved as JPEG format. Another advantage of this algorithm is that it solves the confusion of resampling factor estimation after the JPEG image is resampled. It is worth noting that the disadvantage of this method is that it requires high compression quality of the last JPEG and some resampling factors cannot be detected. In our future work, we aim to achieve blind image forensics in more complex situations.

Acknowledgments. This work was supported in part by the National Natural Science Foundation of China (61702332, 61672354, 61562007), Research Fund of Guangxi Key Lab of Multi-source Information Mining & Security (MIMS16-03), the Guangxi Natural Science Foundation (2017GXNSFAA198222), the Guangxi Collaborative Innovation Center of Multi-source Information Integration and Intelligent Processing. The authors would like to thank the anonymous reviewers for their helpful comments.

References

1. Haouzia, A., Noumeir, R.: Methods for image authentication: a survey. Multimed. Tools Appl. **39**(1), 1–46 (2008)
2. Sreenivas, K., Kamkshi, P.V.: Fragile watermarking schemes for image authentication: a survey. Multimed. Tools Appl. **9**(7), 1193–1218 (2018)
3. Qazi, T., Hayat, K., Khan, S.U.: Survey on blind image forgery detection. IET Image Process. **28**(2), 660–670 (2013)
4. Birajdar, G.K., Mankar, V.H.: Digital image forgery detection using passive techniques: a survey. Digit. Investig. **10**(3), 226–245 (2013)
5. Singh, A.K.: Improved hybrid algorithm for robust and imperceptible multiple watermarking using digital images. Multimed. Tools Appl. **76**(6), 8881–8900 (2017)
6. Kim, C., Shin, D., Leng, L., Yang, C.N.: Lossless data hiding for absolute moment block truncation coding using histogram modification. J. Real-Time Image Process. **14**(1), 101–114 (2018)
7. Lin, O., Yan, H., Huang, Z.: An ID-based linearly homomorphic signature scheme and its application in blockchain. IEEE Access. **6**, 20632–20640 (2018)
8. Alpar, O., Harel, J., Krejcar, O.: Online signature verification by spectrogram analysis. Appl. Intell. **48**(5), 1189–1199 (2018)
9. Huang, Y., Cao, L., Zhang, J., Pan, L., Liu, Y.: Exploring feature coupling and model coupling for image source identification. IEEE Trans. Inf. Forensic Secur. **13**(12), 3108–3121 (2018)
10. Sameer, V., Sugumaran, S., Naskar, R.: K-unknown models detection through clustering in blind source camera identification. IET Image Process. **12**(7), 1204–1213 (2018)
11. Qiao, T., Zhu, A., Retraint, F.: Exposing image resampling forgery by using linear parametric model. Multimed. Tools Appl. **77**(2), 1501–1523 (2018)
12. Hilal, A.: Image re-sampling detection through a novel interpolation kernel. Forensic Sci. Int. **287**, 25–35 (2018)
13. Birajdar, G.K., Mankar, V.H.: Blind method for rescaling detection and rescale factor estimation in digital images using periodic properties of interpolation. AEU-Int. J. Electron. Commun. **68**(7), 644–652 (2014)
14. Shin, H.J., Jeon, J.J., Eom, I.K.: Color filter array pattern identification using variance of color difference image. J. Electron. Imaging **26**(4), 1501–1523 (2018)
15. Chang, T.Y., Tai, S.C., Lin, G.S.: A passive multi-purpose scheme based on periodicity analysis of CFA artifacts for image forensics. J. Vis. Commun. Image Represent. **25**(6), 1289–1298 (2014)

16. Yao, H., Wang, S., Zhang, X.: Detecting image splicing based on noise level inconsistency. Multimed. Tools Appl. **76**(10), 12457–12479 (2017)
17. Yao, H., Cao, F., Tang, Z.: Expose noise level inconsistency incorporating the inhomogeneity scoring strategy. Multimed. Tools Appl. **77**(14), 18139–18161 (2018)
18. Yao, H., Wang, S., Zhao, Y.: Detecting image forgery using perspective constraints. IEEE Signal Process. Lett. **19**(3), 123–126 (2012)
19. Bhardwaj, D., Pankajakshan, V.: Image overlay text detection based on JPEG truncation error analysis. IEEE Signal Process. Lett. **23**(8), 1027–1031 (2016)
20. Yang, J., Zhu, G., Huang, J.: Estimating JPEG compression history of bitmaps based on factor histogram. Digit. Signal Prog. **41**, 90–97 (2015)
21. Thanh, H.T., Cogranne, R., Retraint, F.: JPEG quantization step estimation and its applications to digital image forensics. IEEE Trans. Inf. Forensic Secur. **12**(1), 123–133 (2017)
22. Li, B., Ng, T.T., Li, X.: Revealing the trace of high-quality JPEG compression through quantization noise analysis. IEEE Trans. Inf. Forensic Secur. **10**(3), 558–573 (2015)
23. Luo, W., Huang, J., Qiu, G.: JPEG error analysis and its applications to digital image forensics. IEEE Trans. Inf. Forensic Secur. **5**(3), 480–491 (2010)
24. Taimori, A., Razzazi, F., Behrad, A.: Quantization-unaware double JPEG compression detection. J. Math. Imaging Vis. **54**(3), 269–286 (2016)
25. Yang, J., Xie, J., Zhu, G.: An effective method for detecting double JPEG compression with the same quantization matrix. IEEE Trans. Inf. Forensic Secur. **9**(11), 1933–1942 (2014)
26. Pasquini, C., Boato, G., Perez-Gonzalez, F.: Multiple JPEG compression detection by means of Benford-Fourier coefficients. In: 2014 IEEE International Workshop on Information Forensics and Security (WIFS), pp. 113–118 (2014)
27. Milani, S., Tagliasacchi, M., Tubaro, S.: Discriminating multiple JPEG compression using first digit features. In: 2012 IEEE International Conference on Acoustics, Speech and Signal Processing, pp. 2253–2256 (2012)
28. Farid, H.: Exposing digital forgeries from JPEG ghosts. IEEE Trans. Inf. Forensic Secur. **4**(1), 154–160 (2009)
29. Wang, W., Dong, J., Tan, T.: Exploring DCT coefficient quantization effects for local tampering detection. IEEE Trans. Inf. Forensic Secur. **9**(10), 1653–1666 (2014)
30. Iakovidou, C., Zampoglou, M., Papadopoulos, S.: Content-aware detection of JPEG grid inconsistencies for intuitive image forensics. J. Vis. Commun. Image Represent. **26**(4), 155–170 (2018)
31. Chen, Z., Zhao, Y., Ni, R.: Detection of operation chain: JPEG-Resampling-JPEG. Signal Process.-Image Commun. **57**, 8–20 (2017)
32. Mahdian, B., Saic, S.: Blind authentication using periodic properties of interpolation. IEEE Trans. Inf. Forensic Secur. **3**(3), 529–538 (2008)
33. Schaefer, G., Stich, M.: UCID - An uncompressed colour image database. In: 2014 Conference on Storage and Retrieval Methods and Applications for Multimedia, pp. 472–480 (2004)

Arbitrary Style Transfer of Facial Image Based on Feed-Forward Network and Its Application in Aesthetic QR Code

Shanqing Zhang[1], Shengqi Su[1], Li Li[1(✉)], Jianfeng Lu[1], Ching-Chun Chang[2], and Qili Zhou[1]

[1] School of Computer Science and Technology, Hangzhou Dianzi University, Hangzhou, China
lili2008@hdu.edu.cn
[2] Department of Computer Science, University of Warwick, Coventry CV4 7AL, UK

Abstract. QR code has become essential in daily-life because of the popularity of mobile devices. The visual effect of a conventional QR code is not ideal. Consequently, many good aesthetic algorithms have been proposed. However, both the decoding rate and visual effect of a QR code cannot be guaranteed simultaneously when facial image serves as the background. We propose an arbitrary style transfer of facial image based on feed-forward network as a preprocessing algorithm for an aesthetic QR code. The deep characteristics of content image and style image are unified in the same layer of convolutional neural networks in our style transfer network. Styles are changed. The result of style transfer is restricted with semantic segmentation result, color uniform regularization of facial image and repeating restriction similarity constraints. Experimental results show that both the decoding rate and visual effect of a QR code are guaranteed when our method is used in background preprocessing.

Keywords: Arbitrary style transfer · Feed-forward network · Facial image · Aesthetic QR code

1 Introduction

With the economic development in recent years, smart mobile devices have become more and more popular. The QR code has the advantages of being produced conveniently, having large storage capacity, strong error correction capability and anti-rotation capability. Consequently, QR code is used extensively as an interface between smart mobile devices and the Internet. Today, QR codes have become an integral part of life. An ordinary QR code with visual-unpleasant appearance consists of monotonic black/white encoding modules, which cannot be recognized by human eyes. Recently, visual optimization of QR codes has attracted extensive attention in academia.

Consequently, many good aesthetic algorithms have been proposed. However, most aesthetic QR algorithms are focused on improving the similarities to the original background, but neglect low decoding rate. We found that most aesthetic QR algorithms are suitable for flat texture or uniformly colored anime style images. We use the

C.-N. Yang et al. (Eds.): SICBS 2018, AISC 895, pp. 235–250, 2020.
https://doi.org/10.1007/978-3-030-16946-6_19

scanning error analysis method [1] to analyze the effects of different background images. In Fig. 1, the red parts indicate errors and the green parts indicate correctly-identified values. There are fewer errors in the sampling results matrix of the simple texture image. According to scanning principles, if there are errors in the matrix of sampling results, they need to be corrected by performing data correction. When the percentage of the generated errors exceeds a certain threshold, error correction may fail, which results in a scanning failure. Preprocessing the background is therefore used to increase the decoding rate. Common preprocessing algorithms enhance contrast, remove texture, etc. The visual effect is impacted although satisfactory decoding rate is obtained. It is critical to obtain suitable preprocessing algorithms for facial image with a good visual effect and decoding rate.

Fig. 1. Error analysis of the intermediate results of scanning different QR codes using the ZXing library

We introduce convolutional neural networks (CNN) based style transfer into preprocessing. Style transfer removes details from facial image, thereby improving the decoding rate, achieving both reality and aesthetics. However, the current style transfer cannot be used in facial image because style distortion and non-unified image contents. The corresponding CNN is required for every style image training in conventional image stylization network.

We referenced the network structure of Chen et al. [2]. The image contents and deep characteristics of style images are unified into one layer of CNN, style change is performed, which resulted in faster speed of arbitrary style transfer of images. We further introduced semantic segmentation result to mitigate the impact on texture and color by the background. Color uniform regularization of facial image was introduced and the maximum times in any patch was constrained, greatly improving the quality of facial style images. The experimental results showed that our method yields both a high decoding rate and good visual effect.

2 Related Work

In this section, we review some techniques primarily concerning two topics, aesthetic QR code and image style transfer.

2.1 Aesthetic QR Code

Recently, more and more researches have focused on aesthetic QR code. Current aesthetic algorithms include two categories.

The first category uses the error correction encoding mechanism of RS code [3–5]. It can highlight some important areas of image so that the region of interest (ROI) can be completely displayed. The experimental results show that there are almost no noise pixels on ROI. The main drawback is that the replacement area is limited by error correction capacity. In addition, a processed image is covered with a lot of the black and white points and the visual effect of salient regions is poor. The results are shown in Fig. 2(a).

(a) The result of [4] (b) The result of [6] (c) The result of [7] (d) The result of [9]

Fig. 2. The aesthetic QR codes of different algorithms

The second category uses the XOR characteristic of RS code [6–8] to beautify the QR code. In our previous work, we proposed a novel algorithm based on the XOR mechanism of hybrid basis vector matrices and background image synthetic strategy [9]. The hybrid basis vector matrices include the reverse basis vector matrix (RBVM) and positive basis vector matrix (PBVM). First, RBVM and PBVM are established by Gauss Jordan elimination method according to the characteristics of the RS code. Then, the modification of the parity area of the QR code can be performed with the XOR operation of RBVM. And the XOR operation of PBVM is used to change the data area of the QR code. So a QR code can be modified to be very close to the background image without affecting the error correction ability. Finally, in order to further decrease the difference between a QR code and the background image, a new synthesis strategy is adopted to achieve better aesthetic effect. The experimental results show that this method yield better visual effect without sacrificing recognition rate. The results are shown in Fig. 2(d).

At present, it is difficult for the existing QR code beautification algorithms to address both the decoding rate and the aesthetics of a QR code, especially for facial images.

2.2 Style Transfer

Recently, image style transfer has become a hot topic in the AI field. It is related to texture synthesis. Texture synthesis methods are divided into two categories, i.e., parametric [10–12] and non-parametric [13–15] method. Parametric methods of texture synthesis aim to represent textures through proper statistical models. Non-parametric methods aim to build a perfect image rather than build rich models to understand textures. Li and Wand [16] showed that representations of image content and style were separable by various CNN convolutional layers. Moreover, representation provided the possibility for image decoupling and recombining. Gatys et al. [17] formulated style transfer as an optimization problem that combines texture synthesis with content reconstruction. By matching the global statistics of deep features, the parametric methods can be used to preserve the content of image and the overall visual effect of an artwork.

Based on the existing image style transfer approaches with CNN, we propose an arbitrary style transfer method of facial image based on single-layer activations in a prerained CNN. Similar to Chen's work [2], we reconstruct the complete best matching activation target in the activation space. More importantly, a visually appealing style transfer can be achieved by using our method, which is able to directly adapt to the facial area. We adjust the network of the style image transfer according to the special requirement of facial image as the content image. First, unlike the existing style transfer methods [16, 17], our method does not use pixel-level loss, instead, our method uses loss on the activations. Second, semantic segmentation result and repetitive restriction can help achieve arbitrary style transfer during semantic matching between images. Finally, we use a global constraint based on YUV color space of an image to avoid the color-overmatch problem, which greatly improves the photorealism of the transfer results.

3 A New Method of Style Transfer

The network architecture of our method is shown in Fig. 3. The inputs of feed-forward network are facial image (i.e., content image) and style image. DeepLab [18] is used to obtain semantic segmentation result before content images are inputted into the network. Every patch activated by content images is replaced by the corresponding patch of style transfer images with global balance and semantic segmentation result. Facial image with high fidelity is acquired through a uniform style transfer is obtained. The VGG-19 [19] model is used in this study.

3.1 The Activation Based on Semantic Segmentation Result

In neural algorithm of style transfer, CNN activation is used as the feature of an image. Experimental results show that the interference on facial image from background is the primary factor contributing to infidelity in the process of style transfer. We defined an activation function based on semantic segmentation to solve this problem.

DeepLab [18] is used to generate semantic segmentation result for a facial image. The content image is divided into face and background. The activation of content image with semantic segmentation result is the input of arbitrary style transfer network. The enhanced activation of content image $F_{seg}\mathrm{C}$ is

$$F_{seg}\mathrm{C} = F(C)J(C). \tag{1}$$

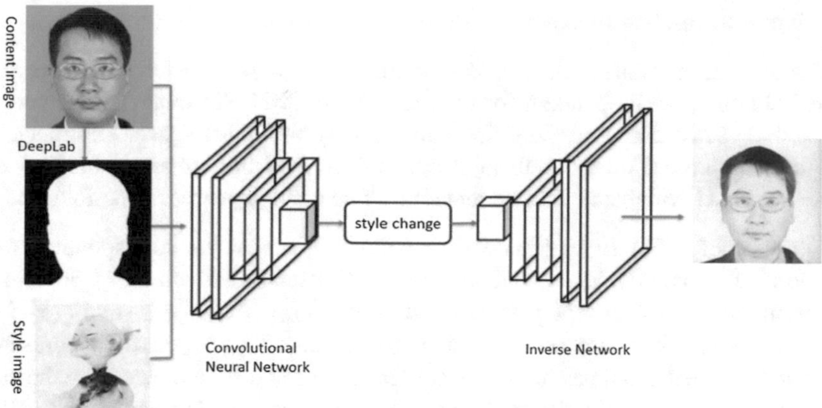

Fig. 3. Network architecture of arbitrary style transfer of a facial image based on feed-forward network

$F(C)$ is the activation of a layer CNN of content image C, $J(C)$ is the semantic segmentation result of the corresponding content image. Downsampling operation matches the content image activation size of every CNN layer.

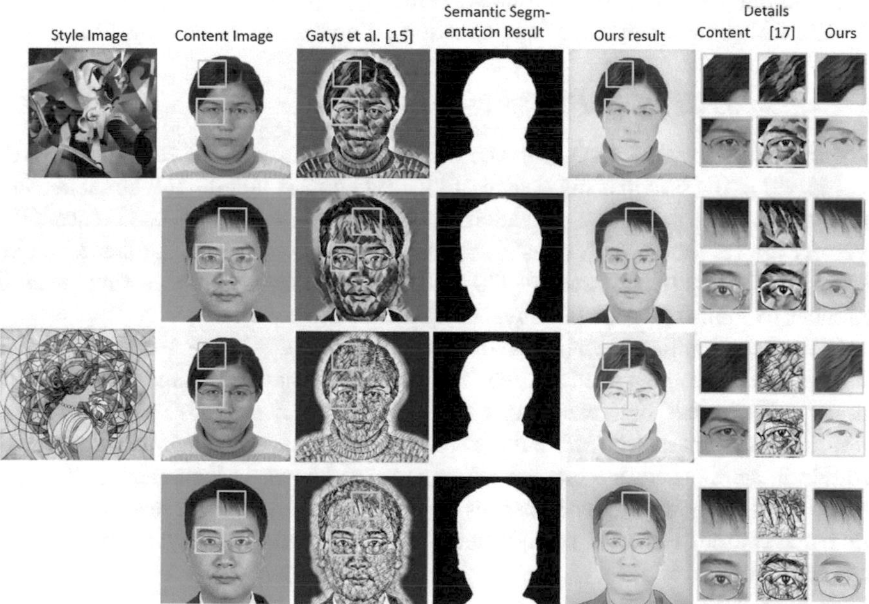

Fig. 4. The contrast of the parameter style transfer network [17] with our non-parametric feed-forward network style transfer

3.2 Non-parametric Image Style Transfer Network

Let C and S denote content image and style image of a style transfer, respectively. $F(\cdot)$ is the full convolution operation for the pre-trained CNN. An image is mapped to an intermediate activation space in CNN from its original image space. The image style change is performed on the activation space $F(S)$ and the enhanced content image activation F_{seg}C with semantic segmentation. The change process is as follows:

(1) Pre-trained VGG-19 [19] network is used to generate the corresponding activations of a content image C and style image S. Patches with size r * r are extracted from the activations. The patches are marked as $f_{seg\_i}(C), i \in n_C$ and $f_j(S), j \in n_S$, respectively. The size of the content image is h * w, n_C and n_S represent the number of extracted patches of the content image and style image from the current layer. The extracted patches should have sufficient overlap and include all activated channels.
(2) The activation of the content image with semantic segmentation result is correlated to the best matching patch of the style image.

$$f_i^{st}(C,S) = \underset{f_j(S),j=1,\ldots,n_S}{\operatorname{argmax}} \frac{\langle f_{seg\_i}(C), f_j(S) \rangle}{\left\| f_{seg\_i}(C) \right\| \cdot \left\| f_j(S) \right\|}. \quad (2)$$

(3) The activation patch of the content image $f_{seg\_i}(C)$ is replaced by the best matching activation patch of the style image $f_i^{st}(C,S)$.
(4) The full content activation is re-constructed and is represented by $F^{st}(C,S)$.

3.3 Repeating Restriction Similarity Constraints

The enhanced activation patch of the content image with semantic segmentation result is correlated to the best matching patch of the style image, and the full target activation is reconstructed. A part of the activation patch of the style image is used repeatedly. It adversely affects the uniform transfer of the image style. It cannot produce rich style transfer of the image. Inspired by [20, 21], we introduced the repeating restriction parameter to limit the usage of activation patch of the style image.

As its name implies, repeating restriction similarity constraints addresses similarity of image area. It is completed in Step 2 of the non-parametric image style transfer in Sect. 3.2. When the best matching style patch with a patch of content image is determined, a count matrix with initial zero is introduced. When an activation patch of the style image is used, the corresponding matrix element is incremented. The patch cannot be used any more after it reaches the threshold and the next new best matching style patch is used. The style expands more uniformly and image style transfer is richer when the threshold is 5.

3.4 The Color Uniform Regularization of Facial Image

Image realistic regularization [22] uses an affine function to widen color difference. It is not suitable for facial image. And regularization in [22] on RGB space requires heavy computation for style transfer. In order to solve these problems, we propose the color uniform regularization of facial image. We directly employ the semantic segmentation result from DeepLab in Sect. 3.1, with the scope to local transformation being controlled. Local transformation is performed in channel Y on YUV space of the content image. The style transfer is faster. Importantly, we decrease the range of the brightness field to weaken the impact from the brightness difference in the facial area.

Color uniform regularization of facial image is defined as follows:

$$L_{color}(I, C) = V_Y(I)^T M_Y(C) V_Y(I), \tag{3}$$

where content image C has N pixels. $V_Y(I)$ the vectorized version (N * 1) of the style transfer image I in channel Y. The matrix $M_Y(C)$ (N * N) is a representation of a standard linear system which can minimize a locally affine function of the content image C in Y channel [22].

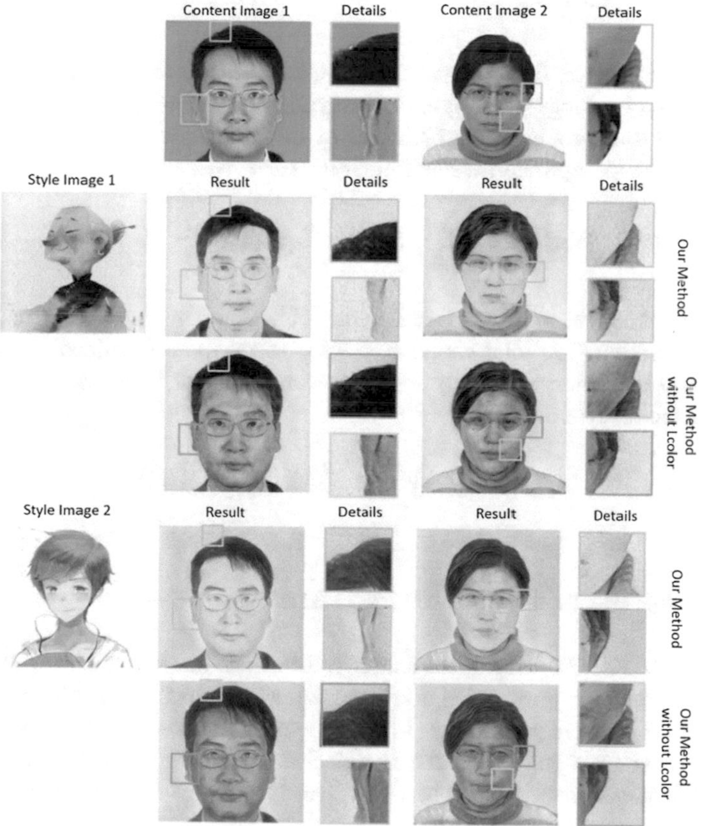

Fig. 5. The contrast of our feed-forward network with the same network without color uniform regularization

3.5 Loss Function

Different from [16, 17], we use the squared-error in our loss function. The enhanced image activation F_{seg}C with semantic segmentation result is redefined. The full activation is re-constructed under the repeating restriction similarity constraints. We pioneer to calculate the total loss in YUV space and add the color uniform regularization of facial image in the loss function. Our loss function is defined as follows:

$$I_{style}(C,S) = \underset{I \in \mathbb{R}^{h*w*d}}{\operatorname{argmin}} \left\| F(I) - F^{ST}(C,S) \right\|_F^2 + \alpha L_{TV}(I) - \beta L_{color}(I,C), \quad (4)$$

Where $\|\cdot\|_F$ is the Frobenius norm, L_{color} is the color uniform regularization of the facial image, $L_{TV}(\cdot)$ is the total variation regularization term [23] that is widely used in image deblurring;

$$L_{TV}(I) = \sum_{i=1}^{h-1}\sum_{j=1}^{w}\sum_{k=1}^{d}\left(I_{i+1,j,k} - I_{i,j,k}\right)^2 + \sum_{i=1}^{h}\sum_{j=1}^{w-1}\sum_{k=1}^{d}\left(I_{i,j+1,k} - I_{i,j,k}\right)^2, \quad (5)$$

where $F(\cdot)$ is activation from the pre-trained CNN. Therefore, the final solution of (4) can be obtained by subgradient-based optimization. Based on [2], an inverse network can be trained to approximate an optimum of the loss function in (4) for any activations. We define the optimal inverse function as follows:

$$\underset{g}{\operatorname{arginf}}\ \mathbb{E}_H\left[\|F(g(H)) - H\|_F^2 + \alpha L_{TV}(g(H)) - \beta L_{color}(H)\right], \quad (6)$$

where f represents a deterministic function and H is a random variable corresponding to the target activation. The total variation regularization term is defined in the previous sections. And the color uniformity regularization of facial image uses the following function in the reverse network:

$$L_{color}(H) = V_Y(g(H))^T M_Y(H) V_Y(g(H)). \quad (7)$$

4 The Implementation of Style Transfer

A pre-trained VGG-19 is used as the feature extractor of our feed-forward network. The training of the inverse network [2] is referenced, Microsoft COCO (MSCOCO) [24] is employed, two training periods are done. The open sources in DeepLab v3+ [18] of encoder-decoder are used. We use the Torch7 framework to implement our method. All results in this paper are generated between 3 and 8 s on NVIDIA GeForce GTX 1060 6 GB GPU.

Target Layer. The effect of different layers of VGG-19 for our method is shown in Fig. 6. The deeper layer of VGG-19 is selected, the more obvious is the texture of the style transfer image. A better style transfer result is in layer Relu3_1, the original image

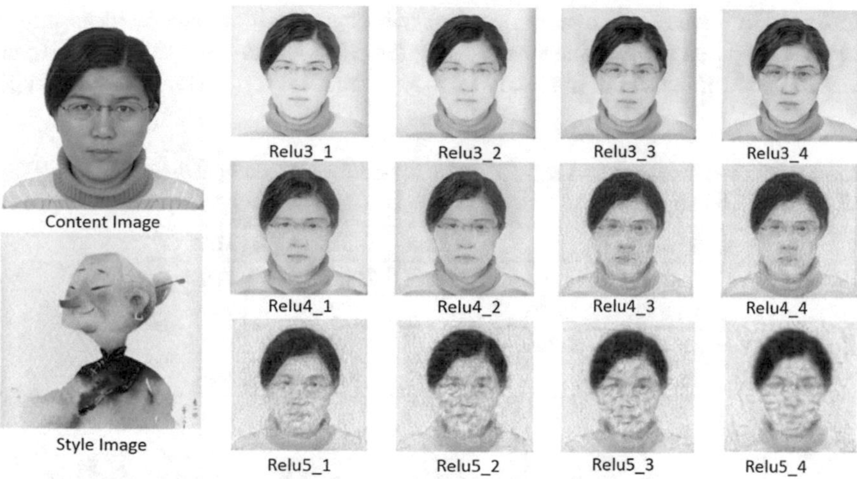

Fig. 6. The effect of arbitrary style transfer network results in different layers of VGG-19

is preserved and texture of style image is added. More image style transfer experiments are performed in layer Relu3_1. According to the naming convention of VGG-19, "Relu X_1" refers to the first ReLU layer after the $(X - 1)_{th}$ maxpooling layer.

Style-Matching Tuning. Image style is directly affected by the patch size and patch stride of convolution in CNN. More texture of styled image is preserved whereas more special structures of the content image are lost with big patch size and patch stride. We therefore set patch size as 3 * 3 and patch stride as 1.

Fig. 7. The effect of arbitrary style transfer network results in different patch size

Fig. 8. The effect of arbitrary style transfer network results in different patch stride

Computation Time. The time needed for style transfer with related work is shown in Table 1. Compared with parameter style transfer [16, 17], only one layer of pre-training VGG-19 network is used in our study with less time in every iteration and fast speed.

Table 1. Different methods stylize 555 * 555 image average computation time on NVIDIA GeForce GTX 1060 6 GB GPU

Method	N.Iters.	Time/Iter. (s)	Total (s)
Gatys et al. [17]	800	0.208	116.4
Chen and Schmidt [2]	1	2.132	2.13
Li and Wand [16]	200	0.988	197.6
Our method	1	4.256	4.26

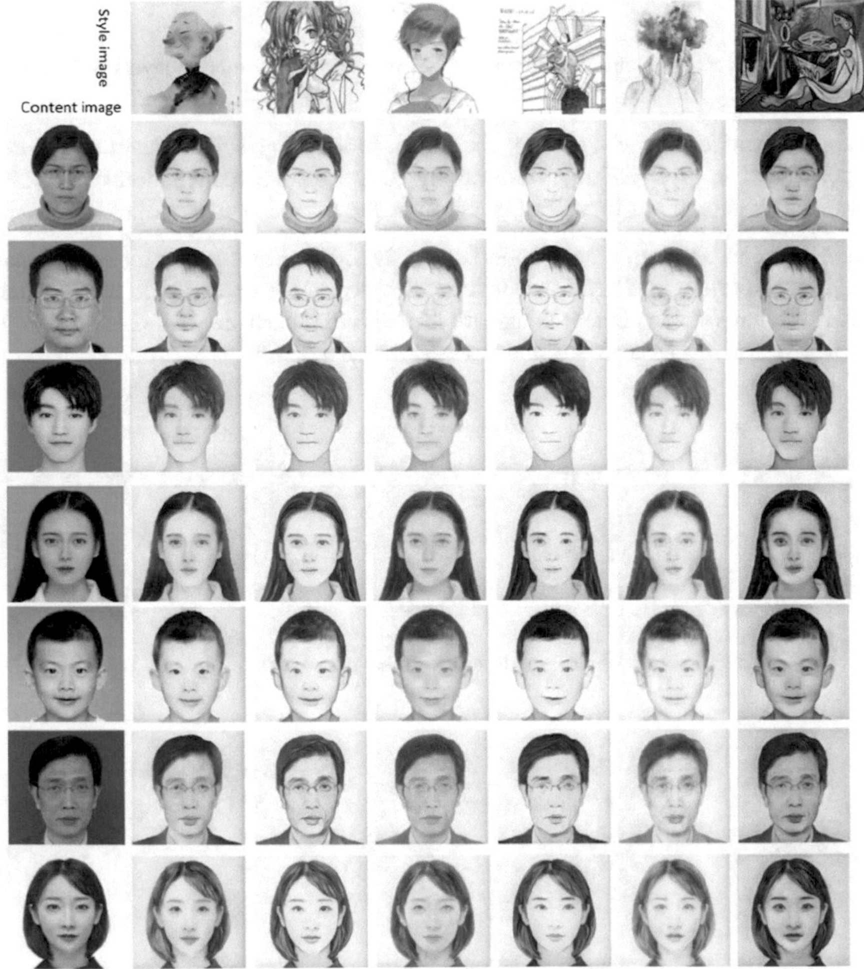

Fig. 9. The results for different facial images in different style images using our network

Our style transfer effect is slightly better (Fig. 11) than [2], but it takes more time. Our total loss function is added color field transfer of visual model, and the style transfer is performed with semantic segmentation result and repeated restriction similarity constraints.

5 Experimental Results of Style Transfer

Experimental results support the effectiveness of our method. Parameters are adjusted as described above. Patch size is 3 * 3, patch stride is 1, $\alpha = 10^{-6}$ and $\beta = 10^{2}$.

The Effectiveness of Style Transfer of Facial Image. Our purpose is to obtain more artistic style transfer images for facial images via aesthetic QR code. The style transfer should be both artistic and realistic, and a high decoding rate of style transfer QR code should be achieved. The ideal style transfer results are achieved for different facial images in different style images in our network (Fig. 8). Details of facial images are well preserved by the new network (Fig. 9). Compared with [21] and [2], some results are shown in Fig. 10. Our network obviously improves the effect of facial image style transfer.

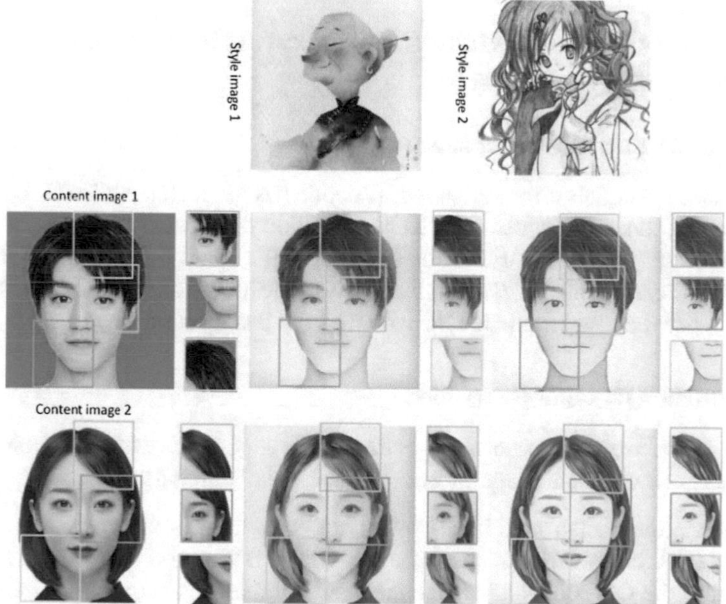

Fig. 10. Details of facial images (such as hair, facial features, and outlines) using our method

Adaptability to Multi-Style Transfer. We propose a feed-forward network for arbitrary style transfer for facial image in order to obtain high fidelity facial image contents and artistic facial style. We need only one network, which supports arbitrary

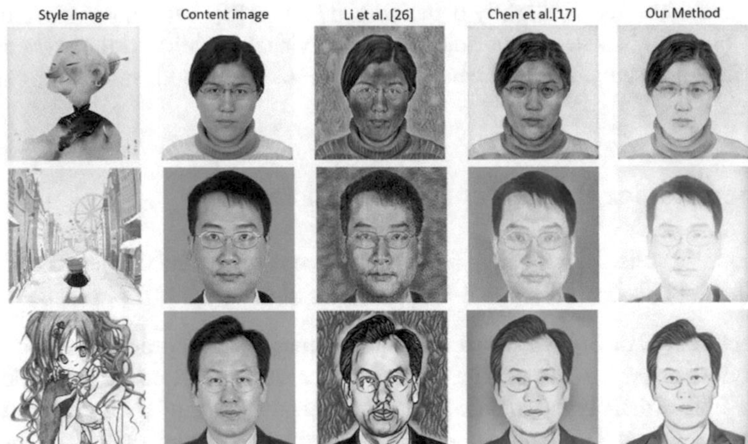

Fig. 11. The results generated by different style transfer method

style transfer of facial image contents and avoids different networks for style image training in previous studies.

6 Aesthetic QR Code Method Based on Arbitrary Stylized Network of Facial Image

6.1 RS Code Encoding Mechanism

QR code uses RS code's error correction. RS code has the features of XOR operation in that a new RS code is obtained with XOR operation of two different RS codes [8]. In [9], we proposed a new algorithm for aesthetic QR code by using the XOR characteristics of RS code with PBVM and RBVM to reduce as much as possible the visual differences between the aesthetic QR code and the background image.

6.2 Aesthetic QR Code Generation

Combining [9], we propose a new aesthetic QR algorithm with the arbitrary style transfer network of facial image in Fig. 12. The details include:

(1) Generate the background image using our style transfer network.
(2) The XOR characteristics of RS code with PBVM and RBVM is applied to beautify the QR code, and reduce the visual differences between the aesthetic QR code and background image as much as possible.
(3) Embed QR code into the background image.

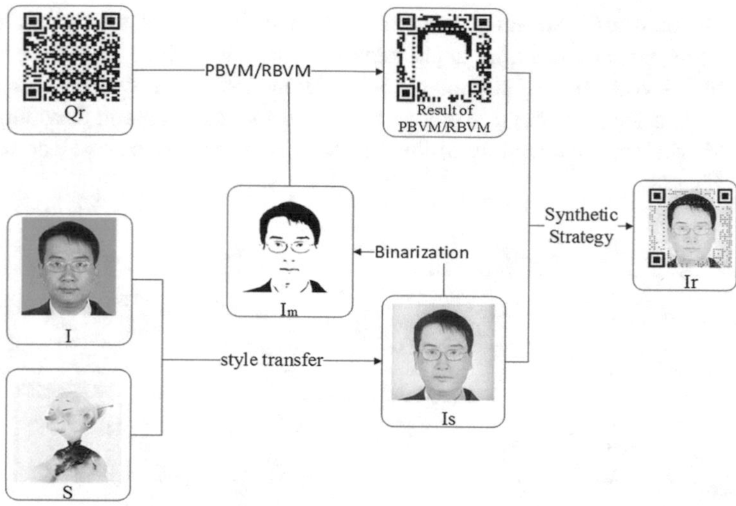

Fig. 12. Flowchart of an aesthetic QR algorithm with arbitrary style transfer network

6.3 Experimental Results

Visual Effect. A collection of 100 images with different styles serves as a dataset in our experiment. All images in the dataset are used as background images to generate aesthetic QR codes. The selected results generated by our method are shown in Fig. 13. The QR version is 5 and the error correction level is L. QR code with stylized image background has more artistic and diverse visual effect. Notice that the additional points that need to be added to the face area are significantly less in the style transfer QR code.

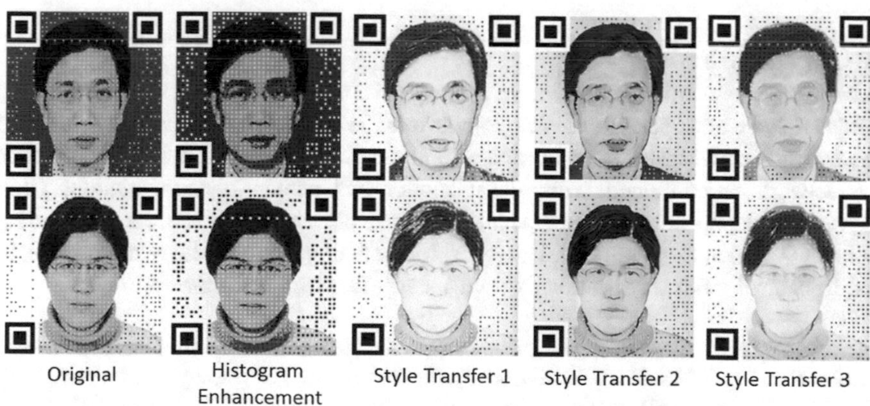

Fig. 13. The generation of aesthetic QR codes with different background image

Scanning Robustness. Scanning robustness of the QR codes was evaluated from various aspects, such as scanning angle variation, scale variation, brightness variation, and coverage. A collection of 100 images with different styles serves as a dataset in our experiment. And the same device (iPhone 8) is used for scanning and decoding tests. It is evaluated whether the decoding of the QR code is successful or not. The results are shown in Fig. 14.

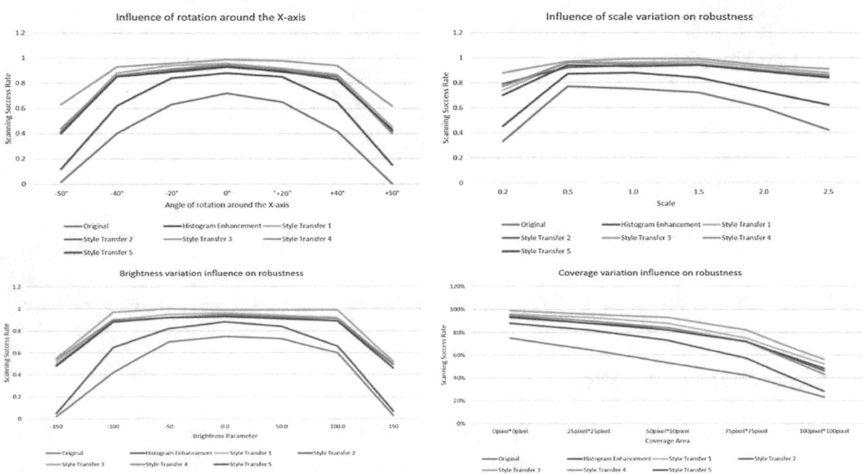

Fig. 14. The results of scanning robustness

Experimental results show that our method not only ensures code usability, but also has the effect of QR code beautification.

7 Conclusion

In this study, we propose an arbitrary style transfer network of facial image based on feed-forward network in order to obtain high fidelity images for aesthetic QR codes. The network is combined with visual model and semantic segmentation. Global brightness balance is used, good visual effect is obtained, and the decoding rate of aesthetic QR code is guaranteed. However, for a small portion of the facial images, it is hard to control color distribution. We will study this further in the future.

Acknowledgments. This work was mainly supported by National Natural Science Foundation of China (No. 61370218, No. GG19F020033), Public Welfare Technology and Industry Project of Zhejiang Provincial Science Technology Department (No. 2016C31081, No. LGG18F020013, No. LGG19F020016).

References

1. Xu, M., et al.: ART-UP: a novel method for generating scanning-robust aesthetic QR codes (2018)
2. Chen, T.Q., Schmidt, M.: Fast patch-based style transfer of arbitrary style (2016)
3. Ono, S., Morinaga, K., Nakayama, S.: Two-dimensional barcode decoration based on real-coded genetic algorithm. In: Evolutionary Computation, pp. 1068–1073. IEEE (2008)
4. Samretwit, D., Wakahara, T.: Measurement of reading characteristics of multiplexed image in QR code. In: International Conference on Intelligent Networking & Collaborative Systems, pp. 552–557. IEEE Computer Society (2011)
5. Li, L., Qiu, J., Lu, J., Chang, C.C.: An aesthetic QR code solution based on error correction mechanism. J. Syst. Softw. S016412121500148X (2016)
6. Fujita, K., Kuribayashi, M., Morii, M.: Expansion of image displayable area in design QR code and its applications. In: Proceedings of the Forum on Information Technology Papers, vol. 10, no. 4, pp. 517–520 (2011)
7. Lin, S.S., Hu, M.C., Lee, C.H., Lee, T.Y.: Efficient QR code beautification with high quality visual content. IEEE Trans. Multimed. **17**(9), 1515–1524 (2015)
8. Cox, R.: QArt Codes. https://research.swtch.com/qart. Accessed 20 Jan 2017
9. Li, L., Li, Y., Wang, B., Lu, J., Zhang, S., Yuan, W., Wang, S., Chang, C.C.: A new aesthetic QR code algorithm based on salient region detection and SPBVM (2017)
10. Bonet, J.S.D.: Multiresolution sampling procedure for analysis and synthesis of texture images. In: Conference on Computer Graphics and Interactive Techniques, pp. 361–368. ACM Press/Addison-Wesley Publishing Co. (2017)
11. Heeger, D.J., Bergen, J.R.: Pyramid-based texture analysis/synthesis. In: International Conference on Image Processing. Proceedings, pp. 229–238 (1995)
12. Portilla, J., Simoncelli, E.P.: A parametric texture model based on joint statistics of complex wavelet coefficients. Int. J. Comput. Vision **40**(1), 49–70 (2000)
13. Ashikhmin, N.: Fast texture transfer. Comput. Graph. Appl. IEEE **23**(4), 38–43 (2003)
14. Efros, A,A., Freeman, W.T.: Image quilting for texture synthesis and transfer. In: Conference on Computer Graphics and Interactive Techniques, pp. 341–346. ACM (2001)
15. Kwatra, V., Essa, I., Turk, G., et al.: Graphcut textures: image and video synthesis using graph cuts. In: ACM SIGGRAPH, pp. 277–286. ACM (2003)
16. Li, C., Wand, M.: Combining Markov random fields and convolutional neural networks for image synthesis, pp. 2479–2486 (2016)
17. Gatys, L.A., Ecker, A.S., Bethge, M.: A neural algorithm of artistic style. Computer Science (2015)
18. Chen, L.C., Zhu, Y., Papandreou, G., Schroff, F., Adam, H.: Encoder-decoder with atrous separable convolution for semantic image segmentation (2018)
19. Simonyan, K., Zisserman, A.: Very deep convolutional networks for large-scale image recognition. arXiv preprint arXiv:1409.1556 (2014)
20. Gu, S., Chen, C., Liao, J., Yuan, L.: Arbitrary style transfer with deep feature reshuffle (2018)
21. Lu, J., Lu, J., Lu, J.: StyLit: illumination-guided example-based stylization of 3D renderings. Acm Trans. Graph. **35**(4), 92 (2016). Krizhevsky, A., Sutskever, I., Hinton, G.E.: ImageNet classification with deep convolutional neural networks. In: International Conference on Neural Information Processing Systems, pp. 1097–1105. Curran Associates Inc. (2012)

22. Luan, F., Paris, S., Shechtman, E., Bala, K.: Deep photo style transfer, pp. 6997–7005 (2017)
23. Johnson, J., Alahi, A., Li, F.F.: Perceptual losses for real-time style transfer and super-resolution, pp. 694–711 (2016)
24. Lin, T.Y., Maire, M., Belongie, S., Hays, J., Perona, P., Ramanan, D., et al.: Microsoft COCO: common objects in context, vol. 8693, pp. 740–755 (2014)

Aesthetic QR Code Authentication Based on Directed Periodic Texture Pattern

Li Li[1], Min He[1], Jier Yu[1], Jianfeng Lu[1(✉)], Qili Zhou[1], Xiaoqing Feng[2], and Chin-Chen Chang[3]

[1] Hangzhou Dianzi University, Hangzhou, China
jflu@hdu.edu.cn
[2] Zhejiang University of Finance and Economics, Hangzhou, China
[3] Feng Chia University, Taichung, Taiwan

Abstract. More and more people use mobile phones to buy goods through scanning the printed aesthetic QR code. QR code has become an important payment tool in today's business. However, there is an inevitable risk in payment processing. In particular, it is difficult to detect whether an aesthetic QR code has been tampered by the attacker. Therefore, it is extremely important to carry out security certification for aesthetic QR code. We propose an algorithm based on the combination of directional periodic texture pattern and aesthetic QR code. Firstly, the aesthetic QR code is generated based on the Positive Basis Vector Matrix (PBVM), whereas the watermark is encoded into 4-bit Gray code and quantized into angles. Then, a periodic texture pattern is generated using a random matrix, and a directed periodic texture pattern is obtained by rotating the texture pattern according to the quantization angle. Finally, the new texture pattern and the aesthetic QR code are fused to obtain an authentication aesthetic QR code. The experiments verified that the aesthetic QR code can be correctly decoded and the watermark can be properly extracted. By utilizing the locating position patterns of the QR code, the additional watermark reference block is not required, therefore, the capacity of the watermark is enhanced. Moreover, the proposed scheme realizes the anti-counterfeiting authentication of aesthetic QR code, and maintains a better visual quality.

Keywords: Aesthetic QR code · Watermark · Authentication

1 Introduction

With the rapid development of the Internet, mobile phones are replacing personal computers due to their small size and portability. Mobile phones can capture, store and share images more quickly. In the marketplace, many applications are accompanied with QR code scanning, for example, website jumps, payment, and sales promotion. QR code stores a large amount of data and yet is independent of the database. It also has strong error correction capability. Furthermore, it can be decoded anywhere by using a smart phone. Therefore, it is widely used in some security sensitive applications, such as mobile payments [1, 2].

C.-N. Yang et al. (Eds.): SICBS 2018, AISC 895, pp. 251–266, 2020.
https://doi.org/10.1007/978-3-030-16946-6_20

Since there is no anti-counterfeiting function in traditional QR code, the attacker can easily tamper with a QR code. More and more researchers worked on improving the visual effects of the traditional QR code. As a result, various algorithms of aesthetic QR cod were proposed [3, 4]. At the same time, for enhancing the security certification of aesthetic QR code, some security authentication methods for aesthetic QR code were presented. For example, Lu et al. [5] proposed an algorithm based on visual cryptography scheme (VCS) combined with aesthetic QR code, which utilizes the visual cryptography divides an original QR code into two shadows, then two shadows are separately embedded into the same background image, and uses the XOR operation to fuse the embedding results with the same carrier QR code. This method can correctly verify the security of the QR code in the mobile payment process. Li et al. [6] presented a multi-safety authentication for aesthetic QR code for shared bicycles, it combines aesthetic QR code with color visual passwords. Two methods in above have good visual quality, but are fragile to print attack.

Print-cam [7] is the process of printing an image and then taking it photos using a mobile phone. During this process, there are some common attacks such as distortion, rotation, scaling, translation, and cropping. In addition, depending on the accuracies of printers, different types of noises will be imposed to the image. These distortions increase the difficulty of extracting the watermark.

However, not many schemes were proposed to protect watermark against print-cam. Katayama et al. [8] firstly proposed a print-cam robust watermark that is based on a sinusoidal watermark pattern and frame synchronization. This method needs to use watermark template positioning, which increases the complexity of watermark extraction. Then, Kim et al. [9] presented watermark extraction using autocorrelation function. They embed pseudo-random vectors by repeating tiling, and then detect the resulting peaks by autocorrelation functions. The method was tested on a digital camera. However, manual interaction must be applied during extracting the watermark. In order to improve the robustness of the watermark in the print-cam process, Pramila et al. [10] modified Chou's methods [11] using autocorrelation functions and directed periodic patterns. The patterns rotated at the angle generated by the watermark message encoding. This method is robust against $\pm 10^{\circ}$ of the tilt of the optical axis, but the embedding capacity is low. At the same time, the method combines the spatial masking and the background luminance components in the JND (just noticeable difference) model. Instead of selecting only one of them, they took more image details into account and made the image look better.

Inspired by directed periodic patterns [10], the watermark is embedded into the aesthetic QR code based on the positive basis vector matrix using the directional periodic texture pattern. During detecting the watermark, the position patterns of the aesthetic QR code are utilized, and no additional watermark reference block is needed. As a result, this scheme not only improves the capacity of watermark, but is also robust against print attack. Moreover, it realizes the QR code anti-counterfeiting authentication.

The rest of this paper is organized as follows. Aesthetic QR code generation and watermark embedding are introduced in Sect. 2. In Sect. 3, the extraction and detection watermark are described in detailed. In Sect. 4, we show the experimental results. Finally, conclusion and future work are given in Sect. 5.

2 Watermark Embedding

In Sect. 2, the watermark embedding process is described in detail. Firstly, an aesthetic QR code is generated based on the PBVM. Then, the aesthetic QR code is preprocessed and the JND value is computed. In order to maintain visual quality, the embedding strength is controlled by JND. Secondly, the watermark message is encoded using Hamming (32, 6) error correction coding and 4-bit Gray coding. Each Gray code sequence is converted to a rotation angle. The periodic texture pattern is rotated according to the angle in order to get a new texture pattern. Finally, the JND and new texture pattern is fused with the aesthetic QR code. The overview of the watermark embedding process is shown in Fig. 1.

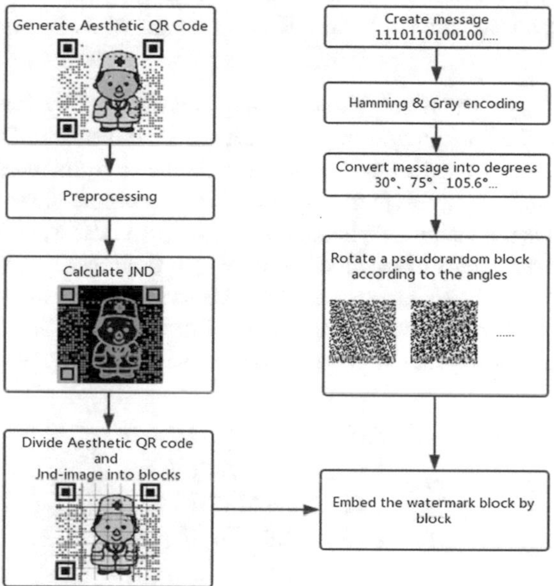

Fig. 1. The embedding process.

2.1 The Generating Algorithm of Aesthetic QR Code

Image Preprocessing. To increase the decoding rate of the QR code, image preprocessing methods are applied on the input background image. The flowchart is shown in Fig. 2.

Firstly, the background image I_0 is converted from RGB color space to Lab color space. Then the image's brightness channel is stretched to improve the contrast. In the new image I_1, the pixels are stretched to both sides according to the corresponding ratio. The pre-processed image is converted into gray scale image I_2 and compressed to the same size as the original QR code image for subsequent operations.

Fig. 2. The flowchart of image preprocess.

Construction of QR Code Based on PBVM. A series of RS codes obtained by Gauss Jordan elimination method formulates a matrix. The matrix is called Positive Basis Vector Matrix (PBVM). When a bit of RS code needs to be modified, special row in the PBVM is selected to execute XOR operation so that the bit is flipped. Based on this principle, we can construction a basic QR code image which is similar to the background image.

The Fusion Strategy. The basic QR code image constructed by the PBVM are fused with preprocessed the background image to get the final aesthetic QR code.

The fusion strategy used in this paper is referenced from literature [12]. In the Eq. (1), T_i represents gray average of gray block. T_0 represents binary threshold, which is obtained from Otsu's method. N_i is the module of QR code, $N_i = 1$ represents white block and $N_i = 0$ represents black block. Where Q = 0 represents that the background image is completely replaced; Q = −1 or 1 represents the central region of the specified module is replaced by the corresponding region of the QR code, and the other region is replaced by corresponding region of background image. Finally, aesthetic QR code is obtained.

$$Q = \begin{cases} 0, & T_i < T_0 \cap N_i = 0, \\ -1, & T_i > T_0 \cap N_i = 0, \\ 1, & T_i < T_0 \cap N_i = 1, \\ 0, & T_i > T_0 \cap N_i = 1. \end{cases} \tag{1}$$

2.2 The Fusion of Aesthetic QR Code and Texture Pattern

Step 1: *Pretreatment for aesthetic QR code.* Aesthetic QR code I is changed from the RGB color space to YUV color space, and the Y-channel is preprocessed using the Wiener filter, denoted as I_w. The Wiener filter minimizes the mean square error between the aesthetic QR code I and the filtered QR code I_w. It is an adaptive filter and has a significant effect on removing Gaussian noise. In order to break the periodicity in the aesthetic QR code I, a part of Wiener filter is replaced with the original image, and the I_p is generated. The replace method is shown in Eq. (2).

$$I_p(x,y) = \begin{cases} I_{w(x,y)}, & when\ r(x,y) = 1\ and \\ & \left(I(x,y) - I_{w(x,y)} > 0\right), \\ I(x,y), & otherwise. \end{cases} \tag{2}$$

where r is random matrix with the same size as the aesthetic QR code and contains only 0 and 1, the sum of the probabilities of 0 and 1 is 1. By controlling the probability of 1 in the matrix, more image details are available to maintain visual quality.

Step 2: *Calculated the JND value.* To ensure the robustness of the print-cam process, calculating the JND value of the aesthetic QR code I_p to measure the sensitivity of people to different areas of distortion. In paper [11], the JND is calculated as follows:

$$\mathrm{JND}(x,y) = \lambda_1 \times (f_1(bg(x,y), mg(x,y)) + \lambda_2) + f_2(bg(x,y)), \tag{3}$$

$$f_1(bg(x,y), mg(x,y)) = bg(x,y) \times \alpha(bg(x,y)) + \beta(bg(x,y)), \tag{4}$$

$$f_2(bg(x,y)) = \begin{cases} T_0 \times \left(1 - \sqrt{\left(\frac{bg(x,y)}{127}\right)}\right) + 3, & \textit{for } bg(x,y) \leq 127, \\ \gamma \times (bg(x,y) - 127) + 3, & \textit{for } bg(x,y) > 127, \end{cases} \tag{5}$$

$$\alpha(bg(x,y) = bg(x,y) \times 0.0001 + 0.115), \tag{6}$$

$$\beta(bg(x,y)) = \lambda - bg(x,y) \times 0.1. \tag{7}$$

where f_1 represents the spatial masking component, $bg(x, y)$ is the average background luminance, $mg(x, y)$ represents the maximum weighted average of luminance differences. $\alpha(x, y)$ is the slope of the f_1, $\beta(x, y)$ represents the intersection with the visibility threshold axis. f_2 is the visibility threshold. The value of T_0, γ and λ to be 17, 3/128 and 1/2. λ_1 and λ_2 are scaling factors. The values are $\lambda_1 = 2.0$ and $\lambda_2 = 3.0$.

Step 3: *Transform watermark message into angle.* The watermark is embedded in the aesthetic QR code by dividing the QR code into blocks. The capacity of embedded is the total bits of all blocks. In our experiments, the aesthetic QR code is divided into non-overlapping 8 × 8 blocks, each block are 64 × 64 pixels. In order to guarantee the decoding rate of the aesthetic QR code, the non-data area of the QR code is used. A center size of 256 × 256 pixels of QR code I_p is divided into 4 × 4 blocks, which is expressed as $I_k(k = 1 \ldots 16)$. 4 bits are embedded in each block and the capacity of watermark is 64 bits. The JND image is divided into block according to the QR code block, denoted as $J_k(k = 1 \ldots 16)$. The block diagram is shown in Fig. 3.

Fig. 3. Block diagram of aesthetic QR code.

The watermark message S = $\{S_1, S_2, \ldots, S_n\}$ is encoded using Hamming (32, 6) error correction coding and 4-bit Gray coding. Firstly, Hamming (32, 6) codes can detect up to two-bit errors. 4 bits per block is divided according to the errors sequence. Secondly, each 4 bits is encoded into 4-bit Gray coding. Each Gray code sequence is converted to a rotation angle, which is assigned a value between 0 and 180°. The angle is quantified according to the number of embedded bits, and then allocated to each Gray coded sequence. If each block is embedded with 4 bits, the quantization angle is given in Table 1. The quantization step size of the angle is calculated by the Eq. (8).

$$\alpha = \frac{180^\circ}{2^m}. \tag{8}$$

where m represents the bits embedded in per block.

Because the adjacent quantization step size of the Gray code is only 1 bit, when the rotation angle of the texture is misinterpreted, the Hamming code and the Gray code can ensure that the watermark message is correctly extracted.

Table 1. Quantizing the rotation angles.

Gray code	Quantization angle
0000	0°–11.25°
0001	11.25°–22.5°
0011	22.5°–33.75°
0010	33.75°–45°
0110	45°–56.25°
0111	56.25°–67.5°
0101	67.5°–78.75°

Step 4: *Generate the texture patterns.* The texture pattern rotates according to the encoded message. Texture patterns W is generated by tiling a small pseudo-random matrix, the pseudo-random pattern w(*x*, *y*) ($1 \leq x \leq 28$, $1 \leq y \leq 7$) is a random matrix containing only {−1, 1} of size 28 × 7.

The periodicity of texture pattern W is shown in Eqs. (9) and (10).

$$\mathrm{W}(\mathrm{x} + hp_0, \mathrm{y}) = \mathrm{w}(\mathrm{x}, \mathrm{y}); h, p_0 > 1, \tag{9}$$

$$\mathrm{W}(\mathrm{x}, \mathrm{y} + vp_1) = \mathrm{w}(\mathrm{x}, \mathrm{y}); v, p_1 > 1. \tag{10}$$

where p_0 and p_1 represent the number of repetitions of the periodicity, h and v denote the number of repetitions in both horizontal and vertical directions.

The texture pattern W is 128 × 128 pixels. Each texture is rotated in turn by the angle and cut the size to 64 × 64.

A directed periodic texture pattern is calculated by the Eq. (11).

$$W^{\theta} = W\left(u^{'}, v^{'}\right) + \varepsilon = \prod\left\{\begin{bmatrix} cos\theta & -sin\theta \\ sin\theta & cos\theta \end{bmatrix} \times W(u, v)\right\}. \tag{11}$$

where $\prod$ defines bilinear interpolation and ε represents the error. θ defines the rotation angle.

Step 5: *The fusion of watermark.* The JND image block J_k and directed periodic texture pattern block $W_k^{\theta_k}$ are fused with aesthetic QR code blocks $I_k(k = 1\ldots16)$. The watermark fusion algorithm is shown in Eq. (12).

$$I_k^{'}(x, y) = I_k(x, y) + W_k^{\theta_k}(x, y)(\delta_1 J_k(x, y) + \delta_2(1 - J_k(x, y))). \tag{12}$$

where $I_k^{'}$ is the kth watermark block of the aesthetic QR code, the parameters δ_1 and δ_2 determine the embedded strength. The watermark strength of $\delta_1 = 40, \delta_2 = 4$ and $\delta_1 = 50, \delta_2 = 5$ arc demonstrated in Fig. 4.

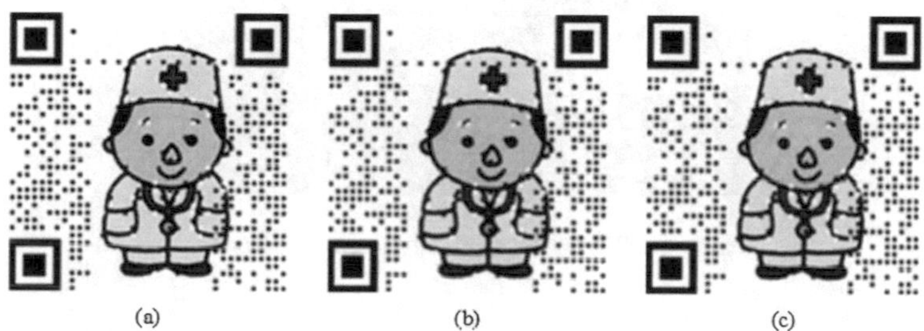

Fig. 4. (a) Original image (b) $\delta_1 = 40, \delta_2 = 4$ (c) $\delta_1 = 50, \delta_2 = 5$.

3 Watermark Extraction and Message Detection

Figure 5 shows the flow of the watermark extraction process. An aesthetic QR code captured by mobile phone and the captured image is corrected using QR code positioning information. The Y-channel of the captured image is preprocessed and then divided into blocks. These blocks are processed by calculating the autocorrelation

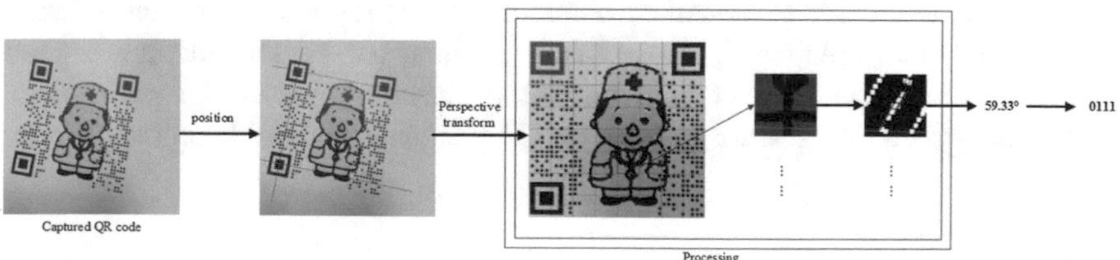

Fig. 5. The watermark extraction process.

function of each block and enhancing the autocorrelation peak. Then, lines are searched from the autocorrelation image using Hough Transform in order to obtain angles. Finally, the watermark is decoded by the angles using Gray and Hamming codes.

3.1 Correction of QR Code

In the print-cam process, the common distortion are rotation, scaling, translation, and cropping. In order to extract message, the aesthetic QR code Y_q captured by the mobile phone should be projected to a new viewing plane by perspective transformation. Firstly, the three position patterns of QR code rapidly locate the center point coordinates. Then, the four control points of the target image calculated by the four vertex coordinates. Finally, geometric distortion correction is accomplished by perspective transformation. The corrected image is denoted as Y_p. The positioned and corrected image are illustrated in Fig. 6.

Fig. 6. QR code positioning and correction.

In addition, the image Y_p is down-sampled by using bilinear interpolation, denoted as $\mathrm{Y}_p^{'}$. Then, the image Y^* is obtained by adjusting the brightness and contrast.

3.2 Calculation of Autocorrelation Function

The center size of the 256×256 of the image Y^* is chosen, and then divided into 4×4 blocks, which is expressed as $Y_k^*(k = 1\ldots16)$. For each block, the median filter removes outliers and noise from the image is to be calculated by the Eq. (13).

$$\widetilde{W}_k(x, y) = Y_k^*(x, y) - h_m(k) * Y_k^*(x, y). \tag{13}$$

where $Y_k^*(x, y)$ is the kth watermarked block and h_m represents the median filtering.

Autocorrelation function is a common method for spatial frequency texture description. The periodicity of autocorrelation function reflects the periodicity of repeated occurrence of texture primitives. The autocorrelation function of regular texture has peaks and valleys, which can be used to detect the arrangement of texture elements. The median filtered image $\widetilde{W}_k$ calculates the autocorrelation function to get Y_R^*. The autocorrelation function is shown in Eq. (14).

$$R_{\widetilde{W}_k \widetilde{W}_k}(u,v) = \sum_x \sum_y \widetilde{W}_k(x,y)\widetilde{W}_k(x+u,y+v). \tag{14}$$

The resulting values of the autocorrelation is normalized between 0 and 1.

$$R^*_{\widetilde{W}_k \widetilde{W}_k}(u,v) = \frac{\left|R_{\widetilde{W}_k \widetilde{W}_k}(u,v)\right|}{\max\left(R_{\widetilde{W}_k \widetilde{W}_k}(u,v)\right)}. \tag{15}$$

In order to enhance the peak detection of autocorrelation function, rotationally symmetric Laplacian of Gaussian filtering operation is performed in Eq. (16).

$$R^{**}_{\widetilde{W}_k \widetilde{W}_k}(u,v) = \frac{\partial^2}{\partial(u,v)^2} h^T * \left(\frac{\partial^2}{\partial(u,v)^2} h \times R^*_{\widetilde{W}_k \widetilde{W}_k}(u,v)\right). \tag{16}$$

The morphological operation usually uses a structuring element for probing and expanding the shapes containing in the input image. A structuring element $\mathrm{G}(u,v)$ is generated by the Eq. (17).

$$\mathrm{G}(u,v) = \begin{cases} 1, if\ M(u,v) \times R^{**}_{\widetilde{W}_k \widetilde{W}_k}(u,v) \geq \gamma, \\ 0,\ if\ \ M(u,v) \times R^{**}_{\widetilde{W}_k \widetilde{W}_k}(u,v) < \gamma. \end{cases} \tag{17}$$

where $M(u,v)$ denotes masking operation and γ is a threshold. Threshold operation is used to find the location of the maximum value in the autocorrelation function. γ is mainly used to control sufficient peaks for line detection.

3.3 Watermark Message Extraction

Lines are detected from the structuring element using Hough Transform. Each quantified angle represents each block of watermark message, as show in Table 1. Finally, the watermark message is decoded by Hamming (32, 6) code. Then, the message of all the blocks is connected to obtain the watermark. The flow chart of message detection is illustrated in Fig. 7.

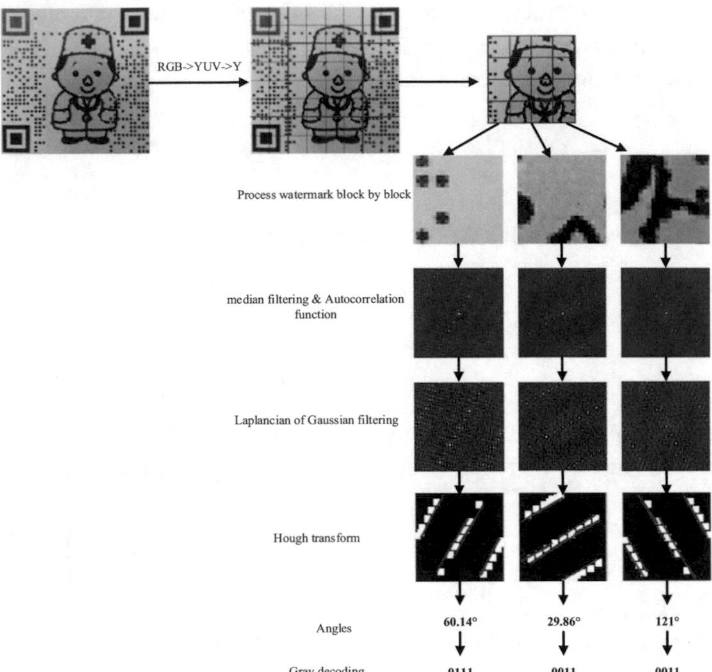

Fig. 7. The message extraction.

4 Experiments

4.1 Subjective Evaluation

The system was implemented by MATLAB. In the experiments, all the test images are of size 512 × 512 and each aesthetic QR code is embedded with different watermarks. We utilized HP Color LaserJet 1025 PCL 6 printer. The physical size of the printed images is 6 × 6 cm. The watermark images are printed on A4 paper, and all experiments are carried out in a normal indoor lighting environment. Part of the test images, watermark images and the print-cam images are given in Table 2.

The quality of the watermark images are evaluated by subjective testing. The strengths of watermark images are $\delta_1 = 40, \delta_2 = 4$ and $\delta_1 = 50, \delta_2 = 5$. Ten persons are invited to evaluate the original images and watermarked ones. The five-point scale is rated on images: (5: imperceptible, 4: perceptible but not objectionable, 3: slightly objectionable, 2: objectionable, 1: very objectionable). The results of subjective evaluation are shown in Table 3.

Table 2. The test images.

Test images	Watermark images(δ_1=40, δ_2=4)	Print-cam images

Table 3. Subjective evaluation scores of watermark images.

Strength	$\delta_1 = 40, \delta_2 = 4$	$\delta_1 = 50, \delta_2 = 5$
image1	4.81	4.66
image2	4.85	4.78
image3	4.84	4.67
image4	4.78	3.93
image5	4.97	4.75
image6	4.95	4.58

4.2 Objective Evaluation

For image quality analysis, objective evaluation has measured by SSIM (structural similarity index) [13]. Given two aesthetic QR codes *x* and *y*, the structural similarity of the two images can be found as follows:

$$\mathrm{SSIM}(x, y) = \frac{(2u_x u_y + c_1)(2\sigma_{xy} + c_2)}{\left(u_x^2 + u_y^2 + c_1\right)\left(\sigma_x^2 + \sigma_y^2 + c_2\right)}. \tag{18}$$

where u_x the average of x, u_y the average of y, σ_x^2 the variance of x, σ_y^2 the variance of y, σ_{xy} the covariance of x and y, $c_1 = (k_1 L)^2$, $c_2 = (k_2 L)^2$ two variables to stabilize the division with weak denominator. L the dynamic range of the pixel-values, $k_1 = 0.01$, $k_2 = 0.03$ by default.

The value of SSIM is between -1 and 1, where value 1 indicates that two images are the same. The experiment result shows our algorithm has good visual quality when the watermark strength is lower than $\delta_1 = 70, \delta_2 = 7$, while the SSIM is greater than 0.8. The obtained results are shown in Table 4.

Table 4. Image quality by SSIM.

Strength	$\delta_1 = 40, \delta_2 = 4$	$\delta_1 = 50, \delta_2 = 5$	$\delta_1 = 60, \delta_2 = 6$	$\delta_1 = 70, \delta_2 = 7$
image1	0.9311	0.9122	0.9010	0.8882
image2	0.9724	0.9641	0.9554	0.9477
image3	0.9623	0.9438	0.9230	0.9105
image4	0.9206	0.9020	0.8879	0.8814
image5	0.9747	0.9677	0.9631	0.9576
image6	0.9785	0.9720	0.9624	0.9543

4.3 Robustness Against Distance Variations

To make sure the aesthetic QR code decoded correctly, the range of the mobile phone from A4 paper is constantly changed. These experiments are tested with Xiaomi 8 mobile phone, a 1200 megapixel CMOS camera. Wide-angle lens is 28 mm and aperture F1.8. Test images are printed on an A4 paper, using HP Color LaserJet 1025 PCL 6 printer.

Fig. 8. Examples of captured images at different distances.

Figure 8 shows the captured image using mobile phone. The captured image is segmented into blocks and the average precision of the rotation angle of each block is calculated. The accuracy of angle detection is determined by the relative error between the calculated average angle and the actual angle. The smaller the error, the higher the accuracy of the test results. In the experiment, the watermark is correctly extracted at a watermark strength of $\delta_1 = 50, \delta_2 = 5$ and a distance between 8 and 18. The experimental results are illustrated in Fig. 9.

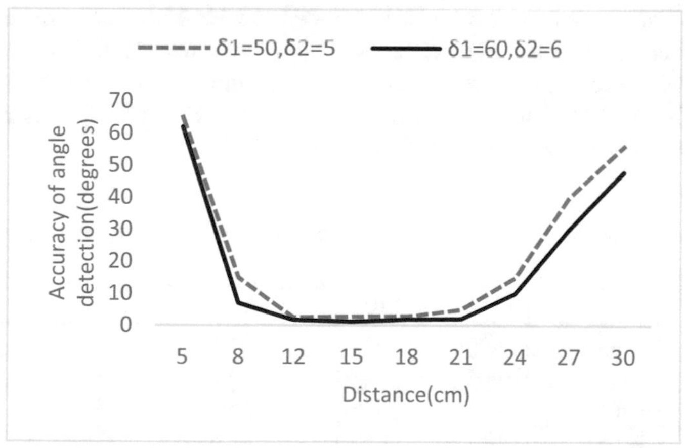

Fig. 9. Effect of distance on angle accuracy.

Fig. 10. Examples of captured images at different angles.

4.4 Robustness Against Angle Variations

For testing robustness at various angles, different angles of the test image are taken at the same distance, and the watermark bits extracted correctly is calculated. The captured images at various angles are shown in Fig. 10. If each block is embedded with 4 bits, the correct extraction result of the watermark is given in Table 5 when the strength is $\delta_1 = 50, \delta_2 = 5$. Table 6 shows the experimental results generated by embedding 5 bits into each block. Compared with [10], our algorithm utilizing the locating position patterns of the QR code, the additional watermark reference block is not required. The proposed algorithm not only improves the embedding capacity of the watermark, but also increases the angle range of the captured image.

The test results show that when the aesthetic QR code is taken vertically, the watermark strength of $\delta_1 = 50, \delta_2 = 5$ is sufficient to extract the watermark correctly from all the images. When each block is embedded with 4 bits, the quantization range of the angle is larger than 5 bits per block. It is shown that embedding more bits into each block will reduce the degree of quantification, and therefore the robustness of the watermark is reduced. The angle of shooting for all printed images is robust to 15 degree tilt when 4 bits are embedded each block.

Table 5. Robustness to attacks with 64 bit message. (Sixteen 4 bit message segments embedded.)

Angle	0°	5°	10°	15°	20°	25°
image1	64/64	64/64	64/64	64/64	63/64	62/64
image2	64/64	64/64	64/64	63/64	62/64	61/64
image3	64/64	64/64	64/64	64/64	64/64	63/64
image4	64/64	64/64	64/64	64/64	63/64	62/64
image5	64/64	64/64	64/64	61/64	60/64	59/64
image6	64/64	64/64	64/64	64/64	64/64	63/64

Table 6. Robustness to attacks with 80 bit message. (Sixteen 5 bit message segments embedded.)

Angle	0°	5°	10°	15°	20°	25°
image1	80/80	80/80	80/80	75/80	78/80	76/80
image2	80/80	80/80	80/80	77/80	79/80	75/80
image3	80/80	80/80	80/80	78/80	80/80	79/80
image4	80/80	80/80	80/80	80/80	79/80	80/80
image5	80/80	80/80	80/80	76/80	78/80	79/80
image6	80/80	80/80	80/80	80/80	79/80	78/80

5 Conclusion

A novel authentication algorithm for aesthetic QR code based on positive basis vector matrix and directed texture pattern is proposed. This method realizes dual security authentications of aesthetic QR code and maintains a better visual quality. By utilizing the locating position patterns of the QR code, the additional watermark reference block is not required. Moreover, the proposed scheme not only enhances the watermark capacity, but also improves the watermark robustness against the print-cam process.

In the future, we hope to improve the robustness and imperceptibility of the watermark. The unused QR code region during the embedding process deserves further investigations.

Acknowledgments. This work was mainly supported by National Natural Science Foundation of China (No. 61370218, No. GG19F020033), Public Welfare Technology and Industry Project of Zhejiang Provincial Science Technology Department (No. 2016C31081, No. LGG18F020013, No. LGG19F020016).

References

1. Suryotrisongko, H., Sugiharsono, Setiawan, B.: A novel mobile payment scheme based on secure quick response payment with minimal infrastructure for cooperative enterprise in developing countries. Proc. Soc. Behav. Sci. **65**, 906–912 (2012)
2. Ortiz-Yepes, D.A.: A review of technical approaches to realizing near-field communication mobile payments. IEEE Secur. Priv. **14**(4), 54–62 (2016)
3. Ahmed, Q., Munib, S., Mirza, M.T., et al.: Smart phone based online medicine authentication using print-cam robust watermarking. In: 13th International Conference on Frontiers of Information Technology, pp. 222–227. IEEE, Islamabad (2015)
4. Kuribayashi, M., Morii, M.: Aesthetic QR code based on modified systematic encoding function. IEICE Trans. Inf. Syst. **E100D**, 42–51 (2017)
5. Lu, J., Yang, Z., Li, L., et al.: Multiple schemes for mobile payment authentication using QR code and visual cryptography. Mob. Inf. Syst. **9**, 1–12 (2016)
6. Li, L., Zhang, S., Yang, Z., Lu, J., Chang, C.C.: Novel schemes for bike-share service authentication using aesthetic QR code and color visual cryptography. In: Sun, X., Chao, H. C., You, X., Bertino, E. (eds.) Cloud Computing and Security, ICCCS 2017. LNCS, vol. 10603, pp. 837–842. Springer, Cham (2017)

7. Pramila, A., Keskinarkaus, A., Seppänen, T.: Increasing the capturing angle in print-cam robust watermarking. J. Syst. Softw. **135**, 205–215 (2018)
8. Katayama, A., Nakamura, T., Yamamuro, M., Sonehara, N.: New highspeed frame detection method: side trace algorithm (STA) for i-appli on cellular phones to detect watermarks. In: Proceedings of the 3rd International Conference on Mobile and Ubiquitous Multimedia, College Park, Maryland, USA, vol. 83, pp. 109–116 (2004)
9. Kim, W., Jang, H.J., Kim, G.Y.: Transmission rate prediction of VBR motion image using the kalman filter. In: International Conference on Computational Science and Its Applications, vol. 3981, pp. 106–113. Springer, Heidelberg (2006)
10. Pramila, A., Keskinarkaus, A., Seppänen, T.: Toward an interactive poster using digital watermarking and a mobile phone camera. SIViP **6**(2), 211–222 (2016)
11. Chou, D.-H., Li, Y.-C.: A perceptually tuned subband image coder based on the measure of just-noticeable-distortion profile. IEEE Trans. Circ. Syst. Video Technol. ACM **5**(6), 467–476 (1995)
12. Li, L., et al.: A new aesthetic QR code algorithm based on salient region detection and SPBVM. In: Peng, S.L., Wang, S.J., Balas, V., Zhao, M. (eds.) Security with Intelligent Computing and Big-data Services, SICBS 2017. AISC, vol. 733, pp. 20–32. Springer, Cham (2018)
13. Hore, A., Ziou, D.: Image quality metrics: PSNR vs. SSIM. In: International Conference on Pattern Recognition, pp. 2366–2369. IEEE (2010)

A Heterogeneous Multiprocessor Independent Task Scheduling Algorithm Based on Improved PSO

Xiaohui Cheng and Fei Dai(✉)

Guangxi Key Laboratory of Embedded Technology and Intelligent System, College of Information Science and Engineering, Guilin University of Technology, Jiangan Road No. 12, Guilin 541000, China
{cxiaohui,102016453}@glut.edu.cn

Abstract. The independent task scheduling problem of heterogeneous multi-processors belongs to the NP-hard problem. The emergence of evolutionary algorithms provides a new idea for solving this problem. Particle swarm optimization (PSO) is a kind of intelligent evolutionary algorithm and it could be used to solve scheduling problem. We firstly discretized the representation of particle swarm optimization algorithm and made it suitable for the scheduling problem of heterogeneous multiprocessors. Then, the PSO algorithm was introduced into heterogeneous multiprocessors independent task scheduling problem by modeling method. In order to overcome particle swarm optimization algorithm's problem that is easy to fall into local optimum and premature convergence. We proposed a heterogeneous multiprocessor independent task scheduling algorithm based on improved PSO by improving the update operation of particle swarm optimization algorithm and transformed it into crossover and mutation operation of genetic algorithm. The experimental results show that the improved PSO scheduling algorithm can overcome the premature defects of PSO algorithm and the makespan of proposed IPSO is smaller than PSO.

Keywords: Task scheduling · Independent tasks · Particle swarm optimization · Heterogeneous multiprocessors

1 Introduction

Today, multiprocessor task scheduling challenges scholars due to the problem of efficiently assigning a great deal of tasks in very short execution time, usually limited to the order of few minutes, or even seconds [1]. Indeed, a good task scheduling algorithm will benefit a lot for improving the performance of multiprocessors system. Therefore, efficiently assigning and mapping tasks to various processors become a critical issue. Traditional homogeneous task scheduling problem is one of the NP-hard problems [2], the heterogeneity makes it even harder. As a result, it is necessary for researchers to solve this problem.

The problem of task scheduling could be transformed into the problem of mapping tasks to different processors. The problem of mapping the tasks to multiple processors can be differentiated in terms of static mapping, dynamic mapping, independent task

C.-N. Yang et al. (Eds.): SICBS 2018, AISC 895, pp. 267–279, 2020.
https://doi.org/10.1007/978-3-030-16946-6_21

set, dependent task set, flow-shop task set, homogeneous processors, heterogeneous processors, or various qualitative parameters used for performance measure [3]. Static mapping, independent task set, and heterogeneous processor are the main research aspects of this paper. On the basis of it, we built a heterogeneous multiprocessor independent task scheduling model and implemented our algorithm on this scheduling model.

In this paper, we focus on the independent task scheduling of the heterogeneous multiprocessors and improved a particle swarm optimization (PSO) algorithm by using genetic algorithm (GA) ideas. The experiment results show our improved algorithm outperform PSO scheduling algorithm.

The rest of the manuscript is organized as follows. The next section presents the definitions of independent task scheduling model, the assumptions and objectives. The main related work of independent task scheduling for heterogeneous multiprocessors are reviewed in Sect. 3. The details about our improved PSO algorithm are described in Sect. 4. The experiments and analysis are presented in Sect. 5. Finally, Sect. 6 introduces the main conclusions of our research and discusses the future work.

2 Heterogeneous Multiprocessor Independent Task Scheduling Model

This section introduces system model for heterogeneous multiprocessor independent task scheduling and its objectives.

2.1 System Model

Heterogeneous multiprocessor system consists of a set of m heterogeneous processors having different processing elements. A task will get the resources only from the processor allocated to it. There is a set $T(T_1, T_2, \ldots, T_n)$ of n independent and simultaneously available tasks and a set $P(P_1, P_2, \ldots, P_m)$ of m various processors. C $(C_1, C_2, \ldots, C_m)$ represents processing time of task $T_i(i = 1, 2, \ldots, n)$ (Fig. 1).

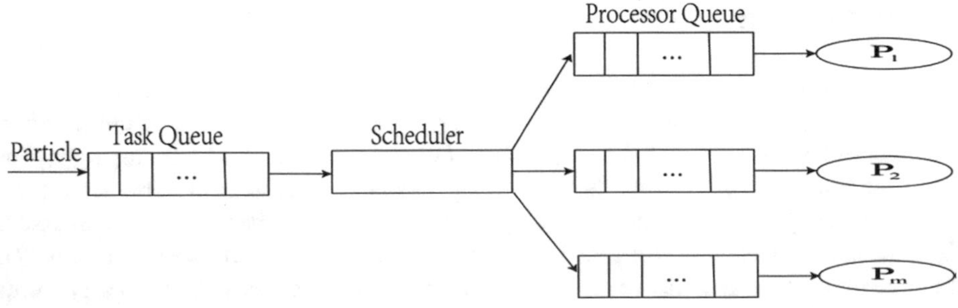

Fig. 1. Particle swarm scheduling model under heterogeneous multiprocessor

2.2 Assumptions

Following assumptions are made while modeling the problem:

1. Processors are heterogeneous (each with different computing speed) and endlessly available from start-to-end time.
2. Each processor can only process one task at a time.
3. All tasks are mutually independent and they do not have executed sequence and cannot be preempted.
4. All set-up times are included in the execution time and are independent from the logical order of tasks.

2.3 Objectives

There are various qualitative parameters available in order to judge the performance of scheduling algorithms. Here, maximum span among all processors is taken to evaluate the performance of scheduling algorithm. Span for processor i named as $Span_i$ could be defined as:

$$Span_{i \in \{1,2,3\ldots,m\}} = \sum\nolimits_{j=1}^{n} C_j M_{ij} \tag{1}$$

where $M_{ij} = 1$ if task T_j is assigned to processor P_i, otherwise $M_{ij} = 0$.

The makespan is stated as the maximum completion time for all processors. Maximum span can be defined as:

$$MakeSpan = Maximum\left[Span_{i \in \{1,2,3\ldots,n\}}\right] \tag{2}$$

The objective of our problem is that it maps the task set T to the set of processors P while minimizing the makespan having considered the constraints.

3 Related Works

In heterogeneous multi-core scheduling problems, it can be divided into two types: independent tasks and dependent task scheduling. At present, there are many research results for task-dependent scheduling [4–13], but the number of research results in independent task scheduling is relatively small. As a result, this domain requires further research and relevant research results should be combined with practice. In the field of independent task scheduling, some of the research results in recent years are as follows: Gogos et al. [14] proposed a heterogeneous multiprocessor independent task algorithm that uses heuristic and column pricing to achieve shorter scheduling lengths than other scheduling algorithms. Braun, Siegel et al. [15] compared the effects of seven static heuristics in heterogeneous environment independent task scheduling. The experimental results showed that the genetic algorithm performs optimally. Iturriaga, Nesmachnow et al. [1] implemented a parallel random search scheduling method for independent task scheduling in CPU/GPU heterogeneous computing systems. Experimental results showed that the method has short execution time and high solution quality. Dorronsoro and Pinele [16] innovatively combined genetic algorithms with machine learning algorithms and introduced them into independent task scheduling problems to solve this problem. Compared with the other two heuristic algorithms, introducing machine learning into genetic algorithm could improve accuracy of scheduling greatly. Zhou, Jiang and Fang [17] proposed a minimum and earliest completion time algorithm for the independent task of heterogeneous environment. The algorithm introduced Min-min algorithm to solve k tasks with earliest completion time, and first dealt with the most costly scheduling. Compared with the Min-min algorithm, the proposed algorithm is better. At present, some scholars have applied the PSO algorithm to the independent task scheduling problem in multi-core processors and proposed corresponding methods from various aspects [3, 18–22].

In summary, the research on independent task scheduling has yet to be deepened, especially the independent task scheduling problem of heterogeneous multi-cores processor. According to the independent task scheduling problem of heterogeneous multiprocessors, firstly, the heterogeneous multiprocessor independent task scheduling model is constructed. Secondly, referring to the characteristics of the particle swarm optimization algorithm, we introduce the crossover and variation of genetic algorithms in the update operation of PSO algorithm. Then, an improved particle swarm heterogeneous multiprocessor independent task scheduling algorithm based on this model is proposed. We hope that it will bring new inspiration and help to industry and scientific research by solving such problems.

4 PSO Algorithm

Particle swarm optimization is inspired by interactions involved in the collective social behavior of animals. For example, a group of migratory birds maintain a certain formation flying between each other. The PSO algorithm focuses on the motion of a group of examples to cover the solution space to find different possible solutions for approximate optimal solutions.

Assuming that search space of the problem is an n-dimensional space, the position and velocity vectors of the first particle can be expressed as follows:

$$X_i = [x_{i1}, x_{i2}, \ldots, x_{in}]$$

$$V_i = [v_{i1}, v_{i2}, \ldots, v_{in}]$$

During each iteration, the particle continuously adjusts its position and updates it by tracking the position and velocity extremum. The optimal solution found by the first particle itself is marked as the individual optimal solution $\mathrm{pbest_i} = [\mathrm{pbest_{i1}}, \mathrm{pbest_{i2}}, \cdots, \mathrm{pbest_{in}}]$, and the other is the optimal solution found by the whole population at present, which is called the global optimal solution, denoted as $\mathrm{gbest_i} = [\mathrm{gbest_{i1}}, \mathrm{gbest_{i2}}, \cdots, \mathrm{gbest_{in}}]$. In addition, the optimal solution of all the neighbors of the particle is the local optimal solution.

In the original particle swarm optimization algorithm, the particle position and velocity variation formula [5] are as follows:

$$V_{is}(t+1) = V_{is}(t) + c_1 r_1(t)[p_{is}(t) - X_{is}(t)] + c_2 r_2(t)[p_{gs}(t) - x_{is}(t)] \quad (3)$$

$$X_{is}(t+1) = X_{is}(t) + V_{is}(t+1) \quad (4)$$

In above equations, i = [1, m], s = [1, S]. c1, c2 are a non-negative integer and are called learning factors. r1, r2 are independent random number between [0, 1]. The V_{max} is the maximum speed of particle, and it is constant and set by users. The m represents the population size and the t represents iteration number.

5 Based on Improved PSO Scheduling Algorithm

The improved particle swarm optimization algorithm proposed in this paper defines each particle as the potential solution of the problem, that is, after initializing the generated particle, each particle obtains a scheduling length and a scheduling sequence.

The innovation of the improved particle swarm optimization scheduling algorithm proposed in this paper is that it mainly transforms the update operation of the particle swarm optimization algorithm and introduces the crossover and mutation operations of the genetic algorithm into the particle swarm optimization algorithm. In the update operation of the algorithm, the main function of the crossover operation is that it exchanges the same type but different sizes tasks of the two processors, and then calculates their scheduling sequence and the scheduling length. The function of the mutation operation is that it chooses two processors in the processor list randomly, one processor reduces a task and the other processor adds the task that former processor's lost, and then calculates the scheduling length and running time of the two processors.

Pseudo code of update algorithm (the core of this algorithm) can be seen as follows (Tables 1, 2 and 3):

The flow chart of the improved algorithm is as follows (Fig. 2):

Algorithm detailed calculation process is as follows:

Step 1: m specified different processing frequencies heterogeneous processor and n tasks to be processed are Initialized (n > m). Also, p particles are randomly generated.

Step 2: After generating p particles, each particle will generate a random scheduling. Firstly, the task is assigned to the processor randomly and equally, the random scheduling sequence is obtained, the scheduling length of the particle is calculated, and the local optimal scheduling is set. When each particle calculates its scheduling length and obtains their scheduling sequence, the global optimal scheduling length and its scheduling sequence are obtained.

Step 3: Particle update operation is performed. Here is the core of the algorithm. Firstly, the scheduling sequence and the local scheduling sequence are run a1 times crossover operation. After the crossover operation, the optimal local scheduling sequence can be obtained between the particles. Then a2 times cross operation is performed on the scheduling sequence and the globally optimal scheduling sequence.

Table 1. Update algorithm

```
Input:
parameters for scheduling algorithm
Output:
Makespan, LocalBestParticle
----------------------------------------
for i=1 to a1
Crossoveroperator(Schedule,LocalBestSchedule);
end for
for j=1 to a2
CrossOverOperator(Schedule,GlobalBestSchedule);
end for
for k=1 to b
MutationOperator();
end for
CalculateMakespan
SetLocalBestParticle
```

After the cross operation is completed, the scheduling sequence obtains the global optimal scheduling sequence. Finally, b times mutation operation is carried out.

Step 4: After the update operation, the scheduling length is calculated, the global optimal particle is set, and the scheduling length is outputted.

Step 5: Judge whether the stop condition is satisfied or not, otherwise repeat the first step.

Table 2. CrossOverOperator algorithm

Input:

Schedule, LocalBestSchedule/GlobeBestSchedule

Output:

New Schedule

1: Choose a randomly a processor ID random_processor_id;

2: Get the processor to move from the chosen processor of best particle;

3: Get a random task from the shortlisted tasks

4: Look for this task in processor_schedule. That is, check which processor currently has the task in processor_schedule

5: Get the task to lose from processor_map

6: Get a random task from the shortlisted tasks

7: Swap the tasks between the processors in processor_schedule

Table 3. MutationOperater algorithm

Input:

Output:

1: Get two random processors, random_p1_id, random_p2_id

2: Compare run time to select which processor will gain/lose a task

3: Select which process will be moved from losing p to gaining p

4: One processor lose a task from its task_list and the other get this task and add it in its task_list

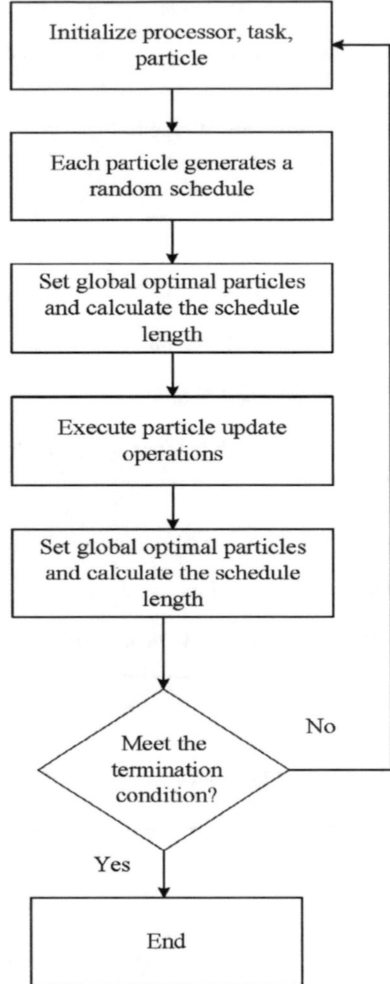

Fig. 2. Algorithm flowchart

6 Experiments and Analysis

6.1 Environment of the Experiments

The experimental environment in this paper is shown in the following table (Table 4):

In this experimental environment, the algorithm parameters are set as follows: cognitive factor a1 = 5, social factor a2 = 5, inertial factor b = 10, particle number = 20, iteration number = 1000. The processing speed of processor is randomly generated from 1000 to 300000 instructions per millisecond. Communication rate between processors is randomly set from 10 to 1000 bytes per millisecond. Average instruction of task is set from 100 to 100000000 instructions randomly. Data amount of task is generated from 0 to 100000 bytes randomly.

Table 4. Experimental environment setup

CPU	AMD Athlon(tm) II X4 645 Processor 2.3 GHz
Memory	4G
Operating system	Windows 7
Development platform	Visual Studio 2013
Program language	C++

6.2 Experimental Result

In order to verify the effectiveness of the algorithm, this paper uses two groups of experiments to test the performance of the algorithm: The first group is set to execute a fixed number of tasks but under the system environment of different number of cores. The second group, as control group, is set to execute different number of tasks and under the system environment of a fixed number of cores. The first group of experiments includes experiments 1 to 3, and the experiment setup is as follows: the number of fixed tasks is 1000, and the experiments are carried out under the environment of 10, 15 and 20 heterogeneous processor cores respectively. In contrast, the second group of experiments includes experiments 4 to 6, the experimental setup is as follows: the fixed number of heterogeneous processor cores is 10, and the performance of the algorithm is tested under the task number of 350, 600, 850. The following experimental results are the average of the results obtained after the algorithm is run 20 times.

In Table 5, the experiment 1 shows every 50 iterations' makespan of PSO and the improved PSO scheduling algorithm in the system environment of processor core number of 10, task number of 1000. The experiment 2 displays every 50 iterations' makespan of PSO and the improved PSO scheduling algorithm in the system environment of processor core number of 15, task number of 1000. The experiment 3 demonstrates every 50 iterations' makespan of PSO and the improved PSO scheduling algorithm in the system environment of processor core number of 20, task number of 1000.

The makespan of the first group of experiments after 1000 iterations is shown below:

In contrast, we set up a second group to test the performance of our proposed algorithm. Table 6 depicts the experiment results of second group.

From Table 6, Experiment 4 reveals comparison of experimental results between PSO and improved PSO scheduling algorithm in the system environment of heterogeneous processor core number of 10 and a task number of 350. Experiment 5 exhibits comparison of experimental results between PSO and improved PSO scheduling algorithm in the system environment of heterogeneous processor core number of 10 and a task number of 600. Experiment 6 displays comparison of experimental results between PSO and improved PSO scheduling algorithm in the system environment of heterogeneous processor core number of 10 and a task number of 850.

The makespan of the second group of experiments after 1000 iterations is shown in the following figure:

From the experimental results of Figs. 3 and 4, it can be seen that the proposed IPSO heterogeneous multiprocessor independent task scheduling algorithm can obtain

Table 5. First group of experiments

Iteration	Experiment 1		Experiment 2		Experiment 3	
	PSO makespan	IPSO makespan	PSO makespan	IPSO makespan	PSO makespan	IPSO makespan
0	1599.269	1601.006	1768.758	1767.658	1367.156	1368.866
1	1518.013	1593.036	1645.268	1611.598	1246.121	1272.121
50	308.924	284.924	225.033	210.033	190.003	167.003
100	299.409	282.409	223.476	208.446	188.852	165.852
150	293.463	280.463	217.459	207.459	188.344	165.344
200	292.646	279.083	216.198	206.691	172.760	164.660
250	292.036	278.636	216.836	205.836	172.722	164.022
300	292.036	278.094	216.386	205.386	171.693	163.693
350	291.831	277.831	216.002	205.002	171.612	163.512
400	291.790	277.790	215.774	204.774	171.612	163.404
450	291.790	277.790	215.588	204.588	171.612	163.155
500	291.790	277.790	215.411	204.411	170.227	163.138
550	291.747	277.747	215.353	204.353	170.227	163.102
600	291.663	277.663	215.310	204.310	170.227	163.027
650	291.663	277.661	214.308	204.308	170.217	163.027
700	290.663	277.634	214.292	204.292	170.185	163.008
750	290.663	277.633	214.260	204.258	170.166	163.007
800	290.663	277.609	214.260	204.258	170.125	163.006
850	290.646	277.567	214.260	204.197	170.106	163.001
900	290.646	277.466	214.260	204.190	170.106	162.864
950	290.646	277.466	214.260	204.160	170.106	162.853
1000	290.646	277.466	214.198	204.099	170.106	162.423

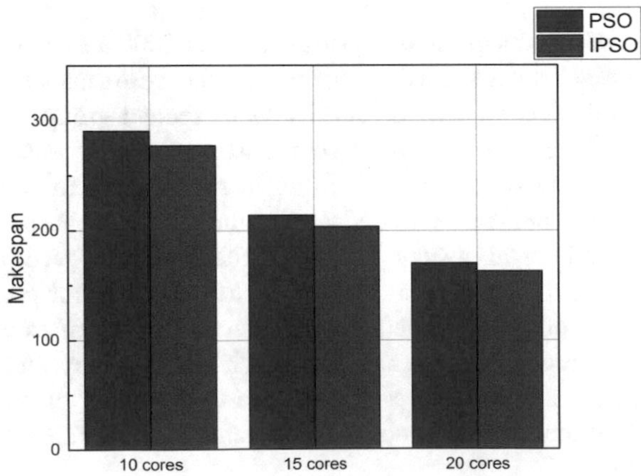

Fig. 3. Two algorithms' makespan diagram in 10, 15, 20 processors after 1000 iterations

Table 6. Second group of experiments

Iteration	Experiment 4		Experiment 5		Experiment 6	
	PSO makespan	IPSO makespan	PSO makespan	IPSO makespan	PSO makespan	IPSO makespan
0	529.489	528.723	985.158	987.559	1426.258	1423.381
1	418.692	414.790	863.159	869.120	1335.365	1310.822
50	138.576	101.024	200.568	173.732	302.924	242.301
100	112.684	99.959	183.526	171.944	259.409	239.831
150	112.684	99.794	183.326	171.331	250.463	238.073
200	110.583	99.721	185.326	170.902	257.083	237.369
250	111.583	99.681	183.232	170.774	257.036	236.953
300	111.583	99.608	183.232	170.701	257.036	236.705
350	111.583	99.467	183.232	170.628	256.831	236.691
400	110.883	99.374	182.126	170.588	256.790	236.675
450	110.883	99.374	182.126	170.520	256.790	236.597
500	110.883	99.311	181.626	170.446	256.790	236.532
550	110.883	99.299	181.626	170.421	255.747	236.408
600	110.883	99.274	181.428	170.406	255.663	236.339
650	110.883	99.263	181.426	170.366	255.663	236.334
700	110.883	99.263	181.126	170.223	255.663	236.294
750	110.879	99.258	180.126	170.212	255.649	236.279
800	110.879	99.258	180.126	170.210	255.649	236.237
850	110.579	99.258	180.126	170.179	255.646	236.237
900	110.579	99.248	180.126	170.122	255.646	236.237
950	110.179	99.248	180.326	170.110	255.646	236.191
1000	110.179	99.238	182.326	170.104	255.646	236.107

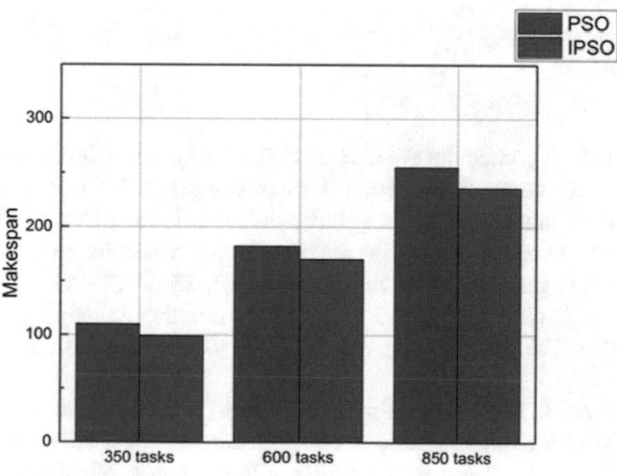

Fig. 4. Two algorithms' makespan diagram in 350, 600, 850 tasks after 1000 iterations

shorter scheduling length than the PSO scheduling algorithm after 1000 iterations, regardless of whether in the system condition of various number of processors and fixed number of tasks or different number of tasks and fixed number of processors.

It can be seen from Table 5 that both IPSO and PSO algorithms' makespan are getting smaller and smaller after each iteration under the conditions of different numbers of processors and 1000 tasks, but IPSO on average obtains lower scheduling length. From Table 6, we can see that both IPSO and PSO algorithm's makespan decrease as the iteration goes, but overall IPSO can obtain lower makespan than PSO under the condition of various tasks' number and 10 processors. Considering the experimental results of above experimental tables and images, we can draw the conclusion that IPSO algorithm has lower scheduling length and faster convergence speed.

7 Conclusions

In this paper, a heterogeneous multi-core independent scheduling model is established, and the assumptions and algorithm objectives are given. On above basis, a heterogeneous multi-core independent task scheduling algorithm based on improved PSO is proposed. The experimental results show that the proposed improved PSO heterogeneous multi-core independent task scheduling algorithm is better than the PSO algorithm under the scheduling model, which can jump out of the local optimal solution faster and avoid premature occurrence. In the next work, we plan to introduce methods such as machine learning to optimize the algorithm further.

Acknowledgments. As the research of the thesis is sponsored by National Natural Science Foundation of China (No: 61662017, No: 61262075), Key R & D projects of Guangxi Science and Technology Program (AB17195042), Guangxi Science and Technology Development Special Science and Technology Major Project (No: AA18118009), Guangxi Key Laboratory Fund of Embedded Technology and Intelligent System, we would like to extend our sincere gratitude to them.

References

1. Iturriaga, S., et al.: A parallel local search in CPU/GPU for scheduling independent tasks on large heterogeneous computing systems. J. Supercomput. **71**(2), 648–672 (2014)
2. Sahni, S.K.: Algorithms for scheduling independent tasks. J. ACM **23**(1), 116–127 (1976)
3. Shriya, S., et al.: Directed search-based PSO algorithm and its application to scheduling independent task in multiprocessor environment **404**, 23–31 (2016)
4. Yi, J., et al.: Reliability-guaranteed task assignment and scheduling for heterogeneous multiprocessors considering timing constraint. J. Signal Process. Syst. **81**(3), 359–375 (2014)
5. Kumar, N., Vidyarthi, D.P.: A novel hybrid PSO–GA meta-heuristic for scheduling of DAG with communication on multiprocessor systems. Eng. Comput. **32**(1), 35–47 (2015)
6. Xu, Y., Li, K., Hu, J., et al.: A genetic algorithm for task scheduling on heterogeneous computing systems using multiple priority queues. Inf. Sci. **270**(6), 255–287 (2014)

7. Ayari, R., et al.: ImGA: an improved genetic algorithm for partitioned scheduling on heterogeneous multi-core systems. Des. Autom. Embed. Syst. **22**(1–2), 183–197 (2018)
8. Jiang, Y., et al.: DRSCRO: a metaheuristic algorithm for task scheduling on heterogeneous systems. Math. Probl. Eng. **2015**, 1–20 (2015)
9. Prescilla, K., Immanuel Selvakumar, A.: Modified Binary Particle Swarm optimization algorithm application to real-time task assignment in heterogeneous multiprocessor. Microprocess. Microsyst. **37**(6–7), 583–589 (2013)
10. Xie, G., et al.: Mixed real-time scheduling of multiple DAGs-based applications on heterogeneous multi-core processors. Microprocess. Microsyst. **47**, 93–103 (2016)
11. Xu, C., Li, T.: Chemical reaction optimization for task mapping in heterogeneous embedded multiprocessor systems. Adv. Mater. Res. **712–715**, 2604–2610 (2013)
12. Xu, Y., et al.: A DAG scheduling scheme on heterogeneous computing systems using double molecular structure-based chemical reaction optimization. J. Parallel Distrib. Comput. **73**(9), 1306–1322 (2013)
13. Rzadca, K., Seredynski, F.: Heterogeneous multiprocessor scheduling with differential evolution. In: IEEE Congress on Evolutionary Computation (2005)
14. Gogos, C., et al.: Scheduling independent tasks on heterogeneous processors using heuristics and Column Pricing. Future Gener. Comput. Syst. **60**, 48–66 (2016)
15. Braun, T.D., et al.: A comparison of eleven static heuristics for mapping a class of independent tasks onto heterogeneous distributed computing systems. J. Parallel Distrib. Comput. **61**(6), 810–837 (2001)
16. Dorronsoro, B., Pinel, F.: Combining machine learning and genetic algorithms to solve the independent tasks scheduling problem. In: IEEE International Conference on Cybernetics (2017)
17. Zhou, Y., Jiang, C., Fang, Y.: Research on independent task scheduling algorithm in heterogeneous environment. Comput. Sci. **35**(8), 90–92+97 (2008)
18. Omidi, A., Rahmani, A.M.: Multiprocessor independent tasks scheduling using a novel heuristic PSO algorithm. In: IEEE International Conference on Computer Science and Information Technology, pp. 369–373. IEEE (2009)
19. Zhang, W., et al.: Energy-aware real-time task scheduling for heterogeneous multiprocessors with particle swarm optimization algorithm. In: Mathematical Problems in Engineering, pp. 1–9 (2014)
20. Sarathambekai, S., Umamaheswari, K.: Intelligent discrete particle swarm optimization for multiprocessor task scheduling problem. J. Algorithms Comput. Technol. **11**(1), 58–67 (2016)
21. Chen, J., Pan, Q.: Improved particle swarm optimization algorithm for solving independent task scheduling problem. Microelectron. Comput. **34**(6), 214–215 (2008)
22. Wang, Y., Wang, N., Yang, C., et al.: A discrete particle swarm optimization algorithm for task assignment problem. J. Cent. South Univ. (Sci. Technol.) **39**(3), 571–576 (2008)

Cross-Domain Text Sentiment Classification Based on Wasserstein Distance

Guoyong Cai, Qiang Lin(✉), and Nannan Chen

Guangxi Key Lab of Trusted Software,
Guilin University of Electronic Technology, Guilin, China
ccgycai@gmail.com, firejohnny@outlook.com,
1728792152@qq.com

Abstract. Text sentiment analysis is mainly to detect the sentiment polarity implicit in text data. Most existing supervised learning algorithms are difficult to solve the domain adaptation problem in text sentiment analysis. The key of cross-domain text sentiment analysis is how to extract the domain shared features of different domains in the deep feature space. The proposed method uses denosing autoencoder to extract the deeper shared features with better robustness. In addition, Wasserstein distance-based domain adversarial and orthogonal constraints are combined for better extracting the deep shared features of the different domain. Finally, the deep shared features are used for cross domain sentiment classification. The experimental results on the real data sets show that the proposed method can better adapt to domain differences and achieve higher accuracy.

Keywords: Cross-domain · Wasserstein distance · Domain adversarial · Text sentiment analysis

1 Introduction

Traditional text sentiment analysis researches usually assume that the training data and the target data are from the same space and with the same distribution. With such data, supervised learning algorithm can train an appropriate sentiment polarity classifier and get qualified classification results. However, it requires a large amount of high-quality manual annotation data to train the classifier. In practical applications, manual annotation data requires experts to understand the data so that it is expensive and labor-intensive to obtain a large amount of annotation data. Text sentiment analysis is a task which is sensitive to domains cause that the sentiment expression features of different domains are different. The classification model trained in one domain usually cannot be directly applied to other domains [1]. In order to reasonably and effectively apply the sentiment classification model trained in one domain to other domains, namely maximizing the domain adaptation ability of the sentiment classification model and reducing the annotation cost of the new domain, cross-domain sentiment analysis [2] has emerged and received increasing attention from the academic community.

C.-N. Yang et al. (Eds.): SICBS 2018, AISC 895, pp. 280–291, 2020.
https://doi.org/10.1007/978-3-030-16946-6_22

At present, there are two main methods in cross-domain text sentiment analysis: instance-based cross-domain sentiment analysis and feature-based cross-domain sentiment analysis [3].

The instance-based cross-domain sentiment analysis realizes the knowledge transfer from the source domain to the target domain by finding the connection between the source domain and the target domain, and applies the transfer knowledge to the target domain for classification task. Pan et al. [2] proposed a spectral feature alignment algorithm to construct the relationship between different domains. Through the spectral clustering algorithm, the domain independent words with high correlation are clustered under one cluster, and the domain co-occurrence words are also clustered in one for the purpose of the transfer learning of domain knowledge. However, the defect of instance-based method is that it has neither extracted the text feature nor learned the language features of text (such as the context information in the text). When the spatial distributions of the source and target domains are different, the model is less robust.

Feature-based cross-domain sentiment analysis constructs a unified feature representation of cross-domain data by searching the correlation or the co-occurrence features between the source and the target domains. Glorot et al. [4] applied Stacked Denoising Autoencoder (SDA) to cross-domain sentiment analysis. They first extracted words or phrases with higher frequency as their respective feature set from the source domain and target domain, and then reconstruct features of the source domain and the target domain through SDA, so that the features of the source domain and the target domain can show the same distribution in the potential space. Finally, they trained the classifier with the source domain feature and applied it to the target domain to realize the cross-domain text sentiment analysis task. Inspired by Goodfellow's generative adversarial nets [5], Ajakan et al. [6] integrated the source domain data with the target domain data, and then labeled the source domain data and the target domain data with domain labels. Through maximizing the cross entropy loss, the weights of network were updated and achieved the purpose of domain adaptation. Next, training the feature extractor with the domain adversarial of the source domain and the target domain, and training sentiment classifier with the deep features of the source domain and the target domain adversarial simultaneously. Finally, with the deep features extracted from target domain by adversarial, they can predict the sentiment polarity of the text of the target domain and thereby realizing cross-domain text sentiment analysis. However, Arjovsky et al. [7] mathematically proved that the method based on adversarial which proposed by Goodfellow et al. proposed would cause instability in the generative adversarial nets training, and the generated samples are also in the lack of lacked diversity. In order to alleviate these problems, Arjovsky et al. [8] proposed another function Wasserstein distance instead of the original cross entropy function. Linear regression combined with weight cropping is used to alleviate the problem that the generative and adversarial process is not easy to converge and the samples are diversity lacked. In the cross-domain classification task of images, Bousmalis et al. [9] proposed Domain Separation Networks (DSN) to separate the domain's own private features and the shared features between domains, and thus achieved domain feature separation, and then, they optimized the similarity loss function to train the feature extraction network, finally trained the classifier through the labeled image data, and achieved good results. This paper improves DSN and applies it to cross-domain text sentiment analysis.

The above research can achieve shared feature extraction to a certain extent by means of the shared network structure. But Salzmann et al. [10] found that the features extracted by the shared feature extraction network contain a large number of private domain features since the sentiment classifier is trained by the shared features of the source domain and the corresponding sentiment labels. In order to obtain better domain shared features and improve the accuracy of cross-domain text sentiment classification, by means of DSN network, this paper proposes a domain separation model based on Wasserstein distance (W-DSN) which combining denosing autoencoder and domain adversarial method based on Wasserstein distance together.

2 Wasserstein Distance Based Domain Adversarial Model

2.1 Task Description

The task of cross-domain text sentiment analysis in this paper refers to training the sentiment classifier with the sentiment-labeled data in the source domain, and achieving sentiment classification in the target domain data. Suppose that there are a source domain data set $\mathcal{D}_S$ and a target domain data set D_T, in which x^{s} is the text data with label and x^{t} is the text data without label. N_s and N_l denote the number of x^{s} and x^{t} respectively. That is $\mathrm{x}^{\mathrm{s}} = \left\{\left(\mathrm{x}_i^{\mathrm{s}}\right)\right\}_{i=1}^{N_s}(\mathrm{x}^{\mathrm{s}} \sim \mathcal{D}_S)$ and $\mathrm{x}^{\mathrm{t}} = \left\{\left(\mathrm{x}_{\mathrm{i}}^{\mathrm{t}}\right)\right\}_{i=1}^{N_l}(\mathrm{x}^{\mathrm{t}} \sim \mathcal{D}_T)$. Obviously, it is easy to use the deep features of the source domain to train the source domain sentiment classifier via the sentiment labels in the source domain. But for the target domain, it is difficult to obtain a classifier since there is no label in the target domain. In order to apply the source domain private deep features and labels to the target domain for the text sentiment classification task, it is necessary to firstly assume that the sentiment classifier is shared by the source domain and the target domain, that is, the distributions over the source domain private features and the target domain private features are same in the feature space. The work of this paper is to train a sentiment classifier with the texts and labels in the source domain data set and then combined with the target domain data set to do the adversarial training. Finally, the sentiment classifier can effectively classify the text data in the target domain.

2.2 W-DSN Description

This section will introduce the cross-domain text sentiment classification network (W-DSN) based on Wasserstein distance. It consists of six parts and the overall structure is show in Fig. 1. Specially, it includes: (1) $\mathrm{E}_{\mathrm{p}}^{\mathrm{s}}(\mathrm{x}^{\mathrm{s}})$, the private encoder of the source domain as Fig. 1(A) shows. (2) $\mathrm{E}_{\mathrm{p}}^{\mathrm{t}}(\mathrm{x}^{\mathrm{t}})$, the private encoder of the target domain as Fig. 1(C) shows. (3) shared encoder $\mathrm{E}(x)$ as Fig. 1(B) shows. (4) sentiment classifier $\mathrm{G}(\mathrm{h}_{\mathrm{c}}^{\mathrm{s}})$ as Fig. 1(D) shows. (5) domain adversarial net $\mathrm{D}_a\left(\mathrm{h}_{\mathrm{c}}^{\mathrm{s+t}}\right)$ as Fig. 1(E) shows. (6) shared decoder $\mathrm{D}_c\left(\mathrm{h}_{\mathrm{p}} + \mathrm{h}_{\mathrm{c}}\right)$ as Fig. 1(F) shows. (7) L_{senti}, L_{domain}, L_{rcon} denote the loss functions.

Firstly, the deep feature representation of text data in different domains is extracted by DAE, and then the domain adaptation is realized by adversarial and orthogonal

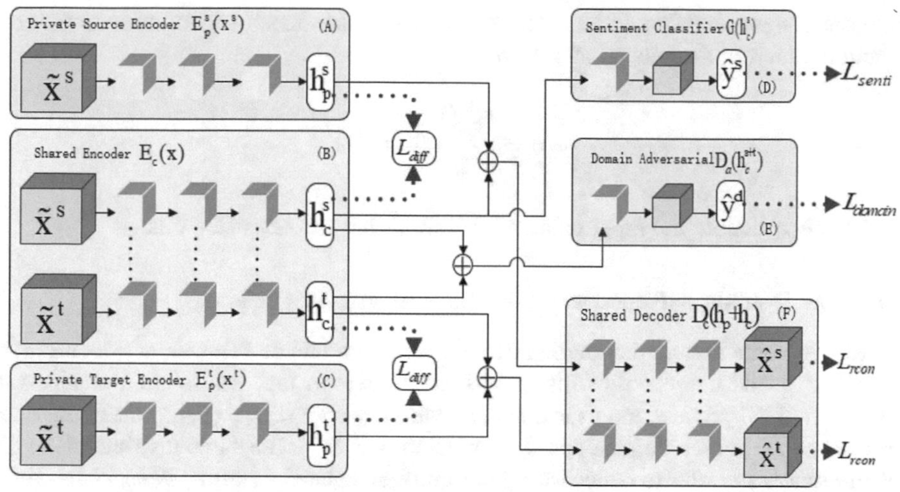

Fig. 1. Cross-domain Text Sentiment Classification method based on domain separation

constraint. Finally, the source domain features are used to train the sentiment classifier and the text sentiment of the target domain can be predicted by the classifier trained in the source domain. The details of the proposed model will be described in Sects. 2.3–2.7.

2.3 Deep Features Extraction with Denosing Autoencoder

Denosing autoencoder (DAE) firstly adds noise to the raw input and then extracts the main feature by encoder and decode the main feature by the decoder. Then, it can reconstruct the raw input. Because of the noise added to the raw input, DAE will own better generation ability than the standard autoencoder. W-DSN utilizes the DAE to extract the deep feature of the source domain text data and the target domain text data so that they can be projected into the same feature space.

Specifically, W-DSN uses a three-layer fully connected network to construct a private encoder of the source domain (as shown in Fig. 1(A)), a private encoder of the target domain (as shown in Fig. 1(C)) and a shared encoder (as shown in Fig. 1(B)) to extract shared features from source domain and target domain text data. Firstly, source domain text data x^s and target domain data x^t are artificially noised as $\tilde{x}^s$ and $\tilde{x}^t$. And then, take $\tilde{x}^s$ into private encoder $E_p^s(x^s)$ of the source domain and then the deep feature representation $h_p^s = E_p^s(\tilde{x}^s)$ of the source domain is obtained. The deep feature representation $h_c^t = E_p^t(\tilde{x}^t)$ of the target domain is obtained in the same way. In addition, W-DSN takes $\tilde{x}^s$ and $\tilde{x}^t$ into the shared encoder to get the shared features $h_c^s = E_c(\tilde{x}^s)$ and $h_c^s = E_c(\tilde{x}^t)$.

After that, a shared decoder (Fig. 1(F)) which is also consisted of a three-layer fully connected network is used to reconstruct the source domain deep feature representation, the target domain deep features representation and the domain shared feature representation. By minimizing the reconstruction loss function L_{rcon}, deep features of different domain texts could be captured. W-DSN takes mean square error as the loss function which is formulated as follows.

$$L_{rcon} = \sum_{i=1}^{N_s+N_t} \|x_i - \hat{x}_i\|^2 \tag{1}$$

Where x_i denote the input data and $\hat{x}_i$ denote the reconstructed data.

2.4 The Domain Adversarial Based on Wasserstein Distance

In order to make feature distribution of the source domain and the target domain more consistent in the deep feature space, W-DSN draws on the idea of the domain confrontation [5, 8] process and uses domain adversarial net (Fig. 1(E)) to make the deep features of the source domain and the target domain with the same distribution in the feature space. In order to better learn the sentiment-related features, W-DSN takes the Wasserstein distance as the adversarial-loss instead of cross entropy. The Wasserstein distance can be used to evaluate the distance between two fields. Even if two domains are absolutely different, Wasserstein distance can offer gradient for extracting features. But for cross entropy, the gradient equals to zero under such a condition. The Wasserstein distance between the two domains needs to be calculated as:

$$W_p(P, Q) = \inf_{\gamma \sim \Pi(P,Q)} \mathbb{E}_{(x,y)\sim\gamma}(\|x - y\|) \tag{2}$$

Where inf denote the low bound of the function, $\prod(P, Q)$ denote all the possible joint distribution of P and Q, thus the marginal distributions of each one in $\prod(P, Q)$ are P and Q. For each possible joint distribution γ, the distance of samples is calculated as $\|x - y\|$. Since Wasserstein can not be solved directly, Arjovsky [7] converts formula (2) to formula (3) according to the duality of Kantorovich-Rubinstein.

$$W_p(P, Q) = \sup_{\|f\|_L \le K} \mathbb{E}_{x\sim P}[f(x)] - \mathbb{E}_{y\sim Q}[f(y)] \tag{3}$$

Where $\|f\|$ is the half-norm of Lipschitz, sup means the upper bound, and $\|f\|$ satisfies with the Lipschitz continuity [5]. Under such a condition, W-DSN takes formula (3) as the domain adversarial loss function, in other words, it takes Wasserstein distance as the domain adversarial loss function. So in Fig. 1, Eq. (4) is obtained.

$$L_{w-d} = \max\left(\frac{1}{N_s}\sum_{i=1}^{N_s} f(h_i^s) - \frac{1}{N_t}\sum_{i=1}^{N_t} f(h_i^t)\right) \tag{4}$$

Where $f(\cdot)$ denote linear function. In order to satisfy with the Lipschitz continuity [11] condition, we cut the gradient to the scale of [c, −c] as Arjovsky [8] does. But Gulrajani [12] point out that such cut will result in Gradient disappearance or gradient explosion. So they use gradient regularization for the condition of Lipschitz continuity. Gradient regularization is to regularize the gradient value of the network before updating the network weight, that is, to regularize all the gradient values and then update the W-DSN model, and this process is called GP-DSN. The regularization of W-SDN is formulated as Eq. (5):

$$L_{gp} = \left(\|\nabla_x f(\mathrm{x})\|_p - c\right)^2 \tag{5}$$

Where c is a hyper parameter. So the loss function L_{domain} of GP-DSN equals to L_{gp-d}.

$$L_{gp-d} = \frac{1}{N_s}\sum_{i=1}^{N_s} f(\mathrm{x}_i^s) - \frac{1}{N_t}\sum_{i=1}^{N_t} f(\mathrm{x}_i^t) + \alpha\left(\|\nabla_x f(\mathrm{x})\|_p - c\right)^2 \tag{6}$$

Where α is the balance coefficient.

2.5 Domain Separation Based on Orthogonal Constraint

Although the shared deep features are extracted by the encoder of the domain adversarial training are already similar, the deep features still contain a large number of private domain features. In order to eliminate the private domain features as much as possible, we take advantage of the orthogonal constraint [10] to separate the private features and the shared feature as much as possible for the purpose of domain separation, and combine with the domain adversarial process to make the shared features' distribution more similar. Orthogonal constraint push the product of two eigenvectors as close as possible to zero in order to make as much as possible differences between the private and shared features. The Orthogonal loss of W-DSN is formulated as follows.

$$L_{diff} = \left\|\mathrm{h}_c^{sT}\mathrm{h}_p^s\right\|_F^2 + \left\|\mathrm{h}_c^{tT}\mathrm{h}_p^t\right\|_F^2 \tag{7}$$

Where h_c^s and h_c^t denote the domain shared deep feature representation, h_p^s denote the private deep feature representation of the source domain, h_p^t denote the private deep feature representation of the target domain and $\|\cdot\|_F^2$ is the square of Frobenius norm. Minimizing the loss could make the domain shared deep feature representation h_c^s and h_c^t more similar by optimizing the adversarial loss and the orthogonal constraint loss.

2.6 Sentiment Classifier

After the process of domain adversarial net and orthogonal constraint, the more similar deep feature representation h_c^s and h_c^t are obtained in the feature space. W-DSN training the sentiment classifier with the source domain shared deep features as shown in Fig. 1 (D) and applying the trained sentiment classifier to the target domain for text sentiment classification task. After the domain confrontation network and orthogonal constraints are processed, the deep shared feature representation the more consistent distribution in the feature space are obtained. W-DSN performs sentiment classification by logistic regression. The loss function of the classifier is show in formula (6).

$$L_{senti} = -\sum_{i=0}^{N_s} y^s \log(\hat{y}^s) \tag{8}$$

Where y^s denote the real sentiment label and $\hat{y}^s$ denote the predicted sentiment label. The weights are updated by minimizing the sentiment classification loss L_{senti}.

2.7 Object Function

All loss functions in the domain separation model are introduced in Sects. 2.3–2.7, where the loss functions L_{senti}, L_{domain} and L_{rcon} need to be minimized, and the loss function L_{domain} needs to be maximized. Suppose that the weights of the network need to be updated as $\{\theta\} = \{\theta_{p-enc}, \theta_{s-enc}, \theta_{dec}, \theta_s, \theta_d\}$, of which the element respectively denote the private feature encoder, domain shared feature encoder, decoder, sentiment classifier and the weight of adversarial net. In order to unify the training process, W-DSN applies the gradient reversal layer (GRL) which is proposed by Ganin et al. [13] between the feature extraction layer and the domain adversarial net. In the forward propagation of the W-DSN, no operation is performed by GRL. But in the back propagation, L_{domain} is multiplied by a hyperparameter $-\lambda$ to minimize the overall loss function. Therefore, the final objective function of W-DSN is shown in Eq. (9) as follows.

$$\min L(\theta) = \frac{1}{N^s + N^t}(L_{senti} + \beta L_{diff} + \gamma L_{recon} - \eta L_{domain}) \tag{9}$$

Where β, γ and η are hyperparameters. The update of $L(\theta)$ is formulated (10) as follows where μ is learning rate.

$$\begin{aligned}
&\theta_{s-enc} \leftarrow \theta_{s-enc} - \mu(\frac{\partial L_{senti}}{\partial \theta_{s-enc}} + \beta \frac{\partial L_{diff}}{\partial \theta_{s-enc}} - \eta \frac{\partial L_{domain}}{\partial \theta_{s-enc}}) \\
&\theta_{p-enc} \leftarrow \theta_{p-enc} - \mu(\frac{\partial L_{diff}}{\partial \theta_{p-enc}}) \\
&\theta_{dec} \leftarrow \theta_{dec} - \mu(\frac{\partial L_{rcon}}{\partial \theta_{dec}}) \\
&\theta_s \leftarrow \theta_s - \mu(\frac{\partial L_{sent}}{\partial \theta_s}) \\
&\theta_d \leftarrow \theta_d - \mu(\frac{\partial L_{domain}}{\partial \theta_d})
\end{aligned} \tag{10}$$

3 Experimental Result

3.1 Datasets

This paper uses the Amazon product review dataset provided by Glorot et al. [4] to evaluate the proposed model. This data set contains comments on specific products in four different areas, including books, DVD disk, electronics, and kitchen appliances. Each of these areas contains 2,000 labeled reviews (1000 positive reviews and 1000 negative reviews) and a number of unlabeled reviews. In this paper, each dataset of four different domain is used as the source domain dataset for one time, and the other three datasets are used as the target domain dataset at this time. The target domain data is divided into training set and test set (1600 is used for training, 400 used for testing). The statistics of the four data sets are show in Table 1:

Table 1. Dataset statistics

	Book	DVD	Electronics	Kitchen
Pos	1000	1000	1000	1000
Neg	1000	1000	1000	1000
Unlabel	6000	34741	13153	16785
Vocabulary	171760	739346	237346	278187

In the text representation, all the data of the source domain and the target domain are processed with uni-grams and bi-grams statistics, and we removed the stop words with the english sentiment analysis stop words table provided by Baidu, and then the tf-idf of the first 5,000 words is used as the representation of the text data. The length of the dictionary for the final statistics is shown in vocabulary in Table 1.

3.2 Parameter Settings

W-DSN uses a three-layer fully-connected network to construct an encoder and a decoder. The number of neurons in the three-layer fully-connected network in the encoder is set to 1000, 500, and 200, respectively, and that in decoder is set to 200, 500, and 1000, respectively. The fully connected network in both the decoder and the encoder use relu as the activation function. This paper uses the RMSprop [14] to optimize network weights, and its learning rate is set to 0.001. The three hyper parameters β, γ and λ involved in the objective function are set to 1. All the experiments are finished on Tesla P100-PCIE GPU woks station. The operate system is Linux, the development environment are Python 2.7, Tensorflow 1.3.0, and the development tool is PyCharm.

3.3 Compared Methods

In order to prove the effectiveness of the proposed method for the prediction of sentiment polarity in the target domain, the method proposed is compared with the best method proposed in the existing research. At the same time, in order to understand the characteristics of the proposed model more intuitively, the paper reduce features into two dimensions after the data is extracted by different algorithms, and then visualizes the data.

We compare the method proposed in this paper with the following methods:

(1) LR: A logistic regression classifier is constructed using a three-layer fully connected neural network. And then trained in the source domain and tested on the target domain.
(2) SDA: Glorot et al. proposed Stack Denosing Autoencoder (SDAE) in [4]. The method uses a multi-layer MLP to construct a SDAE for main feature extraction, and then trains the classifier.
(3) mSDA: Chen et al. [15] proposed an improved model based on SDA. This method uses SDAE for main feature extraction and was time saving.
(4) DANN: Ajakan [6] proposed domain adversarial net which combines mSDA to extract features firstly and then making the domain deep features over the same distribution via domain adversarial. Training the sentiment classifier with the deep features of the source domain and do sentiment classification task on the target domain.
(5) **W-DSN**: W-DSN is our proposed model. It first extracts the deep features of different domains with DAE, and then achieves domain adaptation with adversarial and orthogonal in the adversarial net. Finally training the sentiment classifier with the deep features of the source domain and do sentiment classification task on the target domain as DANN did.
(6) GP-DSN: Gradient Regulation Domain Separation (GP-DSN) is an optimized version of W-DSN. It regularized the gradient before the network weights are updated, and thus accelerate the convergence of the loss function.

3.4 Experimental Results and Analysis

Figure 2 shows the performance of W-DSN, GP-DSN and the compared methods in different domains. As shown in Fig. 2, our proposed method is superior to existing methods on accuracy of sentiment classification.

As shown in Fig. 2(a), LR only uses the source domain data to do cross-domain text sentiment analysis, but without using any cross domain related methods, so it doesn't achieve higher accuracy. This proves that training sentiment classifier on the source domain cannot adopt the features of the target domain. SDA and mSDA have better performance than LR, which proves that DAE can extract the shared feature between the source domain and the target domain and so that the sentiment classifier trained by these features achieves better performance. Further more, the performance of DANN is better than SDA and mSDA. It indicates that the domain adversarial has better generalization ability than SDA and mSDA. And domain adversarial has a positive role in cross-domain analysis. W-DSN not only use the structure of DAE, but also use processing on domain adversarial. The accuracy of W-DSN is 4.0% higher than DANN. It indicates that combining Wasserstein distance based domain adversarial

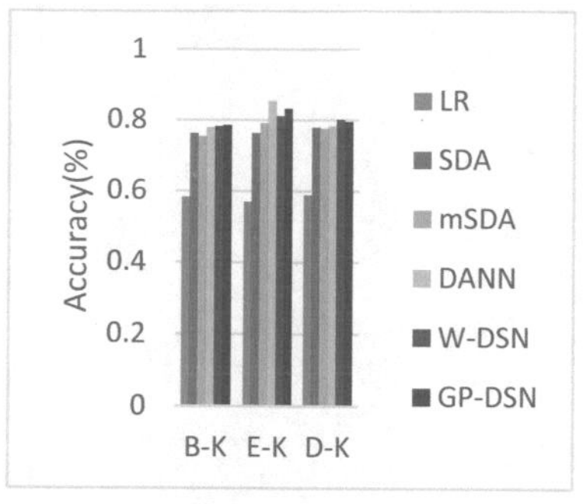

(a) DVD is the target domain

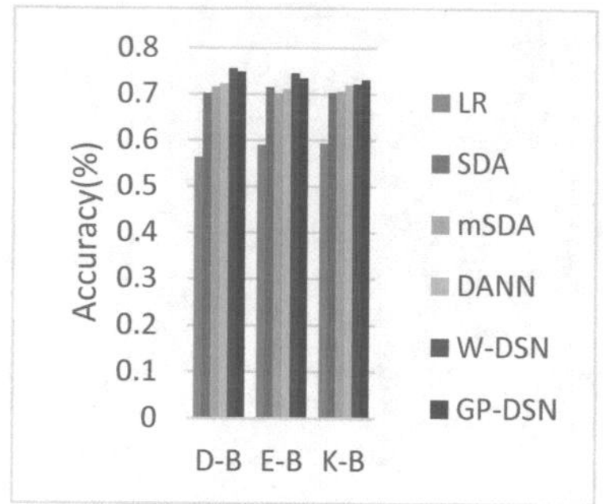

(b) electronics is the target domain

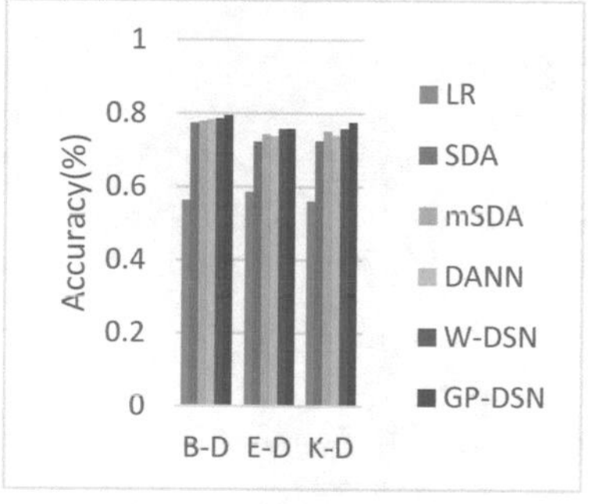

(c) kitchen is the target domain

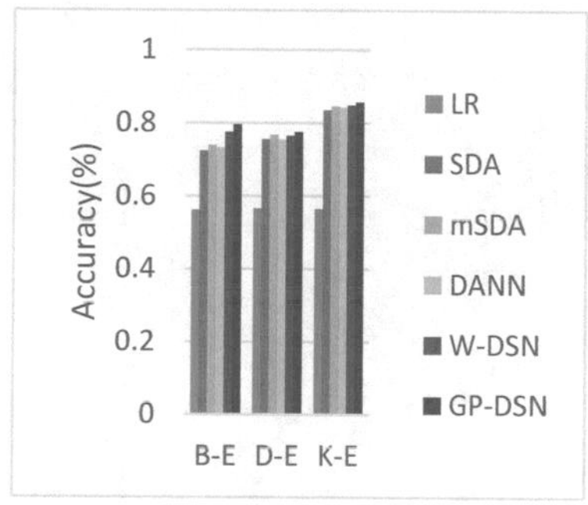

(d) book is the target domain

Fig. 2. The accuracy of cross-domain sentiment classification in four different domains.

with orthogonal constraint is helpful to cross-domain text sentiment analysis. Figure 2 shows that GP-DSN has the best performance, which indicates the positive effect of gradient regularization in cross-domain modelling.

3.5 Data Distribution

In this section, the effectiveness of models will be illustrated by visualizing the distribution of data features. The adaptation of data will also be discussed. Taking book as the source domain data, dvd as the target domain data, and then we use T-SNE [16] to reduce the dimension of data to 2D and visualize the distribution of data features as shown in Fig. 3.

Figure 3(a) shows the feature distribution of the original data, and it can be seen that the feature distribution of the source domain data and the target domain data is inconsistent. Therefore, it is intractable to use the classifier trained to the source domain on the classification task on the target. Figure 3(b) is the data feature distribution obtained from DANN. Compared Fig. 3(a) with Fig. 3(b), it can be found that the feature distribution of the source domain data and the target domain data is confused after domain adversarial training. However, the data feature distribution obtained from DANN has a narrow coverage. Figure 3(c) and (d) respectively show the feature

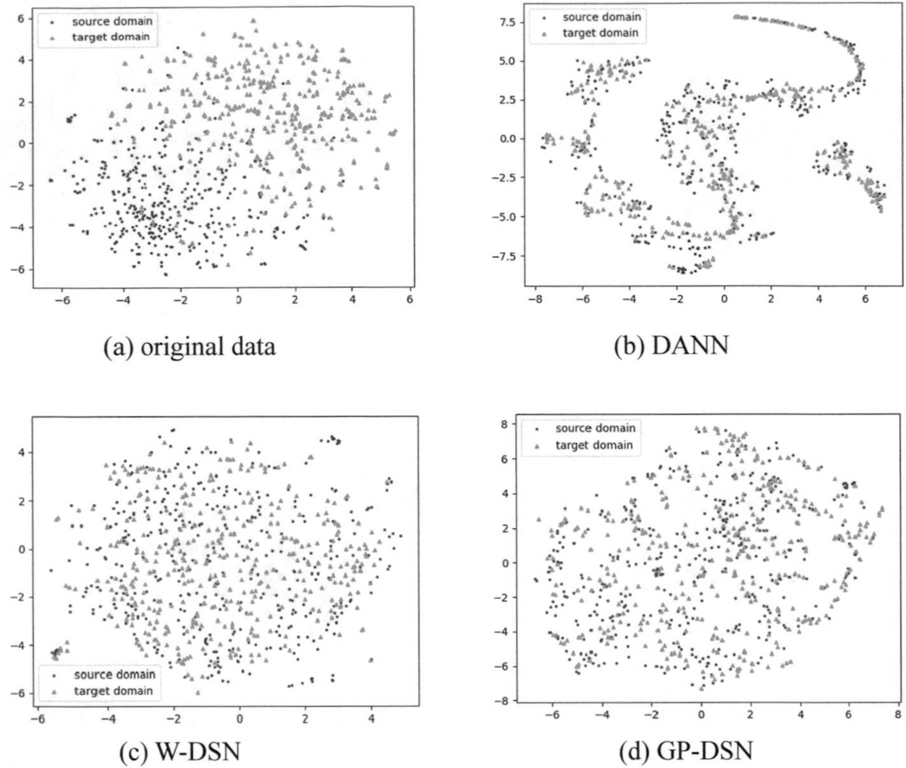

(a) original data

(b) DANN

(c) W-DSN

(d) GP-DSN

Fig. 3. Feature distribution

distribution of W-DSN and GP-DSN. It can be found that W-DSN and GP-DSN perform better in the confused of feature distribution, and the coverage is more complete, which leads to more for features are reserved for sentiment classification, thereby, W-DSN and GP-DSN gets more robust performance.

4 Conclusion

In recent years, cross-domain sentiment analysis has become an increasingly important research hotspot. This paper proposes two cross-domain sentiment analysis models based on Wasserstein distance with domain adversarial model, which better solves the problem of domain adaptation and improves the accuracy of sentiment classification. That is, the proposed method uses DAE to extract the private features and domain shared features, and uses adversarial and orthogonal constraint to achieve domain adaptation, thus that the features distribution of the source domain and the target domain becomes more similar in the feature space, and more sentiment features are reserved. The sentiment classifier is trained with the features of the source domain and applied to sentiment classification on the target domain. The effectiveness of the proposed methods have been evaluated on four benchmark datasets, and the experiment results show that the proposed method is superior to the existing methods, indicating that the proposed method can better solve cross-domain sentiment classification and domain adaptation problem. Although the methods have achieved good results, it does not consider the context of the sentence. The loss of semantic in sentence is still very serious. Therefore, our future plan is how to integrate the semantic information into the feature extraction process for better performance on cross-domain sentiment classification.

References

1. Tan, S., Cheng, X., Wang, Y., Xu, H.: Adapting naive bayes to domain adaptation for sentiment analysis. In: European Conference on Information Retrieval, pp. 337–349. Springer, Heidelberg (2009)
2. Pan, S.J., Ni, X., Sun, J.-T., Yang, Q., Chen, Z.: Cross-domain sentiment classification via spectral feature alignment. In: Proceedings of the 19th International Conference on World Wide Web, pp. 751–760. ACM (2010)
3. Pan, W., Zhong, E., Yang, Q.: Transfer learning for text mining. In: Mining Text Data, pp. 223–257. Springer, Boston (2012)
4. Glorot, X., Bordes, A., Bengio, Y.: Domain adaptation for large-scale sentiment classification: a deep learning approach. In: Proceedings of the 28th International Conference on Machine Learning (ICML-11), pp. 513–520 (2011)
5. Goodfellow, I., Pouget-Abadie, J., Mirza, M., Xu, B., Warde-Farley, D., Ozair, S., Courville, A., Bengio, Y.: Generative adversarial nets. In: Advances in Neural Information Processing Systems, pp. 2672–2680 (2014)
6. Ajakan, H., Germain, P., Larochelle, H., Laviolette, F., Marchand, M.: Domain-adversarial neural networks. arXiv preprint arXiv:1412.4446 (2014)
7. Arjovsky, M., Bottou, L.: Towards principled methods for training generative adversarial networks. arXiv preprint arXiv:1701.04862 (2017)
8. Arjovsky, M., Chintala, S., Bottou, L.: Wasserstein generative adversarial networks. In: International Conference on Machine Learning, pp. 214–223 (2017)
9. Bousmalis, K., Trigeorgis, G., Silberman, N., Krishnan, D., Erhan, D.: Domain separation networks. In: Advances in Neural Information Processing Systems, pp. 343–351 (2016)
10. Salzmann, M., Ek, C.H., Urtasun, R., Darrell, T.: Factorized orthogonal latent spaces. In: Proceedings of the Thirteenth International Conference on Artificial Intelligence and Statistics, pp. 701–708 (2010)
11. Heinonen, J.: Lectures on Lipschitz analysis. No. 100. University of Jyväskylä (2005)
12. Gulrajani, I., Ahmed, F., Arjovsky, M., Dumoulin, V., Courville, A.C.: Improved training of Wasserstein GANs. In: Advances in Neural Information Processing Systems, pp. 5767–5777 (2017)
13. Ganin, Y., Lempitsky, V.: Unsupervised domain adaptation by backpropagation. arXiv preprint arXiv:1409.7495 (2014)
14. Ruder, S.: An overview of gradient descent optimization algorithms. arXiv preprint arXiv: 1609.04747 (2016)
15. Chen, M., Xu, Z., Weinberger, K., Sha, F.: Marginalized denoising autoencoders for domain adaptation. arXiv preprint arXiv:1206.4683 (2012)
16. van der Maaten, L., Hinton, G.: Visualizing data using t-SNE. J. Mach. Learn. Res. **9**, 2579–2605 (2008)

WebRTC-Based On-Site Operation and Maintenance Adaptive Video Streaming Rate Control Strategy

Chuang Liu(✉), Sujie Shao, Shaoyong Guo, and Xuesong Qiu

Beijing University of Posts and Telecommunications, Beijing 100876, China
bupt_lc@163.com

Abstract. At present, the on-site operation and maintenance of the power communication network lacks efficient real-time interaction means of operation and maintenance data, and it is impossible to realize real-time decision-making and accurate implementation, which indicates a moderate efficiency and quality of operation and maintenance. The application of WebRTC and wearable operation and maintenance technology, which establishes a multi-party video call based on P2P connection between the wearable terminal and the operation and maintenance platform, can also realize active on-site operation and maintenance of multi-party coordination and auxiliary decision-making. However, because of the complicated power communication network status of the on-site operation and maintenance, the video transmission quality fluctuates greatly, which restricts the efficient implementation of wearable operation and maintenance. To this end, for wearable on-site operation and maintenance, this paper proposes a WebRTC-based adaptive video streaming transmission rate control strategy, which supports adaptive dynamic adjustment of video streaming data transmission rate with network link quality changing. Simulation shows that the proposed transmission rate control strategy can effectively reduce the network delay and packet loss rate, even under various network environments. Therefore, this strategy can effectively adapt to the complex and variable on-site operation and maintenance of the power communication network.

Keywords: WebRTC · Rate control · Power communication network

1 Introduction

With the development of Internet technology, smart wearable devices has been increasingly used in on-site operation and maintenance of power communication network [1]. With WebRTC [2] (Web Real-Time Communication) and wearable operation and maintenance technology, a P2P-based video call between the wearable terminal and the operation and maintenance platform can be established to realize remote operation of on-site operation and maintenance. However, the complex environment on operation and maintenance site and the lability of the mobile network channel usually lead to the low performance of WebRTC-based video call and the high rate of network delay and packet loss [3]. Therefore, the study of WebRTC-based on-site operation and maintenance adaptive video streaming rate control strategy is of great significance.

C.-N. Yang et al. (Eds.): SICBS 2018, AISC 895, pp. 292–303, 2020.
https://doi.org/10.1007/978-3-030-16946-6_23

In view of the above problems, there are two existing solutions: video streaming rate control and network transmission control. Literature [4] proposes an adaptive method combining video coding and video transmission control systems on heterogeneous networks, which can calculate the coding rate to encode the video and then transmit it. Literature [5] proposes an improved rate control algorithm. The algorithm introduces a content complexity factor to make the rate allocation more accurate, which can more accurately adjust the quantization parameter to be used in the current frame according to the historical information of the encoded frame. Literature [6] proposes a novel RTC congestion control algorithm. In the idea of using Kalman filter to estimate the end-to-end one-way delayed variation of the packet from the sender to the receiver, the algorithm can to some extent adapt video streaming transmission rate according to network conditions. However, these methods do not consider the transmission rate control of the video stream in the complex and variable network environment. The topic on how to adjust the video streaming rate to realize the adaptive transmission of video stream with the network bandwidth is untouched.

Based on the above analysis, this paper proposes a WebRTC-based on-site operation and maintenance adaptive video streaming rate control strategy AVRCS (Adaptive Video Rate Control Strategy). Firstly, at the receiver of the video stream, a new dynamic threshold adjustment is proposed, which calculates the current network link quality threshold as well as the overload detection signal, and the transmission rate is adjusted based on the latter. The result of the calculation is then returned to the sender of the video stream. At the same time, at the sender of the video stream, a new transmission rate is calculated based on the packet loss rate and the transmission rate of the previous frame data. Finally, based on the results calculated by the sender and the receiver, the actual transmission rate of the next frame of data is calculated.

This paper is divided into five sections. The second section introduces the video communication architecture of multi-point remote collaborative operation and maintenance. The third section presents the WebRTC-based on-site operation and maintenance adaptive video streaming rate control strategy. The forth section is simulation data analysis, which verifies the rationality of the AVRCS. The final section gives the conclusion.

2 Multi-point Remote Collaborative Video Communication Architecture

Power communication field wearable multi-point remote collaborative operation and maintenance communication architecture, which is based on Openfire and WebRTC, consists of three parts: intelligent wearable operation and maintenance terminal, intelligent operation and maintenance platform, and instant communication system for wearable multi-point remote cooperative operation and maintenance. As shown in Fig. 1.

Intelligent wearable operation and maintenance terminal mainly refers to wearable AR (Augmented Reality) glasses. Augmented reality technology is used to support real-time guidance of on-site operation and maintenance operations through information interaction between dispatch manager and technical experts. The intelligent operation and maintenance platform mainly implements support functions, for example, operation

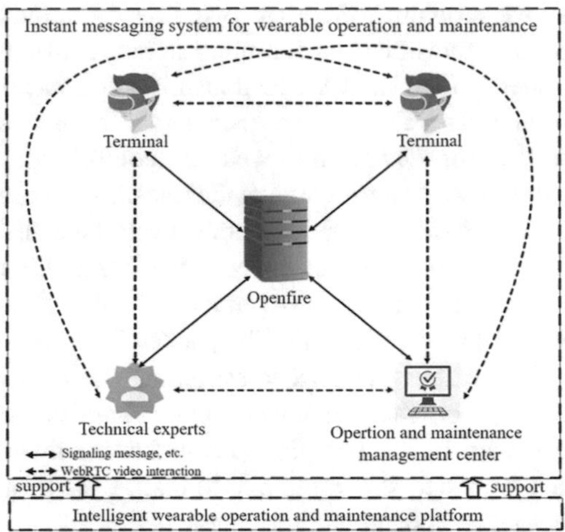

Fig. 1. Remote collaborative video communication architecture for wearable operation and maintenance.

and maintenance work order management and scheduling. It supports dispatchers to confirm and feedback the work order delivery, information support and implementation results. Also, it helps technical experts to guide on-site operation and maintenance according to work order information. Real-time communication system needs support remote collaborative video communication among operation staff, dispatchers and technical experts. This paper uses the Openfire server registration session and its maintained characteristics to build the above instant messaging system, and a point-to-point video call based on WebRTC is established. The instant messaging system mainly includes two parts: WebRTC-based point-to-point video stream communication and Openfire-based signaling message data communication. The former uses WebRTC technology to directly transmit a video stream after establishing a point-to-point connection through signaling messaging. The latter implements core functions such as connection management, message routing, and messaging through the Openfire server.

In the process of establishing a point-to-point video connection, both sides pass metadata through the Openfire server. Metadata mainly includes two types: metadata of network communication connections and metadata of communication content. The former includes IP address, NAT network address translation and firewall, etc. The latter contains commands to open/close dialogs, metadata (encoding format, media type and bandwidth) of media files, etc. The separate transmission of video stream and the signaling message stream effectively takes the advantage of Openfire and WebRTC. Not only is a reliable and controllable instant messaging system established, but also the system has the characteristics of easy deployment, low latency, low server pressure, and high efficiency.

3 Adaptive Transmission Rate Control Strategy

3.1 Video Streaming Rate Control Framework

Under the communication architecture of the multi-point remote video collaboration for wearable operation and maintenance shown in Fig. 1, the control of the WebRTC video streaming rate is the core of the video rate control of the instant communication system. The real-time data transmission protocol used by WebRTC is RTP (Real-time Transport Protocol), which is also a commonly used protocol for streaming media transmission. However, RTP only guarantees real-time data transmission, flow control or congestion control is uninvolved. Thus, RTCP (Real-time Transport Control Protocol) is needed to implement congestion control strategies under different scenarios and requirements. This paper proposes an adaptive video streaming rate control strategy. The workflow of this strategy is shown in Fig. 2. The sender sends the RTP packet and accepts the Receiver Estimated Maximum Bitrate (Receiver Estimated Maximum Bitrate) feedback message. The sender adjusts the data transmission rate according to the feedback message to achieve the purpose of sensing the network quality and controlling the video rate.

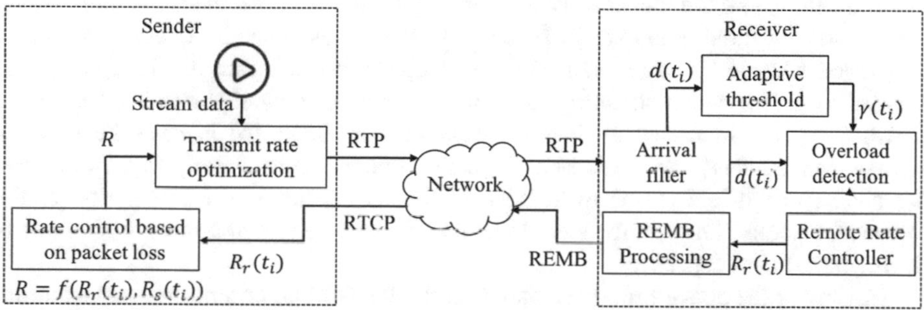

Fig. 2. WebRTC video rate control process.

3.2 Video Streaming Rate Control Strategy

The Specific Steps of the Receiver Are as Follows

Arrival Time Filtering. The purpose of the arrival time filtering module is to calculate the one-way delay gradient estimate $d(t_i)$, as shown in Fig. 3. At the data receiver, based on the one-way delay gradient $g_d(t_i)$, which is defined as Eq. (1), the Kalman filter is used to calculate the estimated value $d(t_i)$ of the one-way delay gradient.

$$g_d(t_i) = (t_i - t_{i-1}) - (t_i^s - t_{i-1}^s) \tag{1}$$

Where, t_i^s and t_{i-1}^s are the start times of the video data transmission of the i and $i-1$ frames respectively, and t_i and t_{i-1} are the cutoff times of all the video data of the i and $i-1$ frames respectively.

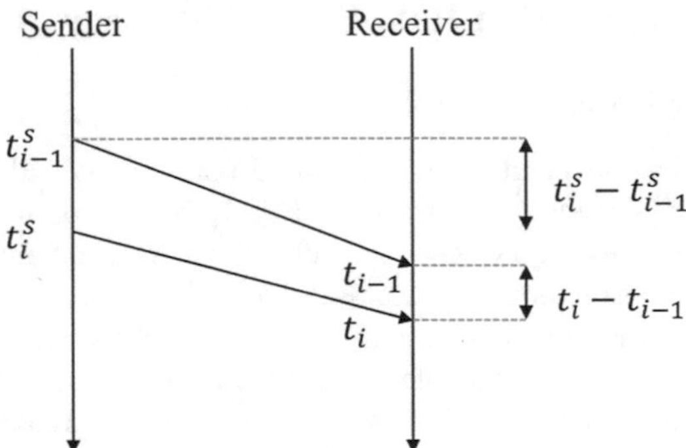

Fig. 3. One-way delay gradient calculation.

Adaptive Threshold. The goal of the adaptive threshold module is to adjust the sensitivity of the algorithm to the delay gradient based on the network condition. Reference [7] proposed a dynamic equation change threshold $\gamma(t_i)$ to achieve real-time monitoring of the current link load. The algorithm performs well in a network environment with a delay of less than 200 ms, but there are two defects. (a) In a network environment where the link quality is poor and the delay is tolerated, the initial value $\gamma(t_0)$ of the threshold is too small. As a result, the link is generally overloaded during the period when the session starts, which results in a lower rate of video streaming transmission. The video streaming transmission rate is gradually stabilized until the threshold $\gamma(t_i)$ is increased to fit the current network state. (b) In case the network environment suddenly deteriorates and the link quality drops sharply in a short time, the algorithm will always be overloaded in the future, the adjustment period of the threshold response link quality is too long, and the video streaming transmission rate is of low adjustment capability.

In view of the above problems, based on the RRTCC algorithm, this paper further dynamically adjusts the threshold initial value $\gamma(t_0)$ and threshold $\gamma(t_i)$ to achieve reasonable delay tolerance. Firstly, in a network environment with poor link quality, increase the threshold initial value. Define $\gamma(t_0)$ as:

$$\gamma(t_0) = \max(\gamma^0(t_0), \partial \frac{d_{t_0}}{d_{t_0}^0} \gamma^0(t_0)) \tag{2}$$

Where, $\gamma^0(t_0)$ is the threshold initial value used by the RRTCC algorithm, and dt_0^0 is the one-way delay gradient estimation mean of the RRTCC algorithm in a network environment with a delay of less than 200 ms, and dt_0 is the one-way delay gradient estimation value of the current network, and ∂ is a dynamic adjustment factor.

Secondly, in the case of complex and variable network link quality, the threshold $\gamma(t_i)$ needs to be dynamically adjusted, as shown in Eq. (3).

$$\gamma(t_i) = \begin{cases} \gamma(t_{i-1}) + k_u(1+\theta)(d(t_i) - \gamma(t_{i-1})), & d(t_i) > \gamma(t_{i-1}) \\ \gamma(t_{i-1}) + k_s\theta(d(t_i) - \gamma(t_{i-1})), & 0 < d(t_i) < \gamma(t_{i-1}) \\ \min(\gamma(t_{i-1})/2, \gamma(t_0)), & d(t_i) < 0 \end{cases} \tag{3}$$

Where, k_u and k_s represent the speed at which the threshold increases or decreases respectively, k_u is slightly greater than 1, while $0 < k_s < 1$.

The adaptive threshold module calculates a new threshold $\gamma(t_i)$ based on the relative magnitude of the estimate $d(t_i)$ of the one-way delay gradient and the threshold $\gamma(t_{i-1})$ of the previous time. When $d(t_i) > \gamma(t_{i-1})$, the quality of the network link is degraded, accompanied with overloaded link, Thus, the "addition and fast increase" phase starts. The threshold is adjusted to a state slightly larger than the estimated value, and the algorithm is guided to enter the "micro-down" phase to increase the network delay tolerance. At the same time, it effectively avoids the link to be always overloaded, and the adjustment period of the threshold response link quality is shorter; When $d(t_i) < \gamma(t_{i-1})$ and $d(t_i) > 0$, the link quality is relatively stable, and the "subtraction and micro-drop" phase starts to maintain the current network steady state; When $d(t_i) < 0$, the current link quality is good, accompanied with low bandwidth utilization, and the "multiply and fast down" phase starts. Then the current threshold is rapidly decreased, which orders the sender to send more video frame data to increase network bandwidth utilization.

Overload Detection. The purpose of the overload detection module is to dynamically trigger the overload detection signal. Each time a video stream is received, the overload detection triggers the state drive signal S based on the relative magnitude of the one-way delay gradient estimate $d(t_i)$, the threshold $\gamma(t_i)$ and the current state hold time T_{keep}. Signal S has three states: overuse, underuse and normal. The overuse signal indicates that the current network is congested, resulting in a large video stream delay and a high packet loss rate. The underuse signal indicates that the current network waits for fewer transmission queues and the available bandwidth resources are abundant. The normal signal is a state between the two.

Overload detection process is as follows: When $d(t_i) > \gamma(t_{i-1})$ and $T_{keep} > T_s$, the overload detection module triggers the overuse signal. T_s indicates the lower limit of the current state holding time. If $T_{keep} < T_s$, the overload detection signal S will not be triggered. Similarly, when $d(t_i) < 0$, and $T_{keep} > T_s$, the underuse signal is triggered; when $0 < d(t_i) < \gamma(t_i)$ and $T_{keep} > T_s$, the normal signal is triggered. In order to meet the reasonable delay tolerance requirements, Tt_i should adapt to the current network link status. Tt_i is as shown in Eq. (4):

$$T_{t_i} = \max(\beta d(t_i), T_{t_i}^0) \tag{4}$$

$T_{t_i}^0$ is the RRTCC algorithm trigger signal time threshold, β is the dynamic adjustment factor of T_{t_i}, and T_{t_i} changes dynamically with $d(t_i)$.

Remote Rate Control. The receiving rate $R_r(t_i)$ of the receiver is calculated based on the state driving signal S, as shown in Eq. (5). When S is overuse signal, the receiving rate is lowered to balance the delay; when S is underuse signal, the receiving rate is increased to improve the bandwidth utilization.

$$R_r(t_i) = \begin{cases} \min\{(1-\lambda_1)R_r^0(t_i), (1-\lambda_2)R_r(t_{i-1})\}, & S = overuse \\ \phi(R_r(t_{i-1})), & S = normal \\ (1+\lambda_2)R_r(t_{i-1}), & S = underuse \end{cases} \tag{5}$$

Where, $R_r^0(t_i)$ is the average receiving rate in the last 500 ms, and λ_1 and λ_2 are the receiving rate reduction factor and the growth factor respectively.

In addition, according to the active bandwidth detection method, the receiving rate $R_r(t_i)$ is further adjusted when the signal S is in the normal state, and the corresponding incremental detection is performed according to the rate change trend [8, 9]. According to the queuing delay $d(t_i)$, the receiving rate $R_r(t_i)$ is adjusted in three cases, as shown in Eqs. (6) and (7).

$$\phi(R_r(t_{i-1})) = \begin{cases} \min\{(1-\lambda_1 P_{d\gamma})R_r^0(t_i), (1-\lambda_2 P_{d\gamma})R_r(t_{i-1})\}, & \frac{3}{4}\gamma(t_i) < d(t_i) \le \gamma(t_i) \\ R_r(t_{i-1}), & \frac{1}{2}\gamma(t_i) \le d(t_i) \le \frac{3}{4}\gamma(t_i) \\ (1+\lambda_2(1-\frac{2d(t_i)}{\gamma(t_i)}))R_r(t_{i-1}), & 0 \le d(t_i) < \frac{1}{2}\gamma(t_i) \end{cases} \quad (6)$$

$$P_{d\gamma} = \frac{4d(t_i) - 3\gamma(t_i)}{2\gamma(t_i)} \quad (7)$$

When the queuing delay takes the lower limit, the receiving rate increases, and the growth factor decreases as the queuing delay increases. When the queuing delay takes the upper limit, the receiving rate decreases, and the decreasing factor increases as the queuing delay increases. When the queued delay value is the intermediate area, the receiving rate is the same as the receiving rate $R_r(t_{i-1})$ of the time t_{i-1}.

REMB Processing. The receive rate $R_r(t_i)$ is sent to the video sender via the REMB message along with the RTCP packet. Under normal circumstances, the REMB message is sent every 1 s. Once the rate $R_r(t_i) < 0.97R_r(t_{i-1})$, the reception rate $R_r(t_i)$ is attenuated by 3% or more, the REMB message will be sent immediately.

The Specific Steps of the Sender Are as Follows

The transmission rate $R_r(t_i)$ is calculated according to the packet loss rate $PLR(t_i)$ of the video stream containing the RTCP packet [10], as shown in Eq. (8):

$$R_s(t_i^r) = \begin{cases} R_s(t_{i-1}^r)(1-\omega_1 PLR(t_i)), & PLR(t_i) > b_u \\ \varphi(R_s(t_{i-1}^r)), & b_l \le PLR(t_i) \le b_u \\ (1+\omega_2)(R_s(t_{i-1}^r) + R_0), & PLR(t_i) < b_l \end{cases} \quad (8)$$

Where b_u, b_l are the thresholds for $PLR(t_i)$ in the $R_r(t_i)$ staged optimization process, $b_u = 0.1$, $b_l = 0.02$, $b_m = (b_u + b_l)/2$. R_0 is the transmission rate detection bandwidth. ω_1, ω_2 is the reduction and growth factor of the transmission rate respectively. $\omega_1 \in [0.1, 1]$, $\omega_2 \in [0.01, 0.1]$. When $PLR(t_i) > b_u$, the transmission rate increases; When $PLR(t_i) < b_l$, the transmission rate is reduced; When $b_l \le PLR(t_i) \le b_u$, adjust the transmission rate according to Eq. (9).

$$\varphi(R_s(t_{i-1}^r)) = \begin{cases} R_s(t_{i-1}^r) - \frac{PLR(t_i) - b_m}{2(b_u - b_m)} R_0, & b_m < PLR(t_i) \le b_u \\ R_s(t_{i-1}^r) + \frac{b_m - PLR(t_i)}{2(b_m - b_l)} R_0, & b_l \le PLR(t_i) \le b_m \end{cases} \quad (9)$$

When $b_m < PLR(t_i) \leq b_u$, according to the value of the packet loss rate $PLR(t_i)$, $0.5R_0$ is used as the reference for speed detection, and the amount of decrease increases with the increase of the packet loss rate. When $b_l \leq PLR(t_i) \leq b_m$, according to the value of the packet loss rate $PLR(t_i)$, the speed increase detection is performed with $0.5R_0$ as the reference quantity, and the increase amount decreases as the packet loss rate increases.

Calculate the Actual Transmission Rate of the Video Stream
Based on $R_r(t_i)$ and $R_s(t_i^r)$, calculate the actual transmission rate R of the video stream at the transmitting end, as shown in Eq. (10):

$$R = f(R_s(t_i^r), R_r(t_i)) = \min\{R_s(t_i^r), R_r(t_i)\} \tag{10}$$

The video stream is coded according to the transmission rate R, and the transmission rate of the current data packet is controlled [11, 12], and the adaptation and matching of the video stream transmission rate to the network link is realized.

4 Simulation

In this paper, the WebRTC module in Chromium browser is modified to realize the proposed data transmission mechanism. Meanwhile, with the official API of WebRTC, the real-time streaming media server is built as an experimental platform [13] to simulate the performance of AVRCS and RRTCC algorithms. The video session is simultaneously started in a wired network of 100M bandwidth and a mobile 4G network environment, and the graph of the transmission rate fluctuation is recorded. The results are shown in Figs. 4 and 5. Then, when the video call is in a stable condition, the average value and standard deviation of the transmission rate are respectively calculated. The results are shown in Table 1.

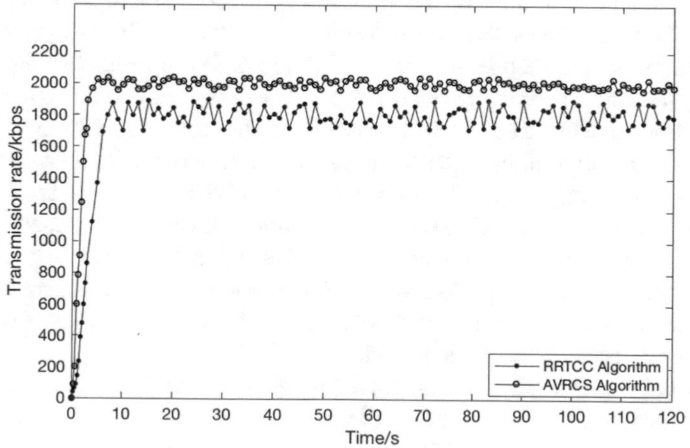

Fig. 4. Transmission rate under wired network

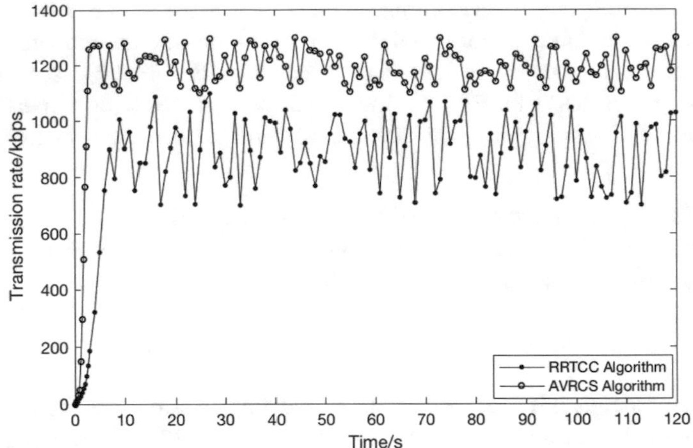

Fig. 5. Transmission rate under mobile 4G network

Table 1. Average and standard deviation of transmission rate.

	Average value/kbps		Standard deviation	
Network scenario	Wired network	Mobile 4G	Wired network	Mobile 4G
RRTCC	1804	896	57	114
AVRCS	1999	1195	25	58

Figure 4 shows the transmission rate of the AVRCS algorithm and the RRTCC algorithm in a wired network environment. During establishing the session connection, the transmission rate increases rapidly. After the transmission rate is stable, it can be seen from Table 1 that the average transmission rate with the AVRCS algorithm is 10.8% higher than that of the RRTCC algorithm, and the standard deviation is reduced by 56.1%. Figure 5 shows the transmission rate of the AVRCS algorithm and the RRTCC algorithm in a mobile network environment. During establishing the session connection, the link is generally overloaded due to the small initial threshold in the RRTCC algorithm. Therefore, the transmission rate increases significantly slower than that of the AVRCS algorithm. After the transmission rate is stable, it can be seen from Table 1 that the average transmission rate of the AVRCS algorithm is 36.4% higher than that of the RRTCC algorithm, and the standard deviation is reduced by 49.1%. The experimental results show that the AVRCS algorithm effectively reduces the transmission rate adjustment period, which not only improves the transmission rate of the video stream, but also reduces the fluctuation of the video stream transmission rate, especially in the case of mobile networks.

Frames Per Second (FPS) is an important indicator of the smoothness of video playback, which is the main factor that affects the user experience of video viewing. In this paper, frames per second is used to compare the smoothness of the video. In the case of wired network and mobile network, two kinds of algorithms are used to transmit a standard video sequence with 15 frames per second, and the frames per second is separately checked at the receiver. The results are shown in Figs. 6 and 7. Then, the average value and standard deviation of the video frames per second were respectively calculated, and the results are shown in Table 2.

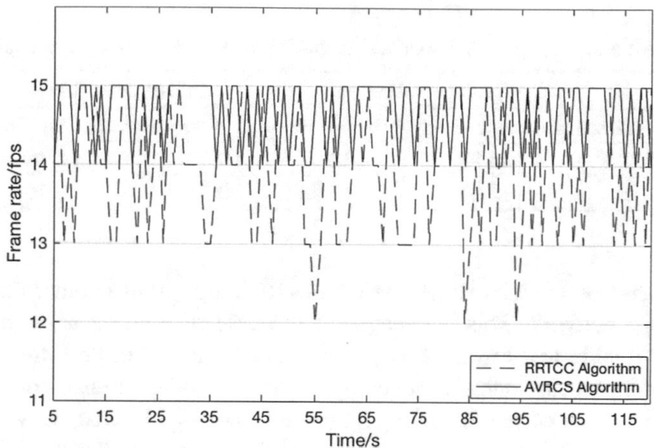

Fig. 6. Video frames per second under wired network

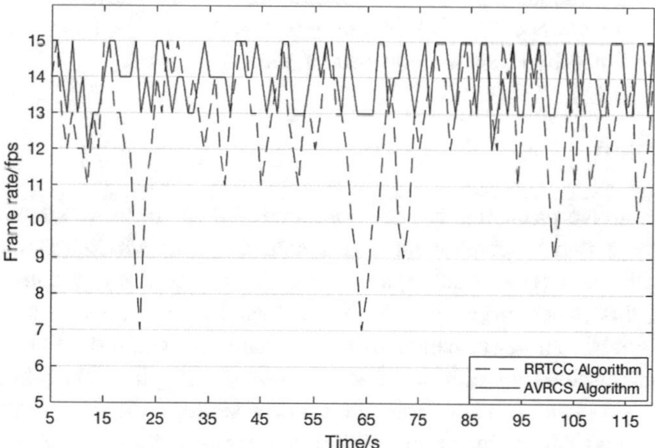

Fig. 7. Video frames per second under mobile 4G network

It can be seen from Table 2 that in the case of mobile networks, the average frames per second of the video received by the AVRCS algorithm is 10.1% higher than that of the RRTCC algorithm, and its standard deviation of the frames per second is reduced by 47.5%. While in the case of a wired network, the average frames per second of the video received by the AVRCS algorithm is 4.9% higher than that of the RRTCC algorithm. Meanwhile, its standard deviation of frames per second decreases by 40.4%. The experimental results show that the AVRCS algorithm increases the stability of the receiving frames per second of the video stream, and the picture quality of the video is also improved.

Table 2. Average and standard deviation of video frames per second.

	Average value/fps		Standard deviation	
Network scenario	Wired network	Mobile 4G	Wired network	Mobile 4G
RRTCC	13.98	12.67	0.89	1.78
AVRCS	14.67	13.96	0.53	0.94

The experimental results show that the AVRCS algorithm is more effective in the case of mobile network. This is because in the case of a wired network with better network link quality and lower latency, the RRTCC algorithm performs well with its higher transmission rate and the generally higher receiving frames per second. As a result, the advantage of the AVRCS algorithm is not highlighted. However, when it comes to mobile network, the adaptive threshold strategy of the AVRCS algorithm effectively reduces the transmission rate adjustment period, which greatly improves not only the network bandwidth utilization but also the transmission rate of the video stream. Therefore, the adaptive video streaming rate control strategy proposed in this paper can effectively cope with different network environments, so that to improve the smoothness and stabilize transmission rate of video playback.

5 Conclusion

In view of the above problems, like the complexity of the network status in the power communication network operation and maintenance site, the video transmission quality fluctuation with the network quality, and low efficiency of the wearable operation and maintenance, this paper proposes a WebRTC-based adaptive video streaming transmission rate control strategy, which supports adaptive dynamic adjustment of video streaming transmission rate with network link quality changing. The simulation results show that the proposed transmission rate control strategy can effectively reduce the network delay and packet loss rate, even under various network environments. It is demonstrated that this strategy can effectively adapt to the complex and variable on-site operation and maintenance of the power communication network.

References

1. 徐丽红,陈端云.VR技术在电力通信网中的应用和研究. 中国新通信 **19**(21), 112 (2017)
2. Johnston, A., Yoakum, J., Singh, K.: Taking on webRTC in an enterprise. IEEE Commun. Mag. **51**(4), 48–54 (2013)

3. Singh, V., Lozano, A.A., Ott, J.: Performance analysis of receive-side real-time congestion control for WebRTC. In: 2013 20th International Packet Video Workshop, San Jose, CA, pp. 1–8 (2013)
4. Cheng, B., Yang, J., Wang, S., Chen, J.: Adaptive video transmission control system based on reinforcement learning approach over heterogeneous networks. IEEE Trans. Autom. Sci. Eng. **12**(3), 1104–1113 (2015)
5. 韩峥,唐昆,崔慧娟.基于H.264的码率控制算法. 清华大学学报(自然科学版) (01), 59–61 (2008)
6. Carlucci, G., De Cicco, L., Holmer, S., Mascolo, S.: Congestion control for web real-time communication. IEEE/ACM Trans. Networking **25**(5), 2629–2642 (2017)
7. Lozano, A.A., Singh, V., Ott, J.: Performance analysis of topologies for Web-based Real-Time Communication (WebRTC) (2013)
8. Fan, W., Li, J., Tang, N., Yu, W.: Incremental detection of inconsistencies in distributed data. In: 2012 IEEE 28th International Conference on Data Engineering, Washington, D.C., pp. 318–329 (2012)
9. Gupta, A., Choudhary, A.: Real-time lane detection using spatio-temporal incremental clustering. In: 2017 IEEE 20th International Conference on Intelligent Transportation Systems (ITSC), Yokohama, pp. 1–6 (2017)
10. Carlucci, G., Holmer, S., et al.: Analysis and design of the google congestion control for web real-time communication (WebRTC). In: International Conference on Multimedia Systems. ACM, Article no. 13 (2016)
11. 闫素英. 基于WebRTC的无线实时通信QoS-QoE评估与预测. 北京交通大学 (2016)
12. Bandung, Y., Subekti, L.B., Tanjung, D., Chrysostomou, C.: QoS analysis for WebRTC videoconference on bandwidth-limited network. In: 2017 20th International Symposium on Wireless Personal Multimedia Communications (WPMC), Bali, pp. 547–553 (2017)
13. Bakar, G., Kirmizioglu, R.A., Tekalp, A.M.: Motion-based adaptive streaming in WebRTC using spatio-temporal scalable VP9 video coding. In: GLOBECOM 2017 - 2017 IEEE Global Communications Conference, Singapore, pp. 1–6 (2017)

Evaluation of Enzymatic Extract with Lipase Activity of Yarrowia Lipolytica. An Application of Data Mining for the Food Industry Wastewater Treatment

Heidy Posso Mendoza[1(✉)], Rosangela Pérez Salinas[1], Arnulfo Tarón Dunoyer[1], Claudia Carvajal Tatis[2], W. B. Morgado-Gamero[2], Margarita Castillo Ramírez[3], and Alexander Parody[4]

[1] Department of Bacteriology, Universidad Metropolitana, Barranquilla, Colombia
Heidy_posso@unimetro.edu.co, Rosyperez55@gmail.com, petetetaron@yahoo.com.mx

[2] Deparment of Exact and Natural Sciences, Universidad de la Costa, Barranquilla, Colombia
{ccarvaja4,wmorgadol}@cuc.edu.co

[3] Barranquilla Air Quality Monitoring Network EPA-Barranquilla Verde, Barranquilla, Colombia
mcastilloramirez87@gmail.com

[4] Engineering Faculty, Universidad Libre, Barranquilla, Colombia
alexandere.parodym@unilibre.edu.co

Abstract. The object of this research was to obtain the Crude Enzymatic Extract (CEE) of *Yarrowia lipolytica* ATCC 9773, in the medium of 30% Water of Sales (SW) applying a biologically treatment to three different concentrations yeast inoculum food wastewater, collected from cheese and whey production. It was evaluated the behavior of the inoculum in a suitable medium that stimulates lipids biodegradation. The standard liquid-liquid partition method SM 5520 B was used to quantify fat and oil removal for each concentration of yeast, before treatment and post treatment. The Industrial Fat effluent was characterized by physical chemical patterns, and two treatments were evaluated; Treatment 1 consisted of pH 5.0 and treatment 2 with a pH of 6.5, both with the following characteristics; Concentration of inoculum 8% 12% and 16% at 27 °C temperature and evaluation time 32 h. The best results (2.702 mg/L fat and 83% degradation oil) were found to be pH 5.0, 16% concentration and 27 °C, BOD5, and COD decreased by 43.07% and 44.35%, respectively during the 32 h; For pH 6.5, 8% concentration at 32 h and at room temperature, degraded 2.177 mg/L fat and oil (67% degradation); The BOD5, and COD decreased by 37.93% and 39.19%, in the same time span. The treatment at pH 5.0 inoculum concentration of 16% was effective in removing 83% of the volume of fats and oil in the effluent, representing a useful tool for the wastewater treatment.

Keywords: Crude enzymatic extract · Wastewater treatment · Biodegradation · Yeast inoculum · Lipases · *Yarrowia lipolytica*

C.-N. Yang et al. (Eds.): SICBS 2018, AISC 895, pp. 304–313, 2020.
https://doi.org/10.1007/978-3-030-16946-6_24

1 Introduction

Industrial development and unsustainable population growth, have increased the need to develop integrated water resources management, as a requirement for environmental preservation and for economic development [1]. According to the World Health Organization WHO, 2 million ton of wastewater (industrial, agricultural and domestic) are generated, without previous treatment, has been discharged into the environment, causing a negative impact related to high organic and inorganic load decomposition, that generates toxic products during the lipid peroxidation process [2]. These peroxides cause cellular damage in animals [3], in addition, this process could spread unpleasant odors [4] and microorganism in the air [5], studies suggest adverse health effects from exposure to bioaerosols affecting the health of the nearby communities [6]; constituting a public health problem [4–6].

As a measure has been implemented strategies as chemical or biotechnological remediation processes. Biodegradation is the biological treatment foundation that is applied to unfold undesirable chemicals such as fats, oils, proteins, nutrients or other substrates present in wastewater [7]. Microorganisms can play an important role in the cycle of organic matter, as effective biodegradation agents catalyzing pollutants transformation in innocuous components by enzymes [8, 9]. *Yarrowia lipolytica* is a strictly aerobic yeast, widely used in bioremediation processes due to its specific metabolic pathways and the ability to alter its cell surface, thus allowing efficient degradation of hydrophobic substrates such as n-alkanes, fatty acids, fats and oils, in addition to being used in industrial applications, due to its ability to produce a broad spectrum of products, such as: Organic acids and/or extracellular enzymes [10].

In order to remove Food industry wastewater pollutants, many studies have proposed to apply lipases, which are enzymes that catalyze the fats and oils hydrolysis by enzymatic bioremediation, giving rise to substances simpler and easier to degrade by other microorganisms. These enzymes have a great importance for their multiple applications because as degradation of substrates with high fat content, as well as, esterification reactions in Food, Pharmaceutical and Cosmetic industry. Lipases have been isolated from many species of animals, plants and microorganisms, however, microbial lipases are more versatile with interesting features such as stability in organic solvents, activity under various conditions, high substrate specificity and selectivity [11–14]. Due to this characteristics, they are very useful to treat wastewater, using microorganisms (lipolytica ones) with bioconverting (or biodegrading) ability in products with less impact on the ecosystem [2]. They also have been used in biofuels production [15, 16]. Lipases industrial production could be expensive, this situation has limited their use. However, this research had focused on evaluating the crude enzymatic extract with lipase activity of *Yarrowia lipolytica* ATCC 9773, using operating conditions that facilitate bioremediation processes in food industry effluents, as well as reducing production costs and becoming an economically viable biotechnological process that does not cause damage to the environment.

2 Materials and Methods

2.1 Strain and Culture Conditions

The strain used in this study was *Yarrowia lipolytica* ATCC 9773. The strain was cultivated in potato dextrose agar (PDA) and olive oil as a lipid source for 3 days at 27 °C and 37 °C, testing for the optimum temperature of growth. For recognizing the microscopic morphology strain was used and the inoculum was adjusted by turbidity at MacFarlan scale of 3. The inoculum consisted in a suspension of mature *Y. lipolytica* spores which were obtained from cultures incubated at 25 °C by 5 days; for biomass removing from the agar surface, cells were isolated adding NaCl 0.9% and they were stirred with a magnetic stirrer. The inoculum was done by duplicate samples with their respective controls.

2.2 Crude Enzymatic Extract (CEE)

Yarrowia lipolytica biomass was obtained and suspended in sterile 0.9% NaCl, then it was inoculated into a salt water culture medium (SW 30%) containing 5% sodium chloride, supplemented with Yeast extract 0.5%, as well as olive oil 1% and 0.1% Triton X-100 as inductor and emulsifier, respectively. Subsequently to 200 mL of the SW medium, 10 mL of the spore solution were added in 1 L Erlenmeyer; the solutions were tempered to 27 ° C for 8 h in a shaker at 200 rpm (Fig. 3). After 8 h the fungal biomass of Yarrowia lipolytica was separated from the supernatant by centrifugation at 5000 rpm for 10 min. The supernatant obtained was the crude enzymatic extract which was filtered on cellulose acetate membranes of 0.45 and 0.22 mm. CEE suspension viability was determined by measuring the absorbance at 620 nm, in a Spectronic 20D spectrophotometer, which being greater than or equal to 0.5 guarantees an approximate amount of viable enzymes of 2.5 × 106/mL [17]. CEE suspension was done by duplicate samples with their respective controls.

2.3 Chemical and Physical Wastewater Evaluation

Wastewater was collected from an industrial company which produce dairy products, 2 L of effluent was used to perform the initial physicochemical parameters before treatment such as pH, Temperature, Chemical Oxygen Demand COD, Biochemical Oxygen Demand BOD, Total Oils and Fats, Water Hardness, Total Proteins, Total Solids and Nutrients [18, 19]. The rest of the collected effluent was used to design the two treatments to be evaluated in this research.

2.4 Enzymatic Activity Evaluation of the of the Lipase Fraction in Different Treatments

For treatment 1, 3 L of wastewater were used by adjusting the pH to 5.0, after, solutions were prepared with one liter of the effluent added with CEE, obtaining different CEE concentrations: 8%, 12%, and 16% (CEE) solutions. For each different concentrations of effluent –CEE, were made 5 replicates of 200 mL. For treatment 2, were used 3 L of effluent by adjusting the pH to 6.5, after, it was prepared solutions with one liter of the effluent added with CEE, obtaining solutions of 8%, 12%, y 16% of CEE. For each different concentrations of effluent –CEE, were made 5 replicates of 200 mL. The oil concentration was determined every 8, 16, 24, and 32 h. The basis of the calculation of the capacity shall be made taking into account the quantity of reduced fat. After 32 h, the oil and grease concentration were determined in the two treatments, the best two parameters were selected according to the results obtained to perform a post-treatment physical-chemical characterization, in order to observe the performance and performance Total of *Yarrowia lipolytica* ATTC9773 strain in the fatty effluent.

2.5 Data Analysis

The statistical treatment of the data focused on the generation of simple linear regression models in the case of the study of the enzymatic activity of the lipase and in the case of the relationship between the concentrations, pH and the level of removal, simple ANOVAs were applied, the combinations of pH and concentration were converted into treatments, in addition the statistical differences detected were explained by means of confidence intervals for the mean with 95% confidence. Figure 1 shows the procedures that had been applied in the laboratory for obtaining and evaluating the crude enzymatic extract.

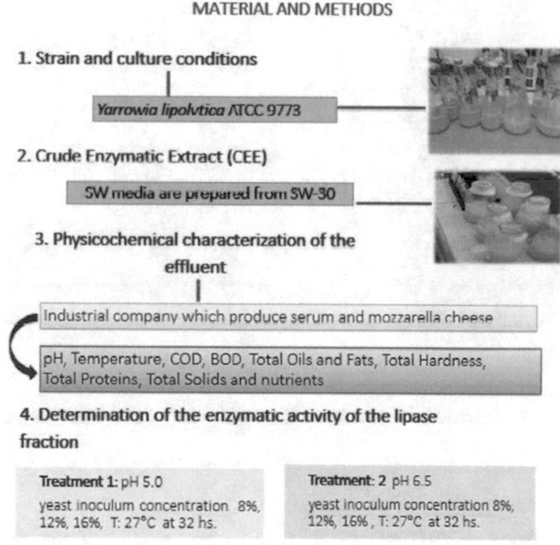

Fig. 1. Materials and methods

3 Results and Discussion

3.1 Chemical and Physical Wastewater Evaluation

Table 1 presents the results of Chemical and physical wastewater evaluation previous treatments with CEE.

Table 1. Chemical physical characterization of the effluent previous CEE treatments

Parameter	Results	Units
Biochemical Oxygen Demand	17320	mg O_2/L
Chemical Oxygen demand	53150	mg O_2/L
Fats, oils and grease	3260	mg/L
pH	8,08	pH units
Total Solids	21325	mg/L
Temperature	21,5	°C
Total Phosphorous	<0,075	mg P/L
Total Hardness	490	mg $CaCO_3$/L
Proteins	2,3	%

According to the Removal percentage BOD and COD parameter post CEE treatments, the best treatments correspond to pH: 5 concentration 16% and pH: 6.5 concentration 8% (Fig. 2).

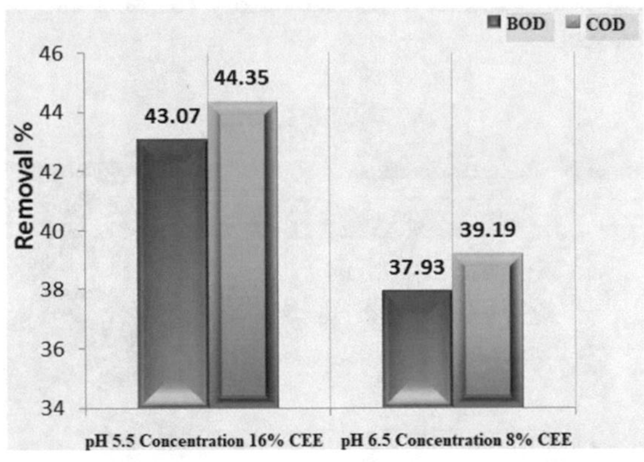

Fig. 2. BOD and COD Removal percentage, pH 5.0 concentration 16% and pH 6.5 concentration 8% at 32 h.

3.2 Enzymatic Activity Evaluation of the Lipase Fraction in Different Treatments

Figure 3 shows a correlation coefficient close to 1 which means that variables in each experimental group have been controlled; relationship between time and fat concentration is an inverse correlation. Fat descent coefficients were −351,7 al 8%, −349,7 al 12%, y −355,7 al 16% a pH 5.0, the highest one was presented at concentration 16%.

Figure 4 shows that in case of pH 6.5 the highest fat descent coefficient −184,6 at concentration 8%.

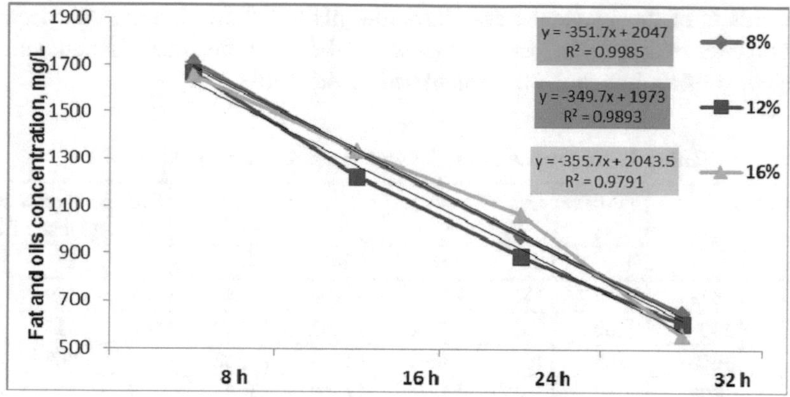

Fig. 3. Lipase enzymatic activity evaluation *Yarrowia lipolytica* ATCC 9773 pH 5.0

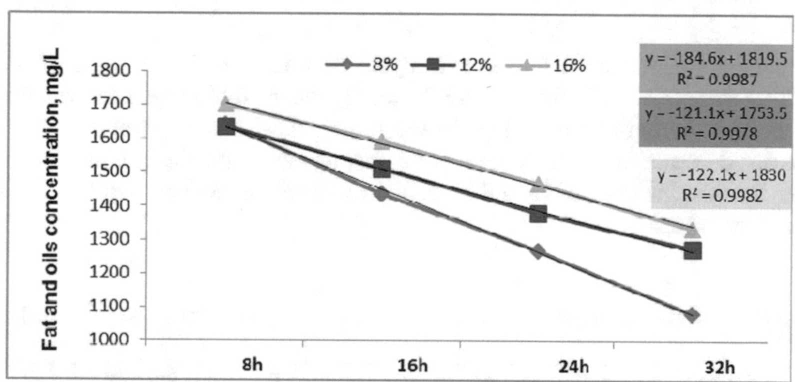

Fig. 4. Lipase enzymatic activity evaluation *Yarrowia lipolytica* ATCC 9773 pH 5.0

Table 2 shows contrasting treatments multifactorial ANOVA analysis, related to pH and concentration, pH 5.0 treatments had demonstrated a higher efficiency than pH 6.5 treatments. It is important to emphasize that we work with the absolute values.

Table 2. Fat descent coefficient multifactorial ANOVA Analysis

Source	SS	df	MS	F	Prob>F
Columns	66003.1	1	66003.1	99.06	0.0006
Error	2665.2	4	666.3		
Total	68668.2	5			

Regardless of the enzymatic concentration, pH 5.0 shows the greatest decrease in fat and oil after treatment. In this same sense the delta or difference of fat and oils in the interregnums of each record were calculated in the Table 3.

Table 3. Difference in the fat and oils removal intervals of 8 h

Concentration		TIME (hours)					Difference between interrregnums of 8 h		
		0	8	16	24	32	1	2	3
pH 5,0	**8%**	3260	1711	1327	978	655	384	349	323
	12%	3260	1666	1226	891	612	440	335	279
	16%	3260	1654	1337	1068	558	317	269	241
pH 6,5	**8%**	3260	1642	1438	1269	1083	204	169	186
	12%	3260	1636	1511	1380	1276	125	131	104
	16%	3260	1702	1592	1469	1336	110	123	133

According to data to a one-way analysis of variance, the analysis of variance (ANOVA) presented in Table 4, there is no significant difference between the treatments in the different concentrations in the same pH, also concentration 8% presents the greatest decrease in the interregnum of measurement of the fat and oil content in the effluent, for this, the delta of fat and oil content was calculated between one record and another, which means every 8 h.

Table 4. Difference in fat concentration Treatments multifactorial ANOVA Analysis

Source	SS	df	MS	F	Prob>F
Columns	168866.4	5	33773.3	19.04	2.48192e-05
Error	21288.7	12	1774.1		
Total	190155.1	17			

Multi-comparis chart of the Matlab R 2014[a] was made. Figure 5 shows that the group 6 corresponding to concentration 16% at pH 6.5 presents significantly different average in comparison with three groups which are in red corresponding to pH 5.0 treatment.

Finally, it is important to mention that when comparing the concentration of 8% in pH 6.5 in this, the greatest decrease in fat and oil is registered with respect to those of 12 and 16% in the same pH, coinciding with the same records in pH 5.0. The following is illustrated below (Fig. 6):

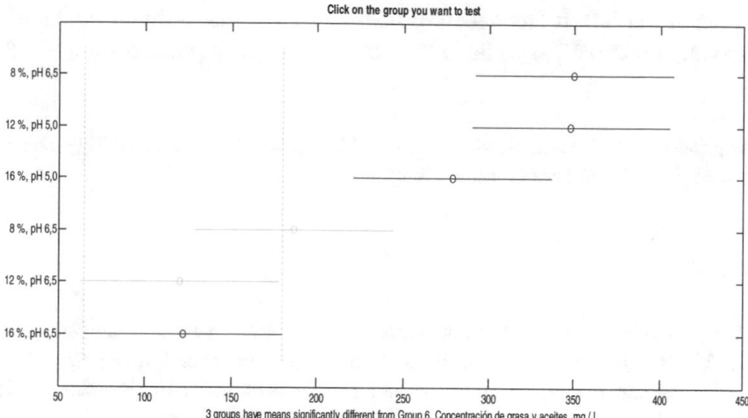

Fig. 5. Significance and differences related to concentration of fats and oils mg/L comparison with treatment 16% concentration pH 6.5

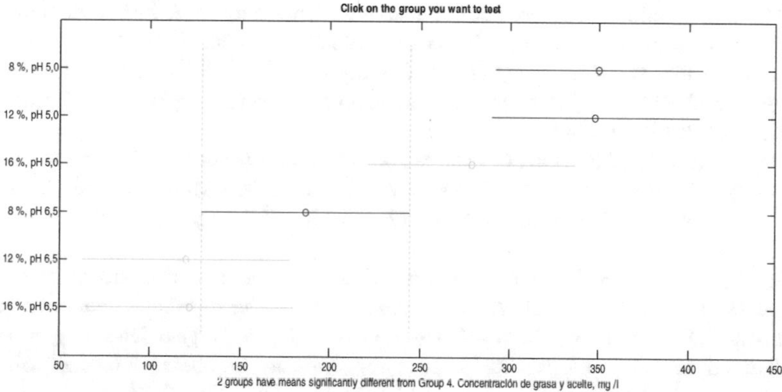

Fig. 6. Significance and differences related to concentration of fats and oils mg/L in comparison with treatment 8% concentration pH 6.5

4 Conclusion

In the research conducted, it was found that the treatments at pH 5.0 and pH 6.5 are: practical, economical, and efficient, observing that in the treatment at pH 5.0 the removal of fats and oils, were excellent with 83%. The lipolytic capacity shown by the CEE of Yarrowia lipolytica ATCC 9773 is influenced by the parameters (PH, concentration of inoculum 8%, 12% and 16%) to which it is subjected having a maximum lipolytic capacity in the 32 h of culture with a temperature of 27 °C and a pH of 5.0 and an inoculum concentration of 16%. The lipolytic capacity of the CEE of *Yarrowia lipolytica* ATCC 9773 was demonstrated, this yeast is useful in future treatment processes in fatty effluents and/or industrial fat traps with efficient biological microorganisms. In conclusion, statistically the best results are obtained with an 8%

concentration at pH 5.0. It has been reported that most microorganisms have a higher enzyme production at pH 7.0 while in this study the highest production was achieved at pH 5.0.

Acknowledgments. This research was supported by grants from Bacteriology Department of Universidad Metropolitana Barranquilla, Colombia.

References

1. MohdKhairul-Nizam, M.Z.: Bioremediation of oil from domestic wastewater using mixed culture: effects of inoculum concentration and agitation speed [dissertation]. Malaysia: University Pahang Faculty of chemical and Natural Resources Engineering (2008). http://iportal.ump.edu.my/lib/item?id=chamo:30963&theme=UMP2
2. Abass, O.A., Ahmad, T.J., Suleyman, A.M., Mohamed, I.A.K., Md. Zahangir, A.: Removal of oil and grease as emerging pollutants of concern (EPC) in wastewater stream. Eng. J. **12**(4), 161–169 (2011). https://doi.org/10.31436/iiumej.v12i4.218
3. Lemus, G.R.: Biodegradation and environmental impact of lipid-rich wastes under aerobic composting conditions [dissertation]. Vancouver: University of British Columbia Department of chemical and Biological Engineering (2003). 10.14288/1.0058966
4. Van Der Walle, N.: Über synthetische Wirkung bakterieller lipasen. Cbl Bakt Parasitenk Inktionskr **70**, 369–373 (1927). http://www.scielo.org.co/scielo.php?script=sci_arttext&pid=S0121-40042012000300001
5. Morgado Gamero, W.B., et al.: Concentrations and size distributions of fungal bioaerosols in a municipal landfill. In: Tan, Y., Shi, Y., Tang, Q. (eds.) Data Mining and Big Data. LNCS, vol. 10943, pp. 244–253. Springer, Cham (2018). https://doi.org/10.1007/978-3-319-93803-5_23
6. Morgado Gamero, W.B., et al.: Hospital admission and risk assessment associated to exposure of fungal bioaerosols at a municipal landfill using statistical models. In: Yin, H., Camacho, D., Novais, P., Tallón-Ballesteros, A. (eds.) Intelligent Data Engineering and Automated Learning IDEAL 2018. Lecture Notes in Computer Science, vol. 11315. Springer, Cham (2018). https://doi.org/10.1007/978-3-030-03496-2_24
7. Kempka, A.P., Lipke, N.R., Pinheiro, T.L.F., Menoncin, S., Treichel, H., Freire, D.M.G.: Response surface method to optimize the production and characterization of lipase from Penicillium verrucosum in solid-state fermentation. Bioprocess Biosyst. Eng. **31**(2), 119–125 (2008). https://doi.org/10.1007/s00449-007-0154-8
8. Ferrer, P., Montesinos, J.L., Valero, F., Sola, C.: Production of native and recombinant lipases by Candida rugosa. Appl. Biochem. Biotechnol. **95**(3), 221–256 (2001). https://doi.org/10.1385/ABAB:95:3:221
9. Contesini, F.J., da Silva, V.C.F., Maciel, R.F., de Lima, R.J., Barros, F.F.C., Carvalho, P.D.: Response surface analysis for the production of an enantioselective lipase from Aspergillus niger by solid state fermentation. J. Microbiol. **47**(5), 563–571 (2009). https://doi.org/10.1007/s12275-008-0279-8
10. Colla, L.M., Rizzardi, J., Pinto, M.H., Reinehr, C.O., Bertolin, T.E., Vieira Costa, J.A.: Simultaneous production of lipases and biosurfactants by submerged and solid-state bioprocesses. Bioresour. Technol. **101**(21), 8308–8314 (2010). https://doi.org/10.1016/j.biortech.2010.05.086

11. Burkert, J.F.M., Maugeri, F., Rodrigues, M.I.: Optimization of extracellular lipase production by Geotrichum sp. using factorial design. Bioresour. Technol. **91**(1), 77–84 (2004). https://doi.org/10.1016/S0960-8524(03)00152-4
12. Saatci, Y., Arslan, E.I., Konar, V.: Removal of total lipids and fatty acids from sunflower oil factory effluent by UASB reactor. Biores. Technol. **87**(3), 269–272 (2001). https://doi.org/10.1016/S0960-8524(02)00255-9
13. Di-Giulio, R.: Indices of oxidative stress as biomarkers for environmental contamination. In: Mayes, M.A., Baeeon, M.B. (eds.) En Aquatic Toxicology and Risk Assesment, vol. 14, pp. 15–31. American Society for Testing and Materials, Philadelphia (1991). https://doi.org/10.1520/STP23561S
14. Cirne, D.G., Paloumet, X., Björnsson, L., Alves, M.M., Mattiasson, B.: Anaerobic digestión of lipid-rich waste: Effects of lipid concentration. Renew. Ener. **32**(6), 965–975 (2007). https://doi.org/10.1016/j.renene.2006.04.003
15. Ayadi, I., Kamoun, O., Trigui-Lahiani, H., et al.: J. Ind. Microbiol. Biotechnol. **43**, 901 (2016). https://doi.org/10.1007/s10295-016-1772-4
16. Li, Z.J., Qiao, K., Liu, N., et al.: J. Ind. Microbiol. Biotechnol. **44**, 605 (2017). https://doi.org/10.1007/s10295-016-1864-1
17. Kirk, T.K., Schultz, E., Connors, W.J., et al.: Arch. Microbiol. **117**, 277 (1978). https://doi.org/10.1007/BF00738547
18. Viloria, A., Campo Urbina, M., Gómez Rodríguez, L., Parody Muñoz, A.: Predicting of behavior of escherichia coli resistance to imipenem and meropenem, using a simple mathematical model regression. Indian J. Sci. Technol. 9(46) (2016). https://doi.org/10.17485/ijst/2016/v9i46/107379
19. Carrero, C., et al.: Effect of vitamin a, zinc and multivitamin supplementation on the nutritional status and retinol serum values in school-age children. In: Tan, Y., Shi, Y., Tang, Q. (eds.) Data Mining and Big Data. DMBD 2018. Lecture Notes in Computer Science, vol. 10943. Springer, Cham (2018)

Load-Balancing-Based Reliable Mapping Algorithm for Virtual Resources in Power Communication Network

Nie Junhao[1(✉)], Qi Feng[1], Li Wenjing[2], and Zhang Zhe[2]

[1] Beijing University of Posts and Telecommunications, Beijing, China
njhnjhnjhao@foxmail.com

[2] State Grid Information and Communication Industry Group Co. Ltd., Beijing, China

Abstract. Based on the service characteristics of power communication network and the reliability requirements of virtual mapping, this paper designs a reliable mapping algorithm for virtual resources based on load balancing with network virtualization technology. The algorithm sorts according to services request priorities, adopting the idea of redundant backup primary and secondary road co-mapping, combining conditions of links constraints, bandwidth constraints, as well as non-intersecting of the main and auxiliary roads, and then maps requests to achieve the goal of load balancing. The simulation results show that the proposed algorithm has a significant effect on improving mapping success rate and stability of the virtual network. It can also alleviate the reliability problem of power communication network when fault occurs.

Keywords: Load balancing · Power communication network · Resource mapping · Reliability

1 Introduction

With the development and application of smart grids, the power distribution and utilization communication network plays an increasingly important role as the connection between the backbone transmission network and the local network of users in the power communication network. As the access layer network of the power backbone network, it covers smart grids at all levels, smart meters, charging piles and some communication terminals. There are a wide variety of devices which carry different functions or services, and the requirements for communication quality and methods are also different. Such a large number of nodes and widely distributed network characteristics make the development of smart grids put forward higher requirements and challenges for the reliability of information transmission in power distribution and utilization communication network.

Network virtualization technology can integrate abstract network resources. Mapping of virtual resources can shield the differences of power distribution communication network infrastructures and facilitate the design of efficient and reliable resource mapping algorithms to achieve optimal scheduling of heterogeneous resources.

C.-N. Yang et al. (Eds.): SICBS 2018, AISC 895, pp. 314–325, 2020.
https://doi.org/10.1007/978-3-030-16946-6_25

Currently, there are two types of virtual mapping architectures which are widely recognized in network virtualization technology with three layers [1, 2] and four layers [3] respectively. This paper adopts three-layer network architecture which contains infrastructure layer, virtual network layer and service layer from bottom to top. At present, the study on resource virtualization of distribution network is still in its infancy. In terms of virtualization, the resource mapping scheme [4] did not take the problem of reduced reliability into account when a line fails. Literature [5] studied the resource-based resource allocation method and allocates resources according to QoS requirements of services, but this method does not consider the impact of the underlying network service reliability. There are also some literatures that took reliability into account. Such as, literature [6] proposed an algorithm based on perceived node reliability and path sharing, realizing network resource overhead reduction through the target of network reliability. Literature [7] proposed a redundant path method to ensure network reliability. Literature [8] proposed a resource optimization allocation mechanism for power wireless virtual private networks, which abstracts physical wireless resources by establishing a network model to achieve resource sharing.

Considering the lack of existing resources in the comprehensive consideration of resource virtualization and network reliability. In order to meet the needs of power communication networks, this paper proposes a reliable mapping scheme for virtual resources based on load balancing. The scheme adopts the idea of redundant backup paths and maps the primary and secondary paths together. The optimization model of the main and auxiliary roads is constructed with the goals of load balancing and resource saving, which can effectively reduce the excessive waste of resources in the mapping process and improve the network reliability.

2 System Model

The power distribution and utilization communication network mainly includes distribution automation services, electricity consumption information collection system, and intelligent electricity services. From a business perspective, the intelligent distribution service communication network is mainly divided into backbone communication network and access layer communication network which can guarantee the timely, safe and reliable transmission of massive data with its high bandwidth and transmission rate. So, we assume that the infrastructure resources of the system model are limited but can meet business needs, and only considering their common attributes, namely the bandwidth of communication medium.

Intelligent distribution network is the smallest and with largest number of devices network in the power grid system which owns a wide range of nodes and complex structure. Therefore, in terms of faults, this paper only considers single point fault, verifying the proposed algorithm can improve network reliability effectively. The virtual network provisioning layer is an abstraction of the infrastructure layer. We abstract the device resources and communication media of the infrastructure layer into nodes and links.

Based on the above analysis, we will construct the virtual resource mapping model of the electric communication network into three layers which are infrastructure layer, virtual network layer and service layer from bottom to top, as shown in Fig. 1. In the virtual network, infrastructure layer is deployed by the intelligent distribution grid infrastructure provider, which aggregates different communication entities in the communication network. Virtual network layer serves as the abstraction of the infrastructure layer resources and is the core layer of the virtual network mapping. Service layer interacts with users and is responsible for managing user requests, sorting according to service request priorities and performing virtual network mapping in turn.

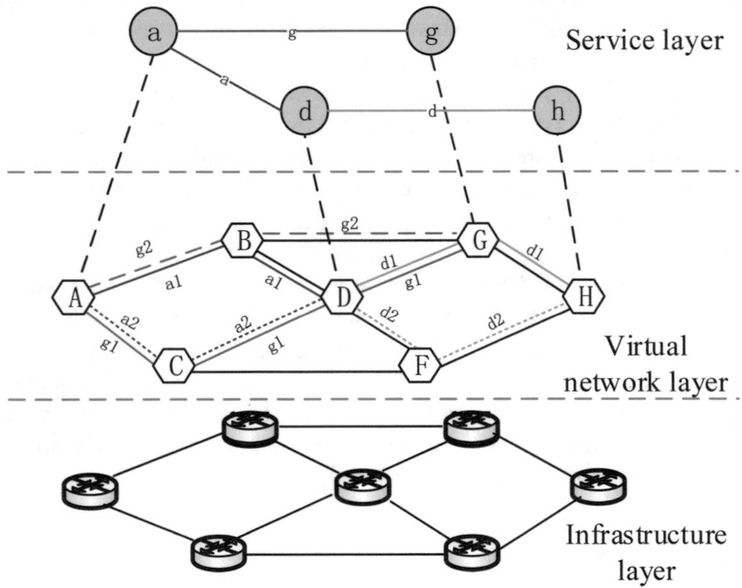

Fig. 1. Reliability mapping model of virtual network based on load balancing

In this paper, the idea of redundant backup primary and secondary road co-mapping is used to improve the reliability of virtual network. Firstly, infrastructure is abstracted into a virtual network. When service layer request occurs, the link will be selected with smallest load variance and then the remaining bandwidth can meet request and lastly, its hop count should meet the limit as the main path mapping. Then, the link that satisfies the network resource-saving and meets the disjoint relationship between the primary and secondary roads is selected as the secondary route. For example, virtual networks a1 and a2, where a1 is the main path, and auxiliary channels a2 and a1 do not intersect each other. When the network node B fails, the network can actively switch the cost-link transmission data, which improves the network reliability.

2.1 Problem Description

Undirected weighted graph $G^V = (N^V, L^V)$ is used to represent the virtual network layer, N^V is the set of nodes in virtual network providing layer, and L^V is the set of links in virtual network providing layer, and the weight is remaining bandwidth of each link. Each virtual network layer link $l_x^V(i,j) \in L^V$, $B_{LMax}(\mathrm{x})$ is total bandwidth of the link, and $B_L(x)$ is remaining bandwidth. Undirected weighted graph $G^S = (N^S, L^S)$ represents a virtual request, where N^V represents a set of nodes in virtual request network node, L^S represents a set of links in virtual request network, and the weight is requested bandwidth of each virtual link. The required bandwidth size for each virtual link $l_x^S \in L^S$ is $b(x)$.

Node Mapping. Virtual node mapping refers to find a physical node that meets its node capability constraints for each virtual node in virtual network request. According to the distribution of physical nodes of infrastructure layer, expression of virtual node mapping is:

$$N^S \rightarrow N^V \tag{1}$$

Each virtual node needs to be mapped to a different physical node. Each physical node need to be in one-to-one correspondence. The specific constraint is as follows:

$$\forall n_i^S = \forall n_u^V, \forall x_i^u = \{0,1\} \tag{2}$$

The variable x_i^u indicates mapping relationship between virtual node n_u^V and physical node n_i^S. In the mapping process, if virtual node n_u^V is mapped to physical node n_i^S, variable takes a value of *1*, otherwise it is *0*.

Primary Link Mapping. In the process of mapping primary and secondary links, load balancing is used as target, load ratio of each physical link is calculated to obtain minimum load variance of entire network. The main path mapping objective function formula is as follows:

$$D = \frac{\sum_{l \in E^V} \delta(l)(R^L(l) - E(R^L(l)))^2}{\sum_{l \in L^V} \delta(l)} \tag{3}$$

Where D is variance of the load rate of link mapped to virtual network providing layer, which is used to measure the load balancing degree of virtual network providing layer. If the variance is smaller, load is more balanced. $\delta(l)$ is a mapping parameter, $\delta(l) = 1$ when there is a virtual link mapped to this physical link, otherwise $\delta(l) = 0$. $R^L(l)$ is load rate of virtual network provider layer link, and its calculation formula is as follows:

$$R^L(l) = \frac{B_{LMax}(l) - B_L(l)}{B_{LMax}(l)} \tag{4}$$

The link remaining bandwidth $B_L(l)$ of virtual network providing layer is calculated by $B_L(l) = B_L^{'}(l) - b(m)$. Where $B_L^{'}(l)$ represents remaining bandwidth of the link when the virtual link is not mapped.

The sum of the virtual request bandwidths of each link mapping in mapping process should not be greater than remaining bandwidth of the link, which is $\sum b(l) < B_L(l)$

The primary and secondary road mapping links do not intersect each other. Constraints are as follows:

$$\sum_{m \in E^S} \varphi_m(l) b(m) \leq (\alpha + \beta) * R_E^{'}(l), \forall l \in E^V \tag{5}$$

When the link has a primary route mapping, $\alpha = 1$, otherwise, $\beta = 1$, $\alpha + \beta = 1$.

In addition, virtual requests generally have delay requirements, so the length of mapping primary and secondary routes should be less than the upper limit of the tolerable path length, and the path length constraint is:

$$length(l) \leq L(l) \tag{6}$$

$length(l)$ is the physical path length, and $L(l)$ request the maximum number of hops tolerated.

Secondary Link Mapping. The auxiliary road mapping aims at resource saving. This model can save mapping resources in the secondary link mapping phase and solve the problem of excessive waste of resources in the resource mapping process. In the secondary link mapping process, the higher the mapping overlap rate of secondary link, the greater the resource resolution. Therefore, the objective function of secondary link mapping model is as follows:

$$B = \sum_{l \in E^S} [(1 - \eta_l) B_{Max}^S(l) - B^{Vj}(x_{\max}^l)], j \in J \tag{7}$$

Where $B^{Vj}(x_{\mathrm{m}}^l)$ is the required bandwidth of virtual request l with the largest bandwidth requirement in the virtual request link set mapped on the secondary link x_{m}^l. η_l is the percentage of primary link in the link l of the infrastructure layer.

The bandwidth constraint needs to be met in the secondary link mapping process as follows:

$$\sum_{j \in J} B^{Vj}(m^j) \leq (1 - \eta) B^{'S}(x), \forall x \in L^S \tag{8}$$

$\varphi_{m^j}^{'}(x) = 1$ when the request link m^j has a secondary link mapping at the infrastructure layer, and otherwise $\varphi_{m^j}^{'}(x) = 0$. At the same time, non-intersecting of the main and auxiliary roads as well as the bandwidth constraint should be satisfied.

2.2 Algorithm Design

In this section, a reliable algorithm for virtual resource mapping based on load balancing is designed according to the characteristics of the distribution network. The algorithm uses load balancing as the main path target and uses resource saving as the secondary path target to map the service request. Each virtual network request is mapped in order of priority.

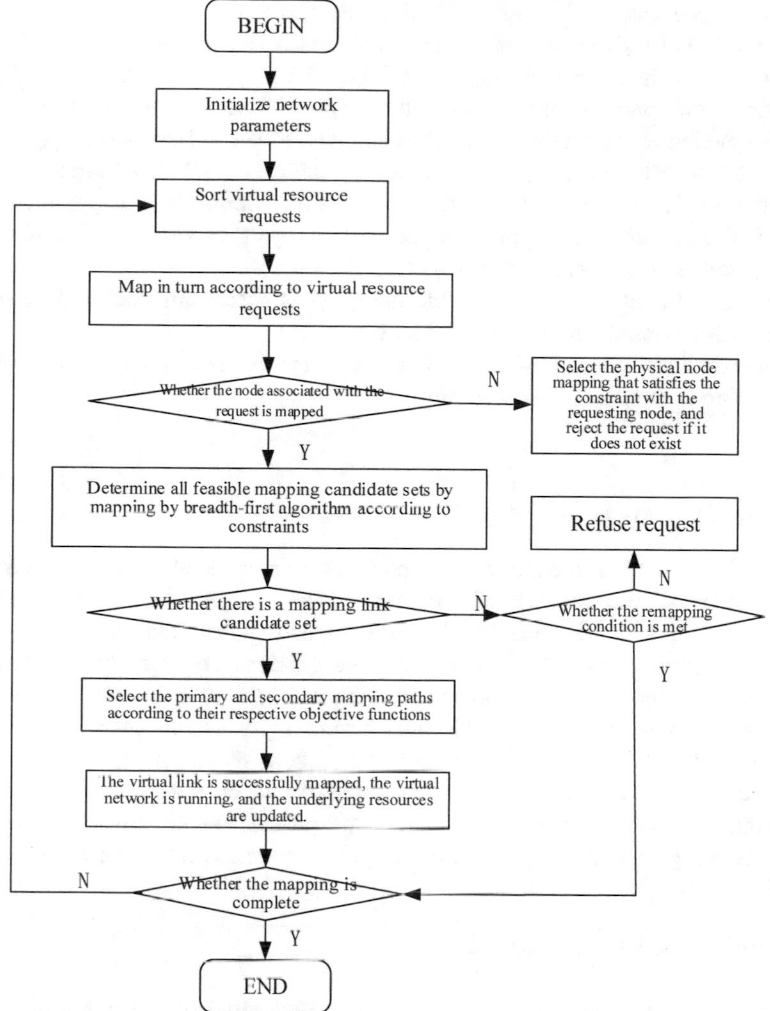

Fig. 2. Reliable virtual network mapping algorithm flow based on load balancing

Algorithm flow shown in Fig. 2 is depicted as follows:

(1) Initialize network resources. Import the distribution network resource data and set parameters.
(2) Sort according to the priority of network request, and meet the principle of high priority request-first mapping.
(3) Node mapping, select physical node that meets the node requirements for mapping, if it does not exist, reject request and return (2).
(4) Find the top network request in the current queue. According to the constraints such as bandwidth and delay, network resources are traversed by breadth-first algorithm to find the link candidate set that satisfies the condition.

(5) If there is a link candidate set, calculate the entire network load variance of each sub-link that satisfies mapping condition, and select the link with smallest load variance as the main path map. It is determined by whether there is a link that satisfies the condition of auxiliary path mapping objective function. If yes, the link with the largest remaining resource is selected as secondary path mapping; if no, secondary path mapping is discarded. If there is no link candidate set, it is determined by whether link resource satisfies the continuation mapping condition, and if it is satisfied, skipping the current mapping and returns (2); if resource mapping condition is not met, rejecting the request.
(6) Update network resources after the mapping is successful. Includes virtual link remaining bandwidth, queues, and so on.
(7) Find whether the queue is mapped or not. If the queue is empty, virtual resource mapping is ended. Otherwise returns (2)

2.3 Algorithm Analysis

The main idea of this algorithm is to implement network load balancing, and realize mapping of virtual resources while improving network stability and reliability. For each virtual request, link and its related nodes are mapped together, and the optimal solution is selected in the solution set to maximize the efficiency of network mapping. The primary and secondary road mappings are targeted at load balancing, and the main purpose is to improve stability of the link. Secondary road improves the stability of virtual network with the goal of resource conservation. When primary link fails, it can be switched to the secondary channel immediately, which reduces the impact of the single point of failure on the entire network, and provides an effective solution for the reliable mapping of virtual resources in the power communication network.

3 Simulation Environment

This simulation uses a network generation tool to randomly generate a 20-node, 31-link rule map for simulating an electrical communication network environment. In order to simulate the actual situation of the power distribution communication network, we divide the network request service into five categories, communication protection, video surveillance, data acquisition, distribution automation and intelligent power. We assign the proportion of the five types' service requests with 5%, 10%, 20%, 30%, and 35%, respectively, according to the security, delay, priority, and bandwidth requirements of the service request. What's more, the distribution of service request is stable. Assume that virtual network layer link bandwidth is an integer and randomly generate requests, including the initial node, the destination node, and the required bandwidth. It is also assumed that the virtual requests are all within the upper limit of the tolerable path length. This experiment is to compare the effectiveness of the algorithm to the virtual link. The request tolerable path length is uniformly set to 8, and the network bandwidth remains same. The link load variance, link mapping success rate, and link failure impact on the service are used as evaluation indicators. This experiment compares and analyzes the load-balancing virtual link reliable mapping algorithm (LRMA) with the Shortest path resource mapping algorithm (SPRMA) and the Greedy algorithm (GA). The shortest path algorithm allocates resources to the virtual network request with the shortest path as the target, and the greedy algorithm aims to allocate the maximum bandwidth resource to the virtual network request. In order to ensure the validity of the experimental data, simulation results are average values obtained after multiple simulations.

4 Simulation Results

In order to evaluate the correctness and effectiveness of the algorithm, in the simulation experiment, we set virtual network request mapping success rate, network load variance, and the impact of single point of failure on the network request as an evaluation index, comparing the algorithm proposed in this paper with SPRMA and GA. the simulation structure is shown in Figs. 3, 4, 5 and 6.

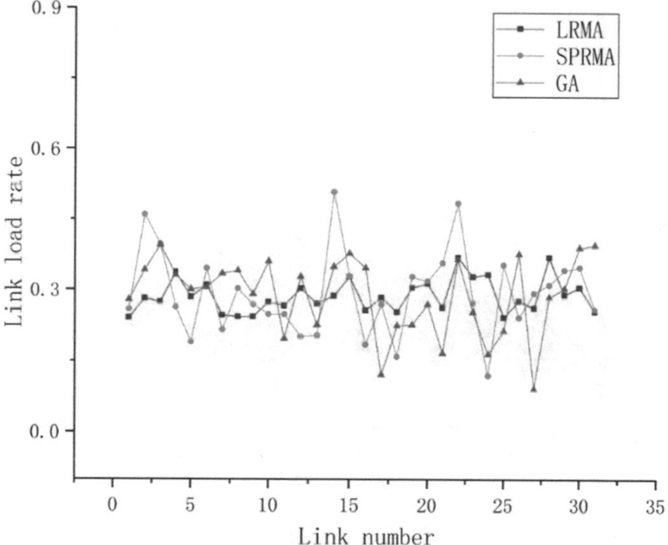

Fig. 3. 40 service request link load rate

Figure 3 shows the load ratio of each link in network after 40 times of service request. The red line indicates reliable mapping algorithm based on load balancing (LRMA), gray line is shortest path mapping algorithm (SPRMA), and blue line is greedy algorithm (GA). The link load rate trend of three algorithms is shown above. It can be seen from figure that the link load rate of LRMA algorithm proposed in this paper fluctuates slightly above and below 0.3, while the other two algorithms have larger fluctuations, the stability is worse than the LRMA algorithm. The reason is that the LRMA fully considers the underlying resources of the network in the mapping process, and maps with load balancing as target. As the number of requests increases, link load is relatively balanced. Even if the network resources are scarce, link load rate can be stabilized. The SPRMA algorithm targets the shortest path and does not consider the underlying physical resources. When the number of requests is small, the proportion of total number of links occupied by the mapped links is relatively small, which has little effect on network load balancing. But when the number of requests is large, nodes with more links in network will be used multiple times and mapped multiple times, so that a certain number of links will be overloaded while some links have sufficient remaining bandwidth and the network load is unbalanced. The GA algorithm targets the maximum bandwidth, when mapping, the underlying resources are considered less. In the mapping process, the high-bandwidth links are repeatedly mapped. When the number of requests is large, the same situation as the SPRMA algorithm occurs, and the link load in network will be in an unbalanced state.

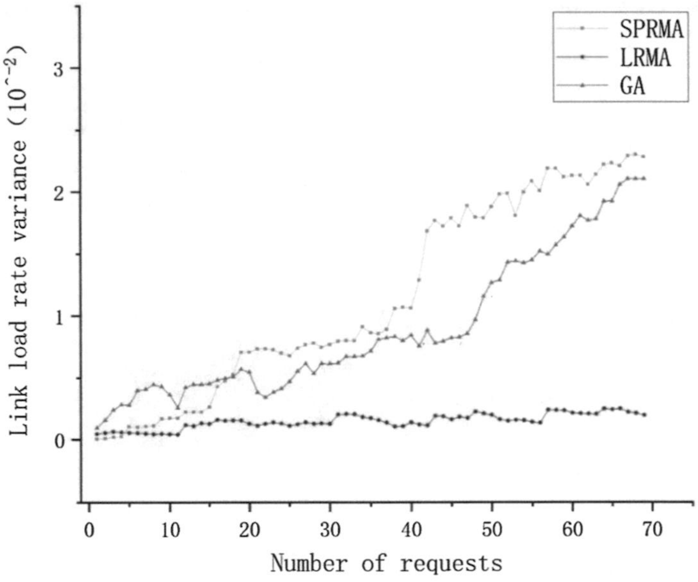

Fig. 4. Link load variance

Figure 4 depicts the network load variance as the number of service requests increases. Red indicates the load-balancing-based reliable mapping algorithm LRMA, gray is the shortest path SPRMA algorithm, and blue is the greedy algorithm targeting the maximum bandwidth. It can be seen from figure that the load variance of LRMA algorithm proposed in this paper fluctuates slightly, and the whole tends to be stable. While the SPTMA and GA algorithms fluctuate greatly, the overall trend is upward, and SPTMA grows slightly faster than GA. The reason is that SPTMA and GA will have multiple requests mapped on the same link, resulting in unbalanced network load and overall service carrying capacity. The SPTMA algorithm targets the shortest path and does not consider the underlying resources when mapping. As the number of requests increases, the link load rate gap between the three algorithms gradually increases. When the number of requests reaches a certain number, SPTMA and GA will prematurely full load, the link load variance of the two algorithms will be stable. The load variance has a certain probability of a small decrease when the request link does not include the full load. SPTMA algorithm targets load balancing and has a small link load variance, which is consistent with the expected effect.

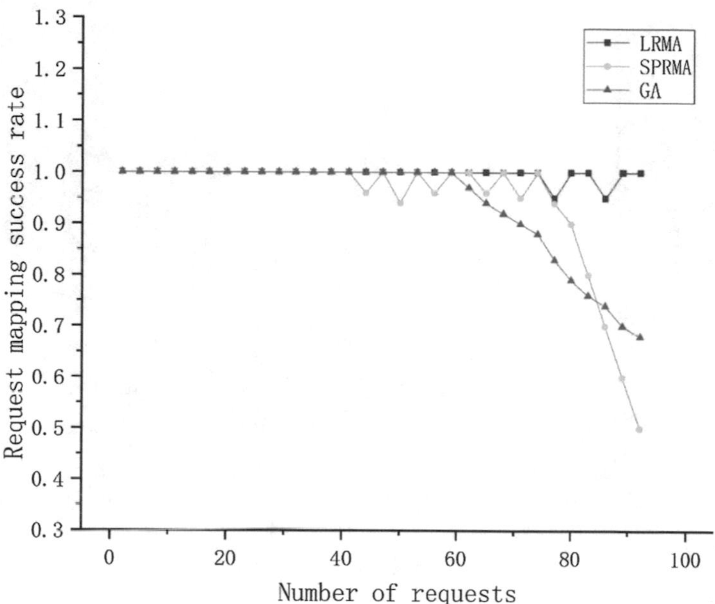

Fig. 5. Request mapping success rate

Figure 5 depicts the acceptance rate of three algorithms to virtual requests. The red line indicates LRMA algorithm designed in this paper, gray is SPRMA algorithm, blue is GA algorithm. As the number of requests increases, the trend of request mapping success rate of three algorithms is shown in the figure. It can be seen that SPRMA and GA algorithms have a significant linear decline when the number of requests is greater than 60, while LRMA fluctuates slightly and is generally stable. It is fully explained that LRMA algorithm proposed in this paper performs well under the condition of lack of bandwidth resources. The main reason is that LRMA algorithm aims at load balancing, reducing the local link pressure of network, improving the network service carrying capacity as a whole, and avoids the excessive load rate of a small part of the link, which affects the service request success rate of overall network. In addition, this algorithm designs a network resource update mechanism. When the number of requests is large, the resource mapping with less bandwidth requirement is less affected. SPRMA algorithm fails to consider the underlying resources. It may occur that multiple requests are mapped to the same link too early. After a small number of links are mapped multiple times, the load ratio is too large, and the mapping success rate fluctuation occurs earlier. GA algorithm selects mapping link with the goal of requesting the remaining maximum bandwidth of the link, and considers some of underlying resources to some extent. When the mapping success rate fluctuates too much, the increase of the full-load link leads to a decrease in the mapping success rate, although compared to SPRMA, the algorithm has improved, but the advantage is not great.

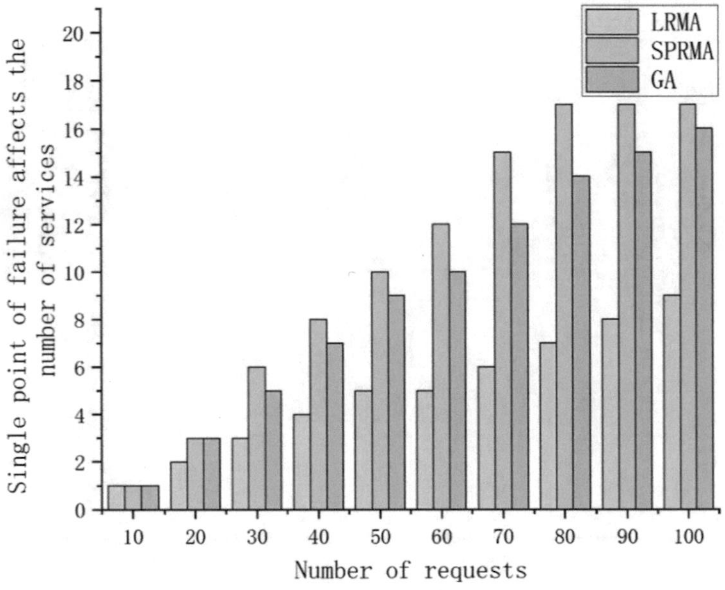

Fig. 6. The impact of single point failure on the network

Network reliability is also an important algorithm evaluation index. The degree of impact on the number of services when a network node fails reflects reliability of mapping network. As shown in Fig. 6, yellow indicates LRMA algorithm, green indicates SPRMA shortest path algorithm, purple represents GA greedy algorithm. As the number of requests increases, it can be seen that the number of services affected by the single-point failure of three algorithms all obtained an increase. Specifically, the increase of LRMA algorithm is smallest, while the increase of SPTMA and GA are significant and their difference is not big. This is because LRMA algorithm has a load balancing goal, underlying resource allocation is relatively uniform, and the single point of failure has less impact on service request. Both LRMA and GA algorithm can be used to concentrate the multi-path request on one link, causing some virtual link load to increase, the link load distribution to be uneven, and the number of services affected by the failure to increase. When the number of service requests reaches upper limit of algorithm mapping link, the number of services affected by single-point faults will tend to be stable. In addition, the design algorithm of this paper adopts the idea of redundant backup. When a single node in the request link fails, it can switch secondary transmission, which also reduces the impact of single-node failure on services.

5 Conclusion

In order to solve the problems existing in the power distribution communication network, such as complex network structure, various types of underlying devices, low resource utilization, and difficulty in management and maintenance of heterogeneous networks, we propose a load-balancing-based reliable mapping algorithm. It is designed by using network virtualization technology based on the three-layer virtual network mapping architecture. The algorithm aims at load balancing and equalization, using redundant backup primary and secondary road co-mapping to improve network reliability. Finally, simulation results prove that the proposed algorithm can improve the stability of link load to the greatest extent, reducing the impact of single point of failure on network services, meeting the requirements of reliability, stability and flexibility of virtual mapping in power communication network.

Acknowledgment. This work is supported by Science and Technology Project from Headquarters of State Grid Corporation of China: "Key technology development and application demonstration of high-confidence intelligent sensing and interactive integrated service system (52110118008H)".

References

1. Guo, T., Wang, N., Moessner, K., Tafazolli, R.: Shared backup network provision for virtual network embedding. In: Proceedings of International Conference on Communications (ICC), June 2011
2. Kim, Y.D., Kim, I., Choi, J., et al.: Adaptive modulation for MIMO systems with V-BLAST detection. In: VTC 2003-Spring, Jeju [s.n.], pp. 1074–1078 (2003)
3. Jarray, A., Karmouch, A.: Periodical Auctioning for QoS Aware Virtual Network Embedding. In: Proceedings of IEEE 20th International Workshop Quality of Service (IWQoS), June 2012
4. Chowdhury, N.M.K., Rahman, M.R., Boutaba, R.: ViNEYard: virtual network embedding algorithms with coordinated node and link mapping. ACM/IEEE Trans. Netw. **20**(1), 206–209 (2012)
5. Butt, N.F., Chowdhury, M., Boutaba, R.: Topology-awareness and reoptimization mechanism for virtual network embedding. In: Proceedings of the Networking 2010, Chennai, pp. 27–39. Springer, Heidelberg (2010)
6. Akkaya, K., Uluagac, A.S., Aydeger, A.: Software defined networking for wireless local networks in smart grid. In: 2015 IEEE 40th Local Computer Networks Conference Workshops (LCN Workshops), Clearwater Beach, FL, pp. 826–831 (2015)
7. Li, Y., Ni, W., Li, Y., Zheng, X.: Availability analysis of permanent dedicated path protection in WDM mesh networks. In: Asia Communications and Photonics Conference and Exhibition, Shanghai, pp. 369–370 (2010)
8. Meng, L., Sun, K., Wei, L., et al.: A virtual resource optimization allocation mechanism for power wireless private networks. Electron. Inf. **39**(07), 1711–1718 (2017)

Security of Network, Cloud and IoT

Enhancing Network Intrusion Detection System Method (NIDS) Using Mutual Information (RF-CIFE)

Nyiribakwe Dominique[1(✉)] and Zhuo Ma[2(✉)]

[1] Department of Computer Science and Technology, Xidian University, Xi'an, China
nyiribakwedom@gmail.com

[2] Department of Cyber Engineering, Xidian University, Xi'an, China
mazhuo@mail.xidian.edu.cn

Abstract. Most modern real word activities use an Internet where network traffic is exponentially increased. The attackers try different techniques and attempts for compromising and make unauthorized access to the network traffic of various network aspects. Intrusion detection systems (IDSs) used to detect both known and unknown/new attacks within the network system. Now a days researchers, security experts have implemented many different algorithms and mechanisms in order to enhance security measures. In this paper, we applied Random forest (RF) combined with conditional infomax feature extraction (CIFE) named as (RF-CIFE) for improving an Intrusion detection system model. In the experiment, four classifiers used are Support Vector Machine (SVM), C5.0, Multilayer Perceptron Neural Network (MLP) and Random Forest Algorithm. The conduction performance results using KDD Cup99 dataset prove that the combination of RF-CIFE with each classifier outperforms better in term of accuracy, detection rate, precision, false alarm rate and error rate.

Keywords: Network Intrusion Detection Systems (NIDS) · Conditional Infomax Feature Extraction (CIFE) · Mutual information theory · Support Vector Machine (SVM) · C5.0 · Multilayer Perceptron Neural Network (MLP) · Random Forest Algorithm

1 Introduction

The growth of current network technology fetch major challenges in the protection and prevention of system and network resources in all areas and organizations like finances, health, education, government and many other entities. Network and information security is the process of taking physical and software preventative measures to protect the underlying networking infrastructure from unauthorized access, misuse, malfunction, modification, destruction, or improper disclosure, thereby creating a secure platform for computers, users,

C.-N. Yang et al. (Eds.): SICBS 2018, AISC 895, pp. 329–342, 2020.
https://doi.org/10.1007/978-3-030-16946-6_26

the large amount of traffic data across the network and programs to perform their permitted critical functions within a secure environment [1,8]. Because The intruders attempt to compromise the confidentiality, integrity and availability of the computer and network resources, a flexible and effective defense system is required against various types of network attacks and ensuring the safety of network equipment, with capable of analyzing large amounts of traffic to exact detect many variety attacks. There are various measures and solutions proposed in order to maintain computer systems and networks secure like firewalls, virus scanners and encryption mechanisms are existing but they are not sufficient to protect network data and its resource [6,20,22]. Network Intrusion Detection Systems (NIDSs) are a common security defense which identify malicious attack behaviors by analyzing the network traffic of key nodes of a network or system audit log, and then send intrusion alerts to system administrators in real time. There are two approaches for implementation of intrusion detection system are:

- Misuse based system: In Misuse based IDS, the recent activity is compared to known signature or known intrusion scenario that can be specific pattern or sequence of data or events. Thus, the disadvantage is that can detect only known attacks for which they have a defined signature.
- Anomaly based system: In anomaly-based IDS, intrusions are detected by analyzing some deviations from the normal behavior or normal pattern. advantage of anomaly detection system is that they can detect previously unknown attacks [22].

Various researches have been implemented using machine learning, data mining and deep learning for grouping the detected normal and attack instances by clustering based on its similarities. And with classification techniques which construct the classified model depends on data. It helps to classify the new data into one predefined classes depends on the attribute values [4,6]. In this paper, we propose the use of ensemble mutual information theory technique, combined with various machine learning techniques. The model is capable of detecting and classifying all network traffic categories to facilitate NIDS operations in modern network. This paper has five sections as follows: Sect. 2 explores Background and Related work. Section 3 describes Our proposed work. Section 4 we present Our Experiment and Evaluation Results. Finally, Sect. 5 Conclusion.

2 Background and Related Work

2.1 Background

Mutual Information. Measure the amounts of information that one random variable contains about another, and the reduction in the uncertainty of one random variable due to the knowledge of the other variable. By the Consideration of two random variables X and Y with a joint probability mass function $p(x, y)$ and marginal probability mass functions $p(x)$ and $p(y)$. The mutual information

$I(X;Y)$ is the relative entropy between the joint distribution and the product distribution $p(x)p(y)$ [9,10,14,20].

$$I(X;Y) = \sum_{x,y} p(x,y) log \frac{p(X,Y)}{p(X)p(Y)} \tag{1}$$

Support Vector Machine (SVM). Is a machine learning algorithm and supervised learning method that analyzes data for classification and regression analysis separates the data into normal and abnormal categorials with the help of decision boundary and performs classification by finding the hyperplane that maximizes the margin between the two classes [1,12,20]. The hyperplane is decided by a subset of training samples, namely support vectors. SVM classifier is able to train samples into separate classes divided by maximum margin and categorize new test samples with high accuracy. Regularization is taken into accounts in SVM to avoid over-fitting as well. Moreover, SVM is efficient since it is formulated as the quadratic optimization problem (no local minimum). In addition to being a linear classifier, SVM performs well on non-linear classification by using kernel tricks [5,6,16] (Fig. 1).

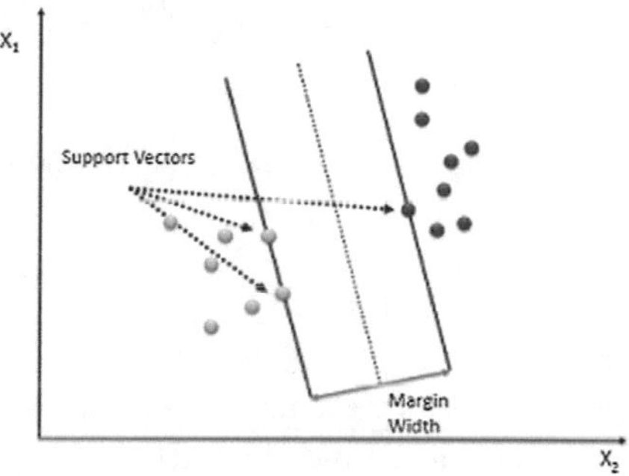

Fig. 1. Support vector machine decision boundary

C5.0. Is an algorithm widely used as a decision tree method in machine learning. This type of decision tree model is based on entropy and information gain, It Gives a binary tree or multi branches tree. C5.0 method is a new updated decision tree based on C4.5 method with many new functions [2]. C5.0 method performs faster than ordinary methods and has a good detection accuracy. The detailed advantages of C5.0 method are given below.

C5.0 method is faster and efficiency. C5.0 method can handle the noise and missing data. C5.0 method can handle the over fitting and error pruning problem. C5.0 method can handle the relevant attributes. Both continuous and categorical features can be implemented in a convenient way [5].

Multilayer Perceptron Neural Network. A multilayer perceptron (MLP) is a class of feedforward deep learning and artificial neural network. It composed of more than one perceptron, with an input layer which receive the signal, an output layer that makes a decision or prediction about the input, and in between those two, an arbitrary number of hidden layers that are trues computational engine of the MLP. MLPs with one hidden layer are capable of approximation any continuous function [4,19] (Fig. 2).

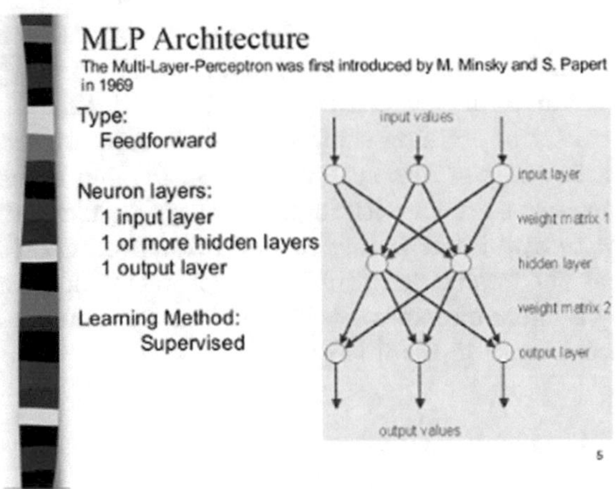

Fig. 2. Multilayer Perceptron Neural Network architecture

Random Forest Algorithm. Random forest Algorithm is an ensemble classifier, which constructs a group of independent and non-identical decision trees based on the idea of randomization. Random forest can be defined as $h(x, \theta k), k = 1, ...L$ in which θk is a kind of mutual independent random vector parameter, and x is the input data. Each decision tree uses a random vector as a parameter, randomly selects the feature of samples, and randomly selects the subset of the sample data set as the training set. The construction algorithm of random forest is as follows: k suggests the number of decision tree in the random forest, n indicates the number of sample in training data-set that each decision tree corresponds to, M refers to the feature number of sample, m represents the number of features when carrying out segmentation on a single node of a decision tree, $m \ll M$ [11,17].

2.2 Related Work

IDS is among the main powerful topics in network and information security field, where a lot of researchers are interested, and applied several different techniques for improving network intrusion detection systems (NIDS) performance.

Teng, Wu, Zhu, Teng, and Zhang [7] proposed an adaptive mechanism of collaboration intrusion detection model CAIDM based on 2 class SVMs and DT, the author built three intrusion detectors according to TCP, UDP, ICMP protocols and application layer protocols. The result revealed that the CAIDM is more accurate and efficient than detector system with a set of single type SVM.

Tao, Sun, and Sun [6] proposed two step optimizations of SVM and on GA. The results proved that the fitness function used in SVM achieved 94,53% of false positive rate and an error rate of 2.4 %.

Yuan and Hogrefe [5] proposed two layers multi-class detection method, the author used C5.0 to identify the most important features and Naives Bayes classifier for final output results, the Experiment results of TLMD had superior detection accuracy compare with single SVM and C5.O.

Aung [11] proposed an analysis of Random forest Algorithm (RFA) based on NIDS, the author used K-means algorithm to generate heterogenous dataset to nearly homogenous dataset, and then applied Random forest algorithm for classifying intrusions and normal traffic. The experiment showed that using RFA based on k – means had less efficient time and suitable accuracy than single RFA.

Karatas [4] proposed Neural network-based intrusion detection system with different functions, the author focus on the performance of the selected training functions in application area of Intrusion detection system, the error rate evaluation metric used to evaluate all functions.

Ali, AL Mohammed, Ismail, and Zolkipli [3] proposed new intrusion detection system based on fast learning network and particle swarm optimization PSO-FLN, the authors' experiment indicated that the accuracy increased when increasing the number of hidden neuron's in hidden layer.

3 Our Proposed Methodology

3.1 Random Forest for Feature Importance

Feature importance measurement Ensemble methods/feature importance evaluation. For each feature, the classifier produces a statistical measurement (and the corresponding standard deviation) for how important the feature was for predicting the target variable. The basic use of this information is to create a "feature ranking" among the features (from high importance value to low). A RF model [17,18] is made of an ensemble of trees, each of which is grown from a bootstrap sample of the n data points. For each tree, the selected samples from the bag (B), the remaining ones constitute the out-of-bag (B). Let B stand for the set of bags over the ensemble and B be the set of corresponding out-of-bag. We have $|B| = |B| = T$, the number of trees in the forest.

In order to compute feature importance, Breiman [18] proposes a permutation test procedure based on accuracy aimed at finding a subset of most relevant features for a prediction task. For each variable X_j, there is one permutation test per tree in the forest. For an out-of-bag sample B_k corresponding to the kth tree of the ensemble, one considers the original values of the variable X_j and a random permutation d_j of its values on Bk. The difference in prediction error using the permuted and original variable is recorded and averaged over all the out-of-bag s in the forest. The higher this index, the more important the variable is because it corresponds to a stronger increase of the classification error when permuting it. The importance measure J_a of the variable X_j is then defined as:

$$\mathbf{J}_a(x_j) = \frac{1}{T} \sum_{B_k \in B} \frac{1}{|B_k|} (h_k^{d_j}(i) \neq y_i - I(h_k(i) \neq y_i)) \tag{2}$$

Where y_i is the true class label of the out-of-bag example i, I is an indicator function, $h_k(i)$ is the class label of the example i as predicted by the tree estimated on the bag B_k, $h_k^{d_j}(i)$ is the predicted class label from the same tree while the values of the variable X_j have been permuted on B_k. Such a permutation does not change the tree but potentially changes the prediction on the out-of-bag example since its Jth dimension is modified after the permutation. Since the predictors with the original variable h_k and the permuted variable $h_k^{d_j}$ are individual decision trees, the sum over the various trees where this variable is present represents the ensemble behavior, respectively from the original variable values and its various permutations [15,17].

3.2 Conditional Infomax Feature Extraction

In NIDS, feature selection and extraction used to describe the samples, so that the classification can be efficient and robust in a feature space of much lower dimension [13]. Dimensionality reduction of the raw input variable space is an essential preprocessing step in the classification process. In general, it is desirable to keep the dimensionality of the input features as small as possible to reduce the computational cost of training a classifier as well as its complexity.

Some studies [9,10,14] show that in contrast to minimize the feature redundancy, the conditional redundancy between unselected features and already selected features given class labels should also be maximized. In other words, as long as the feature redundancy given class labels is stronger than the intra-feature redundancy, the feature selection will be affected negatively. A typical feature selection under this argument is Conditional Infomax Feature Extraction (CIFE) [9,14], in which the feature score for a new selected feature X_k is:

$$\mathbf{J}_{cife}(X_k) = I(X_k;Y) - \sum_{X_j \in S} I(X_j;X_k) + \sum_{X_j \in S} I(X_j;X_k)|Y) \tag{3}$$

This states that the selected features X are independent and class-conditionally independent given the selected feature X_k under consideration.

3.3 Proposed RF-CIFE Mechanism

The main ideas of adapting our proposed mutual information (RF-CIFE) into IDS system is to optimize the performance measures and Minimize the random traffic packets' errors form different resources across Network Intrusion Detection systems (NIDS). CIFE find a transformation that maximizes mutual information of class labels and extracted features. After applying KDD CUP 99 features' normalization and transformation as described in Sect. 4, features dimensionality increased and became highly correlated, we used RF to select the most relevant features according to the prediction labels as shown in Fig. 3 below:

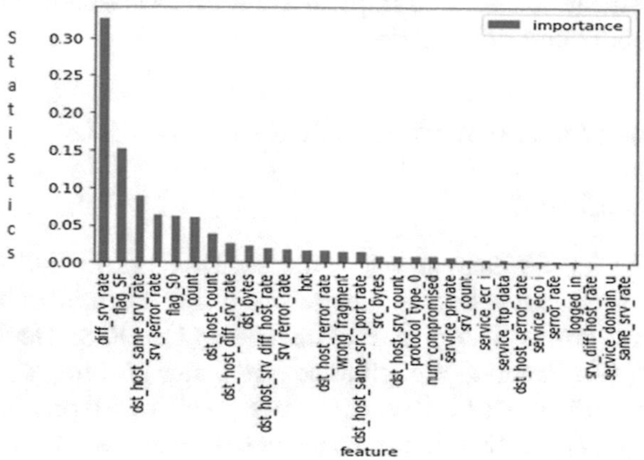

Fig. 3. Feature importance given by Random forest

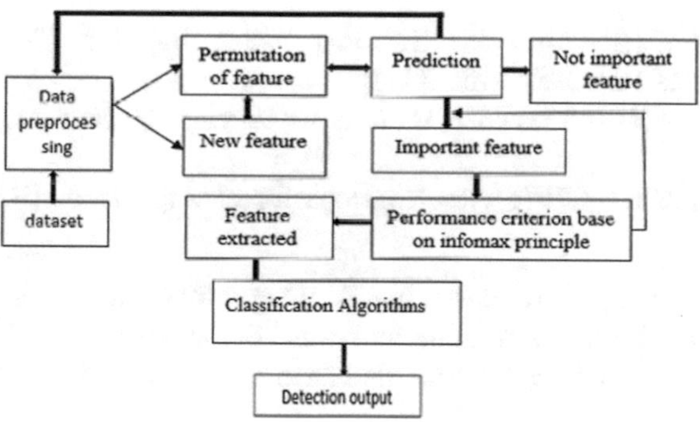

Fig. 4. RF-CIFE model architecture

Figure 4 shows the proposed RF-CIFE architecture of enhancing NIDS performance and minimization of the variables' error. This architecture consists of three main parts:

- Feature importance based on Random forest: There is a loop between dataset, permutation of feature, new feature (temporal place) and prediction fields. Where entered data ($x_1 ... x_n \in R^n$) will generate the permutation of feature based on model's prediction (y), if the permuting values of x_i increase the

model error, x_i will ranked as important. Whereas, Unimportant, if x_i keeps the model error unchanged.
- Conditional Informax principle measure the relevance and redundancy of features, after, it will extract and rank the most maximized features by reducing the class-relevant.
- For the Classification stage we implemented four classifiers to evaluate effective of the model described in Sect. 2.

4 Experimentation and Evaluation Results

4.1 Experimentation

Dataset. KDD Cup 99 dataset was built by DARPA'98 IDS evaluation program, contains 41 features and is labeled as either normal or an attack, The attacks types are grouped into four categories: (1). DOS: Denial of service – e.g. syn flooding (2). Probing: Surveillance and other probing, e.g. port scanning (3). U2R: unauthorized access to local super user (root) privileges, e.g. buffer overflow attacks. (4). R2L: unauthorized access from a remote machine, e.g. password guessing [23], KDD Cup 99 dataset benchmark used for large number of researchers in IDS building models.

In order to build an effective and efficiency Network intrusion detection system model, we proposed (RF-CIFE) for feature importance and extraction combined with 4 algorithms as our classifiers:

- Combination of RF-CIFE with Support vector machine (SVM)
- Combination of RF-CIFE with C5.0
- Combination of RF-CIFE with Multilayer Perceptron Neural Network (MLP-ANN)
- Combination of RF-CIFE with Random forest Algorithm (RFA).

Data Preprocessing. After assessing KDD cup 99 dataset, we observed that it contains categorical and continuous features. Before implementing this dataset, the categorical features were transformed into numerical using OneHotEncoder() function.

For data standardization, we used MinMaxScaler.Fit-transform() function to lie features between a given minimum and maximum value, often between [0, 1] range. As mentioned with the previous researchers, this dataset contains the duplicate values where drop-duplicates () function is used to drop all duplicates rows for reducing the computational cost and improving the classification performance of the model. We combined all attacks into one class as 'attacks' and other class for 'normal'.

Feature Importance and Extraction. Random forest (RF) is used for feature importance, it ranked all features, and Only thirty features were selected which provides more information based on predicted class labels. We implemented CIFE to extract all features given by RF, and eight features selected for the final classification and prediction of our model evaluation results.

KDD cup 99 0nly 10% of train and test dataset was used in all these operations and then, we compared each operation with other existing IDS methods with the following evaluation metrics: Accuracy, Detection rate (DR), Precision, FPR and Error Rate (Misclassification Rate).

- **Accuracy:** is used to evaluate the overall performance of the system

$$\text{Accuracy (Acc)} = \frac{TP + TN}{TP + FP + FN + TN}$$

- **DR:** is used to evaluate the system's performance with respect to its malware traffic detection

$$\text{Detection Rate (Dr)} = \frac{TP}{TP + FN}$$

- **Precision:** is used to evaluate the data instances predicted as positive that are actually positive.

$$\text{Precision (Pr)} = \frac{TP}{TP + FP}$$

- **False Alarm Rate:** is used to evaluate misclassifications of normal traffic.

$$\text{False Alarm Rate (FAR)} = \frac{FP}{FP+TN}$$

- **Error Rate:** is used to evaluate to the proportion of the samples that the classier incorrectly predicts in all samples.

$$\text{Error Rate} = \frac{FP+FN}{TP+TN+FP+FN}$$

Where

(**TP**): True positives are the cases when the actual class of the data point was 1(True) and the predicted is also 1(True).
(**TN**): True negatives are the cases when the actual class of the data point was 0(False) and the predicted is also 0(False).
(**FP**): False positives are the cases when the actual class of the data point was 0(False) and the predicted is 1(True). False is because the model has predicted incorrectly and positive because the class predicted was a positive one.
(**FN**): False negatives are the cases when the actual class of the data point was 1(True) and the predicted is 0(False). False is because the model has predicted incorrectly and negative because the class predicted was a negative one.

4.2 Evaluation Results

We conducted our experiment using anaconda which is a popular python tool. The performance and comparison results of our proposed model and other methods are shown in the tables below:

Table 1. Comparison of RF-CIFE with SVM and other methods

Methods	Acc	Dr	Pr	FAR	Error rate
TLMD [5]	93.32	-	-	-	-
CAIDM [7]	89.02	-	-	-	0.121
GA-SVM [6]	-	94.53	-	-	0.024
RF-CIFE with SVM	**96.3**	**96**	**95**	**0.023**	**0.011**

Table 1: show the experiment results obtained for 10% train and test datasets using RF-CIFE with SVM, the evaluation metrics show that the combination of RF-CIFE with SVM outperform different existing methods in all evaluation metrics experimented.

Table 2. Evaluation Results of RF-CIFE with C5.0

Methods	Acc	Dr	Pr	FAR	Error rate
C5.0 [5]	92.77	-	-	-	-
RF-CIFE with C5.0	**99**	**99.1**	**99**	**0.011**	**0.009**

Table 2: show the performance and evaluation metrics of proposed RF-CIFE with C5.0 using 10% train and test datasets and comparison with C5.0 algorithm and higher accuracy.

Table 3: show the performance of RF-CIFE with MLP-NN using 10% train and test datasets, we realized that by increasing the number of neuron's in hidden

Table 3. Comparison Results of RF-CIFE with MLP-NN and other methods

Methods	nbes hidden layer	Acc	Dr	Pr	FAR	Error rate
DAE-IDS [25]		94.7	95.6	-	-	-
HAST-IDS [24]		98.7	97.83	-	0.4	-
Trainlm function of ANN [4]		-	-	-	-	0.025
RF-CIFE with MLP-NN	13	**97.9**	**99.2**	**98**	**0.04**	**0.02**
	50	**98**	**99.1**	**98**	**0.03**	**0.019**
	150	**98.2**	**99.1**	**98**	**0.03**	**0.02**
	250	**98.3**	**99.2**	**98.2**	**0.03**	**0.016**

Table 4. Evaluation Results of RF-CIFE with Random forest Algorithm (RFA)

Methods	Acc	Dr	Pr	FAR	Error rate
k-mean with RFA [11]	-	-	-	-	0.02
RF-CIFE with RFA	**99.2**	**99.7**	**99**	**0.01**	**0.007**

layer, results of accuracy, detection rate, precision increased, and the false alarm and error rate results' decreasing compare with other methods.

Table 4: show the performance and evaluation metrics of proposed RF-CIFE with RFA using 10% train and test datasets and the reduction Error rate from 0.02% of existing RFA algorithm to the 0.007% for the combination of RF-CIFE with RFA.

Graphs Comparison of All Classifiers at Each Evaluation Metric

See Figs. 5, 6, 7, 8 and 9.

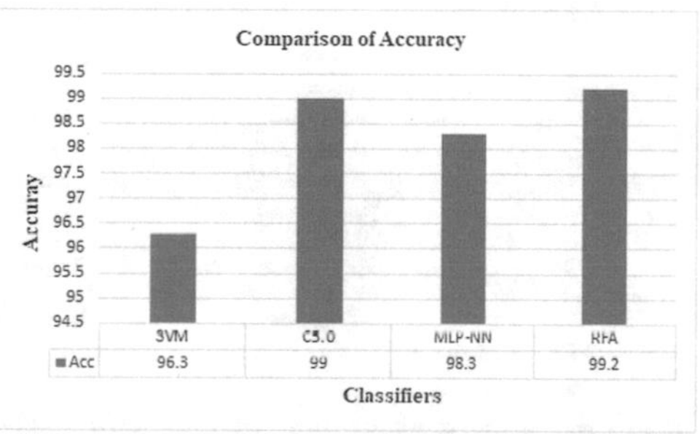

Fig. 5. Describe the comparison in term of Accuracy. It shows that combination of the proposed model RF-CIFE with Random Forest Algorithm has higher accuracy compare to the other combination algorithms experimented, and RF-CIFE with SVM has the lowest.

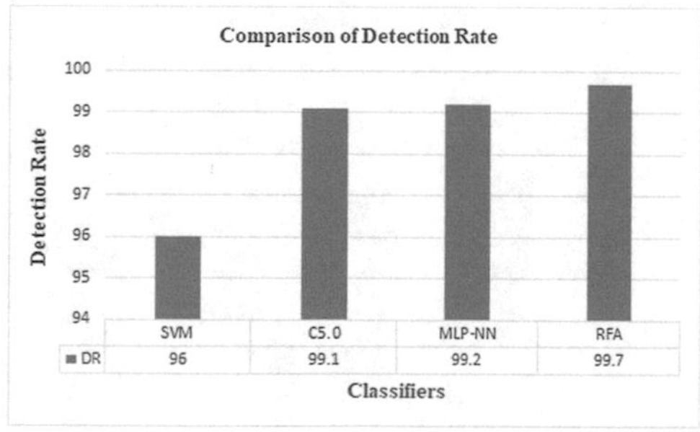

Fig. 6. Describe the comparison in term of Detection Rate. Where experiment results for combination of the proposed model RF-CIFE with C5.0, MLP-NN and Random Forest Algorithm are up to 99% and RF-CIFE with SVM has 96% Which validate the usefulness of this model.

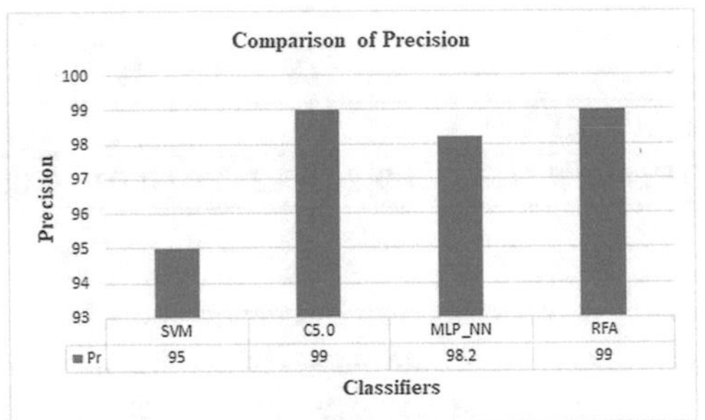

Fig. 7. Compare four classifiers in term of precision. Experiment results describe the correct positive predictions out of all positive predictions. The used classifiers have high precision which indicates the low false positive rate of the proposed model.

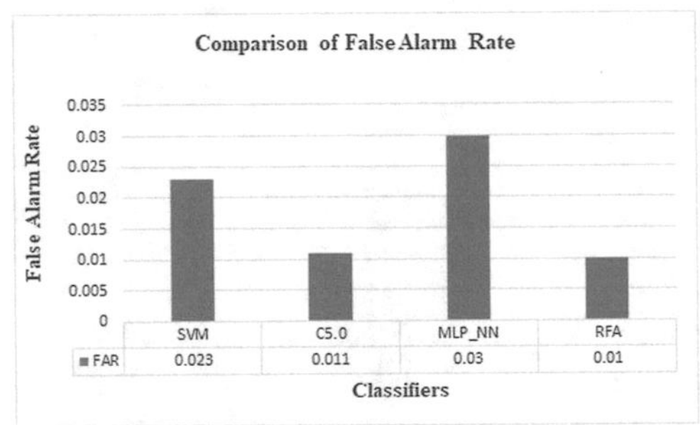

Fig. 8. Describe the minimization of False Alarm rate from RF-CIFE combine with four classification algorithms. It indicates the low false classification error of the model. MLP-NN produces higher False Alarm rate of 0.03% where as other classifiers are around 0% False alarm rate.

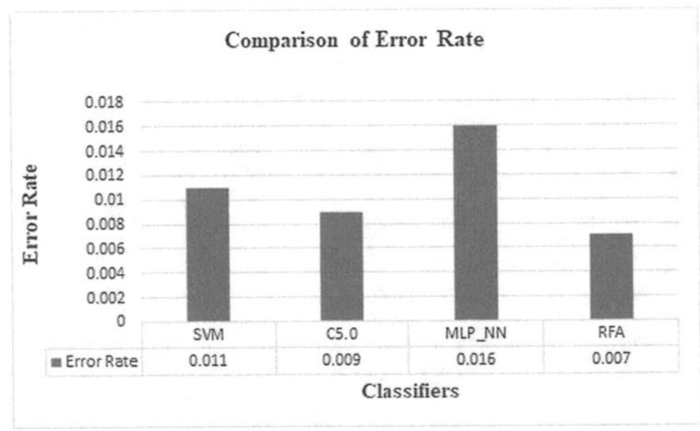

Fig. 9. Show the proportion classification of how the combination of RF-CIFE with each classification's algorithm get wrong. The experiment results show that the proposed model with all used classifiers produces the reduction of classification error around 0%.

5 Conclusion

In this paper, we proposed a mutual information model for an efficient and effectiveness of NIDS. Our approach applied Random forest (RF) for feature importance and CIFE for extracting features based on the corresponding mutual information Maximization. Using 10% train and test KDD cup 99 benchmark; we demonstrated how the combination of RF-CIFE with different algorithms improve the capabilities of an Intrusion Detection System (IDS). The experimental results reveal that the proposed method increase an Accuracy, Detection rate, Precision and provides low False Alarm rate and Error rate. We also compared our evaluation metric results, and the contribution of our approach for different implemented algorithms.

References

1. Wankhade, A., Chandrasekaran, K.: Distributed-intrusion detection system using combination of ant colony optimization (ACO) and support vector machine (SVM). In: 2016 International Conference on Micro-Electronics Telecommunication Engineering, pp. 646–651 (2016)
2. Pandya, R., Pandya, J.: C5.0 algorithm to improved decision tree with feature selection and reduced error pruning. Int. J. Comput. Appl. **117**, 18–21 (2015)
3. Ali, M.H., AL Mohammed, B.A.D., Ismail, M.A.B., Zolkipli, M.F.: A new intrusion detection system based on Fast Learning Network and Particle swarm optimization. IEEE Access **6**, 20255–20261 (2018)
4. Karatas, G.: Neural network based intrusion detection systems with different training functions, pp. 1–6 (2018)
5. Yuan, Y., Hogrefe, D.: Two layers multi-class detection method for network intrusion detection system. In: IEEE Symposium on Computers and Communications (ISCC), p. 7 (2017)
6. Tao, P., Sun, Z., Sun, Z.: An improved intrusion detection algorithm based on GA and SVM. IEEE Access **6**, 13624–13631 (2018)
7. Teng, S., Wu, N., Zhu, H., Teng, L., Zhang, W.: SVM-DT-based adaptive and collaborative intrusion detection. IEEE/CAA J. Autom. Sin. **5**(1), 108–118 (2018)
8. https://www.sans.org/network-security
9. Nie, S., Gao, T., Ji, Q.: An information theoretic feature selection framework based on integer programming. In: Proceedings of International Conference on Pattern Recognition, pp. 3573–3578 (2017)
10. Torkkola, K.: Feature extraction by non-parametric mutual information maximization. J. Mach. Learn. Res. **3**, 1415–1438 (2003)
11. Aung, Y.Y.: An analysis of random forest algorithm based network intrusion detection system, pp. 127–132 (2017)
12. CS229 Lecture notes: Support Vector Machines by Andrew Ng
13. Etemad, K., Chellappa, R.: Discriminant analysis for recognition of human face images. J. Opt. Soc. Am. **14**(8), 1724–1733 (1997)
14. Lin, D., Tang, X.: Conditional infomax learning: an integrated framework for feature extraction and fusion. In: Computer Vision - ECCV 2006. LNCS, vol. 3951, pp. 68–82. Springer, Heidelberg (2006)

15. Paul, J., Verleysen, M., Dupont, P.: Identification of statistically significant features from random forests. In: ECML Workshop Solving Complex Machine Learning Problems with Ensemble Methods (2013)
16. Zhuo, X., Zhang, J., Woo, S.: Network intrusion detection using word embeddings (2017)
17. Ren, Q., Cheng, H., Han, H.: Research on machine learning framework based on random forest algorithm, vol. 1820, p. 080020 (2017)
18. Breiman, L.: Random forests. Mach. Learn. **45**(1), 5–32 (2001)
19. https://deeplearning4j.org/multilayerperceptron#mlp
20. Xu, B., Chen, S., Zhang, H.: Incremental k-NN SCM method in intrusion detection (2017)
21. Cover, T.M., Thomas, J.A. (eds.): Elements of Information Theory II, 2nd edn, pp. 38–48. Wiley, New York (2006)
22. Samrin, R., Vasumathi, D.: Review on anomaly based network intrusion detection system. In: 2017 International Conference on Electrical, Electronics, Communication, Computer, and Optimization Techniques (ICEECCOT), pp. 141–147 (2017)
23. http://kdd.ics.uci.edu/databases/kddcup99/kddcup99.html
24. Wang, W., Sheng, Y., Wang, J., Zeng, X., Ye, X., Huang, Y., Zhu, M.: HAST-IDS: learning hierarchical spatial-temporal features using deep neural networks to improve intrusion detection. IEEE Access **6**(c), 1792–1806 (2017)
25. Farahnakian, F., Heikkonen, J.: A deep auto-encoder based approach for intrusion detection system. In: International Conference on Advanced Communication Technology, ICACT 2018, pp. 178–183, February 2018

You Are What You Search: Attribute Inference Attacks Through Web Search Queries

Tianyu Du[1(✉)], Tao Tao[2], Bijing Liu[3], Xueqi Jin[4], Jinfeng Li[1], and Shouling Ji[1,5]

[1] Zhejiang University, Hangzhou, China
{zjradty,lijinfeng_0713,sji}@zju.edu.cn
[2] State Grid Hangzhou Power Supply Company, Hangzhou, China
taotao980925@aliyun.com
[3] NARI Group Corporation, Beijing, China
beginjing@163.com
[4] State Grid Zhejiang Electric Power Co., LTD., Hangzhou, China
jxqoffice@163.com
[5] Alibaba-Zhejiang University Joint Research Institute of Frontier Technologies, Hangzhou, China

Abstract. Most, if not all, existing attribute inference attacks leverage users' social friendship information and/or social behavioral information to infer the attributes of a target user. In this paper, we study this problem in a novel angle. Specifically, we study whether a user's private attributes (e.g., age, gender, and education) can be inferred based on his or her query history, which is the first such attempt to the best of our knowledge. We present a thorough description of our query-based attribute inference attack and experimentally evaluate our method on a real-world dataset provided by Sogou. Experimental results show that our method can achieve 70.21% for the precision, 68.82% for the recall, and 69.50% for the F1-score on average. When predicting users' gender, the proposed method has precision of 84.56%. This suggests that query records indeed disclose a significant amount of information about users.

Keywords: Attribute inference · Privacy · Query classification · Ensemble learning

1 Introduction

Datasets that contain personal information gradually become public to support research and for other purposes, which is prone to posing privacy risks. For example, users' friends and behavior could disclose users' personal information [9,12], which many people may view as sensitive. An attribute inference attack is a process wherein attackers infer users' private attributes such as age, gender, education or even political orientation and religious belief from public information.

C.-N. Yang et al. (Eds.): SICBS 2018, AISC 895, pp. 343–358, 2020.
https://doi.org/10.1007/978-3-030-16946-6_27

Once users' attributes have been inferred, they can be used for security-sensitive activities like spear phishing [1] and backup authentication [11]. Attackers can be anyone who are interested in users' attributes, like advertisers and hackers. In addition to the privacy leakage, attackers who know the attributes about a person could link users across different sites [2] or with offline records [5] to complete individual profiles, which may cause more severe security risks. Previous attribute inference techniques can be roughly divided into two categories. One is based on social friends, since people connected by social ties tend to share similar attributes. The other is based on behavior, because people who have similar behaviors are inclined to share the same interests and attributes.

Intuitively, characteristics and interests of a user can be disclosed subconsciously through query records. For example, men are more interested in military and cars than women. People who have high education are inclined to get information about society and economy. People in the 19–23 age range often search more about college life and social topics. Based on this conjecture, we may leverage query records to conduct an attribute inference attack. A crucial issue for this kind of attack is to effectively classify massive queries that are short, ambiguous and noisy, which is a challenging problem. First, queries can be as simple as a single character, or as complex as a piece of code segment. Second, some queries are easy to understand, while others may contain multiple meanings. Furthermore, some words may just have the meaning defined in the dictionary, whereas others may have some special meanings on the Internet. Finally, the meanings of queries may also change over time.

Our Work. In this paper, our goal is to leverage queries to infer user attributes, which is the first such attempt to the best of our knowledge. We depart from prior work in the following way: rather than analyzing the risks of attribute inference using social friends information or behavior information, we adopt a combined and in-depth feature representation method that effectively interprets the typically short, ambiguous and noisy query records. We comprehensively evaluate our method for inferring age, gender and education using a dataset with 100,000 users provided by Sogou, one of the main search engines in China. From the experimental results, our method can achieve 70.21% for the precision, 68.82% for the recall, and 69.5% for the F1-score on average. When predicting users' gender, the proposed method has precision of 84.56%. These results imply that an attacker can use users' queries to infer their attributes at high accuracy.

Our Contributions. In conclusion, our key contributions are as follows:

1. We propose a query-based attribute inference framework that exploits users' query records to conduct attribute inference. Our results have serious implications for user privacy – private attributes can be inferred from users' query records.
2. Our method has two distinguishing features: (i) it combines three different features; and (ii) it combines three different kinds of classifiers. Moreover, we analyze the impacts of different feature representation methods and feature selection methods on the given dataset, respectively.

3. Leveraging a real world dataset (100,000 users, 12,608,914 queries), we conduct evaluations to examine the performance. Experimental results demonstrate that our method achieves a satisfying inference performance. For instance, it can successfully infer the age, gender, education for 61%, 84%, 63% of the users, respectively. Our best result for averaged precision, recall and F-score is 70.21%, 68.82%, 69.50%, respectively. This suggests that the query records carries much information about users. Therefore, users' private information suffer from serious risks.

2 Related Work

2.1 Attribute Inference

Previous works that attempt to infer the attributes of the users were mostly based on users' social relationships and behavioral records. He et al. [12] used Bayes network to analyze the relations among people in social network. Lindamood et al. [14] revised Naive Bayes (NB) classifier to integrate social links and other attributes of users to conduct attribute inference. Thomas et al. [16] used information about users' friends and wall posts to infer attributes such as religious views, political views and gender based on multi-label classification. Fang et al. [8] leveraged the correlations between user attributes to improve the performance of attribute inference. Weinsberg et al. [18] studied the inference of gender using rating scores and found that Logistic Regression (LG) classifier outperforms Support Vector Machine (SVM) classifier and NB classifier. Chaabane et al. [6] correlated the music with Wikipedia pages and used topic model to find the latent correlations between music. They identified that users that like similar music are inclined to have same attributes. Gong et al. [9] integrated social friends and user behavior to infer the missing attributes of targeted users. These studies are orthogonal to ours, since they leveraged social network, user behaviors and other information rather than queries.

2.2 Query Classification

Query classification is a difficult problem due to the ambiguous, short and noisy features of web search queries. There are massive algorithms for web search query and short text classification. Cao et al. [5] used neighboring queries and their corresponding clicked URLs in search sessions as context information to solve the problem of query classification based on Conditional Random Field (CRF) models. Beitzel et al. [4] examined three approaches to query classifications and found that combining three techniques has the best performance. The three approaches they used are matching against a list of manually labeled queries, supervised learning of classifiers and mining of selectional preference rules from large unlabeled query logs. Beitzel et al. [3] found that training classifiers from manually classified queries outperforms the bridged classifier using document taxonomy. Hu et al. [13] leveraged Wikipedia to discover massive intent concepts for query classification. Specifically, the intent of any input query is identified through mapping the query into the Wikipedia representation space. Since we focus on Chinese query classification, some extensively used knowledge bases such as Probase, WordNet, Freebase, and YAGO are inapplicable to our problem.

2.3 Ensemble Learning

Ensemble learning is a popular and powerful technique using multiple learning algorithms to get better predictive performance [21]. Intuitively, some classifiers may perform better than others on certain classes. By combining the classification results, different classifiers supplement each other, making the results more robust than single classifier. Many researchers focus on what kinds of base model should be used and how to combine these models [7,15,17,20]. Dietterich [7] reviewed some ensemble methods and explained why ensembles can often perform better than any single classifier. Miskin et al. [15] applied ensemble methods for blind source separation and deconvolution of images. Xia et al. [20] combined features and classification algorithms to produce a more accurate sentiment classification. Verbaeten et al. [17] used ensemble methods to find out noisy training samples in classification tasks. More precisely, they used dozens of filter techniques based on ensemble methods to identify mislabeled training samples and remove them.

3 Methodology

In this section, we first discuss the overall approach to conduct query data-based attribute inference, and then detail the three feature representation methods and one ensemble method.

3.1 Overall Approach

First, we apply text segmentation to divide queries into meaningful units. Second, we use three feature representation methods to express the queries from different aspects. Specifically, We use the Term Frequency - Inverse Document Frequency (TF-IDF) method to obtain word weight vectors from pre-processed data and use term frequency to select the most important feature. Furthermore, we use the Latent Dirichlet Allocation (LDA) model to get the latent user-topic distribution, and use the word2vec method to get word embedding as other features. Finally, we train several classifiers using benchmark machine learning techniques (e.g., Support Vector Machines (SVMs)) to classify these features. To overcome the drawbacks of using a single classifier, we adopt an ensemble method to take advantage of multiple classifiers and achieve better performance.

3.2 Query Representation

We use three different methods to represent queries and get three different kinds of features. The result of TF-IDF is the weight vector of each document. The result of LDA is the topic distribution of each document. The result of word2vec is word embedding, and we take the average of word embeddings in one document as the document embedding. These results are used as the input features of classifiers.

TF-IDF. TF-IDF is evolved from IDF [6]. The intuition is that a frequently used term in different documents is less important and should be assigned with less weight. The basic idea of TF-IDF is that the importance of a term for a given document can be evaluated by term frequency and inverse document frequency. Therefore, the formula of TF-IDF used for term weighting is:

$$w_{i,j} = tf_{i,j} \times log(\frac{N}{df_i}) \tag{1}$$

where $w_{i,j}$ is the weight for term i in document j, $tf_{i,j}$ is the term frequency of term i in document j, df_i is the document frequency of term i and N is the number of documents. However, the tremendous amount of the vocabulary in the dataset makes it computationally expensive to weight all terms. Therefore, we use a term frequency threshold to filter out the trivial terms and use the remaining terms as the input features of classifiers.

Latent Dirichlet Allocation (LDA). For many problems that need a semantic understanding of short text, inferring latent topics is an important task. Conventional topic modeling techniques such as LDA have been studied extensively for various tasks in information retrieval and text mining to extract latent topics from document corpus [10]. The fundamental idea of LDA is that each document is a multinomial distribution over topics and each topic is a multinomial distribution over words. Before we describe the employed LDA model, we briefly introduce its conventional terminology.

- **Word:** an item from the vocabulary of size W.
- **Document:** a sequence of N words represented by $d = \{w_1, w_2, \cdots, w_N\}$ where w_n is the n-th word in the sequence.
- **Corpus:** a collection of M documents represented by $D = \{d_1, d_2, \cdots, d_M\}$ where d_m is the n-th document in the corpus.
- **Topic:** denoted by $Z = \{z_1, z_2, \cdots, z_K\}$ with the assumption that there are K topics in the corpus.

Assume the topic distribution of a document is θ, and word distribution of a topic is ϕ. Then the generative process of each document in a corpus is as follows:

1. Choose the number of word N, $N \sim Poisson(\xi)$
2. Choose θ, $\theta \sim Dir(\alpha)$
3. For $w_n \in d$:
   (a) Choose a topic z_n, $z_n \sim Multinomial(\theta)$

(b) Choose a word w_n from $p(w_n|z_n, \beta)$, a multinomial probability ϕ^{z_n}.

There are several popular solutions for the LDA model, such as Gibbs Sampling and the Expectation Maximization (EM) algorithm. In this paper, the LDA model is trained by the scikit-learn Python package.

Word2vec. Given two words semantically related while rarely co-occur in short texts, the LDA model cannot learn the semantic relatedness of them. Furthermore, people will also use their background knowledge to understand short texts rather than just use the content words. The recent advances in word embedding provide powerful methods for learning semantic relations, which can be employed to improve the topic modeling for short texts.

Word2vec is a kind of word embedding techniques released by Google in 2013 [4]. It takes a text document as input and outputs word vectors, which can be used as features in many natural language processing (NLP) problems. It has two main model architectures: the Continuous Bag-of-Words (CBOW) model and the skip-gram model. CBOW uses the context to predict the current word, and skip-gram uses the current word to predict the context. Since word2vec has been shown to be well suited for conducting Chinese document classification [8], we apply it for query classification in this paper. In particular, we train the word2vec model using the gensim Python package.

3.3 Feature Selection

Term Frequency. Term frequency is the number of words in the collection of documents. We compute the term frequency for each term in corpus and remove terms whose frequency is less than predetermined threshold. The basic idea is that rare terms are non-informative and uninfluential. Removing rare terms can reduce the dimension of the feature space. Term frequency is the simplest technique for feature selection. However, it is usually considered as an "ad hoc" for selecting features.

χ^2 statistic (CHI2). The χ^2 statistic measures the independence between t and c. It is defined to be:

$$\chi^2(t,c) = \frac{N \times (AD - BC)^2}{(A + C) \times (B + D) \times (A + B) \times (C + D)} \tag{2}$$

where A is the number of times t and c co-occur, B is the number of times t occurs without c, C is the number of times c occurs without t, D is the number of times neither c nor t occurs. The χ^2 statistic will be zero if t and c are independent. However, it only considers whether term t occurs in a document regardless of its times. It makes this method exaggerate the importance of low-frequency terms, so it is usually combined with other factors such as term frequency to overcome its drawbacks.

3.4 Stacked Generalization

In this paper, we use stacked generalization or stacking proposed by Wolpert [19] to combine multiple classifiers, which uses the weights learned from a validation dataset to combine the results of different base classifiers where each one may be somewhat biased. The basic idea is to learn a level-1 classifier based on the results of level-0 classifiers.

Level-0 Generalizers. For a data set $D = (\boldsymbol{x_i}, y_i), i = 1, 2, \cdots, n$ where $\boldsymbol{x_i}$ is a feature vector of the i-th example and y_i is corresponding label, we splits D into K sub-dataset $D_1, D_2, \cdots, D_K$ and perform K-fold cross-validation. At each k-th fold, $D_{-k} = D - D_k$ is used as training part and D_k is used as test part. Then, N algorithms $L_1, L_2, \cdots, L_N$ are applied to the D_{-k} to obtain N level-0 classifiers $C_1, C_2, \cdots, C_N$. The results of the N level-0 classifiers on each example in D_k with its corresponding label form a level-1 sub-dataset MD_k. After cross-validation process, the union $MD = \cup MD_k, k = 1, 2, \cdots, K$ makes up the full level-1 data set. It is worth noting that all the N learning algorithms should be trained on the whole data set D to get the final level-0 classifiers $C_1, C_2, \cdots, C_N$.

Level-1 Generalizers. After we get the level-0 classifiers, we use another learning algorithm L_M to get the level-1 classifier C_M. To classify a new example, the results of all the level-0 classifiers form a level-1 vector of N dimension. Then, the vector is assigned with a label by the level-1 classifier as the final classification result.

4 Experiments

In this section, we first introduce the dataset, the evaluation settings and metrics. Then we perform several experiments to demonstrate the effectiveness of our method. Furthermore, we investigate the impacts of some parameters, and then compare the performance of different feature representation methods. Afterwards, we investigate the performance of different classifiers and verify the effectiveness of ensemble learning.

4.1 Dataset

We use the dataset provided by Sogou[1], which contains 100,000 users' records. Each record is corresponding to one user's data and contains the age, gender and education labels as well as the queries made by that user. Each query is a combination of numbers, punctuation, URLs, acronyms, codes and words.

Table 1 describes the meaning of numbers in each attribute. Figure 1(a) shows the frequency of query length. The query's length vary from one character to more than thirty characters, but most queries' length are limited to ten characters.

Intuitively, one user's age and education can be correlated. For example, a 16 years old user can hardly get education higher than high school. Furthermore, the elder users tend to have higher education. To verify our conjecture, we plot

[1] http://www.datafountain.cn/data/science/player/competition/detail/description/239.

Table 1. Dataset description

Label	1	2	3	4	5	6
Age	0–18	19–23	24–30	31–40	41–50	51–999
Gender	Male	Female	-	-	-	-
Education	Ph.D	M.S.	B.S.	High school	Middel school	Primary school

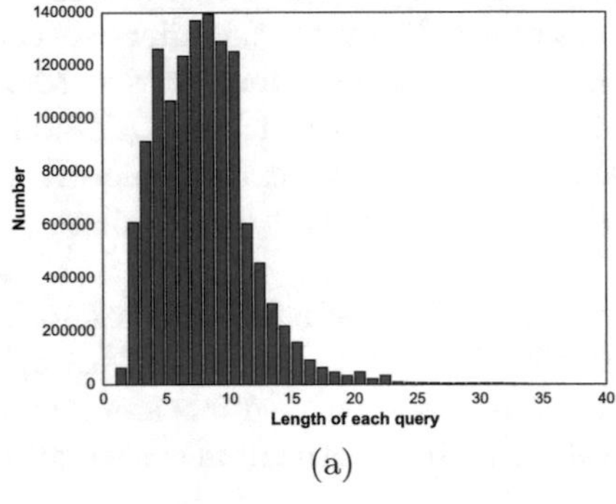

(a)

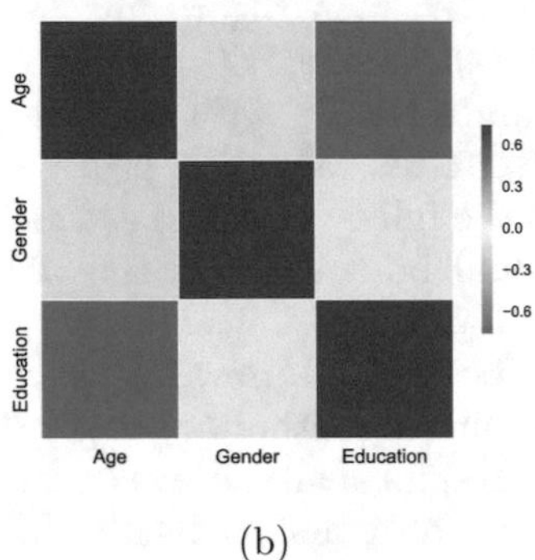

(b)

Fig. 1. (a) Frequency of query length. (b) Correlation of three attributes.

the correlation between different attributes in Fig. 1(b). In Fig. 1(b), the depth of color represent the degree of correlation, and the red means there is positive correlation between two attributes, while the blue means there is negative correlation between them. Apparently, Fig. 1(b) verifies our conjecture.

Figure 2(a) shows the histogram of the number of users in each class in terms of age and gender, and Fig. 2(b) shows the histogram of the number of users in each class in terms of education and gender. Apparently, the numbers of users in different classes are significantly imbalanced. In Fig. 2(a), the largest group "0–18" contributes about 40.64% of the age label in the whole dataset, while the smallest one "51–99" contributes about 0.20%. In Fig. 2(b), the largest group "middle school" contributes about 41.64% of the education label in the whole dataset, while the smallest one "Ph.D" contributes about 0.40%.

4.2 Evaluation Settings and Metrics

Preprocessing. Notice that there are some "0" labels in the dataset, which means the user's attribute is unknown. For preciseness, we remove the record with at least one "0" label in preprocessing. Then, we adopt jieba[2] Python package to segment the queries into words and give them part-of-speech (POS) tags. The preprocessed data is obtained after removing stop words and punctuation characters. Therefore, short queries which contain stop words are not considered

[2] https://pypi.python.org/pypi/jieba/.

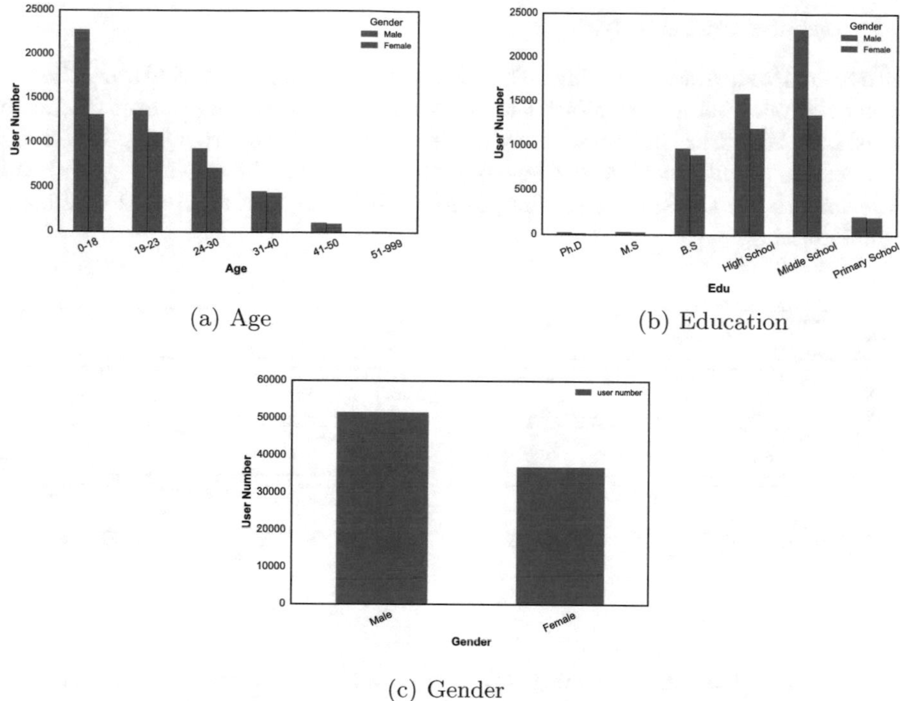

Fig. 2. The number of users in each age/education/gender category.

in this evaluation. We omit the details about how to select the stop words as it is not the keystone.

Evaluation Settings. In our following experiments, we randomly and uniformly sample 10% of the users in the whole dataset. We then remove their attribute labels and take them as the testing dataset to evaluate our method. We take the remaining 90% of the users as the training dataset.

Evaluation Metrics. The performance of a classifier is usually measured by precision, recall and F1-score, which are standard measures in machine learning evaluation. The definitions of precision, recall and F1-score are described as follows:

$$Precision = \frac{A}{B}$$

$$Recall = \frac{A}{C}$$

$$F1 - score = \frac{2 \cdot Precision \cdot Recall}{Precision \cdot Recall}$$

where the A denotes the sum of users correctly classified as c_i, B denotes the sum of users classified as c_i, C denotes the sum of users whose true class is c_i.

4.3 Results and Analysis

Effect of Parameter Tuning. To select the most important features, there is an crucial parameter that affect the performance of classifiers - term frequency threshold. We filter out words that don't reach the term frequency threshold. The results of different word frequency threshold are showed in Fig. 3(a). All experiments use the same classifier: SVM with linear kernel, C = 0.1 and other default setting.

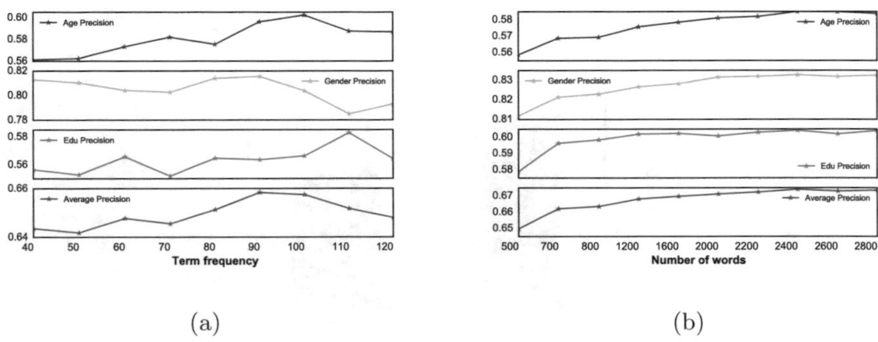

Fig. 3. Effect of parameter tuning. (a) The values of precision with different term frequency thresholds. (b) The values of precision got by the χ^2 algorithm with different number of words.

From Fig. 3(a), we can see that average precision have a peak value when word frequency threshold is 90. As the threshold of term frequency increases, the precision increases at the beginning, and then tends to decrease. The reason is that we need a certain amount of words to get the characteristics of the user. However, too many words may introduce some noise, while too few words can not capture the whole information of users' characteristics. Both of them can reduce the precision of classifiers. Therefore, in the following experiments we choose threshold = 90 in the following experiments.

We have tried another feature selection method - χ^2 test, and the results are shown in Fig. 3(b). To our surprise, this method performs better than term frequency in off-line evaluation but performs worse in online evaluation. We think it may caused by the different distribution of attributes. Therefore, we still use term frequency in the following experiments in terms of generalization ability.

Another parameter is the number of components kept by PCA. Intuitively, more components will contain more information of original data and therefore achieve better performance. However, it is not worth to keep all components because some components only contain trivial information. Figure 4(a) shows the performance of SVM classifier by varying the number of components kept by PCA. The value of precision increases initially, and then decreases. The performance of component = 100 and component = 150 have similar performance, but component = 150 will cost more computation time. Therefore, in the following experiments we only use the component = 100 for classification.

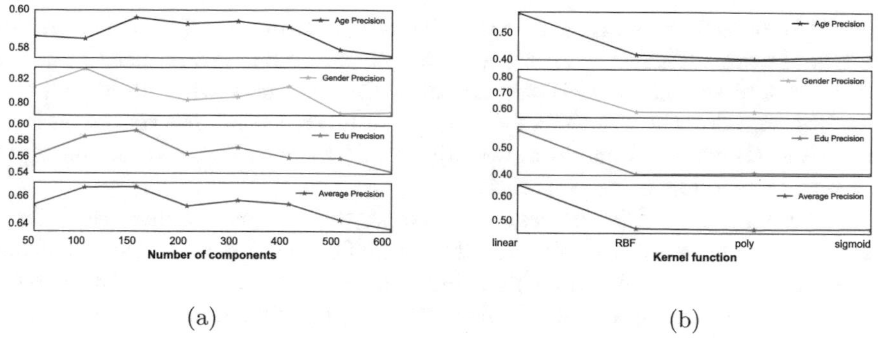

Fig. 4. (a) The values of precision got by the PCA algorithm with different number of components. (b) The values of precision got by the SVMs with different kernel functions.

Comparison of Feature Representation Methods. Table 2 shows the results of different feature representation methods by using the SVM classifier with linear kernel. It is clear that no single kind of features produces better classification results than the combination of all the features. However, traditional "bag-of-words" based methods like TF-IDF can still achieve high precision despite its simplicity. Furthermore, combining three different features performs the best and single LDA performs the worst. The poor performance of LDA is due to the fact that it leverages word co-occurrence to get topics from a corpus of documents. Despite its effectiveness on many problems, it shows bad performance when it comes to short texts such as queries due to the scarce word co-occurrence. Therefore, in the following experiments, we use the combination of three kinds of features as the input data for classifiers.

Table 2. Comparison of different feature representation methods

Feature	AgePre	GenPre	EduPre	AvgPre
TF-IDF	59.00%	83.55%	61.06%	67.87%
word2vec	57.34%	83.75%	55.74%	65.61%
LDA	56.45%	78.68%	53.89%	63.01%
TF-IDF+word2vec	61.54%	83.96%	63.28%	69.59%
TF-IDF+LDA	60.08%	83.94%	62.19%	68.74%
TF-IDF+word2vec+LDA	61.84%	84.27%	63.97%	70.03%

Comparison of SVMs with Different Kernel Functions. To our knowledge, the kernel function can effect the performance of SVM classifier. Therefore, we do several experiments for the sake of finding the SVM classifier with the most appropriate kernel function. Before classification, we use TF-IDF to get the weight of each word, and use word frequency threshold of 90 times to filter out trivial words. Then, we use PCA with 200 components to reduce dimension.

Figure 4(b) shows the performance of SVM classifiers with different kernel functions. Apparently, the SVM classifier with linear kernel greatly outperforms other SVM classifiers. Therefore, we only use SVM with linear kernel for classification in the following experiments.

Comparison of Classifiers. We now study the performance of different classifiers including SVM classifier, MNB classifier and LR classifier. We choose these classifiers as level-0 classifiers for computational reasons. Furthermore, to illustrate the benefits of combining different classifiers, we run the stacked generalization algorithm described in Sect. 3. The used level-1 classifier is a SVM classifier with an RBF kernel and other default settings. Given three level-0 classifiers, we can totally obtain seven ensemble classifiers. Figure 5 show the precision, recall and F1-score performance of the three single classifiers and four ensemble classifiers on the testing dataset, respectively.

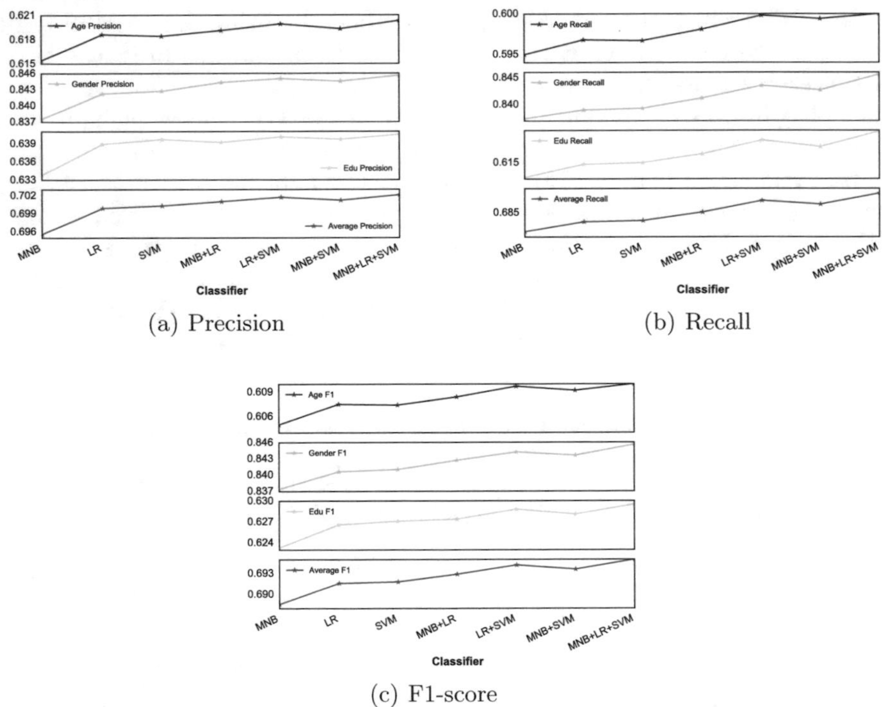

(a) Precision

(b) Recall

(c) F1-score

Fig. 5. The values of precision, recall and F1-score got by the different classifiers and their combination.

We find several interesting observations from Fig. 5. First, the SVM classifier obtains the highest precision, recall and F1-score among the three classifiers. Second, under all the metrics, ensemble classifiers perform better than any single classifier, which verifies the effectiveness of ensemble learning. Specifically, the ensemble classifier of the three level-0 classifiers performs better than the ensemble classifiers of any two level-0 classifiers, which achieves 70.21% for the precision, 68.82% for the recall, and 69.5% for the F-score on average. When predicting users' gender, the ensemble classifier of the three level-0 classifiers has precision of 84.56%. This finding verifies our conjecture that different kinds of classifiers complement each other, and their combination will enhance the performance. We list the specific value for further comparison in Tables 3, 4 and 5.

Table 3. Comparison of classifiers by precision

Model	AgePre	GenPre	EduPre	AvgPre
MNB	0.6154	0.8375	0.6338	0.6956
LR	0.6186	0.8422	0.6389	0.6999
SVM	0.6184	0.8427	0.6397	0.7003
MNB+LR	0.6191	0.8443	0.6392	0.7010
LR+SVM	0.6199	0.8450	0.6401	0.7017
MNB+SVM	0.6193	0.8445	0.6397	0.7012
MNB+SVM+LR	0.6203	0.8456	0.6405	0.7021

Table 4. Comparison of classifiers by recall

Model	AgeRecall	GenRecall	EduRecall	AvgRecall
MNB	0.5931	0.8162	0.6134	0.6742
LR	0.5968	0.8321	0.6191	0.6827
SVM	0.5967	0.8328	0.6197	0.6831
MNB+LR	0.5981	0.8359	0.6195	0.6845
LR+SVM	0.5998	0.8370	0.6210	0.6859
MNB+SVM	0.5994	0.8365	0.6201	0.6853
MNB+SVM+LR	0.6000	0.8379	0.6217	0.6865

Table 5. Comparison of classifiers by F1-score.

Model	AgeF1	GenF1	EduF1	AvgF1
MNB	0.6040	0.8267	0.6234	0.6847
LR	0.6075	0.8371	0.6288	0.6912
SVM	0.6074	0.8377	0.6295	0.6915
MNB+LR	0.6084	0.8401	0.6292	0.6926
LR+SVM	0.6097	0.8410	0.6304	0.6937
MNB+SVM	0.6092	0.8405	0.6297	0.6931
MNB+SVM+LR	0.6100	0.8417	0.6310	0.6942

5 Discussion

Limitations. The possibility that there may be more than one people using the same account will interfere classifiers and harm the attack performance. For example, different family members will search different information, causing the query record reflect different people's characteristics.

Other Methods to Improve Performance. The key problem is how to automatically find the exact means behind the queries that can represent the users' characteristics. Correctly categorizing the queries has the potential to bring major gains in accuracy. One possible feature representation method is to enrich the queries with external knowledge such as knowledge base (e.g., WordNet), and then map queries to intermediate objects and finally maps them to target categories. However, there is no available knowledge base that can be exploited to categorize Chinese short texts. Furthermore, investigating more classifiers and fully understanding the limitations and advantages of each classifier may also improve the performance.

6 Conclusion

In this paper, we present a query data based inference attack wherein adversaries exploit users' query records to disclose their attributes. In the attack, we first use a combined feature representation method based on TF-IDF, LDA and word2vec to obtain features from users' query records and then employ stacked generalization to combine multiple classification results and build an ensemble classifier. We evaluate the performance of our method on a real-world dataset provided by Sogou and the experimental results demonstrate that our query-based inference attack can effectively predict users' attributes. Based on our research, we can imagine that the development of more sophisticated query-based techniques will pose a great threat to user privacy.

A few fascinating directions for future work include representing features, eliminating the noise in queries, leveraging unsupervised methods, as well as defending against our attribute inference attacks.

Acknowledgments. This work was partly supported by NSFC under No. 61772466, the Zhejiang Provincial Natural Science Foundation for Distinguished Young Scholars under No. R19F020013, the Technology Project of State Grid Zhejiang Electric Power co. LTD under NO. 5211HZ17000J, the Provincial Key Research and Development Program of Zhejiang, China under No. 2017C01055, the Fundamental Research Funds for the Central Universities, and the Alibaba-ZJU Joint Research Institute of Frontier Technologies.

References

1. Spear Phishing Attacks. http://www.microsoft.com/protect/yourself/phishing/spear.mspx
2. Bartunov, S., Korshunov, A., Park, S.T., Ryu, W., Lee, H.: Joint link-attribute user identity resolution in online social networks. In: Proceedings of the Workshop on Social Network Mining and Analysis in the 6th International Conference on Knowledge Discovery and Data Mining. ACM (2012)
3. Beitzel, S.M., Jensen, E.C., Chowdhury, A., Frieder, O.: Varying approaches to topical web query classification. In: Proceedings of the 30th Annual International ACM SIGIR Conference on Research and Development in Information Retrieval, pp. 783–784. ACM (2007)
4. Beitzel, S.M., Jensen, E.C., Frieder, O., Grossman, D., Lewis, D.D., Chowdhury, A., Kolcz, A.: Automatic web query classification using labeled and unlabeled training data. In: Proceedings of the 28th Annual International ACM SIGIR Conference on Research and Development in Information Retrieval, pp. 581–582. ACM (2005)
5. Cao, H., Hu, D.H., Shen, D., Jiang, D., Sun, J.T., Chen, E., Yang, Q.: Context-aware query classification. In: Proceedings of the 32nd International ACM SIGIR Conference on Research and Development in Information Retrieval, pp. 3–10. ACM (2009)
6. Chaabane, A., Acs, G., Kaafar, M.A.: You are what you like! Information leakage through users' interests. In: Network and Distributed System Security Symposium, pp. 1–14 (2012)
7. Dietterich, T.G.: Ensemble methods in machine learning. In: International Workshop on Multiple Classifier Systems, pp. 1–15. Springer (2000)
8. Fang, Q., Sang, J., Xu, C., Hossain, M.S.: Relational user attribute inference in social media. IEEE Trans. Multimed. **17**(7), 1031–1044 (2015)
9. Gong, N.Z., Liu, B.: You are who you know and how you behave: attribute inference attacks via users' social friends and behaviors. In: USENIX Security Symposium, pp. 979–995 (2016)
10. Griffiths, T.L., Steyvers, M.: Finding scientific topics. Proc. Natl. Acad. Sci. **101**(suppl 1), 5228–5235 (2004)
11. Gupta, P., Gottipati, S., Jiang, J., Gao, D.: Your love is public now: questioning the use of personal information in authentication. In: Proceedings of the 8th ACM SIGSAC Symposium on Information, Computer and Communications Security, pp. 49–60. ACM (2013)
12. He, J., Chu, W.W., Liu, Z.V.: Inferring privacy information from social networks. In: International Conference on Intelligence and Security Informatics, pp. 154–165. Springer (2006)
13. Hu, J., Wang, G., Lochovsky, F., Sun, J.T., Chen, Z.: Understanding user's query intent with wikipedia. In: Proceedings of the 18th International Conference on World Wide Web, pp. 471–480. ACM (2009)
14. Lindamood, J., Heatherly, R., Kantarcioglu, M., Thuraisingham, B.: Inferring private information using social network data. In: Proceedings of the 18th International Conference on World Wide Web, WWW 2009, vol. 10, p. 1145 (2009). https://doi.org/10.1145/1526709.1526899
15. Miskin, J., MacKay, D.J.: Ensemble learning for blind image separation and deconvolution. In: Advances in Independent Component Analysis, pp. 123–141. Springer (2000)

16. Thomas, K., Grier, C., Nicol, D.M.: unFriendly: multi-party privacy risks in social networks. In: International Symposium on Privacy Enhancing Technologies Symposium. LNCS, vol. 6205, pp. 236–252. Springer (2010)
17. Verbaeten, S., Van Assche, A.: Ensemble methods for noise elimination in classification problems. In: International Workshop on Multiple Classifier Systems, pp. 317–325. Springer (2003)
18. Weinsberg, U., Bhagat, S., Ioannidis, S., Taft, N.: BlurMe: inferring and obfuscating user gender based on ratings. In: Proceedings of the 6th ACM Conference on Recommender Systems, pp. 195–202. ACM (2012)
19. Wolpert, D.H.: Stacked generalization. Neural Netw. **5**(2), 241–259 (1992)
20. Xia, R., Zong, C., Li, S.: Ensemble of feature sets and classification algorithms for sentiment classification. Inf. Sci. **181**(6), 1138–1152 (2011)
21. Zhou, Z.H.: Ensemble Methods: Foundations and Algorithms. Chapman and Hall/CRC, New York (2012)

New Publicly Verifiable Cloud Data Deletion Scheme with Efficient Tracking

Changsong Yang[1,2] and Xiaoling Tao[1,2](✉)

[1] State Key Laboratory of Integrated Service Networks (ISN), Xidian University, Xi'an 710071, China
csyang02@163.com

[2] Guangxi Colleges and Universities Key Laboratory of Cloud Computing and Complex Systems, Guilin University of Electronic Technology, Guilin 541004, China
txl@guet.edu.cn

Abstract. Cloud storage service, one of the most important services in the cloud computing, can offer high-quality storage service for tenants. By employing the cloud storage service, the resource-constraint data owners can outsource their data to the remote cloud server to reduce the heavy storage burden. Due to the attractive advantages, there are an increasing number of data owners prefer to embrace the cloud storage service. However, the data owners will lose the direct control over their outsourced data, and they can not directly execute operations over their outsourced data, such as data deletion operation. That will make outsourced data deletion become a crucial security problem: the selfish cloud server may not honestly perform the data deletion operation for economic interests, and then returns error results to mislead the data owners. Although a series of solutions have been proposed to solve this problem, most of them can be described as "one-bit-return" protocol: the storage server removes the data and then returns a one-bit deletion reply, and the data owners have to trust the deletion reply because they can not verify it conveniently.

In this paper, we put forward a new publicly verifiable data deletion scheme with efficient tracking for cloud storage. In our novel scheme, the remote cloud server executes the data deletion operation and returns a deletion proof. If the remote cloud server maliciously reserves the data backups, the data owner can easily detect the malicious data reservation by verifying the returned deletion proof. Besides, we utilize the primitive of Merkle Hash Tree to solve the problem of public verifiability in data deletion. Furthermore, our scheme can realize efficient data leakage source tracking when the data are leaked, which can prevent the data owner and the cloud server from exposing the data designedly to slander each other. Finally, we prove that our novel scheme can satisfy the desired security properties.

Keywords: Cloud storage · Data deletion · Public verification · Efficient tracking · Merkle Hash Tree

C.-N. Yang et al. (Eds.): SICBS 2018, AISC 895, pp. 359–372, 2020.
https://doi.org/10.1007/978-3-030-16946-6_28

1 Introduction

Cloud computing is a very promising Internet-based computing paradigm, which has got significant development recently years [25,35]. With large-scale resources, cloud computing can provide plenty of attractive services, for example, data storage and sharing service [13,14,27], outsourcing service [3–5], verifiable databases service [6,7,18], and so on. These services have been applied widely, especially cloud storage service. With enormous storage sources, the cloud storage service provider can ubiquitously offer on-demand cloud storage service [26,32]. Thanks to plenty of attractive advantages, there are an increasing number of resource-constraint data owners prefer to embrace the cloud storage service, and outsource their data to the remote cloud server to reduce the heavy storage overhead.

Despite tremendous benefits, the cloud storage suffers from some new security threats inevitably, for example, secure and verifiable outsourced data deletion. Although unlinking can delete the link of the file efficiently, the contents still remain in the disk, and the attacker can utilize a forensic tool to scan the disk to recover the contents easily [9]. In order to remove the contents of the file, plenty of researchers suggest that it should reach data deletion by overwriting [1,8,12,24,30]. In these schemes, they overwrite the storage disk with random data. Although overwriting can theoretically delete the data, it is not efficient to overwrite the disk, especially in distributed storage system. Besides, the attacker who equipped with advanced tools can recover the deleted data with the physical remanence left on the storage medium [10].

To make the data deletion operation more secure and efficient, cryptography technique is utilized to delete the data instead of protecting them [2,16,23]. These schemes encrypt the data before storing, then they reach data deletion by destroying the decryption key to make the ciphertext unavailable. Although they can realize data deletion by efficiently removing a very short decryption key, most of them can not reach verifiability, and the data owner has to trust the returned deletion result. To give the data owner the ability to verify the deletion result conveniently, some solutions have been proposed [11,19,29,31]. These schemes return a proof to the data owner after deleting the data, and the data owner can check the data deletion outcome by verifying the returned proof. However, when the data are leaked, these schemes can not judge who leaks the data since both the data owner and the remote cloud server can obtain the same ciphertext. Therefore, the malicious one may expose the data designedly to slander the other one.

Although various solutions have been proposed to solve the problem of data deletion, there are still some inherent limitations. Firstly, a lot of existing solutions can not reach verifiability, and the data owner has to trust the returned deletion result. However, the selfish cloud server may not honestly erase the data, and then it may return an error result to maliciously cheat the data owner. Secondly, although some existing schemes can realize verifiability, these schemes can not satisfy the property of traceability. When the data are leaked, they can not trace the data leakage source. To the best of our knowledge, it seems that there is not research work on the publicly verifiable data deletion scheme that supports data leakage source tracking simultaneously. Therefore, we put forward a novel publicly verifiable outsourced data deletion scheme with efficient tracking.

1.1 Our Contributions

In this paper, we put forward a novel publicly verifiable cloud data deletion scheme with efficient data leakage source tracking. The main contributions of this paper are as follow:

- We propose a new Merkle Hash Tree-based publicly verifiable outsourced data deletion scheme with efficient tracking. To be specific, after executing the data deletion operation, the cloud server can utilize the Merkle Hash Tree to generate a deletion proof. If the cloud server does not honestly execute the data deletion operation, the proposed scheme can enable the data owner to detect the dishonest behavior by verifying the proof.
- The proposed scheme can satisfy the property of traceability, which is different from the most of the previous works. If the data are leaked, the proposed scheme can trace the data leakage source, which can prevent the dishonest data owner and the malicious cloud server from exposing the data designedly to slander each other. Moreover, the proposed scheme is also very efficient in communication as well as computation.

1.2 Related Work

The problem of secure and verifiable digital data deletion is very important, and it has attracted a large number of researchers' attentions. Garfinkel and Shelat [9] emphasize that deletion by unlinking is not secure because the contents still remain in the physical medium, and the attacker can easily recover the contents by utilizing a tool to scan the disk. Therefore, it is meaningful to study more secure and verifiable data deletion methods.

In order to erase the contents of the file, and offer the data owner the ability to verify the deletion result, Paul and Saxena [20] proposed a proof of erasability (PoE) scheme for cloud storage. They suggest that the physical disk should be overwritten with random patterns. Then the same patterns should be returned as a proof, which will be used to verify the result of the deletion operation. Besides, Perito and Tsudik [21] proposed a proof of secure erasure (PoSE-s) scheme, which works essentially the same way as the PoE scheme in [20]. But there is an additional assumption of bounded storage in scheme [21], which is different from the scheme [20]. In 2016, Luo et al. [15] proposed a permutation-based data deletion scheme. They assume that the cloud server maintains the latest version of the data, and all the backups will be consistent. Then they delete the data by updating them with random data. Finally, the data owner can efficiently check the deletion result by a challenge-response protocol.

Perlman et al. [22] utilize a trusted third party to solve the problem of secure data erasure. Firstly, they encrypt the data with a data key. After that, the trusted third party will further utilize a control key to encrypt the data key.

Then the data will be deleted by destroying the control key. In 2016, Hao et al. [11] presented a scheme to delete secret data with public verifiability. They encrypt the data and then store the private key in the trusted platform module's (TPM) protected memory. Finally, they reach data deletion by deleting the private key. In 2018, Yang et al. [33] put forward a blockchain-based publicly verifiable data deletion scheme for cloud storage. They employ the blockchain to reach public verifiability without any trusted third party, which is different from a lot of the previous works. Recently, Yang et al. [34] put forward a verifiable data transfer and deletion scheme for cloud storage. Their scheme can give the data owner the ability to migrate the data between two different cloud servers, and then delete the transferred data from the original cloud simultaneously.

1.3 Organization

The remainder of this paper is organized as follows: We show the preliminaries in Sect. 2. In Sect. 3, we describe the problem statement, including the system model, the security challenges and the design goals. In Sect. 4, we present our new scheme in detail. A brief analysis of the proposed scheme, and the comparison with a recent scheme are presented in Sect. 5. Finally, we will conclude this paper in Sect. 6.

2 Preliminaries

2.1 Merkle Hash Tree

As a specific binary tree, Merkle Hash Tree (MHT) is always utilized to authenticate digital data [17,28]. In the MHT, each node maintains a hash value. To be

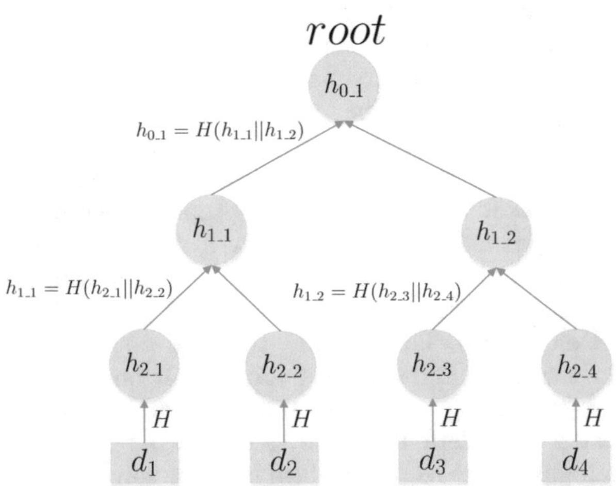

Fig. 1. An example of MHT

specific, every leaf node keeps a hash value of the authenticated data, and each internal node stores a hash value of the concatenation of its left child and right child. We will give a simple example to make the description of the MHT more clear. We assume that $D = \{d_1, d_2, d_3, d_4\}$ is a data set, which will be authenticated with the MHT, as illustrated in Fig. 1. In the leaf nodes, $h_{2\_i} = H(d_i)$, where $i \in [1, 4]$, H is a one-way collision-resistant hash function. Besides, every internal node maintains a hash value of the concatenation of its two child nodes. Finally, the public key signature technique is utilized to compute a signature on the root node of the MHT. After that, the verifier can authenticate any subset with the verification object, which is a set of all sibling nodes on the path from the authenticated leaf node to the root node. For instance, in order to authenticate the data d_3, the verification object may contain h_{1_1}, h_{2_4}, and the signature of h_{0_1}.

3 Problem Statement

3.1 System Model

In the following, we formalize the system model of our scheme, which involves three entities: the data owner O, the cloud server S and the trusted agency TA, as illustrated in Fig. 2.

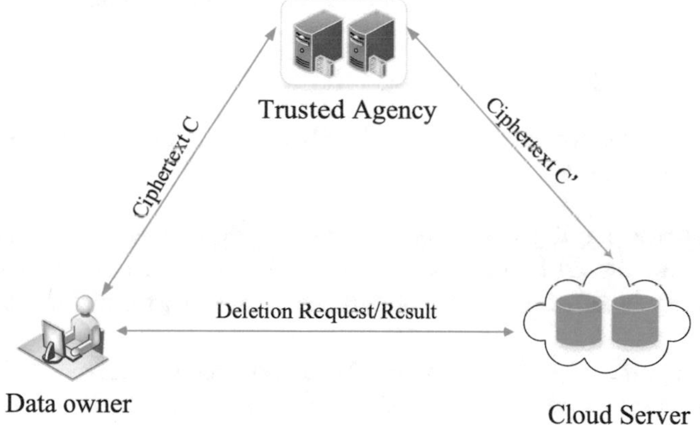

Fig. 2. The system model

- **The Data Owner O.** The data owner O refers to a resource-constraint entity, who is willing to outsource his data to the cloud server. When O will not need the data anymore, he wants to permanently remove the outsourced data and verify the deletion result.
- **The Cloud Server S.** The cloud server S is an entity which has large-scale resources. With enormous resources, S can maintain a large amount of data for O. When O will not need the data anymore, S will delete the data and generate a deletion proof for O.

- **The Trusted Agency** TA**.** The trusted agency TA refers to an entity, which is absolutely righteous. That is to say, the TA will always behave honestly, and he will never collude with O or S to cheat the other maliciously. Therefore, both S and O fully trust the TA unconditionally.

3.2 Security Threats

We mainly consider the following three security challenges in our scheme.

- **Data privacy disclosure.** Outsourced data privacy disclosure is a very serious security threat. On one hand, the attackers try to dig the privacy from the data. On the other hand, the cloud server may move the data to other subcontractors, or share the data with other corporators for economic interests.
- **Data corruption.** The manager performs erroneous operations, software or hardware malfunctions all may cause data loss. Besides, the attackers may modify the data illegally, or delete some data arbitrarily, which will also lead to data corruption.
- **Malicious data reservation.** The data owner prefers to delete the outsourced data when he will not need them anymore. However, the selfish cloud server may not execute the deletion operation honestly for saving computation overhead, or for digging privacy information from the malicious reservation.

3.3 Design Goals

On the basis of the security threats described above, our scheme should realize the following four design goals.

- **Data confidentiality.** The sensitive information which contained in the outsourced data should be protected. That is to say, it is necessary to use cryptography tools to encrypt the outsourced data before storing them on the cloud server.
- **Data integrity.** In order to prevent the outsourced data from being polluted, the data owner should be given the ability to verify the data integrity and availability. If the outsourced data are polluted, the data owner should be able to detect the malicious manipulation.
- **Verifiable data deletion.** To enable the remote cloud server to honestly remove the outsourced data, the data owner should be given the ability to verify the deletion result. If the cloud server reserves the data maliciously, it can not compute a valid proof efficiently to convince the data owner.
- **Accountable traceability.** In order to prevent the data owner and the cloud server from slandering each other, it should reach accountable traceability. To be specific, when the data are leaked, we can trace the data leakage source precisely.

4 Our Construction

4.1 High Description

The proposed scheme can realize publicly verifiable cloud data deletion with efficient tracking, and Fig. 3 describes the main processes of the proposed scheme. Firstly, O encrypts the outsourced file to protect the privacy, and then sends the ephemeral ciphertext to TA. Then TA further encrypts the received ephemeral ciphertext. And then TA sends the final ciphertext to S, and S maintains the final ciphertext. After that, O can remove the local backups. Later, O downloads the ciphertext to decrypt to get the plaintext when he wants the file. Finally, when O will not need the outsourced file anymore, he is willing to send a deletion command to delete the data from S. Upon receiving the deletion request, S deletes the related ciphertext and returns a deletion proof to O. After that, O can check the deletion result by verifying the proof. In our scheme, we utilize MHT to realize public verifiability, and the verification process does not depend on any trusted third party at all.

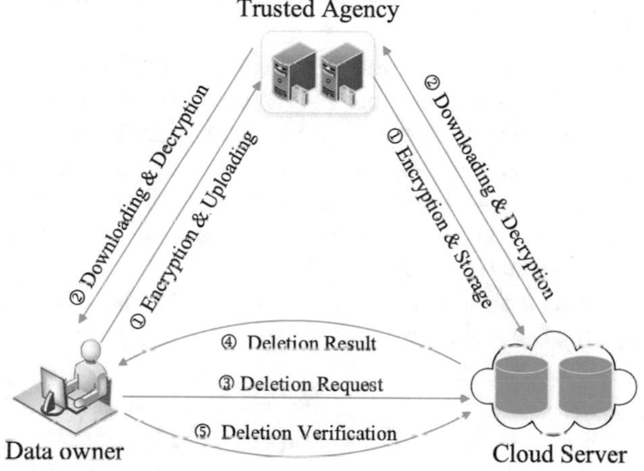

Fig. 3. The main processes of our scheme

4.2 The Concrete Construction

In the following, we will present our novel scheme in detail. For simplicity, we can directly assume that O has passed the authentication and has become a legal user of S. Then O can set a unique identity id, which is so secret that only O and TA know. Moreover, we assume that O, S, TA respectively have an ECDSA key pair (pk_o, sk_o), (pk_s, sk_s) and (pk_t, sk_t). $H_1(\cdot)$ and $H_2(\cdot)$ are two one-way collision resistant hash functions. Without loss of generality, we can assume that O wants to outsource the file F, whose name is n_f, and n_f is kept secret and unique.

- *Encrypt*. In order to protect the sensitive information, O must encrypt the file F before sending it to TA. Then TA should further encrypt the received ephemeral ciphertext to obtain the final ciphertext, and then sends the final ciphertext to S.
  - O firstly encrypts the file F: $C_o = Enc_{k_o}(F)$, where $k_o = H_1(sk_o||id||n_f)$, Enc is an IND-CPA secure traditional symmetric encryption algorithm. Then O computes a file tag $tag_f = H_1(n_f)$ and a hash value $h_o = H_1(C_o||id||tag_f)$. Therefore, O obtains the ephemeral ciphertext $C_{f_o} = (C_o, tag_f, h_o)$, and then sends C_{f_o} to TA.
  - Upon receiving the ephemeral ciphertext C_{f_o}, TA checks that whether the equation $h_o \stackrel{?}{=} H_1(C_o||id||tag_f)$ holds. If $h_o \neq H_1(C_o||id||tag_f)$, TA aborts and returns failure; otherwise, TA further encrypts C_{f_o} as $C_t = Enc_{k_t}(C_{f_o})$, and computes a hash value $h_t = H_1(C_t||id||tag_f)$, where $k_t = H_1(sk_t||id||tag_f)$. Finally, TA obtains the final ciphertext $C_{f_t} = (C_t, tag_f, h_t)$ and sends C_{f_t} to S.
- *StoreCheck*. After storing the outsourced data successfully, S should generate a related storage proof to persuade TA and O that he has stored the outsourced data honestly (m is assumed to be the number of the leaf nodes in the MHT).
  - Upon receipt of C_{f_t}, S stores it on the physical medium. Here we can take $m = 8$ for example, and C_{f_t} is assumed to be stored in the leaf node 6. Then S constructs and publishes the MHT as Fig. 4. After that, S computes a signature on the hash value of the root node as $sig_r = Sign_{sk_s}(h_{0\_1})$, where $Sign$ is an ECDSA signature generation algorithm. Finally, S sends the storage proof $\lambda = (sig_r, \phi)$ to TA, where Φ is the verification object $(h_{1\_1}, h_{2\_4}, h_{3\_5})$.
  - Then TA checks that whether S stores the file honestly. To be specific, TA computes the following equations:

$$\begin{aligned} h'_{3\_6} &= H_2(C_{f_t}); \\ h'_{2\_3} &= H_2(h_{3\_5}||h'_{3\_6}); \\ h'_{1\_2} &= H_2(h'_{2\_3}||h_{2\_4}); \\ h'_{0\_1} &= H_2(h_{1\_1}||h'_{1\_2}); \end{aligned}$$

    Then TA checks that whether h'_{0_1} equals h_{0_1}. If $h'_{0\_1} = h_{0\_1}$, TA verifies the signature sig_r. If sig_r is invalid, TA aborts and returns failure; otherwise, TA sends the storage proof λ to O to inform him that F has been stored honestly. Finally, O will store (h_o, id, n_f, λ) and delete the local backups of F.
- *Decrypt*. After S storing the outsourced data honestly, O will not maintain the local backups anymore. Therefore, when O wants the outsourced file, he will download the ciphertext, and then he can obtain the related plaintext by decrypting the corresponding ciphertext.

- To download the ciphertext, O firstly computes an ECDSA signature sig_d as $sig_d = Sign_{sk_o}(download||tag_f||T_d)$, where T_d is a timestamp. Then O can generate a download request $R_d = (download, tag_f, T_d, sig_d)$ and send it to S. Upon receiving R_d, S checks that whether R_d is valid. If R_d is not valid, S aborts and returns failure; otherwise, S sends $C_{f_t} = (C_t, tag_f, h_t)$ along with R_d to TA.
- Upon receipt of C_{f_t} and R_d, TA firstly verifies the validity of R_d. If R_d is invalid, TA aborts and returns failure; otherwise, TA checks that whether the equation $h_t \stackrel{?}{=} H_1(C_t||id||tag_f)$ holds. If the equation does not hold, TA aborts and returns failure; otherwise, TA decrypts C_t to obtain C_{f_o} as $C_{f_o} = Dec_{k_t}(C_t)$, where Dec is a traditional symmetric decryption algorithm, and $k_t = H_1(sk_t||id||tag_f)$. Finally, TA sends $C_{f_o} = (C_o, tag_f, h_o)$ to O.
- Upon receiving the ciphertext $C'_{f_o} = (C'_o, tag'_f, h'_o)$, O verifies that whether the equation $h_o \stackrel{?}{=} H_1(C'_o||id||tag_f)$ holds. If the verification passes (that is to say $C_o = C'_o$), O decrypts C_o to obtain the plaintext: $F = Dec_{k_o}(C_o)$, where $k_o = H_1(sk_o||id||n_f)$.

– *Delete*. When O will not need the file F anymore, he prefers to require S to delete the file and return a deletion evidence.
  - In order to delete the outsourced file, O should firstly compute a signature: $sig_e = Sign_{sk_o}(delete||tag_f||T_e)$, where T_e is a timestamp. Then O generates a deletion request $R_e = (delete, tag_f, T_e, sig_e)$, and sends R_e to S to delete the outsourced file.
  - Upon receipt of R_e, S checks that whether R_e is valid. If R_e is not valid, S aborts and returns failure; otherwise, S deletes the file and generates a signature $sig_s = Sign_{sk_s}(delete||tag_f||T_e||R_e)$. Then S utilizes sig_s to overwrite C_{f_t} to reconstruct the MHT, as illustrated in Fig. 5. Besides, S computes a new signature: $sig^*_r = Sign_{sk_s}(h^*_{0\_1})$. Finally, S sends the deletion evidence $\tau = (R_e, sig_s, sig^*_r, \Phi)$ to O, where $\Phi = (h_{1\_1}, h_{2\_4}, h_{3\_5})$.

– *DelCheck*. After deleting the outsourced data, O can check the deletion result by verifying τ. Firstly, O checks that whether the signature sig_s is valid. If sig_s is not valid, O aborts and returns failure; otherwise, O computes the following equations:

$$h^{*'}_{3\_6} = H_2(sig_s);$$
$$h^{*'}_{2\_3} = H_2(h_{3\_5}||h^{*'}_{3\_6});$$
$$h^{*'}_{1\_2} = H_2(h^{*'}_{2\_3}||h_{2\_4});$$
$$h^{*'}_{0\_1} = H_2(h_{1\_1}||h^{*'}_{1\_2});$$

After that, O checks that whether $h^{*'}_{0\_1}$ equals $h^*_{0\_1}$. If $h^{*'}_{0\_1} = h^*_{0\_1}$, O aborts and returns failure; otherwise, O verifies the signature sig^*_r. If and only if sig^*_r is a valid signature on $h^*_{0\_1}$ can the verification pass, and O believes that the deletion proof is valid.

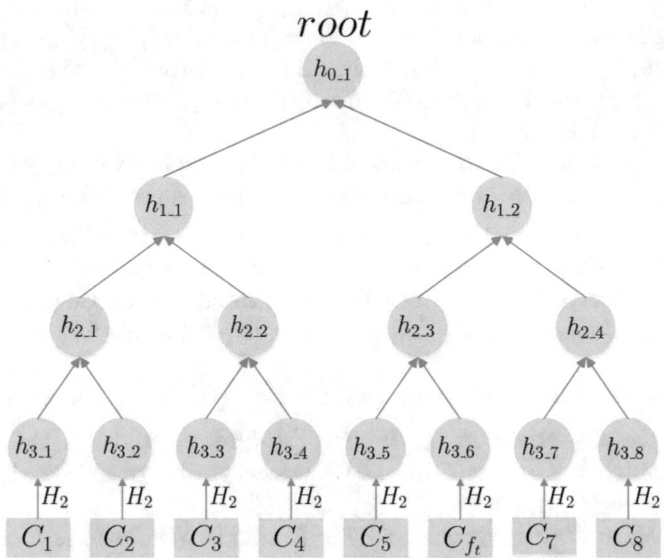

Fig. 4. The MHT for storage proofs

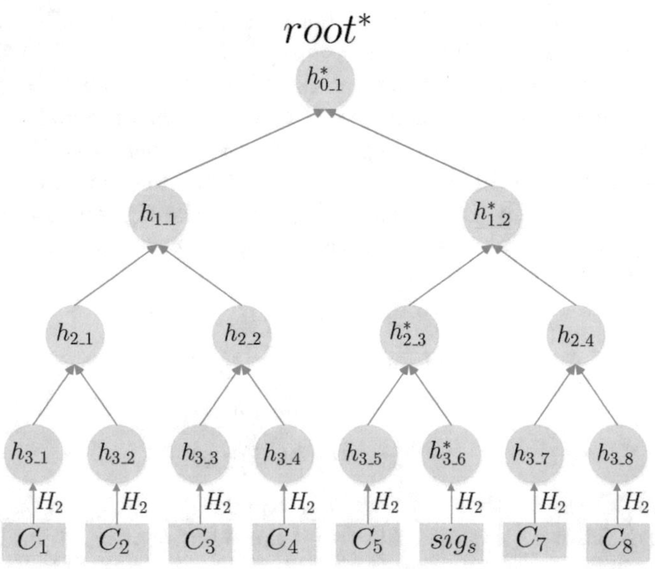

Fig. 5. The MHT for deletion proofs

5 Analysis of Our Scheme

5.1 Security Analysis

In the following, we will analyze the security properties of the proposed scheme.

The Proposed Scheme Satisfies the Property of Data Confidentiality. In the proposed scheme, the data owner firstly utilizes the IND-CPA secure AES encryption algorithm to encrypt the outsourced file to obtain the ephemeral ciphertext. Then the trusted agency further uses the IND-CPA secure AES encryption algorithm to encrypt the ephemeral ciphertext. Besides, both of them maintain the encryption/decryption keys secretly. Therefore, the attacker can not obtain the encryption/decryption keys. That is to say, the attacker can not get the plaintext. Therefore, the proposed scheme can satisfy the property of data confidentiality

The Proposed Scheme Satisfies the Property of Data Integrity. In the decryption process, S will send C_{f_t} and R_d to TA, where $C_{f_t} = (C_t, tag_f, h_t)$ and $R_d = (download, tag_f, T_d, sig_d)$. Upon receiving C_{f_t} and R_d, TA will check R_d and C_{f_t}. If R_d is invalid, TA will abort; otherwise, TA will check C_{f_t}. To be specific, TA checks that whether the equation $h_t \stackrel{?}{=} H_1(C_t||id||tag_f)$ holds. The id is maintained secretly by O and TA. Therefore, S can not falsify C_t into C_t' to make the equation $h_t \stackrel{?}{=} H_1(C_t'||id||tag_f)$ hold. That is, only if C_t is correct and intact can the equation hold. Therefore, TA always can detect the malicious operation if S falsifies C_t. Secondly, O receives $C_{f_o} = (C_o, tag_f, h_o)$ from TA. As TA is impartial and it never behaves maliciously. Therefore, TA will not falsifies C_{f_o}. That is, the proposed scheme satisfies the property of data integrity.

The Proposed Scheme Satisfies the Property of Public Verifiability. Our scheme can reach publicly verifiability. After deleting, S generates a proof τ to prove that he has executed the deletion honestly, where $\tau = (R_e, sig_s, sig_r^*, \Phi)$, and $\Phi = (h_{1\_1}, h_{2\_4}, h_{3\_5})$. Then any verifier who given τ can verify the deletion result. Firstly, the verifier checks that whether R_e is valid. If R_e is valid, it means that O has required to delete the file; otherwise, the verifier aborts. Secondly, the verifier checks the validity of sig_s. If sig_s is invalid, the verifier aborts and returns failure; otherwise, the verifier uses sig_s and Φ to compute $h_{0\_1}^*$. Finally, the verifier checks that whether sig_r^* is a valid signature on $h_{0\_1}^*$. Only if all the verifications pass will the verifier trust that S has deleted the data honestly. As described above, all the verification processes do not involve any private information, and anyone who given τ can verify the result. Therefore, our scheme can satisfy the property of public verifiability.

The Proposed Scheme Satisfies the Property of Leakage Source Traceability. When the data are leaked, our scheme can trace the leakage source precisely. In our solution, the data owner O owns the plaintext of the outsourced file F, and O encrypts F to obtain C_{f_o}. The trusted agency TA further encrypts C_{f_o} to obtain the final ciphertext C_{f_t}. Then the cloud server S maintains C_{f_t}. Besides, O can not access to the final ciphertext C_{f_t}, and S can not access to

F and C_{f_o}. That is, only TA and O can obtain C_{f_o}, and only TA and S can obtain C_{f_t}. However, TA is absolutely impartial, O and S fully trust that TA will never expose the data. Therefore, on one hand, if C_{f_o} is exposed, the data leakage source must be O. Further, the dishonest O can not expose C_{f_o} to successfully slander that S does not honestly remove the data. On the other hand, if C_{f_t} is leaked, the data leakage source must be S. Therefore, if S does not honestly erase C_{f_t} and resulting in data leakage, S can not deny his malicious behavior. In other word, our scheme can realize data leakage source tracking.

5.2 Comparison

We present the comparison between our scheme and scheme [11] in Table 1, and we introduce some marks firstly. We denote by $\mathcal{M}$ a multiplication in cyclic multiplicative group $\mathbb{G}$, $\mathcal{E}$ a data encryption calculation, resp., $\mathcal{D}$ a data decryption calculation. Besides, we denote by $\mathcal{H}$ an operation of computing a hash value, by $\mathcal{S}$ a signature generation calculation (resp., by $\mathcal{V}$ a calculation of verifying a signature). Further, we assume that there are $m = 2^n$ leaf nodes in the MHT, and we will omit the ordinary data transmission operations for simplicity.

Table 1. Comparison between two schemes

Scheme	Scheme [11]	Our proposed scheme
Computational model	Amortized model	Amortized model
Trusted third party	Yes	Yes
Public verifiability	Yes	Yes
Leakage source traceability	No	Yes
Computation **(Delete)**	$1\mathcal{S}$	$3\mathcal{S} + 1\mathcal{V} + (n+1)\mathcal{H}$
Computation **(Verify)**	$1\mathcal{V}$	$2\mathcal{V} + (n+1)\mathcal{H}$

6 Conclusion

In this paper, we put forward a novel publicly verifiable outsourced data deletion scheme with efficient tracking. In cloud storage, there is a trust problem: both the data owner O and the cloud server S do not fully trust each other. In our novel scheme, we utilize the primitive of MHT to solve this trust problem. The cloud server S can utilize the MHT to generate a deletion proof. If S does not remove the data honestly, O will detect the S's malicious behavior by verifying the proof. Besides, our scheme can support data leakage source tracking when the data are leaked. That is, both O and S can not reserve and expose the data maliciously to slander each other successfully.

Acknowledgement. This work was supported by the Open Projects of State Key Laboratory of Integrated Service Networks (ISN) of Xidian University (No. ISN19-13), the National Natural Science Foundation of Guangxi (No. 2016GXNSFAA380098) and the Science and Technology Program of Guangxi (No. AB17195045).

References

1. Bauer, S., Priyantha, N.B.: Secure data deletion for Linux file systems. In: Proceedings of the 10th Conference on USENIX Security Symposium, vol. 174, pp. 153–164. USENIX Association, Berkeley (2001)
2. Boneh, D., Lipton, R.: A revocable backup system. In: Proceedings of the 6th Conference on USENIX Security Symposium, vol. 6, pp. 91–96. USENIX Association, Berkeley (1996)
3. Chen, X., Li, J., Ma, J., Tang, Q., Lou, W.: New algorithms for secure outsourcing of modular exponentiations. IEEE Trans. Parallel Distrib. Syst. **25**(9), 2386–2396 (2014)
4. Chen, X., Li, J., Huang, X., Li, J., Xiang, Y., Wong, D.: Secure outsourced attribute-based signatures. IEEE Trans. Parallel Distrib. Syst. **25**(12), 3285–3294 (2014)
5. Chen, X., Huang, X., Li, J., Ma, J., Lou, W., Wong, D.: New algorithms for secure outsourcing of large-scale systems of linear equations. IEEE Trans. Inf. Forensics Secur. **10**(1), 69–78 (2015)
6. Chen, X.F., Li, J., Weng, J., Ma, J., Lou, W.: Verifiable computation over large database with incremental updates. IEEE Trans. Comput. **65**(10), 3184–3195 (2016)
7. Chen, X., Li, J., Huang, X., Ma, J., Lou, W.: New publicly verifiable databases with efficient updates. IEEE Trans. Dependable Secur. Comput. **12**(5), 546–556 (2015)
8. Diesburg, S., Wang, A.: A survey of confidential data storage and deletion methods. ACM Comput. Surv. **43**(1), 1–37 (2010)
9. Garfinkel, S., Shelat, A.: Remembrance of data passed: a study of disk sanitization practices. IEEE Secur. Priv. **99**(1), 17–27 (2003)
10. Gutmann, P.: Secure deletion of data from magnetic and solid-state memory. In: Proceedings of the Sixth USENIX Security Symposium, vol. 14, pp. 77–89. USENIX Association, Berkeley (1996)
11. Hao, F., Clarke, D., Zorzo, A.: Deleting secret data with public verifiability. IEEE Trans. Dependable Secur. Comput. **13**(6), 617–629 (2016)
12. Hughes, G., Coughlin, T., Commins, D.: Disposal of disk and tape data by secure sanitization. IEEE Secur. Priv. **7**(4), 17–27 (2009)
13. Li, J., Zhang, Y., Chen, X., Xiang, Y.: Secure attribute-based data sharing for resource-limited users in cloud computing. Comput. Secur. **72**, 1–12 (2018)
14. Li, Y., Gai, K., Qiu, L., Qiu, M., Zhao, H.: Intelligent cryptography approach for secure distributed big data storage in cloud computing. Inf. Sci. **387**, 103–115 (2017)
15. Luo, Y., Xu, M., Fu, S., Wang, D.: Enabling assured deletion in the cloud storage by overwriting. In: Proceedings of the 4th ACM International Workshop on Security in Cloud Computing, pp. 17–23. ACM, New York (2016)
16. Lee, J., Yi, S., Heo, J., Park, H., Shin, S.Y., Cho, Y.: An efficient secure deletion scheme for flash file systems. J. Inf. Sci. Eng. **26**(1), 27–38 (2010)
17. Merkle, R.C.: Protocols for public key cryptosystems. In: Proceedings of the 1980 IEEE Symposium on Security and Privacy (SP), pp. 122–134. IEEE Computer Society, Washington, DC (1980)
18. Miao, M., Wang, J., Ma, J., Susilo, W.: Publicly verifiable databases with efficient insertion/deletion operations. J. Comput. Syst. Sci. **86**, 49–58 (2017)

19. Ni, J., Lin, X., Zhang, K., Yu, Y., Shen, X.: Secure outsourced data transfer with integrity verification in cloud storage. In: Proceedings of the 2016 IEEE/CIC International Conference on Communications in China, ICCC 2016, pp. 1–6. IEEE Computer Society, Washington, DC (2016)
20. Paul, M., Saxena, A.: Proof of erasability for ensuring comprehensive data deletion in cloud computing. In: Meghanathan, N., Boumerdassi, S., Chaki, N., Nagamalai, D. (eds.) Recent Trends in Network Security and Applications, CNSA 2010. CCIS, vol. 89, pp. 340–348. Springer, Heidelberg (2010)
21. Perito, D., Tsudik, G.: Secure code update for embedded devices via proofs of secure erasure. In: Gritzalis, D., Preneel, B., Theoharidou, M. (eds.) Computer Security C ESORICS 2010, ESORICS 2010. LNCS, vol. 6345, pp. 643–662. Springer, Heidelberg (2010)
22. Perlman, R.: File system design with assured delete. In: Proceedings of the Third IEEE International Security in Storage Workshop (SISW 2005), pp. 83–88. IEEE Computer Society, Washington, DC (2005)
23. Rahumed, A., Chen, H.C., Tang, Y., Lee, P.P., Lui, J.C.: A secure cloud backup system with assured deletion and version control. In: Proceedings of the 2011 40th International Conference on Parallel Processing Workshops (ICPPW), pp. 160–167. IEEE Computer Society, Washington, DC (2011)
24. Reardon, J., Basin, D., Capkun, S.: SoK: secure data deletion. In: Proceedings of the 2013 IEEE Symposium on Security and Privacy (SP), pp. 301–315. IEEE Computer Society, Washington, DC (2013)
25. Stergiou, C., Psannis, K.E., Kim, B.G., Gupta, B.: Secure integration of IoT and cloud computing. Future Gener. Comput. Syst. **78**, 964–975 (2018)
26. Tang, Y., Lee, P., Lui, J., Perlman, R.: Secure overlay cloud storage with access control and assured deletion. IEEE Trans. Dependable Secur. Comput. **9**(6), 903–916 (2012)
27. Tian, H., Chen, Y., Chang, C., Jiang, H., Huang, Y., Chen, Y., Liu, J.: Dynamic-hash-table based public auditing for secure cloud storage. IEEE Trans. Serv. Comput. **10**(5), 701–714 (2017)
28. Wang, J., Chen, X., Huang, X., You, I., Xiang, Y.: Verifiable auditing for outsourced database in cloud computing. IEEE Trans. Comput. **64**(11), 3293–3303 (2015)
29. Wang, Y., Tao, X., Ni, J., Yu, Y.: Data integrity checking with reliable data transfer for secure cloud storage. Int. J. Web Grid Serv. **14**(1), 106–121 (2018)
30. Wright, C., Kleiman, D., Sundhar, R.S.S.: Overwriting hard drive data: the great wiping controversy. In: Sekar, R., Pujari, A.K. (eds.) Information Systems Security, ICISS 2008. LNCS, vol. 5352, pp. 243–257. Springer, Heidelberg (2008)
31. Xue, L., Ni, J., Li, Y., Shen, J.: Provable data transfer from provable data possession and deletion in cloud storage. Comput. Stand. Interfaces **54**, 46–54 (2017)
32. Yang, C., Ye, J.: Secure and efficient fine-grained data access control scheme in cloud computing1. J. High Speed Netw. **21**(4), 259–271 (2015)
33. Yang, C., Chen, X., Xiang, Y.: Blockchain-based publicly verifiable data deletion scheme for cloud storage. J. Netw. Comput. Appl. **103**, 185–193 (2018)
34. Yang, C., Wang, J., Tao, X., Chen, X.: Publicly verifiable data transfer and deletion scheme for cloud storage. In: Naccache, D., Xu, S., Qing, S., Samarati, P., Blanc, G., Lu, R., Zhang, Z., Meddahi, A. (eds.) ICICS 2018. LNCS, vol. 11149, pp. 445–458. Springer, Heidelberg (2018)
35. Yang, C., Huang, Q., Li, Z., Liu, K., Hu, F.: Big Data and cloud computing: innovation opportunities and challenges. Int. J. Digit. Earth **10**(1), 13–53 (2017)

A Transient Grid Security Control Algorithm Based on EMS System

Yang Su[1(✉)], Song Liu[1], Zhihong Liang[2], Zhizhong Qiao[3], and Xiaodong Li[3]

[1] China Southern Power Grid Co., Ltd., Guangzhou 510670, China
suyang@csg.cn

[2] Dingxin Information Technology Co., Ltd., Guangzhou 510627, China

[3] NARI Information & Communication Technology Co., Ltd., Nanjing 210033, China

Abstract. The transient security control aims to effectively improve the stability of the grid system and prevent large-scale blackouts in the system. In this paper, based on the actual and effective grid data filtered from the EMS system, a transient grid security control algorithm is proposed which is used to solve the optimal power flow. This algorithm is decomposed into two algorithms (i.e., OPF algorithm and optimal control algorithm). In the iteration, the optimal control on the OPF operating point is used to obtain the active output limit of the relevant unit, and the differential equation constraint is transformed into the inequality constraint of the control variable, that is, the inequality constraint related to the number of faults is added, and then solve the problem alternately and finally get the solution. After considering the transient stability constraints, the algorithm proposed is feasible and effective, and has the advantages of reducing the computational burden and the scale of the problem.

Keywords: Grid security control algorithm · Transient stability constraints · EMS system

1 Introduction

At present, due to the rapid development of China's economy, the increase in power is difficult to meet the needs of national economic growth. It is difficult to solve the problem of stable operation of the system through network construction, increase of spare capacity and investment of new devices in a short period of time. How to use various control and adjustment methods to improve the stable operation level of the system is even more important. Since the transient stability of the power system is closely related to the operating state of the system, the fault state, the structure and parameters of the power grid, and the functions of various safety automatic devices in the system, and the current scale of the power system is increasing, The study of transient stability of power systems [1] is a very complex problem. Research in this field has attracted a large number of electric workers and has become an enduring research field [2] in power system research.

C.-N. Yang et al. (Eds.): SICBS 2018, AISC 895, pp. 373–382, 2020.
https://doi.org/10.1007/978-3-030-16946-6_29

Since the French scholars first proposed the optimal trend model [3], the academic community has conducted a lot of researches on the problem. The model can take into account various equations and inequality constraints, and can organically combine the economics and safety of the power system [4–7], which has aroused widespread concern in the power engineering community. In the following decades, the optimal trend has been greatly developed, not only considering different optimization objectives and different constraints, but also presenting various effective algorithms, making the optimal trend more mature and widely used. Power production practices have produced huge economic and social benefits. With the continuous maturity of the grid transient safety control parallel computing technology and the energy management system EMS (Energy Management System [8]), the actual grid data obtained from the EMS system is the basis for the power flow analysis and dynamic security analysis, and the dynamic security analysis program is smooth. The development provided the necessary conditions.

This paper proposes a specific method for correctly and efficiently processing EMS data, so as to screen out the effective information of EMS data and quickly extract the topology structure of the backbone network, which provides the necessary conditions for the smooth development of the dynamic security analysis program. Then the transient grid security control algorithm is used to solve the optimal power flow problem, and the inequality constraints related to the number of faults are added, and then solved alternately. The method proposed in this chapter solves the problem that the calculation burden is too heavy and the problem scale is too large after considering the transient stability constraint.

The remaining parts of the paper are organized as follows: In Sect. 2, some preliminaries (i.e., grid transient stability, energy management system and OPF algorithm) are introduced. And in Sect. 3, transient grid security control algorithm is proposed which is decomposed into two algorithms (i.e., OPF algorithm and optimal control algorithm). In the iteration, the optimal control on the OPF operating point is used to obtain the active output limit of the relevant unit, and the differential equation constraint is transformed into the inequality constraint of the control variable. In Sect. 4, we demonstrate that the algorithm proposed is feasible and effective by experimental results. At last, Sect. 5 is dedicated for conclusion.

2 Preliminaries

2.1 Grid Transient Stability

Transient stability refers to whether the power system can reach a new steady state operation state or return to the original state after a sudden large disturbance in a certain stable point operation state [9]. The so-called large disturbance here generally refers to a short circuit fault, a sudden disconnection of a heavy load line or a generator. If the system can still reach steady state operation after a large disturbance, the system is transiently stable under such operating conditions. On the other hand, if the system is not disturbed enough to reach an acceptable stable operating state, but there is always relative motion between the rotors of the generators, the power, current and voltage of the system are constantly oscillating, so that the entire system cannot continue to run. It is said that the system cannot maintain transient stability under such operating conditions. Obviously, the transient stability of a system is related to the original operation mode of the system and the interference mode [10]. That is to say, the same system is transiently stable under certain operation modes and some disturbances, but in another operation mode and under another disturbance it may be unstable. Therefore, when analyzing the transient stability of a system, the initial operation mode of the system must first be determined in conjunction with the actual situation of the system.

2.2 EMS

EMS (Energy Management System) is a comprehensive automation system for modern power systems based on computers. It consists of three functions: SCADA, PAS and DTS. It organically links data acquisition and monitoring, automatic power generation control and network analysis to improve the automation of isolated "island status" to a unified "management system level", and improve the power system from empirical scheduling to analytical Dispatching has comprehensively improved the level of safety, economy, quality and environmental protection of the power system. EMS is mainly for power generation and transmission systems, and is used in dispatch centers for large-scale power grids and provincial power grids.

The Supervisory Control and Data Acquisition (SCADA) system [11] is a monitoring and data acquisition system. It was proposed by Bonneville Electronics in the United States in the 1987. The first SCADA system appeared in the United States and is located in the Dittmer Control Center in Vancouver. The main equipment of the system is the data terminal of the system, namely RTU (Remote Terminal Unit [12]). Its main function is to complete telemetry, remote signaling, remote control and remote adjustment tasks. In order to overcome the error caused by various factors of the grid measurement data, the power flow does not converge, and the fast algorithm application for power angle and voltage stability calculation is realized. After long-term efforts, power system scholars have developed a real-time network analysis (NA). The Advanced Application Software [13] (PAS) includes a series of basic algorithms such as network topology, external network equivalents, observability analysis, and state estimation. The real-time collected raw data becomes reliable and mature data, thus solving real-time problems. The problem of power flow calculation. In the late 1970s, the Dispatcher Training Simulator [14] (DTS) began to be nested into the EMS system, training the dispatcher with the actual grid scenario, which greatly played the role of DTS and improved the dispatcher's ability to operate normally and handle accidents. The actual grid data obtained from the EMS system is the basis for power flow analysis and dynamic safety analysis.

2.3 OPF

In the context of the increasing scale of distributed power and non-traditional load access distribution networks, it is necessary to analyze the impact of their access on the operational characteristics of the distribution network. The calculation of the distribution network optimization trend is an important part of the research.

The optimized power flow calculation of the distribution network refers to determining the control variables of the system under the conditions given by the network structure and parameters of the distribution network, so that a given objective function describing the operational efficiency of the system is optimized while satisfying the system [15, 16]. Operational and security constraints. It can be described as a simple mathematical form:

$$\begin{aligned} & min f(x,u) \\ & g(x,u) = 0 \\ & \underline{h} \leq h(x,u) \leq \bar{h}, \end{aligned} \tag{1}$$

here, x is the state variable of the system and u is the control variable for optimizing the power flow; $f(x,u)$ is a given objective function used to describe the operational benefits of the system. $g(x,u)$ is an equality constraint that the system needs to meet. $h(x,u)$ is the inequality constraint that the system needs to meet.

The established OPF mathematical model is based on the minimum network loss, considering the constraints of non-traditional loads, adding gears to the optimized variables, compensating the compensation capacity of the capacitor bank, the position of the voltage regulator, and the charge and discharge state of the energy storage battery and many more. The conventional OPF model can be described as a minimum cost problem that satisfies some equality constraints and inequality constraints. In this paper, the total cost of power generation fuel is used as the objective function, and the fuel characteristics of the unit adopt a quadratic function relationship:

$$F = \sum_{i \in S_G} f_i(P_{gi}), \; f_i(P_{gi}) = a_i + b_i P_{gi} + c_i P_{gi}^2, \tag{2}$$

here, P_{gi} is the active output of the first generator, a_i, b_i, c_i respectively represent the fuel cost coefficient of the *i-th* generator. S_G represents a collection of adjustable generators.

(1) Equality Constraint

Polar equations of flow:

$$P_{gi} - P_{li} - V_i^2 G_{ii} - V_i \sum_{j \in I} V_j (G_{ij} \cos\theta_{ij} + B_{ij} \sin\theta_{ij}) = 0 \tag{3}$$

$$Q_{ri} - Q_{ri} + V_i^2 B_{ii} - V_i \sum_{j \in I} V_j (G_{ij} \sin\theta_{ij} - B_{ij} \cos\theta_{ij}) = 0. \tag{4}$$

Here, I is the set of nodes associated with node i. P_{gi} and Q_{ri} are the active and reactive power of the node i, P_{li} and Q_{li} are respectively the active and reactive loads of the node i; G_{ij}, B_{ij} is the mutual admittance between the nodes i, j.

(2) Inequality Constraint

The inequality constraint mainly includes the upper and lower limits of the active power output [17], and the upper and lower limits of the adjustable reactive power output.

$$t_{cl\eta} \leq t_{cr\eta}, \ L(i,j) \in N_c, \tag{5}$$

here, $t_{cr\eta}$ is the critical cut-off time of fault $L(i,j)$; $t_{cl\eta}$ is the actual fault cut-off time of fault $L(i,j)$; $L(i,j)$ is a three-phase short-circuit fault on the side busbar; $t_{cl\eta}$ is the time to eliminate the fault by cutting off the line (i,j); N_c is the set of faults to be analyzed. The perturbation quadratic function $t_{cr\eta} = f_\eta(P_\eta)$ represents the relationship between the critical cut-off time $t_{cr\eta}$ and active power flow P_η of line (i,j). Therefore, the constraint of the critical cutting time of the safe operation of the system can be transformed into the line active power flow constraint: $P_\eta \leq P_\eta^{\text{lim}}$.

3 A Transient Grid Security Control Algorithm Based on EMS System

3.1 Processing Grid Data in EMS System

Before the power flow distribution is calculated, the power loss in the network is unknown. Therefore, the active power P of at least one node in the network cannot be given. This node assumes the active power balance of the system, so it is called the balance node. In addition, a node must be selected, specifying its voltage phase to be zero. As a reference for calculating the voltage phase of each node, this node is called the reference node. The voltage amplitude of the reference node is also given. For the convenience of calculation, the balance node and the reference node are often selected as one node, which is customarily called a balance node. There is only one balance node, its voltage amplitude and phase have been given, and its active power and reactive power are to be determined.

It is reasonable to choose the main FM power plant as the balance node, but it can also be selected according to other principles when calculating the power flow. For example, in order to improve the convergence of the admittance matrix method, the generator with the most outlets can be selected as the balance node.

In the EMS data, we determine the basic principle of the balance point: the balance node is dynamically searched in each round of power flow calculation, and one of the largest capacity of the commissioning unit is selected from the generators of the existing power plants.

3.2 A Transient Grid Security Control Algorithm

Combining the OPF algorithm described above with the optimal control algorithm, the optimal control on the OPF operating point in the iteration is used to obtain the active output limit of the relevant unit, and the differential equation constraint is transformed into the inequality constraint of the control variable, that is, the number of faults and the number of faults are added. The related inequality constraints are then solved alternately to get the solution. In preventive control, the objective function can be selected as the control cost or the purchase cost. The specific forms in different operating modes are roughly the same. We use the system to generate the minimum fuel cost as the objective function, such as

$$min\ F = \sum_{\imath \in S_G} f_\imath(P_{g\imath}), \tag{6}$$

here, $f_\imath(P_{g\imath}) = a_\imath + b_\imath P_{g\imath} + c_\imath P_{g\imath}^2$ is the power generation cost of the unit's fuel characteristics using a quadratic function. $a_\imath$, $b_\imath$, $c_\imath$ is the cost factor of the *ı-th* generator; S_G is a collection of adjustable generators considered.

(1) **Steady State Constraints**

Steady-state constraints include equality constraints and inequality constraints, where the equality constraint is the node flow equation of the system before the failure. In this paper, the power flow equation in polar form is used as the equality constraint, and its expression is

$$P_g - P_L - P(V, \theta) = 0 \tag{7}$$

$$Q_\tau - Q_L - Q(V, \theta) = 0 \tag{8}$$

In the formula, P_g, Q_τ represent the bus active and reactive power vectors, respectively; P_L, Q_L indicate the bus active and reactive load vectors, respectively. Inequality constraints are divided into inequality constraints on variables and inequality constraints on functions. Common variable inequality constraints include:

$$P_{gi}^{min} < P_{gi} < P_{gi}^{max}, \ i \in S_G \tag{9}$$

$$Q_{\tau i}^{min} < Q_{\tau i} < Q_{\tau i}^{max}, \ i \in S_R \tag{10}$$

$$V_i^{min} < V_i < V_i^{max}, \ i \in S_N \tag{11}$$

The above three formulas respectively represent the upper and lower limits of the active power output; the upper and lower limits of the adjustable reactive power output, and the upper and lower limits of the power saving voltage modulus. S_G, S_R and S_N respectively represent a collection of adjustable generators considered, a set of adjustable reactive power supplies and a set of nodes.

(2) **Transient Stability Constraint**

For the convenience of description and without loss of generality, the generator adopts the Golden Model, regardless of the dynamic process of the prime mover and its governor, and the load adopts a constant impedance and constant power hybrid model. The fault modes considered are all three-phase permanent faults at the beginning of the line, and the relay protection operates correctly and the fault is removed. The performance index function is as follows:

$$J(\delta(t_f)) \equiv \phi(\delta(t_f)) = \max_{\forall\tau,j}\left\{\left[\delta_\tau(t_f) - \delta_j(t_f)\right]^2\right\} - \rho^2. \tag{12}$$

Here, the fault removal time t_{cl} and the length of the study period t_f are fixed, which is independent of the operating mode and parameters u.

Rule 1: if $\max\limits_{\forall\tau,j,\ \forall t\in[t_0,t_{\min}]}\{[\delta_\tau(t) - \delta_j(t)]^2\} \geq \xi^2$, the time domain simulation stop at the first time $\max\limits_{\forall\tau,j}\{[\delta_\tau(t) - \delta_j(t)]^2\}$ cross the square of the threshold value ξ^2. The first time it crosses the threshold value is t_f.

Rule 2: if $\rho^2 \leq \max\limits_{\forall\tau,j,\ \forall t\in[t_0,t_{\min}]}\left\{\left[\delta_\tau(t) - \delta_j(t)\right]^2\right\} < \xi^2$, the time domain simulation will continue to $t_{\max}$ and t_f is the max time of $\max\limits_{\forall\tau,j,\ \forall t\in[t_0,t_{\min}]}\left\{\left[\delta_\tau(t) - \delta_j(t)\right]^2\right\}$.

Rule 3: if $\max\limits_{\forall\tau,j,\ \forall t\in[t_0,t_{\min}]}\left\{\left[\delta_\tau(t) - \delta_j(t)\right]^2\right\} < \rho^2$, the time domain simulation is $[t_0,\ t_{\max}]$ and $t_f = t_{\max}$.

If the generator active output P is selected as the control variable u, the transient stability constraint equation can be expressed as

$$J = J(P) \leq 0 \tag{13}$$

In the calculation process, *flag* and variable step factor α represent the number of iterations for the control variable, parameter ε is the convergence allowable error, and $\rho(\rho > 0)$ is the transient stability constraint threshold.

(1) Set $flag = 0,\ k = 0$
(2) Do power flow analysis and identify the relevant units S_t of corresponding to the faulty line and then perform fault simulation. Next we can calculate transient stability $J^k(P^k)$.
(3) If $J^k(P^k) \leq 0$, $\max\left|\Delta P_\tau^{k-1}\right| < \varepsilon$, then the optimal control results are calculated: if $P \leftarrow P^k$ and get output limits $P_{\lim}$, then turn (6); if $P^k \leftarrow P^{k-1} + \alpha\Delta P^{k-1}$, turn (2).
(4) If $flag = 0$, we can get the conjugate equation in $[0,\ t_f]$, turn (5); if $P \leftarrow P^k$, we can calculate output limits $P_{\lim}$ and turn (6).
(5) $P^{k+1} = P^k + \Delta P^k$, set $k \leftarrow k+1, \alpha^k \leftarrow 1, flag \leftarrow 1$ and turn (2).
(6) End.

4 Experiment Evaluation

The no. 1 generator is selected as the balancing machine, and all the generators in the calculation example adopt the classical model, the load uses the 40% constant impedance, 60% constant power mixing model, and the time period of transient stability study is taken as $t_f = 4.0$ s. The iterative convergence accuracy of the optimal control process requires that the error of the active power output of the generator is not greater than $\varepsilon = 0.1$ MW. The threshold value of transient stability is $\rho = 180°$. In practical application, the threshold value of transient stability can be selected according to the actual operation. The active output upper limit, lower limit, power generation cost, OPF solution and total system power generation cost of each generator are shown in Table 1. The upper limit of all bus voltages in the system is 1.1 *p.u.*, and the lower limit is 0.95 *p.u.*

Table 1. Generator data and apparent power of OPF generator under basic conditions.

Dynamo	Output ceiling/MW	Output floor/MW	OPF power output/MVA	Total expense/ $/h
1	1200	400	881.64 + j5.094	61699.52
2	900	600	654.89 + j44.630	61699.52
3	800	300	637.05 + j127.987	61699.52
4	650	200	569.64 + j56.370	61699.52
5	350	100	244.12 + j0.3333	61699.52

The routine *OPF* is first calculated to serve as the initial solution point for *OTS*. At the initial operating point of *OPF*, the entire network fault scan is performed with the actual fault cut-off time $t_{cl} = 0.16$ s to filter out the set of hazardous faults Θ. As shown in Table 2, the symbol $L(m,n)$ indicates that the bus has a three-phase short-circuit fault at the moment m, and the fault is eliminated by cutting the line (m,n) at 0.16 s. The relevant units of $L(22,21)$ are No. 2 and No. 3, the relevant units of $L(25,2)$ are No. 4 and No. 5, and the relevant unit of $L(21,22)$ are No. 1. Therefore, three hazardous faults are divided into two layers. Its effective faults are $L(22,21)$ and $L(25,2)$.

Table 2. Serious failure analysis results before adjustment.

Malfunction number	Malfunction	t_{cl}/s	t_{ci}/s	Time allowance/s
1	$L\,(22,21)$	0.16	0.1068	−0.0532
2	$L\,(25,2)$	0.16	0.1396	−0.0204
3	$L\,(21,22)$	0.16	0.1229	−0.0271

The details on each iteration step of the alternate solution are listed in the Table 3. In the first step of the iteration, the operating cost of the whole network is significantly improved by considering the transient stability constraint, but the anti-disturbance capability of the system still cannot meet the requirements, so it needs to continue to adjust. In the subsequent iterations, the active output of the relevant units continues to decrease, and the stability level of the system is gradually increased. After 5 iterations, the effective faults changed from a hazardous fault to a non-hazardous fault.

Table 3. Algorithm convergence process.

Alternate number	$L\,(22,21)$	$L\,(25,2)$	Stability of $L\,(22,21)$	Stability of $L\,(25,2)$	Total fee
0	——	——	No	No	61699.52
1	1118.80	13388.05	No	No	61792.75
2	1117.12	13361.38	No	No	61800.56
3	1115.76	13261.03	No	Yes	61808.92
4	1113.80	13248.0	No	Yes	61814.01
5	1113.78	13248.0	Yes	Yes	61814.34

The results of the calculation are shown in Table 4. To further verify the effect of the calculation, the system was re-scanned after adjustment, and the result of corresponding to the fault set are showed in Table 5. The stability of the system is guaranteed with minimum cost increase.

Table 4. Optimum active power output and total cost of generator.

Dynamo	Optimal output under transient stability constraints/MVA	Total cost ($/h)
1	1024.39	61814.34
2	594.89	61814.34
3	672.46	61814.34
4	666.51	61814.34
5	601.41	61814.34

Table 5. The result of serious fault analysis after adjustment.

Fault number	Fault	t_{cl}/s	t_{cr}/s	Time margin/s
1	$L\,(22,21)$	0.16	0.1602	0.002
2	$L\,(25,2)$	0.16	0.1600	0
3	$L\,(21,22)$	0.16	0.1818	0.0218

5 Conclusion

On the transient time scale, the main dynamic performances in engineering include the voltage performance on important nodes in the transient process and the power oscillation characteristics on the tie line. According to the characteristics of the actual system stable operation, engineering researchers make different definitions, and the specific form of the algorithm also changes. Transient stability constraints are power angle stability constraints. Power system power angle stability refers to the ability of each synchronous generator to maintain synchronous operation and to accept steady state over time after a large disturbance of the system under normal operating conditions. This paper establishes a model and method for considering multiple expected fault. The method can be solved by an effective nonlinear programming method. Numerical simulation results confirm the validity of the mathematical model and method.

References

1. Duncan Glover, J., Mulukutla, S.S.: Power system analysis and design, pp. 99–110. China Machine Press (2014)
2. Duan, J., Wang, C., Xu, H., et al.: Distributed control of inverter-interfaced microgrids based on consensus algorithm with improved transient performance. IEEE Trans. Smart Grid **99**, 1 (2017)
3. Peter, C., Yang, K.L.: Criterion for the optimal solution of an inventory model with a linear trend in demand. J. Inf. Optim. Sci. **20**(2), 235–248 (1999)
4. Wang, C., Xie, H., Bie, Z., et al.: Reliability evaluation of AC/DC hybrid power grid considering transient security constraints In: IEEE Conference on Automation Science and Engineering, pp. 1237–1242. IEEE (2018)
5. Singh, S.N., David, A.K.: Towards dynamic security-constrained congestion management in open power market. IEEE Power Eng. Rev. **20**(8), 45–47 (2000)
6. Fouad, A.: Dynamic security assessment practices in North America. IEEE Trans Power Syst. **3**(3), 1310–1321 (1988)
7. Fu, S.T., Chen, J.L., Hu, J.X., et al.: Implementation of an on-line dynamic security assessment program for the central China power system. Control Eng. Pract. **6**, 1517–1524 (1998)
8. Solanki, B.V., Bhattacharya, K., Cañizares, C.A.: A sustainable energy management system for isolated microgrids. IEEE Trans. Sustain. Energ. **8**(4), 1507–1517 (2017)
9. Yuanhuan, F.U., Yinhong, L.I., Xuan, H.E., et al.: Corrected transient analysis model of doubly fed induction generator with crowbar protection under grid fault. Proc. CSEE **37**(16), 4591–4600 (2017)
10. Rei, A.M., Leite, D., Silva, A.M., Jardim, J.L., et al.: Static and dynamic aspects in bulk power system reliability evaluations. IEEE Trans. Power Syst. **15**(1), 189–195 (2000)
11. Gaushell, D.J., Darlington, H.T.: Supervisory control and data acquisition. Proc. IEEE **75** (12), 1645–1658 (1987)
12. Shuai, L.L., Tian, G.Y., Xiao-Lu, L.I., et al.: Vertical integration study on power grid model of remote terminal unit and monitor center. Jiangxi Electr. Power **134**(6), 2860–2863 (2012)
13. Criel, M., Godefroid, M., Deckers, B., et al.: Evaluation of the Red Blood Cell advanced software application on the CellaVision DM96. Int. J. Lab. Hematol. **38**(4), 366–374 (2016)
14. Podmore, R., Giri, J.C., Gorenberg, M.P., et al.: An advanced dispatcher training simulator. IEEE Power Eng. Rev. **PER-2**(1), 19–20 (1982)
15. Wang, D., Jiang, Y., Qiu, C., et al.: Research on the on-line dynamic security assessment system and application of Jiangsu power grid. Electric Power Eng. Technol. **1**, 120–125 (2017)
16. Hsiao-Dong, C., Cheng-Shang, W., Hua, L.: Development of BCU classifiers for on-line dynamic contingency screening of electric power systems. Power Syst. **14**(2), 660–666 (1999)
17. Yi, K.K., Choo, J.B., Yoon, S.H., et al.: Development of wide area measurement and dynamic security assessment systems in Korea. In: Proceedings of Power Engineering Society Summer Meeting, Vanouver, Canada, vol. 3, pp. 1498–1499 (2001)

Security Analysis of Bioinformatics WEB Application

Tao Tao[1], Yuan Chen[2(✉)], Bijing Liu[3], Xueqi Jin[4], Mingyuan Yan[5], and Shouling Ji[2,6]

[1] State Grid Hangzhou Power Supply Company, Hangzhou, China
taotao980925@aliyun.com
[2] Zhejiang University, Hangzhou, China
{chenyuan, sji}@zju.edu.cn
[3] NARI Group Corporation, Beijing, China
[4] State Grid Zhejiang Electric Power Co., Ltd., Hangzhou, China
[5] University of North Georgia, Dahlonega, Georgia
[6] Alibaba-Zhejiang University Joint Research Institute of Frontier Technologies, Hangzhou, China

Abstract. Bioinformatics is a subject that focuses on developing methods and software tools, especially web applications, to analyze, understand and utilize biological data. This scientific field attracts large research interest and has been developed rapidly in most aspects but not on security. The lack of security awareness of researchers and insufficient maintenance are the main reasons for security vulnerabilities of bioinformatics web application, such as SQL injection, XSS and file leakage, etc. In the paper, we perform security analysis for website URLs extracted from PubMed abstracts, which contains more than 20,000 URLs. The analysis includes server version CVE matching, HTTPS security evaluation, git leakage detection, and small-scale manual penetration testing. The result shows that the most commonly used server version is outdated and vulnerable. Particularly, only one-fourth HTTPS domains are secure based on our testing, which only count for 7.6% in the entire testing websites. Discovered vulnerabilities are reported to website manager by email and we receive positive feedbacks.

Keywords: Bioinformatics · WEB Security · Git leakage

1 Introduction

In this section, we first introduce the background includes Bioinformatics, data source PubMed, WEB Security and some web security issues. After that, the motivation and problem under investigation are presented.

1.1 Bioinformatics WEB Application

Bioinformatics is an interdisciplinary field of biology, computer science, mathematics, information engineering, and statistics [1]. The research focuses on analyzing and interpreting biological data, in which computer programming plays a significant role.

C.-N. Yang et al. (Eds.): SICBS 2018, AISC 895, pp. 383–397, 2020.
https://doi.org/10.1007/978-3-030-16946-6_30

For example, The Basic Local Alignment Search Tool (BLAST) [2] is a sequence similarity search program, which can be used for inferring functional and evolutionary relationships between sequences as well as helping identify gene families.

In pursuit of better user experience and usability, many researchers develop a web interface. Users can use these applications thorough browser to utilize the software or query the database, so that save time and resources by avoiding downloading, compiling and installation. Take BLAST as an example, NCBI provides a public interface at https://www.ncbi.nlm.nih.gov/BLAST/, which has been cited by more than a thousand papers.

1.2 PubMed

From where can we get bioinformatics research papers? PubMed provides an answer. It includes more than 28 million citations for biomedical literatures. XML format data can be retrieved from NCBI FTP[1]. Though no full text is available on PubMed, it's universal for authors to provide a URL link to their website in paper abstract if their paper is releasing a new bioinformatics database or software. Therefore, we are interested in extracting URLs from PubMed abstracts, and perform security analysis on these websites.

1.3 Web Security of Bioinformatics Applications

Most web applications contain severe security vulnerabilities, which allow hackers access, manipulate data or even make remote code execution without authorization. Even for experienced programmers, identifying potential bugs and security vulnerabilities is a challenging task, let alone for biological researchers.

Furthermore, the develop-publish-graduate pattern in this field also make it worse for bioinformatics web application security. These applications become unmaintained after the developer which typically students graduate. As more and more software vulnerabilities expose nowadays, it is unquestionably not safe to stick to an old version. Even if the code which web application developers write is secure, the underlying software and operating system may be outdated and vulnerable for n-day attacks.

Based on these observations, we speculate that the security status may be even worse for bioinformatics web applications.

1.4 Web Security Issues

SQL Injection

If user provided data concatenates into a SQL query without sanitization, SQL injection occurs. By exploiting this vulnerability, attackers can do arbitrary operations like query, insert, or delete data in the database, leading to data get stolen, lost, corrupted, altered, even the server may be taken over.

[1] ftp://ftp.ncbi.nlm.nih.gov/pubmed/baseline.

Command Injection

Just like SQL injection, command injection occurs when unsafe user-supplied data being passed to a system shell command. Command injection is much more sensitive for web applications, because it allows arbitrary file read which can be leveraged to achieve database connection secrets leakage. Other sites running on the same host will also be affected if no proper isolation or access control made.

Cross Site Scripting (XSS)

Different from server-side code execution, the attack code of Cross Site Scripting (XSS) runs on users' browsers. According to HackerOne, in 2017, XSS was the most common attack type discovered by hackers [3]. Stealing user login credentials, making requests without user approval, and installing malicious software on user's computer are made viable by XSS.

XSS can be classified into two types, reflected XSS and stored XSS. These two types can be differentiated by checking whether server stores attack vector. In reflected XSS, victim clicks attacker crafted URL resulting in the execution of the malicious script embedded in the URL. This type of URL is easy to detect and most recent browsers provide protection techniques to make it much harder to exploit. In stored XSS, also known as persistent XSS, the attacker's payload is stored on the server. And all visiting users will run the payload. For example, a job submitting form allow submitting a job title and description, which are also shown in the result page. Stored XSS occurs when no adequate sanitization performed before outputting the page, affecting all users who visit this page.

1.5 Motivation and Objective

Find existing vulnerabilities and potential security risk inside a web application is essential. Although research proposed web applications may not implement a user system, neither process sensitive data like financial or e-commerce applications, which makes attacks like XSS less profitable, it's still a significant issue affecting the reputation of research institute once being attacked. More importantly, it is possible that these vulnerable applications being used or further developed to deal with sensitive human-related data, like medicine or health advisory service. Discovering vulnerabilities of bioinformatics web application and fix them is meaningful and necessary. Unfortunately, no existing literature on this topic can be found.

Therefore, this work tries to figure out whether or not bioinformatics web application is secure, and answering the following questions:

- Are these websites still available after paper publication?
- What and which type of server they are running?
- Are these server program outdated or with known CVEs?
- How many of them deploy HTTPS?
- Is HTTPS deployed securely or vulnerable for existing attacks?
- Are these web applications vulnerable to typical web security issues?

The contribution of this work can be summarized as follows:

- As far as our knowledge, we are the first to explore the current security situation of bioinformatics web applications.
- Our tests towards more than 20,000 URLs proved that most bioinformatics web applications are running outdated and vulnerable server version, and HTTPS deployment is far from satisfaction.
- We conducted git leakage detection and small-scale manual penetration test, reported discovered vulnerabilities to contact email addresses, and received positive feedbacks.

The remaining contents are organized as follows. In Sect. 2, we present the used method in detail. Our findings and discussion are given in Sect. 3. Section 3.6 shows a case study as a realistic attack example, from git leak to root privilege. Then we summarize and propose future work in Sect. 4.

2 Method

In this section, we will introduce the whole procedure of our analysis. The source of our data will be described in Sect. 2.1, then we perform accessibility analysis, server version detection, HTTPS security rating and git leakage in the following sections. Section 2.7 ends with reporting discovered security issues to contact email addresses.

2.1 Data Source

DaTo [4] is a database of biological software and database. It extracts URL from abstracts from all PubMed papers and uses text mining to locate the tool name. Then, it leverages E-link API provided by NCBI to fetch the citation list between papers, to provide a user-friendly interface for search, map, statistics and network functions. Figure 1 shows the workflow of DaTo.

This work uses DaTo dataset updated in 2017 July. One of us also committed to the development of DaTo, so we can access the dataset. We believe DaTo covers most bioinformatics web applications and outperforms other datasets. DaTo provides 19 data fields, including `pmid`, `lat`, `lng`, `name`, `description`, `url`, `urlstatus`, `country`, `has_mesh`, `abstract`, `date`, `mesh_term`, `jid`, `journal_short`, `country_long`, `state`, `city`. In these data fields, pmid, url, urlstatus, and date are used in this work. Pmid means PubMed ID, url is the URL extracted from the paper abstract, urlstatus shows whether this website can be accessed or not. The date shows the publication time of a paper.

DaTo dataset comprised of 23452 possible bioinformatics tools and databases. After inspecting the data, we found that some URL items are paper publication URL, DOI URL, copyright statement or FTP addresses. Since these URLs are not related to bioinformatics web application, so we excluded these data items from the dataset. After this pre-processing step, 22786 entries retained.

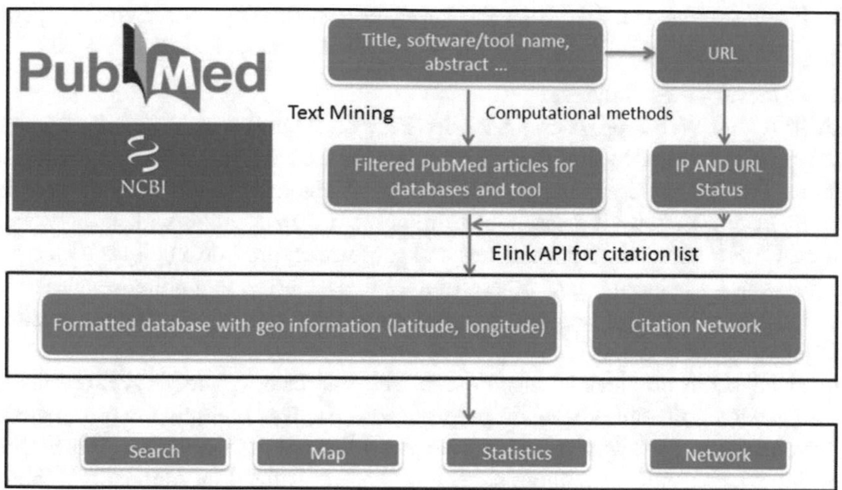

Fig. 1. Data processing workflow of DaTo (From [4])

2.2 Accessibility Analysis

If a web application is not accessible because the server is down, domain is expired, or return 404 errors, its security is out of the question. So, before performing any security analysis, we need to determine whether these applications are still available and online.

DNS Resolution. When a user visits a website, DNS (Domain Name System) provides translation from domain to IP address. If the DNS resolution failed to return an IP address, most likely the domain has expired, and the website is unquestionably not accessible. In this step, we query system default DNS server for A-type DNS response of these domain names. If `NXDOMAIN`, `NoNameservers`, `Timeout` or any other errors occurred more than three times, we marked it as a failure.

Ping Test. Ping is a simple tool using ICMP network protocol for determining whether the host is online. Packet loss rate and network delay can be obtained by pinging target host. In this step, IP addresses collected from domain name resolution are being tested using Ping, we mark it as a failure if it fails to respond three times.

2.3 Server Software Security Analysis

Server Version Retrieval

In HTTP protocol, the server version information is included in the HTTP response message `Server` header field. Censys.io [5] is a search engine for network host, which has collected more than three billion ipv4 addresses using Zmap.

In this step, we query IP information from Censys.io API to get the web server information. Here is an example:

```
131.204.46.201 Apache/2.4.6 (CentOS) OpenSSL/1.0.2k-fips PHP/
5.6.30
```

As it appears, we can learn that the target host is running on a CentOS operating system, using Apache HTTP Server 2.4.6, OpenSSL 1.0.2k, and PHP 5.6.30.

CVE Vulnerabilities Retrieval
OWASP (Open Web Application Security Project) Top Ten aims to raise awareness about application security by identifying some of the most critical risks facing organizations [9]. Using Component with Known Vulnerabilities is one of the major concerns listed. Thereby it's essential to match software version with known vulnerabilities. CVE (Common Vulnerabilities and Exposures) provide a unique ID for public security issues, including ID, description and at least one public reference link. However, risk rate and fix suggestions are not included in the CVE official website, yet NVD provides these data.

cvedetails.com provides an easy-to-use web interface to query CVE entries for a specific vendor, software or specific program version. By combining exploit data from exploit-db.com, cvedetails.com also provides public exploit information. It is helpful to determine whether a CVE is more severe so that get it fixed as soon as possible.

To automatically match software version with its corresponding CVE information, we program to crawl the data provided by cvedetails.com. Starting from searching software name, fetching its version list, and get CVE info for each version. CVE info obtained includes CVE ID, risk rating, description, and public exploits count.

2.4 HTTPS Support Test and Security Rating

HTTPS is an extension of the Hypertext Transfer Protocol (HTTP) for secure communication. It leverages TLS layer to provide a secure channel for HTTP. The HTTPS certificate is used to verify the identity of connection host, which makes MITM (Man in The Middle) attack much harder towards HTTPS connections compared with HTTP connections. Modern browsers also employ a stricter security policy for HTTPS sites, for example if an HTTPS website load insecure HTTP scripts, the browser will show a warning indicating insecurity.

HTTPS Support Test
This step is implemented in Python script, using Request library to try to make https connections. Using parameter `verify=False`, making TLS connection regardless of certificate errors, gives us a list of domains which support HTTPS connection while parameter `verify=True` is leveraged to filter websites deployed correct and trusted HTTPS certificates from the aforementioned list.

HTTPS Security Rating
SSL is a complex and hybrid protocol, allowing numerous features at different connection stages. Therefore, it is challenging to identify the HTTPS security level of a target website. To address this and make our test result more persuasive, we use HTTPS rating API provided by Qualys SSL Labs. Concretely, we use ssllabs-scan to obtain the rating result. The command is as follows:

```
./ssllabs-scan -usecache=true -verbosity=debug --grade -hos-
tfile https_verified.txt > ssltest_output.txt
```

The procedure of generating HTTPS security rating proposed by Qualys SSL Labs HTTPS is as follows:

1. Checking whether the certificate is valid and trusted. If not, give grade M for Certificate Mismatch or grade T for Certificate Untrusted and skip the following checks.
2. Producing a weighted centesimal score based on three dimensions: supported protocols – 30%, supported key exchange methods – 30%, and supported encryption methods – 40%. In particular, if a score for one dimension is zero, the total score is set to zero.
3. Using the following rule to transform the score to grade A to F:
   A: [80,100], B: [65,80), C: [50,65), D: [35,50), E: [20,35), F: [0,20)
4. Some special rules may be applied. Such as reducing from A to A- if unwanted features are present, and using A+ for unusual secure configuration.

We refer [6] for detailed evaluation standard of the rating process.

2.5 Git Leakage Detection

Git is a famous version control tool, especially for managing source code. Some web applications may leverage Git to control the version, such as using webhook to automatically pull the latest code files, so that developers can get rid of trivial work like copy files to the server. But, Git has its drawback if HTTP server wrongly configured, which is Git leakage.

Git leakage means that the .git folder is accessible via GET HTTP request. Attackers can download the entire or partial .git folder using the knowledge of Git internals like the structure of .git/index and zlib compression. Attackers can use tools like GitHack[2] to conduct an attack.

For simple detection propose, we only need to access .git/HEAD. If the server responds with code 200 in the HTTP response and the file content match the desired file format, we consider it has a Git leakage vulnerability.

Although Git leakage will make the source code public, there is no actual harm for open-source websites. In order to exclude open-source websites, we filter those public repositories. We fetch the branch ref file, such as .git/refs/heads/master for branch master (typically listed in .git/HEAD file), to get the latest commit ID. Then we query GitHub API based on the commit ID to filter those open-source websites. The remaining domains will be kept for further vulnerability notification.

2.6 Manual Penetration Test

Although there are numerous existing automatically or semi-automatically vulnerability black-box scanners, we choose not to use these tools due to ethical and legal consideration. Using scanners may put tremendous pressure on the target server, or

[2] https://github.com/lijiejie/GitHack.

even cause a deny of service attack. So, in this step, we choose a small number of websites to test their common vulnerabilities from the following aspects manually:

- Dangerous File Upload: Try to upload a harmless file such as phpinfo.php through a file uploading form, and guess the uploaded location.
- Sensitive File Leakage: For a result page, access the folder page to check if directory listing is enabled. For a dynamic file read page, such as with parameter `?file=job_id.txt`, change the parameter to check if system file `/etc/passwd` can be accessed.
- XSS Attack: Add ">>>" to GET parameters, to check if ">>>" is embedded in the response page. If only ">>>" is present, the website has taken defense against XSS into consideration.

Notably, for websites providing source code download and written in PHP, we use RIPS-0.55 to audit the source code. RIPS is an open-source PHP static analysis software to detect security vulnerabilities such as XSS, SQL Injection, Local File Inclusion, and others.

2.7 Vulnerability Report

After found and confirmed the vulnerability, we decide to report the problem we found to the website owner. According to paper [7], e-mail notifications is the best and economical method to send the alert letter.

We collect the contact email from the web pages manually, such as from the bottom of the index page or contact page, if not present, we seek to use the email address of the paper corresponding author.

"WEB Security issue about {{domain}}" is used as email subject, and in the email body, we provided necessary information about the vulnerability we found such as vulnerability type, url and impact, besides we also provided fix suggestion and reference urls.

3 Result and Discussion

3.1 Accessibility Analysis

After data cleaning, there are 22786 URL data items in DaTo dataset, which contains 12276 domains and 86 IP addresses. DNS lookup found that 1619 (13.2%) domains cannot resolve, which counts for 2214 (9.7%) URLs.

The length of IP list from DNS lookup plus IP address is 8010.

Moreover, the Ping test showed that only 4063 (50.7%) IP replied to our Ping request. Though this does not mean only half hosts were online at that moment, because some hosts may be configured not to answer ICMP messages.

3.2 Server Software Security Analysis

The query from Censys.io returned 5891 data items, in which 3500 (59.4%) successfully matched to known software version. The primary reasons of mismatching are rare software which has no CVE information, and WAF or CDN deployed which hide the Server field.

Figure 2 and Table 1 shows the most frequent Server version and matched CVEs.

Table 1. Top 10 frequent server software version and corresponding CVEs counts

Server software version	Counts	CVE counts	High-risk CVE counts	Public exploits count
Apache/2.2.15	480	28	6	2
Apache/2.4.6	285	21	1	1
Apache/2.2.22	277	15	5	0
Apache/2.4.7	266	20	1	1
Apache/2.2.3	242	50	10	2
Microsoft-IIS/7.5	174	5	3	1
Apache/2.4.10	172	17	4	0
Apache/2.4.18	168	14	4	0
OpenSSL/1.0.1e	154	68	13	2
Apache-Coyote/1.1	149	0	0	0

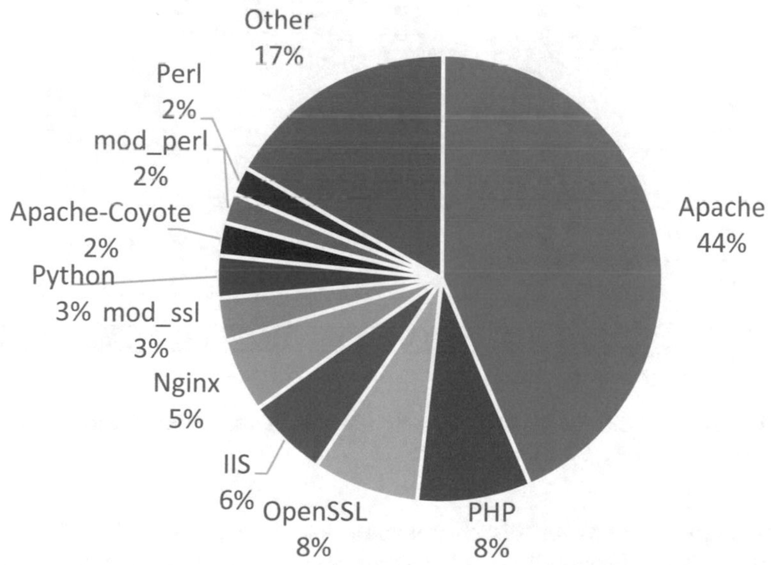

Fig. 2. Frequency distribution of server software

Here we define those CVEs whose CVSS score greater than 7 are high-risk CVEs. Public exploits are exploits that are uploaded to exploit-db.com to exploit the CVE. As we can see from this result, it's evident that a large proportion of servers are running outdated server version software, and there're public exploits attackers can make use of.

In 5891 IP addresses, 3109 (52.8%) of them successfully matched to CVE entries, in which 2991 (50.8%) have at least one high-risk CVE, and 2028 (34.4%) have at least one CVE that corresponding public exploit is available.

It is necessary to note that even high-risk CVEs may not be easy to make use. A specific vulnerability only exists under some particular requirements and environments. Take heart-bleed (CVE-2014-0160) as an example, this vulnerability can lead to leakage of users' cookies and passwords. Yet, if the server has not configured HTTPS, this CVE has no impact to this server, so attackers cannot exploit this vulnerability. Besides, the public exploit may merely be a POC (Proof of Concept), and there is a considerable gap between CVE and real impact. Nevertheless, more CVEs indicate less security and more threats.

Figure 3 shows the geographical distribution of the average count of public exploits per country. The darker, the problem is more severe. We can see developed countries like the United States and European countries are less vulnerable, by contrast, developing countries are more likely to use the vulnerable software.

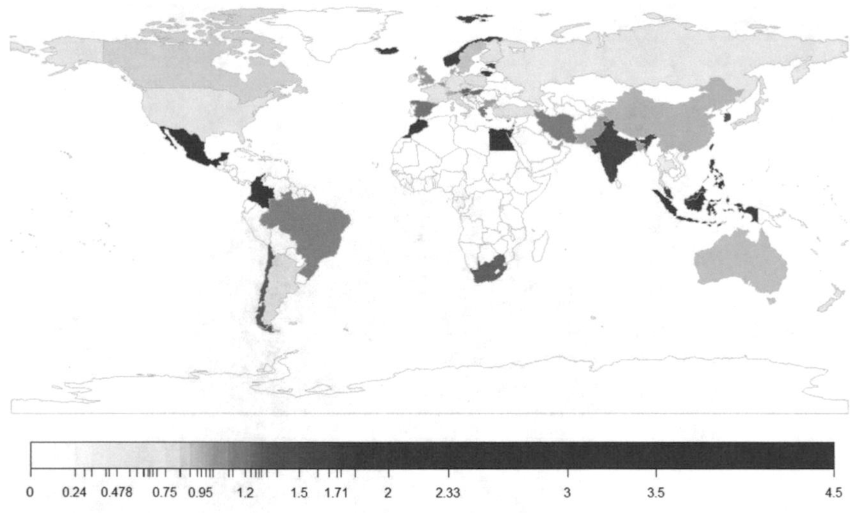

Fig. 3. Geographical distribution of vulnerable bioinformatics web applications. The value is average number of public exploits per country.

Possible explanations for this phenomenon may include the difference introduced by economic level. Economically advanced countries are also technology developed countries, so their developers' security awareness is higher than those in developing countries. Besides, newer applications developed also contribute to less vulnerable applications exploitable.

3.3 HTTPS Support Test and Security Rating

After DNS lookup, we have 10687 resolvable domains and IPs. HTTPS connection without verifying certificate proved that 5278 (49.4%) of them support HTTPS connection, which takes about half proportion. Meanwhile, supporting HTTPS and having a valid and trusted HTTPS certificate only accounts for only one fourth, which are 2727 (25.5%). This data shows that in this field, HTTPS is far from being adopted extensively.

Calling security rating API by ssllab-scan provided by Quals SSL Lab to test the aforementioned 2727 domains/IPs, we collected 1346 data items. The main reasons for missing data are network fluctuation and target server rejection of being tested. We plot the distribution of returned grade data in Fig. 1. In which A and A+ applications count for 60.3% (811), which is a desirable result. However, they only account for 7.6% of entire domains/IPs.

To sum up, only one-fourth of bioinformatics web applications deployed valid HTTPS certificate, and only 7.6% applications deployed secure HTTPS configuration (Fig. 4).

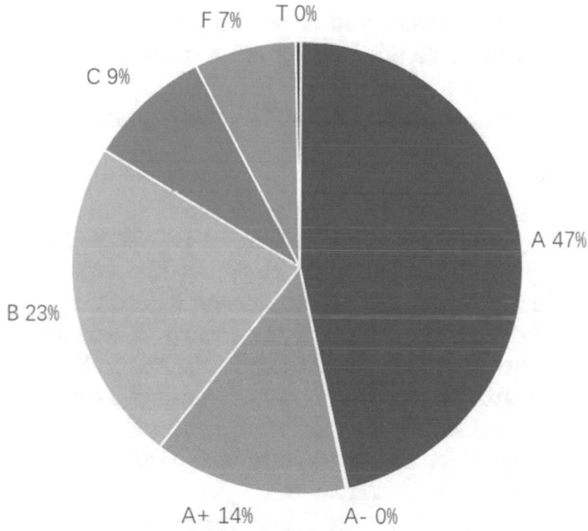

Fig. 4. HTTPS security grade distribution

3.4 Git Leakage Detection and Manual Penetration Test Result

From 18236 distinct URLs, we access each one's corresponding .git/HEAD file, and we found 134 (0.73%) of them have git leakage issue. After querying commit ID of branch ref file, we filtered some open-source projects to get a list of 91 URLs. Then, we manually visited each one to get the contact email addresses and ignore those static websites (only providing description or software download functionality). After this step, we have 55 contact email addresses to send the notification letter.

Manual penetration test result is limited to our technical level and ethical consideration. Nevertheless, we still get some insights proving the existence of typical types of WEB vulnerabilities. Table 2 summarize our findings of manual test.

Table 2. Findings of manual penetration test

Vulnerability type	Count
Cross Site Scripting (XSS)	9
Dangerous file upload	2
Remote command execution	1
Arbitrary file reading	1
Directory listing	1

3.5 Vulnerability Report

We send email to 66 email addresses of 55 websites in sum. If rejection letter received, we try to find another possible email address to send. Then, we got nine replies in total, in which 4 of them finished the repair, one finished part of the work but still need further amendment, one responder said he would consider it (XSS defense) in new version development, and two replied saying not a security threat due to open-source (Git leak issue).

All replies are positive about our work and expressed gratitude to us. Here is a letter sample:

```
Dear ***(redacted),
  thank you very much for pointing out this issue.
  For safety reason, we should deal with it as soon as
possible. I will contact the current Website maintainer
and server administrator to figure out a solution.
  I appreciate your work in Bioinformatics Security and
active involvement in improving security levels. As a
very inexperienced Web-developer by then (even until
now), I often ignored this aspect as long as the code
worked. I hope starting from this issue I will learn more
knowledge about Web security and contribute to building a
better bioinformatical environment.
  Thank you again!
Best regards,
***(redacted)
```

3.6 Case Study – from Git Leak to Root Privilege

In this section, we will show how a bioinformatics web application can be compromised to get root privilege of the target host. We denote the target website as `vuln.com`, denote username as `user`, project name as `name`, our server IP as `our_ip`, and mask some data fields for ethical consideration.

First, from the result of Git leak scan, we found http://vuln.com/.git/config is present, which tells us the remote origin is `git@gitlab.com:user/name-server.git`. By visiting corresponding GitLab project page, we found this website source code is open-source under MIT license.

The repository has more than two hundred commits and contains more than 200 MB files. Most files are committed three years ago (2015), and the last commit is authored in October 2017. Besides, this repository has no README file nor informative description text and has 0 stars. So, this repository can be viewed as not in actively developing and less-concerned.

By searching dangerous functions like shell_exec and exec, we found numerous calling to these dangerous functions with $PARAM. This coding style is rather dangerous, just like concatenating variables to SQL statements, if no sanitization is assured before the function call for any involved variable. Here is a code example extracted from `align.php`:

```
$results_dir = $_REQUEST['results_dir'];
$pairwise = shell_exec("cd $results_dir;echo -n `grep
PAIRWISE $PAR_FILE`");
```

As you can see, the variable results_dir is directly passed to shell_exec calling. So, the attacker can easily exploit this remote code execution vulnerability, just like visit this URL below:

```
http://vuln.com/align.php?results_dir=|whoami#
```

But here the command execution result is not shown in the response page, so we need to send the execution result to our server. Here is the modified proof of concept:

```
http://vuln.com/align.php?results_dir=|curl our_ip:6666
--data `id|base64 -w0`#
```

This exploit leverage the curl utility which typically embedded in the system to send the base64 encoded command output to our server via HTTP POST body. (Our server is running `nc -lp 6666` to receive data). The base64 encoding is introduced to avoid interference of special characters in the command output (Fig. 5).

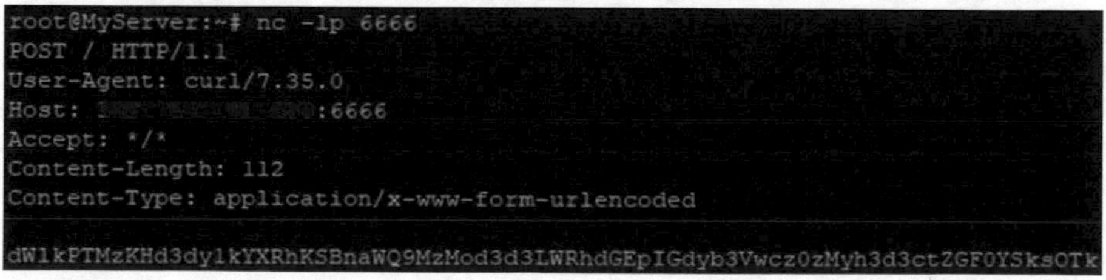

```
root@MyServer:~# nc -lp 6666
POST / HTTP/1.1
User-Agent: curl/7.35.0
Host: [illegible]:6666
Accept: */*
Content-Length: 112
Content-Type: application/x-www-form-urlencoded

```
dWlkPTMzKHd3dy1kYXRhKSBnaWQ9MzMod3d3LWRhdGEpIGdyb3Vwcz0zMyh3d3ctZGF0YSksOTk

Fig. 5. Our server successfully receives the command execution result from the vulnerable host.

Base64-decoding the string gives us: (partially marked):

```
uid=33(www-data) gid=33(www-data) groups=33(www-data),999
(gitlab-www),1000(***)
```

So, attackers can do anything from reading system files, to delete all files owned by www-data. But this is not the end, just like matching HTTP Server version to CVE, we can also check the Linux kernel version to find whether privilege escalation vulnerabilities are present. So, we do cat/etc/issue, and uname -a to get the knowledge of the system. And the corresponding results are:

```
Ubuntu 14.04.5 LTS \n \l
Linux *** 3.13.0-73-generic #116-Ubuntu SMP Fri Dec 4
15:31:30 UTC 2015 x86_64 x86_64 x86_64 GNU/Linux
```

It is clear that stick to an old version kernel is definitely not safe. This kernel version is vulnerable to Dirty COW (CVE-2016-5195). Linux kernel 2.x through 4.x before 4.8.3 are all affected by the Dirty COW [8], leaving attacker an easy way to gain root privilege because of public exploit script (https://www.exploit-db.com/exploits/40839/).

Due to ethical considerations, we did not perform any escalation attempt so that we cannot be certain about root privilege acquisition, it is possible that maybe this server has been patched for this CVE. But, by searching CVE list for this kernel version, there are many more CVEs which malicious attacker can use. (cvedetails.com has 174 CVEs for kernel version 3.13.) So it is definitely a severe threat to the whole system that needs to be fixed as soon as possible.

4 Conclusion and Future Work

In this paper, we evaluated bioinformatics web applications' security from multiple dimensions and found several worrisome facts, such as three-fourths of them still doesn't support HTTPS, and about half of them are using server software with high-risk CVEs, and even 34.4% of them are at risk of public exploit scripts.

After git leakage detection and manual penetration test, we reported security issues to website owners and corresponding authors and received positive responses. This proved our inspection of problem reason, which is the developers' inadequacy of security awareness.

Bioinformatics security is a research field that far from being studied. As fast development of bioinformatics and medicine, more and more related web application or APPs will be developed and commercialized. Security and privacy are critical requirements for these advancements and should be addressed before it's too late. How to better apply security scanner to bioinformatics area, how to get authorization from website owner and how to do a better vulnerability notification campaign are left to future work.

Acknowledgement. We would like to thank the anonymous reviewers for their valuable suggestions for improving this paper. We are also grateful to Yincong Zhou, Dahui Hu and Prof. Ming Chen of The Group of Bioinformatics of Zhejiang University for their work about DaTo and contribution to this work.

This work was partly supported by NSFC under No. 61772466, the Zhejiang Provincial Natural Science Foundation for Distinguished Young Scholars under No. R19F020013, the Provincial Key Research and Development Program of Zhejiang, China under No. 2017C01055, the Fundamental Research Funds for the Central Universities, and the Alibaba-ZJU Joint Research Institute of Frontier Technologies. Technology Project of State Grid Zhejiang Electric Power co. LTD under NO. 5211HZ17000J.

References

1. Bioinformatics Wikipedia. https://en.wikipedia.org/wiki/Bioinformatics. Accessed 12 Oct 2018
2. Johnson, M., et al.: NCBI BLAST: a better web interface. Nucleic Acids Res. **36**(2), W5–W9 (2008)
3. Ranger, S. At $30,000 for a flaw, bug bounties are big and getting bigger – ZDNet. http://www.zdnet.com/article/at-30000-for-a-flaw-bug-bounties-are-big-and-getting-bigger/. Accessed 12 Oct 2018
4. Li, Q., Zhou, Y., et al.: DaTo: an atlas of biological databases and tools. J. Integr. Bioinform. **13**(4), 297 (2016)
5. About Us – Censys. https://censys.io/about. Accessed 12 Oct 2018
6. SSL Server Rating Guide. https://github.com/ssllabs/research/wiki/SSL-Server-Rating-Guide. Accessed 12 Oct 2018
7. Stock, B., Pellegrino, G., Li, F., et al.: Didn't you hear me? - towards more successful web vulnerability notifications. In: Network and Distributed System Security Symposium (2018)
8. CVE-2016-5195 in Ubuntu. https://people.canonical.com/~ubuntu-security/cve/2016/CVE-2016-5195.html. Accessed 12 Oct 2018
9. OWASP Wikipedia. https://en.wikipedia.org/wiki/OWASP. Accessed 12 Oct 2018

Dynamic Network Configuration: An Effective Defensive Protocol for Public Blockchain

Zhengwei Jiang[1(✉)], Chenyang Lv[2(✉)], Bo Zhang[3(✉)], Chao Zhang[4(✉)], Wei Lu[4(✉)], and Shouling Ji[2(✉)]

[1] State Grid Zhejiang Electric Power CO., LTD., Hangzhou, China
jiang_ng_zhengwei@zj.sgcc.com.cn

[2] College of Computer Science and Technology, Zhejiang University, Hangzhou, China
{puppet,sji}@zju.edu.cn

[3] NARI Group Corporation, Beijing, China
zhangbo7@sgepri.sgcc.com.cn

[4] State Grid Hangzhou Power Supply Company, Hangzhou, China
13757118405@139.com, 48618828@qq.com

Abstract. To earning unfair profits, adversaries can attack the legitimate nodes in the Bitcoin network with *selfish mining*, the *eclipse attack* and the *information delaying attack*. In this paper, we study the patterns of the above attacks and then present a scheme to enhance the security of the Bitcoin network. First, we propose a new structure for the Bitcoin network called double-layer dynamic network, which improves the defense capability of Bitcoin against several attacks. Second, we design a new structure for the blocks in the Bitcoin network, which provides a way to store the IP addresses in the blocks. Third, we present a novel network protocol named dynamic network configuration for public blockchain. Our protocol pushes the updating of IP addresses in the blocks and changes the construction of the network periodically. From theoretical analysis and simulated evaluation, we find that under our protocol, the Bitcoin network can defend against the *selfish mining*, the *eclipse attack* and the *information delaying attack* effectively.

Keywords: Bitcoin · Defensive protocol · P2P network

1 Introduction

Bitcoin [1] has received much more attention and trust than any other digital cryptocurrency recently. One of the key reasons for the success of Bitcoin is that it uses a distributed database called blockchain rather than a central bank. Most nodes in the Bitcoin network work on the longest blockchain. The blockchain implements a hash-based Proof of Work (PoW) mechanism and uses blocks to store transactions. By using the PoW mechanism, the blockchain ensures that

C.-N. Yang et al. (Eds.): SICBS 2018, AISC 895, pp. 398–413, 2020.
https://doi.org/10.1007/978-3-030-16946-6_31

the recorded transactions are unmodifiable. Therefore, Bitcoin provides more reliable currencies than other digital cryptocurrencies.

The Bitcoin network has an open network environment. It allows nodes to join and leave the network freely. While the open network environment also brings several security risks. Adversaries can join the network and gain unfair profits by attacking legitimate nodes using typical attacks such as *double spending* and *selfish mining*. Many researches explain the attacks on the Bitcoin network [2,3,5,9,12–14].

To defend against the above attacks, a few research has been conducted [4,15]. However, to our knowledge, there is no research that can defend against multiple intractable attacks for the Bitcoin network.

Therefore, in this paper, we introduce a novel defensive protocol to enhance the security of the Bitcoin network and defend against *selfish mining*, the *eclipse attack*, and the *information delaying attack*. The contributions of our work in this paper can be summarized as follows:

- We analyze the vulnerabilities of the Bitcoin network by discussing the implementations of *selfish mining*, *eclipse attack* and *information delaying attack*. Based on the analysis, we propose a novel protocol named Dynamic Network Configuration (DNC) and a new structure of the overlay network called Double-layer Dynamic Network (DDN) to defend against above mentioned attacks.
- We analyze the defensive capability of the DNC protocol both theoretically and experimentally. Our protocol is very effective against *selfish mining*, *eclipse attack* and *information delaying attack* in the Bitcoin network.

2 Background

In this section, we introduce the basic concepts of the Bitcoin network to facilitate our discussion. A complete explanation of Bitcoin can be found in [6].

Nodes. The nodes in the Bitcoin network are identified by IP addresses and can trade with other nodes. Each node has at most 8 outgoing TCP connections and 117 incoming TCP connections. We model the Bitcoin network by a graph $G = (V, E)$, where $V = \{A_1, A_2, ...\}$ and $E = \{e_{ij} | A_i, A_j \in V\}$ characterize the set of nodes and the set of connections among the nodes respectively. Let $outaddr(A_k)$ be the set of peers that are connected by the outgoing connections of node A_k, and $inaddr(A_k)$ be the set of peers that connect to the node A_k by their outgoing connections. $tried(A_k)$ denotes the set of IP addresses in the *tried table* of node A_k. Let $|V| = n$, i.e., the number of nodes in Bitcoin is n.

Inv Messages. Nodes in the Bitcoin network use the *inv* messages to reduce the cost of spreading messages. The *inv* message only contains the type and the hash of the entire message. Hence, the size of an *inv* message is much smaller than that of a complete message. The *inv* messages are saved in the First In First Out (FIFO) buffer of each node.

The New Table. Each node maintains a *new table* which contains the IP addresses of a node that it has not yet established successful connections with.

The Tried Table. Each node in Bitcoin maintains a *tried table* which contains 4,096 IP addresses. When a node has successfully established outgoing or incoming connections with peers, it stores the IP addresses in its *tried table*. Each node keeps the timestamps of the latest successful connections to the peers.

Transactions. Bitcoin uses the transaction-to-transaction payments to implement the trades among the nodes. Transactions are the messages to transfer the ownerships of coins between the nodes in Bitcoin.

Blocks. The construction of a block has two parts: the transaction data part and the block header. Transactions are stored as the leaf nodes of the merkle trees in the transaction data part of blocks. After hash computing, the merkle roots will be stored in the block headers. The block header also stores the hash of the previous block's header. Due to the high computational cost of hash computing, it is usually assumed that the messages on the blocks are inalterable.

Miners. Miners are a kind of nodes that have computing power. Miners collect transactions and mine blocks to make the blockchain get longer. A miner mines a block means that the miner calculates the hash of the previous block header that satisfies the requirements and obtains a new block. The miner spreads the new block in the Bitcoin network and will be rewarded with coins.

Blockchain. The Bitcoin network uses a decentralized database called blockchain, where transactions are stored in the blocks of the blockchain. Since the block header stores the hash of the previous block's header, blocks are logically organized as a chain. All the nodes in the Bitcoin network reach a consensus on the longest blockchain by downloading from the first block to the latest block. Then, they keep working on the longest blockchain. The height of a block is the number of blocks between this block and the first Bitcoin block in the blockchain.

3 Vulnerability Analysis

3.1 Selfish Mining

Eyal and Sirer showed that a mining pool controlling more than 33% of the computing power in Bitcoin can increase the mining advantages by withholding its mined blocks [5]. The pool releases the hidden blocks to get rewards when new blocks are found by other miners.

From this perspective, we consider the nodes of adversaries with the following strategy: nodes of adversaries are randomly distributed in the Bitcoin network. Their outgoing connections connect to legitimate IP addresses randomly. Meanwhile, the nodes of adversaries mine blocks cooperatively. When a node of adversaries mines a new block, the node transmits the block to other nodes of adversaries and they withhold the block. When legitimate miners mine another block and spread it in the Bitcoin network, sooner or later one or more nodes of adversaries will receive the block. Then, these nodes inform the rest of adversaries' nodes and all they spread the withheld block in order to make other nodes work on adversaries' hidden block. We explore how many nodes will receive the hidden block first and work on the adversaries' block eventually through this strategy via simulation. The result is shown in Fig. 1.

From Fig. 1, even if adversaries control 10% of all the nodes in the Bitcoin network, there are nearly 60% nodes receiving the hidden block of adversaries first and working on the hidden block under this strategy.

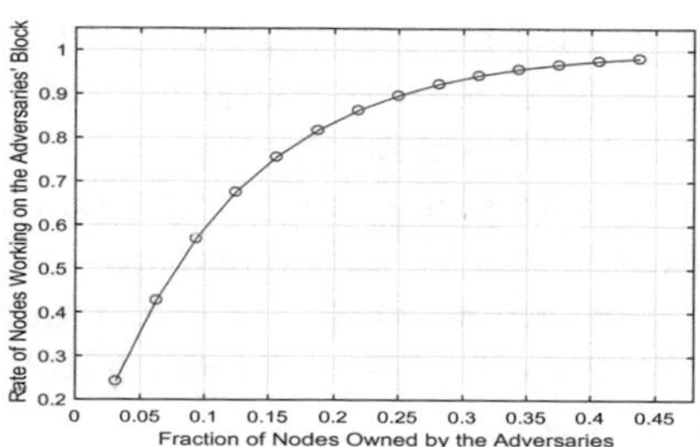

Fig. 1. The number of nodes eventually working on the adversaries' hidden block versus the number of adversaries' nodes.

The result indicates that if adversaries have enough nodes, most nodes will work on their hidden block. The hidden block will be the member of the longest blockchain with much higher probability than the blocks mined by legitimate miners. Thus, adversaries can accumulate mining advantages by always withholding the blocks and releasing one of the hidden blocks when legitimate nodes publish a new block. We come to a conclusion that adversaries can do *selfish mining* without the risk of losing the rewards under this strategy.

3.2 Eclipse Attack

Heilman et al. showed that by monopolizing the outgoing connections and incoming connections of victims and controlling the messages that victims receive and send, adversaries can utilize the computing power of victims for their own use [2]. We call this kind of attack the *eclipse attack*.

There are two main vulnerabilities utilized by *eclipse attack*. (1) Adversaries can attack the *tried table* and *new table* to improve the success rate of monopolizing the victims' outgoing connections. As shown in [2], when the resources are enough, adversaries can monopolize the outgoing connections of victims with a quite high probability. (2) It is difficult for victims to realize that they have already been attacked by adversaries. This is because victims cannot be conscious of that their input and output messages are filtered by adversaries.

3.3 Information Delaying Attack

Gervais et al. showed that by sending victims *inv* messages and ignoring the requests for the entire message, adversaries can delay the victims for receiving specific messages for nearly 20 min [3]. Because of such latency, adversaries can implement double-spending attack more easily and improve the reward of *selfish mining*. We call such attack the *information delaying attack*.

Specifically, *information delaying attack* utilizes the vulnerability that each node in the Bitcoin network has to request the sender of the *inv* message for the entire message in sequence. However, the Bitcoin network does not punish the senders who do not respond the requests. We deduce the reasons as follows: first, there are many reasons that receivers do not receive the entire message. For example, the senders do not receive the requests of the receivers, the entire message is lost during the transmission and so on. Second, adversaries may utilize the punishment mechanism if there is one to harm the profits of legitimate senders. Thus, it is improper to punish the overtime senders, we consider modifying the advertisement-based request management system to defend against the *information delaying attack*.

Based on the aforementioned discussion, we design DNC to patch the vulnerabilities and improve the construction of the Bitcoin network.

4 Dynamic Network Configuration

In this section, we give the details of our DNC protocol. The main idea of DNC is as follows: Different from the traditional design that nodes maintain the long term connections, DNC is designed to build a dynamic structure of a communication network for Bitcoin. Nodes periodically reconnect to different peers on their own initiative. We introduce the primary modules of DNC below:

(1) **Preprocessing Module.** To implement DNC, first, leveraging the Bitcoin network, we logically construct DDN, which contains High Layer Dynamic Network (HLDN) and a Low Layer Dynamic Network (LLDN). Second, we design a new block structure to store the IP addresses for the nodes in HLDN.
(2) **Outgoing Connection Maintenance Module.** We redesign the rules for determining outgoing connections to ensure the defensive capability of the DNC protocol and the structure of DDN. Each node in the Bitcoin network has to follow the rules when the replacement period starts, when a node connects to peers and when a node disconnects from peers.
(3) **Substitutive Peer Selection Module.** The nodes in HLDN are updated after a fixed replacement period. By using the substitutive peer selection module, each node in HLDN will choose a substitutive peer and announce its IP address in the Bitcoin network periodically.

To start the DNC protocol, the community of Bitcoin run the preprocessing module. Then, each node in Bitcoin runs the outgoing connection maintenance module and the substitutive peer selection module to maintain DNC.

4.1 Preprocessing Module

In DNC, we first construct a DDN for the Bitcoin network to transmit information. To implement DDN, we first select a small part of nodes from Bitcoin network to construct HLDN. The remaining nodes form the sub-network called LLDN. The illustration of DDN is shown in Fig. 2.

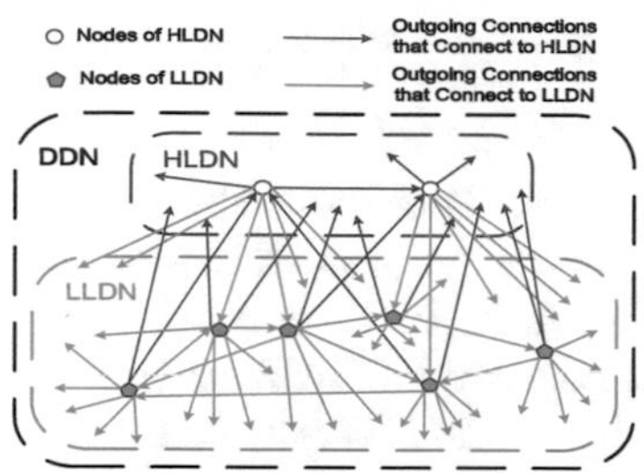

Fig. 2. The illustration of DDN.

Let $V_H = \{A_i, A_j, A_k...\}$ and $V_L = \{A_m, A_n, A_p, ...\}$ characterize the set of the nodes in HLDN and in LLDN respectively. Let $|V_H| = n_H$ and $|V_L| = n_L$.

HLDN has the following properties. First, the number of nodes in HLDN is small. Second, the IP addresses of the nodes in HLDN will be stored in the blocks with a new structure. Therefore, each node can obtain the IP addresses of all the nodes in HLDN from the latest block. Third, each node in DDN is only allowed to use two outgoing connections to connect to HLDN. Fourth, under the DNC protocol, the nodes in HLDN can transmit the entire messages directly without sending *inv* messages first and waiting for the requests from the receivers. Therefore, the nodes in HLDN will expedite the spread speed of messages. Once a node receives an unsolicited message from an incoming connection, it checks whether the IP address that sends the message is in the latest block. If yes, the node receives the message and checks the content; it abandons the message otherwise.

The nature of DDN is indicated by the mechanism that each node in HLDN will select a substitutive peer after a period of replacement (Sect. 4.3).

After constructing DDN, the Bitcoin network also has to implement our new construction of blocks. The IP addresses of the nodes in HLDN will form a merkle tree in the address part of the new blocks and will be stored in the leaf nodes of the merkle tree. Miners obtain a merkle root by hash computing and store the root in the block header. Thus, the IP addresses of the nodes in HLDN will influence the hash computation of the block header. In addition, miners cannot mine the block without the IP addresses of the nodes in HLDN. In particular, The new block structure is shown in Fig. 3.

Note that the new block structure can be implemented without modifying the block structure of the existing blockchain. Since existing blocks are not influenced by the later blocks, the blocks with the new structure can still store the hash

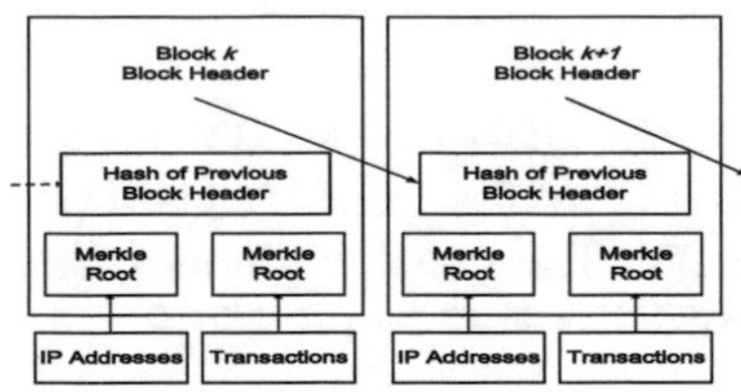

Fig. 3. The new block structure.

headers of the existing blocks. Therefore, the Bitcoin network can implement the new block structure whenever the need arises.

Because of the new block structure, each node that receives a new block can get the IP addresses of the nodes in HLDN from the table, which we call the *global table*. Further, since the nodes in HLDN only select the connected IP addresses to store in blocks and the IP addresses in the *global table* are replaced periodically, the *global table* in the latest block provides more reliable IP addresses than the *tried table* and *new table* of each node.

4.2 Outgoing Connection Maintenance Module

Since all the nodes are logically divided into two parts in DDN: the nodes in HLDN and the nodes in LLDN. We design the rules of determining the outgoing connections for each node in DDN to ensure the performance and the defensive capability of DNC.

There are two rules of constructing outgoing connections for each node in DDN: the first is that two outgoing connections have to connect to the IP addresses in the latest *global table*, the second is that the remaining outgoing connections have to connect to the IP addresses which are not in the latest *global table*. In other words, we limit the number of the outgoing connections for each node to connect to HLDN is two. Under the DNC protocol, each node has to maintain the rules when the node receives blocks containing the updated *global table*, when the node disconnects from other peers, and when the node connects to others.

From the above rules, all the outgoing connections are required to connect to IP addresses. This makes DDN more compact, makes messages spread more quickly and increases the difficulty of attacks on transmitting messages. Now, we describe the details of the outgoing connection maintenance module. Let T_G be the set of IP addresses in the *global table* of the latest block. Specifically,

(1) For a node A_k in DDN, if $|outaddr(A_k) \cap T_G|$ is greater than 2, A_k will randomly select an IP address from $(outaddr(A_k) \cap T_G)$ and disconnect from it until $|outaddr(A_k) \cap T_G|$ is equal to 2.

(2) If $|outaddr(A_k) \cap T_G|$ is less than 2, A_k should connect to a random IP address from $(T_G - outaddr(A_k))$ by an unoccupied outgoing connection until $|outaddr(A_k) \cap T_G|$ is equal to 2. If A_k does not have any unoccupied outgoing connection, A_k will randomly select an IP address from $(outaddr(A_k) - T_G)$ and disconnect from that IP address to get an unoccupied outgoing connection.

(3) After the above steps, A_k has satisfied the requirement that $|outaddr(A_k) \cap T_G|$ is equal to two. If A_k still has unoccupied outgoing connections, A_k will randomly select IP addresses from $(tried - T_G - outaddr(A_k))$ with a bias towards addresses with fresher timestamps. Then, A_k will try to connect to them until there is no unoccupied outgoing connections.

4.3 Substitutive Peer Selection Module

Each node in HLDN will select a substitutive peer after a fixed replacement period, which is implemented by the substitutive peer selection module. Note that, if a node A_n in LLDN runs the module and announces the substitutive IP address, all the nodes in the Bitcoin network can check the *global table* to realize that A_n is not the member of HLDN and ignore its replacement message. We give the details of the module below.

The replacement period is counted by the number of blocks. For example, if the period is counted by 10 blocks, every IP address of the nodes in HLDN will be stored in the *global table* for 10 blocks. Then, the IP address in the *global table* will be replaced by the substitutive IP address selected by itself. Each node in LLDN has approximately the same probability to be selected as the member of HLDN (the node that has more outgoing connections can have a higher probability, since more nodes may select it as the substitutive IP address). Specifically,

(1) When the replacement period starts, each node in HLDN will obtain an unoccupied outgoing connection first. For a node A_k in HLDN, If A_k does not have any unoccupied outgoing connection, A_k will randomly select an occupied outgoing connection connecting to LLDN and disconnect it.

(2) Then each node in HLDN will try to randomly select a substitutive IP address from its incoming connections. The substitutive IP address I_i of A_k should satisfy the following requirements: (i) $I_i \notin T_G$; (ii) $I_i \notin outaddr(k)$; (iii) I_i has not been chosen by other nodes in HLDN. A_k uses an unoccupied outgoing connection to connect to I_i. If they establish the connection successfully, A_k will spread the message in the whole network to announce that I_i will be the substitutive IP address of A_k. Then, I_i will be stored in the *global table* of the subsequent blocks.

(3) If the current incoming connections of A_k do not satisfy the requirements or A_k fails too many times to connect to the IP address from $inaddr(A_k)$, A_k will randomly select an appropriate IP address from $tried(A_k)$. After establishing the connection successfully, A_k announces the substitutive IP address.

In the substitutive peer selection module, DNC requires that each node in HLDN has to connect to the substitutive peer with the outgoing connection. There are two reasons for this design. First, two nodes can reach a consensus on the longest blockchain by establishing a connection and exchanging the version messages, which prevents that the substitutive peer works on a shorter fork. Second, the DNC protocol can preferentially guarantees that the nodes in HLDN will have the incoming connections with a higher probability. Since a node in HLDN can directly send an entire message to its peers, it will transmit messages faster than the nodes in LLDN. Without the incoming connections, the nodes in HLDN cannot receive messages or spread the messages from the peers, which wastes the transmittability of HLDN.

We compare the difference between the current Bitcoin network and that under the DNC protocol. The Bitcoin network at present requires nodes to establish long-range connections. Therefore, the construction of the whole network is static. Our protocol requires nodes to reconnect to different peers periodically by the outgoing connection maintenance module and the substitutive peer selection module. Therefore, the network's construction is dynamic. Since the nodes in DDN connect to the peers randomly and frequently, the distribution of the nodes in DDN will be more balance than the present distribution. In this way, the DNC protocol can improve the security of the Bitcoin network.

5 Defensive Capability Analysis

5.1 Selfish Mining

We first analyze the defensive capability of DNC against *selfish mining.*

Theorem 1. *Legitimate nodes will detect the number of selfish miners' blocks periodically under the DNC protocol.*

Proof. Because of the replacement of nodes in HLDN and the maintenance of outgoing connections, nodes will disconnect from peers and connect to other peers periodically and randomly. If *selfish miners* are connected to other legitimate nodes or are connected by others, they will exchange the version messages containing the length of their blockchains to establish the connection. Then the hidden blocks of *selfish miners* will be exposed to the legitimate nodes, followed by, legitimate nodes can detect the number of selfish miners' blocks periodically.

Because of the same reasons mentioned above, the legitimate nodes whose outgoing connections are connected to *selfish miners* will disconnect from *selfish miners* gradually. If *selfish miners* refuse to establish incoming connections, the number of their incoming connections will be decreased as time goes on. Eventually, *selfish miners* will have no incoming connections.

Then, we consider a worse case that adversaries can forge the version messages to conceal the length of the hidden blockchain and establish connections with legitimate nodes successfully. There is another countermeasure to defend against *selfish mining* as shown in the following theorem.

Theorem 2. *Under the DNC protocol, the mining advantage of selfish miners will be reset when the replacement period starts.*

Proof. When the replacement period starts, the nodes of HLDN will randomly choose the substitutive nodes. All the miners have to wait for the replacement messages of the nodes in HLDN. Adversaries cannot forge the replacement messages or forecast the substitutive IP addresses of all the nodes in HLDN. Without the correct IP addresses, *selfish miners* cannot construct the correct merkle tree, nor calculate the hash of the block header. Thus, *selfish miners* cannot selfishly mine the block containing the updated *global table*. When all the replacement messages are spread in the Bitcoin network, both legitimate miners and *selfish miners* start to mine the subsequent block at the same time. Therefore, the mining advantage of *selfish miners* accumulated before will be reset.

Since *selfish miners* cannot mine the subsequent block and it is unnecessary to wait for the legitimate miners to catch up. If *selfish miners* have hidden blocks and their hidden blockchain is coming up to the block which is going to update the *global table*, *selfish miners* will spread the hidden blocks to get rewards.

By adjusting the number of blocks that each fixed replacement period has, the DNC protocol limits the maximal length of the hidden blockchain. The optimal defensive capability against *selfish mining* is to set the period counted by one block. In other words, the Bitcoin network will update the *global table* when a miner mines a block. In practice, the community of Bitcoin can set an appropriate number of the blocks that one replacement period has. Considering the computing power that adversaries may have and the probability of mining a block by adversaries, the periodic time can be set loosely.

5.2 Eclipse Attack

As we discussed in Sect. 3.2, there are two main vulnerabilities of the Bitcoin network that are utilized by the *eclipse attack*. In DNC, we provide novel ways to patch these two vulnerabilities.

For the first vulnerability, we explain how DNC lowers the success rate that adversaries can monopolize all the outgoing connections of victims.

For each node under the DNC protocol, two outgoing connections connect to HLDN and six outgoing connections connect to LLDN (it is the default setting that each node has eight outgoing connections). When a victim restarts, it will randomly select two IP addresses from the *global table* and randomly select six IP addresses from the *tried table* to connect.

Let H_a be the number of adversaries' nodes in HLDN, and H_n be the number of legitimate nodes in HLDN. Let $P_h(H_a, H_n)$ be the probability that both two outgoing connections connect to the adversaries' nodes in HLDN. Then, we have

$$P_h(H_a, H_n) = \left\{ \frac{C_{H_a}^2}{C_{H_a+H_n}^2} \right\}. \tag{1}$$

Let r be the number of IP addresses that have been rejected so far, and t be the difference between the timestamp of the IP address and the current time.

$p(r,t)$ denotes the function counting the probability that a node uses this IP address rather than reject it. Then, we have

$$p(r,t) = min\left\{1, \frac{1.2^r}{1+t}\right\}. \tag{2}$$

Let t_a be the difference between the timestamp of adversaries' IP addresses and the current time, and t_n be the difference between the timestamp of legitimate IP addresses and the current time. The adversaries' IP addresses occupy α fraction of all the IP addresses in the *tried table*. $F(i,\alpha,t_a,t_n)$ denotes the function counting the probability that the ith node rejects an IP address. Then, we have

$$F(i,\alpha,t_a,t_n) = [1 - p(i-1,t_a)] \cdot \alpha + [1 - p(i-1,t_n)] \cdot (1-\alpha). \tag{3}$$

Let $P_l(k,\alpha,t_a,t_n)$ be the probability that a node rejects an IP address $(k-1)^{th}$ times and at the kth time the node connects to the IP address of adversaries in LLDN. We have

$$P_l(k,\alpha,t_a,t_n) = \alpha \cdot p(k-1,t_a) \cdot \prod_{i=1}^{k-1} \cdot F(i,\alpha,t_a,t_n). \tag{4}$$

$P_m(H_a,H_n,k,\alpha,t_a,t_n)$ denotes the probability that adversaries monopolize all the outgoing connections of a victim. We have

$$P_m(H_a,H_n,k,\alpha,t_a,t_n) = P_h(H_a,H_n) \cdot \prod_{i=1}^{6} P_l(k_i,\alpha,t_a,t_n). \tag{5}$$

We can learn from $P_m(H_a,H_n,k,\alpha,t_a,t_n)$ that adversaries can improve $P_l(k,\alpha,t_a,t_n)$ with the *eclipse attack*. However, it is difficult for adversaries to improve $P_h(H_a,H_n)$. Let P_{Ha} be the proportion of the number of adversaries' nodes in HLDN to the number of all the nodes in HLDN. P_{Aa} denotes the proportion of the number of adversaries' nodes to the number of nodes in the network. Since each node in HLDN selects the substitutive IP address randomly, we conjecture that P_{Ha} should be nearly equal to P_{Aa}.

The growth rate of $P_h(H_a,H_n)$ is pretty low. Even if P_{Aa} is 40%, the expected value of $P_h(H_a,H_n)$ is only 16%. Considering the success rate of monopolizing six outgoing connections that connect to LLDN, the expected value of the final success rate P_m is lower than 16%. Therefore, our protocol significantly lowers the success rate of monopolizing a victim's all the outgoing connections.

We also do simulated experiments to verify the defensive capability of the DNC protocol against monopolizing all the eight outgoing connections by the *eclipse attack*. In the simulation, the number of nodes in the Bitcoin network is 12,288. Each node maintains its *tried table* that can store 4,096 IP addresses at most. We use different fractions of nodes owned by adversaries to attack legitimate nodes and test the success rate of monopolizing all the eight outgoing connections when the timestamps difference of adversaries' nodes and legitimate nodes is 2 h. In the simulation, *num* is the number of nodes owned by adversaries. Let $P_{AinTried}$ be the average proportion of victims' *tried tables* that occupied by adversaries. In addition, $P_{success}$ is the success rate that adversaries monopolize all the outgoing connections of victims.

The results are shown in Fig. 4 and Table 1, which confirm that DNC can significantly lower the success rate of monopolizing all the eight outgoing connections of a victim. Even if adversaries own 40.625% of all the nodes in the Bitcoin network, their success rate of monopolizing all the eight outgoing connections of a victim is less than 16.5%. We can further learn from Table 1 that although adversaries can still occupy the most items in the *tried table* of a victim, it is much harder for them to increase the number of their IP addresses in the *global table*.

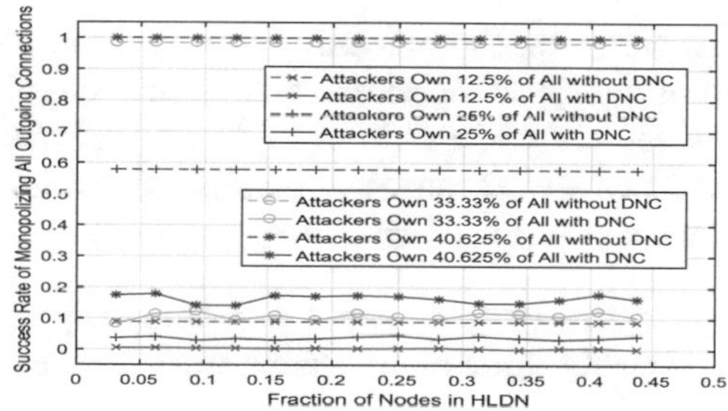

Fig. 4. The success rate of monopolizing all the eight outgoing connections of a victim by the *eclipse attack* with and without the DNC protocol.

Because each node in HLDN randomly selects the substitutive IP address, adversaries' nodes in HLDN should be only influenced by the total number of adversaries' nodes. We can learn from Table 1 that P_{Ha} is nearly equal to P_{Aa}. We can also find that P_{HtoAll} does not affect P_{Ha}. These results confirm our conjecture.

Table 1 also indicates that an effective way for adversaries to improve $P_h(H_a, H_n)$ is to add more nodes to the network, which requires more resources. However, if adversaries put lots of computing resources into the Bitcoin network, they can earn more coins even by following the rules of Bitcoin. If adversaries attack the Bitcoin network, Bitcoin may lose the trust of users. Then, coins may lose their value. As a result, adversaries may waste their computing resources. Thus, the more resources adversaries put into Bitcoin, the less likely they will attack the Bitcoin network.

The above analysis shows that DNC can significantly lower the success rate of monopolizing all the eight outgoing connections. Next, we consider the worst case that all the outgoing connections of a victim have connected to the IP addresses of adversaries. We have the following lemma and theorem to show that the DNC protocol can help victims detect the *eclipse attack*.

Table 1. Simulation results of the eclipse attack.

P_{Aa}	num	P_{HtoAll}	P_{Ha}	$P_{AinTried}$	$P_{success}$	P_{Aa}	num	P_{HtoAll}	P_{Ha}	$P_{AinTried}$	$P_{success}$
12.5%	1,536	0		37.49%	8.82%	25%	3,072	0		74.99%	57.75%
12.5%	1,536	4/32	12.53%	38.18%	0.36%	25%	3,072	4/32	25.01%	75.85%	3.37%
12.5%	1,536	8/32	12.44%	38.65%	0.36%	25%	3,072	8/32	24.93%	76.04%	4.32%
12.5%	1,536	12/32	12.99%	40.47%	0.36%	25%	3,072	12/32	25.49%	77.37%	3.25%
33.3%	4,096	0		99.00%	98.39%	40.625%	4,992	0		100%	100%
33.3%	4,096	4/32	33.24%	99.59%	9.38%	40.625%	4,992	4/32	40.50%	99.94%	14.06%
33.3%	4,096	8/32	33.16%	99.71%	10.33%	40.625%	4,992	8/32	40.46%	99.95%	15.99%
33.3%	4,096	12/32	33.81%	99.77%	10.70%	40.625%	4,992	12/32	41.07%	99.97%	15.62%

Lemma 1. *If an adversary monopolizes all the incoming connections of a victim, the adversary cannot transmit blocks to the victim after the global table updates.*

Proof. If a victim receives a block containing the updated *global table*, in order to follow the rules of outgoing connections, the victim may disconnect from the IP addresses of the adversary and randomly select IP addresses from the *global table* or *tried table* to reconnect. Thus, it is possible that this victim connects to legitimate nodes. As a conclusion, adversaries cannot transmit blocks to victims.

Lemma 1 implies that an adversary cannot continuously utilize the computing power of a victim to selfishly mine or implement other attacks.

Theorem 3. *An adversary cannot monopolize the outgoing connections of a victim for a long time under the DNC protocol.*

Proof. Consider whether an adversary can monopolize all the incoming connections of a victim:

(i) If not, the victim can receive blocks and replacement messages from the Bitcoin network under the DNC protocol. Then, there is a certain probability that the victim's outgoing connections disconnect from the IP addresses of the adversary and connect to legitimate nodes.
(ii) If yes, according to Lemma 1, an adversary cannot transmit blocks to a victim. When the adversary do not transmit blocks, it is easy for the victim to figure out the abnormal condition. Since the victim cannot increase the incoming connections of different IP addresses on its own, it will change the IP addresses of outgoing connections. The victim can obtain IP addresses from the *global table* in the previous blocks or ask *DNS seeds* for IP addresses and connect to them.

We can learn from Theorem 3 that there is a dilemmatic circumstance when an adversary monopolize all the incoming connections of a victim. No matter the adversary sends the block with the updated *global table* or not, the victim will try to reconnect to different IP addresses with the outgoing connections. In other words, the victim can recover from the condition that all its outgoing connections are monopolized by the adversary under the DNC protocol.

5.3 Information Delaying Attack

DNC stipulates that nodes of HLDN can spread an entire message without spreading *inv* message first. Only if there is no outgoing connections of the legitimate nodes in HLDN connecting to victims, can adversaries delay the delivery of blocks and transactions successfully. Therefore, DNC lowers the success rate of the *information delaying attack*. Let $P_{dl}(H_n, n_L)$ be the success possibility of the *information delaying attack* to nodes in LLDN. Then, we have

$$P_{dl}(H_n, n_L) = \left\{ \frac{C^6_{n_L - 1}}{C^6_{n_L}} \right\}^{H_n}. \tag{6}$$

$P_{dh}(H_n, H_a)$ denotes the success possibility of the *information delaying attack* to the nodes in HLDN. Then, we have

$$P_{dh}(H_n, H_a) = \left\{ \frac{C^2_{H_n + H_a - 1}}{C^2_{H_n + H_a}} \right\}^{H_n}. \tag{7}$$

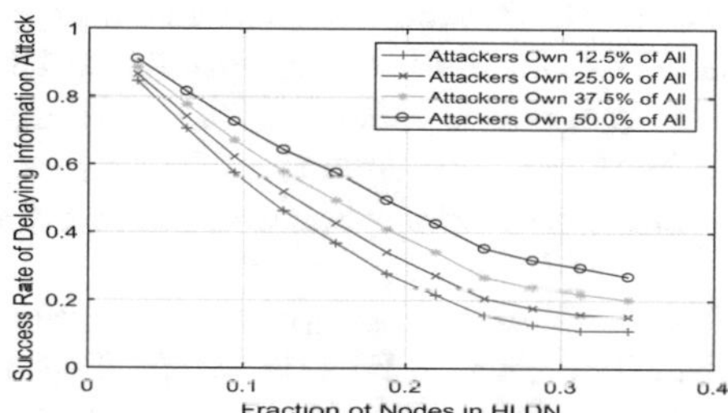

Fig. 5. The success rate of the *information delaying attack* is influenced by P_{HtoAll} and the number of nodes which are owned by adversaries.

We perform a simulation to evaluate the influence of DNC on the success rate of *information delaying attack*. The result is shown in Fig. 5, from which we can learn the following observation: if there are more nodes in HLDN, the success rate of the *information delaying attack* decreases. This is because there are more legitimate nodes in HLDN to deliver messages without sending *inv* messages. We also figure out that if adversaries own more nodes, P_{Ha} will increase. Then, there will be less legitimate nodes in HLDN.

6 Related Work

A number of works analyzed the security of Bitcoin [7,8,10,11]. Eyal and Sirer showed that if a mining pool withholds blocks on purpose, it can earn rewards in exceed of it is supposed to get [5]. These studies indicate that Bitcoin's countermeasures are not very effective to defend against *selfish mining*.

Karame et al. indicated that the measures from Bitcoin developers are not always effective in defending against double-spending [9]. There are some works [2,12–14] showed that *eclipse attack* can isolate the victims from other peers in a peer-to-peer network. Gervais et al. showed that adversaries can delay specific messages delivery to victims [3]. These demonstrate the importance of timely transmitting information in Bitcoin, which is a motivation of our research.

Ruffing et al. introduced a cryptographic primitive for preventing double-spending [4]. Schrijvers et al. introduced a game-theoretic model for reward functions of mining [16]. Bag et al. presented a new scheme for the PoW mechanism [15]. Different from the above researches, our study focused on improving the structure of the Bitcoin network. Our protocol is designed to enhance the security of transmission and to effectively defend against multiple attacks.

7 Conclusion

In this paper, we introduced a novel defensive protocol called Dynamic Network Configuration (DNC) to defend against *selfish mining*, the *eclipse attack* and the *information delaying attack* for the Bitcoin network. At first, we analyzed the vulnerabilities of the present Bitcoin network utilized by the aforementioned attacks. Based on our analysis, we presented the DNC protocol and then analyzed its defensive capability. From theoretical analysis and simulated experiments, we made the conclusions that the DNC protocol is effective against *selfish miners*, *eclipse attack* and *information delaying attack*.

Acknowledgement. This work was partly supported by NSFC under No. 61772466, the Zhejiang Provincial Natural Science Foundation for Distinguished Young Scholars under No. R19F020013, the Provincial Key Research and Development Program of Zhejiang, China under No. 2017C01055, the Fundamental Research Funds for the Central Universities, and the Alibaba-ZJU Joint Research Institute of Frontier Technologies. Technology Project of State Grid Zhejiang Electric Power co. LTD under NO. 5211HZ17000J.

References

1. Nakamoto, S.: Bitcoin: a peer-to-peer electronic cash system (2008)
2. Heilman, E., Kendler, A., Zohar, A., et al.: Eclipse attacks on bitcoin's peer-to-peer network. In: USENIX, pp. 129–144 (2015)
3. Gervais, A., Ritzdorf, H., Karame, G.O., et al.: Tampering with the delivery of blocks and transactions in bitcoin. In: CCS, pp. 692–705 (2015)
4. Ruffing, T., Kate, A., Schröder, D.: Liar, liar, coins on fire!: penalizing equivocation by loss of bitcoins. In: CCS, pp. 219–230 (2015)
5. Eyal, I., Sirer, E.G.: Majority is not enough: bitcoin mining is vulnerable. In: Financial Cryptography, pp. 436–454 (2014)
6. Bitcoin Developer Guide. https://bitcoin.org/en/developer-guide
7. Decker, C., Wattenhofer, R.: Information propagation in the bitcoin network (2013)
8. Bonneau, J., Miller, A., Clark, J., et al.: Sok: research perspectives and challenges for bitcoin and cryptocurrencies. In: S&P, pp. 104–121 (2015)

9. Karame, G.O., Androulaki, E., Capkun, S.: Double-spending fast payments in bitcoin. In: CCS, pp. 906–917 (2012)
10. Courtois, N.T., Bahack, L.: On subversive miner strategies and block withholding attack in bitcoin digital currency. arXiv preprint arXiv:1402.1718 (2014)
11. Gervais, A., Karame, G., Capkun, S., et al.: Is bitcoin a decentralized currency? S&P **12**(3), 54–60 (2014)
12. Singh, A., Ngan, T.W., Druschel, P., et al.: Eclipse attacks on overlay networks: threats and defenses. In: INFOCOM, pp. 1–12 (2006)
13. Sit, E., Morris, R.: Security considerations for peer-to-peer distributed hash tables. In: IPTPS, vol. 2429, pp. 261–269 (2002)
14. Castro, M., Druschel, P., Ganesh, A., et al.: Secure routing for structured peer-to-peer overlay networks. OSDI **36**(SI), 299–314 (2002)
15. Bag, S., Ruj, S., Sakurai, K.: Bitcoin block withholding attack: analysis and mitigation. TIFS **12**(8), 1967–1978 (2017)
16. Schrijvers, O., Bonneau, J., Dan, B., et al.: Incentive compatibility of bitcoin mining pool reward functions. In: Financial Cryptography, pp. 477–498 (2016)

Survival Model for WiFi Usage Forecasting in National Formosa University

Jutarat Kositnitikul and Ji-Han Jiang(✉)

Department of Computer Science and Information Engineering, National Formosa University, Huwei, Taiwan
tanyong250438@gmail.com, jhjiang@nfu.edu.tw

Abstract. This paper presents the effectiveness of adopting survival analysis approach to predict the WiFi usage in the future and the understanding of covariance affect WiFi usage such as date, time, and user, by introducing dataset of WiFi usage historical. The study took place in National Formosa University in Taiwan. Survival analysis is the analysis of data involving times to event of interest. There are three survival analysis methods implemented in this paper which are Kaplan-Meier estimator, Cox Proportional Hazards Model, and Random Survival Forest. The result was shown that survival analysis approach gains a satisfy prediction result. This approach can be adapted for improving WiFi network organization in any organization by understanding the connection of covariance and accomplishing an effective decision.

Keywords: WiFi prediction · Survival analysis · Kaplan Meier survival · Cox Proportional Hazards · Random Survival Forest

1 Introduction

In this day and age, digital technology influences our daily life. The Internet may be combined with other traditional basic needs such as food, shelter, clothing, and medicine because the internet made an individual's life very simple and more productive. The easiest way to connect to the internet is using a WiFi network. WiFi is the name of a wireless networking technology that provides high-speed internet and network connection. It provides any person to connect to the network anytime and anywhere [1]. WiFi provides service in private homes, businesses, as well as in public places for example, in the college. There is a WiFi network connectivity that is easily accessible by all students and staffs providing internet services for free and helping students in studies by allowing them to search and study on the internet. Moreover, the survey presents 40% of respondents chose Wi-Fi as their number one daily essential [2].

In the interest of the high amount of WiFi users nowadays, WiFi data visualization will become more beneficial in order to improve user experience and support effectively WiFi network managing in an organization. Thus the purpose is to study on the WiFi usage data and present a WiFi usage predictive model for determining the possible factors influence WiFi usage. The dataset collected from access points in the college which takes place in all building at National Formosa University in Taiwan, e.g. classroom, laboratory, library, gymnasium, dormitory, and other places.

C.-N. Yang et al. (Eds.): SICBS 2018, AISC 895, pp. 414–422, 2020.
https://doi.org/10.1007/978-3-030-16946-6_32

In the era of data analysis, survival analysis is provided time-to-event approach [3]. The goal is to predict the time an event occurs and shows the relationship of survival time result to explain the elements and correlative of elements in historical data [4]. For example, the event of interest is a heart attack, then the patients are observed over the study time period. When they have a heart attack or have nothing happen until the end of the study is recorded, and there are many factors can be considered, such as gender, treatment, age, and other factors. From this historical data, we can predict time to event for a new patient.

In general, there are two points of the observation period consist of a beginning point and an ending point [5]. Time can be evaluated in days, weeks, or years. And also the event can be death, the occurrence of a disease, divorce, drop off from the school, got a benefit from stock, and other events. Originally, survival analysis is commonly used in medical and biological sciences, then become more widely used in other fields such as engineering, economic, education, etc., but no specific in WiFi prediction. In addition, survival analysis has many different regression modeling strategies. The contribution of the research is to present the use of three different survival models implemented with WiFi usage historical data, which is consist of Kaplan-Meier estimator, Cox Proportional Hazards Model, and Random Survival Forest.

2 Methodology

In this section, the research methods implemented for forecasting the survival rate of WiFi usage are presented. The three methods consist of Kaplan Meier Estimator, Cox Proportional Hazards model, and Random Forest Tree model, which are provided by R package.

2.1 Kaplan Meier Estimator Model (KM)

A very widely used method of estimation is calculating and plotting a Kaplan–Meier curve which is a non-parametric method of estimating the survival function. The non-parametric method is a simple method which does not make any distributional hypothesis. Moreover, it is very useful for summarizing survival data and making simple comparisons. Nonetheless, the non-parametric method cannot deal with more complex situations and difficult to interpret that yields inaccurate estimates.

The KM estimator of survival function is represented as follows;

$$\widehat{S}(t) = \Pi_{j:t_j \leq t} \frac{e_j - m_j}{e_j} \tag{1}$$

Where m_j is the number of variables recognizes the event at time t_j and e_j is the number of variables that have not qualified the event and also the risk at time t_j [6].

2.2 Cox Proportional Hazards Model (Cox PH)

Cox proportional hazards regression model is a popular regression model for survival analysis. It is a semi-parametric method for considering the effect of variables to the time of the specified event takes to happen. In Particular, more than one variables can be combined in a Cox proportional hazards model to modify for the effects of each variable. The variables are given in hazard form as follows;

$$h(t; x_1, \ldots, x_p) = h_0(t)exp(\beta_1 X_1 + \beta_2 X_2 + \ldots + \beta_p X_p) \tag{2}$$

Where $exp(\beta_p)$ is Hazard ratio. The formula shows the event rate at time t conditional on survival until time t or later. Where h(t) is the expected hazard at time t, h0 (t) is the baseline hazard and represents the hazard when all of the independent variables (X1, X2, Xp) are equal to zero [7]. Hazard ratio equal 1 suggests lack of association, a hazard ratio greater than 1 means an increased risk, and a hazard ratio lower than 1 suggests a smaller risk.

2.3 Random Survival Forest Model (RSF)

A random forest is a machine learning that can forecast the risk of a nonparametric variable in survival analysis. The predictor is calculated by combining the results of survival trees. There are several general strategy steps. Step 1, draw G bootstrap samples. Step 2, expand a survival tree based on data of each bootstrap samples g = 1,…, G and consider some following rules: (a) At each tree node selects a subset of the predictor variables. (b) Among all binary splits defined by the predictor variables selected in (a), find the best split into two subsets (the daughter nodes) according to find a suitable benchmark for right censored data. (c) Repeat (a)–(b) circularly on each daughter node until a stopping benchmark is met. Step 3, combine information from the terminal nodes, which is node has no further split, from the G survival trees to gather a risk prediction set [8].

In this paper consider the implementation of random forest trees from R package, which is randomSurvivalForest [8].

Random survival forest is the set of results that constructed by aggregating tree-based Nelson-Aalen estimators. The ensemble survival function from random survival forest is represented as follow:

$$\widehat{S}^{rsf}(t|x) = exp\left(-\frac{1}{G}\sum_{g=1}^{G} \widehat{H}_g(t|x)\right) \tag{3}$$

In each terminal node of a tree, hazard function is estimated using the Nelson-Aalen.

$$\widehat{H}_g(t|x) = \int_0^t \frac{\underline{N}_g(ds, x)}{\underline{Y}_g^*(s, x)} \tag{4}$$

$$\underline{N_i}(s) = \upsilon(\underline{T_i} \leq s, \Delta_i = 1); \underline{Y_i}(s) = \upsilon(\underline{T_i} > s) \tag{5}$$

let υ_g note that the gth survival tree and let $\upsilon_g(x)$ denote the terminal node of subjects in the gth bootstrap sample with predictor x. $\underline{T_i}$ is the minimum of the survival time T_i, and Δ_i is the status (censoring) value indicating a WiFi used ($\Delta_i = 1$) or was right-censored ($\Delta_i = 1$) [9].

2.4 Data

The dataset consists of one-month history of WiFi networks usage in buildings at the National Formosa University in Taiwan such as classroom, laboratory, library, gymnasium, dormitory, and other places between April 20, 2018, and March 21, 2018. The data includes day, month, time (hour), user, and location of each WiFi login event. Each row represents for each hour in each date and collected in a different location. A total of data is 33,792 observations, where 8477 observations show the usage WiFi event. A status is considered to be censored if that place has people using the WiFi at that time.

The survival is observed every hour of each day. Besides, the event of WiFi usage at less once per hour is represented as 1 (event occurrence in survival model), and 0 represents the event not being observed in that hour. The variables below are included in the analysis to inspect their effect with survival possibility.

- User: student and other
- Location: 22 places in NFU
- Day: Monday, Tuesday, Wednesday, Thursday, Friday, Saturday, and Sunday
- Month: April, March

Note that the train data set includes 80% of total data (27,033 observations) collected randomly, and the remaining 20% is the test dataset (6,759 observations).

3 Evaluation Result

The dataset used in the analysis is presented in this section, and also the result of the evaluation. R software is used to prepare data and do the analysis [10]. In this part, the results of KM, Cox PH, and RSF are presented, following by methods comparisons and predictive performances.

3.1 Kaplan Meier Survival

The results from the KM estimator presents the median survival rate for data in each hour using the command **survfit** in Rstudio [11]. Note that the survival curve always falls down from survival rate equal to 1 until zero. The meaning is the more graph falls down, the more event occurs. The KM curve in Figs. 1 and 2 present how survival rates correlate with times, and covariates. Both of them show that the survival rate slightly dropped in the same length before 8 a.m., but more quickly jumped after that period. Hence, the result can be considered that from 0 a.m. until 8 a.m. the school may not be a popular place because the number of WiFi usage event slowly went up.

In addition, Fig. 1 shows the comparison between survival rates and days. For instance, the survival rate fell to a low on Wednesday consequently means that people mostly use WiFi on this day especially after 8 a.m. On the contrary, WiFi is rarely used on Saturday and Sunday. Similarly, Fig. 2 Shows WiFi usages in each location. The graph shows that the stadium has the lowest WiFi usage following by the swimming pool. Note that surv refers to survival rate, and time indicates time unit in an hour.

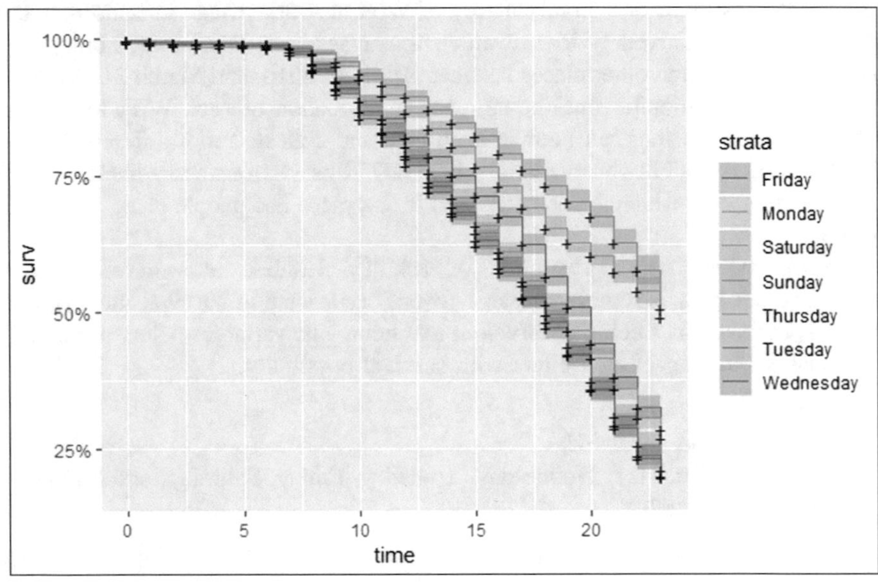

Fig. 1. Comparison of days using Kaplan Meier curve

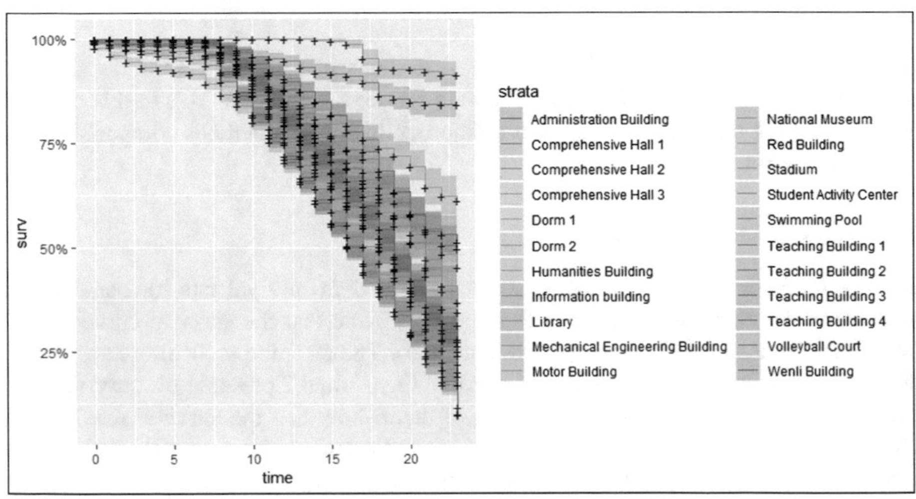

Fig. 2. Comparison of locations using the Kaplan Meier curve

3.2 Cox Proportional Hazards Model

Cox PH model presents the effect of all covariates on the survival rate of the dataset. The method was operated in R; an open source software using the command **coxph** in Rstudio [11]. Table 1 presents the top 10 hazard ratio from 28 results. A positive coefficient represents a worse prediction and a negative coefficient indicates a protective effect of the variable with which it is associated. In addition, the highest hazard ratio means that the variable has the largest effect on a survival rate. From the results, the coefficient starts from −3.22 to 0.93 and hazard ratio starts from 0.03 to 2.53. The hazard ratio greater than 1 means an increased risk, and a hazard ratio lower than 1 suggests a low effect on a WiFi usage event.

The highest value for the User variable represents the WiFi user, showing that students have high event hazard and low survival rate. The meaning is students use WiFi more than other users. In addition, a stadium from the location variable has the lowest hazard ratio which is only 0.03. Therefore, the high survival rate has a low effect on WiFi usage event. The result of both KM and Cox PH have the same figure according to how more or less meaningful of all variables effect to survival rate.

Table 1. The top 10 hazard ratio from Cox model

Element	Coefficient (coef)	Hazard ratios (exp(coef))
User: student	0.93	2.52
Day: Tuesday	0.25	1.28
Day: Wednesday	0.23	1.26
Day: Thursday	0.19	1.21
Day: Monday	0.10	1.10
Location: Comprehensive Hall 2	0.07	1.07
Location: Wenli Building	0.04	1.04
Location: Dorm 1	−0.04	0.95
Location: Library	−0.06	0.94
Location: Student Activity Center	−0.20	0.81
Location: Dorm 2	−0.52	0.59

3.3 Random Survival Forest

Results from fitted an RSF of 1000 survival trees built using the command **rfsrc** in the randomForestSRC package, Rstudio [12]. To identify the most impactful covariates in explaining the survival of WiFi usage in NFU, **vimp** command in R was used to find the measure of variable importance [13]. They indicate that location, user, and day are the most important covariates strongly associated with WiFi usage event as shown in Fig. 3 and also the error rate compared to the number of trees is presented.

The user that was strongly associated with WiFi usage in the presence of Cox PH show up among other covariates but do not appear to be the highest rank in RSF result. It can consume that excluding covariates in RSF cannot be compared to the Cox PH model due to the Cox PH is more specifically considering every elements.

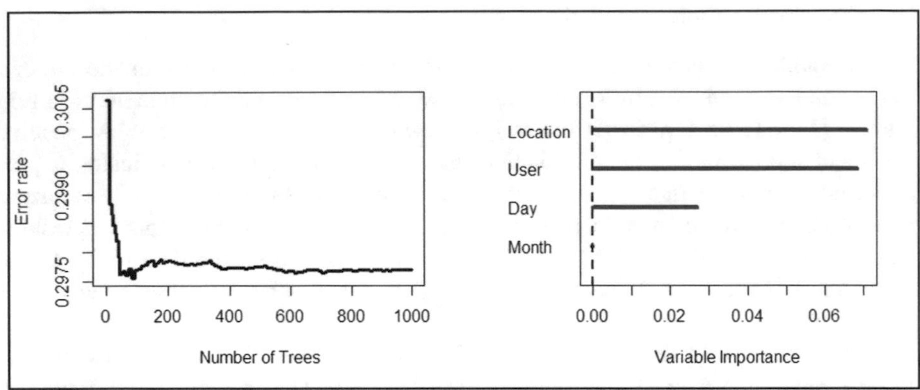

Fig. 3. The prediction error rate vs. number of trees and variable importance

On the other hand, random survival forest provides an importance score for each class of elements in order to present the general aspect of covariates. For the result, Location, User, Day, and Month were got 0.0714, 0.0682, 0.0271, and 0.0002 important scores respectively. Thus, if there is some difference in location data such as the location changed from Monday to Sunday, it will substantially affect the survival rate compared to other elements.

Table 2. Interaction of covariates in RSF

Pair of variable	Paired	Additive	Difference
1. Location: User	0.1611	0.1771	−0.0159
2. Location: Day	0.1272	0.1348	−0.0076
3. Location: Month	0.0987	0.0964	0.0023
4. User: Day	0.1230	0.1274	−0.0044
5. User: Month	0.0902	0.0890	0.0011
6. Day: Month	0.0461	0.0469	−0.0008

Table 2 shows the interaction between each variable. The most interesting value is Difference which compared paired and additive. If a Difference is a positive value it means pair is more powerful than staying alone. Note that RSF model computation take times and the number of variables has a limit.

3.4 Comparison and Predictive Performance

The comparison of the three survival curve is shown in Fig. 4. The **Surv** refers to survival rate, and **Time** unit is an hour. It is slightly different in the curves, thus three of these survival curve have similar results.

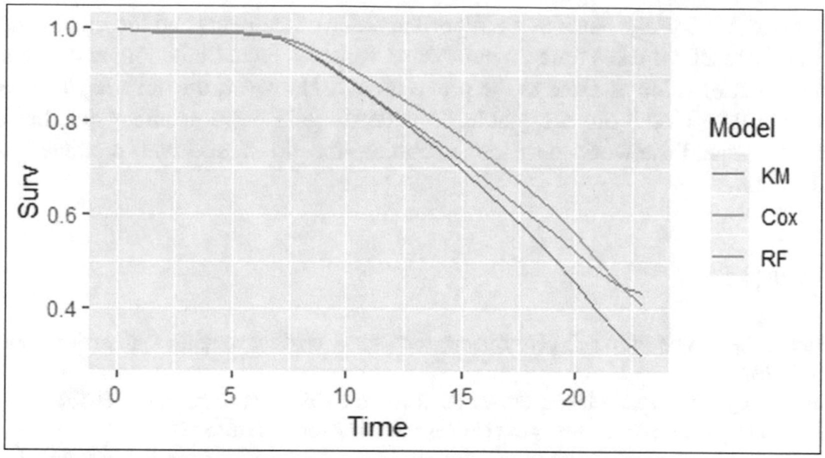

Fig. 4. Comparison of the survival curves

In practice, the KM estimator gives the probability of survival rate at a certain time point on a population level. So, it is impossible to get KM estimates for individual events. In order to get predicted survival probabilities needed to use some model (e.g. Cox PH, accelerated failure time model, neural network etc.). Package **pred** in R was used for this analysis. The error rate of Cox PH model is 29.74% and 30.86% for RSF model. The survival probability of Cox PH model starts from 0 to 1, and the boundary is 0.5. Probability greater than 0.36 were treated as 1 and the rest were treated as 0. The 72% accuracy on the test set is acceptable. On the contrary, the predicted values from RSF model is not in the same range with Cox PH results. The number starts at 0 to more than 10. Therefore, the cut off value decided to be 10 and gain more 73% accuracy. Note that for other experiments different thresholds could be a better option.

4 Conclusion

In this study, the predictive result and accuracy from three survival analysis models are presented based on consideration of the development of WiFi network organization. The survival curve of all this three method went down at the same rate. Nevertheless, the analysis result of this three methods is totally different.

First, the result from the Kaplan Meier survival method shows a superficial pattern of WiFi usage with survival curve. Although this method is easy to implement but difficult to interpret because of survival rate is given for population level at a certain time not individual. Thus, the survival rate for each event cannot be predicted. Second, Cox Proportional Hazards Model method mainly provide a correlation value between variables, that is be more understandable. On the other hand, the distribution of the outcome is unknown leads to be hard to interpret, and survival predicted value cannot assume whether the event will occur or not, so the suitable threshold needed. Third, the output of the Random Survival Forest gives more useful value to understand the condition of WiFi usage, such as an important value of elements and the interaction. By the way, this method takes times, a number of variables have limitation, and also needs a suitable cut of value as same as the previous one. However, the accuracy of Cox PH model and RSF model are acceptable. In order to gain more accuracy for the future work, the neural network may be a choice for do a survival predicted result classification.

References

1. Pan, D.: Analysis of Wi-Fi performance data for a Wi-Fi throughput prediction approach. KTH (2017)
2. iPass Corporate: iPass Mobile Professional Report 2016. iPass company (2016)
3. Kartsonaki, C.: Survival Analysis. University of Oxford, Oxford (2016)
4. Mills, M.: Introducing Survival and Event History Analysis. SAGE Publications, London (2011)
5. Kleinbaum, D.G.: Survival Analysis, a Self Learning Text. Springer, New York (1996)
6. Smith, T., Smith, B.: Kaplan Meier and Cox proportional hazards modeling: hands on survival analysis. SAS® Users Group International Proc. Seattle, Washington (2003)
7. Louzada, F., Cancho, V.G., Oliveira, M.R., Yiqi, B.: Modeling time to default on a personal loan portfolio in presence of disproportionate hazard rates. J. Stat. Appl. Pro. **3**(3), 295–305 (2014)
8. Ishwaran, H., Kogalur, U.B., Blackstone, E.H., Lauer, M.S.: Random survival forests. Cleveland Clinic, Columbia University (2018)
9. Mogensen, U.B., Ishwaran, H., Gerds, A.: Evaluating random forests for survival analysis using prediction error curves. Department of Biostatistics, University of Copenhagen (2012)
10. R Core Team: R: a language and environment for statistical computing. R Foundation for Statistical Computing, Vienna, Austria (2014)
11. Therneau, T.M., Lumley, T.: http://CRAN.R-project.org/package=survival (core) (2009)
12. Ishwaran, H., Kogalur, U.B.: http://CRAN.R-project.org/package=randomForestSRC (2014)
13. Ehrlinger, J.: ggRandomForests: Exploring Random Forest Survival. Microsoft (2016)

An Auditing Scheme for Cloud-Based Checkout Systems

Tao-Ku Chang(✉) and Cheng-Yen Lu

Department of Computer Science and Information Engineering, National Dong Hwa University, Hualien, Taiwan
tkchang@gms.ndhu.edu.tw

Abstract. The goal of this paper is to design and implement a security mechanism for cloud-based checkout systems based on chain-hashing scheme. Many cloud-based checkout systems are developed for merchants. However, storing transaction data in cloud storage is associated with serious security risks. The repudiation problem exists between merchants and service providers. We need a scheme that enables the service provider to prove its innocence and the merchant to prove its guilt. The proof of innocence is also called auditing. This paper designs a secure cloud-based checkout system. We use chain hashing to design auditing scheme for checkout systems.

Keywords: Cloud security · E-commerce · Auditing

1 Introduction

Cloud technologies are applied to many commercial application systems. Cloud computing has given birth to the checkout system as Software as a Service. Cloud-based checkout systems have many advantages over traditional checkout systems. Users do not need to construct computer facility, maintain and purchasing specific hardware. It can be accessed from the internet using internet browser wherever there is internet connection. The cloud-based checkout systems are independent of platform and operating system limitations. They are made compatible with a wide range of hardware. The user data including transactions and inventory information are stored in cloud storages and the system does not run locally as well so there is no requirement of installation.

However, there are substantial amount of existing security issues concerning storing business data or even commercial secrets in cloud storage: (1) Cloud service providers could lost data due to possible system setting errors, computing mistakes, or server crashes; (2) The service provider is likely to then restore the files using a backup of an early version of the files and their associated digital signatures, and can then deny the user's latest version of files lost. This is called a roll-back attack [1] or replay attack [2]; (3) Outside attackers could access or even tamper the data from the server illegally. Employees of cloud service providers could inadvertently allow the external attacks. Even though cloud service providers constructed a perfect security system, these types of security flaws are still inevitable. Therefore, the concern diminishes many people's

C.-N. Yang et al. (Eds.): SICBS 2018, AISC 895, pp. 423–437, 2020.
https://doi.org/10.1007/978-3-030-16946-6_33

willingness to store important data or even vital commercial secrets in the cloud. Also, none of today's cloud-based checkout services provide security guarantees in their service-level agreements (SLAs).

This research proposed a mutual auditing mechanism for cloud-based checkout systems. Client and service providers exchange attestations for every transactions. These attestations are chain hashed and will be used in auditing to ensure the accuracy of the transaction information in cloud storage. Once the transaction information uploaded to the cloud appears to be damaged or tampered, merchants and cloud service providers could clarify and prove the origin of the mistakes. It can guarantee mutual nonrepudiation between the clients and service providers.

The remainder of this paper is organized as follows: Sect. 2 gives an overview of related work and technologies, Sect. 3 presents the proposed mechanism, and conclusions about the work described in the paper are drawn in Sect. 4.

2 Related Works

We need a scheme that enables the service provider to prove its innocence and the users to prove its guilt. The proof of innocence is also called auditing [3]. To solve the mutual auditing issue between cloud service providers and users, some systems consider cloud storage untrusted and provide a way to detect violations of integrity, write-serializability, and freshness. SUNDR is a network file system designed to store data securely on untrusted servers [4]. SUNDR lets clients detect any attempt at unauthorized file modification by malicious server operators or users. SUNDR's protocol achieves a property called fork consistency, which guarantees that clients can detect any integrity or consistency failures as long as they see each other's file modifications. Clients maintain a version structure list which stores a list of version structure in the storage server. When user u performs a file system operation, u's client acquires the global lock and downloads the latest version structure for each user and group. SUNDR dealt with the violation detect problem using the so-called "forking" semantics [5] which is also employed in [6–8]. These solutions guarantee integrity, and by adding some extra out-of-band communication among the clients can also be used to achieve a related notion of consistency.

Popa et al. proposed a system called CloudProof that contains a protocol to detect and prove violations of four desirable security properties of cloud storage: confidentiality, integrity, write serializability, and read freshness (denoted as CIWF) [9]. Users can detect the violations of CIWF of the cloud storage system and the service provider can disprove false accusations made by users; that is, the users of CloudProof cannot frame the cloud. Its framework supports the mutual nonrepudiation guarantee. These proofs are based on attestations, which are signed messages that bind the users to the requests they make and the service provider to maintaining the data in a certain state. The protocol proposed in CloudProof is able to maintain the mutual nonrepudiation property of cloud storage. Hwang et al. also proposed C&L scheme [10] which does not require multiple client devices to exchange any messages and still can guarantee nonrepudiation and maintain the CIWF requirements. It also proposed how to apply the hash tree [11] to remove accumulated attestations.

3 The Proposed Auditing Scheme

The goal of this paper is to makes it possible to detect violations of security properties and guarantee mutual nonrepudiation between the merchants and service providers of the cloud-based checkout system. The client device and service provider exchange attestations for every transaction. These attestations are chain hashed so that the client device of the merchant only has to store the last attestation that contains the last chained hash. The service provider keeps all the attestations for use when proofs are required.

The proposed scheme involves the following entities: a merchant (M), and service provider (P), which have the following public and private key pairs: (pri(M), pub(M)) and (pri(P), pub(P)), respectively. The scheme comprises four phases: a registration phase, a login phase, a transaction phase, and an auditing and clearing phase. The symbol descriptions of the following figures are shown in Table 1.

Table 1. The symbol descriptions

Symbol	Description
M	The devices of a merchant
P	Cloud service provider
CompanyID	Merchant ID (account of a system manager)
CompanyName	Merchant name
CompanyPWD	Password of a merchant
pub(.)	Public key
StaffID	Staff's ID of a merchant
StaffPWD	Staff's Password of a merchant
AA	Administrative authority
OP	Operation (add, delete, modify, query)
Description	Other information
TID	Transaction ID
TD	Transaction data (Transaction detail)
LSN	Local sequence number
CH_i	Hash chain $CH_{i-1} = h(Q_{i-1}, CH_{i-2})_{pri(P)}, CH_0 = h(h(TID)_{pri(M)})_{pri(P)}$
TDHC	Hash chain of the transaction detail $TDHC_i = h(TDHC_{i-1}, h(TD_i)), i = 2, \ldots, n. TDHC_1 = h(h(TD_1))$
$h()_{pri(.)}$	Hash function
L_i	Inform message

3.1 Registration Phase

Merchant has to register an account and then registers employees' accounts. Figure 1 shows the following steps and message exchanges involved in the registration phase:

(1) M sends a register service request, RSR_i = {CompanyID, CompanyName, CompanyPWD, Phone, Email, pub(M)}, to P for registration.

(2) After registration, M sends a login service request, LSR_i = {CompanyID, CompanyPWD, [CompanyID, CompanyPWD]$_{pri(M)}$}, to P for login.
(3) After login, M adds the data {OP, CompanyID, StaffID, StaffPWD, AA, Description, [OP, CompanyID, StaffID, StaffPWD, AA, Description]$_{pri(M)}$} for employee's registration.

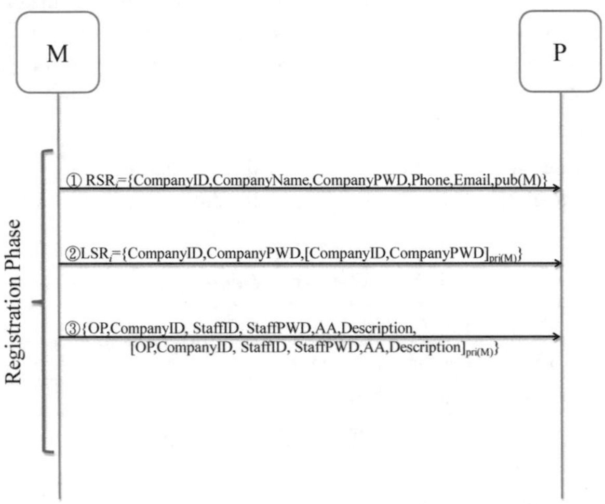

Fig. 1. The message exchange of the registration phase

3.2 Login Phase

Merchant has to login the checkout system and then employees login with his/her account and password before starting the checkout service. Figure 2 shows the steps and message exchanges in the login phase:

(1) M sends LSR_i, where LSR_i = {CompanyID, CompanyPWD, [CompanyID, CompanyPWD]$_{pri(M)}$}, to P for login.
(2) P receives LSR_i. It verify the signature and check whether the account and password are correct. When login process is successful, P sends {pub(P)} to M.
(3) M stores {pub(P)} for verification. Employee who has an account logins for checkout services. M sends $SLSR_i$ (Staff Login Service Request), where $SLSR_i$ = {StaffID, StaffPWD, [StaffID, StaffPWD]$_{pri(M)}$}, to P.
(4) P receives $SLSR_i$. It verifies the signature and check whether employee's account and password are correct. When login process is successful, P sends TDM_i, where TDM_i (Tansaction ID Delivery Message) = {TID}, to M. TID is a non-duplicate transaction ID. Its format is {StaffID + TIME}. For example, TID will be {A610201810120708} when the checkout system assigns a TID to an employee whose ID is "A610" at 07:08 p.m. on October 12, 2018.

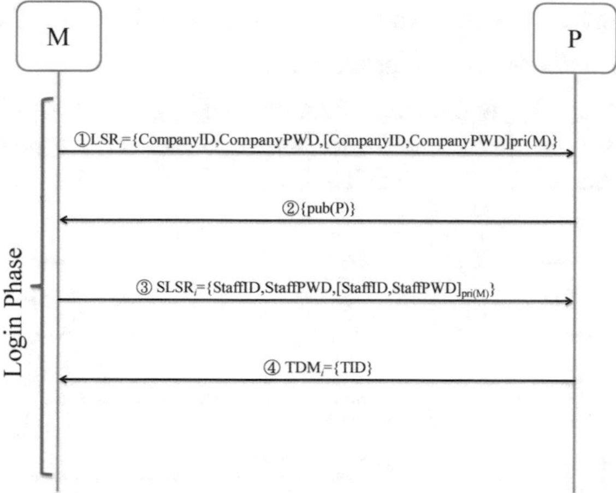

Fig. 2. The message exchange of the login phase

3.3 Transaction Phase

The following steps (depicted in Fig. 3) are involved when a customer intends to check out.

(1) When a customer checks out, the client device of M produces transaction data, TD and makes a TDHC (Transaction Data Hash Chain), where $\mathrm{TDHC}_i = \mathrm{h}(\mathrm{TDHC}_{i-1}, \mathrm{h}(\mathrm{TD}_i)), i = 2, \ldots, n$, $\mathrm{TDHC}_1 = \mathrm{h}(\mathrm{h}(\mathrm{TD}_1))$. TDHC contains each hash vale of TD_i and is stored on the client device for future auditing data stored on the service provider.

(2) The device of M sends a request message Q_i, where $\mathrm{Q}_i = \{\mathrm{TD}, \mathrm{TID}, \mathrm{LSN}, [\mathrm{TD}, \mathrm{TID}, \mathrm{LSN}]_{\mathrm{pri(M)}}\}$, to P. Local sequence number (LSN), which is an increasing number, is bound with TID and maintained by the merchant. TID and LSN enable to make sure that the legality of a sequence of messages between the merchant and the service provider in the chain-hashing.

(3) When P receives Q_i, it verifies if this is a valid request message by checking the digital signature and then checks if LSN is continuous. P sends a response message R_i, where $\mathrm{R}_i = \left\{\mathrm{Q}_i, \mathrm{h}(\mathrm{Q}_i, \mathrm{CH}_{i-1})_{\mathrm{pri(P)}}, \left[\mathrm{Q}_i, \mathrm{h}(\mathrm{Q}_i, \mathrm{CH}_{i-1})_{\mathrm{pri(P)}}\right]_{\mathrm{pri(P)}}\right\}$, to the device of M. Note that CH_{i-1} is $i-1^{\mathrm{th}}$ chain hash and $\mathrm{CH}_{i-1} = \mathrm{h}(\mathrm{Q}_{i-1}, \mathrm{CH}_{i-2})_{\mathrm{pri(P)}}$, $\mathrm{CH}_0 = \mathrm{h}(\mathrm{h}(\mathrm{TID})_{\mathrm{pri(M)}})_{\mathrm{pri(P)}}$.

(4) When the device of M receives R_i, it verifies if its signature is valid. Then it sends reply-response message RR_i to P, where $\mathrm{RR}_i = \left\{\mathrm{R}_i, [\mathrm{R}_i]_{\mathrm{pri(M)}}\right\}$. Note that the service provider keeps all RR_i for when proofs are required. These reply-response messages are not deleted until auditing and clearing phase is completed.

(5) When P receives message from M, it stores TD and RR_i to the database. Finally, it sends an acknowledgement message ACK_i, where $ACK_i = \left\{ RR_i, [RR_i]_{pri(P)} \right\}$, to M. M receives ACK_i and increments the value of LSN by 1. It indicates that this transaction process is completed. M just need to keep the newest ACK_i as the attestation and ACK_{i-1} will be deleted.

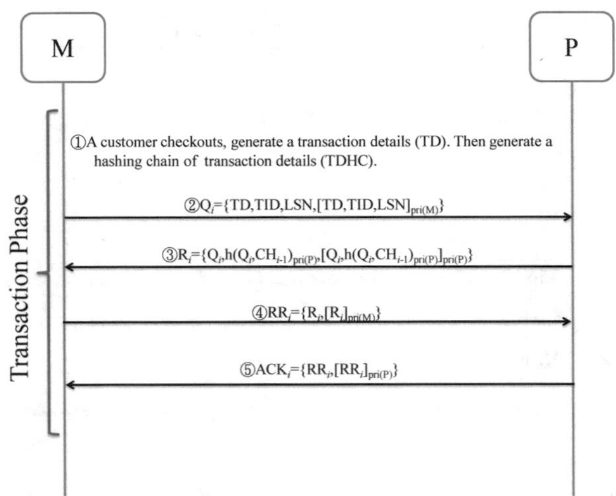

Fig. 3. The message exchange of the transaction phase

3.4 Auditing and Clearing Phase

Over a period of time, it is possible for a client device of the merchant to transfer a huge number of transaction data, which could result in an excessive overhead for service providers to store the server-side attestations or when an employee intends to sign out. We need a way to audit these attestations and discard them. The details of the message exchanges in the auditing and clearing phase illustrated in Fig. 4 are as follows:

(1) When an employee of M signs out or intends to clear attestations, the device of M sends ASR_i, where ASR_i = {TID, TDHC, [TID, TDHC]$_{pri(M)}$} to P for the request of auditing and clearing.
(2) P receives ASR_i from M. It verifies if the signature is valid. It calculates $TDHC'_i$ according to all transaction data that belong to the same TID in ASR_i. If $TDHC_i$ in ASR_i and $TDHC'_i$ are the same, P sends an acknowledgement message, {L_i}, to the client device of M.
(3) When the device of M receives the acknowledgement message, it replies a request of auditing server-side attestation and transaction ($ASAR_i$), where $ASAR_i = \left\{ \text{TID}, L_i, [\text{TID}, L_i]_{pri(M)} \right\}$, to P.
(4) When P receives $ASAR_i$ from M, it sends all of transaction data TD_s with the same TID and RR_s to M for auditing.

(5) M verifies all RR_s from P if the signature of each transaction is valid. If all RR_s are correct, M sends a request of clearing attestation (ACR_i), where $ACR_i = \{TID, L_i, [TID, L_i]_{pri(M)}\}$, to P.

(6) When P receives ACR_i, it verifies if the signature is valid. P deletes $\{RR_1, RR_2, \cdots, RR_i\}$ and the authority of TID. Finally, it sends $\{L_i\}$ to M for replying that the process of auditing and clearing is complete. The client device of M will log out automatically. It must apply a new TID, then the checkout service could be continue.

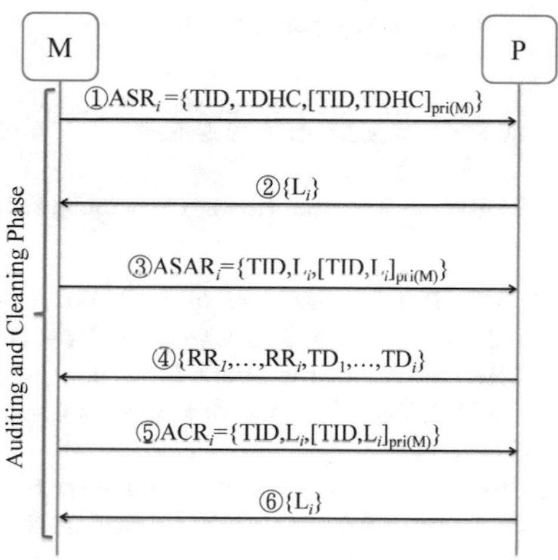

Fig. 4. The message exchange of the auditing and clearing phase

When the auditing is unsuccessful, this mechanism enables the service provider to prove its innocence and the users to prove its guilt. The details of the message exchanges when the auditing and clearing is unsuccessful are illustrated in Fig. 5.

(1) When an employee of M signs out or intends to clear attestations, the device of M sends ASR_i, where $ASR_i = \{TID, TDHC, [TID, TDHC]_{pri(M)}\}$ to P for the request of auditing and clearing.

(2) P receives ASR_i from M. It verifies if the signature is valid. It calculates $TDHC'_i$ according to all transaction data that belong to the same TID in ASR_i. If $TDHC_i$ in ASR_i and $TDHC'_i$ are not the same, P sends an error message, $\{L_i\}$, to the client device of M. $\{L_i\}$ contains auditing failure announcement and which transaction data is incorrect.

(3) The device of M receives the error message of auditing, but it still reply a request of auditing server-side attestation and transaction ($ASAR_i$), where $ASAR_i = \{TID, L_i, [TID, L_i]_{pri(M)}\}$, to P.

(4) When P receives $ASAR_i$ from M, it sends all of transaction data TD_s with the same TID and RR_s to M for auditing.

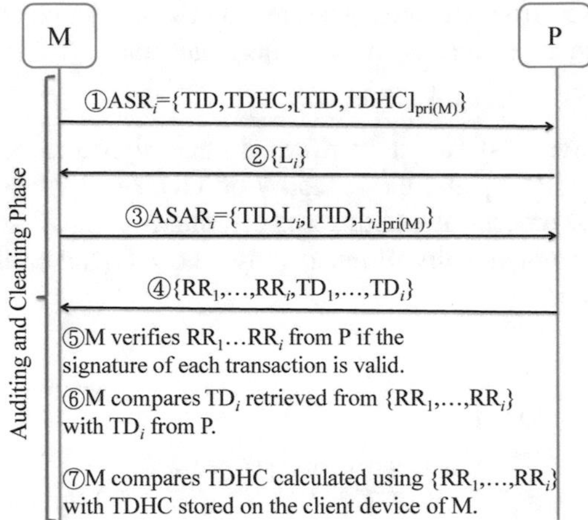

Fig. 5. The message exchange of the auditing failure

(5) M verifies all RR_s from P if the signature of each transaction is valid. If one of signatures is incorrect, that can prove service provider's guilt.
(6) If all signatures in RR_s are valid, the client of M compares TD_i retrieved from $\{RR_1, \cdots, RR_i\}$ with TD_i from P. If one of comparisons is inconsistent, that can prove service provider's guilt.
(7) If all attestations from P are correct (step5 and step 6), M compares TDHC calculated using $\{RR_1, \cdots, RR_i\}$ with TDHC stored on the client device of M. If one of comparisons is inconsistent, that can prove the merchant's guilt.

4 Security Analysis and Implementation

In this study we made no assumption about the honesty of the merchant and service provider. In this mechanism, TID is used to identify each employee when logging in the checkout system even though the same client device is used by the same employee. LSN is to ensure data serializability and freshness. Furthermore, LSN is maintained by the merchant, based on TID and LSN, cloud service providers could not launch roll-back attack to avoid the accountability. The required security and privacy features were implemented as follows:

(1) Merchant counterfeits TDHC and ACK: The merchant can modify TDHC and ask compensation for inconsistent TD modified by the service provider. RR stored in the service provider is signed by the merchant; the merchant cannot repudiate its signature. The service provider can compare TD from RR with counterfeit TD from the merchant to prove its innocence. Moreover, RR includes the newest LSN, the merchant cannot add or delete one TD for compensation. Even though it modifies ACK and LSN, the signature from service provider will be invalid and result in the inability of hash chain CH to work correctly.

(2) Cloud service provider counterfeits RR and TD: The service provider can counterfeit RR and TD for avoid its accountability due to bugs, crashes, operator errors, or misconfigurations. In this scheme, the merchant keep the newest ACK, which can prove the number of transactions, where $\mathrm{ACK}_i = \left\{\mathrm{RR}_i, [\mathrm{RR}_i]_{\mathrm{pri(P)}}\right\}$, as the attestation. If the service provider counterfeits RR or recovers TD, the signature from the merchant will be invalid.

According to the proposed scheme shown in Sect. 3, we implement a system to prove that this scheme can work smoothly. We first define the formats of the transaction data, transaction data hash chain, request, response, reply-response, and ACK messages. These messages were represented as XML documents in our implementation. Figures 6, 7, 8, 9, 10 and 11 show examples of transaction data, transaction data hash chain, request, response, and reply-response, and ACK messages in the chain-hashing on transaction phase, respectively.

```
<Transaction>
  <ReceiptNum>AA12345678</ReceiptNum>
  <Date>2018/10/10 16:30:06</Date>
  <CompanyID>test</CompanyID>
  <StaffID>cchaha</StaffID>
  <TotalPrice>25</TotalPrice>
  <Detail>
    <Good>
      <Barcode>4710782171879</Barcode>
      <Price>25</Price>
      <Amount>1</Amount>
    </Good>
  </Detail>
</Transaction>
```

Fig. 6. The example of TD (Transaction Data)

```
<TDHashChain>
  <RootHash>XHfGEwNAeONXQlOgdLULLNQiVbaiLs17JRDZ2sVxsls=</RootHash>
  <TDFile>
    <TransactionData>
      <Name>TDcchaha20181010162956_1</Name>
      <Hash>XHfGEwNAeONXQlOgdLULLNQiVbaiLs17JRDZ2sVxsls=</Hash>
    </TransactionData>
  </TDFile>
</TDHashChain>
```

Fig. 7. The example of TDHC (Transaction Data Hash Chain)

```xml
<Request>
  <Transaction_data>
    <Transaction>
      <ReceiptNum>AA12345678</ReceiptNum>
      <Date>2018/1010 16:30:06</Date>
      <CompanyID>test</CompanyID>
      <StaffID>cchaha</StaffID>
      <TotalPrice>25</TotalPrice>
      <Detail>
        <Good>
          <Barcode>4710782171879</Barcode>
          <Price>25</Price>
          <Amount>1</Amount>
        </Good>
      </Detail>
    </Transaction>
  </Transaction_data>
  <TID>cchaha20181010162956</TID>
  <LSN>1</LSN>
  <Signature xmlns="http://www.w3.org/2000/09/xmldsig#">
    <SignedInfo>
      <CanonicalizationMethod
       Algorithm="http://www.w3.org/TR/2001/REC-xml-c14n-20010315" />
      <SignatureMethod
       Algorithm="http://www.w3.org/2000/09/xmldsig#rsa-sha1" />
      <Reference URI="">
        <Transforms>
          <Transform Algorithm=
           "http://www.w3.org/2000/09/xmldsig#enveloped-signature" />
        </Transforms>
        <DigestMethod
          Algorithm="http://www.w3.org/2000/09/xmldsig#sha1" />
        <DigestValue>Yt5LW6FRBrth4pMiOLxnaBndMUI=</DigestValue>
      </Reference>
    </SignedInfo>
    <SignatureValue>
      b2s21f7DihrMq7gc9J8K50rpvzjArtXY7uPJZVieZcGraO38wnLwj8TK5irLWfwdmro0YVent0k+OHM42ZNlFu7
      rAv-
      VrzdtWIART9mlYrjW0xZTG5SBljb9Fxg6CT4+Adg4sEeKu1Gab7stoD6mn+wsSf/CePq2x1+Siz/d9/5Y=</Sig
      natureValue>
  </Signature>
</Request>
```


Fig. 8. The example of $Q_i = \left\{ \text{TD, TID, LSN, } [\text{TD, TID, LSN}]_{\text{pri(M)}} \right\}$

```
<Response>
  <Request>
    <Transaction_data>
      <Transaction>
        <ReceiptNum>AA12345678</ReceiptNum>
        <Date>2018/10/10 16:30:06</Date>
        <CompanyID>test</CompanyID>
        <StaffID>cchaha</StaffID>
        <TotalPrice>25</TotalPrice>
        <Detail>
          <Good>
            <Barcode>4710782171879</Barcode>
            <Price>25</Price>
            <Amount>1</Amount>
          </Good>
        </Detail>
      </Transaction>
    </Transaction_data>
    <TID>cchaha20181010162956</TID>
    <LSN>1</LSN>
    <Signature xmlns="http://www.w3.org/2000/09/xmldsig#">
      <SignedInfo>
        <CanonicalizationMethod
         Algorithm="http://www.w3.org/TR/2001/REC-xml-c14n-20010315" />
        <SignatureMethod
         Algorithm="http://www.w3.org/2000/09/xmldsig#rsa-sha1" />
        <Reference URI="">
          <Transforms>
            <Transform Algorithm=
             "http://www.w3.org/2000/09/xmldsig#enveloped-signature" />
          </Transforms>
          <DigestMethod
           Algorithm="http://www.w3.org/2000/09/xmldsig#sha1" />
          <DigestValue>Yt5LW6FRBrth4pMiOLxnaBndMUI=</DigestValue>
        </Reference>
      </SignedInfo>
      <SignatureValue>
         b2s21f7DihrMq7gc9J8K50rpvzjArtXY7uPJZVieZcGraO38wnLwj8TK5irLWfwdmro0YVent0k+OHM
         42ZNlFu7rAvVrzdtWIART9mlYrjW0xZTG5SBljb9Fxg6CT4+Adg4sEeKu1Gab7stoD6mn+wsSf/CePq
         2x1+Siz/d9/5Y=</SignatureValue>
    </Signature>
  </Request>
  <ChaingHash>EcZHnXVswpssB2IbZ4gERQfsI3yZtTbTHVLBJpPd1rc=</ChaingHash>
  <Signature xmlns="http://www.w3.org/2000/09/xmldsig#">
    <SignedInfo>
      <CanonicalizationMethod
       Algorithm="http://www.w3.org/TR/2001/REC-xml-c14n-20010315" />
      <SignatureMethod
        Algorithm="http://www.w3.org/2000/09/xmldsig#rsa-sha1" />
      <Reference URI="">
        <Transforms>
          <Transform Algorithm=
           "http://www.w3.org/2000/09/xmldsig#enveloped-signature" />
        </Transforms>
        <DigestMethod
         Algorithm="http://www.w3.org/2000/09/xmldsig#sha1" />
        <DigestValue>5NIb2kCLEcyRNS64RV+7z2Y2d1Y=</DigestValue>
      </Reference>
    </SignedInfo>
    <SignatureValue>
      tBKn6Yd22NjaJfGeZRj0M8rWj1Ws6HvzJiz14wUNtvCFbvN3ZixVS/2jbDtZvWHWL2NxGSjAlJNIUrLRgS
      ivRjra+SZdmnCT9TTMlbkkXkh4/z7I++z8u3wZATOiivQvKRYeEypGNidctQDyz38mKIdwqgDqcb8yHKGZ
      ABRd0Ww=</SignatureValue>
  </Signature>
</Response>
```


Fig. 9. The example of $R_i = \left\{Q_i,\ h(Q_i,\ CH_{i-1})_{pri(P)},\ [Q_i,\ h(Q_i, CH_{i-1})_{pri(P)}]_{pri(P)}\right\}$

```xml
<ResponseOfResponse>
  <Response>
    <Request>
      <Transaction_data>
        <Transaction>
          <ReceiptNum>AA12345678</ReceiptNum>
          <Date>2018/10/10 16:30:06</Date>
          <CompanyID>test</CompanyID>
          <StaffID>cchaha</StaffID>
          <TotalPrice>25</TotalPrice>
          <Detail>
            <Good>
              <Barcode>4710782171879</Barcode>
              <Price>25</Price>
              <Amount>1</Amount>
            </Good>
          </Detail>
        </Transaction>
      </Transaction_data>
      <TID>cchaha20181010162956</TID>
      <LSN>1</LSN>
      <Signature xmlns="http://www.w3.org/2000/09/xmldsig#">
        <SignedInfo>
          <CanonicalizationMethod Algorithm="http://www.w3.org/TR/2001/REC-xml-c14n-20010315" />
          <SignatureMethod Algorithm="http://www.w3.org/2000/09/xmldsig#rsa-sha1" />
          <Reference URI="">
            <Transforms>
              <Transform Algorithm="http://www.w3.org/2000/09/xmldsig#enveloped-signature" />
            </Transforms>
            <DigestMethod Algorithm="http://www.w3.org/2000/09/xmldsig#sha1" />
            <DigestValue>Yt5LW6FRBrth4pMiOLxnaBndMUI=</DigestValue>
          </Reference>
        </SignedInfo>
        <SignatureValue>
          b2s21f7DihrMq7gc9J8K50rpvzjArtXY7uPJZVieZcGraO38wnLwj8TK5irLWfwdmro0YVent0k+OHM42ZNlFu
          7rAvVrzdtWIART9mlYrjW0xZTG5SBljb9Fxg6CT4+Adg4sEeKu1Gab7stoD6mn+wsSf/CePq2x1+Siz/d9/5Y=
        </SignatureValue>
      </Signature>
    </Request>
    <ChaingHash>EcZHnXVswpssB2IbZ4gERQfsI3yZtTbTHVLBJpPd1rc=</ChaingHash>
    <Signature xmlns="http://www.w3.org/2000/09/xmldsig#">
      <SignedInfo>
        <CanonicalizationMethod Algorithm="http://www.w3.org/TR/2001/REC-xml-c14n-20010315" />
        <SignatureMethod Algorithm="http://www.w3.org/2000/09/xmldsig#rsa-sha1" />
        <Reference URI="">
          <Transforms>
            <Transform Algorithm="http://www.w3.org/2000/09/xmldsig#enveloped-signature" />
          </Transforms>
          <DigestMethod Algorithm="http://www.w3.org/2000/09/xmldsig#sha1" />
          <DigestValue>5NIb2kCLEcyRNS64RV+7z2Y2d1Y=</DigestValue>
        </Reference>
      </SignedInfo>
      <SignatureVal-
ue>tBKn6Yd22NjaJfGeZRj0M8rWj1Ws6HvzJiz14wUNtvCFbvN3ZixVS/2jbDtZvWHWL2NxGSjAlJNIUrLRgSivRjra+SZdm
nCT9TTMlbkkXkh4/z7I++z8u3wZATOiivQvKRYeEypGNidctQDyz38mKIdwqgDqcb8yHKGZABRd0Ww=</SignatureValue>
    </Signature>
  </Response>
  <Signature xmlns="http://www.w3.org/2000/09/xmldsig#">
    <SignedInfo>
      <CanonicalizationMethod Algorithm="http://www.w3.org/TR/2001/REC-xml-c14n-20010315" />
      <SignatureMethod Algorithm="http://www.w3.org/2000/09/xmldsig#rsa-sha1" />
      <Reference URI="">
        <Transforms>
          <Transform Algorithm="http://www.w3.org/2000/09/xmldsig#enveloped-signature" />
    </Transforms>
        <DigestMethod Algorithm="http://www.w3.org/2000/09/xmldsig#sha1" />
        <DigestValue>0Cqo4wo5gvYEFLEVvyWWe71Qq8A=</DigestValue>
      </Reference>
    </SignedInfo>
    <SignatureValue>
      hBaILSvt6BSEha85o+vvuPrMlvBrciBI3rGzCtFwwsP44yIXq3WVkIsJz+/fD3QvGXGjAjTbk6nEADqyY31tbQxErq
      w19V6xuMeILChZRBCRXvgS42w+VywquylYvGb+a7bTfsCe5bq4gOGu31JWv62/huRV6iUJ0lpHob8GWA0=</Signat
      ureValue>
  </Signature>
</ResponseOfResponse>
```


Fig. 10. The example of $RR_i = \left\{ R_i, [R_i]_{pri(M)} \right\}$

```xml
<Acknowledgement>
  <ResponseOfResponse>
    <Response>
      <Request>
        <Transaction_data>
          <Transaction>
            <ReceiptNum>AA12345678</ReceiptNum>
            <Date>2018/10/10 16:30:06</Date>
            <CompanyID>test</CompanyID>
            <StaffID>cchaha</StaffID>
            <TotalPrice>25</TotalPrice>
            <Detail>
              <Good>
                <Barcode>4710782171879</Barcode>
                <Price>25</Price>
                <Amount>1</Amount>
              </Good>
            </Detail>
          </Transaction>
        </Transaction_data>
        <TID>cchaha20181010162956</TID>
        <LSN>1</LSN>
        <Signature xmlns="http://www.w3.org/2000/09/xmldsig#">
          <SignedInfo>
            <CanonicalizationMethod
             Algorithm="http://www.w3.org/TR/2001/REC-xml-c14n-20010315" />
            <SignatureMethod Algorithm="http://www.w3.org/2000/09/xmldsig#rsa-sha1" />
            <Reference URI="">
              <Transforms>
                <Transform
                 Algorithm="http://www.w3.org/2000/09/xmldsig#enveloped-signature" />
              </Transforms>
              <DigestMethod Algorithm="http://www.w3.org/2000/09/xmldsig#sha1" />
              <DigestValue>Yt5LW6FRBrth4pMiOLxnaBndMUI=</DigestValue>
            </Reference>
          </SignedInfo>
          <SignatureValue>
           b2s21f7DihrMq7gc9J8K50rpvzjArtXY7uPJZVieZcGraO38wnLwj8TK5irLWfwdmro0YVent0k+OHM42ZN
           lFu7rAvVrzdtWIART9mlYrjW0xZTG5SBljb9Fxg6CT4+Adg4sEeKu1Gab7stoD6mn+wsSf/CePq2x1+Siz/
           d9/5Y-</SignatureValue>
        </Signature>
      </Request>
      <ChaingHash>EcZHnXVswpssB2IbZ4gERQfsI3yZtTbTHVLBJpPdlrc=</ChaingHash>
      <Signature xmlns="http://www.w3.org/2000/09/xmldsig#">
        <SignedInfo>
          <CanonicalizationMethod
           Algorithm="http://www.w3.org/TR/2001/REC-xml-c14n-20010315" />
          <SignatureMethod Algorithm="http://www.w3.org/2000/09/xmldsig#rsa-sha1" />
          <Reference URI="">
            <Transforms>
              <Transform Algorithm="http://www.w3.org/2000/09/xmldsig#enveloped-signature" />
            </Transforms>
            <DigestMethod Algorithm="http://www.w3.org/2000/09/xmldsig#sha1" />
            <DigestValue>5NIb2kCLEcyRNS64RV+7z2Y2d1Y=</DigestValue>
          </Reference>
        </SignedInfo>
        <SignatureValue>
          tBKn6Yd22NjaJfGeZRj0M8rWj1Ws6HvzJiz14wUNtvCFbvN3ZixVS/2jbDtZvWHWL2NxGSjAlJNIUrLRgSiv
          Rjra+SZdmnCT9TTMlbkkXkh4/z7I++z8u3wZATOiivQvKRYeEypGNidctQDyz38mKIdwqgDqcb8yHKGZABRd
          0Ww=</SignatureValue>
      </Signature>
    </Response>
    <Signature xmlns="http://www.w3.org/2000/09/xmldsig#">
      <SignedInfo>
        <CanonicalizationMethod Algorithm="http://www.w3.org/TR/2001/REC-xml-c14n-20010315" />
        <SignatureMethod Algorithm="http://www.w3.org/2000/09/xmldsig#rsa-sha1" />
        <Reference URI="">
          <Transforms>
            <Transform Algorithm="http://www.w3.org/2000/09/xmldsig#enveloped-signature" />
          </Transforms>
          <DigestMethod Algorithm="http://www.w3.org/2000/09/xmldsig#sha1" />
          <DigestValue>0Cqo4wo5gvYEFLEVvyWWe71Qq8A=</DigestValue>
        </Reference>
      </SignedInfo>
      <SignatureValue>
       hBaILSvt6BSEha85o+vvuPrMlvBrciBI3rGzCtFwwsP44yIXq3WVkIsJz+/fD3QvGXGjAjTbk6nEADqyY31tbQx
       Erqw19V6xuMeILChZRBCRXvgS42w+VywquylYvGb+a7bTfsCe5bq4gOGu31JWv62/huRV6iUJ0lpHob8GWA0=</
       SignatureValue>
    </Signature>
  </ResponseOfResponse>
```


Fig. 11. The example of $\mathrm{ACK}_i = \left\{\mathrm{RR}_i, [\mathrm{RR}_i]_{\mathrm{pri(P)}}\right\}$

```
    <AttestationChainHash>g38vAxDwIRZ5CjZ41NyKWd6UBDdCilwIcFYE5b7p5eY=</AttestationChainHash>
    <Signature xmlns="http://www.w3.org/2000/09/xmldsig#">
      <SignedInfo>
        <CanonicalizationMethod Algorithm="http://www.w3.org/TR/2001/REC-xml-c14n-20010315" />
        <SignatureMethod Algorithm="http://www.w3.org/2000/09/xmldsig#rsa-sha1" />
        <Reference URI="">
          <Transforms>
            <Transform Algorithm="http://www.w3.org/2000/09/xmldsig#enveloped-signature" />
          </Transforms>
          <DigestMethod Algorithm="http://www.w3.org/2000/09/xmldsig#sha1" />
          <DigestValue>sM5iNIWEvhpwOc3eBGEWOd2YT8U=</DigestValue>
        </Reference>
      </SignedInfo>
      <SignatureValue>
       HRQqYh3bzO8k+xtSywnB9fsDisfu59Ssk+LQvGTz5Dz2dZMQnmGEAJWav72xjmKVuSjP23YvO7vDtdtGfFGE/ODPw
       +hSxNp93AUiTG0+a3A4RacCjngwZpKoEg0GELkNkhc4/IV9ntBBpZj4cWBqin21V3N7a7jghqVPWlGF/+w=</Sign
       atureValue>
    </Signature>
  </Acknowledgement>
```


Fig. 11. (*continued*)

5 Conclusion

This research proposed a mutual auditing mechanism for a cloud-based checkout system. It provides and ensures the evidence nonrepudiation and continuity between merchants and cloud service providers through exchanging attestations from every transaction and hashing of these attestations as a chain. Merchants only need to retain the last attestation and TDHC instead of all transaction details for auditing. Furthermore, cloud service providers do not need to worry about merchants forging transaction data to get reimbursement and result in the loss of reputation and business. The proposed scheme can applied to a mobile checkout system and combine mobile payment in the future.

References

1. Feng, J., Chen, Y., Summerville, D., Ku, W.-S., Su, Z.: Enhancing cloud storage security against roll-back attacks with a new fair multi-party non-repudiation protocol. In: 8th IEEE Consumer Communications and Networking Conference (CCNC), pp. 521–522 (2011)
2. Shraer, A., Keidar, I., Cachin, C., Michalevsky, Y., Cidon, A., Shaket, D.: Venus: verification for untrusted cloud storage. In: 17th ACM Cloud Computing Security Workshop (CCSW) (2010)
3. Shah, M.A., Baker, M., Mogul, J., Swaminathan, R: Auditing to keep online storage services honest. In: USENIX HotOS XI: 11st Workshop on Hot Topics in Operating Systems (2007)
4. Li, J., Krohn, M., Mazières, D., Shasha, D.: Secure untrusted data repository (SUNDR). In: USENIX 6th Symposium on Operating Systems Design and Implementation (OSDI 2004) (2004)
5. Mazières, D., Shasha, D.: Building secure file systems out of Byzantine storage. In: 21st Annual ACM Symposium on Principles of Distributed Computing (PODC), pp. 108–117 (2002)
6. Cachin, C., Shelat, A., Shraer, A.: Efficient fork-linearizable access to untrusted shared memory. In: ACM 26th PODC, pp. 129–138 (2007)
7. Majuntje, M., Dobre, D., Serafini, M., Suri, N.: Abortable fork-linearizable storage. In: ACM 13th OPODIS, pp. 255–269 (2009)

8. Cachin, C., Shelat, A., Shraer, A.: Integrity protection for revision control. In: ACM 7th ACNS, pp. 382–399 (2009)
9. Popa, R.A., Lorch, J.R., Molnar, D., Wang, H., Zhuang, L.: Enabling security in cloud storage SLAs with CloudProof. In: USENIX Annual Technical Conference (USENIXATC 2011), p. 31 (2011)
10. Hwang, G.-H., Peng, J.-Z., Huang, W.-S.: A mutual nonrepudiation protocol for cloud storage with interchangeable accesses of a single account from multiple devices. In: 12th IEEE International Conference on Trust, Security and Privacy in Computing and Communications (IEEE TrustCom 2013) (2013)
11. Merkle, R.C.: A digital signature based on a conventional encryption function. In: A Conference on the Theory and Applications of Cryptographic Techniques on Advances in Cryptology (CRYPTO 1987), pp. 369–378 (1988)

Implementation of an IP Management and Risk Assessment System Based on PageRank

Chia-Ling Hou(✉), Cheng-Chung Kuo(✉), I-Hsien Liu, and Chu-Sing Yang(✉)

Institute of Computer and Communication Engineering, Department of Electrical Engineering, National Cheng Kung University, Tainan, Taiwan
{clhou, csyang}@mail.ncku.edu.tw, jjguo@cryptolab.tw, ihliu@twisc.ncku.edu.tw

Abstract. Recently, network technology had brought a variety of attacks on the Internet, unfortunately, no one is safe in this trend. Network managers try to find the attackers and search for the suspicious behaviors in the network connections to defend their services. Intrusion Detection System (IDS) can help network managers to find out the network attacks, but for some special cases, IDS has its limitation. Proposed system integrated network-based IDS (NIDS) and host-based IDS (HIDS) to detect the suspicious behavior and assess the risk value of each IP. This research is dedicated to separating attacks and suspicious behaviors analysis by network-based IDS and host-based IDS. Furthermore, the proposed system will also find the relations among suspicious IP by using the modified PageRank algorithm and correlate the events to estimate the risk for each IP. The ranking of each IP represent the risk level and network managers can protect the hosts by the ranking. The experiment results show that the proposed system can achieve the goal of managing attack and tracking the suspicious ones. It can help users to take appropriate action in time.

Keywords: NIDS · HIDS · Risk assessment

1 Introduction

As computer technology advances, modern network makes people's lives more convenient than before. People cannot get rid of these changes in their lives. Recently, leakage of personal data, ransomware virus, and the attack of botnet emerge in endlessly such as WannaCry raging users [1]. Moreover, in 2018, the famous message board has been attacked by flooding flows [2].

Intrusion Detection System are proposed to against network attacks. IDS are mainly divided into 2 categories, Network based IDS and Host based IDS. NIDS has several ways to distinguish malicious connection such as monitoring packet payloads, number of flows and so on to detect attacks. The common way for packet monitoring is deep packet inspection. Deep Packet Inspection (DPI) is a method which focus on network packets detection and identification. It can also classified to a signature-based detection on NIDS. DPI is a method that has the lowest false positive and is most direct.

C.-N. Yang et al. (Eds.): SICBS 2018, AISC 895, pp. 438–450, 2020.
https://doi.org/10.1007/978-3-030-16946-6_34

Since DPI will check the content of every packets, it can discover more signatures and strings than other detection if DPI database and comparison is as new as possible. However, since it will inspect every packets in time, it consumes too much resource to achieve the goal. The flow detection is more common way in NIDS. On the other hand, HIDS can find out the abnormal behavior in host immediately. HIDS usually use log analysis or file monitoring detection to prevent the attack.

However, it is no longer safe to only use NIDS or HIDS. Because of the limitation about flows detection, NIDS can't accurately discover web-based attacks like SQL injection, cross-site scripting (XSS) and phishing attacks. These attacks can only be found out by HIDS to monitor logs like apache log. Generally, network attacks can't be defensed by single solution or system any more. Thus, this study designed an IP management and risk assessment system (IPMRAS) by combining these two systems. IPMRAS divides data into two groups, attack IPs and suspicious IPs. IPMRAS puts attack IPs into black list and calculating the risk of suspicious IPs to reveal the degree of danger for each IP.

The reminder of this paper is organized as follows. Section 2 introduces the background knowledge and researches related to risk assessment system. Section 3 presents the design idea and the implementation detail of IPMRAS. Section 4 provided the experiments result for proposed IPMRAS. Finally, the last Section concluded this paper and give the future work of this system.

2 Related Works

2.1 Intrusion Detection System

Nowadays, most of attacks invade services of company and government agencies through well-known ports such as website etc. These attacks can't prevent by setting port or IP, so it has to get other information to block the attack behavior. According to the above description, the concept of intrusion detection system (IDS) [3] are proposed. IDS can remind managers whether there are any attack and high risk event by detecting flows attack features and abnormal behaviors. IDS can be classified by data type, detection, deployment (source) to Fig. 1 in [3–5]. According to different data type and application environment, IDS deployment can be usually classified into host-based IDS (HIDS) [6] and network-based IDS (NIDS) [7]. HIDS will focus on protecting critical resource like secret document, personal information. Besides, it will monitor port change, user behavior, system log and application log to prevent any malicious event. In addition, NIDS will detect network flows in local network environment. It can monitor whether there is any intrusion or malicious event by using packets or other information. Both systems have the advantage in different aspects.

2.2 Link Analysis

Since social network become more and more convenient, people want to know the relation and importance among different website. In social network, link analysis is an important method to understanding the link information. It can conquer various

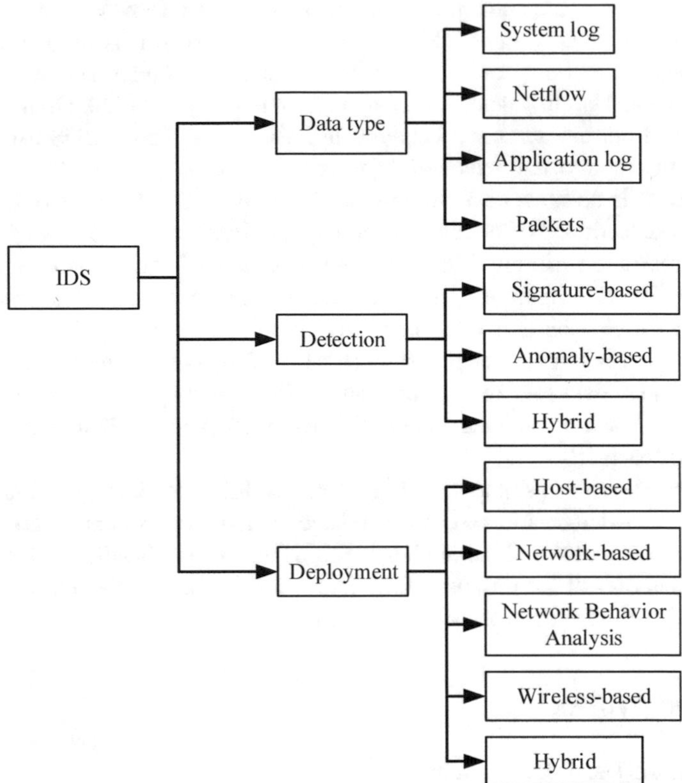

Fig. 1. IDS classification [3–5]

problem in social networks like website relation and also calculate the popularity of website. Take blog for example. Some famous bloggers will introduce popular shops or restaurants and give the hyperlink beside the article. Once bloggers' website citing other website, the relation between them will be closer than other non-siting one. By using link analysis, it can transform this siting relation into degree. In addition, research [8] discover that data mining is actually confluence of multiple disciplines. Generally, the most famous link analysis method used by Google is PageRank. PageRank [9] is proposed to calculate the relation and importance among different website. Actually, it is a method that quantifies the citing relation which is called devoting value among the website. Since the citing actions is similar to the IP connection in the network, it will use PageRank to figure out the close relationship among these IPs and to keep an eye on those IPs to prevent the potential malicious behavior.

2.3 Risk Assessment

Risk assessment is a significant procedure to confirm the security of system. According to ISO 31000:2018 [10], risk assessment can evolve many different risk assessment methods through risk identification, risk analysis and risk evaluation. Risk identification is to determine the target value in the environment. Besides, risk analysis is the way to observe the influence variable, and it will use some ways to confirm the variety such as event replaying. Furthermore, risk evaluation is to execute the analysis solution and do some prevention to protect the environment.

Previously, risk assessment often use hierarchy to judge the level of the risk. Although it is simple to use hierarchy, it can't have a good consideration in general situation. In [11], first mission is to find the goal of the risk assessment and execute the data preprocessing. Then, observe which object will increase risk level and set the quantification mechanism of these risk parameters. After that, set up a threshold to aware users. Finally, when some risk value is out of range, the system will start the protection, and launch the alarm to users.

Since the growth of network technology, attack techniques also become more severe than before. In 2017, the event of LinkedIn data leakage, more than seven hundred million account was used to spread Trojan by spam robots. In 2016, hacker heist over eighty million by installing malware to ATM. Security incidents are endless. In research [12], it proposes the attack tree and attack graph to replay the attack, and also to trace the attack source and path. Additionally, it combine risk assessment to clarify the vulnerability through intrusion and malicious behaviors in order to prevent any attack.

3 System Design

This paper designs an IP management and risk assessment system (IPMRAS) to manage data from NIDS, HIDS and assess IP risk. This paper uses supervise learning NIDS [13] to collect the netflow information then analyze netflow and compare with another netflow network-based intrusion detection system [14] in order to ensure the objectiveness. Netflow network-based intrusion detection system will analyze the attack in time detection and space detection. For example, some attacks will cause a large amount of connection in short time, but usually attackers will reduce the connection frequency for avoiding being detected. Therefore, it will find out strange connection behavior when extending time section. Additionally, supervised learning XGBoost NIDS is going to improve the accuracy of netflow network-based intrusion detection system by using machine learning. It will collect all the feature to build model and use model to monitor the real traffic. Also, the HIDS [15] used in this paper will focus on port monitoring, log analysis and file monitoring. Service providers won't have port change frequently like web server. If port change occur, port monitoring will detect whether there is any changing port action in time. Since port number won't change frequently, port monitoring will detect whether there is any changing port action in time. The log analysis will focus on system log and apache log to detect the abnormal behaviors. HIDS uses signature-based detection to complete the mission, and file monitoring will detect essential folder and dictionary including the modification of sensitive information. After detection, HIDS will output the detection result to alert users.

3.1 System Architecture

Figure 2 is the architecture of IPMRAS. The HIDS will analyze the abnormal behavior of the machine. The monitoring aspect are port monitoring, log analysis and file monitoring. In addition, router transform flows into netflow, and NIDS analyze these netflow by using supervise learning method. After analysis is complete, NIDS will give every flows a corresponding label to distinguish different connection type. There are five types of label which are normal, horizontal scan, vertical scan, flooding flows and brute force. The system will also append the result of another NIDS which monitor different pattern such as threshold to estimate the event type.

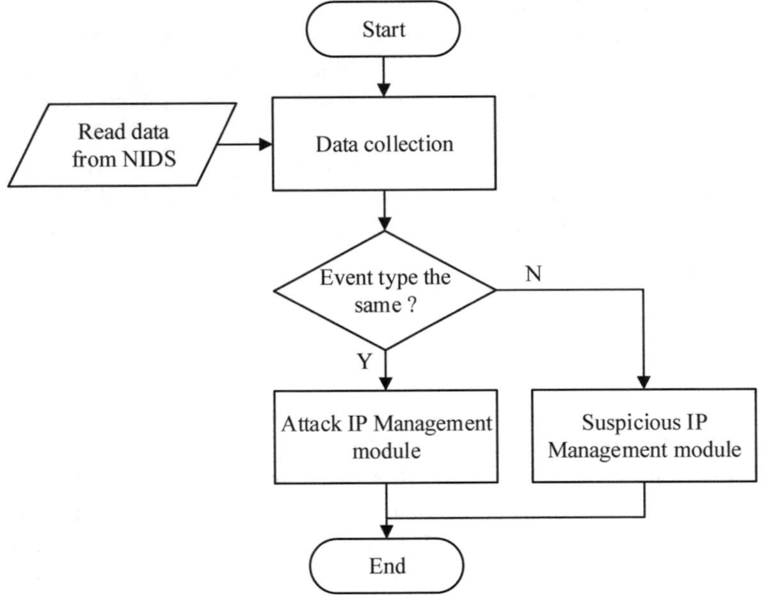

Fig. 2. System Architecture

IPMRAS first will read data of HIDS and NIDS data and compare the feature of supervise learning NIDS label and pattern monitoring NIDS label whether the label are the same. If the label are the same, this flow is certainly an attack. If they are not, then the flow is defined as suspicious connection. The proposed system establish different mechanism to deal with attack and suspicious connections. The attack IP management module will handle attack IP, and suspicious IP management module will analyze the risk of suspicious IP. The detail of each module will introduce in the following.

3.2 Attack IP Management Module

Figure 3 showed the flow chart of the attack IP management module. This module will arrange attack IP to black list and manage black list in the same time. It will collect the same label data from supervised learning NIDS. This module will execute IP comparison to compare with attack IP in HIDS and append the missing-attack IP into black list. Also, attack IP management module accumulate flow information such as duration, packet per second and append the IP latest appearing time. In addition, the black list data will generate the black list relation graph to show the attack relation in the environment.

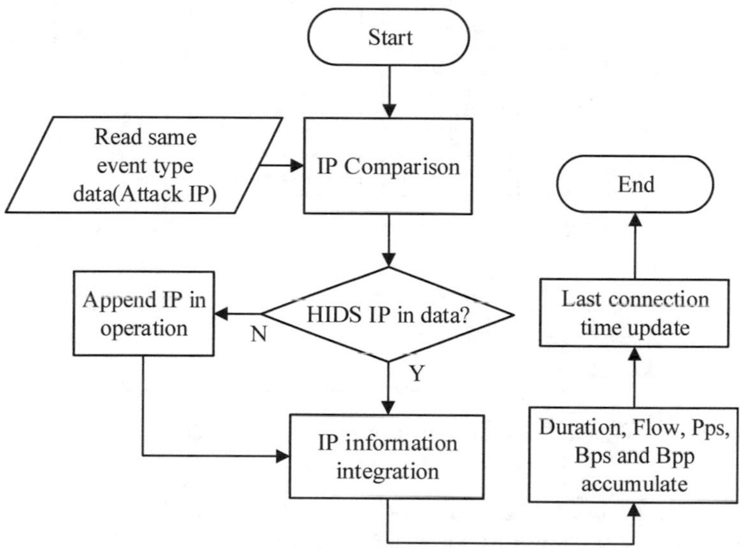

Fig. 3. Attack IP management module flow chart

3.3 Suspicious IP Management Module

In Fig. 4, the suspicious IP management module is going to collect suspicious IP which has different event type label from two NIDS mentioning above. The paper want to know the risk that suspicious IP may transform into attack IP, so this module will estimate suspicious IP risk in behavior and event correlation in the environment.

The behavior estimation is to divide suspicious IP connection into three types, horizontal scan, vertical scan and flooding flows/brute force. Each attack type has specific pattern. By observing each event type pattern, the algorithm will quantify the harm to become figure. Horizontal scan is an attack that attackers want to know whether there is any IP opening specific service like HTTP, HTTPS etc. If some opening-service servers are scanned by attackers, they will respond and it will make attackers know where they can start intrusion or to attack the vulnerability. Horizontal scan usually use one source IP to connect several destination IPs' same port. If source IP scan as inclusive IP as possible, it will make as many host as possible to become victims.

Additionally, vertical scan is the attack that attackers want to assault specific host. Attackers want to know every services opening on the host and use these services and port to find possible weakness. Vertical scan usually use same source IP to connect

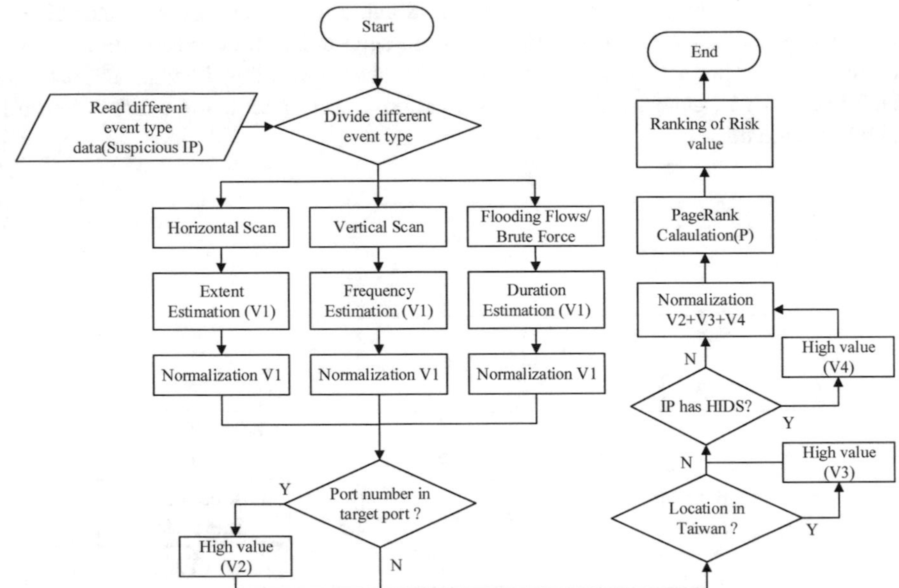

Fig. 4. Suspicious IP management module flow chart

same destination IP, but the scanning port will change dramatically to find out which ports are opened. It is harmful when attackers scan rapidly. Therefore, the frequent is the element to judge the damage of vertical scan.

Flooding flow is an attack that use same source IP to connect same port of same destination IP numerously. Generally, normal network behavior will only have limited connections. Taking HTTP for example, HTTP protocol has keep-alive function to avoid the connection being terminated in order to wait for the client resource delivering. However, in abnormal condition, it will start new connections continuously such as malware connecting external network etc. Apart from this, the same behavior will also occur in brute force because attackers usually want to crack the password quickly. Attacker will use the accelerated cracking tool to achieve the goal. Thus, if the attack continues as long as possible, it will cause huge loss. According to the pattern above, it can use horizontal scan extent, vertical scan frequent and flooding flows/brute force duration to estimate the corresponding harm (V1) of every connections.

Besides, the module will use target port to find out whether the suspicious connection aim at some specific service port. Usually, users won't change service port number in Table 1 like port 80, 443, 22, 3389 etc. Attackers can easily execute attack or intrusion through these target port. Hence, the algorithm will assign a high value (V2) to describe the risk if the suspicious connection used the target port to connect with destination IP. Moreover, the location of IP is also a term to estimate the risk. Attackers are around the world. Because of the special geographic location in Asia and close relationship America, some hackers attack Taiwan in order to improve their skills and to test their malware. Thus, it will assign a high value (V3) to those foreign connection IP to represent the risk of them. Furthermore, if the host has HIDS to pass attack detail to proposed system, it can know the attack immediately and warn users to protect host in time. The host which HIDS deploy is usually the critical server in the environment. Hence, the risk of these servers are higher than other machine. The module will also assign a high value (V4) to the connection which includes important server IP to represent the risk of it. Finally, it will normalize the value to decrease the system load and to compare risk easily.

Table 1. Services with port number

Service	Port number	Application
HTTP	80	Transmission protocol, plaintext
HTTPS	443	Transmission protocol, chipertext
SSH	22	Remote connection protocol, Linux…
RDP	3389	Remote connection protocol, Windows
SMTP	25	Transmission protocol, mail

In event correlation, the module will estimate the IP relation to find out the IP which establish connection most commonly in this section. It use modified PageRank to complete the relation quantification. Traditional PageRank is used to know the importance of every website through the citing relation among the social network. The value of each website is calculated by the being-cited number of it. The citing relation seems like the IP connection in the environment, so it is used to calculate the IP relation by using modified PageRank in this paper. Initially, the modified PageRank has the same procedure with traditional PageRank. It will transform the IP relation into matrix H by using formula (1). c is the relation of every IP, and i is the total number of suspicious IP. For instance, if IP A has connection with IP B and C, only c_{12} and c_{13} will be one.

$$H = \begin{bmatrix} c_{11} & c_{12} & \cdot & \cdot & \cdot & c_{1i} \\ c_{21} & c_{22} & \cdot & \cdot & \cdot & c_{2i} \\ \cdot & \cdot & \cdot & \cdot & \cdot & \cdot \\ \cdot & \cdot & \cdot & \cdot & \cdot & \cdot \\ \cdot & \cdot & \cdot & \cdot & \cdot & \cdot \\ c_{i1} & c_{i2} & \cdot & \cdot & \cdot & c_{ii} \end{bmatrix}, \forall i > 0 \wedge i \in N \tag{1}$$

$$PR = \begin{bmatrix} \frac{1}{i} & \frac{1}{i} & \cdot & \cdot & \cdot & \frac{1}{i} \end{bmatrix}^T, \forall i > 0 \tag{2}$$

$$PR_{t+1} = H * PR_t, \forall t > 0 \tag{3}$$

$$\|PR_{t+1} - PR_t\| < \varepsilon, \varepsilon \in (0, 1] \tag{4}$$

Additionally, the value of each IP is earned by the being-connected relation both in modified PageRank and traditional PageRank. In modified PageRank, it has to display the importance of every IP. The value of each IP is also earned by being-connected relation. The only different between traditional PageRank and modified one is the original value. If IP Z doesn't have any being-connected IPs, the modified PageRank will give it an original value which is $(total\ suspicious\ event\ numbers\ of\ IPZ) * \frac{1}{i}$. Since IP and website have different circumstance, the original value of every IP can't be the same. For example, IP E launch thousands of suspicious events, and IP R launch only ten events. These two IP don't have any being-connected relation. Obviously, these two IP can't have same original value in calculation, so modified PageRank revises this part of the method to suit corresponding situation. Then, the method initializes PageRank value of every IP in formula (2). After that, it uses formula (3), iteration calculation, to find out the PageRank value. In order to know the convergence of the calculation, modified PageRank uses formula (4) to ensure accomplishment of the PageRank calculation. Finally, the PageRank value is represented each IP importance in the environment. After the calculation above, the suspicious IP management module will arrange the result into risk value and output risk ranking.

4 Experiment and Evaluation

The proposed system IPMRAS was deployed in the campus network in Fig. 5. The experiment is composed by EE router, NIDS, Web Server (including HIDS) and IPMRAS. By using IPMRAS, it can collect data from NIDS and HIDS. The data will first separate into attack IP and suspicious IP through the event type from supervise learning NIDS and pattern monitoring NIDS. If the event type is the same, the IP is an attack one. Otherwise, it is a suspicious IP.

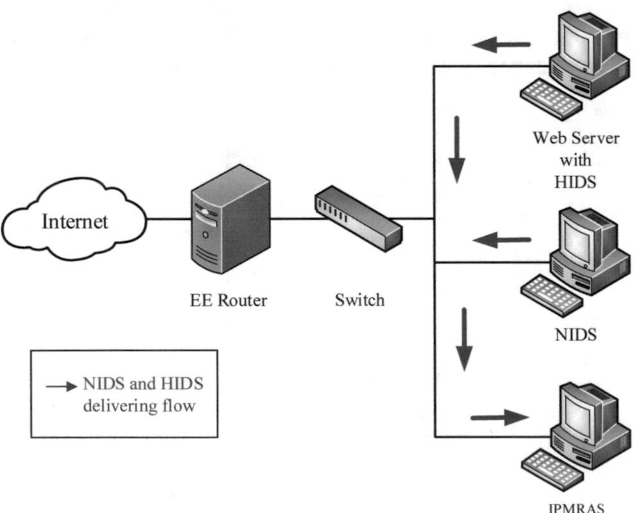

Fig. 5. Experiment environment

Table 2. Black list IP and corresponding information

Black list IP	Duration	Flow	Packet per second	Bytes per second	Bytes per packet	Recent appearing time
201.***.***.126	77727.85	256	0	256	12288	2018-07-14 13:34:27
212.***.***.78	265.25	256	651	297730	13312	2018-07-14 13:25:22
192. ***.***.84	224054.05	742	0	742	34132	2018-07-14 13:34:35
31. ***.***.74	1690.312	270	4	4871	299	2018-07-14 13:34:24
27. ***.***.34	655.868	194	4	5051	147	2018-07-14 13:34:26
185. ***.***.251	67694.3	514	2429	917211	23644	2018-07-14 13:34:35
5. ***.***.27	28682.5	150	391	147902	6900	2018-07-14 13:34:19

The attack IP management module will put attack IP into black list like Table 2 and also accumulate flow information such as duration, flow, packet per second etc. Besides, it will take black list to compare with HIDS attack record. The module will append the attack IP which is not in NIDS result into black list to prevent the missing attack. To ensure mechanism working, the research output an attack IP relation graph in Fig. 6. The big nodes are the attackers who launch many attacks to the little nodes which is represented the attack condition. It can show that lots of NCKU EE IPs are being attacked or connected by the attackers, so the black list IP is really harmful to the environment. By using the graph, it can show the attackers path and remind users about the situation.

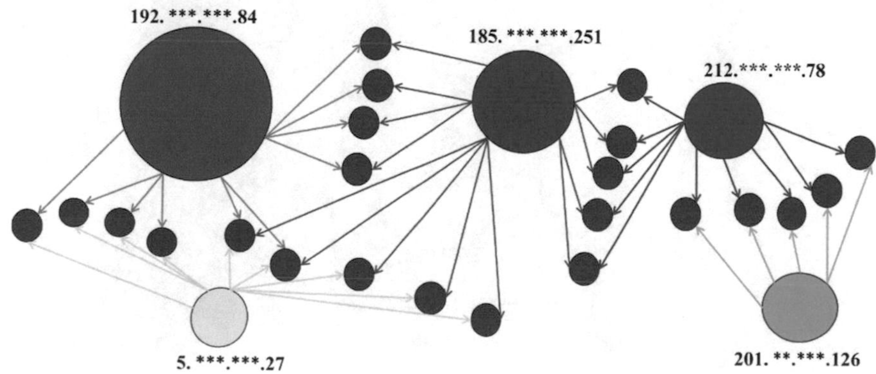

Fig. 6. Attack IP relation graph

The suspicious IP management module will divide different event type IP into three group, horizontal scan, vertical scan and flooding flow/brute force. According to the different patterns of each event type, it can use horizontal scan extent, vertical scan frequent and flooding flows/brute force duration to estimate the harm. Also, it will use the target port monitoring, location monitoring and HIDS availability to calculate the risk of suspicious IP. Besides, the relation among suspicious IP is also a significant feature to decide the risk. Thus, the module use modified PageRank to find out the suspicious IP relation degree. To sum it up, the final result of suspicious IP show in Table 3. Since external suspicious IP can't be found, the research only show the NCKU EE suspicious IP in Fig. 7 to ensure the suspicious IP risk assessment operation. The big nodes are the suspicious IP which connect to other in many times. It can display the suspicious relation among these IPs to show users the situation.

Table 3. Suspicious IP and corresponding information

Suspicious IP	Ranking	Risk	Behavior	PageRank
80.***.***.187	1	1.19	0.19	1.0
103.***.***.76	2	1.07	0.31	0.76
140.***.***.xxx	3	1.00	1.0	0.003
101.***.***.230	4	0.72	0.25	0.47
185.***.***.1	5	0.32	0.06	0.26
202.***.***.42	6	0.30	0.05	0.26
5.***.***.59	7	0.26	0.04	0.22

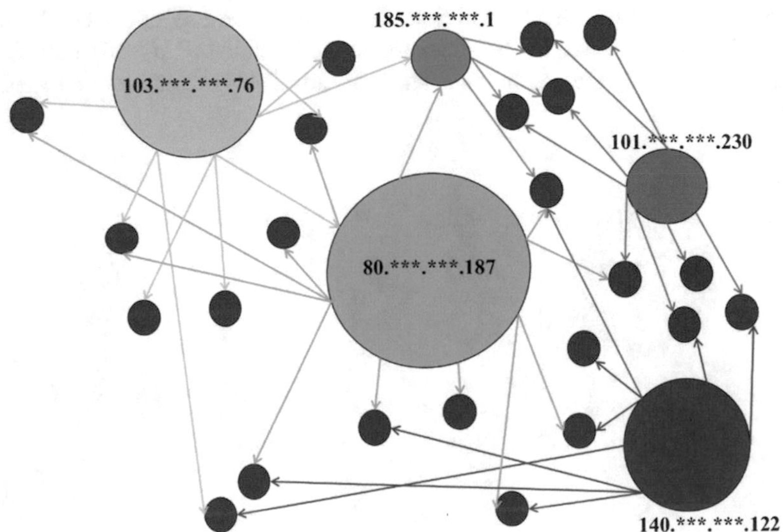

Fig. 7. Suspicious IP relation graph

5 Conclusion and Future Work

The paper proposed IPMRAS to analyze the attack and risk level in the campus environment. First, the system compared the feature of NIDS with different monitoring methods. It will separate the same label result of event type into attack IP and the different one into suspicious IP. Then, attack IP management module will make attack IP become black list. Also, attack IP will compare with HIDS attack record and append the missing-attack IP into black list to warn users about the situation.

On the other hand, the suspicious IP management module will divide suspicious flows into three event types. It will use different pattern of event type to estimate the risk of every IP. Then, it will execute target port monitoring, location monitoring and HIDS availability. Besides, in order to know the relation among the suspicious IP, the module will use modified PageRank to calculate the relation of every IP. The highest one means it launches as many suspicious connection as possible. The module then uses all the value to find out the risk of every suspicious IP and output the risk ranking.

Network really changes people's lives, but every coin has its two sides. No one can be safe and sound in the trend of explosive network attack. Also, network attacks can't be defensed by single solution or system any more. Because of the limitation about flows detection, NIDS can't accurately discover web-based attacks. These attacks can be found out by HIDS to monitor logs like apache log. Thus, IPMRAS combines two systems, HIDS and NIDS, to capture most of the attacks. However, only using HIDS and NIDS to analysis attack is still not enough since network technology grows rapidly. In the future, it can also mix other data source like honeypot to improve the accuracy of detection and risk assessment. The more information it collect, the better result it will show to keep user's safety.

References

1. Ehrenfeld, J.M.: Wannacry, cybersecurity and health information technology: a time to act. J. Med. Syst. **41**(7), 104 (2017)
2. Narayanaswamy, K., Burns, B., Manthena, V.R.R.: Protecting against distributed network flood attacks. U.S. Patent No. 8,789,173, 22 July 2014
3. Sabahi, F., Movaghar, A.: Intrusion detection: a survey. In: 2008 Third International Conference on Systems and Networks Communications (2008)
4. Lazarevic, A., Kumar, V., Srivastava, J.: Intrusion detection: a survey. In: Kumar, V., Srivastava, J., Lazarevic, A. (eds.) Managing Cyber Threats, pp. 19–78. Springer, Boston (2005)
5. Liao, H., Lin, C.R., Lin, Y., Tung, K.: Intrusion detection system: a comprehensive review. J. Netw. Comput. Appl. **36**(1), 16–24 (2013)
6. Vokorokos, L., Balaz, A.: Host-based intrusion detection system. In: 2010 IEEE 14th International Conference on Intelligent Engineering Systems (2010)
7. Vigna, G., Kemmerer, R.: NetSTAT: a network-based intrusion detection system. J. Comput. Secur. **7**(1), 37–71 (1999)
8. Data Mining: Concepts and Techniques. Liacs.leidenuniv.nl (2018). http://liacs.leidenuniv.nl/~bakkerem2/dbdm2007/05_dbdm2007_Data%20Mining.pdf. Accessed 10 Oct 2018

9. Page, L., et al.: The PageRank citation ranking: bringing order to the web. Stanford InfoLab (1999)
10. ISO 31000:2018. Iso.org (2018). https://www.iso.org/obp/ui/#iso:std:iso:31000:ed-2:v1:en. Accessed 3 Oct 2018
11. Li, S., et al.: An improved information security risk assessments method for cyber-physical-social computing and networking. IEEE Access **6**, 10311–10319 (2018)
12. Codetta-Raiteri, D., Portinale, L.: Decision networks for security risk assessment of critical infrastructures. ACM Trans. Internet Technol. (TOIT) **18**(3), 29 (2018)
13. Tseng, D., Yang, C.: A NetFlow based malicious traffic detection research using XGBoost. Cheng Kung University, Institute of Computer and Communication Engineering (2018)
14. Kuo, C., Yang, C.: Design and implementation of a network intrusion detection system based on NetFlow. Cheng Kung University, Institute of Computer and Communication Engineering (2015)
15. Yao, S., Yang, C.: Design and implementation of a host-based intrusion detection system for Linux-based web server on signature-based approach. Cheng Kung University, Institute of Computer and Communication Engineering (2018)

Integrated Forensic Tool for Network Attacks

Chia-Mei Chen[1(✉)], Gu-Hsin Lai[2], and Zheng-Xun Tsai[1]

[1] Department of Information Management, National Sun Yat-sen University, Kaohsiung, Taiwan
cchen@mail.nsysu.edu.tw

[2] Department of Technology Crime Investigation, Taiwan Police College, Taipei, Taiwan

Abstract. With the proliferation of cyber-attacks, Digital Forensic, also known as Computer Forensic, becomes more important to collect and analyze the seamless tracks that hackers leave. Through data acquisition, collection, preservation, analysis, examination and report generation, internet administrators are able to trace hackers and make sure of the loss. However, digital forensic is difficult since hackers tend to clean up the evidence of their existence, the complication of devices and log formats further increase the challenges. Existing digital forensic tools usually support some of the tasks in the forensic process instead of the comprehensive one. To make things harder for forensic investigators, these tools typically do not support each other. In order to ease the burden for investigators and make digital forensic available for general users, this re-search proposes an integrated system that can facilitate evidence acquisition, testing, analysis, and reporting in an integrated manner. This proposed system is expected to enhance the efficiency of digital forensic.

Keywords: Digital forensics · Computer forensics · Tools integration

1 Introduction

As digital transmission has become common in the business and daily communication, digital products usually store lots of confidential data including credit card number, contacts, personal credential, and business information. Once the digital products have been compromised, they might cause serious damage and financial loss. An increasing number of cybercrimes occurred worldwide [1]. The banking industry, government, academic institutes, and technology industry are the primary targets as well [2–8]. Attackers may apply various ways to perform attacks. It would consume a lot of time for the investigators to exercise the process of digital forensics, as the evidences need to be collected and analyzed from multiple sources and multiple forensics tools are acquired to fulfill the needs.

Digital forensics [9], also known as computer forensics, is to find the evidences in a cybercrime to keep track of the criminals. Digital forensics process consists of the following stages: acquisition, collection, preservation, analysis, examination, and reporting [10]. Many tools were developed for different stages. Each one has pros and cons; finding a suitable one is not easy among some with similar functionality. For instance, TCP View in Sysinternals developed by Microsoft and CurrPorts by NirSoft

C.-N. Yang et al. (Eds.): SICBS 2018, AISC 895, pp. 451–455, 2020.
https://doi.org/10.1007/978-3-030-16946-6_35

both perform the evidence collection of network states. The main functions of both tools are nearly the same, but some additional supporting features vary. Therefore, choosing proper tools is not an easy job.

This research proposes an automatic forensic system which integrates multiple existing tools and conduct evidence collection from multiple sources and preliminary analysis. The purpose of this research is to shorten the investigation time and hence reduce the incident damage.

2 Related Work

According to National Institute of Justice (NIJ) [10], handling electronic crime scene involves securing and evaluating the scene, documenting the scene, evidence collection, and packaging, transportation, and storage of digital evidence. Specifically, investigators should follow below steps in electronic crime forensics: (1) Evidence collection; (2) Evidence testing; (3) Evidence analysis; and (4) Reporting.

A variety of forensic tools have been developed to serve the growing needs in Digital Forensic. Popular tools include TCP View [12], Cports [13], Process Explorer [14], Procmon [15], TCP Flow [16], Wireshark [11], and Wmic [17]. These tools differ in the processing and output format.

Although these forensic tools are useful for its distinctive purposes, digital forensic requires a more integrated one to cope with the rapidly growth in cyber-attacks that exhaust investigators. According to Global Information Assurance Certification (GIAC) [18], integrity is needed in digital forensic and electronic data collection. GIAC [18] recommends that (1) use existing tool instead of writing your own program, (2) acquire information from live systems, (3) bit stream imaging, (4) do NOT add, delete, modify or filter collected information, (5) hash all evidence, (6) keep records of all behavior, and (7) back up the evidence.

3 System Design

According to the previous two chapters, there are two kind of problems. First, the diversity of forensics tools makes user hard to pick up proper tools, and evidences formats generated by different tools may also delay the analysis progress, making forensics harder. Second, common users may not comprehend those digital evidences and investigators may also need times to analyze them. To solve the problems, we propose an automatic forensics system, the system integrate some common use forensics tools to collect and analyze digital evidences, then generate report for users and investigators.

To enhance the system intelligence, the proposed integrated forensics system adopts configuration files which include the signatures of malware, network traffic, or anomalous behaviors. Based on NIST forensics analysis policy, the actions of evidence collection and analysis are separated into two stages. In the collection stage, the proposed system collects the digital evidences such as network connection, process, Windows registry and Windows event logs; the analysis stage performs the evidence analysis and correlation searching for suspicious events, processes, and network traffic and generates the report.

3.1 Collection Stage

To maintain the integrity of the digital evidence, the proposed system is not installed into the victim's computer. During the collection stage, the proposed system executes the digital tools and collects the digital evidences automatically. The following types of evidences are collected: (1) network evidences including the connecting ports and the process information which performs the connection; (2) process evidences including the running processes, the program path, the process information; (3) registry evidence [19]; and (4) event logs. The corresponding tools applied for the collection include CurrPorts, Procmon, Autoruns, and wevtutil.

3.2 Analysis Stage

Without the analysis, the evidences would not tell the story of the attack incident. This study proposes an integrated forensic system which performs the evidence collection and analysis. The analysis and correlation on the multiple evidences starts from network analysis as most malware perform certain anomalous network behaviors. Any connection of the process in the evidences which matches with an indicator from the configuration files will be marked for further investigation and will be correlated with the process analysis to see if there is any suspicious process.

Malware exhibit certain anomalous behaviors which diverse from the benign but attempting to behave like a normal process. For example, malicious programs may disguise as a system program such as svchost.exe. By comparing the parent process of each system process, the proposed system can identify such disguised malware. The path of the executable is essential evidence clue as well, as some malware might hide in hidden folders, such as %temp% and %appdata%.

The analysis on registry data mainly focuses on the processes running at the startup time, as malware attempt to execute automatically once the victim machine starts. The anomaly behaviors from malware infection leave traces on different places. By correlating the evidences from various sources, the system will produce the report which could speed up the incident response process.

4 Evaluation

Due to the urgency of the incident investigation, organizations in lack of expertise might not be able to handle the incidents promptly and incur huge loss. The proposed system aims for providing an efficient way to investigate the incidents for the above purpose. The purpose of this evaluation has to validate if the proposed system could identify the attacks automatically and correctly.

In this experiment, eight malware samples were selected to validate if the pro-posed system could identify the attacks automatically. The criteria of the malware selection is to cover various types of malware including coin miners, ransomware, Trojan, and backdoor. In real scenarios, the victims might discover the compromised machines after certain damage has been done. Therefore, the incident response and evidence collection might be performed after it. The experiment attempted to simulate the real cases and therefore the evidence data were collected in different time frames to see how efficiently the system can discover the attacks.

The experimental results in Table 1 demonstrate that the proposed system could identify the attacks automatically at different infection times, except ransomware. Some ransomware stopped the attack after the files were encrypted; therefore, no anomalies could be found afterwards. This experiment proves that the proposed system could identify the attacks automatically and efficiently.

Table 1. Experimental results of detecting various malware

Type of malware	Detection after 30 min	Detection after 24 h
Backdoor	O	O
Coin miner	O	O
Ransomware	O	X
Trojan	O	O

5 Conclusion

The promptness of incident response is crucial for damage control. Organizations might be short of the professional staff to perform the task. The commercial forensic tools are expensive and require expertise to interpret the reports. An automatic forensics tool helps the victims identify the attacks efficiently and reduces the time and efforts on collecting and analyzing the evidences. The proposed automatic forensics system integrates the available tools from the open source as well as the operating system vendor to perform evidence acquisition, analysis and correlation, and reporting in an integrated manner. The experimental results show that the proposed system provides an efficient way to investigate the incidents.

Advanced attacks might need professional experts to comprehend the evidences deeply. The future research could enhance the system by including expertise knowledge base and collect memory evidence into consideration.

References

1. Asia Times Staff: Taiwanese under Siege from Blitz of Chinese Cyberattacks. Asia Times, 6 April 2018. http://www.atimes.com/article/taiwanese-siege-blitz-chinese-cyberattacks/. Accessed 27 May 2018
2. Shen, T.A.: Hacker Group Targeting at Taiwan, The List of Security Incidents in Taiwan Financial Industry. iThome, 13 October 2017. https://www.ithome.com.tw/news/117386. Accessed 27 May 2018
3. Huang, Y.F.: The Hacking Procedures of ATM in Taiwan First Bank. iThome, 25 July 2016. https://www.ithome.com.tw/news/107294. Accessed 25 May 2018

4. Huang, Y.F.: More Hacking Details About Security Incidents of Far Eastern International Bank Are Revealed. iThome, 23 October 2017. https://www.ithome.com.tw/news/117397. Accessed 27 May 2018
5. TANet Computer Emergency Response Team. TWCERT. https://twcert.org.tw/subpages/cert/cert_taiwan_details.aspx?id=9. Accessed 29 May 2018
6. TACERT: Case Study: Linux Server in Campus Infected with Miner Malware, May 2018. https://portal.cert.tanet.edu.tw/docs/pdf/2018052502050404129385033631215.pdf. Accessed 29 May 2018
7. TACERT: Case Study: Website Hijacked by Coinhive Miner, 1 2017. https://portal.cert.tanet.edu.tw/docs/pdf/2018012305014949449974152724093.pdf. Accessed 29 May 2018
8. TACERT: Case Study: WannaCry Spread All Over the Computers in Campus, June 2015. https://portal.cert.tanet.edu.tw/docs/pdf/2017062004063434305283221149085.pdf. Accessed 29 May 2018
9. Baryamureeba, V., Tushabe, F.: The enhanced digital investigation process model. In: The Digital Forensic Research Conference, Baltimore, MD (2004)
10. Ademu, I.O., Imafidon, C.O., Preston, D.S.: A new approach of digital forensic model for digital forensic investigation. Int. J. Adv. Comput. Sci. Appl. **2**(12), 175–178 (2011)
11. Wireshark User Guide. https://www.wireshark.org/docs/man-pages/wireshark.html. Accessed 7 Nov 2017
12. Russinovich, M.: TCP view. https://docs.microsoft.com/en-us/sysinternals/downloads/tcpview. Accessed 5 May 2018
13. CurrPorts. http://www.nirsoft.net/utils/cports.html. Accessed 8 Oct 2017
14. Russinovich, M.: Process Explorer. https://docs.microsoft.com/en-us/sysinternals/downloads/process-explorer. Accessed 5 Nov 2017
15. Process Monitor. https://docs.microsoft.com/en-us/sysinternals/downloads/procmon. Accessed 5 Nov 2017
16. Garfinkel, S.L.: TCP Flow. https://github.com/simsong/tcpflow/wiki/tcpflow-%E2%80%94-A-tcp-ip-session-reassembler. Accessed 4 May 2018
17. Windows Management Instrument Console. https://msdn.microsoft.com/zh-tw/library/aa394531(v=vs.85).aspx. Accessed 6 Nov 2017
18. Chisholm, C., Groman, J.: Integrating forensic investigation methodology into eDiscovery. In: GIAC (GCFA) Gold Certification (2010)
19. Talebi, J., Dehghantanha, A., Mahmoud, R.: Introducing and analysis of the Windows 8 event log for forensic purposes. In: Garain, U., Shafait, F. (eds.) Computational Forensics, vol. 8915, pp. 145–162. Springer, Cham (2015)

Cryptanalysis of Anonymous Three Factor-Based Authentication Schemes for Multi-server Environment

Jiaqing Mo(✉), Hang Chen, and Wei Shen

School of Computer Science and Software, Zhaoqing University, Zhaoqing 526061, China
mojiaqing@126.com

Abstract. Cryptanalyzing the security weaknesses of authentication protocols is extremely important to propose countermeasures and develop a truly secure protocol. Over last few years, many three factor-based authentication schemes with key agreement have been proposed for multi-server environment. In 2017, Ali and Pal developed a three-factor authentication scheme in multi-server environment using elliptic curve cryptography (ECC) to remedy the security flaws in Li et al.'s scheme and claimed their improved version can withstand the passive and active attacks. In this paper, we prove that Ali-Pal's scheme is subject to offline password guessing attack, replay attack, and known session-specific temporary information (KSSTI) attack. In the same year, Feng et al. examined Kumari et al.'s biometrics-based authentication scheme for multi-server environment and found that their scheme was vulnerable to several attacks. To fix these weaknesses, Feng et al. proposed an enhanced three-factor authentication scheme with key distribution for mobile multi-server environment and claimed that their scheme can satisfy the security and functional requirements. However, we show that Feng et al.'s scheme fails to resist offline password guessing attack, and suffers from replay attack. In addition to point out the security defects, we put forward countermeasures to eliminate the security risks and secure the three factor-based authentication schemes for multi-server environment.

Keywords: Authentication · Three-factor security · Offline password guessing attack · Multi-server environment

1 Introduction

Thanks to significant advances in network, communication, software and hardware technologies, lots of services such as online-shopping, electronic bank, and online-chat are becoming more and more popular, and play an important role in people's daily life. Accordingly, as the network scale becomes larger, the multi-server located in different positions, which can address the weaknesses of limited storage and computation capability in the single server, has been developed as a scalable platform. The typical architecture of multi-server is showed in Fig. 1. Because of the openness of the net-

C.-N. Yang et al. (Eds.): SICBS 2018, AISC 895, pp. 456–468, 2020.
https://doi.org/10.1007/978-3-030-16946-6_36

work, an attacker can eavesdrop, delete, insert, modify or intercept the messages in the public channel and launch various attacks. Therefore, it is essential to design a truly secure authentication scheme for multi-server environment.

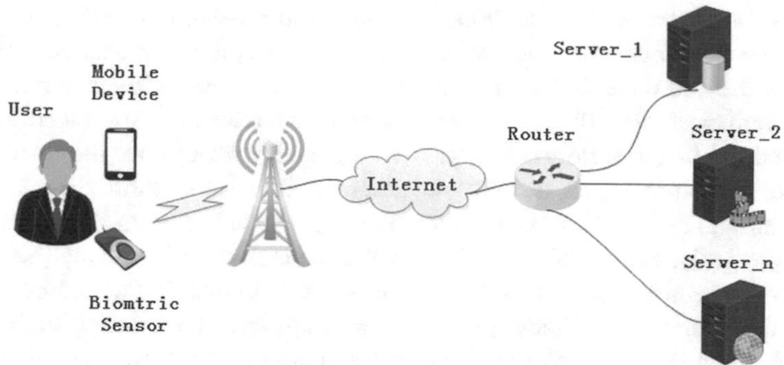

Fig. 1. The typical architecture of multi-server environment

The password based authentication scheme, where the password acts as the secret information to authenticate the user, has become the most convenient way to solve security problems. However, the password can be easily forgotten or may be guessed and stolen by the attacker. To solve this problem, the smart card, which can store the information issued to the user and have some computation capability, is used in authentication schemes. Thus, the password and the smart card are used as two factors in the authentication schemes for multi-server environment [1–9]. However, most of the two factor-based authentication schemes for multi-server are vulnerable to various kinds of attacks.

To address these security defects, the biometrics like iris and fingerprint, are introduced into the authentication protocols. Compared with the password and smart card, biometrics have many advantages as follows [10]:

(1) It is difficult to forge the biometrics;
(2) It is difficult to lose or forget the biometrics;
(3) It is difficult to copy, guess or break the biometric key.

In the three factor-based authentication schemes, the user can gain access only if he submits all the three factors (i.e. password, smart card, biometrics) to prove his identity. Hence, many three factor-based authentication schemes [11–24] were proposed to improve the security of the authentication schemes in multi-server environment. Here, we briefly introduce the schemes closest to this paper.

In 2013, Pippal et al. [25] presented a three-factor authentication scheme for multi-server to resist all types of attacks. However, Wei et al. [26] found that Pippal et al.'s scheme was subject to offline password guessing attack, user impersonation attack, and privileged insider attack. Later on, Guo et al. [27] also found that Pippal et al.'s scheme cannot withstand impersonation attack and offline password guessing attack, and proposed an improved version. Unfortunately, Ali and Pal [28] examined Guo et al.'s scheme and have revealed various attacks, and suggested an improved three factor-based authentication scheme. After that, Li et al. [29] revealed that Pippal et al.'s scheme suffered from impersonation, smart card stolen, and insider attacks. As a remedy, they put forward another three factor-based authentication scheme to overcome these vulnerabilities. Subsequently, Ali and Pal [23] analyzed the scheme of Li et al. and found that their scheme cannot withstand password guessing, user impersonation, insider, smart card theft attacks. To overcome these weaknesses, they have given an enhanced three factor-based authentication scheme for multi-server environment. Nevertheless, we will demonstrate that serious threats of their protocol have been overlooked: (1) It suffers from offline password guessing attack once the smart card and biometrics are compromised, that is to say, their scheme cannot fulfill the goal of three-factor security; (2) It suffers from replay attack; (3) It suffers from KSSTI attack.

More recently, Feng et al. [24] showed that Kumari et al.'s scheme [30] failed to meet user anonymity requirement and was vulnerable to man in the middle attack. To address these drawbacks, they proposed an improved three-factor authentication scheme for mobile multi-server environment and argued that their scheme can satisfy the security and functional requirements. In this work, we prove that Feng et al.'s scheme cannot provide truly three-factor security along with user impersonation attack.

As two cases studies, we demonstrate that most of the three factor-based authentication schemes like [23, 24] cannot provide truly three-factor security for multi-server environment. In particular, we propose countermeasures to thwart this security loophole and improve the security of similar protocols.

The remainder of the paper is organized as follows: We review Ali-Pal's scheme in Sect. 2, and describe the security threat in Sect. 3; Feng et al.'s scheme is reviewed in Sect. 4 and cryptanalysis is made in Sect. 5. Section 6 proposes countermeasures to fix the discovered threats. Finally, we conclude this paper in Sect. 7.

2 Review of Ali-Pal's Scheme [23]

In this section, we briefly review Ali-Pal's scheme for multi-server environment. Their scheme consists of six phases: initialization, server registration, user registration, login, authentication and key agreement, password change. For better presentation, we list the notations used in Ali-Pal's scheme as closely as possible (Table 1).

2.1 Initialization Phase

RC chooses a random number x as a secret parameter and an elliptic curve point P as a public parameter.

2.2 Server Registration Phase

The server S_j selects a free identity SID_j and delivers it to RC via the secure channel. On receipt of SID_j, RC computes $MK = h(SID_j \| x)$ and sends it to S_j via a secure channel.

Table 1. Notations

Notation	Description
U_i	i^{th} user
RC	The registration center
S_j	j^{th} server
$\mathcal{A}$	The adversary
UID_i	The identity of user U_i
SID_j	The identity of user S_j
PW_i	The password of user U_i
B_i	U_i's biometrics
$\oplus$	The bitwise XOR operation
$\|\|$	The concatenation operation
h()	A secure one-way hash function
H()	Bio-hash function
$E_k()/D_k()$	The symmetric encryption/decryption function with key k

2.3 User Registration Phase

In this phase, user U_i registers at RC.

(1) $U_i \rightarrow RC$: $\{UID_i, B_i\}$ (via a secure channel)
(2) $RC \rightarrow U_i$: smart card = $\{D_i, DID_i, P, E_k/D_k, h(), H()\}$ (via a secure channel) $DID_i = E_{h(x)}(UID_i \,||\, R)$, $A_i = H(B_i)P$, $C_i = h(UID_i \,||\, x)P$, $D_i = A_i + C_i$, where R is RC's secret random number.
(3) U_i computes $V_i = h(UID_i \,||\, PW_i \,||\, H(B_i))$ and writes it to the smart card.

2.4 Login Phase

(1) U_i inserts the smart card to the reader, inputs his identity UID_i^* and password PW_i^*, and imprints biometric B_i^*. Then the smart card computers $V_i^* = h(UID_i^*||PW_i^*||H(B_i^*))$ and checks whether $V_i^* = V_i$ holds. If it holds, it continues the session. Otherwise, it aborts the session.
(2) $U_i \rightarrow RC$: $\{DID_i, D_i, M_2, M_3\}$
$M_1 = R_C P$, $M_2 = H(B_i)P + M_1$, $M_3 = h(UID_i \,||\, M_1 \,||\, SID_j \,||\, D_i)$, where R_C is the random number chosen by U_i.

2.5 Authentication and Key Agreement Phase

(1) $RC \rightarrow S_j$: $\{M_5, M_6, DID_i^{new}, E_i\}$

After receiving the login message, RC decrypts DID_i and obtains (UID_i, R), then computes $A_i^* = D_i - h(UID_i||x)P, M_1^* = M_2 - A_i^*, M_3^* = h(UID_i||M_1^*||SID_j||D_i)$, and checks whether $M_3^* = M_3$ holds. If it is true, RC selects a random number R_{SC}, computes $DID_i^{new} = E_{h(x)}(UID_i || R_{SC})$, $E_i = E_{h(SIDj||x)}\ (UID_i || M_1^* || H(B_i))$, $M_4 = R_{SC}P$, $M_5 = M_4 + H(B_i)P$, $M_6 = h(M_4 || DID_i^{new} || UID_i || SID_j)$ and sends $\{M_5, M_6, DID_i^{new}, E_i\}$ to S_j.

(2) $S_j \rightarrow U_i$: $\{DID_i^{new}, M_9, M_8, M_5\}$

S_j computes $D_{h(SIDj\ ||\ x)}\ (E_i) = (UID_i\ ||\ M_1\ ||\ H(B_i))$, $M_4^* = M_5 - H(B_i)P$, $M_6^* = h(M_4^* || DID_i^{new} || UID_i || SID_j)$, and checks $M_6^* \stackrel{?}{=} M_6$. If it is true, S_j selects a random number R_S and computes $M_7 = R_SP$, $M_8 = M_7 + M_1$, $M_9 = h(DID_i^{new} || M_7 || M_4 || H(B_i)P)$ and sends $\{DID_i^{new}, M_9, M_8, M_5\}$ to U_i.

(3) $U_i \rightarrow S_j$: $\{L_i\}$

U_i computes $M_7* = M_8 - M_1, M_4^* = M_5 - H(B_i)P, M_9^* = h(DID_i^{new} || M_7 || M_4 || H(B_i)P)$, and checks whether $M_9^* = M_9$ holds. If it is true, U_i computes $sk = h(R_CP\ ||\ R_{SC}P\ ||\ R_SP)$ and $L_i = skP + h(UID_i\ ||\ H(B_i))P$, and sends $\{L_i\}$ to S_j.

(4) S_j computes $sk^* = h(R_CP\ ||\ R_{SC}P\ ||\ R_SP)$, $L_i^* = sk^*P + h(UIDi\ ||\ H(B_i))P$, and checks whether $L_i^* = L_i$ holds. If it is true, S_j believes that it shares the same session key sk with U_i.

2.6 Password Change Phase

U_i inserts smart card into the card reader, inputs UID_i^*, PW_i^* and imprints B_i^*, the card reader computes $V_i^* = h(UID_i^* || PW_i^* || H(B_i^*))$ and checks whether $V_i^* = V_i$ holds. If it is true, the card reader asks for input password PW_i^{new}, then computes $V_i^{new} = h(UID_i || PW_i^{*new} || H(B_i))$ and replaces V_i with V_i^{new} in the smart card.

3 Cryptanalysis of Ali-Pal's Scheme

In this section, the security weaknesses of Ali-Pal's scheme will be pointed out. Specifically, their scheme is subject to offline password guessing attack, replay attack, KSSTI attack, which make their scheme unsuitable for practical deployment. Before presenting the cryptanalysis of Ali-Pal's scheme, we define the adversarial model as follows [31–33]:

(1) The adversary can capture the messages exchanged between the related parties in the authentication session.
(2) The adversary can somehow obtain the U_i's smart card and reveal the secret parameters by using side-channel attack [34, 35].
(3) The adversary is able to enumerate offline all the possible candidate $D_{ID}*D_{PW}$ in polynomial time, where D_{ID} and D_{PW} represent the identity and password space, respectively.

3.1 Offline Password Guessing Attack

In Ali-Pal's scheme, they claimed that their scheme can withstand password and identity guessing attacks. However, we prove that their scheme is prone to offline password attack.

In the three-factor authentication scheme, even the adversary has learned two elements of the three factors (password, smart card, biometrics), he/she cannot carry out attacks. Unfortunately, if the lost/stolen smart card of U_i is obtained by the dedicated attacker $\mathcal{A}$, and the biometric characteristics (e.g., fingerprint) of U_i is accidentally obtained by $\mathcal{A}$ without the awareness of the owner, $\mathcal{A}$ will perform the following procedure to launch offline password guessing attack.

Step 1: $\mathcal{A}$ extracts the content $\{V_i, D_i, DID_i, P, E_k/D_k, h(), H()\}$ from the smart card using the analysis methods reported in [34, 35].

Step 2: $\mathcal{A}$ chooses a pair (UID_i^*, PW_i^*) from the dictionary space of D_{ID} and D_{PW}, respectively.

Step 3: $\mathcal{A}$ computes $V_i^* = h(UID_i^* || PW_i^* || H(B_i^*))$, where B_i^* is U_i's biometrics.

Step 4: $\mathcal{A}$ verifies the correctness of UID_i^* and PW_i^* by checking whether $V_i^* = V_i$ holds. If it is successful, $\mathcal{A}$ has found the correct pair (UID_i^*, PW_i^*). Otherwise, $\mathcal{A}$ repeats step 1–4 until $V_i^* = V_i$.

Let $|D_{ID}|$ and $|D_{PW}|$ be the size of D_{ID} and D_{PW}, respectively. The time complexity of above procedure is $O(|D_{ID}| * |D_{PW}| * T_h * T_H)$, where T_h, T_H denote the running time of the hash function $h()$, the Bio-hash function $H()$, respectively. Thus, the time needed for $\mathcal{A}$ to perform the above procedure is linear to $|D_{ID}| * |D_{PW}|$. According to [37], D_{ID} and D_{PW} is rather limited in practice with $|D_{PW}| \le |D_{ID}| \le 10^6$, so the time for carrying out an offline password guessing attack can be shortened to seconds in the normal computer. Consequently, Ali-Pal's scheme fails to provide three-factor security.

3.2 Replay Attack

It is highly essential for a cryptography protocol to resist the replay attack. Unfortunately, Ali-Pal's scheme is subject to replay attack owing to lack of timestamp. If $\mathcal{A}$ captures the messages like $\{M_5, M_6, DID_i^{new}, E_i\}$ during the authentication and key agreement phase and replays it to the RC, it is clear that RC will perform the subsequent procedure because RC has no mechanism to check the freshness of the message, the computation resource is wasted. Therefore, Ali-Pal's scheme is inherently vulnerable to replay attack.

3.3 KSSTI Attack

For an authentication scheme with key agreement, $\mathcal{A}$ is unable to compute the session key even though the ephemeral information is compromised. In Ali-Pal's scheme, the session key $sk = h(R_CP || R_{SC}P || R_SP)$, where R_C, R_{SC}, and R_S are the random numbers chosen by the user U_i, the remote center RC, the server S_j, respectively; and P is the public parameter. If the ephemeral information R_C, R_{SC}, and R_S are compromised by $\mathcal{A}$, he can compute the session key $sk = h(R_CP || R_{SC}P || R_SP)$ easily. Therefore, Ali-Pal's scheme is subject to KSSTI attack.

4 Review of Feng et al.'s Scheme [24]

In this session, we will concisely review Feng et al.'s scheme presented in 2017. Their scheme is an improvement over Kumari et al.'s scheme [30] and involves initialization phase, authentication and session key agreement phase, password updating phase. The notations in the Feng et al.'s scheme are almost the same as those in Sect. 2.

4.1 Initialization

The RC chooses a random number $s \in F_p$, computes $PKr = sP$, where F_p is a finite field with order p, and s is the secret master key. Then RC generates the public parameters $\{PKr, P, E_p, h(), H()\}$, E_p is an elliptic curve defined on F_p.

4.2 User Registration

(1) $U_i \rightarrow RC$: $\{ID_i, BPW_i\}$ (via a secure channel)

$BPW_i = H(BIO_i \,||\, PW_i)$, where BIO_i is the biometrics of U_i.

(2) $RC \rightarrow U_i$: smart card (via a secure channel)

RC computes $B_i = h(ID_i \,||\, s) \oplus BPW_i$, and stores $\{B_i, P, h(), H()\}$ into the smart card.

(3) U_i inserts $C_i = H(ID_i \,||\, PW_i \,||\, BIO_i)$ into smart card.

4.3 Cloud Server Registration

(1) $CS_j \rightarrow RC$: $\{SIDj\}$ (via a secure channel)

CS_j represents the jth cloud server, and SID_j is the identity of CS_j.

(2) $RC \rightarrow CS_j$: $\{D_j = h(SID_j \,||\, s)\}$(via a secure channel)

4.4 Authentication and Session Key Agreement

(1) $U_i \rightarrow CS_j$:$\{RID_i, E_i, \alpha\}$

$C_i = H(ID_i \,||\, PW_i \,||\, BIO_i)$, $BPW_i = H(BIO_i \,||\, PW_i)$, $A_i = BPW_i \oplus B_i$, $E_i = eP$, $E_i^* = ePKr = ((E_i^*)_x, (E_i^*)_y)$, $RID_i = ID_i \oplus h((E_i^*)_x, 1)$, $\alpha = h(ID_i||SID_j||E_i||(E_i^*)_y||A_i)$, where e is a random nonce chosen by the smart card.

(2) $CS_j \rightarrow U_i$: $\{RID_i, E_i, \alpha, RID_j, F_j, \beta\}$

CS_j chooses a random number f, computes á$F_j = fP, F_j^* = fPKr = ((F_j^*)_x, (F_j^*)_y)$, $RID_j = SID_j \oplus h((F_j^*)_x, 1)$, $\beta = h(RID_i|| E_i|| \alpha ||F_j||(F_j^*)_y|| SID_j|| D_j)$.

(3) $RC \rightarrow CS_j$:{CID_i,γ, $UIDj$,δ}

RC computes $F_r^* = sF_j = ((F_r^*)_x, (F_r^*)_y), SID_j = RID_j \oplus h((F_r^*)_x, 1)$, $D_j = h(SID_j \,||\, s)$, and checks whether $\beta = h(RID_i || E_i || \alpha || F_j || \left(F_j^*\right)_y || SID_j || D_j\right)$ holds. If it is true, then RC computes $E_r^* = sE_j = ((E_r^*)_x, (E_r^*)_y), ID_i = RID_i \oplus h((E_i^*)_x, 1)$, $A_i = h(ID_i \,||\, s)$, and checks whether $\alpha = h(ID_i||SID_j||E_i||(E_i^*)_y||A_i)$ holds. If it holds, RC computes $CID_i = ID_i \oplus h((F_r^*)_y, 2), CID_j = SID_j \oplus h((E_r^*)_y, 2), \gamma = h(CID_i|| ID_i|| SID_j||(F_r^*)_y|| D_j)$, $\delta = h(UID_j|| ID_i|| SID_j||(E_r^*)_x|| A_i)$.

(4) $CS_j \rightarrow U_i$: {$UIDi$, δ, $UIDj$, η}

CS_j computes $ID_i = CID_i \oplus h((F_j^*)_y, 2)$, and checks whether $\gamma = h(CID_i|| ID_i|| SID_j||\left(F_j^*\right)_x|| D_j\right)$ holds. If it is true, CS_j computes $sk_{ji} = h(ID_i \,||\, SID_j \,||\, fE_i)$, $\eta = h(UID_i \,||\, \delta \,||\, F_j \,||\, ID_i \,||\, SID_j \,||\, sk_{ji})$.

(5) $U_i \rightarrow CS_j$: {ρ}

U_i computes $SID_j = UID_j \oplus h((E_i^*)_y, 2)$, and checks whether $\delta = h(UID_j|| ID_i|| SID_j||\left(E_r^*\right)_x|| A_i\right)$ holds. If it is true, U_i computes $sk_{ij} = h(ID_i \,||\, SID_j \,||\, eF_j)$, and checks whether $\eta = h(UID_i \,||\, \delta \,||\, F_j \,||\, ID_i \,||\, SID_j \,||\, sk_{ij})$ holds. If it holds, U_i computes $\rho = h(UID_i \,||\, \delta \,||\, F_j \,||\, \eta \,||\, ID_i \,||\, SID_j \,||\, sk_{ij})$.

(6) CS_j checks whether $\rho = h(UID_i \,||\, \delta \,||\, F_j \,||\, \eta \,||\, ID_i \,||\, SID_j \,||\, sk_{ij})$ holds. If it is true, CS_j accepts sk_{ij} as the session key.

4.5 Password Updating

(1) $U_i \rightarrow$ smart card: $\{ID_i, PW_i, BIO_i, PW_i^{new}\}$

PW_i^{new}: U_i's new password.

(2) The smart card checks whether $C_i = H(ID_i \,||\, PW_i \,||\, BIO_i)$ holds. If it holds, the smart card aborts this session. Otherwise, the smart card replaces B_i and C_i with $(Bi \oplus H(PW_i|| BIO_i) \oplus H(PW_i^{new}|| BIO_i))$ and $H(ID_i|| PW_i^{new}|| BIO_i)$, respectively.

5 Cryptanalysis of Feng et al.'s Scheme

In this section, we point out that Feng et al.'s scheme cannot prevent offline password guessing attack and suffers from replay attack which make their scheme impractical.

5.1 Offline Password Guessing Attack

Suppose that $\mathcal{A}$ compromised U_i's smart card and biometrics BIO_i^* as illustrated in Sect. 3.1, we address the scenario of offline password guessing attack as follows.

Step 1: $\mathcal{A}$ extracts the secret information $\{C_i, B_i, P, h(), H()\}$, where $C_i = H(ID_i||PW_i||BIO_i^*)$.

Step 2: $\mathcal{A}$ selects a pair (ID_i^*, PW_i^*) from the identity space of D_{ID} and the password space D_{PW}.

Step 3: $\mathcal{A}$ computes $C_i^* = H(ID_i^*||PW_i^*||BIO_i^*)$, and checks whether $C_i^* = C_i$ holds, where C_i is revealed from U_i's smart card. If it is true, $\mathcal{A}$ finds the correct pair of (ID_i, PW_i). Otherwise, $\mathcal{A}$ repeats step 1–3 until the correct pair (ID_i, PW_i) is found.

With the right ID_i and PW_i of the legal user U_i, $\mathcal{A}$ could impersonate U_i to initiate a session with cloud server CS_j easily.

In this regard, Feng et al.'s scheme suffers from offline password guessing attack and fails to provide three-factor secrecy as they proclaimed.

5.2 Replay Attack

In Feng et al.'s scheme, suppose that $\mathcal{A}$ captures message $\{RID_i, E_i, \alpha\}$ and retransmits it to CS_j, CS_j will execute the subsequent procedure defined by the protocol because their scheme doesn't provide timestamp mechanism to check the freshness of the message. Similarly, if $\mathcal{A}$ intercepts message $\{RID_i, E_i, \alpha, RID_j, F_j, \beta\}$ and replays it to RC, RC will not be aware of the replay action resulting in waste of computing resource.

In general, any cryptography scheme using a random number in each party is incapable of implementing explicit authentication while withstanding replay attack, because if $\mathcal{A}$ replays the message to impersonate the legal user, the cloud server has no mechanism to verify whether the replayed message is fresh. Furthermore, if the server maintains a table for the received messages, their scheme becomes impractical with the number of received messages increase rapidly.

6 Countermeasures

In this section, we brief countermeasures to overcome the common security weaknesses found in schemes of Ali-Pal [23] and Feng et al. [24].

6.1 Countermeasure to Offline Password Guessing Attack

Both Ali-Pal's scheme and Feng et al.'s scheme are discovered vulnerable to offline password guessing attack leading to failure of providing truly three-factor security. In these two schemes, the verifier $V_i = h(UID_i || PW_i || H(B_i))$ of Ali-Pal's scheme and $C_i = H(ID_i || PW_i || BIO_i)$ of Feng et al.'s scheme are stored in the smart card. Thus, if $\mathcal{A}$ somehow obtains the smart card and the biometrics of U_i, $\mathcal{A}$ could extract the secret parameters stored in smart card and conduct offline password guessing attack as we mentioned earlier.

To hinder this security threat, one of the feasible ways is to exploit "fuzzy verifier" mechanism [38]. Take Ali-Pal's scheme as an example, we modify V_i in the user registration phase as $V_i = h(h(UID_i || PW_i || H(B_i)) \bmod m)$ and store it in the smart card, where m is used to determine the space of (ID_i, PW_i) pair. If the smart card and the biometrics are compromised by $\mathcal{A}$, there are $\frac{|D_{ID}|*|D_{PW}|}{m}$ candidates (ID_i, PW_i) to thwart $\mathcal{A}$. For example, suppose that ID_i and PW_i are made up of numbers, i.e. $|D_{PW}| = |D_{ID}| = 10^8$, and we set m = 2^8, so there exist $\frac{|D_{ID}|*|D_{PW}|}{m} = \frac{10^8*10^8}{2^8} \approx 2^{45}$ candidates. As a result, it is difficult for $\mathcal{A}$ to perform the offline password guessing attack. One may doubt that if the attacker accidentally inputs an incorrect pair (ID_i^*, PW_i^*) which satisfies $V_i = h(h(ID_i^* || PW_i^* || H(B_i)) mod\, m)$, but (ID_i^*, PW_i^*) is not equal to (ID_i, PW_i)? The probability is $\frac{1}{2^8}$. Further, if the user is required to input his old/new passwords twice and $h()$ responds like a random oracle, the probability will be reduced to $\left(\frac{1}{2^8}\right)^2 = \frac{1}{2^{16}}$. Thus, the fuzzy verifier can be used in the authentication scheme to frustrate the adversary to conduct the offline password guessing attack.

6.2 Countermeasure to Replay Attack

In the cryptography protocol, it is quite hard to avoid replay attack and resisting replay attack has become a challenging issue. The purpose of the replay attack is to imitate as a legal protocol participant. Generally speaking, there are two techniques to avert this attack, i.e. random number and timestamp, the basic idea of these techniques is to identify the freshness of the received messages. If an authentication scheme only adopts random number approach to guarantee the freshness of the received messages without timestamp, it should maintain a database to save all the received messages, which leads to their scheme becomes undesirable. On the other hand, if the authentication scheme adopts timestamp and uses the hash function to protect the timestamp, when $\mathcal{A}$ intends to retransmit the previous messages, the receiver will detect this attack instantly by verifying the timestamp and the corresponding hash value. Here, we take Ali-Pal's scheme as an example and propose countermeasure to thwart replay attack and modify their scheme as follows.

(1) In step 2 of the login phase, we alter M_3 as $M_3 = h(UID_i \,||\, M_1 \,||\, SID_j \,||\, D_i \,||\, T_1)$, and revise the transmitted message as $\{DID_i, D_i, M_2, M_3, T_1\}$, where T_1 is the current timestamp.

(2) In step 1 of authentication and key agreement phase, we modify M_3^* as $M_3^* = h(UID_i || M_1^* || SID_j || D_i || T_1')$, where T_1' is timestamp in the message $\{DID_i, D_i, M_2, M_3, T_1'\}$ from U_i via the public channel.

We consider the following two cases:

Case 1: If $\mathcal{A}$ alters T_1' and the verification $T_R - T_1' \geq \Delta t$ is false, RC will abort this session. T_R is the timestamp when RC received the message.

Case 2: If $\mathcal{A}$ alters T_1' and the verification $T_R - T_1' \geq \Delta t$ is true, RC computes $M_3^* = h(UID_i || M_1^* || SID_j || D_i || T_1')$ and checks whether $M_3^* = M_3$ holds. It is noted that this verification will fail because M_3 involved the original timestamp T_1 in the intercepted message $\{DID_i, D_i, M_2, M_3, T_1\}$.

Therefore, the countermeasure we presented can effectively obviate the replay attack in schemes of Ali-Pal [23] and Feng et al. [24].

7 Conclusion

A large number of three factor-based authentication with key agreement schemes have been presented for multi-server environment. However, most of them suffer from offline password guessing attack which leads to failure of achieving truly three-factor security. Very recently, Ali-Pal and Feng et al. proposed two anonymous three factor-based authentication schemes for multi-server environment to overcome the security defects in previous schemes. However, via careful examination, we prove that both of them still fail to provide three-factor security once the smart card and the biometrics are compromised. Meanwhile, these two schemes are susceptible to replay attack. We also put forward countermeasures to fix the security weaknesses. Our results highlight the importance of a robust security model in the design of three factor-based authentication schemes for multi-server environment and the similar kinds.

Acknowledgements. This work was partially supported by the National Natural Science Foundation of China (Project No. 61672007), Science and Technology Innovation Guidance Project 2017 (Project No. 201704030605).

References

1. Liao, Y.P., Wang, S.S.: A secure dynamic ID based remote user authentication scheme for multi-server environment. Comput. Stan. Interfaces **31**, 24–29 (2009)
2. Liao, Y.P., Wang, S.S.: Improvement of the secure dynamic ID based remote user authentication scheme for multi-server environment. Comput. Stan. Interfaces **31**, 1118–1123 (2009)
3. Sood, S.K., Sarje, A.K., Singh, K.: A secure dynamic identity based authentication protocol for multi-server architecture. J. Network Comput. Appl. **34**, 609–618 (2011)
4. Li, X., Xiong, Y., Ma, J., Wang, W.: An efficient and security dynamic identity based authentication protocol for multi-server architecture using smart cards. J. Network Comput. Appl. **35**, 763–769 (2012)
5. Han, W.: Weaknesses of a dynamic identity based authentication protocol for multi-server architecture. arXiv preprint arXiv:1201.0883 (2012)
6. Xue, K., Hong, P., Ma, C.: A lightweight dynamic pseudonym identity based authentication and key agreement protocol without verification tables for multi-server architecture. J. Comput. Syst. Sci. **80**, 195–206 (2014)
7. Wang, D., Ma, C.-g., Gu, D.-l., Cui, Z.-s.: Cryptanalysis of two dynamic id-based remote user authentication schemes for multi-server architecture. In: International Conference on Network and System Security, pp. 462–475. Springer (2012)
8. Xie, Q., Wong, D.S., Wang, G., Tan, X., Chen, K., Fang, L.: Provably secure dynamic ID-based anonymous two-factor authenticated key exchange protocol with extended security model. IEEE Trans. Inf. Forensics Secur. **12**, 1382–1392 (2017)
9. Chuang, M.-C., Chen, M.C.: An anonymous multi-server authenticated key agreement scheme based on trust computing using smart cards and biometrics. Expert Syst. Appl. **41**, 1411–1418 (2014)
10. Li, C.-T., Hwang, M.-S.: An efficient biometrics-based remote user authentication scheme using smart cards. J. Network Comput. Appl. **33**, 1–5 (2010)

11. Yang, D., Yang, B.: A biometric password-based multi-server authentication scheme with smart card. In: 2010 International Conference on Computer Design and Applications (ICCDA), pp. V5-554–V555-559. IEEE (2010)
12. Yoon, E.-J., Yoo, K.-Y.: Robust biometrics-based multi-server authentication with key agreement scheme for smart cards on elliptic curve cryptosystem. J. Supercomput. **63**, 235–255 (2013)
13. He, D.: Security flaws in a biometrics-based multi-server authentication with key agreement scheme. IACR Cryptology ePrint Archive 2011, 365 (2011)
14. Kim, H., Jeon, W., Lee, K., Lee, Y., Won, D.: Cryptanalysis and improvement of a biometrics-based multi-server authentication with key agreement scheme. In: International Conference on Computational Science and Its Applications, pp. 391–406. Springer (2012)
15. Mishra, D., Das, A.K., Mukhopadhyay, S.: A secure user anonymity-preserving biometric-based multi-server authenticated key agreement scheme using smart cards. Expert Syst. Appl. **41**, 8129–8143 (2014)
16. Lin, H., Wen, F., Du, C.: An improved anonymous multi-server authenticated key agreement scheme using smart cards and biometrics. Wireless Pers. Commun. **84**, 2351–2362 (2015)
17. Lu, Y., Li, L., Yang, X., Yang, Y.: Robust biometrics based authentication and key agreement scheme for multi-server environments using smart cards. PLoS ONE **10**, e0126323 (2015)
18. Wang, C., Zhang, X., Zheng, Z.: Cryptanalysis and improvement of a biometric-based multi-server authentication and key agreement scheme. PLoS ONE **11**, e0149173 (2016)
19. He, D., Wang, D.: Robust biometrics-based authentication scheme for multiserver environment. IEEE Syst. J. **9**, 816–823 (2015)
20. Jiang, P., Wen, Q., Li, W., Jin, Z., Zhang, H.: An anonymous and efficient remote biometrics user authentication scheme in a multi server environment. Frontiers Comput. Sci. **9**, 142–156 (2015)
21. Odelu, V., Das, A.K., Goswami, A.: A secure biometrics-based multi-server authentication protocol using smart cards. IEEE Trans. Inf. Forensics Secur. **10**, 1953–1966 (2015)
22. Reddy, A.G., Yoon, E.-J., Das, A.K., Odelu, V., Yoo, K.-Y.: Design of mutually authenticated key agreement protocol resistant to impersonation attacks for multi-server environment. IEEE Access **5**, 3622–3639 (2017)
23. Ali, R., Pal, A.K.: An efficient three factor-based authentication scheme in multiserver environment using ECC. Int. J. Commun Syst **31**, e3484 (2017)
24. Feng, Q., He, D., Zeadally, S., Wang, H.: Anonymous biometrics-based authentication scheme with key distribution for mobile multi-server environment. Future Gener. Comput. Syst. **84**, 239–251 (2017)
25. Pippal, R.S., Jaidhar, C., Tapaswi, S.: Robust smart card authentication scheme for multi-server architecture. Wireless Pers. Commun. **72**, 729–745 (2013)
26. Wei, J., Liu, W., Hu, X.: Cryptanalysis and improvement of a robust smart card authentication scheme for multi-server architecture. Wireless Pers. Commun. **77**, 2255–2269 (2014)
27. Guo, D., Wen, F.: Analysis and improvement of a robust smart card based-authentication scheme for multi-server architecture. Wireless Pers. Commun. **78**, 475–490 (2014)
28. Ali, R., Pal, A.K.: Three-factor-based confidentiality-preserving remote user authentication scheme in multi-server environment. Arab. J. Sci. Eng. **42**, 3655–3672 (2017)
29. Li, X., Niu, J., Kumari, S., Liao, J., Liang, W.: An enhancement of a smart card authentication scheme for multi-server architecture. Wireless Pers. Commun. **80**, 175–192 (2015)

30. Kumari, S., Li, X., Wu, F., Das, A.K., Choo, K.-K.R., Shen, J.: Design of a provably secure biometrics-based multi-cloud-server authentication scheme. Future Gener. Comput. Syst. **68**, 320–330 (2017)
31. Wang, D., He, D., Wang, P., Chu, C.-H.: Anonymous two-factor authentication in distributed systems: certain goals are beyond attainment. IEEE Tran. Dependable Secure Comput. 1 (2015)
32. Wang, D., Wang, P.: Two birds with one stone: two-factor authentication with security beyond conventional bound. IEEE Trans. Dependable Secure Comput. (2016)
33. Wang, D., Gu, Q., Cheng, H., Wang, P.: The request for better measurement: a comparative evaluation of two-factor authentication schemes. In: Proceedings of the 11th ACM on Asia Conference on Computer and Communications Security, pp. 475–486. ACM (2016)
34. Kocher, P., Jaffe, J., Jun, B.: Differential power analysis. In: Annual International Cryptology Conference, pp. 388–397. Springer (1999)
35. Messerges, T.S., Dabbish, E.A., Sloan, R.H.: Examining smart-card security under the threat of power analysis attacks. IEEE Trans. Comput. **51**, 541–552 (2002)
36. Islam, S.H.: Design and analysis of an improved smartcard-based remote user password authentication scheme. Int. J. Commun Syst **29**, 1708–1719 (2016)
37. Wang, D., Wang, P.: Understanding security failures of two-factor authentication schemes for real-time applications in hierarchical wireless sensor networks. Ad Hoc Netw. **20**, 1–15 (2014)
38. Ma, C.G., Wang, D., Zhao, S.D.: Security flaws in two improved remote user authentication schemes using smart cards. Int. J. Commun Syst **27**, 2215–2227 (2014)

LWE-Based Single-Server Block Private Information Retrieval Protocol

Shuai Liu(✉) and Bin Hu

Information Science and Technology Institute, Zhengzhou, China
sssshuai1993@163.com

Abstract. The appearance of fully homomorphic encryption (FHE) scheme induces new ways to construct the single-server private information retrieval protocol. At PKC 2015, Hiromasa et al. proposed the first FHE scheme that encrypts matrices and supports homomorphic matrix addition and multiplication. (hereafter, referred to as HAO15 scheme). Motivated by their work, we construct a LWE-based single-server block private information retrieval protocol. To get almost optimal communication cost, we adopt the homomorphic-ciphertext compression technique proposed by Naehrig et al. And as an intermediate product, we give a homomorphic algorithm, with no need for the secret key, to check the equality between diagonal matrices that are encrypted under HAO15 scheme.

Keywords: Block private information retrieval · Fully homomorphic encryption · Homomorphic-Ciphertext compression

1 Introduction

Private information retrieval (PIR) protocol allows a user to retrieve the i-th bit in an u-bit database, without revealing to others any information about which bit is retrieved. A trivial solution is that the user retrieves the entire database, but it induces enormous communication cost. Block private information retrieval (BPIR) is a natural and more practical extension of PIR in which, instead of retrieving only a single bit, the user retrieves a data block including multiple bits from the database server. Communication complexity and computation complexity are two important indicators to evaluate a PIR or BPIR protocol.

In 1995, Chor et al. [1] first came up with the notion of PIR in a multi-server setting, where the user retrieves information from multiple database servers, each has a copy of the same database. To protect the user's privacy in the multi-server setting, they must make sure that there is no collusion between all servers. If only single database server is used, the communication complexity will be $\Omega(u)$ in the information-theoretic sense. In [1], they have also shown that we can convert any PIR protocol to a BPIR protocol. So, all the time the developments of PIR and BPIR are synchronous.

In 1997, Ambainis [2] constructed a multi-server PIR protocol with communication complexity of $O(u^{1/(2c-1)})$, where c is the number of database servers. After that, there were varieties of improved schemes [3–5], but the communication complexity of those schemes is still unacceptable.

C.-N. Yang et al. (Eds.): SICBS 2018, AISC 895, pp. 469–480, 2020.
https://doi.org/10.1007/978-3-030-16946-6_37

In [6], Kushilevitz and Ostrovsky used the quadratic residue computational assumption to construct a single-server PIR with communication complexity of $O(u^{\varepsilon})(\varepsilon > 0)$. Subsequently, researchers proposed some single-server PIR protocols that are respectively based on the Φ-hinding assumption [7], discrete logarithm problem [8], one-way trapdoor permutations [9] and so on. The communication complexity of most those protocols is also $O(u^{\varepsilon})(\varepsilon > 0)$.

As the development of FHE, researchers began to adopt new ways to construct the PIR protocol by using FHE schemes. In a nutshell, a fully homomorphic encryption scheme is an encryption scheme that allows evaluation of arbitrarily complex programs on encrypted data. In 2009, Gentry [10] proposed the first FHE scheme based on ideal lattices and used the scheme to construct a PIR protocol roughly. In 2011, Brakerski and Vaikuntanathan [11] presented a FHE scheme that is based solely on the (standard) learning with errors (LWE) assumption. The scheme has very short ciphertexts and they therefore use it to construct an asymptotically efficient LWE-based single-server private information retrieval protocol. The communication complexity of the protocol (in the public-key model) is $\lambda \cdot \text{polylog}(\lambda) + \log u$ bits per single-bit query (here, λ is a security parameter).

In 2013, Xun et al. [12] constructed a simple and universal PIR protocol model that can be instantiated with any FHE scheme, called YKPB-PIR protocol model. They also gave an instance with a variant of Dijk et al.'s [13] somewhat homomorphic encryption scheme. Furthermore, they extended the PIR protocol to a BPIR protocol. Up to now, most of the PIR protocols based on FHE schemes adopt YKPB-PIR protocol model. The performance of this type of protocols mainly depends on the improvement of FHE schemes.

Almost all of previous BPIR protocols are obtained by extending the pre-existing PIR protocols, and it's significant to study how to design efficient BPIR protocols directly.

1.1 Our Results

Motivated by recent work of Hiromasa et al., we construct a LWE-based single-server block private information retrieval protocol. In 2015, Hiromasa, Abe, and Okamoto proposed a variant [14] of GSW-FHE [15], which can encrypt matrices and supports homomorphic matrix addition and multiplication. This is a natural extension of packed FHE and thus supports more complicated homomorphic operations. We store the data block in the plaintext matrix of HAO15 scheme and give a homomorphic algorithm, with no need for the secret key, to check the equality between diagonal matrices that are encrypted under HAO15 scheme.

To provide almost optimal communication cost, we adopt the homomorphic-ciphertext compression technique proposed by Naehrig et al. [16], when the user sends the query to the database server. Then the communication complexity of our protocol can reach $O(\log v)$ bits every query, where v is the number of data blocks in the database.

1.2 Organization

The rest of the paper is organized as follows. In Sect. 2, we give the definition of single-server BPIR and introduce the HAO15 scheme. In Sect. 3 we describe how to encrypt an index under HAO15 scheme and give an algorithm to check the equality homomorphically between encrypted indices. In Sect. 4 we construct a LWE-based single-server block private information retrieval protocol and analyze the correctness, security and performance of the protocol. And we make a conclusion in Sect. 5.

2 Preliminaries

For a nonnegative integer n, we let $[n] = \{1, \cdots, n\}$. We let $\mathbb{Z}$ denote the set of integers. Vectors are written by using bold lower-case letters, e.g., $\boldsymbol{x}$. Matrices are written by using bold capital letters, e.g., $\boldsymbol{X}$, and the i-th column vector of a matrix is denoted by $\boldsymbol{x}_i$. And we let $\boldsymbol{I}_n$ denote the $n \times n$ identity matrix.

2.1 Definitions of Single-Server BPIR

We define the single-server block private information retrieval in the public-key setting. In this setting, there is a public key associated with the user (who holds the respective secret key). This public key is independent of the query and of the database, and can be generated and sent before the interaction begins, and may be used many times. Thus, the size of the public key is not counted towards communication complexity of the protocol. We formalize this by an efficient setup procedure that runs before the protocol starts and generates this public key.

We first initialize all parameters. Let λ be the security parameter. Assume that the server has access to a database $DB \in \{0, 1\}^u$, and DB is composed of v data blocks $B_1, B_2, \cdots, B_v$, each of which includes w bits (so we have $u = vw$). We can represent the database by $DB = B_1|B_2|\cdots|B_v$ with $B_i = (b_{i,1}, b_{i,2}, \cdots, b_{i,w})\ (i = 1, \cdots, v)$. A single-server BPIR protocol in the public-key setting consists of four polynomial-time computable algorithms (BPIR.Setup, BPIR.Query, BPIR.Response, BPIR.Decode) as follows:

0. **Setup.** The user runs the setup algorithm

$$(params, setupstate) \leftarrow \text{BPIR.Setup}(1^\lambda).$$

It then sends the public set of parameters *params* (the public key) to the server and keeps the secret state *setupstate* private.

1. **Query.** The user wants to receive the i-th data block from the server, and it runs

$$(query, qstate) \leftarrow \text{BPIR.Query}(1^\lambda, setupstate, i).$$

The query message *query* is then sent to the server and *qstate* is a query-specific secret information that is kept private.

2. **Answer.** The server has access to a database $DB = B_1|B_2|\cdots|B_v$. Upon receiving the query message *query* from the user, it runs the "answering" algorithm

$$resp \leftarrow \text{BPIR.Response}(1^\lambda, DB, params, query).$$

And it sends the response *resp* back to the user.
3. **Decode.** The user decodes the response *resp* by running

$$x \leftarrow \text{BPIR.Decode}(1^\lambda, setupstate, qstate, resp).$$

Then x is the output of the protocol.

We note that while in general a multi-round interactive protocol is required for each database query, the protocol we present are of the simple form of a query message followed by a response message. Hence, we choose to present the simple syntax above.

The communication complexity of the protocol is defined to be $|query| + |resp|$. Namely, the number of bits being exchanged to retrieve a single data block (excluding the setup phase).

Correctness and security are defined as follows.

Definition 1 (Correctness Definition). A single-server BPIR protocol is correct if for any security parameter λ, any database $DB = B_1|B_2|\cdots|B_v$ with any size, and any index $i \in [v]$, it holds that $B_i = \text{BPIR.Decode}(1^\lambda, setupstate, qstate, resp)$, where $(params, setupstate) \leftarrow \text{BPIR.Setup}(1^\lambda)$, $(query, qstate) \leftarrow \text{BPIR.Query}(1^\lambda, setupstate, i)$, $resp \leftarrow \text{BPIR.Response}(1^\lambda, DB, params, query)$.

The security of the single-server BPIR protocol can be defined with a game between an adversary (the server) $\mathcal{A}$ and a challenger $\mathcal{C}$ as follows:

1. The adversary $\mathcal{A}$ chooses two different indices $i, j \in [v]$ and sends them to $\mathcal{C}$.
2. Let $\beta_0 = i$ and $\beta_1 = j$. The challenger $\mathcal{C}$ chooses a random bit $b \in \{0, 1\}$, runs $(query, qstate) \leftarrow \text{BPIR.Query}(1^\lambda, setupstate, \beta_b)$, and then sends *query* back to $\mathcal{A}$.
3. The adversary $\mathcal{A}$ can experiment with *query* in arbitrary nonblack-box way, and finally outputs $b' \in \{0, 1\}$.

The adversary $\mathcal{A}$ wins the game if $b' = b$ and loses otherwise. We define the adversary's advantage in this game to be

$$\text{Adv}_{\mathcal{A}}(\lambda) = |\Pr(b' = b) - 1/2|.$$

Definition 2 (Security Definition). Asingle-server BPIR protocol is semantically secure if for any PPT adversary $\mathcal{A}$, we have that $\text{Adv}_{\mathcal{A}}(\lambda)$ is a negligible function.

2.2 Matrix GSW-FHE

Making a natural extension of packed FHE, Hiromasa et al. translated GSW-FHE scheme [15] to be able to encrypt a matrix and homomorphically compute matrix addition and multiplication [14]. In this section, we will introduce their FHE scheme (HAO15 scheme) in detail.

Let λ be the security parameter and r be the number of bits to be encrypted, which defines the message space $\{0,1\}^{r\times r}$. The HAO15 scheme is parameterized by an integer dimension n, an integer modulus q, and an error distribution χ over $\mathbb{Z}$ which we assume to be subgaussian. Let $l = \lceil \log q \rceil$, $m = O((n+r)\log q)$, $N = (n+r)\cdot l$. The ciphertext space is $\mathbb{Z}_q^{(n+r)\times N}$. Let $\boldsymbol{g}^T = (1, 2, 2^2, \cdots, 2^{l-1})$ and $\boldsymbol{G} = \boldsymbol{g}^T \otimes \boldsymbol{I}_{n+r}$.

The HAO15 scheme consists of the following algorithms:

- HAO15.KeyGen$(1^\lambda, r)$: Set the parameters n, q, m, l, N, and χ as described above. Sample a uniformly random matrix $\boldsymbol{A} \leftarrow \mathbb{Z}_q^{n\times m}$, secret key matrix $\boldsymbol{S}' \leftarrow \chi^{r\times n}$, and noise matrix $\boldsymbol{E} \leftarrow \chi^{r\times m}$. Set $\boldsymbol{S} = [\boldsymbol{I}_r || -\boldsymbol{S}'] \in \mathbb{Z}_q^{r\times(n+r)}$. We denote the i-th row of $\boldsymbol{S}$ by $\boldsymbol{s}_i^T$. Let

$$\boldsymbol{B} = \begin{pmatrix} \boldsymbol{S}'\boldsymbol{A} + \boldsymbol{E} \\ \boldsymbol{A} \end{pmatrix} \in \mathbb{Z}_q^{(n+r)\times m}.$$

Let $\boldsymbol{M}_{(i,j)} \in \{0,1\}^{r\times r} (i,j = 1, \cdots, r)$ be the matrix with 1 in the (i, j)-th position and 0 in the others. For all $i,j = 1, \cdots, r$, sample uniformly random matrices $\boldsymbol{R}_{(i,j)} \leftarrow \{0,1\}^{m\times N}$, and set

$$\boldsymbol{P}_{(i,j)} = \boldsymbol{B}\boldsymbol{R}_{(i,j)} + \begin{pmatrix} \boldsymbol{M}_{(i,j)}\boldsymbol{S} \\ \boldsymbol{0} \end{pmatrix} \boldsymbol{G} \in \mathbb{Z}_q^{(n+r)\times N}.$$

Output pk $= (\{\boldsymbol{P}_{(i,j)}\}_{i,j\in[r]}, \boldsymbol{B})$ and sk $= \boldsymbol{S}$.

- HAO15.SecEnc$_{\text{sk}}(\boldsymbol{M} \in \{0,1\}^{r\times r})$: Sample a random matrix $\boldsymbol{A}' \leftarrow \mathbb{Z}_q^{n\times N}$ and $\boldsymbol{E} \leftarrow \chi^{r\times N}$, and output the ciphertext

$$\boldsymbol{C} = \left[\begin{pmatrix} \boldsymbol{S}'\boldsymbol{A}' + \boldsymbol{E} \\ \boldsymbol{A}' \end{pmatrix} + \begin{pmatrix} \boldsymbol{M}\boldsymbol{S} \\ \boldsymbol{0} \end{pmatrix} \boldsymbol{G} \right]_q \in \mathbb{Z}_q^{(n+r)\times N}.$$

- HAO15.PubEnc$_{\text{pk}}(\boldsymbol{M} \in \{0,1\}^{r\times r})$: Sample a random matrix $\boldsymbol{R} \leftarrow \{0,1\}^{m\times N}$, and output the ciphertext

$$\boldsymbol{C} = \boldsymbol{B}\boldsymbol{R} + \sum_{i,j\in[r]:\boldsymbol{M}[i,j]=1} \boldsymbol{P}_{(i,j)} \in \mathbb{Z}_q^{(n+r)\times N},$$

where $\boldsymbol{M}[i,j]$ is the (i, j)-th element of $\boldsymbol{M}$.

- HAO15.Dec$_{\text{sk}}(\boldsymbol{C})$: Output the matrix $\boldsymbol{M} = (\lfloor \langle \boldsymbol{s}_i, \boldsymbol{c}_{jl-1} \rangle \rceil_2)_{i,j\in[r]} \in \{0,1\}^{r\times r}$.
- $\boldsymbol{C}_1 \oplus \boldsymbol{C}_2$: Output $\boldsymbol{C}_{add} = \boldsymbol{C}_1 + \boldsymbol{C}_2$ as the result of homomorphic addition between the input ciphertexts.

- $\boldsymbol{C}_1 \odot \boldsymbol{C}_2$: Output $\boldsymbol{C}_{mult} = \boldsymbol{C}_1\boldsymbol{G}^{-1}(\boldsymbol{C}_2)$ as the result of homomorphic multiplication between the input ciphertexts, where $\boldsymbol{G}^{-1}(\boldsymbol{C})$ is the function that outputs a matrix $\boldsymbol{X}'$ such that $\boldsymbol{G}\boldsymbol{X}' = \boldsymbol{C} \bmod q$.

3 Homomorphic Comparison Between Encrypted Indices

Assume that the user wants to retrieve the i-th data block in the database $DB = B_1|B_2|\cdots|B_v$, we write the index $i \in [v]$ in the binary representation, denoted as $i = a_1a_2\cdots a_t$(where $t = \lceil \log(v+1) \rceil$). In this section, we describe how the binary index is encrypted under HAO15 scheme and then give a homomorphic algorithm to check the equality between encrypted indices.

3.1 Encrypt an Index Under HAO15 Scheme

In our BPIR protocol, we encrypt binary index $i = a_1a_2\cdots a_t$ under HAO15 scheme. Let $k = \lceil t/r \rceil$ and store all bits $a_1, a_2, \cdots, a_t$ in the diagonal of plaintext matrices $\boldsymbol{M}_1^i, \boldsymbol{M}_2^i, \cdots, \boldsymbol{M}_k^i$(with 0 in other positions except for diagonal) in sequence. We fill up the diagonal of $\boldsymbol{M}_k^i$ with zero behind binary index. Then encrypt $\boldsymbol{M}_1^i, \boldsymbol{M}_2^i, \cdots, \boldsymbol{M}_k^i$ under HAO15 scheme and get corresponding ciphertexts $\boldsymbol{C}_1^i, \boldsymbol{C}_2^i, \cdots, \boldsymbol{C}_k^i$.

For the convenience of description, we define the following function:

- $\text{IndexEnc}_{\text{pk}}(i)$: Input an index $i \in [v]$, output a ciphertext $\bar{i} = \{\boldsymbol{C}_1^i, \boldsymbol{C}_2^i, \cdots, \boldsymbol{C}_k^i\}$, where $\boldsymbol{C}_1^i, \boldsymbol{C}_2^i, \cdots, \boldsymbol{C}_k^i$ are obtained as described above.

3.2 A Homomorphic Comparison Algorithm

We first introduce a permutation algorithm of HAO15 scheme. Plaintext slots in packed FHE correspond to diagonal entries of plaintext matrices in the HAO15 scheme. It is easy to see that we can correctly compute homomorphic slot-wise addition and multiplication. In many applications, we may want to permute plaintext slots. This can be achieved by multiplying the encryptions of a permutation and its inverse from left and right.

- $\text{SwitchKeyGen}(\text{pk}, \sigma)$: Given a public key pk of the HAO15 scheme and a permutation σ, let $\boldsymbol{P}_\sigma \in \{0,1\}^{r\times r}$ be a matrix corresponding to σ, and generate

$$\boldsymbol{W}_\sigma \leftarrow \text{HAO15.PubEnc}_{\text{pk}}(\boldsymbol{P}_\sigma),$$
$$\boldsymbol{W}_{\sigma^{-1}} \leftarrow \text{HAO15.PubEnc}_{\text{pk}}(\boldsymbol{P}_\sigma^T).$$

Output the switch key $\text{spk}_\sigma = (\boldsymbol{W}_\sigma, \boldsymbol{W}_{\sigma^{-1}})$.

– $\mathrm{SlotSwitch}_{\mathrm{spk}_\sigma}(\boldsymbol{C})$: Take as input a switch key spk_σ and a ciphertext $\boldsymbol{C}$, output

$$\boldsymbol{C}_\sigma \leftarrow \boldsymbol{W}_\sigma \odot \boldsymbol{C} \odot \boldsymbol{W}_{\sigma^{-1}}.$$

For arbitrary two indices $i, j \in [v]$, encrypt them under the HAO15 scheme as described in Sect. 3.1, then get $\bar{i} = \mathrm{IndexEnc}_{\mathrm{pk}}(i)$ and $\bar{j} = \mathrm{IndexEnc}_{\mathrm{pk}}(j)$. How can we check the equality homomorphically between indices i and j when we only get access to ciphertexts $\bar{i}, \bar{j}$ and public key pk? We certainly can't get the test result in clear, but we can get the encryption of the test result under HAO15 scheme. We give the homomorphic comparison algorithm as follows.

– CompKeyGen(pk): Let $\sigma_h (h = 0, 1, \cdots, r-1)$ be the cyclic permutation that maps the first element to the $((1+h) \bmod r)$-th element, compute

$$\mathrm{spk}_{\sigma_h} = \mathrm{SwitchKeyGen}(\mathrm{pk}, \sigma_h),$$

output the compare key $\mathrm{cpk} = \{\mathrm{spk}_{\sigma_h} | h = 0, 1, \cdots, r-1\}$.
– $\mathrm{Eq\_Test}_{\mathrm{cpk}}(\bar{i}, \bar{j})$: Given $\bar{i} = \{\boldsymbol{C}_1^i, \boldsymbol{C}_2^i, \cdots, \boldsymbol{C}_k^i\}$ and $\bar{j} = \{\boldsymbol{C}_1^j, \boldsymbol{C}_2^j, \cdots, \boldsymbol{C}_k^j\}$ that are corresponding to the index $i \in [v]$ and $j \in [v]$ separately. Set

$$\boldsymbol{C}^* \leftarrow \bigodot_{d=1}^{k} (\boldsymbol{C}_d^i \oplus \boldsymbol{C}_d^j \oplus \boldsymbol{C}_I),$$

where $\boldsymbol{C}_I$ is a ciphertext that encrypts the identity matrix $\boldsymbol{I}$ under the HAO15 scheme. Output a ciphertext $\boldsymbol{C} \in \mathbb{Z}_q^{(n+r) \times N}$ encrypting an identity matrix if $i = j$ and a zero matrix otherwise, as

$$\boldsymbol{C} \leftarrow \bigodot_{h=0}^{r-1} \mathrm{SlotSwitch}_{\mathrm{spk}_{\sigma_h}}(\boldsymbol{C}^*).$$

Theorem 1. The algorithm $\mathrm{Eq\_Test}_{\mathrm{cpk}}(\bar{i}, \bar{j})$ outputs a ciphertext $\boldsymbol{C}$ encrypting an identity matrix if $i = j$ and a zero matrix otherwise, where cpk = CompKeyGen(pk).

Proof. Given the functions cpk = CompKeyGen(pk), $\boldsymbol{C}^* \leftarrow \bigodot_{d=1}^{k} (\boldsymbol{C}_d^i \oplus \boldsymbol{C}_d^j \oplus \boldsymbol{C}_I)$, and $\boldsymbol{C} \leftarrow \bigodot_{h=0}^{r-1} \mathrm{SlotSwitch}_{\mathrm{spk}_{\sigma_h}}(\boldsymbol{C}^*)$. We prove the two cases when $i = j$ and $i \neq j$ separately.

When $i = j$, it's evident that $\boldsymbol{C}^*$ encrypts the identity matrix according to the property of homomorphic matrices addition and multiplication operations. Observe $\boldsymbol{C} \leftarrow \bigodot_{h=0}^{r-1} \mathrm{SlotSwitch}_{\mathrm{spk}_{\sigma_h}}(\boldsymbol{C}^*)$, it's easy to see that the ciphertext $\boldsymbol{C}$ also encrypts the identity matrix.

When $i \neq j$, there is at least a zero in the diagonal of the plaintext matrix corresponding to $\boldsymbol{C}^*$. Through $\boldsymbol{C} \leftarrow \bigodot_{h=0}^{r-1} \mathrm{SlotSwitch}_{\mathrm{spk}_{\sigma_h}}(\boldsymbol{C}^*)$, the zero is diffused to every position in the diagonal. So $\boldsymbol{C}$ encrypts a zero matrix.

4 LWE-Based Single-Server BPIR Protocol

We describe how to construct a LWE-based single-server BPIR protocol by using HAO15 scheme. Specially, to reduce the communication complexity, we use a homomorphic-ciphertext compression technique proposed by Naehrig et al. [16] and first used in constructing a private information retrieval protocol in [11]. In Sect. 4.1, we present our BPIR protocol whose correctness, security and performance are discussed in Sect. 4.2.

4.1 Our BPIR Protocol

We have given the definition of the single-server BPIR protocol in Sect. 2.1. Here, we will instantiate the components based on the HAO15 scheme.

In our BPIR protocol, the user sends a query that encrypts a binary index with HAO15 scheme to the server. Unfortunately, ciphertext expansion (i.e. the ciphertext size divided by the plaintext size) of current FHE scheme is prohibitive, and this induces big communication complexity. To solve this issue, we adopt the way proposed in [16] to instead send the index encrypted with a block encryption scheme $\mathrm{BE} = (\mathrm{BE.KeyGen}, \mathrm{BE.Enc}, \mathrm{BE.Dec})$. The server then encrypts the ciphertexts with the HAO15 scheme and the user's public key and homomorphically decrypts them before they are processed. By this way, the communication is lowered to the optimal size plus a costly one-time setup that consists of sending the public key of the HAO15 scheme and an encryption of the block cipher secret key.

After previous preparation, we give our BPIR protocol formally as follows:

0. **Setup.** The protocol begins in an off-line setup phase that does not depend on the index to be queried nor on the contents of the database. Let λ be the security parameter and $r = \lceil \sqrt{w} \rceil$, the user generates key $(\mathrm{pk}, \mathrm{sk})$ of the HAO15 scheme and key s of the block encryption scheme BE:

$$\begin{aligned}(\mathrm{pk}, \mathrm{sk}) =& \mathrm{HAO15.KeyGen}(1^\lambda, r),\\ s =& \mathrm{BE.KeyGen}(1^\lambda).\end{aligned}$$

Then encrypt the secret key s under HAO15 scheme

$$\bar{s} = \mathrm{HAO15.SecEnc}_{\mathrm{sk}}(\mathrm{s}).$$

It thus obtains a public set of parameters $params = \{\mathrm{pk}, \bar{s}\}$ (the public key) that is sent to the server, and a secret state $setupstate = \{\mathrm{sk}, s\}$ that is kept private.

Upon receiving *params*, the server computes $\bar{j} = \text{IndexEnc}_{\text{pk}}(j)$ for every $j \in [v]$. For all $j \in [v]$, store all bits $b_{j,1}, b_{j,2}, \cdots, b_{j,w}$ of the data block B_j in a plaintext matrix $\boldsymbol{M}_j$ (fill up it with 0 in remainder positions) in sequence, set

$$\boldsymbol{C}_j = \text{HAO15.PubEnc}_{\text{pk}}(\boldsymbol{M}_j).$$

And the server runs

$$\text{cpk} = \text{CompKeyGen(pk)}.$$

Then the server gets a set of parameters $\{\bar{j}, \boldsymbol{C}_j, \text{cpk}\}$.

Once the setup phase is complete, the user and server can run the remainder of the protocol an unbounded number of times.

1. **Query.** When the user wants to retrieve the i-th data block in the database $DB = B_1|B_2|\cdots|B_v$, it encrypts the binary index $i = a_1a_2\cdots a_t$ with the block encryption scheme BE

$$query = \text{BE.Enc}(s, i = a_1a_2\cdots a_t).$$

The query message *query* is then sent to the server.

2. **Answer.** Getting the query message *query*, the server first encrypts *query* under HAO15 scheme

$$\overline{query} = \text{IndexEnc}_{\text{pk}}(query).$$

Decrypt *query* homomorphically as

$$\bar{i} = \text{Evaluate}(C_{\text{BE.Dec}}, (\overline{query}, \bar{s}), \text{pk}).$$

Then compute

$$\boldsymbol{C} = \bigoplus_{j \in [v]} (\text{Eq\_Test}_{\text{cpk}}(\bar{i}, \bar{j}) \otimes \boldsymbol{C}_j).$$

The response $\boldsymbol{C}$ is then sent back to the user.

3. **Decode.** Upon receiving the response $\boldsymbol{C}$, the user decrypt the ciphertext $\boldsymbol{C}$ under the secret key sk:

$$\boldsymbol{M} = \text{HAO15.Dec}_{\text{sk}}(\boldsymbol{C})$$

The user can get the i-th data block B_i by extracting the top w bits of the matrix $\boldsymbol{M}$.

It's easy to transform our BPIR protocol to a standard PIR protocol. But the performance of the PIR protocol is not better than previous schemes.

4.2 Correctness, Security and Performance

In this section, we will discuss the correctness, security, and performance of our BPIR protocol. Correctness holds as the following theorem.

Theorem 2 (Correctness). In the LWE-based single-server BPIR protocol presented in Sect. 4.1, the user can get the i-th data block B_i by extracting the top w bits of the matrix $\boldsymbol{M}$, where $\boldsymbol{M} = \text{HAO15.Dec}_{\text{sk}}(\boldsymbol{C})$ and $\boldsymbol{C} = \bigoplus_{j\in[v]} (\text{Eq\_Test}_{\text{cpk}}(\bar{i},\bar{j}) \otimes \boldsymbol{C}_j)$.

Proof. From Theorem 1, $\text{Eq\_Test}_{\text{cpk}}(\bar{i},\bar{j})$ is a ciphertext encrypting an identity matrix if $i = j$ and a zero matrix otherwise. So $\boldsymbol{C} = \bigoplus_{j\in[v]} (\text{Eq\_Test}_{\text{cpk}}(\bar{i},\bar{j}) \otimes \boldsymbol{C}_j)$ is a ciphertext that encrypts the same plaintext matrix $\boldsymbol{M}_i$ as $\boldsymbol{C}_i$. And hence $\boldsymbol{M} = \text{HAO15.Dec}_{\text{sk}}(\boldsymbol{C})$ is the matrix $\boldsymbol{M}_i$. Because for all $j \in [v]$, we store all bits $b_{j,1}, b_{j,2}, \cdots, b_{j,w}$ of the data block B_j in a plaintext matrix $\boldsymbol{M}_j$(fill up it with 0 in remainder positions) in sequence, the user can get the i-th data block B_i by extracting the top w bits of the matrix $\boldsymbol{M}$.

Before we analyze the security of our protocol, we first introduce the Lemma 1.

Lemma 1 ([14]). Let $\boldsymbol{B}, \boldsymbol{M}_{(i,j)}, \boldsymbol{R}_{(i,j)}, \boldsymbol{P}_{(i,j)} (i,j = 1, \cdots, r)$ be the matrices generated in HAO15.KeyGen, and $\boldsymbol{R}$ be the matrix generated in HAO15.PubEnc. For every $i,j = 1, \cdots, r$, if the HAO15 scheme is circular secure and DLWE holds, then the joint distribution $(\boldsymbol{B}, \boldsymbol{B}\boldsymbol{R}_{(i,j)}, \boldsymbol{P}_{(i,j)}, \boldsymbol{B}\boldsymbol{R})$ is computationally indistinguishable from uniform over $\mathbb{Z}_q^{(n+r)\times m} \times \mathbb{Z}_q^{(n+r)\times N} \times \mathbb{Z}_q^{(n+r)\times N} \times \mathbb{Z}_q^{(n+r)\times N}$.

Our single-server BPIR protocol is constructed based on HAO15 scheme and a block encryption scheme, so we have Theorem 3.

Theorem 3 (Security). Our LWE-based single-server BPIR protocol can be semantically secure by assuming that HAO15 scheme is circular secure and DLWE holds.

Proof. The security of our protocol is based on security of HAO15 scheme and the block encryption scheme, so we can get the conclusion directly from Lemma 1.

Here, we evaluate the performance of our protocol by analyzing the two important indicators communication complexity and computation complexity.

It's easy to see that the communication complexity of our LWE-based single-server BPIR protocol is $|query| + |\boldsymbol{C}| = \log v + (n + \sqrt{w})^2 \cdot \log^2 q$.

As for computation complexity, we instantiate the block encryption scheme with SIMON-64/128 scheme which is a lightweight block cipher proposed by Beaulieu et al. [17] in 2013. Then the procedure to decrypt the query message *query* homomorphically needs $\frac{22t}{r}$ homomorphic multiplications, which is negligible. The function $\boldsymbol{C} = \bigoplus_{j\in[v]} (\text{Eq\_Test}_{\text{cpk}}(\bar{i},\bar{j}) \otimes \boldsymbol{C}_j)$ needs $vkr = vt$ homomorphic multiplications, which is sequentialized absolutely and therefore induces the lower noise in the ciphertexts (the noise terms in ciphertexts of HAO15 scheme grow asymmetrically). So $O(vt)$ homomorphic multiplications is needed in every retrieval.

5 Conclusion

In this paper, we have presented a new way to construct the BPIR protocol. Almost all of previous BPIR protocols derive from the standard PIR protocols that only support single-bit retrieval. We take advantage of the packed FHE scheme (HAO15 scheme) to construct a BPIR protocol directly. Our LWE-based single-server BPIR protocol has lower communication complexity than existing BPIR protocols because we just send a single ciphertext back to the user.

There are many interesting works that are worth further study. And we plan to optimize the HAO15 scheme itself to obtain higher efficiency.

References

1. Chor, B., Goldreich, O., Kushilevitz, E., et al.: Private information retrieval. In: Symposium on Foundations of Computer Science. IEEE Computer Society, p. 41 (1995)
2. Ambainis, A.: Upper bound on communication complexity of private information retrieval. In: International Colloquium on Automata, Languages and Programming, pp. 401–407. Springer-Verlag (1997)
3. Beimel, A., Ishai, Y.: Information-theoretic private information retrieval: a unified construction. In: International Colloquium on Automata, Languages and Programming, pp. 912–926. Springer-Verlag (2001)
4. Itoh, T.: Efficient private information retrieval. Tech. Report IEICE ISEC **98**(1), 11–20 (1998)
5. Ishai, Y., Kushilevitz, E.: Improved upper bounds on information-theoretic private information retrieval (Extended Abstract). In: ACM Symposium on Theory of Computing, pp. 79–88. ACM (1999)
6. Kushilevitz, E., Ostrovsky, R.: Replication is not needed: single database, computationally-private information retrieval. In: Symposium on Foundations of Computer Science, pp. 364–373. IEEE (2002)
7. Cachin, C., Micali, S., Stadler, M.: Computationally private information retrieval with polylogarithmic communication. In: Advances in Cryptology — EUROCRYPT 1999, pp. 402–414. Springer, Heidelberg (1999)
8. Wang, S., Agrawal, D., Abbadi, A.E.: Generalizing PIR for practical private retrieval of public data. In: Lecture Notes in Computer Science, vol. 6166, pp. 1–16 (2010)
9. Kushilevitz, E., Ostrovsky, R.: One-way trapdoor permutations are sufficient for non-trivial single-server private information retrieval. Proc. Eurocrypt. **1807**, 104–121 (2000)
10. Gentry, C.: A Fully Homomorphic Encryption Scheme. Stanford University, Stanford (2009)
11. Brakerski, Z., Vaikuntanathan, V.: Efficient fully homomorphic encryption from (Standard) LWE. In: Foundations of Computer Science, pp. 97–106. IEEE (2011)
12. Yi, X., Kaosar, M.G., Paulet, R., et al.: Single-database private information retrieval from fully homomorphic encryption. IEEE Trans. Knowl. Data Eng. **25**(5), 1125–1134 (2013)
13. Dijk, M.V., Gentry, C., Halevi, S., et al.: Fully homomorphic encryption over the integers. In: International Conference on Theory and Applications of Cryptographic Techniques, pp. 24–43. Springer-Verlag (2010)
14. Hiromasa, R., Abe, M., Okamoto, T.: Packing messages and optimizing bootstrapping in GSW-FHE. In: Public-Key Cryptography – PKC 2015, pp. 73–82. Springer, Heidelberg (2015)

15. Gentry, C., Sahai, A., Waters, B.: Homomorphic encryption from learning with errors: conceptually-simpler, asymptotically-faster, attribute-based. In: Cryptology Conference, pp. 75–92. Springer, Heidelberg (2013)
16. Naehrig, M., Lauter, K., Vaikuntanathan, V.: Can homomorphic encryption be practical? In: ACM Cloud Computing Security Workshop, CCSW 2011, Chicago, Il, Usa, October. DBLP, pp. 113–124 (2011)
17. Beaulieu, R., Treatman-Clark, S., Shors, D., et al.: The SIMON and SPECK lightweight block ciphers, pp. 1–6. IEEE (2015)

Design and Implementation of an Automatic Scanning Tool of SQL Injection Vulnerability Based on Web Crawler

Xiaochun Lei[1(✉)], Jiashi Qu[2], Gang Yao[3], Junyan Chen[4], and Xin Shen[5]

[1] School of Computer Science and Information Security, Guilin University of Electronic Technology, Guilin Guangxi 541004, China
glleixiaochun@qq.com
[2] Guangxi Cooperative Innovation Center of Cloud Computing and Big Data, Guilin University of Electronic Technology, Guilin Guangxi 541004, China
[3] Guangxi Colleges and Universities Key Laboratory of Cloud Computing and Complex Systems, Guilin University of Electronic Technology, Guilin Guangxi 541004, China
[4] Guangxi Colleges and Universities Key Laboratory of Intelligent Processing of Computer Images and Graphics, Guilin University of Electronic Technology, Guilin Guangxi 541004, China
[5] Guangxi Key Lab of Trusted Software, Guilin University of Electronic Technology, Guilin Guangxi 541004, China

Abstract. An automatic detection tool for SQL injection vulnerability based on web crawler is designed and implemented. By studying the characteristics of various web application vulnerabilities, the causes and detection methods of SQL injection vulnerabilities are analyzed in detail. In addition, functions such as URL (Uniform Resource Locator) optimization and similarity determination are added to each module's characteristics, so that the vulnerabilities can be scanned more accurately and quickly. The tool can automatically explore the target based on web crawler framework. After testing, it is proved that the scanning tool can effectively detect potential SQL injection security vulnerabilities in a website.

Keywords: Web crawler · SQL injection · Automatic scanning

1 Introduction

In the short span of several decades since the formation of the Internet, the number of websites has exploded. At the same time, with the rapid development, more and more security vulnerabilities are endangering the security of websites. Common security problems include SQL injection vulnerability, XSS (Cross Site Scripting) vulnerability detection, weak login, etc. The ten key web application security risks released by the Open Web Application Security Project (OWASP) in 2017 ranked first injection class attacks represented by SQL injection. In December 1998 in hacker magazine "Phrack", Jeff Forristal published a series of SQL injection problems with the Microsoft SQL

C.-N. Yang et al. (Eds.): SICBS 2018, AISC 895, pp. 481–488, 2020.
https://doi.org/10.1007/978-3-030-16946-6_38

server, this was the earliest article about SQL injection attacks. Chris Anley defines SQL injection as "an attacker is able to insert a series of SQL statements into a 'query' by manipulating data input into an application" [1]. Therefore, SQL injection vulnerabilities can easily exist in web applications with database connections. Talk's information leak in 2015 exposed sensitive information of about 150,000 people, using SQL injection technology. Therefore, many vulnerability scanning tools have the function of SQL injection vulnerability detection. The scanning tool designed in this paper can automatically scan and detect SQL injection vulnerabilities based on the active framework of web crawler.

2 Design and Implementation of Network Crawler Module

2.1 The Design and Workflow of the Crawler Module

The crawler is the core component of the system. It visits all web content from the target site and extracts all the content of the URLs and web pages for the purpose of scanning vulnerabilities. The general crawler's working principle is as follows: start downloading the web page from the initial URL set by a person and find out all URLs from the web page; add the newly acquired URL to the URL queue; extract a URL and repeat the previous process until all URLs are downloaded [2].

The crawling process of this system:

(1) After the user enters the initial URL, he chooses the corresponding checking strategy and various parameters (such as crawling strategy, depth, etc.).
(2) The scanner checks the validity of the input URL and runs the crawler according to the parameters if it is valid, otherwise it prompts the user to re-enter the URL.
(3) The crawler module performs URL crawling operations according to the corresponding parameters and strategies (depth first, breadth first, random priority, default by random priority strategy).
(4) The URL optimization module filters and removes the initial results.
(5) The URL optimization results are put into the cache queue to wait for storage in the database.
(6) The crawler module checks whether the depth of the current web page reaches the set depth value, and if it reaches, it ends the crawler operation, otherwise the depth adds one and continues to run the crawler until the end.
(7) The crawler module ends, prompting the user system to run to the scanning module.

2.2 Regular Expression Filtering

The effective optimization of URL results by web crawler can lighten the burden on the database and scan module. Regular expressions are a logical formula for a string operation. Regular tables are usually used to retrieve and replace texts that conform to a certain pattern (rule), and make a string of characters by forming a regular string expression. A filter is used [3] because the essence of vulnerability scanning tools is to simulate attacks on websites with interactive functions, so that some links crawled by crawlers, such as .jpg, .gif, .doc, .zip, .mp4 and so on, can be directly removed from meaningless web pages, and the method is simple and effective.

The crawler uses a random crawling strategy, and when the depth is set to 3 it crawls a school site 10 times to get the average value. The effective URLs are compared before and after using regular expressions, as shown in Table 1.

Table 1. Data control table before and after URL filtering

	Before using regular expressions	After using regular expressions
Valid URL number	2893	2668

The results show that the filtration efficiency is (2893-2268)/2893 ≈ 7.78%.

In fact, the calculated values on different websites are often unequal. There are certain differences according to the type and quantity of website resources, and the pictures or resources on the campus website are few, which can serve as a reference for the minimal filtering effect, so it can be judged that regular expression is effective for URL filtering.

2.3 Bloom Filter

In addition to filtering meaningless web pages, crawlers are likely to find duplicated URLs. How to prevent duplicate downloading of the same page is one aspect that the crawler needs to consider. If it is not handled well, it will seriously slow down the crawler speed, and it may also lead to crawling into a dead circle. Generally speaking, to determine whether an element exists in a collection, we need to save the elements and make comparisons. But when the data is large, the storage space will become larger and larger, and the efficiency will be lower and lower. An efficient way to remove duplicate URLs is to use the Bloom filter.

The Bloom filter was proposed by Howard Bloom in 1970. He only used a series of bits to save data, and it could detect whether an element existed in the collection, so the algorithm had good space utilization [4]. The advantage of the Bloom filter is that the space efficiency and query time are much more than that of the general algorithm. In the occupied space, the Bloom filter only needs the size of the hash table 1/8–1/4 to solve the same problem; more importantly, in terms of time complexity, the search time of the Bloom filter is constant and does not increase with filter slow-down [5].

Assuming the set is {url1, url2, url3} and K is 2, look for url_x as an element in the collection, as shown in Fig. 1.

Because the mapping relation of hash function is many to one, the hash value of different strings may be the same. And because of the realization principle of the Bloom filter, 1 of the bits from different URL hashes in the Bloom filter vector can be identified as another unknown URL by the hash combination, which makes it difficult to get from the filter. To reduce the defects of a URL and the erroneous recognition rate which is difficult to get rid of in the process of removing duplicate URLs, a lot of work around the data structure has been done since the advent of this data structure, and a lot of improved algorithms based on the Bloom filter have been proposed, such as

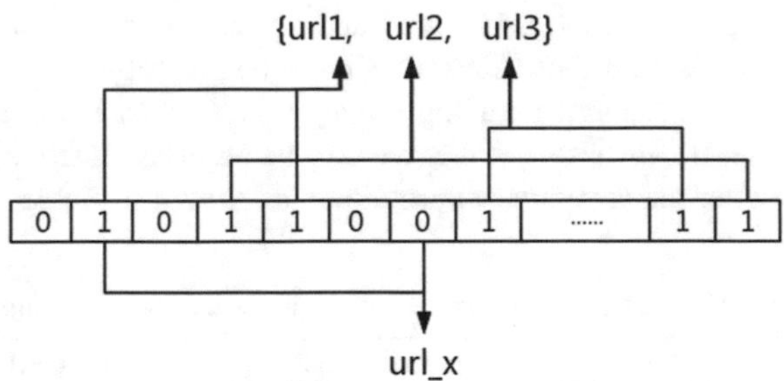

Fig. 1. Remove duplicate URLs

counting compression type, splitting type, spectral type and so on. The current performance and effect are basically satisfied for this system, so we don't delve into it here [5].

The most important advantage of the Bloom filter for duplicate URL removal is memory saving. For example, if there are 1 million URLs, each URL assumes 50 bytes of storage space. If the hash table is used to store it, assuming that the hash table's loading factor is 0.5, then the storage space is 100 million bytes. If we use the Bloom filter to store it, and suppose the error rate is not greater than ε, according to the common formula, the most suitable size of digit group m is [5]:

$$m \geq 1.44 \times n \times \log_2 \frac{1}{8} \tag{1}$$

Assuming that ε is 1/100000, the size of the occupied array of one million URLs can be $2.39*10^7$ bits, and only the 1/33 occupied by the hash table is converted. This shows that the spatial efficiency of the Bloom filter algorithm is very high.

3 Design and Implementation of SQL Injection in the Detection Module

3.1 Detection Design and Process of SQL Injection Detection Module

SQL injection applies some grammatical features of the SQL language to attack the server. The main injection principles are string concatenation of SQL commands, parcel passing by single quotes, and annotation and so on. The goal of SQL injection is to attack the server databasc first, to control the website directly by obtaining the user name and password of the administrator in the data path, or to directly operate the server directly through high authority database users.

For SQL injection vulnerabilities, the main checking methods are script sensitive character filtering, matching and filtering based on attack features, and autonomous scanning of vulnerabilities.

When the system performs SQL injection detection, it first obtains the URL target to be detected from the database and stores the target in the list. The system control module will be divided according to the current computer performance and network condition, and each URL -generated thread is detected in the thread pool, respectively. The process is as follows:

(1) Obtaining URL information from the database.
(2) Corresponding function detects whether there is WAF (Web Application Firewall), if there is, then continues to judge whether there is a filter script, otherwise the fingerprint information is obtained directly.
(3) If a filter script is found, try to bypass it, otherwise start the heuristic injection test directly.
(4) If it can be simply injected, fingerprint information is obtained, otherwise the formal SQL injection test is carried out.
(5) If a vulnerability is found to obtain fingerprint information, end reading the next URL.

When the system carries on the formal injection detection, first it will distinguish the DBMS (Database Management System) type, according to the page error information or bool type test. When you get the DBMS type, you'll use payload to test one by one, which will detect vulnerabilities in six ways:

(1) Boolean-based blind SQL injection
(2) Error-based SQL injection
(3) UNION query SQL injection
(4) Stacked queries SQL injection
(5) Time-based blind SQL injection
(6) Inline-query

After the target detection is completed, the main thread is returned and the result of scanning is saved. When there are loopholes in the website target, it will try to use advanced query statements to test the detailed information of the server and save the final results in the database.

3.2 Web Page Stability and URL Parameters

The results of the crawler scanning may contain a lot of useless URLs. In order to ensure the stability and process of the system, and to improve the efficiency of the system, the stability of the web page should be checked before checking the loopholes.

First, the SQL injection module will determine the survival of the web page. If each URL crawling through the link test can be accessed, it is judged that the target host is saved after the survival state and then used for blind injection operation. Within a shorter interval, two grasping operations are performed on the same web page and the summary value is calculated. If the two values are equal, the web page is defined as stable.

URL injection parameters are randomly replaced by parameter conditions in the injection statements, and the conditions triggered by change detection vulner abilities are compared. If different pages are accessed by changing parameters, this parameter can be used as parameter conditions for SQL injection.

For the web page stability test, we can filter out the useless links effectively, reduce the resource cost of the scanner, and improve the efficiency of the scanner.

3.3 Similarity Judgment of Web Pages

To get the injected results, the system first needs to compare the similarity of web pages. Before reading, the system will read the relevant settings. If regular expressions or special string patterns are used for comparison, the result will be returned directly. If not, then the page is stable. For web page similarity judgment, the system uses the low collision rate of MD5 (Message-Digest Algorithm) encryption. The system will save the MD5 value of the web page and compare the current MD5 value with the MD5 value of the previous web page, and judge its similarity through threshold setting.

4 System Test

The functional modules are developed in Python language, the UI is designed with QT Designer, and integrated with PyQt4 open source toolkit.

4.1 Detection Function Test of Web Crawler

We carried out crawling tests on three websites: a news portal, a technology forum, and a school webpage. The crawling depth of the news portal website was set to 2, for the technical forum it was set to 3, and for the home page of the campus it was set to 5. The crawling total is the total number of URLs crawled by the crawler module. The optimization result is the number of URLs left after the URL filter and the de-weighting algorithm. The crawling results are shown in Table 2.

Table 2. Web crawler performance comparison table

Site type	Depth	Crawling total	Optimization results	Common web crawler (mean)
News portal	2	38117	9735	33000About
Technical Forum	3	10258	2974	7400About
Home page of the campus	5	16883	3864	9000About

From Table 2 we can see that the ability of our web crawler is better than the average level of ordinary crawlers. After the corresponding optimization, the number of URLs is greatly reduced and the optimization efficiency is:

$$\text{News portal}: \ (38117 - 9735)/38117 \approx 74.46\%$$
$$\text{Technical Forum}: \ (10258 - 2974)/10258 \approx 71.01\%$$
$$\text{Home page of the campus}: \ (16883 - 3864)/16883 \approx 77.11\%$$

From the above calculation, we can see that the crawler of this system has greatly improved the URL optimization compared with the traditional crawler. It can exclude more than 70% of similar or useless web pages, which can reduce the number of web objects that need to be scanned and achieve a great improvement in system efficiency.

4.2 SQL Vulnerability Detection Test

The results of SQL injection test on the website of a Marine College and the website of our university are shown in Table 3.

Table 3. SQL injection detection

Site	The number of test URLs	The depth of web crawler	Total time (second)	The number of detected vulnerabilities
a Marine College website	14	5	124	2
Website of our university	3864	5	18000	24

The contents of the system can be displayed as: the target station with vulnerabilities, the date of detection, the result of detection, the operating system used by the target server, the database used by the target server, and the payload statement. After detecting, we found that there are SQL injection vulnerabilities in the websites of our digital campus, the school news page, the School United Front, the Foreign Languages Institute, the National Defense College, the Marine Information Institute, the Art and Design Institute, and the Mathematical Modeling Society. There are 24 websites with vulnerabilities. Using an online SQL injection detection tool, 22 websites with vulnerabilities were detected among websites of our university, which took 206 s. The efficiency of this tool is far less than that of the commercial software, but the result of vulnerability detection is better.

5 Conclusions

The test shows that the scanning tools described in this paper can effectively tap potential SQL injection vulnerabilities in a website, help website developers and maintenance personnel quickly find and repair vulnerabilities in a website, and the designed process models and collected test data during the development and testing process can also be developed in the future.

How to find the appropriate injection point in static or pseudo static web pages is a difficult point which can be further studied, and other types of vulnerabilities can be added to improve the scope of vulnerability detection. Some improved algorithms such as MD5-based URL duplicate elimination tree [7] and Gap-Weighted String Subsequence [8] are used to achieve higher efficiency in Similarity judgment of web pages. In addition, the overall speed of the system is not ideal. How to optimize the detection of concurrency, adjust the multi-threading method, extend application scope [9] are also an important research topic for the future.

Acknowledgment. This work is supported by Guangxi Colleges and Universities Key Laboratory of cloud computing and complex systems (Nos. 14103,15208) Guangxi Colleges and Universities Key Laboratory of cloud computing and complex systems (No. YD16303), Guangxi Key Lab of Trusted Software(No. kx201320), Guangxi Colleges and Universities Key Laboratory of Intelligent Processing of Computer Images and Graphics (No. GIIP201509).

References

1. Anley, C.: Advanced SQL injection in SQL server applications. An Ngs software Insight Security Research Publication (2002)
2. Zhou, L.Z., Lin, L.: Survey on the research of focused crawling technique. Comput. Appl. **09**, 1965–1969 (2005)
3. Jun, M.: Research on application of Information collection engine based on regular expression technology. University of Electronic Science and Technology of China (2006)
4. Pan, H., Hai-hong, E., Song, M.: The bloom filter applies in data deduplication. Software **36** (12), 166–170 (2015)
5. Huang, E.B.: A method for URL duplicate removal based on bloom Filter. Mod. Comput. **14**, 7–10 (2013)
6. Gol, D., Shah, N.: Detection of web application vulnerability based on RUP model. In: Recent Advances in Electronics & Computer Engineering, pp. 96–100. IEEE (2016)
7. Yan, L., Ding, B., Yao, Z., et al.: Design and optimisation of md5 duplicate elimination tree-based network crawler. Comput. Appl. Softw. **2**, 325–329 (2015)
8. Mcwhirter, P.R., Kifayat, K., Shi, Q., et al.: SQL Injection Attack classification through the feature extraction of SQL query strings using a Gap-Weighted String Subsequence Kernel. J. Inf. Secur. Appl. **40**, 199–216 (2018)
9. Qiuhong, P., Zhanqi, C., Linzhang, W.: Static detection approach for SQL injection vulnerability in android applications. J. Front. Comput. Sci. Technol. (2018)

Malware Detection Based on Opcode Sequence and ResNet

Xuetao Zhang, Meng Sun, Jiabao Wang, and Jinshuang Wang(✉)

Department of Cyberspace Security, Army Engineering University of PLA, Nanjing 210000, China
zxtzhangxuetao@163.com, siyezhishuang@163.com

Abstract. Nowadays, it is challenging for traditional static malware detection method to keep pace with the rapid development of malware variants, therefore machine learning based malware detection approaches begin to flourish. Typically, operation codes disassembled from binary programs were sent to classifiers e.g. SVM and KNN for classification recognition. However, this feature extraction method does not make full use of sequence relations between opcodes, at the same time, the classification model still has less dimensions and lower matching ability. Therefore, a malware detection model based on residual network was proposed in this paper. Firstly, the model extracts the opcode sequences using the disassembler. To improve the vector's expressibility of opcodes, Word2Vec strategy was used in the representation of opcodes, and word vector representations of opcodes were also optimized in the process of training iteration. Unfortunately, the overlapping opcode matrix and convolution operation results in information redundancies. To overcome this problem, a method of downsampling to organize opcode sequences into opcode matrix was adopted, which can effectively control the time and space complexity. In order to improve the classification ability of the model, a classifier with more layers and cross-layer connection was proposed to match malicious code in more dimensions based on ResNet. The experiment shows that the malware classification accuracy in this paper is 98.2%. At the same time, the processing time consumption comparing with traditional classifiers is still negligible.

Keywords: Opcode · N-gram · ResNet · Word2vec

1 Introduction

In recent years, in order to overcome the challenges brought by the malware variants to the static analysis method, the study of machine learning based malware analysis has attracted extensive attentions. Assaleh et al. [2] used n-gram to process the program code for further classification in 2004. Shabtai et al. proposed a method that use opcodes instead of binary codes [3] to achieve higher recognition accuracy. Muazzam Siddiqui et al. used data mining technology [4] against assembly instructions, and obtained the accuracy rate of 0.98 on their data sets. Santos et al. also used opcode frequency as a feature [5], and used SVM classifier to identify malicious codes. Jahan et al. used markov model to process the extracted features [6], which were characterized by the frequency of opcodes too. To keep the information of the original

C.-N. Yang et al. (Eds.): SICBS 2018, AISC 895, pp. 489–502, 2020.
https://doi.org/10.1007/978-3-030-16946-6_39

program, operation code is directly obtained by disassembling. That means the scheme of using the operation code as the data source of malware detection is widely adopted.

The n-gram model was widely used to divide the sequence of opcode to short opcode sequences. The occurrence times of each segment are sent to classifiers as features. Most of the operation code fragments extracted in the previous work were based on 2-gram or 3-gram, which is not good at representing the sequence information of malicious programs. In order to further mine opcode sequence, how to introduce the behavior characteristics of malicious code in the case of guaranteed cost is a key problem [7]. For example, a single opcode cannot carry any sequence information, and a short opcode sequence only carries very limited sequence information. Some of the behavior logic of malicious apps exists in the permutation and combination of opcode sequences. Meanwhile, most related work use K-Means, SVM, or feedforward neural network with relatively simple structures. These networks do not have sufficient parameters to carry the information in the opcode sequence. Therefore, new classification models should be studied.

Due to the limitations of short operation code fragments in describing malware semantic hierarchy information, instead of using the operation code fragment frequency as characteristics, this paper uses operation code sequence as the features directly, i.e. variable step length n-gram model was used to generate operation code sequence.

Experiments show that the window size between 30 and 50 is good for preserving opcode sequence information. In order to reduce data redundancy generated by the n-gram model and the convolution operation, a downsampling method was proposed to compress the training data while keeping the accuracy.

The main ideas of this paper are as follows: The word vector concept of NLP was used to represent opcodes, by incorporating the parameter update into the training iteration process, as a result, the semantic relation of opcodes can be kept in vector space. By using convolution as basic operation unit, and a cross-layer connection inspired by ideas of ResNet was added, which enables the matching of multiple aspects of the malicious program structures, and the expressiveness of classification models were enhanced.

2 Malware Detection Model Based on ResNet

Feature extraction, model structure, and optimization strategy will be presented in order below.

2.1 Convolution Operation with N-Gram Model

Text Convolution and Data Redundancy of the N-Gram Model. The classifier structure used in traditional malware detection based on machine learning is relatively simple, which can't match deep information existed in the opcode sequence [8]. According to Yoon Kim's method of using convolutional neural network for text classification [9], convolution was adopted as the basic arithmetic unit in this paper. In order to illustrate the data redundancy generated by convolution operation and the organization of n-gram data, a simple convolutional neural network is taken as an example to explain this opinion. As shown in Fig. 1, each row of the left opcode matrix is generated by one shift of the n-gram model, which serves as the data source of the convolution operation. The width of convolution kernel is consistent with the width of matrix, and the depth is different.

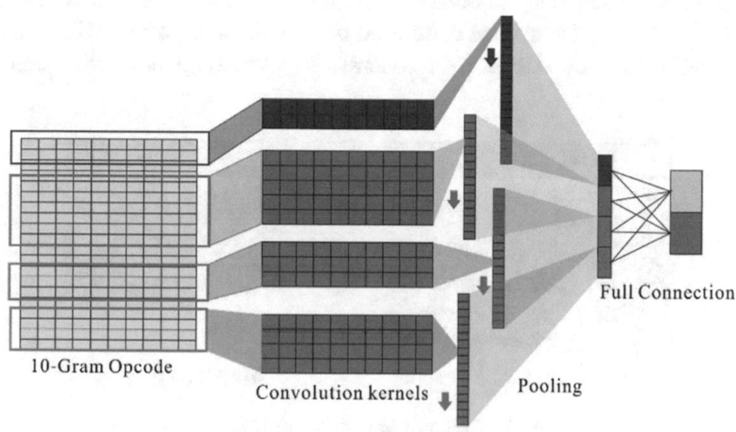

Fig. 1. Convolution diagram

The n-gram model is used to organize the assembly instruction, which can dig the relationship between each opcode in instruction sequence. However, excessive context connection will appear when using it to process the instruction matrix, resulting in a significant decline in information representation ability, as shown in the following figure.

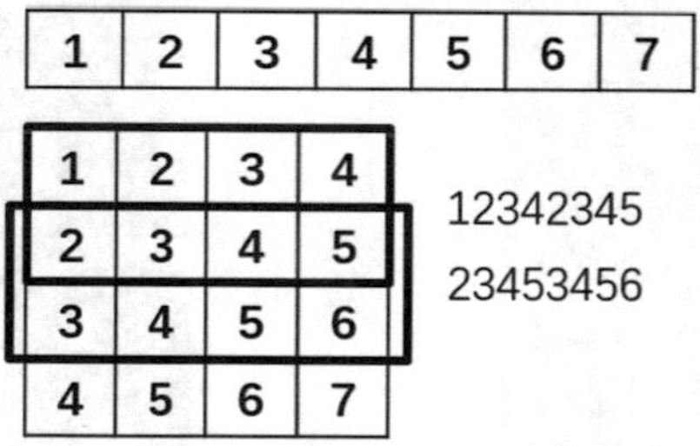

1	2	3	4	5	6	7

1	2	3	4
2	3	4	5
3	4	5	6
4	5	6	7

12342345

23453456

Fig. 2. Convolution traversal schematic, each number in the figure represents an opcode.

As shown in Fig. 2, the original opcode sequence is at the top, and the opcode matrix generated by n-gram model was shown below in the figure. After the convolution kernel moves twice, the suborder ‘234’ occurs three times. In other words, the context of instruction segment can be traversed multiple times by the convolution kernel. This is because multiple repeated sub-sequences are produced during the organization of the data, which resulted in a significant increase of the data source and increases the calculation amount of data to preprocess, reduces the efficiency of classification. In order to eliminate the redundancy of information, this paper proposed a down sampling method to process the opcode matrix.

N-Gram Down Sampling. In the n-gram model, a window size usually specified to slide on the original sequence to obtain subsequences. The effect of different step sizes is shown on the Fig. 3: a sequence composed of five opcodes, which will generate three sub-sequences when step size is 1, or generate two sub-sequences with step size 2.

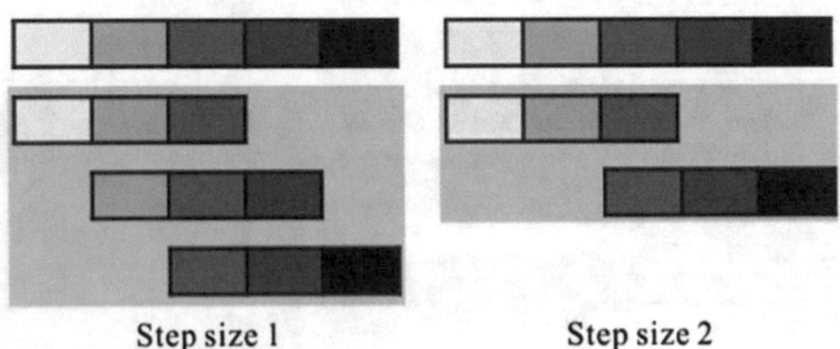

Fig. 3. Convolution traversal schematic

In this work, traditional n-gram model is improved by increasing the sliding step of the window. Take Fig. 4 as an example.

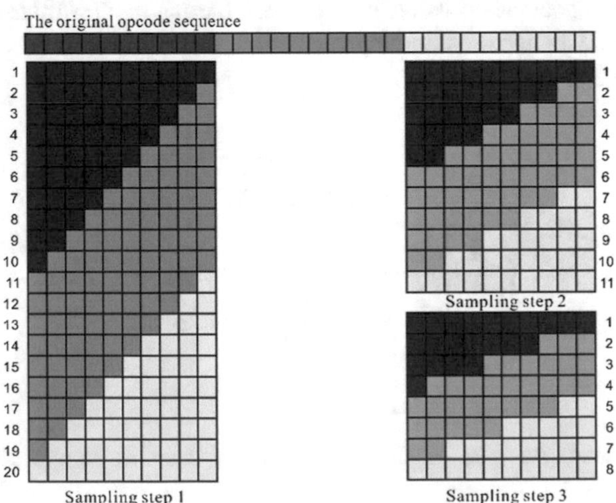

Fig. 4. Down sampling schematic

In Fig. 4, the original opcode sequence is shown above. The opcode matrix generated by 2-gram and step size 1 is shown in the left figure, and the size of the matrix is 10 × 20. The upper right is the opcode matrix generated by 2-gram and step size 2, the matric size is 10 × 11. The bottom right step size increases to 3, and the size of the opcode matrix is 10 × 8. It can be seen that the opcode matrix will be compressed to different degrees with different step sizes. It should be pointed out that this data compression method will cause loss of connection information. The test is carried out to select the appropriate step length. To reduce time and space complexity as possible and keep recognition accuracy, the compression ratio should be selected in a reasonable level, which can only remove redundant information and ensure data integrity.

2.2 Feature Extraction

The feature extraction process is shown in Fig. 5. Firstly, the malicious program is processed with the disassembler to obtain .asm file. Opcodes were extracted from it using the regular matching method. After saving, the n-gram model was used to sample the opcode sequence based on a certain step size. Then a dictionary was constructed to map the opcode into a vector, which was finally sent to the classifier.

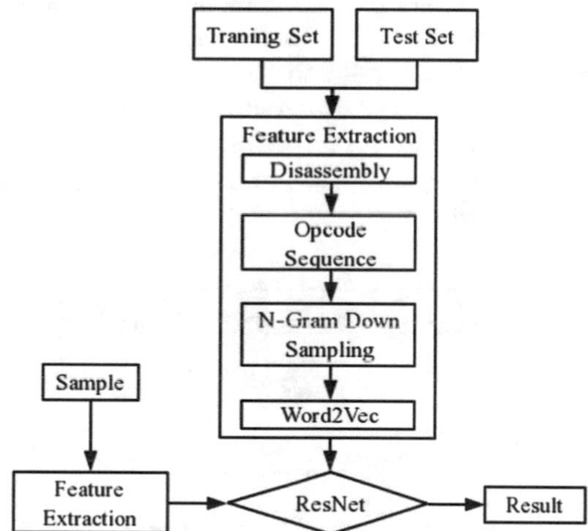

Fig. 5. Testing process. This diagram describes the whole malware detection process based on the opcode sequence.

According to relevant works' verification, operators in assembly instructions contain little information. So mainstream data sources always discard operators [5]. Therefore, when preprocessing, we also abandoned operands and only retain opcodes.

2.3 Word2Vec

Lee et al. proposed a strategy named instruction2vec [10], which mapped the opcode instruction to a vector, and applied it to software vulnerability mining. The data organized by this strategy obtained with a 0.96 accuracy on the Text-CNN. We enhanced this method through adding word vector into the parameter update range of the whole network, so can the opcode be measured at the semantic level through the distance relation between vectors. To enhance the model's ability of describing the opcode sequence. The effectiveness of this strategy is verified by experiments.

2.4 Classifier Based on ResNet

In 2016, He et al. published a paper on CVPR [11], proposed the ResNet structure. Relevant research points out that with the number of network layers increasing, the problem of gradient disappearance and gradient explosion becomes increasingly prominent. When the number of network layers exceeds a certain scale, the relevant performance of the model will deteriorate. ResNet aimed at this problem, improved the traditional network and added shortcut connection structure to ensure the gradient can be successfully returned in a deep neural network. At the same time, residual network can be regarded as the integration of different depth network.

As shown in Fig. 6, *f1*, *f2* and *f3* are operation units such as convolution in the model, and there are shortcut connections between each layer. That is, data can flow through different combinations of layers, and the whole model structure can be regarded as a combination of classifiers based on different levels, this can help the model measure different levels of information based on the original data. In this paper, the structure is used to match different levels of information contained in programs such as program statement, function and behavior.

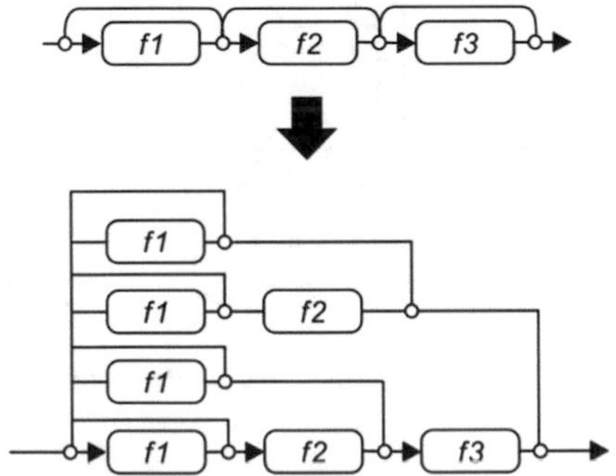

Fig. 6. ResNet structure diagram.

Taking Fig. 7 as an example. After a convolution layer, the description granularity of the program can be regarded as statement level. After two convolution layers, it can be understood as function level. Adding cross-layer connections can match structures such as statement calls to functions. As the number of model layer increases, the model's description dimensions are also increasing. The detailed structure of the network is as follows.

The classification network is shown in Fig. 8. The opcode sequence matrix is firstly mapped to vector space. And then send to the convolution unit. There are three residual blocks in this classifier, each contains two convolutional layers. Regularization strategies are added after convolutional layers to overcome the over-fitting problem in the model. A single convolutional layer is added on shortcut connection to ensure that the data dimension is 128, which is convenient for adding the convolved data after BN (Batch Normalization) [12]. Relu is used as linear activation function in this network. There is a max pooling layer after convolutional. Finally, the target category is output through the full connection. According to the idea of ResNet, shortcut connections were added between three residual blocks, so that the information of different granularity can be combined to enhance the description ability of the model.

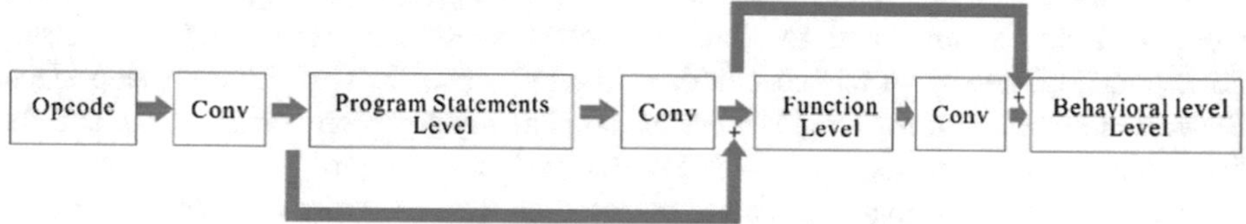

Fig. 7. Schematic diagram of program structure.

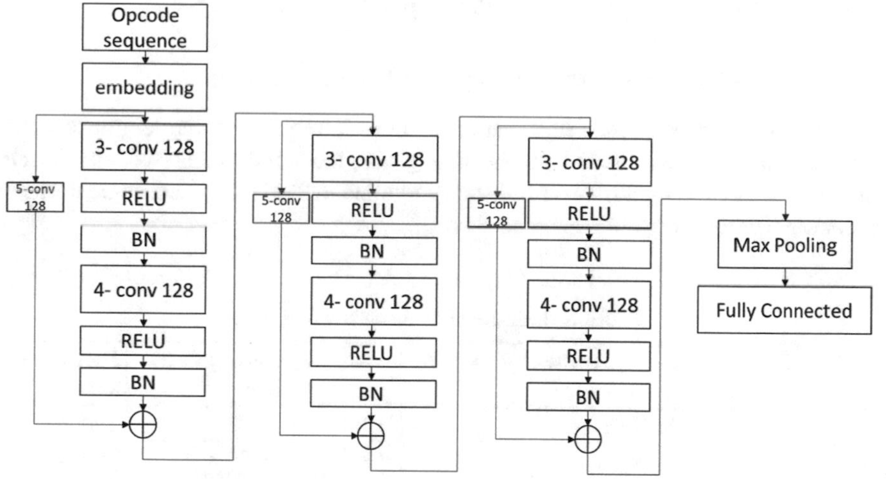

Fig. 8. Classifier structure.

3 Experimental Verification

The experimental scheme is designed as follows, including data set description and projects which need to be tested.

3.1 Kaggle Malware Classification Competition Data Set

The malware data set came from the malicious code classification competition held by Microsoft in Kaggle 2015 [13]. All samples in this data set are malicious code based on Windows platform, including training set, test set and labels. There are nine different types samples in it, and each sample contains two files, which one is binary file (excluding PE header), and another is generated by IDA disassembler. In the data extraction phase, regular matching is used to extract opcodes from asm file and organize them into opcode sequences. The test set is divided into 10% from the total sample. Each sample has a corresponding label.

3.2 Test Scheme

To verify the effectiveness of the downsampling strategy on reducing data volume and reducing training time, and to find the best compression step length. Four-step downsampling step length is specified, to exclude the effect of different step sizes. Besides, different window lengths were tested. The network convergence time and the final classification accuracy is recorded under each step length.

To verify the performance of classifier based on ResNet, a text convolutional neural network was selected as a contrast classifier. Different window lengths were used in the experiment to eliminate the influence of different window sizes on the classification accuracy. Besides, in order to verify the effect of the Word2Vec policy, we compared the result using and not using the word2vec policy on the same model.

3.3 Step Size Test of Sequence Length and Downsampling

In order to keep the sequence information of opcode sequences, the length of opcode sequence was first investigated. In the first place, the window size N was selected, which was 5, 10, 20 and 30. Test results are recorded in Table 1. It should be noted that the data in Table 1 were not sampled down.

Table 1. Classification accuracy and time

Window size (N)	Model	Downsampling step size	Accuracy	Convergence time	Dictionary size
5	Text-CNN	1	0.741	1852 s	190
10	Text-CNN	1	0.901	3050 s	190
20	Text-CNN	1	0.983	6836 s	189
30	Text-CNN	1	0.994	11455 s	190

Then, under the same conditions, the step size of the downsampling was increased to 2 and 10, and the experimental results and relevant parameters were shown in Tables 2 and 3.

Table 2. Precision and convergence time when step length of downsampling is 2

Window size (N)	Model	Downsampling step size	Convergence time	Dictionary size	Accuracy
5	Text-CNN	2	3010 s	190	0.717
10	Text-CNN	2	3246 s	190	0.886
20	Text-CNN	2	3782 s	190	0.979
30	Text-CNN	2	4170 s	190	0.976
40	Text-CNN	2	4692 s	190	0.998
50	Text-CNN	2	5197 s	190	0.999

Table 3. Precision and convergence time when step length of downsampling is 10

Window size (N)	Model	Downsampling step size	Convergence time	Dictionary size	Accuracy
5	Text-CNN	10	798 s	190	0.675
10	Text-CNN	10	792 s	190	0. 03
20	Text-CNN	10	939 s	190	0.903
30	Text-CNN	10	970 s	190	0.934
40	Text-CNN	10	1097 s	190	0.976
50	Text-CNN	10	1142 s	190	0.982

The above three groups of experimental data showed that in the case of fixed down sampling step, the accuracy has an obviously rise with the growth of window length. The Opcode description ability achieved the best result when sequence length located between 30 to 40. Although the value of N increased again, the classification accuracy will not have obviously improved but the data quantity will continue to rise, which lead to an increasing time consumption.

Comparing the data between Tables 2 and 3, opcode sequence retains a large amount of original connection information when the step size is 2, and the model has a good classification performance. The optimal classification result is close to 100%, but the time complexity is relatively high. The experimental group in Table 3 used 10 as the down sampling step size, so the connection information was further eliminated. The final classification accuracy was only slightly lower than before, but the time complexity was reduced by nearly 5 times. It shows that the data compression method proposed in this paper can ensure the classification accuracy while removing most of the redundant information to improve the classification efficiency of the model.

The convergence time and classification accuracy with different downsampling step size (i.e. data in Tables 2 and 3) were plotted in Figs. 9 and 10.

Under the same downsampling step size with Fig. 9, the classification accuracy is plotted in Fig. 9.

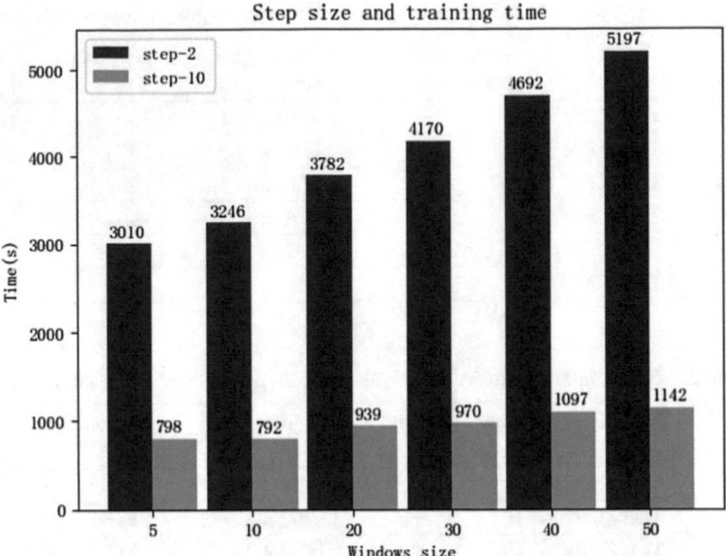

Fig. 9. Network convergence time comparison.

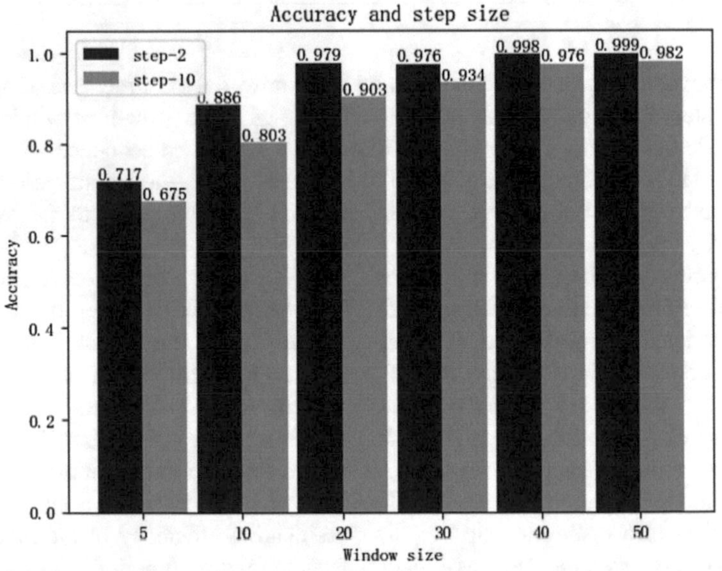

Fig. 10. Precision comparison under different step-down sampling steps.

As can be seen from Fig. 9, under the condition of different downsampling step sizes, the training time of the network varies greatly, but the accuracy differs little. These results can prove that the data compression method proposed in this paper can efficiently remove the redundant information than the original n-gram data organization mode, reducing the network training time while ensuring the accuracy.

3.4 Classifier Evaluation

To test the classification ability of our model, and verify the validity of the Word2Vec strategy at the same time, the experiment was divided in three rounds. We only used Text-CNN in the first round. The second round use Word2Vec strategy for only once mapping with the RES. Word vectors were not updated in the process of training. In the third round, the parameters of word vector were included into the updated range. Two rounds of experimental results were compared to verify whether the network structure proposed in this work improved the classification ability.

Text-CNN Without Word2Vec Policy. The selected window size include 10, 20, 30, 40, 50. The step size of downsampling is 20. The network convergence time and classification accuracy are recorded in Table 4.

Table 4. Sample classification accuracy (Text-CNN)

Window size (N)	Model	Downsampling step size	Convergence time	Dictionary size
10	Text-CNN	20	190	0.790
20	Text-CNN	20	189	0.832
30	Text-CNN	20	190	0.829
40	Text-CNN	20	189	0.840
50	Text-CNN	20	190	0.853

The purpose of this round experiment was to compare the detection effect with the model we tested in the next round.

RES Without Word2Vec Policy. All parameters are the same in this round. The network convergence time and classification accuracy are recorded in Table 5.

Table 5. Sample classification accuracy (RES)

Window size (N)	Model	Downsampling step size	Convergence time	Dictionary size
10	RES	20	190	0.800
20	RES	20	189	0.868
30	RES	20	190	0.867
40	RES	20	189	0.872
50	RES	20	190	0.878

According a simple comparison between the classification accuracy of the network and Text-CNN network using the same parameters, we can see that the ResNet structure proposed in this paper has increased the classification accuracy by several percentage points.

Use the Word2Vec policy. To keep other parameters of the network unchanged, we only add Word2Vec strategy on the basis of the above experiment. The experimental results were recorded in Table 6.

Table 6. Sample classification accuracy (RES+Word2Vec)

Window size (N)	Model	Downsampling step size	Convergence time	Dictionary size
10	Our model	20	190	0.824
20	Our model	20	189	0.838
30	Our model	20	190	0.899
40	Our model	20	189	0.922
50	Our model	20	190	0.900

By comparing the data in Tables 4 and 5, it can be proved that our proposed model can effectively improve the detection effect. And by comparing the data in Tables 5 and 6, it can be concluded that the Word2Vec strategy is applicable to describe opcode. It should be noted that Word2Vec is mostly used in natural language processing. Our experimental results show that this strategy can improve the classification performance of the model, but the principle needs further exploration.

Comparing with relevant works, for example, in the similar work [10], the trainable Word2Vec strategy is not used, so the mapping is only done once during model initialization. Such a strategy cannot explore the semantic features in opcode. The experimental results show that the trainable word vector can improve the model's ability of describing opcodes to a certain extent. In this work, the classification accuracy was improved by several percentage points compared with that without the strategy.

3.5 Classifier and Word Vector Result Analysis

We selected data from the Text-CNN, our model with Word2Vec strategy, our model without Word2Vec strategy. The data of the three classifiers were plotted in Fig. 11.

Figure 11 shows that the accuracy of our model has a better accuracy baseline than Text-CNN by 2–3%. Besides, the accuracy of the model is further improved after applying Word2Vec. This comparison proves that the residual network proposed in this paper has a better program description ability than the original CNN network and the Word2Vec strategy is also applicable to describe opcode.

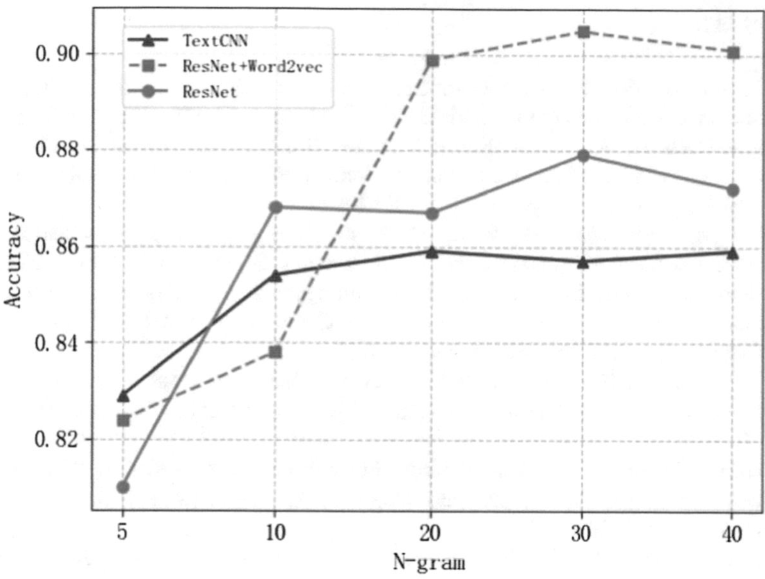

Fig. 11. Classifier performance comparison

4 Conclusion

In this paper, a new classifier based on residual network was proposed. In order to resolve the problem of information redundancy caused by convolutional n-gram model, we design a novel down sampling method with variable steps to get opcode subsequences based on n-gram. The experimental result shows that this data compression method can effectively eliminate redundant information and maintains good performance at the same time. On the side of network structure, we proposed a classifier model based on ResNet, which has powerful program description ability by means of cross-layer connection. Experimental results show that it can improve accuracy obviously. By referring to the relevant knowledge in NLP, the trainable word vector was introduced into the processing to represent opcodes, which further improves the accuracy. Through the combination of the above strategies, our malware detection system has higher precision and lower time complexity than the original Text-CNN with the same data set.

Acknowledgements. This work is supported by the Natural Science Foundation of Jiangsu Province for Excellent Young Scholars (BK20180080).

References

1. Li, J., Sun, L., Yan, Q., et al.: Significant permission identification for machine-learning-based android malware detection. IEEE Trans. Industr. Inf. **14**(7), 3216–3225 (2018)
2. Abou-Assaleh, T., Cercone, N., Keselj, V., et al.: N-gram-based detection of new malicious code. In: Proceedings of the International Computer Software and Applications Conference. COMPSAC 2004, vol. 2, pp. 41–42. IEEE (2004)
3. Shabtai, A., Moskovitch, R., Feher, C., et al.: Detecting unknown malicious code by applying classification techniques on Opcode patterns. Secur. Inform. **1**(1), 1–22 (2012)
4. Siddiqui, M., Wang, M.C., Lee, J.: Data mining methods for malware detection using instruction sequences. In: Iasted International Conference on Artificial Intelligence and Applications, pp. 358–363. ACTA Press (2008)
5. Santos, I., Brezo, F., Ugarte-Pedrero, X., et al.: Opcode sequences as representation of executables for data-mining-based unknown malware detection. Inf. Sci. **231**(9), 64–82 (2013)
6. Divandari, H., Pechaz, B., Jahan, M.V.: Malware detection using Markov Blanket based on Opcode sequences. In: International Congress on Technology, Communication and Knowledge. IEEE (2016)
7. Kang, B.J., Yerima, S.Y., Mclaughlin, K., et al.: N-Opcode Analysis for Android Malware Classification and Categorization, 1–7 (2016)
8. O'Kane, P., Sezer, S., Mclaughlin, K., et al.: SVM training phase reduction using dataset feature filtering for malware detection. IEEE Trans. Inf. Forensics Secur. **8**(3), 500–509 (2013)
9. Kim, Y.: Convolutional Neural Networks for Sentence Classification. Eprint Arxiv (2014)
10. Lee, Y.J., Choi, S.-H., Kim, C., Lim, S.-H., Park, K.-W.: Learning binary code with deep learning to detect software weakness (2017)
11. He, K., Zhang, X., Ren, S., et al.: Deep residual learning for image recognition, pp. 770–778 (2015)
12. Rasmus, A., Valpola, H., Honkala, M., et al.: Semi-supervised learning with ladder networks. Comput. Sci. **9 Suppl 1**(1), 1–9 (2015)
13. Microsoft Malware. https://www.kaggle.com/c/malware-classification

Design of Elderly Care System Integrated with SLAM Algorithm

Jun-Yan Chen and Long Huang(✉)

Guilin University of Electronic Technology, Guilin, China
1016121532@qq.com

Abstract. Currently, China has entered the aging society, and the problems of elderly care have been focused by the whole society. In order to help the young people to take care of the elderly, a care service system integrated with SLAM algorithm was designed. This system includes four parts: ZigBee networking equipments, robot, cloud server and Andriod APP. It makes the robot based on SLAM algorithm achieve indoor navigation and map building. And it provides convenience for the disabled people, monitors the environment security information in real time and shows the physical data of the elderly, which can help young people who work outside master the physical healthy conditions of the elderly who are at home.

Keywords: Elderly care · SLAM algorithm · Robot · ZigBee

1 Introduction

According to the statistics, at the end of 2017, there were 241 million people over 60 years of age in China, accounting for 17.3% of the total population. China has completely entered an aging society and has become the country with the largest number of elderly people in the world [1, 4]. The implementation of China's "Family Planning" policy has led to more families with only one child. Nowadays, an average of 1 young people needs to take care of 1.5 elderly people. The large number of elderly people has become a burden for young people, and it has attracted highly attention in society.

With the rapid development of Internet of Things technology, creative IoT products continue to emerge [7]. The IoT products that serve the elderly are produced more and more, but they are only a single-scale and independent, have not yet formed a complete service system.

In response to the above problems, a remote elderly care system based on ZigBee network has been designed, which deeply integrates multiple IoT products into a complete system. The robot in the system can collect environmental information at home and receive people's voice to control appliances. It can also be controlled and monitored by the remote APP. Using ZigBee networking technology to establish a LAN of a coordinator and multiple terminals can control the doors, windows and lighting equipments as well as collect sensors data. Each terminal wirelessly transmits data to the coordinator, and the coordinator connects to the intelligent router through the serial port. Then the intelligent router uploads the data to the cloud server [5]. For the elderly, the wristband worn on the hands of the elderly can detect the body data

C.-N. Yang et al. (Eds.): SICBS 2018, AISC 895, pp. 503–513, 2020.
https://doi.org/10.1007/978-3-030-16946-6_40

such as heart rate and blood pressure [2, 3] and then upload it to the cloud server through the intelligent router for real-time analysis. At the same time, a fall alarm function is added to the wristband to prevent an emergency from occurring. Andriod APP of mobile phone uses mobile network to access cloud server to obtain relevant data. APP has the functions of displaying the home environment and the elderly's body information, accepting emergency alarm reminders, and issuing commands to remotely control multiple IoT technology products in the home [4]. Using ZigBee networking technology to integrate fragmented technology products into a whole can help young people care for the elderly who are at home, which has a high market value [5, 6].

2 System Overall Frame Design

The system consists of four parts: ZigBee network equipments, intelligent router, cloud servers and Android APP. The design frame diagram of system is shown in Fig. 1. Utilizing ZigBee network technology can reconstruct the equipments. The ZigBee central node named coordinator is connected to the intelligent router through the USB to serial port. The data is uploaded to the cloud server by using the intelligent router data gateway platform. The APP accesses the cloud server to obtain corresponding information.

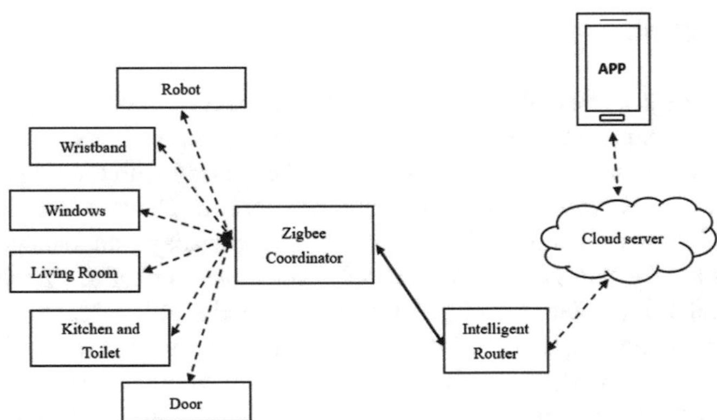

Fig. 1. The design frame diagram of system

3 System Design Implementation Plan

3.1 Hardware Design Implementation

The hardware design of this system mainly consists of two parts: the robot and the wristband worn by the elderly.

(1) Robot: The robot used to serve the elderly uses a STM32F103 chip as the main control chip. The main modules are voice recognition module, ZigBee communication module, sensors module and motors driving module. STM32F103 Chip has a wealth of pin resources. It is mainly used for driving the motors and provides +5v for kinds of sensors. Various sensors include infrared obstacle avoidance sensor, flame sensor, DHT11 temperature and humidity sensor, MQ-2 smoke sensor, MQ-7 CO toxic gas sensor, ultrasonic sensor. In order to make full use of the pin resources of STM32F103, an external circuit board is designed to lead the I/O port pins for sensors and other peripherals like Wi-Fi camera module and audio power amplifier module. In addition, a control circuit board with STM32F101 as the core is specially designed for the control of the robot arms. The control panel draws eight PWN waves, which are used for driving the six steering gears of the robot arms to control the robot arms to take objects. The schematic diagram of the robot motherboard design is shown in Fig. 2.

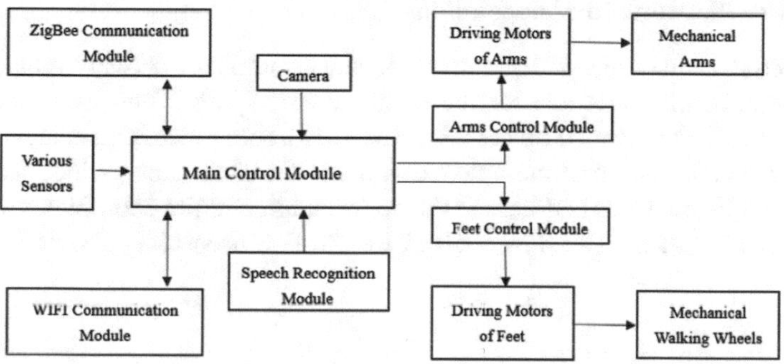

Fig. 2. The design schematic diagram of robot

(2) Smart Wristband: The master chip of wristband worn by the elderly is the CC2530 chip of TI Company. The wristband needs to be powered by a small piccc of aluminum battery. The wristband circuit board is equipped with a heart rate sensor, blood pressure sensor, and three-axis acceleration sensor, which heart rate sensor and blood pressure sensor are installed on the inner side of the hand ring contacting with the skin. They are close to the old man's skin to achieve accurate measurement, such as real-time feedback when the elderly have unsafe changes in heart rate and blood pressure. The three-axis acceleration sensor is installed on the surface of the wristband to detect the movement of the elderly in real time. When the three-axis acceleration sensor suddenly detects a large acceleration in the vertical direction, it can be judged that the elderly accidentally fall down, then the wristband will send out an alarm message to young people [6]. The design frame of the smart wristband is shown in Fig. 3.

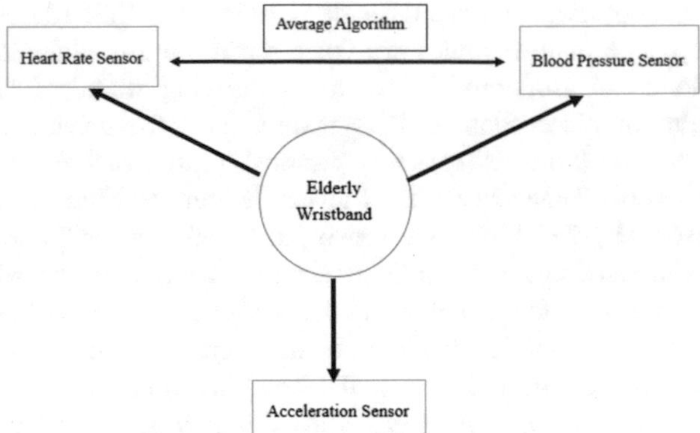

Fig. 3. The design diagram of intelligent wristband

3.2 ZigBee Network Implementation

As the central node of the ZigBee network, the coordinator actively creates and initializes the network and waits for the child nodes to join. Then it is receiving and forwarding the data uploaded by the terminals. It is a data processing center. In order to avoid message congestion of multiple terminals in the same time period, each terminal needs to send its own short address to the coordinator, and the coordinator allocates an address space for them. The flow chart of the ZigBee network is shown in Fig. 4.

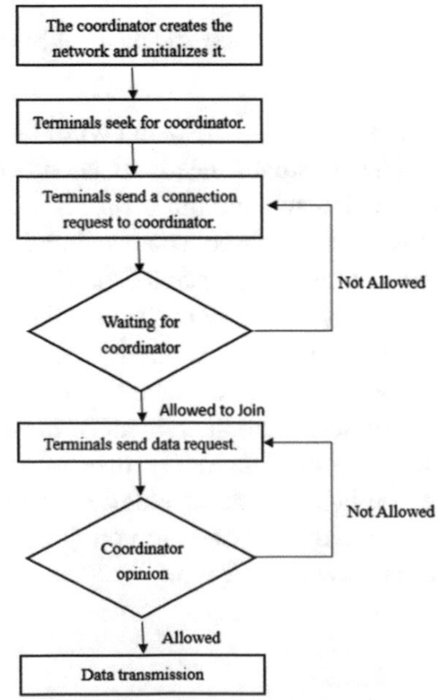

Fig. 4. The flow chart of ZigBee network

(1) The first terminal - Robot: The robot is equipped with a Wi-Fi camera to achieve remote monitoring and a voice recognition module to achieve voice control of the home equipments. The robot uses serial communication to connect with the ZigBee terminals for data transmission. The robot converts the voice into instructions and uploads it to the coordinator, then the coordinator sends it to other terminals. It means to use the robot as one of the terminals to control other terminals.
(2) The second terminal - Wristband: The Wristband is designed by ZigBee technology. It is equipped with body sensors to detect heart rate and blood pressure in real time so as to master the physical conditions. A three-axis acceleration is used for detecting the physical movement of the elderly. The wristband uploads the data to the coordinator. Then the coordinator uploads it to the cloud server through the intelligent router.
(3) Other terminals - Home equipments: Utilizing the I/O pin functions of ZigBee, the equipments in the home can be transformed. The terminal of the living room can remotely switch the air conditioners, the lamps and the TV. The terminal of the windows is equipped with the rain drop sensor, which can automatically close the window when it rains. It can also remotely control the switch of the windows and the curtains. The terminal of the kitchen and toilet is equipped with MQ-2 sensor, MQ-7 sensor and flame sensor to monitor the safety situation. The infrared pyroelectric body sensor can automatically switch the light whether there someone is using the bathroom or not. It can also remotely switch electric water heaters and exhaust fans. The terminal of the door can open the door lock remotely if the key is forgotten. It can sound the alarm when the door lock is vandalized. These terminals can be controlled by the human voice recognition of the robot so that the elderly can enjoy the effects of relaxing life.

3.3 Mobile Implementation

The system aims to help more young people who work outside care for the elderly and grasp the situation in the room. Mobile terminal APP can monitor the safe information of the home and display the data of the elderly's body so that the young man can master the elderly's physical conditions and remotely observer the elderly's living conditions. In addition, the APP can achieve the operation of robots and home equipments, which is convenient for the elderly with mobility problems. The mobile terminal APP of the system is divided into three parts: status module, communication module and setting module. APP is rich in functions and simple in operation. The design framework of APP is shown in Fig. 5.

(1) Status module: It is including remotely robot controlling, environmental monitoring, household electrical appliances controlling and camera monitoring. The robot controlling refers to remotely controlling the robot to walk under the camera page. The environmental monitoring refers to displaying the real-time data by historical polygon charts collected by the sensors in the home. The household electrical appliances controlling refers to using the buttons of APP to switch air conditioners, lamps, windows, door and other household electrical appliances.

The camera monitoring can open the Wi-Fi camera on the robot to observe the living conditions of the elderly.

(2) Communication module: It is including common contacts, emergency alarm phone and voice interaction. In the common contacts section, we can add an acquaintance phone number to contact. In the emergency alarm phone section, we can enter alarm phone numbers like 110, 120 in advance so that it is a key to seek help when we are in trouble. In the voice interaction section, we can use iFlytek Voice Input to send voice commands to control the home equipments remotely.

(3) Setting module: It is including login/registration, personal data settings, opinions feedback and exit the software. In addition, mobile notification status bar can display the alarm information what the APP has received.

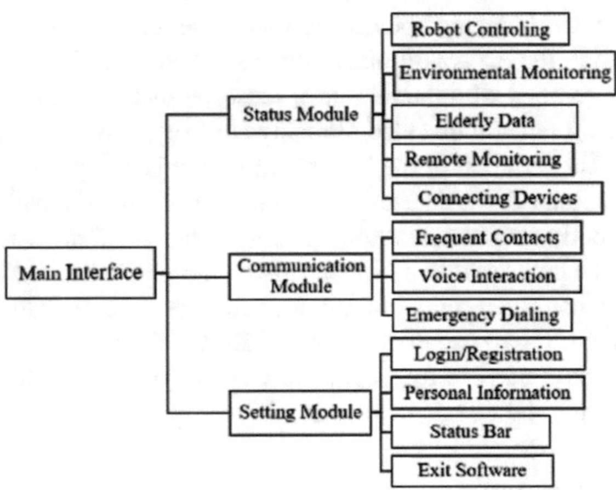

Fig. 5. The design framework of APP

3.4 SLAM Algorithm Implementation

The sensor used by the robot to implement the SLAM algorithm is the Kinect sensor of Microsoft. The Kinect [16] sensor is mainly made of infrared projector, RGB color camera and infrared depth camera. It works by projecting infrared structured light onto an object and the reflected light by the surface scattering enters the infrared camera [14]. Then according to the shape and position of the collected infrared spot, we can calculate the distance between objects in the field of view. Kinect sensor can acquire both depth images and RGB color images, in which depth images contain the spatial distance information of the current scene.

There have two parts to help the robot based on SLAM algorithm [9, 12] implement indoor [13] navigation and map building. One is learning state, in an unknown environment, the robot continuously detects the environment through Kinect sensor and records environmental characteristics. At the same time, it calculates the corresponding behavior pattern by recording odometer information. We use environment features and behavior patterns as state nodes to build the topological map [8]. Another part is navigation state. The robot compares the path nodes that has been learned before. That is, environmental features detected and preserved by previous visual systems. We can achieve condition positioning and select the corresponding behavior patterns to achieve navigation. Flow chart of SLAM algorithm is shown in Fig. 6.

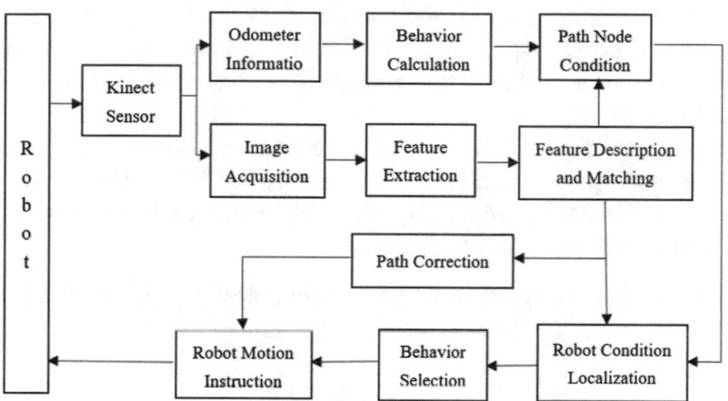

Fig. 6. Flow chart of SLAM algorithm

3.4.1 Learning State

In the learning state, we put the robot in a strange environment and let it extract environmental feature point information by Kinect sensor. The robot performs feature matching to record the current environmental information and builds the map [11]. The ORB [15] (Oriented FAST and Rotated BRIEF) is an algorithm for fast feature point extraction and description. The feature point extraction is performed by the oFAST algorithm improved by the FAST [16] algorithm. The feature point description is performed by the rBRIEF algorithm with the addition of rotation invariance improved by the BRIEF algorithm.

The feature points extracted by oFAST algorithm are determined by comparing the gray values of the circular pixels and the central pixels.

$$\mathrm{N} = \sum\nolimits_{x\forall(circle(p))} |I(x) - I(p)| > \varepsilon_d \tag{1}$$

The ORB algorithm proposes the use of moment method to determine the direction of FAST feature points. The definition of distance is as follows: m_{10}

$$m_{pq} = \sum\nolimits_{x,y\in r} x^p y^q I(x,y) \tag{2}$$

We can find the centroid of the moment:

$$\mathrm{C} = \left(\frac{m_{10}}{m_{00}}, \frac{m_{01}}{m_{00}}\right) \tag{3}$$

It assume that the corner coordinates are O. Then the angle θ of the vector is the direction of the feature point. The calculation formula as follows:

$$\theta = \arctan(\frac{m_{01}}{m_{00}} / \frac{m_{10}}{m_{00}}) = \arctan(m_{01}/m_{10}) \tag{4}$$

In the above formulas, (x, y) is the coordinate of the FAST corner point, and the circular radius is r, (x, y) ranges from [−r, r], and C is the center of mass of the circle. The I(x, y) is an image grayscale expression. The arctan() represents the arctangent function of the value interval (−π, π].

Feature point description refers to the description of image features by detecting and extracting good points by FAST algorithm. rBRIEF [19] algorithm implements feature point description by comparing the pixel value of the points. The rBRIEF algorithm is described in detail as follows.

(1) We assume that the coordinates of n points are (x_i, y_i), and the point pair is constructed as a 2 × n matrix.

$$\mathrm{M} = \begin{bmatrix} x_1 & \cdots & x_n \\ y_1 & \cdots & y_n \end{bmatrix} \tag{5}$$

(2) The M matrix is processed by the mathematical method of rotation transformation. The transformed M rotation matrix uses the main direction of the corner.

$$R_n = \begin{bmatrix} \cos\theta & \sin\theta \\ -\sin\theta & \cos\theta \end{bmatrix} \tag{6}$$

(3) We make the M matrix into a directed form, and the direction can be expressed as a formula (7).

$$M_\theta = \arctan(m_0, m_{10}) \tag{7}$$

(4) Finally, the result of the obtained descriptor is expressed as a formula (8).

$$g_n(P, \theta) = f_n(P) | \{x_i, y_i\} \varepsilon M_\theta \tag{8}$$

3.4.2 Navigation State

After using the ORB algorithm to complete the learning stage task, then we let the robot achieve the autonomous cruise task, in which includes robot walking direction, state positioning, interference eliminating and path correcting.

We can judge the direction of the robot by using the robot odometer information to calculate the changes of the direction angle Δt within the state node interval time. We stipulate that the angular velocity of the robot during walking is w = 0.2 rad/s, and the left turn direction is the positive direction.

① If $\Delta t > 5°$, the robot should turn left.
② If $\Delta t < -5°$, the robot should turn right.
③ If $-5° < \Delta t < 5°$, the robot should go straight.

The state of the robot is determined by comparing and matching the current scene with the memory road sign in the learning map. If the match is successful, the next step is to decide the direction of the robot's movement.

During the robot learning stage and navigation state, in order to avoid mismatching of image features [10] acquired by the robot, we use K-nearest neighbor algorithm to remove mismatch. It can improve the matching accuracy and meet the need of robot's navigation task.

In the process of robot navigation [17], the robot's driving path and learning path will be deviant because of factors such as offset from the starting position or environmental dynamics. So, We need to add a correction mechanism to the robot's driving process to ensure the accuracy of robot navigation.

If $X_t < X_d$ *and* $X_t < 0$, then the robot turns left.
If $X_t > X_d$ *and* $X_t > 0$, then the robot turns right.
X_t is the horizontal coordinate of the road sign in the input scene. X_d is the transverse coordinate of the corresponding feature points in the database.

4 System Test Results

The test of this system need the 700 mAH lithium batteries as supplied power. After the system starts, the ZigBee network is successful, and various parts of the system operate smoothly under the net. The system can last for about 60 h, and there is no network disconnection during the test. It is showing the stability of the network. Table 1 is the test and result of each module in system.

Table 1. System module test and results

Test module	Test results
Robot module	The robot can walk front, back, left and right through the APP to avoid obstacles. The robot can achieve autonomous cruise. The users can use Mandarin to control the home devices by robot's voice recognition. The camera on robot is clearly observed under the mobile network
Wristband module	The APP can display the value of tester's heart rate and blood pressure, and the wristband sends an alarm message to the APP when the tester suddenly falls to the ground
Living-room module	The value of temperature and humidity can be collected. The air conditioners can be controlled. When the smoke sensor senses a toxic gas, the buzzer is activated and reported to the mobile terminal

(*continued*)

Table 1. (*continued*)

Test module	Test results
Window module	The windows and curtains can be controlled by the stepper motors. The photosensitive sensor and the raindrop sensor can detect the amount of rain and whether rain or not
Kitchen and toilet module	A small tank of liquefied petroleum gas deliberately leaks a little gas, and the sensor of the lampblack machine senses the poisonous gas. Then the lampblack machine automatically drakes and exhausts it. The infrared thermo release human body sensor is used for judging whether the person is using the bathroom, and the lamps can be switched automatically
Door module	By receiving control instructions sent from the mobile terminal, the door's functions of opening and closing can be executed

5 Summary and Prospect

As the number of elderly people increases year by year, the issue of caring for the elderly has become an important issue for every family. The system can achieve remote operation of robot and household equipments, and provide convenience for the elderly with mobility problems. At the same time, the APP can monitor the environmental safety information of the home in real time and display the physical data of the elderly. Young people can grasp the physical conditions of the elderly and remotely observe the living conditions of the elderly in the home. Due to ZigBee network has the advantages of convenient self-organizing network, fast transmission speed and low power consumption, which can greatly improve the battery life. This system can bring practical effects to every family. It can give the elderly who stay alone at home more security. It allows young people working outside the home to keep abreast of the elderly in their home. Not only to reduce the pressure of young people, but also to take care of the elderly and promote the inheritance of traditional culture of respecting old-age endowment.

Although the system has achieved most of the functions, but still need further improvement and perfection. The main manifestations are as follows: (1) Adding robot arms to grasp objects for the elderly more intelligently; (2) Camera combined with face recognition algorithm, which can identify whether strangers come in home illegally and strengthen security; (3) The robot has an indoor patrol function, and it can detect the danger and call the police in time. The above functions will continue to improve in the later period.

Acknowledgements. This work is supported by Guangxi Cooperative Innovation Center of cloud computing and Big Data under Grant no. YD16515 and Student's Platform for Innovation and Entrepreneurship Training Program under Grant no. 201810595038.

References

1. Fu, W., Jiang, D., Xiong, P.: Design and implementation of android-based intelligent monitoring system for the elderly. Software **38**(7), 10–13 (2017)
2. Geng, F., Yin, J., Zhang, H.: Design of human body ECG signal monitoring system based on intelligent terminal. J. Transduct. Technol. **27**(3), 289–292 (2014)
3. Ma, B.: Design of elderly monitoring and tracking system based on ZigBee technology. Shandong Ind. Technol. **23**, 136 (2016)
4. Qiu, M.: Intelligent community home care system based on IoT ZigBee technology. J. Zhangjiakou Vocat. Tech. Coll. **33**(2), 71–75 (2014)
5. Xu, L., Huang, C.: Research and implementation of empty nest elderly home care system based on Internet of Things. Wirel. Interconnect Technol. **13**, 45–46 (2016)
6. Zhao, Y., Yu, Z.: Research and design of elderly care system based on cloud architecture. Sci. Technol. Vis. **19**, 138–140 (2015)
7. Zhu, Z.: Progress and trend of sensor network and Internet of Things. Microcomput. Appl. **26**(1), 1–3 (2010)
8. Hua, G., Hasegawa, O.: A robust visual-feature extraction method for simultaneous localization and mapping in public outdoor environment. J. Adv. Comput. Intell. Intell. Inform. **19**(1), 11–22 (2015)
9. Strasdat, H., Montiel, J., Davison, A.: Scale drift—aware large scale monocular SLAM. In: Proceedings of Robotics: Science and Systems (RSS), pp. 101–108 (2010)
10. Bostanci, E., Kanwal, N., Bostanci, B., Guzel, M.S.: A fuzzy brute force matching method for binary image features. arXiv preprint arXiv:1704.06018 (2017)
11. Cadena, C., Carlone, L., Carrillo, H., et al.: Past, present, and future of simultaneous localization and mapping: toward the robust-perception age. IEEE Trans. Robot. **32**(6), 1309–1332 (2016)
12. Mur-Artal, R., Tards, J.D.: ORB-SLAM2: an open-source SLAM system for monocular, stereo, and RGB-D cameras. IEEE Trans. Robot. **33**(5), 1255–1262 (2017)
13. Henry, P., Krainin, M., Herbst, E., et al.: RGB-D mapping: using depth cameras for dense 3D modeling of indoor environments. In: Experimental Robotics, pp. 647–663. Springer, Heidelberg (2014)
14. Davison, A.J., Reid, I.D., Molton, N.D., et al.: MonoSLAM: Real-time single camera SLAM. IEEE Trans. Pattern Anal. Mach. Intell. **29**(6), 1052 (2007)
15. Mur-Artal, R., Montiel, M.M., Tardos, J.D.: ORB-SLAM: a versatile and accurate monocular SLAM system. IEEE Trans. Robot. **31**, 1147–1163 (2015)
16. Hartmann, J., Dariush, F., Marek, L., Kluessendorff, J.H., Maehle, E.: Real-time visual SLAM using FastSLAM and the microsoft kinect camera. In: Proceedings of ROBITIK (2012)
17. Shim, V.A., Tian, B., Yuan, M.L., Tang, H.J., Li, H.Z.: Direction-driven navigation using cognitive map for mobile robots. In: Proceedings of IEEE/RSJ International Conference on Intelligent Robots and Systems (2014)
18. Zhang, S., Xie, L., Adams, M.: An efficient data association approach to simultaneous localization and map building. Int. J. Robot. Res. **24**, 49–60 (2005)
19. Calonder, M., Lepetit, V., Strecha, C.: BRIEF: binary robust independent elementary features. In: European Conference on Computer Vision (2010)

A Fog-Based Collusion Detection System

Po-Yang Hsiung, Chih Hung Li, Shih Hung Chang, and Bo-Chao Cheng(✉)

Department of Communications Engineering, National Chung Cheng University, 168 University Road, Min-Hsiung, Chia-Yi 62145, Taiwan
paul100dtj@gmail.com, zlsh09830123@gmail.com, will168891688916889@gmail.com, bcheng@ccu.edu.tw

Abstract. The threat of malicious software (malware) programs on users' privacy is now worse than ever. Some malwares can even collude to increase their permissions through sharing, making it easier to breach users' privacy. However, detecting malware on a mobile phone consumes a considerable amount of a mobile phone's energy. While the advancements of cloud technology make it possible to transmit malware data to the cloud for analysis, transmitting large amounts of data contributes to the energy consumption problem of mobile phones. Therefore, this study proposes a fog-based computing technique that targets intelligent collusion attacks and intelligently controls the collection and transmission behavior to reduce the amount of data that is transmitted to lower the energy consumption of mobile phones. This study conducts experiments using an official application and compares it with other methods. The experimental results show that the proposed method transmits less data than other methods and saves the energy consumption of mobile phones.

Keywords: Mobile malware detection · Collusion attack · Fog-based computing · Energy consumption

1 Introduction

In the modern society, mobile device usage is growing exponentially, and advances in technology have made mobile applications more convenient and diverse. Although applications can be downloaded from official markets or third-party stores, users often have limited knowledge about the applications they download. In addition, official markets or third-party stores often fail to filter malware. In recent years, researchers have found that users' private information has been leaked because many applications contain malicious advertisements, and two or more applications may obtain more permissions through sharing their permissions to conduct collusion attacks, thus threatening user privacy [1, 2].

Previously prevalent collusion attack detection technologies are static analysis and dynamic behavior detection [3, 4]. Static analysis refers to performing detections in non-runtime environments—such as permission detection and feature detection—analyzing application code, and identifying coding defects and potential malicious code. Dynamic behavior detection refers to performing behavior analysis while the host system is running, usually in a sandbox environment, such as behavior detection or

C.-N. Yang et al. (Eds.): SICBS 2018, AISC 895, pp. 514–525, 2020.
https://doi.org/10.1007/978-3-030-16946-6_41

battery monitoring, to observe the impact of the application on the overall performance of the host. However, performing detection on a mobile device consumes a considerable amount of energy from the mobile phone and tends to be ineffective. To overcome this problem, research in recent years has proposed cloud malware detection technology. The core of this technology is to transmit detection and analysis to the cloud and use the data of the application on the mobile device to conduct the analysis in the cloud, thus reducing the energy consumed by the mobile phone and overcoming the limitation of mobile device resources.

However, using cloud technology to detect a large number of applications consumes a huge amount of the mobile device's energy and often results in exhausting the battery. For example, when collecting information about all applications on the mobile phone and transmitting the data to the cloud for analysis, as the user's number of applications increases, the amount of data also increases, thus increasing the energy consumption. By using cloud technology to analyze the data, the mobile phone consumes more energy to collect and transmit the data, thus also exhausting the battery. This research therefore focuses on how to use the framework of fog computing to carry out mobile device collusion attack detection more effectively. We set up a fog computing environment and combine it with the "Anubis" method, which can monitor and control the data transmission of mobile phones, as shown in Fig. 1. This study aims to reduce the consumption caused by data transmission in the fog computing environment, uses the permission-based detection method to compare the permissions and judgments of the application against the conspiracy attack, and compares the probability of collusion and permission information. The use of "Anubis" allows the mobile terminal to know how to collect the necessary information and when to transmit the data, thus reducing the energy consumed by the mobile phone for collecting and transmitting data, and setting up an energy-saving and efficient collusion detection mechanism.

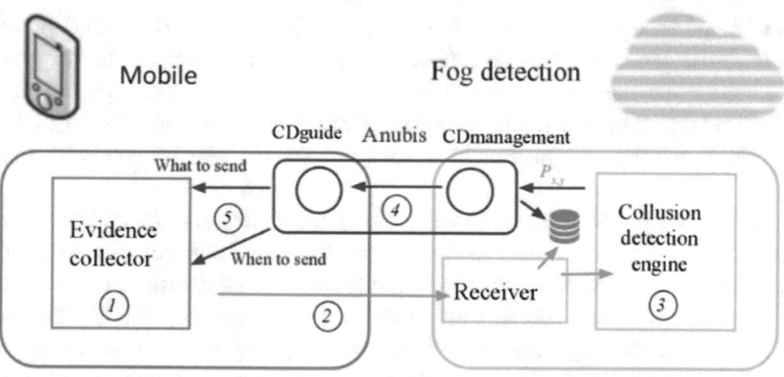

Fig. 1. Anubis concept diagram.

The remainder of this paper is structured as follows. Section 2 introduces research on using cloud technology to detect malicious software and collusion attacks. Section 3 presents a detailed description of the "Anubis" method proposed in this paper. Section 4 presents the experimental results and validations, and Sect. 5 presents the conclusion and future direction of the research.

2 Related Work

Portokalidis et al. [5] proposed a solution called Paranoid Android for smart phone security, which tracks and records the programs executed in the phone and transmits a copy of the tracking file to the remote server. In the virtual environment, the program is re-executed according to the original mobile phone process and checked securely. Paranoid Android consists of three parts: (1) smart phone, (2) remote server, and (3) network/proxy. First, the program known as Tracer is run on the smartphone. Tracer records the operation of other programs in the mobile phone (such as system call and asynchronous signal) and encrypts them in HMAC mode at the storage. Tracer connects to the network through the proxy on the mobile terminal, which sends the data to the remote server. After receiving the data, the remote server starts the virtual machine to execute the copy program in the mobile phone. At this time, the server can detect the malicious software.

Although their proposed method provided a new way to detect mobile device malware and overcame the problem of performing detection on a mobile phone, this method has some drawbacks. To upload the detection analysis to the remote server, the mobile terminal collects and transmits the data of all programs. Such huge data transmission results in increased battery consumption, making it impossible to analyze multiple applications before the device is exhausted.

Burguera et al. [6] developed a cloud technology method called Crowdroid to detect mobile malware. The Crowdroid method is a behavior-based detection method that consists of an application installed on a mobile device and a cloud server. The mobile terminal sends the data executed by the detected application, and the cloud server receives the data and forms a behavior data set for each application. After collecting enough data, K means, a method similar to cluster, is used to divide each dataset into two sets: one is benign program behavior and the other is malicious behavior.

As it is not feasible to implement the security mechanism on the mobile phone, this method performs the entire analysis process on the cloud and combines the behavior detection method with the K means algorithm to classify the datasets. The experimental results showed that the proposed method could successfully distinguish between benign and malicious programs. However, analyzing the behavior of the application using behavior detection is time-consuming. To collect these behavioral data, it is necessary to execute the application on a mobile phone or sandbox, which increases the time consumption and raises the risk of a malicious attack during the detection process.

Xu et al. [7] examined collusion attacks of applications. A collusion attack is mainly conducted through the internal communication mechanism of the mobile phone's Inter-Component Communication (ICC). Mobile phones use ICC to share user data and control or send data from other programs over the network, making mobile devices more vulnerable. The authors developed a model called AppHolmes, which analyzes the application's two-way links to detect whether the application faces the possibility of a collusion attack. Their proposed method comprises data extraction, ICC analysis, and data matching to determine the possibility of a collusion attack. The experiment analyzed a number of third-party software programs, which were downloaded from Baidu and Pea Pod. The results of the experiment showed that 30% of the software programs had the possibility of a collusion attack, and 5 to 6 applications communicated with each other. The experiment also analyzed the impact of collusion attacks on mobile phone energy consumption, CPU, and memory, factors that should not be ignored. However, the authors failed to address the impact of application collusion attacks on mobile phone batteries.

Kashefi et al. [8] compared malware with predetermined rules to determine which applications obtain too many permissions. The paper was mainly for a single application of "Kirin". However, when two or more malwares attempt to gain access through sharing, namely launching collusion attacks, these applications can have too many privileges to steal user data, meaning that it is not enough to merely compare the rules of a single application. Their paper therefore proposed a mechanism to determine how to detect an application with too many permissions. This mechanism detected not only a single application, but also whether a group of applications gains excessive permissions through collusion attacks. To detect different collusion attacks, the authors added some new rules to improve the detection mechanism. In their experiment, different types of applications were detected. While the experimental results showed that social, communication, and audio-visual software have a significantly higher possibility of a collision attack, the study used a permission-based detection method to compare predetermined rules, which only considered the behavior of application permissions and failed to take the possibility of sharing between applications into account.

3 Approach

In this paper, the mobile phone transmits the data to the fog end to analyze whether a collusion attack exists between the applications. We use ifogsim [9], which was proposed in 2016, simulate the mobile phone and the fog end environment, measure the mobile phone transmission data at the fog end, and observe the effect of the amount of transmission on the delay and energy consumption. Using the edge deployment strategy of the paper in the fog end environment and placing the module near the edge of the network and the cloud, the execution can be controlled, thus reducing the transmission delay and transmission consumption, and avoiding network congestion caused by increased data transmission.

This paper proposes a fog-based collusion attack detection architecture. For the detection mechanism of a collusion attack, combined with the algorithm "Anubis", the intelligent control allows the mobile phone to capture only the information and data necessary for detecting the application and allows the mobile phone to successfully transmit data in the case of limited energy. The proposed "Anubis" can reduce the data collection and transmission, thus lowering the battery consumption. To help understand the algorithm Anubis, we define notations with brief descriptions in Table 1.

Table 1. Comparison of relevant research.

Symbol	Description
D_j^i	The evidence of App_i collected at T_i
R	Detection rules of a collusion attack
$\Theta_{p,q}$	The evidence after matching action rule at T_p $to T_q$
$Z_{p,q}$	The evidence after matching action rule at T_p $to T_q$
$P_{x,y}$	Collusion probability (pairs of apps)
$\alpha_{p,q}$	The evidence set R, $\Theta \cup Z = \alpha$
ω_i	The evidence set identifying a collusion attack $i, \omega_i \subseteq R$
E_i	The residual energy at time i
E_m	The reserve energy for up to time m (e.g., when the user charge his/her smartphone usually)
μ	Energy consumption for collecting and transmitting evidence

Our fog-based collusion attack detection architecture is divided into three parts:

- Mobile phone: This part comprises four steps: (1) Capture, (2) Encoding, (3) Storage, and (4) Transmission. Androguard is used to retrieve the permissions of the application on the mobile phone, collect the permission data of the application, encode and compress the data, store and ensure that the attacker cannot be stolen, and transfer the processed data to the server of the fog computing for detection.
- Collusion Attack Detection Mechanism: Detection technology is a permission-based approach to the fog computing architecture. By comparing the data transmitted from the mobile phone and referring to various rules for judging the attack, we use Eq. (1) to calculate the probability of each group of applications for collusion attacks. The obtained results will be passed to the main algorithm of this paper, Anubis, for analysis.

$$P^i_{x,y} = \frac{|(\Theta^x_{p,q} \cap \omega i) \cup (\Theta^y_{p,q} \cup \omega i)|}{|\omega i|} + \frac{|(Z^x_{p,q} \cap \omega i) \cup (Z^y_{p,q} \cap \omega i)|}{|\omega i|} \tag{1}$$

- Anubis: First, the probability of a collusion attack and related information are obtained from fog computing, the predetermined rules are subtracted from the application's permission, and the data necessary for judging the application collusion is obtained. Then, using Eq. (2), the necessary data can be transmitted several times before the mobile phone needs to be charged.

$$n = (E_i - E_m)/\mu \tag{2}$$

The following section provides detailed algorithms. Table 2 presents the steps on the mobile side. The mobile phone stage comprises four steps: (1) Capture, (2) Encoding, (3) Storage, and (4) Transmission.

Table 2. Mobile device terminal.

Algorithm 1. Mobile Side

Input: The evidence of App

Output: Collusion App's data

1. Extractor.collect(D_j^i);
2. Facts = Extractor_collect(D_j^i);
3. Encode.(Facts);
4. Storage.deposit (Encode.(Facts));
5. Transmitter.send(Encode.(Facts), fog);

Table 3. Collusion attack detection mechanism.

Algorithm 2. Collusion Attack Detection Mechanism

Input: Encode.(Facts)

Output: $P_{s,t}$

1. D = Receiver.receive(Encode.(Facts));
2. (α, ω_i)= Detection.match(D,R);
3. Ps,t = formula_1(α, ωi) ; // use Eq. (1)
4. return Ps,t;

Table 3 shows the detection mechanism, comparing the permissions of the application with the rules for judging the attack. Equation (1) is a simple function of the main algorithm Anubis. The formula can be used to determine how many times the mobile phone can transmit the necessary data to the fog computing detection mechanism.

The previous step obtains the collusion probability P of each pair of applications. Further analysis is performed to establish how many times the necessary evidence γ can be transmitted before charging is required, using Eq. (2), to avoid battery exhaustion when transmitting data. The algorithm proposed in this study determines the possibility of collusion attacks in each pair of applications. After conducting the analysis, it is possible to obtain the data necessary for the collusion and to ensure that the necessary data can be transmitted several times before the next charge (Table 4). We conduct experiments to prove the Anubis method, detect the collusion to obtain the necessary data, determine how many times it can be transmitted, and compare it with other detection methods.

Table 4. Anubis.

Algorithm 3. Anubis

```
CDmanagement  /* at the fog computing side */
   Input: ω, α, Px, y
   Output: γ, n, analysis_result
while (Px, y > 0%){
   if (ω ==α) {
         analysis_result = 1;
         γ = 0; }
   else {
         analysis_result = 0;
         γ = ω – α;
         n = CalculateTimes (); // use Eq. (2). }
   sendToCDguide(n);
}
-------------------------------------------------------------------
CDguide /* at the mobile device side */
while (1){
   n = recievingFromCDmanagement()
  for (i = 0; i< n; n++){
        Extractor_collect(γ);
        TransmitData2Fog(γ);}
}
```

3.1 Example

We assume there are three groups: i, j, and k; each group has 16 apps for detection, two apps are paired for analysis, and each pair has the same result. We compare the rules for two types of collision attacks, (ω_1, ω_2). We also assume that the phone has 20% power now, the detection mechanism will consume 16% of the power before the next charge.

$$\omega_1 = \Theta_1, \Theta_2, Z_1$$
$$\omega_2 = \Theta_2, \Theta_3, Z_{12}, \Theta \cup Z = \omega;$$

The following examples illustrate the operation of the entire algorithm. The first step is to take the evidence α of the three sets of programs on the mobile phone. Here, one pair is used for analysis, and m is the comparison. The evidence is shown as follows:

$$\mathrm{m}(D^i_{p,q}, R) = \Theta_1, \Theta_3, Z_1 = \alpha^i_{p,q}$$

$$\mathrm{m}(D^j_{p,q}, R) = \Theta_1, \Theta_3, Z_2 = \alpha^j_{p,q}$$

$$\mathrm{m}(D^k_{p,q}, R) = \Theta_2, \Theta_3, Z_2 = \alpha^k_{p,q}$$

In the second step, after the collection, the detection mechanism compares the evidence of the three sets of application programs i, j, and k with the rules of the collusion attack ω1, ω2, and uses Eq. (1) to find the possibility of a collusion, P (Table 5).

Table 5. The possibility of collusion.

	ω_1	ω_2
$Appi_1, i_2$	2/3	2/3
$Appj_1, j_2$	2/3	1
$Appk_1, k_2$	1	1

In the third step, Anubis is judged to be the evidence γ lacking in the conspiracy attack and finds the necessary evidence to be transmitted several times using a limited amount of charge. After finding the probability of P, the rule ω of ω − α = γ is used in Anubis to subtract the evidence α. Table 6 shows the necessary data for the collusion of a pair of decision applications.

Table 6. γ, the parameter necessary for determining a collusion attack.

	ω_1	ω_2
$Appi_1, i_2$	Θ_2	Θ_2
$Appj_1, j_2$	Z_1	$\varnothing$
$Appk_1, k_2$	$\varnothing$	$\varnothing$

We use Eq. (2) to make several calculations before the next charge. N can transmit three sets of necessary data, as shown in Table 7.

Table 7. n, number of times that data can be transmitted.

	ω_1	ω_2
$Appi_{1-16}$	10	10
$Appj_{1-16}$	10	0
$Appk_{1-16}$	0	0

4 Experiment

This section draws a comparison with other detection platforms from the example, targeting the data transmission amount of each detection method, the execution time of the detection, and the impact on the battery consumption of the mobile phone after the detection. For the experiment, the research is validated using the languages and devices shown in Table 8.

To evaluate the proposed method Anubis, 80 different applications were downloaded from Google Play and third-party APP Store apps and Pea Pods for analysis. As mobile users download an average of 16 APPs per person, we analyze 16 APPs for each integration. In this experiment, the performance of detecting collusion attacks in different experimental environments was discussed separately. In the experiment, three methods of collusion detection are discussed: the fog computing architecture of Anubis, cloud-based Paranoid, and the detection by Mobile on phones.

Table 8. The simulation environment in the experiment.

SW/HW	Description
Processor	Intel Core i7-4790
RAM	8 GB
Operating System	Microsoft Windows 7
Programming Language	Python, Java
Mobile battery	4000 mAh
Data	Download from Google Play and 3rd party app markets

As shown in Fig. 2, after the proposed method Anubis performs fog detection, 40% of the experimented applications have a possibility of collusion attacks. The experimental diagram also shows that Anubis in the fog uses less energy than cloud technology (Paranoid) and mobile phone detection (Mobile). There are two reasons for this. First, while Anubis and Paranoid collect and pass the data to the fog/cloud for analysis, in the Mobile method, when the number of applications reaches 64, the mobile phone is exhausted because the analysis is conducted on the mobile phone. Second, as Paranoid

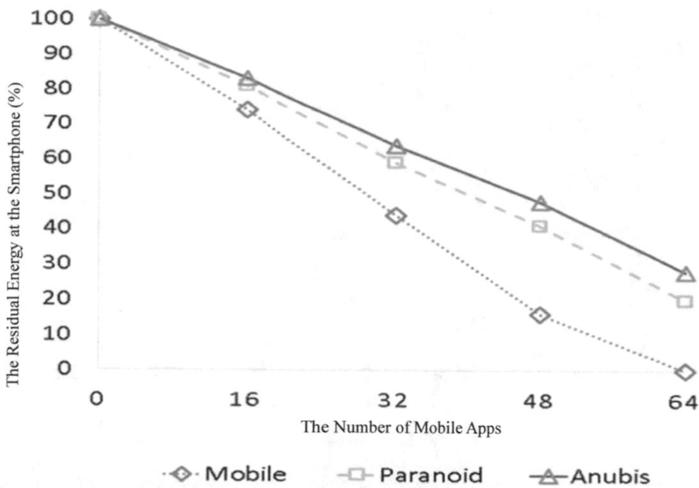

Fig. 2. Relationship between the number of applications and the remaining battery capacity of the mobile phone for each detection.

analyzes all the data, the mobile phone consumes too much energy when transmitting data. Figure 2 shows the power status of the simulated mobile phone after detection. The power consumed by the experiment is affected by the data volume and execution time.

Figures 3 and 4 compare the two cloud technologies Anubis and Paranoid, which are different from Mobile's analysis on mobile phones. Figure 3 shows that, as the number of detection applications increases, fewer data are collected by the fog end Anubis than by Paranoid; the average amount of data per 16 applications is about 55% less, thus reducing the amount transmitted for analysis. This is because Anubis can extract the necessary data and identify the application's more dangerous permissions. These dangerous authorities can more accurately determine whether a collusion exists and pass the necessary information to the fog to detect. A reduced amount of data transferred can be more effectively analyzed to determine whether a collusion attack is present, thus avoiding the energy consumption caused by transmitting too much data.

As shown in Fig. 4, as the number of applications increases, the analysis time of Anubis is still less than that of Paranoid, with an average analysis time of 45% less for every 16 applications. The reason is the same as Fig. 3. As Anubis analyzes the necessary data and transmits it to the fog for analysis, it can more quickly detect which applications have collusion attacks, thus reducing the analysis time and notifying users of malicious software more quickly. Therefore, the problem of exhausting the mobile phone's battery during the detection process is avoided. Taking the necessary information and transmitting it to the fog end analysis can reduce the amount of mobile phone to cloud transmission and achieve a more efficient and energy-saving collusion attack detection mechanism.

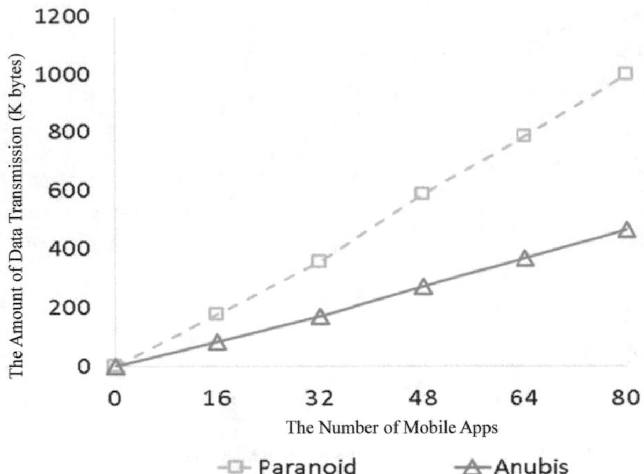

Fig. 3. Relationship between the number of applications and the amount of data transmission for each detection.

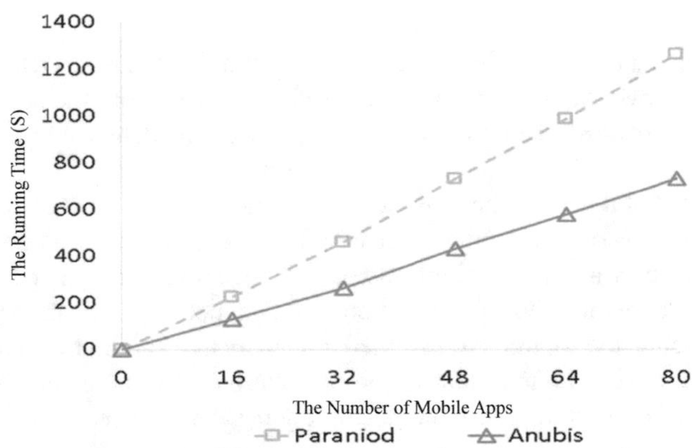

Fig. 4. Relationship between the number of applications and the time for analysis for each detection.

5 Conclusion and Future Work

The objective of this research is to prevent the mobile phone from consuming too much battery power when detecting whether an application is a collusion attack. When the phone detects multiple applications, it tends to capture too much data, which will largely consume the phone's power. In the past, as all data were transmitted once or regularly, the battery was exhausted halfway through the detection. Therefore, this paper proposed the Anubis' algorithm to allow the mobile device to transmit only the necessary data before the next charge, which it combined with intelligent control to effectively detect collusion attacks.

Although this is a practical detection method, it is a permission-based method, and the analysis result is affected by predetermined rules. As the number of applications increases, the predetermined rule becomes more complex. Our future research will focus on improving the detection technology to achieve a higher detection rate and accuracy. We also aim to optimize the algorithm to improve the intelligence of the entire mechanism, more effectively control the data capture and transmission, make the detection mechanism more efficient, and reduce the battery consumption of the mobile phone.

Acknowledgement. This work was supported by the Ministry of Science and Technology (MOST), Taiwan, under MOST Grants: 107-2221-E-194-022 – & 105-2221-E-194-014-MY2.

References

1. Peng, S., Yu, S., Yang, A.: Smartphone malware and its propagation modeling: a survey. IEEE Commun. Surv. Tutor. **16**(2), 925–941 (2014)
2. Wang, D., Cheng, H., He, D., Wang, P.: On the challenges in designing identity-based privacy-preserving authentication schemes for mobile devices. IEEE Syst. J. **12**(1), 916–925 (2018)
3. Bhandari, S., Jaballah, W.B., Jain, V., Laxmi, V., Zemmari, A., Gaur, M.S., Mosbah, M., Conti, M.: Android inter-app communication threats and detection techniques. Comput. Secur. **70**, 392–421 (2017)
4. Chen, H., Su, J., Qiao, L., Zhang, Y., Xin, Q.: Malware collusion attack against machine learning based methods: issues and countermeasures. In: ICCCS, vol. 5, pp. 465–477 (2018)
5. Portokalidis, G., Homburg, P., Anagnostakis, K., Bos, H.: Paranoid android: versatile protection for smartphones. In: ACSAC 2010, pp. 347–356 (2010)
6. Burguera, I., Zurutuza, U., Nadjm-Tehrani, S.: Crowdroid: behavior-based malware detection system for android. In: SPSM@CCS 2011, pp. 15–26 (2011)
7. Xu, M., Ma, Y., Liu, X., Lin, F.X., Liu, Y.: AppHolmes: detecting and characterizing app collusion among third-party android markets. In: WWW 2017, pp. 143–152 (2017)
8. Kashefi, I., Kassiri, M., Salleh, M.: Preventing collusion attack in android. Int. Arab J. Inf. Technol. **12**(6A), 719–727 (2015)
9. Gupta, H., Dastjerdi, A.D., Ghosh, S.K., Buyya, R.: iFogSim: a toolkit for modeling and simulation of resource management techniques in the Internet of Things, edge and fog computing environments. Softw. Pract. Exper. **47**(9), 1275–1296 (2017)

A Real-Time Online Security Situation Prediction Algorithm for Power Network Based on Adaboost and SVM

Haizhu Wang[1(✉)], Wenxin Guo[1], Ruifeng Zhao[1], Bo Zhou[2], and Chao Hu[2]

[1] Electric Power Dispatching and Control Center of Guangdong Power Grid Co., Ltd., Guangzhou 510600, China
mmyzwhz_23@qq.com

[2] NARI Information & Communication Technology Co., Ltd., Nanjing 210033, China

Abstract. The power network has a great impact on the national economy, and power accidents will cause great losses. Therefore, strengthening the mastery and control of the online security situation of the power network timely has become a topic of widespread concern. The traditional power network online security situation prediction algorithms have low accuracy and efficiency. In this paper, Adaboost and SVM are combined to predict real-time online security situation of power network, and an experimental analysis is carried out. Compared with the traditional methods, this method has certain improvement in the correctness and efficiency of the algorithm.

Keywords: Power network · Adaboost · SVM · Security situation prediction

1 Introduction

The power network is the lifeblood of the national economy. With the development of the economy, the power network is also constantly developing. With the continuous growth of the power network and the enrichment of control components, the dynamic characteristics of the power grid become more complex and changeable, and a series of security problems caused by transient stability appear. The impact of security incidents on power networks is also growing. In this context, online security analysis and control of power networks have attracted more and more attention [1–5]. This is an important mean to ensure the normal operation of power systems and prevent the destruction of potential unstable factors.

The online security and stability control decision are to search for the optimization measures to eliminate various security and stability problems for the expected fault set and the set of candidate control measures. The set of candidate control measures includes control objects, control actions, control ranges, priorities, and cost functions. Currently, it mainly relies on manual offline setting, which is difficult to adapt to changes in the external environment. In order to solve this shortcoming, we must adopt a streamlined approach to online security control. The flow of online security control is

C.-N. Yang et al. (Eds.): SICBS 2018, AISC 895, pp. 526–534, 2020.
https://doi.org/10.1007/978-3-030-16946-6_42

as follows: (1) We collect and input data; (2) The network security situation predicts in real time and outputs the predicted result; (3) Record/report the security situation results. If there is a security problem, skip to the next step, otherwise we skip to step (1). (4) Online security control. We can see that in the process of online security control, it has an important position to predict the network security situation in real time and output the prediction result.

The security situation of the power network is the calculation process and has the following requirements:

(1) Intelligent: Automatic prediction based on the security situation of input data, without excessive expert domain knowledge, greatly reducing the amount of manual work;
(2) High efficiency: It can get the predicted results in a short time and report the records, which is convenient for the next adjustment work;
(3) High accuracy: the prediction result can be obtained with a high accuracy, and the artificial waste caused by the wrong prediction can be reduced;
(4) Strong flexibility: The model used to predict the online security situation prediction of the power network can be flexibly adjusted and designed according to requirements.

It is a meaningful work to predict and mitigate the impact of power network accidents through real-time online security situation prediction analysis. In order to be able to detect problems and control in time to reduce the impact of power network accidents, people have conducted research and achieved some results. Such as BP neural network [6], radial basis neural network [7], various improved algorithms of neural network [8], support vector machine [9], improved generalized regression neural network [10] and so on. The above methods have been studied to some extent for the prediction of the security situation of the power network, but there are defects that are difficult to overcome:

(1) On the one hand, the training parameters of BPNN, RBFNN and SVM are difficult to determine, resulting in strong fluctuations in prediction results, large prediction errors, and reduced NSSF accuracy;
(2) On the other hand, although the literature [8] uses genetic algorithms to improve the accuracy, the time complexity is too high, which affects its application and promotion.

Based on the above research status and industrial demand, an efficient and high-accuracy online security situation prediction algorithm needs to be proposed. This paper proposes a real-time online security situation prediction algorithm based on Adaboost and SVM, which is divided into two parts: First, the labeled data is used to train the Adaboost algorithm based on the SVM weak classifier and obtain a fitted prediction model. Then use the prediction algorithm to perform real-time predictive analysis of the security posture of the power network and determine its security status.

The structure of this paper is as follows: First, the preliminary knowledge of Adaboost and SVM is introduced in the Sect. 2. Then in the Sect. 3, the real-time online security situation prediction algorithm for power network based on Adaboost and SVM is introduced. In the Sect. 4, we show the experimental results on the correctness and efficiency of the algorithm. And in the last chapter, we analyze and summarize the experimental results and look forward to the future research directions.

2 Preliminary Knowledge

2.1 Adaboost Algorithm

The Adaboost algorithm [11] was first proposed by Schapire and Freund. It is currently the most representative combination classification algorithm, and it is the most widely used and has a good theoretical basis. The Adaboost algorithm has the advantages of high speed, high flexibility and high detection rate. The core idea is to repeat training for the same training set, get different weak classifiers, and then combine these weak classifiers to form a strong classifier according to their respective speech rights. The classification effect of a single weak classifier is not very strong, but once multiple weak classifiers are combined, it will achieve good results. In theory, the greater the number of weak classifiers, the higher the accuracy of the combined strong classifier.

The flow of the Adaboost algorithm is shown in Algorithm 1. After the T round iteration, we get a strong classifier composed of T weak classifiers based on the right of speech.

Algorithm 1. Process of Adaboost algorithm

1. Initialize the weight vector;
2. Train weak classifiers;
3. Calculate the weight error of the weak classifier and calculate the discourse weight of the weak classifier;
4. The weight is updated, skip to the second step and start the next iteration;
5. Combine a strong classifier.

However, we know that the traditional Adaboost is to solve the problem of the two classifications. In reality, the prediction problem of our power network security situation cannot be limited to two categories, and there are multiple classifications. The multi-category classification methods based on AdaBoost mainly include:

(1) AdaBoost. M1 and AdaBoost. M2 [12]: It directly generalizes the Adaboost algorithm to multiple types of problems, but it has higher requirements for weak classifiers;
(2) AdaBoost. MH [13]: Directly convert k-type problems into k two-class problems. Because of the conversion to two types of problems, the accuracy requirements of such weak classifiers can be easily satisfied. However, when the algorithm is relatively large, the algorithm has a complicated calculation process and a large amount of calculation.
(3) AdaBoost SAMME [14]: Different from the above two methods, the Adaboost algorithm is directly extended to multiple classes, and the correct rate requirement of the weak classifier is reduced from 1/2 to 1/k.

In this paper, considering the multi-classification and computational complexity, we will adopt a third multi-classification method to predict the security situation of the power network in real time.

2.2 Support Vector Machine

Support Vector Machine (SVM) [15] was proposed by Cortes and Vapnik in 1995 and soon became the mainstream technology for machine learning, and soon became the mainstream of machine learning. The main idea of SVM is: (1) For the linear separability case, finding a hyperplane causes the samples to be divided into two categories with the largest interval; (2) For the case of linear inseparability, the linear indivisible samples of low-dimensional input space are transformed into high-dimensional feature spaces by linear mapping algorithm to make them linearly separable.

SVM solutions are usually designed with the help of convex optimization techniques [16] and are generally designed for two-class tasks. SVM can also be used to solve nonlinear problems after the introduction of the kernel method. There are three general SVMs:

(1) Hard-spaced support vector machine (linear separable support vector machine): When the training data is linearly separable, a linear separable support vector machine can be obtained by hard-stitching.
(2) Soft-interval support vector machine: When the training data is approximately linearly separable, a linear support vector machine can be obtained by the maximum soft-spacing.
(3) Nonlinear support vector machine: When the training data is linearly inseparable, a nonlinear support vector machine can be obtained by the kernel method and the soft interval maximum chemistry.

3 Method Description

In order to predict the online security situation of the power network in real time, we must first have a prediction model. As we all know, the classification problem in machine learning is often affected by the data dimension, and some outliers may occur. The Adaboost algorithm training process only selects functions that are known to improve the predictive power of the model, reduces the dimensions and potentially improves execution time because there is no need to calculate unrelated features. In addition, the real-time online security situation data of the power network is likely to have some outliers, and the AdaBoost algorithm is not sensitive to noisy data and outliers, so on some issues, it may not be easy compared to other learning algorithms. Under the influence of overfitting.

The learning effect of the individual learning algorithm used in the AdaBoost algorithm may be weak, but as long as the performance is better than the random guess SVM as the weak classifier, the final model can be proved to converge to a powerful learner. Therefore, we use the Adaboost algorithm and the SVM algorithm to predict the security situation of the power network. The algorithm is divided into two parts.

The first part is the training of the model. The tagged data is input into the model to train the model. After the error rate is less than the threshold, the model is used as the real-time security situation online prediction of the power network, otherwise iteratively Continue to train the model.

The flow of this algorithm is shown in Fig. 1. First, the data is prepared and the data is preprocessed. Then, one of the processed data is put into the model to train, and after training, the trained model is predicted with the remaining data. After obtaining the prediction result, the accuracy analysis is performed. If the error is greater than the threshold, iteratively continues training. If the error is less than the threshold, the final model we need is obtained. Then you can put in the trained model.

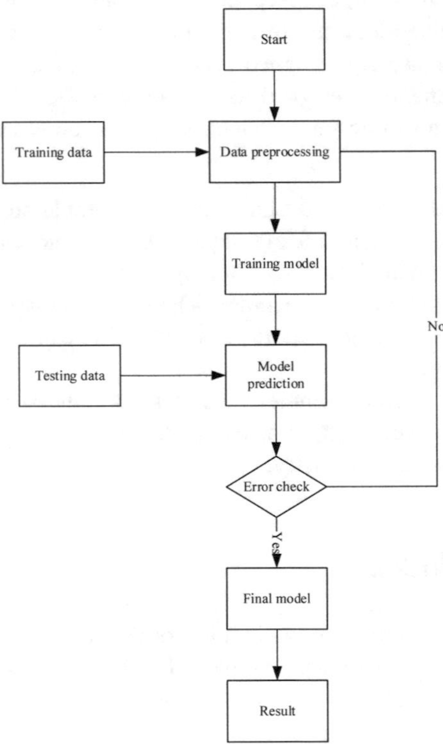

Fig. 1. Process of our algorithm.

In the algorithm flow of this algorithm, the training of the model is the most important step. The algorithm is based on the following specific steps:

(1) Given N labeled samples $\{(x_1, y_1), (x_2, y_2), \cdots, (x_N, y_N)\}$, where $x_i \in X$ (X is the data sample space) and $y_i \in Y$ (Y is the category space) where $i = 1, 2, \cdots, N$. The data set is assigned, with about 85% of the samples being used as training samples and the rest as test samples, and the number of trainings T is known.

(2) Initialization weight vector $D_1 = (w_1^1, w_2^1, \cdots, w_N^1)$, where $w_i^1 = \frac{1}{N}$, represents the weight of the i th sample in the first round, where $i = 1, 2, \cdots, N$, the data set is distributed according to D_t.

(3) SVM is classified as a weak classifier, and the decision tree h_t of the t round is constructed according to the SVM algorithm.

(4) Use the generated weak classifier h_t to calculate the weight error on the training data set:

$$\varepsilon_t = \sum\nolimits_{i=1}^{N} w_i^t |h_t(x_i) - y_i|; \tag{1}$$

It can be seen that the better the classification effect of the weak classifier is, the smaller the weight error ε_t is, and vice versa. According to the weight error, use the formula:

$$\alpha_t = \frac{1}{2} \ln[(1 - \varepsilon_t)/\varepsilon_t]. \tag{2}$$

Calculate the weak classifier's discourse weight α_t of the round. The smaller the weight error, the larger the discourse weight, and vice versa.

(5) The number of iterations is increased once. If the number of iterations is greater than T, the next step is continued. Otherwise, the weight vector of the next round is updated and jumps to step (3). The weight update formula is as follows:

$$\begin{aligned} D_{t+1} &= (w_1^{t+1}, w_2^{t+1}, \cdots, w_N^{t+1}) \\ w_i^{t+1} &= \frac{w_i^t \exp(-\alpha_t h_t(x_i) y_i)}{Z_t} \end{aligned} \tag{3}$$

Where $Z_t = \sum_{i=1}^{N} w_i^t \exp(-\alpha_t h_t(x_i) y_i)$ is the normalized constant, and $i = 1, 2, \cdots, N$.

(6) Combine weak classifiers above random selection into strong classifiers according to a certain voice weight α_t:

$$H(x) = sign \sum_{t=1}^{T} \alpha_t h_t(x) \tag{4}$$

(7) Calculate the error rate of the model. If it is greater than the threshold τ, jump to the first step and re-train. If it is less than the threshold τ, the training ends and the model is obtained.

4 Experiment and Analysis

In this section, we use the intrusion detection data published by HoneyNet [17] as the experimental data source, and compare the SVM, Adaboost, BPNN and the algorithm from the correctness and efficiency of the algorithm. We randomly selected 85% of them as the training sample set, and the rest were used as test sample sets for experimental verification. Next, we conduct comparative experiments on the two aspects of correctness and efficiency and analyze the results.

4.1 Accuracy Analysis

First, we analyze the correctness of SVM [9], Adaboost [11], BPNN [6] and our algorithm. Table 1 (Fig. 2) shows the experimental results of ten experiments. We can clearly see that the line of BPNN algorithm is the lowest under other algorithms, that is, its accuracy is lower. The algorithm accuracy of Adaboost and SVM is slightly better than BPNN algorithm. The line of our algorithm is on top of other algorithms, which means that our algorithm is more accurate than other algorithms.

Table 1. Accuracy of four methods (%).

	1	2	3	4	5	6	7	8	9	10
SVM	81.8	74.0	79.1	79.7	74.1	76.0	78.8	80.9	75.7	74.0
Adaboost	84.5	84.0	78.3	82.6	75.1	76.7	79.4	83.6	84.0	79.7
BPNN	70.7	69.1	74.9	65.8	73.2	68.1	66.0	72.9	64.5	67.2
Ours	90.7	87.2	87.6	90.0	88.6	84.4	90.9	90.7	85.4	85.0

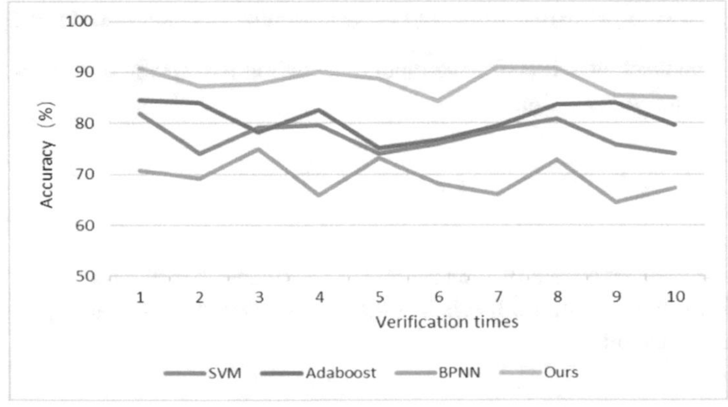

Fig. 2. The accuracy comparison among SVM, Adaboost, BPNN algorithms and ours.

4.2 Efficiency Analysis

We continue to compare the efficiency of SVM, Adaboost, BPNN and our algorithms. From Table 2 and Fig. 3, we can clearly see that BPNN takes the longest time to make predictions. The algorithm efficiency of SVM and Adaboost algorithm is better than BPNN. Our algorithm takes the shortest time in predicting results. Because the real-time security situation prediction of the power network requires the algorithm model to quickly give a judgment result, the algorithm is superior in efficiency to the other three algorithms.

Table 2. Time consumption of three methods (s)

	1	2	3	4	5	6	7	8	9	10
SVM	20	27	30	29	27	26	24	31	31	30
Adaboost	35	42	41	32	43	36	42	34	33	35
BPNN	67	70	55	71	64	58	60	57	65	66
Ours	15	18	21	16	13	25	15	20	19	23

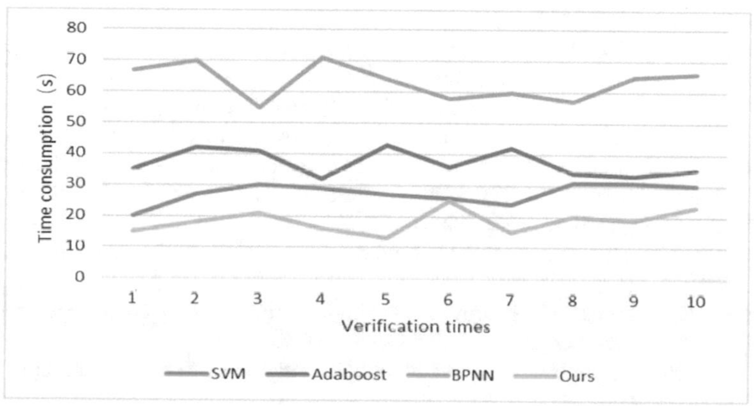

Fig. 3. Comparison of time consumption of SVM, Adaboost, BPNN and our method

5 Conclusion

We proposed a real-time online security situation analysis algorithm for power networks based on Adaboost and SVM. We combine the advantages of the Adaboost algorithm and the SVM algorithm to predict the cyber security posture, and conducted an experimental analysis, we compared SVM, Adaboost, BPNN and our algorithm from two aspects of accuracy and efficiency. The experimental results show that our algorithm is not only more accurate than other algorithms, but also takes less time. Next, we can consider how to control and adjust it to achieve intelligent and flexible processing in the true sense. In the case of obtaining the security situation of the power network.

References

1. Zhang, B.-M.: Concept extension and prospects for modern energy control centers. Autom. Electr. Power Syst. **27**(15), 1–6 (2003)
2. Sun, H.-B., Xie, K., Jiang, W.-Y., et al.: Automatic operator for power systems: principle and prototype. Autom. Electr. Power Syst. **31**(16), 1–6 (2007)
3. Huang, T.-E., Sun, H.-B., Guo, Q.-L., et al.: Knowledge management and security early warning based on big simulation data in power grid operation. Power Syst. Technol. **39**(11), 3080–3087 (2015)
4. Sun, H.-B., Huang, T.-E., Guo, Q.-L., et al.: Power grid intelligent security early warning technology based on big simulation data. South. Power Syst. Technol. **10**(3), 42–46 (2016)
5. Huang, T.-E., Sun, H.-B., Guo, Q.-L., et al.: Distributed security feature selection online based on big data in power system operation. Autom. Electr. Power Syst. **40**(4), 32–40 (2016)
6. Chen, T., Gong, Z.-H., Hu, N.: A predicting model of network situation based on proved BP. Electron. Commer. China **2009**(3), 93–99 (2009)
7. Ren, W., Jiang, H.X., Sun, Y.F.: Network security prediction based on RBF neural network. Comput. Eng. Appl. **42**(31), 136–138 (2006)
8. Lin, Z., Chen, G., Guo, W., et al.: PSO-BPNN-based prediction of network security situation. In: International Conference on Innovative Computing Information and Control. IEEE Computer Society (2008)
9. Zhang, X., Hu, C.-Z., Liu, S.-H.: Research on network attack situation forecast technique based on support vector machine. Comput. Eng. **33**(11), 10–12 (2007)
10. Wang, Y.-F., Shen, H.-Y.: Network Security Situation forecast based on improved general regression neural network. J. North China Electr. Power Univ. **38**(3), 91–95 (2011)
11. Freund, Y., Schapire, R.-E.: A decision-theoretic generalization of on-line learning and an application to boosting. In: European Conference on Computational Learning Theory, pp. 23–37. Springer, Heidelberg (1995)
12. Freund, Y., Robert E.-S.: Experiments with a new boosting algorithm. In: ICML, vol. 96 (1996)
13. Schapire, R.-E, Singer, Y.: Improved boosting algorithms using confidence-rated predictions, pp. 80–91. Kluwer Academic Publishers (1998)
14. Hastie, T., Rosset, S., Zhu, J., et al.: Multi-class AdaBoost. Stat. Interface **2**(3), 349–360 (2009)
15. Cortes, C., Vapnik, V.: Support-vector networks. Mach. Learn. **20**(3), 273–297 (1995)
16. Boyd, S., Vandenberghe, L.: Convex Optimization. Cambridge University Press, Cambridge (2004)
17. HoneyNet Homepage (2002). http://www.honeynet.org/paper/stats/

A Pi-Based Beehive IoT System Design

Yi-Liang Chen, Hung-Yu Chien(✉), Ting-Hsuan Hsu, Yi-Jhen Jing, Chun-Yu Lin, and Yi-Chun Lin

Department of Information Management, National Chi Nan University, Nantou, Taiwan
hychien@ncnu.edu.tw

Abstract. Applying Internet-of-Things (IoT) technologies in various agriculture challenges and ecosystem challenges not only can reduce the man efforts but also improve the productivity and the efficiency. Among many agriculture or ecosystem challenges, monitoring bees is one of the most interesting and imperative ones, as bees play a critical role in both the ecosystem and the agriculture and their habitats are under very serious pressures. Even though there are several commercial beehive monitoring systems on the market, localization and customization of such systems is inevitable, due to various environments, climates, box designs, or various bee-keeping practices. In this article, we introduce the challenges here in Taiwan and the design to cope these challenges. Based on some low-cost components (like raspberry pi, various sensors, and communication facilities) on the market, we design our bee monitoring system which can monitor various environment data (like temperature, humidity, and GPS) and bee data (like bee sound and infra-red images). The prototype has been tested in the field, and we are evaluating its effectiveness.

Keywords: Infra-red · Beehive · Thermal camera · Raspberry pi · Internet of Things

1 Introduction

IoT is one of the most popular buzzwords recently, and various IoT news and applications dazzle us daily in the media. Various applications attract the attention from the academia, industry and consumers. These applications have the great potential to improve the quality, the efficiency, or the productivity in various domains.

Bee keeping is one of the potential IoT applications, and it does deserve our attraction for several critical reasons. Due to the big demand of honey, there have been many food safety issues happening on the honey supply on the market. The nature environments and the nature habitats for bees deteriorate globally. The improper practice of using pesticide and herbicide. The possible new viruses and possible parasites threats the colonies of bees. The global climate change also has the potential to impact bee colonies. The Colony Collapse Disorder (CCD) on bee colonies frightens people around the globe. Many institutes and communities start to make various efforts and seek possible solutions to cope with this challenge. IoT systems that monitor the environments and collect data from both the environments and the bee colonies seem to be a very promising solution.

C.-N. Yang et al. (Eds.): SICBS 2018, AISC 895, pp. 535–543, 2020.
https://doi.org/10.1007/978-3-030-16946-6_43

Bee keeping is a very skillful profession. Both the knowledge and the experiences matter a lot for a successful bee keeping business. An experienced and knowledgeable bee keeper could have much better productivity than a naïve beginner. However, it is very difficult to effectively and quickly pass on the knowledge and the experiences to beginners. Luckily, the application of new technologies and IoT systems could reduce the labors and could enhance the productivity.

There are several commercial or open-source bee monitoring systems. The functions of these products and packages include environment/beehives data collection, bee sound analysis, images capturing, and so on. These tools and packages are quite handy. However, we find that every region has its own environment conditions and the bee-keeping practice in different regions has its own uniqueness and different challenges; for example, the bee boxes might have different size and different designs. The ventilation designs of the boxes could be quite different. The suitable places for bee boxes could be quite different: the boxes could be placed in sunny location, while some regions preferring shady location; this affect how the power/battery design. The weighting system (scale system) for measuring the weight of a beehive is highly dependent on the structure of the bee boxes. All these factors call for the customization and localization of bee monitoring systems. We, therefore, aim at designing beehive monitoring systems that fits local environments and local bee-keeping practices, using low-cost and open-source components and tools. The design process and the experiences could help similar-interest communities design their own systems.

Related Works
The tangible resources on the Internet and the open and resourceful hardware and software support contribute to the fast evolution and development of various IoT systems. Bee-keeping is one of the potential IoT applications, and it could benefit from the technologies to reduce the efforts/man power, to improve both the productivity and the quality, and to reduce the cost.

Many countries like Taiwan, the farmers and the bee keepers usually own small business, and they cannot afford expensive solutions. Fortunately, there are many low-cost devices and open-source platforms available. This facilitates us a great potential to build powerful but cheap solutions to improving the agriculture practice or the bee-keeping practices.

Regarding micro controllers and tiny computers, Arduino [1, 2] and Raspberry pi [3] are two popular platforms for building various IoT platforms. We adopt popular sensors like temperature, humidity, sound recorders, GPS, NoIR (No infra-red filter camera) [17], and the FLIR thermal camera [18].

There are several publications and projects related to our work. Sanbo [4] focuses on the hardware design and the wireless networks integration to facilitate efficient and intelligent watering/irrigation system in the field. Chen and Chien [11], inspired by the DotPiDot project [5], design their green-house IoT system, which collects the green-house environment data, controls the switches, and analyze the Normalized Difference Vegetation Index (NDVI) [10] values of the plants. DotPiDot [5] integrates Tornado [7], RethinkDB [6] database, Raspberry Pi [3], various sensors, fans and light actuators to design a greenhouse demonstration system that can collect and display the conditions in a real-time. Chien et al. [14] explore the potential application of thermal cameras [13] and NoIR cameras [12] in agriculture IoTs.

Aumann and Emanetoglu [15] present a radar microphone for studying incidental and deliberate insect sounds; It was specifically designed to record the sounds coming from honey bees inside a beehive.

Lofaro introduces [16] the "Smart Hive" project that aims at collect beehive data remotely and the systems will be made open-source. Sentinel apiary program [20] is one another open project that monitors honey bee health in real-time using monthly disease assessments of Varroa and Nosema loads. Heiseet al. [17] apply the computational auditory scene analysis (CASA) to detect the bee buzz in the field.

APIVOX SMART MONITOR [18] is an exclusive application for beekeepers, that will allow you to monitor the status of colonies of bees and to forecast possible problems.

Zacepins et al. [19] design an automatic bee colony weighting system.

Bromenshenk [21] discusses the potential of applying the Infrared technology on the bee management. Beehacker [22] surveys the acoustic analysis of bee behaviors. Seidle [23] shares his design of bee monitoring system. BuzzBox [24] is an open source beehive monitoring system; the kit collects beehive data and sends them to the cloud from which users can browse the data. We bought one Buzzbox kit as a reference.

The rest of this paper is organized as follows. Section 2 introduces the requirements from the local bee-keeping practices. Section 3 discusses our system design. Section 4 shows some results from the field testing and evaluations. Section 5 states our conclusions and future work.

2 The Requirements from the Local Practices

Figure 1 depicts the process of this study. We first did some site visits and interviews with the bee keepers. Based on the information, we identify the concerns and the challenges, and a survey of the publications and the products. Then, we design the system, implement it, and then perform the site testing and evaluation. This phase is repeated several times to adjust the system design according to the site environment. Finally, we focus on the data collection and develop some automatic data analysis.

Now we discuss the requirements and challenges, according to the local environments and the local practices. In addition to the common monitoring like temperature, humidity, and luminosity, we discuss some other concerns for Taiwan bee-keeping practices.

The climate in Taiwan is usually hot and humid in summer and it is windy in the northern regions of the island in winter. Therefore, most of the bee keepers place their bee boxes in some shady area to keep away from the scorching sun in summer and to avoid strong winds. This affects the power supply design and simple small solar panel might not be the first choice. These sites do not have any power support in the nearby.

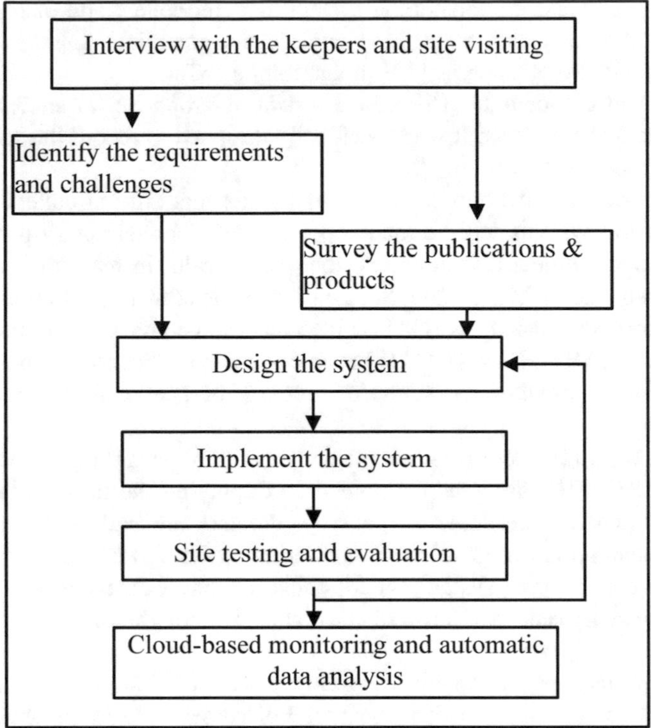

Fig. 1. The process of this study

The bee-keeping here is very labor-intensive and requires lots of knowledge and experiences. The knowledge and experiences about bee swarming, new queen and missing queen are very helpful.

The population is quite dense, and the patches of plants that can attract bees are usually small. Therefore, the average number of boxes one keeper has is usually less than 100. They work diligently and examine each box frequently. Therefore, the weighting system or the scale for hive is not popular or desirable, as an experienced and diligent keeper estimate the weight manually on a daily basis.

Varroa and Nosema are two bothering issues that weaken the bee colonies here. Early detection could be quite helpful to take proper actions. Apis cerana (中國蜂) colonies are vulnerable to the A. cerana Sacbrood virus (AcSBV or CSBV; or called 東方蜂囊狀幼蟲病毒in Chinese).

Due to the demand of honey is larger than the supply here, the price of honey rises very quickly these years, and the theft of beehives happen more and more these years.

The locations of beehives are usually in rural areas, and wifi communications are difficult or inaccessible.

In the following, we summarize these challenges.

1. Monitoring the environment data (like temperature, humidity, and luminosity) of the hives.
2. Detection of possible theft of hives. GPS might be a solution for this challenge.
3. Detection of bee swarming, new queens, and missing queens.
4. Detection of Varroa, Nosema, and AcSBV.
5. Easy access of the data of the hives, and the solutions should be cheap.

3 The System Design and Implementation

3.1 The System Architecture

The whole system architecture is depicted in Fig. 2. The system consists of three entities in three layers. Each bee box is installed with our integrated pi-based sensors, and it transmits its data to a wifi/4G gateway. The gateway provides the wifi connection to nearby Pis, and forwards to data to a remote server via the 4G connection. This design facilitates the server to collect the data like sounds and images. Other wireless communication like LoRa cannot support the required bandwidth. The 4G connection is not free; Luckily, there is a promotion here, and we can access the connection with barely no cost; we still need to constantly survey the other options and the prices, and another option might be better in the future. The server manages the databases, the web pages, and the data analysis and alarming.

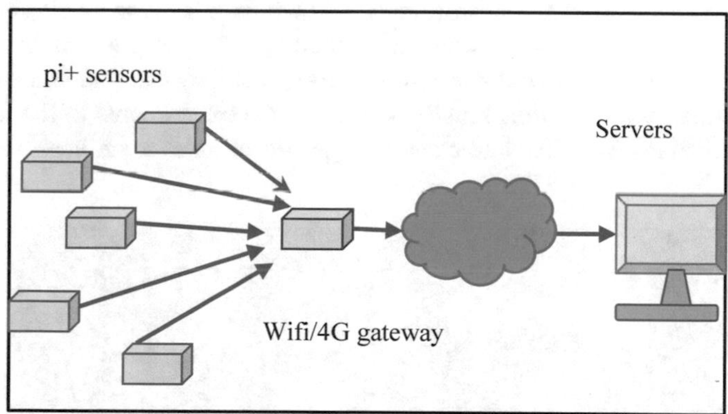

Fig. 2. The system architecture

3.2 The Components and the Functions

The integrated pi-based sensors consist of temperature, humidity, GPS, sounds, and thermal images. The sounds are used for analyzing the status of the hive; the status set include the normal active state, the swarming state, the new queen state, and the missing queen state. The thermal images are collected for the possible detection of the virus infection and parasite infection.

To integrate the various sensors with an affordable size, we adopt the GrovePi board [25] which has embedded several popular I/O ports with modest size. Due the bee box site is usually shady and rainy is quite popular in summer, we use a power usage device to estimate the required power supply. Figure 3 depicts the integrated sensors and the power usage detector. Based on the power usage data, we find that a 10000 mAh mobile battery can supports 27–33 h. So we have several batteries.

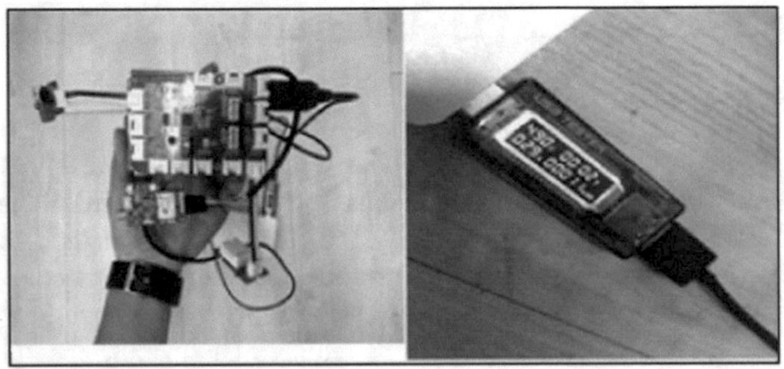

Fig. 3. The integrated sensors and the power detector

One another challenge comes out is that there are many ants attacking our hives when we first experiment the system in our campus. Figure 4 shows the initial bee box stand we designed. The PVC foots of the stands will be placed in small buckets which hold water to deter the attacks from ants. Finally, both designs fail in our campus experiments, because the ant colonies are so strong that they can find their ways to pass through the small water barrier. Finally, we move the experiments to the keepers' site where they regularly keep the site clean and get rid of ants in the nearby.

Fig. 4. Two bee box stand designs

Now our system works well and the data has been constantly collected.

Figure 5 shows some of the web pages and functions. In the top figure, the system shows the site in the google map, and users can click on the site to examine the details. The middle-right figure shows the detailed data of the site, and the middle-left one shows the data in chart format. The bottom figure shows the sound in the Audacity tool [26].

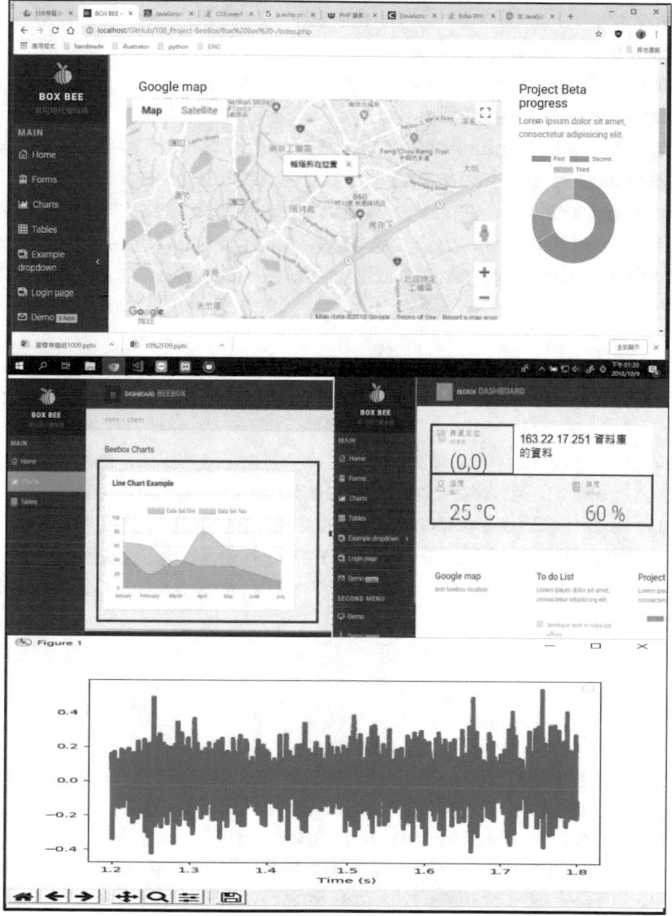

Fig. 5. The integrated sensors and the power detector

4 Evaluations and Discussions

The system has been tested and evaluated. Up to now, there are still some experiments going on and some functions still under being developed. Now we give some discussions and observations as follows.

The 4G connection now can happily support our requirements of video/image transmission. In the promotion period, it does not cost us a lot, but we need to continue the survey and determine the proper options as the technologies advance and the market changes.

The data collected can be shared to users via web pages, and the chart can easily show the trend. The GPS data can be used as a source for possible theft, but the device is so visible that the thief might take off the device when they notice that the device might disclose their actions.

Posting the alarms on the LINE or the FB could be a good function to instantly notify the users and the community.

The battery and the plan of power supply should be re-evaluated and designed so that the process is more convenient.

The sound data is under analyzed. Now we study the acoustic analysis papers and reports, and are evaluating the procedures for automatic sound analysis.

The thermal camera is so sensitive and fragile in the field. After the installation, it blacks out after some trials. Because it is the most expensive component in this project, we are pondering whether we should still have this component or seek other possible substitutes.

5 Conclusions

In this article, we have introduced the local environments, the system design and the problems for remote beehive monitoring. Up to now, we have successfully build the prototype and the data collected has been under investigation. But, we still have several challenges to solve. Based on this experiment, we believe that customized beehive IoT monitoring system could be a very helpful tool for both bee research and bee keeping practice.

References

1. Wiki, Arduino. https://en.wikipedia.org/wiki/Arduino. Accessed 07 Apr 2018
2. Arduino project. https://www.arduino.cc/. Accessed 07 Apr 2018
3. Raspberry pi. https://www.raspberrypi.org/. Accessed 07 Apr 2018
4. Sanbo, L.: Application of the internet of things technology in precision agriculture irrigation systems. In: 2012 IEEE International Conference on Computer Science & Service System (CSSS), pp. 1009–1013. IEEE (2012)
5. DotPiDot, Greenhouse Pi (2016). https://hackaday.io/project/12762-greenhouse-pi. Accessed 07 Apr 2018
6. RethinkDB. http://www.rethinkdb.com/docs/sharding-and-replication/. Accessed 07 Apr 2018
7. Tornado. http://www.tornadoweb.org/en/stable/. Accessed 07 Apr 2018
8. Splunk. https://en.wikipedia.org/wiki/Splunk. Accessed 07 Apr 2018
9. WebSock. https://developer.mozilla.org/enUS/docs/Web/API/WebSockets_API. Accessed 07 Apr 2018
10. NDVI and NRG. https://publiclab.org/wiki/ndvi. Accessed 07 Apr 2018

11. Chen, Y.-J., Chien, H.-Y.: IoT-based green house system with splunk data analysis. In: 2017 IEEE 8th International Conference on Awareness Science and Technology (iCAST 2017), Taichung, Taiwan, 8–10 November 2017
12. Pi NoIR Camera. https://www.raspberrypi.org/products/pi-noir-camera-v2/. Accessed 07 Apr 2018
13. FLIR and Lepton Thermal camera. https://lepton.flir.com/getting-started/lepton-quick-start-raspberry-pi/. Accessed 07 Apr 2018
14. Chien, H.-Y., Tseng, Y.-M., Hung, R.-W.: Some study of applying infra-red in agriculture IoT. In: 2018 IEEE International Conference on Awareness Science and Technology (IEEE ICAST 2018), Fukuoka, Japan, 18–21 September 2018. IEEE press (2018)
15. Aumann, H.M., Emanetoglu, N.W.: The radar microphone: a new way of monitoring honey bee sounds. In: 2016 IEEE SENSORS, Orlando, FL, USA, 30 October–3 November. IEEE (2016)
16. Lofaro, D.M.: The honey bee initiative - smart hive. In: 2017 14th International Conference on Ubiquitous Robots and Ambient Intelligence (URAI), Jeju, Korea, 28 June–1 July 2017
17. Heise, D., Miller-Struttmann, N., Galen, C., Schul, J.: Acoustic detection of bees in the field using CASA with focal templates. In: 2017 IEEE Sensors Applications Symposium (SAS), Glassboro, NJ, USA, 13–15 March. IEEE (2017)
18. APIVOX SMART MONITOR. https://www.apivoxauditor.com/. Accessed 07 Apr 2018
19. Zacepins, A., Pecka, A., Osadcuks, V., Kviesis, A., Engel, S.: Solution for automated bee colony weight monitoring. Agron. Res. **15**(2), 585–593 (2017)
20. Sentinel Apiary Program. https://beeinformed.org/programs/sentinel/. Accessed 07 Apr 2018
21. Bromenshenk, J.: Infrared: the next generation in colony management. Bee Culture, December 2015
22. Beehacker: Acoustic analysis of bee behavior – Part 1, http://www.beehacker.com/wp/?page_id=189. Accessed 07 Apr 2018
23. Seidle, N.: The Internet of bees: adding sensors to monitor hive health. https://makezine.com/projects/bees-sensors-monitor-hive-health/. Accessed 07 Apr 2018
24. BuzzBox Sensor Kit: The open source beehive. https://www.kickstarter.com/projects/181034265/buzzbox-advanced-beehive-sensor-and-smartphone-app. Accessed 07 Apr 2018
25. The GrovePi. https://www.dexterindustries.com/grovepi/. Accessed 07 Apr 2018
26. The Audacity tool. https://www.audacityteam.org/download/Accessed 07 Apr 2018

PLVSNs: A Privacy-Preserving Location-Sharing System for Vehicular Social Networks

Chang Xu[1], Xuan Xie[1], Liehuang Zhu[1(✉)], Chuan Zhang[1], and Yining Liu[2]

[1] School of Computer Science and Technology, Beijing Institute of Technology, Beijing, China
liehuangz@bit.edu.cn

[2] Guilin University of Electronic Technology, Guilin, China

Abstract. Combining Vehicular Ad Hoc Networks (VANET) and mobile online social networks (mOSNs), Vehicular Social Networks (VSNs) can provide more complex and rich services, including location sharing services, which brings more convenient communication and information sharing. However, location sharing in VSNs demands advanced privacy-preserving schemes to protect vehicles' social relationships and location information. Previous solutions employ Location Based Server (LBS) to manage users' location information. However, the server may collect the messages it has received to get more information of vehicles. Moreover, the threshold distances privacy is not properly protected, which may disclose users' identities. In order to overcome the shortage, we propose PLVSNs, a privacy-enhanced location-sharing scheme in VSNs. It takes advantage of vehicles' increasingly superior computing power, allowing vehicles and SNS to implement the distance comparison without LBS. Furthermore, we introduce a privacy enhanced distance comparison protocol, which provides a more efficient comparison between the actual distance and the threshold distance while not disclosing vehicles' privacy. We also show that the new construction is secure. Finally, we implement PLVSNs and the experimental results demonstrate the efficiency of our proposed scheme.

Keywords: Privacy-preservation · Location-sharing · Vehicular Social Networks

1 Introduction

The continuing advances of Dedicated Short Range Communications (DSRC) [6] and mobile computing [1,4,18,21] have greatly promoted Vehicular Ad Hoc Networks (VANET) [5,15], where vehicles can build communications with other vehicles, Roadside Unit (RSU), and remote servers through a direct or indirect way. Combining VANET and mobile online social networks (mOSNs), Vehicular Social Networks (VSNs) have attracted attention from industries and academia,

C.-N. Yang et al. (Eds.): SICBS 2018, AISC 895, pp. 544–557, 2020.
https://doi.org/10.1007/978-3-030-16946-6_44

which significantly improves the driving experiences of users and elevated the development of Intelligent Transport System (ITS).

VSNs can provide various location based services (LBSs) [9,15,19]. In LBSs, users update their location information to obtain different types of custom services, such as road surface conditions detection, traffic data sensing, collaborative driving, and so on [2,13,16]. However, privacy concern is still a major challenge. To enjoy location based services, vehicles need to broadcast their locations periodically. The messages, nevertheless, are usually sensitive and need strict protection. For example, vehicles' locations can be associated with their personal attributes, such as friend lists, attitudes towards friends/strangers, and disclosing this information will put users' social relationships at risk.

To address the privacy preserving issue in mOSNs, a number of research works regarding location privacy have been proposed. In 2012, Wei et al. presented a system called MobiShare [17]. Then, inspired by this work, Liu et al. provided a N-Mobishare system [10,11]. Li et al. [8] proposed MobiShare+ in 2014. In 2016, BMobishare was proposed, which employs Bloom Filter to protect users' privacy [14]. A more secure system was put forwarded by Li et al. in 2017 [7].

However, these research works are still not perfect. Firstly, they adopt Location Based Server (LBS) to realize the management of vehicles' position. The location based server is assumed to be honest but curious, and it may collect messages to guess more sensitive information of vehicles. Second, the threshold distances set for different target are also sensitive, and adversaries can identify a vehicle through these data. Previous schemes either ignore this threat or adopt a protection algorithm with large computation and communication consume.

Our Contributions: Motivated by these issues, we propose a privacy-enhanced location-sharing scheme in VSNs, namely PLVSNs. Our contributions are described as follows.

- Previous research works employ LBS to manage vehicles' geographical positions and calculate distances between vehicles. The server may collect messages sent by vehicles to guess more sensitive information of vehicle users. Our scheme takes advantage of the increasingly powerful computing power of the vehicles, removes LBS from the architecture.
- The threshold distances set by users can release users' attitude toward their friends/strangers, an adversary may collect this information to infer more sensitive information of users. In PLVSNs, a more efficient privacy enhanced comparison protocol is defined to prevent vehicles from knowing the threshold distances other vehicles set for them and prevent social network server from knowing the distances between vehicles.

The remainder of this paper is organized as follows. The building blocks including the privacy enhanced distance comparison protocol are proposed in Sect. 2. In Sect. 3, we formalize the problem by describing the models, the threat model and design goals of the system. Section 4 introduces the proposed PLVSNs scheme. The security analysis and performance analysis is provided in Sect. 5.

Some related works are reviewed in Sect. 6. Finally, we draw our conclusion in Sect. 7.

2 Building Blocks

To provide location based services while protecting users' privacy, PLVSNs make use of building blocks as blow.

2.1 Broadcast Encryption

PLVSNs use a broadcast encryption (BE) scheme for vehicles to send messages directly [12]. The BE scheme consists of four algorithms: (Setup, Extraction, Encryption, Decryption) which work as follows:

- Setup: Take a security parameter 1^k as input and output a master key mk.
- Extraction: Take mk and user's identifier ID as input and output user's public key and private key pair k_{pub}, k_{pri}.
- Encryption: Take a set of users U, a message m and master key mk as input and output the ciphertext c.
- Decryption: Take ciphertext c and a user's private key k_{pri} as input, output a message m or error message.

2.2 Paillier Encryption

Paillier cryptosystem is a widely used homomorphic cryptosystem [3]. The description of Paillier encryption is as follows:

- Setup: Let $n = pq$, where p and q are large prime numbers. Choose a random number g which is multiples of N. $\lambda = lcm(p-1, q-1)$. The public key is (N, g) and the private key is λ.
- Encryption: Take a message m as input and the ciphertext is $c = g^m r^N \bmod N^2$, r is random number, $r \in Z_N^*$, $c \in Z_N^{2*}$, $m \in Z_N$.
- Decryption: Take ciphertext c as input and output a message $m = L\left(c^\lambda \bmod N^2\right) * \left(L\left(g^\lambda \bmod N^2\right)\right)^{-1} \bmod N$.

Paillier cryptosystem has homomorphic properties as below:

- $E(m_1, r_1)E(m_2, r_2) \bmod n^2 = E(m_1 + m_2, r_1 + r_2), m \bmod n^2$
- $E(m_1, r_1)^{m_2} \bmod n^2 = E\left(m_1 m_2, r_1 m_2\right) \bmod n^2$

2.3 Privacy Enhanced Distance Comparison Protocol

To provide location sharing services, PLVSNs need to compare two values: the distance between two vehicles and the corresponding threshold distance. We propose a privacy enhanced distance comparison protocol (Protocol 1) based on [20] to provide this comparison while protecting the privacy of vehicles.

Assume that $d_{threshold}$ represents the threshold distance, l_{actual} represents the actual distance, $d_{threshold}$ and l_{actual} are integers. E_1 is Paillier encryption using SNS's public key Pk_1, and D_1 is Paillier decryption using SNS's private key.

Protocol 1. Privacy enhanced distance comparison protocol

Input: Threshold distance $d_{threshold}$; Actual distance l_{actual}

Output: $d_{threshold} > l_{actual}$ as TRUE or FALSE

1: SNS sends $E_1(0, r_1)$, $E_1(d_{threshold}, r_2)$ and Pk_1 to vehicles.

2: Each vehicle randomly picks two arbitrary positive numbers δ_1, δ_2. For comparing computation, vehicles compute

$$\begin{aligned} e_1 =& E_1(\delta_1(d_{threshold} - l_{actual}) + \delta_2, \delta_1(r_2 + r_4) + r_3) \\ =& E_1(\delta_1(d_{threshold} - l_{actual}), \delta_1(r_2 + r_4)) E_1(\delta_2, r_3) \\ =& E_1(d_{threshold} - l_{actual}, r_2 + r_4)^{\delta_1} E_1(\delta_2, r_3) \\ =& (E_1(d_{threshold}, r_2) E_1(-l_{actual}, r_4))^{\delta_1} E_1(\delta_2, r_3) \end{aligned} \tag{1}$$

$$\begin{aligned} e_2 =& E_1(\delta_2, \delta_1 r_1 + r_2) \\ =& E_1(0, r_1)^{\delta_1} E_1(\delta_2, r_2) \end{aligned} \tag{2}$$

3: SNS compare $d_1 = D_1(e_1)$ and $d_2 = D_1(e_2)$, iff $d_1 > d_2$, then $d_{threshold} > l_{actual}$, SNS outputs TRUE, otherwise SNS outputs FALSE.

3 System Models and Design Goals

We present the system model of PLVSNs in this section. The security goals are also identified and listed for the proposed scheme.

3.1 System Architecture

The system architecture is depicted in Fig. 1 where three main entities are involved.

Vehicles. The vehicles of VSNs can communicate with each other and roadside unit (RSU) directly. They can request for locations of friends and strangers. We assume that vehicles know their own locations from GPS.

RSU. RSU can communicate with vehicles and SNS directly, it is the transport intermediary between vehicles and SNS.

Social Network Server (SNS). SNS manages vehicles' information, including profiles and social relationships. It participates in the location request process. SNS communicates with RSU directly.

In our system, locations of vehicles should be protected from SNS. Three types of queries are allowed for vehicles to submit: (1) requests for nearby friends' locations, (2) requests for nearby strangers' locations, and (3) requests for specific friends' locations.

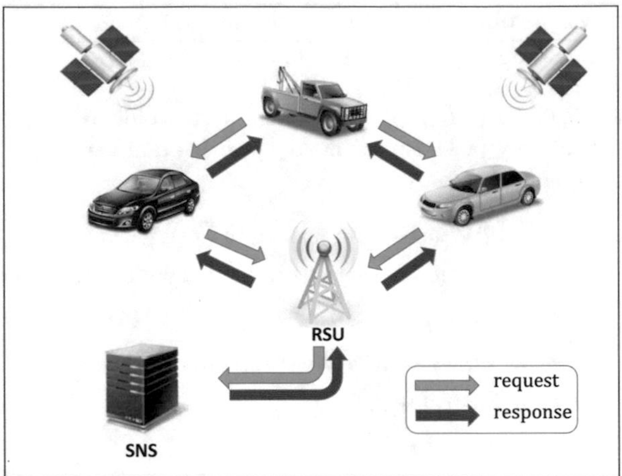

Fig. 1. System architecture

3.2 System Workflows

Before exploring our proposed scheme, an overview of the mechanism which mainly includes three main workflows given as follows.

1. *User registration.* Vehicles must register for LBSs at SNS. The registration process requires vehicles submitting their profiles, including their pseudo-identities and individual preferences. Moreover, vehicles need to prove the authenticity. SNS maintains a database and processes all the information.
2. *Updates.* When exchanging the pseudo-identities with other vehicles or change social relationships, vehicles update their information to SNS. SNS maintains the new identity & pseudo-identity mapping relationship, new social relationship and threshold distances of vehicles.
3. *Location query.* There are three type of queries supported in the proposed system, which are described as follows.
   - *Nearby friends' locations query.* When a vehicle wants to know the locations of nearby friends, he/she submits a query for nearby friends' locations to vehicles around him/her. The requester will receive the location information if he/she satisfies the access control policies set by their friends.
   - *Nearby strangers' locations query.* When a vehicle wants to know the locations of nearby strangers, he/she submits a query for nearby strangers' locations to vehicles around him/her. The requester will receive the location information if he/she satisfies the access control policies set by strangers nearby.
   - *Specific friends' locations query.* When a vehicle wants to know the locations of specific friends, he/she submits a query for specific friends' locations to vehicles around him/her. The requester will receive the location information if he/she satisfies the access control policies set by their specific friends.

3.3 Threat Model

Our system consists of vehicles, RSU and SNS, different trust assumptions over them are listed as below:

- Vehicles are assumed to be dishonest, they may try to use authorized operations to access the location information they do not have the permission to access.
- RSU is supposed to be honest but curious, that means it is supposed to follow the proposed protocol general but try to obtain other vehicles' information as possible.
- SNS is also considered to be honest but curious. It will follow the scheme and try to get more information of vehicles, such as their locations.
- We assume that RSU and SNS will not be compromised by an adversary at the same time. That means RSU and SNS will not collude with each other. SNS can not collect the messages sent by vehicles.

3.4 Security Goals

Using the threat model defined previously, the security goals for our proposed location-sharing system is defined as below:

- The system should protect vehicles' location information from SNS and unauthorized vehicles. Vehicles' locations cannot be leaked to friends or strangers who do not satisfy the predefined access policy.
- RSU transports information between vehicles and SNS, it should not know vehicles' identity information or social network information.
- SNS provides social relationships related services and should not be able to determine (directly or indirectly) the vehicle locations.

4 System Design

In this section, the details of our proposed system will be described. The summarize of notations used in this paper is given in Table 1.

Each vehicle has an identifier ID, several PIDs and a public-private key pair (pk_v, sk_v) which can be updated. They will exchange their PIDs at appropriate occasions to protect their privacy. RSU manages a broadcast set and vehicles within the service area are members of this set. We assume the location sharing request is within the serve area of RSU. RSU has a public-private key pair (pk_r, sk_r). SNS also has a public-private key pair (pk_s, sk_s). All vehicles know the public key pk_r and pk_s.

4.1 Initialization

PLVSNs use symmetric encryption and broadcast encryption for vehicles to communicate with each other directly. When a vehicle reaches the service range of RSU, he/she has to prove his/her legitimacy, then a symmetric key k will send to him/her using broadcast encryption. Vehicles use this key to send request messages to each other in the request phase.

Table 1. Summary of notations

Symbol	Description
ID	A vehicle's social network identifier
PID	A vehicle's pseudo-identifier
df	Threshold distance for a friend
ds	Threshold distance for strangers
$Flist$	A vehicle's friend list
ts	Time stamp
T	A certain time period
(pk_r, sk_r)	RSU's public-private key pair
(pk_s, sk_s)	SNS's public-private key pair
(x, y)	Location of a vehicle
SE	Symmetric encryption algorithm
SD	Symmetric decryption algorithm
E	Paillier encryption algorithm
D	Paillier decryption algorithm

4.2 Registration

Each vehicle needs to register for the location sharing services at SNS before they enjoy it. For user A with an identifier ID, the registration process is described as follows:

Step 1. A sends a registration message in the form of $(C_{pk_s}(ID, PID, Sig(ID), Sig(PID), pk_A, Flist, (df_{A,1}, df_{A,2}, ..., ds)), ts, Sig(ts))$ to RSU, where C_{pk_s} is RSA encryption using SNS's public key, pk_A is A's public key, $(df_{A,1}, df_{A,2}, ...)$ are the threshold distances for A's friends $(1, 2, ...)$ set by A, ds is the threshold distance for A's strangers.

Step 2. RSU checks the time stamp and signature, if the signature is valid, RSU forwards the message to SNS.

Step 3. When SNS receives the registration message, it checks the signature. SNS keeps a record as $(ID, PID, pk_A, Flist, (df_{A,1}, df_{A,2}, ..., ds))$ in the database.

4.3 Update

When A changes his/her information in $(PID, pk_A, Flist, (df_{A,1}, df_{A,2}, ..., ds))$, or after a certain time period, he/she needs to update his/her information. The detailed steps are as follows:

Step 1. A sends a update message in the form of $(C_{pk_s}(ID, PID, Sig(ID), Sig(PID), pk_A, Flist, (df_{A,1}, df_{A,2}, ..., ds)), ts, Sig(ts))$ to RSU, where PID is A's pseudo-identifier in use, pk_A is A's new public key, $Flist$, $(df_{A,1}, df_{A,2}, ..., ds)$ are A's new relationship with friends/strangers and his/her tendency to share location information.

Step 2. RSU receives the update message, checks the time stamp signature and forwards the message to SNS.

Step 3. When SNS receives the update message, it checks the signature. SNS finds the record with same ID and updates the new information of A in its database.

4.4 Request for Nearby Friends' Locations

When vehicle A wants to find nearby friends' locations, there are five steps as follows.

Step 1. When A intends to know his/her friends' locations within a certain region, he/she submits a query for nearby friends' locations in the form of $SE_k(PID, (x, y), l, f, pk_A)$ to vehicles around him/her, where k is the symmetric key, (x, y) is A's current location, l is the distance within which A wants to find his/her friends, f represents the request type. pk_A is A's public key. RSU also receives this request message and makes a record in its database.

Step 2. When vehicles $v_1, v_2, \ldots$ around A receive the queries for nearby friends' locations, they check the distances between them and A. If v_1 is within the request area, he/she will send a message $(C_{pk_r}(PID_1, af, PID))$ to RSU, where PID_1 is v_1's pseudo identity, PID is A's pseudo identity, af represents the message type. Waiting for a certain time T to collect all response messages, RSU forwards the response messages to SNS in the form $(C_{pk_s}((PID_1, PID_2, \ldots), af, PID))$, where $(PID_1, PID_2, \ldots)$ are pseudo identities of users who response for the request.

Step 3. After receiving the response messages, SNS checks in its database, screens out the vehicles who is not A's friends. SNS finds out the threshold distances set by A's friends for him/her.

Step 4. SNS and A's friends execute our proposed privacy enhanced distance comparison protocol. Find A's friends $(v_a, v_b, \ldots)$ who are willing to share their locations with A.

Step 5. Vehicles $(v_a, v_b, \ldots)$ use A's public key to encrypt their identities and locations and send these messages to A.

4.5 Request for Nearby Strangers' Locations

Similar to request for nearby friends' locations, when vehicle A wants to find nearby strangers' locations, there are five steps as follows.

Step 1. When A intends to find strangers' locations within a certain area, he/she submits a query for nearby strangers' locations in the form of $(SE_k(PID, (x, y), l, s, pk_A))$ to vehicles around him/her, where k is the symmetric key, (x, y) is A's current location, l is the distance within which A wants to find strangers, s represents the request type. pk_A is A's public key. RSU also receives this request message and makes a record in its database.

Step 2. When vehicles $v_1, v_2, \ldots$ around A receive the queries for nearby strangers' locations, they check the distances between them and A. If v_1 is within the request area, he/she will send a message $(C_{pk_r}(PID_1, as, PID))$ to RSU, where PID_1 is v_1's pseudo identity, PID is A's pseudo identity, as represents the message type. Waiting for a certain time T to collect all response messages, RSU forwards the response messages to SNS in the form $(C_{pk_s}((PID_1, PID_2, \ldots), as, PID))$, where $(PID_1, PID_2, \ldots)$ are pseudo identities of users who response for the request.

Step 3. After receiving the response messages, SNS checks in its database, screens out the vehicles who is A's friends. SNS finds out the threshold distances for strangers set by these vehicles.

Step 4. SNS and A's strangers execute our proposed privacy enhanced distance comparison protocol. Find strangers $(v_c, v_d, \ldots)$ who are willing to share their locations with A.

Step 5. Vehicles $(v_c, v_d, \ldots)$ use A's public key to encrypt their identities and locations and send these messages to A.

4.6 Request for Specific Friends' Locations

When vehicle A wants to find locations of specific friends nearby, five steps below will be performed.

Step 1. When A intends to obtain specific friends' locations within a certain area, he/she submits a query for specific friends' locations in the form of $(SE_k(PID, (x, y), l, sf, pk_A, C_{pk_s}(ID, (ID_{f1}, ID_{f2}, \ldots))))$ to nearby vehicles and RSU, where k is symmetric key, (x, y) is A's current location, l is the distance within which A wants to find his/her specific friends, sf is the request type, pk_A is A's public key and $(ID_{f1}, ID_{f2}, \ldots)$ are A's specific friends' identities. RSU forwards $(PID, C_{pk_s}(ID, (ID_{f1}, ID_{f2}, \ldots)))$ to SNS. SNS finds pseudo identities $(PID_{f1}, PID_{f2}, \ldots)$ corresponding to $(ID_{f1}, ID_{f2}, \ldots)$ in its database.

Step 2. When vehicles $v_1, v_2, \ldots$ around A receive the queries for specific friends' locations, they check the distances between them and A. If v_1 is within the request area, he/she will send a message $(C_{pk_r}(PID_1, asf, PID))$ to RSU, where PID_1 is v_1's pseudo identity, PID is A's pseudo identity, asf represents the message type. Waiting for a certain time T to collect all response messages, RSU forwards the response messages to SNS in the form $(C_{pk_s}((PID_1, PID_2, \ldots), asf, PID))$, where $(PID_1, PID_2, \ldots)$ are pseudo identities of users who response for the request.

Step 3. After receiving the response messages, SNS checks in $(PID_{f1}, PID_{f2}, \ldots)$ who response for the request, assume that they are $(PID_{F1}, PID_{F2}, \ldots)$, SNS finds out the threshold distances set by $(PID_{F1}, PID_{F2}, \ldots)$ for him/her.

Step 4. SNS and $(PID_{F1}, PID_{F2}, \ldots)$ execute our proposed privacy enhanced distance comparison protocol. Find A's specific friends $(v_a, v_b, \ldots)$ who are willing to share their locations with A.

Step 5. Vehicles $(v_a, v_b, \ldots)$ use A's public key to encrypt their identities and locations and send these messages to A.

5 Security Analysis

PLVSNs assume that RSU and SNS are both honest but curious. That means they follow the scheme, but try to know more private information of vehicles. We assume that they will not collude with each other and being controlled by the adversary at the same time. Researches on VANET guarantee the pseudo identities of vehicles can not be tracked. The security analysis is provided below.

Threat 1: Vehicles are dishonest and unauthorized vehicles may want to find target vehicles' locations illegally.

PLVSNs use threshold distances to implement access control, vehicles can set different threshold distances for different targets. Since the protocol is performed honestly by RSU and SNS, vehicles can only get the location information they are permitted to access.

In our system, vehicles communicate with each other using pseudo identities when they query for the locations of friends/strangers, these query messages include some sensitive information, such as vehicles' locations. For the adversary outside RSU's service, PLVSNs use broadcast encryption to prevent them from knowing the information. For malicious vehicles inside the broadcast set, the mapping relationship between vehicles and their pseudo identities are well protected in VANET and malicious users cannot identify a vehicle from his/her request messages.

Threat 2: RSU in our system can receive all messages sent by vehicles and SNS, and it may collect these messages to get sensitive information of an target vehicle.

In registration and update phase, PLVSNs use asymmetric encryption to protect vehicles' sensitive information, RSU can not get any personal information of vehicles.

In the request phase, RSU can not identify the target vehicle from request messages, since VANET protects the mapping relationship between vehicles and their pseudo identities. RSU can receive the final response message from target vehicles' friends/strangers, it may collect these response messages and guess the social relationship of vehicles. Note that, vehicles have a wide range of activities and each vehicle can only request for a finite interval, that means only a few friends/strangers will receive the request messages. Moreover, not all these friends/strangers will reply. Furthermore, dummy vehicles can be added into the response set. Social network privacy is protected in our system.

Threat 3: SNS manages users' social network information, it may want to know the location information of an target vehicle. Vehicles do not submit their locations in registration and update phase in PLVSNs. We have assumed that RSU and SNS will not collude with each other, SNS can not get locations of specific vehicles from their request message through RSU.

The distances between vehicles and the threshold distances are also sensitive, and our proposed protocol uses homomorphic encryption to protect distances between vehicles from SNS and protect threshold distances from vehicles.

6 Experimental Evaluation

In this section, we implement the PLVSNs system and present an evaluation of the real time performance.

6.1 Implementation

PLVSNs employ four main encryption algorithms: digital signature, symmetric encryption, asymmetric encryption, and homomorphic encryption. We use RSA for asymmetric encryption and signature, AES for symmetric encryption and Paillier for homomorphic encryption. The request area is set as a circle region with radius of 1.5 km. The threshold distance is set to an optional set of values. Python is employed to perform the proposed algorithms. We use package *phe* to implement homomorphic encryption, *pycrypto* for symmetric encryption and asymmetric encryption. The simulation is implemented on an Intel Xeon E3 with 8 GB memory.

6.2 Evaluation

In PLVSNs, the digital signature technology is only used in the registration and update phases, and any secure signature algorithm can be used including RSA, so we mainly observe the system performance in each request phase.

When a vehicle wants to know locations of his/her friends/strangers, he/she sends a request to vehicles around him/her. SNS chooses vehicle's friends or strangers from the vehicles which response, then executes the privacy enhanced distance comparison protocol. PLVSNs's system execution time is related to the number of friends/strangers who response for the request. Thus, we observe the time spent for the request process and the time spent for privacy enhanced distance comparison protocol against different number of friends/strangers response. Each experiment is performed 10 times and we choose the average values. Figures 2 and 3 show the results respectively.

Figure 2 shows that the time consumption of the system increases with the number of friends/strangers responses. Figure 3 shows that the execution time of privacy enhanced distance comparison protocol also increases with the number of friends/strangers responses. The time of other processes outside the comparison algorithm is basically unchanged, and takes up a small part of the whole process. Note that our simulation assumes that the comparison between the actual distances and the threshold distances is carried out in sequence, while in practice, this process can be carried out by each vehicle and SNS separately, which can be processed in parallel, thus the system running time will be greatly reduced.

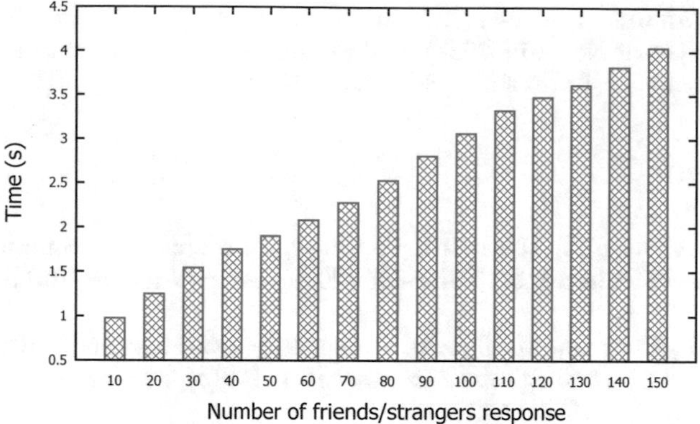

Fig. 2. Entire request process

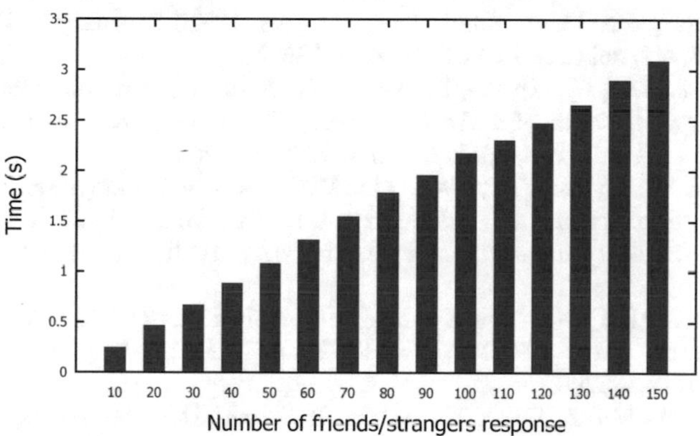

Fig. 3. Privacy enhanced distance comparison protocol process

7 Conclusion

In this paper, we make a research on protecting location privacy and social network privacy in VSNs. We present a scheme called PLVSNs, a privacy enhanced location sharing system. By using DSRC and mobile computing technologies, the system allows users to request directly to nearby vehicles, thus enabling PLVSNs to remove LBS from the architecture. Vehicles complete the location sharing process with RSU and SNS. We also propose a new privacy enhanced distance comparison protocol, which protects identity privacy, location privacy and social network privacy while providing efficient comparison. We show that PLVSNs is secure under the stronger security model. The simulations demonstrates the efficiency of our system.

Acknowledgement. This research is supported by National Natural Science Foundation of China (Grant Nos. 61402037), and Graduate Technological Innovation Project of Beijing Institute of Technology (No. 1320012211801).

References

1. Abbas, N., Zhang, Y., Taherkordi, A., Skeie, T.: Mobile edge computing: a survey. IEEE Internet Things J. **5**(1), 450–465 (2018). https://doi.org/10.1109/JIOT.2017.2750180
2. Barnard, J., Huisman, M., Drevin, G.: Security awareness and the use of location-based services, technologies and games. IJESDF **9**(4), 293–325 (2017). https://doi.org/10.1504/IJESDF.2017.10008010
3. Damgård, I., Jurik, M.: A generalisation, a simplification and some applications of Paillier's probabilistic public-key system. In: Public Key Cryptography, Proceedings of the 4th International Workshop on Practice and Theory in Public Key Cryptography, PKC 2001, Cheju Island, Korea, 13–15 February 2001, pp. 119–136 (2001). https://doi.org/10.1007/3-540-44586-2_9
4. Hoang, D.T., Lee, C., Niyato, D., Wang, P.: A survey of mobile cloud computing: architecture, applications, and approaches. Wirel. Commun. Mob. Comput. **13**(18), 1587–1611 (2013). https://doi.org/10.1002/wcm.1203
5. Hu, W., Xue, K., Hong, P., Wu, C.: ATCS: a novel anonymous and traceable communication scheme for vehicular ad hoc networks. I. J. Network Security **13**(2), 71–78 (2011). http://ijns.femto.com.tw/contents/ijns-v13-n2/ijns-2011-v13-n2-p71-78.pdf
6. Kenney, J.B.: Dedicated short-range communications (DSRC) standards in the united states. Proc. IEEE **99**(7), 1162–1182 (2011). https://doi.org/10.1109/JPROC.2011.2132790
7. Li, J., Yan, H., Liu, Z., Chen, X., Huang, X., Wong, D.S.: Location-sharing systems with enhanced privacy in mobile online social networks. IEEE Syst. J. **11**(2), 439–448 (2017). https://doi.org/10.1109/JSYST.2015.2415835
8. Li, J., Li, J., Chen, X., Liu, Z., Jia, C.: {MobiShare}+: security improved system for location sharing in mobile online social networks. J. Internet Serv. Inf. Secur. **4**(1), 25–36 (2014)
9. Liao, D., Li, H., Sun, G., Zhang, M., Chang, V.: Location and trajectory privacy preservation in 5G-enabled vehicle social network services. J. Netw. Comput. Appl. **110**, 108–118 (2018). https://doi.org/10.1016/j.jnca.2018.02.002
10. Liu, Z., Li, J., Chen, X., Li, J., Jia, C.: New privacy-preserving location sharing system for mobile online social networks. In: Eighth International Conference on P2P, Parallel, Grid, Cloud and Internet Computing, 3PGCIC 2013, Compiegne, France, 28–30 October 2013, pp. 214–218 (2013). https://doi.org/10.1109/3PGCIC.2013.38
11. Liu, Z., Luo, D., Li, J., Chen, X., Jia, C.: N-Mobishare: new privacy-preserving location-sharing system for mobile online social networks. Int. J. Comput. Math. **93**(2), 384–400 (2016). https://doi.org/10.1080/00207160.2014.917179
12. Phan, D.H., Pointcheval, D., Shahandashti, S.F., Strefler, M.: Adaptive CCA broadcast encryption with constant-size secret keys and ciphertexts. Int. J. Inf. Sec. **12**(4), 251–265 (2013). https://doi.org/10.1007/s10207-013-0190-0
13. Shaham, S., Ding, M., Liu, B., Lin, Z., Li, J.: Privacy preservation in location-based services: a novel metric and attack model. CoRR abs/1805.06104 (2018)

14. Shen, N., Yang, J., Yuan, K., Fu, C., Jia, C.: An efficient and privacy-preserving location sharing mechanism. Comput. Stand. Interfaces **44**, 102–109 (2016)
15. Sun, G., Zhang, Y., Liao, D., Yu, H., Du, X., Guizani, M.: Bus-trajectory-based street-centric routing for message delivery in urban vehicular ad hoc networks. IEEE Trans. Veh. Technol. **67**(8), 7550–7563 (2018). https://doi.org/10.1109/TVT.2018.2828651
16. Usman, M., Asghar, M.R., Ansari, I.S., Granelli, F., Qaraqe, K.A.: Technologies and solutions for location-based services in smart cities: past, present, and future. IEEE Access **6**, 22240–22248 (2018). https://doi.org/10.1109/ACCESS.2018.2826041
17. Wei, W., Xu, F., Li, Q.: Mobishare: flexible privacy-preserving location sharing in mobile online social networks. In: Proceedings - IEEE INFOCOM, vol. 131, no. 5, pp. 2616–2620 (2012)
18. Xue, K., Hong, P.: A dynamic secure group sharing framework in public cloud computing. IEEE Trans. Cloud Comput. **2**(4), 459–470 (2014). https://doi.org/10.1109/TCC.2014.2366152
19. Zhang, C., Zhu, L., Xu, C.: INBAR: a new interest-based routing framework in vehicular social networks. In: 2017 International Conference on Security, Pattern Analysis, and Cybernetics (SPAC), pp. 479–484. IEEE (2017)
20. Zhang, L., Li, X., Liu, Y., Jung, T.: Verifiable private multi-party computation: ranging and ranking. In: Proceedings of the IEEE INFOCOM 2013, Turin, Italy, 14–19 April 2013, pp. 605–609 (2013). https://doi.org/10.1109/INFCOM.2013.6566844
21. Zhang, Y., Chen, X., Li, J., Wong, D.S., Li, H., You, I.: Ensuring attribute privacy protection and fast decryption for outsourced data security in mobile cloud computing. Inf. Sci. **379**, 42–61 (2017). https://doi.org/10.1016/j.ins.2016.04.015

Cryptography and Cryptosystem

Witness-Based Searchable Encryption with Aggregative Trapdoor

Xin Xie[1,2], Yu-Chi Chen[2,3(✉)], Jun-Rui Wang[2], and Yingjie Wu[1]

[1] College of Mathematics and Computer Science, Fuzhou University, Fuzhou 350002, People's Republic of China

[2] Department of Computer Science and Engineering, Yuan Ze University, Taoyuan 320, Taiwan
wycchen@saturn.yzu.edu.tw

[3] Innovation Center for Big-Data and Digital Convergence, Yuan Ze University, Taoyuan 320, Taiwan

Abstract. The well-known open problem in public key encryption with keyword search is how to avoid internal adversaries as the server. Implicitly, the internal attack is implemented as follows. Upon receiving a trapdoor, the probability polynomial time internal adversary can always act as a sender to produce each ciphertext for each keyword if keyword space is bounded by a polynomial of the security parameter. Then, the adversary runs the test algorithm for the trapdoor and all produced ciphertext, and then infer the correct keyword. To overcome this problem, the original framework must be changed slightly. A fundamental goal is creates a secure *bridge* between the sender and receiver. It not only keeps testability of the server, but also avoids imitating a sender. Witness-based searchable encryption (WBSE) is a manner to realize the design goal. In this paper, we formalize an abstracted notion, witness-based searchable encryption with aggregative trapdoor. Under the notion, we present a nearly optimal solution for WBSE under the barrier with trapdoor size proportional to n (the number of senders). Comparing with the existing scheme with trapdoor size $O(n)$, the proposed scheme is based on bilinear map, and offers size only in n.

Keywords: Public key encryption · Keyword search · Searchable encryption · Witness · Aggregative trapdoor

1 Introduction

Security issues, trends, and solutions of cloud storage systems has been discussed and considered in the last decade, since the ideas of cloud computing are gradually mature, and mobile phones and their APPs have been widely used.

This work supported in part by the Innovation Center for Big-Data and Digital Convergence, Yuan Ze University, and Ministry of Science and Technology of Taiwan, under grant MOST 106-2218-E-115-008-MY3.

C.-N. Yang et al. (Eds.): SICBS 2018, AISC 895, pp. 561–573, 2020.
https://doi.org/10.1007/978-3-030-16946-6_45

There are plenty of systems providing such the services (i.e., Dropbox, iCloud, and Google Drive). For protecting confidentiality of sensitive data, encryption is a straight way against attackers, but indeed cannot offer efficient search over encrypted data.

Public key encryption with keyword search (PEKS) is a notion to address the above security goal with the sender-receiver scenario. PEKS allows that given a trapdoor of keyword T_w (produced by a receiver), the server can search keyword ciphertexts $C_{w'}$ (sent by a sender). If the test algorithm considers that a pair of ciphertext and trapdoor matches (implicitly denotes $\mathsf{Test}(T_w, C_{w'}) = 1$ if $w = w'$), this concludes the ciphertext and trapdoor are exactly generated from the same keyword. Upon searching, the receiver will obtain the encrypted data, referred to as an original ciphertext of a message from the server. As a result, the PEKS systems [17] can reach encrypted data sharing very easily, and many schemes depend on pairing [1,9,11,16,20,22]. In the literature, Boneh et al. [6] introduced the notion of PEKS, and defined the security models for ciphertext security, and proposed the first PEKS scheme. Recap the security defined by [6]. It relies on a secure channel to the server, and thus only the server can run Test. Some follow-up works got rid of using any secure channel by adding server's public key for constructing designate server PEKS schemes (dPEKS) [4,13,18,19]. In fact, dPEKS can withstand outside adversaries to run Test without any security channel.

Related Work. The open problem in PEKS or dPEKS is how to exactly avoid internal adversaries $\mathcal{A}$ as the server. Implicitly, the internal attack is implemented as follows. Upon receiving a trapdoor T_w, the probability polynomial time $\mathcal{A}$ can always act as a sender to produce a ciphertext $C_{w'_i}$ for each keyword $w'_i \in \mathbb{K}$ if $|\mathbb{K}|$ is bounded by a polynomial of λ where λ is the security parameter. Then, $\mathcal{A}$ runs $\mathsf{Test}(T_w, C_{w'_i})$ and then infer the correct w. To overcome this problem, the framework of PEKS or dPEKS must be changed slightly. There is a idea to keep testability of the server, but to avoid imitating a sender. It creates a secure *bridge* between the sender and receiver. Typically, two simple manners implement this idea.

- (PEKS with authenticated senders [14]) The sender must be authenticated by the receiver before generating ciphertext.
- (Witness-based PEKS [17]) The sender holds witness, and the receiver retrieves the corresponding instance and then use it to produce the trapdoor of keyword.

However, these manners are realized but unavoidably with additional round and communication complexity.

Contributions. In this paper, we address witness-based PEKS (WBSE, for short) for improving efficiency. Our starting point is to observe that the existing WBSE scheme of Ma et al. [17] relies on trapdoor size proportional to n where n is the number of senders. However, the factor n is a barrier on size of trapdoor in

WBSE, since the receiver must obtain all distinct instances (produced by distinct senders) from the server or public board. Shrinking size from $O(n)$ to n is our goal in this work. We obtain a nearly optimal solution under the barrier. We use bilinear map as the building block and construct a new WBSE scheme. Our main technique is to pack n different and individual trapdoors together, and thus create a degree $n-1$ polynomial f where every point x_i induces $f(x_i) = s$ and x_i is a Diffie-Hellman session key produced from a witness and receiver's public key. At a high level, the trapdoor is composed of n encodings of n coefficients of f and one encoding of a keyword[1], and in particular the keyword ciphertext is of encodings of a point x_i and its powers $x_i^2, ..., x_i^n$. Finally, bilinear map can support evaluation of Test over the encodings of f and x_i. Summarizing the result, the trapdoor contains f, and the ciphertext does x_i. This suffices to achieve the size of trapdoor is exactly n with constant overheads.

Organization. The rest of the article is arranged as follows. In Sect. 2, we introduce some definitions and tools that will be used throughout this paper. In Sect. 3, we propose an improved WBSE scheme and analysis of correctness. In Sect. 4, We compare the performance of our scheme with Ma et al.'s scheme in the literature. Finally, Sect. 5 concludes the paper and our future work. The notations are located in Table 1.

Table 1. Notations

Symbol	Description
λ	Security parameters
R	Witness relation function
m	Keywords
$\mathcal{M}$	Keyword space
$\mathcal{WT}$	Witness space
w	Witness produced by the data sender
t	Instance produced by the data sender
C_w	Ciphertext of keyword
T_w	Trapdoor of keyword

2 Preliminaries

In this section, we briefly introduce some preliminaries. We denote the security parameter and probabilistic polynomial time by λ and PPT. The following assumptions are produced by using λ. Moreover, we have to define the

[1] Keep *encoding* implicitly. Intuitively, we say that encoding converts an input x to a group element with some additional randomness.

negligible function. It is a function that for every polynomial time function $p(\lambda)$ and all sufficiently large λ, $\mathsf{negl}(\lambda) \leq \frac{1}{p(\lambda)}$ holds.

2.1 Hash Function

A hash function is a pair of probabilistic polynomial time algorithms $\prod = (Gen, H)$ that satisfy the following two conditions [15]:

(1) Gen is a probabilistic algorithm that outputs a key z with a security parameter 1^λ as input.
(2) There is a polynomial l such that H takes the key z and a string of strings $x \in \{0,1\}^*$ as input and outputs a string of characters $H_z(x) \in \{0,1\}^{l(\lambda)}$.

2.2 Asymmetric Bilinear Pairing Groups

As we knew, Boneh and Franklin [7] proposed an efficient IBE scheme which used bilinear maps on elliptic curves. Eisenträger et al. [10] improved the efficiency of operations of elliptic curves and speed up the Weil and Tate pairings. Formally, a bilinear map is denoted by $e : \mathbb{G}_1 \times \mathbb{G}_2 \rightarrow \mathbb{G}_T$ where $\mathbb{G}_1$ and $\mathbb{G}_2$ are cyclic multiplicative groups of prime order q. Let g and h be generators of $\mathbb{G}_1$ and $\mathbb{G}_2$ respectively. We define an asymmetric bilinear group with tuple $(q, \mathbb{G}_1, \mathbb{G}_2, \mathbb{G}_T, e)$ and satisfy the following properties.

(1) Bilinear: For any integers $x \in \mathbb{G}_1, y \in \mathbb{G}_2$ and $a, b \in \mathbb{Z}_q^*$, the equation $e(x^a, y^b) = e(x, y)^{ab}$ is held.
(2) Non-degeneracy: $e(g, h)$ generates $\mathbb{G}_T$

We use the description $(q, \mathbb{G}_1, \mathbb{G}_2, \mathbb{G}_T, e)$ to illustrate the output of algorithm $\mathcal{G}_{abpg}(1^\lambda)$.

2.3 Mathematical Assumptions

Assumption 1 (Decisional Diffie-Hellman (DDH) [2,5]**).** *$\mathbb{G}$ is a cyclic group of order q, and g is an element of $\mathbb{G}$. $a, b, c \in \mathbb{Z}_q^*$ are uniformly randomly chosen. If there is no probabilistic polynomial-time adversary $\mathcal{A}$ can distinguish between tuples (g, g^a, g^b, g^{ab}) and (g, g^a, g^b, g^c), then the DDH assumption holds in $\mathbb{G}$. Formally, we state it as*

$$|\Pr[\mathcal{A}(1^\lambda, q, \mathbb{G}, g^a, g^b, g^{ab}) = 1] - \Pr[\mathcal{A}(1^\lambda, q, \mathbb{G}, g^a, g^b, g^c) = 1]| \leq \mathsf{negl}(\lambda).$$

Assumption 2 (Symmetric External Diffie-Hellman (SXDH) [8,21]**).** *Given two distributions $((q, \mathbb{G}_1, \mathbb{G}_2, \mathbb{G}_T, e, g_1, g_2), g_1^u, g_1^v, Y_{uv})$ and $((q, \mathbb{G}_1, \mathbb{G}_2, \mathbb{G}_T, e, g_1, g_2), g_1^u, g_1^v, Y_r)$ where g_1, g_2 are elements of $\mathbb{G}_1, \mathbb{G}_2$ respectively, and $u, v, r \xleftarrow{\$} \mathbb{Z}_q^*$, $Y_{uv} = g_1^{uv}, Y_r = g_1^r$, the external Diffie-Hellman assumption is that for a PPT algorithm $\mathcal{A}$, the advantage of $\mathcal{A}$ is defined as:*

$$\begin{aligned}&|\Pr[\mathcal{A}(1^\lambda, (q, \mathbb{G}_1, \mathbb{G}_2, \mathbb{G}_T, e, g_1, g_2), g_1^u, g_1^v, g_1^{uv}) = 1] \\ &- \Pr[\mathcal{A}(1^\lambda, (q, \mathbb{G}_1, \mathbb{G}_2, \mathbb{G}_T, e, g_1, g_2), g_1^u, g_1^v, g_1^r) = 1]| \leq \mathsf{negl}(\lambda).\end{aligned}$$

However, the symmetric external Diffie-Hellman assumption holds if DDH is also intractable in $\mathbb{G}_2$.

Assumption 3 (Inverse Computational Diffie-Hellman(InvCDH) [3]). *Given* (g, g^a), *for* g *as a generator of* $\mathbb{G}$ *and* $a \in \mathbb{Z}_q^*$, *the InvCDH assumption is hard to compute* $g^{a^{-1}}$ *as* $\Pr[\mathcal{A}(1^\lambda, q, \mathbb{G}_1, g, g^a) \rightarrow g^{a^{-1}}] \leq \mathsf{negl}(\lambda)$ *for all PPT adversaries* $\mathcal{A}$.

2.4 Witness-Based Searchable Encryption with Aggregative Trapdoor

Definition 1. *The searchable encryption algorithm based on witness consists of the following polynomial time algorithms, where the witness relation* $R(w, t) = 1$ *is defined on NP-language [12]* $\mathcal{L}$ *of for any polynomial algorithms* A_w, A'_w, $w \in \mathcal{WT}$, *where* $\mathcal{WT}$ *denotes the witness space.* $\Pr[A_w(1^\lambda, w) = t] = 1$ *and* $\Pr[A'_w(1^\lambda, t) = w] < \mathsf{negl}(\lambda)$, *where* λ *is a security parameter. For simplicity, the algorithm of instance generation is denoted by* $\mathsf{InsGen}(w) = t$.

- $\mathsf{KeyGen}(1^\lambda)$: The algorithm takes input as a security parameter 1^λ and generates a public/private key pair (pk, sk).
- $\mathsf{WBSE}(pk, m; w)$: For public key pk and a keyword $m \in \mathcal{M}$ with a random witness w $\in \mathcal{WT}$, produces a ciphertext $C = (t, d)$ where $t \in \mathcal{L}$ and (w, t) satisfies the witness relation R.
- $\mathsf{Trapdoor}(sk, m; t_1, t_2 ..., t_n)$: Given a private key sk, a keyword m and any instances $t \in \mathcal{L}$, produces a trapdoor $T = (h_k, t_d)$ where h_k and t_d are calculated from instance t.
- $\mathsf{Test}(C, T)$: Given the a searchable ciphertext $C = (t, d)$ and a trapdoor $T = (h_k, t_d)$, check if the instance t contained in the trapdoor is consistent with the t of the searchable ciphertext denoting that the keyword test on C and T can be executed, and if both contain keywords that are the same outputs 1, or a bit 0 otherwise.

Remark 1. *Analog to encryption is random, it should be safe for random selection of witnesses* w *because it is based on the fact that each time the encryption is finished, it is discarded, which is reasonable.*

Remark 2. *Under the witness relation* $R(w, t)$, *the framework works properly. The premise is that the instance* t *contained in the trapdoor is consistent with the instance* t *contained in the ciphertext. Therefore, the adversary cannot produce a valid ciphertext to guess its keyword because it is almost impossible to derive from* t *to* w. *That is to resist the essence of keyword guessing in our framework.*

Remark 3. *As with the PEKS system, the receiver and sender do not need to communicate or share a secret value throughout the process. Therefore, WBSE is an asymmetric encryption.*

Definition 2. *The correctness of the* $\mathcal{WBSE}$ = (KeyGen, WBSE, Trapdoor, Test) *satisfies the following three conditions. Where for any* $\lambda \in \mathcal{N}$, KeyGen(1^λ) *outputs the* (pk, sk), $m, m'(m \neq m') \in \mathcal{M}$, $t, t'(t \neq t') \in \mathcal{L}$ *and* $w, w'(w \neq w') \in \mathcal{WT}$ *satisfy* $R(w,t) = 1$ *and* $R(w',t') = 1$. *Case 1: When* $R(w,t) = 1$ *and the WBSE's keywords is the same as Trapdoor's keywords, the Test algorithm always outputs 1. Case 2: When* $R(w,t') \neq 1$, *regardless of whether the keywords are equivalent, the Test algorithm always outputs 0. Case 3: When* $R(w,t) = 1$, *but the WBSE's keywords is different form the Trapdoor's keywords, the probability that the Test algorithm outputs 1 is negligible. Figure 1 illustrates the system framework.*

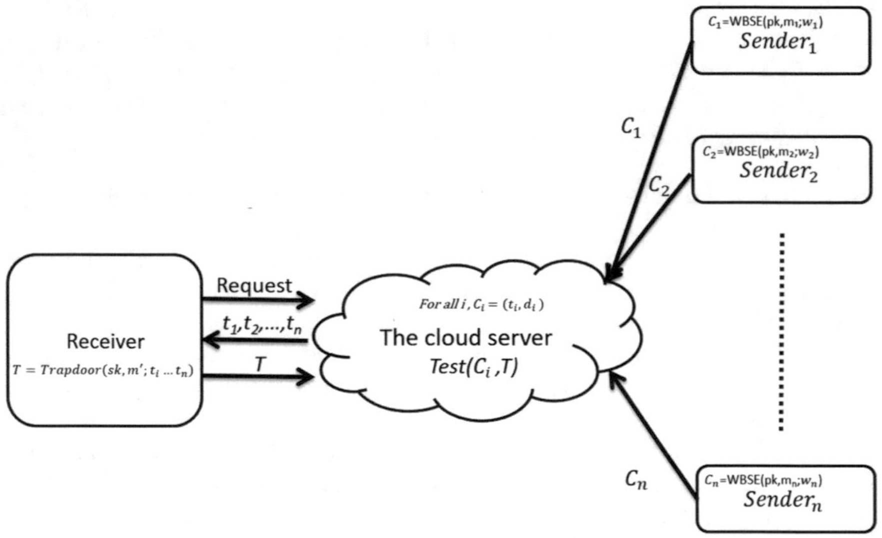

Fig. 1. Sketch of the scheme.

3 The Proposed Scheme of WBSE with Aggregative Trapdoor

3.1 Construction

There are three groups $\mathbb{G}_1, \mathbb{G}_2, \mathbb{G}_T$ of prime order q. The bilinear map is specified by $e : \mathbb{G}_1 \times \mathbb{G}_2 \rightarrow \mathbb{G}_T$. $H(\cdot)$ be a cryptographic hash function defined on $\{0,1\}^* \rightarrow \mathbb{Z}_q^*$. We choose a NP-language $\mathcal{L}$ for the witness relation and $\mathrm{w} \in \mathcal{WT}$, where $\mathcal{WT}$ denotes the witness space, $\mathcal{WT} = \mathbb{Z}_q^*$ and the instance t is defined as follows:

$$R(\mathrm{w}, t) = 1 \wedge \mathrm{w} \in \mathcal{WT} \wedge t = g^w.$$

- KeyGen(1^λ): It takes the security parameter λ as input, and then returns the description of the groups $\mathbb{G}_1, \mathbb{G}_2, \mathbb{G}_T$ of order q, the bilinear map e, the generator $g \in \mathbb{G}_1$, and the public/secret key pair where $sk = (\mu, \alpha)$ (μ and α are uniformly chosen) and $pk = (q, \mathbb{G}_1, \mathbb{G}_2, \mathbb{G}_T, e, g, A = g^\alpha, U = g^\mu, H(\cdot))$.
- WBSE($pk, m; w$): It products an instance t for a witness $w \in \mathcal{WT}$, and then computes $x = H(A^w)$. Then, it chooses $r \in \mathbb{Z}_q^*$ uniformly at random and computes L_u sequence(n is a large enough number):

$$L_u = [U^r, U^{r \cdot x}, U^{r \cdot x^2}, ..., U^{r \cdot x^n}].$$

For a keyword m, it obtains the hash value $H(m)$ and compute $\nu = A^r \cdot g^{H(m) \cdot r}$. Finally, it returns the $\mathcal{WBSE}$ ciphertext as

$$C = (t, \nu, L_u).$$

- Trapdoor($sk, m, t_1, t_2, ..., t_n$): It firstly fetches all the instance t_i and computes L_x as follows

$$L_t = [t_1, t_2, ..., t_k] = [g^{w_1}, g^{w_2}, ..., g^{w_k}].$$
$$L_x = [x'_1, x'_2, ..., x'_k].$$

where $x'_i = H((g^{w_i})^\alpha)$. Then, it chooses a_k, $s \in \mathbb{Z}_q^*$ uniformly at random and constructs k-degree F function:

$$F(x) = \prod_{i=1}^{k}(x - x'_i) + s = a_k \cdot x^k + a_{k-1} \cdot x^{k-1} + ... + a_0.$$

Finally, it chooses a generator $h \subset \mathbb{G}_2$ and returns a trapdoor $T = (h_k, t_d)$ by setting

$$h_k = (h^{a_k}, h^{a_{k-1}}, ..., h^{a_0}).$$
$$t_d = h^{\frac{u \cdot s}{\alpha + H(m)}}.$$

- Test(C, T): Check if the searchable ciphertext and trapdoor enjoy the same keyword m by computing the following equation:

$$\prod_{i=0}^{k} e(U^{r \cdot x^i}, h^{a_i}) = e(\nu, t_d). \tag{1}$$

If yes, outputs 1 denoting that C and T contain the same keyword, and 0 otherwise.

3.2 Correctness

Expend the scheme:

$$\begin{aligned}
&\prod_{i=0}^{k} e(U^{r\cdot x^i}, h^{a_i})\\
&= e(U^r, h^{a_0}) \cdot e(U^{r\cdot x}, h^{a_1}) \cdot e(U^{r\cdot x^2}, h^{a_2}) \cdots e(U^{r\cdot x^k}, h^{a_k})\\
&= e(g^{u\cdot r}, h^{a_0}) \cdot e(g^{u\cdot r\cdot x}, h^{a_1}) \cdot e(g^{u\cdot r\cdot x^2}, h^{a_2}) \cdots e(g^{u\cdot r\cdot x^k}, h^{a_k})\\
&= e(g,h)^{a_0\cdot u\cdot r} \cdot e(g,h)^{a_1\cdot x\cdot u\cdot r} \cdot e(g,h)^{a_2\cdot x^2\cdot u\cdot r} \cdots e(g,h)^{a_k\cdot x^k\cdot u\cdot r}\\
&= e(g,h)^{u\cdot r\cdot(a_0+a_1\cdot x+a_2\cdot x^2+a_k\cdot x^k)}.
\end{aligned}$$

Note that if the instance t is generated by the witness w, then the trapdoor function $F(x) = s$, if the instance t of the ciphertext is equal to the instance t that was brought into the F function calculation (In short, the instance t that the F function brings into the calculation is the instance t provided by the ciphertext). So for the left side of Eq. 1, we conclude that

$$\prod_{i=0}^{k} e(U^{r\cdot x^i}, h^{a_i}) = e(g,h)^{u\cdot r\cdot s},$$

and for the right side

$$e(\nu, t_d) = e(g^{a\cdot r+H(m)\cdot r}, h^{\frac{u\cdot s}{\alpha+H(m')}}).$$

If $m = m'$, the right side of Eq. 1 is equal to $e(g,h)^{u\cdot r\cdot s}$, $\mathsf{Test}(C,T)$ will outputs 1.

If $m \neq m'$ and Eq. 1 still holds, then we know $H(m) = H(m')$. Due to the collision resistant property of $H(\cdot)$, $\Pr[\mathsf{Test}(C,T) = 1]$ in this case is negligible.

3.3 Security Analysis

We postpone the security definition in Appendix. Here we highlight the security of the scheme in an informal way with respect to ciphertext security, trapdoor security, and trapdoor unforgeability. To simplify the argument, directly use the viewpoint of the internal adversary for ciphertext and trapdoor security, and consider the sender as the adversary for trapdoor unforgeability.

- (Ciphertext and trapdoor security) To distinguish ciphertext, $\mathcal{A}$ faces $g^{\alpha r+H(m)r}$, and $g^{\alpha r}$ is a SXDH challenge. By standard hybrid argument, we can pay negligible security loss to move it to $g^{c+H(m)r}$ where c is an element in $\mathbb{G}_1$. Moreover, to distinguish trapdoor, $\mathcal{A}$ faces $h^{\frac{us}{\alpha+H(m)}}$, and h^{us} is a SXDH challenge. By the same procedure, we complete the analysis.
- (Trapdoor unforgeability) To forge a trapdoor, we assume the sender who holds the witness as the most powerful adversary, and the forgery must include $h^{\frac{1}{\alpha}}$. It meets the hardness given h, h^{α} to find the inverse $h^{\frac{1}{\alpha}}$. Therefore, the trapdoor unforgeability holds as well.

We will provide the details of security analysis in the full version.

4 Comparison and Experiment

We improve efficiency of the WBSE scheme proposed by Ma et al., and the security of the scheme is also held with that described in Ma et al. The following compares the size and time complexity with Ma et al.'s WBSE scheme.

Table 2. Size comparison.

	$\|pk\|$	$\|C\|$	$\|T\|$
MMSY	$3\|\mathbb{G}\|$	$9\|\mathbb{G}\|$	$4\|\mathbb{G}\| \cdot n$
Ours	$\|\mathbb{G}_1\| + \|\mathbb{G}_2\|$	$n \cdot \|\mathbb{G}_1\|$	$\|\mathbb{G}_2\| + \|\mathbb{G}_2\| \cdot n$

$|pk|$: The length of public key. $|C|$: The length of the ciphertext. $|T|$: The length of trapdoor. Our related works are constructed in pairing groups: $\mathbb{G}_1 \times \mathbb{G}_2 \rightarrow \mathbb{G}_T$, while MMSY scheme's instantiation is implemented in discrete logarithm group $\mathbb{G}$.

Table 3. Computation complexity comparison.

	Encryption	Trapdoor	Times
MMSY	$12E$	$4E \cdot n$	3
Ours	$n \cdot E$	$E{+}E \cdot n$	3

P: A pairing operation. E: An exponentiation operation. n: the number of tested ciphertexts.

We compare all the comparisons under a single keyword. Table 2 shows the size of the public key, ciphertext and trapdoor. Table 3 shows the computational time complexity of the encryption algorithm and the trapdoor algorithm. From Tables 2 to 3, the cost of the trapdoor size has been improved. In reality, the number of searchable ciphertexts increases exponentially with the increase of users. For a single keyword query, in the MMSY scheme, the trapdoor is multiplied by a multiple of n, which is a waste of time and space. In practical applications, the receiver may test all ciphertexts. Although the Test algorithm in our scheme will use the two bilinear pairing, the calculation cost is higher, but we noticed that Trapdoor algorithm only needs to be calculated once and does not have to be recalculated like the MMSY scheme. In short, our structure is not only suitable for security-critical applications, but also against keyword guessing attacks.

5 Conclusion

In this paper, we address witness-based searchable encryption and present a nearly optimal solution under the barrier with trapdoor size proportional to n (the number of senders) by constructing a new scheme. The main technique is simple, and packs n different and individual trapdoors together. For packing, we create a degree $n-1$ polynomial f where every point x_i induces $f(x_i) = s$ and x_i is a Diffie-Hellman session key from a witness and receiver's public key. In particular, we use only one randomness for masking instead of n, and thus reduce size from cn to n for some constant c. For keyword search, we need one more component involving the keyword in a trapdoor. Finally, this suffices to achieve the size of trapdoor is exactly n plus a constant. Our future works will attempt to achieve size of polylog or constant.

A Security Definition of WBSE

A.1 Ciphertext Security

About $\mathcal{WBSE}$ ciphertext security, a ciphertext $\mathsf{WBSE}(pk, m; w)$ does not reveal any information about m unless the trapdoor T_w for (m, t) is available under the witness relation $R(w, t)$, where adversary can access to the trapdoor oracle for any keyword-instance pair (m, t).

Experiment $\boldsymbol{Exp}^{\textbf{WB-IND-CCA}}_{\mathcal{WBSE},\mathcal{A}}(\boldsymbol{\lambda})$. Let λ be the security parameter and $\mathcal{A}$ be the adversary against the ciphertext security of witness-based searchable encryption, defined as witness-based indistinguishable encryption under chosen ciphertext attack.

1. **Steup:** The challenger runs the $\mathsf{KeyGen}(1^\lambda)$ algorithm to generate a public/private key pair (pk, sk). Secondly, it generates $WtList = \{w_1, w_2, ..., w_n\} \xleftarrow{\$} \mathcal{WT}$, then use $\mathsf{InsGen}(w_i)$ to generate instance list $InsList = \{t_1, t_2, ..., t_n\}$. It gives $(pk, InsList)$ to $\mathcal{A}$.
2. **Phase I:** The adversary $\mathcal{A}$ can adaptively ask the trapdoor T_w for any keyword m and instance $t \in InsList$. It returns the trapdoor $T_w = (h_k, t_d)$.
3. **Challege:** The adversary $\mathcal{A}$ sends the challenger two keywords (m_0, m_1) and a instance $t^* \in InsList$. The only restriction is that the adversary can not ask the previously trapdoors T_{w_0} of keyword-instance pair (m_0, t^*) or T_{w_1} of keyword-instance pair (m_1, t^*). The challenger picks a random $b \in \{0, 1\}$ and choose the witness $w^* \in WtList$ according to t^* under the relation of $R(w^*, t^*) = 1$. Finally, the challenger runs $\mathsf{WBSE}(pk, m_b; w^*)$ to generate ciphertext c^* and return c^* to $\mathcal{A}$.
4. **Phase II:** The adversary $\mathcal{A}$ can continue to ask for trapdoor T_w for its chosen keyword m and instance t as long as (m, t) not equal to (m_0, t^*) or (m_1, t^*).
5. **Guess:** In the end, the adversary $\mathcal{A}$ must guess $b^{'}$, $\mathcal{A}$ win the game if $b = b^{'}$, indicating that the experiment outputs is 1, 0 otherwise. In other words, the

adversary wins the game if he can correctly guess whether he was given the WBSE for the m_0 or m_1. We define $\mathcal{A}$'s advantage as

$$Adv_{\mathcal{WBSE},\mathcal{A}}^{WB-IND-CCA} = \left| Pr\left[Exp_{\mathcal{WBSE},\mathcal{A}}^{WB-IND-CCA}(\lambda) = 1 \right] - \frac{1}{2} \right|.$$

Definition 3 *($\mathcal{WBSE}$-WB-IND-CCA). We say that $\mathcal{WBSE}$ = (KeyGen, WBSE, Trapdoor, Test) is witness-based indistinguishable under chosen ciphertext attack if for all probabilistic polynomial time adversary $\mathcal{A}$, $Adv_{\mathcal{WBSE},\mathcal{A}}^{WB-IND-CCA}$ is a negligible function.*

A.2 Trapdooor Security

Regarding $\mathcal{WBSE}$ trapdoor security, a trapdoor $\mathsf{Trapdoor}(sk, m, t)$ does not reveal any information about m under the witness relation $R(w, t)$, where adversary can access to the trapdoor oracle for any keyword-instance pair (m, t) without returning the trapdoor containing the challenge instance t^*.

Experiment $Exp_{\mathcal{WBSE},\mathcal{A}}^{WB-IND-TD}(\lambda)$. Let λ be the security parameter and $\mathcal{A}$ be the adversary against the trapdoor security of witness-based searchable encryption, defined as witness-based indistinguishable trapdoor.

1. **Steup:** The challenger runs the $\mathsf{KeyGen}(1^\lambda)$ algorithm to generate a public/private key pair (pk, sk). Secondly, it generates $WtList = \{w_1, w_2, ..., w_n\} \xleftarrow{\$} \mathcal{WT}$, then use $\mathsf{InsGen}(w_i)$ to generate instance list $InsList = \{t_1, t_2, ..., t_n\}$. Besides, We choose randomly a witness w^* and compute its instance t^*. Assume that $InsList^* = InsList \bigcup t^*$. It gives $(pk, InsList^*)$ to $\mathcal{A}$.
2. **Phase I:** The adversary $\mathcal{A}$ can adaptively ask the trapdoor T_w for any keyword m and instance $t \in InsList$ (hence $t \neq t^*$). It returns the trapdoor $T_w = (h_k, t_d)$.
3. **Challege:** The adversary $\mathcal{A}$ sends the challenger two keywords (m_0, m_1). The challenger picks a random $b \in \{0, 1\}$, generates the challenge trapdoor ${T_w}^* = (h_k^*, t_d^*)$ by running the $\mathsf{Trapdoor}(sk, m_b, t^*)$ algorithm and return ${T_w}^*$ to $\mathcal{A}$.
4. **Phase II:** The adversary $\mathcal{A}$ continue to ask for the trapdoor oracle the same as Phase I.
5. **Guess:** In the end, the adversary $\mathcal{A}$ must guess b', $\mathcal{A}$ win the game if $b = b'$, indicating that the experiment outputs is 1, 0 otherwise. In other words, the adversary wins the game if he can correctly guess whether he was given the Trapdoor for the m_0 or m_1. We define $\mathcal{A}$'s advantage as

$$Adv_{\mathcal{WBSE},\mathcal{A}}^{WB-IND-TD} = \left| Pr\left[Exp_{\mathcal{WBSE},\mathcal{A}}^{WB-IND-TD}(\lambda) = 1 \right] - \frac{1}{2} \right|.$$

Definition 4 *($\mathcal{WBSE}$-WB-IND-TD). We say that $\mathcal{WBSE}$ = (KeyGen, WBSE, Trapdoor, Test) is witness-based indistinguishable trapdoor if for all probabilistic polynomial time adversary $\mathcal{A}$, $Adv_{\mathcal{WBSE},\mathcal{A}}^{WB-IND-TD}$ is a negligible function.*

A.3 Trapdoor Unforgeability

Regarding $\mathcal{WBSE}$ trapdoor unforgeability, given the public key pk generated by the receiver, the adversary is allowed to choose the keyword-instance pair (m^*, t^*) to generate the trapdoor $T_w{}^*$, but the adversary cannot output a meaningful trapdoor except the keyword-instance pair he has previously asked. The "meaningful" trapdoor as follows: Given a trapdoor that generates an existing instance of the available ciphertext as input (the witness of the instance is unknown), we say that the trapdoor is meaningful for its testing of the available ciphertext. Otherwise, given a trapdoor which is generated on an instance produced by itself (the witness of the instance is known), we say that the trapdoor is meaningless, and because of the witness relation, any others can generate the ciphertext of its chosen keyword to do the trapdoor test.

Experiment $\boldsymbol{Exp_{\mathcal{WBSE},\mathcal{A}}^{EUFT-CIA}(\lambda)}$. Let λ be the security parameter and $\mathcal{A}$ be the adversary against the trapdoor unforgeability of witness-based searchable encryption. The concept of a meaningful trapdoor security is called as the existence of an unforgeable trapdoor against chosen instance attack (EUFT-CIA).

1. **Steup:** The challenger runs the $\mathsf{KeyGen}(1^\lambda)$ algorithm to generate a public/private key pair (pk, sk). Secondly, it generates $WtList = \{w_1, w_2, ..., w_n\} \xleftarrow{\$} \mathcal{WTWT}$, then use $\mathsf{InsGen}(w_i)$ to generate instance list $InsList = \{t_1, t_2, ..., t_n\}$. It gives $(pk, InsList)$ to $\mathcal{A}$.
2. **Phase:** The adversary $\mathcal{A}$ can adaptively ask the trapdoor T_w for any keyword m and instance $t \in InsList$. It returns the trapdoor $T_w = (h_k, t_d)$.
3. **Challege:** The adversary $\mathcal{A}$ outputs a challenge forged trapdoor $T_w{}^*$ for its chosen (m^*, t^*): $((m^*, t^*), T_w{}^*) = ((m^*, t^*), (h_k^*, t_d^*))$ and (h_k^*, t_d^*) does not appear before.
4. **Guess:** The adversary $\mathcal{A}$ wins the game if $\mathsf{Trapdoor}(sk, m^*, t^*) = T_w{}^*$, indicating that the experiment outputs is 1, 0 otherwise.

$$Adv_{\mathcal{WBSE},\mathcal{A}}^{TDUnforgeability} = \left| Pr\left[Exp_{\mathcal{WBSE},\mathcal{A}}^{EUFT-CIA}(\lambda) = 1 \right] \right|.$$

Definition 5 *($\mathcal{WBSE}$ – Trapdoor Unforgeability). We say that $\mathcal{WBSE}$ = (KeyGen, WBSE, Trapdoor, Test) is witness-based indistinguishable trapdoor if for all probabilistic polynomial time adversary $\mathcal{A}$, $Adv_{\mathcal{WBSE},\mathcal{A}}^{TDUnforgeability}$ is a negligible function.*

References

1. Baek, J., Safavi-Naini, R., Susilo, W.: Public key encryption with keyword search revisited. In: International Conference on Computational Science and Its Applications, pp. 1249–1259. Springer (2008)
2. Ballard, L., Kamara, S., Monrose, F.: Achieving efficient conjunctive keyword searches over encrypted data. In: International Conference on Information and Communications Security, pp. 414–426. Springer (2005)

3. Behnia, R., Heng, S.-H., Tan, S.-Y.: On the security of a certificateless short signature scheme. Malays. J. Math. Sci. **9**, 103–113 (2015)
4. BingJian, W., TzungHer, C., FuhGwo, J.: Security improvement against malicious server's attack for a dpeks scheme. Int. J. Technol. Des. Educ. **1**, 350–353 (2011)
5. Boneh, D.: The decision Diffie-Hellman problem. In: International Algorithmic Number Theory Symposium, pp. 48–63. Springer (1998)
6. Boneh, D., Di Crescenzo, G., Ostrovsky, R., Persiano, G.: Public key encryption with keyword search. In: International Conference on the Theory and Applications of Cryptographic Techniques, pp. 506–522. Springer (2004)
7. Boneh, D., Franklin, M.: Identity-based encryption from the Weil pairing. In: Annual International Cryptology Conference, pp. 213–229. Springer (2001)
8. Chen, J., Lim, H.W., Ling, S., Wang, H., Wee, H.: Shorter IBE and signatures via asymmetric pairings. In: International Conference on Pairing-Based Cryptography, pp. 122–140. Springer (2012)
9. Cheng, L., Jin, Z., Wen, O., Zhang, H.: A novel privacy preserving keyword searching for cloud storage. In: 2013 Eleventh Annual International Conference on Privacy, Security and Trust (PST), pp. 77–81. IEEE (2013)
10. Eisenträger, K., Lauter, K., Montgomery, P.L.: Fast elliptic curve arithmetic and improved Weil pairing evaluation. In: Cryptographers' Track at the RSA Conference, pp. 343–354. Springer (2003)
11. Fang, L., Susilo, W., Ge, C., Wang, J.: Public key encryption with keyword search secure against keyword guessing attacks without random oracle. Inf. Sci. **238**, 221–241 (2013)
12. Garg, S., Gentry, C., Sahai, A., Waters, B.: Witness encryption and its applications. In: Proceedings of the Forty-Fifth Annual ACM Symposium on Theory of Computing, pp. 467–476. ACM (2013)
13. Hu, C., Liu, P.: A secure searchable public key encryption scheme with a designated tester against keyword guessing attacks and its extension. In: International Conference on Computer Science, Environment, Ecoinformatics, and Education, pp. 131–136. Springer (2011)
14. Huang, Q., Li, H.: An efficient public-key searchable encryption scheme secure against inside keyword guessing attacks. Inf. Sci. **403**, 1–14 (2017)
15. Lindell, Y., Katz, J.: Introduction to Modern Cryptography. Chapman and Hall/CRC, Boca Raton (2014)
16. Liu, Q., Wang, G., Wu, J.: An efficient privacy preserving keyword search scheme in cloud computing. In: International Conference on Computational Science and Engineering, CSE 2009, vol. 2, pp. 715–720. IEEE (2009)
17. Ma, S., Yi, M., Susilo, W., Yang, B.: Witness-based searchable encryption. Inf. Sci. **453**, 364–378 (2018)
18. Rhee, H.S., Park, J.H., Susilo, W., Lee, D.H.: Trapdoor security in a searchable public-key encryption scheme with a designated tester. J. Syst. Softw. **83**(5), 763–771 (2010)
19. Shao, Z.-Y., Yang, B.: On security against the server in designated tester public key encryption with keyword search. Inf. Process. Lett. **115**(12), 957–961 (2015)
20. Yang, Y., Liu, X., Zheng, X., Rong, C., Guo, W.: Efficient traceable authorization search system for secure cloud storage. IEEE Trans. Cloud Comput. (2018)
21. Zhao, Q., Zeng, Q., Liu, X., Xu, H.: Simulation-based security of function-hiding inner product encryption. Sci. China Inf. Sci. **61**(4), 048102 (2018)
22. Zhou, Y., Zhao, X., Liu, S., Long, X., Luo, W.: A time-aware searchable encryption scheme for EHRs. Digit. Commun. Netw. (2018)

Robust/Recover Provable Data Possession Protocol

Chao Feng[1,4(✉)], Honghong Wang[2], Wenbo Wan[3], Qinghua Li[1], and Fangzhou Xu[1]

[1] Department of Physics, School of Electronic and Information Engineering, Qilu University of Technology (Shandong Academy of Sciences), Jinan 250353, People's Republic of China
cfeng@qlu.edu.cn

[2] School of Electrical Engineering and Automation, Qilu University of Technology (Shandong Academy of Sciences), Jinan 250353, People's Republic of China

[3] School of Information Science and Engineering, Shandong Normal University, Jinan 250358, People's Republic of China

[4] Institute of Automation, Shandong Academy of Sciences, Jinan 250101, People's Republic of China

Abstract. Provable data possession (PDP) allows a client that has stored data at remote server to verify that the server correctly possesses the original data. A long-standing problem is how to reduce I/O cost. Through the integration of Online-code and PDP, a challenge/check protocol that can verifies the possession is proposed. The protocol generates probabilistic proofs of possession by sampling tiny sets of data, which obviously reduces I/O cost. Meanwhile, the protocol can recover corrupted data. The authors formalize this notion in the Robust/Recover (RR) provable data possession guarantee. Briefly speaking, the client maintains a constant amount of metadata to verify the proof. The challenge/check protocol transmits a constant amount of data, which reduces communication complexity. The authors give a detailed analysis of this protocol and build a simulation to evaluate practicability in reliability, space overhead, computation complexity, and communication complexity.

Keywords: Data outsourcing · Provable data possession · Online-code · Robust/Recover

1 Introduction

Outsourcing of data allows the data owner (client) with limited resources uploads its data to a remote server (e.g. Cloud Storage Service Providers), which is supposed to correctly stores the data and make it available to the client on demand. The main advantages of data outsourcing include reduced costs from savings in storage, maintenance and as well as increased availability. Unfortunately, the server is not trusted. Therefore, the client should verify if their data has been tampered or deleted. How to verify the availability of data has turned into a critical issue in outsourcing data services.

C.-N. Yang et al. (Eds.): SICBS 2018, AISC 895, pp. 574–587, 2020.
https://doi.org/10.1007/978-3-030-16946-6_46

Early work concentrates on data authentication, i.e., how to efficiently verify that the server returns correct and complete results in response to its clients' queries [1, 2]. The following work focused on outsourcing encrypted data and associated difficult problems mainly having to do with efficient querying over encrypted domain [3–6]. Ateniese et al. [7] firstly proposed a model for provable data possession (PDP). In this model, data (represented as a file F) is preprocessed by the client, and the client produced metadata used for verification purposes. The file F and metadata were transferred to a server, and the client deleted the local copy of the file and stored the metadata locally. By randomly chosen a set of blocks from the file F, the client computed a challenge, and then the server computed a proof corresponding to the challenge, the client kept some secret information to verify server's proof later. The authors presented several variations of their scheme under different cryptographic assumptions.

Juels et al. [8] proposed a model for proofs of retrievability (PORs), which focusing on static archival storage of large files. The effectiveness rests largely on preprocessing steps before sending a file to the server: "sentinel" blocks are randomly inserted to detect corruption, which was encrypted to hide these sentinels, and error-correcting codes were used to recover from corruption. As expected, the error-correcting codes improved the error-resiliency of their system. Unfortunately, the number of queries was limited.

Curtmola et al. [9] proposed a robust model for provable data possession, which can recovered the corrupted data, Unfortunately, the server should retrieved a set of raw data with high I/O cost. Inspired by Juels [8], Bowers et al. [10] proposed an optimized application of Forward Error Correction (FEC) codes. The concept played a more prominent role as the remote storage community moves from the theoretical to the practical. The same pairing of erasure coding with data checking had been used by remote storage systems that distribute or replicate data among many servers [11]. Indeed, the use of FEC codes in the frameworks of [10] was not optimal and sacrificed the performance. Meanwhile, spot checking was probabilistic in nature and could not detect the corruption of small parts of the data (e.g., 1 byte).

In this paper, we propose a robust/recover model for provable data possession (RRPDP). We focus on the provable data possession (PDP) [7] framework, as being representative for remote data checking based on spot checking. More importantly, PDP allows easy and immediate integration with online codes to improve the data possession guarantee. In the model, a file F is first encoded using online code and PDP is applied on the encoded file $\widetilde{F}$ (instead of F). The PDP framework provides the ability to detect if the server corrupts a fraction of $\widetilde{F}$. Through the integration of online codes and PDP, a server can provides a robust/recover data possession guarantee.

The contribution is two-fold. Firstly, we propose a robust provable data possession that provides proof of possession, which can detects corrupted data with high probability. Secondly, through the careful integration of online codes and PDP, the server can recovers a tiny fraction of the file that has been deleted or corrupted.

Comparison with related works [7, 9, 12] was showed in Table 1. The important performance parameters of a PDP scheme include: Computation complexity, Communication complexity and Block access complexity. The features of our scheme (RRPDP) was listed in Table 1.

Table 1. Comparison with related work.

	[7]	[9]	[12]	RRPDP
Data possession	Yes	Yes	No	Yes
Client computation	$O(n)$	$O(c)$	$O(1)$	$O(c)$
Server computation	$O(1)$	$O(c)$	$O(c)$	$O(c)$
Communication complexity	$O(c)$	$O(c)$	$O(1)$	$O(c)$
Decoding complexity	–	$O(n\ln(n))$	–	$O(n)$
Robustness	No	Yes	No	Yes
Recoverability	No	Yes	No	Yes

n: the number of data block in F; c: the number of requested data block.

The rest of the paper is organized as follows. In Sect. 2, we describe the spot checking algorithm and the online codes. Section 3 introduces main construction and the notations used throughout the paper. In Sect. 4, we discuss its correctness, security, and the main results of our experiments. Conclusions and future work are presented in Sect. 5.

2 Spot Checking and Online Codes

In this section, we first describe the spot checking algorithm. We can prove data possession with high probability by verifying a small amount of blocks, which obviously reduces I/O cost. After formalizing the spot checking algorithm, an overview of online-code is given. We refer the reader to [13] for a more detailed exposition. In the end, a formally definitions of the robust/recover provable data possession is given.

2.1 Spot Checking

For an effective solution, the amount of block accesses at the server should be minimized, for the server may be involved in concurrent interactions with many clients. To improve the performance, our construction introduces a technique that allows the client makes a challenge by randomly choosing a subset of blocks, which named as "spot checking". As a result, we can prove data possession with high probability based on accessing a fraction of the file, which obviously improves the performance of scheme. Once the server deletes a fraction of the file, the client can detects server misconduct with high probability by verifying a small amount of blocks that independently of the total number of file blocks.

For example, a file with $n = 100000$ blocks. If the server (S) has deleted 1% of the blocks, then the client (C) can detects server misconduct with probability more than 99% by asking proof of possession for only 1000 randomly selected blocks.

2.2 Online Codes

We informally describe the online-code, which was firstly proposed by Maymounkov [13]. Online-code is parameterized by the two variables, k and δ, which defines the relations between the complexity and performance of the online-code. According to the author's suggestion, $k = 3$ and $\delta = 0.005$ [13]. The code algorithm consists of three phases.

Preprocess. Firstly, n data blocks are translated into a composite message by appending some auxiliary blocks. Each auxiliary block is the exclusive-OR of some number of the original message blocks.

The Encoding Process. From the composite message, we generate encoding blocks of size $(1 + k\delta)n$. Encoding blocks are named as *check blocks* below. A check block is the exclusive-OR of i data blocks, which are selected uniformly and independently from an ordered set of all composite blocks; i is the degree of one check block. The degree is chosen randomly according to a appropriate probability distribution $p = (p_1, p_2, \cdots p_L)$, such that degree i is chosen with probability p_i, where $\sum_1^L p_i = 1$. L is a constant.

The Decoding Process. For each check block e, all of whose data blocks are recovered, except for one. We call this data block m_x. We have $m_x = e \oplus m_1 \oplus \cdots \oplus m_{i-1}$, where $m_1, \ldots, m_{i-1}$ are the recovered data blocks that are corresponding to e. Apply this step until no more data blocks can be decoded. Upon receiving a certain number of check blocks some fraction of the composite message can be recovered. The composite message can be used to recover the original message. For more details in [13] (Fig. 1).

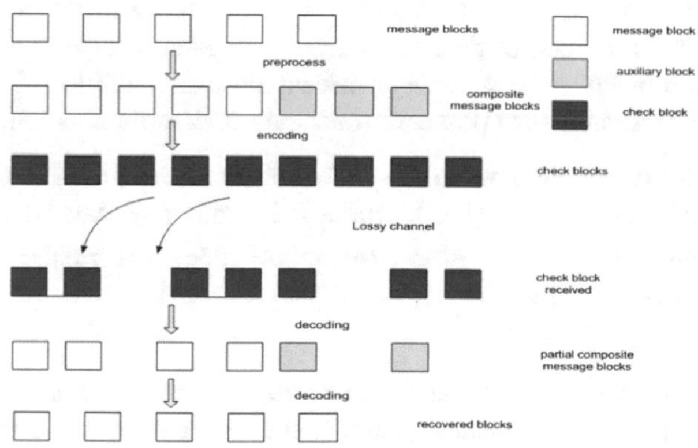

Fig. 1. High level view of online codes

The main result of online-code is described as follows, which useful for our construction, especially, for robustness.

Theorem 1. For a file F, and parameters k and δ, the probability that can recovers a $1 - \delta$ fraction of the original message from check blocks is $(\frac{\delta}{2})^{k+1}$.

2.3 RRPDP

In this paper, we propose a robust/recover version of provable data possession, which adapts from [7].

Definition 1. Robust/Recover provable data possession (RRPDP). An RRPDP scheme is a collection of six polynomial time algorithms (Encode, KeyGen, TagBlock, ChalGen, GenProof, CheckProof) such that:

Encode(F). Takes an original file F as input, and generate the encoded file $\widetilde{F}$.

KeyGen(1^λ). Takes the security parameter λ as input, output $(\mathrm{pk}, \mathrm{sk})$ such that $\mathrm{pk} = (N, g)$ and $\mathrm{sk} = (e, d, \upsilon)$.

TagBlock($\mathrm{pk}, \mathrm{sk}, m_i$). It takes as inputs a public key pk, a secret key sk and a file block m_i, and returns the verification metadata (T_{m_i}, W_i).

ChalGen(c, k_1). Takes c and k_1 as input, output a challenge set_{chal} corresponding to randomly chosen blocks of size c.

GenProof($c, \widetilde{F}$). Takes c and the encoded file $\tilde{F}$ as inputs, output $Ver = (T, \rho)$.

CheckProof($\mathrm{pk}, \mathrm{sk}, set_{chal}, Ver$). Run by the client to validate a proof of possession. It takes as inputs a public key pk, a secret key sk, a challenge set_{chal}, and a proof of possession $Ver = (T, \rho)$. It returns whether Ver is a correct proof of possession for the blocks determined by set_{chal}.

3 Construction

A robust/recover provable data possession scheme incorporates mechanisms for mitigating arbitrary amounts of data corruption. We define a notion of mitigation that includes the ability to both efficiently detects data corruption and recovers the corrupted blocks. When the corruption of data is detected, the client can recovers data in time. That is to say, a robust/recover scheme ensures that no data will lost. The definition of the robustness of provable data possession scheme is described as follows.

Definition 2. Robustness/Recoverability. A robust/recover provable data possession scheme RRPDP is a two-tuples (P,O), where P is a provable data possession scheme for a file $\widetilde{F}$, and O is a coding procedure(online code) that yield $\widetilde{F}$ applied on F. A provable data possession schemes is robust/recover, if it has the following two properties:

1. The client can detect w.h.p if the server corrupts more than a φ fraction of $\widetilde{F}$;
2. The client can recover the data F w.h.p if the server corrupt at most a φ fraction of $\widetilde{F}$.

The implementation of robust data possession guarantee for F can be described as follows: the file F is encoded using online code and PDP is then applied on the encoded file $\widetilde{F}$ (instead of F). When combined with an appropriate online code, a PDP scheme can provides a robust data possession guarantee for the encoded file $\widetilde{F}$. Compared with other rateless erasure codes, the decoding complexity of online codes is $O(n)$; meanwhile, the decoding complexity of R-S is $O(n^2)$.

3.1 Notations

In this section, we introduce some notations used throughout the work. λ is the security parameter, and κ is the seed of a pseudo-random permutation. Let $p = 2p^{'} + 1$ and $q = 2q^{'} + 1$ be safe primes and let $N = pq$ be an **RSA** modulus. A file F is consist of a finite ordered collection of n data blocks: $F = (m_1, m_2, \ldots, m_n)$. Let g be a generator of QR_N, which is the unique cyclic subgroup of $\mathbb{Z}_N^*$ of order $p^{'}q^{'}$. All exponentiations are performed modulo N (Table 2).

Table 2. Parameters.

Symbol	Suggested value	Description
λ	512	Security parameter
s		The seed of cryptographic hash functions
N	1024 bits	RSA module $N = pq$
p'	512 bits	Random prime
q'	512 bits	Random prime
p		RSA random prime $p = 2p' + 1$
q		RSA random prime $q = 2q^{'} + 1$
b	256 KB	Data block
φ	0.01	The ratio of corruption
Θ	0.005	The ratio of unrecoverable data
δ	.0005	Precoding parameter: Control redundancy
ε	0.01	Encoding parameter: Control redundancy
k	3	Precoding parameter
average	8	The average degree of check block
n		The number of data block of F
n'		The number of data block of $\widetilde{F}$

3.2 Algorithm

The Code Process. In this work, the file F is described as a $m \times n$ matrix. The jth column of matrix $\mathbf{b}_j$ is the jth data block $\mathbf{b}_j = (b_{1,j}, b_{2,j} \ldots b_{m,j})$, b_{ij} is the sub data block, $m = \lceil b/(1024-1) \rceil$, each element of matrix is less than N.

$$F = (\mathbf{b}_1, \mathbf{b}_2, \ldots, \mathbf{b}_n) = \begin{pmatrix} b_{11} & \cdots & b_{1n} \\ \vdots & \ddots & \vdots \\ b_{m1} & \cdots & b_{mn} \end{pmatrix} \quad (1)$$

As described above, the code algorithm consists of three phases. In preprocess phase, the client generates $n\delta k$ auxiliary data block from n original data block. Then, constructs a matrix $Aux_{n \times n\delta k}$, such that the sum of value in the same line is κ, and the sum of value in the same column is $1/\delta$. Secondly, the client generates another matrix $B_{n \times n'} = [I_{n \times n} | Aux_{n \times n\delta k}]$, and outputs a encoded file $F' = FB$. In the end, the client computes a vector $\mathbf{e} = F'\mathbf{u}$, where $\mathbf{u}$ is a random vector.

PDP. The PDP protocol can be divided into four steps. Firstly, the client preprocess a tag for each data block $\mathbf{b}_i$ of the file $\widetilde{F}$ and then store the file $\widetilde{F}$ and its tags $(T_{\mathbf{b}_i}, W_i)$ with a server S. Secondly, the client C generates a challenge by randomly choose a subset of file. Thirdly, the server S generates a proof of possession. Finally, the client C verifies the validity of the proof. It is worth noting that the client C stores on the server S a file $\widetilde{F}$ (instead of F), which is a finite ordered collection of $(1 + k\delta)n$ blocks: $\widetilde{F} = (m_1, m_2, \ldots, m_{(1+k\delta)n})$.

Table 3. The key generation of PDP algorithm.

Key generation
$\text{KeyGen}(1^\lambda)$. Take security parameter λ as input, output $\text{pk} = (\text{N,g})$ and $\text{sk} = (\text{e},d,v)$. Based on DSS-16, one generate two distinct safe primes p and q, and compute $p = 2p'+1$, $q = 2q'+1$, $N = pq$ be an **RSA** modulus. Secondly, choose randomly $a \overset{R}{\leftarrow} ¢^*_N$ such that $\gcd(a \pm 1, N) = 1$, and compute $g = a^2$. In the end, compute e and d such that $ed \equiv 1 \bmod (p'q')$, where e is a secret prime such that $e > \lambda$ and $d > \lambda$, g is a generator of QR_N, and $v \overset{R}{\leftarrow} \{1^\lambda\}$ .Output $\text{pk} = (\text{N,g})$ and $\text{sk} = (\text{e},d,v)$.

Table 3 describes the key generation process of PDP algorithm. In order to generate a couple of key with security guarantee, p and q are two safe primes.

Table 4 describes the pre-process stage of PDP algorithm, which is divided into two steps. The first step is the tag generation process, and the second step generates a challenge of chosen data block.

Table 4. The pre-process of PDP algorithm

Pre-process
TagBlock(pk,sk,**b**,i). Generate a tag, i is the indices of the chosen data block. One compute $W_i = v \| i$, and $T_{i,\mathbf{b}} = (h(W_i) \cdot g^{\mathbf{b}_i})^d \bmod N$. In order to compute the tag efficiently, we preprocess the order of generator $ord(g)$, then compute $\mathbf{b}_i^{'} = \mathbf{b}_i \bmod (org(g))$ .Output W_i and $T_{i,\mathbf{b}_i}$.
ChalGen. One generate set_{chal}=(c,k,g^s), then send it to a server. c is the number of the chosen data block, s is the seed of cryptographic hash functions, and k is the seed of PRF. Compute g^s=$g^s \bmod N$, and $s \stackrel{R}{\leftarrow} Z_N^*$.

Table 5 describes the proof-check process stage of PDP algorithm, which is divided into two steps. The first step is the generation of proof, and the second step is the checking of proof.

Table 5. The proof-check of PDP algorithm

Proof-check
GenProof. Take the public key $\mathrm{pk} = (\mathrm{N,g})$, the chosen subset of encoded file $\widetilde{F} = (\mathbf{b}_1^{'}, \mathbf{b}_2^{'}, \ldots, \mathbf{b}_{n\delta k}^{'})$, and the tags $\sum = (T_{1,\mathbf{b}_1^{'}}, T_{2,\mathbf{b}_2^{'}}, \ldots T_{n\delta k, \mathbf{b}_{n\delta k}^{'}})$ as inputs, generate the proof. S1: For $1 \le j \le \mathrm{c}$, compute $T = \prod_1^c T_{i_j, \mathbf{b}_{i_j}^{'}}$ S2: Compute $\rho = H\left(\mathrm{g}_s^{\sum_c^r b_j} \right) \bmod N$ S3: Output $VER = (T, \rho)$.
CheckProof. Take $VER = (T, \rho)$, and pk, sk, set_{chal} as inputs, the client verify the validity of the proof. S1:Compute $\eta = T^e$ S2:if $H(\eta^s \bmod N) = \rho$, correct else error

4 Analysis

4.1 Correctness

The scheme is based on the KEA1 assumption which was introduced by Damgard in [14]. In particular, Bellare and Palacio [15] provided a formulation of KEA1, and we adapt to work in the RSA setting.

Theorem 2. According to the KEA1-assumptions [14], the scheme can correctly verifies the validity of the server's proof by checking T and ρ.

Proof: The client pre-computes a tag $T_{\mathbf{b}_i} = (W_i \cdot g^{h(\mathbf{b}_i)})^d \bmod N$ for one data block $\mathbf{b}_i$ of the file $\widetilde{F}$, then stores the file $\widetilde{F}$ and its tags $(T_{\mathbf{b}_i}, W_i)$ at a server S. The client generates a challenge set_{chal} corresponding to a randomly selected block set of size c, and then the server S generates a proof $Ver = (T, \rho)$ in response to the challenge. In the end, the client C verifies the validity of proof. According to the KEA1-assumptions, the server S generates $i_j = \pi_{k_1}(j)$, and then compute $T = \prod_{j=1}^{c} T_{\mathbf{b}_{i_j}}$, $\rho = \prod_{j=1}^{c} W_{i_j} g^{h(\mathbf{b}_{i_1}) + \ldots + h(\mathbf{b}_{i_c})} \bmod N$. The proof is $Ver = (T, \rho)$. The client C has a secret key e, he/her can computes $\eta = T^e$.

4.2 Security

In this subsection, the security analysis of RRPDP algorithm was given. Intuitively speaking, an adversary cannot generate a correct proof without possessing the blocks corresponding to a selected challenge. Using a game, the security of our algorithm was intuitively described as follows.

Inception Phase: the client sends a public key pk to an adversary, and keeps the corresponding sk secret.

Generate metadata: the adversary can arbitrarily selects ℓ block $\mathbf{b}_i$ $(c \leq \ell \leq n)$, and request the metadata $T_{\mathbf{b}_i}$ corresponding to $\mathbf{b}_i$.

Generate a challenge: the client generates a challenge set_{chal}, and requests the adversary to provide a proof of chosen blocks $\mathbf{b}_1, \mathbf{b}_2, \ldots, \mathbf{b}_c$ determined by set_{chal}.

Generate a proof: the adversary computes a proof of possession for the chosen blocks $\mathbf{b}_1, \mathbf{b}_2, \ldots, \mathbf{b}_c$, and sends it to client.

Proof-check: the client computes CheckProof, if its output is "correct", the adversary wins the Game.

In our security definition, if the adversary is able to win the Game, means that the adversary can correctly executing GenProof without private key sk, or means that the adversary can guessing all the chosen blocks. The probability of generating a correct secret key sk corresponding to the given public key is negligible. On the other hands, the probability of guessing all the chosen blocks is less than $n!/((n-c)! \cdot c!)$. We refers the reader to [7] for a more generic and extraction-based security definition for PDP.

4.3 Experimental

We measure the performance based on our implementation of RRPDP in Linux. All experiments are conducted on an Intel 2.2 GHz systems. Algorithms use the NTL version 5.5.2 with a modulus N of size 1024 bits, and one data block of size 24 KB. For comparison, we also implemented the scheme of Ateniese et al. [7], which named as E-PDP.

Preprocess Time. In preparing a file for outsourced storage, the client firstly generates its tags for verify. In this experiment, preprocess time is the time of metadata generation, which does not include the time of loading data to the client and the time of transferring metadata to disk. Figure 2 shows that the pre-processing time as a function of file size for RRPDP and E-PDP. Compared with E-PDP, RRPDP exhibits slower pre-processing performance. In order to generate the per-block tags, RRPDP performs an exponentiation on each file block (including the auxiliary block) of file $\widetilde{F}$ instead of F.

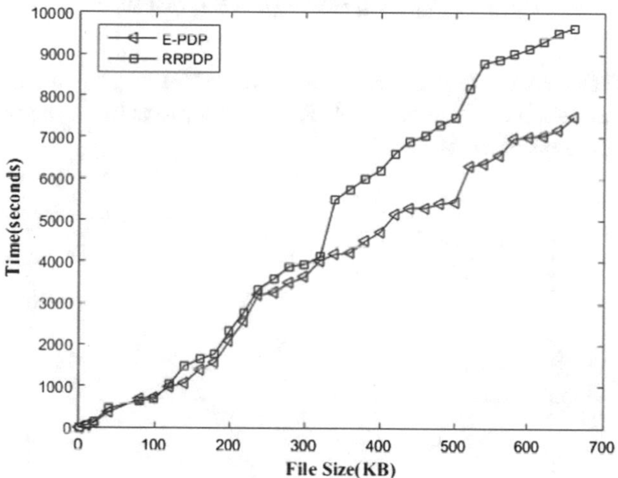

Fig. 2. The comparison of preprocess time

Challenged Time. Figure 3 shows the comparison of challenged time of E-PDP and RRPDP. Note the logarithmic scale. Challenged time includes the time to access the memory blocks that contain file data in cache. We restrict this experiment to blocks of 768 KB, which was suggested by E-PDP [7]. The experimental result shows that the efficiency of E-PDP is slightly higher than RRPDP. According to the result, it can be conclude that the integration of online code and PDP can makes the scheme robust/recover, which also sacrifices the performance of scheme, especially, the challenged time.

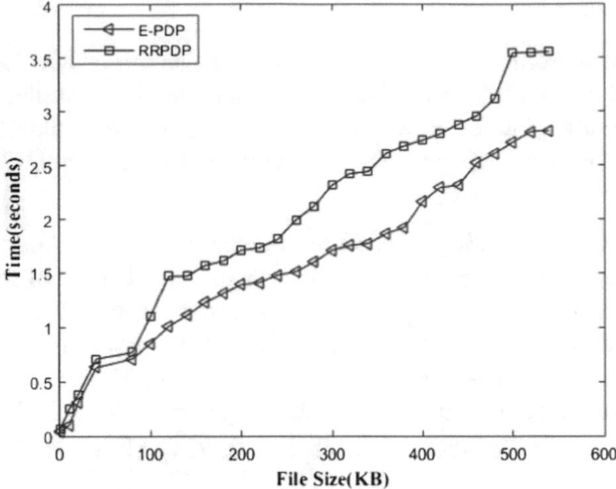

Fig. 3. The comparison of challenged time

The Trade-Off Between Pre-processing and Challenge. Figure 4 shows that the best balance occurs at data block size of 20–30 KB. In order to achieve the best balance, the suggested size of data block is 24 KB.

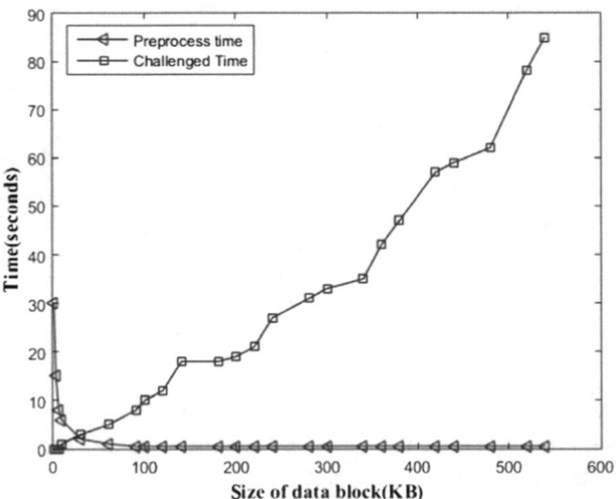

Fig. 4. The trade-off between pre-processing and challenge

Robustness. The robustness of the scheme we proposed including two aspects:

1. The detecting probability of corrupted data. φ: the lower limit of corrupted data. The scheme allows the server to correctly proves possession of selected blocks of $\widetilde{F}$. Spot checking greatly reduces the workload on the server, while still achieving

detection of server misbehavior with high probability. Assume S deletes $t = n\varphi$ blocks out of the n-block file $\widetilde{F}$. Let c be the number of different blocks for which C asks proof in one challenge. Let X be a discrete random variable, which is defined to be the number of data blocks chosen by C. Compute P_X, the probability that at least one of the blocks picked by C matches one of the blocks deleted by S. We have

$$P_X = P\{X \geq 1\} = 1 - P\{X = 0\} = \prod_{i=0}^{c-1} 1 - \frac{n-i-t}{n-i} \tag{2}$$

Since

$$\frac{n-i-t}{n-i} \geq \frac{n-i+1-t}{n-i-1} \tag{3}$$

We have

$$1 - \left(\frac{n-t}{n}\right)^c \leq P_X \leq 1 - \left(\frac{n-c+1-t}{n-c+1}\right)^c \tag{4}$$

2. The recovery probability of corrupted data. In this work, c blocks were randomly chosen from the encoded files $\widetilde{F}$. We assume that the ratio of corrupted data block is $\varphi(\varphi = 0.01)$. According to the previous parameters ($k = 3$, $\delta = 0.005$) and Theorem 1, we conclude that the probability that data can be restored is more than 0.9999 (Fig. 5).

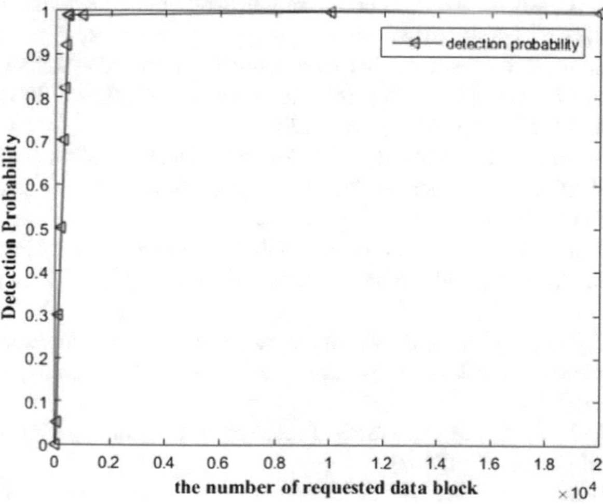

Fig. 5. The detecting probability of corrupted data

5 Conclusions

In this paper, a robust/recover provable data possession protocol was proposed. The main contribution of our scheme is the careful integration of online-code and PDP. By introduce the online codes, the protocol can verify data possession without access the whole data file, and can also recovery corrupted data. Experiments show that our scheme is practical to verify possession of large data sets. Given the practical importance of the online-code and the emergence of storage outsourcing service, we believe the study of robust/recover provable data possession to be important and motivated.

Acknowledgment. The authors would like to thank G. Ateniese, K. Bowers, Guomin Yang and Haiying Liu for sharing their deep insights about delegated computation and KEA1-assumption related matters. This work was supported in part by the National Natural Science Foundation of China (Grant No. 61701270) and Cooperation Foundation for The Youth Doctors of QiLu University of Technology (Shandong Academy of Sciences) (Grant No. 2017BSHZ008).

References

1. Devanbu, P., Gertz, M., Martel, C., et al.: Authentic third-party data publication. J. Comput. Secur. **11**(4), 291–314 (2003)
2. Mykletun, E., Narasimha, M., Tsudik, G.: Authentication and integrity in outsourced databases. ACM Trans. Storage **2**(2), 107–318 (2006)
3. Xue, L., Ni, J., Li, Y., Shen, J.: Provable data transfer from provable data possession and deletion in cloud storage. Comput. Stand. Interfaces **11**(Part 1), 46–54 (2017)
4. Clementine, G., Chen, R., Willy, S., Thomas, P.: Dynamic provable data possession protocols with public verifiability and data privacy. In: ISPEC 2017. LNCS, Melbourne, vol. 10701, pp. 485–505. Springer (2017)
5. Fan, X., Yang, G., Mu, Y., Yu, Y.: On indistinguishability in remote data integrity checking. Comput. J. **58**(4), 823–830 (2015)
6. Golle, P., Staddon, J., Waters, L.B.: Secure conjunctive keyword search over encrypted data. In: Proceedings of the Second International Conference ACNS 2004. LNCS, Yellow Mountain, vol. 3089, pp. 31–45. Springer (2004)
7. Ateniese, G., Burns, R., Curtmola, R., et al.: Provable data possession at untrusted stores. In: Proceedings of ACM Conference on Computer and Communications Security, pp. 598–609. ACM, New York (2007)
8. Juels, A., Kaliski, S.B.: PoRs: proofs of retrievability for large files. In: Proceedings of ACM Conference on Computer and Communications Security, pp. 584–597. ACM, New York (2007)
9. Curtmola, R., Khan, O., Burns, R.: Robust remote data checking. In: Proceedings of the 4th ACM International Workshop on Storage Security and Survivability, pp. 63–68. ACM, New York (2008)
10. Bowers, K., Juels, A., Oprea, A.: Proofs of retrievability: theory and implementation. ePrint Archive Report (2008/175) (2008)
11. Kotla, R., Alvisi, L., Dahlin, M.: SafeStore: a durable and practical storage system. In: USENIX Annual Technical Conference, Santa Clara, pp. 129–142 (2007)

12. Golle, P., Jarecki, S., Mironov, I.: Cryptographic primitives enforcing communication and storage complexity. In: Proceedings of the 6th International Conference on Financial Cryptography, Melbourne, pp. 120–135. Springer (2002)
13. Maymounkov, P.: Online codes. New York University, TR2003-883 (2003)
14. Damgard, I.: Towards practical public key systems secure against chosen ciphertext attacks. In: Proceedings of the 11th Annual International Cryptology Conference on Advances in Cryptology. LNCS, Santa Barbara, vol. 740, pp. 445–456. Springer (1992)
15. Bellare, M., Palacio, A.: The knowledge-of-exponent assumptions and 3-round zero-knowledge protocols. In: Proceedings of the CRYPTO 2004, LNCS, Santa Barbara, vol. 3152, pp. 273–289. Springer (2004)

A Provably Secure Scalable Revocable Identity-Based Signature Scheme Without Bilinear Pairings

Changji Wang[1,2], Hui Huang[3(✉)], and Yuan Yuan[4]

[1] School of Information Science and Technology, Guangdong University of Foreign Studies, Guangzhou 510006, China
[2] Shanghai Key Laboratory of Integrated Administration Technologies for Information Security, Shanghai 200240, China
[3] School of Statistics and Mathematics, Guangdong University of Finance and Economics, Guangzhou 510320, China
xxxhuanghui@163.com
[4] School of Mathematics and Statistics, Guangdong University of Foreign Studies, Guangzhou 510006, China

Abstract. Revocation functionality is essential for the practical deployment of identity-based cryptosystems because a user's private key may be corrupted by hacking or the period of a contract expires. Many researchers are focusing on revocable identity-based encryption scheme, while revocable identity-based signature scheme has received limited concentration. Recently, several revocable identity-based signature schemes have been proposed. However, these schemes are not scalable and are vulnerable to signing key exposure attack. In this paper, we first refine the security model of revocable identity-based signature scheme by considering the signing key exposure attack. Then, we propose a scalable revocable identity-based signature scheme with signing key exposure resistance. Finally, we prove the proposed scheme is existentially unforgeable against adaptively chosen message and identity attacks under the standard discrete logarithm assumption in the random oracle model.

Keywords: Revocable identity-based encryption · Revocable identity-based signature · KUNode algorithm · Random oracle model

1 Introduction

The concept of identity-based cryptography was first proposed by [1] to simplify certificate management in tradition public key infrastructure. In an identity-based cryptosystem, an entity's public key can be publicly computed from arbitrary strings that uniquely identifies the entity, such as a complete name or an email address. A trusted third party named as private key generator (PKG) generates the private key for the entity and sends it to the entity through a

C.-N. Yang et al. (Eds.): SICBS 2018, AISC 895, pp. 588–597, 2020.
https://doi.org/10.1007/978-3-030-16946-6_47

secure channel. Since Boneh and Franklin proposed the first practical and secure identity-based encryption (IBE) scheme [2], various IBE schemes, identity-based signature (IBS) schemes, identity-based key establishment protocols have been proposed and widely applied [3].

Key revocation mechanism is necessary for the practical deployment of any public key cryptosystems. For example, a user is no longer qualified for the public key, or a user's private key has been corrupted by hacking. In these cases, it is crucial for the cryptosystems to revoke the misbehaving or compromised user. In the tradition certificate setting, two popular solutions have been proposed, i.e., certificate revocation list and online certificate status protocol [4]. In the identity-based setting, however, key revocation is non-trivial because a user's identity is itself a public key, one can not simply change his public key, as this changes his identity as well.

Boneh and Franklin proposed a naive revocation method for IBE scheme (BF-RIBE for short) [2], which requires all users, regardless of whether their private keys have been exposed or not, have to regularly get in contact with the PKG, prove their identities and get new private keys. The PKG must be online for all such transactions, and a secure channel must be established between the PKG and each user to transmit the private key. Obliviously, this will lead to huge computation and communication overhead for the PKG that are linearly increased in the number of non-revoked users. Tseng and Tasi improved BF-RIBE scheme and proposed a RIBE scheme with a public channel (TT-RIBE for short) [13], where each user's decryption key consists of a fixed initial private key and a time update key, and the time update key is changed along with time period. The PKG periodically generates new time update keys and sends them to the non-revoked users via a public channel. Obviously, TT-RIBE scheme is still not scalable.

Boldyreva et al. proposed the first selective-ID secure, scalable RIBE scheme [6] by combining the complete subtree method [5] with a fuzzy IBE scheme [15], where the PKG is only required to perform logarithmic work in the number of users and no secure channel is required between users and the PKG. Then, Libert and Vergnaud [17] proposed the first adaptive-ID secure, scalable RIBE scheme by combining the complete subtree method with a black-box accountable authority IBE scheme [11]. Later, Seo and Emura considered the decryption key exposure attack in the scalable RIBE schemes, and proposed an adaptive-ID secure, scalable RIBE scheme with decryption key exposure resistance [12]. Recently, Wang et al. proposed an adaptive-ID secure, scalable RIBE scheme with constant size public system parameters and decryption key exposure resilience [19].

Although RIBE scheme has attracted researchers' attention in recent years, revocable identity-based signature (RIBS) scheme has received limited concentration. There are only a few RIBS schemes in the literature [7,8]. Existing RIBS schemes adopt the idea of TT-RIBE scheme, where each user's signing key consists of a fixed initial private key and a time update key, and the time update key is changed along with time period. For non-revoked users, the PKG periodically generates new time update keys and sends them to the non-revoked users via a public channel. Thus, they are not scalable because the PKG has to generate time update key for all non-revoked users in each time period. Furthermore, exiting security model for RIBS scheme is a natural extension of the security model for the ordinary IBS scheme. It allows an adversary to obtain any private keys of a chosen identity, the only one restriction is that if the adversary obtains a private key of the challenge identity ID^*, then ID^* should be revoked before the challenge time T^*. However, it does not consider signing key exposure attack, wherein an adversary may obtain a signer's private key d_{ID^*} from a compromised signing key $sk_{\mathsf{ID}^*,\mathsf{T}}$, thus the adversary can forge signatures by combining it with subsequent updated keys. It is important and challenging task to construct a scalable RIBS scheme with signing key exposure resistance.

In this paper, we first refine the security model of RIBS scheme by considering the signing key exposure attack. Then, we propose a scalable RIBS scheme with signing key exposure resistance by combining the complete subtree method with Galindo and Garcia's IBS scheme [9]. Finally, we proved the proposed RIBS scheme is existentially unforgeable against chosen message and identity attacks under the standard discrete logarithm assumption in the random oracle model.

The rest of the paper is organized as follows. In Sect. 2, we introduce some preliminary works necessary for our RIBS scheme. In Sect. 3, we give formal syntax and security definitions for the RIBS scheme. In Sect. 4, we describe our RIBS scheme without bilinear pairings. In Sect. 5, we present efficiency analysis and security proofs of our RIBS scheme. Finally, we conclude the paper in Sect. 6.

2 Preliminaries

We denote by $x \xleftarrow{\$} \mathbf{S}$ the operation of picking an element x uniformly at random from the set $\mathbf{S}$.

2.1 Complexity Assumptions

Let $\mathbf{G}_p$ be a prime p order group with a generator g. The Discrete Logarithm Problem (DLP) in $\mathbf{G}_p$ is defined as: given $g^x \in \mathbf{G}_p$ for unknown $x \xleftarrow{\$} \mathbf{Z}_p^*$, to compute x.

An adversary $\mathcal{A}$ has advantage ε in solving the DLP in $\mathbf{G}_p$ if

$$\Pr[\mathcal{A}(g, g^x) = x] \geq \varepsilon$$

where the probability is over the random choice of x in $\mathbf{Z}_p^*$, and the random bits consumed by $\mathcal{A}$. The DLP assumption in $\mathbf{G}_p$ holds if all probabilistic polynomial time (PPT) adversaries have at most a negligible advantage in solving the DLP.

2.2 KUNode Algorithm

Denote by Path(η) the set of nodes on the path from a leaf node η to the root node of the binary tree $\mathbb{T}$, by ζ_L and ζ_R the left and right child of a non-leaf node ζ, respectively. The KUNode algorithm takes as input a binary tree $\mathbb{T}$, revocation list $\mathbf{RL}$, and time period T, it outputs a set of nodes. Each user is assigned to a leaf node. If a user (assigned to η) is revoked on time period T, then $(\eta, \mathsf{T}) \in \mathbf{RL}$. The algorithm is described as follows [19].

$\mathsf{KUNode}(\mathbb{T}, \mathbf{RL}, \mathsf{T})$
  $\mathbf{X}, \mathbf{Y} \leftarrow \emptyset$
  $\forall(\eta_i, \mathsf{T}_i) \in \mathbf{RL}$,
    If $\mathsf{T}_i \leq \mathsf{T}$, then add $\text{Path}(\eta_i)$ to $\mathbf{X}$
  $\forall x \in \mathbf{X}$
    If $x_L \notin \mathbf{X}$, then add x_L to $\mathbf{Y}$
    If $x_R \notin \mathbf{X}$, then add x_R to $\mathbf{Y}$
  If $\mathbf{Y} = \emptyset$, then add `root` to $\mathbf{Y}$
  Output $\mathbf{Y}$

Upon registration, the PKG assigns a leaf node η of $\mathbb{T}$ to the user, and provides the user with a set of distinct private keys, wherein each private key is associated with a node on $\text{Path}(\eta)$. At time period T, the PKG broadcasts key updates for a set $\mathbf{Y} \subset \mathbb{T}$ of nodes which contains no ancestors of revoked users and precisely one ancestor of any non-revoked user.

3 Syntax and Security Definitions of RIBS Scheme

We denote by $\mathbf{I}$, $\mathbf{T}$ and $\mathbf{M}$ the identity space, the time space and the message space, respectively.

3.1 Syntax Definition of RIBS Scheme

A RIBS scheme consists of the following seven polynomial time algorithms:

- **Setup**: The setup algorithm takes as input a security parameter κ and a maximal number of users N. It outputs the public parameter mpk, the master secret key msk, the initial revocation list $\mathbf{RL} = \emptyset$, and an initial state st_0.
- **KeyGen**: The key generation algorithm takes as input the public parameter mpk, the master secret key msk, an identity $\mathsf{ID} \in \mathbf{I}$ and a state st. It outputs a private key d_{ID} of ID and an updated state st.
- **KeyUp**: The key update algorithm takes as input the public parameter mpk, the master secret key msk, the key update time $\mathsf{T} \in \mathbf{T}$, the current revocation list $\mathbf{RL}$, and state st. It outputs the update key ku_{T}.
- **SKG**: The signing key generation algorithm takes as input the public parameter mpk, a user's private key d_{ID} and the update key ku_{T}. It outputs a signing key $sk_{\mathsf{ID},\mathsf{T}}$ that is valid on time period T or a special symbol $\perp$ indicating that ID was revoked.
- **Sign**: The signing algorithm takes as input the public parameter mpk, a message m, and the signer's signing key $sk_{\mathsf{ID},\mathsf{T}}$. It outputs the signature $\sigma_{\mathsf{ID},\mathsf{T}}$. For simplicity and without loss of generality, we assume that $\sigma_{\mathsf{ID},\mathsf{T}}$ implicitly contains ID and T.

– **Verify**: The verification algorithm takes as input the public parameter mpk and a pair of message and signature $(m, \sigma_{\mathsf{ID},\mathsf{T}})$. It outputs 1 if $\sigma_{\mathsf{ID},\mathsf{T}}$ is valid signature on m for identity ID signed on time period T. Otherwise, it outputs 0.
– **Revoke**: The key revocation algorithm takes as input an identity ID to be revoked, a revocation time period T, the revocation list $\mathbf{RL}$, and the state st. It outputs an updated revocation list $\mathbf{RL}$.

The consistency condition requires that for a message $m \in \mathbf{M}$ and corresponding signature $\sigma_{\mathsf{ID},\mathsf{T}}$ which is signed by a non-revoked user with identity ID on time period T, the algorithm $\mathbf{Verify}(mpk, m, \sigma_{\mathsf{ID},\mathsf{T}})$ must return 1.

3.2 Security Definition of RIBS Scheme

The existentially unforgeable against adaptively chosen message and identity attacks (EUF-ID-CMA) for RIBS scheme is defined via the following game interacting between a forger $\mathcal{F}$ and a challenger $\mathcal{C}$.

– **Setup**: The challenger $\mathcal{C}$ runs $\mathbf{Setup}(1^{\kappa}, N) \to (msk, mpk)$. Then, $\mathcal{C}$ gives mpk to $\mathcal{F}$, while keeps msk secret.
– **Queries**: The forger $\mathcal{F}$ is allowed to issue the following queries adaptively.
  • *KeyGen Oracle*: $\mathcal{F}$ asks for the private key of any identity $\mathsf{ID} \in \mathbf{I}$, and $\mathcal{C}$ forwards the private key d_{ID} and update state st to $\mathcal{F}$ by running $\mathbf{KeyGen}(mpk, msk, \mathsf{ID}, \mathsf{st}) \to (d_{\mathsf{ID}}, \mathsf{st})$.
  • *KeyUp Oracle*: $\mathcal{F}$ asks for the update key on any time period $\mathsf{T} \in \mathbf{T}$, and $\mathcal{C}$ forwards the update key ku_{T} to $\mathcal{F}$ by running $\mathbf{KeyUp}(mpk, msk, \mathsf{T}, \mathbf{RL}, \mathsf{st}) \to ku_{\mathsf{T}}$.
  • *SKG Oracle*: $\mathcal{F}$ asks for the signing key of any identity $\mathsf{ID} \in \mathbf{I}$ on any time period $\mathsf{T} \in \mathbf{T}$, and $\mathcal{C}$ forwards the signing key $sk_{\mathsf{ID},\mathsf{T}}$ to $\mathcal{F}$ by sequentially running $\mathbf{KeyGen}(mpk, msk, \mathsf{ID}, \mathsf{st}) \to d_{\mathsf{ID}}$, $\mathbf{KeyUp}(mpk, msk, \mathsf{T}, \mathbf{RL}, \mathsf{st}) \to ku_{\mathsf{T}}$ and $\mathbf{SKG}(mpk, d_{\mathsf{ID}}, ku_{\mathsf{T}}) \to sk_{\mathsf{ID},\mathsf{T}}$.
  • *Sign Oracle*: $\mathcal{F}$ asks for the signature for any message $m \in \mathbf{M}$ of any identity $\mathsf{ID} \in \mathbf{I}$ on any time period $\mathsf{T} \in \mathbf{T}$, and $\mathcal{C}$ forwards the signature $\sigma_{\mathsf{ID},\mathsf{T}}$ to $\mathcal{F}$ by sequentially running $\mathbf{KeyGen}(mpk, msk, \mathsf{ID}, \mathsf{st}) \to d_{\mathsf{ID}}$, $\mathbf{KeyUp}(mpk, msk, \mathsf{T}, \mathbf{RL}, \mathsf{st}) \to ku_{\mathsf{T}}$, $\mathbf{SKG}(mpk, d_{\mathsf{ID}}, ku_{\mathsf{T}}) \to sk_{\mathsf{ID},\mathsf{T}}$ and $\mathbf{Sign}(mpk, m, sk_{\mathsf{ID},\mathsf{T}}) \to \sigma_{\mathsf{ID},\mathsf{T}}$.
  • *Revoke Oracle*: $\mathcal{F}$ asks for the revocation of any identity $\mathsf{ID} \in \mathbf{I}$ on any time period $\mathsf{T} \in \mathbf{T}$, and $\mathcal{C}$ forwards the updated revocation list $\mathbf{RL}$ to $\mathcal{F}$ by running $\mathbf{Revoke}(mpk, \mathsf{ID}, \mathsf{T}, \mathbf{RL}, \mathsf{st}) \to \mathbf{RL}$.

  The forger $\mathcal{F}$ is allowed to query above oracles with the following restrictions:
  • *SKG Oracle* cannot be queried on time period T before *KeyUp Oracle* was queried.
  • *KeyUp Oracle* and *Revoke Oracle* can be queried on time period which is greater than or equal to the time period of all previous queries.
  • *Revoke Oracle* cannot be queried on time period T if *KeyUp Oracle* was queried on T.

- **Forge**: At the end of the game, $\mathcal{F}$ outputs a forgery of a pair of message and signature $(\hat{m}, \sigma_{\hat{\mathsf{ID}},\hat{\mathsf{T}}})$ on behalf of an identity $\hat{\mathsf{ID}}$ on a time period $\hat{\mathsf{T}}$.

We say that $\mathcal{F}$ wins the experiment provided that the following conditions hold:

- **Verify**$(mpk, \hat{\mathsf{ID}}, \hat{\mathsf{T}}, \hat{m}, \sigma_{\hat{\mathsf{ID}},\hat{\mathsf{T}}}) = 1$.
- If **KeyGen**$(\hat{\mathsf{ID}})$ was queried, then **Revoke**$(\hat{\mathsf{ID}}, \mathsf{T})$ must be queried for $\mathsf{T} \leq \hat{\mathsf{T}}$.
- *SKG Oracle* cannot be queried with respect to $(\hat{\mathsf{ID}}, \hat{\mathsf{T}})$.
- *Sign Oracle* cannot be queried with respect to $(\hat{\mathsf{ID}}, \hat{\mathsf{T}}, \hat{m})$.

The forger's advantage ε in the above game is defined by the probability that $\mathcal{F}$ outputs a valid forgery. If there does not exist any PPT forger who has non-negligible advantage in the above game, we say that a RIBS scheme is EUF-ID-CMA secure.

4 Our RIBS Scheme

The proposed RIBS scheme is described as follows.

- **Setup**$(1^\kappa, N)$: The PKG performs the following steps.
  1. Choose a prime p order group $\mathbf{G}_p$ with a generator g, where $2^\kappa \leq p < 2^{\kappa+1}$.
  2. Choose three cryptographic hash functions $H_1 : \mathbf{G}_p \times \mathbf{I} \to \mathbf{Z}_p$, $H_2 : \mathbf{G}_p \times \mathbf{T} \to \mathbf{Z}_p$ and $H_3 : \mathbf{M} \times \mathbf{G}_p \times \mathbf{I} \times \mathbf{T} \to \mathbf{Z}_p$.
  3. Choose $x_1, x_2, x_3 \xleftarrow{\$} \mathbf{Z}_p$, compute $h_1 = g^{x_1}$, $h_2 = g^{x_2}$ and $h_3 = g^{x_3}$.
  4. Set the master secret key $msk = \langle x_1, x_2, x_3 \rangle$, the revocation list $\mathbf{RL} = \emptyset$, and initial state $\mathsf{st} = \mathbf{BT}$, where $\mathbf{BT}$ is a binary tree with N leaves.
  5. Publish the public parameters $mpk = \langle \mathbf{G}_p, g, p, h_1, h_2, h_3, H_1, H_2, H_3 \rangle$.
- **KeyGen**$(mpk, msk, \mathsf{ID}, \mathsf{st})$: The PKG chooses an unassigned leaf node η from $\mathbf{BT}$ at random, and stores ID in the node η. For each node $\theta \in \mathsf{Path}(\eta)$, the PKG performs the following steps.
  1. Recall c_θ if it was defined. Otherwise, choose $c_\theta \xleftarrow{\$} \mathbf{Z}_p$ and store $(c_\theta, \tilde{c}_\theta = x_3 - c_\theta)$ in the node θ.
  2. Choose $r_{1,\theta} \xleftarrow{\$} \mathbf{Z}_p$, compute $R_{1,\theta} = g^{r_{1,\theta}}$ and $y_{1,\theta} = r_{1,\theta} + x_1 \cdot H_1(R_{1,\theta}, \mathsf{ID}) + c_\theta$, and set $d_{\mathsf{ID},\theta} = (R_{1,\theta}, y_{1,\theta})$.
  3. Finally, the PKG assigns the user with identity ID the private key $d_{\mathsf{ID}} = \{(\theta, d_{\mathsf{ID},\theta})\}_{\theta \in \mathsf{Path}(\eta)}$.
- **KeyUp**$(mpk, msk, \mathbf{T}, \mathbf{RL}, \mathsf{st})$: The PKG parses $\mathsf{st} = \mathbf{BT}$. For each node $\theta \in \mathsf{KUNode}(\mathbf{BT}, \mathbf{RL}, \mathsf{T})$, the PKG performs the following steps.
  1. Retrieve $\tilde{c}_\theta$ (note that $\tilde{c}_\theta$ is always pre-defined in the **KeyGen** algorithm).
  2. Choose $r_{2,\theta} \xleftarrow{\$} \mathbf{Z}_p$, compute $R_{2,\theta} = g^{r_{2,\theta}}$ and $y_{2,\theta} = r_{2,\theta} + x_2 \cdot H_2(R_{2,\theta}, \mathsf{T}) + \tilde{c}_\theta$, and set $\mathsf{ku}_{\mathsf{T},\theta} = (R_{2,\theta}, y_{2,\theta})$.
  3. Return $ku_{\mathsf{T}} = \{(\theta, ku_{\mathsf{T},\theta})\}_{\theta \in \mathsf{KUNode}(\mathbf{BT},\mathbf{RL},\mathsf{T})}$.

- **SKG**$(mpk, d_{\mathsf{ID}}, ku_{\mathsf{T}})$: The PKG parses $ku_{\mathsf{T}} = \{(\theta, R_{2,\theta}, y_{2,\theta})\}_{\theta \in \mathsf{KUNode}(\mathbf{BT},\mathbf{RL},\mathsf{T})}$ and $d_{\mathsf{ID}} = \{(\theta, R_{1,\theta}, y_{1,\theta})\}_{\theta \in \mathsf{Path}(\eta)}$. If $\mathsf{Path}(\eta) \cap \mathsf{KUNode}(\mathbf{BT}, \mathbf{RL}, \mathsf{T}) = \emptyset$, it returns $\perp$. Otherwise, the PKG performs as follows.
  1. Choose $r_d \xleftarrow{\$} \mathbf{Z}_p$.
  2. Compute $h_1' = h_1^{r_d}$, $h_2' = h_2^{r_d}$, $h_3' = h_3^{r_d}$, $R_{1,\theta}' = R_{1,\theta}^{r_d}$, $R_{2,\theta}' = R_{2,\theta}^{r_d}$, $D_\theta = r_d(y_{1,\theta} + y_{2,\theta})$ for all $\theta \in \mathsf{Path}(\eta) \cap \mathsf{KUNode}(\mathbf{BT}, \mathbf{RL}, \mathsf{T})$.
  3. Return $sk_{\mathsf{ID},\mathsf{T}} = \langle h_1', h_2', h_3', \{D_\theta, R_{1,\theta}, R_{2,\theta}, R_{1,\theta}', R_{2,\theta}'\}_{\theta \in \mathsf{Path}(\eta) \cap \mathsf{KUNode}(\mathbf{BT},\mathbf{RL},\mathsf{T})}\rangle$.
- **Sign**$(mpk, sk_{\mathsf{ID},\mathsf{T}}, m)$: To sign a message $m \in \{0,1\}^*$ on the time period T, the signer generates a signature $\sigma_{\mathsf{ID},\mathsf{T}}$ on a message m using his signing key $sk_{\mathsf{ID},\mathsf{T}}$ by performing the following steps.
  1. Choose $a \xleftarrow{\$} \mathbf{Z}_p$.
  2. Compute $A = g^a$ and $b_\theta = a + D_\theta \cdot H_3(m, A, \mathsf{ID}, \mathsf{T})$ for $\theta \in \mathsf{Path}(\eta) \cap \mathsf{KUNode}(\mathbf{BT}, \mathbf{RL}, \mathsf{T})$.
  3. Return the signature $\sigma_{\mathsf{ID},\mathsf{T}}$, where

$$\sigma_{\mathsf{ID},\mathsf{T}} = \langle A, h_1', h_2', h_3', \{b_\theta, R_{1,\theta}, R_{2,\theta}, R_{1,\theta}', R_{2,\theta}'\}_{\theta \in \mathsf{Path}(\eta) \cap \mathsf{KUNode}(\mathbf{BT},\mathbf{RL},\mathsf{T})}\rangle$$

- **Verify**$(mpk, \sigma_{\mathsf{ID},\mathsf{T}}, m)$: Upon receiving a signature $\sigma_{\mathsf{ID},\mathsf{T}}$ on message m for ID on the time period T, a verifier checks the equation

$$g^{b_\theta} = A \cdot (R_{1,\theta}' R_{2,\theta}' h_1'^{H_1(\mathsf{ID}, R_{1,\theta})} h_2'^{H_2(\mathsf{T}, R_{2,\theta})} h_3')^{H_3(m, A, \mathsf{ID}, \mathsf{T})}$$

  holds or not. If it holds, the algorithm outputs 1; otherwise, the algorithm outputs 0.
- **Revoke**$(mpk, \mathsf{ID}, \mathsf{T}, \mathbf{RL}, \mathsf{st})$: Let η be the leaf node associated with ID. The PKG updates the revocation list by $\mathbf{RL} \leftarrow \mathbf{RL} \cup \{(\eta, \mathsf{T})\}$ and returns the updated revocation list.

5 Efficiency Analysis and Security Proof

We denote by P a computation of the bilinear pairing $\hat{e}(\mathbf{G}_1, \mathbf{G}_1) \rightarrow \mathbf{G}_T$, by M a scalar multiplication in $\mathbf{G}_1$, by H a map-to-point hash function, by E an exponentiation in $\mathbf{G}_p$, by N the maximum number of users in the system, and by r the number of revoked users. Compared to scalar multiplication in $\mathbf{G}_1$, map-to-point hash function, and exponentiation in $\mathbf{G}_p$, bilinear pairing is considered as the most expensive computational operations. The relative computation cost of a bilinear pairing is approximately twenty times higher than that of the scalar multiplication over elliptic curve group [21].

As shown in the Table 1, our RIBS scheme is more efficient and practical than existing RIBS schemes [7,8] in terms of verification cost and scalability.

Theorem 1. *Let $\mathcal{F}$ be an $(\epsilon, q_{H_1}, q_{H_2}, q_{H_3})$-forger against RIBS in the EUF-ID-CMA model, where q_{H_1}, q_{H_2} and q_{H_3} are denoted by the upper bound of the number of queries on H_1-oracle, H_2-oracle and H_3-oracle, respectively. If H_1, H_2 and H_3 are modeled as random oracles, we can construct either*

Table 1. Comparison of RIBS schemes.

	Scalability	Update cost	Sign cost	Verify cost	Intractable problems
[7]	No	$O(N-r)$	$2M$	$2P+M+2H$	CDHP
[8]	No	$O(N-r)$	E	$3E$	DLP
Our scheme	Yes	$O(r\log(N/r))$	E	$4E$	DLP

- *Algorithm $\mathcal{R}$ which ε-breaks the DLP, where*

$$\varepsilon \geq \frac{\epsilon^2}{q^4 q_{H_3}}$$

- *Algorithm $\mathcal{R}_1$ which ε_1-breaks the DLP, where*

$$\varepsilon_1 \geq \frac{\epsilon^2}{q^4(q_{H_3}+q_{H_2})^2} + \frac{\epsilon^2}{q^4(q_{H_3}+q_{H_2})^6}$$

- *Algorithm $\mathcal{R}_2$ which ε_2-breaks the DLP, where*

$$\varepsilon_2 \geq \frac{\epsilon^2}{q^4(q_{H_3}+q_{H_1})^2} + \frac{\epsilon^2}{q^4(q_{H_3}+q_{H_1})^6}$$

- *Algorithm $\mathcal{R}_3$ which ε_3-breaks the DLP, where*

$$\varepsilon_3 \geq \frac{4\epsilon^2}{q^4(q_{H_3}+q_{H_1}+q_{H_2})^2} + \frac{2\epsilon^2}{q^4(q_{H_3}+q_{H_1}+q_{H_2})^6}$$

Proof. We will give detailed security proof in the full version.

6 Conclusions

In this paper, we first refined the security model for revocable identity-based signature scheme by considering a realistic threat, called signing key exposure. Then, we proposed the first scalable revocable identity-based signature scheme without bilinear pairings in the new security model by combining the lightweight Galindo and Garcia's identity-based signature scheme with the complete tree method. Finally, we proved our proposed revocable identity-based signature scheme is existentially unforgeable against adaptively chosen message and identity attacks under the standard discrete logarithm assumption in the random oracle model.

Acknowledgments. This research is funded by Science and Technology Program of Guangzhou (Grant No. 201707010358) and Opening Project of Shanghai Key Laboratory of Integrated Administration Technologies for Information Security (Grant No. AGK201707).

References

1. Shamir, A.: Identity-based cryptosystems and signature schemes. In: Blakley, G.R., Chaum, D. (eds.) Advances in Cryptology - CRYPTO 1984. Lecture Notes in Computer Science, vol. 196, pp. 47–53. Springer, Heidelberg (1985)
2. Boneh, D., Franklin, M.: Identity-based encryption from the Weil pairing. In: Kilian, J. (ed.) Advances in Cryptology - CRYPTO 2001. Lecture Notes in Computer Science, vol. 2139, pp. 213–229. Springer, Heidelberg (2001)
3. Chen, L.Q.: An interpretation of identity-based cryptography. In: Aldini, A., Gorrieri, R. (eds.) Foundations of Security Analysis and Design IV. Lecture Notes in Computer Science, vol. 4677, pp. 183–208. Springer, Heidelberg (2007)
4. Zhang, P.F.: Tradeoffs in certificate revocation schemes. Comput. Commun. Rev. **33**(2), 103–112 (2003)
5. Naor, D., Naor, M., Lotspiech, J.: Revocation and tracing schemes for stateless receivers. In: Kilian, J. (ed.) Advances in Cryptology - CRYPTO 2001. Lecture Notes in Computer Science, vol. 2139, pp. 41–62. Springer, Heidelberg (2001)
6. Boldyreva, A., Goyal, V., Kumar, V.: Identity-based encryption with efficient revocation. In: Proceedings of the 15th ACM Conference on Computer and Communications Security - CCS 2008, pp. 417–426. ACM, New York (2008)
7. Wu, T.Y., Tsai, T.T., Tseng, Y.M.: Revocable ID-based signature scheme with batch verifications. In: 8th International Conference on Intelligent Information Hiding and Multimedia Signal Processing, pp. 49–54. IEEE (2012)
8. Sun, Y.X., Zhang, F.T., Shen, L.M., Deng, R.: Revocable identity-based signature without pairing. In: 5th International Conference on Intelligent Networking and Collaborative Systems, pp. 363–365. IEEE (2013)
9. Galindo, D., Garcia, F.D.: A Schnorr-like lightweight identity-based signature scheme. In: Preneel, B. (ed.) Progress in Cryptology - AFRICACRYPT 2009. Lecture Notes in Computer Science, vol. 5580, pp. 135–148. Springer, Heidelberg (2009)
10. Chatterjee, S., Kamath, C., Kumar, V.: Galindo-Garcia identity-based signature revisited. In: Kwon, T., Lee, M.K., Kwon, D. (eds.) International Conference on Information Security and Cryptology - ICISC 2012. Lecture Notes in Computer Science, vol. 7839, pp. 456–471. Springer, Heidelberg (2013)
11. Libert, B., Vergnaud, D.: Towards black-box accountable authority IBE with short ciphertexts and private keys. In: Jarecki, S., Tsudik, G. (eds.) Public Key Cryptography - PKC 2009. Lecture Notes in Computer Science, vol. 5443, pp. 235–255. Springer, Heidelberg (2009)
12. Seo, J.H., Emura, K.: Revocable identity-based encryption revisited: security model and construction. In: Kurosawa, K., Hanaoka, G. (eds.) Public Key Cryptography - PKC 2013. Lecture Notes in Computer Science, vol. 7778, pp. 216–234. Springer, Heidelberg (2013)
13. Tseng, Y.M., Tsai, T.T.: Efficient revocable ID-based encryption with a public channel. Comput. J. **55**(4), 475–486 (2012)
14. Boneh, D., Boyen, X.: Efficient selective-ID secure identity-based encryption without random oracles. In: Cachin, C., Camenisch, J.L. (eds.) Advances in Cryptology - EUROCRYPT 2004. Lecture Notes in Computer Science, vol. 3027, pp. 223–238. Springer, Heidelberg (2004)
15. Sahai, A., Waters, B.: Fuzzy identity-based encryption. In: Cramer, R. (ed.) Advances in Cryptology - EUROCRYPT 2005. Lecture Notes in Computer Science, vol. 3494, pp. 457–473. Springer, Heidelberg (2004)

16. Bellare, M., Neven, G.: Multi-signatures in the plain public-key model and a general forking lemma. In: Proceedings of the 13th ACM Conference on Computer and Communications Security - CCS 2006, pp. 390–399. ACM, New York (2006)
17. Libert, B., Vergnaud, D.: Adaptive-ID secure revocable identity-based encryption. In: Fischlin, M. (ed.) Topics in Cryptology - CT-RSA 2009. Lecture Notes in Computer Science, vol. 5473, pp. 1–15. Springer, Heidelberg (2009)
18. Lee, K., Lee, D.H., Park, J.H.: Efficient revocable identity-based encryption via subset difference methods. IACR Cryptology ePrint Archive, Report 2014/132 (2014). http://eprint.iacr.org/2014/132
19. Wang, C.J., Li, Y., Xia, X.N., Zheng, K.J.: An efficient and provable secure revocable identity-based encryption scheme. PLoS ONE **9**(9), 1–11 (2014)
20. Boldyreva, A., Palacio, A., Warinschi, B.: Secure proxy signature schemes for delegation of signing rights. J. Cryptology **25**(1), 57–115 (2012)
21. Chen, L., Cheng, Z., Smart, N.P.: Identity-based key agreement protocols from pairings. Int. J. Inf. Secur. **6**(4), 213–241 (2007)

Efficient Computation Method of Participants' Weights in Shamir's Secret Sharing

Long Li[1], Tianlong Gu[1], Liang Chang[1], and Jingjing Li[2(✉)]

[1] Guangxi Key Laboratory of Trusted Software, Guilin University of Electronic Technology, Guilin 541004, China

[2] School of Information and Communication, Guilin University of Electronic Technology, Guilin 541004, China

1402101004@mails.guet.edu.cn

Abstract. Shamir's secret sharing is an important means to realize data protection. Since participants in a specific weighted secret sharing scheme have different weights, these weights need to be computed and allocated in advance. In [15], a weight calculation method is proposed based on Karnaugh map, but this method has certain application bottlenecks and the algorithm efficiency is not efficient enough. To solve the above problems, this paper proposes a novel weight calculation method based on ordered binary decision diagrams. The new method can calculate weights for any number of participants, and the algorithm has lower space-time complexity. Theoretical analysis shows that the proposed scheme is feasible and effective.

Keywords: Secret sharing · Weight computation · Karnaugh map · Ordered binary decision diagram

1 Introduction

Secret sharing (SS) is one of the most important primitives in cryptography, which is used to keep the confidentiality of information. Its basic idea is to split a secret s into multiple shares in an appropriate way, and s can be reconstructed with partial or full shares. In reality, different share is distributed to different participants, and s can be reconstructed by eligible participant combination with the help of correct algorithm.

The concept of SS was proposed by Shamir [1] and Blakley [2] in 1979 separately. Besides, the concept of weighted SS (WSS) is proposed simultaneously in Shamir's publication. In view of its research significance and application value, SS has received extensive attention and many different types of SS schemes have been proposed. McEliece and Sarwate [3] construct a SS scheme by using error correction code. Asmuth and Bloom [4] propose a threshold SS scheme based on Chinese Remainder Theorem (CRT). Benaloh [5] first puts forward the concept of homomorphic SS. Ito et al. [6] propose the concept of general SS (GSS), and construct a GSS with monotone access structures by introducing the cumulative array technique. In a GSS, any qualified shareholder combination can reconstruct the secret, while any unqualified shareholder combination cannot reconstruct the secret.

C.-N. Yang et al. (Eds.): SICBS 2018, AISC 895, pp. 598–605, 2020.
https://doi.org/10.1007/978-3-030-16946-6_48

With the continuous development of information technology and the increasingly complex network scene, especially the rise of cloud computing, distributed storage, big data and other technologies, SS shows a broader application perspective. Because the above-mentioned simple SS schemes can no longer meet the demand, the majority of scientific researchers have conducted in-depth research, and proposed a number of more sophisticated SS schemes [7–9]. Aiming at dishonest distributors and participants, Chor et al. [10] proposed the concept of verifiable SS (VSS) for the first time. By adding an interactive authentication algorithm, the scheme allows shareholders to authenticate the authenticity of their secret shares, thus preventing dishonest secret dealer (SD) from cheating participants. In this VSS, only one secret can be shared at a time, while exceptional cases, such as multiple participants sharing multiple secrets, may be occur in practical applications. Therefore, He et al. [11] propose the concept of multistage SS (MSS). But the general implementation proposed in this paper cannot defense collision attack and can be used only once. Kumar et al. [12] propose a dynamic SS scheme for the data storage service of cloud computing. Besides, the scheme is secure and publicly verifiable. Based on Birkhoff interpolation, Traverso et al. [13] present a dynamic and verifiable hierarchical SS scheme. Without secret reconstruction, the scheme allows the change of shareholders, shares and access policies, and each shareholder can verify its share. To minimize the communication cost between participants during the secret reconstruction procedure, Huang et al. [14] prove a tight lower bound on the communication bandwidth, and construct schemes achieving the bound based on Shamir's SS.

Through analyzing the above literatures, it can be found that most of the research works focus on the secret splitting and reconstruction in SS, and these literatures split the secret based on assumed user shares. Obviously, the research on how to determine the secret shares held by users is few. According to the literatures we have inquired, only Harn et al. [15] propose a user share allocation scheme based on Karnaugh map. But the solution cannot be applied to situations where there are large numbers of users [16]. In addition, the spatial complexity of the Karnaugh map is proportional to the number of users, resulting in a large time and space cost of the relevant algorithms. Therefore, to overcome above defects, this paper proposes a novel share allocation scheme based on ordered binary decision diagram (OBDD). The new scheme can allocate secret shares among any number of users, and can reduce the time and space complexity of the algorithm and improve the allocation efficiency by subgraph isomorphism and structure reduction.

2 Preliminaries

2.1 Secret Sharing

Shamir's (*t*, *n*) SS. Shamir's (*t*, *n*) SS contains two parts: share generation and secret reconstruction.

(1) Share generation. *SD* select a finite field F_q, where $q \geq n$, then randomly select n distinct non-zero elements $x_1, x_2, \ldots, x_n$ and $t - 1$ elements $a_1, a_2, \ldots, a_{t-1}$ from

F_q. Construct a $(t-1)$-degree polynomial $f(x) = s + a_1x + a_2x^2 + \ldots + a_{t-1}x^{t-1}$, compute sub-secrets $f(x_i)\,(1 \le i \le n)$ and send $(x_i, f(x_i))$ to share holder p_i $(1 \le i \le n)$.

(2) Secret reconstruction. Without loss of generality, assuming the shareholders participated in share reconstruction are $\{p_1, p_2, \ldots, p_t\}$. $p_i(1 \le i \le t)$ provide sub-secrets $(x_i, f(x_i))\,(1 \le i \le t)$, and then recover polynomial $f(x) = \sum_{i=1}^{t} f(x_i) \prod_{j=1, j \ne i}^{t} \frac{x - x_j}{x_i - x_j}$ by means of Lagrange interpolation. Finally, secret s can be figured out by calculating $f(0)$.

Weighted SS (WSS). AWSS scheme can be represented by a triple (*k*, *n*, *w*), where *k* is the threshold value, *n* is the total number of users participating in SS, and *w* represents the user weight allocation function. Suppose the weight of participant $p_i(1 \le i \le n)$ is $w_i(1 \le i \le n)$. The specific meaning of WSS is that for a certain user combination P', if $\sum_{p_i \in P'} w_i \ge k, P'$ can recover the secret; otherwise, the secret recovery cannot be recovered.

Access Structure. Given a set of users $P = \{p_1, p_2, \ldots, p_n\}$, and define a set Γ, which is composed of subsets of *P*. If any subset contained in Γ can calculate the shared secret, Γ is called an access structure of *P*. If a subset is contained in Γ, it is called as authorization subset; otherwise it is called as unauthorized subset.

2.2 OBDD

OBDD. Each OBDD is a directed acyclic graph, which can be used to represent a Boolean function *f*.

An OBDD is composed of nodes and edges. A single *root* node is used to represent *f*. Each non-terminal node *u* can be represented by a tuple <*id*, *i*, *low*, *high*>. In the above tuple, *id* is the sequence number of *u*, *i* is the sequence number of the variable contained in *u*, and *low* (*high*) is the sequence number of *u*'s left (right) child node. The left (right) child is reached only when the value of the variable contained in *u* is 0 (1). Two terminal nodes are used to represent Boolean constants 0 and 1 respectively, and these two nodes have no edge.

Shannon's Expansion Theorem. Shannon's expansion theorem is the construction principle of OBDDs, which can be explicitly described by the following equation:

$$f(x_1, \ldots, x_i, \ldots, x_n) = x_i \cdot f(x_1, \ldots, 1, \ldots, x_n) + x_i' \cdot f(x_1, \ldots, 0, \ldots, x_n)$$

A unique OBDD will be finally generated if all variables, i.e., $x_i(1 \le i \le n)$, are expanded according to Shannon's expansion theorem. Obviously, different variable order will construct different OBDD. Therefore, when constructing an OBDD structure, the variable order must be pre-defined.

3 Weight Computation Method for Shamir's (*t*, *n*) WSS

An OBDD-based weight computation method for Shamir's (*t*, *n*) WSS will be proposed in this chapter. The basic idea of the mechanism is as follows: *t* can be calculated by comparing the weights of all authorized user combinations and unauthorized user combinations, and the weight of each user can be further obtained.

3.1 Construction of OBDD Access Structure

In practical applications, SS policies are often described by natural language or Boolean functions. In this section, an OBDD access structure will be proposed to depict SS policies, and the process used to generate OBDD Access structure will be explained. The basic components of OBDD access structure are nodes and edges. The structure of nodes is defined as follows:

```
struct node {
  unsigned short index; // number of variable
  unsigned short id; // number of node
  struct node *low; // 0-branch child
  struct node *high; // 1-branch child
  double val; //value of terminal node, 0 or 1
};
```

For any SS policy, after converting it into a Boolean function, the following algorithm can be used to construct the OBDD access structure.

Assume that there are *n* participants, denoted as $X = \{x_1, x_2, ..., x_n\}$, the SS policy is expressed by Boolean function $f(x_1, x_2, ..., x_n)$, and the variable order in OBDD is $\pi: x_1 < x_2 < ... < x_n$. Based on Shannon's expansion theorem, OBDD can be constructed by the following recursive algorithm (Table 1).

In an OBDD access structure, a node represents a user, its 1-branch indicates that the user participates in the secret reconstruction, and the 0-branch means the opposite. On a certain path between root and terminal node [1] (root->[1]), if variables appear strictly according to the variable order π, the path is called a valid path. Correspondingly, paths between root and terminal node [0] (root->[0]) are called invalid paths. Obviously, a valid path represents an authorized user combination (or a positive access instance), and an invalid path represents an unauthorized user combination (or a negative access instance).

3.2 Construction of Sub-OBDDs

According to the conditions of the secret reconstruction and the meaning of nodes and edges in OBDD access structure, it can be seen that all users who appear on a certain valid path and whose edge value is 1 can reconstruct secret together, and all users who appear on a certain invalid path and whose edge value is 1 cannot reconstruct secret. Therefore, a qualified sub-OBDD and an unqualified sub-OBDD can be constructed

Table 1. Algorithm used to obtain the OBDD corresponding to a Boolean function

Inputs: A Boolean function f and the maximum index of variables n
Output: The OBDD representation of f with the variable ordering π: $x_1<x_2<\ldots<x_n$

```
node* Construct-step(char *f, int i);
node* Construct(char *f) {
  int i = 1;
  node *u;
  Empty the computed-table;
  return (u = Construct-step(f, i));
}
node* Construct-step(char *f, int i) {
  static int id=2;
  node *u, *v0, *v1;
  if ((i > n)&( *f== "0")) return [0];
  else if ((i > n)&( *f== "1")) return [1];
  else {
    v0=Construct-step(f|x_i=0, i+1);
    v1=Construct-step(f|x_i=1, i+1);
    if computed-table entry (v0, v1,u) exists return u;
    Generate node u= < i, ++id, v0, v1>;
    Store (v0, v1,u) in computed-table;
    return u;
  }
}
```

based on OBDD access structure. A qualified sub-OBDD contains only all valid paths between root and $\boxed{1}$, representing all positive access instances. An unqualified sub-OBDD contains only invalid paths between all root and $\boxed{0}$, representing all negative access instances.

Construct Qualified Sub-OBDD. The rules for constructing qualified sub-OBDD based on OBDD access structure are as follows:

(1) Pruning rule: starting from terminal node $\boxed{0}$, cut off nodes and edges irrelevant to the valid path from the bottom up.
(2) Deletion rule 1: for node u, if $u.low = u.high$, connect all edges pointing to u to $u.low$, and then delete u.
(3) Deletion rule 2: for node u, if $u.low = 1$, connect all edges pointing to u to terminal node $\boxed{1}$, and then delete u.
(4) Merging rule: for node u and v, if $(u.i = v.i) \wedge (u.low = v.low) \wedge (u.high = v.high)$, connect all edges pointing to u to v, and then delete u.

The above rules need to be repeated until the nodes and edges in the structure no longer change.

Construct Unqualified Sub-OBDD. The rules for constructing unqualified sub-OBDD based on OBDD access structure are as follows:

(1) Pruning rule: starting from terminal node $\boxed{0}$, cut off nodes and edges irrelevant to the valid path from the bottom up.
(2) Deletion rule 1: for node *u*, if *u.low* = *u.high*, delete *u.low*.
(3) Deletion rule 2: for node *u*, if *u.low* = 0, connect all edges pointing to *u* to terminal node $\boxed{0}$, and then delete *u*.
(4) Merging rule: for node *u* and *v*, if $(u.i = v.i) \wedge (u.low = v.low) \wedge (u.high = v.high)$, connect all edges pointing to *u* to *v*, and then delete *u*.

The above rules need to be repeated until the nodes and edges in the structure no longer change.

3.3 Extract User Combinations from Sub-OBDDs

Before extracting user combinations, the following definitions must be first given.

Definition (Minimum qualified user combination): the smallest user combination that can recover the shared secret, denoted as *Min*. The set composed of all *Min* is denoted as *MIN*.

Definition (Maximal unqualified user combination): the biggest user combination that cannot recover the shared secret, denoted as *Max*. The set composed of all *Max* is denoted as *MAX*.

Since sub-OBDDs represent access instances, they can be used to extract *MIN* and *MAX* respectively.

Extract a *MIN* from a Qualified Sub-OBDD. The steps of extracting a *MIN* from a qualified sub-OBDD are as follows:

Step 1, obtain all valid paths by traversing the qualified sub-OBDD. Assuming these valid paths are denoted as $V = \{V_0, V_1, \ldots, V_{s-1}\}$.
Step 2, read (x_j, $edge_j$) and store x_j in Min_i if $edge_j = 1$. After all tuples contained in valid path V_i are checked, the minimum authorized user composition corresponding to V_i is obtained, i.e., $Min_i = \{x_j | (x_j, edge_j) \in V_i \,\text{and}\, edge_j = 1\}$.
Step 3, obtain the *MIN* according to *V*, i.e., $MIN = \{Min_0, Min_1, \ldots, Min_{s-1}\}$.

Extract a *MAX* from an Unqualified Sub-OBDD. The steps of extracting a *MAX* from an unqualified sub-OBDD are as follows:

Step 1, obtain all invalid paths by traversing the unqualified sub-OBDD. Assuming these invalid paths are denoted as $F = \{F_0, F_1, \ldots, F_{f-1}\}$.
Step 2, read (x_j, $edge_j$) and store x_j in Max_i if $edge_j = 1$. After all tuples contained in invalid path F_i are checked, the maximum unauthorized user composition corresponding to F_i is obtained, i.e., $Max_i = \{x_j | (x_j, edge_j) \in F_i \,\text{and}\, edge_j = 1\}$.
Step 3, obtain the *MAX* according to *F*, i.e., $MAX = \{Max_0, Max_1, \ldots, Max_{s-1}\}$.

3.4 Calculate Each Participant's Weight

Before calculation, the following notations are first defined: the weight of user x_i is s_i, and $s_i > 0$; for Min_i and Max_j, $|Min_i|$ and $|Max_j|$ are used to represent the weights owned by all users contained in each combination.

In a WSS, any authorized user combination can recover the secret, while any unauthorized user combination cannot recover the secret. Assuming that Shamir (t, n) WSS is adopted, it must satisfy (1) $\forall Min_i \in MIN$, $|Min_i| \geq t$; (2) $\forall Max_j \in MAX$, $|Max_j| < t$. Therefore, the following polynomials can be used to express the conditions that the weights calculation needs to be satisfied.

$$|Min_i| > |Max_j|, \text{ where } Min_i \in MIN \text{ and } Max_j \in MAX.$$

Furthermore, the specific weights of each participant can be obtained by solving the above polynomials, and t can be set as the minimum value of $|Min_i|$.

4 Analysis and Comparison

Before making quantitative comparisons, it is assumed that n users participate in SS. Compared with Karnaugh map-based weight computation method, the newly proposed OBDD-based weight computation method has the following advantages:

(1) When more than 6 variables is contained [16], the automatic construction and simplification of Karnaugh map will become very complicated. But the construction and reduction of OBDD are not limited by the number of variables.

(2) The spatial complexity of Karnaugh map is 2^n, that is, it grows exponentially with the number of users; while the existence of subgraph isomorphism in OBDD makes its spatial complexity significantly reduced, and 2^n will be reached only in the worst case.

Obviously, in large-scale distributed systems, especially when there are a large number of users or the SS policies are complex, it is complicated if Karnaugh map-based weight computation method is adopted to compute users' weights, and even the computation cannot be completed. By contrast, the OBDD-based weight computation method can complete the weight computation in any case, and has better efficiency performance.

5 Conclusion and Future Work

In this paper, a new weight computation scheme is proposed based on OBDD. By comparing with Karnaugh map-based weight computation scheme, the effectiveness of the new scheme is proved. In the follow-up work, actual comparison experiment will be conducted to analyze the spatial and temporal complexity of schemes quantitatively. Besides, with the advantage that OBDDs can perform computations directly, dynamic weight computation scheme that supports policy update will be further studied.

Acknowledgements. This work was supported in part by the Natural Science Foundation of China (U1711263, U1501252, 11603041), in part by the Key Research and Development Program of Guangxi (AC16380014, AA17202048, AA17202033), and in part by the Natural Science Foundation of Guangxi Province (2017GX NSFAA198283).

References

1. Shamir, A.: How to share a secret. Commun. ACM **22**(11), 612–613 (1979)
2. Blakley, G.R.: Safeguarding cryptographic keys. In: Proceedings of AFIPS 1979 National Computer Conference, pp. 313–317. AFIPS Press, Montvale (1979)
3. McEliece, R.J., Sarwate, D.V.: On sharing secrets and reed solomon codes. Commun. ACM **24**(8), 583–584 (1981)
4. Asmuth, A., Bloom, J.: A modular approach to key safeguarding. IEEE Trans. Inf. Theory **30**(2), 208–210 (1983)
5. Benaloh, J.C.: Secret sharing homomorphisms: keeping shares of a secret. In: Proceedings of CRYPTO 1986, pp. 412–417. Springer, Berlin (1986)
6. Ito, M., Saito, A., Nishizeki, T.: Secret sharing scheme realizing general access structure. In: Proceedings of the IEEE Global Telecommunications Conference, pp. 99–102. IEEE Press, Globecom (1987)
7. Hsu, C.F., Cheng, Q., Tang, X., Zeng, B.: An ideal multi-secret sharing scheme based on MSP. Inf. Sci. **181**(7), 1403–1409 (2011)
8. Liu, Y., Zhang, F., Zhang, J.: Attacks to some verifiable multi-secret sharing schemes and two improved schemes. Inf. Sci. **329**(1), 524–539 (2016)
9. Lu, H.C., Fu, H.L.: New bounds on the average information rate of secret-sharing schemes for graph-based weighted threshold access structures. Inf. Sci. **240**(11), 83–94 (2013)
10. Chor, B., Goldwasser, S., Micali, S., Awerbuch, B.: Verifiable secret sharing and achieving simultaneity in the presence of faults. In: Proceedings of 26 IEEE Symposium on Foundations of Computer Science, Portland, pp. 383–395. IEEE Press (1985)
11. He, J., Dawson, E.: Multistage secret sharing based on one-way function. Electron. Lett. **30** (19), 1591–1592 (1994)
12. Kumar, P.S., Ashok, M.S., Subramanian, R.: A publicly verifiable dynamic secret sharing protocol for secure and dependable data storage in cloud computing. Int. J. Cloud Appl. Comput. **2**(3), 1–25 (2017)
13. Traverso, G,. Demirel, D., Buchmann, J.: Dynamic and verifiable hierarchical secret sharing. In: Proceedings on International Conference on Information Theoretic Security, pp. 24–43. Springer, Cham (2016)
14. Huang, W., Langberg, M., Kliewer, J., Bruck, J.: Communication efficient secret sharing. IEEE Trans. Inf. Theory **62**(12), 7195–7206 (2016)
15. Harn, L., Hsu, C., Zhang, M., He, T., Zhang, M.: Realizing secret sharing with general access structure. Inf. Sci. **367–368**, 209–220 (2016)
16. Prasad, V.C.: Generalized Karnaugh map method for Boolean functions of many variables. IETE J. Educ. **58**(1), 1–9 (2017)

An Identity-Set-Based Provable Data Possession Scheme

Changlu Lin[1,2(✉)], Fucai Luo[1,3,4], Jinglong Luo[1,2], and Yali Liu[5]

[1] College of Mathematics and Informatics, Fujian Normal University, Fuzhou 350117, China
cllin@fjnu.edu.cn

[2] Fujian Provincial Key Laboratory of Network Security and Cryptology, Fujian Normal University, Fuzhou 350007, China

[3] School of Cyber Security, University of Chinese Academy of Sciences, Beijing 100093, China

[4] State Key Laboratory of Information Security, Institute of Information Engineering, Chinese Academy of Sciences, Beijing 100093, China
luofucai@iie.ac.cn

[5] College of Computer Science and Technology, Jiangsu Normal University, Xuzhou 221116, China
liuyali@jsnu.edu.cn

Abstract. Provable Data Possession (PDP) scheme is a cryptographic protocol that allows the users to check the availability and integrity of outsourced data on cloud storage servers (CSS) which are not completely trusted. Most of the PDP schemes are publicly verifiable, while private verification is necessary in some applications to prevent the disclosure of any relevant information. In this work, we consider the scenario of allowing the cloud user to determine whether the clients can use the data or not through controlling their ability to verify the proof correctly, and propose an identity-set-based PDP (ISB-PDP) scheme. Our ISB-PDP scheme is not only proved to be secure under the hardness of the computational Diffie-Hellman (CDH) problem, but also select/cut some clients (verifiers) *dynamically* to check the proof correctly according to the needs of the cloud user.

Keywords: Cloud storage server · Provable Data Possession (PDP) · Identity-set-based · CDH problem

1 Introduction

In recent years, outsourcing data to remote cloud storage server (CSS) is a growing trend for numerous customers and organizations, since it alleviates the burden of local data storage and maintenance. Essentially, it takes data storage as a service, which allows universal data access with independent geographical locations. Thus, it avoids the capital expenditure on hardware, software, and personnel maintenance, and so on. However, due to the fact that the cloud

C.-N. Yang et al. (Eds.): SICBS 2018, AISC 895, pp. 606–618, 2020.
https://doi.org/10.1007/978-3-030-16946-6_49

users (data owners) are no longer physically possess their sensitive data, it raises new formidable and challenging tasks related to data confidentiality, integrity and availability protection in CSS. Moreover, the remote CSS are not trusted completely it might delete or modify some data and it conceal their improper behaviors from the cloud users for their own benefits (might just for saving storage space). It might not store all data quickly as required by the cloud users, i.e., place it on hardware or other off-line media and thus using less fast storage. Furthermore, the cloud infrastructures are subject to wide range of internal and external security threats. As a result, many users are still hesitate to use cloud storage due to these concerns. Hence, it is a crucial demand of cloud users and clients to get strong evidences that the CSS still possess the outsourced data not being tampered or partially deleted over time. However, it is impractical for the cloud users and clients to download all stored data in order to validate its integrity, since this would require immense communication overheads across the network. Therefore, it is essential to design an efficient mechanism to verify the integrity of outsourced data with minimum computation, communication and storage overheads.

Related Works. In order to solve the above issues, the cryptography and security researchers have proposed two basic approaches "Proofs of Retrievability (PORs)" [4,7,9,17–19] and "Provable Data Possession (PDP)" [1,5,6,8,10–16] schemes under different systems and security models. The PDP scheme is a "challenge-response" protocol that the cloud users pre-process the file to generate some metadata and signatures which will be sent later to CSS. Then, upon receiving the challenge (query a random set of outsourced data) from a verifier (client), the CSS calculates a proof as the response. After that, the verifier checks the PDP (proof) and outputs "success" if it passes the verification, and outputs "failure" otherwise. The PORs protocol is a variant of PDP, but has stronger security than the regular PDP notion. With PoRs protocol, the CSS can not only prove to a verifier that he is actually storing all of data, but also prove that the users can retrieve them at any time. The benefit of a PORs protocol over simple transmission of file is efficiency, the response can be highly compact (tens of bytes), and the verifier can complete the proof by using a small fraction of the file, as a stand alone tool for testing file retrievability against a single server, a PORs protocol is of limited value. However, detecting that a file is corrupted is not helpful if the file is irretrievable and the client has no recourse. Thus, PORs protocols are mainly useful in environments where file is distributed across multiple systems, such as independent storage services. Moreover, most PORs protocols are less efficient than their PDP counterparts, based on an observation made by Barsoum *et al.* [2]. While most of the PDP schemes are probability proof techniques (a few of PDP schemes provide deterministic proof in which all blocks of file have been checked) for verifying availability and integrity of outsourced data without downloading the whole data.

Considering the role of the verifiers in the model, all of the above PDP schemes can be divided into two categories: *the private verification* and *the public verification.* Most of the PDP schemes are public verifiability while some

PDP schemes [1,14] provide both the private verification and the public verification. However, the public verification is undesirable in many circumstances, while the private verification is necessary in some applications to prevent the disclosure of relevant information. For instance, the cloud users (data owners) will be restricted to access the Internet, e.g., on the ocean-going vessel, or on the warfare. In these situations, the cloud users cannot perform the remote data availability and integrity checking, while the cloud users hope that some limited people can verify the PDP correctly. For instance, the delegable provable data possession was proposed in [10] for the remote data in the clouds. The proxy provable data possession was proposed by wang [13] in Public Clouds. The provable multiple-replica dynamic data possession was proposed by Hou *et al.* [6] to support big data storage in cloud computing. Wang *et al.* [14] and Wang *et al.* [12] proposed the identity-based remote data possession checking in public clouds. Nevertheless, these PDP schemes only considered minor clients (verifiers) that they cannot satisfy the case of multiple clients. In particular, they are not available in the following scenario:

For data stored in CSS, such as some installation packages, some softwares, which are owned by a company or an organization treated as a cloud user, and are downloaded to be used by the clients (data users of the company or organization) frequently. The clients must be able to verify the data integrity before downloading it. Otherwise, it has a risk due to the nondeterminacy of the data integrity, thus, the clients will not use the data if it cannot been verified correctly. It turns out that verifying the PDP correctly means the data is safely used, vise versa. Therefore, under the circumstance, the cloud user determines whether the clients can use the data or not through controlling their ability to verify the PDP correctly.

To address this issue, traditional methods usually require the cloud users (data owner) to deliver decryption keys to the clients (authorized data users). These methods, however, usually involve complicated key management and huge overhead on cloud users. Ren *et al.* [8] proposed the attribute-based PDP in public cloud storage which utilizes attribute-based signature to construct the homomorphic authenticator. In their scheme, the homomorphic authenticator contains an attribute strategy, only the clients, who satisfied the strategy, can check the PDP (proof) correctly. In spite of the cloud user can adaptively choose some clients to verify the PDP in Ren *et al.* [8], in which some clients may be authorized to use some data so as to verifying the corresponding PDP, some other clients may be authorized to use other data so as to verifying the corresponding PDP. However, the scheme in Ren *et al.* [8] is inefficiency, in case of the number of clients are enormous, and only minority of clients are forbidden to use some data result in failing to verify the corresponding PDP. Since the clients are inefficiently revoked or cut from verifying the PDP correctly in Ren *et al.* [8]. Specially, in some cases, the cloud user hopes just a certain some specified people can verify, in which Ren *et al.* [8] can achieve efficiently, but fail to efficiently realize when the cloud user hopes all people except for a certain some designated people.

Our Contributions. To solve the above-mentioned problems, we consider designing an efficient system in which the cloud user can dynamically select/cut the verifiers (clients) to check the PDP. Therefore, we propose an identity-set-based provable data possession (ISB-PDP) scheme that can dynamically select/cut some clients to check the PDP correctly according to the needs of the cloud users. Furthermore, we prove that the proposed ISB-PDP is secure under the security of SBE (for Identity-Set-Based Broadcast Encryption) in Zhu *et al.*'s [20], and the hardness assumption of the computational Diffie-Hellman (CDH) problem.

1.1 Organization

The rest of the paper is organized as follows. The system model, the bilinear pairings and the hardness assumption will be given in Sect. 2. In Sect. 3, we provide the definition of our ISB-PDP. In Sect. 4, we first recall the identity-set-based encryption (SBE), and then present our ISB-PDP scheme. We analyze the security of the proposed ISB-PDP scheme in Sect. 5. Finally, Sect. 6 concludes the paper.

2 Preliminaries

2.1 The System Model

For the sake of simplicity, this paper considers a cloud storage server, while it can be extended to multi-cloud storage servers as well. The proposed ISB-PDP involves three different entities as illustrated in Fig. 1:

- The cloud user, who has a huge number of data files to be stored in clouds and is entitled to access and manipulate stored data, he/she can be a company and an organization, etc., who has a large number of clients (data users).
- The clients, who are data users (authorized users) but are not entitled to access and manipulate stored data, while they maybe can and have to check the availability and integrity of the data before using it (some of them maybe revoked by cloud user so that they will fail to check).
- The cloud storage server (CSS), which provides data storage services and has enough storage space and significant computation resources but is manipulated by manpower such as the CSS providers.

The paper assumes the CSS is untrusted, since it is manipulated by manpower such as the CSS provider who is omitted in our paper. The paper also assumes the cloud user generates the private keys of the clients and the corresponding public keys (trusted third party can be entrusted to do this work), then the cloud user entitles or revokes the permission of the clients to verify the PDP by operating our proposed ISB-PDP. Specially, the cloud user generates the private keys of the clients and the corresponding public keys, then and computes the block-tag pairs of data files. Finally, the cloud user posts the public keys in the public cloud, and sends the private keys to the clients and deliveries the block-tag pairs to the CSS. The clients obtain a PDP (proof) after querying the CSS, then the clients use their private keys and the public keys to check the PDP.

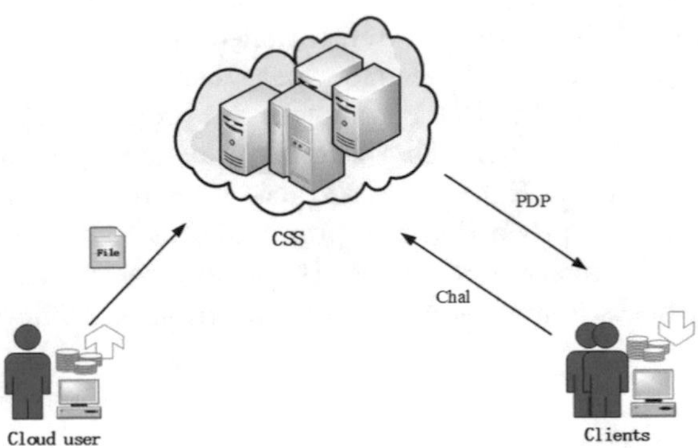

Fig. 1. The PDP model

2.2 Bilinear Pairings

Let $\mathbb{S} = (p, \mathbb{G}_1, \mathbb{G}_2, \mathbb{G}_T, e)$ be a bilinear map group system, where $\mathbb{G}_1$, $\mathbb{G}_2$ and $\mathbb{G}_T$ are multiplicative cyclic groups of prime order p, and $e : \mathbb{G}_1 \times \mathbb{G}_2 \longrightarrow \mathbb{G}_T$ is a bilinear map. The bilinear maps e has the following properties: for any $G \in \mathbb{G}_1$, $H \in \mathbb{G}_2$ and all $a, b \in \mathbb{Z}_p$, we have

- bilinearity: $e(aG, bH) = e(G, H)^{ab}$;
- non-degeneracy: $e(G, H) \neq 1$ unless G or $H = 1$;
- computability: $e(G, H)$ is efficiently computable.

Definition 1 (CDH problem). *Given g, g^x, $h \in \mathbb{G}$ for some group $\mathbb{G}$ and $x \in \mathbb{Z}_p$, it is to compute h^x.*

3 Definition of ISB-PDP

In this section, we present the formal definition of the ISB-PDP scheme.

Definition 2 (ISB-PDP). *An IBS-PDP protocol is a collection of two algorithms **(SBE-KeyGen, TagGen)** and two interactive proof system **(GenProof, CheckProof)**, which are described in detail below.*

***SBE-KeyGen**($\mathbb{S}$). Input a bilinear map group system $\mathbb{S}$, it outputs the system public parameters params, the public keys pk and the secret keys sk.*

***TagGen**(sk, pk, F). Input the private keys sk and the public keys pk, the block m_i (the cloud user splits file F into n blocks, i.e., $F = (m_1, m_2, \cdots, m_n)$, where $m_i \in \mathbb{Z}_p^*$, it outputs the tuple $\{\Phi_i, (m_i, T_{m_i})\}$, where Φ_i denotes the $i-th$ record of metadata, (m_i, T_{m_i}) denotes the $i-th$ block-tag pair. All the metadata $\{\Phi_i\}$ denote as Φ and all the block-tag pairs $\{(m_i, T_{m_i}), i \in [1, n]\}$ denote as Σ.*

***GenProof**($pk, chal, \Sigma$). Input the public keys pk, a chal from the query of a client c_k and Σ, it outputs a PDP Γ as the response to the query.*

***CheckProof**($pk, sk_k, chal, \Gamma$). Input the public keys pk, the private key sk_k, the chal and the PDP Γ, the client c_k outputs "success" or "failure".*

4 Our Proposed ISB-PDP Scheme

Our ISB-PDP scheme is comprised by *SBE*, *KeyGen*, *TagGen* and *GenProof*, where the *SBE* is brought from Zhu *et al.*'s [20] some work. The cloud user first generates the private keys of clients and outputs ciphertext by implementing the *SBE*, then sets the ciphertext as the public parameters. After that, the cloud user computes the block-tag pairs by running *KeyGen* and *TagGen*. Finally, the client decrypts the ciphertext and executes *GenProof*. In the following, we present them detail by detail in this section.

4.1 The Identity-Set-Based Broadcast Encryption (SBE)

Here we first review the formal definition of identity-set-based broadcast encryption (*SBE*) with key encapsulation mechanism [3], and present the full construction of identity-set-based broadcast encryption in Zhu *et al.*'s [20] work with minor modification on description to meet the needs of our ISB-PDP scheme. For the sake of simplicity, the game-based security definition model, the definition and construction of **aggregation functions** that are **ZerosAggr** and **PolesAggr**, the secure analysis, the performance evaluation and experimental results in [20] will not be presented in this paper. Nevertheless, the security of the proposed ISB-PDP scheme is based on the security of *SBE*, and we will give the details in our security analysis.

For ease of reading, we present some notations which will be used in the following. Let $\mathbf{U} = \{ID_1, \cdots, ID_n\}$ denote the set of clients chosen by the cloud user, $\mathbf{R} = \{ID_1, \cdots, ID_m\}$ be a subset of $\mathbf{U}$ which is used in **aggregation functions**, and $\mathbf{S} = \{ID_1, \cdots, ID_t\} \subset \mathbf{U}$ denote the set of selected/cut clients. We have $|\mathbf{U}| = n$, $|\mathbf{R}| = m$, $|\mathbf{S}| = t$, and we require $t < m < n$. The identity-set-based broadcast encryption is made up of four algorithms, shown as follows:

Setup$(\mathbb{S})$. Takes as input a bilinear map group system $\mathbb{S}$. It outputs a public key mpk and a master secret key msk, where mpk contains a list of clients' profiles pp.

KeyGen(msk, ID_k). Takes as input msk and a client's identity ID_k. It outputs the client's secret key sk_k and adds a client's profile pp_k to pp, i.e., $\mathrm{pp} = \mathrm{pp} \cup \{pp_k\}$.

Encrypt$(mpk, \mathbf{S}, mode)$. Takes as input mpk, a set of clients' identities $\mathbf{S}$, and a mode of operation $mode$, where the $mode$ belongs to one of two modes in $\{c \in \mathbf{S}, c \notin \mathbf{S}\}$. It outputs a ciphertext $\mathcal{C}$ and a random session key ek, where $(\mathbf{S}, mode)$ is included in $\mathcal{C}$.

Decrypt$(mpk, sk_k, \mathcal{C})$. Takes as input mpk, a ciphertext $\mathcal{C}$, and a client's secret key sk_k. If this client satisfies the access mode $mode$, then the algorithm can decrypt the ciphertext $\mathcal{C}$ and return a session key ek.

In *SBE*, the *client's profile* includes the identity of this client and a public parameter generated in registry. The *SBE* makes use of these profiles to realize encryption and decryption for a subset of clients, and the clients can verify the

PDP if they can decrypt the ciphertext in our ISB-PDP. As a group-oriented cryptosystem, we employ a list of profiles to realize the management of memberships. According to different operations, the set of clients $\mathbf{S}$ will be used in two cases:

Select-mode ($c \in \mathbf{S}$). Used to specify multiple verifiers, where $\mathbf{S}$ denotes a set of specified clients, such that the client $c \in \mathbf{S}$ will be authorized to decrypt the message, then he/she can verify the PDP.

Cut-mode ($c \notin \mathbf{S}$). Used to revoke multiple verifiers, where $\mathbf{S}$ denotes a set of revoked clients, such that the client $c \notin \mathbf{S}$ will be authorized to decrypt the message, then he/she can verify the PDP.

Consider all possible mpk from $Setup(\mathbb{S}) \to (mpk, msk)$, a valid ciphertext C from $Encrypt(mpk, \mathbf{S}, mode) \to (\mathcal{C}, ek)$ and $KeyGen(msk, ID_k) \to sk_k$. If the client's identity ID_k satisfies the operation mode $(\mathbf{S}, mode)$ in $\mathcal{C}$, then the decryption algorithm will retrieve the session key ek, i.e.,

$$\Pr\begin{bmatrix} Decrypt(mpk, sk_k, \mathcal{C}) = ek : \\ mode(ID_k, \mathbf{S}) = 1 \end{bmatrix} = 1,$$

where $mode(ID_k, \mathbf{S}) = 1$ denotes the boolean judgment over $mode := \{c \in \mathbf{S}, c \notin \mathbf{S}\}$ for a certain ID_k and a set of clients' identities $\mathbf{S}$.

Remark 1. In Zhu *et al.*'s [20] work, there are three modes: Select model, All-model and Cut model. While in our paper, we omit the All-model, since if we do not embed the *SBE* in our PDP scheme, it is also public verification. And the main ingredients of our work is to consider making a system in which the cloud user can select/cut the verifiers (clients) to check the PDP, thus we will not consider public verification. Therefore, it makes no sense to consider the All-model in our ISB-PDP.

4.2 The Full Construction of Identity-Set-Based Broadcast Encryption (SBE)

Setup($\mathbb{S}$). Chooses two elements $G \xleftarrow{R} \mathbb{G}_1$, $H \xleftarrow{R} \mathbb{G}_2$, and two exponents $\gamma, \epsilon \xleftarrow{R} \mathbb{Z}_p^*$, and then set $R = e(G, H)^\epsilon$ and sets $G_k = G^{\gamma^k}$ for $k \in [1, m]$. The master key is outputted as $msk = (\gamma, \epsilon, G, G^\epsilon)$ and the public key is $mpk = \{\mathbb{S}, H, R, \{G_k\}_{k \in [1,m]}, pp = \phi\}$.

KeyGen(msk, ID_k). Given a client's identity ID_k, defines $x_k = hash(ID_k)$ and computers the k-th client's secret key

$$sk_k = G^{\frac{\epsilon \cdot x_k}{\gamma + x_k}}, \qquad \text{and} \qquad H_k = H^{\frac{\epsilon}{\gamma + x_k}},$$

where, $pp_k = (ID_k, H_k)$ is appended to pp, i.e., $\text{pp} = \text{pp} \cup \{pp_k\}$.

Encrypt$(mpk, \mathbf{S}, mode)$. Picks a random $s \xleftarrow{R} \mathbb{Z}_p^*$ and executes the following process:

- Case $mode := (c \in \mathbf{S})$: invokes **PolesAggr**$(mpk, \mathbf{S}) \rightarrow H_{\mathbf{S}} = H^{\epsilon \prod_{e_i \in \mathbf{S}} \frac{1}{\gamma + x_i}}$, then computes

$$C_1 = H^s, \qquad \text{and} \qquad C_2 = (H_{\mathbf{S}})^s.$$

- Case $mode := (c \notin \mathbf{S})$: invokes **ZerosAggr**$(mpk, \mathbf{S}) \rightarrow G_{\mathbf{S}} = G^{\gamma \cdot \prod_{e_i \in \mathbf{S}} (\gamma + x_i)}$, then computes

$$C_1 = H^s, \qquad \text{and} \qquad C_2 = (G_{\mathbf{S}})^s.$$

Finally, the ciphertext is published as $C = (\mathbf{S}, mode, C_1, C_2)$. The corresponding session key is $ek = R^s$.

Decrypt$(mpk, sk_k, \mathcal{C})$. Chooses one action from two following cases according to the $mode$ in C:

- Case $mode := (c \in \mathbf{S})$: checks whether ID_k is a member of $\mathbf{S}$, that is, $ID_k \in \mathbf{S}$. If true, it sets $\mathbf{S}_- = \mathbf{S} \setminus \{ID_k\}$ and invokes **ZerosAggr** $(mpk, \mathbf{S}_-) \rightarrow G_{\mathbf{S}_-} = G^{\gamma \cdot \prod_{e_i \in \mathbf{S}_-} (\gamma + x_i)}$. Next it retrieves the session key

$$ek' = e(sk_k, C_1) \cdot e(G_{\mathbf{S}_-}, C_2). \tag{1}$$

- Case $mode := (c \notin \mathbf{S})$: checks whether ID_k satisfies the relation $ID_k \notin \mathbf{S}$. If true, it sets $\mathbf{S}_+ = \mathbf{S} \cup \{ID_k\}$ and invokes **PolesAggr**$(mpk, \mathbf{S}_+) \rightarrow H_{\mathbf{S}_+} = H^{\epsilon \prod_{e_i \in \mathbf{S}_+} \frac{1}{(\gamma + x_i)}}$. Next, it also retrieves the session key

$$ek' = e(sk_k, C_1) \cdot e(C_2, H_{\mathbf{S}_+}). \tag{2}$$

4.3 The Concrete ISB-PDP Scheme

First we define the following cryptographic hash function which are used in our scheme:

$$\begin{aligned}
&h(\cdot) : \mathbb{G}_T \longrightarrow \mathbb{Z}_p^*,\\
&H(\cdot) : \{0,1\}^* \longrightarrow \mathbb{G}_1,\\
&f : \mathbb{Z}_p^* \times \{1, 2, \cdots, n\} \longrightarrow \mathbb{Z}_p^*,\\
&\pi : \mathbb{Z}_p^* \times \{1, 2, \cdots, n\} \longrightarrow \{1, 2, \cdots, n\}.
\end{aligned}$$

Where, f is a pseudo-random function and π is a pseudo-random permutation. The parameters $\{h, H, f, \pi\}$ are made public. Our concrete ISB-PDP scheme is constructed as follows.

SBE-KeyGen$(\mathbb{S})$. First, the cloud user generates the private keys of clients, master keys msk, public keys mpk and the corresponding ciphertext $C = (\mathbf{S}, mode, C_1, C_2)$ according to the two modes by implementing the SBE, then computes a generator $g \in \mathbb{G}_2$, selects a random element $u \in \mathbb{G}_1$, and computes $v \longleftarrow g^{\epsilon}$, where $\epsilon \in msk$. Then, the secret keys are $sk = (msk, ek)$ and the public parameters are $pk = \{mpk, C_1, C_2, h, H, f, \pi, u, v, g, p\}$.

TagGen(sk, pk, F). The cloud user splits file F into n blocks, *i.e.*, $F = (m_1, m_2, \cdots, m_n)$, where $m_i \in \mathbb{Z}_p^*$ of which $i \in [1, n]$. For every block m_i, the cloud user performs the procedures as follows:

(1) The cloud user calculates $h(ek)$, and generates $u\|i$, then, computes $H(u\|i)$.

(2) The cloud user calculates

$$T_{m_i} = (H(u\|i)u^{h(ek)}u^{m_i})^{\epsilon}.$$

(3) The cloud user outputs (m_i, T_{m_i}).

After the procedures are performed n times, all the block tags are generated by the cloud user, then, the cloud user denotes the collection of all block-tag pairs $\{(m_i, T_{m_i}), i \in [1, n]\}$ as $\Sigma = \{(m_1, T_{m_1}), (m_2, T_{m_2}), \cdots, (m_n, T_{m_n})\}$, and sends the block-tag pairs collection Σ to the CSS, after that, the CSS stores the block-tag pairs collection Σ. At the same time, the cloud user deletes the block-tag pairs collection Σ from its local storage but stores the private key sk and the corresponding metadata Φ. Next, we assume a client c_k whose private key is sk_k queries the CSS.

GenProof$(\pi, f, chal, \Sigma)$. Upon receiving a query from the client c_k which is $chal = (c, k_1, k_2)$, where $c \in [1, n]$, $k_1 \in \mathbb{Z}_p^*$ and $k_2 \in \mathbb{Z}_p^*$, the CSS performs the procedures as follows:

(1) For $1 \leq j \leq c$, the CSS calculates the indexes and coefficients of the blocks for which the proof is generated:

$$i_j = \pi_{k_1}(j), v_j = f_{k_2}(j).$$

(2) The CSS calculates

$$T = \prod_{j=1}^{c} T_{m_{i_j}}^{v_j}, \mu = \sum_{j=1}^{c} v_j m_{i_j}.$$

(3) The CSS outputs $\Gamma = (\mu, T)$, and sends Γ to the client c_k as the response to the query.

CheckProof$(pk, sk_k, chal, \Gamma)$. Upon receiving the response Γ from the CSS, the client c_k performs the procedures as follows:

(1) For $1 \leq j \leq c$, the client c_k calculates the indexes and coefficients of the blocks for which the proof is generated: $i_j = \pi_{k_1}(j), v_j = f_{k_2}(j)$ in the same way with the CSS.

(2) According to decryption procedures of the SBE, the client c_k computes the Eq. (1) $ek' = e(sk_k, C_1) \cdot e(G_{\mathbf{S}_-}, C_2)$ if it is in the **Select-model**. Otherwise, the client c_k computes the Eq. (2) $ek' = e(sk_k, C_1) \cdot e(C_2, H_{\mathbf{S}_+})$ in the **Cut-model**. If it does not satisfy both models, the client c_k declares termination. Otherwise, the client c_k obtains $h(ek')$.

(3) With the value $h(ek')$, the client c_k checks whether the following equation holds:

$$e(T, g) \stackrel{?}{=} e(\prod_{j=1}^{c} H(u\|i_j)^{v_j} u^{h(ek')v_j}, v) \cdot e(u^{\mu}, v). \tag{3}$$

(4) If it holds, then the client c_k outputs "success". Otherwise, the client c_k outputs "failure".

5 The Security Analysis of the Proposed ISB-PDP Scheme

In this section, we present a formal security analysis for our proposed ISB-PDP scheme depending on the security of the SBE and the hardness of the computational Diffie-Hellman (CDH) problem. For the sake of simplicity, the security proof of the SBE in Zhu *et al.*'s [20] will not be reviewed in our paper.

Theorem 1. *In our proposed ISB-PDP scheme, we assume the SBE is secure (which has been proved in [20]). Then there is no client (verifier) c_i can verify the PDP correctly, if the client belongs to neither the* **Select-model** *nor* **Cut-model**, *i.e., the client $c_i \notin \mathbf{S}$ in* **Select-model**, *and $c_i \in \mathbf{S}$ in* **Cut-model**.

Proof. According to procedures of the proposed IBS-PDP scheme, the client (verifier) c_i has to figure the value ek' out by the Eqs. (1) or (2) correctly. Otherwise, he/she can't calculate the value $h(ek')$ correctly so as to fail to verify the Eq. (3). In Zhu *et al.*'s [20] security proof, the client c_i who does not satisfy both model **Select-model** and **Cut-model** can't decrypt the ciphertext C, i.e., the client c_i can't obtain the value ek. Therefore, according to the security proof of the SBE and the procedures of our proposed ISB-PDP scheme, the client (verifier) c_i can't verify the Eq. (3) correctly if the client $c_i \notin \mathbf{S}$ in **Select-model**, and $c_i \in \mathbf{S}$ in **Cut-model**, which completes the proof.

Theorem 2. *If the CDH problem is hard in bilinear groups, then the malicious CSS can't pass the checking procedures, i.e., the verification Eq. (3) in $CheckProof$ will fail to pass, if the original blocks of file even only one of the blocks has been modified or incomplete. Except by responding with the correctly computed proof $\Gamma = (\mu, T)$, where $\mu = \sum_{j=1}^{c} v_j m_{i_j}$.*

Proof. Since the proof is simple, we just present a simple analysis proof, rather than a formal adversary-challenge model proof. First, it is infeasible for the malicious CSS to pass the checking procedures using blocks at different indices even if the cloud user uses the same secret key ϵ with all tags of blocks, since the hash value $H(u\|i)$ and the block index are embedded into the block tag which specified the tag so as to prevent using the tag to obtain a proof for the different block. Second, the value $u^{h(ek)}$ of tags is fixed and independent which is computed by the cloud user beforehand and will be computed by verifiers afterward, it is useful for the verifiers but useless for the malicious CSS. Then, the goal of the malicious CSS is to generate a proof which is not computed correctly but can pass the verification process. Let $\Gamma' = (\mu', T')$ be the malicious CSS's response. While let $\Gamma = (\mu, T)$ be the expected response from the honest CSS, where $T = \prod_{j=1}^{c} T_{m_{i_j}}^{v_j}$ and $\mu = \sum_{j=1}^{c} v_j m_{i_j}$. In the following, the CDH problem will be solved by the algorithm $\mathscr{B}$ (which plays the role of a verifier) if the malicious CSS passes the verification process. Given the CDH problem: given g, g^{ϵ}, $h \in \mathbb{G}$ for some group $\mathbb{G}$, while $\epsilon \in \mathbb{Z}_p$ is unknown, the algorithm $\mathscr{B}$ expects to compute h^ϵ. We assume $\mathscr{B}$ can compute all values except ϵ in our ISB-PDP scheme. $\mathscr{B}$ sets $u = g^\alpha h^\beta$ for $\alpha, \beta \in \mathbb{Z}_p^*$ beforehand, while all other values remain unchanged in our ISB-PDP scheme.

If the malicious CSS's responses $\Gamma' = (\mu', T')$ and passes the verification process, i.e. $\Gamma' = (\mu', T')$ satisfies the Eq. (3), thus, $\mathscr{B}$ computes

$$e(T', g) = e(\prod_{j=1}^{c} H(u\|i_j)^{v_j} u^{h(ek')v_j}, \upsilon) \cdot e(u^{\mu'}, \upsilon). \quad (4)$$

At the same time, the expected proof $\Gamma = (\mu, T)$ also satisfies the Eq. (3), then, $\mathscr{B}$ computes

$$e(T, g) = e(\prod_{j=1}^{c} H(u\|i_j)^{v_j} u^{h(ek')v_j}, \upsilon) \cdot e(u^{\mu}, \upsilon). \quad (5)$$

Since $\Gamma' \neq \Gamma$, then, there will be three cases.

(1) $T' = T$, $\mu \neq \mu'$. $\mathscr{B}$ gets $e(T', g) = e(T, g)$ and $T' \cdot T^{-1} = 1$, then according to the Eqs. (4) and (5), $\mathscr{B}$ defines $\Delta\mu = \mu' - \mu \neq 0$, dividing the Eq. (5) by Eq. (4) to obtain

$$e(T' \cdot T^{-1}, g) = e(u^{\Delta\mu}, \upsilon). \quad (6)$$

Then, $\mathscr{B}$ transforms Eq. (6) above to obtain

$$e(T' \cdot T^{-1}, g) = e(u^{\epsilon\Delta\mu}, g). \quad (7)$$

Since $u = g^\alpha h^\beta$, $\mathscr{B}$ computes h^ϵ according to the Eq. (7) as follows:

$$\begin{aligned} (g^\alpha h^\beta)^{\epsilon\Delta\mu} &= T' \cdot T^{-1} \\ h^{\beta\epsilon\Delta\mu} &= 1 \cdot \upsilon^{-\alpha\Delta\mu} \\ h^\epsilon &= \upsilon^{-\frac{\alpha}{\beta}}. \end{aligned} \quad (8)$$

Hence, the given CDH problem is solved by $\mathscr{B}$.

(2) $T' \neq T$, $\mu \neq \mu'$. The same to the above procedures, $\mathscr{B}$ divides the Eq. (5) by Eq. (4) result in $e(T' \cdot T^{-1}, g) = e(u^{\Delta\mu}, \upsilon)$, then, computes h^ϵ as follows:

$$\begin{aligned} (g^\alpha h^\beta)^{\epsilon\Delta\mu} &= T' \cdot T^{-1} \\ h^{\beta\epsilon\Delta\mu} &= T' \cdot T^{-1} \cdot \upsilon^{-\alpha\Delta\mu} \\ h^\epsilon &= (T' \cdot T^{-1} \cdot \upsilon^{-\alpha\Delta\mu})^{\frac{1}{\beta\Delta\mu}}. \end{aligned} \quad (9)$$

Hence, the given CDH problem is also solved by $\mathscr{B}$.

(3) $T' \neq T$, $\mu = \mu'$. In this case, $\mathscr{B}$ gets $e(T', g) = e(T, g)$ according to the Eqs. (4) and (5), then $\mathscr{B}$ gets $e(T' \cdot T^{-1}, g) = 1$ resulting in $T' = T$ which contradicts the precondition. In other word, this case is non-existent.

From the above, we have that there is no malicious CSS that can pass the checking procedures except by responding with correctly computed proof in our ISB-PDP scheme, i.e., the malicious CSS cannot pass the checking procedures except that the data blocks that are queried by the client are intact. This completes the proof.

6 Conclusion

In this paper, we propose a new PDP scheme called ISB-PDP. To the best of our knowledge, our proposed ISB-PDP is the first PDP scheme which is not only proved to be secure under the hardness of the computational Diffie-Hellman (CDH) problem, but also can dynamically select/cut some verifiers to check the PDP correctly according to the needs of the cloud user. In other words, the cloud user can determine whether the clients could use the data or not through controlling their ability to verify the PDP correctly, this may produces some economic effectiveness.

Acknowledgements. We are very grateful to the anonymous referees, who pointed out several inaccuracies and suggested improvements in the presentation of the paper. This work was supported by the National Natural Science Foundation of China (Nos. 61572132, U1705264, 61672030, and 61702237), the Natural Science Foundation of Jiangsu Province, China (No. BK20150241), and the Special Foundation of Promoting Science and Technology Innovation of Xuzhou City, China (No. KC18005).

References

1. Ateniese, G., Burns, R.C., Curtmola, R., Herring, J., Kissner, L., Peterson, Z.N.J., Song, D.X.: Provable data possession at untrusted stores. In: Proceedings of the 2007 ACM Conference on Computer and Communications Security, CCS 2007, Alexandria, Virginia, USA, 28–31 October 2007, pp. 598–609 (2007)
2. Barsoum, A.F., Hasan, M.A.: Integrity verification of multiple data copies over untrusted cloud servers. In: 12th IEEE/ACM International Symposium on Cluster, Cloud and Grid Computing, CCGrid 2012, Ottawa, Canada, 13–16 May 2012, pp. 829–834 (2012)
3. Boneh, D., Gentry, C., Waters, B.: Collusion resistant broadcast encryption with short ciphertexts and private keys. In: Advances in Cryptology - CRYPTO 2005: Proceedings of the 25th Annual International Cryptology Conference, Santa Barbara, California, USA, 14–18 August 2005, pp. 258–275 (2005)
4. Bowers, K.D., Juels, A., Oprea, A.: Proofs of retrievability: theory and implementation. In: Proceedings of the First ACM Cloud Computing Security Workshop, CCSW 2009, Chicago, IL, USA, 13 November 2009, pp. 43–54 (2009)
5. Curtmola, R., Khan, O., Burns, R.C., Ateniese, G.: MR-PDP: multiple-replica provable data possession. In: 28th IEEE International Conference on Distributed Computing Systems (ICDCS 2008), Beijing, China, 17–20 June 2008, pp. 411–420 (2008)
6. Hou, H., Yu, J., Hao, R.: Provable multiple-replica dynamic data possession for big data storage in cloud computing. I. J. Network Security **20**(3), 575–584 (2018)
7. Juels, A., Kaliski Jr., B.S.: PORs: proofs of retrievability for large files. In: Proceedings of the 2007 ACM Conference on Computer and Communications Security, CCS 2007, Alexandria, Virginia, USA, 28–31 October 2007, pp. 584–597 (2007)
8. Ren, Y., Yang, Z., Wang, J., Fang, L.: Attributed based provable data possession in public cloud storage. In: 2014 Tenth International Conference on Intelligent Information Hiding and Multimedia Signal Processing, IIH-MSP 2014, Kitakyushu, Japan, 27–29 August 2014, pp. 710–713 (2014)

9. Shacham, H., Waters, B.: Compact proofs of retrievability. In: Advances in Cryptology - ASIACRYPT 2008, Proceedings of the 14th International Conference on the Theory and Application of Cryptology and Information Security, Melbourne, Australia, 7–11 December 2008, pp. 90–107 (2008)
10. Shen, S., Tzeng, W.: Delegable provable data possession for remote data in the clouds. In: Proceedings of the Information and Communications Security - 13th International Conference, ICICS 2011, Beijing, China, 23–26 November 2011, pp. 93–111 (2011)
11. Wang, C., Chow, S.S.M., Wang, Q., Ren, K., Lou, W.: Privacy-preserving public auditing for secure cloud storage. IEEE Trans. Comput. **62**(2), 362–375 (2013)
12. Wang, F., Xu, L., Wang, H., Chen, Z.: Identity-based non-repudiable dynamic provable data possession in cloud storage. Comput. Electr. Eng. **69**, 521–533 (2018)
13. Wang, H.: Proxy provable data possession in public clouds. IEEE Trans. Serv. Comput. **6**(4), 551–559 (2013)
14. Wang, H., Wu, Q., Qin, B., Domingo-Ferrer, J.: Identity-based remote data possession checking in public clouds. IET Inf. Secur. **8**(2), 114–121 (2014)
15. Wang, Q., Wang, C., Li, J., Ren, K., Lou, W.: Enabling public verifiability and data dynamics for storage security in cloud computing. In: Computer Security - ESORICS 2009, Proceedings of the 14th European Symposium on Research in Computer Security, Saint-Malo, France, 21–23 September 2009, pp. 355–370 (2009)
16. Yu, Y., Au, M.H., Ateniese, G., Huang, X., Susilo, W., Dai, Y., Min, G.: Identity-based remote data integrity checking with perfect data privacy preserving for cloud storage. IEEE Trans. Inf. Forensics Secur. **12**(4), 767–778 (2017)
17. Yuan, J., Yu, S.: Proofs of retrievability with public verifiability and constant communication cost in cloud. In: Proceedings of the 2013 International Workshop on Security in Cloud Computing, SCC@ASIACCS 2013, Hangzhou, China, 8 May 2013, pp. 19–26 (2013)
18. Zheng, Q., Xu, S.: Fair and dynamic proofs of retrievability. In: Proceedings of the First ACM Conference on Data and Application Security and Privacy, CODASPY 2011, San Antonio, TX, USA, 21–23 February 2011, pp. 237–248 (2011)
19. Zhu, Y., Wang, H., Hu, Z., Ahn, G., Hu, H.: Zero-knowledge proofs of retrievability. Sci. China Inf. Sci. **54**(8), 1608–1617 (2011)
20. Zhu, Y., Wang, X., Ma, D., Guo, R.: Identity-set-based broadcast encryption supporting "cut-or-select" with short ciphertext. In: Proceedings of the 10th ACM Symposium on Information, Computer and Communications Security, ASIA CCS 2015, Singapore, 14–17 April 2015, pp. 191–202 (2015)

(2, 2) Threshold Robust Visual Secret Sharing Scheme for QR Code Based on Pad Codewords

Longdan Tan, Yuliang Lu, Xuehu Yan(✉), Lintao Liu, and Jinrui Chen

National University of Defense Technology, Hefei 230037, China
publictiger@126.com

Abstract. Quick response (QR) code has been used widely because of its advantages, such as, fast reading, error correction, and encoding multiple types of data. More and more researches on the combination of QR and visual cryptography have emerged, but there are few robust visual secret sharing scheme (VSS) for QR code. In this paper, we propose a robust $(2, 2)$ threshold VSS for QR code based on pad codewords. We utilize the bug of no checking on the pad codewords when encoding the QR code, we encode initial shadows shared with secret image as pad codewords of cover QR code images to generate QR code shadows. In this way, the decoding results of shadow QR code images are identical with those of cover QR code images, no doubt arises when shadow images are exposed to the public channel. While a certain degree of damage and noise appearing in the shadow images, the secret can still be revealed, so the scheme is robust, which is important for practice. The experimental results demonstrate the effectiveness and robustness of our scheme.

Keywords: Quick response (QR) code · Visual secret sharing scheme (VSS) · Pad codewords · Robustness

1 Introduction

Quick response (QR) code is using popularly in many fields of work and life recently. QR codes can encode various types of data and correct errors. In particular, the use of error correction codewords makes the QR code robust to some degree damage or noise.

The threshold-based visual secret sharing scheme (VSS) is a scheme that sharing a binary secret image into n different shadow images looking like random noise. No information can be obtained from less than k shadow images. While, k or more shadow images can be stacked to reveal partial information about the secret image without any computations or cryptographic knowledge. However,

Supported by the National Natural Science Foundation of China (Grant Number: 61602491) and the Key Program of the National University of Defense Technology (Grant Number: ZK-17-02-07).

C.-N. Yang et al. (Eds.): SICBS 2018, AISC 895, pp. 619–628, 2020.
https://doi.org/10.1007/978-3-030-16946-6_50

the threshold-based VSS requires the use of codebook, and there is also the problem of pixel expansion.

Kafri and Keren [14–16] proposed random grid (RG)-based VSS which can solve the problem of the threshold-based VSS well. When the secret image is sharing, n shadow images of the same size with the original secret image are generated. Then, k or more shadow images are superimposed by OR-based VSS (OVSS) [13] based on RG to reveal the original secret image. However, when more shadow images are superimposed to reconstruct the secret image, the image will become darker.

XOR-based VSS (XVSS) by performing XORing operation on the shadow images to reveal secret image can solve the problem of RG-based OVSS. It requires a small amount of computation, but in this way, better image relative difference can be achieved [10,14].

Based on the characteristics of QR code and VSS, there has been a lot of researches on combination of them in recent years.

Lin et al. [7] proposed distributed secret sharing approach based on QR code to detect cheater. Wang et al. [12] also proposed a scheme to prevent cheating by embedding QR codes into the best regions of given shares. Chow et al. [5] proposed a $(n, n)(n \geq 3)$ threshold VSS utilizing QR code error correction mechanism. Two-level information management scheme was proposed in [9] that makes the decoding of shadow images not feasible if appropriate angle and distance for scanning are not found. Wan et al. inserted the secret image into a continuous area considering the theory of the error correction of QR code and VSS [11]. Flexible access structures and the characteristic of security are taken into account based on QR code applications in VSS with high sharing efficiency in [3] by Chen et al. Chow et al. used symmetric keys to encrypt initial shadow images and then embedded them into cover QR codes to protect the shadow images [4]. Huang et al. [6] proposed a Sudoku-based secret sharing approach using QR code to prevent cheating. The problem of robustness is important for secret image recovery in practice, but it not be mentioned in their schemes.

In this article, we propose a robust VSS for QR Code based on pad codewords. In our scheme, we better utilize the bug of no checking on the pad codewords when encoding the QR code, robust VSS can be achieved. We substitute the bits of initial shadow images for the pad codewords of cover QR codes to generate QR code shadows. In this way, the scheme keeps the error correction abilities of QR code shadow images, so the scheme is robust to the conventional image attacks and the QR code shadow images can be decoded by standard QR code decoder. Experimental results will display the effectiveness of our scheme.

The rest of the paper is organized as follows. QR codes and $(2, 2)$ threshold VSS are introduced in Sect. 2. The secret image sharing, recovering phase of the scheme are described in Sect. 3. Section 4 demonstrates the experimental results and comparison. Finally, Sect. 5 concludes this paper.

2 Preliminaries

2.1 QR Codes

QR code [8] is a matrix symbol consisting of black and white square modules distributed in a square pattern. Forty versions denoted as $V(1 \sim 40)$ of QR code are introduced in standard [1], and each version has four error correction level denoted as $E(L, M, Q, H)$, the error correction ability are corresponding to $L \sim 7\%$, $M \sim 15\%$, $Q \sim 25\%$, $H \sim 30\%$.

A QR code symbol [1] includes functional patterns and encoding regions. The functional patterns include alignment, unique finding patterns, timing and separation. The encoding region is consisting of version information, data and error correction codewords and format information. Different version and error correction level have different data and error correction codewords. The QR code of higher version with lower error correction level has more data capacity.

The data codewords and error correction codes generated by the data codewords are distributed in corresponding error correction blocks. A QR code has one or more error correction blocks, which is depended on the version and error correction level.

When the input data is encoding, firstly, the data are converted into a bit stream, mode indicators are inserted as necessary at the beginning of new mode segment to change mode, then a terminator is added at the end of the segment. The modes, such as numeric mode, alphanumeric mode, 8-bit byte mode, Kanji mode and so on, are defined based on the assignments and character values related with the default ECI. The resulting bit stream is split into 8-bit codewords. Pad codewords are added as necessary to fill the total number of data codewords. Corresponding error correction codewords are generated by data codewords.

After encoding all the codewords, one of eight data mask patterns is selected to mask the encoding area to balance the black and white modules and avoid the confusion functional pattern with encoding area. Functional patterns are generated relied on the version. The encoding method of input data is important for our scheme.

2.2 (2,2) Threshold RG-Based VSS

In RG-based VSS [2], S denotes the value of a pixel in secret images. SC_n denotes the value of a pixel in shadow images. Here white pixel denoted as 0, black pixel denoted as 1. The sharing phase of (2, 2) threshold RG-based VSS is described in the following steps.

Step 1: Generating one RG SC_1 Randomly.

Step 2: Calculating the value of SC_2 according to Eq. (1).

$$SC_2(i,j) = \begin{cases} \overline{SC_1(i,j)} & if\ S = 1 \\ SC_1(i,j) & if\ S = 0 \end{cases} \tag{1}$$

All pixels in two shadow images are generated according to the above steps, so two shadow images have the same size with secret image look like random noise images, no information of secret image can be revealed in any shadow images. However, stacking the two shadow images can reveal the original secret image although there are some image contrast losses, and performing XORing operation on the two shadow images with lightweight computation can reveal the original secret image losslessly. Figure 1 shows the example of (2, 2) threshold VSS of recovery by stacking.

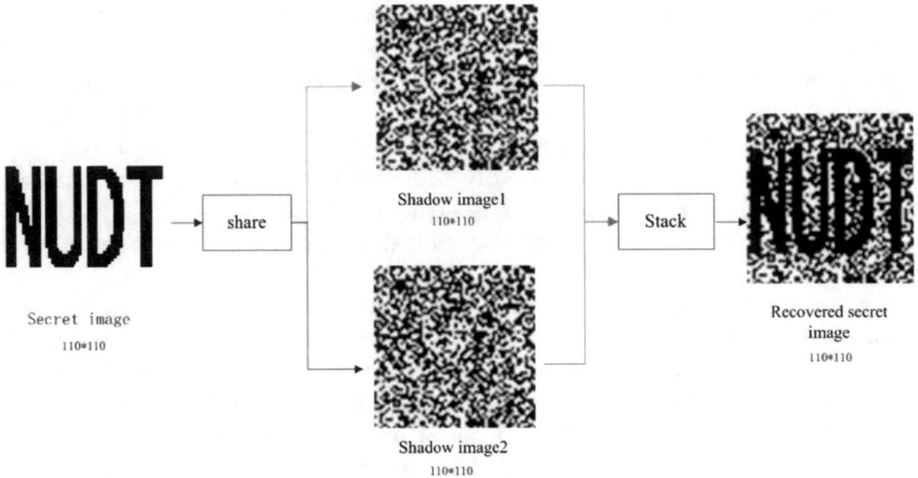

Fig. 1. performing ORing on the two shadow images in (2, 2) threshold VSS

In this paper, we use the RG-based (2, 2) threshold VSS generation phase to generate the initial shadow images. The recovery way is performing ORing and XORing operation respectively on the shadow images to reveal the original secret image.

3 Proposed (2, 2) Threshold Robust VSS for QR Code Based on Pad Codewords

3.1 The Main Idea

In this section, we propose a (2, 2) threshold RG-based VSS based on QR code to generate robust shadow images. The secret image and cover QR code images are binary images, (2, 2) threshold RG-based VSS is used to generate two initial shadow images. The shadow images are binary images looking like random noise. Then the two shadow images are converted to two bit streams, which are converted to decimal data streams added at the back of the decoding results of cover QR code to encode the QR code shadow images. In the process of encoding the

QR code shadow images, the mode indicators of coded segment of the decimal data streams are replaced by terminator sequence 0000 and corresponding error correction codewords are changed. The decimal data streams are encoded as pad codewords, since there is no checking on the pad codewords, the decoding results of QR code shadows are identical with those of cover QR images, and the error correction abilities are preserved.

The size of secret image is $m * n$. It is converted to a binary bit stream with the length of $m * n$. The $m * n$ bits are divided into groups of ten binary number equivalent, each group is converted to three bits decimal digit. If the total number of the last group is less than ten, we will convert it to three bits decimal digit too. If the decimal digit less than three bits, pad 0 on the left of the result. The length of converted result of $m * n$ pixels is $\lceil m * n/10 \rceil * 3$. C denotes the data capacity of QR code having the same version and error correction level as the cover QR code in numeric mode. The modes of last mode segment of the decoding result of the cover QR code images are not numeric mode. M denotes the length of the decoding result of the cover images which is equivalent to that of its corresponding numeric mode, the size of embeddable secret image has to satisfy the condition in the Eq. (2).

$$\lceil m * n/10 \rceil * 3 + \lceil M \rceil \leq C \tag{2}$$

The QR code shadow images generation architecture of our scheme is illustrated in Fig. 2.

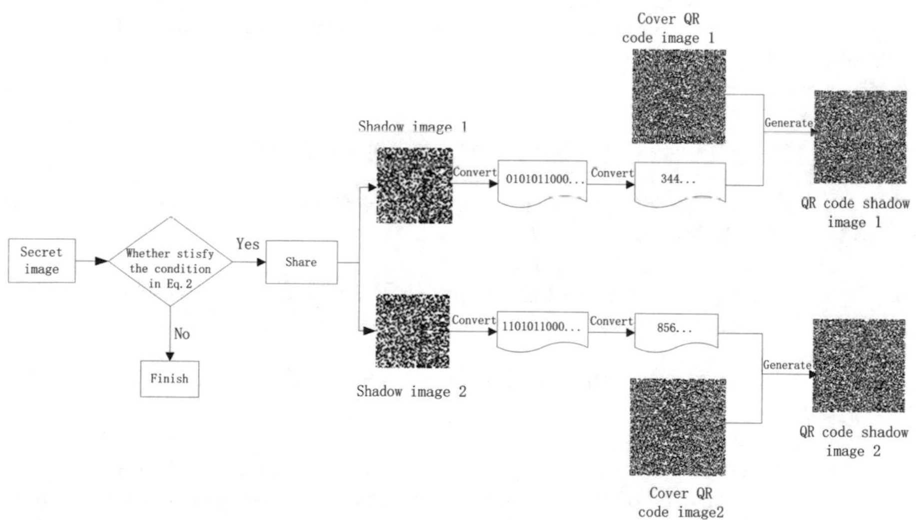

Fig. 2. The QR code shadow images generation architecture of our scheme

When we recover the secret image, we find the position of the terminator of the QR code shadows, and then turn it into the mode indicator of numeric mode, change corresponding error correction codewords, so we can decode the changed QR code, We extract the messages behind the decoding result of the QR code shadows, and then convert them to binary streams. We perform ORing and XORing operation on the bit streams respectively. The resulting bit streams are arranged in new binary images to create the original secret images.

The details of sharing algorithm steps are described in Algorithm 1. One secret image S and two cover QR code images are the input and two QR code shadow images are the output in Algorithm 1. When the secret image is sharing, we should make sure the size of the secret image and cover QR code images satisfies the condition in the Eq. (2). This conversions in steps 3–13 enable the number of ten bit binary numbers to be represented by three bits decimal digit, but there may be some deviations. After modifying the mode indicator with terminators in step 6, the codewords added behind the terminator are treated as pad codewords, so the shadows can be decode by a standard QR code, and the decoding contents are identical with the cover QR codes, which will not raise doubts.

Algorithm 1. (2, 2) threshold robust VSS for QR code based on pad codewords

Require: one $m * n$ binary secret image S; two binary cover QR code images C_1, C_2;
Ensure: two QR code shadows SC_1, SC_2.

1: Share a secret image using (2, 2) threshold RG-based VSS to generate two noise-like shadow images S_1, S_2.
2: Convert S_1 and S_2 to binary bit streams B_1, B_2 with the same size of $m * n$. The $m * n$ bits are divided into g groups of ten binary number equivalent, the ten binary number in group i of B_1, B_2 are converted to three bits decimal digits K_i, T_i respectively.
3: **for** $i = 1$ to g **do**
4:     **if** $K_i, T_i \leq 99$ **then**
5:         pad 0 on the left of the result to gain three decimal digits K'_i, T'_i.
6:     **else if** $K_i > 999$ and $T_i > 999$ **then**
7:         randomly replace one of their highest five digits to 0, then converted K_i, T_i to three bits decimal digits K'_i, T'_i.
8:     **else if** $K_i > 999$ or $T_i > 999$ **then**
9:         Exchange one of highest five digits between K_i and T_i or convert it to 0 to make K_i, T_i can be represented by three bits decimal digits K'_i, T'_i.
10:     **else**
11:         Convert K_i, T_i to three bits decimal digit K'_i, T'_i.
12:     **end if**
13: **end for**
14: Decode the cover QR code of C_1, C_2 image to get the messages D_1, D_2.
15: Add K', T' at the end of D_1, D_2 respectively. Input the message and the mode indicators of coded segment of the decimal data streams are replaced with terminator sequence 0000, corresponding error correction codewords are changed to generate shadow images SC_1, SC_2
16: Output the n QR code shadow images SC_1, SC_2.

3.2 The Recovery Phase

The corresponding recovery algorithm steps are described in Algorithm 2.

Algorithm 2. Secret image recovery of our proposed scheme

Require: two shadow images SC_1, SC_2;
Ensure: the reconstructed secret QR code image $SOR, SXOR$.
1: Decode QR code shadow SC_1, SC_2 to get its decoding messages A_1, A_2.
2: Locate the terminator of SC_1, SC_2 based on the length of A_1, A_2, and modify the terminator to mode indicator of numeric mode, and change corresponding error correction codewords respectively to get two new QR code Q_1, Q_2.
3: Decode QR code Q_1, Q_2, extract the messages behind the decoding results of the QR code shadows, convert them to binary streams L_1, L_2.
4: Perform ORing and XORing operation on the bit streams respectively to get two bit streams.
5: Arrange the two bit streams into two $m * n$ images to reconstruct two secret image $SOR, SXOR$.
6: Output the reconstructed secret image $SOR, SXOR$.

In step 2, we change the terminator to mode indicator of numeric mode, so the messages containing initial shadow images can be decoded. The secret images can be reconstructed in two ways in steps 4–5.

4 Experiments and Analysis

4.1 Image Illustration

Experimental results in python language are given to display the effectiveness of the proposed scheme. The size of the secret image is $70 * 70$. All the QR codes used in the experiments are version 27 with error correction level H. Figure 3 shows our $(2, 2)$ threshold robust VSS for QR code. Figure 3(a) is the secret image S. Figure 3(b)–(c) are cover images C_1, C_2. Shadow images SC_1, SC_2 are shown in Fig. 3(d)–(e), Fig. 3(f)–(g) are images captured from a computer screen, P_1, P_2. Figure 3(h)–(i), reS_1 and reS_2 are the results of modifying the four bits of the terminators with the mode indicators of numeric mode and changing corresponding error correction codewords in Fig. 3(d)–(e). Decoding P_1, P_2 by a standard QR code decoder to get all the codewords, and then modifying the four bits of the terminators with the mode indicators of numeric mode and changing corresponding error correction codewords to generate new QR codes with the same pixels of reS_1, reS_2. SOR in Fig. 3(j) is the reconstructed secret image created by performing ORing operation. $SXOR$ in Fig. 3(k) is the reconstructed secret image created by performing XORing operation. Figure 3(l) shows the difference between S and $SXOR$.

As Fig. 3, two recovery methods can be realized. The reason for using screen shots is that secret images can still be revealed when there are a lot of noises added at the same time, indicating that the algorithm is applicable to the actual robust requirements.

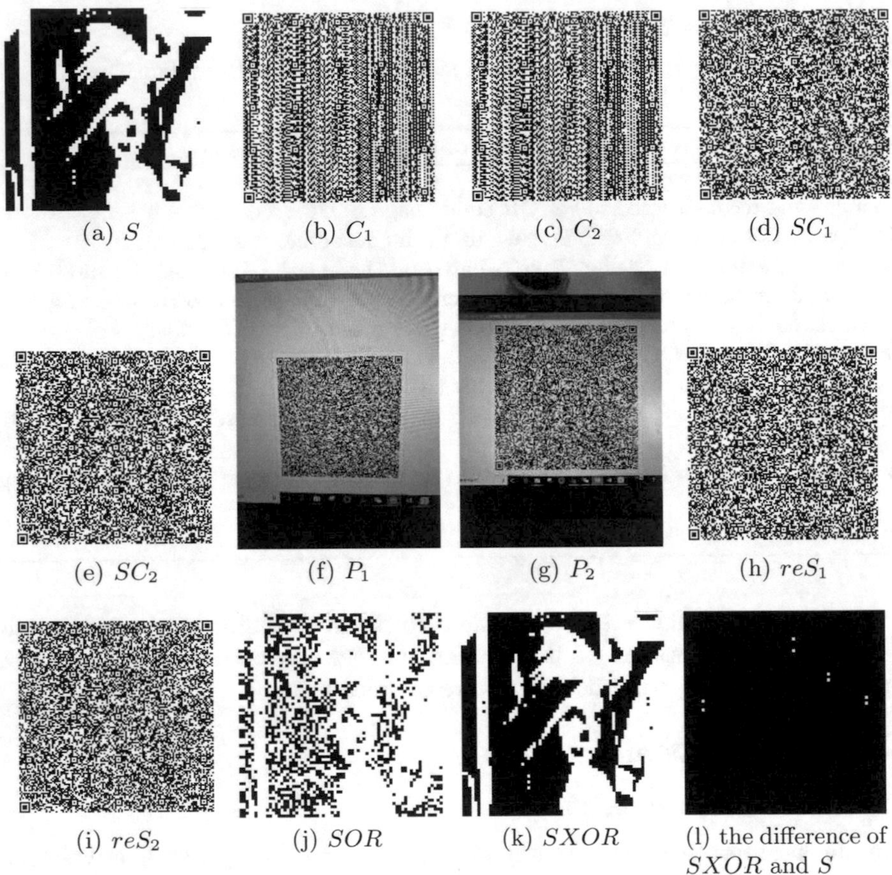

(a) S (b) C_1 (c) C_2 (d) SC_1

(e) SC_2 (f) P_1 (g) P_2 (h) reS_1

(i) reS_2 (j) SOR (k) $SXOR$ (l) the difference of $SXOR$ and S

Fig. 3. Our (2, 2) threshold robust VSS based on QR code with version 27 and level of error correction H.

4.2 Comparison and Analyses

In Fig. 3, the last mode segment of decoding result of the cover images is not numeric mode, the size of embeddable secret image is $70 * 70$, the length of the input messages in 8-bit byte mode is 8, converted to numeric mode, the length is equivalent to $\lceil 8 * 1501/625 \rceil = 20$, the data capacity of the cover QR code in numeric mode is 1501, while in 8-bit byte mode is 625. $\lceil 70 * 70/10 * 3 \rceil + 20 < 1501$, so the condition in the Eq. (2) is satisfied and the secret image can be shared in the cover QR code images. If mode segment of decoding result of the cover images is numeric mode, we can use lager secret image as long as it satisfies the condition of $\lceil m * n/10 \rceil * 3 + M \leq C$, but the value of character count indicator, not the value of terminators should be modified.

Since the initial shadow images are encoded as pad codewords of cover QR code, and there is no checking on the codewords when encoding, so the decoding results of QR code shadows are the same as the cover QR images, the error correction capacity is also preserved. From Fig. 3(h)–(i), we can find that when camera and screen noises appeared simultaneously, as long as the shadow images can be decoded correctly when added with noise, the secret image can be recovered, so the scheme is robust to noise and damage to certain a degree. But the scheme in [5] and [11] can not be robust to conventional attack.

The secret image can be reconstructed almost losslessly. When converting ten binary number to three bits decimal digit, which is large than 999, the value may be different from the initial ten binary number, but for the whole image, the probability is very small, and Fig. 3(l) just shows the probability of the conversion.

The visual secret sharing scheme can be applied on color QR codes. For example, there are three planes in a eight colors QR code, each plane is a binary QR code, so our scheme can be applied on the binary QR code.

5 Conclusion

This paper investigates a robust (2, 2) threshold VSS for QR code. The scheme utilizes the bug of no checking on the pad codewords when encoding the QR code, modifies the mode indicator with the terminator and encodes the bits encoding of shadow images as pad codewords to generate QR code shadows. The proposed approach is robust since the error correct capacity of QR code is preserved. There are two ways to reveal the secret image, by performing OR and XOR operation respectively. As testing in Fig. 3, the decoding results of shadow images are identical with those of cover QR code images, thus the scheme is robust to camera and screen noise. Different threshold schemes with different versions will be our future work.

Acknowledgment. The authors would like to thank the anonymous reviewers for their valuable comments. This work is supported by the National Natural Science Foundation of China (Grant Number: 61602491) and Key Program of National University of Defense Technology.

References

1. Information technology-automatic identification and data capture techniques-bar code symbology-QR code. ISO/IEC
2. Chen, T.-H., Tsao, K.-H.: Threshold visual secret sharing by random grids. J. Syst. Softw. **84**(7), 1197–1208 (2011)
3. Cheng, Y., Fu, Z., Yu, B.: Improved visual secret sharing scheme for QR code applications. IEEE Trans. Inf. Forensics Secur. **PP**(99), 1 (2018)
4. Chow, Y.W., Susilo, W., Tonien, J., Vlahu-Gjorgievska, E., Yang, G.: Cooperative secret sharing using QR codes and symmetric keys. Symmetry **10**(4), 95 (2018)
5. Chow, Y.W., Susilo, W., Yang, G., Phillips, J.G., Pranata, I., Barmawi, A.M.: Exploiting the error correction mechanism in QR codes for secret sharing. In: Australasian Conference on Information Security and Privacy (2016)

6. Huang, P.C., Chang, C.C., Li, Y.H.: Sudoku-based secret sharing approach with cheater prevention using QR code. Multimedia Tools Appl. (1), 1–20 (2018)
7. Lin, P.Y.: Distributed secret sharing approach with cheater prevention based on QR code. IEEE Trans. Industr. Inform. **12**(1), 384–392 (2016)
8. Samretwit, D., Wakahara, T.: Measurement of reading characteristics of multiplexed image in QR code. In: International Conference on Intelligent NETWORKING and Collaborative Systems (2011)
9. Tkachenko, I., Puech, W., Destruel, C., Strauss, O., Gaudin, J.M., Guichard, C.: Two-level QR code for private message sharing and document authentication. IEEE Trans. Inf. Forensics Secur. **11**(3), 571–583 (2015)
10. Tuyls, P., Hollmann, H.D., Van Lint, J.H., Tolhuizen, L.M.G.M.: XOR-based visual cryptography schemes. Des. Codes Crypt. **37**, 169–186 (2005)
11. Wan, S., Lu, Y., Yan, X., Liu, L.: Visual secret sharing scheme with (k,n) threshold based on QR codes. In: International Conference on Mobile Ad-Hoc and Sensor Networks (2017)
12. Wang, G., Liu, F., Yan, W.Q.: 2D barcodes for visual cryptography. Multimedia Tools Appl. **75**(2), 1223–1241 (2016). https://doi.org/10.1007/s11042-014-2365-8
13. Wu, X., Sun, W.: Random grid-based visual secret sharing with abilities of OR and XOR decryptions. J. Vis. Commun. Image Represent. **24**(1), 48–62 (2013)
14. Yan, X., Liu, X., Yang, C.-N.: An enhanced threshold visual secret sharing based on random grids. J. Real-Time Image Proc. **14**(1), 61–73 (2018). https://doi.org/10.1007/s11554-015-0540-4
15. Yan, X., Lu, Y.: Participants increasing for threshold random grids-based visual secret sharing. J. Real-Time Image Proc. **14**(1), 13–24 (2018). https://doi.org/10.1007/s11554-016-0639-2
16. Yan, X., Wang, S., Niu, X.: Threshold construction from specific cases in visual cryptography without the pixel expansion. Sig. Proc. **105**, 389–398 (2014)

Hybrid Chain Based Hierarchical Name Resolution Service in Named Data Network

Zhuo Yu[1(✉)], Aiqiang Dong[1(✉)], Xin Wei[2(✉)], Shaoyong Guo[2(✉)], and Yong Yan[3(✉)]

[1] Research and Development Division, Beijing China-Power Information Technology Co., Ltd., Beijing, China
2227375991@qq.com

[2] State Key Laboratory of Networking and Switching Technology, Beijing University of Posts and Telecommunications, Beijing, China
vaisy@bupt.edu.cn

[3] "Da Yun Wu Yi" Power Application Technology Laboratory, State Grid Zhejiang Electric Power Research Institute, Hangzhou, China
55682381@qq.com

Abstract. Named Data Networking (NDN) is a promising framework which advocates ubiquitous in-network caching to enhance content delivery. The framework provides mechanisms for users to request content restricted by public key of its source or the cryptographic digest of the content object to avoid receiving fake content. However, it does not provide any mechanisms to learn this critical information. Blockchain is a distributed ledger technology which could supply enhanced security, easy extensibility and scriptable programmability. Till now, the technology has attracted lots of attentions in network management research. However, each type of blockchain has its drawbacks, which could hurt the efficiency or credibility of system. This paper combines different types of blockchain, leverages a hybrid chain based Hierarchical Name Resolution Service (HNRS) for NDN. The design could provide a credible, efficient, flexible HNRS for NDN, thus supply an important support for NDN to build a credible, efficient, flexible benign ecological environment.

Keywords: Blockchain · Data authentication · Future internet architecture · Name Data Network · Name Resolution Service

1 Introduction

Name Data Network (NDN, or CCN simplified by Content Centric Networking) is a proposed architecture which fetch content by content names. As in [1], NDN requires that not only can content be retrieved by name, but that the content so retrieved is valid – users can securely retrieve desired content by name, and authenticate the result regardless of where it comes from. The requirement means that users who request content must either have the public key of the user who publish the content, or know the digest of the content before requesting for it. However, NDN does not provide any solution to this proverbial "chicken-and-egg" problem [2]. Therefore, it is necessary to

C.-N. Yang et al. (Eds.): SICBS 2018, AISC 895, pp. 629–637, 2020.
https://doi.org/10.1007/978-3-030-16946-6_51

research name resolution mechanism for users retrieving desired content securely in NDN.

There have been a lot of research on name resolution. It is proposed in [3] that globally distinct DNS servers self-organize form a flat peer-to-peer network to provide fast lookups, automatically reconfigures and thwarts distributed denial of service attacks. It is proved that Distributed Hash Table (DHT) based DNS is more robust while traditional DNS has higher performance in [4, 5]. In [6], the difference between DNS and name resolution in NDN is discussed, and requirement to the latter which including dedicated name cache servers, multiple independent trust anchors at the top and robustness, and robustness is proposed. In [1], a hierarchical naming scheme based resolution service which could provide a system for the registration, storage and distribution of security information associated with namespaces in CCN is designed. As proved in [7], it is safe to store NDN data with hash chain. At the same time, a technology using hash chain to record transactions called blockchain attract the interest of researchers. In [8, 9], it is proposed that blockchain can be used to secure Internet sources since the technology can supply a framework with the elimination of any PKI-like root of trust, a verifiable and distributed transaction history log, multi-signature based authorizations for enhanced security, easy extensibility and scriptable programmability. However, blockchain can be divided into public chain and permissioned chain, and different blockchain has different features. In [10, 11], a public chain based name resolution is proposed. In [12], disadvantages of the public chain including unpredictable long-term status and uncertain costs are pointed out. Learn from above, we design a Hierarchical Name Resolution Service (HNRS) by combining public chain with permissioned chain for NDN. To the best of our knowledge, it is the first study that combines different types of blockchain for name resolution.

In our design, namespace can be divided into one global namespace and lots of local namespaces. Since permissioned chain can get trust by chaining up to public chain or other permissioned chains, we can build a hierarchical trust structure by taking public chain as root of trust. As it is easily for a permissioned chain to anchor in another public chain, the solution can reduce the risk of the decline of a specific public chain. In addition, a permissioned chain could design own name scheme autonomously, which could easily compatible with traditional programs. More than programmability, traceability and credibility of the traditional public chain design, hybrid chain based design could improve the efficiency, portability and flexibility.

The remainder of this paper is organized as follows. Section 2 briefly introduces name resolution process in NDN and how blockchain operates. Section 3 provides an overview of our design and operations in the design. Section 4 describes our implementation. Section 5 make a discussion for hybrid chain based application in communication and conclude the paper.

2 Preliminaries

2.1 Name Resolution Service in NDN

In NDN architectures, users who request data called consumers while who produce data called producers. Each peers in NDN including consumers and producers act as a forwarder. Packets in NDN can be divided into Data and Interest, and each packets will include a hierarchically structured name. To fetch desired Data packets, consumers send out Interest packets. Forwarders will forward Interest packets towards available data replicas by performing a longest prefix match on the name in their Forwarding Information Base (FIB). If a Data packet whose name exactly matches that in the Interest, the Data will be return back along the path toke by Interest.

Since Data may come from everywhere, it is imperative to verify the packet. Each packet will obtain a signature when it is created. Any changes to the content will make the signature invalid. The key locator field in signature field names the signing key to be used for the signature verification While a key locater in NDN is simply another Data containing the public key bits and the corresponding signature, which in turn contains its own key. These key locators form a certification chain up to a trust anchor key that a consumer must have obtained and trust a Priori [6].

In fact, fetch the Data by its names called name resolution. To find corresponding Data efficiently, forwarders need efficiently synchronize update FIB. Ensure the data is valid need trace trusted key credibly. As above, a NRS should synchronize and update the address and key of users efficiently and credibly.

2.2 How Blockchain Works

A blockchain is a distributed data structure that is replicated and shared among the members of a network [13]. Data will be encapsulated into block, and each block will contains the hash of previous block. By the way, a chain likes Fig. 1 forms. Once attend the network, members must synchronize with the whole network to update its chain, otherwise it will be abandoned by the network. The process that members synchronize with others called consensus. By consensus, chains obtained by different users tend to be consistent, so it is difficult to be corrupt or extinguished.

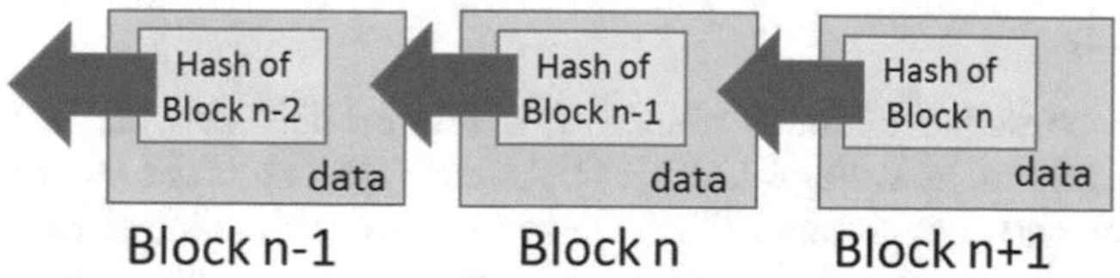

Fig. 1. Blockchain structure.

When a new member attends public chain, a pair of key will be generated by asymmetric encryption. If the public key can be viewed as account, the private key can be considered as the password of the account. The process can be operated offline, because it relies on fixed algorithms instead of approval of anyone. As the name suggests, everyone can attend to public chain while only permissioned nodes can attend to permissioned chain.

In public chain, the common way to reach consensus is Prove of Work (PoW) and Prove of Stake (PoS) or their derivative. In PoW, nodes who encapsulate the data need solve a puzzle then broadcast the block. A special transaction will be record in the block: the network will generate some fees like points to the node, and these points of public chain has derived currency significance. As nodes will choose the longest chain to synchronize, so the first node solve puzzle will get a fee as incentives. If a nodes want change a data without traces, it need solve puzzles in very fast speed to produce longer chain thus be synchronized by others. As it is difficult, we think data in public chain which adopt PoW is credible. In PoW, the one who solve the problem fastest get incentives while the one who have most assets in the chain get incentives in PoS. The reason why choose these methods is that, the one who paid more energy to solve puzzles or the one who have more asset in the chain always wants to keep the chain credible for his assets.

In permissioned chain, the process is simplified by restricting who can amend the database to a set of known entities. Nodes record data follow the rules not because they are getting paid but because other people in the network, who know their identities, hold them accountable [12]. Without worrying about trust, the operation efficiency in blockchain is improved. More importantly, operate costs is reduced by getting rid of incentives.

Smart contract means scripts stores in blockchain. When a smart contract be triggered, it will execute independently and automatically in a prescribed manner on every node in the network according to the data that was included in the triggering [13].

By blockchain, a system which could synchronize update data credibly and automatically can be implemented.

3 Hierarchical Name Resolution Service Design

In this section, we will present our own design for name resolution service in NDN and introduce its naming mechanism and name resolution mechanism.

3.1 Overview

Similar to the structure of today's Internet, a future NDN network is likely to be a network of autonomous systems (AS) [14]. As each AS is a profit seeking entities and use an interior gateway protocol (IGP) for optimizing their internal routes, we take a permissioned chain to manage one AS. Each AS coordinate with other AS using the Border Gateway Protocol (BGP), while the border gateway can exchange identify information with their neighbors. In this architecture, network can be divided into hierarchical ASes. For entities with business relationship, they can build a permissioned chain to cooperate with each other, thus form an AS, each member of the AS can be a Border Gateway (BG). As everyone can trust an information in blockchain, permission chain can get trusted by public chain.

To build an AS can be trusted by anyone, entity first need get a global namespace in public chain. To be verified, message recorded in public chain must contain its identity and accessible address of Border Gateway (BG). As the position of the message in public chain is uniqueness and tamper-resistant, it can be a namespace for AS. In this way, the public chain anchors the permissioned chain. Since permissioned chain can be customized, traditional system can be migrated to permissioned chain easily without hurt the efficiency of origin system.

The hierarchical name can be illustrated as Fig. 2. As one AS can be divided into several ASes, permissioned chain can anchor in higher level permissioned chain. For example, a permissioned chain managing second namespace called 'market/fruit' is valid only when it is chain up to the chain managing first namespace called 'market' while the latter need anchor in public chain because it manages a first namespace. Each AS has a namespace and chain up to another chain until public chain. Long-term status of a special public chain is unpredictable, while this design can be migrated to another public chain easily by update the top name. In this way, we construct a flexible name system for NDN.

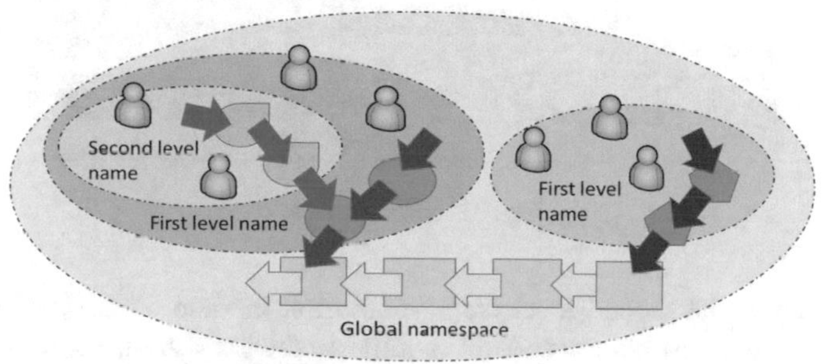

Fig. 2. Hierarchical name structure.

Consensus between ASes reaching by synchronizing the public chain, while consensus within AS reaching consensus by consistency algorithm. Because it is slow for public chain to reach consensus, we only ensure identity of entities and access it by public chain. Since the public chain can be trusted by everyone, entity can trust the information from other ASes. As mentioned in Sect. 2. A, packet will contain a signature field including a public key field which names the signing key to be used for the signature verification. Since the public key field can be the certificate issued by the entity, others can verify the certificate by the public key of the entity thus verify the data by the public key in the certificate. Finally, we get a credible name structure.

As above, by combining permissioned chain and public chain, we can construct an efficient, flexible, and credible NRS for NDN.

3.2 Setup

The NRS design for the system contains public chain based first level name resolution and permissioned chain based name resolution. As demonstrated in Fig. 3, one first get an account by account management module, and publish name transactions to public chain through data process module. Information extraction module synchronize with public chain and filter out information about name, resolving name in first level can be completed by name resolution module according to the information. Furthermore, if the one want to cooperate with others thus form an AS, he can build a permissioned chain. By member management module, joining and leaving of member can be executed in consensus. By contract management module, smart contract can be published by members, and be executed automatically. Name registration and name resolution in multilevel is developed as smart contracts.

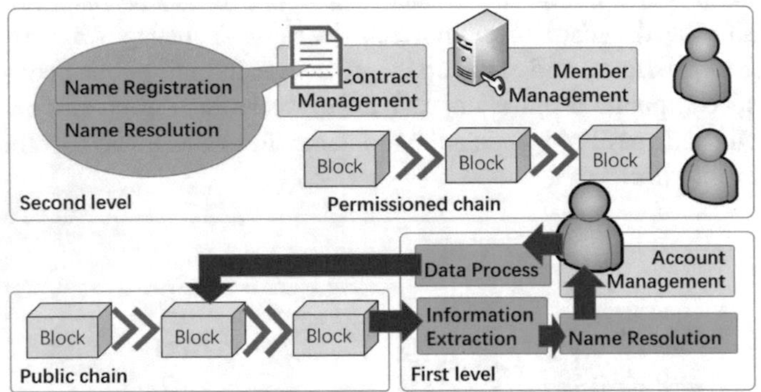

Fig. 3. NRS service.

Just like construct second level name resolution module based on the module in first level, one can construct lower level module in the way. And, nodes trust others in the same permissioned chain, so only several nodes in each AS install the name resolution module for upper level is OK.

3.3 Namespace and Naming

NDN use hierarchical name structure. In our architecture, each AS corresponds to a namespace. By appending names to namespace, new name forms. For example:

'447447.169' is a first level name.
'447447.169/' is a first level name zone.
'447447.169/x' is a second level name.

And so on for each name and namespace. Nodes in NDN publishes data its own namespace in the form of data packets.

To get a first level name, one must register an account in public chain. And he need to publish a transaction to public chain which contains his own information, the transaction will be encapsulated into block by miner. As we mentioned in Sect. 3. A, position of the transaction is immutable and unique, and it can be a first level name. In the way, a first level AS is born with a first level namespace. Everyone can choose to believe the AS or not by the information exposed in public chain.

After getting a first level name, the one who register the name in public chain can build a permissioned chain by issuing certificate to nodes that he can trust. As each node in the AS can get a name by him, second name is born. The formation process of third level namespace is similar to second level name. It is noteworthy that each namespace has been authorized by its parent namespace.

Because the operation of name is not only registered and resolution, but also update and transfer, we design a service for name resolution in public chain called ODIN [15], which could encapsulate message submitted by nodes into transaction. The message can be divided into three types: Register, update and transfer. In register or update information, nodes need submit information to be verified, while in transfer information, the destination address should be the next owner of the name. According to the format of information, the ODIN message will be encapsulated in a transaction which meets the format of transaction in public chain. Similar to name service in public chain, permissioned chain also need provide name service by smart contract.

3.4 Name Resolution

As we mentioned in Sect. 2. A, communication in NDN is implemented through name based forwarding. The process is detailed in this part. Since name resolution within AS is the same as tradition network, we just introduce operations beyond an AS.

We would take the example in Sect. 3. A to illustrate the work of name resolution. When a forwarder in 'market' received an Interest for 'market/fruit/apple', it will run the longest match algorithm in its FIB and forward it to 'market/fruit', while when it receives an Interest for 'farm/fruit/apple', it will forward it to 'farm' according to the accessible address in the message recorded in public chain (If it not synchronizes with public chain, it can forward it to the BG of its AS). When a consumer belongs to AS called 'market' receive a data called 'farm/fruit/apple', it will find a certificate in the data. The certificate contains the public key of 'farm/fruit', and the certificate has been signed by 'farm'. Consumer can verify the certificate by the public key of 'farm' in public chain, then verify the signature by the public key in the certificate. In this way, communication across ASes can be completed.

4 Implementation

To verify and exercise the proposed design, we have developed a prototypes of the hybrid chain based HNRS system.

First, we take bitcoin as the public chain, and write a java based tool for publishing name message in Bitcoin and synchronizing by API applied by bitcoin. The work has been implemented in [15], user can register, trade and query namespace by synchronizing with blockchain in the web. And we register a first level name in bitcoin. Then we build a fabric [16] based permissioned chain for managing second level name, and write smart contract for permissioned chain to manage namespace in its own AS. Finally, we build a NDN by ccn-lite [17], and revise the code of forwarder to support the parse for hybrid chain based name. To simulate communications cross ASes, we build a simulate network as Fig. 4.

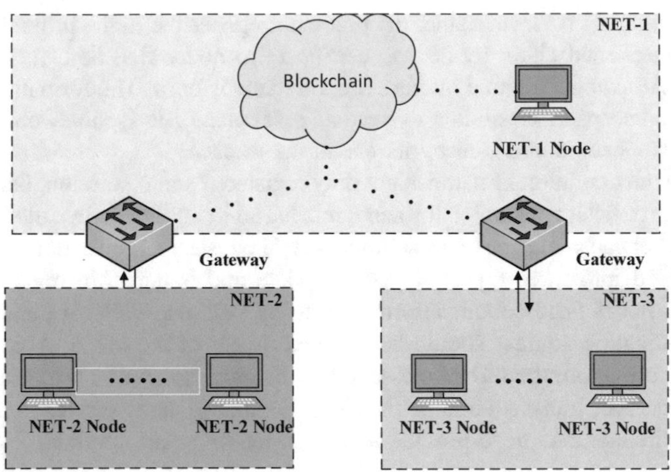

Fig. 4. Simulate network.

Then, we call the public network as NET-1, and set a node in NET-1 to synchronizing with blockchain, and filter out the naming information from blocks in public chain. Then we build NET-2 and NET-3 as ASes. Both NET-2 and NET-3 contains one gateway and several nodes. These gateways act as BG to manage the name information in its own AS and keep connectivity with NET-1. When a node wants publish contents, it will run the smart contract in the permissioned chain of the AS to register a name for its data. Communications among ASes can be completed as Sect. 3 demonstrated.

As lots of data is published in public chain, the cost of synchronizing with blockchain is enormous. In the simulation, we use NET-1 node to synchronize with public chain as a name resolve server. When others need resolve name in global namespace, they just query it in NET-1 node. When they find the address in bitcoin of the record, then go to bitcoin to query the record by website [18].

5 Discussions Conclusion

In this paper, we propose a hybrid chain based HNRS for NDN and describe it. By hybrid chain, users can verify contents easily and credibly. Different from other blockchain based solution, we introduce permissioned chain to reconcile the risk and inefficiency of public chain. The design inherits the security of public chain and the customizability of permissioned chain, provide a credible, efficient, flexible HNRS for NDN.

By the powerful HNRS, users can trust data and service supplied in NDN. Predictively, the research will provide an important support for NDN to build a sustainable development of benign ecological environment.

Acknowledgement. This work was supported in part by State Grid Corporation of Science and Technology Project (Grant: 52110417000G).

References

1. Smetters, D., Jacobson, V.: Securing network content. Technical report, PARC, October 2009
2. Mahadevan, P., et al.: CCN-KRS: a key resolution service for CCN. In: Proceedings of the 1st International Conference on Information-Centric Networking, Paris, France, pp. 97–106. ACM, (2014)
3. Ramasubramanian, V.: The design and implementation of a next generation name service for the internet. ACM SIGCOMM Comput. Commun. Rev. **34**(4), 331–342 (2004)
4. Cox, R., Muthitacharoen, A., Morris, R.T.: Serving DNS using a peer-to-peer lookup service. In: LNCS, pp. 155–165. Springer, Berlin (2002)
5. Pappas, V., Massey, D., Terzis, A., Zhang, L.: A comparative study of the DNS design with DHT-based alternatives. In: Proceedings IEEE INFOCOM 2006. 25th IEEE International Conference on Computer Communications, Barcelona, Catalunya, Spain, pp. 1–13 (2006)
6. Afanasyev, A., Jiang, X., Yu, Y., Tan, J., Xia, Y., Mankin, A., Zhang, L.: NDNS: a DNS-like name service for NDN. In: Proceedings of 2017 26th International Conference on Computer Communication and Networks (ICCCN), Vancouver, Canada, pp. 1–9 (2017)
7. Refaei, T., Horvath, M., Schumaker, M., Hager, C.: Data authentication for NDN using hash chains. In: Proceedings of 2015 IEEE Symposium on Computers and Communication (ISCC), Golden Bay, Larnaca, Cyprus, pp. 982–987 (2015)
8. Bozic, N., Pujolle, G., Secci, S.: A tutorial on blockchain and applications to secure network control-planes. In: 2016 3rd Smart Cloud Networks & Systems (SCNS), Dubai, UAE, pp. 1–8 (2016)
9. Hari, A., Lakshman, T.V.: The Internet blockchain: a distributed, tamper-resistant transaction framework for the Internet. In: ACM Workshop on Hot Topics in Networks, Atlanta, Georgia, USA, pp. 204 210. ACM (2016)
10. Fotiou, N., Polyzos, G.C.: Decentralized name-based security for content distribution using blockchains. In: Proceedings of the IEEE Conference on Computer Communications Workshops (INFOCOM WKSHPS), San Francisco, CA, USA, pp. 415–420 (2016)
11. Benshoof, B., Rosen, A., Bourgeois, A.G., Harrison, R.W.: Distributed decentralized domain name service. In: 2016 IEEE International Parallel and Distributed Processing Symposium Workshops (IPDPSW), Chicago, Illinois, USA, pp. 1279–1287 (2016)
12. Peck, M.E.: Blockchain world - Do you need a blockchain? This chart will tell you if the technology can solve your problem. IEEE Spectr. **54**(10), 38–60 (2017)
13. Christidis, K., Devetsikiotis, M.: Blockchains and smart contracts for the Internet of Things. IEEE Access **4**(4), 2292–2303 (2016)
14. Pacifici, V., Dan, G.: Coordinated selfish distributed caching for peering content-centric networks. IEEE/ACM Trans. Netw. **24**(6), 3690–3701 (2016)
15. PPkpub project: Blockchain based communication project. https://github.com/ppkpub/javatool. Accessed Dec 2016
16. Hyperledger/Fabric: Blockchain fabric incubator code. https://github.com/hyperledger/fabric. Accessed Mar 2017
17. Ccn-lite project: Implementation of the CCNx and NDN protocols. https://github.com/cn-uofbasel/ccn-lite. Accessed Mar 2016
18. Blockchain.info: Bitcoin block information query online. https://blockchian.info. Accessed Dec 2016
19. Le Bon, G.: The Crowd: A Study of the Popular Mind. Fischer, Frankfurt (1897)

A Secret Sharing Scheme of Dynamic Threshold with Cheaters Detection and Identification

Yong-Zhong He[1], Xue-Jiao Xu[1,2], Dao-Shun Wang[2(✉)], Shun-Dong Li[3], Xiang-Hui Zhao[4], and Ching-Nung Yang[5]

[1] Beijing Key Laboratory of Security and Privacy in Intelligent Transportation, School of Computer and Information Technology, Beijing Jiaotong University, Beijing, China

[2] Department of Computer Science and Technology, Tsinghua University, Beijing, China

daoshun@mail.tsinghua.edu.cn

[3] School of Computer Science, Shaanxi Normal University, Xi'an, China

[4] China Information Technology Security Evaluation Center, Beijing, China

[5] Department of Computer Science and Information Engineering, National Dong Hwa University, Hualien, Taiwan

Abstract. There are many schemes to detect and identify cheaters when the number of participants is exactly equal to t in secret sharing schemes. However, most of them need dealers or redundant information to detect the dishonest participants when participants are greater than threshold t. Harn *et al.* proposed a dynamic threshold secret reconstruction scheme, which the threshold can be improved to k during reconstruction. Less than $k-1$ participants cannot recover the shared secret. Their scheme uses the symmetric polynomial to resist the external adversary. However, each participant needs to hold a polynomial. Thus, the complexity of storage of the scheme is high. Furthermore, the scheme cannot detect and identify the cheaters who present falsified information during secret reconstruction. In this paper, we propose a secret sharing scheme with dynamic threshold k that can identify up to $k-1$ dishonest participants who have no legal share information. The scheme does not need the help of the dealer to detect cheaters. The dealer uses the traditional polynomial to distribute a secret to each participant. In the secret reconstruction phase, the share information of each participant is transformed and then presents to other participants who verify it using the public key of the sender. If the presented information is falsified, it will fail the verification and the related participant is identified as a cheater. Otherwise, the secret can be correctly reconstructed. Meanwhile, our scheme ensures that the external adversary who eavesdrops the reconstruction messages cannot recover the secret. The trade-off is that our scheme has additional computation overhead.

Keywords: Cheater detection · Dishonest participants · Secret sharing

C.-N. Yang et al. (Eds.): SICBS 2018, AISC 895, pp. 638–649, 2020.
https://doi.org/10.1007/978-3-030-16946-6_52

1 Introduction

Shamir proposed the (t, n) secret sharing scheme based on the Lagrange polynomial in 1979 [1]. In the scheme, a secret S is divided into n shares and distributed among a set of n shareholders through the secret access, such that any t or more than t shares can reconstruct the secret S by the Lagrange polynomial, and fewer than t shares cannot reconstruct the secret S. In the process of the secret recovery, the shares are showed by the participants, which may be eavesdropped and tampered by the external adversary. For this reason, Chor *et al.* [2] first proposed the concept of verifiable secret sharing in 1985. In 1989, Laih *et al.* [3] proposed the dynamic threshold scheme, aiming to resist the information leakage or ensure the security of the scheme changed by increasing the number of participants. After that, resisting external adversary attacks, a dynamic verifiable secret sharing scheme [4–7] and dynamic multi-secret sharing scheme [8–11] based on RSA assumption, discrete logarithm hypothesis, elliptic curve and bilinear mapping were proposed successively to resist the information eavesdropped and tampered.

Yuan *et al.* [7] proposed the threshold changeable secret sharing schemes using a two-variable one-way function, and the scheme can resist collusion attacks and could store the information security by the combiner. By encrypting the share with RSA and changing the threshold by the combiner, the scheme can verify the correctness of the share information by the participants, improving the traditional scheme. However, it needs an extra combiner, which should be safe and credible. Wei *et al.* [12] proposed a scheme based on the bilinear maps, which does not depend on any secure channel between the dealer and the participants. For all the schemes above, in the reconstruction of the secret, they only consider the scenario that exact t shareholders reconstruct the secret. However, when there are more than t shareholders reconstructing the secret, the scheme is not secure under the attacks by dishonest participants. In order to prevent an external adversary, Harn *et al.* [13] proposed a new encryption verification method based on the symmetry of binary symmetric polynomials.

Harn *et al.* proposed a dynamic threshold secret reconstruction scheme with the threshold that can be improved to k during reconstruction. Less than $k - 1$ participants cannot recover the shared secret. Their scheme uses the symmetric polynomial to resist the external adversary. However, each participant needs to hold a polynomial. Thus, the complexity of storage of the scheme is high. Furthermore, the scheme cannot detect and identify the cheaters who present falsified information during secret reconstruction. Traverso *et al.* [15] proposed the dynamic and verifiable hierarchical secret sharing scheme based on Birkhoff interpolation where shareholders are divided into disjoint levels and the power of a shareholder to reconstruct the secret depending on the level. It can achieve the verifiable of the share. However, it does not consider the situation that the dishonest participants do not show their own shares but still recover the secret correctly.

In this paper, we propose a new method to detect the dishonest participants who use the fake information in reconstruction stage. Our scheme depends on univariate polynomial and ElGamal encryption to distribute the secret and verify the share. Because in reconstruction stage, the share of each participant is sent in ciphertext, no external adversary can obtain the share information of the participants. If the dishonest participants present false share in ciphertext, other participants receiving the message can verify it based on its signature, so that the cheater can be identified.

The rest of this paper is organized as follows. Section 2 introduces the ElGamal signature algorithm and Harn *et al.* scheme [14]. Section 3 presents our scheme and discusses the security and correctness of the scheme. Section 4 compares the scheme in attack resistance and performance with Harn *et al.* scheme. The conclusion is in Sect. 5.

2 Preliminary Work

In this section, we introduce the preliminary work of the ElGamal signature algorithm and Harn *et al.* scheme [14].

2.1 ElGamal Signature Algorithm

ElGamal algorithm [16] can be used for both data encryption and digital signature, and its security depends on the computation of discrete logarithm problems in finite fields. The algorithm is described as follows.

Data encryption
Step1: Random selection of a large prime number p, two random numbers g and x, where ($g < p$, $x < p$).
Step2: Compute $y = g^x \bmod p$, x is the private key, y is the public key, g and p is the public information.
Step3: Encryption algorithm: the sender selects the random number v, $1 < v < p-1$, v and $p-1$ are coprime. The sender uses the plaintext M to compute $d = g^v \bmod p$, $m = y^v M \bmod p$, and gets the ciphertext (d, m).
Step4: Decryption algorithm: the receiver receives the ciphertext (d, m) and computes the plaintext by $M = m/d^x \pmod{p}$.
As:
$M = m/d^x \pmod{p} \equiv y^v M/(g^v)^x \pmod{p} \equiv g^{xv} M/g^{vx} \pmod{p} \equiv M \pmod{p}$.

Digital signature
Step1: The initial process: Let p is a large prime. The integer g is a generator of Z_p^*. The private key of the sender is $x(1 < x < p-1)$. The corresponding public key is $y = g^x \bmod p$. **Step2:** The signature process: For the message to be signed m, the sender picks a random number v, $1 < v < p-1$, v and $p-1$ are coprime. Computes $r = g^v \bmod p$ and $s = (m - xr)v^{-1} \bmod (p-1)$, and let the (r, s) as the Signature of the message, and sends the m and (r, s) to receiver. **Step3:** The verification process: The receiver receives the m and (r, s), and verifies the equation $g^m = y^r r^s \bmod p$. If it is satisfied, the signature verification is OK, otherwise failed. As: $g^{vs} = g^{(m-xr)} \bmod p \Rightarrow (g^v)^s = g^m (g^x)^{\,r} \bmod p \Rightarrow g^m = y^r r^s \bmod p$.

2.2 Harn *et al.* Scheme

In the Harn *et al.* [14] scheme, in the secret distribution stage, the secret is a linear combination of the polynomial. The dealer uses the characteristic of symmetric polynomials to form the symmetric key to encrypt the sub-secret information. It can achieve the verifiable secret sharing without interactivity. In the secret reconstruction stage, each participant calculates the sub-secret by using the share. Then, the participant uses the symmetric key to encrypt the sub-secret and transmit the information to the other participants. Therefore, it has a problem that the dishonest participants not show their sub-secret can still get the secret without noticing.

Harn's scheme is as follows:

Initialization
Input: p is a large prime. Z_p is a finite field. S is the secret, where $S \in Z_p$. $P = \{P_1, P_2 \cdots, P_n\}$ represents a set of participants. t is the threshold, n is the number of participants. The identity value $\{v_1, v_2, \cdots, v_n\}$, $v_i \notin \{0,1\}$ is the value of the identity of the participants. **Output:** Reconstruct the secret S.

Secret distribution
Step1: The dealer selects a $t-1$ degree symmetric polynomial, where t is the threshold, $f(x,y)=a_{0,0}+a_{1,0}x+a_{0,1}y+\cdots+a_{t-1,0}x^{t-1}+a_{0,t-1}y^{t-1} \bmod p$, where $a_{i,j}\in Z_p$, $a_{i,j}=a_{j,i}$, $\forall i,j\in[0,t-1]$, and the secret $S\in Z_p$ satisfies that: $S=f(0,0)+bf(1,1)$, where $b\in Z_p$ and public the information b. **Step2:** The participant P_i chooses the corresponding n random numbers $\{v_1,v_2,\cdots,v_n\}$, $v_i\notin\{0,1\}$ as the value of the identity of the participants. The dealer computes $s_i(y)=f(v_i,y) \bmod p$, $i=1,2,\cdots n$ and sends the share to the participant P_i $(1<i<n)$ secretly.

Secret reconstruction
Step1: When k $(t\le k\le \frac{t(t+1)}{2}+1\le n)$ participants want to recover the secret, each participant P_i $(1<i<k)$ retrieves the public information b, identity value and his own shares to compute the sub-secret: $$w_i=s_i(0)\prod_{l=1,l\ne i}^{k}\frac{-v_l}{v_i-v_l}+bs_i(1)\prod_{l=1,l\ne i}^{k}\frac{1-v_l}{v_i-v_l} \bmod p.$$ **Step2:** A symmetric encryption key $k_{ij}=k_{ji}$ is built between two participants P_i and P_j by computing $k_{ij}=f(v_i,v_j)=f(v_j,v_i)=k_{ji}$, $i=1,2,\cdots k$, $j=1,2,\cdots k$, $i\ne j$, which can use to encrypt the sub-secret. Then each participant sends the encrypted sub-secret to the other participant P_j, $i\ne j$. After receiving the information each participant decrypts the encrypted information and computes the secret $S=\sum_{i=1}^{k}w_i \bmod p$. So we can reconstruct the secret S.

In the Harn *et al.* scheme, the secret is a linear combination of the polynomial, w_i is a linear combination, ensuring that other participants are not able to obtain any information about the secret. Through binary symmetric polynomials, a pair of symmetric encryption keys is formed between every two participants to encrypt the sub-secret information preventing external adversary from eavesdropping and tampering with the transmission of information. However, when the $k(t\le k\le \frac{t(t+1)}{2}+1)$ participants want to recover the secret, the dishonest participant can deduce the i participant's share information according to the sub-secret information sent by the $(2t\le i\le k)$ participant. Because of $w_i=s_i(0)\prod_{l=1,l\ne i}^{k}\frac{-v_l}{v_i-v_l}+bs_i(1)\prod_{l=1,l\ne i}^{k}\frac{1-v_l}{v_i-v_l} \bmod p$ we know $\prod_{l=1,l\ne i}^{k}\frac{-v_l}{v_i-v_l}$, $\prod_{l=1,l\ne i}^{k}\frac{1-v_l}{v_i-v_l}$, where $s_i(1)$ is a polynomial with an unknown parameter of t. One weakness of the scheme is that a dishonest participant can get other participants' shares to recover the secret. In addition, because that the scheme must meet the requirement $(t \leq k \leq \frac{t(t+1)}{2} + 1 \leq n)$ for reconstructing the secret in the reconstruction process.

3 Our Scheme

In this section, we present our scheme in details. Our scheme uses the improved ElGamal signature to detect the dishonest participants. We use the ElGamal to encrypt the transmission information that the external adversary cannot eavesdrop and tamper the information. We use fewer shares than Harn *et al.* scheme to identify the dishonest participants and the external adversaries.

3.1 Assumptions and Security Model

The security model for our scheme is described below.

- The dealer is trusted and the communication channel between the dealer and participants is secure and authenticated.
- Participants include the honest participant, the dishonest participants and the external adversaries.

Honest participants: the honest participants have the secret share information and are strict compliance with the requirements of the scheme, ensuring the confidentiality of their own information and the participants do not collusion with any participants in the secret scheme.

Dishonest participants: the dishonest participants are the illegal passive participants. They use the fake information to take part in the secret reconstruction phase but do not have a legal identity.

External adversaries: the external adversary will eavesdrop and tamper the information exchanged between any participants.

3.2 The Algorithm

The Algorithm are presented in three stages: the initialization stage, the secret distribution stage, and the secret reconstruction stage. Based on the security model, in the (t, n) secret sharing scheme, t is the threshold, n is the number of participants $(2 \leq t \leq n)$, S is the secret that the participants want to recover. P_i is the $i(1 \leq i \leq n)$ participant. v_i is unique identity of each participant. In the reconstruction stage, there are k participants to reconstruct the shared secret S. The scheme can ensure that any $k < t$ or more than $k - 1$ participants cannot reconstruct the secret S, and only exact k participants can reconstruct the secret S. In the secret reconstruction stage, the share information of each participant is encrypted and then presented to other participants who verify the share using the public key of the sender. If the presented information is falsified, it will fail the verification and the related participant is identified as cheater.

Otherwise, the secret can be correctly reconstructed. Furthermore, the dishonest participants cannot decrypt the ciphertext to get the sub-secret because the dishonest participants do not have the legal identity security information to decrypt.

Initialization
Input: p is a large prime. Z_p is a finite field. S is the key secret, where $S \in Z_p$. $P = \{P_1, P_2 \cdots, P_n\}$ represents a set of participants. t is the threshold, n is the number of participants. The integer g is a random generator of Z_p, which can obtain the identity value of each participant s_i. What's more, the identity value is the public key of each participant P_i $(i = 1, 2, \cdots, n)$. k $(t < k \le n)$ is number of the reconstruction participants. **Output:** reconstruct and verify the correctness of the secret S.

Secret distribution
Step1: The participant P_i $(i = 1, 2, \cdots, n)$ randomly selects a non-zero random number x_i as its private key value and computes $s_i = g^{x_i} \notin \{0,1\}$ as the identity value of the participant P_i, which is the public key and public the value. Meanwhile, the dealer should ensure the uniqueness of public key values. Otherwise, the participant should reselect the private key and compute the corresponding public key. **Step2:** The dealer randomly selects t integers that are not fully zero $a_i\,(i = 0, 1, 2, \cdots, t-1)$, where $a_i \in Z_p$, $a_{t-1} \neq 0$, to construct the $t-1$ degree polynomial $f(x) = a_0 + a_1 x + a_2 x^2 + \cdots + a_{t-1} x^{t-1} \bmod p$, and the secret $S \in Z_p$ satisfies that $S = f(0) + bf(1)$, where $b \in Z_p$ and public the value b. **Step3:** The dealer uses the public key s_i to compute the share $y_i = f(s_i) \bmod p$, and sends it to the participant P_i, $i = 1, 2, \cdots, n$. The participant P_i needs to keep the private key x_i and the share y_i secret.

Secret reconstruction
Verification steps: **Step1:** The participant P_i $(i = 1, 2, \cdots, k)$ $(t < k)$ who want to reconstruct the secret publish their own identify values (public key) s_i $(i = 1, 2, \cdots, k)$. Then each participant P_i uses the formula $w_i = f(s_i)\prod_{l=1,l\neq i}^{k}\frac{0-s_l}{s_i - s_l} + bf(s_i)\prod_{l=1,l\neq i}^{k}\frac{1-s_l}{s_i - s_l} \bmod p$ to compute the sub-secret w_i. **Step2:** The participant P_i uses the ElGamal signature algorithm to encrypt and sign the sub-secret w_i. When the signature of information verified successfully, the participants can decrypt the encrypted information to recover the secret S. The signature rule is as follows. The participant P_i picks a random number v_i, $1 < v_i < p - 1$, v_i and $p - 1$ are coprime. Using the sub-secret w_i to compute $m_{ij} = s_j^{v_i} w_i \bmod p$, we can obtain the ciphertext m_{ij}. The P_i computes $d_i = g^{v_i} \bmod p$ and $r_i = (m_{ij} - x_i d_i) v_i^{-1} \bmod (p-1)$ to obtain the signature of information (d_i, r_i). The participant P_i sends the (d_i, r_i) and the m_{ij} $(i \neq j)$ to the participant P_j $(i \neq j)$, so does the participant P_j $(i \neq j)$. **Step3:** The participant P_j $(i \neq j)$ receives the (d_i, r_i) and the m_{ij} from the participant P_i, verify the equation $g^{m_{ij}} = s_i^{d_i} d_i^{r_i} \bmod p$. If it is not true, it means the sub-secret has been tampered and the participant is dishonest. The participant P_i also verifies the information received from the participant P_j. When it is correct, the participant P_i computes $w_i = m_{ij} / d_i^{x_j} \pmod p$ to get the sub-secret w_i. The dishonest participant does not have the privacy key, so it cannot decrypt the ciphertext m_{ij} and it cannot obtain the sub-secret from other participants. **Reconstruction steps:** After computing the sub-secret, each participant can use the formula to compute the secret $S = \sum_{i=1}^{k} w_i \bmod p$. Then we can reconstruct the secret S.

3.3 Correctness Analysis

Theorem 1. Any *t* or more than *t* shares of the participants can reconstruct the secret *S*, and fewer than *t* shares cannot reconstruct the secret *S*.

- Firstly, we will prove that any *t* or more than *t* shares of the participants can reconstruct the secret *S*.

Because of the equation $S = \sum_{i=1}^{k} w_i \bmod p$, if the number of w_i is k, we can reconstruct the secret, where $t \le k \le n$. Otherwise, if the number of w_i is few than k, we cannot reconstruct the secret.

Because of $w_i = f(s_i) \prod\limits_{l=1, l \ne i}^{k} \frac{0 - s_l}{s_i - s_l} + bf(s_i) \prod\limits_{l=1, l \ne i}^{k} \frac{1 - s_l}{s_i - s_l} \bmod p$, we can get the following formula.

$$\begin{aligned}\sum_{i=1}^{k} w_i \bmod p &= \sum_{i=1}^{k} \left(f(s_i) \prod_{l=1, l \ne i}^{k} \frac{0 - s_l}{s_i - s_l} + bf(s_i) \prod_{l=1, l \ne i}^{k} \frac{1 - s_l}{s_i - s_l} \bmod p \right) \\ &= \sum_{i=1}^{k} f(s_i) \prod_{l=1, l \ne i}^{k} \frac{0 - s_l}{s_i - s_l} + b \sum_{i=1}^{k} f(s_i) \prod_{l=1, l \ne i}^{k} \frac{1 - s_l}{s_i - s_l} \bmod p \\ &= f(0) + bf(1) \\ &= S \end{aligned} \tag{1}$$

- Then we will prove that the fewer than the t shares cannot reconstruct the secret S. Suppose that, the participant P_j, $(1 \le j \le k)$ cannot show their sub-secret w_j. When recovery the secret, other participants compute the following formula.

$$\begin{aligned}\sum_{i=1}^{k-1} w_i \bmod p &= \sum_{i=1}^{k-1} \left(f(s_i) \prod_{l=1, l \ne i}^{k} \frac{0 - s_l}{s_i - s_l} + bf(s_i) \prod_{l=1, l \ne i}^{k} \frac{1 - s_l}{s_i - s_l} \bmod p \right) \\ &= \sum_{i=1}^{k-1} f(s_i) \prod_{l=1, l \ne i}^{k} \frac{0 - s_l}{s_i - s_l} + b \sum_{i=1}^{k-1} f(s_i) \prod_{l=1, l \ne i}^{k} \frac{1 - s_l}{s_i - s_l} \bmod p \\ &\ne (0) + bf(1) \bmod p \\ &\ne S \end{aligned} \tag{2}$$

Fewer than t shares cannot reconstruct the secret, so the participant P_j cannot reconstruct the secret.

3.4 Security Analysis

Theorem 2. When $k(t \le k \le n)$ participants recover the secret, they are able to resist the collusion attacks of the $k - 1$ participants. So the set of the $k - 1$ participants cannot recover the secret, and each participant cannot obtain any shares information according to the received information from other participants.

- Because each participant reconstructs the polynomial of degree $k - 1$, only when $k(t \le k \le n)$ participants reconstruct the secret honestly, the secret can be reconstructed to resist the collusion attack of $k - 1$ participants.
- In the reconstruction process, when the sub-secret is sent to the other participants to recover the secret, the participant may use Lagrange interpolation to infer the sub-secret information, which received from other participants. However, because the

sub-secret is once a time, the sub-secret cannot use in the next time. From the formula (2) we can know that the dishonest participants cannot obtain the sub-secret information held by other participants.

4 Discussions

Compared with the Harn's scheme, our scheme can resist the dishonest and external adversary attack to guarantee the security of secret information. Moreover, we can detect which one is the dishonest participant in the reconstruction phase.

4.1 Security Provisions

In our scheme, participants can verify the share information provided by other participants to decide if there is cheating and who is the cheater.

Suppose that the participant P_i is dishonest. In the secret reconstruction step, when the participant P_i sends the false information m'_{ij} and (d'_i, r'_i) to the participant $P_j(i \neq j)$, the participant $P_j(i \neq j)$ computes the $g^{m'_{ij}}$ using the public key to verify the equation $g^{m_{ij}} = s_i^{d_i} d_i^{r_i} \bmod p$. As follows.

$$s_i^{d_i} d_i^{r_i} \bmod p = (g^{x_i})^{d_i} d_i^{(m_{ij}-x_i d_i) v_i^{-1}} = (g^{x_i})^{d_i} g^{v_i (m_{ij}-x_i d_i) v_i^{-1}} = (g^{x_i})^{d_i} g^{(m_{ij}-x_i d_i)} = g^{m_{ij}} \quad (3)$$

Therefore, the participant P_j can verify that P_i is the dishonest participant (Table 1).

Table 1. Comparison table

Schemes	Resist external adversary	Share security	Collusion attack	Verify the cheater
Harn's scheme	Resist eavesdrop and tamper behavior	Share can be inferred	Exist $j(t \leq j \leq k)$ participant collusion attack	NO
Our scheme	Resist eavesdrop and tamper behavior	No share can be inferred	Resist $j(t \leq j \leq k)$ participant collusion attack	YES

Even if the participant P_i obtains other participant's information, the dishonest participant does not have the illegal identity information to decrypt the ciphertext to calculate the sub-secret using the equation $w_i = m_{ij} / d_i^{x_j} (\text{mod} p)$. Therefore, the dishonest participants cannot recover the secret.

4.2 Performance

Because our scheme uses public key signature and univariate polynomial, the storage needed by participants is smaller than Harn's scheme. Other differences are illustrated in Table 2.

Table 2. Comparison table

	The share of the participants	Transmission encryption	Key management	Number of participants
Harn's scheme	Univariate polynomial coefficients	Polynomial encryption	Store a pair of keys between any two participants	$k \le 1 + \frac{t(t+1)}{2} \le n$
Our scheme	Share and privacy key	Improved ElGamal Encryption	Only need to store own private key and decrypt all the ciphertext received	$t < k \le n$

As can be seen from the previous table, because of the boundary requirements for the number of the reconstruction participants k, Harn's scheme should meet $(t \le k \le \frac{t(t+1)}{2} + 1 \le n)$. In our scheme, because that we construct the univariate polynomial, the relationship is $t < k \le n$. The scheme ensures that the number of participants n increases linearly with the increase of reconstruction participants k. The drawbacks of our scheme is that it causes additional computation overhead and space overhead in the verification.

5 Conclusion

In this paper, we propose a (t, n) secret sharing scheme of dynamic threshold that can identify up to $k-1$ dishonest participants who have no legal share information. Any set of the $k-1$ participants cannot recover the secret, and the external passive adversary cannot recover the secret. During the distribution phase, the participants only need to store their private key x_i and the share y_i distributed from the dealer. In the reconstruction phase, participants will firstly verify the identity of other participants and the correctness of sub-secret information. If the verification is correct, the participants can recover the secret using the Lagrange polynomial. The ElGamal encryption scheme is used to encrypt the information before transmission, ensuring that sub-secret is not eavesdropped and tampered by the external adversary in the transmission. ElGamal signature is used to identify dishonest participants who present false share information.

Acknowledgment. This work was supported in part by the National key R&D program of China under Grant 2017YFC0820102, the National Natural Science Foundation of China under Grant U1536102, and Grant U1536116, in part by China Mobile Research Fund Project (MCM20170407), and State Administration of Press, Publication, Radio, Film and Television (SAPPRFT) Key Laboratory of Digital Content Anti-Counterfeiting and Security Forensics.

References

1. Shamir, A.: How to share a secret. Commun. ACM **22**(11), 612–613 (1979)
2. Chor, B., Goldwasser, S., Micali, S., Awerbuch, B.: Verifiable secret sharing and achieving simultaneity in the presence of faults. In: 26th Annual Symposium on Foundations of Computer Science, pp. 383–395. IEEE (1985)
3. Laih, C.S., Harn, L., Lee, J.Y., Hwang, T.: Dynamic threshold scheme based on the definition of cross-product in an N-dimensional linear space. In: Advances in Cryptology, pp. 286–298. Springer, New York (1989)
4. Min, Z., Huang, T.L., Min, Z.: A RSA keys sharing scheme based on dynamic threshold secret sharing algorithm for WMNs. In: International Conference on Intelligent Computing and Integrated Systems, pp. 160–163 (2010)
5. Wang, F., Zhou, Y.S., Li, D.F.: Dynamic threshold changeable multi-policy secret sharing scheme. Secur. Commun. Netw. **8**(11), 1002–1008 (2015)
6. Tadayon, M.H., Khanmohammadi, H., Arabi, S.: An attack on a dynamic multi-secret sharing scheme and enhancing its security. In: Electrical Engineering, pp. 1–5 (2013)
7. Yuan, L., Li, M., Guo, C., Choo, K.R., Ren, Y.: Novel threshold changeable secret sharing schemes based on polynomial interpolation. PLoS ONE **11**(10), 1–19 (2016)
8. Qu, J., Zou, L., Zhang, J.: A practical dynamic multi-secret sharing scheme. In: IEEE International Conference on Information Theory and Information Security, pp. 629–631. IEEE (2010)
9. Harn, L.: Secure secret reconstruction and multi-secret sharing schemes with unconditional security. Secur. Commun. Netw. **7**(3), 567–573 (2014)
10. Tadayon, M.H., Khanmohammadi, H., Sayad Haghighi, M.: Dynamic and verifiable multi-secret sharing scheme based on Hermite interpolation and bilinear maps. IET Inf. Secur. **9** (4), 234–239 (2014)
11. Pilaram, H., Eghlidos, T.: A lattice-based changeable threshold multi-secret sharing scheme and its application to threshold cryptography. Scientia Iranica **24**(3), 1448–1457 (2017)
12. Wei, C., Xiang, L., Bai, Y., Gao, X.: A new dynamic threshold secret sharing scheme from bilinear maps. In: International Conference on Parallel Processing Workshops, pp. 19–22. IEEE Computer Society (2007)
13. Harn, L., Hsu, C.F.: A novel threshold cryptography with membership authentication and key establishment. Wirel. Pers. Commun. **97**(11), 1–8 (2017)
14. Harn, L., Hsu, C.F.: Dynamic threshold secret reconstruction and its application to the threshold cryptography. Inf. Process. Lett. **115**(11), 851–857 (2015)
15. Traverso, G., Demirel, D., Buchmann, J.: Dynamic and verifiable hierarchical secret sharing. In: Information Theoretic Security. Springer (2016)
16. Harn, L., Lin, C.: Detection and identification of cheaters in (t, n) secret sharing scheme. Des. Codes Cryptogr. **52**(1), 15–24 (2009)

Game-Based Security Proofs for Secret Sharing Schemes

Zhe Xia[1(✉)], Zhen Yang[1], Shengwu Xiong[1], and Ching-Fang Hsu[2]

[1] School of Computer Science, Wuhan University of Technology, Wuhan 430071, China
{xiazhe,zhenyang,xiongsw}@whut.edu.cn

[2] Computer School, Central China Normal University, Wuhan 430079, China
cherryjingfang@gmail.com

Abstract. Secret sharing schemes allow the secret to be shared among a group of parties, so that a quorum of these parties can work together to recover the secret, but less number of parties cannot learn any information of the secret. In the literature, secret sharing schemes are normally analysed using heuristic arguments rather than strict security proofs. However, such a method may overlook some security flaws, especially when it is used to analyse the secrecy property. In this paper, we illustrate this issue using some concrete examples. We show that in two existing secret sharing schemes, the secrecy property was originally conjectured to be satisfied, but the adversary still can employ some security flaws to violate this property. We then introduce a game-based model that can be used to formally analyse the secrecy property in secret sharing schemes. We prove that our model captures the definition of the secrecy property. And as an example, we show how our method can be used to analyse Shamir secret sharing scheme. Note that our method might find applications in other secret sharing schemes as well.

1 Introduction

Secret sharing schemes [2,4,14,16] are important techniques in modern cryptography and information security. They allow the secret to be shared among different parties, so that it can be recovered only if more than a quorum of these parties work together, but no information of the secret is revealed otherwise. Either to learn the secret or destroy it, the adversary needs to compromise multiple parties instead of a single one, making her tasks more complicated compared with the traditional storage methods. Therefore, secret sharing schemes are useful tools that enhance both the secrecy and availability of sensitive information. Moreover, secret sharing schemes are fundamental building blocks for various cryptographic protocols, such as group authentication and key exchange [10,13], attribute-based encryptions [15], threshold cryptosystems [7,17], multi-party computations [3,6], etc.

C.-N. Yang et al. (Eds.): SICBS 2018, AISC 895, pp. 650–660, 2020.
https://doi.org/10.1007/978-3-030-16946-6_53

In the literature, a number of secret sharing schemes have been proposed over the last few decades, and generally speaking, these schemes can be classified into the following three categories based on their mathematical structures:

- *Schemes based on polynomial interpolation* [16]. In a (t, n) threshold scheme in this category, the dealer selects a random polynomial with degree at most $t-1$, and sets the secret as the polynomial's constant coefficient. Then, n points that satisfy this polynomial are calculated by the dealer, and each participating party will receive one of these points as her share. In the reconstruction phase, every party reveals her point. If more than t points are known, the polynomial can be uniquely reconstructed through polynomial interpolation, e.g. Lagrange interpolation, and the secret can be recovered. However, the knowledge of $t-1$ points can interpolate into infinite many polynomials with degree $t-1$, hence no information of the secret is leaked. Shamir secret sharing is an example that uses polynomial interpolation, and it is widely used in designing cryptographic protocols.
- *Schemes based on hyperplane geometry* [4]. Suppose the dealer would like to share the secret among n parties, such that at least t of them can recover the secret. The dealer first randomly selects a point P in the t-dimensional space, and then computes n random hyperplanes that pass through this point. Each party will receive one of these hyperplanes as her share. In the reconstruction phase, if t of these hyperplanes are revealed, they will intersect at the point P, and the secret can be recovered. However, if only $t-1$ of these hyperplanes are revealed, they only intersect at a line in the t-dimensional space, and there are infinite many points in this line, hence no information of the secret is leaked. It is widely known that secret sharing schemes based on polynomial interpolation can be considered as special cases of those that are based on hyperplane geometry [12].
- *Schemes based on Chinese Remainder Theorem* [2,14]. To share the secret among n parties so that t of them can recover the secret, n different primes are selected by the dealer. The relationship between the secret and these primes needs to satisfy two requirements: (1) the secret is larger than the product of any $t-1$ of these primes; and (2) the secret is smaller than the product of any t of these primes. Each party then receives a share as the secret being moduloed by the corresponding prime. Therefore, thanks to the Chinese Remainder Theorem, t shares can recover the secret without ambiguity, but $t-1$ of these shares cannot uniquely determine a value. An interesting feature of schemes in this category is that the shares can be randomised without affecting their functionalities, and this feature has some interesting applications in cryptography [8].

To capture the security properties in secret sharing schemes, we denote the set of n parties as $\mathcal{P} = \{P_1, P_2, \ldots, P_n\}$, and let $\mathcal{K}, \mathcal{S}$ be the secret set and the share set, respectively. Let Γ be a collection of authorised subsets of $2^{\mathcal{P}}$, called the *access structure*. In the share distribution phase, to share a secret $s \in \mathcal{K}$, each party $P_i \in \mathcal{P}$ receives a share $\mathsf{sh}_i \in \mathcal{S}$ from the dealer. In the secret reconstruction phase, any authorised subset $\mathcal{A} \in \Gamma$ of parties can use their shares to recover the secret. But any non-authorised subset $\mathcal{B} \notin \Gamma$ of parties can learn no information about the secret.

These two requirements can be formalised using the entropy $\mathsf{H}(\cdot)$ of random variables in information theory. Denote S as the random variable associated to the secret, SH_i as the random variable associated to P_i's share, and $\mathsf{SH}_{\mathcal{A}}$ as the vector of random variables associated to the shares belonging to the parties in the subset $\mathcal{A} \subseteq \mathcal{P}$. The secret sharing scheme should satisfy the following two security properties:

- *Correctness:* Given the subset of shares $\{\mathsf{sh}_i\}_{P_i \in \mathcal{A}}$, we have $\mathsf{H}(\mathsf{S}|\mathsf{SH}_{\mathcal{A}}) = 0$ for any subset $\mathcal{A} \in \Gamma$.
- *Secrecy:* Given the subset of shares $\{\mathsf{sh}_i\}_{P_i \in \mathcal{B}}$, we have $\mathsf{H}(\mathsf{S}|\mathsf{SH}_{\mathcal{B}}) = \mathsf{H}(\mathsf{S})$ for any subset $\mathcal{B} \notin \Gamma$.

Moreover, for any unconditional secure secret sharing scheme that achieves the above properties, Brickell [5] has given the lower bounds regarding the length of each share, that the equation $\mathsf{H}(\mathsf{SH}_i) \geq \mathsf{H}(\mathsf{S})$ needs to hold for every party $P_i \in \mathcal{P}$. In other words, the length of each share has to be equal or larger than the length of the secret.

When analysing the security properties in secret sharing schemes, it is relatively easy to justify the correctness property. Because we only need to prove that any subset of t honest shareholders can correctly recover the secret. However, it is much more challenging to prove the secrecy property. To the best of our knowledge, most of the existing works only justify the secrecy property using heuristic arguments rather than strict security proofs.

In this paper, we raise the point that heuristic arguments are not adequate to ensure the secrecy property in secret sharing schemes. We illustrate this using two concrete examples: one is a common misuse of Shamir secret sharing and the other is a modification of Shamir secret sharing scheme. Both these schemes are originally thought to satisfy the secrecy property. However, we show that they fail to achieve what they claim. To formally analyse the secrecy property in secret sharing schemes, we introduce a game-based model. We prove that our model captures the definition of the secrecy property. We then show how our method can be used to analyse Shamir secret sharing as an example. We note that our method is very general, and it can find applications in many other secret sharing schemes as well.

The rest of the paper is organised as follows. In Sect. 2, we describe the models and assumptions for secret sharing schemes. In Sect. 3, we present two examples in which the secrecy property can be violated, which is contrasting to the original intuitions. In Sect. 4, we introduce our game-based method to formally analyse the secrecy property in secret sharing schemes, and we show how this method can be used to analyse Shamir secret sharing. Finally, we conclude in Sect. 5.

2 Models and Assumptions

System Model. The players include a trusted dealer $\mathcal{D}$, n participating parties $\mathcal{P} = \{P_1, P_2, \ldots, P_n\}$ and some insider or outsider adversaries. It is assumed that all these players have unlimited computational resources. Among these parties, it is assumed that at least t of them are honest, where $t > n/2$. Note that this setting prevents the dishonest parties from learning the secret even if they all collude. Here, the word "dishonest" means honest-but-curious. That is, these dishonest parties will follow the protocol, but they may try to learn information that they are not authorised to access.

Communication Model. It is assumed that there exists a secure channel between the dealer and every party, so that the shares can be securely distributed to these parties. Moreover, it is assumed that every player is connected to a common authenticated broadcast channel $\mathcal{C}$. Any message sent through $\mathcal{C}$ can be heard by the other players. The adversary can neither modify messages sent by an honest player through $\mathcal{C}$, nor she can prevent honest players from receiving messages from $\mathcal{C}$. Note that both these channels are standard assumptions that are widely used in existing secret sharing schemes, and they can be implemented by standard cryptographic primitives.

Adversary Model. Two types of adversaries are considered in secret sharing schemes:

- *Inside adversary* $\mathcal{A}_I$ is assumed to be able to control at most $t-1$ parties. If a party is controlled, her internal states, e.g. the share generated by the dealer, will be learnt by the inside adversary. $\mathcal{A}_I$'s objective is to learn some meaningful information of the secret.
- *Outside adversary* $\mathcal{A}_O$ is an attacker who does not own any valid share. But $\mathcal{A}_O$ can participate in the secret reconstruction phase, impersonate to be a legitimate party, and try to learn the secret after the other parties having revealed their shares.

3 Two Insecure Secret Sharing Schemes

3.1 A Common Misuse of Shamir Secret Sharing

In the rest of this paper, we assume that all equations are modulo p unless otherwise stated. In some of the existing works, Shamir secret sharing is incorrectly described as follows:

- **Share distribution phase:** to share a secret s, the dealer randomly selects a polynomial $f(x) = a_0 + a_1 x + \cdots + a_{t-1}x^{t-1}$ over $\mathbb{Z}_p$ **with degree** $t-1$, such that $a_0 = s$. Then the dealer sends the shares $s_i = f(x_i)$ to each shareholder through some secure channel. Here, the set of integers $\{x_1, x_2, \ldots, x_n\}$ are public parameters associate with the participating parties that are pairwise different.
- **Secret reconstruction phase:** any subset $\mathcal{A}$ of these shareholders can reconstruct the polynomial $f(x)$ using Lagrange interpolation, if $|\mathcal{A}| \geq t$:

$$f(x) = \sum_{i \in \mathcal{A}} (s_i \cdot \prod_{j \in \mathcal{A}, j \neq i} \frac{x - x_j}{x_i - x_j})$$

  and the constant coefficient of $f(x)$ is the secret.

To see why the above scheme satisfies the correctness property. Firstly, since $j \neq i$ ensures that $x_i - x_j \neq 0$. Hence the above expression is always well defined. Then denote $\mathcal{L}_i(x) = \prod_{j \in \mathcal{A}, j \neq i} \frac{x - x_j}{x_i - x_j}$ as the Lagrange coefficient and $|\mathcal{A}| = k$. In one aspect, for all $\sigma \neq i$, the iteration of x_j will meet x_σ at some point, and we have:

$$\mathcal{L}_{i \neq \sigma}(x_\sigma) = \prod_{j=1, j \neq i}^{k} \frac{x_\sigma - x_j}{x_i - x_j} = \frac{x_\sigma - x_1}{x_i - x_1} \cdots \frac{x_\sigma - x_\sigma}{x_i - x_\sigma} \cdots \frac{x_\sigma - x_k}{x_i - x_k} = 0$$

In the other aspect, if $\sigma = i$, we have:

$$\mathcal{L}_{i = \sigma}(x_\sigma) = \prod_{j=1, j \neq i}^{k} \frac{x_\sigma - x_j}{x_i - x_j} = 1$$

Therefore, we have

$$\mathcal{L}_i(x_\sigma) = \begin{cases} 1 \text{ if } \sigma = i \\ 0 \text{ if } \sigma \neq i \end{cases}$$

And this implies that for every point x_δ, where $\delta \in \{1, 2, \ldots, n\}$, we have $f(x_\delta) = 0 + \cdots + s_\delta + \cdots + 0 = s_\delta$, showing that the polynomial interpolation always works, hence the scheme satisfies the correctness property.

To justify that the above scheme enjoys the secrecy property. The common arguments are that since the polynomial $f(x)$ has degree $t-1$, it contains t unknown coefficients. But the inside adversary $\mathcal{A}_I$ is assumed to only corrupt at most $t-1$ of these parties, $\mathcal{A}_I$ cannot solve the system of equations, and hence $\mathcal{A}_I$ cannot obtain any information of the secret.

We now show that the above arguments regarding the secrecy property is inadequate, using the attack introduced by Ghodosi et al. [9]. The key point here is that the dealer randomly selects a polynomial $f(x)$ with degree $t-1$. In this case, the length of each party's share still satisfies the lower bounds given by Brickell [5], and there are still more unknown coefficients than the number of polynomials obtained by the inside adversary $\mathcal{A}_I$. However, if $\mathcal{A}_I$ knows that the coefficient with the highest degree is not 0, then she can use the $t-1$ controlled parties to preclude a possible value for the secret using the following strategy.

Denote $f(x) = a_0 + a_1 x + \cdots + a_{t-1} x^{t-1}$ with $a_{t-1} \neq 0$. Then, $t-1$ colluded parties can interpolate a $t-2$ degree polynomial $g(x) = b_0 + b_1 x + \cdots + b_{t-2} x^{t-2}$, such that $f(x_i) = g(x_i)$ for $1 \leq i \leq t-1$. This leads the system of equations:

$$\begin{cases} (a_0 - b_0) + (a_1 - b_1)x_1 + \cdots + (a_{t-2} - b_{t-2}){x_1}^{t-2} + a_{t-1}{x_1}^{t-1} & = 0 \\ (a_0 - b_0) + (a_1 - b_1)x_2 + \cdots + (a_{t-2} - b_{t-2}){x_2}^{t-2} + a_{t-1}{x_2}^{t-1} & = 0 \\ & \vdots \\ (a_0 - b_0) + (a_1 - b_1)x_{t-1} + \cdots + (a_{t-2} - b_{t-2}){x_{t-1}}^{t-2} + a_{t-1}{x_{t-1}}^{t-1} & = 0 \end{cases}$$

By contradiction, if we assume that $a_0 = b_0$, then the above system of equations with $t-1$ equations and $t-1$ unknown values $\{a_1, a_2, \ldots, a_{t-1}\}$ will have a unique solution. This is because the determinant of a Vandermonde matrix is not 0.

Hence, the solution must be $a_1 = b_1, a_2 = b_2, \ldots, a_{t-2} = b_{t-2}$, and $a_{t-1} = 0$. This contradicts the assumption that $a_{t-1} \neq 0$. Therefore, the inside adversary $\mathcal{A}_I$ can preclude b_0 as a possible value of the secret, and the above scheme fails to achieve the secrecy property.

3.2 A Modification of Shamir Secret Sharing

Note that Shamir secret sharing allows more than t parties to participate in the secret reconstruction phase. Therefore, the outside adversary $\mathcal{A}_O$ who impersonates a legitimate party can learn the secret as follows. $\mathcal{A}_O$ simply releases a random value as her share, and meanwhile, $\mathcal{A}_O$ learns the other parties' shares. Because only t shares are needed to recover the secret, $\mathcal{A}_O$ can recover the secret using the other parties' shares. A straightforward method to prevent this attack is to authenticate each party in the secret reconstruction phase. But in [11], Harn tries to address this problem without using an additional authentication algorithm. Harn's scheme works as follows:

- **Share distribution phase.**
  1. The dealer $\mathcal{D}$ selects k random polynomials $f_l(x)$ over $\mathbb{Z}_p$ for $l = 1, 2, \ldots, k$, with degree at most $t-1$ each. Here, p is a prime that satisfies $p > n$, where n is the number of participating parties.
  2. Then, $\mathcal{D}$ generates the shares $\mathsf{sh}_i = f_l(x_i)$ for $i = 1, 2, \ldots, n$, and sends each share to the corresponding party through the secure channel. The integers $\{x_1, x_2, \ldots, x_n\}$ are publicly known and pairwise different.
  3. To share the secret $s \in \mathbb{Z}_p$, the dealer finds integers $w_l, d_l \in \mathbb{Z}_p$ for $l = 1, 2, \ldots, k$, such that $s = \sum_{l=1}^{k} d_l f_l(w_l)$. The values w_l need to be pairwise different, and the intersection of the two sets $\{x_1, x_2, \ldots, x_n\}$ and $\{w_1, w_2, \ldots, w_k\}$ needs to be empty. The dealer $\mathcal{D}$ makes these integers w_l, d_l publicly known for $l = 1, 2, \ldots, k$.
- **Secret reconstruction phase.**
  1. Suppose u parties participate in the secret reconstruction phase, where $t \leq u \leq n$. Each party P_i uses her share sh_i and the value u to compute the token c_i as:
  $$c_i = \sum_{l=1}^{k} d_l f_l(x_i) \prod_{v=1, v\neq i}^{u} \frac{w_l - x_v}{x_i - x_v}$$
  And then, P_i sends the token c_i to the authenticated broadcast channel.
  2. After receiving all the tokens c_i for $i = 1, 2, \ldots, u$, every party can compute the secret as $s = \sum_{i=1}^{u} c_i$.

In the security analysis, we need to show that the scheme achieves both the correctness and the secrecy properties. It is straightforward to see that the correctness property is satisfied, because we have:

$$
\begin{aligned}
s = \sum_{i=1}^{u} c_i &= \sum_{i=1}^{u}\sum_{l=1}^{k}(d_l f_l(x_i) \prod_{v=1, v\neq i}^{u} \frac{w_l - x_v}{x_i - x_v}) \\
&= \sum_{l=1}^{k}(d_l \sum_{i=1}^{u}(f_l(x_i) \prod_{v=1, v\neq i}^{u} \frac{w_l - x_v}{x_i - x_v})) \\
&= \sum_{l=1}^{k} d_l f_l(w_l)
\end{aligned}
$$

Moreover, Harn has claimed that if $kt > n-1$, then the secrecy property also holds. The arguments are as follows: considering the worst case where n players are involved to recover the secret and the outside adversary $\mathcal{A}_O$ is among these players. Then $\mathcal{A}_O$ can obtain at most $n-1$ equations from the $n-1$ revealed tokens. But since the number of unknown coefficients kt is larger than the number of obtained equations, $\mathcal{A}_O$ cannot recover the polynomials, hence she cannot learn any information of the secret.

However, it has been shown by Ahmadian et al. [1] recently that the above arguments are incorrect. The outside adversary $\mathcal{A}_O$ can use a clever method, called the linear subspace attack, to construct a linear subspace spanned by the revealed tokens and then compute an additional token without recovering the polynomials. The sum of the revealed tokens and this additional token allows $\mathcal{A}_O$ to recover the secret. Therefore, Harn's secret sharing scheme also fails to achieve what it claims. The readers who are interested in this attack can refer to the paper in [1] for more details.

4 Game-Based Security Proofs

Both the above two examples demonstrate that intuitions might be faulty sometimes. In this section, we introduce a game-based model that can be used to formally prove the secrecy property in secret sharing schemes. We present step-by-step why our model captures the secrecy property. And then, we use an example to demonstrate its usage by analysing Shamir secret sharing scheme.

4.1 A Game-Based Method

Original Definition of Secrecy: Let S and Γ be the random variables denoting the value of the secret to be shared and the shares possessed by the adversary, respectively. The secrecy property is formalised as follows:

Definition 4.1 (Secrecy). *Given a subset $\mathcal{B}$ of shares ($|\mathcal{B}| < t$), the adversary learns no information of the secret. Formally,*

$$\mathsf{Pr}[\mathsf{S} = s | \Gamma = \mathcal{B}] = Pr[\mathsf{S} = s]$$

Another Definition of Secrecy: Then, we give another definition of the secrecy property, and we prove that our new definition captures the original definition.

Theorem 4.2. *If the probability distribution of the subset $\mathcal{B}$ ($|\mathcal{B}| < t$) does not depend on the shared secret, then the secrecy property holds. Formally, for every $s, s' \in \mathbb{Z}_p$, if we have*

$$Pr[\Gamma = \mathcal{B}|\mathsf{S} = s] = Pr[\Gamma = \mathcal{B}|\mathsf{S} = s']$$

then the definition 4.1 holds.

Proof. For every $s, s' \in \mathbb{Z}_p$, if we have $\Pr[\Gamma = \mathcal{B}|\mathsf{S} = s] = \Pr[\Gamma = \mathcal{B}|\mathsf{S} = s']$, then for any $s'' \in \mathbb{Z}_p$, we have $\Pr[\Gamma = \mathcal{B}|\mathsf{S} = s''] = 1/|p|$. Using Bayes' Theorem, we have:

$$\begin{aligned}
\Pr[\mathsf{S} = s|\Gamma = \mathcal{B}] &= \frac{\Pr[\Gamma = \mathcal{B}|\mathsf{S} = s] \cdot \Pr[\mathsf{S} = s]}{\Pr[\Gamma = \mathcal{B}]} \\
&= \frac{\Pr[\Gamma = \mathcal{B}|\mathsf{S} = s] \cdot \Pr[\mathsf{S} = s]}{\sum_{s'' \in \mathsf{S}} \Pr[\Gamma = \mathcal{B}|\mathsf{S} = s''] \cdot \Pr[\mathsf{S} = s'']} \\
&= \frac{1/|p| \cdot \Pr[\mathsf{S} = s]}{1/|p| \cdot \sum_{s'' \in S} \Pr[\mathsf{S} = s'']} \\
&= \Pr[\mathsf{S} = s]
\end{aligned}$$

A Game-Based Model for Secrecy: Now, we define the following experiment $\mathcal{G}_{SS}$, and we prove that if the adversary has no advantage in $\mathcal{G}_{SS}$, then the secrecy property holds.

$\underline{\mathcal{G}_{SS}}$:

1. The adversary $\mathcal{A}$ outputs two arbitrary messages $s_0, s_1 \in \mathbb{Z}_p$;
2. The challenger $\mathcal{C}$ chooses a uniform bit $b \in \{0, 1\}$, and run the secret sharing scheme, treating s_b as the secret. Then, a subset $\mathcal{B}$ of shares are given to the adversary $\mathcal{A}$, where $|\mathcal{B}| < t$;
3. The adversary outputs a bit b';
4. The output of the experiment is defined to be 1 if $b' = b$, and 0 otherwise.

Theorem 4.3. *The secret sharing scheme satisfies the secrecy property if in the above game:*

$$Pr[\mathcal{G}_{SS} = 1] = \frac{1}{2}$$

Proof. The equation $\Pr[\mathcal{G}_{SS} = 1] = 1/2$ in this theorem implies that for any $s, s' \in \mathbb{Z}_p$, we have

$$\Pr[\mathsf{S} = s|\Gamma = \mathcal{B}] = \Pr[\mathsf{S} = s'|\Gamma = \mathcal{B}]$$

Moreover, using Bayes' Theorem, we have

$$\Pr[\Gamma = \mathcal{B}|\mathsf{S} = s] = \frac{\Pr[\mathsf{S} = s|\Gamma = \mathcal{B}] \cdot \Pr[\Gamma = \mathcal{B}]}{\Pr[\mathsf{S} = s]}$$

$$\mathsf{Pr}[\Gamma = \mathcal{B}|\mathsf{S} = s'] = \frac{\mathsf{Pr}[\mathsf{S} = s'|\Gamma = \mathcal{B}] \cdot \mathsf{Pr}[\Gamma = \mathcal{B}]}{\mathsf{Pr}[\mathsf{S} = s']}$$

Because the challenge bit b is uniformly chosen, $\mathsf{Pr}[\mathsf{S} = s] = \mathsf{Pr}[\mathsf{S} = s'] = 1/2$. Therefore, we have $\mathsf{Pr}[\Gamma = \mathcal{B}|\mathsf{S} = s] = \mathsf{Pr}[\Gamma = \mathcal{B}|\mathsf{S} = s']$. Note that the last equation is Theorem 4.2, which means that Theorem 4.3 implies Theorem 4.2.

Another Game-Based Model for Secrecy: Finally, we define another experiment $\mathcal{G}'_{SS}$, and we prove that it captures the experiment $\mathcal{G}_{SS}$. Moreover, by transitivity of proofs, if the adversary has no advantage in $\mathcal{G}'_{SS}$, then we can conclude that the secrecy property holds.
$\underline{\mathcal{G}'_{SS}}$:

1. The challenger $\mathcal{C}$ shares the secret $s \in \mathbb{Z}_p$ through the secret sharing scheme.
2. $\mathcal{C}$ chooses a uniform bit $b \in \{0, 1\}$. If $b = 0$, $\mathcal{C}$ sets $\hat{s} = s$. If $b = 1$, $\mathcal{C}$ chooses a uniform $\hat{s}$ within $\mathbb{Z}_p$.
3. Given a subset $\mathcal{B}$ of shares where $|\mathcal{B}| < t$ as well as the value $\hat{s}$, the adversary $\mathcal{A}$ outputs a bit b'. The output of the experiment is defined to be 1 if $b' = b$, and 0 otherwise.

Theorem 4.4. *The secret sharing scheme satisfies the secrecy property if*

$$Pr[\mathcal{G}'_{SS} = 1] = \frac{1}{2}$$

Proof. According to this theorem, $\mathsf{Pr}[\mathcal{G}'_{SS} = 1] = 1/2$ implies that when the secret s has been shared, for any $s' \in \mathbb{Z}_p$, we have

$$\mathsf{Pr}[\mathsf{S} = s|\Gamma = \mathcal{B}] = \mathsf{Pr}[\mathsf{S} = s'|\Gamma = \mathcal{B}]$$

This implies that for any $s', s'' \in \mathbb{Z}_p$, we have $\mathsf{Pr}[\mathsf{S} = s'|\Gamma = \mathcal{B}] = \mathsf{Pr}[\mathsf{S} = s''|\Gamma = \mathcal{B}]$. Note that the last equation is Theorem 4.3, which means that Theorem 4.4 implies Theorem 4.3.

4.2 An Application of This Method

The above theorems prove that if the adversary does not have any advantage in the experiment $\mathcal{G}'_{SS}$, the secret sharing scheme will satisfy the secrecy property. Now, we demonstrate how this game-based model works by analysing the secrecy property of Shamir secret sharing scheme.

We denote Game_0 as the original game that the real secret value $s \in \mathbb{Z}_p$ is shared using Shamir secret sharing, and Game_1 is a game simulated by the simulator that a random value $s' \in \mathbb{Z}_p$ is shared instead. The purpose is to prove that the adversary cannot distinguish between these two games, even if she has unlimited computational resources.

Game_0:

- **In the share distribution phase:** a random polynomial $f(x) = a_0 + a_1 x + \cdots + a_{t-1}x^{t-1}$ over $\mathbb{Z}_p$ with degree at most $t-1$ is selected by the dealer, such that $a_0 = s$. Then, the dealer evaluates $f(x)$ and computes $s_i = f(x_i)$, where $\{x_1, x_2, \ldots, x_n\}$ are public parameters associate with the participating parties that are pairwise different. These s_i values are called the shares, and they are sent to the parties through the secure channel. In this phase, the adversary learns the corrupted parties' internal states $\{s_1, s_2, \ldots, s_{t-1}\}$.
- **In the secret reconstruction phase:** $\mathcal{A}$ is denoted as the set containing the participating parties in the secret reconstruction phase. Each party reveals her share s_i for $i \in \mathcal{A}$, and the polynomial $f(x)$ can be recovered by polynomial interpolation. In this phase, the adversary learns all the revealed shares $\{s_i\}_{i\in\mathcal{A}}$ as well as the polynomial $f(x)$.

Game_1:

- **In the share distribution phase:** the simulator just sends $\{s_1, s_2, \ldots, s_{t-1}\}$ to the adversary.
- **In the secret reconstruction phase:** the simulator randomly selects $s' \in \mathbb{Z}_p$, and uses the set $\{s_1, s_2, \ldots, s_{t-1}, s'\}$ to interpolate a polynomial $f'(x) = b_0 + b_1 x + \cdots + b_{t-1}x^{t-1}$ over $\mathbb{Z}_p$ with degree at most $t-1$. The simulator then evaluates $f'(x)$ and computes the shares $s'_i = f'(x_i)$ for $i \in \mathcal{A}$. The simulator sends $f'(x)$ and $\{s'_i\}_{i\in\mathcal{A}}$ to the adversary.

Indistinguishability Between Game_0 from Game_1. In the share distribution phase of both games, the adversary sees exactly the same set $\{s_1, s_2, \ldots, s_{t-1}\}$. Hence her view of this phase is the same in both games. In the secret reconstruction phase of Game_0, the adversary sees a polynomial $f(x)$ in which all its coefficients are randomly distributed in $\mathbb{Z}_p$, as well as a set of shares $\{s_i\}_{i\in\mathcal{A}}$ that are also randomly distributed in $\mathbb{Z}_p$. Moreover, these shares satisfy the polynomial as $s_i = f(x_i)$ for $i \in \mathcal{A}$. In the secret reconstruction phase of Game_1, the adversary sees a polynomial $f'(x)$. Because s' is a random value in $\mathbb{Z}_p$ and $f'(x)$ is interpolated by the set $\{s_1, s_2, \ldots, s_{t-1}, s'\}$, all coefficients of $f'(x)$ are randomly distributed in $\mathbb{Z}_p$. Moreover, because the shares are computed as $s'_i = f'(x_i)$ for $i \in \mathcal{A}$, they satisfy the polynomial $f'(x)$ and they are also randomly distributed in $\mathbb{Z}_p$. Therefore, the adversary has the same view in this phase as well. Putting everything together, the adversary cannot distinguish Game_0 from Game_1, even if she has unlimited computational resources.

5 Conclusion

In this paper, we have pointed out that heuristic arguments are inadequate to analyse the secret sharing schemes, especially for the secrecy property. To address this problem, we introduce a game-based method that could be used to analyse the secrecy property in a formal way. We have shown how this method could be applied to analyse Shamir secret sharing. And our method might find applications in other secret sharing schemes as well.

Acknowledgement. This work was partially supported by the National Natural Science Foundation of China (Grants No. 61772224) and Natural Science Foundation of Hubei Province (Grant No. 2017CFB303).

References

1. Ahmadian, Z., Jamshidpour, S.: Linear subspace cryptanalysis of harn's secret sharing-based group authentication scheme. IEEE Trans. Inf. Foren. Secur. **13**(2), 502–510 (2018)
2. Asmuth, C., Bloom, J.: A modular approach to key safeguarding. IEEE Trans. Inf. Theory **29**(2), 208–210 (1983)
3. Ben-Or, M., Goldwasser, S., Wigderson, A.: Completeness theorems for non-cryptographic fault-tolerant distributed computation. In: Proceedings of the Twentieth Annual ACM Symposium on Theory of Computing, pp. 1–10. ACM (1988)
4. Blakley, G.R., et al.: Safeguarding cryptographic keys. In: Proceedings of the National Computer Conference, vol. 48, pp. 313–317 (1979)
5. Brickell, E.F.: Some ideal secret sharing schemes. In: Workshop on the Theory and Application of of Cryptographic Techniques, pp. 468–475. Springer (1989)
6. Chaum, D., Crépeau, C., Damgard, I.: Multiparty unconditionally secure protocols. In: Proceedings of the Twentieth Annual ACM Symposium on Theory of Computing, pp. 11–19. ACM (1988)
7. Desmedt, Y.: Threshold cryptosystems. In: International Workshop on the Theory and Application of Cryptographic Techniques, pp. 1–14. Springer (1992)
8. Fuyou, M., Yan, X., Xingfu, W., Badawy, M.: Randomized component and its application to (t, m, n)-group oriented secret sharing. IEEE Transa. Inf. Forensics Secur. **10**(5), 889–899 (2015)
9. Ghodosi, H., Pieprzyk, J., Safavi-Naini, R., Remarks on the multiple assignment secret sharing scheme. In: International Conference on Information and Communications Security, pp. 72–80. Springer (1997)
10. Harn, L.: Group authentication. IEEE Trans. Comput. **62**(9), 1893–1898 (2013)
11. Harn, L.: Secure secret reconstruction and multi-secret sharing schemes with unconditional security. Secur. Commun. Networks **7**(3), 567–573 (2014)
12. Kothari, S.C.: Generalized linear threshold scheme. In: Workshop on the Theory and Application of Cryptographic Techniques, pp. 231–241. Springer (1984)
13. Li, J., Wen, M., Zhang, T.: Group-based authentication and key agreement with dynamic policy updating for MTC in LTE-A networks. IEEE Internet Things J. **3**(3), 408–417 (2016)
14. Mignotte, M.: How to share a secret. In: Workshop on Cryptography, pp. 371–375. Springer (1982)
15. Sahai, A., Waters, B.: Fuzzy identity-based encryption. In: Annual International Conference on the Theory and Applications of Cryptographic Techniques, pp. 457–473. Springer (2005)
16. Shamir, A.: How to share a secret. Commun. ACM **22**(11), 612–613 (1979)
17. Shoup, V.: Practical threshold signatures. In: International Conference on the Theory and Applications of Cryptographic Techniques, pp. 207–220. Springer (2000)

Dress Identification for Camp Security

Jiabao Wang(✉), Yang Li, Yihang Xiong, Zhixuan Zhao, and Dexing Kong

Army Engineering University of PLA, Guanghua Road, Haifu Streat, No. 1, Nanjing 210007, China
jiabao_1108@163.com

Abstract. With the development of artificial intelligence and internet of things, the intelligent embedded devices are becoming more and more popular. So it is urgent to achieve the lightweight application of artificial intelligence. In this paper, a dress identification framework is developed for camp security. The framework has both hardware and software. The hardware is composed of LattePanda board, Arduino chip, USB camera, and buzzer. To achieve the identification, we extract the color histogram feature of moving person from the images captured by USB camera, and identify the military dress by using support vector machine algorithm. The equipment can output sound or light signals by Arduino chip, when it identifies a non-military dress. We implement the identification function based on the OpenCV library. The framework can run in real-time, with a reliable precision.

Keywords: Dress identification · Embedding development · Camp security · Support vector machine · Color histogram

1 Introduction

As a special place, ensuring the safety of military camp is an important task of all countries [7]. The military camp requires strict handling for 24 h, but the manual duty requires human resources and is affected by the personal factors of the sentinel, so it is difficult to guarantee the 100% monitoring. So it requires the use of intelligent surveillance to complete the corresponding guard tasks, and the intelligent guards need to identify the dress of each passing person [12].

The daily management of military camp is also closely related to the dress identification. There are not only military soldiers in the military camp, but also many logistics staff, builders and civilian staff employed in the camp. As there are many sensitive areas in the military camp that are not accessible to non-military personnel, the identification of military and non-military personnel entering the sensitive areas is also one of the important methods to ensure the security of the military camps. Furthermore, the dress identification is also very important to identify attacking targets in combat, such as civilians or soldiers.

C.-N. Yang et al. (Eds.): SICBS 2018, AISC 895, pp. 661–672, 2020.
https://doi.org/10.1007/978-3-030-16946-6_54

1.1 State-of-the-Arts

In view of the troops, the military dress identification has a strong value, but it has only limited application. The representative application is identify individual photographs of military uniform on the Internet, ensuring the supervision of soldier's confidentiality. It's lack of smart equipment for real-time identification of sensitive areas in military camps. At present, the military dress identification in the camp mainly includes the following two ways:

Human Guard. Mostly, human guard is the mainly used method to identify dress, as shown in Fig. 1(left). The advantage is that the person dress can be accurately identified. However, it's dangerous to set a human guard for security. And more, Long time duty will overdraw the guard's energy and affect the guard's judgement and execution. It is a time-consuming and exhausting work.

Video Surveillance. Video surveillance is another dress identification method in actually, as shown in Fig. 1(right). In a special position where the visual field is clear, the camera is installed and a person is arranged in the monitoring room to stare at the screen for person identification. The advantage of this approach is that it replaces the guard's responsibility partially and the security of the guard is guaranteed, but it also requires human labor to verify the dress identification.

Fig. 1. Two identification methods: human guard (left) and video surveillance (right)

1.2 Comparison

Both of them have strong performance in the accuracy of dress identification, and can effectively identify personnel in the relevant sensitive areas. However, both of them cost a lot of manpower. In the modern era of artificial intelligent, the cost of human labor resource is different from the past. The development of a country requires its army to provide strong security guarantee, improve the efficiency of military action, and accelerate the transformation from quantity to quality, intelligence and efficiency. As a result, both of above methods are negative ways.

At present, artificial intelligence technology [14] has achieved a great development, and it's application is very few in the army. This paper develops a new embedded equipment to keep the security of the camp, with artificial intelligence technology.

1.3 Contributions

The design of the equipment has the characteristics of miniaturization, low power consumption, mobility and easy deployment in practical application, and the contributions are as follows:

- In terms of hardware design, we use the LattePanda embedded board as the platform, combined with the Arduino chip, USB camera, buzzer, and other peripherals. It has strong peripheral expansion ability.
- In the aspect of software design, we finish the video data acquisition, processing, analysis and storage, with the Python language and OpenCV library. The program can achieve automatic near real-time dress identification.
- We test the equipment and functions in different conditions. Experimental results show that the developed dress identification function can identify person dress effectively in near real-time.

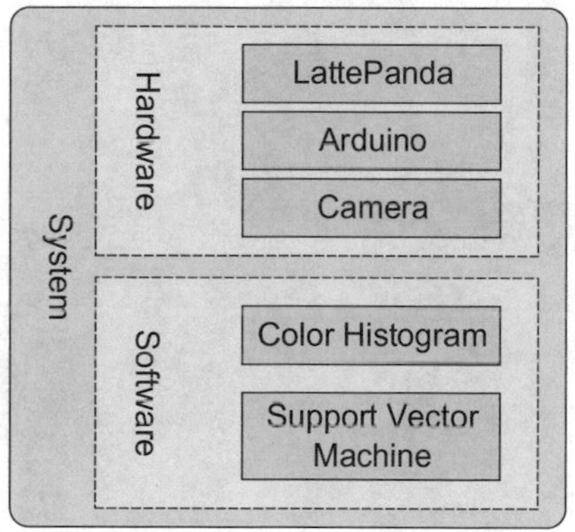

Fig. 2. The designed architecture.

2 Dress Identification

2.1 Architecture

As mentioned above, human guard can achieve accurate identification due to the existence of guards, but it requires a lot of manpower. For video surveillance, due to the existence of camera, guards can be freed, but it still needs to be manually identified when someone comes in. Therefore, intelligently identify the dress of pedestrians can reduce manpower costs, based on the video surveillance.

2.2 Hardwares

To achieve the function of dress identification, we have to select hardware equipments. The embedded development platforms, such as raspberry PI [8], LattePanda [1], are compared. According to requirements and limitations of task, we choose LattePanda as the core development board, because it has stronger computing ability than raspberry PI. Besides, we adopt USB camera as input, and LED light as output.

LattePanda. LattePanda embedded board [1] has surpassed the raspberry PI in terms of hardware specifications. It can run Windows 10 system on the board. It is equipped with 4 cores Intel Z8350 processor, 2G DDR3 memory and 32G eMMC high-speed flash memory. The premium version is added to 4G memory and 64G high-speed flash memory. As for I/O, the TF card slot on the LattePanda board can be used for extended storage. There are USB3.0 and USB2.0 provided for connecting external devices. The standard HDMI interface can connect the monitor directly, the 3.5 mm standard audio interface is compatible with the speaker, and bluetooth 4.0 and wireless network card are all available.

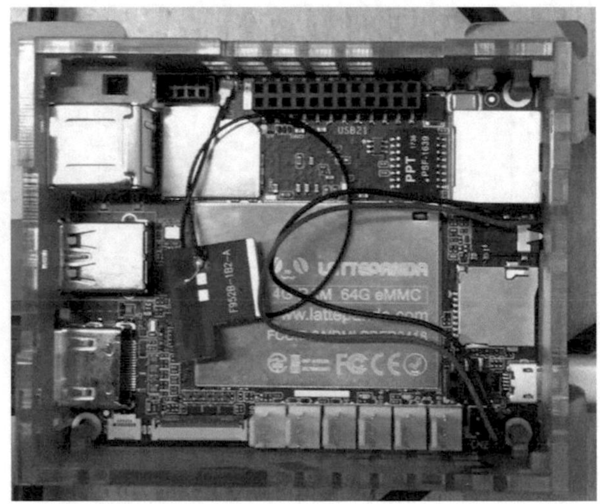

Fig. 3. LattePanda embedded board

This LattePanda used in this paper is showed in Fig. 3. It provides the environment for the compilation, debugging and running of the equipment. At the same time, it inputs external information such as video signals, and outputs the results through Arduino [2] after analysis.

Arduino. Arduino is a convenient and flexible open electronic platform and widely used in internet of things [10]. It can perceive the environment through a variety of sensors, and feedback with controlling lights, motors, and other output devices. The micro-controller on the board can be programmed through Arduino's programming language. The LattePanda has built-in an Arduino coprocessor, ATmega32u4.

In experiments, Arduino is adopted to provide the platform for output results, such as light signals or buzz signals, so as to realize the alarm when abnormal happens.

IO Sensors. In experiments, a USB camera is used as the visible light sensor, shown in Fig. 4(left). It captures the surveillance video and provides the images for intelligent analysis. The LED lamp is used as the information for indicating the state of analysis. The final equipment platform is shown in Fig. 4(right).

In terms of equipment and communication by means of serial port or WIFI, it provides the information transmission between multiple platforms, and has the capability of long distance communication and video transmission.

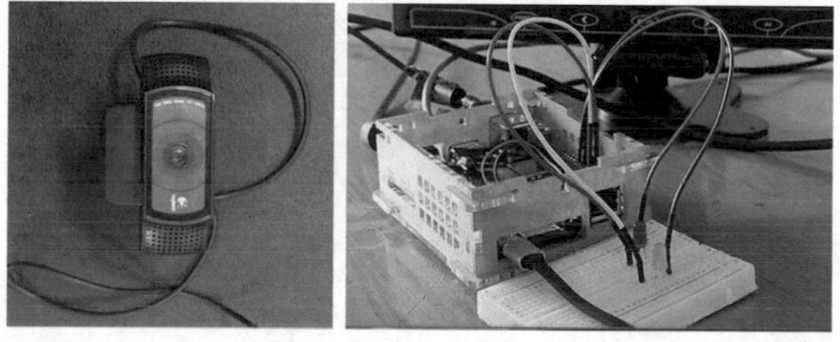

Fig. 4. Camera and integrated platform

2.3 Identification Method

SVM Classifier. In order to realize the recognition function, support vector machine (SVM) [5] is used for classification or identification. SVM is a machine learning method based on statistical learning theory, by seeking structural minimum risk to improve the generalization ability [13]. The basic principle of support vector machine is as follows.

For binary classification, given a set of training data $X = \{(x_i, y_i)\}_{i=1}^{n}$, where $x_i \in R^d$ represents the feature of d dimension, $y_i \in \{+1, -1\}$ is the label, and n is the number of samples. The target is to learn a model to predict the label when given a new sample.

The basic idea of SVM is to construct a super-plane for segmenting the positive samples and negative samples. The best super-plane can be found with solving the following problem:

$$\begin{aligned} \min_{w \in R^d, b \in R} \quad & \frac{1}{2}||w||^2 \\ s.t. \quad \forall i, \quad & y_i(w \cdot x_i + b) \geq 1 \end{aligned} \tag{1}$$

where w and b is the parameters for solving.

To tackle the noise samples, the constraint is relaxed for obtaining a feasible solution. A non-negative parameter is introduced to punish the samples violated of the linear separable principle. So the soft-SVM is introduced as

$$\begin{aligned} \min_{w \in R^d, b \in R, \xi \in R^n} \quad & \frac{1}{2}||w||^2 + C\sum_{i=1}^{n} \xi_i \\ s.t. \quad \forall i, \quad & y_i(w \cdot x_i + b) \geq 1 - \xi_i \\ \forall i, \quad & \xi_i \geq 0 \end{aligned} \tag{2}$$

where C is the punish factor, to balance the margin and the error. The parameter ξ_i is the relax value. When the sample is violate the constraint, the punish value is ξ_i, as shown in Fig. 5. In the figure, circle points and square points are two classes, and the two red points are the punished points.

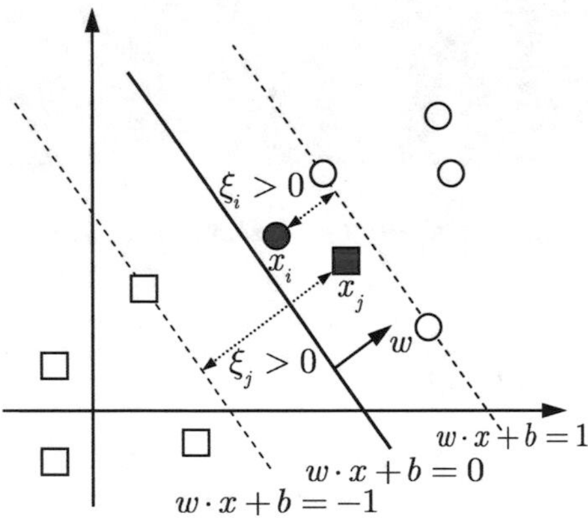

Fig. 5. Soft SVM principle

The solution of SVM can be obtained from the dual problem, which is transformed from the primal problem, according to the duality principle. The dual SVM is

$$\begin{aligned} \max_{a\in R^n} \quad & \sum_{i=1}^{n} a_i - \frac{1}{2}\sum_{i=1}^{n}\sum_{j=1}^{n} a_i a_j y_i y_j x_i \cdot x_j \\ s.t. \quad & \forall i, \ \ 0 \le a_i \le C \\ & \sum\nolimits_{i=1}^{n} a_i y_i = 0 \end{aligned} \tag{3}$$

where α_i is the dual parameter, and corresponding to the constraint $y_i(w \cdot x_i + b) \geq 1 - \xi_i$. So we can obtain the primal parameter as $w = \sum_{i=1}^{n} a_i y_i x_i$.

Generally speaking, it is a binary classification problem for dress identification, and the optimization of SVM can finally be transformed into the solution of a convex quadratic programming problem [11].

To represent a pedestrian image as a point in SVM, we need to transform the image into a feature vector, which is depicted in next subsection.

Feature Representation. Although convolutional neural network can be used for clothes classification, but it is too time-consuming [6]. To speedup the computation of dress identification in LattePanda, color histogram is used for representation. The different of military dress and civil dress is the color of dress. As we known, color feature is widely used in image retrieval [4,9]. The values in the color histogram are obtained by statistical calculation, which describes the quantitative features of the color in the image, and can reflect the statistical distribution of the image color.

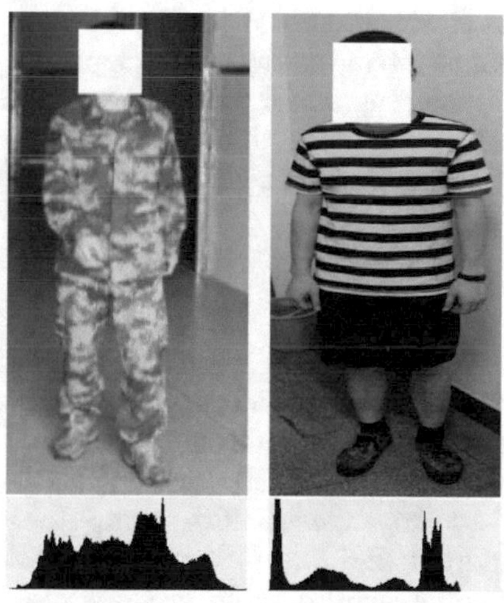

Fig. 6. Color histogram feature

The proportion of different colors in the whole image is described, while the spatial position is ignored. Each image can only give a corresponding histogram, but different images may have the same color distribution and thus have the same histogram. Therefore, histogram and image are one-to-many relationship.

In order to represent an image as a high-dimensional vector, we create the color histogram in the RGB space. For the pixel of r, g, b values, we split the values in each channel with the segment size of 16. The number of color feature bins is 16^3, and each is computed as

$$h(v) = \sum_{x,y} \delta(q[v_r(x,y), v_g(x,y), v_b(x,y)] = v) \tag{4}$$

where x, y are the coordinates of given pixels in an image, and

$$q[v_r, v_g, v_b] = 16 * 16 * (v_r/16) + 16 * (v_g/16) + v_b/16 \tag{5}$$

which represents the quantization of a given pixel of v_r, v_g, v_b values.

A color histogram feature vector extracted from each pedestrian image corresponds to a point in the high-dimensional space of SVM.

2.4 Implementation

In implementation, the hardware and software includes:

- LattePanda embedded development board, CPU main frequency 1.6ghz, memory size of 4GB, high-speed flash memory size of 64GB, including 1 network port, two USB 2.0 interfaces, one USB 3.0 interface, bluetooth 4.0 and wireless network card.
- Auxiliary development board and sensors include Arduino board, USB camera, LED lamp, buzzer, etc.
- Windows 10 operating system, software development using Anaconda integration environment, the programming language Python.
- The communication between LattePanda and Arduino is through COM, and the external control of Arduino is realized through Arduino development language.
- Video and image data are processed by opencv-python visual development package.

The intelligent dress identification acquires video through USB camera, detects moving person, and trains SVM model and predicts dress classes. In order to achieve the near real-time requirement, the video frame must be processed and analyzed fast, so we chose the OpenCV library for speedup.

OpenCV is an open-source, cross-platform computer vision library that runs on Linux, Windows, Android, and Mac OS operating systems. It is composed of a series of C functions and a small number of C++ classes [3]. It also provides interfaces in Python, Ruby, MATLAB and other languages, and implements many common algorithms in image processing and computer vision.

After processing and predicting of the input video, the dress identification outputs results by light signals, with the device of Arduino. We can also combine Arduino with other embedded devices, such as LattePanda, raspberry PI.

3 Deployment

3.1 Operation Process

The whole process of operation are shown in Fig. 7, where includes five steps. The installation, data collection and identification are the mainly three processes.

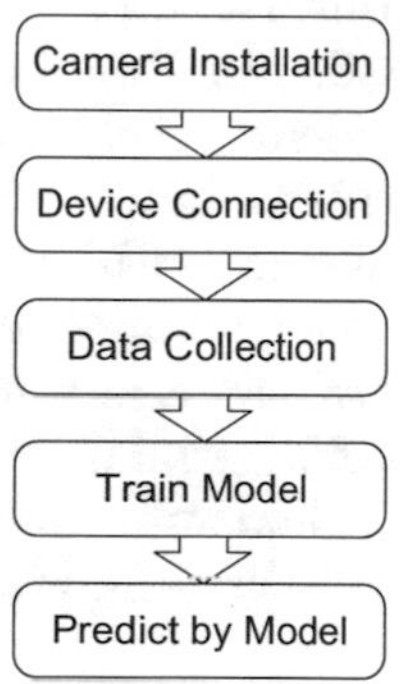

Fig. 7. Flow chart of operation

Equipment Installation. In order to identify the pedestrian's dress more effectively, there are two requirements on the installation of the equipment:

- Since dress identification is based on the pedestrian's appearance, the angle between the camera and the horizontal plane should be kept within a small range to maximize the pedestrian's appearance.
- The camera should be set up in a proper lighting position, because the color features are sensitive to the lighting.

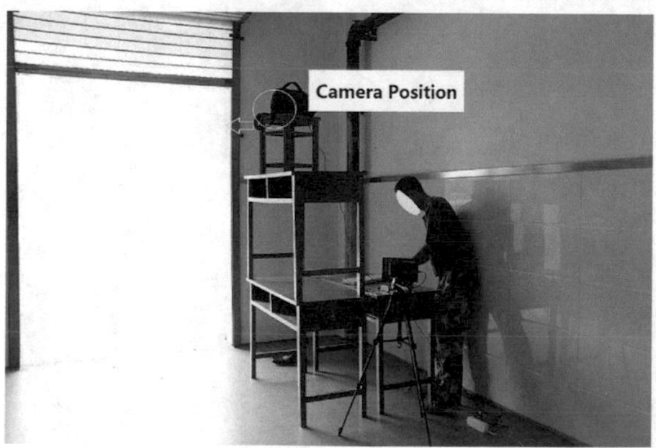

Fig. 8. Equipment deployment

For the intensity of lighting, if the background light is too dim or too strong, the appearance color may be distorted. Figure 8 shows an example of an installation location.

Data Collection. For training the SVM model, a training data set should be established first. The set must contain enough positive and negative samples. For different clothing, such as army, navy, and air force, they all have their own military clothing to meet the needs of military application. Therefore, different military clothing should be considered in the collection of training samples. In order to make the trained model have better adaptability, during the data collection, the dress should have different clothing for the diversity of the samples.

In the early stage of building the training data set, it was found that different surveillance scenes would also have a certain impact on the representation of model features, and the scene color around pedestrians would have a certain impact on the color histogram. Therefore, for the collected data, the data of different scenes should be sampled to increase the adaptability of different scenes.

Intelligent Identification. After data collection, SVM model needs to be trained and saved to memory. In testing, the model can be loaded to do intelligent dress identification. When a person enters surveillance scene, the device will automatically capture the video and analyze the video frames. The SVM model is used for dress identification, and the result signal outputs through Arduino. In experiments, different LED lights are set to represent the military dress and civil dress. Figure 9 shows a set of test results, where the red light represents military dress(left) and the green light represents civil dress(right).

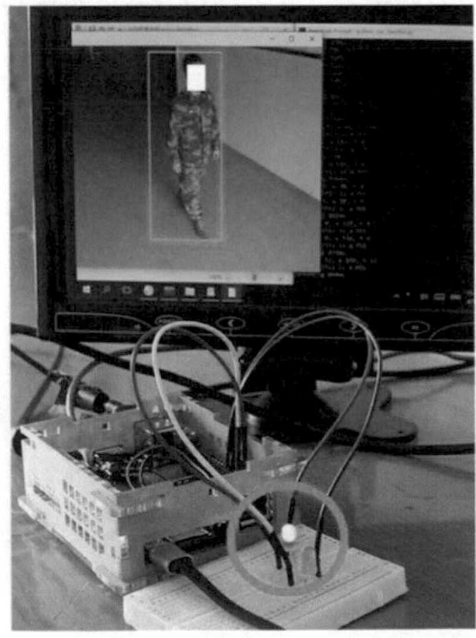

Fig. 9. Identification results. (Best viewing in color)

3.2 Discussion

Intelligent dress identification is used to replace soldier and video monitoring equipment at the camp gate. Its specific functions include three aspects. The first is able to process and analyze captured video to achieve real-time or near-real-time speed. The second is able to identify pedestrians to determine whether they are wearing military or civil clothes. The third is able to output signals through LED lights, buzzers to alarm dangerous.

Furthermore, there are some aspects for practical usage. Firstly, do not expose the LattePanda device to sunlight, and pay attention to waterproof. Secondly, regularly clean out unwanted and expired junk files in the storage device. Thirdly, keep all parts connected well before power on, and make sure to disassemble parts after power cut off.

4 Conclusion and Further Work

In this paper, we develop dress identification for camp security. We design the whole framework based on the LattePanda board, with Arduino chip, USB camera, and buzzer. To achieve the identification, the color histogram feature extracted and used for identifying the dress with support vector machine algorithm. The system is implemented based on the OpenCV library, and can run in real-time, with a reliable precision.

In the future, we can reorganize the functions according to different needs. A variety of sensors can be assembled, such as night-time infrared sensor, temperature sensor, humidity sensor, flame sensor and other sensors. The system can provide intelligent surveillance with comprehensive monitoring and warning functions. Besides, it can be set up on the platform of unmanned aerial vehicle (UAV), and be used to identify the dress in battlefield. At the same time, it can be applied to other branches of the armed forces, for example navy and air force, and also can be extended to the civilian fields such as intelligent identification of parcels in industry.

References

1. https://www.lattepanda.com/
2. https://www.arduino.cc/
3. Bradski, G.R., Kaehler, A.: Learning OpenCV - computer vision with the OpenCV library: software that sees. O'Reilly (2008). http://www.oreilly.de/catalog/9780596516130/index.html
4. Chakravarti, R., Meng, X.: A study of color histogram based image retrieval. In: Sixth International Conference on Information Technology: New Generations, ITNG 2009, Las Vegas, Nevada, USA, 27-29 April 2009, pp. 1323–1328 (2009). https://doi.org/10.1109/ITNG.2009.126
5. Cortes, C., Vapnik, V.: Support-vector networks. Mach. Learn. **20**(3), 273–297 (1995). https://doi.org/10.1007/BF00994018

6. Cychnerski, J., Brzeski, A., Boguszewski, A., Marmolowski, M., Trojanowicz, M.: Clothes detection and classification using convolutional neural networks. In: 22nd IEEE International Conference on Emerging Technologies and Factory Automation, ETFA 2017, Limassol, Cyprus, September 12-15, 2017, pp. 1–8 (2017). https://doi.org/10.1109/ETFA.2017.8247638
7. Kardas, K., Cicekli, N.K.: SVAS: surveillance video analysis system. Expert Syst. Appl. **89**, 343–361 (2017). https://doi.org/10.1016/j.eswa.2017.07.051
8. Kyaw, A.K., Truong, H.P., Joseph, J.: Low-cost computing using raspberry pi 2 model B. JCP **13**(3), 287–299 (2018)
9. Liu, G., Yang, J.: Content-based image retrieval using color difference histogram. Pattern Recogn. **46**(1), 188–198 (2013). https://doi.org/10.1016/j.patcog.2012.06.001
10. Perilla, F.S., Villanueva Jr., G.R., Cacanindin, N.M., Palaoag, T.D.: Fire safety and alert system using arduino sensors with IoT integration. In: Proceedings of the 7th International Conference on Software and Computer Applications, ICSCA 2018, Kuantan, Malaysia, February 08–10, 2018. pp. 199–203 (2018). http://doi.acm.org/10.1145/3185089.3185121
11. Rajendran, A., Li, P., Zhang, C., Deng, Y.: Parallel training of multi-class support vector machines using sequential minimal optimization. In: Proceedings of the 2007 International Conference on Machine Learning; Models, Technologies & Applications, MLMTA 2007, June 25–28, 2007, Las Vegas Nevada, USA, pp. 31–37 (2007)
12. Rego, A., Canovas, A., Jiménez, J.M., Lloret, J.: An intelligent system for video surveillance in IoT environments. IEEE Access **6**, 31580–31598 (2018). https://doi.org/10.1109/ACCESS.2018.2842034
13. Vieira, D.A.G., Takahashi, R.H.C., Palade, V., Vasconcelos, J.A., Caminhas, W.M.: The Q-norm complexity measure and the minimum gradient method: a novel approach to the machine learning structural risk minimization problem. IEEE Trans. Neural Network. **19**(8), 1415–1430 (2008). https://doi.org/10.1109/TNN.2008.2000442
14. Yannakakis, G.N., Togelius, J.: Artificial Intelligence and Games. Springer, Cham (2018). https://doi.org/10.1007/978-3-319-63519-4

ETSB: Energy Trading System Based on Blockchain

Xia Dong[1], Wei Zaoyu[2(✉)], Mao Hua[1], Xu Jing[1], Yang Debo[3], Wang Fanjin[4], and Bi Wei[5]

[1] Tianjin Electric Power Company Economic and Technological Research Institute of State Grid, Tianjing 300010, China
[2] School of Cyberspace Security, Beijing University of Posts and Telecommunications, Beijing 100876, China
weizaoyu2017@bupt.edu.cn
[3] Tianjin Electric Power Company Chengdong Power Supply Branch of State Grid, Tianjing 300250, China
[4] Nankai University, Tianjin 300071, China
[5] Zsbatech Corporation, Beijing 100088, China

Abstract. The emergence of blockchain solves the trust problem of the traditional energy trading mode, making the distributed multi-energy system possible and providing technical support for the new energy trading system. Therefore, a kind of energy trading system based on blockchain is proposed. The ETSB model is established firstly, and the definition of ETSB is expanded. The ETSB is a four-tuples, including node set, data set, consensus mechanism and intelligent contract, which are all newly constructed according to the characteristics of energy trading. On the basis of ETSB architecture, the potential security risks of ETSB are analyzed hierarchically. Finally, a complete security protection mechanism is proposed for those system vulnerabilities.

Keywords: Energy trading mode · Blockchain · ETSB · Four-tuples · Security

1 Introduction

The traditional energy trading mode is generally based on exchange trading [1], supplemented by the over-the-counter trading. In the exchange trading mode, exchanges have to coordinate the planning of trading in a centralized manner to maintain the stability, and therefore they must consume high costs and carry out rigorous safety ratings to keep their credibility as a central institution. In addition. There are also security issues in the exchanges where data may be lost or tampered with and users' privacy is difficult to be guaranteed. As for the OTC (Over-the-counter) trading mode, though it has the advantages of flexibility and convenience, users still have certain security risks due to the information asymmetry of the two parties. Besides, in the traditional energy trading system, because of the lack of a credible interconnection platform, different energy industries are closed relatively and operate independently, which cannot promote a wide range of multi-energy comprehensive utilization. The

C.-N. Yang et al. (Eds.): SICBS 2018, AISC 895, pp. 673–684, 2020.
https://doi.org/10.1007/978-3-030-16946-6_55

energy market needs a new “fair and open” trading mechanism, which is also efficient, accurate, private, trustworthy, and traceable.

The concept of blockchain is proposed by [2], which considers blockchain technology is an open, transparent, decentralized database. The blockchain also began to act as a shared database in practical applications [3]. [4, 5] provides a combination of digital currency and smart grid, using digital currency for smart grid settlement. Multi-signature technology and blockchain are also applied in distributed energy transactions, which is a solution for ensuring transaction security when a third party is not trusted in a distributed smart grid [6]. [7] proposes a new block-chain-based point-to-point energy trading network model, and [8] discusses the efficiency, scale and security of the smart grid with blockchain. [9] also explores the use of blockchain systems based on smart contract to share energy and promote power management. [10] studies the monitoring method of smart grid based on blockchain.

Obviously, the combination of blockchain and energy trading is a new development trend in the energy field, and the two are highly compatible. The credibility provided by blockchain technology can effectively solve the problem of openness and trust in energy transaction data. It also can guarantee the data owner’s intrinsic protection needs for the privacy and confidential information, break through the bottleneck of the traditional energy trading mode, and provide strong technical support for energy trading. However, because the new energy trading system involves multi-energy linkage and special high-value transactions, its security is also in urgent need of attention. When exploring new energy trading modes supported by blockchain, the security of the blockchain ecology must be improved.

In order to apply to energy trading supported by blockchain with higher security requirements, the Energy Trading System Base On Blockchain is proposed. According to the blockchain principle of the literature [11] and the energy trading process of the literature [12], the basic framework of the ETSB model and the definition of ETSB are first established and described in detail through a four-tuples. In the four-tuples, the node set includes the operators node and the users’ node, which form the ETSB network structure together. The data set consists of three elements, namely the energy supply data, the energy transaction data, and the energy transmission data. The consensus mechanism draws on BFT-DPoS to propose a new consensus agreement for energy trading. And finally the role of smart contracts in ETSB is introduced. Based on the architecture of ETSB, this paper analyzes the possible security risks of ETSB from its attack layers, attack surfaces and attack points, and targetedly develop reasonable security risk control measures.

2 ETSB Model

Blockchain is a chronological chain structure which can be regarded as a distributed database of various computer technologies, such as consensus mechanism, encryption algorithm, P2P network, and smart contract. In the ETSB model, the energy blockchain refers to a data sharing system, in which energy nodes use secure and transparent blockchain technology to realize distributed storage of energy transaction data, and then use smart contracts to automatically filter users, conclude transactions, and develop supply and delivery strategy.

The basic framework of the ETSB model supported by the energy blockchain is shown in Fig. 1. An integrated energy trading system generally includes multiple modules such as energy supply system, intelligent transmission network, energy storage system, users and agents. In the ETSB model, the integrated energy trading is divided into three modules: energy supply, energy transmission and energy trading, then the data of complete energy ecological information is carried by the energy blockchain. Specifically, the information flow of an integrated block is to generate a new block by consensus algorithm, then energy supply data, scheduling data, transaction records and other information of each energy source in each period are recorded on the block after the authorization of the energy operator and a timestamp is added to the block. Finally, a time-increasing hash data chain is formed, which ensures the difficult tampering and traceability of energy transaction data, and records the operation process and essential data of energy transactions comprehensively and accurately.

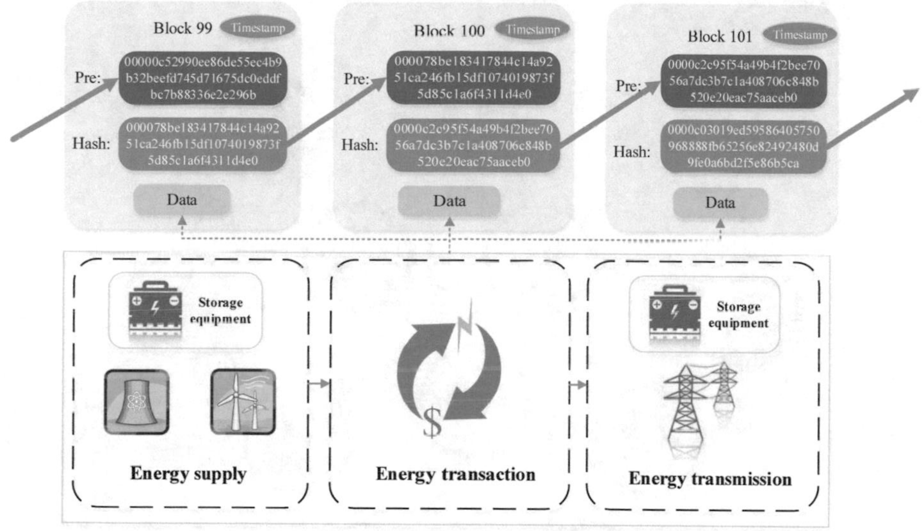

Fig. 1. ETSB basic framework

Definition. ETSB (Energy Trading System Based on Blockchain) is a system model that mainly records the operational data of energy supply, energy transmission and energy transaction based on the energy blockchain. The ETSB model can be represented by a four-tuples $ETSB = (N, D, C, SC)$.

2.1 Node Set

In the node set $N = \{ O, U\}$, O is a finite set of various energy operators, and U is a finite set of energy trading users. The node set is both providers and users of the information in the ETSB, and they constitute the node of the energy blockchain network together. Among them,$O = \{ O_i | 1 \leq i \leq n\}$, n is the number of energy operators participating in data sharing, j is the energy type index, **m** is the total number of energy types, and k is the user type index, which is divided into individuals, general enterprises, and large enterprises. Since the nodes in this model involve multiple energy operators and trading users, energy operators cannot be opened some key data about the energy to external users inevitablely. Therefore, the ESTB chooses to use the consortium blockchain. As shown in Fig. 2. In this mode, the node privilege can affect the nodes' reading and writing of trading data in the energy blockchain. The energy operator node has read and written permissions to the energy blockchain. But the openness of the user node is limited to a certain extent, and it just has partial reading and writing permissions after being authorized by the high-privileged nodes.

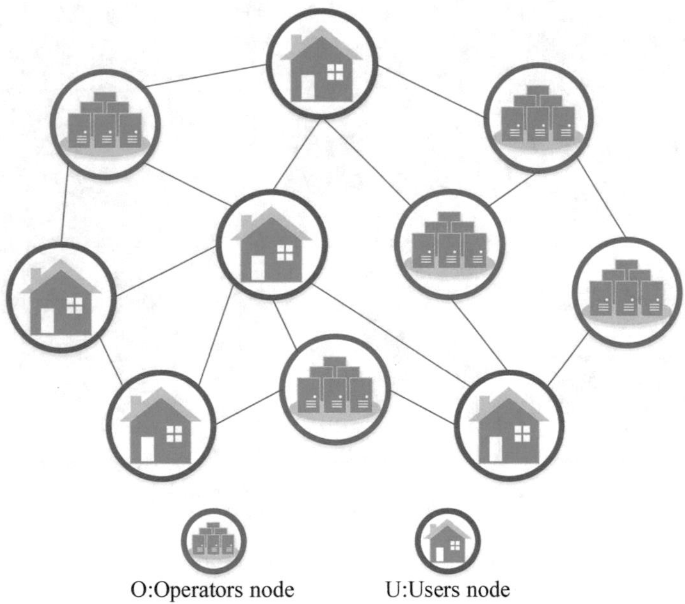

Fig. 2. ETSB network structure

In ETSB, when a new $O_i$node joins, it needs to provide node information such as enterprise name, capacity scale, energy type, and geographic location, so that the node can pass the system authentication, obtain the private key randomly generated by ETSB, get the public key by the private key and irreversible cryptographic algorithm. Whenever a new U_{jk}node joins, it needs to provide basic user information such as user type, identity information, and intended energy type, and obtain its private key and corresponding public key.

2.2 Data Set

The data set of *ESTB* is $D = \{ES, ETS, ETM\}$, *ES*, *ETSB*, and *ETM* represent energy supply data set, energy transaction data set, and energy transmission data set respectively. Each block of the energy blockchain consists of a block header and a block body. The block header encapsulates the timestamp, the hash used for the connection, the related information of the consensus algorithm, and the merkle root. The corresponding data(including energy supply, energy transaction, and energy transmission) in each block generation time will be recorded by the nodes in the block body, so that the energy transaction process in this time interval is completely recorded, ensuring that each energy transaction can be traced back to its supply source and transaction party, and transport stream. Taken the energy supply process as an example, the corresponding block data will include its index number, operating unit ID, energy type, agent, geographic location, quantity-price, and corresponding sales period, as shown in Fig. 3.

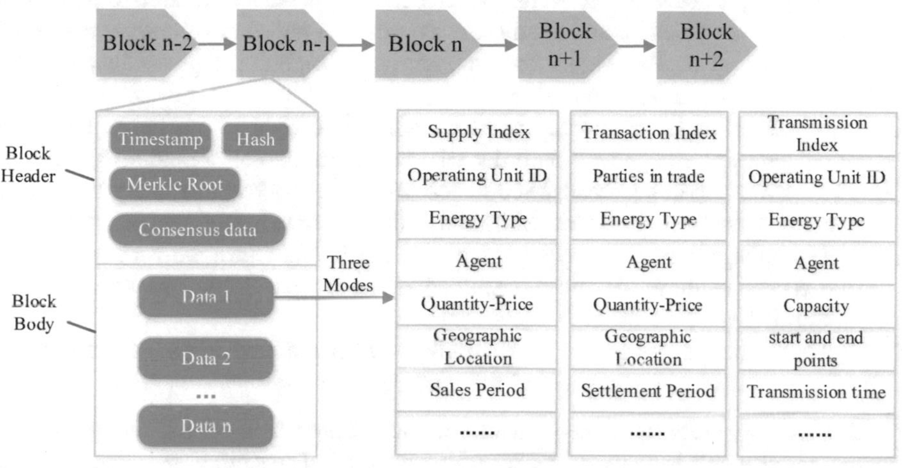

Fig. 3. ETSB block data

2.3 Consensus Mechanism

C represents the consensus mechanism for the energy blockchain. Obviously, the commonly used consensus mechanism PoW for digital currency blockchain is not suitable for energy trading. Its mining mechanism is redundant in ETSB, which not only wastes computing resources, but also makes mining management difficult because of the complexity of the energy nodes. Therefore, when constructing the energy blockchain, we proposes a new algorithm called E-DPoS by drawing on the BFT-DPoS consensus mechanism. The specific process is as follows:

(1) At the beginning of each cycle *T*, all operator nodes will broadcast their weight information $W = \{c, t, p, et, ID, w\}$ to the entire network of ETSB. In the weight information *W*, *c* is the credit rating which is calculated by the operators' inherent scale and transaction success rate. *t* is the trading age of the operator node which

will be cleared when it participates in a witness election. p is the transaction amount of specific operator in each cycle T. et is the energy type coefficient. ID is the operator identification number. w is the weight accumulation of each feature item, and each feature item weight can be referred to Table 1.

Table 1. Weight information

Feature item	Weight	Reference
c	W_c	1.0
t	W_t	1.0
p	W_p	0.4
et	W_{et}	0.2–0.8

(2) The operator nodes whose weight w exceeds the threshold are able to participate in the witness election, and finally select n operator nodes as witnesses, where n is a multiple of 3. The higher the weight w, the higher the probability that the operator node becomes witnesses.
(3) The witness node will obtain the block generation right in this cycle, and these n witnesses blocks generation order will be determined by a random algorithm.
(4) Each witness will be broadcasted to the whole network immediately when generating a new block. Other witnesses will be immediately verified the corresponding signature after receiving the new block, and return the verified block to the current witness. At the moment when more than 2/3 of the witnesses are confirmed, the new block is irreversible.
(5) After all the witness nodes have completed their required block generation, this blocks generation cycle end and the next round of consensus begins.

The E-DPoS consensus mechanism can be adapted well to the energy trading market, and it won't be caused blockchain forks at all. The setting of trading age and weight also make the attacks from malicious nodes more difficult to implement. In practical applications, the feature items and weights of the E-DPoS algorithm can be appropriately adjusted according to the operation.

2.4 Smart Contracts

SC represents the Smart Contract. It links the various modules of the ETSB and plays a role in management and control. According to their specific needs, the nodes can develop a smart contract, and spread it to the ETSB whole nodes through the P2P network. When it detects that external data situation is consistent with the terms of the contract, the smart contract can be executed automatically. In ETSB, only the operator node and the authorized high-privileged user node are able to write smart contracts for the energy blockchain, and other user nodes have only the right to use partial smart contracts.

We only introduce two major smart contracts in ETSB. The first one is the transaction negotiation smart contract, which can retrieve the users' demand information and the energy suppliers' supply information from the energy trading platform, and select suitable trading objects for both parties according to the specific requirements. The second one is the data analysis smart contract, which can retrieve and analyze block data and then develop energy management strategies for the next stage. For example, the operator nodes can extract the energy block transaction data set and the energy transmission data set in a period of time through the smart contract, and then analyze the fluctuation of the energy transaction price under the factors of geographical location, user type, transaction time and the like, thereby achieving goals such as pricing more reasonable, targeting the best user base.

Figure 4 is the basic operation block diagram of ETSB. The energy blockchain is equivalent to an energy trading sharing database. The user nodes log into the energy trading platform through the management APP and grants energy blockchain the right to record transaction data. The energy supply data and energy transmission data also require the operator node to authorize to record. The smart contract can obtain the supply and demand information of both parties of the energy transaction, and accurately locate the appropriate energy supply information for the users, promoting the completion of the transaction. Besides, it can also analyze and summarize all kinds of useful data, and provide energy operators with data support for formulating strategies. The energy blockchain is the information exchange center and data storage center of ETSB, and its secure and non-tamperabe architecture also makes the shared energy data more credible. While the smart contract can be regarded as the lubricant of ETSB, which makes the overall operation scheduling of the system more flexible.

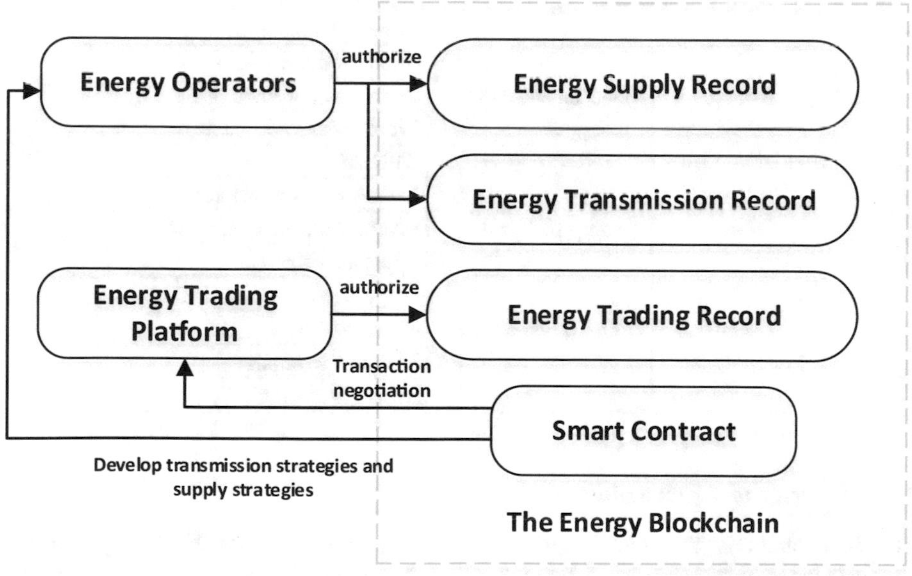

Fig. 4. ETSB basic operation block diagram

3 ETSB Security

In order to improve the security of ETSB, we identify the security risks of ETSB and divide them into attack layer, attack surface and attack point according to the architecture of ETSB, so as to propose a more complete security protection mechanism based on these existing system vulnerabilities. In Table 2, it is shown the security risks of ETSB, in which the attack layer is divided into blockchain, interface, and users.

Table 2. ETSB security risks

Attack layer	Attack aspect	Attack point
Blockchain	Block data	Malicious information
		Resource abusing
	P2P network	Eclipse attacking
	Consensus protocol	Network monopoly
		Alliance centralization
	Smart contract	Reentrancy
		Access control
		Integer overflow
		Unchecked-send
		Bad Randomness
		Denial of Service
		Front-Running
		Time manipulation
		Short Address Attack
Interface	Management APP	Private key storage
		Weak password
		RPC API
		Unknown vulnerability
	Trading platform	oAuth protocol vulnerability
		Payment vulnerability
		Single-sign-on vulnerability
Users	High-privileged users	Phishing attack
		Man-in-the-middle
		Trojan hijacking

3.1 Security of Blockchain

The blockchain is the underlying technical foundation of the ETSB system and supports the regular running of the entire ETSB system. If there is a security problem at the energy blockchain, it will definitely cause the upper level services to be seriously affected. Therefore, the ETSB must establish a favorable blockchain security mechanism.

First of all, data is the key content of ETSB. Due to the non-tamperablity and non-deletablity of blockchain, malicious information and resource abusing attacks will greatly increase the redundancy of block data, reducing the overallrunning speed of energy trading, or even hindering the normal operation of transactions. Therefore, in order to increase the security of the input data from the nodes, the ETSB performs strict sensitive word checking and filtering on the input data from the nodes. The format and digital limit of the writable position of the energy blockchain are provided, and only fixed value option inputs are permitted at some key points.

Secondly, energy trading and network communication are inseparable. If any high-privilege operator node in the P2P network is subjected to traffic theft or hijacking attacks such as eclipse attacks, the node may be isolated from the main network and is spoofed by the attacker to obtain erroneous data orinformation, which will seriously affect the economic benefits of the node, and even cause other user nodes to be deceived. The data transmission of ETSB at the network layer enforces the secure HTTPS protocol. Besides, to prevent all incoming and outgoing connections of the victim node from being monopolized, we open 13 outgoing connections by learning from the Ethereum and control the number of node IDs that each machine can generate, which greatly improves the difficulty of eclipse attacks. In addition, a check code is added in the transmission information to authenticate the P2P network data.

The consensus mechanism is the basis of block generation. Since the improved algorithm based on D-PoS is adopted by ETSB, it also faces some security issues with D-PoS, such as network monopoly and alliance centralization. This problem is more dependent on the credibility of the operators' organization and the improvement of the credit information system.

Smart contracts are the focus of ETSB security. It is because the essence of smart contract is codes, which makes it difficult to guarantee the absence of security vulnerabilities. Therefore, a rigorous security audit is required before smart contracts (Smart contract here is based on solidity development) are written into ETSB.

First, we need care about making external calls. For example, we should avoid using *raw calls* or *contract calls* and using *send()*, *transfer()* instead of *call.value()* when sending important data. When using low-level calling function such as *address.call()*, *address.callcode(), address.delegatecall() and address.send()*, we need to handle the possibility of the command to check the return value is failed to be call. In addition, we should mark the visibility of functions and state variables explicitly and note the distinction between functions and events.

Second, we should examine the code logic of important operations carefully in order to avoid logic traps and standardize the use of exception handling *assert()*, *require()* and *revert()* so as to increase the code robustness of smart contracts.

At the data level, smart contracts should not assume that its initial state contains a zero balance. Note that integer division is to round to down the nearest integer. Besides, in order to prevent integer overflow, it is also necessary to verify the data length of each location, especially the data length in the fallback function. ETSB also locks programs to specific compiler versions to avoid security vulnerabilities due to different compiler versions.

3.2 Security of Interface and Users

The interface includes the management APP and the trading platform website, and the security of the interface is closely related to the security of the user node, which is more biased towards the traditional APP security protection and website security protection.

The security of the management APP is more relevant to the users' personal. For example, if the APP stores the private key related file unencrypted in the system local, the attacker may restore the algorithm logic by means of reverse analysis or other techniques, thereby recovering the user's private key. Such Insecure behaviors will cause the leakage of the users' private key and serious economic losses.

The attack points of the management APP are summarized into four aspects: running environment security, protocol interaction security, data storage security, and function design security. Firstly, when the management APP runs on the users' mobile phone, it will automatically detect the system version, the root environment, and the network agent of the current mobile phone, and provide the function of the APP's own integrity checking. Secondly, in the process of conducting the energy transaction interaction, to prevent information forgery and other attacks, the account of both parties to the transaction will be verified twice, and the function of querying the transaction records is added. In the aspect of data storage security, in order to prevent the loss of the private key, the idea of image information hiding and key distribution can be adopted and we may design a new key storage algorithm to promote the security of key storage. The security of functional design is mainly about transaction password security and transfer security. The management APP needs to improve the detection and restriction of transaction password security level. And in the meanwhile, it should add strict verification and confirmation mechanism to ensure that the transfer address cannot be tampered with.

The protection of the trading platform is similar to the security protection of the website. Online business system needs to undergo strict security auditing and testing. In addition, it is significant to formulate a reasonable security isolation policy, open only necessary service ports, establish a security emergency response mechanism, and conduct security checks regularly. Moreover, because the energy trading platform involves large-scale energy trading and capital flow, therefore, a high-intensity account security system is needed. The specific protective measures are as follows:

(1) To prevent brute force and collision attack, it is asked to limit the login frequency, and perform man-machine identification through verification code;
(2) As for XSS and CSRF vulnerabilities, it is asked to open http only to prevent cookie leakage, and use POST method as much as possible when submitting actions;
(3) Enable multi-factor authentication when the account performs sensitive functions such as payment function, and it is also necessary to prevent vulnerabilities such as "login bypass", "override access" and "password recovery";
(4) The design of business logic must be rigorous, and each piece of business logic code requires a lot of fuzzing and code auditing.

The users layer mainly depend on the users' own security awareness, because most attack methods for users mainly use social engineering to conduct network spoofing. For example, attackers stole the users' login password and transfer address by means of counterfeiting trading platform website domain and page. What the system official needs to do is to provide more security reminders and suggestions, and to find security issues promptly and respond in a timely manner.

4 Conclusion

In order to adapt to the new energy trading mode, it is combined new technologies such as blockchain and smart contracts with energy trading in this paper. It probes the energy trading system of multi-energy and multi-operators with blockchain as the link, and proposes the ETSB model. The model is based on the general operation of energy trading, which is divided into three modules of energy supply, energy trading and energy transmission. The three modules are used as data carriers of the energy blockchain. And the definition and establishment of the basic framework of the ETSB model are proposed, which the ESTB model is represented as a a four-tuples consisting of node set, data set, consensus mechanism, and smart contracts. The node set includes the operator nodes and the user nodes, which together constitute the network structure of the ETSB. The data set includes the supply data set, the transaction data set and the transmission data set, which is the core of the ETSB data sharing. The consensus mechanism is an improved BFT-DPoS algorithm based on the characteristics of the energy trading field, which is both efficient, private, and secure. And the coordination and management of each module is completed on the basis of smart contracts. Finally, the security of ETSB is also analyzed. We systematically study its possible security threats according to the architecture of ETSB, so that we can develop comprehensive protection against potential system vulnerabilities and improve overall system security.

However, there is no uniform standard for the security risk assessment technology of the energy blockchain. Although the list of security risks in the energy blockchain has a clear hierarchical structure, it still needs a large number of actual vulnerability cases to be improved. Secondly, the construction of the energy data set is relatively completed, but the actual application may require more detailed content division, and more accurate and effective adjustments needed to be made in practical application. In addition, to automate the management process of the system, more types of smart contracts should be developed to extend the application of ETSB. Therefore, subsequent research needs to be further optimized in practice, and more complete and stable models and algorithms should be built from practical results to support the new energy trading system.

References

1. Wiyono, D.S., Stein, S., Gerding, E.H.: Novel energy exchange models and a trading agent for community energy market. In: European Energy Market, pp. 1–5. IEEE (2016)
2. Melanie, S.M.: Blockchain: Blueprint for a Neweconomy. OReilly, Sebastopol (2015)
3. Schneider, J., Blostein, A., Lee, B., et al.: Blockchain: putting theory into practice. The Goldman Sachs Group (2016)
4. Mihaylov, M., Jurado, S., Avellana, N., et al.: NRGcoin: virtual currency for trading of renewable energy in smart grids. In: Proceedings of the 11th International Conference on the European Energy Market (EEM), pp. 1–6. IEEE, Krakow (2014)
5. Jurado, S., Jurado, S., Moffaert, K.V., et al.: SCANERGY: a scalable and modular system for energy trading between prosumers. In: International Conference on Autonomous Agents and Multiagent Systems. International Foundation for Autonomous Agents and Multiagent Systems, pp. 1917–1918 (2015)
6. Aitzhan, N.Z., Svetinovic, D.: Security and privacy in decentralized energy trading through multi-signatures, blockchain and anonymous messaging streams. IEEE Trans. Dependable Secure Comput. (99) (2016). https://doi.org/10.1109/tdsc.2016.2616861
7. Sabounchi, M., Wei, J.: Towards resilient networked microgrids: blockchain-enabled peer-to-peer electricity trading mechanism. In: IEEE Conference on Energy Internet and Energy System Integration, pp. 1–5. IEEE (2018)
8. Mylrea, M., Gourisetti, S.N.G.: Blockchain for smart grid resilience: exchanging distributed energy at speed, scale and security. IEEE (2018)
9. Crypto-trading: Blockchain-oriented energy market
10. Gao, J., Asamoah, K.O., Sifah, E.B., et al.: GridMonitoring: secured sovereign blockchain based monitoring on smart grid. IEEE Access **PP**(99), 1 (2018)
11. Nakamoto, S.: Bitcoin: a peer-to-peer electronic cash system. Consulted (2009)
12. Dynamic Energy Trading for Energy Harvesting Communication Networks: A Stochastic Energy Trading Game

An Analysis on Inverted Mirrored Moiré and Ribbon of Band Moiré

Hui-Ying Liu[1], Dao-Shun Wang[1(✉)], Xiang-Hui Zhao[2], Shun-Dong Li[3], and Ching-Nung Yang[4]

[1] Department of Computer Science and Technology, Tsinghua University, Beijing, China
daoshun@mail.tsinghua.edu.cn

[2] China Information Technology Security Evaluation Center, Beijing, China

[3] School of Computer Science, Shaanxi Normal University, Xi'an, China

[4] Department of Computer Science and Information Engineering, National Dong Hwa University, Hualien, Taiwan

Abstract. Band moiré pattern is produced by the superposition of the base layer with the revealing layer that has a different periodic frequency. Band moiré can effectively describe the geometric transformation of moiré, so that we can design more elements and specific moiré in band moiré image, which can be used to engage and decipher the secret image. Hersch et al. put forward a mathematical model to describe the geometric transformation of moiré pattern, and based on this, a method for constructing band moiré pattern was proposed. Furthermore, the inverted mirrored moiré effect is also described for the case that the period of the revealing layer is less than the period of the base layer. However, the principle and limitations of inverted mirrored moiré effect have not been given. In this paper, we introduce the principle of the inverted mirrored moiré effect and specify the conditions of producing inverted mirrored moiré, That is, when the period of the revealing layer is less than the period of the basic layer and more than half of the period of the base layer, the inverted mirrored moiré will be shown. We also noticed the phenomenon of the repeated local moiré pattern with the analysis of the moiré generation. We give the restriction of producing ribbon moiré, when the period of the revealing layer is equal to the period of the base layer, the ribbon moiré will show. The limitations of inverted mirror image or ribbon were analyzed and the results were verified by experiments.

Keywords: Ribbon moiré · Inverted mirrored moiré · Moiré pattern

1 Introduction

When two transparent layers with slightly different spatial frequencies are superimposed together, the two layers are shifted relative to each other within a certain range then a new pattern is produced. This new pattern is called moiré pattern. This visual effect is called a moiré effect. According to the periodicity of the superposition layers, moiré pattern can be divided into periodic moiré [2, 3] and non-periodic moiré [4].

C.-N. Yang et al. (Eds.): SICBS 2018, AISC 895, pp. 685–698, 2020.
https://doi.org/10.1007/978-3-030-16946-6_56

Periodicity refers to the amount occurring within a regular or certain time interval, or the repetition of space or time. Rotating a transparent layer with a periodic pattern by a certain angle and then superimposing it with the layer itself will produce a certain pattern of dark and dark distribution, that is, periodic moiré pattern. For the study of periodic moiré, Bryngdahl [3] used Fourier and vector method to analyze the moiré pattern produced by changing periodic grating pattern, and summarized the moiré effect. Amidror [5] used the Fourier method to analyze the moiré pattern generated by the geometric transformation of the periodic layers. Amidror *et al.* [2] studied the characteristics of periodic moiré patterns such as size and orientation, and pointed out that the Fourier method cannot analyze the moiré generated by the superimposition of all periodic patterns. Periodic moiré is repeated in a certain period or in space or time, so for the purpose of anti-counterfeiting, the periodic moiré effect can make clear use of the magnification effect generated by superposing layers, and generate the corresponding moiré image. Therefore, the study of periodic moiré pattern is helpful in improving the algorithms of anti-counterfeit, image hiding, information hiding and so on. As a typical periodic moiré, we analyze the effect of moiré by analyzing the geometric transformation model of banded moiré. We can freely design different elements, which play a role in the construction of specific moiré image. Hersch *et al.* [1] explained the geometrical principle of banded moiré formation and the construction of banded moiré image. Zhou *et al.* [6] analyzed the moiré patterns produced by grating superposition from Fourier method. It is concluded that the two moiré patterns produced by the superposition of the extended grating can be regarded as a compound transformation of a moiré. Walger *et al.* [7] proposed a method of embedding seven secret messages in a base layer of banded moiré image. Cadarso *et al.* [8] proposed a mini band moiré image that improved the complex high-resolution safety features in terms of copy security. Hersch *et al.* [1] extended the construction of curvilinear moiré by constructing the moiré of band moiré image, and then referred to the mirrored moiré, but they did not discuss the principle and limitations of the inverted mirrored moiré pattern.

This paper is organized as follows. In Sect. 2, we introduced the geometric principle of banded moiré image and the construction of banded moiré image. Section 3 analyzes the principle of the inverted mirrored moiré pattern in detail and the restrictions for the appearance of moiré in the mirror are given, and the inference is verified. In Sect. 4, we give an explanation of the appearance of ribbons and the restriction of ribbons. We also use the experiment to verify the inference.

2 Review the Construction of Band Moiré Image

In this section, we briefly review the restrictions for the formation of band moiré image, the geometric principle of moiré and our motivation for our research of the different moiré pattern. Band moiré pattern can reveal the secret image in the base layer. In the process of adjusting the period of base layer T_b and the period of revealing layer T_r, the moiré pattern will appear band moiré and inverted mirrored moiré pattern. These are different from the phenomenon of the moiré image being erected, which has motivated our research on the principles and restriction of the different effects of band moiré pattern.

2.1 The Parametric Equations Model

The parametric equations model is the easiest and wide-ranging method to predict the moiré geometry after superposing the grating linear [9]. Taking the moiré pattern produced by a linear raster plot as an example can be used to explain the parametric equation model. When two layers are superimposed, the human visual system will automatically select the most prominent moiré pattern [9]. Therefore, as shown in Fig. 1, when the layers of grating m and the layer of grating n are superposed, the most prominent moiré pattern is p-moiré.

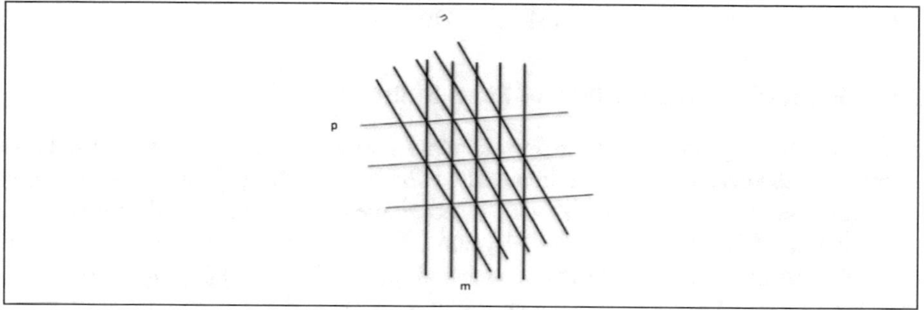

Fig. 1. Grating superposition

Take the moiré in Fig. 1 as an example, the parameter equation model of moiré is expressed: We define T_1 and T_2 as the periods of the base layer and the revealing layer respectively, and θ is the smaller angle between the grating and the vertical direction. Set the straight line grating layer 1 (ignoring the width and strength of the grid) of the linear equation as $x = mT_1$, (m ∈ Z); set the straight line grating layer 2 (ignoring the width and strength of the grid) of the linear equation as $x \cos\theta + y \sin\theta = nT_2$, (n ∈ Z). After superposing of layer 1 and layer 2, the most prominent moiré pattern is p-moiré pattern, which can be presented as $k_1 m + k_2 n = p$, (p ∈ Z), the moiré that satisfies this equation is called (k_1, k_2)-moiré.

Example 1: As shown in Fig. 2, when p = 1, the intersection points of the grating m and n are (1, 0), (2, 1) and so on, and substituted into the moiré equation $k_1 m + k_2 n = p$, we obtain $k_1 = 1, k_2 = -1$. Therefore, p-moiré is called $(1, -1)$-moiré. Simplify the Moiré equation as $x(T_2 - T_1 \cos\theta) - yT_1 \sin\theta = T_1 T_2 p$.

The parametric equation model can be used to calculate the moiré pattern obtained by superimposing multiple grating layers, and the same applies to the curvilinear grating layers. However, it is necessary to know the parametric equation of the superposition layer to derive the equation for the moiré generated. Moreover, the parametric equation model cannot indicate which moiré is the most obvious and most recognizable to the human eyes [9]. In fact, the parametric equation model has been completely included in the model based on Fourier theory.

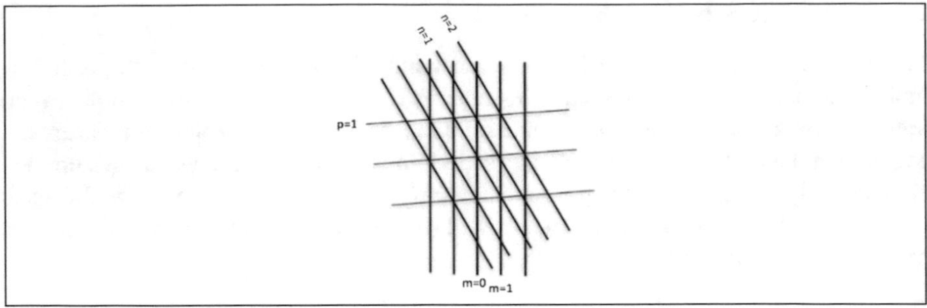

Fig. 2. $(1, -1)$-moiré [9]

2.2 Geometric Transformation of Band Moiré Image

Hersch et al. [1] proposed that in the intensity-based band moiré pattern, the parallelogram of the base band is superimposed by the transparent gratings of the revealing layer, and the moiré image shows an enlarged element of the base band, that is, moiré parallelogram is generated. The parallelogram ABCD on the base band B_0 is mapped to the moiré parallelogram ABEF (as shown in Fig. 3) after the superimposition of the revealed layers. The geometric transformation of the principle and parameters are defined as follows [1].

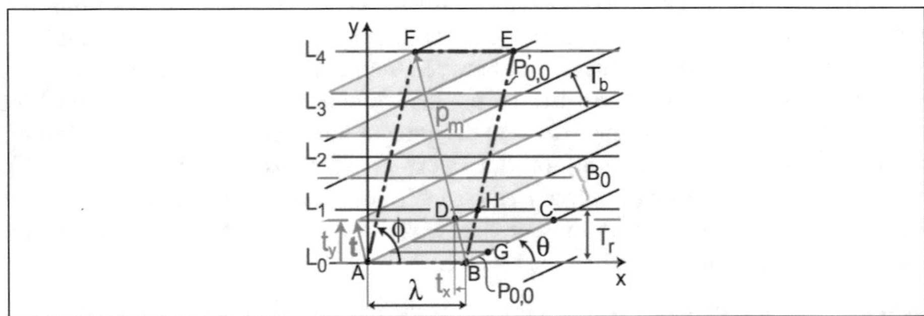

Fig. 3. Geometric transformation [1]

We define the period of revealing layer as T_r, and the width of base band B_0 as T_b, the angle between the basic layer and the revealing layer is θ. The layer that is copied in accordance with the copy vector $t = \left(t_x, t_y\right)$ is called base layer. Point H is the intersection of the revealing layer L_1 and the upper edge lines of the base band B_0, that is, $H = (T_r/\tan\theta, T_r)$. The coordinates of point G are copied according to the copy vector $= \left(t_x, t_y\right)$, to obtain the coordinates of point H, so the coordinates of point G are $G = \left(T_r/\tan\theta - t_x, T_r - t_y\right)$.

During the mapping process, point B is a fixed point, that is, the coordinates of point $B(\lambda, 0)$ are invariant and point G is mapped to point H in the geometric transformation, therefore, we can obtain the geometric transformation $\begin{bmatrix} x' \\ y' \end{bmatrix} = \begin{bmatrix} p & q \\ r & s \end{bmatrix} \begin{bmatrix} x \\ y \end{bmatrix} = \begin{bmatrix} 1 & \frac{t_x}{T_r - t_y} \\ 0 & \frac{T_r}{T_r - t_y} \end{bmatrix} \begin{bmatrix} x \\ y \end{bmatrix}$. The geometric transformation formula shows that the moiré parallelogram enlargement factor is $s = T_r / (T_r - t_y)$. The geometrical transformation equation has no relationship with the angle θ between two layers, but the Moiré directions ϕ is related to θ. Moiré parallelogram angle ϕ meets the formula $\tan\phi = \frac{T_r}{\frac{T_r}{\tan\theta} - \lambda}$. We can obtain the parallelogram direction $\tan\phi = \frac{T_r \cdot \sin\theta}{T_r \cdot \cos\theta - T_b}$ according to the slope $\lambda = \mathrm{T}_b / \sin\theta$ of the base band. According to the enlargement factor s, we can get the moiré parallelogram replication vector is $p_m = \left(t_x + t_y \cdot \frac{t_x}{T_r - t_y}, t_y \cdot \frac{T_r}{T_r - T_y}\right) = \frac{T_r}{T_r - t_y} \cdot t$ and the period of the moiré pattern $T_m = \frac{T_b \cdot T_r}{\sqrt{T_b^2 + T_r^2 - 2 \cdot T_b \cdot T_r \cdot \cos\theta}}$. Thus the vertical replication vector of base band B_0 is $t_y = \frac{H_0' \cdot T_r}{(H_0' + T_r)}$.

According to $\tan\phi = \frac{T_r}{\frac{T_r}{\tan\theta} - \lambda}$, we know that the moiré parallelogram orientation ϕ is determined by period T_r, T_b and the angle θ between two layers. Hersch et al. proposed the geometrical transformation of banded moiré images, constructed the mathematical relationship between the base layer, revealing layer and the layers and the moiré layer. According to the geometric transformation, it is possible to construct a base layer and a revealing layer that can generate a specified moiré, so as to realize the construction of a specified moiré.

2.3 Construction of Band Moiré

When the angle between the base layer S_0 and the reveal layer is 0, the base band is obtained by compressing the secret image S longitudinally, and the basic graphic band is replicated longitudinally to generate a base layer with a single secret. The transparent grating that reveals the layer has continuous sampling points in every base band. After the identification of human vision system, the moiré image with base band magnified will be obtained. Therefore, the decryption of moiré image is a secret image. The introduction and examples of the vertical moiré algorithm for the moiré construction are given below: [10].

Input:	Secrete image S
Output:	Base layer B, revealing layer L
Generating procedure	• The secret image S is compressed longitudinally in a certain proportion to generate the base band B_0 with the period of T_b; • Replicate the base band B_0 longitudinally to generated the secrete layer B; According to the size of the period of basic layer T_b, determine the period of revealing layer T_r and the width of the transparent grating m, then generate the corresponding revealing layer.

Example 2: The size of the secret image S is 720 * 300, and the encrypted information is "NLMY" (as shown in Fig. 4). One can compress the secret image into the base band B_0 of the period $T_b = 30$ in the vertical direction. The base band B_0 replicates 24 times in the vertical direction to generate the basic encryption layer B (as in Fig. 5), the size of the encryption layer is 720 * 720. Define the period of the grating revealing layer L as $T_r = 36$, in which the width of the white transparent aperture is $m = 6$, to generate the revealing layer (as in Fig. 5). Superposing the basic layer and the revealing layer can get the secret image (as shown in Fig. 6).

Fig. 4. Secret image S

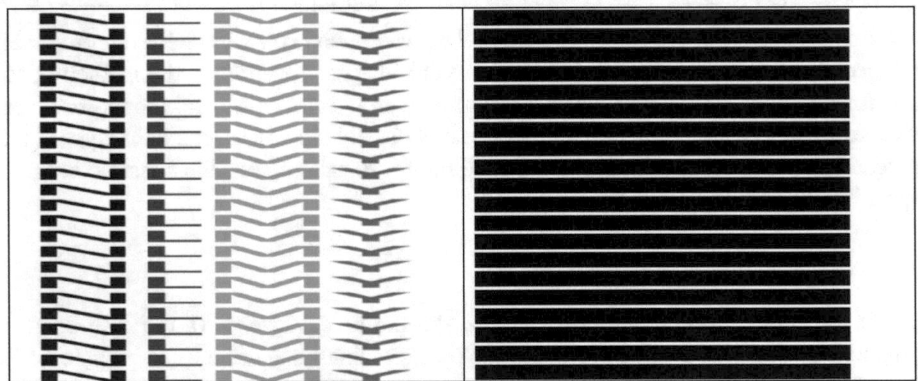

Fig. 5. Base layer B and revealing layer L

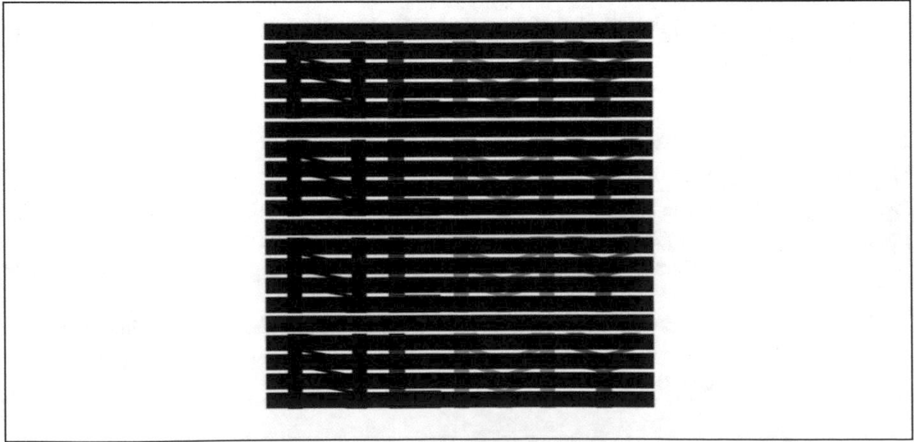

Fig. 6. Superposition

3 Principles and Restrictions of Inverted Mirrored Moiré

In this section, we analyze the inverted mirrored moiré phenomenon of band moiré and give the restriction conditions of the base layer period and the revealing layer period when the inverted mirrored moiré appears.

Moiré is very sensitive to any slight change, so the changing the period of the superposition layers affects the generated moiré effect. In the Band Moiré Image [1], Hersch mentioned that when the periods of the superposition layers change, the inverted mirrored moiré phenomenon appears, but he did not explain the restriction of the layer period.

When the period of the revealing layer T_r gradually reduced to less than the period of the base layer T_b, we can see the inverted mirrored moiré as shown in Fig. 7(a–c). In Fig. 7, the period of the basic layer T_b is consistent with previous period, and the period of the revealing layer T_r of (a)–(c) decreases gradually, generating the moiré effect as follows.

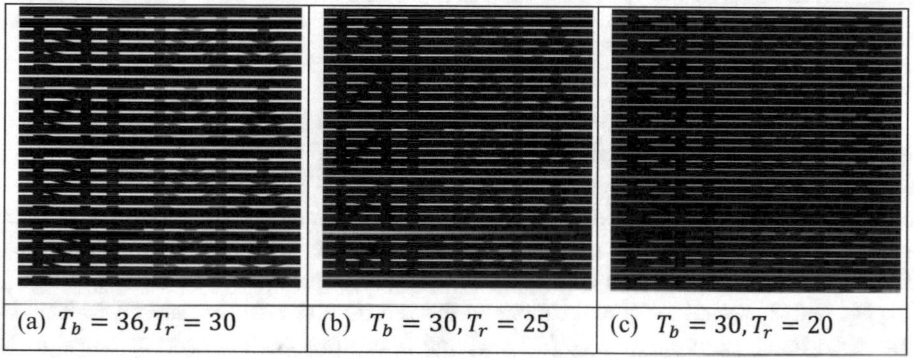

(a) $T_b = 36, T_r = 30$ (b) $T_b = 30, T_r = 25$ (c) $T_b = 30, T_r = 20$

Fig. 7. Inverted mirrored moiré

Therefore, we speculate that the appearance of inverted mirrored moiré is related to the relationship between T_b and T_r. During the decrease of T_r, the pattern of inverted mirrored Moiré decreases, and then disappears. We infer that the appearance of the inverted mirrored moiré is related to the decrease of the revealing layer period, and the inverted mirrored moiré effect ends when another moiré effect appears. Through the theoretical analysis of experiment and geometric transformation, we analyze the relationship between T_b and T_r, and the above speculation is verified in the conclusion. The example is given to illustrate the analysis of the inverted mirrored moiré restriction.

3.1 Inverted Mirrored Moiré Principle

In Band Moiré Images [1], the enlargement factor of moiré is $s = T_r/\left(T_r - t_y\right)$, when the revealing layer period gradually reduces to less than the base layer period T_b, the enlargement factor is negative, the inverted mirrored moiré phenomenon appears. In this section, we take the vertical base band in band moiré as an example to explain the principles and restrictions of inverted mirrored moiré in detail.

When the revealing layer period T_r decreases gradually on the basis of Sect. 2.3, the image starts to appear inverted mirror image. Taking the construction of band moiré as an example, according to the theoretical analysis of the erection effect of band moiré in [2], this paper constructs the generation theory of the inverted mirrored moiré effect of the band moiré.

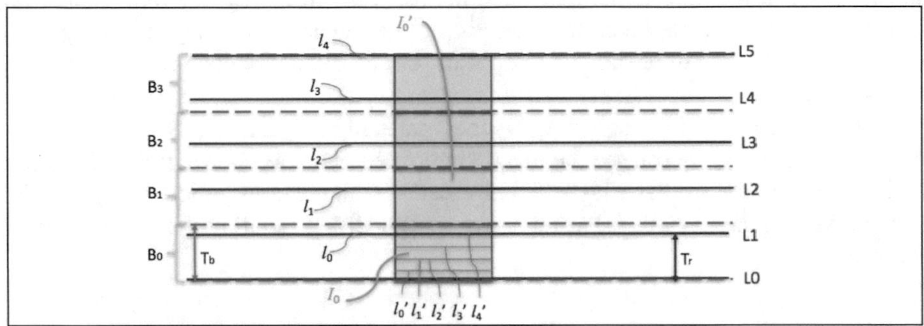

Fig. 8. Principle of mirrored (4 basic bands, 5 revealing bands)

As shown in Fig. 8, the revealing layer is composed of a series of black and white transparent gratings with a period of T_r (4 base bands and 5 revealing bands), which are represented by L_0, L_1, L_2, L_3 and L_4 respectively. B_0 is the base band that contains a secret image. B_0 is the base band that contains a secret image, and it is replicated in the vertical direction along the replication vector into the base bands B_0, B_1, B_2, B_3, forming repeated base layers containing the secret image. The green rectangles in B_0, B_1, B_2, and B_3 are all exact replications of the rectangular L in the base layer. That is, the raster transparent aperture l_0, l_1, l_2, l_3, and l_4 of the revealing layer samples in B_3, B_2, B_1, B_0 of the base layer respectively. Thus, the rectangular l_0', l_1', l_2', l_3', and l_4' in the sampling zone B_0 of the base layer are mapped to the transparent apertures l_4, l_3, l_2, l_1, l_0 of the revealing layer respectively.

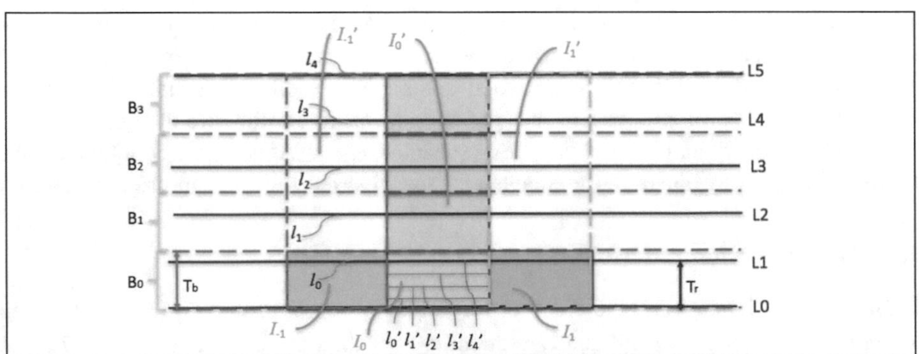

Fig. 9. Inverted mirrored image principle

As shown in Fig. 9, according to the same principle, the rectangular l_{-1} of the base band B_0 will be mapped to l'_{-1}, l_1 *mapped to* l'_1, and so on, the base band B_0 will be mapped to B'_0 of the moiré layer respectively. That is magnified inverted mirrored moiré image of the base band B_0. When the revealing layer is superposed over the base layer, the secret will be decrypted. Through the human's visual system, an inverted mirrored moiré image is produced.

In the restriction of inverted mirrored moiré, we continued to discuss the relationship of period between the base layer and the revealing layer through the principle of geometric transformation, deduced the mirrored moiré formula in vertical direction. The related parameters and variables are defined as follows.

T_r: the period of the revealing layer;
T_b: the period of the base layer, that is, the width of the base band;
n: the number of hidden secrets in the encrypted layer;
m: the width of the transparent aperture in the period of the revealing layer grating;

The mathematical relationship between variables and parameters is as follows.

$$T_b = m + T_r \tag{1}$$

$$T_r = T_b - m = n * m \tag{2}$$

Through the above formula, the vertical inverted mirrored moiré is generated.

3.2 Experimental Results and Verification

Example 3: Suppose that the secret image S has a length and width of 720 * 300, and the encrypted information is "NLMY" (as shown in Fig. 10). The secret image is compressed in the vertical direction into a basic band B_0 with a width $T_b = 36$. The base band B_0 is replicated 20 times in the vertical direction, and generated a basic encrypted layer B (as shown in Fig. 11). The length and width of the encrypted layer are 720 * 720. Suppose that the black and white grating period of the revealing layer is $T_r = 30$, wherein the size of the white transparent aperture is $m = 6$, generating a revealing layer L (as shown in Fig. 11). The base layer B and the revealing layer L are superimposed to obtain the moiré pattern of the inverted mirrored pattern (as shown in Fig. 12).

Fig. 10. Secret image S

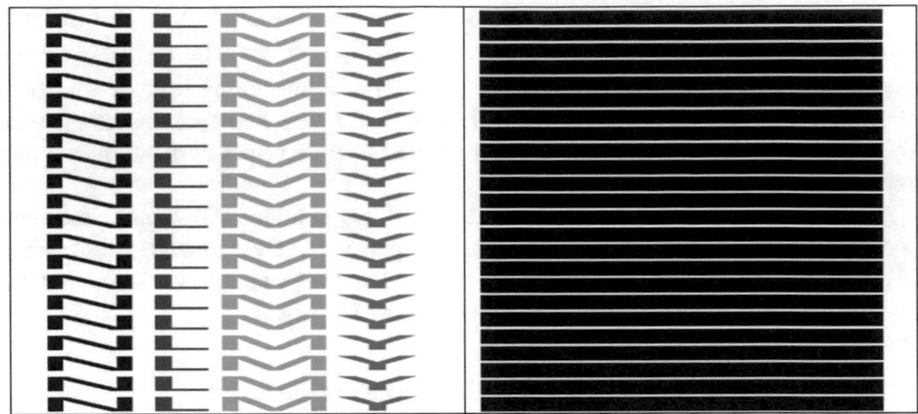

Fig. 11. Base layer B and revealing layer L

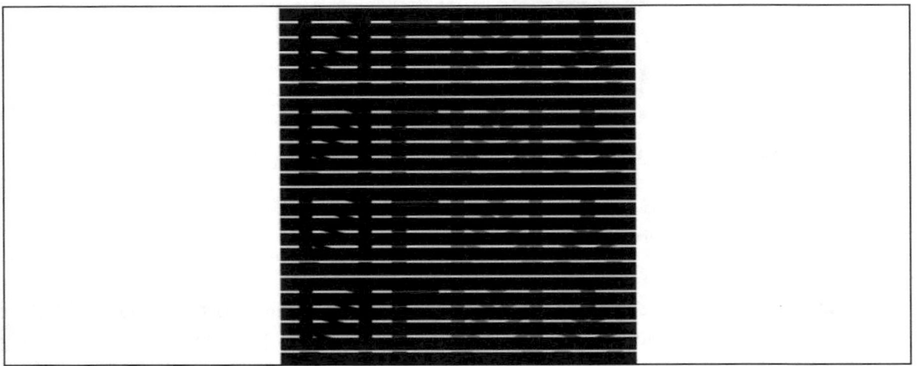

Fig. 12. Inverted mirrored moiré pattern

Through the above principle analysis and experimental verification, the restriction of the appearance of the inverted mirrored moiré is consistent with the previous speculation. The inverted mirrored moiré, which appears when the revealing layer period T_r is less than the base layer period T_b, and it will disappear when other effects appear. When the revealing layer period is less than the base layer period, and disappears when other effects appear the inverted mirrored moiré will appear.

We can notice from Examples 2 and 3 that the relationship between revealing layer period T_r and base layer period T_b is changed. In Example 2, the base layer period $T_b = 30$, the revealing layer period $T_r = 36$, which means T_b is less than T_r. In Example 3, the base layer period $T_b = 36$, the revealing layer period $T_r = 30$, which means T_r is less than T_b. Inverted mirrored moiré appears when $T_r < T_b$. With the decreasing of the revealing layer period, another moiré effect appears and inverted mirrored moiré disappears.

4 Ribbon Principle and Restriction

The ribbon phenomenon refers to the local moiré images that occurs after the superposition of two layers and appears in a certain period circularly. When the relationship between the revealing layer period and the basic layer period continues to change, moiré pattern will appear as a ribbon. In the last section, when the revealing layer period T_r decreases to $T_r = \frac{T_b}{2}$ gradually, the moiré pattern appears as shown in Fig. 13. Therefore, we infer that when the revealing layer period T_r has the equation relation with the basic layer period T_b, the ribbon appears.

Fig. 13. Ribbon

In this section, we analyze the principle of ribbon theoretically, and give the restriction of the base layer period and the revealing layer period when ribbon appears. An example is given to illustrate the analysis of ribbon restrictions. The experimental results verify the correctness of the inference.

4.1 Ribbon Principle

As shown in Fig. 14, the base layer is represented by B_0, B_1, B_2, B_3 and B_4, and the transparent gratings of the revealed layer is represented by L_0', L_1', L_2', L_3' and L_4'. When the base layer period is equal to the revealing layer period, that is $T_b = T_r'$, the transparent aperture of the layer grating l_{10}, l_{11}, l_{12} and l_{13} are sampled for B_3, B_2, B_1, B_0 in the basic layer respectively. The rectangles l_{10}', l_{11}', l_{12}' and l_{13}' in the B_0 sampling zone in the base layer are respectively mapped to the transparent aperture of gratings l_{10}, l_{11}, l_{12} and l_{13} of the revealing layer, that is, the transparent aperture of the revealing layer samples the same region of the base band cyclically. Therefore, the image of the base layer through each of the transparent apertures is the same.

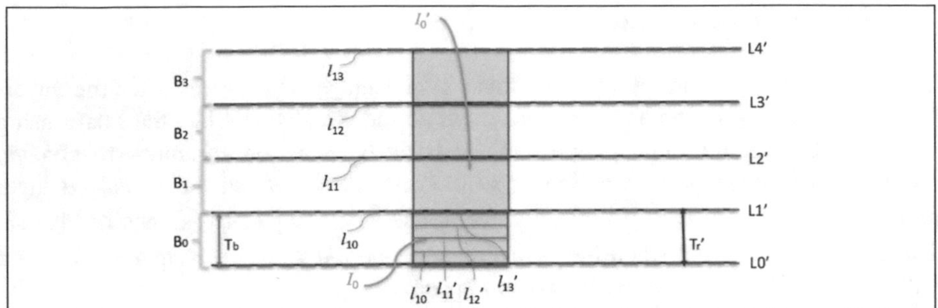

Fig. 14. "Ribbon" geometric principle diagram

Through the above principle analysis, we can deduce the restriction of the ribbon's appearance. When the revealing layer period is equal to the base layer period, that is, when $T_b = T_r$ is present, the ribbon appears, and the same principle can be deduced. When $T_r = \frac{T_b}{2}$, there will also be a ribbon phenomenon. We can continue to infer to all cycles.

4.2 Experimental Verification

Example 4: Let the base layer period $\mathrm{T_b} = 30$, the revealing layer period $T_r' = 30$, the size of the transparent aperture is $m' = 4$. At this time, since the sampling positions are the same in each period, the moiré images generated by the superposition of the two layers after the transparent aperture had sampled the basic band cyclically, that is the "color bands" corresponding to the color of the secret image as shown in Fig. 15.

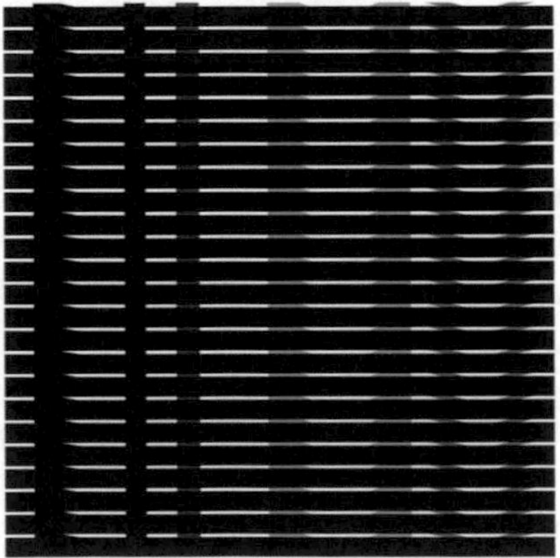

Fig. 15. Ribbon

4.3 Result Comparison

We introduce the geometric principle and restriction of ribbon moiré, and give a detailed explanation of the principle and the restriction of the inverted mirrored moiré mentioned, analyze the principle and restriction for the appearance of the ribbon. The following is the comparison of the experimental results for the three types of moiré phenomena, shown in Table 1.

Table 1. Comparison of experimental results

Moiré	Restriction Conditions	Result
Normal	$T_r > T_b > \frac{T_r}{2}$	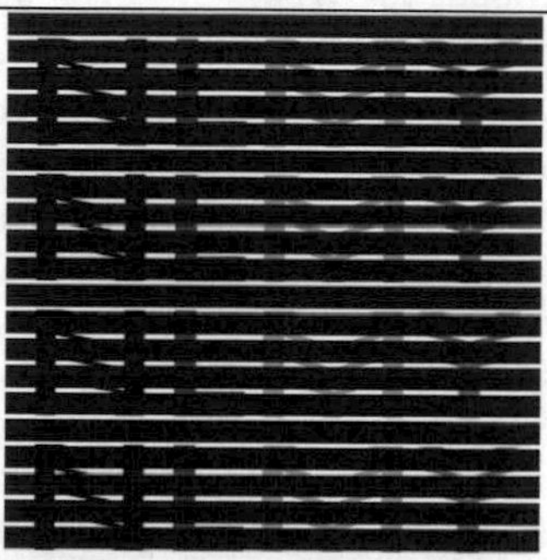
Mirrored	$T_b > T_r > \frac{T_b}{2}$	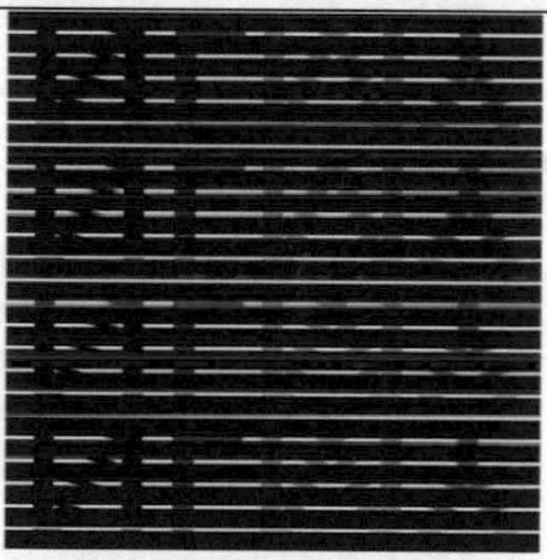
Ribbon	$T_b = nT_r$ 或 $nT_b = T_r$，$(n \in Z)$	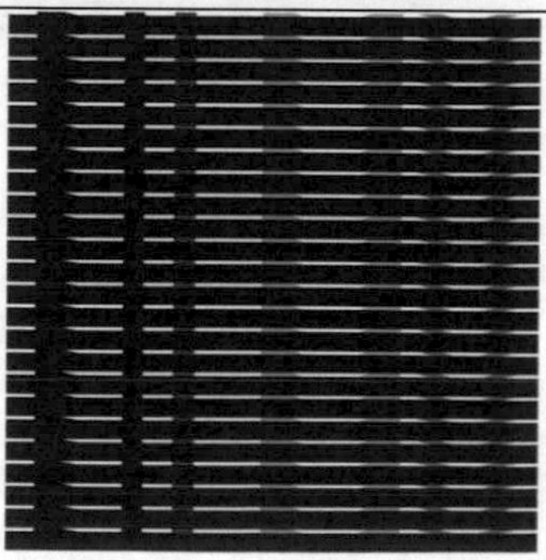

The appearance of three moiré phenomena is continuous. Starting from the superposition layers that produces normal moiré, the revealing layer period gradually decreases from $T_r > T_b > \frac{T_r}{2}$ to $T_r = T_b$, when $T_r = T_b$, moiré appears the ribbon phenomena, and when it continues to decrease to $T_b > T_r > \frac{T_b}{2}$, mirror moiré phenomena appeared.

5 Conclusion

In this paper, based on the Hersch et al. [1] band moiré image, we analyze the principle of the appearance of the inverted mirrored moiré referred by Hersch, and give the restriction of inverted mirrored moiré. At the same time, the principle and restriction of ribbon appearance in moiré are analyzed. We notice that moiré is a visual effect of layers. The band moiré has periodicity, which makes the restriction of different moiré effects presented in this paper also has periodicity. Our future work is to analyze the restrictions and periodicity of inverted mirrored moiré and ribbon, as well as the human eye's effect of moiré effect under different periods.

Acknowledgement. This work was supported in part by the National Natural Science Foundation of China under Grant U1536102, and Grant U1536116, in part by China Mobile Research Fund Project (MCM20170407), and State Administration of Press, Publication, Radio, Film and Television (SAPPRFT) Key Laboratory of Digital Content Anti-Counterfeiting and Security Forensics.

References

1. Hersch, R.D., Chosson, S.: Band moiré images. ACM Trans. Graph. **23**(3), 239–247 (2004)
2. Amidror, I.: The Theory of the Moiré Phenomenon. Kluwer Academic Publishers, Dordrecht, Boston, London (2000)
3. Bryngdahl, O.: Moiré formation and interpretation. J. Opt. Soc. Am. **64**(10), 1287–1294 (1974)
4. Glass, L.: Moiré effect from random dots. Nature **223**(5206), 578–589 (1969)
5. Amidror, I., Hersch, R.D.: Fourier-based analysis and synthesis of moirés in the superposition of geometrically transformed periodic structures. J. Opt. Soc. Am. **15**(5), 117–118 (1998)
6. Zhou, S.L., Fu, Y.Q.: Fourier-based analysis of moiré fringe patterns of superposed gratings in alignment of nanolithography. Opt. Soc. Am. **16**(11), 7869–7880 (2008)
7. Walger, T., Hersch, R.D.: Hiding information in multiple level-line moirés. In: ACM Symposium on Document Engineering, pp. 21–24. ACM (2015)
8. Cadarso, V.J., Chosson, S., Sidler, K.: High-resolution 1D moirés as counterfeit security features. Light Sci. Appl. **2**(7), 86–92 (2013)
9. Amidror, I., Hersch, R.D.: Mathematical moiré models and their limitations. J. Mod. Opt. **57**(1), 23–36 (2010)
10. Xin, M.: Moiré image hiding technology with multi-secret hiding performance. Tsinghua University, Beijing, pp. 5–9 (2015)

Learning and Intelligent Computing

A LSTM-Based Approach to Haze Prediction Using a Self-organizing Single Hidden Layer Scheme

Xiaodong Liu[1], Qi Liu[1(✉)], Yanyun Zou[2], and Qiang Liu[3(✉)]

[1] School of Computing, Edinburgh Napier University, Edinburgh, UK
{x.liu,q.liu}@napier.ac.uk
[2] Jiangsu Collaborative Innovation Center of Atmospheric Environment and Equipment Technology (CICAEET), Nanjing University of Information Science and Technology, Nanjing, China
so_cloudy@163.com
[3] School of Computer, Hunan University of Technology, Zhuzhou, Hunan, China
liuqiang@hut.edu.cn

Abstract. The air quality in urban areas seriously affects the physical and mental health of human beings. And $PM_{2.5}$ (a particulate matter whose diameter is smaller than or equal to 2.5 microns) is the chief culprit causing haze-fog. Since the meteorological data and air pollutes data are typical time series data, it's reasonable to adopt a single hidden-layer LSTMNN (Long Short-Term Memory Neural Network) containing memory capability to implement the prediction. As for deciding the best structure of the neural network, this paper employs a self-organizing algorithm, which uses Information Processing Capability (IPC) to adjust the number of the hidden neurons automatically during a learning phase. In a word, to predict PM2.5 concentration accurately, this paper proposes a Self-organizing Single Hidden-Layer Long Short-Term Memory Neural Network (SSHL-LSTMNN) to predict $PM_{2.5}$ concentration. In the experiment, not only the hourly precise prediction but also the daily longer-term prediction is taken into account. At last, the experimental results reflect that SSHL-LSTMNN performs the best.

Keywords: Haze-fog · $PM_{2.5}$ forecasting · Time series data · Long Short-Term Memory Neural Network · Self-organizing algorithm · Information Processing Capability

1 Introduction

Haze-fog is not only related to meteorological conditions, but also has a non-negligible relationship with human activities. Once the emission exceeds the atmospheric circulation capacity and carrying capacity, the concentration of fine particles will be getting to high. As a result, it is easy to have a large range of haze-fog. The greatest impact of haze-fog is the human health, it's easy to affect the respiratory tract of the body and causes various diseases.

C.-N. Yang et al. (Eds.): SICBS 2018, AISC 895, pp. 701–706, 2020.
https://doi.org/10.1007/978-3-030-16946-6_57

The early $PM_{2.5}$ prediction methods were mainly based on the original statistical methods. Fuller et al. [1] use the average of the pollutants of API (Air Pollution Index), and statistics the linear relationship between the factors and the $PM_{2.5}$ and PM_{10}, so as to realize the forecasting of $PM_{2.5}$. Jian et al. [2] find the correlation between meteorological factors, that is, the humidity is positively related to haze-fog, and the wind speed is negatively related to haze-fog. It is proved that auto-regressive integrated moving average (ARIMA) model can effectively explore the relationship between haze and meteorological factors.

To improve the accuracy of prediction, machine learning methods have been widely used in this field. Some researchers combine statistical methods and machine learning methods to predict. Liu et al. [3] use the comprehensive prediction model to forecast the $PM_{2.5}$ concentration using the autoregressive moving average (ARIMA), artificial neural network (ANNs) model and exponential smoothing method (ESM). Zhu et al. [4] put forward an improved BP (Back Propagation) neural network algorithm, combining the auto-regressive and moving average (ARMA) model with BP (Back Propagation) neural network to predict $PM_{2.5}$ concentration. And some researchers adopt some optimization methods into machine learning methods. Venkadesh et al. [5] combine genetic algorithm and BP neural network to fuse multiple time domain meteorological factors, and determine the duration and resolution of prior input data, and improve the accuracy of prediction. Mishra et al. [6] combine the Principle Component Analysis (PCA) and artificial neural network to get the correlation between meteorology and air pollutants variables, so as to predict the concentration of NO_2 in the air. Li et al. [7] estimate the pollutant concentration using stepwise regression and support vector machine (SVM) based on the existing data from the nearby stations to decrease the costs of air quality monitoring. Mishra et al. [8] use non-meteorological parameters (CO, O_3, NO_2, SO_2, $PM_{2.5}$) and meteorological parameters to make the fusion analysis combining artificial intelligence to forecast haze-fog, it is concluded that compared with the artificial neural network and multilayer perceptron model, NF (Neuro-Fuzzy) model based on the artificial intelligence can better predict the urban haze-fog events in Delhi, India.

According to the above literatures, owing to the typical time series character of air pollutants and meteorological data, this paper uses an improved recurrent neural network, a Self-organizing Single Hidden-Layer Long Short-Term Memory Neural Network (SSHL-LSTMNN) to predict $PM_{2.5}$ concentration.

2 Self-organizing Algorithm

The growing and pruning algorithm is to add or delete hidden layer nodes by using Information Processing Capability (IPC) to achieve self-organizing ability of the single hidden-layer LSTM network during learning phase. By using this algorithm, the number of hidden layer nodes can be adjusted to the best. So the structure of the network is able to satisfy the high precision prediction condition.

The calculation of IPC is divided into two parts: the input IPC and output IPC, which are defined as

$$\begin{cases} S_j^1(t) = \frac{1}{K}\sum_{k=1}^{K} e^{-\mathbf{x}(t-k+1)} \\ S_j^2(t) = I_j(t) \end{cases} \tag{1}$$

where $\mathbf{x}(t-K+1)$ is the input vector at time $(t-K+1)$. The self-organizing algorithm contains three steps: growing step, pruning step and keeping step. The detailed process of each step is described below.

Growing Step. The larger $S_j^1(t)$ and $S_j^2(t)$ are, the information processing ability of the node is more powerful. In this case, if the IPC satisfies the following conditions:

$$\begin{cases} S_j^1(t) = \max \mathbf{S}^1(t) \\ S_j^2(t) = \max \mathbf{S}^2(t) \end{cases} \tag{2}$$

where $\mathbf{S}^1(t) = \left(S_1^1(t), \ldots, S_{m-1}^1(t), S_m^1(t)\right)$ and $\mathbf{S}^2(t) = \left(S_1^2(t), \ldots, S_{m-1}^2(t), S_m^2(t)\right)$ are input IPC vector and output IPC vector of hidden nodes respectively. If the input IPC vector of a node is the maximum value among all input IPC vectors, and the output IPC vector is the maximum value among all output IPC vectors, a new hidden node will be inserted into the hidden layer. And the connection weights of this node will be initialized.

Pruning Step. Like the above step, if the IPC of a node satisfies the following conditions:

$$\begin{cases} S_i^1(t) = \min \mathbf{S}^1(t) \\ S_i^2(t) = \min \mathbf{S}^2(t) \end{cases} \tag{3}$$

If the input IPC vector of a node is the minimum value among all input IPC vectors, and the output IPC vector is the minimum value among all output IPC vectors, the node will be removed. And the connection weights of neighbor nodes will be adjusted.

Keeping Step. If the input IPC as well as output IPC of a hidden node are not the maximum information processing capability or the minimum information strength, the node will be kept.

3 Experiment

In order to predict the concentration of $PM_{2.5}$ comprehensively and accurately, not only the hourly precise prediction but also the daily longer-term prediction is taken into account.

3.1 Data Preprocessing

To predict PM2.5 concentration, this study uses hourly and daily files of Nanjing, which include meteorological data as well as air pollutants. The meteorological data are collected from Meteorological Data Center of China Meteorological Administration, and the data sources of air pollutants are from Environmental Monitoring Stations of

China. Raw data contains 27 factors, such as O_3, NO_2, CO, SO_2, pressure, relative humidity, temperature and so on. And a selection method called Mutual Information (MI) is adopted. This method can calculate that whether there is a relationship between the two variables X and Y, as well as the strength of the relationship. As a result, the MI values of these factors are high: O_3, NO_2, $PM_{2.5}$, pressure, wind speed of instant maximum, wind direction of instant maximum, temperature, wind direction of maximum wind speed, relative humidity, water vapor pressure, minimum relative humidity, horizontal visibility, body temperature.

3.2 Hourly Prediction

To prove that the number of hidden nodes decided by the self-organizing algorithm is the most suitable one, some different numbers of hidden nodes are used for comparison. And Table 1 shows the result of predicting $PM_{2.5}$ concentration after 1, 4, 8 and 12 h.

Table 1. The hourly predicting comparison between different numbers of hidden nodes

Future hours	The number of hidden nodes	RMSE	MAE
1	5	6.622	4.665
	6	6.978	4.865
	7	6.743	4.694
	8	**6.604**	**4.578**
	9	6.705	4.752
	10	6.795	4.693
4	5	13.056	9.866
	6	13.169	9.786
	7	13.174	9.899
	8	**12.866**	**9.611**
	9	13.186	9.779
	10	13.563	10.067
8	5	14.083	10.863
	6	14.208	11.152
	7	14.130	10.983
	8	**13.914**	**10.885**
	9	13.956	10.754
	10	14.103	10.855
12	5	15.383	12.257
	6	15.903	11.855
	7	15.017	11.900
	8	**14.084**	**10.959**
	9	15.021	11.862
	10	14.963	11.786

According to this table, it's obvious that RMSE and MAE are the lowest when the number of hidden nodes is 8, which is determined by the self-organizing algorithm.

3.3 Daily Prediction

Hourly prediction can forecast near-time concentration of $PM_{2.5}$ instead of daily average concentration. So this section uses daily dataset to evaluate the daily prediction performance. After training by the self-organizing algorithm, the number of hidden neurons is set to be 7. And to validate this result, a comparison experiment is operated. The results are shown in Table 2.

Table 2. The daily average predicting comparison between different numbers of hidden nodes

Future days	The number of hidden neurons	RMSE	MAE
1	5	20.509	15.748
	6	18.818	13.519
	7	**18.293**	**13.432**
	8	18.641	13.793
	9	18.573	13.303
	10	18.705	13.529
4	5	23.072	18.621
	6	22.100	17.415
	7	**21.086**	**16.518**
	8	22.439	17.845
	9	21.808	16.890
	10	22.170	17.413
8	5	22.809	18.074
	6	22.233	17.293
	7	**21.674**	**16.158**
	8	22.415	17.586
	9	21.732	16.799
	10	22.125	17.136
12	5	22.983	18.503
	6	21.389	16.848
	7	**20.810**	**15.687**
	8	22.081	17.340
	9	20.951	16.373
	10	21.407	16.837

From Table 2, it's obvious that when the number of hidden nodes is 7, RMSE and MAE perform better than other conditions. So the capability of the growing and pruning algorithm is validated to be effective.

4 Conclusion and Future Work

Because meteorological data and pollutant data are typical time series data, recurrent LSTM network is suitable to apply in this case for its memory capability. And to solve the problem that is difficult to determine the number of hidden nodes, a self-organizing algorithm is adopted. So in this paper, a self-organizing single hidden-layer LSTM neural network (SSHL-LSTMNN) is employed to predict $PM_{2.5}$ concentration.

A self-organizing algorithm using Information Processing Capability (IPC) to determine the number of hidden nodes is illustrated. This method can add or delete nodes during training phase. Then experiments on hourly prediction and daily prediction are put into practice. After learning phase using the self-organizing algorithm, the number of hidden nodes is determined to be 8 and 7 respectively. As a result, the network with 8 hidden nodes in hourly prediction and the one with 7 hidden nodes in daily prediction show the higher prediction accuracy comparing to other number of hidden nodes.

Acknowledgments. This work was supported by Major Program of the National Social Science Fund of China (Grant No. 17ZDA092) and Marie Curie Fellowship (701697-CAR-MSCA-IF-EF-ST).

References

1. Fuller, G.W., Carslaw, D.C., Lodge, H.W.: An empirical approach for the prediction of daily mean PM_{10} concentrations. Atmos. Environ. **36**(9), 1431–1441 (2002)
2. Jian, L., Zhao, Y., Zhu, Y.P., Zhan, M.B., Bertolatti, D.: An application of ARIMA model to predict submicron particle concentrations from meteorological factors at a busy roadside in Hangzhou. China Sci. Total. Environ. **426**(2), 336–345 (2012)
3. Liu, D.J., Li, L.: Application study of comprehensive forecasting model based on entropy weighting method on trend of $PM_{2.5}$ concentration in Guangzhou. China Int. J. Environ. Res. Public Health **12**(6), 7085–7099 (2015)
4. Zhu, H., Lu, X.: The prediction of $PM_{2.5}$ value based on ARMA and improved bp neural network model. In: 2016 International Conference on Intelligent Networking and Collaborative Systems (ICINCS), pp. 515–517 (2016)
5. Venkadesh, S., Hoogenboom, G., Potter, W.: A genetic algorithm to refine input data selection for air temperature prediction using artificial neural networks. Appl. Soft Comput. **13**(5), 2253–2260 (2013)
6. Mishra, D., Goyal, P.: Development of artificial intelligence based NO_2, forecasting models at Taj Mahal. Agra. Atmos. Pollut. Res. **6**(1), 99–106 (2015)
7. Li, M., Wang, W., Wang, Z., Xue, Y.: Prediction of $PM_{2.5}$ concentration based on the similarity in air quality monitoring network. Build. Environ. **137**, 11–17 (2018)
8. Mishra, D., Goyal, P., Upadhyay, A.: Artificial intelligence based approach to forecast $PM_{2.5}$, during haze episodes: a case study of Delhi, India. Atmos. Environ. **102**, 239–248 (2015)

Target Information Fusion Based on Memory Network for Aspect-Level Sentiment Classification

Zhaochuan Wei[1], Jun Peng[1], Xiaodong Cai[1(✉)], and Guangming He[2]

[1] Guilin University of Electronic Technology, Guilin, Guangxi, China
caixiaodong@guet.edu.cn

[2] China Comservice Public Information Industry Co., Ltd., Urumqi, Xinjiang, China

Abstract. Aspect-level sentiment classification is a fine-grained task that provides more complete and deeper analysis results. Attention networks are widely used for aspect-level sentiment classification task. However, when multiple target words in a sentence contain opposite sentiments or the expressions of different targets are similar, the network tends to perform poorly. Our studies find that the method of averaging a sentence or target words weakens the capacity of key words. A target information fusion memory network is proposed to solve this problem in this paper. Firstly, the feature of sentences is extracted through a Bi-LSTM network. Then, the feature of target is extracted incorporated into the sentence feature extracted. Then, the memory information of the specific target is formed by the position coding. Finally, the recurrent attention network is utilized to extract the sentiment expression from the memory. Compare with IAN, the method proposed achieves 1.5% and 1.9% accuracy improvement on SemEvil2014 restaurant dataset and self-defined Chinese mobile phone dataset, respectively. A further extend experiment proves that the proposed method can effectively improve the performance in the case of complex sentences.

Keywords: Aspect-level sentiment classification · Interactive attention neural network · Memory network · Information fusion

1 Introduction

Aspect-level sentiment classification (also mentioned as "target-based" or "target-level" in some papers) is a fine-grained sentiment analysis task. It aims to discriminate the sentiment polarities of specific target words in context, and it can provide more complete and deeper sentiment analysis results [1, 2].

Unlike general sentiment analysis tasks, aspect-level sentiment classification depends not only on the specific context information, but also the feature information of the target [1, 8]. For example, it is difficult to accurately determine the sentiment when the sentence "The processor performance is strong, but the battery is not very durable." is given, because for the targets "processor" and "battery" are positive and negative, respectively.

C.-N. Yang et al. (Eds.): SICBS 2018, AISC 895, pp. 707–713, 2020.
https://doi.org/10.1007/978-3-030-16946-6_58

Aspect-level sentiment classification is a fundamental task in natural language processing and catches many researchers' attention [1, 2]. Traditional approaches mainly focus on designing a set of features such as bag-of-words or sentiment lexicon to train a classifier for aspect-level sentiment classification [3, 4]. These methods need either feature engineering work or extra-linguistic resources. However, feature engineering or extra-linguistic resources is labor intensive and almost reaches its performance bottleneck [8].

With the development of deep learning techniques, some researchers have designed effective neural networks to aspect-level sentiment analysis. In [5], TD-LSTM and TC-LSTM is proposed to extend LSTM by taking the target words into consideration. In [6], the attention mechanism is introduced to be an effective approach. In [7], a deep memory network is proposed for capturing importance information of context words. In [8], an interactive attention networks is proposed which use two attention networks to model the target and context interactively. In [9], to improve the memory network, Bi-LSTM network is used to extract the feature information as the memory, and recurrent attention network is used to implement the sentiment classification. In [10], re-examined the drawbacks of attention mechanism, a CNN network is used to extract the characteristic information of the CPT component to for aspect classification. In [11], approaches that transfer knowledge from document level data is used, which is much less expensive to obtain, to improve the performance of aspect-level sentiment classification. In [12], a way to incorporate inter-aspect dependencies in the task of aspect sentiment analysis is presented and incorporate this relationship by simultaneous classification of all aspects in a sentence along with temporal dependency processing of their corresponding sentence representations using recurrent networks.

It is found that the output of target words and sentences are averaged to obtain the weights of attention in the networks. Firstly, when multiple targets in a sentence contain opposite sentiments, the method of averaging sentence information weakens the characteristic expression. In addition, when the expressions of different targets or emotional words is similar, the method of averaging weakens the expression of key information. This paper proposes a strategy in which memory network with target information integrated is used by combining with the recurrent attention network proposed by [9].

2 Target Information Fusion Memory Network

The architecture of the information fusion network is shown in Fig. 1. It consists of four parts: input embedding layer, Bi-LSTM layer, memory layer, and the recurrent attention layer. For convenience, the sentence is denoted as $S = \{s_1, \ldots, s_i, \ldots, s_n\}$ and the target words is denoted as $T = \{t_1, \ldots, t_i, \ldots, t_m\}$. Where n and m represent the length of the target and sentence, respectively. It is worth noting that sentence S may contain multiple target words. In this case, different target and sentence are distinguished separately.

2.1 Input Embedding Layer and Bi-LSTM Layer

The Word2Vec [14] method is used to map each word into a vector. The vocabulary $L \in R^{|V| \times d}$ can be defined, where, $|V|$ is the size of the vocabulary and d is the dimension of the word vector. Then the input matrix of Bi-LSTM is obtained as $X = [w_s^1, \ldots, w_s^i, \ldots, w_s^n]$. The output of Bi-LSTM network is transmitted to the memory layer for further processing.

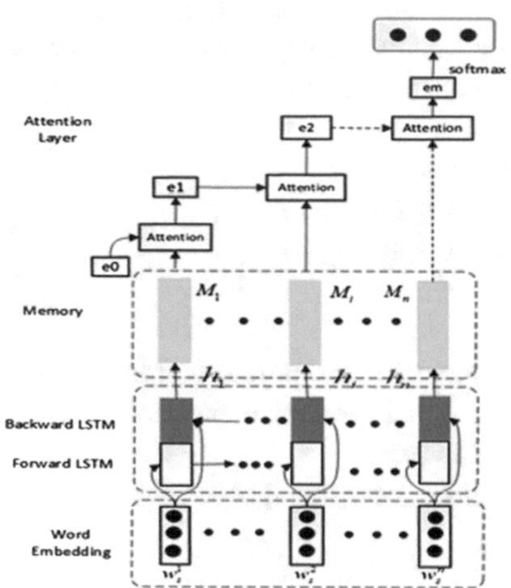

Fig. 1. Model architecture

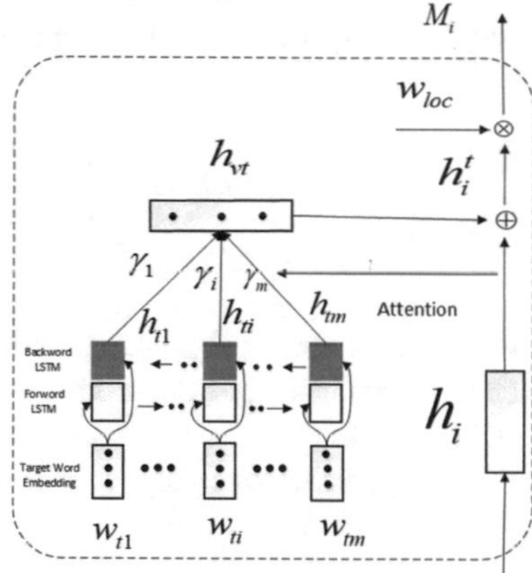

Fig. 2. Internal architecture of memory

2.2 Memory Layer

The above Bi-LSTM network does not consider the relationship between target words and sentence fully. In [8], it is verified that the feature information of the target is crucial in sentiment classification. Therefore, target information fusion memory network is proposed. As shown in Fig. 2, a Bi-LSTM network is used to model the target. Then the attention mechanism is used to model the correlation between target and h_i. Here, h_i represents i^{th} output of the Bi-LSTM network. Finally, the value of h_i and h_{vt} are summed to implement the information fusion.

Specifically, Bi-LSTM network is used to obtain the hidden vector h_{tj} of the target. Target expression h_{vt} is calculate as follows:

$$\gamma_j = h_{tj}^T W_v h_i + b_v \tag{1}$$

$$h_{vt} = \sum_{j=1}^{m} \gamma_j h_{tj} \tag{2}$$

Where, γ_j is a weighted coefficient, and h_{vt} is a vector of target information. Then the sum is calculated to fuse the target information with the sentence:

$$h_i^t = h_i + h_{vt} \tag{3}$$

In target-based sentiment classification, it is generally believed that the position information is important, as described by [9, 10]. Intuitively, the word closer to the target, the greater impact on the aspect-level sentiment classification. For example, in the sentence of "Great food but the service was dreadful!", the sentiment word "great" is closer to the "food" than "service", so a higher weight is assigned when the polarity of "food" is judged. Position coding *w(i)* is calculated as follows:

$$w(i) = \begin{cases} 1 - L(i)(k - i), & 0 < i < k \\ 1, & k \leq i \leq k + m \\ 1 - L(i)(i - k), & k + m \leq i \leq n \\ 0, & others \end{cases} \tag{4}$$

Where, k is the start position of the target, m and n is lengths of the target and sentence, i represent the positions of the words. $L(i)$ represent a position weighting factor and the length of the sentence is used in this paper.

The characteristics of the target information and position weighted are combined to obtain the memory information.

2.3 Recurrent Attention Layer

In order to accurately predict the polarity of the target, recurrent attention mechanism that proposed by [9] is adopted to extract target-specific sentiment information from memory. In the process of obtaining the input of the current memory, the LSTM network is used to obtain the aspect expression in this paper. The mean values taken by [9] weaken the feature expression of the key words, because different words in the target tend to contribute differently [10]. With recurrent attention mechanism, the result is transmited to the softmax to obtain the corresponding sentiment prediction values.

3 Experimental Results and Analysis

3.1 Dataset and Parameter Settings

In order to verify the model, the SemEvil2014's benchmark dataset and self-defined Chinese mobile phone dataset are used. In addition, extended experiments on complex sentences is carried out. The self-defined dataset named dataset 1 contains 1509 positive examples, 521 neutral examples and 1021negative examples for training set. 591 positive examples, 129 neutral examples and 280 negative examples are used for test. The complex sentences are extracted from the benchmark dataset to verify that the model proposed is effective. A dataset 2 is defined as sentences that removes sentences containing only one sentiment word and retains sentences with complex goals and multiple targets from benchmark. A dataset 3 is defined based on dataset 2, it only retains sentences with complex goals and multiple emotional words.

In our experiment, all English word vectors are initialized with GloVe [13]. Chinese word vector is obtained by Word2vec [14]. For words not in the vocabulary list, a uniform distribution obeying U(−0.1, 0.1) is used. The initial learning rate is set to 0.001, and the dropout strategy is used to suppress overfitting [15].

3.2 Benchmark Dataset Experiment Results

As shown in Table 1, the model proposed is achieved 1.5%, 0.9%, and 1.9% improvement over IAN on restaurant, Laptop, and the Chinese dataset, respectively. Compared to the RAM, the results achieved 0.6% and 0.8% improvement in the restaurant and Chinese dataset. It shows that the model proposed can overcome the shortcomings of IAN by using the feature fusion and improve the performance of the IAN network. Different from the RAM model, the characteristics of the target is used to form the memory information of the target. This help the proposed network outperform the RAM model.

Table 1. Accuracy on 3-class for benchmark dataset

	Restaurant	Laptop	Chinese[a]
ATAE-LSTM [6]	0.772	0.687	0.857
MemNet [7]	0.782	0.703	0.863
IAN [8]	0.786	0.721	0.885
RAM [9]	0.795	0.741	0.896
Our model	0.801	0.730	0.904

[a] *The results of self-defined Chinese dataset are provided by using original models of the papers.*

3.3 Extend Experimental Results

By extending the experiment, it is found that the IAN model achieves 78.6% and 88.5% accuracy on the restaurant and Chinese dataset. The accuracy on the dataset 2 is 76.7% and 85.1%, respectively. The accuracy of the sentences contain the opposite sentiment is 66.7% and 81.9%. It proves that the average method used by IAN weakens the ability of feature expression (Table 2).

Table 2. Accuracy on 3-class for extend experiment

Heading level	Model	Restaurant	Chinese
Dataset 2	IAN [8]	0.767	0.851
	RAM [9]	0.784	0.870
	Our model	0.793	0.884
Dataset 3	IAN [8]	0.667	0.819
	RAM [9]	0.696	0.846
	Our model	0.717	0.863

It is found that RAM, IAN and our model reduced the performance of complex sentences. However, a small decline is showed with our model. In dataset 3, the result of proposed model achieves improvement of 5.0% improvement in the dataset of the Restaurant and 4.4% improvement in the dataset of Chinese. Compared to the RAM, our model achieves 2.1% and 1.7% improvement on dataset 3, which indicates that the efficaciousness by introducing the characteristics of the target information in the memory information.

However, the results on Chinese dataset better than English dataset both the degradation on the complex dataset is not as obvious as the English dataset. It due to the fact that the self-defined Chinese dataset comes from the online product reviews, most of the data is relatively clear.

4 Conclusions

This paper proposes a memory network based on target information fusion for aspect-level sentiment classification. By integrating the target information and the location information into the feature expression of sentence, the memory information of the specific target is formed. The recurrent attention mechanism is used to extract the corresponding sentiment expression from the memory information. Using this method, it can effectively overcome the shortcomings of attention networks. The results show that the model can improve the performance of the IAN and RAM networks. However, it does not implement obvious improvement in simple dataset and it is complicated than IAN. Further research can be done on the combination of the IAN and our model.

Acknowledgements. Our work is supported by Key Research and Development Projects of Xinjiang Autonomous Region in 2018 (No. 2018B03022-1 and No. 2018B03022-2), Innovation Project of GUET Graduate Education (No. 2018YJCX38 and No. 2017YJCX38) and Key Research and Guangxi Director Fund of the Key Laboratory of Wireless Broadband Communication and Signal Processing (No. GXKL0614107).

References

1. Zhang, L., Liu, B.: Sentiment analysis and opinion mining. Synth. Lect. Hum. Lang. Technol. **30**(1), 1–167 (2016)
2. Zhang, L., Wang, S., Liu, B.: Deep learning for sentiment analysis: a survey. Wiley Interdiscip. Rev. Data Min. Knowl. Discov. (2018)
3. Ding, X., Liu, B., Yu, P.S.: A holistic lexicon-based approach to opinion mining. In: Proceedings of the 2008 International Conference on Web Search and Data Mining, pp. 231–240 (2008)
4. Perez-Rosas, V., Banea, C., Mihalcea, R.: Learning sentiment lexicons in Spanish. European Language Resources Association (ELRA) (2012)
5. Tang, D., Qin, B., Feng, X., Liu, T.: Effective LSTMs for target-dependent sentiment classification. Comput. Sci. (2015)

6. Wang, Y., Huang, M., Zhu, X., Zhao, L.: Attention-based LSTM for aspect-level sentiment classification. In: Conference on Empirical Methods in Natural Language Processing, pp. 606–615 (2017)
7. Tang, D., Qin, B., Liu, T.: Aspect level sentiment classification with deep memory network. In: Conference on Empirical Methods in Natural Language Processing, pp. 214–224 (2016)
8. Ma, D., Li, S., Zhang, X., Wang, H.: Interactive attention networks for aspect-level sentiment classification. In: Twenty-Sixth International Joint Conference on Artificial Intelligence, pp. 4068–4074 (2017)
9. Chen, P., Sun, Z., Bing, L., Yang, W.: Recurrent attention network on memory for aspect sentiment analysis. In: Conference on Empirical Methods in Natural Language Processing, pp. 452–461 (2017)
10. Li, X., Bing, L., Lam, W., Shi, B.: Transformation networks for target-oriented sentiment classification, pp. 946–956. Association for Computational Linguistics (2018)
11. He, R., Lee, W.S., Ng, H.T., Dahlmeier, D.: Exploiting document knowledge for aspect-level sentiment classification, pp. 579–585. Association for Computational Linguistics (2018)
12. Hazarika, D., Poria, S., Vij, P., Krishnamurthy, G., Cambria, E., Zimmermann, R.: Modeling inter-aspect dependencies for aspect-based sentiment analysis. In: Proceedings of the 2018 Conference of the North American Chapter of the Association for Computational Linguistics: Human Language Technologies, Volume 2 (Short Papers), pp. 266–270 (2018)
13. Pennington, J., Socher, R., Manning, C.: Glove: Global vectors for word representation. In: Conference on Empirical Methods in Natural Language Processing, pp. 1532–1543 (2014)
14. Mikolov, T., Sutskever, I., Chen, K., Corrado, G., Dean, J.: Distributed representations of words and phrases and their compositionality. In: International Conference on Neural Information Processing Systems, pp. 3111–3119. Curran Associates Inc. (2013)
15. Srivastava, N., Hinton, G., Krizhevsky, A., Sutskever, I., Salakhutdinov, R.: Dropout: a simple way to prevent neural networks from overfitting. J. Mach. Learn. Res. **15**(1), 1929–1958 (2014)

Dangerous Objects Detection of X-Ray Images Using Convolution Neural Network

Lekang Zou(✉), Tanaka Yusuke, and Iba Hitoshi

Department of Information & Communication Engineering,
The University of Tokyo, Tokyo, Japan
{suurakukou, iba}@iba.t.u-tokyo.ac.jp,
wing0920@gmail.com

Abstract. Object detection is a popular research direction in the field of traditional visual research. It has a wide range of applications in our real life. Because the deep learning model training for object detection requires a large amount of training data, the research on object detection until now is mostly based on natural images datasets. While natural image object recognition techniques are already very mature, the research on X-ray images is still in its infancy. This paper applies X-ray image dataset for object detection experiments and in particular aims to build an X-ray security system which can be widely used in airports, subway stations and high security conference venues. Our experiments mainly focus on three kinds of dangerous objects: scissor, knife and bottle. The results show that we achieve mean average precision of 86.41% and recall rate of 87.70%.

Keywords: Dangerous object detection · X-ray image · Security system

1 Introduction

In recent years, as population mobility has become more frequent and the number of tourists have increased, it has become particularly important to conduct safety checks on passenger luggage. The X-ray security system is widely used in airports, subway stations, convention centers, government office buildings and logistics industry and is of great significance in protecting the lives of the people. X-ray security system is an efficient security system. When passengers send their luggage to the security inspection machine, the machine scans the luggage using X-ray radiation. Due to the radiopaque nature of X-ray, items in the luggage compartment will be displayed on the screen in real time. This makes it easier for human operators who must review compact, cluttered and highly varying baggage content within limited time-scales.

The published works on raw X-ray images baggage security is still rare. Singh and Singh proposed an explosives detection system for aviation security based on X-ray images [3]. However, threat detection is a complex task and not limited to explosives. Some researchers conducted their research from the perspectives of image enhancement and segmentation [4, 5], but these studies still cannot solve this problem fundamentally. Baştan et al. proposed bag of visual words (BoVW) methods to deal with X-ray images and they achieved quite well performance [6], giving us an idea of adding

C.-N. Yang et al. (Eds.): SICBS 2018, AISC 895, pp. 714–728, 2020.
https://doi.org/10.1007/978-3-030-16946-6_59

extra information in raw X-ray images. Still, this method cannot help build a security system that can directly handle raw X-ray images. A similar method for firearm classification was described in [2], this research was also based on multi-view X-ray images. Jaccard et al. described a method of X-ray image detection of objects in freight containers based on local image window classification using a random forest classifier [7, 8]. However, their researches are based on cargo containers not baggage security. There are also some studies applying X-ray classification tasks to Support Vector Machines (SVM) [9, 10]. Convolution neural network (CNN) was introduced into the field of X-ray images processing by [11], showing state-of-art performance in classification tasks. Riffo et al. conducted experiments on single view grayscale X-ray images [12]. They tested their method for the detection of three kinds of dangerous objects: razor blades, shuriken and handguns. Although this research is based on raw X-ray images, their dataset is a little limited. There is still not enough published work focusing on using raw X-ray images in applications relating to baggage security screening.

Most of previous researches of baggage security screening are based on multi-view X-ray images [1, 2, 6, 11]. In fact, such colored images have been processed for helping human operators distinguish different items because humans' eyes are sensitive to color. However, multi-view X-ray security systems have three shortcomings. Firstly, their devices are heavy and expensive. Furthermore, converting single-view X-ray images to multi-view X-ray images is very time consuming. Since most of the luggage is safe, the workload of converting all the images is tedious and meaningless. Finally, this manual recognition not only requires a wealth of experience but also has high error rates.

This paper has three main contributions. Firstly, we introduce a portable X-ray system for collecting single view X-ray images in place of existing multi view X-ray devices that are heavy and expensive. Secondly, we built a baggage security screening system for detecting dangerous objects based on single view X-ray image using CNN, which can directly handle the baggage information. The most significant contribution of this paper is that we propose a method to synthesize X-ray baggage images and improve the detection performance on raw X-ray baggage images.

Multiview X-ray devices are inconvenient to use in many situations. For example, in 2020 Tokyo Olympics, it is necessary to check a large amount of luggage in a short time. We need more efficient devices to accomplish this task. We introduce a portable X-ray system for X-ray image photography (see Fig. 1). This is a system whose electron acceleration voltage in the X-ray tube can output to 100 kV (Product Name NS-100-L). It is lightweight, compact and easy to carry. We can directly collect single-view X-ray images by using this system. This system contains three devices, a computer for saving X-ray images, a portable X-ray generator and an imaging device. We only need to put the baggage on the imaging device and use the X-ray generator to illuminate the baggage, the X-ray images containing baggage information will be saved in the computer in real time.

The rest of this paper is organized as follows: Sect. 2 discusses the related work of object detection. Section 3 describes our research method and evaluation method. Section 4 shows our experiment results and all these results are discussed in Sect. 5. Section 6 summarizes the paper and discusses future work.

Fig. 1. Portable X-ray device (Product Name NS-100-L). From left to right are computer for saving X-ray images, portable X-ray generator and X-ray imaging device.

2 Related Work

Unlike image classification tasks, object detection tasks not only classify different kinds of object but also locate the positions of multiple objects on the image with anchor. The research process of object detection can be summarized as three stages.

The first stage is traditional object detection method. In traditional method, candidate regions will be selected on the image. Then, extractor will extract features from these regions and features will finally be classified by a trained model. This method has two shortcomings. One is that region selection strategy is based on the sliding window algorithm with high time complexity and redundancy. Another is that the features of the manual design lack robustness for diversity changes.

The second stage is Region with CNN features framework (R-CNN). The region proposal takes use of the low-level information such as edge, color and texture in the image to find the possible position of the object. One of the region proposal methods is Selective Search [13]. In this method, the algorithm gets small-scale feature regions, and then combines these features recursively to get larger scale feature regions. It is similar to the visual perception of humans. Selective search greatly reduces the time complexity of subsequent operations. Inspired by this method, R-CNN [14] generates candidate object regions by using selective search. These candidate object regions are extracted by CNN and finally the SVM is trained to classify the features extracted by CNN. To further improve the performance of R-CNN, some enhancement network models such as SPP-NET [15], Fast R-CNN [16], Faster R-CNN [17] have been proposed.

Unlike R-CNN, YOLO [18] is qualified for object detection task in real time. The third stage represents a network based on an idea of detecting objects by a single shot and converting object detection task to regression problem. YOLO makes use of the whole image as the input and returns the position of the bounding box and the category to which the bounding box belongs at the output layer. This makes the model extremely fast because it is not necessary to recalculate features in the same area as region proposal. There are also some models similarly to YOLO such as YOLOv2 [19], SSD [20]. We show performance comparison of these CNN models based on VOC2007 dataset (see Fig. 2). These results may not apply to X-ray dataset. Considering that R-CNN models cannot meet the real-time detection requirements of real security systems. We chose YOLOv2 to conduct our experiments using X-ray dataset.

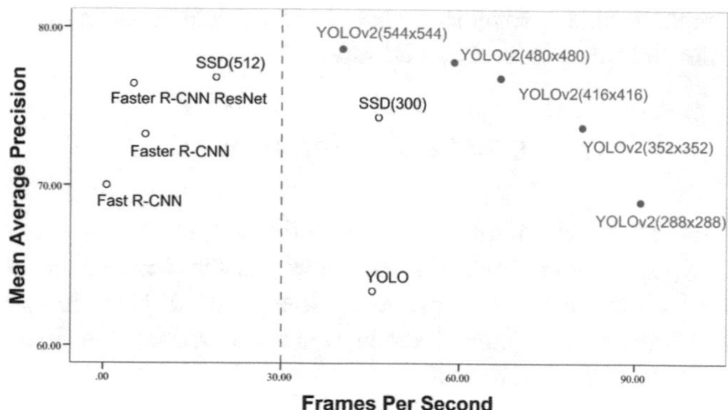

Fig. 2. Performance comparison of object detection CNN models.

3 Research Method

3.1 Raw X-Ray Data Collection

In order to collect enough raw X-ray data, we made use of a mixture of scissors, knives, bottles and some other safe objects as seeds. We have adjusted appropriate illumination distance and light intensity of our X-ray device and collected 1104 raw X-ray images by folding, rotating, overlapping and changing the angle of the objects. These images can be roughly divided into two types. One is an image with only one or several objects, which we call simple image. Another is an image with cluttered scene with the inclusion of non-dangerous objects. These images are very close to the luggage images in reality. We show some examples of our collected raw X-ray data (see Fig. 3).

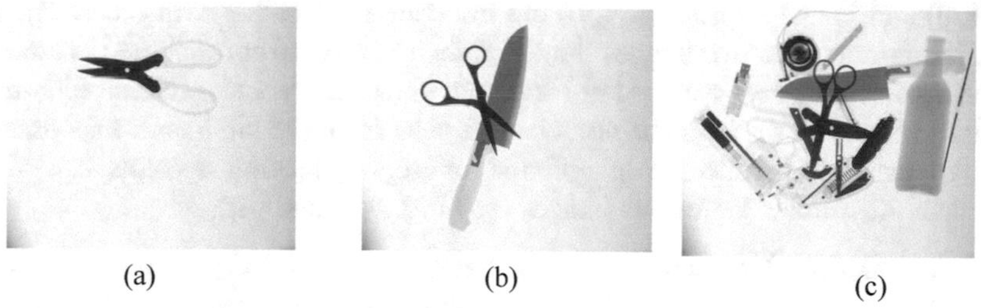

Fig. 3. Examples of raw X-ray data, (a) and (b) are simple images, (c) is complex image.

3.2 Synthesizing X-Ray Data

Beer–Lambert Law. Jaccard, Rogers et al. proposed a method for synthesizing X-ray cargo images [23]. We introduce their method first because our method for synthesizing data is an improvement over their method. Assuming that all X-ray images obey the following Beer–Lambert law:

$$I_{xy} = I_0 \exp(-\int u_{xy}(z)dz), \tag{1}$$

where I_{xy} is the illumination intensity at coordinate (x, y), I_0 is the source beam intensity, and $u_{xy}(z)$ is photon reduction coefficient in the Z-axis direction. When this equation is used for synthesizing cargo X-ray images in [23], the integral part of the equation can be split into two parts of attenuation contributions. One is from the object (O) and the other is from container (C).

$$\begin{aligned} I_{xy} &= I_0 \exp(-\int_O u_{xy}(z)\mathrm{d}z)\exp(-\int_C u_{xy}(z)\mathrm{d}z) \\ &= I_0 O_{xy} C_{xy} \end{aligned} \tag{2}$$

In Eq. (2), we can obtain I_{xy} from the luminance value of the image. Thus, if we can measure I_0 and C_{xy}, we can obtain O_{xy} by computing $I_{xy}/(I_0 C_{xy})$. Then, this O_{xy} can be projected into other X-ray images by multiplication for synthesizing data.

Image Pre-processing. The histogram of each raw X-ray image may vary significantly due to different photographic environment. To remove the impurities in these images, we have properly preprocessed these images and obtained clean, uniform brightness images for synthesizing data. The image pre-processing is divided into two steps. The first step is histogram equalization, which removes the influences of different photographic environment on images. The second step is gaussian filtering, which removes noise and impurities from the image to obtain clean images with clear features.

We show two examples of image pre-processing (see Fig. 4). Both image 1 (see Fig. 4(a)) and image 2 (see Fig. 4(b)) are indistinguishable for each other. The luminance histogram of the two images show that their pixels are concentrated in areas with low luminance (see Fig. 4(e)). After histogram equalization and gaussian filtering, the two images become distinguishable. It can be seen from the luminance histogram (see Fig. 4(f)) even data collected in different photographic environments can also be converted into similar luminance images (see Fig. 4(c) and (d)).

Dangerous Object Extraction. In this stage, we extract dangerous objects from preprocessed images for synthesizing. The extraction process is divided into three steps. First, binarizing the image while changing the threshold. The raw X-ray image is 256-bit grayscale image. We set the luminance value of the pixel whose luminance value exceeds the threshold to 0 and set the luminance value to 255 if it is lower than the threshold. A suitable threshold determines whether the contour of the object can be correctly extracted. Therefore, we need to constantly change the value to find the most appropriate threshold. We show a preprocessed original image and its binarized image with the threshold of 55 (see Fig. 5).

Second, after binarizing the image, we can extract the contour of the object. We show sampling contours extracted with different thresholds (see Fig. 6). The red circles

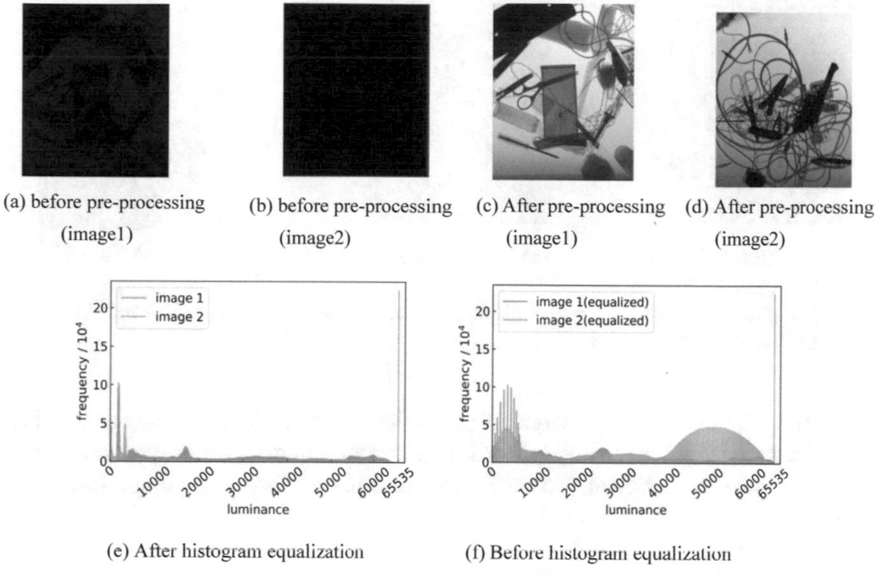

(a) before pre-processing (image1) (b) before pre-processing (image2) (c) After pre-processing (image1) (d) After pre-processing (image2)

(e) After histogram equalization (f) Before histogram equalization

Fig. 4. Examples of image pre-processing.

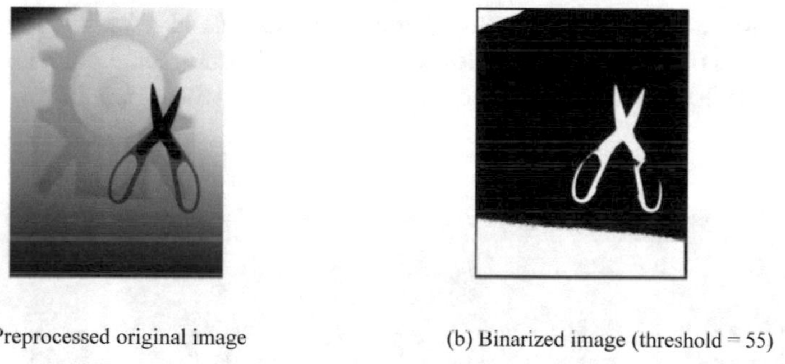

(a) Preprocessed original image (b) Binarized image (threshold = 55)

Fig. 5. Example of binarizing the image.

indicate the parts that were not correctly extracted. The results show that, for a scissor, it is difficult to extract the perfect contour when the threshold is 55 or 65, but when the threshold is 75, the complete contour can be extracted. While extracting the contour, we also separate the object from the image along the outer rectangle of the contour. It should be noted that in order to correct the extracted contour luminance value, it is necessary to divide the extracted contour image by the original image.

Finally, we need to set all the luminance values to 1 in the area which does not belong to contour. This is to keep the background luminance values of the synthesized image constant. In this case, the O_{xy} value of Eq. (2) can be taken as 1. It means that we can ignore the effects of attenuation factors on all objects except those that need to be synthesized. This way, we can extract any object for synthesizing data.

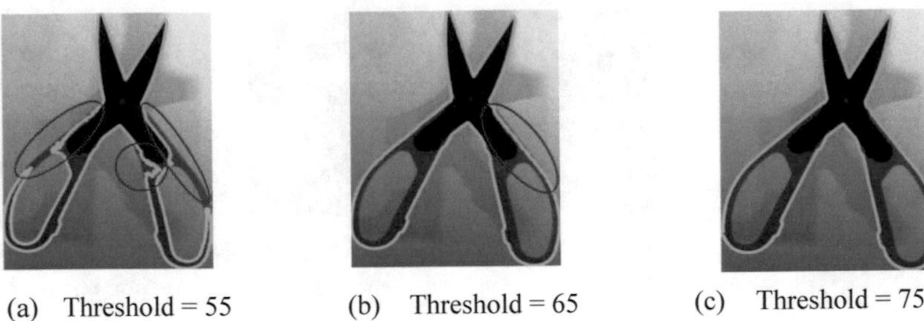

(a) Threshold = 55 (b) Threshold = 65 (c) Threshold = 75

Fig. 6. Sampling contours extracted with different thresholds. The red circles indicate the part that were not correctly extracted.

Synthesizing Data. In the previous step, we have extracted some dangerous objects and randomly selected these objects for image synthesizing. In order to ensure that our synthesized data is sufficiently robust, we perform the following operations on the objects:

1. Randomly enlarging or reducing each object by α times $(0.7 \leq \alpha \leq 1.3)$.
2. Randomly rotating each object by β° $(0 \leq \beta \leq 359)$.
3. Randomly selecting position of each object on the background image.

We show some examples of our synthesized data (see Fig. 7), the object inside the red rectangle is the one we used for synthesizing. These rectangular boxes are used only to mark the objects we synthesized and are not included in the real synthesized images.

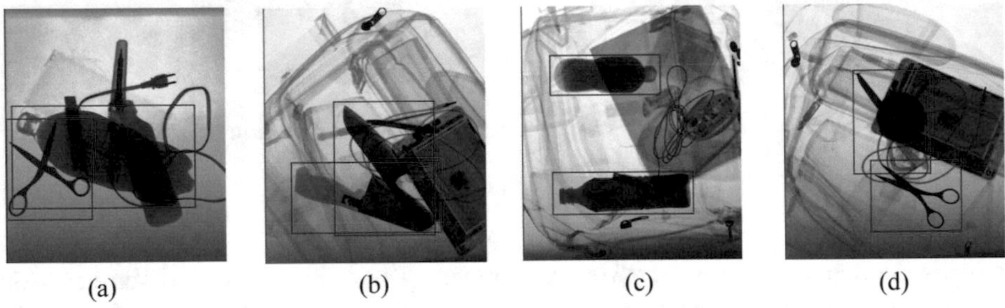

(a) (b) (c) (d)

Fig. 7. Examples of our synthesized data, these rectangular images are not included in the real synthesized images.

3.3 Evaluation Method

The following results make use of two metric measures. The first metric measure is Recall. There are four mathematical definitions: True positive (TP), True negative (TN), False positive (FP) and False negative (FN). TP is the count of the results, which are classified as belonging to the class correctly. TN is the count of the results, which are classified as not belonging to the class correctly. FP is the count of the results, which are classified as belonging to the class incorrectly. FN is the count of the results, which are classified as not belonging to the class, incorrectly. The following figure shows their relationships (see Fig. 8(a)). The mathematical definition of Recall is shown in Eq. (3). This value is used to measure the ability of the object detection model to reject non-related information.

$$\text{Recall} = \frac{TP}{TP + FN} \tag{3}$$

The second metric measure is mean Average Precision (mAP). To demonstrate the calculation of mAP, we first show the definition of Precision in Eq. (4). This value is used to measure the ability of the object detection model to detect relevant information. Recall and Precision are mutually constrained. The confidence value – threshold; is used to make the CNN model detect only the objects which exceed this threshold. By varying the threshold, we can get a Precision-Recall curve (see Fig. 8(b)). The definition of Average Precision (AP) is the area enclosed by the Precision-Recall curve and the X axis.

$$\text{Precision} = \frac{TP}{TP + FP} \tag{4}$$

For continuous Precision-Recall curve, the mathematical definition is Eq. (5).

$$\text{AP} = \int_0^1 PR\, dr \tag{5}$$

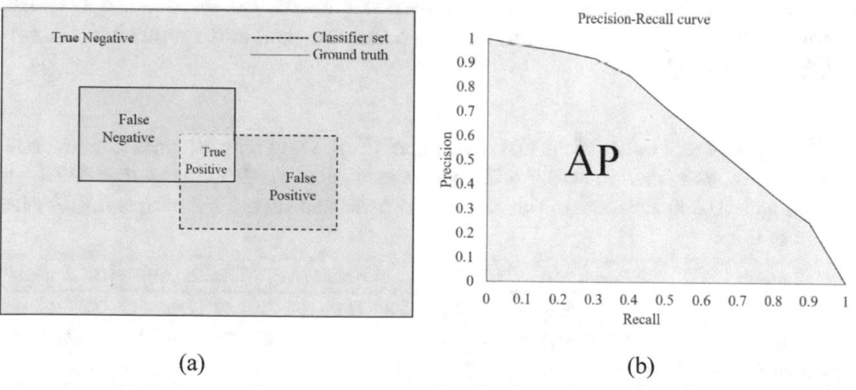

Fig. 8. Definitions of True positive, True negative, False positive and False negative (a) and definitions of AP (b).

For discrete Precision-Recall curve, the mathematical definition is Eq. (6).

$$AP = \sum\nolimits_{i=1}^{n} P(i)\Delta r(i) \tag{6}$$

mAP is the average of multiple classifications of APs.

4 Experiments and Results

The purpose of our experiments is to prove that our synthesized data can improve the performance of detection. In supervised learning, there are currently two major methods that successfully employ CNNs to X-ray images processing. One is training the CNN model from scratch. Another is transfer learning [24]. Considering that our X-ray dataset is small, it is easy to overfit in the case of training from scratch. This time we train YOLOv2 from a pre-trained model for fine-tuning. Because X-ray images are grayscale, we replicate all X-ray images three times to get 3-channel X-ray images as the input to YOLOv2. Our experiments focus on three kinds of dangerous objects detection: scissor, knife and bottle. To clarify the contribution of the synthesized data to the performance, we conduct five types of experiments. We have collected 1104 raw X-ray images (600 $\times$ 497) and synthesized 5017 images (512 $\times$ 614). All the images are labelled in VOC2007 format. For fair comparison, we run twenty times for each type of the experiment. In each run, both training and testing data are randomly selected.

The first experiment is conducted on training only raw X-ray images. We set learning rate to be 0.001 and batch size to be 64. We make use of 662 raw X-ray images for training. Our testing data includes 442 raw X-ray images and 442 synthesized X-ray images. All the results are recorded with the threshold of 0.25. The following is the results of 10,000 iterations based on YOLOv2. Table 1 shows the average test values of the first experiment repeated over 20 times. These results are based on testing both raw data and synthesized data of test dataset. To confirm the contribution of our synthesized data, we tested the raw data and synthesized data in the test dataset separately.

Table 1. Average test values of the first experiment (Exp. 1) running 20 times (training 662 raw images, testing 442 raw images + 442 synthesized images, learning rate = 0.001, batch size = 64, and 10,000 iterations). The results are presented as the Mean ± STDEV with the threshold of 0.25.

YOLOv2 results	Recall (%)	mAP	Bottle (AP)	Knife (AP)	Scissor (AP)
Total test (884)	54.05 ± 3.99	54.72 ± 3.88	41.06 ± 3.95	49.12 ± 4.45	73.97 ± 4.72
Raw data test (442)	84.20 ± 2.78	84.21 ± 1.81	88.63 ± 1.23	78.78 ± 2.88	85.20 ± 2.70
Synthesized data test (442)	30.05 ± 5.51	27.97 ± 5.51	14.37 ± 4.81	19.13 ± 3.69	50.42 ± 10.94

The second experiment is conducted on training only synthesized X-ray images. For fair comparison, the total of training images is the same as the previous experiment. This makes it easier to see the differences between raw X-ray images and synthesized images. We train 662 synthesized images and the proportion of the test dataset is the same as the previous experiment. The other experimental settings are also the same as the previous experiment. Table 2 shows all the results of the second experiment.

Table 2. Average test values of the second experiment (Exp. 2) running 20 times (training 662 synthesized images, testing 442 raw images + 442 synthesized images, learning rate = 0.001, batch size = 64, and 10,000 iterations). The results are presented as the Mean ± STDEV with the threshold of 0.25.

YOLOv2 results	Recall (%)	mAP	Bottle (AP)	Knife (AP)	Scissor (AP)
Total test (884)	67.05 ± 2.46	69.58 ± 2.27	77.28 ± 3.08	66.08 ± 2.21	65.38 ± 3.31
Raw data test (442)	48.55 ± 4.36	52.95 ± 4.65	53.97 ± 7.86	49.57 ± 4.64	55.31 ± 4.06
Synthesized data test (442)	85.25 ± 1.68	85.74 ± 2.00	89.62 ± 0.77	80.57 ± 3.28	87.03 ± 3.62

The third experiment is conducted on training both raw data and synthesized data. The total of training data should be the same as previous two experiments. We keep the amount of raw data and synthesized data in balance to verify the results. The test dataset in this experiment includes 773 raw X-ray images and 773 synthesized X-ray images. The other experimental settings are also the same as the previous two experiments. Table 3 shows all the results of the third experiment.

Table 3. Average test values of the third experiment (Exp. 3) running 20 times (training 331 raw images + 331 synthesized images, testing 773 raw images + 773 synthesized images, learning rate = 0.001, batch size = 64, and 10,000 iterations). The results are presented as the Mean ± STDEV with the threshold of 0.25.

YOLOv2 results	Recall (%)	mAP	Bottle (AP)	Knife (AP)	Scissor (AP)
Total test (1546)	80.15 ± 1.66	80.54 ± 1.31	86.55 ± 1.78	74.06 ± 1.70	81.00 ± 2.06
Raw data test (773)	79.40 ± 2.28	79.96 ± 1.70	85.91 ± 2.44	73.59 ± 2.23	80.36 ± 2.14
Synthesized data test (773)	81.00 ± 1.41	81.92 ± 1.12	87.43 ± 1.16	74.64 ± 1.72	83.68 ± 2.46

To confirm the effects of synthesized images in the training dataset on our experiment results, the rest of the two experiments are conducted by adding synthesized images in the training dataset. We conducted one experiment on the training dataset of 662 raw images and 331 synthesized images, the other experiment is conducted on the training dataset of 662 raw images and 3010 synthesized images. The experimental settings and results are in Tables 4 and 5.

From the practical point of view, raw data test results are more important than synthesized data test results. To compare the mean differences of raw data test results, we show t-test for equality of means of raw data test results in Table 7. We also compare the raw data test results with synthesized data test results for each experiment (see Table 6).

Table 4. Average test values of the fourth experiment (Exp. 4) running 20 times (training 662 raw images + 331 synthesized images, testing 442 raw images + 442 synthesized images, learning rate = 0.001, batch size = 64, and 10,000 iterations). The results are presented as the Mean ± STDEV with the threshold of 0.25.

YOLOv2 results	Recall (%)	mAP	Bottle (AP)	Knife (AP)	Scissor (AP)
Total test (884)	83.10 ± 1.74	83.61 ± 1.19	88.15 ± 0.74	77.37 ± 1.84	85.31 ± 2.55
Raw data test (442)	85.05 ± 2.09	84.91 ± 1.57	89.22 ± 0.72	79.89 ± 2.20	85.62 ± 3.14
Synthesized data test (442)	81.25 ± 2.12	82.60 ± 1.34	87.36 ± 1.14	75.73 ± 2.64	84.70 ± 2.65

Table 5. Average test values of the fifth experiment (Exp. 5) running 20 times (training 662 raw images + 3010 synthesized images, testing 442 raw images + 2007 synthesized images, learning rate = 0.001, batch size = 64, and 25,000 iterations). The results are presented as the Mean ± STDEV.

YOLOv2 results	Recall (%)	mAP	Bottle (AP)	Knife (AP)	Scissor (AP)
Total test (2449)	92.00 ± 0.79	89.97 ± 0.26	90.65 ± 0.10	88.94 ± 0.63	90.32 ± 0.20
Raw data test (442)	87.70 ± 0.01	86.41 ± 1.19	89.29 ± 1.04	81.97 ± 2.35	87.97 ± 1.33
Synthesized data test (2007)	92.90 ± 0.91	90.36 ± 0.17	90.75 ± 0.07	89.65 ± 0.44	90.67 ± 0.14

Table 6. Independent samples test (95% interval confidence) of raw data test results and synthesized data test results for each experiment. The results indicate that the raw data test results are significantly different with synthesized data test results for each experiment.

Comparison of raw data test results and synthesized data test results for each experiment			Levene's test for equality of variances		t-test for equality of means
			F-value	Sig	Sig (p-value)
Exp. 1	mAP	Equal variances assumed	4.862	0.034	0.000
		Equal variances not assumed			0.000
	Recall	Equal variances assumed	1.900	0.176	0.000
		Equal variances not assumed			0.000
Exp. 2	mAP	Equal variances assumed	10.661	0.002	0.000
		Equal variances not assumed			0.000
	Recall	Equal variances assumed	18.708	0.000	0.000
		Equal variances not assumed			0.000
Exp. 3	mAP	Equal variances assumed	0.520	0.475	0.000
		Equal variances not assumed			0.000
	Recall	Equal variances assumed	3.569	0.067	0.0055
		Equal variances not assumed			0.006
Exp. 4	mAP	Equal variances assumed	1.235	0.273	0.000
		Equal variances not assumed			0.000
	Recall	Equal variances assumed	0.149	0.702	0.000
		Equal variances not assumed			0.000
Exp. 5	mAP	Equal variances assumed	27.374	0.000	0.000
		Equal variances not assumed			0.000
	Recall	Equal variances assumed	3.279	0.078	0.000
		Equal variances not assumed			0.000

Table 7. Independent samples test (95% interval confidence) of raw data test results. The results indicate that the raw data test results of Exp. 1 are significantly different from the raw data test results of Exp. 2, Exp. 3 and Exp. 5, but there is no significant difference from the raw data test results of Exp. 4.

Comparison of Raw data test results			Levene's test for equality of variances		t-test for equality of means
			F-value	Sig	Sig (p-value)
Exp. 1 vs Exp. 2	mAP	Equal variances assumed	12.915	0.001	0.000
		Equal variances not assumed			0.000
	Recall	Equal variances assumed	7.653	0.009	0.000
		Equal variances not assumed			0.000
Exp. 1 vs Exp. 3	mAP	Equal variances assumed	0.066	0.798	0.000
		Equal variances not assumed			0.000
	Recall	Equal variances assumed	0.005	0.944	0.000
		Equal variances not assumed			0.000
Exp. 1 vs Exp. 4	mAP	Equal variances assumed	0.074	0.787	0.0975
		Equal variances not assumed			0.0975
	Recall	Equal variances assumed	0.001	0.978	0.141
		Equal variances not assumed			0.141
Exp. 1 vs Exp. 5	mAP	Equal variances assumed	0.761	0.388	0.000
		Equal variances not assumed			0.000
	Recall	Equal variances assumed	1.464	0.234	0.000
		Equal variances not assumed			0.000

5 Discussion

In Exp. 1, the results of synthesized data test (mAP: 27.97, Recall: 30.05) are significantly lower than raw data test results (mAP: 84.21, Recall: 84.20). This is because we only trained raw data in this experiment, the result of which objects in raw data were easier to detect than the objects in synthesized data. Exp. 2 also verified this phenomenon. When only training synthesized data, the results of synthesized data test (mAP: 85.74, Recall: 85.25) are higher than the results of raw data test (mAP: 52.95, Recall: 48.55). The results of these two experiments are not symmetrical. This is because our synthesized data helps improve the performance of detecting raw data. Exp. 3 proved this explanation. When we train the same amount of raw and synthesized data, we found that the results of synthesized data test (mAP: 81.92, Recall: 81.90) are slightly higher than the results of raw data test (mAP: 79.96, Recall: 79.40). The significance of t-test (p-value of mAP: 0.000, p-value of Recall: 0.006) in Table 7 also show that this is not due to random factors. This is because the color of raw data is lighter than synthesized data. It means that the objects in raw data is harder to detect. Since CNN extracts the features layer by layer. If the feature color is light, these features are likely to disappear in the deep layers. In contrast, the feature color of objects in synthesized data is dark. These features are less likely to disappear in the deep layers of CNN, which is more conducive to CNN learning the features of the objects. The results of raw data test in Exp. 3 are lower than the results of raw data test in Exp. 1. This is because the amount of raw training data (training 331 raw images) in Exp. 3 is less than the amount of raw training data (training 662 raw images) in Exp. 1. The results of Exp. 4 and Exp. 5 show the effects of training different amounts of synthesized data (331 synthesized images in Exp. 4 and 3010 synthesized images in Exp. 5) on the performance while training the same amount of raw data (662 raw images). Compared with the results of raw data test in Exp. 1, the results in Exp. 4 of t-test (p-value of mAP: 0.0975 p-value of Recall: 0.141) show that our results are not significant. This result indicates that training a small amount of our synthesized data does not improve the detection performance of raw data. However, when the synthesized data is enough, the results of raw data test in Exp. 5 (mAP: 86.41 Recall: 87.70) shows that our synthesized data can improve the detection performance of raw data (improving 2.2% in mAP, improving 3.5% in Recall). The t-test results (p-value of mAP: 0.000 p-value of Recall: 0.000) also show that both improvements in mAP and Recall are significant.

We show some detection examples of the best performance running in Exp. 5 (see Fig. 9). We can find that some dangerous objects that are difficult for humans to recognize can also be detected by our system. Still, our method has some limitations. The first one is that a small amount of our synthesized data cannot show significant improvement on raw data. This is because the luminance values of the raw data and our synthesized data are not the same, which can be improved by synthesizing data to balance the luminance values. The second limitation is that our security system still can only detect three kinds of dangerous objects. It is necessary to add more kinds of dangerous objects to test. The third limitation is that our detection accuracy is not

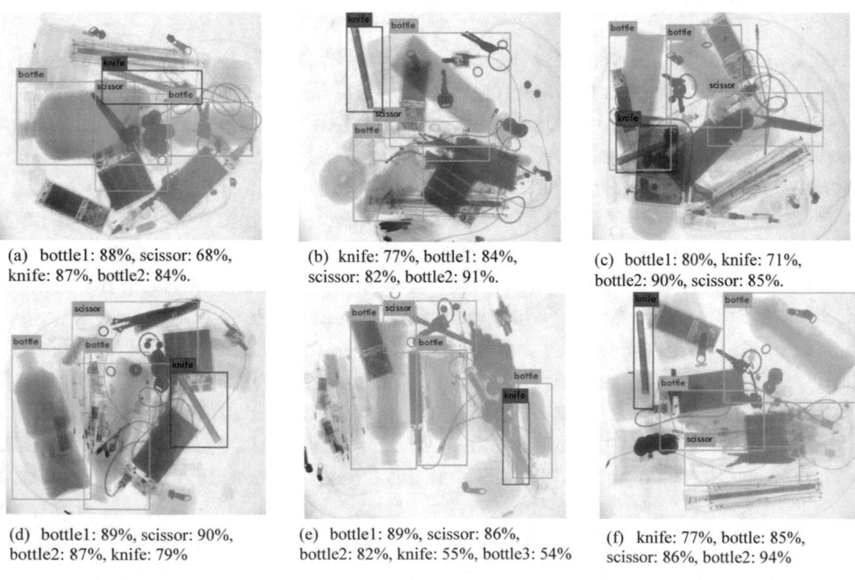

(a) bottle1: 88%, scissor: 68%, knife: 87%, bottle2: 84%.

(b) knife: 77%, bottle1: 84%, scissor: 82%, bottle2: 91%.

(c) bottle1: 80%, knife: 71%, bottle2: 90%, scissor: 85%.

(d) bottle1: 89%, scissor: 90%, bottle2: 87%, knife: 79%

(e) bottle1: 89%, scissor: 86%, bottle2: 82%, knife: 55%, bottle3: 54%

(f) knife: 77%, bottle: 85%, scissor: 86%, bottle2: 94%

Fig. 9. Detection examples of the best performance running on YOLOv2, the value in % refers to the overlapping area of the target.

enough to meet the needs of practical applications. For future research, we will train other advanced CNN models such as DSSD [21] and YOLOv3 [22] to improve the performance.

6 Conclusion

Implications. In this paper, we applied CNN to dangerous objects detection experiments with X-ray image dataset. We empirically focused on three kinds of dangerous objects detection: scissor, knife and bottle and used synthesized X-ray images to improve the detection performance of raw data. We conducted five types of experiments to compare our synthesized data with raw X-ray data and have shown the significance of t-test to verify the contributions of our synthesized data. We have confirmed that our synthesized data can improve the detection performance of raw data. As for future work, we will balance the luminance values of raw data and synthesized data and conduct the experiments on DSSD and YOLOv3 for the sake of improving the detection performance while working on adding other dangerous objects to the experiments.

Acknowledgements. We thank Yokoshima Shin and Nikaido Yoji in T & S Co., Ltd., for providing us with portable X-ray device information and data collection environments.

References

1. Akcay, S., Kundegorski, M., Willcocks, C., Breckon, T.: Using deep convolutional neural network architectures for object classification and detection within X-ray baggage security imagery. IEEE Trans. Inf. Forensics Secur. **13**, 2203–2215 (2018)
2. Turcsany, D, Mouton, A., Breckon, T.P.: Improving feature-based object recognition for X-ray baggage security screening using primed visualwords. In: 2013 IEEE International Conference on Industrial Technology (ICIT), pp. 1140–1145. IEEE (2013)
3. Singh, S., Singh, M.: Explosives detection systems (EDS) for aviation security. Signal Process. **83**, 31–55 (2003)
4. Chen, Z., Zheng, Y., Abidi, B.R., Page, D.L., Abidi, M.A.: A combinational approach to the fusion, de-noising and enhancement of dual-energy X-ray luggage images. In: IEEE Computer Society Conference on Computer Vision and Pattern Recognition-Workshops, 2005. CVPR Workshops, p. 2. IEEE (2005)
5. He, X.P., Han, P., Lu, X.G., Wu, R.B.: A new enhancement technique of x-ray carry-on luggage images based on dwt and fuzzy theory. In: International Conference on Computer Science and Information Technology, 2008, ICCSIT 2008, pp. 855–858. IEEE (2008)
6. Baştan M., Yousefi M.R., Breuel T.M.: Visual words on baggage X-ray images. In: Real, P., Diaz-Pernil, D., Molina-Abril, H., Berciano, A., Kropatsch, W. (eds.) Computer Analysis of Images and Patterns. CAIP. Lecture Notes in Computer Science, vol 6854. Springer, Berlin (2011)
7. Jaccard, N., Rogers, T.W., Griffin, L.D.: Automated detection of cars in transmission X-ray images of freight containers. In: 2014 International Conference on Advanced Video and Signal Based Surveillance (AVSS), pp. 387—392. IEEE (2014)

8. Rogers, T.W., Jaccard, N., Protonotarios, E.D., Ollier, J., Morton, E.J., Griffin, L.D.: Threat image projection (TIP) into X-ray images of cargo containers for training humans and machines. In: 2016 IEEE International Carnahan Conference on Security Technology (ICCST), pp. 1–7. IEEE (2016)
9. Baştan, M.: Multi-view object detection in dual-energy X-ray images. Mach. Vis. Appl. **26**, 1045–1060 (2015)
10. Kundegorski, M.E., Akçay, S., Devereux, M., Mouton, A., Breckon, T.P.: On using feature descriptors as visual words for object detection within X-ray baggage security screening (2016)
11. Akçay, S., Kundegorski, M.E., Devereux, M., Breckon, T.P.: Transfer learning using convolutional neural networks for object classification within X-ray baggage security imagery. IEEE (2016)
12. Riffo, V., Mery, D.: Automated detection of threat objects using adapted implicit shape model. IEEE Trans. Syst. Man Cybern. Syst. **46**, 472–482 (2016)
13. Van de Sande, K.E., Uijlings, J.R., Gevers, T., Smeulders, A.W.: Segmentation as selective search for object recognition. In: 2011 IEEE International Conference on Computer Vision (ICCV), pp. 1879–1886. IEEE (2011)
14. Girshick, R., Donahue, J., Darrell, T., Malik, J.: Rich feature hierarchies for accurate object detection and semantic segmentation. In: Proceedings of the IEEE Conference on Computer Vision and Pattern Recognition, pp. 580–587 (2014)
15. He, K., Zhang, X., Ren, S., Sun, J.: Spatial pyramid pooling in deep convolutional networks for visual recognition. IEEE Trans. Pattern Anal. Mach. Intell. **37**, 1904–1916 (2015)
16. Girshick, R.: Fast R-CNN. In: Proceedings of the IEEE International Conference on Computer Vision, pp. 1440–1448 (2015)
17. Ren, S., He, K., Girshick, R., Sun, J.: Faster R-CNN: towards real-time object detection with region proposal networks. IEEE Trans. Pattern Anal. Mach. Intell. **39**, 1137–1149 (2017)
18. Redmon, J., Divvala, S., Girshick, R., Farhadi, A.: You only look once: unified, real-time object detection. In: Proceedings of the IEEE Conference on Computer Vision and Pattern Recognition, pp. 779–788 (2016)
19. Redmon, J., Farhadi, A.: YOLO9000: better, faster, stronger. arXiv preprint (2017)
20. Liu, W., Anguelov, D., Erhan, D., Szegedy, C., Reed, S., Fu, C.Y., Berg, A.C.: SSD: single shot multibox detector. In: European Conference on Computer Vision, pp. 21–37. Springer, Cham (2016)
21. Fu, C.Y., Liu, W., Ranga, A., Tyagi, A., Berg, A.C.: DSSD: deconvolutional single shot detector. arXiv preprint arXiv:1701.06659 (2017)
22. Redmon, J., Farhadi, A.: YOLOv3: an incremental improvement. arXiv preprint arXiv:1804.02767 (2018)
23. Jaccard, N., Rogers, T.W., Morton, E.J., Griffin, L.D.: Tackling the X-ray cargo inspection challenge using machine learning. In: Anomaly Detection and Imaging with X-Rays (ADIX), vol. 9847, p. 98470N. International Society for Optics and Photonics (2016)
24. Yosinski, J., Clune, J., Bengio, Y., Lipson, H.: How transferable are features in deep neural networks?. In: Advances in Neural Information Processing Systems, pp. 3320–3328 (2014)

An Efficient JPEG Steganalysis Model Based on Deep Learning

Lin Gan[1(✉)], Yang Cheng[1], Yu Yang[1,2], Linfeng Shen[1], and Zhexuan Dong[1]

[1] School of Cyberspace Security, Beijing University of Posts and Telecommunications, Beijing 100876, China
ganlin@bupt.edu.cn

[2] Guizhou Provincial Key Laboratory of Public Big Data, GuiZhou University, Guizhou, China

Abstract. Convolutional neural networks (CNN) have gained an overwhelming advantage in many domains of pattern recognition. CNN's excellent data learning ability and automatic feature extraction ability are urgently needed for image steganalysis research. However, the application of CNN in image steganalysis is still in its infancy, especially in the field of JPEG steganalysis. This paper presents an efficient CNN-based JPEG steganographic analysis model which is called JPEGCNN. According to the pixel neighborhood model, JPEGCNN calculates the pixel residual as a network input with a 3×3 kernel function. In this way, JPEGCNN not only solves the problem that direct analysis of DCT coefficients is greatly affected by image content, but also solves the problem that larger kernel functions such as 5×5 do not effectively capture neighborhood correlation changes. Compared with the JPEG steganographic analysis model HCNN proposed by the predecessors, JPEGCNN is a lightweight structure. The JPEGCNN training parameters are about 60,000, and the number of paramcters is much lower than the number of parameters of the HCNN. At the same time of structural simplification, the simulation results show that JPEGCNN still maintains accuracy close to HCNN.

Keywords: Steganalysis · Convolutional neural network · Transform domain

1 Introduction

Steganalysis analyzes the statistical properties of a digital multimedia documents to determine whether there are extra information hidden in the cover or even the estimates amount of information embedded and the content of the hidden information. The current steganalysis research field generally considers steganalysis as a two-category problem with the goal of distinguishing cover and stego. Under this circumstance, the existing methods mainly construct steganalysis detectors through the following two steps: feature extraction and classification. In the feature extraction step, a series of hand-crafted features are extracted from the image to capture the effects of the embedding operation. The effectiveness of steganalysis depends heavily on feature design. However, this work is complicated due to the lack of an accurate natural image model. At present, the most reliable characteristic design paradigm is to calculate the

C.-N. Yang et al. (Eds.): SICBS 2018, AISC 895, pp. 729–742, 2020.
https://doi.org/10.1007/978-3-030-16946-6_60

noise residuals and then model the residuals using the conditions of adjacent elements or joint probability distribution. With the increasing complexity of steganography, especially in recent years the emergence of adaptive steganography HUGO [1], S-UNIWARD [2], WOW [3], HILL-CMD [4], MiPOD [5], the more complex statistical characteristics of images need to be considered in the design process of the steganalysis domain, and the characteristics are gradually moving towards complexity and high dimensionality. For example, from the early method of describing the correlation between adjacent pixels or coefficients by co-occurrence matrix [6, 7] to the late method of SRM (Spatial Rich Model) [8] and PSRM (Projection Spatial Rich Model) [9], the dimension of feature has developed from several hundred dimensions to tens of thousands of dimensions. In the classification step, classifiers such as SVM or Ensemble classifiers are used to detect stego images. Because the feature extraction and classification steps are separate, they cannot be uniformly optimized, which means that the classifier may not be able to make full use of the information in feature extraction.

In order to solve the aforementioned problems, researchers have introduced deep learning theory into steganalysis in recent years. In 2015, Qian et al. [10] proposed a steganalysis model based on CNN, called Gaussian-Neuron CNN (GNCNN). The model includes an image preprocessing layer, several convolutional layers, and several fully-connected layers. In the image preprocessing layer, the high-pass filter (HPF) is used to enhance the steganography information. The convolution layer is used to extract the image features, and the fully-connected layer is used to classify the images. The model performance is close to the SRM. Pibre et al. [11] proposed that when steganography uses the same embedding key, the CNN-based steganalysis model can achieve better results and has the ability to solve the problem of cover source mismatch. Xu et al. [12] proposed a CNN-based model for steganalysis, called ICNN. Similar to GNCNN, high-pass filtering was used to pre-process the image to enhance the steganalysis information. The difference is that the network structure in the pretreatment unit outside the author's choice, from the CNN structural unit of new success to other more applications. The ICNN has four contributions: (1) The first layer convolutional layer is followed by an absolute activation (ABS) layer to exploit the symmetry of the residual image; (2) The batch normalization (BN) layer is used; (3) The 1×1 size convolutional kernels and global average pooling is used; (4) The tanh function is used in some layers. Experimental results show that the network model has achieved slightly better detection results than the SRM. In 2017, Ye et al. [13] proposed a 10-layer CNN steganography model, called Truncated Linear Unit CNN (TLUCNN). The first layer of the network uses 30 basic linear filters and a Truncated Linear Unit (TLU), achieving significantly better results than the SRM.

According to domain where features are extracted, image steganalysis can be divided into two categories: spatial domain steganalysis and transform domain steganalysis. The aforementioned methods are all focus on spatial steganalysis, and few studies have applied deep learning to transform domain steganalysis. In 2016, Zeng et al. [14] applied the deep learning framework to transform domain steganalysis for the first time and proposed a JPEG steganalysis model with three CNN subnetworks, which is called Hybrid CNN (HCNN). The final experimental results show that the accuracy of HCNN is higher than DCTR [15] and PHARM [16]. At the same time,

Zeng et al. also proved that the two key components of the rich model which are the convolutional kernel for extracting different noise residuals and the threshold quantizer used to reduce the complexity of the model, can't be effectively obtained through deep learning. It is important to recognize these two points for studying deep learning based steganalysis algorithms. Zeng et al.'s findings demonstrate the feasibility of applying a deep learning framework for transform domain steganalysis, which is undoubtedly a huge step forward. However, the HCNN is more complex than the previous deep learning based steganalysis model in spatial domain. The reasons are as follows: (1) The network has two additional steps quantitative and truncated. (2) The feature extraction module contains three paths, whereas the previous model only contains one path. Such a structure leads to an increase in the complexity of the network and the computational overhead. In order to apply the deep learning theory to the transform domain steganalysis and fill in the blanks of previous research, this paper presents an efficient JPEG steganalysis model based on deep learning, called JPEGCNN. Compared with the HCNN, the JPEGCNN has the following advantages: (1) The core of the HPF layer in the pre-processing module is simpler. (2) The network does not require quantification and truncation operation. (3) The network does not require parallel subnetwork structure. These three points make the network proposed in this paper easier to implement and less computational overhead. In this paper, a large number of experiments were conducted on the BOSSBase 1.01 data set. The experimental results show that the JPEGCNN has a good analysis ability for the JPEG steganography algorithm.

The next part of the article is as follows: The second part mainly elaborates the structure of the JPEGCNN. The third part is the experiment. The fourth part is the conclusion and outlooks and the fifth part is acknowledgement.

2 JPEGCNN Model

A well-designed CNN can well simulate the three key steps of steganalysis. Therefore, the development of image steganalysis through CNN is not only realistic but also achievable. The CNN used for steganalysis usually has the following modules: pre-processing module, feature extraction module and classification module. The JPEGCNN also has these three modules.

2.1 Preprocessing Module

The preprocessing module usually has a HPF layer. The HPF layer is a special convolutional layer that is at the forefront of the entire network. In this layer, a pre-defined high-pass filter is usually used for filtering.

In general, high-frequency steganographic noise is a weak signal and is greatly influenced by the image content. Therefore, high-pass filtering can enhance steganographic information and reduce the interference of image content. In addition, the meticulously designed cover and stego are not only visually very similar, but also statistically similar. Based on this, it is not difficult to find that a CNN with a random initialization weight usually does not converge when applied to steganalysis. The HPF layer is crucial for the CNN. If there is no HPF layer, CNN convergence will be much slower or even non-convergent. The HPF layer was first appeared in GNCNN, and the filter of this layer came from SRM, as shown below.

$$\frac{1}{12}\begin{pmatrix} -1 & 2 & -2 & 2 & -1 \\ 2 & -6 & 8 & -6 & 2 \\ -2 & 8 & -12 & 8 & -2 \\ 2 & -6 & 8 & -6 & 2 \\ -1 & 2 & -2 & 2 & -1 \end{pmatrix} \tag{1}$$

HCNN used 25 5 × 5 DCT basic pattern kernels in the HPF layer kernel. The design of the HPF layer kernel is based on DCTR [15]. The convolutional kernel structure is defined as B(k, l) = (Bmn(k, l)), 0 ≤ k, l ≤ 5, 0 ≤ m, n ≤ 5:

$$\begin{aligned} B_{mn}^{(k,l)} &= \frac{W_k W_l}{4} \cos\frac{\pi k(2m+1)}{10} \cos\frac{\pi l(2n+1)}{10} \\ w_0 &= \frac{1}{\sqrt{2}}, w_k = 1 \text{ for k } > 0 \end{aligned} \tag{2}$$

Unlike the 25 5 × 5 DCT basic pattern kernels used by HCNN, the kernel used by the HPF layer in JPEGCNN is still from the SRM and size is 3 × 3. The absolute values of the weights in the kernel are distributed between 0 and 1, as follows.

$$\frac{1}{4}\begin{pmatrix} -1 & 2 & -1 \\ 2 & -4 & 2 \\ -1 & 2 & -1 \end{pmatrix} \tag{3}$$

The use of the filter described above is based on the following reasons: (1) Pixel residuals, rather than quantized DCT coefficients, are more conducive to steganalysis. The steganography algorithm usually has a small modification range for JPEG quantized coefficients, mostly plus or minus one. Therefore, the influence on the statistical characteristics of JPEG quantized coefficients is not significant. Using a filter similar to HCNN, there is no significant advantage in learning the difference between the two types of sample data from the transform domain. For spatial domain, the situation is completely different. Although the quantization coefficient is only changed by 1 unit, the interference of the steganographic operation on the spatial pixels is further amplified by the amplification step of the quantization step, which is advantageous for the steganographic analysis. In addition, SRM has given a neighborhood pixel correlation model. There is a clear model in the spatial domain analysis as a guide. Pixel residuals suppress the interference of image content. However, the correlation of neighborhood pixel correlation in the DCT domain is not modeled. The direct analysis of the DCT coefficients is inevitably affected by the image content, which further demonstrates the rationality of analyzing the residuals in the spatial domain. (2) A reasonably sized filter kernel function is more conducive to steganalysis. Research indicates that for complex images, neighborhood pixel correlation will decrease dramatically as the distance between the boundary and the center increases. The 5 × 5 filter used by GNCNN contains too many pixels with a distance from the center pixel greater than 2, and its ability to capture the neighborhood pixel correlation will decrease. Therefore, JPEGCNN uses a 3 × 3 size filter.

2.2 Feature Extraction Module

After residual is extracted by the preprocessing module, image is input into the feature extraction module composed of several convolutional layers.

Feature extraction module usually consist of multiple convolutional layers. Convolutional layer's input and output are a set of arrays called feature map, and each convolutional layer typically produces feature map in three steps, convolution, non-linear activation, and pooling operation. The first step uses k convolutional kernels for filtering, resulting in k new feature maps. $F^n(X)$ denotes the feature map of the nth layer output. $Weight^n$ denotes the nth layer convolutional kernel and $Bias^n$ denotes the offset. The convolutional layer can be expressed as follow:

$$F^n(X) = P(A^n(F^{n-1}(X) * Weight^n + Bias^n)) \tag{4}$$

In thc formula, $F^0(X) = X$ represents input data and $A^n(\cdot)$ represents the nonlinear activation function. Nonlinear activation function is applied to each input element. Typical activation functions are sigmod, TanH, and ReLU. $P(\cdot)$ represents pooling operation, including average pooling operation and maximum pooling operation. In general, nonlinear activation function and pooling operation are optional in a specific layer, and a convolutional layer can also be set as untrainable.

In convolutional layer, each output featurc map combines the features of multiple input feature maps by convolution. The structure of the convolutional layer involves the concepts of local perception and weight sharing. For local perception, each low-dimensional feature is only calculated from a subset of the inputs, such as the area of the pixel at a given location in the image. The local feature extractor shares the same parameters when applied to different adjacent input positions, which corresponds to the convolution of the image pixel values with the kernel containing the weight parameters. Weight sharing generates a shift-invariant operation, which also reduces the number of free variables, thereby increasing the generalization of the network.

The feature extraction module in HCNN network contains three subnetworks. Each subnetwork contains special structure such as quantization, truncation, ABS and BN. The number of HCNN parameters increases dramatically with the use of subnetwork. One subnetwork contains three convolution layers, and three subnetworks will contain up to nine convolution layers in total. The structure of the subnetwork is shown in Fig. 1. The structure of the whole feature extraction module is shown in Fig. 2.

Unlike HCNN, JPENGCNN does not adopt subnetwork. The reasons for not using subnetworks are as follows: (1) Subnetworks bring a lot of parameters, which greatly increase the computational cost. (2) There is no obvious evidence proves that the use of subnetworks can significantly improve performance. The structure of the feature extraction module is shown in Fig. 3.

The feature extraction module of JPEGCNN has 5 layers. Each of layer contains 16 convolution kernels of size 3 × 3 or 5 × 5. Compared to HCNN, the number of

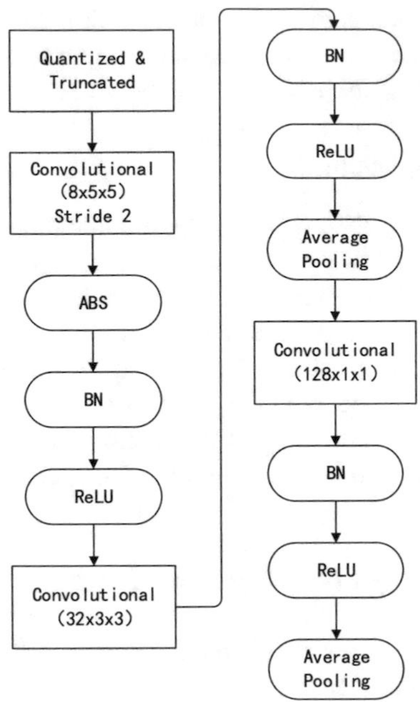

Fig. 1. A type of subnetwork in HCNN

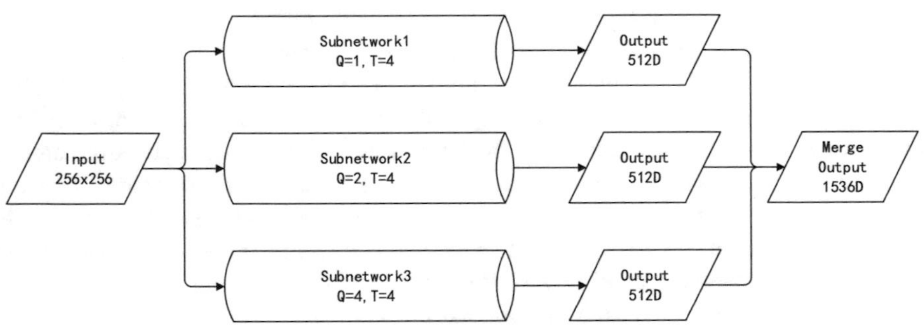

Fig. 2. The structure of the entire feature extraction module in HCNN

convolutional layers in JPEGCNN has been reduced by nearly half. The first convolutional layer of the feature extraction module uses the Gaussian activation function, and the deeper layers use ReLU activation function. Combined with the experimental results of ICNN, it is known that using ReLU as an activation function at a deeper level of CNN is a better choice. The average pooling operation is used at the end of each convolutional layer. The purpose of the pooling operation is to convert low-level feature representation into more useful feature identifier, thereby saving important information and discarding irrelevant detail. In general, deeper feature representation requires information from progressively larger input regions. The role of pooling is to merge information into a small set of local regions while reducing computational

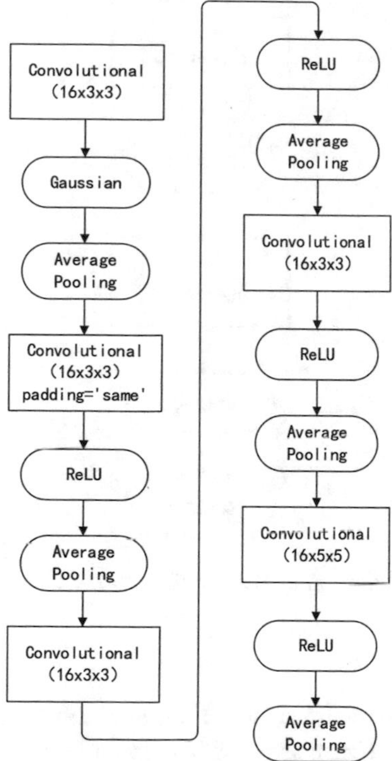

Fig. 3. The feature extraction module of the proposed network in this paper

overhead, which is similar to the purpose of quantifying and truncating operations to aggregate useful features and discarding useless information.

2.3 Classification Module

The features extracted by the feature extraction module are input into the classification module for classification. The classification module finally outputs the category of image which is cover or stego. The classification module usually contains several fully-connected layers and a classifier.

Ye et al. [13] proposed that the fully-connected layer usually contains more training parameters. When the training set is not large enough, it easily leads to overfitting. One solution is to use only one fully-connected layer during training. There is another way to solve the problem of overfitting that is to use the dropout to regularize the fully-connected layer. When using dropout for training, the neuron output in the corresponding layer is set to 0 with a certain probability. This technology can improve CNN's generalization ability to some extent.

The structure of the classification module proposed in this paper is shown in Fig. 4.

The classification module has two 128-dimensional fully-connected layers and a softmax classifier. Dropout layers are added after each fully-connected layer to prevent overfitting. The dropout parameter is set to 0.5.

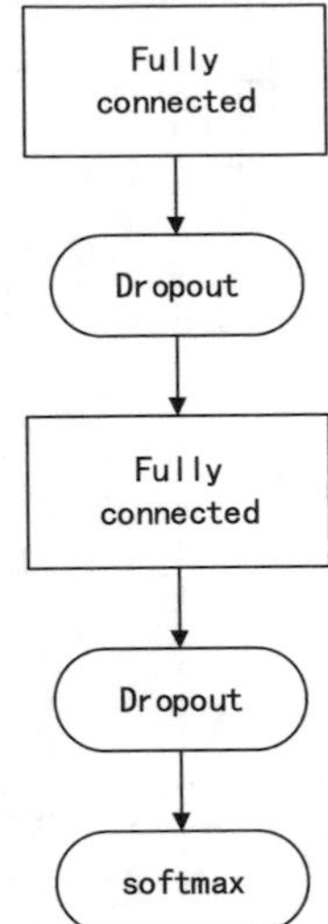

Fig. 4. The classification module of the proposed network in this paper

2.4 The Overall Structure

Combining the above three modules, the structure of JPEGCNN is shown in the Fig. 5. Taking the 256 × 256 size image as input. The input size of each layer is marked in the upper left corner.

3 Experimental Results and Analysis

3.1 The Data Set

The data set used in this paper's experiment is the standardized data set BOSSBase 1.01 [17]. BOSSBase 1.01 contains 10000 images. Each of image's size is 512 × 512 (The number of images corresponding to the MB1 and MB2 algorithms is 9600. Some images were excluded because their maximum embedding rate supported for the MB1 and MB2 algorithms are too small. The rest 9600 images can support embed rates up to 0.8). Due to limitation in computing resource, this paper resized the size of BOSSBase 1.01 images to size 256 × 256 in the experiment. 5 JPEG domain steganography algorithms were used to evaluate the steganalysis ability of the JPEGCNN. These JPEG domain steganography algorithms were Jsteg, nsf5, MB1, MB2, and J-UNIWARD [18].

In experiment, steganography algorithms and cover were used to generate the corresponding stego. These 10,000 cover images and 10000 stego images together formed the data of one experiment (10,000 pairs of cover-stego images). The training set, verification set, and test set used a ratio of 8:1:1, that was, the training set was 8000 pairs of cover-stego images, the verification set was 1000 pairs of cover-stego images, and the test set was 1000 pairs of cover-stego images.

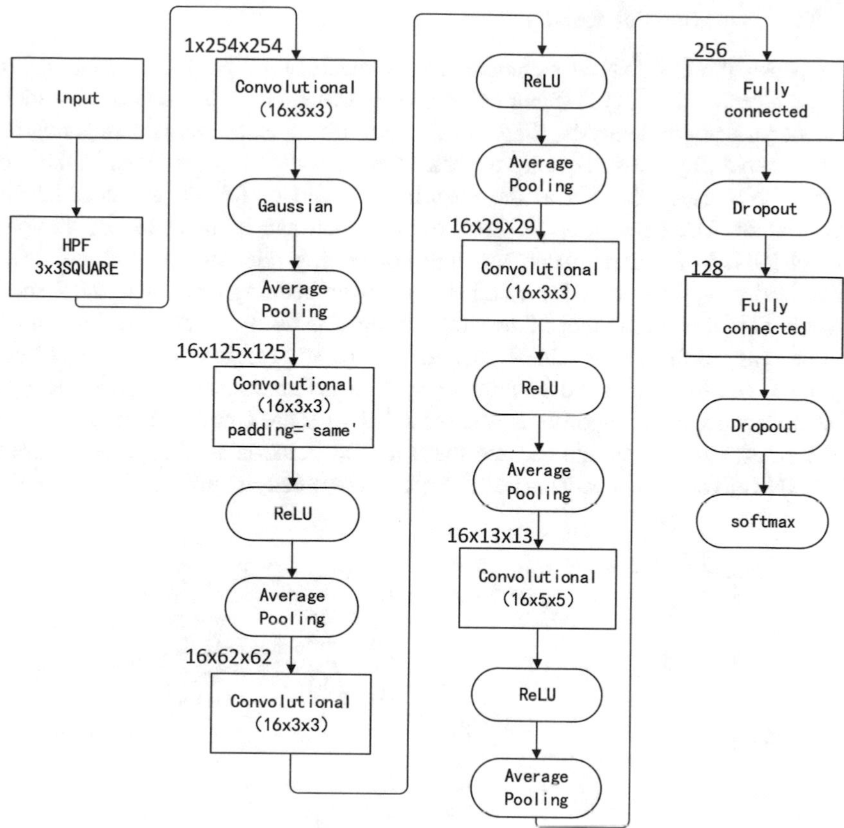

Fig. 5. The overall network structure of JPEGCNN

3.2 Experimental Parameters

The network learning rate in this paper was mostly set to 0.05 (occasionally 0.005 depending on experimental results). Adadelta [19] gradient descent algorithm was used during training. The size of mini-batch was 64. In the pre-processing module, the HPF layer initialization weight has been described in the second section and was set to non-trainable. The parameter of the Gaussian activation function was $\sigma^2 = 0.2$. In the feature extraction module, the convolutional kernel of each layer was initialized using a "Xavier" [20] initializer. The pooling size of each layer was 3×3 and the step size was 2. The total number of parameters of JPEGCNN is 63,212, and the number of parameters that could be trained is 63,202, while the number of parameters of HCNN is one million. JPEGCNN greatly reduces the number of parameters and improves computational efficiency.

3.3 The Experimental Results

In this paper, the GNCNN was compared with the JPEGCNN. The reasons for using GNCNN as reference: (1) Although GNCNN is designed to be used as a model for spatial steganography analysis, it also has a certain frequency domain steganographic analysis capability. Subsequent experimental results also proved this point. (2) JPEGCNN uses GNCNN as design reference and optimized for steganalysis in JPEG domain. The effect of JPEGCNN optimization can be analyzed by comparing with GNCNN. In the experiment, for high embedding rate, such as 0.8 and 1.0, the uninitialized weight was used for training. For low embedding rates such as 0.2 and 0.5, the result weight of high embedding rate training was used for initialization, and then fine-tune was performed on this basis. In this paper, the relationship between the accuracy of verification set and the number of training iterations for J-UNIWARD, Jsteg and nsf5 steganography algorithms was recorded with the embedding rate of 1.0. The result is shown in Fig. 6. We can see that the 200 iteration is enough to ensure that JPEGCNN converges to a better state. Therefore, the iteration number of training is 200.

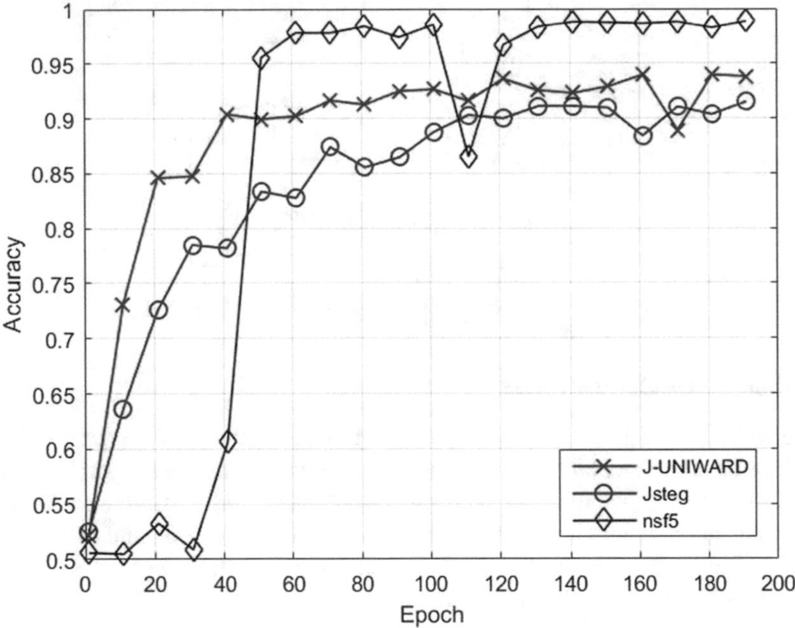

Fig. 6. The relationship between the accuracy and the iteration number

The experimental result is shown in the following Table 1. The first column in the table shows the steganography algorithms used, the second column is payload, the third and fourth columns are the accuracy of the two network JPEGCNN and GNCNN on the test set, respectively. The payload is calculated based on the bpnzac (bits per nonzero AC DCT).

Table 1. The performance of JPEGCNN and GNCNN

Algorithm	Payload	JPEGCNN	GNCNN
Jsteg	0.2	0.8285	0.7865
	0.5	0.9045	0.8375
	1.0	0.9080	0.8590
nsf5	0.2	0.7355	0.6730
	0.5	0.9230	0.8050
	1.0	0.9920	0.9865
MB1	0.2	0.9492	0.8667
	0.5	0.9750	0.9533
	0.8	0.9817	0.9642
MB2	0.2	0.9500	0.9125
	0.5	0.9808	0.9650
	0.8	0.9817	0.9692
J-UNIWARD	0.2	0.7060	0.6545
	0.5	0.8485	0.7675
	1.0	0.9390	0.8980

It can be seen that although GNCNN is designed for spatial steganalysis, GNCNN still has some analytical ability for various JPEG steganography. Compared with GNCNN, JPEGCNN has higher accuracy in JPEG steganalysis. It is also found that JPEGCNN is significantly better than other steganography algorithms in MB1 and MB2 steganography.

The difference between JPEGCNN and GNCNN is mainly reflected in the high pass filter and the activation function of feature extraction module. This paper combines the feature extraction module of JPEGCNN with the preprocessing module of GNCNN to form a reference network between JPEGCNN and GNCNN, which is used to analyze the effect of high pass filter and activation function on the accuracy of JPEGCNN. The reference network is called Reference CNN (RCNN). In order to make the results obvious, JPEGCNN, RCNN, and GNCNN were used to perform 10 steganographic analysis on Jsteg steganography with an embedding rate of 1.0. The results are shown in Tables 2 and 3.

It can be seen that JPEGCNN has an accuracy improvement of about 1.5% compared to RCNN. The accuracy of this part is improved by the stronger capture capability of the 3×3 filter for adjacent pixel correlation. Compared with GNCNN, RCNN has an accuracy increase of about 4%. This part of the accuracy improvement is due to the choice of activation function.

J-UNIWARD is one of the most advanced JPEG steganography algorithms available. In the case of low embedding rate, the author of HCNN used HCNN to analysis J-UNIWARD. In the paper of HCNN, we can see that the analysis accuracy of HCNN is about 0.53 at 0.1 embedding rate and 0.81 at 0.5 embedding rate. In order to compare the performance of JPEGCNN and HCNN, this paper also uses JPEGCNN to perform steganographic analysis on J-UNIWARD in the case of low embedding rate.

The result is shown in Fig. 7. The experimental results show that the accuracy of JPEGCNN is 4–5% higher than that of HCNN in analyzing J-UNIWARD at the same embedding rate.

Table 2. Analysis results of Jsteg steganography by JPEGCNN, RCNN and GNCNN

	JPEGCNN	RCNN	GNCNN
1	0.9080	0.9025	0.8505
2	0.9285	0.8975	0.8590
3	0.8945	0.8700	0.8735
4	0.9090	0.8960	0.8580
5	0.9035	0.9140	0.8630
6	0.9190	0.9140	0.8555
7	0.9255	0.8975	0.8725
8	0.9180	0.8995	0.8410
9	0.9230	0.9005	0.8470
10	0.9265	0.9025	0.8710

Table 3. The average and variance of JPEGCNN, RCNN and GNCNN on Jsteg steganography results

Algorithm	Average	Variance
JPEGCNN	0.9156	0.000115
RCNN	0.8994	0.000133
GNCNN	0.8591	0.000111

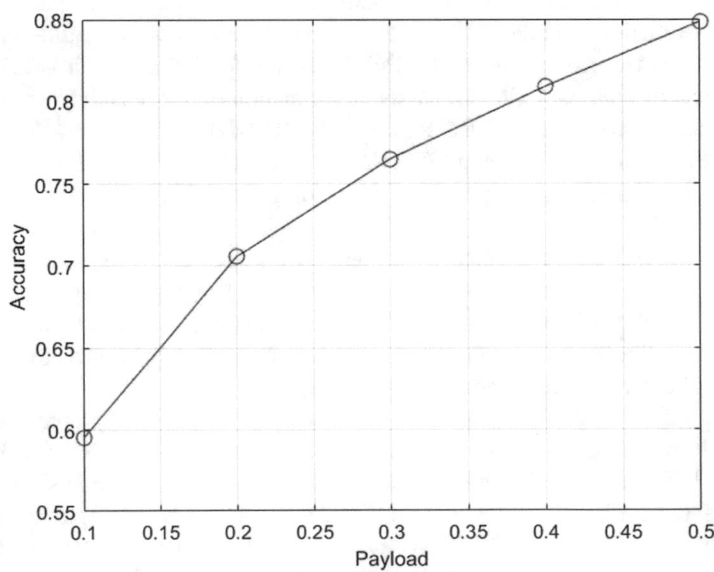

Fig. 7. JPEGCNN analysis accuracy of J-UNIWARD for low embedding rate.

4 Conclusion

For the research gap of applying deep learning to the steganalysis of JPEG domain and the difficulty of artificially designing steganalysis feature, this paper presented an efficient JPEG steganalysis model JPEGCNN and carried on relevant verification and test on BOSSBase 1.01 data set. It was found that the JPEGCNN has better analysis capabilities for JPEG steganography. At the same time, compared to the HCNN, the structure of the JPEGCNN is more simplified and the computational overhead is lower. While simplifying the network structure, the accuracy of classification has maintained a level equivalent to that of previous network.

The future research direction will focus on the following two points: The first is the network JPEGCNN still has room for improvement in the accuracy of steganalysis, and in particular, how can the accuracy rate be further improved at low embedded rates? The second is that the accuracy rate of MB1 and MB2 is significantly superior to other steganography algorithm. The reasons for this are worthy of further research and discussion.

Acknowledgement. This work is supported by the National Key R&D Program of China (No. 2017YFB0802703) and Open Foundation of Guizhou Provincial Key Laboratory of Public Big Data (No. 2018BDKFJJ014).

References

1. Pevný, T., Filler, T., Bas, P.: Using High-Dimensional Image Models to Perform Highly Undetectable Steganography. Springer, Berlin (2010)
2. Holub, V., Fridrich, J.: Designing steganographic distortion using directional filters. In: IEEE International Workshop on Information Forensics and Security (WIFS), vol. 2, pp. 234–239 (2012). https://doi.org/10.1109/WIFS.2012.6412655
3. Holub, V., Fridrich, J.: Digital image steganography using universal distortion. In: Proceedings of the First ACM Workshop on Information Hiding and Multimedia Security, Montpellier, 17–19 June 2013, pp. 59–68 (2013). https://doi.org/10.1145/2482513.2482514
4. Sedighi, V., Cogranne, R., Fridrich, J.: Content-adaptive steganography by minimizing statistical detectability. IEEE Trans. Inf. Forensics Secur. **11**, 221–234 (2016). https://doi.org/10.1109/TIFS.2015.2486744
5. Li, B., Wang, M., Li, X., Tan, S., Huang, J.: A strategy of clustering modification directions in spatial image steganography. IEEE Trans. Inf. Forensics Secur. **10**, 1905–1917 (2015). https://doi.org/10.1109/TIFS.2015.2434600
6. Fridrich, J., et al.: Breaking HUGO—the process discovery. In: Information Hiding-international Conference, vol. 6958, pp. 85–101 (2011). https://doi.org/10.1007/978-3-642-24178-9_7
7. Jan, K., Fridrich, J.: Steganalysis in high dimensions: fusing classifiers built on random subspaces. In: Proceedings of SPIE—The International Society for Optical Engineering, vol. 7880, pp. 181–197 (2011)
8. Fridrich, J., Kodovsky, J.: Rich Models for Steganalysis of Digital Images. IEEE Trans. Inf. Forensics Secur. **7**, 868–882 (2012)

9. Holub, V., Fridrich, J.: Random projections of residuals for digital image steganalysis. IEEE Trans. Inf. Forensics Secur. **8**, 1996–2006 (2013). https://doi.org/10.1109/TIFS.2013.2286682
10. Qian, Y., Dong, J., Wang, W., Tan, T.: Deep learning for steganalysis via convolutional neural networks. In: Proceedings of the SPIE, Media Watermarking, Security, and Forensics 2015, pp. 94 090 J–1–94 090 J–10 (2015)
11. Pibre, L., Jérôme, P., Ienco, D., Chaumont, M.: Deep learning is a good steganalysis tool when embedding key is reused for different images, even if there is a cover source-mismatch. In: Proceedings of the Media Watermarking, Security, and Forensics, Part of IS&T International Symposium on Electronic Imaging (EI 2016), February 2016
12. Xu, G., Wu, H.Z., Shi, Y.Q.: Structural design of convolutional neural networks for steganalysis. IEEE Signal Process. Lett. **23**(5), 708–712 (2016)
13. Ye, J., Ni, J., Yi, Y.: Deep learning hierarchical representations for image steganalysis. IEEE Trans. Inf. Forensics Secur. **12**(11), 2545–2557 (2017)
14. Zeng, J., Tan, S.: Large-scale JPEG steganalysis using hybrid deep-learning framework. IEEE Trans. Inf. Forensics Secur. **13**(5), 1200–1214 (2016)
15. Holub, V., Fridrich, J.: Low-complexity features for JPEG steganalysis using undecimated DCT. IEEE Trans. Inf. Forensics Secur. **10**(2), 219–228 (2015)
16. Holub, V., Fridrich, J.: Phase-aware projection model for steganalysis of JPEG images. In: Proceedings of the SPIE, Electronic Imaging, Media Watermarking, Security, and Forensics XVII, vol. 9409 (2015)
17. Bas, P., Filler, T., Pevný, T.: Break our steganographic system: the ins and outs of organizing boss. In: Information Hiding, pp. 59–70. Springer (2011)
18. Holub, V., Fridrich, J., Denemark, T.: Universal distortion function for steganography in an arbitrary domain. EURASIP J. Inf. Secur. **2014**(1), 1–13 (2014)
19. Zeiler, M.D.: ADADELTA: An adaptive learning rate method. https://arXiv:1212.5701 (2012)
20. Glorot, X., Bengio, Y.: Understanding the difficulty of training deep feedforward neural networks. In: Proceedings of the Aistats, vol. 9, pp. 249–256 (2016)

Cross-Domain Image Steganography Based on GANs

Yaojie Wang[1,2(✉)], Xiaoyuan Yang[1,2], and Jia Liu[1,2]

[1] Engineering University of PAP, Xi'an 710086, China
wangyaojie0313@163.com
[2] Key Laboratory of Network and Information Security of PAP, Xi'an 710086, China

Abstract. According to the embedding method of secret information, steganography can be divided into: cover modification, selection and synthesis. In view of the problem that the cover modification will leave the modification trace, the cover selection is difficult and the load is too low, this paper proposes a cross-domain image steganography scheme based on GANs, which combines with cover synthesis. In the case that the cover type is not given in advance, the proposed scheme is driven by secret message, selects material to build the encrypted carrier, and converts it into cross-domain image, which is mapped to the generative image space for transmission. This scheme is consistent with the idea of coverless information hiding and can effectively resist the detection of steganalysis algorithm. Experiments were carried out on the data set of CelebA, and the results verified the feasibility and security of the scheme.

Keywords: Information hiding · Cover synthesis · Cross-domain · Generative Adversarial Networks

1 Introduction

In Fridrich's groundbreaking work of modern steganography [1], steganographic channel is divided into three categories, cover selection, modification and synthesis. Cover modification is the most common method of traditional information hiding, but it is inevitable to leave some traces of modification on the cover, which makes it difficult to resist the detection based on statistical analysis algorithm. Cover selection method does not modify the cover image, thereby avoiding the threat of the existing steganalysis technology. This method cannot be applied to practical applications because of its low payload [2]. Compared with the former two methods, the cover synthesis method is more suitable. However, this method is only a theoretical conception, rather than a practical steganography, because it is difficult to obtain multiple natural samples [3].

Fortunately, a data-based sampling technique, generative adversarial networks (GANs) [4] have become a new research hot spot in artificial intelligence. The biggest advantage and feature of GANs is the ability to sample real space and generate samples driven by noise, which provides the possibility for cover synthesis. This paper tries to use the message as the driver, selects the object from the image library to synthesize the encrypted cover, and then transforms it into the cross-domain image for transmission.

C.-N. Yang et al. (Eds.): SICBS 2018, AISC 895, pp. 743–755, 2020.
https://doi.org/10.1007/978-3-030-16946-6_61

This method in this paper is consistent with the concept of coverless information hiding [5], which can effectively improve the security of communication.

The remainder of this letter is organized as follows: We detail generative steganography and the derivatives of GANs. Section 3 shows how to build cross-domain image steganography scheme by GANs. Experiment results are demonstrated in Sect. 4. Section 5 concludes this research and details our future work.

2 Related Work

2.1 Generative Steganography

With the support of different technologies, information hiding has made great progress. In particular, the cover synthesis method represented by generative steganography [6] is becoming more and more possible. Compared with the traditional information hiding methods, generative steganography does not embed secret information into the cover images, but directly uses secret information as the driver to "generate/acquire" the encrypted cover. Some methods of generative steganography have appeared in recent years.

Through early texture synthesis techniques, [7, 8] use texture samples and multiple color points generated by secret messages to construct encrypted texture images; [9] increases the embedding capacity by proportional to the size of the stego texture image; Qian et al. [10] proposed a robust steganography algorithm based on texture synthesis. However, these methods are different from the real world in terms of visual effect, while the purpose of information hiding is not to change the original visual effect.

Recently, adversarial training has also been applied to steganography. Volkhonskiy et al. [11] first propose a new model for generating image-like containers based on Deep Convolutional Generative Adversarial Networks (DCGAN [12]). This method can generate a more suitable cover based on the standard steganography algorithm, which is more secure. Similar to [11], Shi et al. [13] introduce WGAN [14] to increase convergence speed and achieve more stable image quality. Tang et al. [15] propose an automatic steganographic distortion learning framework based on GANs, which is composed of a steganographic generative subnetwork and a steganalytic discriminative subnetwork. However, most of these GAN-based steganographic schemes are still the cover modification, and they are not in the true sense of generative steganography.

Since the biggest advantage of GANs is the generation of samples, it is a bold assumption to use GANs to generate a encrypted cover directly from a secret message. If we successfully extract the message from the generated encrypted cover, generative steganography has achieved great success. Some researcher have made a preliminary attempt on this intuitive idea. Ke et al. [16] proposed generative steganography method called GSK in which the secret messages are generated by a generator rather than embedded into the cover, thus resulting in no modifications in the cover. [17] propose a method that using ACGANs [18] to classify the generated samples, and they make the class output information as the secret message.

For further study on generative steganography, we propose a novel method—cross-domain image steganography based on GANs. Its feasibility and safety have been verified through experiments, and our main contributions are as follows:

1. Compared with text-based methods [5] and image-text mapping methods [10], the cross-domain mapping method can effectively resist the detection of steganalysis algorithm while correctly representing secret information.
2. This method of generative steganography abides by Kerckhoffs' principle [19]. All processes can be made public except for training parameters and shared code dictionary. Ideally, without the code dictionary, the extraction of secret information is equivalent to brute force cracking.

2.2 Discovery GAN

As a new framework of generative model, Generative Adversarial Nets, proposed in 2014, is able to generate better synthetic images than previous generative models, which has become one of the most popular research areas. Its idea comes from the two-player zero-sum game in game theory, so the structure of GANs consists of a generator and a discriminator. The generator trics to generate pseudo-real data to cheat the discriminator, while the discriminator tries to distinguish authenticity from synthetic data. By competing with each other, the generator and discriminator are iteratively optimized. The loss function to be optimized is shown in Eq. 1, where $P_{data}(x)$ denotes the true data distribution and $P_z(z)$ denote the noise distribution.

$$\min_G \max_D \rightarrow L(D,G) = \mathrm{E}_{x\sim p_{data}(x)}[\log D(x)] + \mathrm{E}_{z\sim p_{noise}(x)}[\log(1 - D(G(z)))] \quad (1)$$

Recently, many researchers have proposed a large number of improved models [20–22]. In order to avoid the high cost of matching samples and solve the cross-domain relationship in the case of unpaired data, Taeksoo Kim et al. [23] proposed Discovery GAN (DiscoGAN), which better realizes the cross-domain conversion of image style. Its structure is shown in Fig. 1. Unlike pix2pix model [24], which must have constraints on paired data, DiscoGAN can use non-paired data for training and successfully establish cross-domain relationships.

To learn the relationship between different domains, DiscoGAN defines G_{AB} as a function for mapping from domain A to domain B, and similarly defines G_{BA}. Under unsupervised conditions, the rules of the function are uncertain and the mapping rules can be defined arbitrarily. Therefore, we need to add constraints on the loss function, that is, the rules must be one-to-one mapping, and the functions G_{AB} and G_{BA} are mutually reversible. Ideally, the loss function is as follows:

$$G_{BA}[G_{AB}(X_A)] = X_A \text{ and } G_{AB}[G_{BA}(X_B)] = X_B \quad (2)$$

In actual training, this constraint is difficult to achieve. Iterative optimization is used in actual training, and the distance $(G_{BA}[G_{AB}(X_A)];\ X_A)$ can be minimized using any metric function. Again, we also need to minimize the distance $(G_{AB}[G_{BA}(X_B)];\ X_B)$. That is, the generator loss function is expressed as:

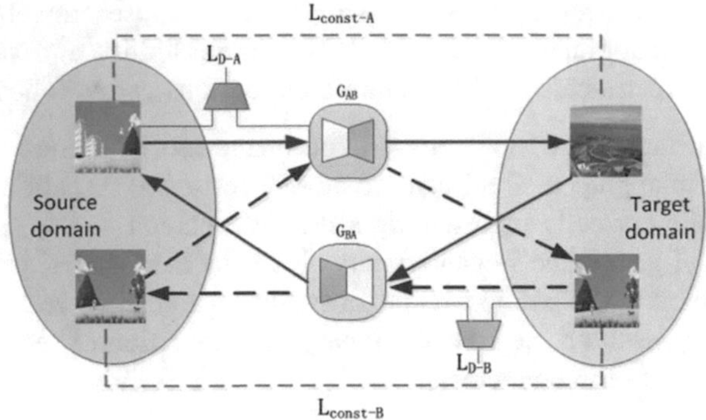

Fig. 1. The structure of DiscoGAN

$$\begin{aligned} &\min \rightarrow E_{X_A \sim P_A}[\log D_B(G_{AB}(X_A))] \\ &\min \rightarrow E_{X_B \sim P_B}[\log D_A(G_{BA}(X_B))] \end{aligned} \tag{3}$$

Both source and target domain objects can be reconstructed in the model, and the parameter of functions G_{AB} and G_{BA} are Shared. At the same time, the generated images G_{AB} and G_{BA} are used as input of the discriminator, and two reconstruction losses of $Lconst\text{-}_A$ and $Lconst\text{-}_B$ were added to solve the problem of schema collapse and one-to-one mapping.

$$\begin{aligned} L_G &= L_{G_{AB}} + L_{G_{BA}} \\ &= L_{G_{AB}} + L_{G_{BA}} + L_{CONST-A} + L_{CONST-B} \end{aligned} \tag{4}$$

$$L_D = L_{D_A} + L_{D_B} \tag{5}$$

The loss of generator is the sum of the loss of the GAN model and the reconstruction loss (Eq. 4). Similarly, the loss of the discriminator is the sum of the loss of the two discriminators (Eq. 5). If the discriminator satisfies max $\rightarrow L_D$ and the generator satisfies min $\rightarrow L_G$ in the training, we can achieve cross-domain images for transmission. This model can effectively establish cross-domain mapping relationship, and there are obvious differences between source domain and target domain, which creates conditions for the realization of generative steganography.

3 Cross-Domain Image Steganography

Considering that DiscoGAN's generator can establish the cross-domain mapping relationship of images, this paper proposes a cross-domain image steganography scheme based on DiscoGAN, which combines with cover synthesis, as shown in Fig. 2.

The scheme of this paper mainly consists of the following parts:

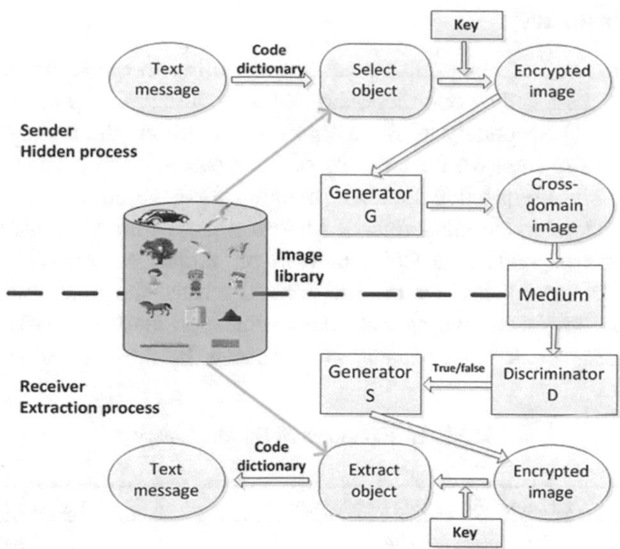

Fig. 2. The structure of Cross-domain image steganography

(1) Build code table dictionary

The purpose of building code dictionary is to transform the text information to the corresponding label images with different semantics, so that both the sender and receiver can use the same dictionary to realize the reversible transformation between label image and text information.

(2) The method of information hiding and extraction

Before communication, the sender and receiver agree in advance to train DiscoGAN with the same parameters and data sets to ensure that they get the same generator and discriminator. The above information is kept strictly confidential by both partners.

In the information hiding, the hidden text information is divided into pieces according to the code dictionary firstly. Then we select the label image from the dictionary and build a relatively reasonable encrypted image, whose state is pseudo-randomly adjusted according to the key. Finally we import the encrypted image into the trained DiscoGAN and generate a cross-domain image through the generator for transmission.

During the extraction, the received cross-domain images are input into the discriminator. According to the mapping relationship, we inversely output the constructed encrypted image. Based on the same key and code dictionary, we adjust the state of encrypted image. Then we decode it into corresponding text information, and realize the extraction of generative steganography.

3.1 Code Dictionary

Considering the need of real communication, according to the Chinese national standard code GB2312-80, the code dictionary should contain all common Chinese characters and national secondary fonts, as well as common phrases and non-Chinese character symbols to improve the capacity of information hiding. The network images based on cloud storage provide sufficient material for generative steganography. This paper selects characteristic label images to form image library. Each label image in image library corresponds to a Chinese word or phrase, as shown in Table 1. The dictionary can be established by the program to ensure the randomness of the dictionary. At the same time, we should change the dictionary regularly to reduce the frequency of using the same dictionary and increase the difficulty of decryption.

Table 1. Examples of the dictionary.

Chinese characters or phrases	Label image
retreat (撤)	
Fujian(福建)	
Rocket army(火箭军)	
Yun-20 large transport aircraft(运-20大型运输机)	
…	…

3.2 Information Hiding

The text information hiding and extraction is the key of information hiding algorithm. The main consideration in hiding is how to select the label image from the image library and build a relatively reasonable encrypted image, Meanwhile, we input them into DiscoGAN generator to generate cross-domain images for transmission. As shown in Fig. 3. The specific hiding scheme is as follows:

- Step 1: According to the principle of forward maximum matching, the text information T is divided into segments Ti based on code dictionary. In order to enable the extraction of the hidden text information correctly, the position of the label image in the encrypted image is determined by corresponding rules.
- Step 2: Under the condition that the position of each label image is unchanged, we can adjust their state according to the key, including color, size, etc. Then we can get a encrypted image with relatively reasonable content, which is denoted as K.
- Step 3: the encrypted image K is input into the pre-trained DiscoGAN, and generate a cross-domain image for transmission through a series of operations.

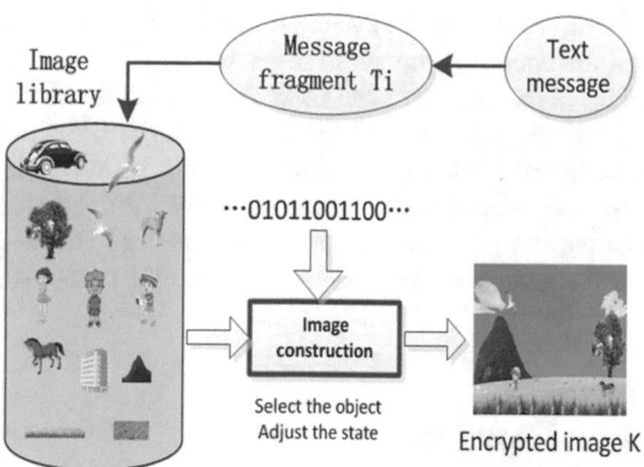

Fig. 3. The structure of the hiding method

3.3 Information Extraction

At the time of extraction, the receiver inputs the cross-domain image to the discriminator, which can only judge the authenticity of the image and cannot directly output the hidden text information. If the cross-domain image is determined to be true, then we input it to the generator. According to the mapping relationship, we can get the encrypted image of the output in reverse. Based on the same key and code dictionary, it is decoded into the corresponding text information to realize the extraction of hidden information. As shown in Fig. 4, the extraction method is as follows:

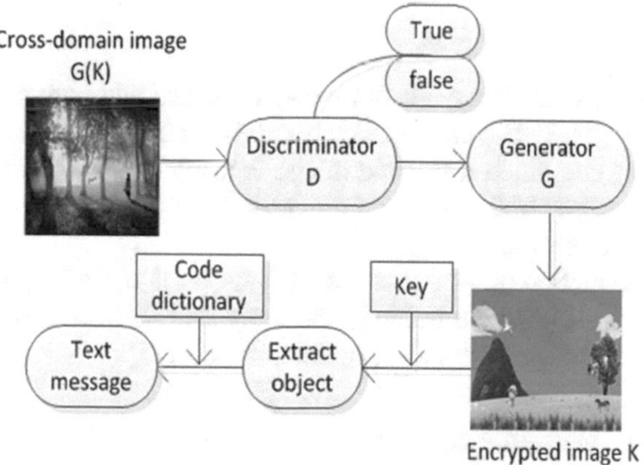

Fig. 4. The structure of extraction algorithm

- Step 1: After receiving the cross-domain image, the receiver will input it into the discriminator that has been trained in advance, and the discriminator outputs the authenticity of the cross-domain image.

- Step 2: If the discriminator determines that it is true, and inputs it into the generator, which generates the encrypted image in reverse.
- Step 3: According to the key, the state of each label image in the encrypted image can be restored with appropriate deviation, but it must still be recognized visually, so that the extraction of hidden information will not be affected.
- Step 4: Compare the code dictionary, we decode it into corresponding text information, and then the text information is sorted according to the corresponding rules, so as to recover the hidden secret information and realize the extraction of hidden information.

4 Experiment and Analysis

In order to verify the performance of the scheme, we used the encrypted image (source domain) to face image (target domain) to train the network model. In the experiment, the random noise was uniformly distributed on (−1, 1); The real sample data set is the CelebA data set (Ziwei Liu and Tang 2015), which contains 200,000 images. The secret information to be hidden is 10 articles randomly selected from the People's Daily official website. The experimental environment is shown in Table 2.

Table 2. Experimental environment

Software platform	Tensorflow v0.12	
Hardware environment	CPU	i7-8250U 3.2 GHz
	RAM	16 GB DDR3 1600 MHz
	GPU	NVIDIA 1080

In the experiments, the data set is first pre-cropped so that both the input and output images are 64 × 64 × 3. The clipped CelebA data set is taken as the target domain B; the pseudo-random combined encrypted image is taken as the source domain A, and is divided into two groups: 12,000 sheets are randomly selected as the training set T_A, and the rest are taken as the test set T_B.

The optimizer in DiscoGAN uses an Adam-based optimization method with a learning rate of 0.0002. At each training, the weight of the discriminator D is updated once, the weight of the generator G is updated twice [25].

4.1 Message Hiding and Extraction

Assuming that each encrypted image has 10 label images, and the hidden text information is randomly selected. According to the code dictionary, the label images are selected from the image library, and we construct a relatively reasonable encrypted image and input it into the pre-trained DiscoGAN generator to generate cross-domain images to realize generative steganography.

Every time we train DiscoGAN network 100 times in the experiment, we carry out a test that generates encrypted image by secret message K and extracts secret message T from encrypted image. The cross-domain image $G(K)$ generated by the encrypted image K as shown in Fig. 5.

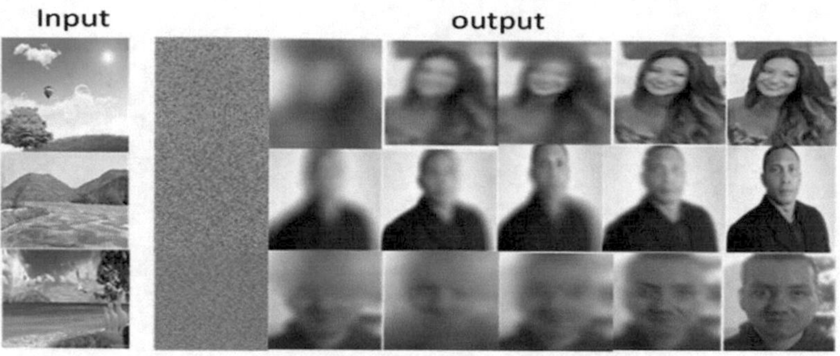

Fig. 5. Example of generated cross-domain images

The relationship between error rate of extracting information and the number of iterations is shown in Fig. 6. After 7,400 training sessions, the average error rate of extracting information was less than 20%. When the training times were more than 9000, the average error rate of information extraction was less than 14.3%. We can reduce the error rate by reducing the number of label images in the encrypted image or reserve error correction position.

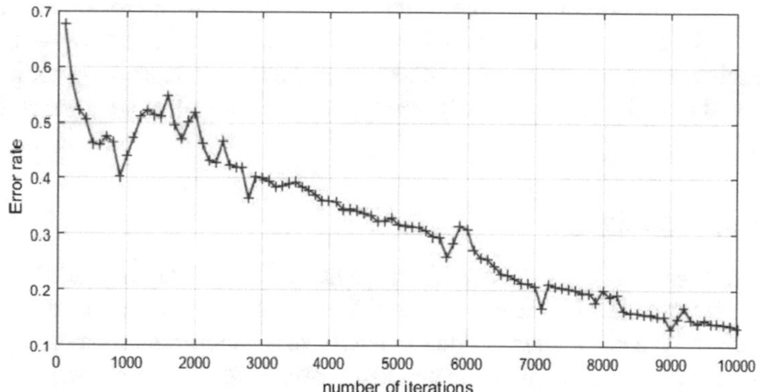

Fig. 6. The relationship between the error rate and the number of iterations

To further verify the feasibility of the scheme, we change the capacity of the hidden information, that is, we change the number of label images in the encrypted images. As shown in Fig. 7. The number of label images is 3, 6, 9 and 12, respectively. In the case of the same number of iterations, the fewer the label images in the encrypted images, the higher the accuracy of information extraction. When the number of label images is less than or equal to 6, after 8,000 training sessions, the extracted information error rate is less than 9%. We only need to add error correction code to ensure the correctness of the decoding. It can also be seen from this that the proposed scheme allows for errors in communication without affecting the correct communication.

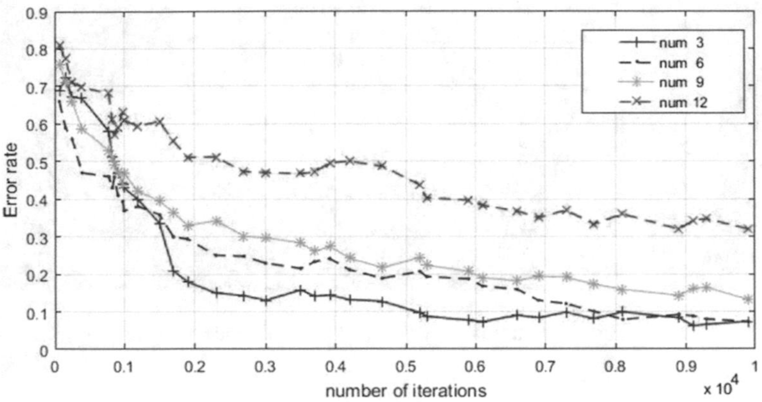

Fig. 7. The relationship between the error rate, the number of label images, and the number of iterations

It can be obtained through experiments that the sender can build a corresponding encrypted image according to the specified label image and convert it into a cross-domain image for transmission. The receiver can correctly extract the hidden label image from the cross-domain image, thus realizing generative steganography. When the number of label images does not exceed a certain threshold, the transmission error rate of the scheme can be effectively reduced, and we do not make any modifications to the cover, which can more effectively resist the detection of steganographic analysis algorithm.

4.2 Hiding Capacity

According to Sect. 3.1, each encrypted image has 10 label images. The corresponding hiding capacity is tested by changing the different word segmentation methods in the code dictionary. We conducted hidden capacity test experiments according to 3 different word segmentation methods. The experimental results are shown in Table 3.

1. We don't use the words segmentation in the dictionary but directly divide the text information into single Chinese character. In the hidden experiment, each word corresponds to a label image. That is, each encrypted image is composed of 10 label images, so the hidden capacity of each encrypted image is 10 words. Since no words segmentation dictionary is used, the number of phrases in the dictionary and the average word length are zero.

Table 3. Experimental results for the hiding capacity test

Average words length of dictionary (Chinese characters/words)	Words numbers of phrases dictionary	The capacity of literature [15] 's method (Chinese characters/image)	The capacity of our method(Chinese characters/encrypted-image)
0	0	1.00	10.00
2	100	1.57	17.42
3	100	1.86	30.11

2. Select words with an average length of 2 to establish a dictionary (100 phrases, average length is 2). This dictionary not only includes label images corresponding to the common words, but also the label image corresponding to the words of different lengths. According to the principle of forward maximum matching, if the secret information contains the phrase in the mapping dictionary, the words are divided into the phrase, otherwise it is divided into a minor phrase or word, and so on. Randomly select 10 text segments, the experimental results show that the average hidden capacity of each image is 17.42 words.
3. Similarly to method 2, a dictionary is established for words with an average length of 3 (100 phrases, average length is 3). We also adopted the principle of forward maximum matching. The experimental results show that the average hiding capacity of each image was 30.11 words.

The experimental results show that establishing a reasonable code dictionary and increasing the average length of the words in the dictionary can improve the information hiding capacity. Theoretically, the hidden capacity of a single label image is the number of Chinese characters in the dictionary. The average information hiding capacity of multiple images is the average length of secret information fragments after word segmentation:

$$\overline{C} = \frac{\sum_{i=1}^{n} C_i}{n} \tag{6}$$

Where:n is the number of secret information fragments, C_i is the length of the i-th secret information fragment.

It can be seen from Table 3 that the scheme of this paper has a large improvement in the hidden capacity. The reason is that each label image corresponds to one keyword, and the average length of the text information corresponding to the encrypted image containing multiple label images is greatly increased, making the capacity of hidden information. Each of the encrypted images in [15] corresponds to one high-frequency keyword, resulting in a relatively small amount of hidden information.

4.3 Security

The security of this paper is based on three aspects: Firstly, the cross-domain images transmitted in public channels are directly generated by the generator without any modification, which belong to cover synthesis, so it can effectively resist the detection of steganographic analysis algorithm. Secondly, based on the DiscoGAN, this scheme does not require any specific matching rules, that is, we can establish cross-domain relations and realize image cross-domain conversion, which sets up a great obstacle for attackers to decipher secret information. Thirdly, the code dictionary and key are kept secret, and the code dictionary is built randomly, which greatly increases the amount of calculation. The algorithm is easy to implement, but it is difficult to decipher in the case of changing the code dictionary periodically. Therefore, compared with the traditional encryption and information hiding methods, the proposed scheme can communicate more concealed and have higher security.

Assume that the attacker suspects that the cross-domain image contains secret information. Because he does not have the same DiscoGAN model as the communicating parties, it is difficult to extract the secret information from the encrypted images by the discriminator. Even if the attacker accidentally extracts the secret information, he cannot decode the secret information into the original text information because he doesn't know the code dictionary and key. Thereby we can ensure the security of the covert communication.

5 Conclusion and Future Work

This paper proposes a cross-domain image steganography scheme based GANs, which expands new ideas for generative steganography. By properly building encrypted images and establishing cross-domain relationships, the security of the scheme is guaranteed. And the scheme is consistent with the idea of coverless information hiding, and can effectively resist the detection of steganalysis algorithms. Theoretical analysis and experiments show that the scheme has a large hidden capacity and good security.

How to construct the code dictionary and reduce the error rate of information extraction rate is the key work for our next step. The future direction is to modify DiscoGAN to handle mixed modes, such as text and images.

Acknowledgment. This work was supported by National Key R&D Program of China (Grant No. 2017YFB0802000), National Natural Science Foundation of China (Grant Nos. 61379152, 61403417)

References

1. Fridrich, J., Kodovsky, J.: Rich models for steganalysis of digital images. IEEE Trans. Inf. Forensics Secur. **7**(3), 868–882 (2012)
2. Fridrich, J.: Steganography in Digital Media: Principles, Algorithms, and Applications. Cambridge University Press, Cambridge (2010)
3. Holub, V., Fridrich, J., Denemark, T.: Universal distortion function for steganography in an arbitrary domain. EURASIP J. Inf. Secur. **2014**(1), 1 (2014)
4. Goodfellow, I., Pouget-Abadie, J., Mirza, M., Xu, B., WardeFarley, D., Ozair, S., Courville, A., Bengio, Y.: Generative adversarial nets. In: Advances in Neural Information Processing Systems, pp. 2672–2680 (2014)

5. Zhou, Z.L., Cao, Y., Sun, X.M.: Coverless information hiding based on bag-of-words model of image. J. Appl. Sci. **34**(5), 527–536 (2016)
6. Liu, M., Liu, J., Zhang, M., Li, T.: Tian generative information hiding method based on generative adversarial networks. J. Appl. Sci. **36**(2), 27–36 (2018)
7. Otori, H., Kuriyama, S.: Data-embeddable texture synthesis. Smart graphics. In: Proceedings of the International Symposium, Sg 2007, Kyoto, Japan, June 25–27, 2007. DBLP, vol. 4569, pp. 146–157. (2007)
8. Otori, H., Kuriyama, S.: Texture synthesis for mobile data communications. IEEE Comput. Graphics Appl. **29**(6), 74–81 (2009)
9. Wu, K.C., Wang, C.M.: Steganography using reversible texture synthesis. IEEE Trans. Image Process. **24**(1), 130 (2015)
10. Qian, Z., Zhou, H., Zhang, W., Zhang, X.: Robust steganography using texture synthesis. In: Processing of the Advances in Intelligent Information Hiding and Multimedia Signal. Springer International Publishing (2017)
11. Volkhonskiy, D., Nazarov, I., Borisenko, B., Burnaev, E.: Steganographic generative adversarial networks (2017)
12. Radford, A., Metz, L., Chintala, S.: Unsupervised representation learning with deep convolutional generative adversarial networks. Comput. Sci.. (2015)
13. Shi, H., Dong, J., Wang, W., Qian, Y., Zhang, X.: Ssgan: secure steganography based on generative adversarial networks (2017)
14. Arjovsky, M., Bottou, L.: Towards Principled Methods for Training Generative Adversarial Networks [DB/OL]. [2017–2-09]. http://arxiv.org/abs/1701.04862
15. Tang, W., Tan, S., Li, B., Huang, J.: Automatic steganographic distortion learning using a generative adversarial network. IEEE Signal Process. Lett. **24**(10), 1547–1551 (2017)
16. Ke, Y., Zhang, M., Liu, J., Su, T., Yang, X.: Generative steganography with kerckhoffs' principle based on generative adversarial networks (2017)
17. Liu, M., Liu, J., Zhang, M., Li, T.: Tian generative information hiding method based on generative adversarial networks. J. Appl. Sci. **36**(2), 27–36 (2018)
18. Odena, A., Olah, C., Shlens, J.: Conditional image synthesis with auxiliary classifier gans (2016)
19. Petitcolas, F.A.P.: Kerckhoffs' Principle, 675–675 (2011)
20. Zhu, J.Y., Park, T., Isola, P., et al.: Unpaired image-to-image translation using cycle-consistent adversarial networks, 2242–2251 (2017)
21. Yi, Z., Zhang, H., Tan, P., et al.: DualGAN: unsupervised dual learning for image-to-image translation, 2868–2876 (2017)
22. Taigman, Y., Polyak, A., Wolf, L.: Unsupervised cross-domain image generation (2016)
23. Kim, T., Cha, M., Kim, H., et al.: Learning to discover cross-domain relations with generative adversarial networks. 한국지능정보시스템학회 학술대회논문집 (2017)
24. Isola, P., Zhu, J.Y., Zhou, T., et al.: Image-to-image translation with conditional adversarial networks, 5967–5976 (2016)
25. Dumoulin, V., Belghazi, I., Poole, B., et al.: Adversarially Learned Inference (2016)

Facial Expression Recognition by Transfer Learning for Small Datasets

Jianjun Li[2(✉)], Siming Huang[2], Xin Zhang[2], Xiaofeng Fu[2], Ching-Chun Chang[3], Zhuo Tang[1], and Zhenxing Luo[1]

[1] Science and Technology on Communication and Information Security Control Laboratory of the 36th Institute of China Electronics Technology Group Corporation, Jiaxing, China

[2] School of Computer Science and Engineering, Hangzhou Dianzi University, Hangzhou, China

{jianjun.li,zhangxin}@hdu.edu.cn

[3] Department of Computer Science, University of Warwick, Coventry, UK

c.chang.2@warwick.ac.uk

Abstract. As a re-identification of facial attributes, facial expression recognition remains a challenging problem and the small datasets further exacerbate the task. Most previous works realize facial expression by fine-tuning the network pre-trained on a related domain. Therefore they have limitations inevitably. In this paper, we propose an optimal Feature Transfer Learning (FTL) algorithm to model the high-level neurons in a unified way. The proposed FTL structure is based on two models by correcting marginal distribution, matching the distribution between domains and optimizing the entire network connection by a parameter sharing method. Evaluation experiments based on three most public datasets of facial expression recognition: CK+, Oulu-CASIA and MMI, show that the proposed method is comparable to or better than most of the state-of-the-art approaches in both recognition accuracy and model size. Furthermore, we also demonstrate that our approach obtains more accurate results than other methods, such as directly fine-tuning a deeper network, training a shallower network from scratch.

Keywords: Facial expression recognition · Transfer learning · Feature transfer

1 Introduction

Nowadays, Deep Convolutional Neural Networks (DCNNs) have made remarkable achievements in image processing, such as segmentation, classification and recognition. However, DCNNs have limitations when applied directly to small datasets, especially for facial expression recognition (FER). One of the main reasons is that small datasets are not well suited for a deep CNN network with a large number of parameters so that a DCNN cannot obtain discriminative

C.-N. Yang et al. (Eds.): SICBS 2018, AISC 895, pp. 756–770, 2020.
https://doi.org/10.1007/978-3-030-16946-6_62

features. Some commonly used methods, such as dropout [22] and small sample CNN [9,25], often do not directly produce satisfied results. However, more recently, transfer learning [1,17,19] has been proven to be an effective method for deep learning tasks related to small datasets by fine-tuning a network that has been pre-trained on a large dataset. This method has achieved great success benefiting from the generality of pre-learned features. Motivated by the observation, some large-scale datasets for face recognition have been used in recent several works [12,17,28] to pre-train CNN networks specially for facial expression recognition purpose and these trained networks are then fine-tuned on small-scale expression datasets.

Although the above strategy performs well, some issues are worth noting. (1) The custom-designed network for face recognition is too big to fit facial expression recognition. It is also strictly limited in the model architecture for sharing model parameters. For example, we cannot randomly remove the layers in the pre-trained network because it will change the receptive fields of the network and thus make the originally pre-turned parameter lose its meaning. (2) The domains used for pre-training and fine-tuning with different distributions interact with each other in a complex and fragile way on successive layers [29]. Therefore, the traditional transfer method leads to optimization difficulties in the splicing of parameters from different networks, which in turn causes performance to drop.

In this paper, we propose a deep learning algorithm for FER task with small datasets by feature transfer learning techniques. First, we fine-tune a pre-trained large face net to make it have a certain classification ability of facial expression. Then, we transfer features trained by the fine-tuned large model into features trained by the small model. During the feature transfer, we supervise the training of the small model using the pre-finetuned face net at the feature level and standardize the features to be transferred by ensuring they obey the standard state distribution. Thus we can ensure to accurately calculate the difference between two models in the same feature space and promote effective transmission of features [7]. Due to the hierarchical structure of the standard DCNN network, the deeper of the layer is, the more abstract of the features are. In previous work [27], the deep layers which are more heavily affected by the task are called task-specific layers. We explore the sharing of the parameters of task-specific layers because we have the same task as the face net after fine-tuning an expression dataset. These layers also have the characteristics that parameter learning is greatly influenced by the front layers and the learning rate is relatively high during back propagation. Therefore, we choose to transfer the layers that learn relatively slowly and have a greater impact on the later layers. Overall, our work provides three main contributions:

(1) Firstly, we propose a novel transfer learning framework for small dataset tasks, aiming to learn more discriminative features with small datasets and shallow networks. Specifically, we utilize a larger network pre-trained on a related domain to guide the training of small models at the feature level.

This makes the network framework more flexible because layers do not have to strictly share parameters from pre-trained networks.

(2) Secondly, inspired by the domain adaption algorithm, we, for the first time, introduce the domain transfer to more complex recognition tasks. In particular, we propose an optimal solution for correcting marginal distribution by minimizing data distribution differences from different networks.

(3) Finally, we share the weights of task-specific layers between the networks with the same task, which greatly ameliorates optimization difficulties caused by splicing parameters of different domains. We find this greatly helps reduce the number of training iterations and thus speed up convergence.

Experiments are conducted on three expression recognition datasets: CK+ [15], Oulu-CASIA [32] and MMI [26], demonstrating the superiority of our method over previous methods. The proposed approach can also be applied to other small datasets for classification tasks.

The remainder of this paper is organized as follows. Related work is briefly introduced in Sect. 2. Section 3 presents our algorithm. Experiments and computational analysis are discussed in Sect. 4. Conclusion is given in Sect. 5.

2 Related Work

Transfer learning [18] aims to train a model in a training dataset and then apply it to a related but different test dataset. Transfer learning techniques have been widely applied in computer vision, especially in the applications with no labels or few labeled datasets. As for the expression recognition task, many prior works [2,16,17] directly utilize or involve the transfer learning method. They usually pre-train a deep network on the source dataset as a feature extractor for the target dataset and then train the classifier on labeled data. Methods [16,17] extend the target dataset by mixing multiple domain related datasets or using data enhancements to fine-tune more parameters of the pre-trained networks. All of these methods belong to shared parametric transfer learning.

Recent research works have focused on the deep feature transfer learning instead of the previous transfer learning method in which only parameters are transferred. Until now, there are no enough and suitable research works available in this area. In the feature transfer learning, network features are transferred from a labeled source dataset into a target domain where there are either no or just few labeled data available. Therefore, the main technical problem with the method is how to match the data distribution in the two domains effectively. Studies [29,30] have found that the learned parameters become increasingly dependent on learned specific tasks as layers of the network go deeper. Inspired by the observation, the Domain Adaptive Neural Network (DaNN) method [3] proposed a simple network structure with feature extraction layers and classifier layers, and added a maximum mean discrepancies (MMD) adaptation layer [5] into the classifier layers to compute the norm of the mean difference between two domains. Deep Domain Confusion (DDC) [24] and Deep Adaptation Network (DAN) [14] extend this method into a deeper network and a multi-kernel MMD multi-layer adaptation respectively. As for expression recognition tasks, we find the study [2,33] have been inspired by the idea of domain transfer learning.

Inspired by the deep feature transfer learning method, we explore effective feature distributions and regression loss between the transfer features. A work related to ours is Facenet2expnet [2], which proposed a two-stage training algorithm. The convolution layers of the network are trained by a pre-trained face net in the first stage while the fully-connected (FC) layers are randomly initialized in the second stage and then the whole network is jointly trained with the label information. This study argues that FC layers are generally to be used in capturing domain-specific semantics and the face net is only used to guide the learning of the convolutional layers. However, due to the change of parameters, the covariate shift [7] of data distribution between the two networks is larger during supervised training. It is difficult to achieve satisfactory results using the proposed distribution function. Also, since the study uses the pre-trained face net to supervise the training of the expression net, the feature information learned does not completely match the task of expression classification and still contains information useful for the face task. Despite fine-tuning the whole network in the second phase, it is difficult to change the fact because fine-tuning with few iterations will be biased toward task-specific layers (i.e, FC layers) with a higher learning rate.

3 The Proposed Approach

In this section, we present a general framework for small dataset recognition tasks. Input images are given as I=$\{X, Y\} = \{x_i, y_i\}_{i=1}^{n}$, with n labeled samples. The proposed network structure is described as follows:

$$O_s = h_{\theta_{s,c}}(f_{\theta_{s,m}}(I)) \tag{1}$$

$$O_t = h_{\theta_{t,c}}(g_{\theta_{t,m}}(I)) \tag{2}$$

where $\theta_{s,c}$, $\theta_{t,c}$ are the parameters of classifier layers h and the parameters $\theta_{s,m}$, $\theta_{t,m}$ correspond to the feature layers f, g of FaceNet and TransferNet, respectively. Our goal is to learn the parameters of feature layers $\theta_{t,m}$ and classifier layers $\theta_{t,c}$ in that they can directly classify expression images into K categories in the expression net.

3.1 Feature Transfer Learning

In the feature transfer learning phase, the main goal is to guide the transfer learning of feature $g_{\theta_{t,m}}(I)$ by minimizing the distance between the feature distributions of the source $f_{\theta_{s,m}}(I)$ and the target $g_{\theta_{t,m}}(I)$. Firstly, we normalize the feature mappings, $f_{\theta_{s,m}}(I)$, based on the mini-batch so that they follow the same data distribution: standard normal distribution:

$$\hat{x}_i \leftarrow \frac{x_i - \mu}{\sqrt{\sigma^2 + \epsilon}} \tag{3}$$

where x_i is the *i-th* element of mini-batch $B = \{x_1, x_2, \dots, x_m\}$, μ is the mean of the mini-batch, σ is a constant for numerical stability and $\hat{x}_i$ is the normalized value. Secondly, we denote the feature mapping of the source and the target of the *l-th* layer as $\hat{x}_{i,s} = f_{\theta_{s,m}}(I)$ and $\hat{x}_{i,t} = g_{\theta_{t,m}}(I)$ respectively. More specifically, we denote the feature output of the *c-ch* channel of the *l-th* layer as $\hat{x}_i^c$.

$$\hat{x}_{i,ave}^c = \{\frac{1}{k*k}\sum_{w=0}^{k-1}\sum_{h=0}^{k-1}\hat{x}_{i,w,h}^c\}_1^{\frac{W+H}{k}} \tag{4}$$

where $\hat{x}_{i,ave}^c$ is the local average response value on the spatial dimension. W, H is the width and height of the feature map and k is the kernel size. Finally, to minimize the distance between the two functions, the loss function L_d is derived as follows:

$$L_d(X, \theta_{s,m}, \theta_{t,m}) = \frac{1}{2}\|f_{\theta_{s,m}}(X), g_{\theta_{s,m}}(X)\|_2^2 \tag{5}$$

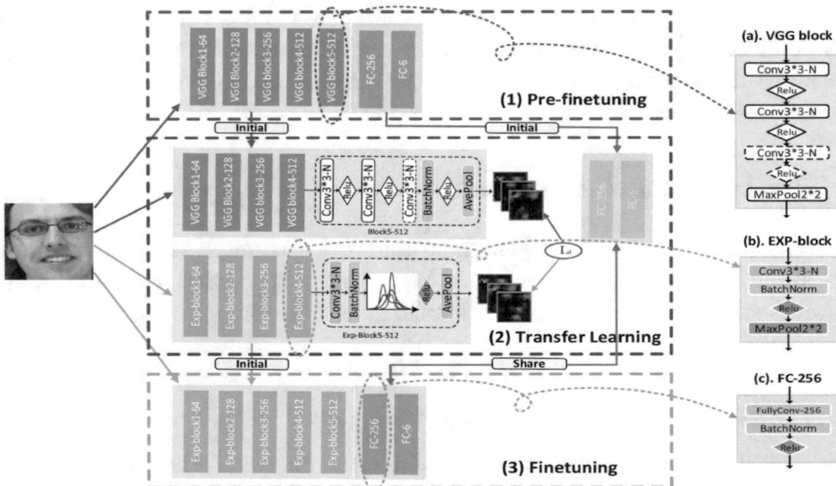

Fig. 1. The proposed CNN network structure. The blue and green part correspond to feature layers of faceNet-V2 and FTL-ExpNet, and the orange part correspond to the classifier layers. The blocks details are shown in the corresponding color part on the right image **(a)**, **(b)** and **(c)**. In pre-finetuning phase, we fine-tune faceNet-V2 on the expression images. In transfer learning phase, the faceNet-V2 is kept unchanged and it only provides the supervision for the FTL-ExpNet based on the same feature distribution. In the following fine-tuning phase, we fine-tune the whole FTL-ExpNet with the shared parameters of classifier layers.

3.2 Training Procedure

The complete proposed structure is illustrated in Fig. 1. In the pre-finetuning phase (shown as the blue box with a dotted line). We preserve and initialize the convolutional layers from faceNet, which is based on VGG16 and trained on the datasets of [19]. The fully connected layers are replaced with FC-512 and FC-6 and initialized from Xavier distributions [4]. Additionally, we add a batch normalization layer in Eq. 4 between the last conventional layer and the last ave-pooling layer in order to unify the data distribution for the next phase. The modified VGG structure (denoted as faceNet-V2) is shown in the blue part of Fig. 1 and the details of the VGG block is shown as the right-hand image **(a)**. We fine-tune the whole faceNet-V2 using a standard supervised loss, cross-entropy loss, as the follows:

$$L_c(X, O_s) = \sum_{j=1}^{k} y_i \log(\hat{y}_i) \tag{6}$$

where y_i and $\hat{y}_i$ are the i-th value of the ground truth label and the i-th output value of softmax separately. In the feature transfer learning phase (shown as the orange box with a dotted line), we freeze the parameters $\theta_{s,m}$ and $\theta_{s,c}$ of facenet-V2. The deep feature of the last average pooling layer is used to provide the supervision of the conventional layers of the feature transfer learning expression net (denoted as FTL-ExpNet). We train the network by minimizing the loss function in Eq. 5 to make the distribution $g_{\theta_{t,m}}(I)$ approach $f_{\theta_{t,m}}(I)$. Our FTL-ExpNet consists of two components (shown as the green box with a dotted line): five conventional units $\theta_{t,m}$ and classifier layers $\theta_{t,m}$. These are shown as the green and orange parts, respectively, in Fig. 1. $\theta_{t,m}$ is available from the feature transfer learning phase. We consider the FTL-ExpNet to have the same expression classification task as the pre-finetuning phase, and the fine-tuned classifier layer (shown as the orange block) $\theta_{s,c}$ is directly applied to the FTL-ExpNet, $\theta_{t,c} = \theta_{s,c}$. Among these conventional Exp-blocks, the first contains two convolutional layers for getting larger receptive fields from the input images, as in VGG16. Also, in order to make the training smoother and obtain the same feature distribution as in the last phase, each Exp-block is followed by a batch normalization layer. For feature transformation calculations, we then connect a non-linear activation function and a max-pooling layer into the first four Exp-blocks, as shown in the right-hand image **(b)** in Fig. 1, and a ReLU and an ave-pooling layer are connected into the last one. The kernel size of all the conventional layers is 3*3. For the pooling layer, its size is 2*2 with stride of 2. The numbers of the output channels are 64, 64, 128, 256, 512, and 512 separately. To improve accuracy, we fine-tune the FTL-ExpNet with the cross-entropy loss in Eq. 6 on the expression dataset, which is the same input as in the previous two phases.

4 Experiments and Analysis

4.1 Implementations

In this section, experiments and detailed comparisons are made between our approach and the state-of-the-art algorithms. To assess the performance of our method, we validate its effectiveness on three widely used FER datasets: CK+ [15], Oulu-CASIA [32] and MMI [26]. All these three datasets have six expressions including anger(An), disgust(Di), fear(Fe), happiness(Ha), sad(Sa) and surprise(Su). In order to compare fairly, the face detector of Dlib-library is used for face detection and the detected faces are accordingly normalized, cropped and resized into 256*256 pixels. The conventional data augmentation method is also used by randomly sampling and horizontal flipping. The batch size is 32 and the momentum is fixed at 0.9. First, we fine-tune faceNet-V2 to make the model have a certain expression classification ability. The learning rate is initalized ass 1e-4 and decreases by 0.1 after 1000 iterations. The total number of training iterations is 2000. Secondly, in the feature transfer learning stage, we use a lower learning rate, 1e-6, because of the large regression loss L_d and it also decreases by 0.1 after 5000 iterations. We train the faceNet-V2 net for 15000 iterations in total. Finally, in order to obtain a higher accuracy, we fine-tune the whole FTL-ExpNet and initialize all parameters by $\theta_{t,m}$ and $\theta_{s,c}$. The learning rate is 1e-4 and it decreases by 0.1 after 1000 iterations (2000 iterations in total). We adopt stochastic gradient descent (SGD) in all training phases and all the experiments are conducted under the deep learning framework Caffe [8]. The facial expression recognition results of our model are shown as Fig. 2. The softmax value of the samples indicates the classification accuracy whether the sample is assigned to the corresponding class. By contrast, it is obvious that some samples have higher softmax values, such as the samples (green boxes) from CK+ dataset and the first one from MMI dataset, comparing to the last two samples from MMI dataset as shown in Fig. 2(b). By observation, we find that the face with beards, glasses and other obstructions can confuse the facial expression and thus reduce the recognition accuracy.

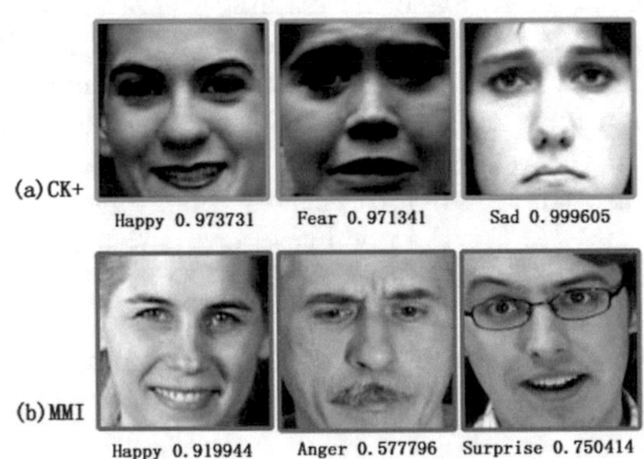

Fig. 2. Facial expression recognition results vs. Recognition accuracies. (a) Images of CK+ dataset: "Happy", "Fear" and "Sad". (b) Images of MMI dataset: "Happy", "Anger" and "Surprise".

4.2 Comparison with Different Models

In this section, a fully comparison has been made between our approach with state-of-the-art methods on three public datasets: CK+, Oulu-CASIA and MMI respectively. Also, the performance analysis has been addressed in Sect. 4.2.

Table 1. The average accuracy in the CK+ dataset.

Method	Average accuracy
BDBN [13]	96.7%
Inception [16]	93.2%
PPDN [33]	97.3%
FN2EN [2]	98.6%
STM-ExpLet [12]	94.2%
DTAGN [9]	97.3%
MSCNN [23]	95.54%
PHRNN-MSCNN [23]	98.50%
DFSN [31]	98.10%
DFSN-I [31]	98.73%
FaceNet Fine-Tune(baseline)	95.5%
Train From Scratch(FTL-ExpNet)	**97.8%**
FTL-ExpNet	**99.3%**

We compare the average accuracy of the proposed approach with both deep learning-based and traditional methods. The fine-tuned VGG-16 FaceNet is considered as the baseline. In order to prove the superiority of the proposed method, the results trained from scratch with our FTL-ExpNet are also included. As shown in Tables 1 and 2, we separate the results into three categories with horizontal lines. The first block shows the experimental results based on static images, which are obtained by using exactly the same data and protocols with ours. The second block shows the state-of-the-art results based on image-sequence. Among these methods, MSCNN and DFSN are single models based on deep learning, while DTAGN, PHRNN-MSCNN and DFSN-I are composite models which combine the temporal features of face key-points. The third block lists several methods proposed in this paper.

CK+ Dataset. CK+ is a representative dataset of FER and there are 327 image sequences with 7 emotion labels including Anger (An), Contempt (Co), Disgust (Di), Fear (Fe), Happiness (Ha), Sadness (Sa) and Surprise (Su). In the experiment, we only use sequences with six of the base emotion labels (excluding contempt) with a total of 309 expression sequences. Like in other works [2,12], we select the top three peak expressions in the last three frames of each expression sequence as expression recognition samples. Thus, there are a total of 927 (309*3) original image samples. All these images are divided into 10 groups by subject ID in an ascending order. Nine groups of them are used for training and the remaining one is for validation. In Table 1, the methods are listed on the left hand and the recognition accuracy of their are listed on the right hand accordingly. The performance of the proposed method significantly outperforms all others including image-sequence based methods, and achieves the accuracy of 99.3% vs. the previous best of 98.6% for six emotions. Figure 3 illustrates the confusion matrix for CK+ dataset. Our proposed method performs well in recognizing expressions of fear, happiness and surprise, but those with pronounced changes in the shape of the mouth while anger are more likely to be confused with disgust.

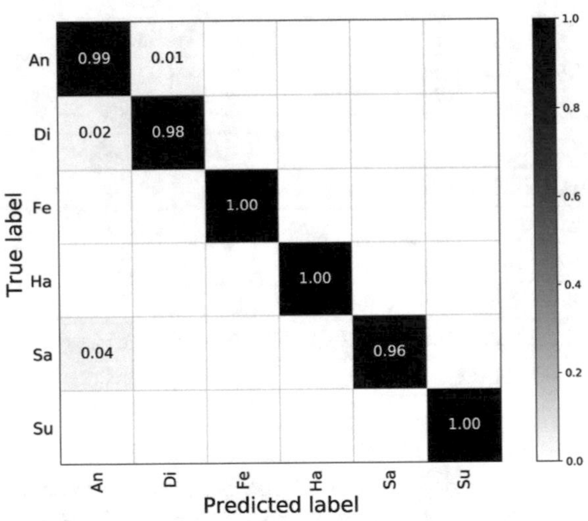

Fig. 3. Confusion matrix of the feature transfer learning method for CK+ dataset.

Oulu-CASIA Dataset. For the further experiment, the Oulu-CASIA dataset is originally composed of 480 image sequences that are taken under different conditions: dark, strong, and weak illumination. Like the CK+ dataset, each sequence begins with a neutral facial expression and ends with the facial expression. In our experiment, only videos with a strong illumination condition are chosen, i.e., 80 subjects with 6 expressions in total. Only the last three apparent emotion frames are used so the total number of expression images is 1440 (480*3). The face images are resized into 256*256 pixels by bilinear interpolation and 10-fold of cross-validations are conducted as for the previous CK+ dataset.

The comparison is illustrated in Table 2 based on the Oulu-CASIA dateset. Although our recognition accuracy is in the runner-up, our baseline is also lower than FN2EN's by almost 2%, i.e. 81.30% vs 83.26%. The accuracy improvement of our algorithm against the baseline is still slightly higher than FN2EN's. And, our experimental results are comparable to most advanced spatio-temporal models (second block). Furthermore, compared to training a model directly from scratch with the same structure, our method is significantly improved with a gain of 11.3%. The confusion matrix of the proposed algorithm is shown as Fig. 4: happiness and sadness are confused in the Oulu-CASIA dataset while anger and disgust are confused in the CK+ dataset.

Table 2. The average accuracy in the Oulu-CASIA dataset.

Method	Average accuracy
PPDN [33]	84.59%
FN2EN [2] (baseline)	87.71% (83.26%)
HOG 3D [10]	70.63%
Atlases [6]	75.52%
STM-ExpLet [12]	74.59%
DTAGN [9]	81.46%
LOMo [20]	82.10%
MSCNN [23]	77.67%
PHRNN-MSCNN [23]	86.25%
DFSN [31]	86.88%
DFSN-I [31]	87.50%
Train From Scratch(FTL-ExpNet)	**74.60%**
FTL-ExpNet (baseline)	**85.90% (81.30%)**

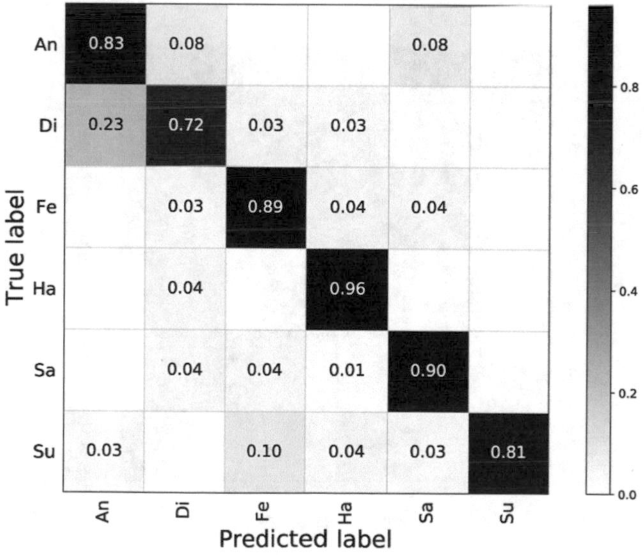

Fig. 4. Confusion matrix of the feature transfer learning method for the Oulu-CASIA dataset.

MMI Dataset. The last experiment is conducted on the MMI dataset, which is composed of 30 subjects of both sexes and ages from 19 to 62. Also, there are 213 sequences labeled with 6 basic expressions, of which 205 sequences captured in a frontal view. Each sequence contains the whole temporal activation patterns (onset → apex → offset) of a single facial expression type. In our experiments, we choose the best 4–6 facial expressions as a sample of facial expressions in each sequence and thus there are a total of 1124 data samples. The chosen images are transformed from color space to gray space to reduce the influence of noise. The experiment is also conducted in 10-fold of cross-validation as the above. Compared to CK+ dataset, MMI is thought to be more challenging in the expressions which are non uniform and wear accessories (e.g., glasses, moustache). Based on the distribution characteristics of expressions in the dataset (onset → apex → offset), MMI is often used for sequence-based expression recognition. Therefore, we mainly compare with sequence-based expression recognition methods including traditional methods and deep learning methods shown in Table 3. We can see that many models based on image-sequence have better precision than static images, so we get the conclusion that the pose, beard and occlusion of the subjects have a great influence on the static image recognition result. However sequence-based features can alleviate this effect. From the confusion matrix as shown in Fig. 5, we find that anger is easily confused with disgust and sadness, while fear, happiness and surprise are also easily confused. We inspect the expressions in the dataset and find that the former three expressions (anger, disgust, sadness) and the latter three expressions (fear, happiness, surprise) have similar appearances in both mouth and eye respectively.

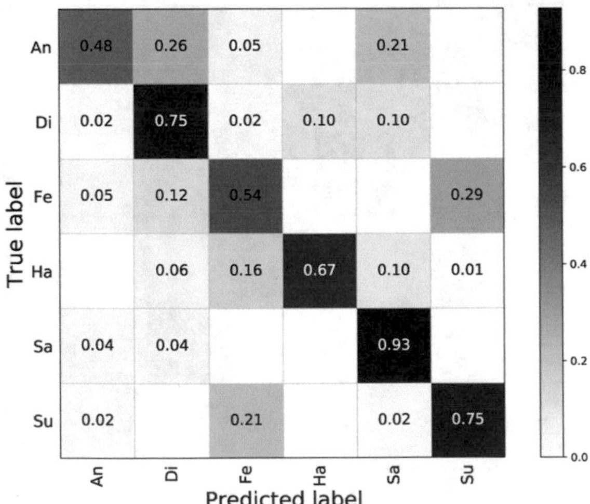

Fig. 5. Confusion matrix of the feature transfer learning method for the MMI dataset.

Table 3. The average accuracy in the MMI dataset.

Method	Average accuracy
HOG 3D [10]	60.89%
3DCNN-DAP [11]	63.40%
3D SIFT [21]	64.39%
DTAGN [9]	70.24%
STM-ExpLet [12]	75.12%
MSCNN [23]	77.05%
PHRNN-MSCNN [23]	81.18%
Train From Scratch(FTL-ExpNet)	**64.50%**
FaceNet Fine-Tune (baseline)	68.20%
FTL-ExpNet	**70.75%**

Comparison on Three Datasets. Analysis has been made based on Tables 1, 2 and 3. We find that the fine-tuned faceNet-V2 is a fairly stable network and suitable for being transferred into facial expression recognition. Also, the proposed method achieves improvements in recognition accuracy compared to other methods that has been verified in three different datasets. Moreover, from Fig. 6, we find that the proposed method has faster convergence rate and more early start point of a high accuracy rate. It also shows that the proposed approach has better capability of deep learning in that the better initial parameters of $\theta_{t,m}$ and $\theta_{t,c}$ can be easily learned in the first two phases. It further proves that the shared parameter of $\theta_{s,c}$ can provide better initial parameters for our network.

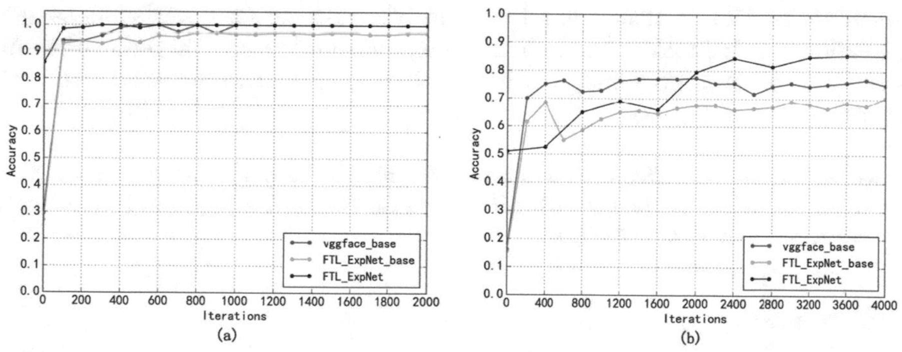

Fig. 6. Comparisons of the validation accuracy for different methods on CK+ **(a)** and Oulu-CASIA dataset **(b)**.

4.3 Comparison on Model Size

The model size is an importanct fact for training and implementation. In Table 4, we compare our model's efficiency with representative works on facial expression recognition. Model PPDN [33] is based on GoogleNet and has model size of 50 M. Iception structure [16] has a model size of 25 M with half parameters of GoogleNet. FaceNet with fine-tuning is based on VGG16 and its model size is around 139 M. By contrast, the proposed network and FN2EN [2] have fairly small-scale parameters and model size. Therefore, the proposed approach gets lightweight network structure and has higher efficiency for facial expression recognition.

Table 4. Comparision of model size.

Method	Parameters
PPDN [33]	50 M
Iception [16]	25 M
FN2EN [2]	11 M
FaceNet Fine-Tune	139 M
FTL-ExpNet	**10 M**

5 Conclusion

In this paper, we have proposed an optimal FTL approach for facial expression recognition of static images based on small datasets. FTL aims to correct marginal distribution, match the distribution between domains and use a parameter sharing method to optimize the entire network connections. An important advantage of FTL is that it is robust and also suitable for other small-sample migration learning tasks. Comprehensive experimental results show that the proposed method is comparable to or better than state-of-the-art methods in both accuracy and model size.

Acknowledgments. This work was supported by the National Natural Science Fund of China (No. 61871170. and No. 61672199) and the National Equipment Development Pre-research Fund: 6140137050202.

References

1. Bodla, N., Zheng, J., Xu, H., Chen, J.-C., Castillo, C., Chellappa, R.: Deep heterogeneous feature fusion for template-based face recognition. In: IEEE Winter Conference on Applications of Computer Vision, pp. 586–595 (2017)
2. Ding, H., Zhou, S.K., Chellappa, R.: Facenet2expnet: regularizing a deep face recognition net for expression recognition. In: Proceedings of the IEEE International Conference on Automatic Face Gesture Recognition, pp. 118–126 (2017)

3. Ghifary, M., Kleijn, W.B., Zhang, M.: Domain adaptive neural networks for object recognition. In: Pacific Rim International Conference on Artificial Intelligence, pp. 898–904. Springer (2014)
4. Glorot, X., Bengio, Y.: Understanding the difficulty of training deep feedforward neural networks. In: Proceedings of the Thirteenth International Conference on Artificial Intelligence and Statistics, pp. 249–256 (2010)
5. Gretton, A., Borgwardt, K.M., Rasch, M.J., Schölkopf, B., Smola, A.: A kernel two-sample test. J. Mach. Learn. Res. **13**, 723–773 (2012)
6. Guo, Y., Zhao, G., Pietikäinen, M.: Dynamic facial expression recognition using longitudinal facial expression atlases. In: European Conference on Computer Vision, pp. 631–644. Springer (2012)
7. Ioffe, S., Szegedy, C.: Batch normalization: accelerating deep network training by reducing internal covariate shift. arXiv preprint arXiv:1502.03167 (2015)
8. Jia, Y., Shelhamer, E., Donahue, J., Karayev, S., Long, J., Girshick, R., Guadarrama, S., Darrell, T.: Caffe: convolutional architecture for fast feature embedding. In: Proceedings of the 22nd ACM international conference on Multimedia, pp. 675–678 (2014)
9. Jung, H., Lee, S., Yim, J., Park, S., Kim, J.: Joint fine-tuning in deep neural networks for facial expression recognition. In: Proceedings of the IEEE International Conference on Computer Vision, pp. 2983–2991 (2015)
10. Klaser, A., Marszałek, M., Schmid, C.: A spatio-temporal descriptor based on 3d-gradients. In: British Machine Vision Conference, pp. 1–10 (2008)
11. Liu, M., Li, S., Shan, S., Wang, R., Chen, X.: Deeply learning deformable facial action parts model for dynamic expression analysis. In: Asian Conference on Computer Vision, pp. 143–157. Springer (2014)
12. Liu, M., Shan, S., Wang, R., Chen, X.: Learning expression lets on spatio-temporal manifold for dynamic facial expression recognition. In: Proceedings of the IEEE Conference on Computer Vision and Pattern Recognition, pp. 1749–1756 (2014)
13. Liu, P., Han, S., Meng, Z., Tong, Y.: Facial expression recognition via a boosted deep belief network. In: Proceedings of the IEEE Conference on Computer Vision and Pattern Recognition, pp. 1805–1812 (2014)
14. Long, M., Cao, Y., Wang, J., Jordan, M.I.: Learning transferable features with deep adaptation networks. arXiv preprint arXiv:1502.02791 (2015)
15. Lucey, P., Cohn, J.F., Kanade, T., Saragih, J., Ambadar, Z., Matthews, I.: The extended cohn-kanade dataset (ck+): a complete dataset for action unit and emotion-specified expression. In: IEEE Computer Society Conference on Computer Vision and Pattern Recognition Workshops, pp. 94–101 (2010)
16. Mollahosseini, A., Chan, D., Mahoor, M.H.: Going deeper in facial expression recognition using deep neural networks. In: IEEE Winter Conference on Applications of Computer Vision, pp. 1–10 (2016)
17. Ng, H.-W., Nguyen, V.D., Vonikakis, V., Winkler, S.: Deep learning for emotion recognition on small datasets using transfer learning. In: Proceedings of the 2015 ACM on International Conference on Multimodal Interaction, pp. 443–449 (2015)
18. Pan, S.J., Yang, Q.: A survey on transfer learning. IEEE Trans. Knowl. Data Eng. **22**(10), 1345–1359 (2010)
19. Parkhi, O.M., Vedaldi, A., Zisserman, A., et al.: Deep face recognition. In: British Machine Vision Conference, p. 6 (2015)
20. Sikka, K., Sharma, G., Bartlett, M.: Lomo: latent ordinal model for facial analysis in videos. In: Computer Vision and Pattern Recognition, pp. 5580–5589 (2016)

21. Sikka, K., Wu, T., Susskind, J., Bartlett, M.: Exploring bag of words architectures in the facial expression domain. In: European Conference on Computer Vision, pp. 250–259. Springer (2012)
22. Srivastava, N., Hinton, G., Krizhevsky, A., Sutskever, I., Salakhutdinov, R.: Dropout: a simple way to prevent neural networks from overfitting. J. Mach. Learn. Res. **15**(1), 1929–1958 (2014)
23. Tang, Y., Zhang, X.M., Wang, H.: Geometric-convolutional feature fusion based on learning propagation for facial expression recognition. IEEE Access **6**, 42532–42540 (2018)
24. Tzeng, E., Hoffman, J., Zhang, N., Saenko, K., Darrell, T.: Deep domain confusion: maximizing for domain invariance. arXiv preprint arXiv:1412.3474 (2014)
25. Urban, G., Geras, K.J., Kahou, S.E., Aslan, O., Wang, S., Caruana, R., Mohamed, A., Philipose, M., Richardson, M.: Do deep convolutional nets really need to be deep and convolutional? arXiv preprint arXiv:1603.05691 (2016)
26. Valstar, M., Pantic, M.: Induced disgust, happiness and surprise: an addition to the MMI facial expression database. In: Proceedings of the 3rd International Workshop on EMOTION (satellite of LREC): Corpora for Research on Emotion and Affect, p. 65 (2010)
27. Vittayakorn, S., Umeda, T., Murasaki, K., Sudo, K., Okatani, T., Yamaguchi, K.: Automatic attribute discovery with neural activations. In: European Conference on Computer Vision, pp. 252–268. Springer (2016)
28. Yan, C., Xie, H., Yang, D., Yin, J., Zhang, Y., Dai, Q.: Supervised hash coding with deep neural network for environment perception of intelligent vehicles. IEEE Trans. Intell. Transp. Syst. **19**(1), 284–295 (2018)
29. Yosinski, J., Clune, J., Bengio, Y., Lipson, H.: How transferable are features in deep neural networks? In: Advances in Neural Information Processing Systems, pp. 3320–3328 (2014)
30. Yosinski, J., Clune, J., Nguyen, A., Fuchs, T., Lipson, H.: Understanding neural networks through deep visualization. arXiv preprint arXiv:1506.06579 (2015)
31. Zhang, K., Huang, Y., Du, Y., Wang, L.: Facial expression recognition based on deep evolutional spatial-temporal networks. IEEE Trans. Image Process. **PP**(99), 1 (2017)
32. Zhao, G., Huang, X., Taini, M., Li, S.Z., PietikäInen, M.: Facial expression recognition from near-infrared videos. Image Vis. Comput. **29**(9), 607–619 (2011)
33. Zhao, X., Liang, X., Liu, L., Li, T., Han, Y., Vasconcelos, N., Yan, S.: Peak-piloted deep network for facial expression recognition. In: European Conference on Computer Vision, pp 425–442. Springer (2016)

A Network Intrusion Detection Model Based on Convolutional Neural Network

Wenwei Tao[1(✉)], Wenzhe Zhang[1], Chao Hu[2], and Chaohui Hu[3]

[1] China Southern Power Grid Co., Ltd., Guangzhou 510670, China
taoww@csg.cn

[2] NARI Information & Communication Technology Co., Ltd., Nanjing 210033, China

[3] Dingxin Information Technology Co., Ltd., Guangzhou 510627, China

Abstract. Intrusion detection is an important research direction in the field of power monitoring network security. The increase of data volume and the diversification of intrusion modes make the traditional detection methods unable to meet the requirements of the current network environment. The emergence of convolutional neural network provides a new way to solve this dilemma. An intrusion detection model based on convolutional neural network is proposed in this paper. The method that converts the flow data into an image is used to represent the flow data in the form of a grayscale image, and use the texture representation in the image to classify the intrusion modes. Through the conversion of traffic data to images, the intrusion detection problem is transformed into image recognition problem, which substitute convolutional neural network technology into the intrusion detection problem. Firstly, the intrusion data set KDD 99 is preprocessed, and generate a two-dimensional image matrix group that meets the requirements. Then, the appropriate model structure for training is selected through comparison experiments. Finally, comparing the trained model with the other machine learning methods is to verify the model about reliability and effectiveness.

Keywords: Convolutional neural network · Image matrix group · Intrusion detection

1 Introduction

In recent years, information security issues have become increasingly prominent, and cybersecurity threats have increased year by year, with an average annual growth rate of about 50%. It has brought immeasurable impact on individuals, businesses and even the country. Last year's ransomware caused at least $55 billion in global losses.

Network intrusion is one of the biggest threats to cyberspace. It refers to any collective attempt to compromise the confidentiality, integrity and availability of hosts or networks. It is the basis of all network damage [1]. The current network intrusion detection situation is not optimistic. For the new intrusion method, the passive situation is often found after the event. Therefore, it is outdated to continue relying on the traditional detection method, and it is impossible to effectively solve the problems of

C.-N. Yang et al. (Eds.): SICBS 2018, AISC 895, pp. 771–783, 2020.
https://doi.org/10.1007/978-3-030-16946-6_63

multiple intrusion methods and unknown intrusion. An efficient network intrusion intelligent detection solution is a top priority.

In recent years, with the rapid development of Convolutional neural network, the application of it which applicate to network intrusion detection has gradually increased. Commonly used models include Deep Belief Network (DBN), Convolutional Neural Network (CNN) and Recurrent Neural Network (RNN).

Nadeem et al. [2] combined neural networks with semi-supervised learning to obtain higher accuracy with a small number of labeled samples. Experiments were carried out using the KDD Cup 99 dataset. The unlabeled data was tracked through the ladder network, and then the tag data was classified by the Deep Belief Network (DBN) to obtain detection accuracy similar to that of supervised learning. Staudemeyer et al. [3] implemented intrusion detection based on LSTM regression neural network. The results show that the LSTM classifier has certain advantages over other strong static classifiers. These advantages are in detecting DoS attacks and Probe attacks, both of which produce unique time series features. In order to compensate for the high false alarm rate, Kim [4] et al. proposed a method of calling language modeling to improve the host intrusion detection system based on LSTM. Agarap et al. [5] replaced Softaax by introducing a Linear Support Vector Machine (SVM) into the final output layer of the GRU model and applied the model to the two classification of intrusion detection. Kolosnjaji et al. [6] implemented a neural network consisting of convolution and feedforward neural structures. A hierarchical feature extraction method is provided that vectorizes n-gram instruction features and convolution features.

This paper presents an intrusion detection model based on Convolutional neural network. By using the flow data visualization method, the data is presented in the form of grayscale graph, and the invasion is classified by the texture feature in the image. Through the conversion of traffic data to image,the intrusion detection problem is transformed into image recognition problem, and convolutional neural network technology is introduced into the problem of intrusion detection. First of all, the intrusion data set KDD 99 is preprocessed to generate a 2d image matrix group that meets the experimental requirements. Then, using the generated data to compare the deep convolutional neural networks of various structures, the time cost and experimental results are taken into consideration, and the appropriate model structure is selected for training. After that, the training model is compared with the mainstream machine learning model to verify the effectiveness of the model.

In the next section, we introduce intrusion detection system simply and the classic model of convolutional neural network. We build intrusion detection model based on convolutional neural network and explore the indicators of the evaluation model in Sect. 3. After that, we analysis experiment about intrusion detection based on convolutional neural network, and compare with other machine learning models to verify the advanced nature of the model in Sect. 4. Finally, we summarize it in Sect. 5.

2 Primaliary Knowledge

2.1 Intrusion Detection System and Technology

Intrusion Detection System (IDS) is a type of software system that applies and detects access to computer operating systems or wired and mobile networks without protocol approval [7]. The intrusion detection system is used to discover various types of abnormal network traffic or unconventional host instructions. It mainly includes some common network attacks, drive attacks, host attacks and malware.

The classification of intrusion detection systems mainly includes detection methods, protection measures and data sources. Data sources are widely used as standards in the field, including network intrusion detection systems, host intrusion detection systems and distributed intrusion detection systems [8, 9].

Intrusion detection technology is mainly used to defend against illegal intrusions from the outside network and internal network, to reduce the maintenance pressure of security personnel, and to take necessary protective measures to avoid the occurrence of hazards when an intrusion occurs. It includes misuse detection, anomaly detection and hybrid detection techniques.

2.2 Convolutional Neural Network

Convolutional Neural Network (CNN) is the first artificial neural network to be successfully applied, and it has excellent performance in the fields of image processing and speech analysis. It utilizes the convolutional feature to effectively reduce the amount of weight and greatly reduce the amount of computation. When the data to be trained is multi-dimensional, this advantage is very obvious, it can reduce the difficulty of manually processing data, and the modeling efficiency is greatly improved. A convolutional network is a multi-layer sensor designed to identify two-dimensional shapes that are not fixed in height, such as translation, scaling, tilting, or other forms of deformation.

Convolutional neural networks are the most widely used deep learning model, as shown in Fig. 1. It reduces the number of parameters that need to be learned to improve the training performance of the BP algorithm through spatial relationships. CNN can learn different levels of features from a large amount of unlabeled data. Therefore, the application prospect of CNN in the field of network intrusion detection is very broad.

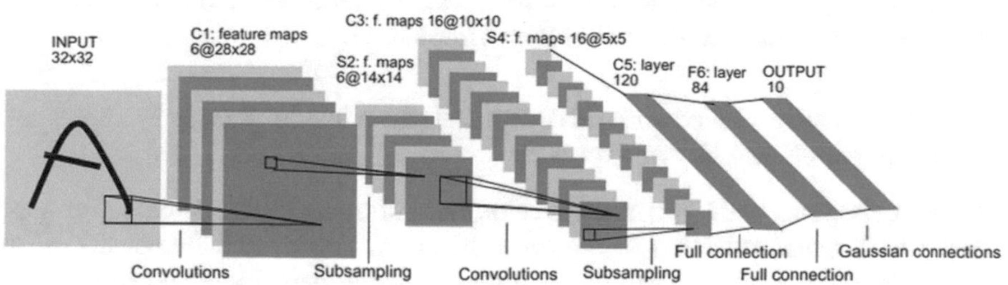

Fig. 1. CNN structure

There are three means for CNN to reduce network training parameters, which are convolution, parameter sharing and pooling [11].

(1) Local receptive field

Suppose we have a 1000 × 1000 pixel image, if there are 10^6 recessive neurons in the hidden layer and they are all fully connected. To train an image, we need to learn 10^{12} parameters. If we want to classify a group of images, it will be astronomical.

Abolishing the full connection is a more common method, so that each recessive neuron only feels the local image area, and then combines to get the information of the whole picture. Assuming that a single hidden layer unit is only connected to a partial image of 10 × 10 pixel size, the resulting parameters are 10^8, which is four orders of magnitude lower than the initial.

(2) Parameter sharing

Although the use of local receptive fields greatly reduces the weight parameters, there are still 10^8 parameters. Therefore, it is necessary to further reduce the parameters. Previously, a recessive neuron was used to connect 10 × 10 pixel images, that is, each neuron has 100 parameters. If the 100 parameters of all neurons are the same, then only 100 parameters are calculated. Intuitively, the 10 × 10 convolution kernel is used to convolve the image. But another problem comes out. A single convolution kernel can only represent one feature and cannot fully reflect the image texture features. Therefore, several convolution kernels are needed. For example, using 100 convolution kernels, the total parameters are only 10^4. Compared with the local receptive field, it is reduced by 4 orders of magnitude.

(3) Pooling

The fewer parameters to learn, the better, so after convolution, we can use pooling (also known as subsampling) to further reduce the parameters. The maximum pooling method is the most commonly used pooling method. The input image is divided into several rectangular regions, and the maximum value is output for each sub-region.

3 Network Intrusion Detection Model Based on Convolutional Neural Network

This paper aims to explore the possibility of convolutional neural networks, and mine the image features of invading samples, and then implement intrusion detection systems based on deep convolutional neural networks. This section will build the structure of the model and introduce the methods and ideas used.

3.1 The Structure of Network Intrusion Detection Model Based on Convolutional Neural Network

This paper proposes a CNN-based intrusion detection model and attempts to apply convolutional neural networks to intrusion detection. The overall framework of the intrusion detection model based on convolutional neural network is shown in Fig. 2, which includes three stages: data preprocessing, model training and intrusion detection.

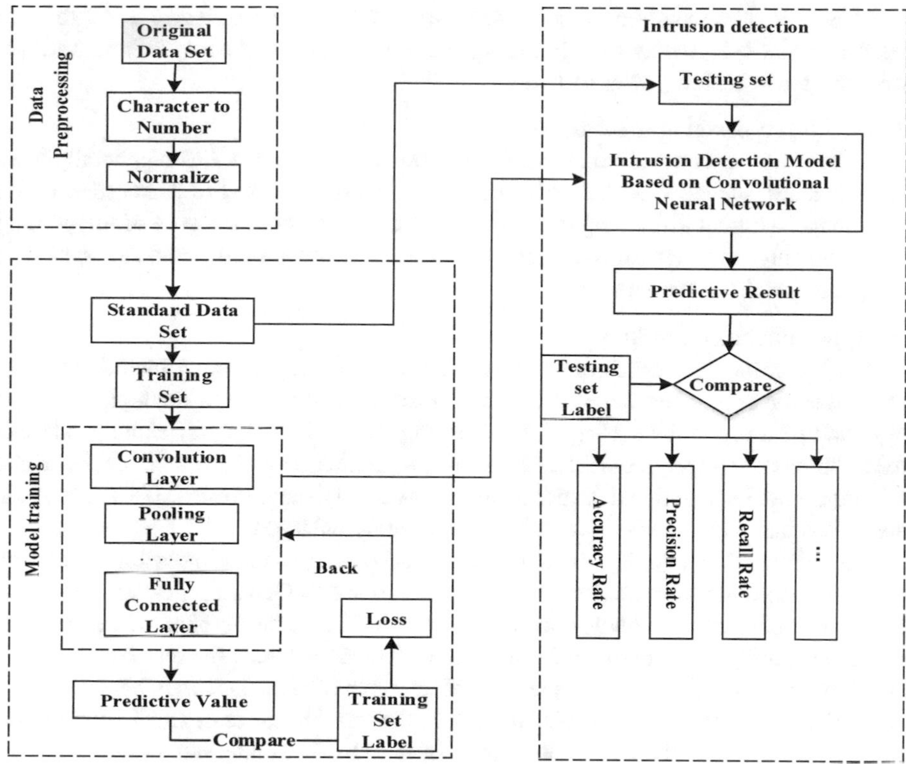

Fig. 2. Overall framework of intrusion detection model based on convolutional neural network

As can be seen from Fig. 2, in the data preprocessing stage, the model uses the KDD 99 data set as the original data set. First, it needs to be digitized to convert the character data in the data sample into a digital type. The data is then normalized and processed into the standard data sets available to the model. In the model training phase, this phase will be adjusted according to the model parameters mentioned in Sect. 3.2, and the model with better prediction effect on the training set will be selected as the detection model. In the intrusion detection phase, we will use the thousands of models trained in the model training phase to test the test set, compare the detection models of different parameters through the indicators introduced in Sect. 3.3, and select the best model that meets the requirements as the final model.

3.2 Parameters of Convolutional Neural Network Model

(1) Size of input data

When using convolutional neural networks to process network data, the data is compressed or extended to $n \times n$ size for ease of computation.

(2) Convolution layer parameter setting

The parameters of the convolution layer mainly include the size of the convolution kernel, the number of convolutions, and the convolution step size [11]. In practice, a

3×3 or 5×5 convolution kernel is generally used, and the step size is set to 1. In order to facilitate hardware storage management, the number of convolution kernels is generally set to a high power of two.

(3) Pool layer parameter setting

Similar to the convolutional layer, the core of the pooled layer generally uses smaller values, such as 2×2, 3×3, and so on. The step size is 1 or 2. In order to lose performance without discarding excessive input characteristics, the core of the pooling layer generally does not exceed three [12]. The commonly used parameter setting is 2×2 core, and the step size is 2.

(4) Hyper-parameter setting

The hyper-parameters of convolutional neural networks generally include learning rate, model layer number, iteration number, batch size, and dropout method. The most important of these is the setting of the learning rate [13]. An ideal learning rate can make the model converge quickly. Otherwise, a gradient explosion will occur and the training cannot be completed. Experiments are generally started from 0.0001, and if the model is found to be slower, it can be appropriately enlarged.

The number of model layers should be set according to the dimension of the data being processed and the problem that you want to solve. Generally, experiment from the small model and gradually increase the depth [14]. The number of iterations is usually set during the experiment. When the difference between the test error rate and the training error rate is small, the current number of iterations can be considered appropriate. Convolutional neural networks in the training process, small batches will perform better, generally select the number in the [16, 128] interval.

Finally, in order to prevent the model from being too strong, dropout is also needed. It is a commonly regularization method that randomly does not activate a certain proportion of hidden layer units in each calculation. This parameter is generally set to 0.5.

3.3 Evaluation Indicators of Network Intrusion Detection Model Based on Convolutional Neural Network

When we evaluate the training effect of the model, there are several commonly used indicators, each of which has its own advantages and disadvantages, and some of them are mutually constrained. It is necessary to comprehensively consider the demand and efficiency, and try to balance the relationship between the two. The following is a summary of these indicators in combination with Table 1.

Table 1. Two-class confusion matrix

Reality	Prediction	
	Positive	Negative
Positive	True Positive (TP)	False Negative (FN)
Negative	False Positive (FP)	True Negative (TN)

True positive (TP) is a positive sample correctly classified by the model; False Negative (FN) is a positive sample misclassified by the model; False Positive (FP) is a negative sample classified by the model; True Negative (TN) is the correct model a negative sample of the classification.

(1) Accuracy rate

The formula for accuracy rate is as follows:

$$Accuracy = \frac{TP + TN}{TP + TN + FP + FN} \tag{1}$$

As can be seen from the above formula, the accuracy rate represents the proportion of the model's correct classification of the actual value of the sample to the data set.

(2) Precision rate

The formula for the precision rate is as follows:

$$Precision = \frac{TP}{TP + FP} \tag{2}$$

Precision rate represents the proportion of true positive samples to positive samples of the model classification.

(3) Recall rate

The formula for the recall rate is as follows:

$$Recall = \frac{TP}{TP + FN} \tag{3}$$

Recall rate shows the proportion of true positive samples in actual samples.

(4) F1 value

$$F1 = \frac{2 \times Precision \times Recall}{Precision + Recall} = \frac{2 \times TP}{2 \times TP + FN + FP} \tag{4}$$

The F1 value is a harmonic average of the accuracy rate and the recall rate, which is equivalent to a comprehensive evaluation index of two values. The larger the value, the better the performance of the model. This article mainly uses the F1 index as the evaluation benchmark, taking into account the accuracy and recall rate considerations.

4 Experiment and Analysis

The main content of this chapter is based on the previous chapter, using the program to implement the intrusion detection model based on convolutional neural network, and evaluate it, then experiment on the model structure and parameters, and select the model with good detection effect.

We will provide a basic strategy for model training based on the model structure discussed above. We use ten percent of KDD 99 data set to train a network model, firstly to train a small convolutional neural network without any regularization, and to achieve

The goal is to set a baseline. In this process, our main problem is overfitting. Then, we reduce the over-fitting phenomenon by increasing the amount of training data and adjusting the hyper-parameters, and improve the network to achieve higher accuracy.

4.1 Selecting the Number of Layer of the CNN Model

Model structure is the basis of training, and it is also difficult to determine. It needs to be tried continuously. This paper adopts the construction strategy from easy to complex, starting from the simplest three-layer convolutional neural network, and increasing the depth and capacity of the model.

An important indicator for judging whether a model is effective is the degree of convergence of the training set loss. If it continues to decrease, it is a better training model. At the same time, it is necessary to compare the accuracy rate of the test set to determine whether it is overfitting. At this stage, the convergence degree of each layer model is mainly compared. The results are shown in Figs. 3 and 4.

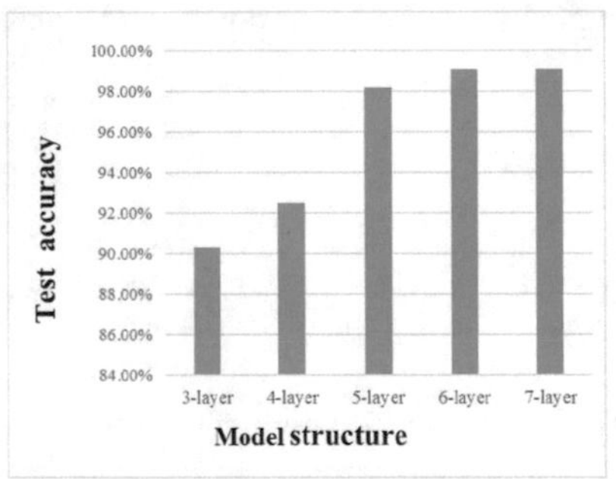

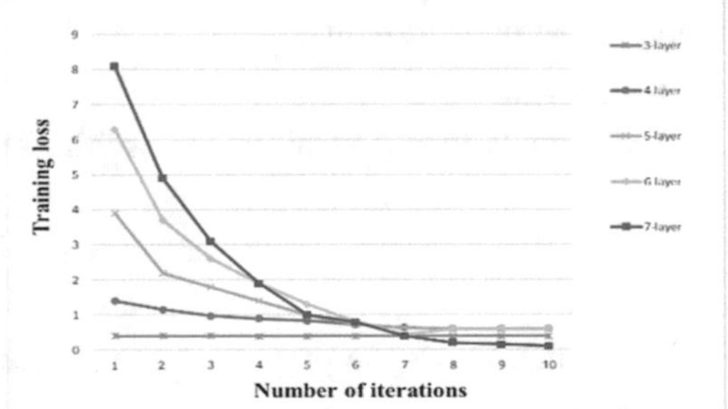

Fig. 3. Training loss of model structure

Fig. 4. Test accuracy of model structure

As can be seen from Figs. 3 and 4, the 6-layer structure and the 7-layer structure have similar convergence efficiency and accuracy. In the model selection, the model with smaller layers tends to be more competitive, because the deployment and training time will be cut back. Therefore, a six-layer structure is used as a detection model in this paper.

Based on the previous information, we will proceed to the structure of the design model. The first layer is the Input Layer, which converts the 121-dimensional flow data into a 11×11 2-dimensional matrix. The second layer is the Convolution Layer, and 32 filters are used to extract the input data. The third layer is the Maxpooling Layer, and subsample the data with a 2×2 convolution kernel, and with a step size of 1. The fourth layer is similar to the second layer except that the number of convolution kernels becomes 64. The fifth layer is the same as the third layer. The sixth layer is the Flatten Layer, which the input multidimensional vector is converted to one dimension. The seventh and the ninth layers are Fully Connected Layer. The eighth and tenth layers are Dropout Layer for regularization to prevent overfitting. The last layer is Softmax Layer for classification. Although there are actually 11 layers, because the Maxpooling Layer, the Flatten Layer, and the Dropout layer have no parameters to operate, the model structure is still called 6-layer.

4.2 Selecting Hyper-Parameters of CNN Model

In the last section, the meaning and members of the hyper-parameters have been introduced. The following experiments are used to compare the effects of various parameters on the performance of the model to determine the optimal solution.

(1) Learning rate

As the most important hyper-parameter, you need to determine it first. Generally, the violent test method is adopted, which multiple learning rate values are selected for testing, and the optimal solution is determined by the loss curve. In this paper, [0.00001, 0.00005, 0.0001, 0.0003] is selected as the comparison interval, and we select the appropriate learning rate from them.

The abscissa is the number of training iterations, and the ordinate is the loss value. The experimental rcsults are shown in Fig. 5.

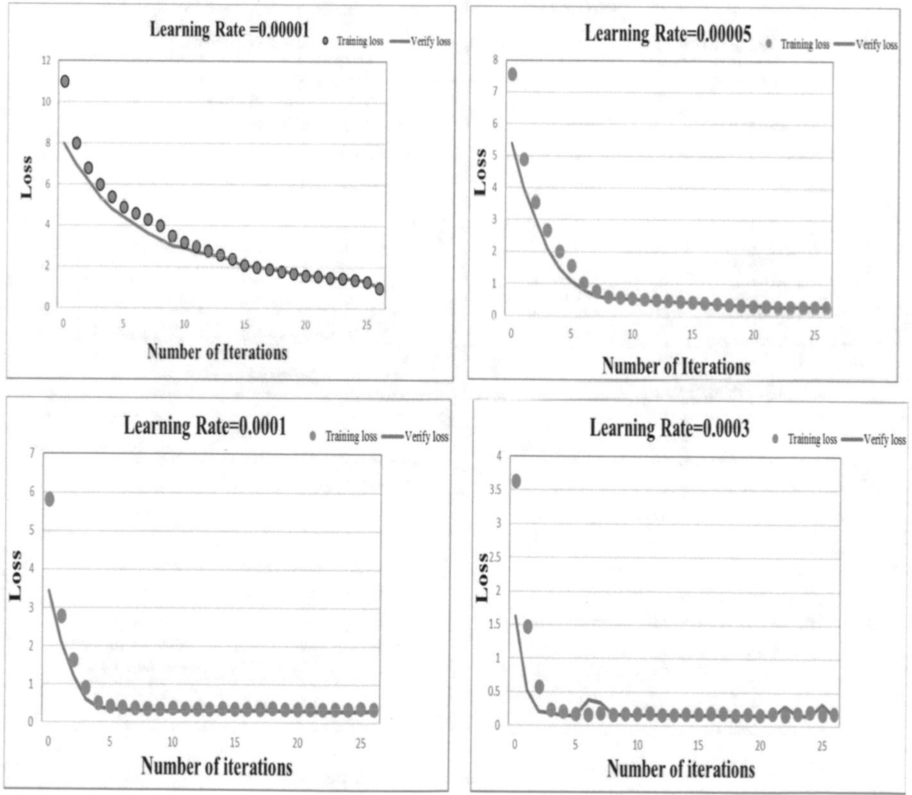

Fig. 5. Learning rate and loss

As can be seen from the above Fig. 5, among the four learning rates, 0.00001 is too small, and after 30 iterations, it still does not converge; 0.0003 is too large, and the curve is uneven. 0.00005 and 0.001 are not much different, but the curve of the former is smoother. Therefore, the learning rate of CNN model in this paper is 0.00005.

(2) Optimizer

The purpose of the optimizer is to optimize the gradient update process. In the convolutional neural network model, the commonly used optimizers are SGD, RMSprop, Adagrad, Adamelta, Adam and Nadam. Under the condition that the learning rate is 0.00005, the performance of each optimizer is shown in Fig. 6.

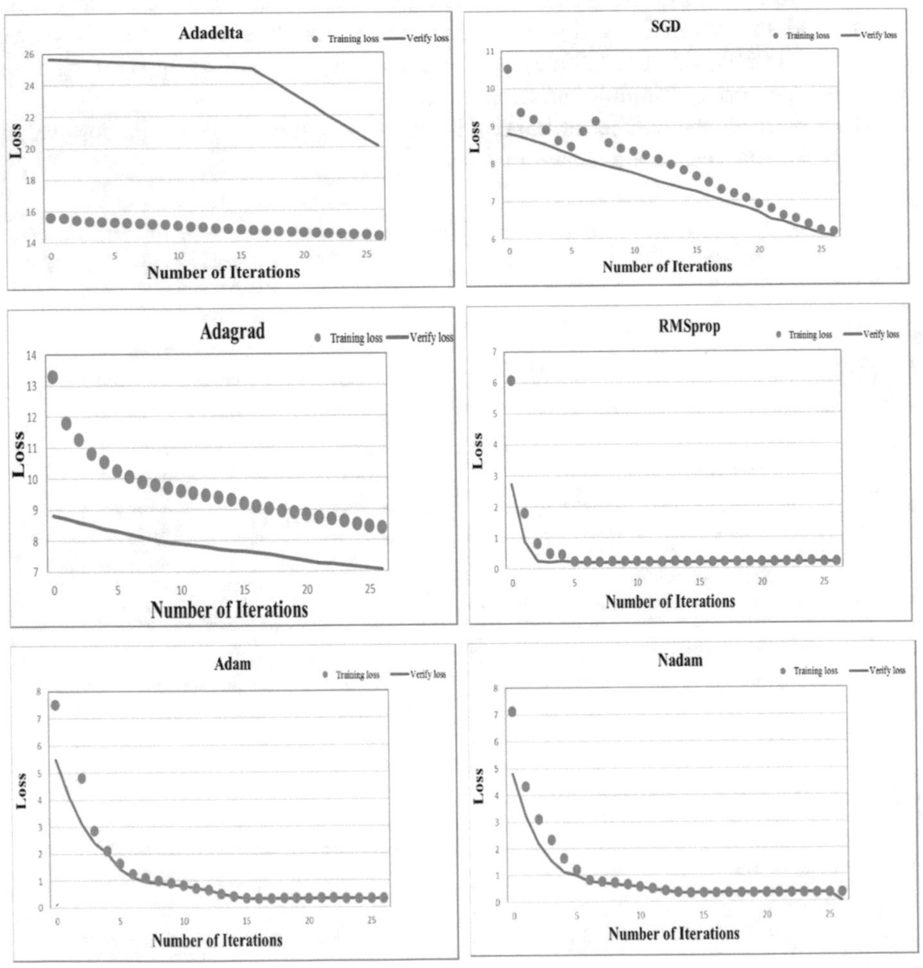

Fig. 6. Optimizer effect comparison

As can be seen from the above Fig. 6, the Adalelta, SGD and Adagrad optimizers perform poorly, and the optimization effect is not satisfactory; while RMSprop, Adam and Nadam have better optimization effects. In this paper, Adam is used as the model optimizer.

(3) Regularization

The purpose of regularization is to reduce the degree of over-fitting of the model and improve the generalization ability of the model. The CNN model regularization mainly uses the L2 and Dropout methods. Here are the following three situations: (1) Only use L2 regularization. (2) Only use Dropout regularization. (3) Combine L2 and Dropout. The specific model performance is shown in the following Fig. 7.

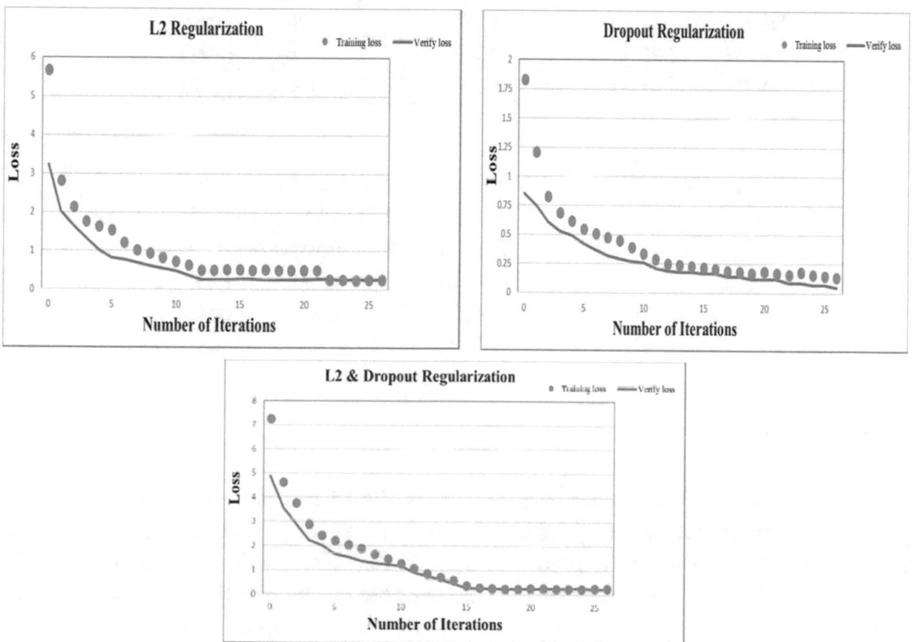

Fig. 7. Regularization method comparison

It can be seen from the Fig. 7 that the simple use of L2 regularization can obtain a better fitting effect after 25 iterations, and the overfitting phenomenon occurs only by the Dropout method. By combining the two methods, both better training results and faster data fitting can be achieved.

4.3 Comparing with Other Machine Learning Algorithms

Through the previous experimental work, the six-layer structure of the model is determined. 0.00005 is selected as the learning rate, Adam is used as the optimization algorithm, and the L2 and Dropout regularization methods are used. In order to reflect the validity and superiority of the model, it also needs to be compared with other machine learning algorithms. This paper mainly selects SVM, decision tree (DT), KNN, LSTM and DBN models for comparison. The training set used is KDD Cup 99 data set. The experimental indicators use accuracy rate, precision rate, recall rate and F1 value. The comparison results are shown in Table 2.

Table 2. The comparison of machine learning algorithms

Algorithms	Accuracy Rate	Precision Rate	Recall Rate	F1 Value
SVM	82.4%	74.0%	80.3%	0.77
KNN	85.2%	89.6%	75.6%	0.82
DT	91.94%	98.9%	86.0%	0.92
DBN	93.49%	92.3%	93.7%	0.93
LSTM	96.93%	98.8%	90.4%	0.95
CNN	97.8%	99.9%	97.7%	0.98

As can be seen from the above Table 2, we can see the processing effect of each machine learning algorithm on the KDD data set. Traditional machine learning algorithms that include SVM, KNN and DT algorithm models have lower experimental indexes; while convolutional neural network algorithms, such as DBN, RNN and CNN, have obtained good experimental results. The CNN used in this paper performs best in the selected six models, which reflects the effectiveness and advancement of the model.

5 Conclusion

Based on the analysis and summary of the results of intrusion detection, this paper proposes an intrusion detection model based on convolutional neural network. We visualize the flow data and explore the metrics that evaluate the performance of the model. After that, we determined the structure of the model and the value of the hyper-parameters. We used the KDD 99 data set for training. Comparing with other machine learning methods under the same index, we obtain better experimental indexes and prove the validity model of the experimental model. Although the performance is good, there are still many problems, we need to continue to research to apply it to the actual network.

References

1. Buczak, A.L., Guven, E.: A survey of data mining and machine learning methods for cyber security intrusion detection. IEEE Commun. Surv. Tutorials **18**(2), 1153–1176 (2016)
2. Nadeem, M., Marshall, O., Singh, S., Fang, X., Yuan, X.: Semi-supervised deep neural network for network intrusion detection. In: KSU Proceedings on Cybersecurity Education, Research and Practice (2016)

3. Staudemeyer, R.C.: Applying long short-term memory recurrent neural networks to intrusion detection. S. Afr. Comput. J. **56**(1), 136–154 (2015)
4. Kim, G., Yi, H., Lee, J., Paek, Y., Yoon, S.: LSTM-Based System-Call Language modeling and robust ensemble method for designing Host-Based Intrusion Detection Systems. arXiv preprint arXiv:1611.01726 (2016)
5. Agarap, A.F.: A Neural Network Architecture Combining GatedRecurrent Unit (GRU)and Support Vector Machine (SVM) for Intrusion Detection in Network Traffic Data. arXiv: 1709.03082 (2017)
6. Kolosnjaji, B., Zarras, A., Webster, G., Eckert, C.: Deep learning for classification of malware system call sequences. In: 29th Australasian Joint Conference. AI 2016: Advances in Artificial Intelligence, Hobart, pp. 137–149 (2016)
7. Dubey, S., Dubey, J.: KBB: a hybrid method for intrusion detection. In: International Conference on Computer, Communication and Control, pp. 1–6 (2015)
8. Geng, M.X.: A survey of network intrusion detection technology. Network Secur. Technol. Appl. **6**, 2830 (2004)
9. Pan, S., Morris, T., Adhikari, U.: Developing a hybrid intrusion detection system using data mining for power systems. IEEE Trans. Smart Grid **6**(6), 3104–3113 (2015)
10. Hinton, G.E.: Deep belief networks. Scholarpedia **4**(6), 5947 (2009)
11. Li, Z., Qin, Z., Huang, K., Yang, X., Ye, S.: Intrusion detection using convolutional neural networks for representation learning. In: Neural Information Processing vol. 10638, pp. 858–866 (2017)
12. Bontemps, L., V., Cao, L., Mcdermott, J., Le-Khac, N. A.: Collective anomaly detection based on long short-term memory recurrent neural networks. In: International Conference on Future Data and Security Engineering, pp. 141–152 (2017)
13. Alom, M.Z., Bontupalli, V.R., Taha, T.M.: Intrusion detection using deep belief networks. In: Aerospace and Electronics Conference, pp. 339–344 (2016)
14. Al-Janabi, S.T.F., Saeed, H. A.: A neural network based anomaly intrusion detection system. In: Developments in E-Systems Engineering, pp. 221–226 (2011)

A Dual-Branch CNN Structure for Deformable Object Detection

Jianjun Li[2(✉)], Kai Zheng[2], Xin Zhang[2], Zhenxing Luo[1], Zhuo Tang[1], Ching-Chun Chang[3], Yuqi Lin[4], and Peiqi Tang[2]

[1] Science and Technology on Communication and Information Security Control Laboratory of the 36th Institute of China Electronics Technology Group Corporation, Jiaxing, China

[2] School of Computer Science and Engineering, Hangzhou Dianzi University, Hangzhou, China
{jianjun.li,zhangxin}@hdu.edu.cn

[3] Department of Computer Science, University of Warwick, Coventry, UK

[4] Yunnan Key Laboratory of Computer Technology Application/Faculty of Information Engineering and Automation, Kunming University of Science and Technology, Kunming, China

Abstract. Object detectors based on CNN are now able to achieve satisfactory accuracy, but their ability to deal with some targets with geometric deformation or occlusion is often poor. This is largely due to the fixed geometric structure of the convolution kernel and the single inflexible network structure. In our work, we use dual branch parallel processing to extract the different features of the target area to coordinate the prediction. To further enhance the performance of the network, this study rebuilds the feature extraction module. Finally, our detector learns to adapt to a variety of different shapes and sizes. The proposed method achieves up to 81.76% mAP on the Pascal VOC2007 dataset and 79.6% mAP on the Pascal VOC2012 dataset.

Keywords: Dual-branch structure · Convolution neural network · Deformable object detection

1 Introduction

Object recognition, such as fingerprint recognition and face recognition, plays an important role in the security field. As we know, that the first step in object recognition is object detection. Therefore, object detection is also essential in the field of security. Object detection is widely used to precisely locate and classify kinds of targets in an image or video. To perform properly, these detectors need to "learn" as many different categories of feature representations as possible. Also, they need to learn the different scales, poses, viewing angles, and even non-rigid body deformations which the same individual may present differently in different images.

C.-N. Yang et al. (Eds.): SICBS 2018, AISC 895, pp. 784–797, 2020.
https://doi.org/10.1007/978-3-030-16946-6_64

In this work, we propose to create an object detector that can view images at the same way that humans do. When a human identifies an object, the first thing to be observed is the overall shape of the target object, which represents the global information about the target. Based on the shape and the scale of the object, a human can preliminarily judge the general characteristics of the object. Next, a human will gradually identify the specific features of the target based on the upper, lower, left, and right sides of the object. This is known as local information. By considering the local and global information combinedly, humans can make precise judgments about the target object. Our objective in this study is to design a detection model that can adapt to different shapes of objects, in other words, obtain both local and global information simultaneously.

We have assessed existing object detectors based on Convolutional Neural Networks (CNN). Fast/Faster R-CNN [5,19] and Region-based Fully Convolutional Networks (R-FCN) [1] are two representative region-based CNN approaches. Fast/Faster R-CNN use a subnetwork to predict the category and bounding box of each region proposal. Unlike Fast/Faster R-CNN, R-FCN proposes the concept of position-sensitive score map and conducts the inference with position-sensitive region of interest pooling (PSRoIPooling). The reason why R-FCN achieves more accuracy is that it focuses on details of the target object and sensitivity of the location. Inspired by their work [1], this study uses position-sensitive RoI pooling to extract the local information of the object and employing Faster R-CNN to extract global information. Besides, a dual branch structure based on Faster R-CNN has been used to rebuild the network in the proposed method, which improves the performance significantly.

In addition, we employ deformable convolution and deformable position-sensitive RoI pooling to enable the detection model to handle the geometric transformations, which is inspired by [2]. These two new modules can greatly enhance the capability of the model to handle different shapes or poses of the same object. Therefore, our model can achieve performance beyond the current level of some detectors for objects with deformation, occlusion and overlap. We have also achieved outstanding results on objects with high similarity, such as chairs and tables.

The remainder of the paper is organized as follows. In Sect. 2, the related works are presented. Section 3 describes the methodology of the proposed approach. Section 4 presents the experimental results. Section 5 concludes our work with a summary.

2 Related Work

With the rapid development of deep convolutional networks [9,10,12,18,23–28], a number of object detection methods have been proposed and made substantial improvements in this field. Some of the recent object detection works are introduced as follows.

R-CNN: Region-based CNN (R-CNN) [6] achieves high accuracy by extracting features via deep neural networks. Following R-CNN, a large number of variants of R-CNN have evolved. In Faster R-CNN [19], Region Proposal Network is introduced to generate proposals and RoIPooling is adopted in the subnetwork on each proposal. Following Faster R-CNN [19], R-FCN [1] proposes the Position-sensitive RoIPooling, which also speeds up the detection when dealing with a large number of proposals. Feature Pyramid Networks (FPN) [15] construct feature pyramids with the inherent multi-scale pyramidal hierarchy of deep convolutional networks. Mask R-CNN [8] further implements a mask predictor by adding an extra branch in parallel based on Faster R-CNN. It also incorporates an RoIAlign layer that removes the harsh quantization of RoIPooling. RetinaNet [16] is an FPN-based single stage detector that employs Focal-Loss to address the class imbalance issue caused by extreme foreground-background ratio.

STN: Spatial Transform Networks (STN) [11] is the first work that learns spatial transformation from image datasets by deep neural networks. STN warps the feature map by means of a global parametric transformation, for example, affine transformation. Such warping is expensive, and it is quite difficult to infer the transformation parameters. STN is effective to solve the classification problem on small-scale images. The inverse STN method [14] is then proposed to address the aforementioned problems by replacing the expensive feature warping with efficient transformation parameter propagation.

DPM: Deformable Part Models [4] is a shallow model that can maximize the classification score by learning the spatial deformation of object parts. Its inference process equals to CNNs [7] when treating the distance transform as a special pooling operation. However, the training process is not end-to-end and involves manual adjustments, including the selection of components and hyper-parameter.

Deformable Convolution: Convolutional neural networks (CNNs) are inherently restricted to model geometric transformations, due to the fixed geometric structure in the module construction. The Deformable Convolution Network [2] indicates that the pixels in a receptive field have different impact on the output response. The pixels on the object contribute greater than others. Therefore, Deformable Convolution Networks enable the receptive field to distinguish objects from background. With this improvement, a convolution filter can autonomously attain the ability to sense the object. It rebuilds the convolution filter, introducing two new modules (Deformable Convolution and Deformable PSRoIPooling) which greatly enhance CNN's capability of modeling geometric transformation. During network training, offsets can be learned by additional convolution layers so that different locations can be sampled more flexibly.

Muti-Branch: Single-branch networks often have limitations, such as fixed receptive field and simple hierarchy, which can result in the network not being able to make full use of feature diversity. However, in recent network structure, like CoupleNet [28] and Mask R-CNN [8], which both adopted multi-branch structure. CoupleNet uses mult-branch to obtain feature maps of different inclination which then merged together. Mask R-CNN uses two-branch to collaboratively perform segmentation and detection of two related tasks, and achieves mutually beneficial results.

Our model draws on the idea of feature fusing that is easily overlooked in many of recent work, continuing the multi-branch structure of CoupleNet [28] and MASK R-CNN [8]. Based on the muti-branch of CoupleNet, we carefully verified its theory and experimental details. Unlike the CoupleNet, we choose the structure of dual-branch while retaining some of the fully-connected layers. With experimental support, we design the Branch B which keep the fully-connected layer (the experimental data in Table 4 proves that our determination's correction). In Branch A, convolution layer is used instead of fully-connected layer. Combining two different branches, our network can not only take into account the amount of model parameters, but also ensure the accuracy. At the same time, we add an additional Deformable module in Branch B, which greatly improved the ability of the Branch to adapt to deformation of object (experiment results are shown in Tables 1 and 2).

3 The Proposed Approach

In this section, we first present the architecture of the proposed detection model and then describe design details.

3.1 Network Architecture

Figure 1 shows the architecture of our model. The proposed detector is mainly composed of two components. The first one is RPN (Region Proposal Network), which is in charge of generating candidate proposals. The second is R-CNN

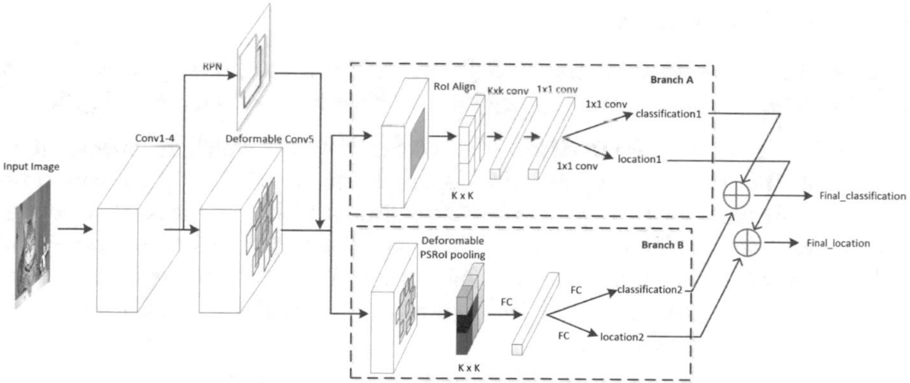

Fig. 1. Our detector: In Feature Extractor, this study uses Deformable Conv5. In R-CNN subnetwork, we employ dual branch (branch A and branch B). Branch A: for the 2000 proposals produced by RPN, we process all the proposals by *RoI Algin*. We replace the last two FC layers by a fully convolution layer. Branch B: for the 2000 proposals produced by RPN, we use Deformable PSRoIPooling and then add one fully connected layer. Finally, we fuse the two feature maps produced by branch A and branch B to predict the final classification and location.

subnetwork which makes the final location and classification. The R-CNN subnetwork involves two separate branches: **(A)** a RoIAlign to gather global feature information for an RoI, and **(B)** Deformable PSRoIPooling to encode details and local feature information of an RoI. Our network is initialized in the same way as the pre-trained ImageNet model, ResNet-101 [9]. To realize the detection task, we remove the last average pooling layer and *FC* (Fully-Connected) layer. All proposals produced by RPN are fed to Branch A and Branch B. Finally, the outputs of the two branches are fused together to do classification and box regression.

3.2 Regional Proposal Networks (RPN)

The RPN is based on sliding-window class-agnostic object detector and it uses features extracted from the 4th stage, following [19]. Specifically, the RPN predefines a set of anchors that are related to scales and aspect ratios. In the proposed approach, three aspect ratios are set to $\{1{:}2, 1{:}1, 2{:}1\}$ and three scales are set to $\{128^2, 256^2, 512^2\}$ separately to cover objects of different shapes, to the extent possible. Besides, a number of proposals are redundant with each other, which is addressed by non-maximum suppression (NMS), and the intersection-over-union (IoU) threshold of NMS is set to 0.7 in our approach. The anchor is assigned with label if its IoU is the highest or exceeds 0.7 with any ground-truth box. On the other hand, if an anchor's IoU is smaller than 0.3 with all the ground-truth boxes, it will be assigned with a negative label. After processing by RPN, each input image outputs 2000 proposals, which is then used in R-CNN subnet.

3.3 R-CNN Subnetwork

In this section, we discuss the R-CNN subnetwork in our approach in four aspects. Also, we provide the topology of our network for readers in Fig. 2.

Maintain Spatial Information. For Faster R-CNN which is presented in Fig. 3, we find that two 1024-d *FC* layers bring a great deal of computations. Meanwhile, the *FC* layer forces the features to be compressed to a one-dimensional vector. The reduction of dimension is not conducive to a detection task that is sensitive to spatial location. Inspired by [21], we replace the two *FC* layers by two *conv (convolution)* layers and use a 1×1 convolution to do classification and location.

Accurate Quantification. Faster R-CNN [19] uses RoIPooling to extract features from each candidate box, and the extracted features are then used in classification and box regression. As discussed in [8], RoIPooling inevitably involves quantization two times. The quantization introduces misalignments between the RoI and the extracted features. To a large extent, it affects the accuracy of localization. Inspired by the work [8], we remove the RoIPooling to reduce the misalignments caused by quantization. Instead, we use bilinear interpolation [11] in RoIAlign to compute the exact values of the input features sampled at four

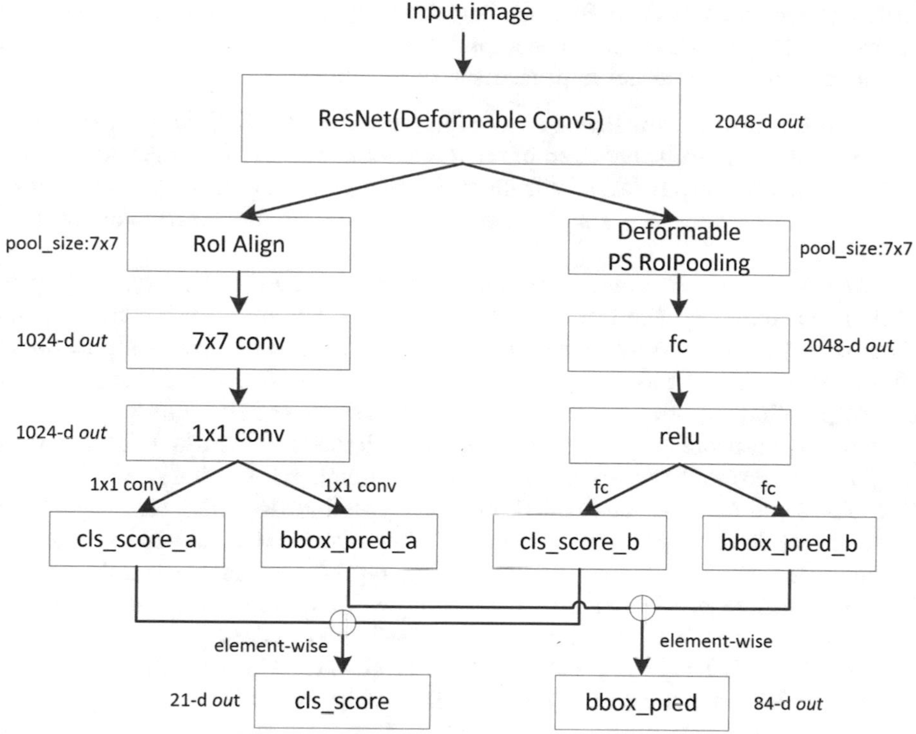

Fig. 2. The Topology of our network.

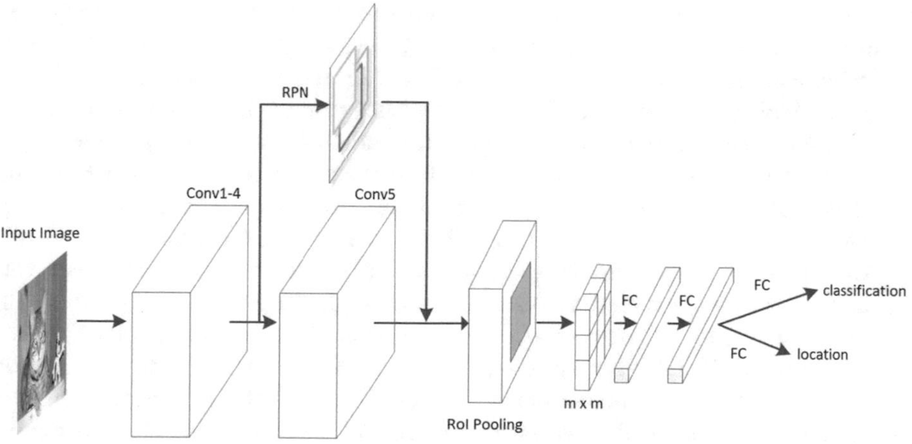

Fig. 3. Faster R-CNN [19]: the whole network can be divided into two parts: RPN and R-CNN subnet. After feature extracting network (like VGG or ResNets), the feature map flows into RPN. Then RPN produces 2000 proposals, which is then used in R-CNN subnet. In R-CNN subnet, each proposal is used to make classification and location regression.

regularly locations in each RoI bin, and then aggregate the output with max or average pooling strategies. The experimental results show that the RoIAlign operator brings substantial improvement to our detector.

Feature Combination. Based on our analysis of Faster R-CNN [19], RoIPooling transforms an arbitrary-sized input rectangular region into specified features (*eg.*, 7×7). The output features from RoIPooling tend to describe overall information of a RoI, while ignoring the details of objects such as internal orientation and hierarchy.

Therefore, we extend an extra branch in parallel like [8] to extract features in detail. Inspired by the work [1,13], we attach a 490-dim 1×1 convolutional layer to construct a position-sensitive score map. Considering computational efficiency, we choose 490-dim ($10 \times 7 \times 7$) rather than 1029-dim ($21 \times 7 \times 7$) for the output channel of our score map, and yet achieve equal accuracy.

For the position-sensitive score map, our detector conducts PSRoIPooling on the 7^2 scores and then vote on the RoI, producing a 10-dim vector for each RoI. The score map can be divided into 7×7 small grids, which can encode the cases of {top-left, top-center, top-right, ... , bottom-right} of an object category. Through the analysis above, PSRoIPooling represents the local and detailed feature information. We then attach two simple *FC* layers to do the classification and regression, in which the dimension is raised to 21 and 84 (4×21). Note that we use a layer rather than a convolutional layer here. The experimental results show that this branch gets lower mAP when using the convolutional layer.

Finally, the detector uses element-wise to fuse class probability and box prediction value, which are produced by Branch A and Branch B, and then outputs the final class and box prediction.

Effective Receptive Field. For the object detection task, we find that the pixels in a receptive field have diffrent impact on the output response, and the pixels on object contribute greater than others. Based on this information, we hope the receptive field to distinguish objects against background. In other words, we utilize the receptive field to sense the target autonomously. Inspired by the work [2], we find that the deformable convolution can make adaptive adjustment based on the scale and shape of the ojbects, and therefore greatly enhances the capability of modeling geometric transformations for the convolution kernel.

Motivated by [2], we add deformable convolution to the last stage of ResNet. Then, the 2D offsets extracted from the preceding feature map via convolutional layers are added to the regular grid sampling locations in the standard convolution. As for PSRoIPooling, an offset is also added to each bin position in the regular bin partition of the previous PSRoIPooling [1,2]. Therefore, our detector can accommodate geometric variations or model geometric transformations in object scale, pose, viewpoint, and part deformation. Meanwhile, the localization capability is improved, especially for non-rigid objects. By using the receptive field, the detector can learn semantic information adaptively and effectively and handle the overlapping and semi-occlusion issues.

4 Experiments

We first evaluate Faster R-CNN [19], and it achieves 77.1% mAP on the VOC2007 test set and 73.8% mAP on the VOC2012 test set. Our detector, by contrast, achieves 81.6% mAP on the VOC2007 test set and 79.6% mAP On the VOC2012 test set, which largely surpasses our baseline model 81.6% vs. 77.1% and 79.6% vs. 73.8%, and other detection models, such as R-FCN [1] (81.6% vs. 79.2% and 79.6% vs. 77.6%). In the meantime, we also compare our detector with SSD [17], D-R-FCN [2] (R-FCN with Deformable Convolution) and Mask R-CNN [8]. The results in Tables 1, 2 and 3 validate the effectiveness of our method over Faster R-CNN. In the meantime, we also compare our model with the SSD method [17]. In the following discussion, we present the training details as well as ablation experiments.

4.1 Implementation Details

Our approach is trained on 2 NVIDIA TITAN X GPUs, where the weight decay is set to 0.0005 and the momentum is set to 0.9. The batch size for each GPU is 2 and each image has 2000/1000 RoIs for training and testing phase. Besides, the learning rate is 0.001 for the first 80k iterations, and decrease to 0.0001 for later 30k iterations. During the training phase, the image scale is randomly sampled from {480, 570, 670, 760, 860}, and the shorter edge of the image is resized to the sampled scale. We adopt online hard example mining (OHEM) [22] techniques. The backbone network in our approach is initialized according to the pre-trained ImageNet [20] unless explicitly noted. Besides, the parameters of stage 2, 3 and 4 in the base model are also adjusted and batch normalization is also fixed to improve the training speed.

Next, a series of ablation experiments are conducted to validate the effectiveness of the proposed approach. All the ablation experiments use single-scale training and testing.

4.2 Ablation Experiments

We conduct experiments on PASCAL VOC 2007 [3], which includes 20 object categories for detailed evaluation of the proposed approach. Table 1 lists the detailed comparison results of Faster R-CNN [19], R-FCN [1], SSD [17], D-R-FCN [2], Mask R-CNN [8] and the proposed method. Specifically, the proposed detector is trained on the union set of VOC 2007 trainval and VOC 2012 trainval, and is evaluated on VOC 2007 test set. To be fair, we only provide the experimental results of a single model without multi-scale testing. All the methods use ResNet-101 as a base network (except SSD512, which uses VGG16). Note that Faster means Faster R-CNN method. D-R-FCN means R-FCN with the Deformable Convolution. Mask R-CNN reimplemented in our experiment is the one without FPN. We find that our method achieves 81.6% mAP, which outperforms R-FCN by 1.6 percent point. The standard mAP (mean Average Precision) score [3] is adopted to evaluate the performance of all methods.

Table 1. Results on PASCAL VOC 2007 test set (trained on VOC 2007 trainval and VOC 2012 trainval). Ours∗: only includes Branch A and Branch B. Ours§: includes Branch A and Branch B with *RoIAlign*. Ours†: includes Branch A and Branch B with *RoIAlign* and *deformable module*.

Method	SSD512	Faster R-CNN	R-FCN	D-R-FCN	Mask R-CNN	Ours∗	Our§	Ours†	Ours
mAP	78.5	77.1	79.2	81.8	80.5	80.0	80.5	81.1	**81.6**
aero	90.0	79.2	78.5	81.2	84.6	81.2	83.3	85.6	83.4
bike	85.3	84.1	85.8	87.5	87.6	87.6	86.1	86.1	88.2
bird	77.7	77.4	80.1	84.3	79.4	78.4	79.0	80.3	80.5
boat	64.2	68.7	70.8	79.4	78.2	74.4	74.6	74.8	75.5
bottle	58.4	59.1	68.4	69.4	68.4	65.5	67.8	68.5	67.3
bus	85.3	86.1	85.1	89.0	88.5	86.9	87.2	88.7	87.4
car	84.4	84.9	86.9	88.7	85.4	87.9	87.9	88.1	88.3
cat	92.5	86.0	88.4	92.4	90.3	88.2	88.3	88.9	88.6
chair	61.3	60.9	65.6	66.5	63.2	64.8	66.7	67.4	**69.1**
cow	83.4	86.4	86.8	89.4	88.4	87.2	87.1	87.2	88.1
table	65.0	72.8	73.1	72.3	70.1	73.9	73.8	74.0	**74.6**
dog	89.8	88.0	88.4	91.3	90.2	89.7	89.5	88.4	88.9
horse	88.5	85.1	88.6	90.1	87.3	87.9	87.8	88.0	88.9
mbike	88.2	83.8	80.4	83.2	84.4	83.1	84.5	85.1	84.0
person	85.5	79.0	80.6	80.4	80.5	83.3	84.2	85.6	**85.8**
plant	54.4	49.4	52.3	57.3	54.3	56.8	57.0	57.3	**57.4**
sheep	82.4	80.9	80.3	85.0	84.9	82.6	82.9	83.3	85.6
sofa	70.7	76.0	80.1	80.6	75.3	78.2	78.5	79.2	**81.2**
train	87.1	78.3	84.7	89.2	88.3	84.3	84.3	86.2	88.4
tv	75.4	76.4	78.5	78.6	79.7	76.9	78.7	80.3	81.2

Besides, we also perform experiments on PASCAL VOC 2012. The models are trained on the union set of VOC 2007 trainval+test and VOC 2012 trainval, and are evaluated on VOC 2012 test set as shown in Table 2. To be fair, we only provided the experimental results of a single model without multi-scale testing. The present comparison on PASCAL VOC 2012 dataset, our method achieves 79.6% mAP, which outperforms R-FCN by 2.0 percent point.

As discussed above, our model can sense the objects of different geometric transformations, which can also adapt to part deformation, occlusion, and partial overlap. As shown in Tables 1 and 2, the proposed model achieves better performance on sofa, person, chair, and table, which validates the effectiveness of our model.

Table 2. Results on PASCAL VOC 2012 test set (trained on VOC 2007 trainval+test and VOC 2012 trainval).

Method	SSD512	Faster R-CNN	R-FCN	D-R-FCN	Mask R-CNN	Ours
mAP	77.7	73.8	77.6	79.4	78.3	**79.6**
aero	88.9	86.5	86.9	86.9	88.4	88.1
bike	84.3	81.6	83.4	86.4	83.2	85.7
bird	76.9	77.2	81.5	78.6	78.3	**81.6**
boat	63.2	58.0	63.8	72.2	67.8	**71.2**
bottle	57.8	51.0	62.4	64.6	62.8	**64.0**
bus	85.0	78.6	81.6	83.4	83.2	83.2
car	83.4	76.6	81.1	84.3	84.3	83.9
cat	91.8	93.2	93.1	94.2	90.3	93.1
chair	60.7	48.6	58.0	60.0	59.3	**61.2**
cow	83.1	80.4	83.8	81.2	80.3	83.0
table	64.0	59.0	60.8	64.3	63.5	**65.6**
dog	88.2	92.1	92.7	93.0	93.9	92.3
horse	88.1	85.3	86.0	90.1	89.3	88.1
mbike	87.9	84.8	84.6	83.2	86.4	85.6
person	84.5	80.7	84.4	84.3	86.3	**87.5**
plant	53.8	48.1	59.0	59.9	58.8	**62.0**
sheep	81.6	77.3	80.8	85.0	80.0	83.0
sofa	69.7	66.5	68.6	71.5	68.4	**71.6**
train	86.5	84.7	86.1	88.4	86.0	87.2
tv	74.4	65.6	72.9	76.4	74.5	74.3

Table 3. Results on the test set of PASCAL VOC 2007.

Method	mAP	
	mAP@0.5 (%)	mAP@0.7 (%)
Faster R-CNN [19]	77.1	61.0
R-FCN [1]	79.2	62.8
D-R-FCN [2]	81.8	68.0
Mask R-CNN [8]	80.5	65.7
Ours	**81.6**	**68.3**

In Table 3 we compare three models' mAP scores with IoU thresholds being set to 0.5 or 0.7. Our detector gets a higher accuracy even at a high IOU threshold, which strongly verifies the efficiency.

For visual comparison, we present detected images in Figs. 4 and 5. In Fig. 4, our model (top) shows great robustness for geometric deformation and semi-occluded objects. For example, the proposed model successfully detects the

Table 4. Results on comparing using fully-connected *(fc)* layer and convolutional *(conv)* layer in Branch B.

Branch B	mAP
fc layer	81.6
conv layer	78.8

Fig. 4. Visualization detection results of our detector (top) and Faster R-CNN (bottom). It is clear that our detector has great robustness for geometric deformation and semi-occluded objects. It verifies our model's capability to learn geometric deformation and the ability to fuse global and local features to make predictions.

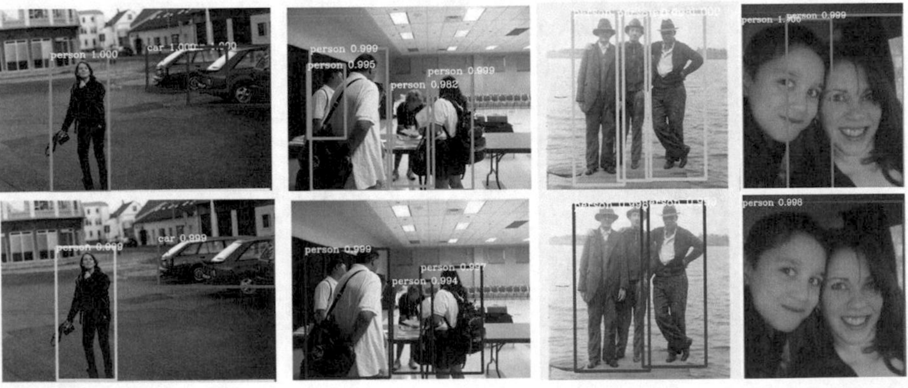

Fig. 5. Comparison for overlapping objects detection. Images in the first row (top) show the detection results by our proposed model. Images in the second row (bottom) are the detection results produced by R-CNN. The comparison demonstrates the superior ability of our model to distinguish overlapping objects.

regions for the sofa and the animal lies on the sofa in the first image, the red bus in the third image and the chair in the fourth image. However, Faster R-CNN fails to capture these objects with geometric deformation or occlusion. The images in Fig. 4 include overlapping objects which should be detected separately. Compared to R-CNN, our model is able to distinguish the overlapping objects.

Demonstrated by the quantitative metric assessment and visual comparison, the proposed model is effective for object detection task. We extract features both globally and locally and combinedly take advantage of deformable convolution and pooling to make our model adaptive to objects with geometric deformation, occlusion and overlapping.

5 Conclusions

In this paper, we present a novel and adaptive object detection model. The proposed approach is able to adaptively adjust geometric variations and model geometric transformations in object scale, pose and part deformation. Especially, we adopts the state-of-the-art image classification methods as backbones, and also add deformable modules into our framework to enhance the capability of transformation modeling. Besides, our method fuses the feature produced by R-FCN and Faster R-CNN, then generates the final feature for accurate prediction. In general, our detector provides a novel idea that utilizes a dual branch which combines global and local feature information to make further classification and location. Benefiting from the architecture, our detector can well adapt to objects with occlusion or deformation, even at a high IoU threshold, which strongly verifies that the approach we propose is robust and efficient. In the future, we will try to optimize the structure and efficiency of the proposed approach to obtain the real-time detection result.

Acknowledgments. This work was supported by the National Natural Science Fund of China (No. 61871170) and the National Equipment Development Pre-research Fund: 6140137050202.

References

1. Dai, J., Li, Y., He, K., Sun, J.: R-FCN: object detection via region-based fully convolutional networks, pp. 379–387 (2016)
2. Dai, J., Qi, H., Xiong, Y., Li, Y., Zhang, G., Hu, H., Wei, Y.: Deformable convolutional networks. In: IEEE International Conference on Computer Vision, ICCV 2017, pp. 764–773 (2017)
3. Everingham, M., Gool, L.J.V., Williams, C.K.I., Winn, J.M., Zisserman, A.: The pascal visual object classes (VOC) challenge. Int. J. Comput. Vis. **88**(2), 303–338 (2010)
4. Felzenszwalb, P.F., Girshick, R.B., McAllester, D.A., Ramanan, D.: Object detection with discriminatively trained part-based models. IEEE Trans. Pattern Anal. Mach. Intell. **32**(9), 1627–1645 (2010)

5. Girshick, R.: Fast R-CNN. In: IEEE International Conference on Computer Vision, pp. 1440–1448 (2015)
6. Girshick, R.B., Donahue, J., Darrell, T., Malik, J.: Rich feature hierarchies for accurate object detection and semantic segmentation. In: 2014 IEEE Conference on Computer Vision and Pattern Recognition, CVPR 2014, Columbus, OH, USA, 23–28 June 2014, pp. 580–587 (2014)
7. Girshick, R.B., Iandola, F.N., Darrell, T., Malik, J.: Deformable part models are convolutional neural networks. In: IEEE Conference on Computer Vision and Pattern Recognition, CVPR 2015, Boston, MA, USA, 7–12 June 2015, pp. 437–446 (2015)
8. He, K., Gkioxari, G., Dollár, P., Girshick, R.B.: Mask R-CNN. In: IEEE International Conference on Computer Vision, ICCV 2017, Venice, Italy, 22–29 October 2017, pp. 2980–2988 (2017)
9. He, K., Zhang, X., Ren, S., Sun, J.: Deep residual learning for image recognition. In: 2016 IEEE Conference on Computer Vision and Pattern Recognition, CVPR 2016, pp. 770–778 (2016)
10. He, K., Zhang, X., Ren, S., Sun, J.: Identity mappings in deep residual networks. In: Computer Vision - ECCV 2016 - 14th European Conference, Amsterdam, The Netherlands, 11–14 October 2016, Proceedings, Part IV, pp. 630–645 (2016)
11. Jaderberg, M., Simonyan, K., Zisserman, A., Kavukcuoglu, K.: Spatial transformer networks. In: Advances in Neural Information Processing Systems 28: Annual Conference on Neural Information Processing Systems 2015, Montreal, Quebec, Canada, 7–12 December 2015, pp. 2017–2025 (2015)
12. Krizhevsky, A., Sutskever, I., Hinton, G.E.: Imagenet classification with deep convolutional neural networks. In: International Conference on Neural Information Processing Systems, pp. 1097–1105 (2012)
13. Li, Z., Peng, C., Yu, G., Zhang, X., Deng, Y., Sun, J.: Light-head R-CNN: in defense of two-stage object detector. CoRR abs/1711.07264 (2017)
14. Lin, C., Lucey, S.: Inverse compositional spatial transformer networks. In: 2017 IEEE Conference on Computer Vision and Pattern Recognition, CVPR 2017, Honolulu, HI, USA, 21–26 July 2017, pp. 2252–2260 (2017)
15. Lin, T., Dollár, P., Girshick, R.B., He, K., Hariharan, B., Belongie, S.J.: Feature pyramid networks for object detection, pp. 936–944 (2017)
16. Lin, T., Goyal, P., Girshick, R.B., He, K., Dollár, P.: Focal loss for dense object detection. In: IEEE International Conference on Computer Vision, ICCV 2017, Venice, Italy, 22–29 October 2017, pp. 2999–3007 (2017)
17. Liu, W., Anguelov, D., Erhan, D., Szegedy, C., Reed, S.E., Fu, C., Berg, A.C.: SSD: single shot multibox detector. In: Computer Vision - ECCV 2016 - 14th European Conference, Amsterdam, The Netherlands, 11–14 October 2016, Proceedings, Part I, pp. 21–37 (2016)
18. Lu, Q., Liu, C., Jiang, Z., Men, A., Yang, B.: G-CNN: object detection via grid convolutional neural network. IEEE Access **5**, 24023–24031 (2017)
19. Ren, S., He, K., Girshick, R., Sun, J.: Faster R-CNN: towards real-time object detection with region proposal networks, pp. 91–99 (2015)
20. Russakovsky, O., Deng, J., Su, H., Krause, J., Satheesh, S., Ma, S., Huang, Z., Karpathy, A., Khosla, A., Bernstein, M.S., Berg, A.C., Li, F.: Imagenet large scale visual recognition challenge. Int. J. Comput. Vis. **115**(3), 211–252 (2015)
21. Shelhamer, E., Long, J., Darrell, T.: Fully convolutional networks for semantic segmentation. IEEE Trans. Pattern Anal. Mach. Intell. **39**(4), 640–651 (2017)

22. Shrivastava, A., Gupta, A., Girshick, R.B.: Training region-based object detectors with online hard example mining. In: 2016 IEEE Conference on Computer Vision and Pattern Recognition, CVPR 2016, Las Vegas, NV, USA, 27–30 June 2016, pp. 761–769 (2016)
23. Simonyan, K., Zisserman, A.: Very deep convolutional networks for large-scale image recognition. CoRR abs/1409.1556 (2014)
24. Szegedy, C., Liu, W., Jia, Y., Sermanet, P., Reed, S.E., Anguelov, D., Erhan, D., Vanhoucke, V., Rabinovich, A.: Going deeper with convolutions. In: IEEE Conference on Computer Vision and Pattern Recognition, CVPR 2015, pp. 1–9 (2015)
25. Xie, S., Girshick, R.B., Dollár, P., Tu, Z., He, K.: Aggregated residual transformations for deep neural networks. In: 2017 IEEE Conference on Computer Vision and Pattern Recognition, CVPR 2017, pp. 5987–5995 (2017)
26. Yan, C., Xie, H., Yang, D., Yin, J., Zhang, Y., Dai, Q.: Supervised hash coding with deep neural network for environment perception of intelligent vehicles. IEEE Trans. Intell. Transp. Syst. **19**(1), 284–295 (2018)
27. Zhang, X., Zhou, X., Lin, M., Sun, J.: Shufflenet: an extremely efficient convolutional neural network for mobile devices. CoRR abs/1707.01083 (2017)
28. Zhu, Y., Zhao, C., Wang, J., Zhao, X., Wu, Y., Lu, H.: Couplenet: coupling global structure with local parts for object detection. In: IEEE International Conference on Computer Vision, ICCV 2017, Venice, Italy, 22–29 October 2017, pp. 4146–4154 (2017)

Initial Sensitivity Optimization Algorithm for Fuzzy-C-Means Based on Particle Swarm Optimization

Zilong Ye[1(✉)], Feng Qi[1], Jingquan Li[2], Yanjun Liu[2], and Han Su[2]

[1] Beijing University of Posts and Telecommunications, No 10, Xitucheng Road, Haidian District, Beijing 100876, China
shandianshi@bupt.edu.cn

[2] State Grid HeBei Electric Power Supply Co. LTD, Shijiazhuang, HeBei, China

Abstract. As a local search algorithm, FCM is sensitive to the initial value. Randomly initializing the centroids or membership matrix will cause the algorithm to fall into local optimum, thus affecting the accuracy and classification results of FCM. In this paper, a fuzzy-C-Means initial sensitivity optimization algorithm, which based on particle swarm, is proposed for the above problems. In the standard PSO algorithm, the Levi flight formula is introduced to simulate global random walk to enhance particle activities and control the balance of local walking and global random walking in the distance formula by a switching parameter, finally coupled with FCM algorithm. In the experimental stage, this paper conducts clustering test and validity analysis on the accuracy and fitness variety of the algorithm through a suite of UCI standard data sets. The experimental results show that compared with the FCM algorithm and the PSO-FCM algorithm, the PSO-LF-FCM algorithm enhances the clustering accuracy and the global search performance in the later iteration of the algorithm, which implies its superior global convergence and optimal solution search ability.

Keywords: Fuzzy-C-Means · Particle Swarm Optimization · Levy Flight

1 Introduction

FCM algorithm is a soft partitioning clustering algorithm that uses membership to determine the degree to which each data point belongs to a certain cluster. However, FCM is sensitive to initial value, and the algorithm uses the gradient operator to iteratively optimize the objective function, which leads to the algorithm easily falling into the local minimum thus affect the clustering result.

Many scholars have incorporated biological algorithms or genetic algorithms into FCM algorithms. Dai et al. [1] proposed an improved clustering algorithm based on optimized artificial fish swarm algorithm and FCM. This algorithm obtains a better initial clustering result by integrating the feature of global optimization and fast convergence in artificial fish swarm algorithm, and then applies FCM algorithm in local search. The algorithm avoids the initial value sensitivity problem of the FCM method, but the algorithm requires numerous parameters and large computational overhead. Compared with other genetic algorithms and bionic algorithms, the particle swarm

C.-N. Yang et al. (Eds.): SICBS 2018, AISC 895, pp. 798–807, 2020.
https://doi.org/10.1007/978-3-030-16946-6_65

algorithm greatly reduces the initial parameters, and brings less expanses in calculation process. Many scholars have used particle swarm optimization to improve FCM in different scenarios, but the particle swarm algorithm itself may cause premature phenomenon caused by its fast converge process. Fu [2] proposed a text clustering method of FCM clustering based on Chaotic Oscillation Particle Swarm Optimization. This method introduces the oscillation link and chaos theory into the standard PSO algorithm to increase the diversity and convergence of the algorithm, but the algorithm only clusters in text. Aspects are applied, lack of generality.

Zhang et al. [3] proposes a dynamic adaptation cuckoo search algorithm that introduces the diversity and concentration of particle swarm into the CS search algorithm. By combining the CS algorithm with the learning parameter mechanism of particle swarm, the new algorithm is more flexible and versatile compared to CS algorithm. Inspired by the idea of [3], this paper combined the mechanism of global random walk in CS algorithm, introduced Levi flight into PSO, and integrated the improved PSO algorithm and FCM in order to solve the defect of FCM algorithm to initial value sensitivity and help PSO in avoiding stagnation. The optimization algorithm not only retains the local search ability, but also introduces a global random walk mechanism to improve the global search ability.

2 The FCM Algorithm

The FCM algorithm is a partition-based clustering algorithm and its goal is to maximize the similarity between elements that parted into the same cluster, while minimalizing the similarity between different clusters. FCM is an improvement of the C-means algorithm. The C-means algorithm divide the data in hard clustering, while FCM uses flexible fuzzy division, which is to determine the degree to which each data point belongs to a certain cluster by membership degree. The clustering algorithm FCM divides n vectors xi(i = 1, 2, … , n) into c fuzzy groups, and finds the centroids of each group, aims to minimize the non-similarity index value function.

The main difference between FCM and other hard clustering algorithms is that FCM uses fuzzy partitioning, so that each data point can have membership to multiple clusters. By relaxing the definition of membership coefficients from strictly 1 or 0, these values can range from any value from 1 to 0 to determine the extent to which it belongs to each group. After normalization, the sum of memberships of a data set is always equal to 1:

$$\sum\nolimits_{i=1}^{c} u_{ij} = 1, \forall j = 1, \ldots, n \tag{1}$$

After applying the algorithm to the fuzzy set, Dunn weights the distance between the sample and the class center by the square of the membership degree. Furthermore, Bezdek introduces fuzzifier m in to the objective function:

$$\mathrm{J} = \sum\nolimits_{i=1}^{k} \sum\nolimits_{j=1}^{n} (u_{ij})^m s_{ij}^2 \tag{2}$$

Where s_{ig} is the distance between dataset element x_i to centroid C_j

To make the formula (2) reach the minimum value, the centroid C_i and the membership degree u_{ij} need to satisfy the following conditions:

$$C_i = \frac{\sum_{j=1}^{n} (u_{ij})^m X_j}{\sum_{j=1}^{n}} u_{ij}^m \tag{3}$$

$$u_{ij} = \frac{1}{\sum_{l=1}^{k} \left(\frac{u_{ij}}{u_{lj}}\right)^{2/(m-1)}} \tag{4}$$

The FCM algorithm can be described as a simple iterative process with the requirements of clustering center and membership degree. The method to optimize the objective function is to update the membership matrix by applying Lagrange Multiplier, and the objective function is Gradient Descent through iteration. The FCM algorithm belongs to the climbing hill algorithm of local search mechanism, which is very sensitive to the initial value selection. Since the initial membership matrix or initial centroid is randomly generated, the algorithm may result in local optimum when the initial value is not selected properly. Besides, the algorithm's accuracy will be affected while multiple extreme points are existed in the date sets.

Aiming at the problems above, we proposed A Fuzzy-C-Means algorithm based on Particle Swarm Optimization and Levy Flight. Using Particle swarm optimization as a global optimization method can make up for the shortcomings of FCM, while the Levi's flight mechanism is further introduced into the particle swarm optimization algorithm, which enhances the particle activity and the global search ability.

3 Fuzzy-C-Means Algorithm Based on Particle Swarm Optimization and Levy Flight

3.1 Particle Swarm Optimization Method Based on Levi Flight

The Particle Swarm Algorithm was proposed by Kennedy and Eberhart in 1995. The algorithm corrects the model of Hepper's simulated bird population in order to make particles fly to the solution space and land at the best solution. Similar to genetic algorithm, the Particle Swarm Algorithm is based on population iteration, however it doesn't apply crossover and mutation used in genetic algorithm, but the particle searches in the solution space to follow the optimal particle.

PSO is advantaged in its simplicity, easy to implement with fewer parameters and requires no gradient information. Its real-number-encoding feature is suitable for dealing with optimization problems. The particle updates its velocity and positions with following equation:

$$v_i^{t+1} = wv_i^t + c_1 r_1 (p_i^t - x_i^t) + c_2 r_2 (p_g^t - x_i^t) \tag{5}$$

$$x_i^{t+1} = x_i^t + v_i^{t+1} \tag{6}$$

c_1 and c_2 are social and cognitive acceleration constants, $r1$ and $r2$ are random numbers between 0 and 1, x_i^t, v_i^t represent the position and velocity of ith particle in tth iteration. whereas p_i^t and p_g^t are the *personal best* and *global best* positions. ω represents the inertia of the particle.

The particle flight direction is determined based on the individual's best position and the global best position during the iterative process. Thus when a particle reaches the local extremum, the particle's velocity will drop rapidly, leading to premature convergence. Here, the randomized Levi flight mechanism in the CS search algorithm is introduced to help PSO algorithm escape the local minima.

The main feature of Levi's flight is the combination of long-distance step and short-distance step. The random walk takes place on a discrete grid rather than on a continuous space, and step-lengths have a probability distribution that is heavy-tailed. Which is an effective ways to find the target. The formula for Levi's flight update particle position is read as:

$$x_{t+1} = x_t + \alpha \otimes Levy(\beta) \tag{7}$$

Where α is the stability parameter, $Levy(\beta)$ is levy flight random walk path, $\otimes$ is matrix dot product operation.

P. Levy determined that the integral form of the symmetric Levy stable distribution as the equation below:

$$Levy(s) = \frac{1}{\pi}\int_0^{+\infty} exp(-\beta|k|^{\lambda})cos(ks)dk \tag{8}$$

$$s = \frac{u}{|v|^{\frac{1}{\beta}}} \tag{9}$$

When $s \geq s_0 > 0$, $s \to \infty$, $Levy(s) \approx \frac{\lambda\beta\Gamma(\lambda)\sin(\frac{\pi\lambda}{2})}{\pi} \cdot \frac{1}{s^{1+\lambda}}$, $\beta = 1$, $u \sim N(0, \sigma^2)$, $v \sim N(0, 1), \sigma = \left\{\frac{\Gamma(1+\beta)\sin(\frac{\pi\beta}{2})}{\beta\Gamma(\frac{1+\beta}{2}) \cdot 2^{(\frac{\beta-1}{2})}}\right\}^{\frac{1}{\beta}}$.

The variance of Levi's flight indicates an exponential relationship with time, *i.e.* $(t) \sim t^{3-\beta}, 1 \leq \beta \leq 3$, which is essentially a Markov chain. The next state/position is only It depends on the current state x_t and the transition probability $Levy(\beta)$ A large probability of its new solution is generated by far-field randomization, and their position is far enough away from the current best solution, which can ensure that the system does not fall into local optimum. Therefore, Levi's flight is superior to Brown's motion in enhancing particle activity.

By introducing Levi flight to update the particle position, Particle Swarm Optimization Method Based on Levi Flight (LF-PSO) adopted different modes to update particles' position: global random walk and local walk. The local walking formula uses formula (5), and the global random walk uses the following equation:

$$x_{t+1} = x_t + \alpha \otimes Levy(\beta) \otimes (p_g^t - x_t) \tag{10}$$

After each iteration, when the current fitness of the particle is less than its personal best fitness, the global random walk is performed with the probability p_a and performs local walking with the probability $1 - p_a$. Besides, when the ratio of the objective function before and after the iteration is smaller than the threshold value, the particles' position is updated by the Levi flight formula to increase the particle activity and avoid premature convergence in early stages and the lack of particle activity in later period iterations. The PSO-LF algorithm flowchart is outlined below (Fig. 1):

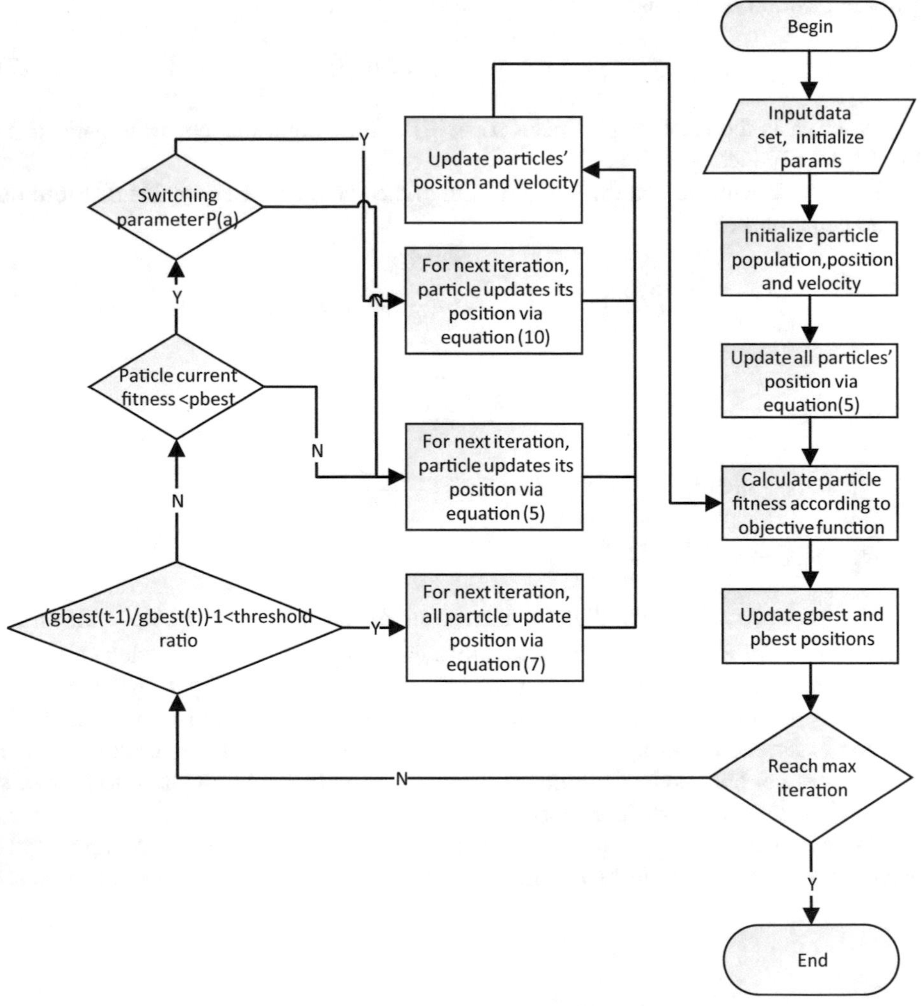

Fig. 1. PSO-LF flowchart

3.2 FCM Clustering Algorithm Based on LF-PSO

3.2.1 Particle Encoding

In this approach, each particle is a n-dimensional candidate solution that can be formally represented as the vector $C = (c_1, c_2, c_i \ldots c_n)$, where c_i represents a possible centroid solution to $i'th$ cluster, each solution is a $1 * d$ shaped vector, d is the dimension of the data set.

3.2.2 Fitness Function

We use the objective function of FCM as the algorithm's fitness function

$$f(C_l) = J = \sum\nolimits_{i=1}^{k} \sum\nolimits_{j=1}^{n} (u_{ij})^m s_{ij}^2 \tag{11}$$

s_{ig} is the distance between dataset element x_i to centroid C_j, u_{ij} is the membership value data set element x_i affiliated to centroid C_j. A lower the fitness value means better clustering results.

3.2.3 Specific Steps

The PSO-LF-FCM algorithm flowchart is outlined below (Fig. 2):

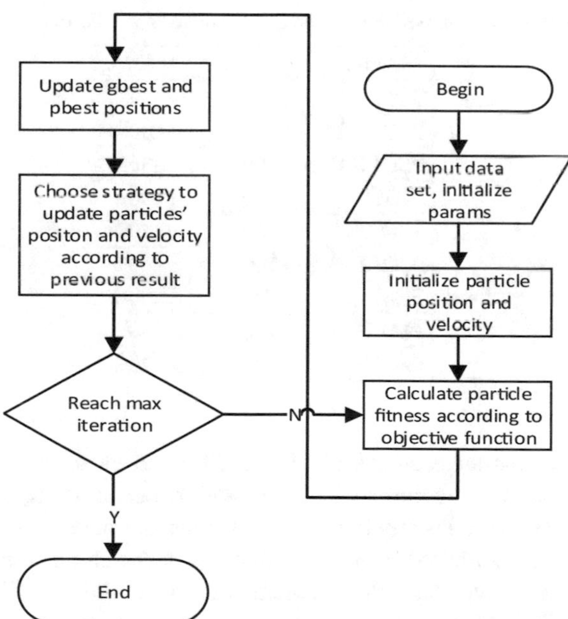

Fig. 2. PSO-LF-FCM flowchart

4 Results and Analysis

In order to verify the validity of the algorithm, we write a verification program based on UCI dataset. The program is implemented in Python 2.7.12 with an Intel(R) Core(TM) i7-6700HQ CPU @2.60 GHz. Experimental results for 10 trials are tabulated and are thereafter analyzed.

In this experiment, fuzzifier m = 2, particles population N = 25, FCM membership matrix minimum deviation value $\varepsilon = 0.001$, inertia weight w is decreased linearly from 0.9 to 0.1 over the course of iterations, learning parameters c_1 and c_2 are chosen as 2, iterations T = 100, switching parameters $p_a = 0.15$, fitness ratio threshold $p = 1\%$. This paper selects data set Iris, Wine and Seed from UCI Machine Learning Repository for the test, the experimental results use accuracy and fitness value as indicators, wherein the accuracy is the algorithm output classification results checked in the official results, which reflects the validity and accuracy of the clustering results. The fitness value is represented as the global particle fitness value variety during the iterative process, reflecting the convergence speed of the algorithm iterative process. Table 1 shows the accuracy FCM, PSO-FCM and PSOLF-FCM algorithms. Figures 3, 4 and 5 are the fitness value variety of the PSO-FCM and PSOLF-FCM algorithms for the Iris, Wine, and Seed data sets under Mahalanobis distance metric.

Table 1. Comparison of various accuracy for UCI data set

	Algorithm	Iteration	Accuracy
Iris	FCM	17	89.30%
	PSO-FCM	100	90.00%
	PSOLF-FCM	100	91.30%
Wine	FCM	30	68.50%
	PSO-FCM	100	76.20%
	PSOLF-FCM	100	80.10%
Seed	FCM	22	89.50%
	PSO-FCM	100	91.30%
	PSOLF-FCM	100	92.40%

Table 1 implies that the accuracy of PSOLF-FCM on most data sets is better than previous algorithms. The traditional FCM algorithm performs unpromising since it adopts gradient operator and is sensitive to initial value and pretend to fall into the local minimum, while the PSO-based FCM algorithm has better global search performance, which can effectively overcome the problem that the traditional FCM algorithm is sensitive to the initial value.

Figures 3, 4, and 5 present that PSOLF-FCM is converge more gradually than the PSO-FCM in early phase, but the particle is more active in the late iterations, and the final fitness value is better, which proves the effectiveness of introducing the Levi flight mechanism. The result also indicates that the algorithm is also effective to avoid the premature convergence due to its improved global search ability.

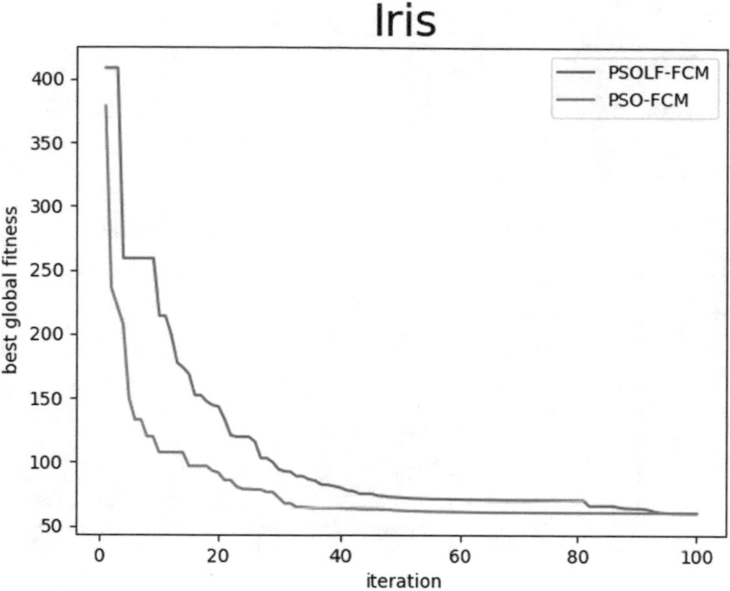

Fig. 3. Fitness variety of algorithms on iris

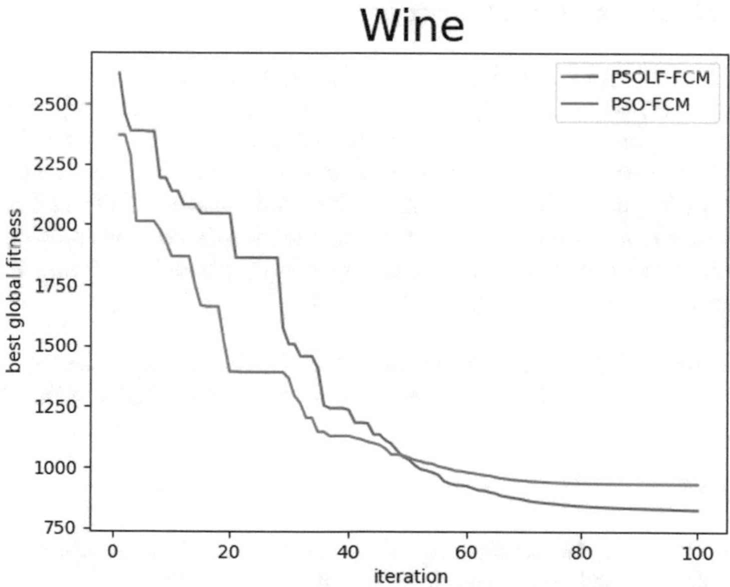

Fig. 4. Fitness variety of algorithms on wine

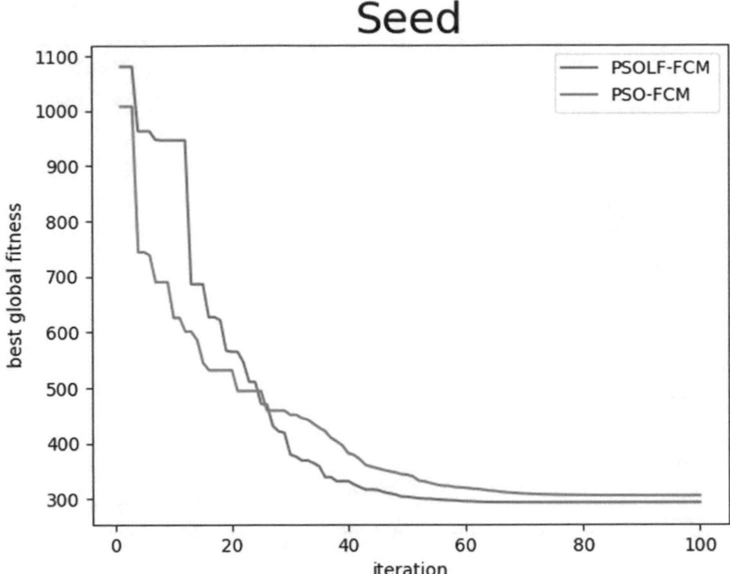

Fig. 5. Fitness variety of algorithms on seed

5 Conclusions and Future Scope

This paper makes an effort to integrate Levi's flight and particle swarm optimization algorithm into FCM clustering algorithm and contrast the accuracy and performance of Fuzzy C-Means upon hybridization with the standard. PSO-LF-FCM utilizes fuzzy membership concepts of FCM and the superior global convergence ability of PSO-LF, thus avoiding stagnation in local optima in the multidimensional fitness landscape in FCM and premature convergence in PSO. Future work will focus on applying dynamic learning mechanisms and analyzing other supervised approaches, aiming at reducing time expanses and improving algorithm stability.

Acknowledgement. This paper is supported by Data Network Security Assessment Strategy and State Analysis Research Information Communication and Security Technology Project.

References

1. Dai, Y., Zhao, L.: An improved, clustering algorithm based on optimized artificial fish swarm algorithm and FCM. Comput. Appl. Softw. **33**(12) (2016)
2. Fu, B.: The text clustering method of FCM clustering based on chaotic oscillation particle swarm optimization. J. Hechi Univ. **35**(02), 74–77 (2015)
3. Zhang, Y., Wang, L., Wi, Q.: Dynamic adaptation cuckoo search algorithm. Control Decis. **29**(04), 617–622 (2014)
4. Sun, W., Meng, b, Wu, X.: Improved clustering algorithm based on cuckoo search. Microelectron. Comput. **35**(08), 16–20 (2018)

5. Zhang, J., Shen, L.: An improved fuzzy c-Means clustering algorithm based on shadowed sets and PSO. Comput. Intell. Neurosci. **2014**, 368628 (2014)
6. Zhou, K., Fu, C., Yang, S.: Fuzziness parameter selection in fuzzy c-means: the perspective of cluster validation. Sci. China (Inf. Sci.) **57**(11), 252–259 (2014)
7. Samadzadegan, F.: Evaluating the potential of particle swarm optimization in clustering of hyperspectral imagery using fuzzy c-means. Asia-Pacific Chemical, Biological & Environmental Engineering Society (APCBEES). In: Proceedings of International Conference on Asia Agriculture and Animal (ICAAA 2011). Asia-Pacific Chemical, Biological & Environmental Engineering Society (APCBEES), p. 7 (2011)
8. Chen, X., Liao, J., Zhao, X., Chen, J.: On the FCM clustering method based on particle swarm optimization with tabu search. J. Hubei Univ. Technol. **28**(02), 45–48 (2013)
9. Yin, H., et al.: Fish swarm algorithm with Levy flight and firefly behavior. Control Theory Appl. **35**(4) (2018)
10. Zhao, Y.: Application of improved cuckoo search in parameter inversion of average elastic moduli of dam and foundation. Pearl River **39**(8) (2018)
11. Zhu, C., Li, L., Guo, J.: Fuzzy clustering image segmentation algorithm based on improved cuckoo search. Comput. Sci. **44**(6) (2017)
12. Silva Filho, T.M., Pimentel, B.A., Souza, R.M.C.R., Oliveira, A.L.I.: Hybrid methods for fuzzy clustering based on fuzzy c-means and improved particle swarm optimization. Expert. Syst. Appl. **42**(17–18), 6315–6328 (2015)
13. Haklı, H., Uğuz, H.: A novel particle swarm optimization algorithm with Levy flight. Appl. Soft Comput. J. **23**, 333–345 (2014)
14. Wang, Y.: Fuzzy C-means clustering algorithm based on particle swarm optimization. Microcomput. Appl. **37**(08), 36–39+44.11 (2018)
15. Wang, J., et al.: Constrained multi-objective particle swarm optimization algorithm based on self-adaptive evolutionary learning. Control Decis. **29**(10), 1765–1770 (2014)

Multi-modal Behavioral Information-Aware Recommendation with Recurrent Neural Networks

Guoyong Cai[1(✉)], Nannan Chen[1], Weidong Gu[1], and Jiao Pan[2]

[1] Guangxi Key Lab of Trusted Software,
Guilin University of Electronic Technology, Guilin, China
ccgycai@guet.edu.cn

[2] Guilin Kaige Information Technology Co. Ltd., Guilin, China

Abstract. Data sparsity is one of the most challenging problems in recommendation systems. In this paper, we tackle this problem by proposing a novel multi-modal behavioral information-aware recommendation method named MIAR which is based on recurrent neural networks and matrix factorization. First, an interaction context-aware sequential prediction model is designed to capture user-item interaction contextual information and behavioral sequence information. Second, an attributed context-aware rating prediction model is proposed to capture attribution contextual information and rating information. Finally, three fusion methods are developed to combine two sub-models. As a result, the MIAR method has several distinguished advantages in terms of mitigating the data sparsity problem. The method can well perceive diverse influences of interaction and attribution contextual information. Meanwhile, a large number of behavioral sequence and rating information can be utilized by the MIAR approach. The proposed algorithm is evaluated on real-world datasets and the experimental results show that MIAR can significantly improve recommendation performance compared to the existing state-of-art recommendation algorithms.

Keywords: Data sparsity · Multi-modal information · Recommendation · Recurrent neural networks

1 Introduction

In recommendation systems, traditional collaborative filtering algorithms usually are suffered by the data sparsity problem, which leads to low recommendation accuracy. In reality, rich contextual information (e.g. time and location) plays an important role in recommendations [1].

Incorporating contextual information into recommendation can mitigate the data sparsity problem and boost the recommendation performance. Usually, contextual information is divided into two different types [2]. One is the attribute context of users and items [3] (e.g. age, sex, occupation, genre, etc.). The other is the user-item interaction context [4] (such as time, location and so on). As contextual information

C.-N. Yang et al. (Eds.): SICBS 2018, AISC 895, pp. 808–823, 2020.
https://doi.org/10.1007/978-3-030-16946-6_66

becomes increasingly important in recommendation systems, many different kinds of context-aware recommendation methods have been developed.

There are various representations methods of contextual information in recommendation, Karatzoglou [5] proposed a context-aware n-dimensional tensor recommendation methods, which treats the context as dimensions of the user-item-context tensor. Rendle [6] proposed a fast context-aware recommendation model with factorization machines, which takes different contexts as the corresponding contextual feature vectors and then fuses the context with feature vectors of user and item. Zhang [3] proposed an attribute context-aware recommendation model with matrix factorization, which defines different types of attribute context as the corresponding bias variables. These works above consider contextual information as dimensions similar to users and items, which captures the common effect of various contexts on users and items but ignore the specific impact of contextual information on each user or item. To deal with this problem, Liu [4] proposed a contextual operating tensor recommendation model, which can perceive specific influences of interaction context on users or items. Nevertheless this method neglect the influence of attribute context of users and items.

There are different methods of incorporating contextual information. Zheng [7] proposed a collaborative filtering method for mobile recommendation on locations and activities, which utilizes GPS trajectory data of users. Levandoski [8] proposed a location context-aware recommendation method which exploits spatial attributions of users and items. Ren [9] proposed an interaction context-aware probabilistic matrix factorization method, which captures geographical information, social information and other information for the POI recommendation. Based on the contextual operating tensors model in [4], Wu [10] proposed three contextual operating tensors for user attribute, item attribute and interaction context, which can capture specific contextual influences on the latent vector of users or items. However, this method is not scalable. Cai [2] proposed a heterogeneous context-aware recommendation model. This model contains two different complicated context-aware sub-models of interaction and attribute context based on tensor factorization and matrix factorization respectively. Then the two sub-models are fused by semi-supervised co-training.

However, the above context-aware recommendation models in [2–10] neglect the significant effects of behavioral sequence information on recommendation process.

In fact, behavioral sequence information also plays an increasingly important role in the recommendation because the real application scenes contain various context information as well as a large amount of behavioral sequence information. Recently, recurrent neural networks are reported to model dependencies of sequence information in some circumstances such as click prediction [11], location prediction [12] and so on. With the increasing importance of behavioral sequence information in recommendation, various kinds of behavioral sequence methods based on the recurrent neural networks have been developed. Wu [13] proposed a real time recommendation model in e-commerce platform, which captures historical sequence information of users. Okura [14] proposed a news recommendation model, which contains a variant of the denoising autoencoder to capture distributed representations of articles and a recurrent neural network sub-model to capture user representations of historical sequences. However, the above works [11–14] do not capture effects of contextual information. To deal with this problem, Liu [15] proposed a context-aware sequential recommendation model (CARNN), which can capture influences of behavioral sequence and interaction contextual information. Whereas, this method ignores effects of ratings and attribution contextual information on recommendation.

In a recommendation procedure, the rating behavior of a user can be divided into two steps [16]. Step1 is that a user choose an item. Step2 is that the user rates on the selected item. Where selecting behavioral sequences of users are implicit feedback information and ratings are explicit feedback information. Therefore, to some extent, fusing implicit and explicit feedback can also mitigate data sparsity problems.

To deal with above mentioned problems of context-aware and behavioral sequence recommendation methods and mitigate the data sparsity, we exploit several types of information and propose a new multi-modal behavioral information-aware recommendation method named MIAR based on recurrent neural networks. Firstly, we construct an interaction context-aware sequential prediction model based on method in [15]. Secondly, we construct an attribution context-aware rating prediction model based on matrix factorization method. Finally, the two sub-models are fused into a unified MIAR model effectively via three fusion strategies.

To summarize, our main contributions of this work are listed as follows:

- Various types of context are perceived by different methods after analyzing their properties of various context.
- After analyzing behavioral process of users, we use matrix factorization to capture rating information of the explicit feedback and choose recurrent neural networks to capture behavioral sequences information of the implicit feedback, which can mitigate the data sparsity problem.
- We propose three fusion strategies from different perspectives, i.e., linear weighted fusion, product fusion and stages fusion, to capture a variety of information jointly and the proposed approaches are compared and analyzed through extensive experiments.

2 The Framework of MIAR

An overall of the proposed MIAR model is depicted in Fig. 1, which contains three steps as follows:

Step1: Constructing Interaction Context-aware Sequential Prediction Model. In this step, a variant of recurrent neural network model is designed to capture influences of correspondingly complicated interaction context information and behavioral sequence information for recommendation.

Step2: Constructing Attribution Context-aware Rating Prediction Model. In this step, a variant of matrix factorization model is applied to capture influences of rating and attribution context information for recommendation.

Step3: Fusing Model. In the final step, interaction context-aware sequential prediction model and attribution context-aware rating prediction model are fused into a unified MIAR model.

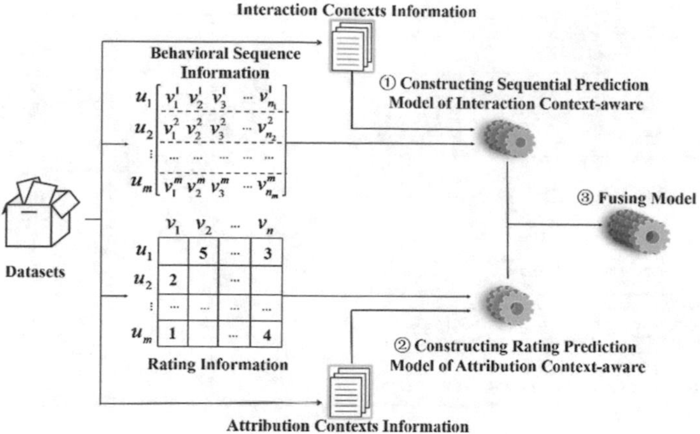

Fig. 1. An overview of the proposed MIAR method.

As shown in Fig. 1, users and items are denoted by $U = \{u_1, u_2, \ldots\}$ and $V = \{v_1, v_2, \ldots\}$. Historical behaviors of user i are denoted by $V^i = \{v_1^i, v_2^i, \ldots\}$ where v_k^i denotes the k-th selected item by user i.

2.1 Constructing Interaction Context-Aware Sequential Prediction Model

With the development of information technology, behavioral sequence information and interaction context information become important in recommendation. Therefore, we construct an interaction context-aware sequential prediction model.

Definition and Notation. We define the input context as specific interaction context (time, location and so on) under user's behaviors. The transition context is defined as time intervals between adjacent behaviors. Usually, the shorter time interval between the adjacent behaviors in historical sequences indicate stronger influence on prediction to next behavior.

As shown in Fig. 2, the left part is an example of interaction context-aware sequential prediction of a user. It consists of input context denoting specific interaction context (weather and location) under user's purchasing behaviors and transition context denoting time intervals between adjacent user's purchasing behaviors.

As shown in Fig. 2, the right part illustrates the model of interaction context-aware sequential prediction, where input context and transition context have influence on predicting next behavior of users. Behavior timestamps of user i selecting items are denoted by $T^i = \{t_1^i, t_2^i, \ldots\}$, where t_k^i denotes a corresponding timestamp of the k-th behavior in the user i's behavioral sequence. Input context of user i's behavior at the timestamp t_k^i is denoted by $c_{IN,k}^i$ which means the specific interaction context under user's behaviors, and the corresponding input context matrix is denoted by $M_{c_{IN,k}^i}$. The transition context $c_{T,k}^i$ denotes time intervals of adjacent user i's behaviors between the timestamp t_k^i and the previous timestamp t_{k-1}^i, and the corresponding transition context

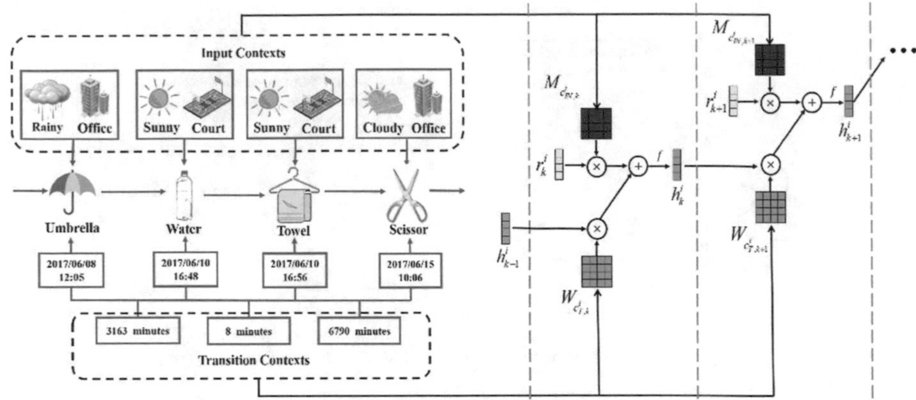

Fig. 2. Behavior scene and interaction context-aware sequential prediction.

matrix is denoted by $W_{c^i_{T,k}}$. The d dimensional latent vector of the corresponding item v^i_k is denoted by r^i_k. Besides,h^i_{k-1} denotes the d dimensional hidden status of user i at the timestamp t^i_{k-1} in the behavior sequence. Furthermore, f denotes the activation function $f(x)$ which is usually selected as a sigmoid function.

Traditional Recurrent Network Networks. The architecture of the traditional recurrent neural networks is shown as following equation:

$$h^i_k = f(r^i_k M + h^i_{k-1} W), \tag{1}$$

where h^i_k denotes d dimensional hidden status of user i at the timestamp t^i_k in the behavior sequence. r^i_k denotes d dimensional latent vector of the corresponding item v^i_k. M denotes $d \times d$ dimensional input matrix to capture inputs of the user current behavior. W denotes $d \times d$ dimensional transition matrix to propagate the hidden status of previous hidden layer. $f(x)$ denotes activation function.

However traditional recurrent neural networks can't well deal with contextual information. Therefore, constructing our proposed interaction context-aware sequential prediction model consists of three steps given as follows: Modeling Input Context, Modeling Transition Context, and Interaction Context-aware Prediction.

Modeling Input Context. In reality, the interaction contextual information (time, location, weather and so on) also plays an important role in recommendation. Therefore, we use the input context $c^i_{IN,k}$ to denote the specific interaction context (such as time, location and so on) under user's behaviors and then model the input context $c^i_{IN,k}$

by replacing input matrix M with input context matrix $M_{c^i_{IN,k}}$ in traditional recurrent neural networks as following:

$$h^i_k = f(r^i_k M_{c^i_{IN,k}} + h^i_{k-1} W), \tag{2}$$

where $M_{c^i_{IN,k}}$ denotes the $d \times d$ dimensional corresponding input context matrix of the input context $c^i_{IN,k}$.

Modeling Transition Context. Diverse lengths of time intervals between adjacent behaviors in historical sequences have different influences on predicting next behavior. Generally speaking, the shorter time interval between the adjacent behaviors in historical sequences suggests stronger influence on prediction to next behavior. Therefore, we use the transition context $c^i_{T,k}$ to denote the time intervals between adjacent behaviors in historical sequences and then model the transition context $c^i_{T,k}$ by replacing transition matrix W with transition context matrix $W_{c^i_{T,k}}$ in traditional recurrent neural networks as following:

$$h^i_k = f(r^i_k M_{c^i_{IN,k}} + h^i_{k-1} W_{c^i_{T,k}}) \tag{3}$$

where $W_{c^i_{T,k}}$ denotes the $d \times d$ dimensional corresponding transition context matrix of the transition context $c^i_{T,k}$ to capture different influences of diverse time intervals.

However in fact, time intervals between adjacent behaviors in historical sequences always are irregular and not constant. There are too many parameters should be learning if we construct transition context matrix $W_{c^i_{T,k}}$ respectively for each transition context $c^i_{T,k}$. Therefore, we divide all possible time intervals into discrete time bins, and use $\lfloor t^i_k - t^i_{k-1} \rfloor$ to denote the time bins which $c^i_{T,k}$ corresponding time interval belong to. Equation (3) can be replaced by Eq. (4) as following:

$$h^i_k = f(r^i_k M_{c^i_{IN,k}} + h^i_{k-1} W_{\lfloor t^i_k - t^i_{k-1} \rfloor}) \tag{4}$$

Interaction Context-Aware Prediction. Given the historical behaviors sequences $V^i = \{v^i_1, v^i_2, \ldots\}$ of user i, the d dimensional hidden status h^i_k of user i at the timestamp t^i_k in the behavior sequence mentioned by Eq. (4), corresponding input context and transition context, the purpose of interaction context-aware sequential prediction model is to predict the next selected item v^i_{k+1} by user i at the timestamp t^i_{k+1} under the input context $c^i_{IN,k+1}$ and transition context $c^i_{T,k+1}$ corresponding to $M_{c^i_{IN,k+1}}$ and $W_{\lfloor t^i_{k+1} - t^i_k \rfloor}$.

The predicting function of this interaction context-aware sequential prediction model $b_1(i,j)$ can be written as following:

$$b_1(i,j) = ranking_{i,k+1,j} = h_k^i W_{\lfloor t_{k+1}^i - t_k^i \rfloor} (r_j M_{c_{IN,k+1}^i})^T, \quad (5)$$

where $ranking_{i,k+1,j}$ denotes the behavioral ranking of next selected item j by user i at the timestamp t_{k+1}^i under corresponding input context and transition context. r_j denotes the d dimensional latent vector of the item j.

This interaction context-aware sequential prediction model can use Bayesian Personalized Ranking (BPR) and Back Propagation through Time (BPTT) for learning parameters [15].

2.2 Constructing Attribution Context-Aware Rating Prediction Model

In a recommendation system, except interaction context and user behavioral sequences information, attribute context and rating information also contain a large amount of preference information.

Therefore, we integrate the attribute context of users and items into the matrix factorization recommendation model as follows.

$$\hat{r}_{i,j} = w_0 + w_i + w_j + \sum\nolimits_{m \in user\_attributes} w_m + \sum\nolimits_{n \in item\_attributes} w_n + v_j^t u_i \quad (6)$$

where $\hat{r}_{i,j}$ denotes the prediction rating that user i gives to item j. w_0 denotes global average bias. w_i and w_j respectively denote user i bias and item j bias. w_m and w_n respectively denote user attribute bias and item attribute bias. u_i and v_j respectively denote the latent vector of user i and the latent vector of item j.

To further capture information, we can not only use w_{in} to connect the user i with the biases w_n of item context (such as genre and so on) but also use w_{jm} to relate the item j to the biases w_m of user context (such as age, sex and so on).

The predicting function of this attribution context-aware rating prediction model $b_2(i,j)$ can be written as following:

$$b_2(i,j) = \hat{r}_{i,j} = w_0 + w_i + w_j + \sum\nolimits_{m \in user\_attributes} w_{jm} + \sum\nolimits_{n \in item\_attributes} w_{in} + v_j^t u_i \quad (17)$$

The objective function of this model $b_2(i,j)$ can define the following equation:

$$\min_{u_*, v_*, w_*} J = \sum_{r_{i,j} \in \Omega} (r_{i,j} - \hat{r}_{i,j})^2 + \lambda(\|u_i\|^2 + \|v_j\|^2 + \sum w_*^2), \quad (8)$$

where Ω denotes the train set, λ denotes regularization factor and w_* denotes all types of bias parameters w_0, w_i, w_j, w_{jm} and w_{in}.

The parameters of the attribution context-aware rating prediction model can be optimized by the following stochastic gradient descent (SGD) as the following equation:

$$\Delta\theta = -\gamma\frac{\partial e_{i,j}^2}{\partial\theta} - \lambda\theta = 2\gamma e_{i,j}\frac{\partial\hat{r}_{i,j}}{\partial\theta} - \lambda\theta,\ \theta = \theta + \Delta\theta \tag{9}$$

where $e_{i,j}^2 = (r_{i,j} - \hat{r}_{i,j})^2$, γ is the learning rate, and θ denotes each parameter u_i, v_j and w_*.

2.3 Fusing the Models

The proposed interaction context-aware sequential prediction model $b_1(i,j)$ can predict the behavioral rankings of the user. The proposed attribution context-aware rating prediction model $b_2(i,j)$ can predict the ratings of the user grading items.

From different aspects, we propose three fusing methods to incorporate $b_1(i,j)$ and $b_2(i,j)$ into a unified model–MIAR.

Linear Weighted Fusion. In various applications, diverse types of context (interaction context and attribution context) or diverse types of information (behavioral sequence information and rating information) have different influence on recommendation.

Therefore in this aspect, we adopt the linear weighted fusion method to fuse $b_1(i,j)$ model and $b_2(i,j)$ model into the MIAR-1 model by a weight coefficient α between 0 and 1 as the following equation.

$$b_{linear}(i,j) = RANK(i,j) = \alpha b_1(i,j) + (1-\alpha)b_2(i,j) \tag{11}$$

where $RANK(i,j)$ denotes the behavioral ranking prediction of selected item j by user .. at the next timestamp in MIAR-1 model.

In the flexibility aspect of model, this linear weighted fusion method can set different weight coefficients α to fusion according to corresponding applications, which can bring more flexibilities to the whole model.

However, we need to empirically set the coefficient α which is often challenging in practice.

Product Fusion. To eliminate the coefficient α, we adopt a product fusion method to construct MIAR-2 model as the following by three steps.

Step1: Normalizing results of model $b_1(i,j)$. Divide the predicted behavioral ranking results of $b_1(i,j)$ by a maximum predicted behavioral ranking of user i from the model $b_1(i,j)$, which leads to the predicted probability $\hat{P}_{i,j}$ of a selection.
Step2: Normalizing results of model $b_2(i,j)$. Divide the predicted rating results of $b_2(i,j)$ by a maximum predicted rating of user i from the model $b_2(i,j)$, which leads to the normalized prediction ratings $\hat{R}_{i,j}$.

Step3: Product fusion. We fuse the predicted probability $\hat{P}_{i,j}$ of a selection and the normalized prediction ratings $\hat{R}_{i,j}$ into our MIAR-2 model $b_{product}(i,j)$ as the following equation:

$$b_{product}(i,j) = RANK(i,j) = \hat{P}_{i,j} \times \hat{R}_{i,j} \tag{12}$$

To sum up, the main advantages of this product fusion method are listed as follows:

- In calculating $RANK(i,j)$, the normalizing of prediction ratings (Step2) can avoid this inaccurate situation that prediction ratings have far more influence on recommendation than behavioral rankings in some applications where the score range for rating items by users is relatively large.
- There are some rating biases that some people tend to give higher ratings for items while some people tend to give lower ratings. In calculating $RANK(i,j)$, the normalizing of prediction ratings (Step2) can also mitigate the inaccuracy brought by above rating biases.
- This proposed product fusion method not only can avoid the predefined parameter in the linear weighted fusion method, but also can translate calculating $RANK(i,j)$ into calculating the mathematical expectation as follows to effectively recommend.

Where $\hat{P}_{i,j}$ denotes the predicted probability of user i selecting item j. Therefore, $1 - \hat{P}_{i,j}$ denotes that user i don't select item j and corresponding ratings can be regarded as zero to some extent. According to the definition of the mathematical expectation, the Eq. (12) can be translated into the Eq. (13) as followings:

$$b_{product}(i,j) = RANK(i,j) = \hat{P}_{i,j} \times \hat{R}_{i,j} = \hat{P}_{i,j} \times \hat{R}_{i,j} + (1 - \hat{P}_{i,j}) \times 0 = E[\hat{R}_{i,j}] \tag{13}$$

where $E[\hat{R}_{i,j}]$ denotes the mathematical expectation for $\hat{R}_{i,j}$.

This model not only mitigates the inaccuracy brought by rating biases but also can be translated into the mathematical expectation to reasonable recommendation.

However to some extent, the adaptability of this product fusion method is less than the linear weighted fusion method when diverse types of context or diverse types of information have different influence on recommendation in some real application scenes.

Stages Fusion. The linear weighted fusion and the product fusion methods proposed above have common patterns that the latent representation of users, items and context are simultaneously contribute to recommendation item list. Besides, if there are several stages to optimize recommendation items list step by step, it will possibly get better recommendation results.

Therefore, we adopt a kind of stages fusion method to construct MIAR-3 model as the following by three stages.

We firstly assume that most users will not want to select those items which are predicted with low ratings for themselves.

Stage1: Learning an initial recommendation items list.

Using $b_1(i,j)$ can obtain $ranking_{i,k+1,j}$ denoted the behavioral ranking prediction of selected item j by user i at the next timestamp t^i_{k+1}, which can generate an initial recommendation items list $\vec{L}$.

Stage2: Predicting user-item ratings.

Using $b_2(i,j)$ can predict user-item ratings which has significant influences on the next Stage3.

Stage3: Optimizing recommendation items list.

According to the above assumption, setting a predicted rating threshold β for user i, if there is an item j that the predicted rating $\hat{r}_{i,j}$ is less than β in the initial recommendation items list $\vec{L}$, the item j will be removed, and another item j_1 that the corresponding $ranking_{i,k+1,j_1}$ is the largest except those items in pre-existing $\vec{L}$ and $\hat{r}_{i,j_1}$ is bigger than β will be appended to the end of this list $\vec{L}$. Therefore, the initial recommendation items list $\vec{L}$ is optimized to form the optimized recommendation items list $\vec{LIST}$.

To some extent, this stages fusion method consists of three stages which are similar to two steps of the user behavioral process proposed in the above introduction by the work [16].

In a user behavioral process, Step1 is that a user firstly should choose a desired item. The above proposed Stage1 can generates the initial recommendation items list by the behavioral ranking prediction, which is similar to Step1.

Step2 is that the user rates the selected item. Similarly, the above proposed Stage2 can obtain the user-item predicted ratings.

In fact, if the predicted rating of an item is too low, the user may not select this item at the next timestamp. Therefore, the above proposed Stage3 can further optimize the recommendation items list by predicting ratings, which may lead to better recommendation results.

3 Experiment

3.1 Experiment Settings

Dataset. Our experiments are conducted on the real dataset, MovieLens-100K, consisting of 100,000 ratings (1–5) from 943 users on 1,682 movies.

The MovieLens-100K dataset contains not only ratings but also rich context information, such as attribute context (the age, sex and occupation of users and the genre of films) and interaction context (the day and hour of timestamps).

According to the timestamps information, we randomly sample the previous 80% ratings of users behavioral sequence information from the original dataset as training set and the remaining 20% ratings of users behavioral sequence information as test set.

On the Movielens-100K dataset, firstly we extract two kinds of input context: seven days in a week (containing 7 input context values), and twenty-four hours in a day (containing 24 input context values). Therefore, there are totally 168 input context values. Besides, we set time intervals discretized by one-day time bins between adjacent behaviors in sequences as transition context. We treat these transition context values where time intervals are larger than thirty days as 1 transition context value and less than a day also as 1 transition context value. So, towards the range of time intervals, there are totally 32 transition context values.

Compared Methods. We compare MIAR model with three traditional collaborative filtering recommendation algorithms and a state-of-the-art context-aware recommendation models.

Traditional Collaborative Filtering Algorithms
UserCF: User-based collaborative filtering algorithm [17].
ItemCF: Item-based collaborative filtering algorithm [18].
SVD++: Matrix factorization based collaborative filtering algorithm [19].
Context-aware Recommendation Models
CARNN: The interaction context-aware sequential recommendation model [15], given by Eq. (5).

Evaluation Metrics. To measure the performance, we use the most popular metric, Recall@k and Precision@k. For the above k we report results with k = 1, 2, 4 and 6 in our experiments.

3.2 Experiment Results and Analysis

In this section, we respectively demonstrate three different types of experiment results by the proposed three diverse MIAR fusion methods depicted in Sect. 2.3. In experiment setting, the MIAR-1 model adopts weight coefficient $\alpha = 0.7$, the MIAR-3 model adopts predicted rating threshold $\beta = 2$ and MIAR-2 does not need hyper parameter.

The Recall@K Performance. The overall recall performance on the MIAR and various compared methods are given by Table 1. Since traditional collaborative filtering algorithms neglect influences of the contextual information on recommendation system, the performance of these traditional collaborative filtering algorithms are usually inferior to the performance of context-aware models.

Table 1. Recall performance on the MIAR-1 and compared methods

Method	Recall@1	Recall@2	Recall@4	Recall@6
UserCF	0.0004	0.0011	0.0024	0.0036
ItemCF	0.0003	0.0006	0.0009	0.0012
SVD++	0.0011	0.0022	0.0046	0.0078
CARNN	0.0042	0.0078	0.0146	0.0226
MIAR-1	**0.0042**	**0.0090**	**0.0183**	**0.0271**
MIAR-2	**0.0044**	**0.0084**	**0.0160**	**0.0239**
MIAR-3	**0.0045**	**0.0089**	**0.0181**	**0.0272**

The recall performance of our proposed MIAR-1 and various compared methods are given in Table 1. Comparing with traditional collaborative filtering methods UserCF, ItemCF, SVD++, our proposed MIAR-1 largely improves recall performance. Besides, comparing with CARNN, the proposed MIAR-1 improves recall performance by 3.7%, 5.2%, 6.4% and 3.4% respectively when k = 1, 2, 4 and 6 correspondingly.

The recall performance of our proposed MIAR-2 and various compared methods are given in Table 1. Comparing with traditional collaborative filtering methods UserCF, ItemCF, SVD++, our proposed MIAR-2 also largely improves recall performance. Besides, comparing with CARNN, the proposed MIAR-2 improves recall performance by 3.6%, 7.6%, 9.6% and 5.7% respectively when k = 1, 2, 4 and 6 correspondingly.

The recall performance of our proposed MIAR-3 and various compared methods are given in Table 1. Comparing with traditional collaborative filtering methods UserCF, ItemCF, SVD++, our proposed MIAR-3 also largely improves recall performance. Besides, comparing with CARNN, the proposed MIAR-3 improves recall performance by 4.5%, 3.9%, 8.4% and 6.9% respectively when k = 1, 2, 4 and 6 correspondingly.

The overall recall performance on the proposed various fusion methods MIAR-1, MIAR-2 and MIAR-3 are given by the following Fig. 3. It seems that MIAR-3 and MIAR-1 have similar recall performance and both of them are a little better than MIAR-2.

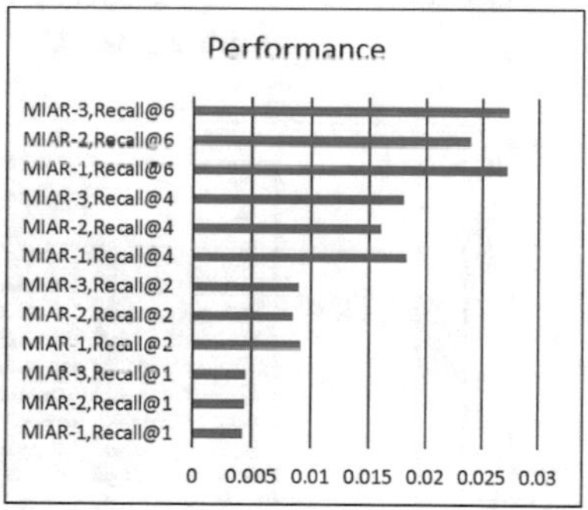

Fig. 3. Recall performance comparison of MIAR-1, 2 and 3.

The Precision@K Performance. Both the recall and precision play an important role in the recommendation system. To measure the combined effect of the recall and precision, we use the context-aware compared method CARNN which has similar recall performance with proposed method MIAR to compare the precision performance.

The precision performance of our proposed MIAR-1 and the context-aware compared method CARNN are given by Fig. 4 and the hyper-parameters settings of CARNN are the same with the hyper-parameters settings of MIAR-1. Comparing with CARNN, the proposed MIAR-1 improves precision performance by 8.3%, 1.8%, 5.6% and 8.8% respectively when k = 1, 2, 4 and 6 correspondingly.

The precision performance of our proposed MIAR-2 and the context-aware compared method CARNN are given by Fig. 5. Comparing with CARNN, the proposed MIAR-2 improves precision performance by 4%, 1.6%, 1.8% and 2% respectively when k = 1, 2, 4 and 6 correspondingly.

The precision performance of our proposed MIAR-3 and the context-aware compared method CARNN are given by Fig. 6. Comparing with CARNN, the proposed MIAR-3 improves precision performance by 2.6%, 2.8%, 3% and 1.2% respectively when k = 1, 2, 4 and 6 correspondingly.

The overall precision performance on the proposed various fusing methods MIAR-1, MIAR-2 and MIAR-3 are given by the following Fig. 7.

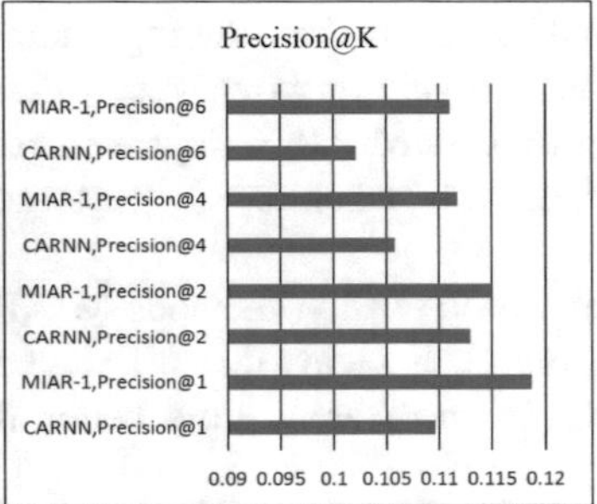

Fig. 4. The Precision@K performance for MIAR-1.

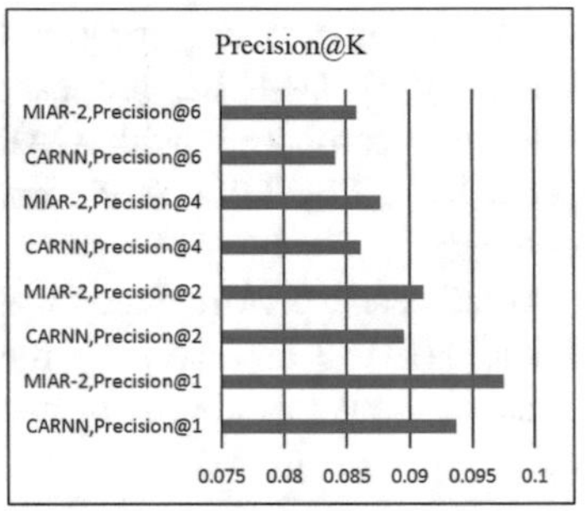

Fig. 5. The Precision@K performance for MIAR-2.

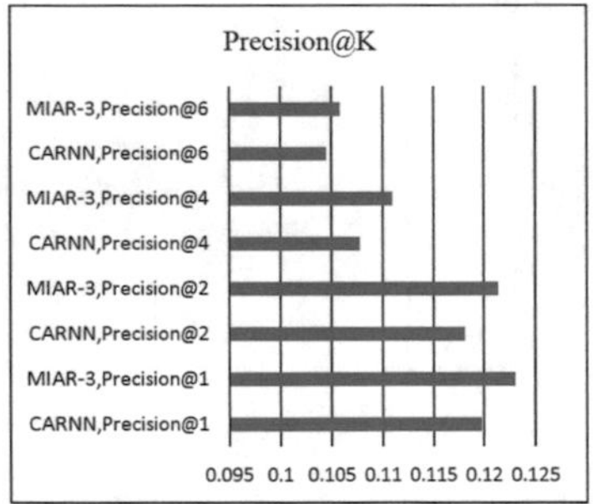

Fig. 6. The Precision@K performance for MIAR-3.

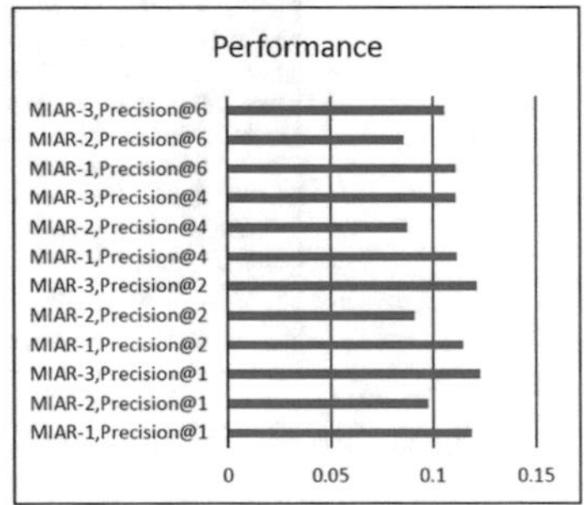

Fig. 7. Precision performance comparison of MIAR-1, 2 and 3.

From the Figs. 3 and 7, these experiment results of the proposed various fusion methods depict that the performance of MIAR-1 by the linear weighted fusing are overall slightly superior to the performance of MIAR-2 by the product fusing. After analyzing the experiment dataset, maybe the above phenomenon is resulted from that the historical sequences information and interaction context information have more influence on Top-N recommendation in the Movielens-100K dataset. Namely this linear weighted fusion method can bring more adaptability to the MIAR-1 model in some application scenes. Besides, the performance of MIAR-3 is similar to the performance of MIAR-1, and also slightly superior to the performance of MIAR-2.

Comparing with others algorithms, MIAR model not only perceives the influence of interaction context and attribution context but also captures behavioral sequences information and rating information, which further alleviates the data sparsity problem and improves recommendation accuracy.

4 Conclusion

As contextual information, implicit and explicit feedbacks information become increasingly important in recommendation systems, incorporating diverse types of context information, behavioral sequence information and rating information into one model should be a well idea to construct multi-modal behavioral information-aware recommendation models.

Therefore in our proposed MIAR model, firstly we construct an interaction context-aware sequential prediction model to perceive influences of interaction context information and behavioral sequence information on the recommendation. Secondly, we construct an attribution context-aware rating prediction model to perceive influences of attribution context information and rating information on the recommendation. Finally, the interaction context-aware sequential prediction model and the attribution context-aware rating prediction model two sub-models are combined effectively by the linear weighted fusion, the product fusion and the stages fusion three different fusion methods to recommend.

The experiment results show that our proposed MIAR model outperforms the various compared collaborative filtering algorithms and a compared state-of-the-art context-aware recommendation models. Therefore, our proposed MIAR model not only perceives the influence of interaction context and attribution context but also captures implicit feedback behavioral sequences information and explicit feedback rating information, which makes a significant progress in alleviating the data sparsity problem to some extent and improves recommendation accuracy.

In the future, we would like to incorporate the attention mechanism into recurrent neural networks to construct a better behavioral context-aware recommendation models and improve recommendation performance.

Acknowledgments. This work is supported by the Chinese National Science Foundation (#61763007), the Guilin Science and Technology Project (20170113-6) and the Guangxi Natural Science Foundation (#2017JJD160017).

References

1. Adomavicius, G., Tuzhilin, A.: Context-aware recommender systems. In: Ricci, F., Rokach, L., Shapira, B., Kantor, P. (eds.) Recommender Systems Handbook, pp. 217–253. Springer, Boston (2011)
2. Cai, G., Gu, W.: Heterogeneous context-aware recommendation algorithm with semi-supervised tensor factorization. In: International Conference on Intelligent Data Engineering and Automated Learning, Guilin, pp. 232–241. Springer (2017)
3. Zhang, M., Tang, J., Zhang, X., et al.: Addressing cold start in recommender systems: a semi-supervised co-training algorithm. In: Proceedings of the 37th International ACM SIGIR Conference on Research & Development in Information Retrieval, pp. 73–82. ACM, New York (2014)
4. Liu, Q., Wu, S., Wang, L.: COT: Contextual Operating Tensor for context-aware recommender systems, pp. 203–209. AAAI, Menlo Park (2015)
5. Karatzoglou, A., Amatriain, X., Baltrunas, L., et al.: Multiverse recommendation: n-dimensional tensor factorization for context-aware collaborative filtering. In: Proceedings of the Fourth ACM Conference on Recommender Systems, pp. 79–86. ACM (2010)
6. Rendle, S., Gantner, Z., Freudenthaler, C., et al.: Fast context-aware recommendations with factorization machines. In: Proceedings of the 34th International ACM SIGIR Conference on Research and Development in Information Retrieval, pp. 635–644. ACM, New York (2011)
7. Zheng, V.W., Cao, B., Zheng, Y., et al.: Collaborative filtering meets mobile recommendation: a user-centered approach, vol. 10, pp. 236–241. AAAI, Menlo Park (2010)
8. Levandoski, J.J., Sarwat, M., Eldawy, A., et al.: Lars: a location-aware recommender system. In: 2012 IEEE 28th International Conference on Data Engineering (ICDE), pp. 450–461. IEEE, Piscataway (2012)
9. Ren, X., Song, M., Haihong, E., et al.: Context-aware probabilistic matrix factorization modeling for point-of-interest recommendation. Neurocomputing **241**, 38–55 (2017)
10. Wu, S., Liu, Q., Wang, L., et al.: Contextual operation for recommender systems. IEEE Trans. Knowl. Data Eng. **28**(8), 2000–2012 (2016)
11. Zhang, Y., Dai, H., Xu, C., et al. Sequential click prediction for sponsored search with recurrent neural networks, pp. 1369–1375. AAAI, Menlo Park (2014)
12. Liu, Q., Wu, S., Wang, L., et al.: Predicting the next location: a recurrent model with spatial and temporal contexts, pp. 194–200. AAAI, Menlo Park (2016)
13. Wu, S., Ren, W., Yu, C., et al.: Personal recommendation using deep recurrent neural networks in NetEase. In: 2016 IEEE 32nd International Conference on Data Engineering (ICDE), pp. 1218–1229. IEEE, Piscataway (2016)
14. Okura, S., Tagami, Y., Ono, S., et al.: Embedding-based news recommendation for millions of users. In: Proceedings of the 23rd ACM SIGKDD International Conference on Knowledge Discovery and Data Mining, pp. 1933–1942. ACM, New York (2017)
15. Liu, Q., Wu, S., Wang, D., et al.: Context-aware sequential recommendation. In: 2016 IEEE 16th International Conference on Data Mining (ICDM), pp. 1053–1058. IEEE, Piscataway (2016)
16. Zhao, X., Niu, Z., Chen, W., et al.: A hybrid approach of topic model and matrix factorization based on two-step recommendation framework. J. Intell. Inf. Syst. **44**(3), 335–353 (2015)
17. Zhao, Z.D., Shang, M.S.: User-based collaborative-filtering recommendation algorithms on hadoop. In: International Conference on Knowledge Discovery and Data Mining, pp. 478–481. IEEE, Piscataway (2010)

18. Sarwar, B., Karypis, G., Konstan, J., et al.: Item-based collaborative filtering recommendation algorithms. In: Proceedings of the 10th International Conference on World Wide Web, pp. 285–295. ACM, New York (2001)
19. Koren, Y.: Factorization meets the neighborhood: a multifaceted collaborative filtering model. In: Proceedings of the 14th ACM SIGKDD International Conference on Knowledge Discovery and Data Mining, pp. 426–434. ACM, New York (2008)

Energy-Efficient Task Caching and Offloading Strategy in Mobile Edge Computing Systems

Qian Chen[1(✉)], Zhoubin Liu[2], Linna Ruan[1], Zixiang Wang[2], Sujie Shao[1], and Feng Qi[1]

[1] Beijing University of Posts and Telecommunications, Beijing 100876, China
chenqian199636@foxmail.com

[2] State Grid Zhejiang Electric Company, Hangzhou, China

Abstract. As the limited processing power and energy of mobile terminals, the QoS of delay-sensitive service cannot be guaranteed. This paper proposes an SDN-based task caching and offloading strategy (SD-TCO) by mobile edge computing technology. The strategy mainly includes two algorithms: SDN-based mobile edge computing network (SD-MEN) task caching algorithm and branch-bounding algorithm based on greedy strategy. The SD-MEN task caching algorithm is used to increase the cache hit ratio by saving the frequently called task results on the edge server, and the branch-bounding algorithm based on greedy strategy is used to offload the task reasonably, which can ensure the QoS of users and minimize energy consumption. Simulation results show that SD-TCO has achieved effective improvement in stable delay and energy consumption.

Keywords: Mobile Edge Computing · SDN · Greedy strategy · Cache

1 Introduction

With the rapid development of mobile communication technologies and Internet of Things, delay-sensitive applications such as augmented reality (AR), virtual reality (VR), and image recognition are also emerging. However, due to the limitation of the computing power and power of the mobile terminals, these applications cannot be widely implemented and promoted. In order to solve this problem, computing task is considered to be offloaded to the cloud, which can overcome the shortcomings of the lack of computing power of mobile terminals. However, when mobile devices are connected to the cloud through a wireless network, a relatively high delay occurs [1], which is not suitable for delay-sensitive tasks. Therefore, task can be offloaded to an edge server deployed near a cell (SCeNB), a macro base station (MBS), etc. by Mobile Edge Computing (MEC), which can provide low latency and high performance computing services for mobile terminal. MEC has two main advantages:

- Compared with local computing [2], MEC can overcome the limitations of mobile device computing power.
- Compared with cloud computing [3], MEC can avoid high latency caused by transmitting large amounts of data to cloud servers.

C.-N. Yang et al. (Eds.): SICBS 2018, AISC 895, pp. 824–837, 2020.
https://doi.org/10.1007/978-3-030-16946-6_67

Therefore, AR, VR and other delay-sensitive, computationally intensive services are more suitable for processing using MEC.

Previous research on MEC mainly involved collaborative computing with the cloud, content caching and task offloading. Among them, caching and offloading are key issues. In order to reduce the delay and energy consumption caused by multiple requests for the same content, Wang et al. [4] proposed to save frequently called content on the edge server, make full use of storage resources of the edge server, and designed a variety of caching strategies to optimize system performance. For the study of task offloading, Chen et al. [5] proposed different tasks offloading strategies by considering multiple servers, multiple users and other factors, and finally achieved the goal of reducing energy consumption. However, computational tasks and cached content are considered independently in all of the above studies.

Tang [6] discussed the application of content caching and computational offloading in a cell network, but did not consider caching task data, just saving important data. Hao [7] proposed to store frequently called task data on the MEC server. Each time user requests the same task, the MEC server performs the calculation, but ignores the limited storage resources and computing power on the MEC server. Due to AR, etc. business involves the processing of images, and the huge amount of data directly affects the number of tasks and computational overhead of cache. In addition, many researchers consider the existence of a hub to complete the collection of information, the scheduling of resources and the offloading of tasks when considering cache and offloading strategies. In response to the above problems, the main contributions of paper include:

- We propose an SDN-based mobile edge computing network (SD-MEN) architecture to control global information by deploying SDN controllers at macro base stations.
- We design the task cache algorithm, which greatly improves the hit probability of the requested task. Considering the limited storage resources of the MEC server, we change the cached content to the execution result of the task.
- We propose a branch and bound method based on greedy strategy to offload tasks and achieve minimum energy consumption while ensuring user QoS. We refer to task caching and task offloading collectively as SDN-based task caching and offloading strategies (SD-TCO).

The paper is organized as follows. The second section introduces system model. In third section, we present problem description. The fourth section presents the solution and algorithm for task caching and task offloading, the fifth section shows and discusses the simulation results. Finally, the last section concludes this paper.

2 System Model

2.1 The Architecture of SD-MEN

As shown in Fig. 1, the architecture of SD-MEN can be divided into three layers, i.e., mobile terminals layer, mobile edge computing layer and SDN control layer. Mobile terminal layer is composed of mobile terminals (MTs) that need to offload tasks, and mobile terminals use CoMP technology to communicate with the SCeNBs of mobile edge computing layer. The mobile edge computing layer consists of SCeNBs and one MBS, and the set of SCeNBs is denoted by $M = \{1, 2, \cdots, m\}$. Both SCeNBs and MBS deploy the MEC server. MBS and SCeNBs use optical fiber communication. MEC server is also connected to SCeNBs and MBS through the optical fiber. In the SD-MEN architecture, in order to reduce system power consumption, the tasks generated by the terminal are allocated to multiple SCENBs for calculation. MBS is mainly responsible for processing tasks in the task cache queue and maintaining and updating multiple information tables with the SDN controller, does not communicate directly with mobile terminal. SDN control layer is mainly composed of an SDN controller, and SDN controller is deployed on the MBS. The key to the SDN technology is to separate the control plane from the data plane by virtualization. In this paper, SDN controller globally controls the SD-MEN based on the OpenFlow protocol by sending a flow table and setting forwarding rules. SDN controller can collect information of all SCeNBs and MBSs within its coverage, and complete resource allocation and task scheduling according to the information.

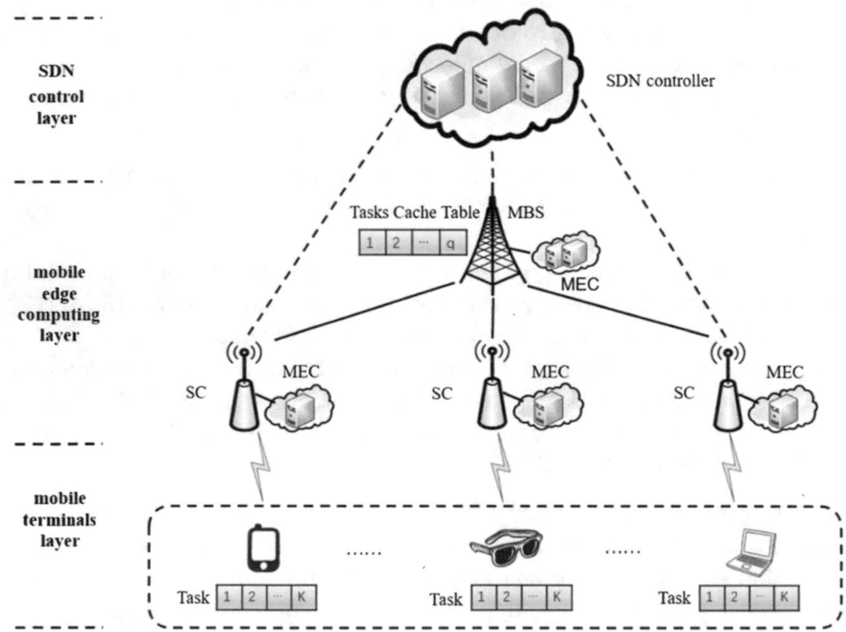

Fig. 1. The architecture of SD-MEN

2.2 Communication Model

In this paper, a widely used task model is used to describe subtasks $q_i = (s_i, c_i)$, which s_i represent the size of the computational task and c_i represent the CPU computation cycle required to complete task. To facilitate the description of the cell site, SCeNB is denoted by sc. It is assumed that a certain task Q requested by mobile device includes multiple subtasks, for example, a typical AR task [8] request includes sub-tasks such as modeling of virtual scenes, fusion of virtual reality signals, tracking of user line of sight changes, etc. Moreover, there is also information communication between different subtasks. In the information communication process, the set N is used to represent n subtasks included in a task requested by the mobile terminal. The communication delay $t_{i,k}$ between the subtasks q_i, q_k can be given as

$$t_{i,k} = \begin{cases} \frac{d_{i,k}}{c}, & q_i, q_k \text{ is assigned to different sc} \\ 0 & q_i, q_k \text{ is assigned to same sc} \end{cases} \tag{1}$$

The communication energy consumption $e_{i,k}$ between subtasks can be given as

$$e_{i,k} = \begin{cases} e_{start} + p_{sc}\frac{d_{i,k}}{c}, & q_i, q_k \text{ is assigned to different sc} \\ 0 & q_i, q_k \text{ is assigned to same sc} \end{cases} \tag{2}$$

where e_{start} indicates the energy consumption generated by the task when the communication is started, p_{sc} is the output power of the sc, $d_{i,k}$ is the amount of data transmitted between the subtasks, and the c denotes the communication transmission rate between the sc_m and sc_n.

The wireless uplink data rate $r_{i,j}$ can be computed by

$$r_{i,j} - B\log_2(1 + \frac{p_i g_{i,j}}{\sigma^2 + I_{i,j}}) \tag{3}$$

where B is channel bandwidth, p_i indicated transmit power of MT_i, σ^2 is noise power of mobile device, and $I_{i,j}$ denotes interference power within the cell. Similar to above case that offloads subtask q_i to sc_j, the transmission delay $t_{i,j}^{tran}$ can be computed by

$$t_{i,j}^{tran} = \frac{s_i}{r_{i,j}} \tag{4}$$

The consumed energy $e_{i,j}^{tran}$ in such case can be given as

$$e_{i,j}^{tran} = p_i t_{i,j}^{tran} \tag{5}$$

2.3 Computation Model

This section mainly introduces the two models involved in the calculation process, MBS-MEC task cache model implemented by MBS and the MEC server connected to it, and SC-MEC task offload model for calculating and offloading tasks.

MBS-MEC Task Caching Model. In this paper, task cache model is built from the global information of SDN controller. This paper processes the information in the SDN controller through the SD-MEN task caching algorithm, filters out the frequently requested tasks, and saves the final execution results on the MEC server. MBS-MEC model can directly return the results of frequently requested tasks to mobile devices, which can effectively reduce latency and reduce energy consumption. If the requested task exists in the task cache table, there is no need to transfer the data. Since the amount of data of the execution result is very small compared to the task itself, the power consumption and delay caused by such data transmission are ignored.

SC-MEC Task Offloading Model. $f_j^{sc}, j \in M$ is used to denote computing power of sc_j. The computational delay $t_{i,j}^{sc\_compute}$ in such case can be given as

$$t_{i,j}^{sc\_compute} = \frac{c_i}{f_j^{sc}} \tag{6}$$

The consumed energy can be given as

$$e_{i,j}^{sc\_compute} = t_i^{sc\_compute} p_{sc} \tag{7}$$

In order to get the delay and energy consumption caused by the task, the delay and energy consumption generated by communication and calculation need to be added separately, which is expressed as

$$t_{i,j}^{sc} = t_{i,j}^{sc\_compute} + t_{i,j}^{tran} + \sum_{k=0,k \neq i}^{n} t_{i,k} \tag{8}$$

$$e_{i,j}^{sc} = e_{i,j}^{tran} + e_{i,j}^{sc\_compute} + \sum_{k=0,k \neq i}^{n} e_{i,k} \tag{9}$$

3 Problem Formulation

3.1 Task Caching

The task cache mainly faces two problems: 1. How to determine cache status of the request task. 2. The criteria and conditions for the task to join the cache. Define a set of cache vectors $X = [x_1, x_2, \cdots, x_n]$ to represent the cache status of a subtask contained in a request task. When $x_i = 0$, the task q_i needs to be calculated and offloaded; When $x_i = 1$, it indicates task q_i locates in the cache, MBS will directly return execution result of this task to the corresponding mobile terminal.

3.2 Task Offloading

When subtask requested to be executed is not in the cache table, task needs to be offloaded. In this paper, task offloading is performed on sc. The task assignment function is represented by a matrix $Y_{nm} = \{y_{i,j} | 1 \leq i \leq n, 1 \leq j \leq m\}$. $y_{i,j} = 1$ denotes task q_i is assigned to sc_j and $y_{i,j} = 0$ denotes task q_i is not assigned to sc_j. According to process of task cache and task offloading, the delay of executing task Q can be obtained

$$T = Max\left\{\sum_{i=0}^{n} y_{i,j}x_i(t_{i,j}^{sc\_compute} + t_{i,j}^{tran} + \sum_{k=0,k\neq i}^{n} t_{i,k})\right\}, \forall j \in M \tag{10}$$

In this process, the energy consumed by processing task Q can be expressed as

$$E = \sum_{i=0}^{n}\sum_{j=0}^{m} y_{i,j}x_i(e_{i,j}^{tran} + e_{i,j}^{sc\_compute} + \sum_{k=0,k\neq i}^{n} e_{i,k}) \tag{11}$$

Our goal is to minimize the energy cost of processing task Q requested by mobile terminal while guaranteeing the quality of service. Finally, the problem can be expressed as follows:

$$\underset{x,y}{Min}\sum_{i=0}^{n}\sum_{j=0}^{m} y_{i,j}x_i(e_{i,j}^{tran} + e_{i,j}^{sc\_compute} + \sum_{k=0,k\neq i}^{n} e_{i,k}) \tag{12}$$

$$C1 : T \leq D$$

$$C2 : \sum_{i=0}^{n} y_{i,j}c_i \leq F_j^{sc}, \forall j \in M$$

$$C3 : \sum_{i=0}^{n} y_{i,j} \leq K, \forall j \in M$$

$$C4 : y_{i,j}, x_i \in \{0, 1\}, \forall i \in N, \forall j \in M$$

$C1$ indicates that processing delay of the task must be less than or equal to the maximum delay that task can tolerate; storage resources and the load that are respectively bound by $C2$ and $C3$, F_j^{sc} is the maximum computing power, and K denotes the maximum number of tasks to be represented; $C4$ describe the range of values.

4 Problem Formulation

Observing the problem (12), we can find that cache vector and the task allocation matrix are the key of the problem, so the problem can be transformed into two sub-problems: 1. Determine cache state of each sub-task and get cache vector. 2. Under the condition of limited resources, select the best task offloading scheme. The approach we will address these two issues is called SDN-based task caching and offloading strategy (SD-TCO).

4.1 Task Caching Scheme

MBS is responsible for maintaining the task history table and the task cache table, which are denoted by I_h and I_c, respectively. Task history table records the most recently requested task and corresponding execution result, and uses M_h to indicate the maximum number of tasks that can be saved; Task cache table records tasks that have been called more than once, using M_c to indicate the maximum number of tasks that can be saved.

It is assumed that task Q requested by the mobile device can be divided into n subtasks, and the number of requests of each subtask follows the Poisson distribution in unit time, which satisfies $P(C = k) = \frac{e^{-\lambda}\lambda^k}{k!}$, where C is the number of requests for the task, and λ denotes the average incidence of tasks per unit time. Vector group X is set to empty, and the number of subtasks is n. For each subtask, task type is first determined according to the Poisson distribution, and then it is judged whether task exists in the task cache table. If it exists, execution result is directly returned by MBS and set x_i to 1, otherwise, continue to judge whether task exists in the task history table. If it exists in the table, it does not need to process task x_i be set to 1, and save task into the task cache table, and update task history table. If neither table records task q_i, set x_i to 0, traverse subtask space, and finally get the vector group $X = [x_1, x_2, \cdots, x_n]$.

Algorithm 1. SD-MEN Task Caching Algorithm

```
Initialize n, M_h, M_c, I_h, I_c, X=∅;
for i=1: n do
 get the task q_i according to the Poisson distribution;
 if q_i ∈ I_c then
   x_i =1;
   sort(I_c) Put q_i in the first place;
 else
   if q_i ∈ I_h then
     x_i =1;
     I_h = I_h \ {q_i};
     if len(I_c) < M_c then
       I_c = I_c ∪ {q_i};
     else
       I_c = I_c ∪ {q_i}
   else
     x_i =0;
     if len(I_h) < M_h then
       I_h = I_h ∪ {q_i};
     else
       I_h = I_h ∪ {q_i};
     end if
   end if
 end if
end for
Output X
```

4.2 Task Offloading Scheme

In this section, the main problem of offloading the task will be solved. Firstly, the objective function is simplified according to vector group X obtained in the previous section, and then the optimal task offloading scheme is obtained by using the branch-bounding algorithm based on greedy strategy.

Problem Simplification: With cached decisions, you can get the set N_{cache}, which represents tasks that already exist in the cache, and the definition set $N_0 = N - N_{cache}$, which represents tasks that need to be offloaded. Observing structure of the problem (12), the energy consumption of a task can be divided into two parts:

- Execution energy consumption of task $(e_{i,j}^{tran} + e_{i,j}^{sc\_compute})$.
- Communication energy consumption between tasks $(e_{i,k})$, represented by matrices $H_{nm} = \{h_{i,j} | 1 \le i \le n_0, 1 \le j \le m\}$ and $C_{nn} = \{c_{i,j} | 1 \le i \le n_0, 1 \le j \le n_0\}$, respectively.

Therefore, problem (12) can be simplified into the following form:

$$\begin{array}{ll} \underset{y}{Min} \sum_{i=0}^{n_0} \sum_{j=0}^{m} y_{i,j} (h_{i,j} + \sum_{k=0, k \ne i}^{n_0} c_{i,k}) \\ s.t. \qquad C1 - C4 \end{array} \tag{13}$$

Branch-Bounding Algorithm Based on Greedy Strategy: Problem (13) can prove to be a combinatorial optimization problem. When the number of subtasks is too large or there are too many sc to choose, the last solved problem becomes an NP-hard problem [9], and time complexity is $O(n_o^m)$. The solution process of the branch and bound method based on greedy strategy can be divided into three steps. First, combined with greedy strategy, the size of the task set N_0 is reduced by the unity method, and then greedy strategy is used to obtain the initial optimal solution of the objective function. Finally, solution space tree is searched and pruned.

(1) The subtask $q_1, q_2 \cdots q_{n_0}$ are regarded as nodes in the graph, and relationship diagram between the subtasks is constructed according to the matrix C_{nn}. If there is information interaction between two subtasks, corresponding task nodes are connected by an undirected edge, and the weight of the edge is Energy consumption for communication between tasks. First, select a pair of subtask nodes with the largest weight in the graph. If there is sc capable of processing the pair of subtasks and satisfy delay and resource constraints, the nodes are combined into one node, ready for next round iterate, otherwise select next pair of nodes with the greatest weight, repeat above process until there is no task pair that meets requirement of the combination, then the iterative process ends.

(2) Define a new collection $F = \varnothing$ to save tasks that have already been assigned. Select the smallest value $h_{i,j}(i \in N_l, j \in M, i \notin F)$ in matrix H_{nm}. If sc_j can process subtasks and meet delay and resource constraints, assign task q_i to F, change the load of sc_j, add the task to set F, and prepare for the next wheel selection, otherwise choose next minimum. Repeat above process until all tasks are added to set F, and an initial energy consumption E_o and an extended set F can be obtained.

(3) Branch and bound method uses a search tree to represent the offloading problem, and each leaf node represents an offloading scheme. The node's bound function is $f(i,j) = f(i-1,s) + h_{i,j} + \sum_{k=0,k\neq i}^{n_0} c_{i,k}, f(0,j) = 0$, and $f(i,j)$ refers to the energy consumed when searching for child node that represents the task being processed. The initial bound is E_o, according to the order of the tasks in the set F as the order of node expansion. The definition set L is used to save the live leaf node, and the first task of sorting the first is taken from set F and $f(1,1), f(1,2), \cdots f(1,m)$ are respectively calculated to determine whether m nodes satisfy the constraint condition. For nodes that do not satisfy the constraint, pruning operation is performed, and remaining nodes are added to set L. If set L is not empty, take a bounded optimal node from L as an extended node, and remove node out of L and then compare bounded value of the node with E_o. If it is bigger than E_o, subtree rooted at the node will be discarded. Otherwise, the above process is repeated. When all the tasks in the collection F are traversed completely, a feasible solution can be obtained, and E_o is recorded and updated until set L is empty, the loop ends, and finally the optimal solution E_o is obtained.

Algorithm 2. Branch and bound method based on greedy strategy

Initialize C_{nm} , N_l , H_{nm} , D , L , Y_{nm} , $F=\varnothing$,update N_l with the unity method;
for size(F)<size(N_l) **do**
  $h_{i,j}$=min$\{H_{nm}\}$,$(i \in N_l, j \in M, i \notin F)$;
  calculate $t_j^{sc} = h_{i,j} + \sum_{k=0,k\neq i}^{n} t_{i,k}$;
  if($t_j^{sc} < D \&\& F_j^{sc} > c_i$) **then**
    $F = F \cup q_i$;
    $F_j^{sc} = F_j^{sc} - c_i$;
  end if
end for
Obtain the initial bound E_o and the extended set F ;
calculate $f(1,1), f(1,2), \cdots f(1,m)$, And add these m nodes to the collection L ;
for size(L)>0 **do**
  $f(i,j) = \min\{L\}$;
  if it is not satisfied constraint condition, $L = L / \{f(i,j)\}$;
  for s=0: M **do**
    $f(i+1,s) = f(i,s) + h_{i+1,s} + \sum_{k=0,k\neq i}^{n_l} c_{i,k} (i \in N_l)$
    if($f(i+1,s) < E_o$) **then**
      $L = L \cup \{f(i+1,s)\}$;
    else
      $L = L / \{f(i+1,s)\}$;
    end if
  end for
  if($i == len(F)$)
    update E_o ;
  end if
end for
Output E_o , **according to the search tree output** Y_{nm} ;

5 Simulation Results and Discussion

In this section, MATLAB simulation tool is used to verify proposed SD-TCO scheme under SD-MEN architecture. In order to verify superiority of the scheme, our proposed scheme will be compared with other two schemes.

Cache+Local: There is a caching strategy, and tasks that are not cached in advance are handled by mobile device itself.

Edge: There is no caching strategy. The task is allocated by the branch-bounding algorithm based on the greedy strategy proposed in this paper, and is processed by the MEC server of the sc.

5.1 Parameter Settings

In the simulation experiment, the location of the mobile device is randomly generated. Suppose there are 5 sc and 1 MBS. The task size s_i requested by the mobile device and the number n of sub-tasks both follow the Poisson distribution, and the number of calculation cycles c_i required for execution is positively related to the task size. The computing power of the MEC server connected to the sc is set to 8 GHz, the number of tasks that can be carried by each MEC server is 3, and the computing power of the mobile device is 0.8 GHz. The transmission bandwidth of the mobile device is 20 MHz, the transmission energy consumption of mobile devices and sc is 0.5 W and 1 W respectively. the channel gain and noise power of the communication are 10^{-5} W, 10^{-9}, respectively. The transmission rate of the fiber $c = 1\,\text{Gbps}$, the amount of information transferred between subtasks $d_{i,k} = 0.02(s_i + s_j)$, and the energy consumed by the task at the start of communication $e_{start} = p_{sc}(s_i + s_j)/c$.

5.2 Performance Analysis

In order to evaluate the SD-TCO performance, we consider the impact of the number of subtasks and size of the task on three different processing schemes.

Figure 2(a) shows the delay comparison of the three schemes when the number of subtasks changes. It can be seen from the figure that the SD-TCO delay has been stable at a small value, which basically does not change with the number of subtasks. Because the caching strategy can effectively reduce the subtasks that need to be processed. The distributed parallel processing of the sc ensures that the delay is in a stable range.

Figure 2(b) shows the energy consumption comparison of the three schemes when the number of subtasks changes. As can be seen from the figure, when the number of subtasks is small, Cache+Local generates the least energy consumption. This is because local processing eliminates energy consumption of data transmission, but as the number of subtasks increases, Cache+Local will increase computing energy consumption due to limitation of local computing power, while SD-TCO is mainly energy generated during the task transmission. Due to the existence of the caching strategy, change in the number of subtasks does not cause a significant change in energy consumption.

Figure 2(c) shows the delay comparison of the three schemes when task size changes. As can be seen from the figure, delays generated by the three schemes will increase as the size of the task increases, but SD-TCO always maintains a tolerable delay. And Edge is very sensitive to changes in the size of the data transmission task. This is because there is no processing of the caching strategy, so that Edge needs to offload all tasks, and delay is also increased.

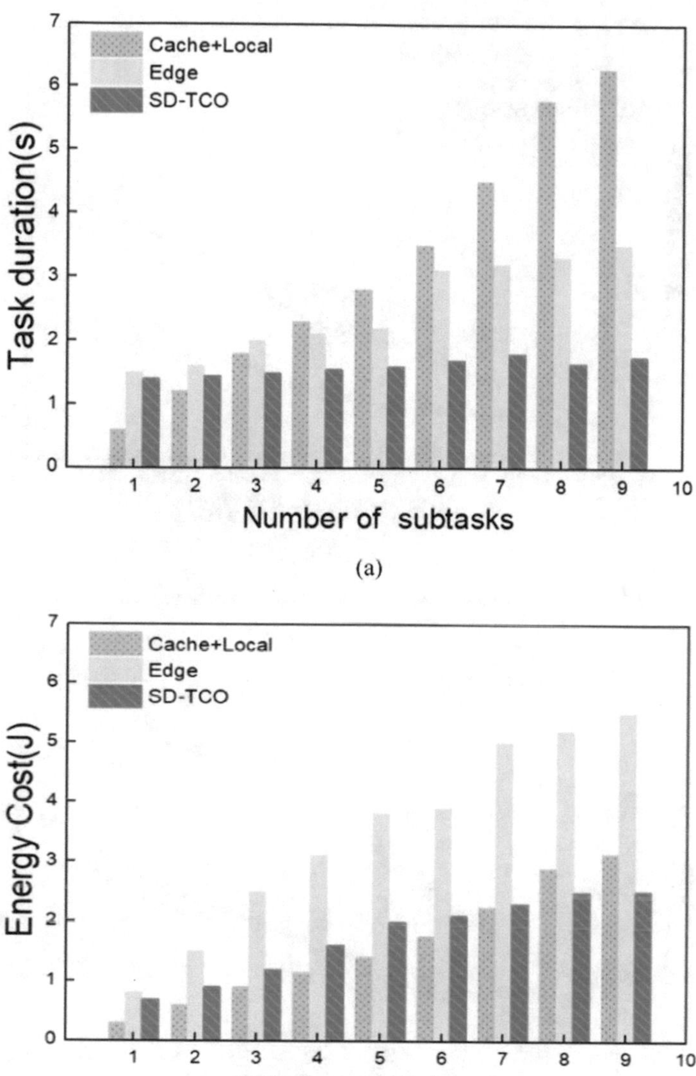

Fig. 2. Program performance comparison

Figure 2(d) shows the energy consumption comparison of the three schemes when the task size changes. As can be seen from the figure, the energy consumption generated by SD-TCO is in a controllable range, but as the amount of task data increases, energy consumption of SD-TCO will gradually exceed that of Cache+Local, due to transmission of large data volume tasks. But it is worth noting that energy consumption generated by the SD-TCO also includes the energy consumed by the sc, and actual energy consumed by mobile terminal is less than the value indicated on the graph.

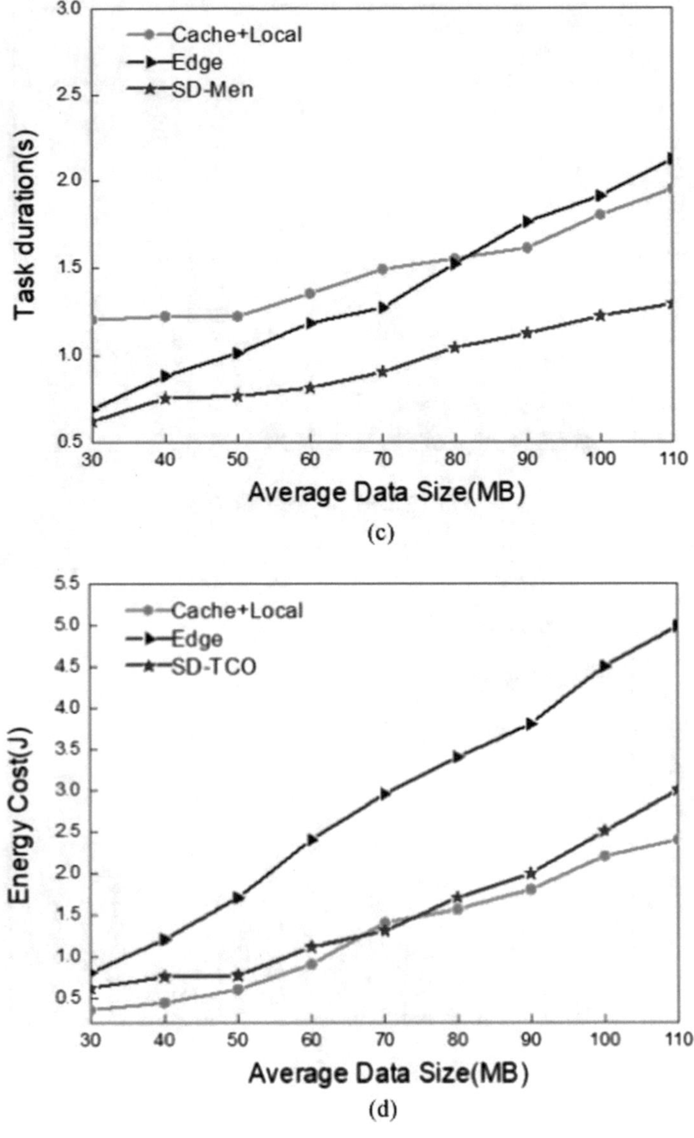

Fig. 2. (*continued*)

6 Conclusions

This paper mainly designs SDN-based task caching and offloading strategy (SD-TCO) based on the proposed SDN-MEN architecture. The user's QoS can be guaranteed under limited constraints and energy consumption can be minimized. The strategy is mainly composed of two algorithms: SD-MEN task caching algorithm and branch-bounding algorithm based on greedy strategy. The experimental results confirm the effectiveness of the strategy in reducing energy consumption and ensuring user QoS. Further research can consider joint cloud for collaborative offloading, research more comprehensive caching strategies, and further reduce energy efficiency costs.

Acknowledgment. This work was supported by State Grid Technology Project 'Edge Computing Research in Smart Grid Application and Security' (Grant: 52110118001H, Contract No: 52110418001B).

References

1. Meng, S., Wang, Y., Miao, Z., Sun, K.: Joint optimization of wireless bandwidth and computing resource in cloudlet-based mobile cloud computing environment. Peer-to-Peer Netw. Appl., 1–11 (2017)
2. Barbera, M., Kosta, S., Mei, A., Stefa, J.: To offload or not to offload? The bandwidth and energy costs of mobile cloud computing. In: Proceedings of IEEE INFOCOM, pp. 1285–1293, April 2013
3. Hou, X., Li, Y., Chen, M., Wu, D., Jin, D., Chen, S.: Vehicular fog computing: a viewpoint of vehicles as the infrastructures. IEEE Trans. Veh. Technol. **65**(6), 3860–3873 (2016)
4. Wang, X., Wang, H., Li, K., Yang, S., Jiang, T.: Serendipity of sharing: Large-scale measurement and analytics for device-to-device (D2D) content sharing in mobile social networks. In: Proceedings of IEEESECON, SanDiego, CA, USA, pp. 1–5, June 2017. https://doi.org/10.1109/sahcn.2017.7964925
5. Chen, X., Jiao, L., Li, W., Fu, X.: Efficient multi-user computation offloading for mobile-edge cloud computing. IEEE/ACM Trans. Netw. **24**(5), 2795–2808 (2016)
6. Tan, Z., Yu, F.R., Li, X., Ji, H., Leung, V.C.M.: Virtual resource allocation for heterogeneous services in full duplex-enabled SCNs with mobile edge computing and caching. IEEE Trans. Veh. Technol. **PP**(99), 1 (2017)
7. Hao, Y., Chen, M., Hu, L., Hossain, M.S., Ghoneim, A.: Energy efficient task caching and offloading for mobile edge computing. IEEE Access **6**, 11365–11373 (2018). https://doi.org/10.1109/ACCESS.2018.2805798
8. Li, G., Liu, Y., Wang, Y.: Evaluation of labelling layout methods in augmented reality. In: 2017 IEEE Virtual Reality (VR), Los Angeles, CA, 2017, pp. 351–352 (2017). https://doi.org/10.1109/vr.2017.7892321
9. Loh, K.H., Golden, B., Wasil, E.: Solving the maximum cardinality bin packing problem with a weight annealing-based algorithm. In: Operations Research and Cyber-Infrastructure, vol. 47, pp. 147–164 (2009)

Emotion Detection in Cross-Lingual Text Based on Bidirectional LSTM

Han Ren[1], Jing Wan[2(✉)], and Yafeng Ren[3]

[1] Laboratory of Language Engineering and Computing, Guangdong University of Foreign Studies, Guangzhou 510420, China
hanren@gdufs.edu.cn

[2] Center for Lexicographical Studies, Guangdong University of Foreign Studies, Guangzhou 510420, China
jingwan@gdufs.edu.cn

[3] Collaborative Innovation Center for Language Research and Services, Guangdong University of Foreign Studies, Guangzhou 510420, China
renyafeng@whu.edu.cn

Abstract. Emotions in cross-lingual text can be expressed in either monolingual or bilingual forms. Current researches have focused on analyzing emotions in monolingual text, whereas such approaches may achieve low performances in the case of identifying emotions in cross-lingual texts, which appear frequently in social media. In this paper, a bidirectional LSTM neural network with emotional knowledge is introduced to detect emotions in cross-lingual texts. This approach also employs the cross-lingual feature and the lexical level feature to analyze texts with multilingual forms and take advantage of emotional knowledge. The evaluation results show that our approach is effective for detecting emotion in cross-lingual texts.

Keywords: Neural network · Bidirectional LSTM · Emotional knowledge · Emotion detection

1 Introduction

The task of emotion detection is to classify emotional expressions in texts. Emotion detection is one of the most important research topics in social computing, and widely used in many natural language processing applications, such as product recommendation, opinion analysis and human-computer conversation [1].

Current research on emotion detection has mainly focused on analyzing emotions in monolingual text. However, such approaches may achieve low performances in the case of identifying emotions in cross-lingual texts, which appear frequently in social media. In cross-lingual texts, there are expressions with more than one language, and the emotions in them may be indicated by emotional texts with either monolingual form or multilingual form. Although knowledge bases such as emotional lexicons contributes to identifying the emotion of a single word or phrase, it is still a challenging task to detect emotions in cross-lingual texts, in which such emotions are typically expressed by language-mixed phrases.

C.-N. Yang et al. (Eds.): SICBS 2018, AISC 895, pp. 838–845, 2020.
https://doi.org/10.1007/978-3-030-16946-6_68

This paper proposes an approach based on neural network models for cross-lingual emotion detection. The approach tackles the problem from two aspects: (1) how does the model find the relation between the text of guest language (i.e., the embedded language) and the text of host language, (2) how does the model know the affective meaning of each word, namely emotion class. Following this idea, the cross-lingual feature and the lexical level feature are employed. The former one aims to find those texts of guest language from cross-lingual texts, while the latter one finds the emotion class of each word by leveraging on emotional knowledge. The system runs as follows: firstly, a word2vec tool is employed to train Chinese and English word vectors; then the proposed model with those two features is utilized to learn feature representation; after that, a softmax classifier is used to classify emotions for cross-lingual texts.

The rest of this paper is organized as follows. Section 2 gives a brief description of related work. Section 3 shows the detail of our system. Section 4 shows experimental results as well as some discussions. Finally, a conclusion is drawn in Sect. 5.

2 Related Work

Current research on emotion detection can be mainly classified into two categories: coarse-grained approaches and fine-grained ones, most of which focus on detecting emotions in monolingual texts [5]. Coarse-grained emotional detection, also known as emotional classification, defines such problem as bi-categorization one, namely judging a text if it holds the emotion *like* or *hate*, *positive* or *negative*. In contrast, fine-grained sentiment classification is devoted to text classification of multiple emotional categories, such as happy, sad, angry, and so on.

There are many statistical methods to detect emotion, including corpus-based statistical models, dictionary-based heuristics as well as user modeling approaches [6–10]. Recently, more researchers focus on modeling emotional texts via neural networks for sentiment and emotion analysis [11, 12]. For example, Tang et al. [13] proposed a joint GRNN and LSTM neural network to learn vector-based document representation, while Zhang et al. [14] compared three neural CRF models for pipeline, joint and collapsed targeted sentiment labeling.

Some researchers also raised their concerns with emotions in cross-lingual text. Lee et al. [2] built a dataset by labeling emotions in cross-lingual texts with happiness, sadness, fear, anger and surprise, and the texts are mixed Chinese and English ones. Wang et al. [3] proposed an attention model with LSTM neural network for feature representation of cross-lingual texts, while Wang et al. [4] proposed a joint factor graph model, in which attribute functions of the factor graph model are utilized to learn both monolingual and bilingual information from each post. Although such approaches and models contribute to analyzing sentiments in real social texts, their performances are not satisfying in contrast to those of emotion detection in monolingual texts, where emotional knowledge has been widely used in monolingual emotion detection tasks. In addition, relations of texts with different languages are not represented distinctly.

3 Our System

3.1 The System

The system in this paper consists of four main modules, i.e. word vector training, data preprocessing, feature extraction and emotion classification. To word vector training, words are vectorized over Wikipedia corpus via the word2vec tool. In the data preprocessing module, we will analyze the data obtained and process the labeled text into the input format of our model. Considering that a text may contain more than one emotion, we build a transformed training and test dataset, each of which contains the text itself and an emotion class that the text holds. To feature extraction, two types of features, namely the cross-lingual feature and the lexical level feature, are computed and combined with the word embeddings for representation learning via a bidirectional LSTM neural network, a.k.a., Bi-LSTM. As to emotion classification, a softmax classifier is built to train and predict emotions.

3.2 Bidirectional LSTM

The LSTM model is a kind of Recurrent Neural Network (RNN). It proposes an improvement for the gradient disappearance problem in the RNN model, replacing the hidden layer nodes in the original RNN model with one memory element. This memory unit consists of memory cell, oblivious gate, input gate, and output gate. The memory cell is responsible for storing historical information, recording and updating historical information through a state parameter. The three-door structure uses the sigmoid function to determine the choice of information and thus acts on the memory cell.

The equations below describe how a layer of memory cell is updated at every timestep t. In these equations, x_t is the input to the memory cell layer at time t, W_i, W_g, W_o, U_i, U_g, U_o and V_o are weight matrices, b_i, b_g and b_o are bias vectors.

$$
\begin{aligned}
i_t &= \sigma(W_i x_t + U_i h_{t-1} + V_i c_{t-1} + b_i) \\
f_t &= 1 - i_t \\
o_t &= \sigma(W_0 x_t + U_0 h_{t-1} + V_0 c_t + b_0) \\
g_t &= tanh(W_g x_t + U_g h_{t-1} + b_g) \\
c_t &= f_t \otimes c_{t-1} + i_t \otimes g_t \\
h_t &= o_t \otimes tanh(c_t)
\end{aligned}
$$

It can be seen that the LSTM model uses the accumulated linear form to process the information of the sequence data, thereby avoiding the problem of gradient disappearance and also learning long-term information, overcoming the shortcomings of the RNN model.

3.3 The Learning Model

Cheng et al. [15] proposed a learning framework for recommendation systems, where raw input features (or feature combination) as well as sentence level feature representation are combined to train a logistic regression classifier. By this approach, the model can learn effective low-dimensional representations for queries in order to tackle the over-generalization problem. Following this idea, our system adopts a similar framework by training the emotion classifier using both emotional keyword features and sentence level feature representation. And the reason is that, emotional keywords serve as important cues for emotion detection, thus it benefit system performance if such emotion knowledge can be considered directly in the learning model. Figure 1 shows the framework of the joint approach of our system.

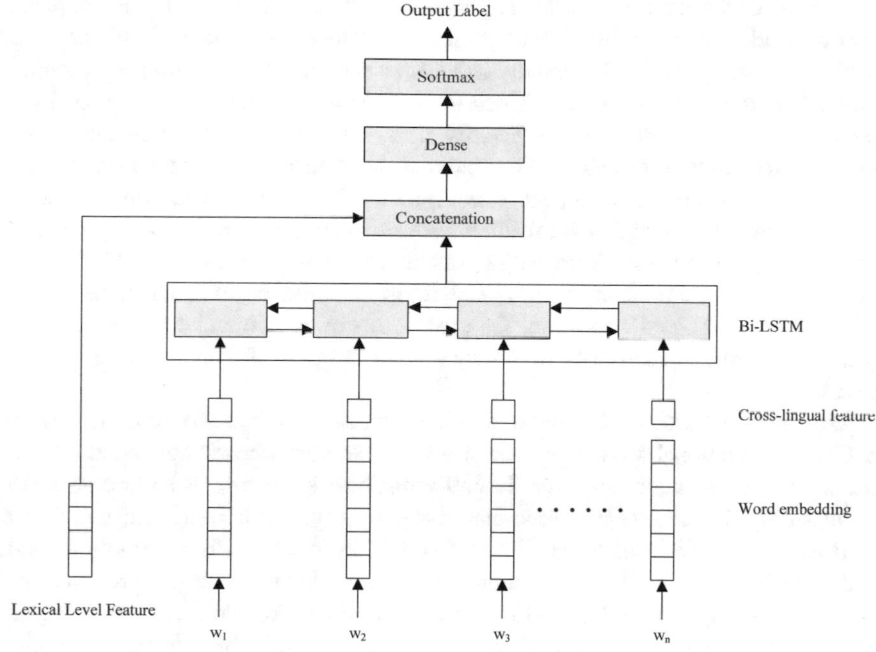

Fig. 1. The learning framework

There are two types of inputs of the joint learning model: emotional keyword features and sentence level feature representation. For emotional keyword features, we build a small emotion lexicon, in which each emotional keyword selected from an emotion knowledge resource also appears at the training and test data. For sentence level feature representation, a Bi-LSTM neural network is utilized. In such model, representation learning is a standard one: first, each Chinese and English word in a sentence is vectorized using the word2vec tool and the corpus comes from Chinese and

English Wikipedia texts; meanwhile, the cross-lingual feature is computed and combined with those vectorized word embeddings; then, sentence level feature representation is learned by a Bi-LSTM neural network; after that, the emotional keyword feature as well as sentence level feature representation are concatenated and put into a softmax classifier for training and prediction.

3.4 Feature Extraction

There are two additional features in this model: the cross-lingual feature and the lexical level feature. The cross-lingual feature is a simple one: if a word is a guest language, the feature value is 1, and if such word is a host language, the value is 0. Although such feature is simple, it can effectively distinguish two types of languages and contribute to investigating the impact of the guest language to emotion classes.

The idea of the lexical level feature comes from Zeng et al. [16] who proposed a neural network framework to classify semantic relations in sentences. In the model, lexical level features defined manually as well as sentence level feature representation via a CNN neural network are combined to train a softmax classifier. Inspired by this approach, we employ two types of features, that is, lexical level features and sentence level ones, for emotion detection. The sentence level feature representation is learned by a Bi-LSTM neural network, since both forward and backward contexts help to detect emotions. For lexical level features, we propose a lexicon-based approach, and the reason is that, emotion knowledge contributes to targeting emotional keywords, which helps detect emotions in emotional texts. It could be of much help to detect emotions in cross-lingual texts, since feature vectors of emotional keywords, especially English keywords, in such texts may very sparse because of their low frequencies of appearance.

Two emotion lexicons of Chinese and English are employed to build the features. The Chinese emotional keywords come from an emotion knowledge called Affective Lexicon Ontology[1], which includes 10,259 emotional keywords in 7 emotion classes. Note that only five classes of words that also appear at the training and test data are picked out. The English emotional keywords come from a well-known emotion lexicon called WordNet-Affect, which is a linguistic resource for the lexical representation of affective knowledge. We also build a small emotional keyword list including 225 words that do not appear in the above lexicons, each of which is labeled with an emotion class. Such data contributes to the recognition of emotional expressions with multilingual form.

Two lexical level features are defined: one is emotional keyword itself, the other is its emotion classes. All the keywords are vectorized to build features to concatenate with sentence level features.

[1] http://ir.dlut.edu.cn/EmotionOntologyDownload.

4 Evaluation Results and Discussions

The data collection in this experiment comes from the data annotated by Lee et al. [2]. The data includes 3,530 posts express emotions with five basic emotions defined, namely happiness, sadness, fear, anger and surprise. Each post may contain multiple emotions, thus the system needs to predict all the emotions that may be contained in each post. The experiment metric is F1 measure.

For systems are set in this experiment. The first one (lstm) employs a standard lstm based model with word embeddings to classify the emotions. The second one (bi-lstm) is very similar with the first one, except that the model is replaced by a standard bi-lstm model. The third one (bi-lstm + f_{cw}) also uses a standard bi-lstm model, but such system integrates each word embedding with its cross-lingual feature. The fourth one (bi-lstm + f_{cw} + f_{lex}) is the model proposed in this paper, which uses all the features including the lexical level feature. We also compare the performance of our approach with two state-of-the-art models. BAN [3] uses attention to capture the informative words from both monolingual, while BLP-BS [4] employs a Bipartite graph based label propagation framework with lexical level bilingual and sentimental information and bilingual context. The following Table 1 shows the experimental results.

Table 1. The official evaluation results of our formal runs

Team	Happiness	Sadness	Anger	Fear	Surprise	Marco-F1
lstm	0.652	0.609	0.667	0.684	0.561	0.635
bi-lstm	0.659	0.612	0.674	0.683	0.565	0.639
bi-lstm + f_{cw}	0.664	0.619	0.683	0.690	0.571	0.645
bi-lstm + f_{cw} + f_{lex}	**0.685**	**0.642**	**0.729**	**0.732**	**0.601**	**0.678**
BAN	0.678	0.634	0.728	0.728	0.594	0.672
BLP-BS	0.638	0.628	0.700	0.693	0.560	0.645

We can see from the experimental results that:

(1) Our approach achieves the best performance over the current state-of-the-art models. Since the other two models do not use emotion knowledge in their systems, it is possible to say that emotion knowledge benefits the performance improvement of learning approaches for emotion detection. The stronger evidence comes from the performance comparison of our approach and the system bi-lstm + f_{cw}. Take a detailed view of the performances of the two systems we can see that, the former one outperforms much than the latter one: there is an increasing 2.1% performance of emotion happiness, an increasing 2.3% of sadness, an increasing 4.6% of anger, an increasing 4.2% of fear and an increasing 3.0% of surprise. It indicates that emotion knowledge contributes to targeting emotional keywords, which help predict emotion classes in cross-lingual texts.

(2) The bi-lstm model outperforms the lstm one in the overall performance, and most emotion detection performances as well: the former one achieves an increasing 0.7% performance of emotion happiness, an increasing 0.3% of sadness, an increasing 0.7% of anger, and an increasing 0.4% of surprise to the latter one. It indicates that future contexts are also important to emotion detection in comparison with past ones, thus bi-lstm models is probably more appropriate for detecting emotions in cross-lingual texts.

(3) Since cross-lingual texts include multilingual forms, it is more appropriate to point out them clearly rather than to acquire such information by self learning. Such conclusion comes from the performance comparison of the system bi-lstm + f_{cw} and the system bi-lstm: the former one achieves an increasing 0.5% performance of emotion happiness, an increasing 0.7% of sadness, an increasing 0.9% of anger, an increasing 0.7% of fear, and an increasing 0.6% of surprise to the latter one.

5 Conclusions

In this paper, we propose a model based on Bi-LSTM neural network to detect emotions in cross-lingual texts, which is a mixed Chinese and English text with emotions. The model employs two kinds of features, that is, the cross-lingual feature and the lexical level feature. Such features contribute to understanding cross-lingual texts from two aspects: one is the relation between the text of guest language and the text of host language, the other is the affective meaning of each word, namely emotion class. Experimental results show that our approach is effective for detecting emotion in cross-lingual texts.

Acknowledgements. This work is supported by Natural Science Foundation of Hainan (618MS086), Special innovation project of Guangdong Education Department (2017KTSCX064), Natural Science Foundation of China (61702121) and Bidding Project of GDUFS Laboratory of Language Engineering and Computing (LEC2016ZBKT002).

References

1. Das, D., Bandyopadhyay, S.: Emotion analysis on social media: natural language processing approaches and applications. In: Agarwal, N. et al. (eds.) Online Collective Action: Dynamics of the Crowd in Social Media, Lecture Notes in Social Networks, pp. 19–37. Springer, Vienna (2014)
2. Lee, S., Wang, Z.Q.: Emotion in code-switching texts: corpus construction and analysis. In: Proceeding of SIGHAN-2015, Beijing (2015)
3. Wang, Z.Q., Zhang, Y., Lee, S., Li, S.S., Zhou., G.D.: A bilingual attention network for code-switched emotion prediction. In: Proceeding of COLING-2016, Osaka, Japan (2016)
4. Wang, Z.Q., Lee, S., Li, S.S., Zhou, G.D.: Emotion analysis in code-switching text with joint factor graph model. IEEE/ACM Trans. Audio Speech Lang. Process. **25**(3), 469–480 (2017)
5. Li, S.S., Lee, S., Liu, H.H., Huang, C.-R.: Implicit emotion classification with the context of emotion related event. J. Chin. Inf. Process. **27**(6), 90–95 (2013)

6. Rao, Y.H., Quan, X.J., Liu, W.Y., Li, Q., Chen, M.L.: Building word-emotion mapping dictionary for online news. In: Proceeding of the 1st International Workshop on Sentiment Discovery from Affective Data (2012)
7. Liu, H.H., Li, S.S., Zhou, G.D., Huang, C.-R., Li, P.F.: Joint modeling of news reader's and comment writer's emotions. In: Proceedings of the 51st Annual Meeting of the Association for Computational Linguistics, Sofia, Bulgaria, pp. 511–515 (2013)
8. Wen, S.Y., Wan, X.J.: Emotion Classification in microblog texts using class sequential rules. In: Proceedings of the Twenty-Eighth AAAI Conference on Artificial Intelligence, Québec, Canada, pp. 187–193 (2014)
9. Ren, H., Ren, Y.F., Li, X., Feng, W.H., Liu, M.F.: Natural logic inference for emotion detection. In: Proceedings of the 16th China National Conference and the 5th International Symposium on Natural Language Processing Based on Naturally Annotated Big Data, Nanjing, China, pp. 424–436 (2016)
10. Ren, Y.F., Wang, R.M., Ji, D.H.: A topic-enhanced word embedding for twitter sentiment classification. Inf. Sci. **369**, 188–198 (2016)
11. Ren, Y.F., Ji, D.H., Ren, H.: Context-augmented convolutional neural networks for twitter sarcasm detection. Neurocomputing **308**, 1–7 (2018)
12. Vo, D.-T., Zhang, Y.: Target-dependent twitter sentiment classification with rich automatic features. In: Proceedings of the Twenty-Fourth International Joint Conference on Artificial Intelligence, Buenos Aires, Argentina, pp. 1347–1353 (2015)
13. Tang, D.Y., Qin, B., Liu, T.: Document Modeling with gated recurrent neural network for sentiment classification. In: Proceedings of the 2015 Conference on Empirical Methods in Natural Language Processing, Lisbon, Portugal, pp. 1422–1432 (2015)
14. Zhang, M.S., Zhang, Y., Vo, D.-T.: Neural networks for open domain targeted sentiment. In: Proceedings of the 2015 Conference on Empirical Methods in Natural Language Processing, Lisbon, Portugal, pp. 612–621 (2015)
15. Cheng, H.-T., Koc, L., Harmsen, J., Shaked, T., Chandra, T., Aradhye, H., Anderson, G., Corrado, G., Chai, W., Ispir, M., Anil, R., Haque, Z., Hong, L.C., Jain, V., Liu, X.B., Shah, H.: Wide & deep learning for recommender systems. In: Proceedings of the 1st Workshop on Deep Learning for Recommender Systems, Boston, USA (2016)
16. Zeng, D.J., Liu, K., Lai, S.W., Zhou, G.Y., Zhao, J.: Relation classification via convolutional deep neural network. In: Proceedings of the 25th International Conference on Computational Linguistics: Technical Papers, Dublin, Ireland, pp. 2335–2344 (2014)

Information Hiding

Sharing More Information in Visual Cryptography Scheme with Block Expansion

Peng Li[1(✉)], Ching-Nung Yang[2], Dan Tang[3], and Jianfeng Ma[1]

[1] Department of Mathematics and Physics, North China Electric Power University, Baoding 071003, Hebei, China
lphit@163.com

[2] Department of CSIE, National Dong Hwa University, Hualien City, Taiwan

[3] School of Software Engineer, Chengdu University of Information Technology, Chengdu 610225, Sichuan, China

Abstract. Visual cryptography scheme (VCS) shares secret image into shadows. Without computing, stacking shadows can reveal the secret image. In this paper, we propose a VCS with block expansion. Each shadow pixel is expanded into a 4-pixel block with extra secret information embedded in. In the revealing process, the secret image can be revealed by stacking shadows. In the meanwhile, we can reveal the extra shared secret information by computation.

Keywords: Visual cryptography · Secret sharing · Secret image sharing · Block expansion

1 Introduction

Visual cryptography scheme (VCS) [1] is a technique to encode a secret image into multiple noise-like shadow images (also called shadows). Only stacking qualified set of shadows can reveal the secret image. Usually, VCS is referred as (k, n)-VCS, where a secret image is shared into n shadows. Each shadow is printed on a transparency. Stacking k transparencies can visually receive the content of the secret image. However, $k - 1$ or less shadows cannot get any information about the secret image.

VCS has the special advantage that the decoding process is not rely on computer resource. In the meanwhile, VCS also has two shortcomings: size expansion of the shadows and degraded visual quality of the revealed image. Many researchers dedicated to improve the performance of VCS [2, 3]. Also, many researchers proposed VCSs with special properties, like sharing multiple secret images [4, 5], meaningful shadows [6, 7]. Actually, the stacking operation on transparencies can be simulated by OR operation on Boolean vectors. In order to further improve the visual quality of the revealed image, XOR-based VCS is proposed [8–10].

Secret information can be also shared by a polynomial. Shamir first introduced secret sharing scheme by polynomial [11]. The secret number is embedded into the constant term of a randomly generated polynomial. Shadow pixels are generated by this polynomial with different input. In the revealing process, the polynomial can be reconstructed by Lagrange interpolation. To reduce the shadow size, Thien and Lin [12] proposed polynomial-based secret image sharing scheme that embed secret image

C.-N. Yang et al. (Eds.): SICBS 2018, AISC 895, pp. 849–857, 2020.
https://doi.org/10.1007/978-3-030-16946-6_69

pixel into all coefficients of a $(k - 1)$-degree polynomial. The shadow size is reduced to $1/k$ times of the secret image. Some PSIS scheme were proposed with different properties, like scalable PSIS [13, 14], PSIS with essential participants [15–18]. Compared with VCS, PSIS can reveal the precise secret image by computation. Some two-in-one secret image sharing schemes were proposed with the advantages of stacking to see and precise decoding [19–23]. The vague secret image can be revealed by stacking shadows. It also can decode the original secret image by computer. Li et al. proposed a two-in-one secret image sharing scheme by replacing shadow pixels with gray pixels [22]. Although it can embed more extra secret information, the shadows need to be saved in disk. It is hard to precisely extract the gray pixels from transparency.

In this paper, we propose a VCS with block expansion. The proposed scheme can share extra secret information in addition to the secret image. Each shadow pixel is expanded into a 4-pixel block. The extra information are embedded into the types of the block. The shadows are still binary image. Therefore, the embedded information is easy to be extracted from transparencies. The rest of this paper is organized as follows. Section 2 reviews some related works. The proposed VCS with block expansion is introduced in Sect. 3. Some experiment results are shown in Sect. 4. Finally, we conclude in Sect. 5.

2 Related Works

In this section, we briefly review visual cryptography scheme (VCS) and polynomial based secret image sharing (PSIS).

2.1 Visual Cryptography Scheme

Noar and Shamir first introduced visual cryptography scheme (VCS) [1]. For threshold (k, n)-VCS, a binary secret image is shared into n shadows. Each shadow is printed on transparency. In revealing process, stacking any k transparencies can visually reveal the secret image. Usually, a VCS is constructed by a pair of $n \times m$ binary matrices, called basis matrices B_0 and B_1, where 1 represent black pixel and 0 represent white pixel. If the secret pixel is black (white resp.), then choose matrix B_1(B_0 resp.) with the columns randomly permuted to generate the shadows. Each row of basis matrix is assigned to a shadow. Since there are m columns in basis matrix, sharing one secret pixel will generate m pixels for each shadow. Therefore, the size of each shadow is m times as the secret image. This is also called size expansion of the shadows. In the revealing process, the stacking operation on any k shadows can be simulated by OR operation on any k rows of the basis matrix. To visually detect the contrast between white and black pixels of the revealed secret image, the OR-ed result by any k rows of B_1 should have more 1s than that of B_0.

Example 1. (2, 2)-VCS with basis matrices

$$B_0 = \begin{bmatrix} 1 & 0 \\ 1 & 0 \end{bmatrix} \text{ and } B_1 = \begin{bmatrix} 1 & 0 \\ 0 & 1 \end{bmatrix} \quad (1)$$

Figure 1 shows an example of (2, 2)-VCS. A secret image (Fig. 1(a)) is shared into two shadows (Fig. 1(b) and (c)). The size of each shadow is two times as the secret image. Stacking two shadows reveal the secret image (Fig. 1(d)). As we can see, the visual quality of the revealed image is degraded than the secret image.

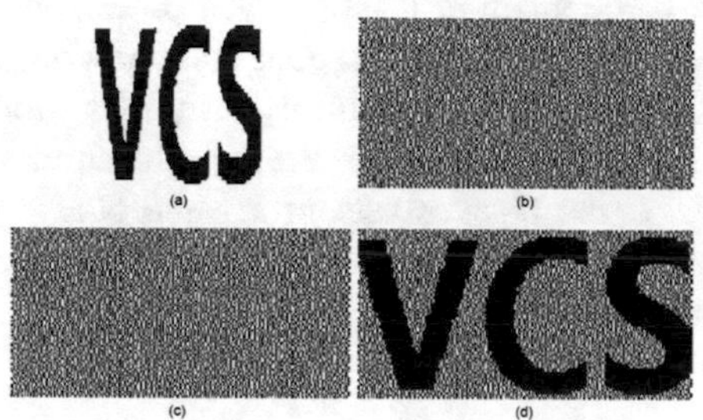

Fig. 1. (2, 2)-VCS. (a) secret image; (b)–(c) two shadows; (d) the revealed image by stacking two shadows.

2.2 Polynomial Based Secret Image Sharing

Polynomial based secret image sharing (PSIS) scheme can also share a secret image into multiple shadows [11, 12]. For a (*k*, *n*)-PSIS scheme, the secret pixels are embedded into the *k* coefficients of a *k*-1 degree polynomial.

$$f(x) = a_0 + a_1x + a_2x^2 + \ldots + a_{k-1}x^{k-1} \bmod p, \quad (2)$$

where $a_0, a_1, \ldots, a_{k\text{-}1}$ are replaced by secret pixels. With different x_i, we can generate the corresponding shadow pixel $f(x_i)$. In the revealing process, with *k* shadow pixels $f(x_1)$, $f(x_2), \ldots, f(x_k)$, we can reconstruct the polynomial $f(x)$, and then get the secret pixels. Since sharing *k* secret pixels generates on pixel for each shadow, the shadow size is 1/*k* of the secret image. In order to deal with grayscale pixels and make sure the result is still a grayscale value, many researchers adopt Galois Field GF(2^8) in the arithmetic.

3 The Proposed VCS with Block Expansion

In this paper, we propose a new (*k*, *n*)-VCS with block expansion. The proposed (*k*, *n*)-VCS can share more information. A secret image is first shared into multiple shadows by traditional VCS. Then each shadow is further operated to embed extra information. The main idea is that each shadow pixel is expanded into a block of four pixels, and the four pixels in each block are also white or black. With different choice of the combination of pixels in each block, we can embed extra information in addition to the original secret image. The extra information embedded in each block is the shadows of extra secret information shared by (k, n)-PSIS scheme.

3.1 The Choice of the Pixels Combination in Each Block

For each shadow generated by traditional (k, n)-VCS, we need to expand each shadow pixel into a four-pixel block. In order to make sure the black area of the revealed image is darker than the revealed white area, a block tends to contain more (less resp.) black pixels corresponding to the black (white resp.) shadow pixel. Let black (white resp.) block corresponds to a block expanded from a black (white resp.) shadow pixel. Suppose a black block can contain at least p black pixels, and a white block should contain at most q black pixels. To satisfy the contrast condition, stacking k black blocks must contain more black pixels than the result by stacking k white blocks. As we know, if the black pixels in k black blocks are all in the same locations, then the stacked block also has p black pixels. If the black pixels in k white blocks are all in the different locations, the stacked block can have kq black pixels. That is, the choice of p, q should satisfy the following condition.

$$p > kq \tag{3}$$

Obviously, kq must smaller than 4. Otherwise, the stacked block of k white block will be totally black.

Table 1. The choices of p and q for (2, 3)-VCS

	$p = 0$	$p = 1$	$p = 2$	$p = 3$	$p = 4$
$q = 0$	No	Yes	Yes	Yes	Yes
$q = 1$	No	No	No	Yes	Yes
$q = 2$	No	No	No	No	No
$q = 3$	No	No	No	No	No
$q = 4$	No	No	No	No	No

For example, for (2, 3)-VCS, the possible choices of p and q are as shown in Table 1.

3.2 Extra Information Shared by Proposed VCS

In order to share more information, the extra information is embedded in each shadow block by choosing different types of pixels of the block. For example, for a block with 1 black pixel, we have four types: [1 0 0 0], [0 1 0 0], [0 0 1 0] and [0 0 0 1]. Table 2 shows all types of shadow block with different number of black pixels.

Table 2. Different types of shadow block.

Amount of black pixels in a block	All the different types	Amount of types
0	[0 0 0 0]	1
1	[1 0 0 0] [0 1 0 0] [0 0 1 0] [0 0 0 1]	4
2	[1 1 0 0] [1 0 1 0] [1 0 0 1] [0 1 1 0] [0 1 0 1] [0 0 1 1]	6
3	[0 1 1 1] [1 0 1 1] [1 1 0 1] [1 1 1 0]	4
4	[1 1 1 1]	1

Let T be the types of candidate types of the block. Then the information amount I that can be embedded in each block is evaluated as follows.

$$I = \log_2(T) \text{ bits} \tag{4}$$

For example, in a (2, 3)-VCS with $p = 3$ and $q = 1$, there are both four types of blocks when a block having 3 and 1 black pixels. Therefore, each black block can embed $\log_2(4) = 2$ bits extra information, and each white block also can embed 2 bits extra information. Let m be the size expansion of (2, 3)-VCS, then the total extra information TI embedded in each shadow image is:

$$TI = 2m|S| \text{ bits}, \tag{5}$$

where $|S|$ denotes the pixel amount of the secret image.

As we know, the extra information embedded in each shadow is a also shadow generated by a (k, n)-PSIS scheme. And the shadow size of (2, 3)-PSIS scheme is 1/2 times of the secret information. Therefore, the total extra secret information TSI shared in the proposed (2, 3)-VCS is evaluated as follows.

$$TSI = 4m|S| \text{ bits} \tag{6}$$

3.3 The Algorithms of Sharing and Revealing Process

The proposed (k, n)-VCS with block expansion has two processes: sharing process and revealing process. In this subsection, we show the details of the algorithms of sharing and revealing process.

Algorithm 1. The sharing process of (k, n)-VCS with block expansion

Input: Value of k, n, p, q, the basis of matrices of (k, n)-VCS, secret image S, and extra secret information SI
Output: n expanded shadows $ES_1, ES_2, \ldots, ES_n$.
(A1-1): Share extra secret information SI by (k, n)-PSIS scheme, and generate n shadows $PS_1, PS_2, \ldots, PS_n$.
(A1-2): Share secret image S by (k, n)-VCS, and generate n shadows $VS_1, VS_2, \ldots, VS_n$.
(A1-3): For each shadow pixel in VS_i, expand it into a 4-pixel block chosen by the information of $PS_i(i=1, 2, \ldots, n)$:
(A1-3a): If the shadow pixel in VS_i is black, then choose a 4-pixel block with p black pixels as the corresponding shadow block in ES_i, the chosen block is corresponding to the information of PS_i.
(A1-3b): If the shadow pixel in VS_i is white, then choose a 4-pixel block with q black pixels as the corresponding shadow block in ES_i, the chosen block is corresponding to the information of PS_i.
(A1-4): Repeat (A1-3) until all shadow pixel in VS_i are expanded.
(A1-5): Finally, we get n expanded shadows $ES_1, ES_2, \ldots, ES_n$.

Algorithm 2. The revealing process of (k, n)-VCS with block expansion

Input: k shadows $ES_1, ES_2, \ldots, ES_k$.
Output: The vague secret image S', and extra secret information SI.
(A2-1): Directly stacking k shadows $ES_1, ES_2, \ldots, ES_k$ to reveal a vague secret image S'.
(A2-2): For each shadow ES_i $(i=1, 2, \ldots, k)$, divide ES_i into multiple 4-pixel blocks.
(A2-3): Read in each block, find out the type of the block, and get the secret information of PS_i corresponding to the type number.
(A2-4): Repeat (A2-3) until all secret information of PS_i is extracted from ES_i. Then we have k shadows PS_i $(i=1, 2, \ldots, k)$.
(A2-5): Decode SI by (k, n)-PSIS with k shadows PS_i $(i=1, 2, \ldots, k)$.

4 Experimental Results

In this section, we conduct two experiments to show the effective of the proposed scheme. The first experiment is a (2, 2)-VCS. The secret image is a print-text image "VCS" with 128 × 128 pixels (Fig. 1(a)). First the secret image is shared into two shadows using basis matrices in Eq. (1). Then each shadow pixel is expanded into a 4-pixel block. If the shadow pixel is black, then the expanded shadow block contain 3 black and 1 white pixels. If the shadow pixel is white, then the expanded shadow block contain 1 black and 3 white pixels. By choosing different types of the block, we embed a shadow of extra secret information into the shadow of VCS. Since there are 4 types block for black block, and 4 types block for white block, expanding each shadow pixel can embed additional 2 bits. Totally, we the extra shared secret information has $128 \times 128 \times 2 \times 2 = 65536$ bits. The final shadows as shown in Fig. 2(b) and (c) have the size 256×512 pixels. Stacking these two shadows we can decode the secret image (Fig. 2(d)).

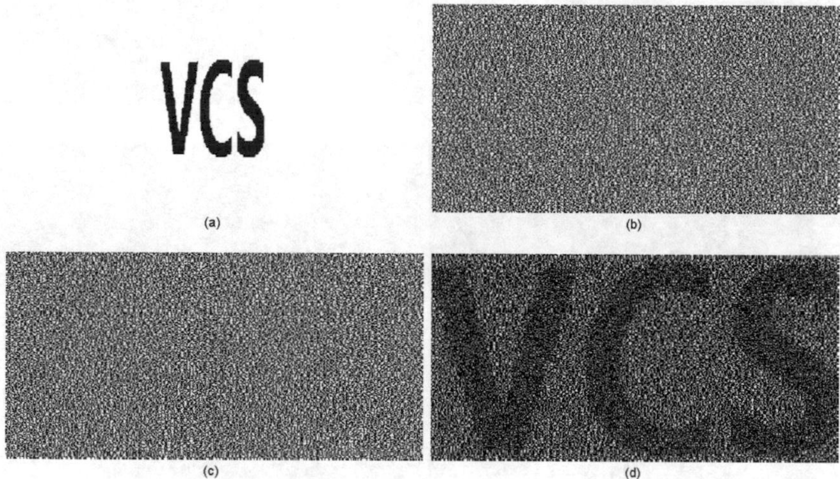

Fig. 2. The experiment of (2, 2)-VCS with block expansion. (a) the secret image; (b)–(c) two shadows; (d) the stacking result by two shadows.

The second experiment is a (2, 3)-VCS. Two basis matrices are chosen as follows.

$$B_0 = \begin{bmatrix} 1 & 0 & 0 \\ 1 & 0 & 0 \\ 1 & 0 & 0 \end{bmatrix} \text{ and } B_1 = \begin{bmatrix} 1 & 0 & 0 \\ 0 & 1 & 0 \\ 0 & 0 & 1 \end{bmatrix} \tag{7}$$

The secret image (Fig. 3(a)) with 128×128 pixels is shared into three shadows, then each shadow pixel is expanded into a 4-pixel block. Since each block can embed 2 bits information, and the size of the shadow is 3 times of the secret image, the total extra shared secret information is $128 \times 128 \times 3 \times 2 = 98304$ bits. 3 generated shadows are shown in Fig. 3(b)–(d). By stacking any two or three shadows, the secret image can be visually revealed. Figure 3(e)–(h) show the revealed image by stacking.

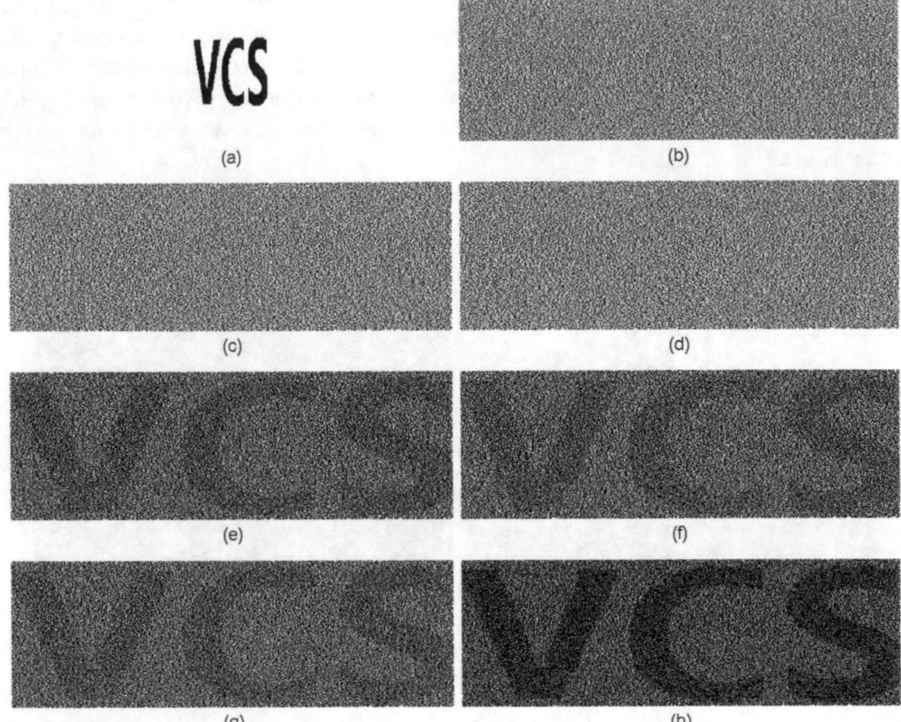

Fig. 3. The experiment of (2, 3)-VCS with block expansion. (a) the secret image; (b)–(d) three shadows; (e) the stacking result by shadow (b) and (c); (f) the stacking result by shadow (b) and (d); (g) the stacking result by shadow (c) and (d); (h) the stacking result by three shadows.

5 Conclusion

In this paper, we proposed a new visual cryptography scheme with block expansion. The proposed scheme can share extra secret information in addition to the secret image. The embedded extra secret information can be easily extracted from shadows. In the revealing process, stacking shadows can reveal the secret image. Then the extra embedded secret information can be extracted by computer.

Acknowledgements. This work was supported by the National Natural Science Foundation of China (No. 61602173).

References

1. Naor, M., Shamir, A.: Visual cryptography. In EUROCRYPT 1994, Lecture Notes in Computer Science, vol. 950, pp. 1–12. Springer, Berlin (1995)
2. Yang, C.N.: New visual secret sharing schemes using probabilistic method. Pattern Recogn. Lett. **25**(4), 481–494 (2004)

3. Cimato, S., Prisco, R., Santis, A.: Probabilistic visual cryptography schemes. Comput. J. **49**(1), 97–107 (2006)
4. Shyu, S.J., Huang, S.Y., Lee, Y.K., Wang, R.Z., Chen, K.: Sharing multiple secrets in visual cryptography. Pattern Recogn. **40**(12), 3633–3651 (2007)
5. Feng, J.B., Wu, H.C., Tsai, C.S., Chang, Y.F., Chu, Y.P.: Visual secret sharing for multiple secrets. Pattern Recogn. **41**(12), 3572–3581 (2008)
6. Liu, F., Wu, C.K.: Embedded extended visual cryptography schemes. IEEE Trans. Inf. Forensics Secur. **6**(2), 307–322 (2011)
7. Wang, D.S., Yi, F., Li, X.B.: On general construction for extended visual cryptography schemes. Pattern Recogn. **42**(11), 3071–3082 (2009)
8. Yang, C.N., Wang, D.S.: Property analysis of XOR-based visual cryptography. IEEE Trans. Circuits Syst. Video Technol. **24**(2), 189–197 (2014)
9. Ou, D., Sun, W., Wu, X.: Non-expansible XOR-based visual cryptography scheme with meaningful shares. Signal Process. **108**, 604–621 (2015)
10. Wu, X., Sun, W.: Extended capabilities for XOR-based visual cryptography. IEEE Trans. Inf. Forensics Secur. **9**(10), 1592–1605 (2017)
11. Shamir, A.: How to share a secret. Commun. Assoc. Comput. Mach. **22**, 612–613 (1979)
12. Thien, C.C., Lin, J.C.: Secret image sharing. Comput. Graph. **26**, 765770 (2002)
13. Yang, C.N., Huang, S.M.: Constructions and properties of k out of n scalable secret image sharing. Opt. Commun. **283**(9), 1750–1762 (2010)
14. Yang, C.N., Chu, Y.Y.: A general (k, n) scalable secret image sharing scheme with the smooth scalability. J. Syst. Softw. **84**(10), 1726–1733 (2011)
15. Li, P., Yang, C.N., Wu, C.C., Kong, Q., Ma, Y.: Essential secret image sharing scheme with different importance of shadows. J. Vis. Commun. Image Represent. **24**(7), 1106–1114 (2013)
16. Yang, C.N., Li, P., Wu, C.C., Cai, S.R.: Reducing shadow size in essential secret image sharing by conjunctive hierarchical approach. Signal Process. Image Commun. **31**, 1–9 (2015)
17. Li, P., Yang, C.N., Zhou, Z.: Essential secret image sharing scheme with the same size of shadows. Digit. Signal Process. **50**, 51–60 (2016)
18. Li, P., Liu, Z., Yang, C.N.: A construction method of (t, k, n)-essential secret image sharing scheme. Signal Process. Image Commun. **65**, 210–220 (2018)
19. Lin, S.J., Lin, J.C.: VCPSS: a two-in-one two-decoding-options image sharing method combining visual cryptography (VC) and polynomial-style sharing (PSS) approaches. Pattern Recogn. **40**, 3652–3666 (2007)
20. Yang, C.N., Ciou, C.B.: Image secret sharing method with two decoding-options: lossless recovery and previewing capability. Image Vis. Comput. **28**, 1600–1610 (2010)
21. Li, P., Ma, P.J., Su, X.H., Yang, C.N.: Improvements of a two-in-one image secret sharing scheme based on gray mixing model. J. Vis. Commun. Image Represent. **23**, 441453 (2012)
22. Li, P., Yang, C.N., Kong, Q., Ma, Y., Liu, Z.: Sharing more information in gray visual cryptography scheme. J. Vis. Commun. Image Represent. **24**, 1380–1393 (2013)
23. Li, P., Yang, C.N., Kong, Q.: A novel two-in-one image secret sharing scheme based on perfect black visual cryptography. J. Real-Time Image Process. **14**(1), 1–10 (2018)

High Capacity Data Hiding Based on AMBTC and Interpolation

Cheonshik Kim[1(✉)], Dongkyoo Shin[1], and Ching-Nung Yang[2]

[1] Department of Computer Engineering, Sejong University, Seoul 05006, Republic of South Korea
mipsan@paran.com, shindk@sejong.ac.kr

[2] Department of Computer Science and Information Engineering, National Dong Hwa University, Hualien 97401, Taiwan, R.O.C.
cnyang@gms.ndhu.edu.tw

Abstract. Absolute Moment Block Truncation Coding (AMBTC) is a compression method with low complexity and simple computation to easily compress a block. Various AMBTC-like methods have been proposed for developing data hiding schemes. All the existing techniques demonstrated steady performance. To further improve performance, in this paper, we present a method to extend the cover image using the neighbor mean interpolation (NMI) method, and improve data hiding performance based on this extended cover image and AMBTC. Experimental results show that our framework is superior in terms of existing BTC-based data hiding function and concealment efficiency.

Keywords: Data hiding · NMI · BTC · AMBTC · DBS

1 Introduction

For the cover image based on spatial domain, data hiding (DH) [1–5] embeds secret bits into the least significant bit (LSB) [6,7] of pixels by flipping the LSBs according an algorithm. The merit of DH based on the spatial domain is to offer high embedding capacity and good image quality. On the other hand, it is vulnerable to image compression, image transformation, and noises like salt & peppers.

Meanwhile, an image based on transform domains derived from various compressing techniques are generally very strong in aspect to attack of image processing. Here, the transform domains are cosine transform (DCT) [8], discrete wavelet transform (DWT) [9], and singular value decomposition (SVD) [10] etc. Block truncation coding (BTC) [11] is very attractive for real-time image compression coding at moderate bit-rate due to its low computation. The absolute moment BTC (AMBTC) [12] proposed by Lema and Mitchell is the most widely used one of various BTC.

The $trios(a, b, BM)$ may be obtained by transforming an image into AMBTC, where (a, b) is two quantization levels and BM (bitmap) is a compressed coefficients. Chuang and Chang [13] proposed a DH that embeds data

C.-N. Yang et al. (Eds.): SICBS 2018, AISC 895, pp. 858–867, 2020.
https://doi.org/10.1007/978-3-030-16946-6_70

using direct bitmap substitution (DBS) for each block. Chen et al. [14] proposed a method to store "1" bit per block using the order of two quantization levels (OTQL). Hong et al. [15] examine the difference values between the bits to be hidden and the bits of the bitmap, and if the number is more than half, flip two quantization levels.

Ou and Sun [16] introduced mixed DH using DBS and OTQL. Bai and Chang [22] used matrix encoding on both the quantization levels and the BM on smooth blocks. Huang et al. [17] also proposed a scheme to embed data using DBS, OTQL, and modification of pixels difference (hidden bits = $log_2 T$) on the two quantization levels. Hong [18] proposed a method of inserting secret bits into *trios* (compressed AMBTC). This method used pixel pair matching (PPM) [19], DBS, and OTQL for DH. In [20], Jung and Yoo proposed a new image interpolation method based on DH (IIDH). They design a new scale-up neighborhood mean interpolation (NMI) with low complexity and high computation speed. Based on the NMI, the original image is expanded to the cover image. A secret data are added to the extended pixels.

In this paper, we propose DH based on AMBTC compressed image, which may provide high EC by using expanded cover images. For experiments, the cover image is made by using combine AMBTC image (generated from the original image) and the expanded image (derived from the original image using NMI) (referred to Sect. 3). For embedding data, the proposed method classified AMBTC blocks into smooth and complex.

The rest of this paper is organized as follows. Section 2 briefly reviews AMBTC and Optimal quantization level adjustment. In Sect. 3, we present embedding algorithm for expanded AMBTC (i.e., $trios(a, b, BM)$). In Sect. 4, the proposed scheme is tested using cover images. A comparison with the Ou and Sun [16], Huang et al. [17], Hong [18], and our proposed scheme is also provided. Finally, this paper concludes in Sect. 5.

2 Related Works

2.1 AMBTC Compression Technique

For AMBTC, the image is divided into non-overlapping blocks sized $m \times m$ and each block is process separately. The total number of pixels in a block is $n = m \times m$. The mean value for each block is calculated as follows:

$$\mu = \frac{1}{n}\sum_{i=1}^{n} x_i \tag{1}$$

where x_i indicates a pixel of ith in a block. The original block can be quantized into a bitmap (BM) composed of $\ddot{b}$ (0 or 1), i.e., if $x_i \geq \mu$, $\ddot{b}_i$ is set to '1', otherwise set to '0'. The pixels in each block are classified into two groups, i.e., 1's pixel group, and 0's pixel group. The number of 1's group denotes p_n and the number of 0's group denotes $n - p_n$.

The mean value in each group, that is, a and b are variables for 0's and 1's groups, where the variables preserve the quantization levels for each block. The two quantization levels are calculated as follows:

$$a = \frac{1}{n - p_n} \sum_{x_i < \mu} x_i \tag{2}$$

$$b = \frac{1}{p_n} \sum_{x_i \geq \mu} x_i \tag{3}$$

where a and b are used to reconstructed AMBTC.

$$\ddot{b}_i = \begin{cases} 1, & \text{if } (x_i \geq \mu) \\ 0, & \text{if } (x_i < \mu) \end{cases} \tag{4}$$

$$x'_i = \begin{cases} a, & \text{if } (\ddot{b}_i = 0) \\ b, & \text{if } (\ddot{b}_i = 1) \end{cases} \tag{5}$$

Equation (4) is used to generate BM of a grayscale image. Equation (5) is applied to restore grayscale image from the compressed image based on AMBTC, where x'_i is a restored grayscale image using the quantization levels. The encoding compressed code, $trio(a, b, BM)$, may be implemented by using Eqs. (2), (3), and (4).

165	175	176	188
175	185	193	166
156	106	100	123
147	167	178	143

(a) Natural image block

1	1	1	1
1	1	1	1
0	0	0	0
0	1	1	0

(b) AMBTC bitmap

176	176	176	176
176	176	176	176
129	129	129	129
129	176	176	129

(c) Reconstructed image block

Fig. 1. An example of AMBTC algorithm.

Figure 1 is an example of AMBTC for encoding and decoding an image. Figure 1(a) shows a block of a natural image, where the block size is 4×4 pixels and average of the block is 158. If we apply Eq. (4) to the natural block, then the bitmap block of Fig. 1(b) come out. The two quantization levels of the block in Fig. 1(c) are obtained $a = 129$ and $b = 177$ using to Eqs. (2) and (3). As a result, the $trio$ representing a form of compression will be (129,177, 1111111100000110). The restoration from $trio$ to the grayscale image (Fig. 1(c)) is simply obtained through the use of Eq. (4).

2.2 Optimal Quantization Level Adjustment

Hong [18] proposed optimal quantization level adjustment to improve quality of stego image. The proposed technique shows to make an image's quality improving some degree. Let n_0 and n_1 denote the number of the unchanged bits and changed, respectively. Re-calculated a' and b' of a stego block are follows:

$$a' = \frac{a_i n_0 + b_i n_i}{n_0 + n_1}, b' = \frac{a_i n_1 + b_i n_0}{n_0 + n_1} \tag{6}$$

3 The Proposed Scheme

For embedding secret data, we apply DBS to smooth blocks. Moreover, the optimal adjustment technique for quantization levels is used for maintaining a good image quality.

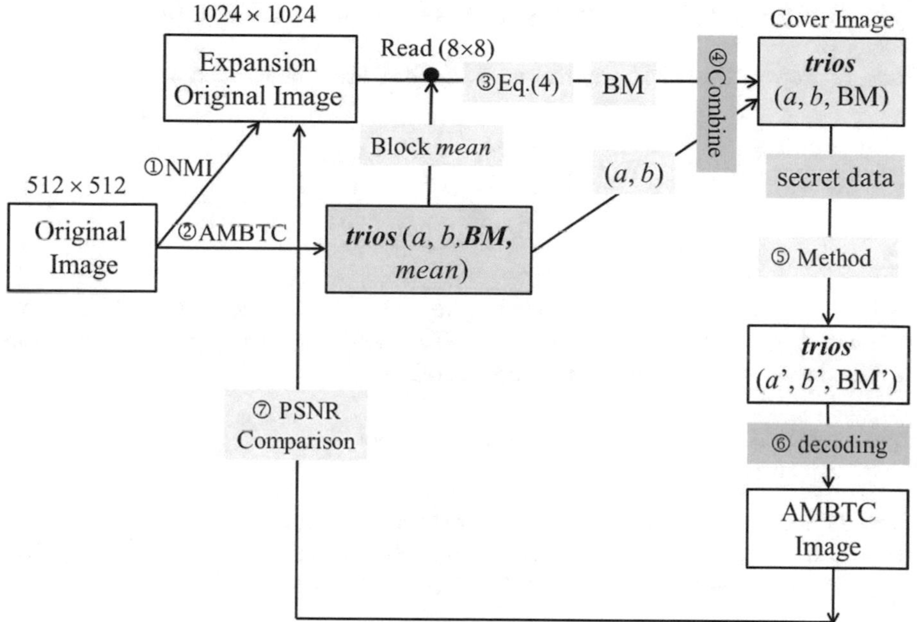

Fig. 2. Diagram of the procedure to generate *trios*.

As shown in Fig. 2, after the original image is converted to *trios*, we apply the proposed schemes to the *trios* to conceal secret data. First, we transform the original image into the extended original image and $trios(a, b, BM, mean)$, respectively①②. The bitmap is generated by comparing the mean values derived from the *trio* with the extended original image using Eq. (4)③. The *trios* for the cover image are done by combining the quantized values of the generated

BM and $trios(a, b)$④. The stego *trios* are generated by applying our proposed scheme to the cover *trios*⑤. The generated *trios* are decoded⑥ and compared with the extended cover image⑦.

3.1 Embedding Procedure

The original cover image is generated from original image by NMI and AMBTC. First, we must decide a threshold T to determine a smooth block. Then, traverse list of $trios(a_i, b_i, BM)$ and fetch the secret bits by the size of BM and replace the secret bits with BM. The detail of our proposed Method follows:

Input: AMBTC codes $trios(a, b, BM)$, secret data κ, block size $m \times m$
Output: AMBTC stego code $trios(a, b, BM)$, threshold T

Step 1: Decide threshold T as optimum value.
Step 2: If $(b - a \leq T)$, extract the number of $m \times m$ bits from κ and replace the secret bits with BM using Eq. (7).
Step 4: If $(b - a \leq T)$, adjust the quantization pair (a_i, b_i) to (a_i', b_i') by using Eq. (6).
Step 5: Repeat Step 2 through Step 4 until all trios have been scanned.

$$B_i = \begin{cases} \kappa_1^{m \times m}, & b - a \leq T \\ no\ action, & b - a > T \end{cases} \tag{7}$$

3.2 Extraction Procedure

Here, we extract secret bits from stego image, i.e., $trios(a, b, BM)$ and threshold T. Then, traverse the list of $trios(a_i, b_i, BM)$ and return BM to secret variable κ. The detail of extraction procedure as follows:

Input: AMBTC codes $trios(a, b, BM)$, threshold T
Output: return secret bits

Step 1: Set $B = trios(a, b, BM)_{i=1}^{N}$ (notes: assign bitmap).
Step 2: If $(b - a \leq T)$, extract BM from the B and assign to κ, i.e., $\kappa = \kappa || BM$.
Step 3: Steps 2–4 are repeated until every *trios* are scanned.

Example 1: We demonstrate the embedding procedure using AMBTC compressed block as shown in Fig. 3. In Fig. 3a, first, $trio(103, 107, BM)$ and secret bits κ are given. For the *trio*, since $b - a = 107 - 103 = 4 \leq T$, it is classified as a smooth block. As shown in Fig. 3b, secret bits κ = '1100 1001 0110 1010' may be embedded into BM using the proposed scheme, which may perform by replacing κ with B directly. B_1 is the appearance after embedding the secret bits. After then, the adjustment for two quantization levels is performed, i.e., $a' = 105$ and $b' = 104$ for B_1. The MSEs of $trio(103, 107, B_1)$ and $trio(105, 104, B_1)$ are 10 and 4.3125, respectively, thus we may know the fact that the adjustment of two quantization levels may be effective.

a. AMBTC encoding – Trio expression

Grayscale image (AMBTC)

103	107	103	107
103	107	107	107
107	103	103	107
107	103	103	107

Encode →

Block **B**, (103,107, **BM**), T=15

0	1	0	1
0	1	1	1
1	0	0	1
1	0	0	1

If $|a-b| < T$*, this block is smooth*

trio $(a, b, \mathbf{B})$ = (103,107, 0101011110011001)

B is smooth block

b. Method

Block **B**, (103,107, **B**), T=15

Stego block $\mathbf{B_1}$,(105,104, $\mathbf{B_1}$)

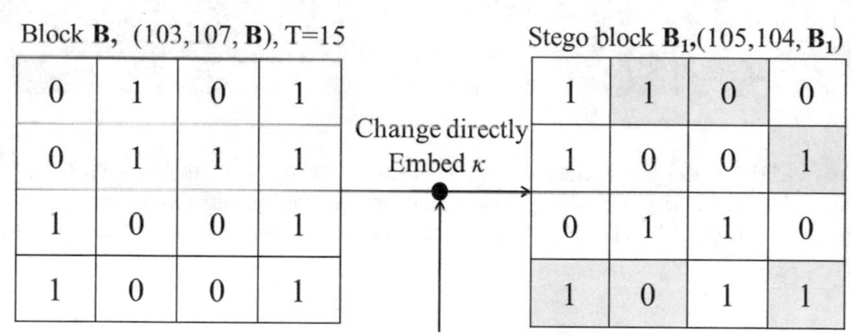

Secret bit κ: **1100 1001 0110 1010**

Replace **B** with secret bits κ

: set $B_1 = \kappa_1^{16}$

Adjustment a and b $\quad a' = \frac{a_i n_0 + b_i n_1}{n_0 + n_1} = 105, b' = \frac{a_i n_1 + b_i n_0}{n_0 + n_1} = 104$

☐ Unchanged pixel. For extraction κ, the *trio*, assign $B = \kappa$

Fig. 3. Examples of the proposed scheme.

4 Experiment and Comparison

In this section, for fairly evaluating analysis of the performance between previously proposed schemes and our proposed scheme, we compare the performance through conducting some experiments using nine numbers of standard 512×512 sized grayscale images [21].

These standard images are transformed into *trios* through the process of Fig. 2. For existing schemes, we experiment using *trios* generated from original grayscale images directly. Through the experiment by comparing the performance, we illustrate the superiority of our proposed scheme.

Figure 4 shows the types of images used in the experiment, i.e., (a) original "Lena" image, (b) expanded original image, (c) AMBTC image obtained from original image by AMBTC algorithm, and (d) the cover image obtained from "the expanded original image" and "AMBTC image" based on the procedure of Fig. 2. The type of two images, Fig. 4(c) and (d), are used to experiment for existing schemes previously proposed and our proposed scheme, respectively.

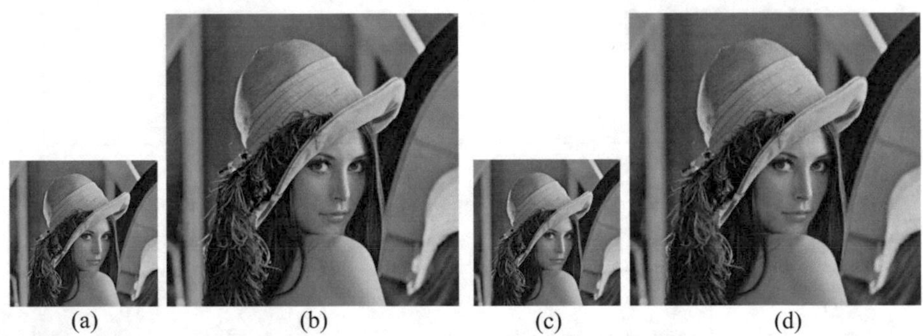

Fig. 4. Original image, expansion original image, compressed image (AMBTC), and cover images: (a) original image, (b) Enlarged original image obtained from the original image by NMI, (c) AMBTC image from original image by AMBTC algorithm, and (d) cover image.

The secret data used in the experiment is randomly generated binary number using pseudo random number generator. After applying the proposed methods, the compressed cover images (*trios*) including secret data are transformed into a grayscale image for comparing with the original image.

An objective evaluation for DH is by measuring the quality and the embedding capacity of stego images. Peak signal to noise ratio (PSNR) is used to measure the quality of an image and it is calculated using mean square error (MSE). The MSE is calculated by averaging the squared intensity differences between the distorted image pixels and the original image pixels. The lower the MSE value of the image, the better the image quality.

Table 1 shows the comparison between the proposed scheme and existing schemes in respect of the experimental results. For EC and PSNR, Ou and

Table 1. The comparison of the proposed scheme (methods) with other schemes.

Images	Huang et al. [17]		Ou and Sun [16]		Hong [18]		Proposed method	
	bits	dB	bits	dB	bits	dB	bits	dB
Bridge	312955	22.6373	253910	20.6783	318500	19.9828	703425	24.8072
Elaine	304231	28.2317	260090	26.2786	338325	25.9399	1001025	28.0364
Boats	303571	25.1200	257090	23.1576	329725	22.2374	879937	27.1110
Goldhill	308401	26.1750	261440	24.1422	333455	24.2783	925825	27.1826
Airplane	288081	26.1662	256760	24.2661	331820	23.0684	909569	28.9018
Lena	295104	27.6240	259940	25.6597	335055	24.9597	953793	29.1137
Peppers	294353	28.0096	257225	26.0507	334835	23.9778	955969	29.4434
Tiffany	294595	28.4624	260045	26.3874	336520	24.5438	975809	29.7136
Zelda	298176	30.2349	262100	28.2456	339490	28.3856	1014593	29.9662
Average	299940	26.9623	258733	24.98513	333081	24.15263	924438	28.2528

Sun's method has the lowest performance. In addition, theirs scheme forces to make sacrifice of a part of PSNR improving EC. Thus, it is not advisable to use this technique. For EC, W Hong's method shows good performance except for our proposed scheme. But, the stego image derived from this method is not appropriate to secret communication when EC is more than 300000 bits, because the PSNR is lower than 30 dB. Therefore, our proposed scheme outperforms the other schemes such as Huang et al. and W Hong in terms of PSNR and EC. In our proposed method, when the threshold is $T \leq 5$ (e.g., Lena image), the existing problems may be enough resolved. As a result, we improved the performance of the two criteria for DH compare to existing schemes.

5 Conclusion

We presented DH based on AMBTC expanded by NMI with low complexity and high computation speed. That is, we used cover images originated from an original image by using AMBTC and NMI. Secret data are embedded into compressed AMBTC images (i.e., $trios(a, b, BM)$) by using our proposed scheme. The performance (EC and PSNR) is superior to existing schemes under the same threshold T. Our proposed scheme can provide sufficient data and image quality, can be a very good choice.

Acknowledgments. This research was supported in part by Ministry of Science and Technology (MOST), under Grant 107-2221-E-259-007-. This research was supported by the Basic Science Research Program through the National Research Foundation of Korea (NRF) funded by the Ministry of Education (2018R1D1A1B07047395), National Research Foundation of Korea (NRF) funded by (2015R1D1A1A01059253), and was supported under the framework of international cooperation program managed by NRF (2016K2A9A2A05005255).

References

1. Bender, W., Gruhl, D., Morimoto, N., Lu, A.: Techniques for data hiding. IBM Syst. J. **35**(3.4), 313–336 (1996)
2. Yang, C.N., Hsu, S.C., Kim, C.: Improving stego image quality in image interpolation based data hiding. Comput. Stand. Interfaces **50**, 209–215 (2017)
3. Kim, C., Shin, D., Yang, C.N., Chou, Y.S.: Improving capacity of Hamming (n, k)+1 stego-code by using optimized Hamming+k. Digit. Signal Process. **78**, 284–293 (2018)
4. Shi, Y.Q., Li, X., Zhang, X., Wu, H.T., Ma, B.: Reversible data hiding: advances in the past two decades. IEEE Access **4**, 3210–3237 (2016)
5. Huang, F., Qu, X., Kim, H.J., Huang, J.: Reversible data hiding in JPEG images. IEEE Trans. Circuits Syst. Video Technol. **26**(9), 1610–1621 (2016)
6. Mielikainen, J.: LSB matching revisited. IEEE Signal Process. Lett. **13**, 285–287 (2006)
7. Chan, C.K., Cheng, L.M.: Hiding data in images by simple LSB substitution. Pattern Recognit. **37**(3), 469–474 (2004)
8. Chang, C.C., Lin, C.C., Tseng, C.S., Tai, W.L.: Reversible hiding in DCT-based compressed images. Inf. Sci. **177**, 2768–2786 (2007)
9. Chang, C.C., Chen, T.S., Chung, L.Z.: A steganographic method based upon JPEG and quantization table modification. Inf. Sci. **141**, 123–138 (2002)
10. Bergman, C., Davidson, J.: Unitary embedding for data hiding with the SVD. In: Proceedings of the SPIE, Security, Steganography, and Watermarking of Multimedia Contents VII, vol. 5681, 21 March 2005. https://doi.org/10.1117/12.587796
11. Delp, E.J., Mitchell, O.R.: Image compression using block truncation coding. IEEE Trans. Commun. **27**(9), 1335–1342 (1979)
12. Hong, W., Chen, T.S., Shiu, C.W.: Lossless steganography for AMBTC compressed images. In: Proceedings of 1st International Congress on Image and Signal Processing, Sanya, China, vol. 2, pp. 13–17 (2008)
13. Chuang, J.C., Chang, C.C.: Using a simple and fast image compression algorithm to hide secret information. Int. J. Comput. Appl. **28**(4), 329–333 (2006)
14. Chen, J., Hong, W., Chen, T.S., Shiu, C.W.: Steganography for BTC compressed images using no distortion technique. Imaging Sci. J. **58**, 177–185 (2010)
15. Hong, W., Chen, J., Chen, T.S., Shiu, C.W.: Steganography for block truncation coding compressed images using hybrid embedding scheme. Int. J. Innov. Comput. Inf. Control **7**, 733–743 (2011)
16. Ou, D., Sun, W.: High payload image steganography with minimum distortion based on absolute moment block truncation coding. Multimed. Tools Appl. **74**, 9117–9139 (2015)
17. Huang, Y.H., Chang, C.C., Chen, Y.H.: Hybrid secret hiding schemes based on absolute moment block truncation coding. Multimed. Tools Appl. **76**, 6159–6174 (2017)
18. Hong, W.: Efficient data hiding based on block truncation coding using pixel pair matching technique. Symmetry **10**(2), 1–18 (2018)
19. Hong, W., Chen, T.S.: A novel data embedding method using adaptive pixel pair matching. IEEE Trans. Inf. Forensics Secur. **7**, 176–184 (2012)
20. Jung, K.H., Yoo, K.Y.: Data hiding method using image interpolation. Comput. Stand. Interfaces **31**(2), 465–470 (2009)

21. Image database ref. http://sipi.usc.edu/database/database.php?volume=misc. Accessed 31 May 2018
22. Bai, J., Chang, C.C.: A high payload steganographic scheme for compressed images with hamming code. Int. J. Netw. Secur. **18**, 1122–1129 (2016)

A Reversible Data Hiding Scheme in Encrypted Image with LSB Compression and MSB Replacement

Wei Zhang[1,2], Qing Zhou[2], Zhenjun Tang[1], Heng Yao[2], and Chuan Qin[1,2(✉)]

[1] Guangxi Key Lab of Multi-source Information Mining and Security, Guangxi Normal University, Guilin 541004, China
wellzhang@yeah.net, tangzj230@163.com

[2] School of Optical-Electrical and Computer Engineering, University of Shanghai for Science and Technology, Shanghai 200093, China
zhouqing19940320@163.com, {hyao,qin}@usst.edu.cn

Abstract. A new scheme of reversible data hiding in encrypted images is proposed in this paper. Content owner first encrypts divided blocks by the specific stream cipher and permutation. During data embedding, data hider embeds secret data into the compressed least significant bits (LSB) and the most significant bits (MSB) of a part of pixels in the smooth region. After the receiver receives the marked, encrypted image, he or she can obtain a directly decrypted image with encryption key. With data-hiding key, the receiver can extract the embedded data from the compressed LSB sequences and MSB layers. If the receiver has both encryption key and data-hiding key, he or she not only can extract the embedded data from the marked, encrypted image, but also can recover original image correctly. Experimental results show our scheme achieves better rate-distortion performance than some reported schemes.

Keywords: Image encryption · Reversible data hiding · Data extraction · Image recovery

1 Introduction

In recent years, the problems of information security for digital images have become a hot issue. Data hiding is a useful technique to protect the secret data [1, 2], and attackers cannot extract the embedded data from the images without data hiding key. Digital image has become a popular cover data for communication in recent years, but original image may be degraded due to data embedding. In some applications, such as military or medical fields, we not only need data hiding to protect the security of embedded data, but also can recover original image losslessly on the receiver side [3], which was called reversible data hiding (RDH).

The technique of reversible data hiding in encrypted image (RDHEI) was extensively studied in recent years, which realized data hiding and protect the privacy of content owner at the same time [4–12]. In [4], a scheme of reversible data hiding in encrypted images was proposed. The authors embedded secret data into images by

C.-N. Yang et al. (Eds.): SICBS 2018, AISC 895, pp. 868–878, 2020.
https://doi.org/10.1007/978-3-030-16946-6_71

modifying a small proportion of encrypted pixels. Based on [4], the scheme in [5] improved the accuracy of data extraction and image recovery by using side match in data hiding phase. To further improve the performances of [4] and [5], Liao et al. [6] proposed a new smoothness function to calculate the complexity of the image blocks according to the absolute mean difference of multiple neighboring pixels. In [7], a high-capacity scheme of reversible data hiding in encrypted images by using MSB prediction was proposed. In [8], low-density parity check (LDPC) matrix was used in data hiding phase to compress LSB sequences for embedding secret data. In the recovery phase, the receiver recovered original image based on data-hiding key and side information. In [9], a novel scheme of reversible data hiding in encrypted images was proposed using distributed source coding (DSC). In data embedding phase, low-density parity check codes was used to embed secret data. The receiver recovered original image and extracted embedded data by distributed source decoding. In [10], the authors compressed LSB sequences by matrix operation and embedded secret data into compressed bits in the data hiding phase. In recovery phase, the receiver extracted the embedded data and recovered original image according to data hiding key and smoothness function. The scheme [11] improved the recovery performance of the scheme [10], in which data hider divided the encrypted image into three channels and embedded the secret data into channels, respectively. The original image was recovered with high embedding rate. In [12], original image was encrypted by the encryption key for privacy protection of image content, and then, each encrypted block was embedded with one secret bit by data-hiding key. Finally, the embedded bits were extracted and original image was recovered successfully.

In this paper, we propose a new scheme of reversible data hiding in encrypted images. Content owner encrypts divided blocks by the specific stream cipher and permutation. Then, the data hider embeds secret data into some encrypted compressed LSB sequences and MSB layers. If the receiver only has encryption key, he can decrypt the marked image to generate a similar image. If the receiver only has data-hiding key, he can extract the embedded data from the marked image. If the receiver has both encryption key and data-hiding key, he can extract embedded data and recover original image without error.

2 Proposed Scheme

In our scheme, four main stages are included, which consist of image encryption, data embedding, image decryption and image recovery. Figure 1 shows the framework of our scheme. In image encryption stage, content owner divides original image into a number of blocks and encrypts the blocks by a specific stream cipher and permutation. In data embedding stage, data hider firstly classifies encrypted blocks into smooth region and complex region by a pre-defined threshold T_1, then collects the smooth blocks and embeds secret data into the compressed LSBs of the collected blocks by matrix operation. Secondly, data hider finds the encrypted blocks which are smaller than or equal to another predefined threshold T_2, and embeds secret data by MSBs substitution. On the receiver side, the receiver can conduct separable operations of

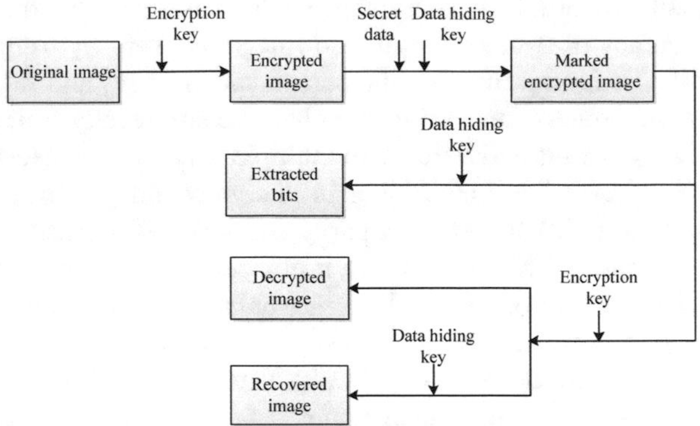

Fig. 1. Framework of the proposed scheme

direct decryption, data extraction and image recovery according to the availability of encryption and data-hiding keys.

2.1 Image Encryption

In this stage, the original image with the size of $M \times N$ is denoted as **I**, and the gray value of each pixel falls into [0, 255], which is represented by 8 bits:

$$b_{i,j,k} = \left\lfloor \frac{I(i,j)}{2^k} \right\rfloor \bmod 2, \quad k = 0, 1, \ldots, 7 \tag{1}$$

$$I(i,j) = \sum_{k=0}^{7} (2^k \times b_{i,j,k}). \tag{2}$$

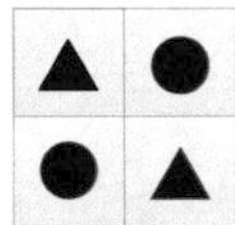

Fig. 2. Four pixels in a block $\mathbf{X}_{u,v}$

Content owner divides original image into a number of non-overlapping blocks sized 2×2. Then the $M \times N/4$ blocks are denoted by $\mathbf{X}_{u,v}$ where (u, v) represents the indices. As shown in Fig. 2, the pixels $X_{u,v}^{(0,0)}$ and $X_{u,v}^{(1,1)}$ are represented by "▲", $X_{u,v}^{(0,1)}$ and $X_{u,v}^{(1,0)}$ by "●". We encrypt 8 bits of circle pixels with pseudo-random bits by:

$$b'_{i,j,k} = b_{i,j,k} \oplus e_{i,j,k} \,. \tag{3}$$

For the triangle pixels, the exclusive-or results of the M least significant bits and pseudo-random bits are calculated:

$$b'_{i,j,p} = b_{i,j,p} \oplus e_{i,j,p}, \quad p = 0,\ 1,\ 2, \ldots,\ M-1. \tag{4}$$

In addition, M denotes the number of LSB layers and will be used in data embedding phase. Here, M is a small positive integer and smaller than 4. Then the other $(8 - M)$ least significant bits are encrypted by another pseudo-random binary sequence $Z_{u,v}$:

$$\begin{cases} \begin{cases} X'^{0,0}_{u,v,p} = \overline{X^{0,0}_{u,v,p}} \\ X'^{1,1}_{u,v,p} = \overline{X^{1,1}_{u,v,p}} \end{cases} \text{if } Z_{u,v} = 1 \\ \begin{cases} X'^{0,0}_{u,v,p} = X^{0,0}_{u,v,p} \\ X'^{1,1}_{u,v,p} = X^{1,1}_{u,v,p} \end{cases} \text{if } Z_{u,v} = 0 \end{cases} \quad p = M,\ M+1, \ldots,\ 7. \tag{5}$$

After all pixels in the image are encrypted, content owner permutes the blocks pseudo-randomly according to the encryption key. Thus, the final encrypted image $\mathbf{I}_e$ is generated.

2.2 Data Embedding

During this stage, data hider embeds secret data into the encrypted image $\mathbf{I}_e$ by two steps. In the first step, data hider divides the encrypted image into a series of 2×2 blocks $\mathbf{E}_{u,v}$ and calculates the absolute difference of two triangle pixels in each block, as illustrated in Fig. 2. In each block $\mathbf{E}_{u,v}$, four pixels are denoted as $E^{(0,0)}_{u,v}$, $E^{(0,1)}_{u,v}$, $E^{(1,0)}_{u,v}$, and $E^{(1,1)}_{u,v}$ in the raster-scanning order.

$$h_{u,v} = \left| 2^M \cdot \left\lfloor E^{(0,0)}_{u,v} / 2^M \right\rfloor - 2^M \cdot \left\lfloor E^{(1,1)}_{u,v} / 2^M \right\rfloor \right|, \tag{6}$$

where $h_{u,v}$ represents the smoothness degree of the block $\mathbf{E}_{u,v}$. Higher value of $h_{u,v}$ means the block $\mathbf{E}_{u,v}$ is more complex. According to the smoothness degrees, data hider classifies the encrypted blocks into smooth region $\mathbf{R}_1$ and complex region $\mathbf{R}_2$ by predefined threshold T_1:

$$\begin{cases} \mathbf{E}_{u,v} \in \mathbf{R}_1, & \text{if } h_{u,v} \le T_1, \\ \mathbf{E}_{u,v} \in \mathbf{R}_2, & \text{if } h_{u,v} > T_1, \end{cases} \tag{7}$$

where $\mathbf{R}_1$ and $\mathbf{R}_2$ represent smooth region and complex region, respectively. Then the smooth blocks are divided into a number of groups, each of which contains g pixels. In each pixel-group, the M least significant bits of the g pixels are collected and denoted by $A(d, 1)$, $A(d, 2)$, …, $A(d, M \cdot g)$, where d is a group index and M is a positive integer. In this way, a matrix **C** sized $(M \cdot g - f) \times M \cdot g$ is generated according to the data hiding key:

$$\mathbf{C} = [\mathbf{I}_{M \cdot g - f}, \ \mathbf{Q}]. \tag{8}$$

The matrix **C** is composed of two parts, the left part is an identity matrix, the right part **Q** sized $(M \times g - f) \times f$ is a pseudo-randomly binary matrix. Here, f is a small positive integer. For each group, the data hider compresses the bits:

$$\begin{bmatrix} A'(d,\ 1) \\ A'(d,\ 2) \\ \vdots \\ A'(d,\ Mg - f) \end{bmatrix} = \mathbf{C} \cdot \begin{bmatrix} A(d,\ 1) \\ A(d,\ 2) \\ \vdots \\ A(d,\ Mg) \end{bmatrix}, \tag{9}$$

where the arithmetic is modulo-2. After calculation, $[A(d, 1), A(d, 2), \ldots, A(d, M \cdot g)]$ are compressed as $(M \cdot g - f)$ bits. So, the compressed f bits in each group are used to accommodate the secret data. In the second step, the data hider also calculates the smoothness of the encrypted blocks by Eq. (6) and classifies the encrypted blocks into smooth region $\mathbf{O}_1$ and complex region $\mathbf{O}_2$ by pre-defined threshold T_2:

$$\begin{cases} \mathbf{E}_{u,v} \in \mathbf{O}_1, & \text{if } h_{u,v} \le T_2, \\ \mathbf{E}_{u,v} \in \mathbf{O}_2, & \text{if } h_{u,v} > T_2. \end{cases} \tag{10}$$

In each encrypted block of smooth region $\mathbf{O}_1$, the data hider collect the triangle pixels $E_{u,v}^{(0,0)}$ and $E_{u,v}^{(1,1)}$ which are used to accommodate the secret data by MSB replacement, and the MSB of two triangle pixels in the same block are replaced by the same secret bit b_s:

$$E'^{(x,y)}_{u,v} = (E^{(x,y)}_{u,v} \bmod 128) + b_s \times 128, \quad (x,y) \in \{(0,0),(1,1)\}. \tag{11}$$

After two steps, the marked image $\mathbf{I}_m$ is generated.

2.3 Data Extraction and Image Recovery

After the receiver obtains the marked image, three cases may occur. If receiver only has the encryption key, he can decrypt the marked image directly. If receiver only has the data hiding key, he can extract the secret data from the marked image directly. If receiver has both the encryption key and data hiding key, he can extract the secret data and recover the original image.

In decryption phase, the receiver first divides the marked image into $M \times N/4$ non-overlapping blocks $\mathbf{\Psi}_{u,v}$ sized 2×2 and classifies them into smooth region $\mathbf{O}_1$ and complex region $\mathbf{O}_2$ by the threshold T_2. Then, the pixels in each block are decomposed into 8 bits and decrypted by using exclusive OR operation according to the encryption key. In addition, all the $M \times N/4$ blocks are inversely permuted to their original locations by the encryption key to generate basic image $\mathbf{I}'_{\mathrm{m}}$. Then, the MSBs of two triangle pixels in smooth region $\mathbf{O}_1$ are needed to recovery. Here, we denote four pixels in each block $\boldsymbol{\Psi}_{u,v}$ in smooth region $\mathbf{O}_1$ by $\Psi_{u,v}^{(0,0)}$, $\Psi_{u,v}^{(0,1)}$, $\Psi_{u,v}^{(1,0)}$ and $\Psi_{u,v}^{(1,1)}$. The estimated gray value of corresponding block can be generated by the following function:

$$\hat{\Psi}_{u,v} = \left\lfloor \Psi_{u,v}^{(0,1)}/2^M \right\rfloor \cdot 2^M + \left\lfloor \Psi_{u,v}^{(1,0)}/2^M \right\rfloor \cdot 2^M, \tag{12}$$

where the estimated gray value is generated from the two neighbors of the same block in the basic image $\mathbf{I}'_{\mathrm{m}}$. Then in each block the receiver calculates:

$$\Psi'^{x,y}_{u,v} = \begin{cases} 128 + \mathrm{mod}(\Psi^{x,y}_{u,v}, 128) & \text{if} \left| 128 + \mathrm{mod}(\Psi^{0,0}_{u,v}, 128) + 128 + \mathrm{mod}(\Psi^{1,1}_{u,v}, 128) - \hat{\Psi}_{u,v} \right|, \\ & < \left| \mathrm{mod}(\Psi^{0,0}_{u,v}, 128) + \mathrm{mod}(\Psi^{1,1}_{u,v}, 128) - \hat{\Psi}_{u,v} \right|, \\ \mathrm{mod}(\Psi^{x,y}_{u,v}, 128) & \text{otherwise,} \end{cases} \tag{13}$$

where $\Psi'^{x,y}_{u,v}$ is the corresponding decrypted pixel, and $(x, y) \in \{(0, 0), (1, 1)\}$. Note that, in each block, the two triangle pixels are replaced by the same secret data. In this way, the decrypted image $\mathbf{I}_{\mathrm{d}}$ is generated.

If receiver only has the data hiding key, he can extract the secret data from the marked image directly. Firstly, the blocks in smooth region $\mathbf{R}_1$ can be collected by the threshold T_1, and the embedded data in compressed LSBs of smooth blocks can be extracted. In the following, the secret data embedded in MSBs of the smooth region $\mathbf{O}_1$ can be also extracted by MSBs replacements according to the data hiding key. Thus, the receiver extracts the embedded data successfully.

If receiver has both encryption key and data hiding key, he can recover the original image and extract the embedded data. Here, after decryption phase, only the LSBs in smooth region $\mathbf{R}_1$ are needed to recover. The pixel groups in smooth region are collected, and the M least significant bits of the g pixels are generated. Then, the receiver can obtain the vector $[A'(d,1), A'(d,2), \ldots, A'(d, M \cdot g - f)]$ from the marked image. However, the original bits vector of each group $[A(d,1), A(d,2), \ldots, A(d, M \cdot g)]$ must be one in the function:

$$\boldsymbol{\alpha} = [A'(d, 1), A'(d, 2), \ldots, A'(d, Mg - f), 0, 0, \ldots, 0] + \beta \cdot \mathbf{H}, \tag{14}$$

$$\mathbf{H} = [\mathbf{Q}', \mathbf{I}_f], \tag{15}$$

where β is an arbitrary binary vector sized $1 \times f$, and $\mathbf{H}$ is a matrix sized $f \times M \cdot g$. There are 2^f possibilities in each pixel group recovery phase. The receiver first puts the elements in each vector $\boldsymbol{\alpha}$ into their original position and decrypts the pixel group to generate decrypted pixel group O_d. Then the total difference of the decrypted pixels $r_{i,j}$ and estimated pixels $\tilde{p}_{i,j}$ in each group can be calculated:

$$\tilde{p}_{i,j} = \frac{p_{i-1,j} + p_{i+1,j} + p_{i,j-1} + p_{i,j+1}}{4}, \tag{16}$$

$$D = \sum_{(i,j) \in O_d} \left| r_{i,j} - \tilde{p}_{i,j} \right|. \tag{17}$$

The 2^f possible values of D correspond to 2^f decrypted pixel group O_d. We choose the smallest D and regard the corresponding pixel group as the original segment because of spatial correlation in natural image. After all pixel groups are recovered, the recovered image is generated.

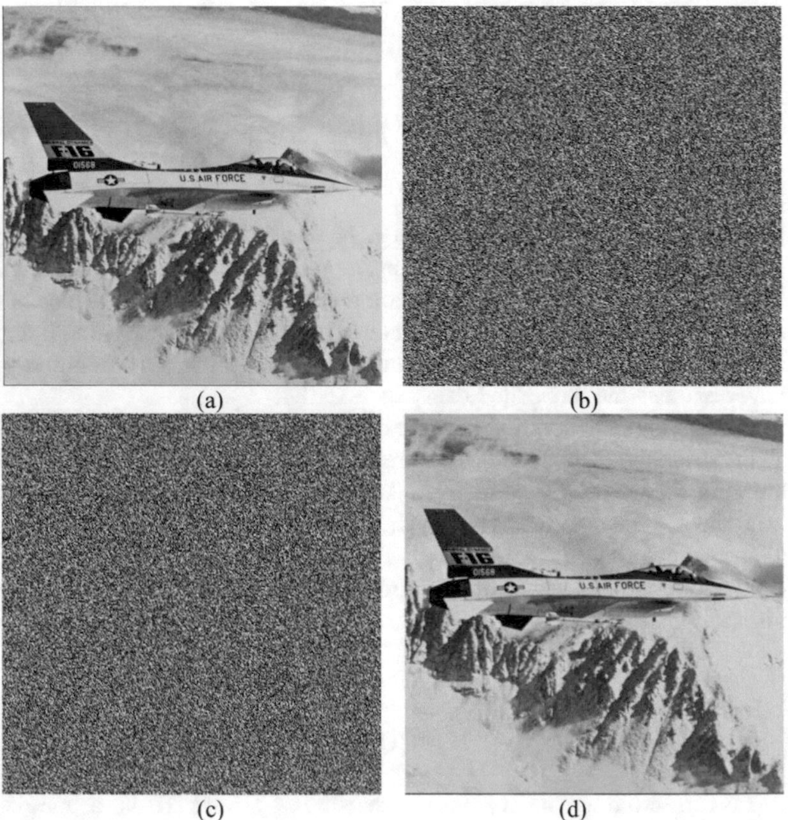

Fig. 3. Results of the proposed scheme for *Airplane*. (a) Original image, (b) Encrypted image, (c) Marked, encrypted image, (d) Directly decrypted image.

3 Experimental Results and Comparisons

In the experiments, the thresholds T_1 and T_2 were set as 40 and 0, respectively. The size of four test images are 512 × 512. The results of our proposed scheme are shown in Fig. 3, the parameters of g, M and f are 1000, 1 and 1, respectively. Figure 3(a) shows the original image *Airplane*. Figures 3(b–c) show the encrypted image and marked image with embedding rate 0.0593 bpp, respectively. Figure 3(d) illustrates the directly decrypted image with PSNR of 54.6 dB.

Figures 4 and 5 show the comparisons between the propose scheme and [6, 10, 11]. The parameter p of method in [6] is 1. Figure 4 shows the PSNR of directly decrypted image with different embedding rate comparisons. Figure 5 illustrates the comparisons of SSIM of directly decrypted image in different embedding rate. The results of SSIM and embedding rate of our scheme are also better. We can observe from Figs. 4 and 5

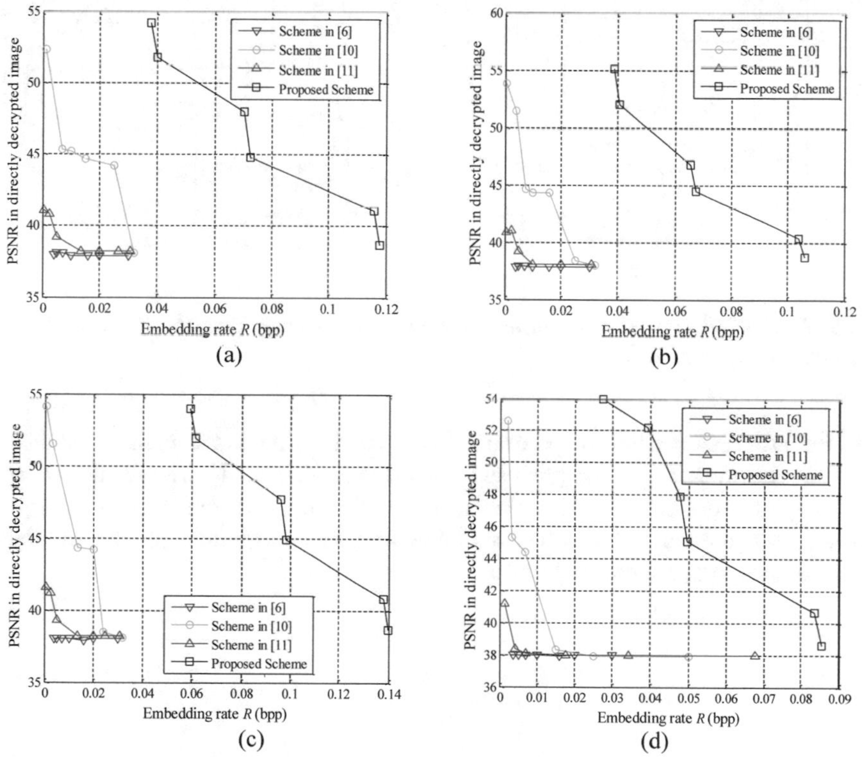

Fig. 4. Rate-PSNR comparisons between the proposed scheme and [6, 10, 11]. (a) *Lena*, (b) *Man*, (c) *Airplane*, (d) *Couple*.

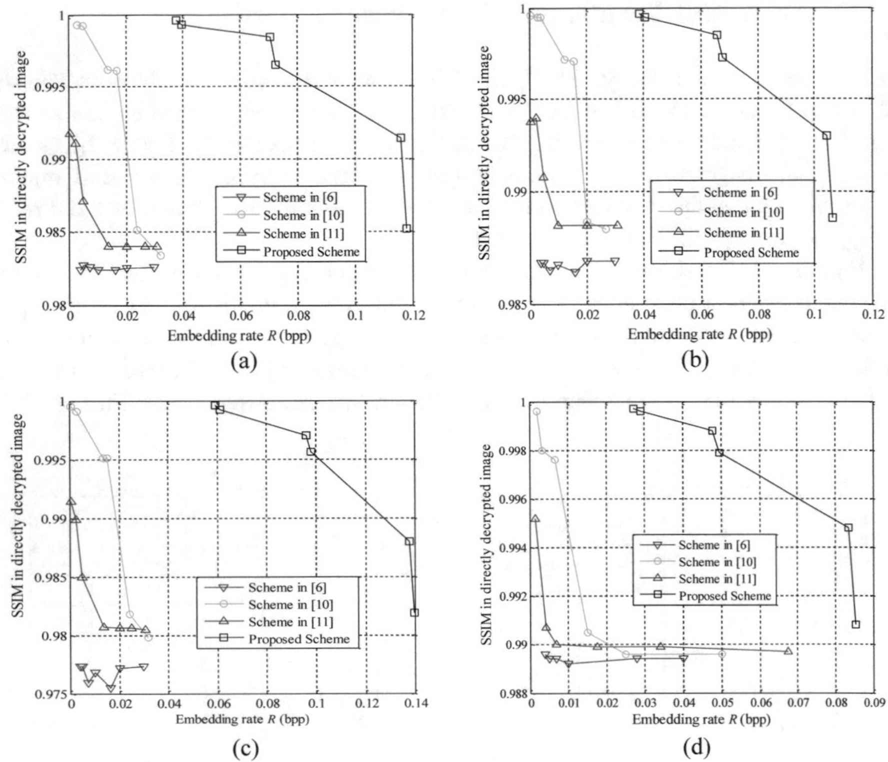

Fig. 5. Rate-SSIM comparisons between the proposed scheme and [6, 10, 11]. (a) *Lena*, (b) *Man*, (c) *Airplane*, (d) *Couple*.

that, the proposed scheme outperforms the other three methods. Figure 6 shows the recovery performances of the proposed scheme and [6, 10, 11]. The results of recovered image for our scheme are better than the other three methods. In summary, the proposed scheme can achieve higher embedding rate and better quality for directly decrypted image and recovered image.

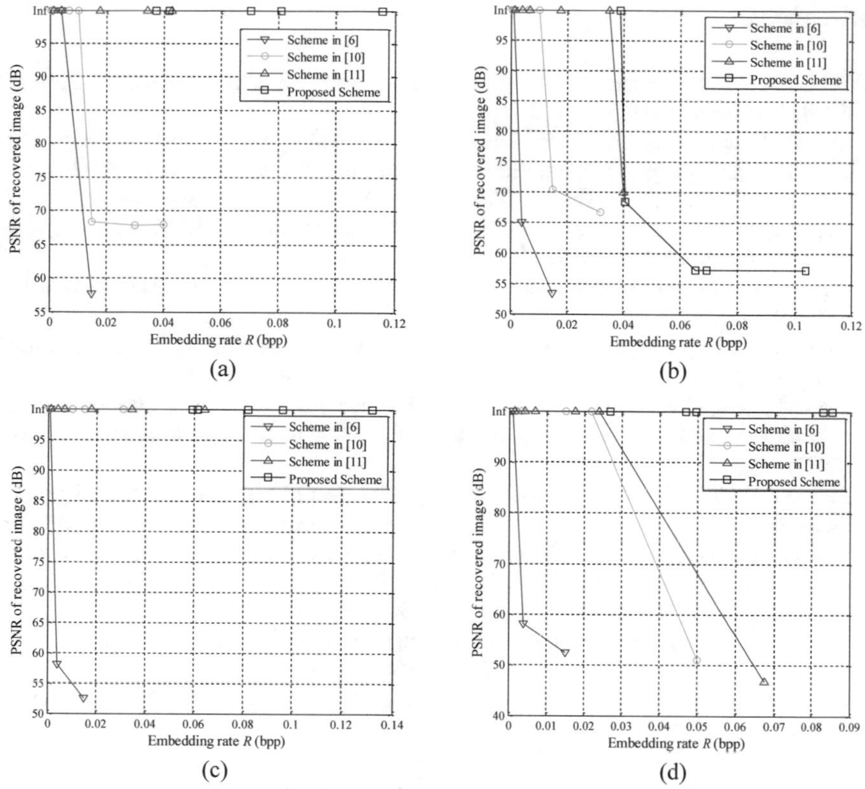

Fig. 6. Recovery comparisons between the proposed scheme and [6, 10, 11]. (a) *Lena*, (b) *Man*, (c) *Airplane*, (d) *Couple*.

4 Conclusions

In this work, we propose a new scheme of reversible data hiding in encrypted images. In the encryption stage, content owner divides original image into blocks and encrypts the blocks by the specific stream cipher and permutation. Data hider embeds secret data into the encrypted image by matrix compression and MSB replacements. If the receiver only has encryption key, he can decrypt the image directly. If the receiver only has data hiding key, he can extract the embedded data. If the receiver has both encryption key and data-hiding key, he can extract the embedded data correctly and recover original image perfectly. Experimental results demonstrate that, our scheme can achieve higher embedding rate and better image quality than some state-of-the-art schemes.

Acknowledgments. This work was supported by National Natural Science Foundation of China (61702332, 61562007, 61672354), Natural Science Foundation of Guangxi (2017GXNSFAA198222), and Research Fund of Guangxi Key Lab of Multi-source Information Mining & Security (MIMS15-03).

References

1. Petitcolas, F.A.P., Anderson, R.J., Kuhn, M.G.: Information hiding — a survey. Proc. IEEE **87**(7), 1062–1078 (1999)
2. Qin, C., Chang, C.C., Chiu, Y.P.: A novel joint data-hiding and compression scheme based on SMVQ and image inpainting. IEEE Trans. Image Process. **23**(3), 969–978 (2014)
3. Qin, C., Chang, C.C., Huang, Y.H., Liao, L.T.: An inpainting-assisted reversible steganographic scheme using a histogram shifting mechanism. IEEE Trans. Circuits Syst. Video Technol. **23**(7), 1109–1118 (2013)
4. Zhang, X.P.: Reversible data hiding in encrypted image. IEEE Signal Process. Lett. **18**(4), 255–258 (2011)
5. Hong, W., Chen, T.S., Wu, H.Y.: An improved reversible data hiding in encrypted images using side match. IEEE Signal Process. Lett. **19**(4), 199–202 (2012)
6. Liao, X., Shu, C.W.: Reversible data hiding in encrypted images based on absolute mean difference of multiple neighboring pixels. J. Vis. Commun. Image Represent. **28**, 21–27 (2015)
7. Puteaux, P., Trinel, D., Puech, W.: High-capacity data hiding in encrypted images using MSB prediction. In: Proceedings of 2016 Sixth International Conference on Image Processing Theory, Tools and Applications (IPTA), Oulu, Finland, pp. 1–6, December 2016
8. Zhang, X.P., Qian, Z.X., Feng, G.R., Ren, Y.L.: Efficient reversible data hiding in encrypted images. J. Vis. Commun. Image Represent. **25**(2), 322–328 (2014)
9. Qian, Z.X., Zhang, X.P.: Reversible data hiding in encrypted image by distributed encoding. IEEE Trans. Circuits Syst. Video Technol. **26**(4), 636–646 (2016)
10. Zhang, X.P.: Separable reversible data hiding in encrypted image. IEEE Trans. Inf. Forensics Secur. **7**(2), 526–532 (2012)
11. Qian, Z.X., Zhang, X.P., Feng, G.R.: Reversible data hiding in encrypted images based on progressive recovery. IEEE Signal Process. Lett. **23**(11), 1672–1676 (2016)
12. Qin, C., Zhang, X.P.: Effective reversible data hiding in encrypted image with privacy protection for image content. J. Vis. Commun. Image Represent. **31**, 154–164 (2015)

Halftone Image Steganography with Flippability Measurement Based on Pixel Pairs

Wanteng Liu, Wei Lu(✉), Xiaolin Yin, Junhong Zhang, and Yuileong Yeung

School of Data and Computer Science, Guangdong Key Laboratory of Information Security Technology, Ministry of Education Key Laboratory of Machine Intelligence and Advanced Computing, Sun Yat-sen University, Guangzhou 510006, China
{liuwt25,yinxl6}@mail2.sysu.edu.cn, luwei3@mail.sysu.edu.cn

Abstract. In recent years, many state-of-the-art steganographic schemes for halftone images have been proposed. Most of them only focus on the visual quality of the marked image but ignore the statistical security. In this paper, a novel steganographic scheme for halftone images is proposed, which focuses on both visual imperceptibility and statistical security. For halftone steganography, embedding messages can only be achieved by flipping pixels. The significant distortion will occur if the pixels are flipped arbitrarily in a halftone image. To select the optimal flipping pixels, we define the flippability measurement based on pixel pairs, which can provide a flippability score for each pixel in halftone images. The higher flippability score of a pixel means that the visual distortion caused by flipping the pixel will be smaller, so the pixels with high flippability scores can be selected as carriers. To achieve high embedding capacity and minimize the embedding distortions, syndrome-trellis code (STC) is employed in the embedding process. The experimental results demonstrate that the proposed scheme can achieve high embedding capacity and realize acceptable statistical security with high visual imperceptibility.

Keywords: Halftone image steganography · Flippability measurement · Pixel pairs · Syndrome-trellis code (STC)

1 Introduction

Steganography, different from the data hiding, aims to embed secret messages into digital covers, which is a practice of covert communication. With the rapid development of digital multimedia, digital images play an important role as carriers for covert communication and the content protection of digital images [2,13,16,26] have received much attention. Nowadays, quite a number of steganography schemes have been proposed for binary images [3,18]. Most of these schemes hide secret information in the complex textural region to minimize the embedding distortions. Halftone image is a specific kind of binary

C.-N. Yang et al. (Eds.): SICBS 2018, AISC 895, pp. 879–890, 2020.
https://doi.org/10.1007/978-3-030-16946-6_72

image, which utilizes scattered black and white pixels to analog grayscale images. Halftone images are generated by digital halftone technique [22], which convert the multi-tone images into two-tone images and the converted two-tone images look like the original multi-tone images when viewed from a distance. Ordered dithering [22] and error diffusion [6] are two main kinds of halftone techniques. Ordered dithering compares the pixel values with some pseudo-random threshold patterns to determine its binary output, which has low computational complexity. Error diffusion spreads the error to surrounding pixels to reduce the visual errors, so error diffusion has the higher visual quality than ordered dithering. In our daily life, halftone images are very common since they are widely used in the printing industry, such as the printing of books, magazines and newspapers. Therefore, steganography on halftone image is significant and meaningful.

In a general way, there are two main kinds of steganography on halftone images. The first kind of schemes is to embed data during the process of halftoning [10,15,21]. Such schemes have the fairly large embedded capacity with negligible visual distortion in halftone images while the original multi-tone images are needed before the schemes can be applied. The second kind of schemes is to embed data directly into the generated halftone images without requiring the original multi-tone images [8,9,11,14,19,24]. Considering the versatility of steganography, our work focuses on the latter schemes.

In recent years, many state-of-the-art steganography schemes on halftone images have been proposed to achieve the visual imperceptibility [9,11,14,24]. In [9], Fu and Au proposed a method called data hiding smart pair toggling (DHSPT) which embedded data into randomly selected locations in the halftone image with intensity and connection selection. As the payload increases, the visual distortion of halftone images will be remarkable. To improve the visual quality of marked halftone images, Lien *et al.* [14] selected the most suitable pair of connected neighboring pixels from a sequence of candidate pairs along Hilbert curve to embed messages. In [11], Guo and Zhang presented a block-based method which defined the Grouping Index Matrix (GIM) and utilized it to embed secret messages by sacrificing some embedding space. There is no doubt that the visual imperceptibility and embedding capacity are important criteria for halftone images steganography. Halftone images have only 1 bit per pixel compared with 8 bits per pixel in grayscale images, so halftone images steganography can only embed messages by flipping pixels. Therefore, the classic halftone steganographic methods, such as DHSPT and GIM, mainly focus on improving the visual quality of marked halftone images. However, with the development of the steganalysis techniques, the security of statistic is also an important criterion for halftone steganography. In [24], Xue *et al.* introduced the concept of dispersion degree (DD) in a local region, which can measure the complexity of the region texture in halftone images. They think that the regions with high dispersion can be changed without incurring obvious noise because it can minimize the amount and the size of the salt-and-pepper clusters. Based on dispersion degree, they proposed a secure halftone image steganography which minimized the distortion of texture structure. Inspired by this, we use the numbers of black pixels in pixel pairs to describe the texture structures of halftone images. Based on the statistical model, we can determine whether a specific pixel should be flipped for embedding messages in the halftone image.

In this paper, we propose a novel steganographic method for halftone images, which can achieve the goals of high embedding capacity, good visual quality and acceptable statistical security. Since halftone images have only 1 bit per pixel, embedding messages in halftone images can only be achieved by flipping pixels. The flipped pixels should be selected carefully otherwise significant distortion will occur if we flip the pixels randomly in a halftone image. Therefore, we propose the flippability measurement based on pixel pairs, which can provide a flippability score for each pixel. With the flippability scores, we are capable of selecting the suitable pixels to flip for embedding information. To achieve high embedding capacity, syndrome-trellis code (STC) is employed in the embedding process, which is a kind of matrix embedding. The experiment results show that the proposed steganographic method can achieve a significant performance compared with the state-of-the-art works.

The rest of this paper is organized as follows. Section 2 develops the proposed flippability measurement for each pixel and the construction of distortion map for STC. In Sect. 3, the proposed steganographic scheme is presented. In Sect. 4, experimental results about visual imperceptibility and the statistical security are presented. Finally, the conclusion of this paper is given in Sect. 5.

2 The Proposed Method

In this section, we present the definition of flippability measurement for each pixel in the halftone image. Based on the flippability measurement, the optimal pixels with high flippability scores will be select to be flipped for embedding messages. To realize the high embedding capacity and minimize the visual distortion, syndrome-trellis code (STC) is employed in embedding process.

2.1 Flippability Measurement

As a specific kind of binary images, halftone images utilize scattered black and white pixels to analog grayscale images. Halftone images have only 1 bit per pixel even though they look like the original grayscale images when viewed from a distance. Therefore, different from the grayscale image steganography, halftone image steganography can only embed information by flipping pixels from black to white and vice versa. The significant distortion will appear if we flip the pixels arbitrarily in a halftone image. Thus, a measurement is needed to determine whether a pixel should be flipped for embedding messages. Obviously, a pixel should be prioritized to be flipped if the visual distortion and the texture destruction is negligible.

Noting that binary images naturally represent the texture [12,20,23], Feng *et al.* [3] exploited the texture model to measure the embedding distortion. They proposed a distortion measurement based on statistics and first extracted the local texture pattern (LTP) as the texture primary. However, the description of halftone image texture is totally different from ordinary binary image because the pixels in the halftone image are discrete and discontinuous. Thus, the method of LTP cannot be used directly for halftone images. Xue *et al.* [24] introduced the concept of dispersion degree (DD) to measure the complexity of the region texture in halftone images. Based on the dispersion property of the pixels in halftone images, we use the number of black pixels in a local region to describe texture structures. In [3], LTP with size 3×3 is employed to describe texture structures of binary images. Inspired by this, we proposed the flippability measurement to evaluate whether a pixel is suitable for flipping.

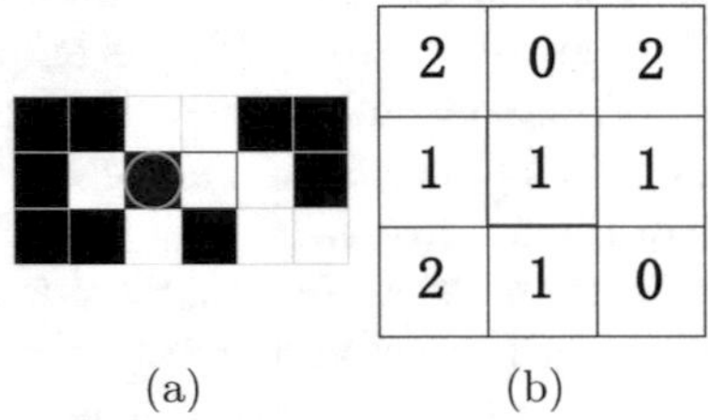

Fig. 1. (a) An example of local texture pattern in a halftone image. The central pixel pair P in the red box is composed of the pixel k in the blue circle and the pixel to the right of it. (b) The number of black pixels corresponding to the pixel pairs of (a).

Given a specific pixel k, we combine it and another pixel on the right side of pixel k into a pixel pair P. As shown in Fig. 1(a), we consider the eight adjacent pixel pairs around the central pixel pair P, which represent the local environment. To describe the texture of halftone images, we focus on the number of black pixels in pixel pairs, which is in the range from 0 to 2. As shown in Fig. 1(b), the 9 pixel pairs can be converted to 9 density values, which represent the number of black pixels in the corresponding pixel pair. From a statistical point of view, we can estimate the number of black pixels in the center pixel pair P from the surrounding pixel pairs. It should be noted that if it is difficult to estimate the number of black pixels in the center pixel pair P, it is also difficult to distinguish whether the pixel k has been flipped or not, which means the pixel k is flippable. Therefore, we can calculate the flippability score for each pixel in the halftone image based on the statistical model.

Let T denotes the neighborhood texture of 3×3 pixel pairs in a local region, which is described by the number of black pixels in the pixel pair. Therefore, T can be defined as:

$$T = \sum_{i=1}^{9} 3^{i-1} N_i \tag{1}$$

where $i = 1, 2, \ldots, 8, 9$ denotes the i-th pixel pairs in the local region and N_i denotes the number of black pixels in the pixel pair i as shown in Fig. 2. It is worth mentioning that the number of black pixels in a pixel pair is in the range

N_6	N_2	N_7
N_5	N_1	N_3
N_9	N_4	N_8

Fig. 2. The local region texture described by N_i, which denotes the number of black pixels in the corresponding pixel pair i.

from 0 to 2. Take Fig. 1(b) as an example, the texture type $T = 3^0 + 3^2 + 3^3 + 3^4 + 2 \times 3^5 + 2 \times 3^6 + 2 \times 3^8 = 15184$. Obviously, the total number of texture types T is $3^9 = 19683$.

Based on the statistical model, the number of black pixels in the surrounding 8 pixel pairs (denoted as S) can be utilized to estimate that in the center pixel pair (denoted as C). It is worth noting that when a specific pixel k is given, there are only two cases of the number of black pixels in the pixel pair P. For instance, if the specific pixel k is black, the pixel to the right of it can be black or white, which means that the number of black pixels in the pixel pair P can only be 1 or 2. Therefore, on the condition that the given pixel K and the number of black pixels in the surrounding pixel pairs S are known, the probability of C is defined as:

$$p_0 = P\{C = 0 \mid S = s, K = k\} \tag{2}$$

$$p_1 = P\{C = 1 \mid S = s, K = k\} \tag{3}$$

$$p_2 = P\{C = 2 \mid S = s, K = k\} \tag{4}$$

where s and k are each a specific example of S and K, respectively. The p_0, p_1 and p_2 indicate the number of black pixels of the pixel pair is 0, 1 and 2, respectively.

Furthermore, we propose the definition of the flippability score in two cases based on the specific pixel k. If the specific pixel k is black, the number of black pixels in this pixel pair can only be 1 or 2 as mentioned before. In this case, the probability of C can only be p_1 and p_2. If the p_1 and p_2 are near, it is difficult to estimate the value of C, which means that it is also difficult to distinguish whether the pixel in the center pixel pair has been flipped or not. According to the information entropy, we define the flippability score $F(k)$ of pixel k as follows:

$$F(k) = -[p_1 \log_2(p_1) + p_2 \log_2(p_2)] \tag{5}$$

and $log_2(\cdot)$ means the logarithms at the base of 2. The flippability score $F(k)$ is high when p_0 and p_1 are near, so it can evaluate whether the pixel k should be flipped. In another case, if the specific pixel k is white, we can similarly define the flippability score $F(k)$ as follows:

$$F(k) = -[p_0 \log_2(p_0) + p_1 \log_2(p_1)] \tag{6}$$

No matter the pixel k is black or white, the range of flippability score $F(k)$ is $[0, 1]$.

2.2 Distortion Map for STC

When the payload is given, matrix embedding is usually employed to minimize the visual distortion caused by embedding. In [5], Filler *et al.* proposed the Syndrome-trellis code (STC), which utilizes the convolutional code and the redundancy of cover carrier to minimize additive distortion in steganography. To achieve better steganographic performance, we use STC in our embedding procedure and a single pixel is used as STC's carrier. For the utilization of STC, the flippability score should be transformed to distortion measurement. Given a specific pixel k, the corresponding flipping distortion is defined as:

$$D(k) = 1 - F(k) \tag{7}$$

where $F(k)$ denotes the flippability score of pixel k. The pixel with higher flippability score has a lower flipping distortion, which means the pixel is more suitable to be flipped. The flipping distortion of all pixels can be combined into a distortion map, which represents the distortions introduced by flipping pixels individually in a halftone image.

3 The Framework of Embedding and Extraction

Based on flippability measurement, the distortion map is proposed to construct the STC's carriers. In this section, the embedding and extraction procedures of the proposed steganographic scheme are presented in detail, whose block diagrams are shown in Figs. 3 and 4, respectively.

3.1 Embedding Procedure

Given a halftone cover image C of size $H \times W$ and the secret message M, the embedding procedure contains the following steps:

1. Calculate the flippability score for each pixel in C and form a flippability map by using Eqs. (5) and (6);
2. Transform the flippability map to a distortion map D by using Eq. (7);
3. Reshape C and D into two one-dimensional vectors, denoted as V_C and V_D;
4. Scramble V_C and V_D into two one-dimensional vectors randomly with the same scrambling seed, donated as V'_C and V'_D;
5. Employ the STC encoder with V'_C, V'_D and M;
6. Obtain a one-dimensional stego vector V_S;
7. Descramble V_S into V'_S and reshape V'_S into the size $H \times W$. The stego image S is obtained.

3.2 Extraction Procedure

The random scrambling seed used in the embedding procedure is known in the extraction procedure. Given a halftone stego image S of size $H \times W$, and the

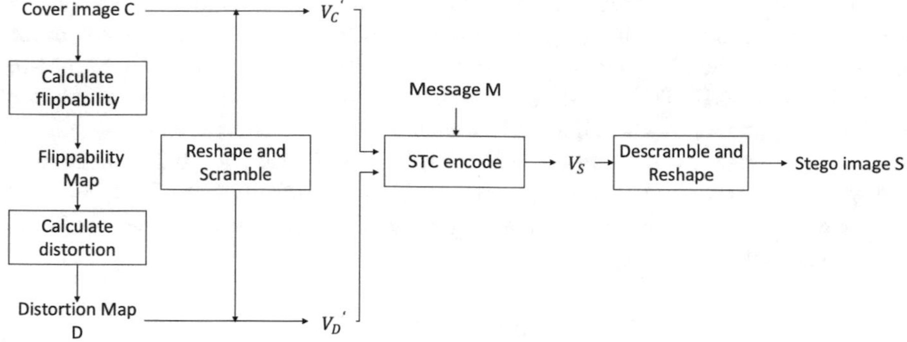

Fig. 3. The framework of embedding procedure.

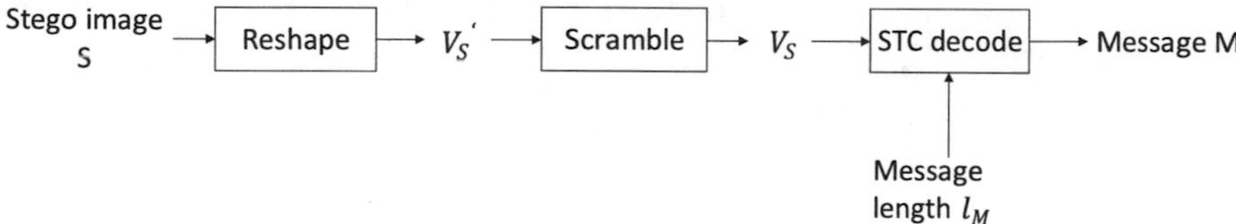

Fig. 4. The framework of extraction procedure.

length of the secret message l_M, the extraction procedure contains the following steps:

1. Reshape S into a one-dimensional vector, denoted as V_S;
2. Scramble V_S into a one-dimensional vector randomly with the same scrambling seed, donated as V_S;
3. Employ the STC encoder with V_S and l_M;
4. Obtain the secret message M.

4 Experiments and Results

In this section, we first introduce the experiment conditions, and then present a set of experiments to demonstrate the high performance and effectiveness of the proposed approach.

4.1 Experiment Conditions

There are 9000 halftone images with size of 256 × 256 used in our experiments compared with DHSPT [9], GIM [11] and DD [24]. These original binary images are from BossBased-1.01 [1]. For our experiments, we converted these original images to halftone images by error diffusion halftone images [7].

To further evaluate the high performance of the proposed method, we conduct some experiments for visual quality comparison, the statistical security comparison and the embedding capacity comparison. Especially, the Peak Signal-to-Noise Ratio (PSNR) [25] presented the distortion by measuring the amount of the flipped pixels and the DRD [17] measured the reciprocal distance on the neighboring pixels do not match well with the evaluation for halftone images. Using the similar standard, we measure the amount and the size of the salt-and-pepper clusters to evaluate the visual quality of halftone images [9]. As in [9], the authors defined five scores as:

$$S_1 = \sum_{i=0}^{4} N_i \tag{8}$$

$$S_2 = \sum_{i=0}^{4} (i+1) N_i \tag{9}$$

$$S_3 = \frac{S_2}{S_1} \tag{10}$$

$$S_4 = \sum_{i=2}^{4} N_i \tag{11}$$

$$S_5 = \sum_{i=0}^{4} i N_i = S_2 - S_1 \tag{12}$$

where N_i is the total number of the locations of the pixels which are modified by steganography and the flipped pixels having i neighbors with same pixel values in the 4-neighborhood. It demonstrates that black flipping pixels in the bright region (class 1) and white flipping pixels in the dark region (class 4) will cause significant salt-and-pepper noise, which reduce the visual quality of halftone images. S_1 gives the total number of these two types of elements. S_2 denotes the total coverage area of clusters class 1 and class 4. Thus, the smaller S_1 and S_2 are, the better visual quality of halftone image is. S_3 gives the average area per cluster. S_4 is the number of class 1 and class 4 elements associated with clusters of size 3 or more, which are useful because clusters of size 1 or 2 are not very visually disturbing. S_5 is a perceptual measure with a linear penalty model. It gives a zero penalty score to isolated black or white pixels which look visually pleasing. In general, algorithms with smaller scores of S_1, S_2, S_3, S_4 and S_5 are better.

Statistical security is another important criterion in steganographic performance evaluation. We conduct some experiments to assess the security of the proposed steganographic scheme. PMMTM-320D [4] is a statistical model employing the dependence between texture structures to describe the embedding distortions on connectivity and smoothness. The number following each feature name is the dimension of the corresponded feature. PMMTM-320D and SVM with an optimized Gaussian kernel are used to construct the steganalyzers. We define P_E the decision error rate to measure the steganographic performance as

$$P_E = \frac{1}{2}(P_{F_p} + P_{F_n}) \tag{13}$$

where P_{F_p} is the probabilities of false positive (detecting cover as stego) and P_{F_n} is the probabilities of false negative (detecting stego as cover).

Also, some experiments are conducted to demonstrate the high embedding capacity of our proposed scheme.

4.2 Results and Comparisons

As shown in Table 1, the visual scores of the proposed schemes are smaller than that of DHSPT, GIM and DD, except S_3. The smaller of the five scores demonstrates that good visual quality for halftone images. The S_3 gives the average area per cluster. Although distortion in halftone image is always exhibited in the form of salt-and-pepper artifacts due to local clusters of pixels, clusters also represent important content in the image sometimes. Thus, the results can further demonstrate that the proposed scheme has good visual imperceptibility.

Table 1. Average scores (S_1 to S_5) of various schemes on the halftone images dataset with 1024 bits embedded.

	S_1	S_2	S_3	S_4	S_5
Proposed	**136.5**	**458.4**	3.345	**118.3**	**304.2**
DHSPT [9]	775.9	2410.8	3.107	468.2	1635.9
GIM [11]	683.6	2198.5	3.216	398.3	1508.4
DD [24]	398.3	1204	**3.022**	289.6	803.2

In addition, Fig. 5 shows the statistical security comparison of different halftone steganographic schemes by using PMMTM-320D. GIM and DHSPT employ "pixel slave" strategy to flip pixels in pairs rather than individually in each block, which inevitably increase destructions between texture. DD flips one pixel by minimizing the change of the cluster composition in each block, but ignores the distortion of the whole image by flipping a single pixel. Based on flippability measurement in the pixel pair, the proposed method provides the flippability score for each pixel in the halftone image, which can be employed to minimize the distortion of flipping a single pixel. As a result, the proposed steganographic scheme has the best performance in statistical security. It is worth mentioning that compared with the limited embedding capacity of the other three methods, the proposed scheme achieves higher embedding capacity.

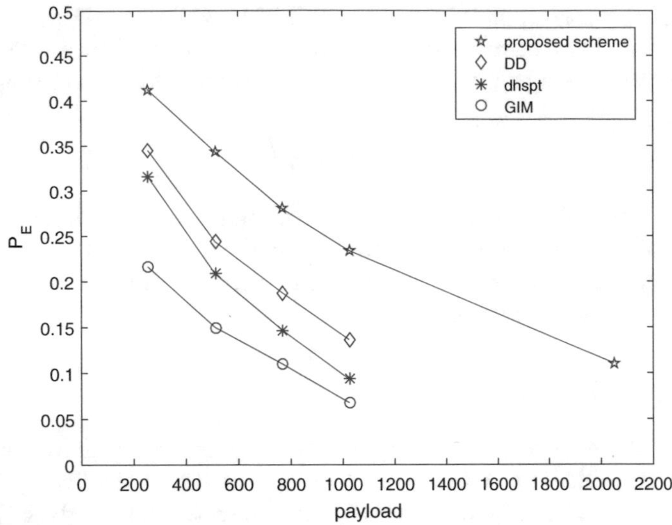

Fig. 5. Statistical security comparison of halftone steganographic schemes using PMMTM [4].

5 Conclusions

In this paper, we have proposed a novel steganographic method with flippability measurement based on pixel pairs for halftone images. By employing the proposed concept of flippability measurement, we can select the optimal flipping pixels to embed messages. To achieve high embedding capacity and minimize the embedding distortions, we utilize syndrome-trellis code (STC), a widely used matrix embedding strategy, in our embedding process. Experimental results demonstrate that the proposed scheme can achieve high embedding capacity and realize acceptable statistical security with high visual imperceptibility.

For halftone steganography, if the embedded capacity is increased, the visual quality and statistical security will drop accordingly. Therefore, we have to make a trade-off between embedded capacity and visual quality in halftone steganography. Herein, we may focus on how to improve the visual quality and statistical security as much as possible without reducing the embedded capacity in the future.

Acknowledgements. This work is supported by the National Natural Science Foundation of China (No. U1736118), the Natural Science Foundation of Guangdong (No. 2016A030313350), the Special Funds for Science and Technology Development of Guangdong (No. 2016KZ010103), the Key Project of Scientific Research Plan of Guangzhou (No. 201804020068), Shanghai Minsheng Science and Technology Support Program (17DZ1205500), Shanghai Sailing Program (17YF1420000), the Fundamental Research Funds for the Central Universities (No. 16lgjc83 and No. 17lgjc45).

References

1. Bas, P., Filler, T., Pevn, Y.T.: Break our steganographic system: the ins and outs of organizing BOSS. J. Am. Stat. Assoc. **96**(454), 488–499 (2011)
2. Chen, J., Lu, W., Fang, Y., Liu, X., Yeung, Y., Xue, Y.: Binary image steganalysis based on local texture pattern. J. Vis. Commun. Image Represent. **55**, 149–156 (2018)
3. Feng, B., Lu, W., Sun, W.: Secure binary image steganography based on minimizing the distortion on the texture. IEEE Trans. Inf. Forensics Secur. **10**(2), 243–255 (2014)
4. Feng, B., Lu, W., Sun, W.: Binary image steganalysis based on pixel mesh markov transition matrix. J. Vis. Commun. Image Represent. **26**(C), 284–295 (2015)
5. Filler, T., Judas, J., Fridrich, J.: Minimizing additive distortion in steganography using syndrome-trellis codes. IEEE Trans. Inf. Forensics Secur. **6**(3), 920–935 (2011)
6. Floyd, R.W.: An adaptive algorithm for spatial grayscale. In: Proceedings of SID International Symposium Digest of Technical Papers, pp. 75–77 (1975)
7. Floyd, R.W., Steinberg, L.: Adaptive algorithm for spatial greyscale. In: Proceedings of SID, pp. 75–77 (1976)
8. Fu, M.S., Au, O.C.: Data hiding watermarking for halftone images. IEEE Trans. Image Process. **11**(4), 477–84 (2002)
9. Fu, M.S., Au, O.C.: Halftone image data hiding with intensity selection and connection selection. Signal Process. Image Commun. **16**(10), 909–930 (2001)
10. Guo, J.M., Liu, Y.F.: Halftone-image security improving using overall minimal-error searching. IEEE Trans. Image Process. **20**, 2800–2812 (2011)
11. Guo, M., Zhang, H.: High capacity data hiding for halftone image authentication. In: International Conference on Digital Forensics and Watermarking, pp. 156–168 (2013)
12. Huang, D., Shan, C., Ardabilian, M., Wang, Y., Chen, L.: Local binary patterns and its application to facial image analysis: a survey. IEEE Trans. Syst. Man Cybern. Part C Appl. Rev. **41**(6), 765–781 (2011)
13. Li, J., Lu, W., Weng, J., Mao, Y., Li, G.: Double JPEG compression detection based on block statistics. Multimed. Tools Appl. **2**, 1–16 (2018)
14. Lien, B.K., Lan, Z.L.: Improved halftone data hiding scheme using hilbert curve neighborhood toggling. In: Seventh International Conference on Intelligent Information Hiding and Multimedia Signal Processing, pp. 73–76 (2011)
15. Lien, B.K., Pei, W.D.: Reversible data hiding for ordered dithered halftone images. In: IEEE International Conference on Image Processing, pp. 4181–4184 (2009)
16. Liu, X., Lu, W., Huang, T., Liu, H., Xue, Y., Yeung, Y.: Scaling factor estimation on jpeg compressed images by cyclostationarity analysis. Multimed. Tools Appl., 1–18 (2018). https://doi.org/10.1007/s11042-018-6411-9
17. Lu, H., Kot, A.C., Shi, Y.Q.: Distance-reciprocal distortion measure for binary document images. IEEE Signal Process. Lett. **11**(2), 228–231 (2004)
18. Lu, W., He, L., Yeung, Y., Xue, Y., Liu, H., Feng, B.: Secure binary image steganography based on fused distortion measurement. IEEE Trans. Circuits Syst. Video Technol. **PP**(99), 1–1 (2018)
19. Mars, J., Au, O.C.: Data hiding by smart pair toggling for halftone images. In: Proceedings of IEEE International Conference on Acoustics, Speech, and Signal Processing, ICASSP 2000, vol. 4, pp. 2318–2321 (2000)

20. Ojala, T., Pietikäinen, M., Mäenpää, T.: Multiresolution gray-scale and rotation invariant texture classification with local binary patterns. IEEE Trans. Pattern Anal. Mach. Intell. **24**(7), 971–987 (2000)
21. Pei, S.C., Guo, J.M.: High-capacity data hiding in halftone images using minimal error bit searching. In: International Conference on Image Processing, vol. 5, pp. 3463–3466 (2004)
22. Ulichney, R.: Digital Halftoning. MIT Press, Cambridge (1987)
23. Wang, B., Li, X.F., Liu, F., Hu, F.Q.: Color text image binarization based on binary texture analysis. Pattern Recognit. Lett. **26**(11), 1650–1657 (2005)
24. Xue, Y., Liu, W., Lu, W., Yeung, Y., Liu, X., Liu, H.: Efficient halftone image steganography based on dispersion degree optimization. J. Real-Time Image Proc., 1–9 (2018). https://doi.org/10.1007/s11554-018-0822-8
25. Yoo, J.C., Ahn, C.W.: Image matching using peak signal-to-noise ratio-based occlusion detection. IET Image Process. **6**(5), 483–495 (2012)
26. Zhang, Q., Lu, W., Wang, R., Li, G.: Digital image splicing detection based on markov features in block DWT domain. Multimed. Tools Appl. **3**, 1–22 (2018)

Reversible Data Hiding in Binary Images by Symmetrical Flipping Degree Histogram Modification

Xiaolin Yin, Wei Lu[(✉)], Wanteng Liu, and Junhong Zhang

School of Data and Computer Science, Guangdong Key Laboratory of Information Security Technology, Ministry of Education Key Laboratory of Machine Intelligence and Advanced Computing, Sun Yat-sen University, Guangzhou 510006, China
{yinxl6,liuwt25,zhangjh65}@mail2.sysu.edu.cn, luwei3@mail.sysu.edu.cn

Abstract. In recent years, reversible data hiding in binary images has become more and more important in the field of data hiding. For the binary property, hiding information in binary images can only be achieved by flipping pixels. But flipping the pixels randomly in a binary image will cause remarkable distortion. The existing methods are capable of reversible data hiding in binary images but the visual quality is not good. In this paper, we propose a novel reversible data hiding method based on symmetrical flipping degree (SFD) histogram modification. First, we define the SFD of the block, which is a visual score measured the distance between the pixels in the block. The smaller SFD is, the better the visual quality the image has. Then the histogram is generated from the SFD, and it utilizes the zero or the minimum point and modifies the pixels with high SFD for its pre-processing. Finally, the appropriate corresponding pixels are selected adaptively to flip for information embedding. Experimental results demonstrate the feasibility of our proposed method and the visual quality is satisfactory.

Keywords: Reversible data hiding · Symmetrical flipping degree (SFD) · Histogram modification · Binary image

1 Introduction

In recent years, multimedia security and multimedia content protection [4,12,14,21] have received much attention. Data hiding [19] is a technique for embedding secret messages into cover media such as images, audios, and videos, which is a practice of digital authentication. With the development of the Internet and digital multimedia, a large number of digital images can spread fast and easily through the Internet. For the purpose of information protection, copyright protection, and integrity authentication, some authentication information should be embedded in digital media by the method of data hiding. So far, a considerable number of data hiding techniques [3,5,7,8,11] have been proposed, which focus

C.-N. Yang et al. (Eds.): SICBS 2018, AISC 895, pp. 891–903, 2020.
https://doi.org/10.1007/978-3-030-16946-6_73

on gray or color images only. However, binary images have different visual representations from gray or color images because binary images have only 1 bit per pixel as compared with 8 bits per gray pixel or 24 bits per color pixel, which means that binary images have fewer bit planes to embed secrets. Therefore, data hiding techniques for grayscale images are not suitable for binary images. In our daily life, binary images are used in a wide range of applications, such as electronic documents, scanned texts, and digital signature. Therefore, data hiding in the binary images is of great significance.

Many data hiding methods [2,6,22] have been proposed in recent years for binary images. In [22], Yang and Kot proposed a blind data hiding method with maintaining the connectivity of pixels in a local region. They imposed three transition criteria to determine the "flippability" of a pixel, and embedded the secret messages in "embeddable" blocks which achieved good performance in visual quality and the embedding capacity. Feng and Lu [6] defined a distortion function to evaluate the impact of flipping considering the pixels cluster and boundary connectivity. With the distortion function, the marked image had good visual quality without degrading the embedding capacity. Cao et al. [2] proposed an edge-adaptive data hiding method by establishing a dense edge-adaptive grid (EAG) along the object contours.

Reversible data hiding (RDH), also known as invertible or lossless data hiding, is a technique that can completely recover both the hidden messages and the cover image. Different from most data hiding methods, RDH can restore the original image after data extraction. This property plays an important role in some sensitive scenarios such as medical imagery [1], law enforcement, and military imagery. The unrecoverable distortion in cover images is not allowed and the exact recovery of the original image is necessary for these applications. In recent years, some reversible data hiding techniques for binary image [9,10,17,20,24] have been released. In [17], Tsai et al. proposed a reversible data hiding mechanism based on pair-wise logical computation (PWLC). The proposed method can recover the host image without utilizing any additional information, while the embedding capacity is limited and the visual quality of the marked image is not good. Ho et al. [9] gathered and quantified statistical data of the occurrence frequencies of various patterns and designed a high-capacity reversible data hiding scheme based on pattern substitution after establishing some pattern exchange relationships. This method can achieve high capacity but the visual quality of the marked image is still not satisfactory. In [20], Xuan et al. proposed a scheme using run-length (RL) histogram modification. They scanned the host image from a specific order to generate a sequence of RL couples, then embed data into black RL histogram. Zhang et al. [24] embedded secret message by modifying the histogram for RDH. They employed the decompression and compression processes of an entropy coder to insert data, which improve the performance of previous RDH schemes especially for larger images. In [10], Kim et al. proposed a n-pairs pattern method which can achieve great embedding capacity by histogram modification.

In this paper, we propose a novel reversible data hiding scheme using symmetrical flipping degree (SFD) histogram modification. In view of the above reversible data hiding methods that can achieve a great payload but poor visual quality, SFD, a visual score measured the distance between the pixels in blocks is proposed to evaluate which pixels are flippable. And because the SFD of the blocks in binary images is symmetrical, we can achieve the reversible process of the message embedding and extraction.

The remainder of this paper is organized as follows. Section 2 describes the SFD histogram construction and modification for reversible data hiding. The proposed reversible data hiding scheme in detail is presented in Sect. 3. Section 4 discusses our experimental results and proves our proposed method achieves better visual quality. Finally, we present the conclusion in Sect. 5.

2 The Proposed Method

In this section, we should solve the following problems: (1) How to find the best flipping pixel to minimize the visual distortion. (2) How to find the precise flipped pixel inside each block in case of its unknown location. We will present the proposed method in details. Firstly, the SFD of the blocks will be defined, and we will explain why the smaller the SFD is, the better the visual quality the image has. Then the histogram is constructed from the SFD. Finally, the appropriate corresponding pixels are selected adaptively to flip for information embedding.

2.1 SFD Definition

Since the visual quality of a binary image is sensible and easily affected by flipped pixels, the proposed reversible data hiding method should ensure that the visual imperceptibility is satisfactory.

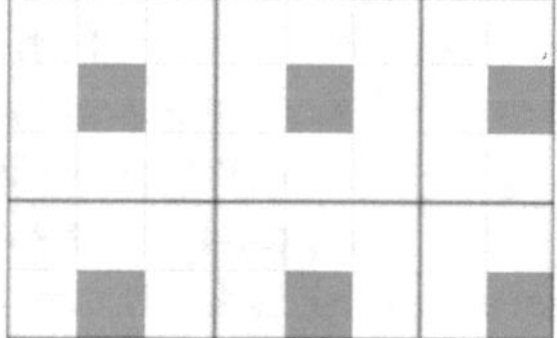

Fig. 1. Binary image divided into non-overlapped 3×3 blocks.

In a binary image, the distance between two pixels is important for the mutual interference perceived by human eyes. Based on observations, for a central pixel, the visual distortion caused by flipping diagonal neighbors flipping is less than horizontal or vertical neighbors. In addition, when the center pixel is mostly the same as the neighbor pixels, flipping this center pixel will cause a large visual distortion.

In [18], the concept of "flippability" is proposed to rank each pixel's flipping priority based on the smoothness and connectivity in 3×3 image blocks. Let the binary image divide into non-overlapped 3×3 blocks, as shown in Fig. 1, and B denotes a 3×3 block, where the center pixel is denoted as b_5. The nine pixels in B are $[\ b_2\ b_1\ b_4;\ b_3\ b_5\ b_7;\ b_6\ b_7\ b_8\ ]$ in Matlab notation, as shown in Fig. 2. The SFD of the block B is defined as:

$$SFD = \sum_{i=1}^{9} w(i) f(b_5, b_i), \qquad f(a,b) = \begin{cases} 1, & a \neq b \\ 0, & a = b \end{cases} \tag{1}$$

Where w is defined as follows:

$$w(i) = \begin{cases} 0, & \text{i} = 5 \\ \dfrac{1}{\sqrt{1 + ((i-5) \bmod 2)}}, & \text{otherwise} \end{cases} \tag{2}$$

<table>
<tr><td>b_2</td><td>b_1</td><td>b_4</td></tr>
<tr><td>b_3</td><td>b_5</td><td>b_7</td></tr>
<tr><td>b_6</td><td>b_9</td><td>b_8</td></tr>
</table>

Fig. 2. The 9 pixels in the 3×3 block B. The center pixel is denoted as b_5.

Actually, SFD describes the number of the neighbor pixels different from the center pixel. In Eq. (2), a larger weight is assigned to the horizontal or vertical neighbors b_1, b_3, b_7, and b_9, because they are closer to the center pixel, and the effect of diagonal neighbors flipping is less than theirs for a central pixel. Obviously, if B is a uniform block (all pixels of the block are black or white), the $SFD = 0$. The greater the SFD the block has, the greater distance to the neighbor pixels it has, which means the center pixel is less similar to the neighbor pixels, and it is considered as the edge point or noise of the image. The examples of the blocks with a great and small SFD is shown in Fig. 3.

2.2 SFD Histogram Construction

Some the reversible data hiding methods [16,25] are proposed based on histogram modification for grayscale images. We propose the SFD histogram and explain why the SFD measurement of the block is symmetrical and provides the feasibility for reversible data hiding in binary images. For a central pixel, it has four diagonal neighbors, two vertical neighbors, and two horizontal neighbors. In Eq. (1), it can be seen that the $w(i)$ of these four vertical or horizontal neighbors

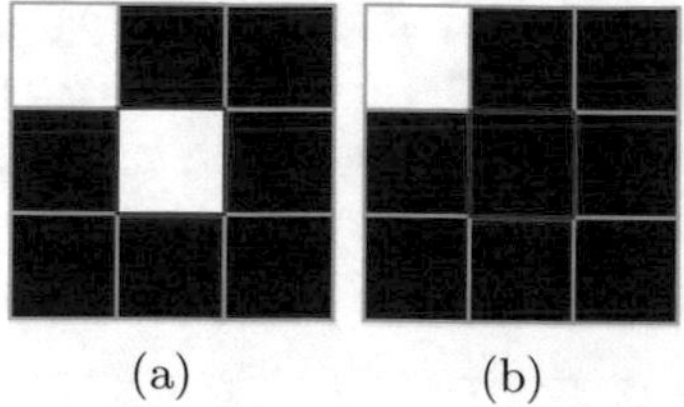

Fig. 3. Examples of the blocks with a great and small SFD. (a) SFD= 6.1213. The center pixel is considered to be the edge point or noise. (b) SFD= 0.7071. The center pixel is considered an important part to ensure the visual quality of this block.

are the same, and they are greater than the $w(i)$ of the four diagonal neighbors. We define that the blocks are called the same patterns if they have the same SFD. An example of two blocks, with the same SFD and different shapes, is shown in Fig. 4.

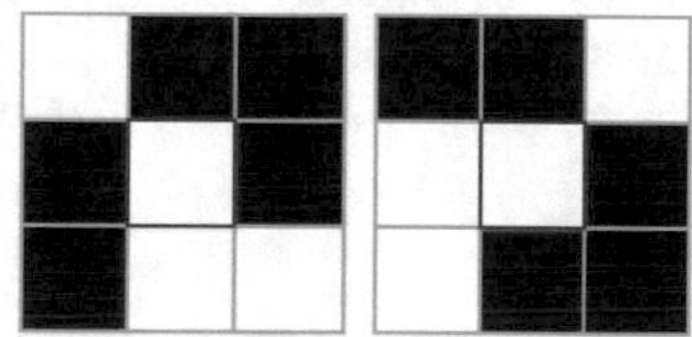

Fig. 4. Example of two blocks with the same SFD = 4.4142 and different shapes.

According to Eq. (1), the number D of the diagonal neighbors different from b_5 in a block, and the number N of the vertical and horizontal neighbors different from b_5 are defined as:

$$D = \sum_{i=1,3,7,9} f(b_5, b_i) \tag{3}$$

$$N = \sum_{i=2,4,6,8} f(b_5, b_i) \tag{4}$$

Since the values of D and N can be 0, 1, 2, 3, and 4 respectively, there are 25 patterns of blocks can be generated. Obviously, when $D = 4$ and $N = 4$, the block has the greatest SFD denoted as SFD_{Max}, which means that its b_5 is completely distinct from the neighbor eight points, as shown in Fig. 5(a). When $D = 0$ and $N = 0$, the block has the smallest $SFD = 0$, which means that its b_5 is completely the same as the neighbor eight points, as shown in Fig. 5(b).

Figure 5 also shows that after we flip the center pixel in a block who has a great SFD, the new SFD of the center pixel will become small. It can be found if a block B in the original image has a $D = d$ and an $N = n$, after flipping b_5 into b_5' in the marked image, the new block B' will have a $D' = (5 - d)$ and an $N' = (5 - n)$. Since D' and N' also take the value of 0, 1, 2, 3 and 4 respectively, the new block B' still belongs to the 25 patterns. Exactly, the new SFD' conforms that $SFD' = SFD_{Max} - SFD$ after flipping b_5. In conclusion, the $SFDs$ in the 25 patterns have a symmetrical relationship, which is in an ascending order by N.

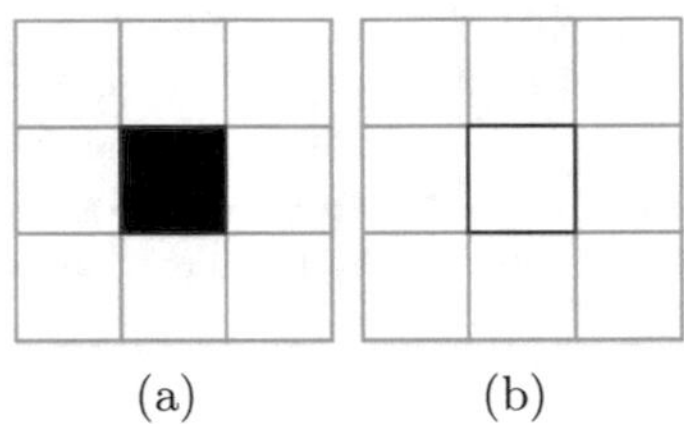

Fig. 5. Example of the SFD's change of flipped the center pixel. (a) $D = 4$, $N = 4$ and $SFD = 6.8284$. (b) $D = 0$, $N = 0$ and $SFD = 0$.

For an $M \times N$ given binary image, there are $\frac{M \times N}{3 \times 3}$ blocks. We first generate its histogram $H(x)$, where the serial number is $x \in i \mid i = 1, 2, \ldots, 25$ and the SFD of each block in the image defined is as $SFD(k)$, where $k = 1, 2, \ldots, \frac{M \times N}{3 \times 3}$. H(x) is defined as:

$$H(x) = \sum_{k=1}^{\frac{M \times N}{3 \times 3}} g(SFD(k), Pat\_SFD(x)), \qquad g(a, b) = \begin{cases} 1, & a = b \\ 0, & a \neq b \end{cases} \tag{5}$$

Where $Pat\_SFD(x)$ is the SFD of the 25 patterns. A SFD histogram of a pattern image is shown in Fig. 6. As we can see, there are 25 patterns in the histogram. For this binary pattern image, the peak point [15] is $H(1)$, and the zero points [15] are $H(5)$, $H(10)$, and $H(16)$.

This symmetrical relationship shows that once the SFD in marked image is known, we may know the original block. In particular, when $x = 13$, $D = 2$, and $N = 2$, the SFD of this block is unchanged after flipping regardless of the state of the original b_5 is '0' or '1'. So $x = 13$ is the center of symmetry in the histogram. Due to the definition of SFD, only the blocks with great SFD could be flipped. In the histogram, we hope to find a minimum point zero $H(\alpha) = 0$ who has no block with $Pat\_SFD(\alpha)$ in the original image. Then set the $H(\beta) = 0$ by flipping the center pixel, where $H(\beta)$ is the symmetry point of $H(\alpha) = 0$ and $\beta = 26 - \alpha$. Before embedding the secret message, the SFD histogram is constructed with an empty point $H(\beta)$.

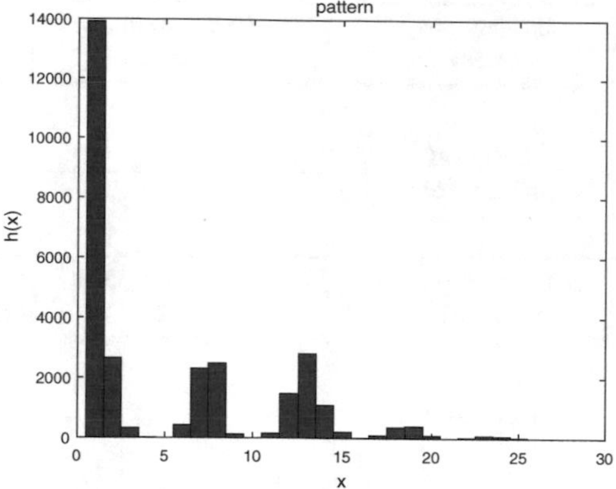

Fig. 6. An SFD histogram of a pattern image.

2.3 Flippable Pixel Selection

After constructing the SFD histogram, we need to consider how to embed the message. For a central pixel, the visual effect of diagonal neighbors flipping is less than horizontal or vertical neighbors. We define the groups in the SFD histogram as:

$$G(n) = \{H(5n-4), H(5n-3), H(5n-2), H(5n-1), H(5n)\} \tag{6}$$

Where $n = 1, 2, 3, 4, 5$. In Sect. 2.2, when the zero point $H(\alpha)$ is in $G(\theta)$, the symmetry point $H(\beta)$ is known in $G(6-\theta)$. At this time, as $H(\beta)$ is equal to 0, another point γ in SFD histogram is chosen to embed the message by flipping the diagonal neighbors of this block. This related point γ must meet the following two conditions:

1. $\gamma > \beta$;
2. $H(\gamma)$ is the greatest one in its group.

The first condition is to ensure the SFD smaller after flipping the diagonal neighbors. The second condition gives the maximum embedding capacity. Finally, in the block with $Pat\_SFD(\gamma)$, counting the number of b_2, b_4, b_6, and b_8 different from b_5 separately, and the top $(\gamma - \beta)$ of them are selected as the message-embedded pixels. As long as the message-embedded pixels is known, the $H(\gamma)$ will be restored.

3 The Framework of Embedding and Extraction

The proposed reversible data hiding method is constructed in this section, and the data embedding and extraction procedures can be illustrated by the block diagrams shown in Figs. 7 and 8, respectively.

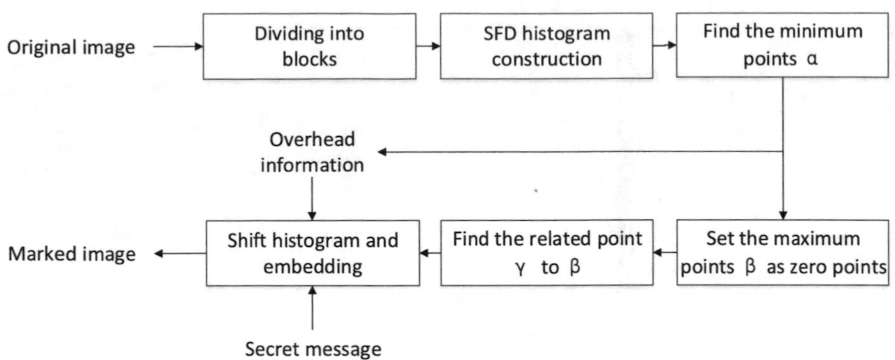

Fig. 7. The framework of embedding procedure.

3.1 Embedding Procedure

For a given binary original image I_O of size $M \times N$, we embed the secret messages sequence S with the length of l_s. The embedding procedure includes the following steps:

1. Divide I_O into non-overlapped 3×3 blocks and calculate the SFD of each block. Generate the SFD histogram $H(x)$ of I_O, where $x = 1, 2, \ldots, 25$;
2. In the histogram $H(x)$, find the minimum point $H(\alpha)$ and the $H(\beta)$ who is symmetry point of α;
3. If the minimum point $H(\alpha) > 0$, all the blocks with $Pat\_SFD(\alpha)$ in I_O need to be recorded as overhead bookkeeping information. Then set $H(\beta) = 0$, and obtain image I'_O;
4. Find the related point γ in the same group as β is;
5. In the block whose center pixel has $Pat\_SFD(\gamma)$, count and select the top $(\gamma - \beta)$ amount of b_2, b_4, b_6, and b_8 different from b_5. Assume that $(\gamma - \beta) = 2$, if the amount of b_2 and b_4 who are different from b_5 is greater then b_6 and b_8, b_2 and b_4 is denoted as the message-embedded pixels;
6. Scan the blocks in the image I'_O, once meet the block with $Pat\_SFD(\gamma)$ and its message-embedded pixels are different from the center pixel. If the to-be-embedded bit of S is '1', the message-embedded pixels need to be flipped. Otherwise, if the to-be-embedded bit of S is '0', the pixels of this block will be remained. Finally the marked image I_M is obtained.

3.2 Extraction Procedure

For a given marked image I_M with the secret message embedded of size $M \times N$, the extracting procedure includes the following steps:

1. Divide I_M into non-overlapped 3×3 blocks and calculate the SFD of each block;

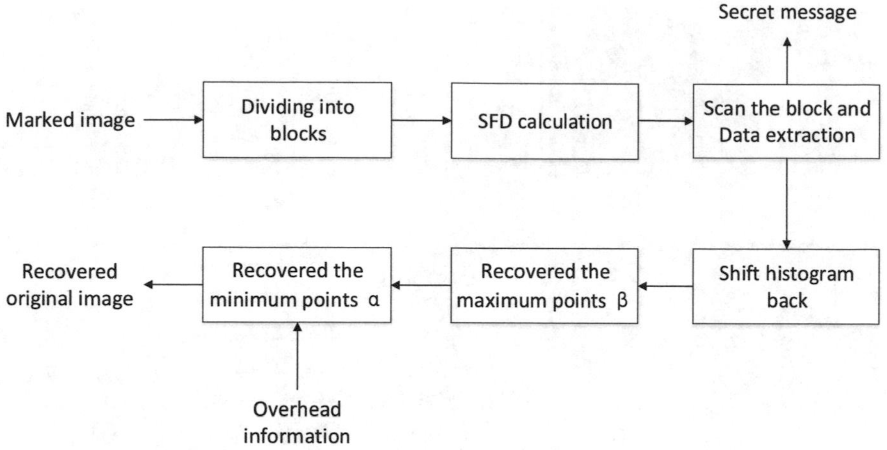

Fig. 8. The framework of extraction procedure.

2. Scan the blocks, once meet the block with $Pat\_SFD(\gamma)$, a bit '0' is extracted. If a block with $Pat\_SFD(\beta)$ is encountered, a bit '1' is extracted. And then flip the message-embedded pixels of this block;
3. According to the overhead bookkeeping information, recover the blocks with $Pat\_SFD(\beta)$ from the blocks with $Pat\_SFD(\alpha)$. Finally the recovered original image I_R is obtained.

4 Experiments and Results

In this section, we first introduce the experiment conditions, and then present a set of experiments to demonstrate the high performance and effectiveness of the proposed approach.

4.1 Experiment Conditions

This paper uses the different types of binary images including pattern, CAD graph, painting, cartoon and word document, which are some common binary image, to compare the performance of n-pair patterns [10]. Some original test image examples are shown in Fig. 9.

To further evaluate the visual quality, two measurements are employed to measure the objective visual quality of a marked image. The Peak Signal-to-Noise Ratio (PSNR) [23], is presented the distortion by measuring the amount of the flipped pixels in the marked image. The DRD [13] is to measure the reciprocal distance on the neighboring pixels to present the distortion caused by flipping pixels.

Some experiments are conducted to demonstrate the high performance and effectiveness of the proposed approach. In [10], Kim et al. proposed n-pair

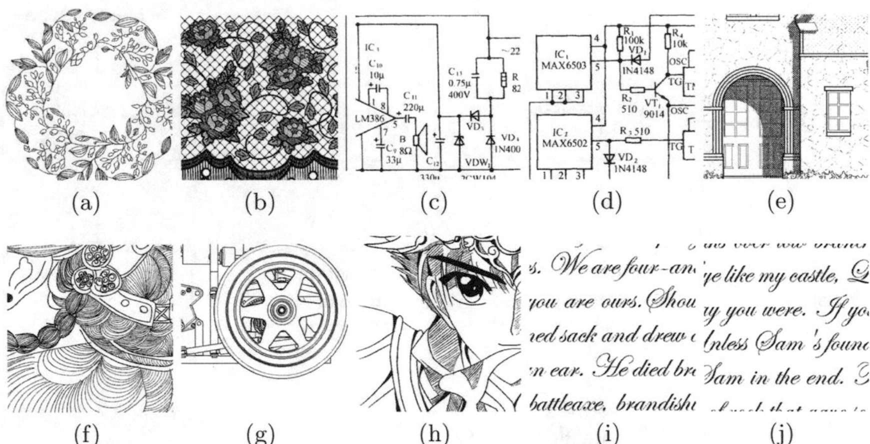

Fig. 9. Original test images (512×512) used for experiments. (a) Pattern 1. (b) Pattern 2. (c) CAD 1. (d) CAD 2. (e) Painting 1. (f) Painting 2. (g) Cartoon 1. (h) Cartoon 2. (i) Word document 1. (j) Word document 2.

Table 1. The objective visual quality and location map comparison of a marked image

Images (512 × 512)	Payload (bits)	n-pair pattern [10]			Proposed method		
		LM	PSNR	DRDM	LM	PSNR	DRDM
Pattern 1	103	41	29.8879	0.0878	**2**	**34.9946**	**0.0303**
Pattern 2	158	140	29.5915	0.0429	**2**	**30.3652**	**0.0427**
CAD 1	138	30	31.0891	0.0901	**0**	**32.3102**	**0.0822**
CAD 2	162	34	30.922	0.0777	**0**	**31.6569**	0.084
Painting 1	76	11	34.9946	0.0158	**2**	**35.6121**	0.0191
Painting 2	171	108	29.4872	0.0469	**1**	**31.3979**	**0.04**
Cartoon 1	73	14	33.7715	0.0311	**0**	**33.9735**	0.0361
Cartoon 2	88	23	34.057	0.0291	**2**	**34.9426**	0.0314
Word document 1	68	2	34.2731	0.0453	**0**	**37.0254**	**0.0292**
Word document 2	75	5	33.5408	0.0574	**0**	**36.0563**	**0.0402**

patterns method, which divides the binary image into pairs of n-length pixel. It tricks the pairs as n-bit binary and converts them to decimal, generates the histogram.

4.2 Visual Qualities Comparison

The objective visual quality comparison of a marked image is shown in Table 1. The results shows that under the same embedding capacity, our proposed method is better than n-pair patterns [10] employed the PSNR measurement in all experimental images. And the pattern and word document marked images

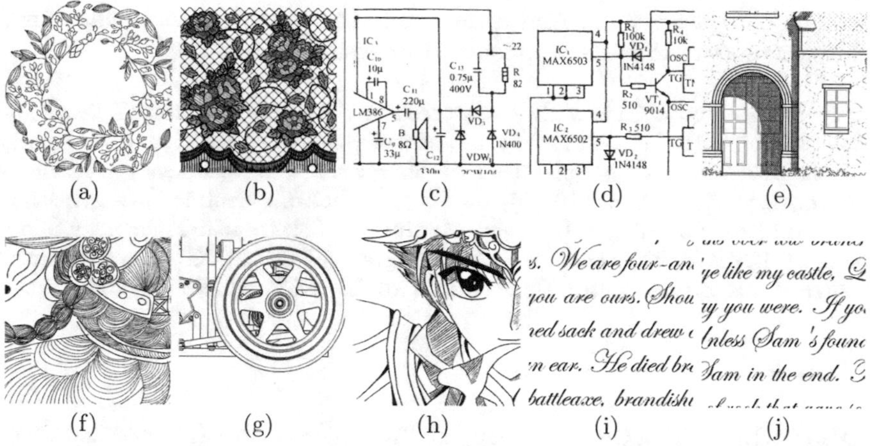

Fig. 10. Marked images (512 × 512) used for experiments. (a) Pattern 1 - 103 bits. (b) Pattern 2 - 158 bits. (c) CAD 1 - 138 bits. (d) CAD 2 - 162 bits. (e) Painting 1 - 76 bits. (f) Painting 2 - 171 bits. (g) Cartoon 1 - 73 bits. (h) Cartoon 2 - 88 bits. (i) Word document 1 - 68 bits. (j) Word document 2 - 75 bits.

present the satisfactory visual quality as the DRDs are also better than n-pair patterns [10].

In [10], the embedding of secret message relies on the modification of the decimal histogram. The image content does not be taken into account when flipping the pixels, resulting in a large impact on the visual quality of the image. For example, when $n = 3$, if it is transformed from "011(3)" to "100(4)", 3-bit pixels are flipped, and the image content will be changed significantly. The marked images including secret messages are shown in Fig. 10.

4.3 Location Map Comparison

In some cases, both two methods need to record some additional information to assist in restoring the image. The overhead bookkeeping information of a marked image, call location map (LM), is shown in Table 1. The result shows that under the same embedding capacity, the two methods have the close value in visual quality but the size of our location maps are much smaller than the other one. In summary, our method achieve image reversible data hiding with less storage space.

5 Conclusions

In this paper, we proposed a novel reversible data hiding scheme in binary image using symmetrical flipping degree (SFD) histogram modification. The SFD is defined to measure the visual quality of an image, and the histogram modification is reversible with the symmetry of SFD. our scheme is achieved with low distortion of visual quality. And it does not require the original image for the extraction process, which is a undoubted reversible method.

Acknowledgements. This work is supported by the National Natural Science Foundation of China (No. U1736118), the Natural Science Foundation of Guangdong (No. 2016A030313350), the Special Funds for Science and Technology Development of Guangdong (No. 2016KZ010103), the Key Project of Scientific Research Plan of Guangzhou (No. 201804020068), Shanghai Minsheng Science and Technology Support Program (17DZ1205500), Shanghai Sailing Program (17YF1420000), the Fundamental Research Funds for the Central Universities (No. 16lgjc83 and No. 17lgjc45).

References

1. Bao, F., Deng, R.H., Ooi, B.C., Yang, Y.: Tailored reversible watermarking schemes for authentication of electronic clinical atlas. IEEE Trans. Inf. Technol. Biomed. **9**(4), 554 (2005)
2. Cao, H., Kot, A.C.: On establishing edge adaptive grid for bilevel image data hiding. IEEE Trans. Inf. Forensics Secur. **8**(9), 1508–1518 (2013)
3. Chen, J., Lu, W., Yeung, Y., Xue, Y., Liu, X., Lin, C., Zhang, Y.: Binary image steganalysis based on distortion level co-occurrence matrix. Comput. Mater. Continua **55**(2), 201–211 (2018)
4. Chen, X., Jian, W., Wei, L., Jiaming, X.: Multi-gait recognition based on attribute discovery. IEEE Trans. Pattern Anal. Mach. Intell. **40**(7), 1697–1710 (2018)
5. Feng, B., Lu, W., Dai, L., Sun, W.: Steganography based on high-dimensional reference table. In: International Workshop on Digital Watermarking, Taipei, Taiwan, pp. 574–587 (2014)
6. Feng, B., Lu, W., Sun, W.: High capacity data hiding scheme for binary images based on minimizing flipping distortion. In: International Workshop on Digital Watermarking, pp. 514–528 (2013)
7. Feng, B., Lu, W., Sun, W.: Secure binary image steganography based on minimizing the distortion on the texture. IEEE Trans. Inf. Forensics Secur. **10**(2), 243–255 (2015)
8. Feng, B., Weng, J., Lu, W., Pei, B.: Steganalysis of content-adaptive binary image data hiding. J. Vis. Commun. Image Represent. **46**, 119–127 (2017)
9. Ho, Y.A., Wu, H.C., Wu, H.C., Chu, Y.P.: High-capacity reversible data hiding in binary images using pattern substitution. Comput. Stan. Interfaces **31**(4), 787–794 (2009)
10. Kim, C., Baek, J., Fisher, P.S.: Lossless data hiding for binary document images using n-pairs pattern. In: International Conference on Information Security and Cryptology, pp. 317–327 (2014)
11. Lin, X., Feng, B., Lu, W., Sun, W.: Content-adaptive residual for steganalysis. In: International Workshop on Digital Watermarking, Taipei, Taiwan, pp. 389–398 (2014)
12. Liu, X., Lu, W., Huang, T., Liu, H., Xue, Y., Yeung, Y.: Scaling factor estimation on JPEG compressed images by cyclostationarity analysis. Multimedia Tools Appl. pp. 1–18 (2018)
13. Lu, H., Kot, A.C., Shi, Y.Q.: Distance-reciprocal distortion measure for binary document images. IEEE Signal Process. Lett. **11**(2), 228–231 (2004)

14. Lu, W., He, L., Yeung, Y., Xue, Y., Liu, H., Feng, B.: Secure binary image steganography based on fused distortion measurement. IEEE Trans. Circuits Syst. Video Technol. **PP**, 1 (2018)
15. Ni, Z., Shi, Y.Q., Ansari, N., Su, W.: Reversible data hiding. IEEE Trans. Circuits Syst. Video Technol. **16**(3), 354–362 (2006)
16. Tai, W.L., Yeh, C.M., Chang, C.C.: Reversible data hiding based on histogram modification of pixel differences. IEEE Trans. Circuits Syst. Video Technol. **19**(6), 906–910 (2009)
17. Tsai, C.L., Chiang, H.F., Fan, K.C., Chung, C.D.: Reversible data hiding and lossless reconstruction of binary images using pair-wise logical computation mechanism. Pattern Recogn. **38**(11), 1993–2006 (2005)
18. Wu, M., Liu, B.: Data hiding in binary image for authentication and annotation. IEEE Trans. Multimedia **6**(4), 528–538 (2004)
19. Wu, M., Tang, E., Lin, B.: Data hiding in digital binary image. In: IEEE International Conference on Multimedia and Expo, pp. 393–396 (2000)
20. Xuan, G., Shi, Y.Q., Chai, P., Tong, X., Teng, J., Li, J.: Reversible binary image data hiding by run-length histogram modification. In: International Conference on Pattern Recognition, Tampa, FL, USA, pp. 1–4 (2008)
21. Xue, Y., Liu, W., Lu, W., Yeung, Y., Liu, X., Liu, H.: Efficient halftone image steganography based on dispersion degree optimization. J. Real Time Image Process. pp. 1–9 (2018)
22. Yang, H., Kot, A.C.: Pattern-based data hiding for binary image authentication by connectivity-preserving. IEEE Trans. Multimedia **9**(3), 475–486 (2007)
23. Yoo, J.C., Ahn, C.W.: Image matching using peak signal-to-noise ratio-based occlusion detection. IET Image Process. **6**(5), 483–495 (2012)
24. Zhang, W., Hu, X., Li, X., Yu, N.: Recursive histogram modification: establishing equivalency between reversible data hiding and lossless data compression. IEEE Trans. Image Process. **22**(7), 2775–2785 (2013)
25. Zhao, Z., Luo, H., Lu, Z.M., Pan, J.S.: Reversible data hiding based on multilevel histogram modification and sequential recovery. AEU Int. J. Electron. Commun. **65**(10), 814–826 (2011)

Binary Image Steganalysis Based on Local Residual Patterns

Ruipeng Li, Lingwen Zeng, Wei Lu(✉), and Junjia Chen

School of Data and Computer Science, Guangdong Key Laboratory of Information Security Technology, Ministry of Education Key Laboratory of Machine Intelligence and Advanced Computing, Sun Yat-sen University, Guangzhou 510006, China
{lirp5,zenglw3,chenjj233}@mail2.sysu.edu.cn, luwei3@mail.sysu.edu.cn

Abstract. A binary image steganalysis scheme based on the statistic model of local residual pattern (LRP) is proposed in this paper. LRP means the pattern of a local area of the residual map of the binary image, which is calculated with the XOR operation. The XOR operation is sensitive to the difference between adjacent pixels, which leads to the emphasis on edge property of the residual map. The neighbouring LRPs of the modified pixel will be affect, which makes the statistic model of LRPs change. Thus the trace of steganography can be detected according to the difference between the statistic models. Finally, the experiments we conducted show that our proposed scheme is effective on binary image steganalysis.

Keywords: Steganalysis · Binary image · Steganography · Machine learning

1 Introduction

Steganography hiding secret messages in some mediums, such as videos, audio or images, is mainly for secret communication. Anyone but the sender and the receiver could not easily notice any difference of mediums which have been embedded secret messages. Presently, steganography mainly choose images as the mediums. Some people would exploit the steganography to transmit criminal information which is embedded in images on the internet. As for steganalysis that is the counter type of steganography, it is used for detecting the existence of hidden messages. Steganalysis is a hot topic of information security. It is significantly important to research steganalysis. In this paper, we propose a novel scheme for binary image steganalysis.

The gray-scale image and binary image are mainly used for embedding secret messages. For the gray-scale image, steganography usually modifies the pixel to embed the messages. As for the binary image, the pixel has only two values 0 or 1. In order to embed message in binary image, we can flip the pixels. At present, the binary images are widely used in the electronic document or digital signature which are around our daily life. Because of the only two kinds of pixels

C.-N. Yang et al. (Eds.): SICBS 2018, AISC 895, pp. 904–913, 2020.
https://doi.org/10.1007/978-3-030-16946-6_74

(0 or 1) in binary image, the modification on binary pixels would be obvious. However, people still put forward many practical steganography methods not only minimizing distortion on human visual system but also preserving the statistical imperceptibility. Feng et al. [9] minimize the embedding distortion on the texture. By extracting the complement, rotation, and mirroring-invariant local texture patterns (crmiLTP), then they measure the flipping distortion by weighting the sum of crmiLTP changes. In [13], Lu et al. proposed a binary steganographic scheme which combines the advantages of flipping distortion measurement and the edge adaptive grid method and the "Connectivity Preserving" criterion. More and more practical binary image steganography methods have been developed in recent years [1,12,16–18], hence there are many challenges in steganalysis research. Steganalysis is a binary classification that whether an image has been embedded secret message. Presently the binary image steganalysis is mainly focused on two steps: the statistical features extraction and the classifier training. For the feature extraction step, many researchers have presented some novel steganalysis schemes. In [8], pixel mesh Markov transition matrix (PMMTM) measures the embedding distortion on texture consistency. In PMMTM, the Markov transition of pixel meshes represent the dependence among texture structures. The final dimensionality-reduced feature set could be formed by shrinking the obtained PMMTM according to its detection performance. In [3], the steganalysis scheme which we call BISLTP primarily utilizes the embedding effect associated with the L-shape pattern-based embedding criterion, the size of this L-shape pattern is 4×3. Finally this scheme gets a 32-dimensional steganalytic feature set. In [4], the changes of distortion scores around the flipped pixels, which are caused by embedding, are used to effectively detect the existence of steganography. In [10], another novel steganalysis scheme LP is propsed, which expands the L-shape size to 5×5 and employs Manhattan distance to measure the inter-pixels correlation. This scheme would get a 8192-dimensional steganalytic feature set. After extracting the diverse features, these features would be trained in the classifier. SVM [2] or ensemble classifier [15] could be used to classify the cover and the stego.

In this paper, we proposed a new steganalysis scheme for binary image. The proposed scheme is based on the statistic model of local residual pattern (LRP), which means the pattern of a local area of the residual map of binary image. The residual map of binary image is calculated with the XOR operation, which is sensitive to the difference between adjacent pixels. An LRP around a residual point describes the property of the neighbourhood of the residual point. Furthermore, the edge is emphasized in the residual. Thus, we can detect the modification on pixels, especially those on the edge, through the change of the statistic model of LRP. The experiments we conducted show that the proposed scheme is effective on binary image steganalysis.

The reminder of this paper is organized as follows. Section 2 would introduce our steganalysis scheme in details. In Sect. 3, we will show our experimental results, including comparing with state-of-the-art steganalysis schemes. After that, we will conclude our paper in Sect. 4.

2 Proposed Scheme

There are three main processes in our proposed schemes: residual extraction, residual pattern counting and feature selection. First, 8 kinds of residual maps of the binary image, which corresponds to 8 directions, are extracted based on the binary image. Then the probabilities of local residual patterns (LRP) are estimated based on 8 different residual maps. Finally, the features (probabilities of LRP) are selected according to fisher criterions which are relevant to the features independently.

2.1 Residual Extraction

In the field of gray-scale image steganalysis, residual extraction is a usual method to catch the trace of steganography on the gray-scale image. e.g., this method is adopted in SPAM [14], SRM [11] and so on. The performances of these schemes reflect the effectiveness of residual extraction. On the one hand, the modification on image pixel only makes small changes, which are invisible for human eyes but affect the noise component instead of the content of the image, and the noise component is emphasized in the residual map. On the other hand, the values of residuals mainly distribute around 0, then the range of residuals can be reduced by quantization, thus the dimensionality of the model can be reduce.

In our proposed method, the process of residual extraction is also adopted. Unlike the residual of the gray-scale image pixel, the residual of the binary image pixel is generated by XOR operation between pixels instead of substraction operation. The meaning of the residual of the gray-scale image pixel is not the same as the meaning of the residual of the binary image pixel. On the one hand, the dimensionality of the model cannot be reduced by residual extraction and quantization. On the other hand, the substraction operation between binary pixels makes no sense, since there are only two discrete for each binary pixel. The residual of binary pixel reflect the property of edges. Figure 1 shows an example of a binary image and some of its residual maps.

There are 8 kinds of residual maps $\mathbf{R}^{\rightarrow}$, $\mathbf{R}^{\leftarrow}$, $\mathbf{R}^{\downarrow}$, $\mathbf{R}^{\uparrow}$, $\mathbf{R}^{\searrow}$, $\mathbf{R}^{\nwarrow}$, $\mathbf{R}^{\swarrow}$, $\mathbf{R}^{\nearrow}$, are extracted in our schemes. e.g., the calculation of $\mathbf{R}^{\rightarrow}$ is expressed as

$$R^{\rightarrow}{}_{i,j} = I_{i,j} \oplus I_{i,j+1} = \begin{cases} 0, & I_{i,j} = I_{i,j+1} \\ 1, & I_{i,j} \neq I_{i,j+1} \end{cases} \tag{1}$$

The value of residual is sensitive to the difference between two adjacent pixels. Thus, the property of edge will be emphasized in the residual map. The edges of different directions will be emphasized by residual maps corresponding to different directions. Figure 1 shows different residuals of a binary image.

The edges are emphasized in our proposed method because the pixels on the edge are frequently selected to flip to embed the message by steganalysis schemes. The changes on the pixels on the edges are not obvious for human eyes. Furthermore, the changes on the flat area causes impact on the connectivity of the image.

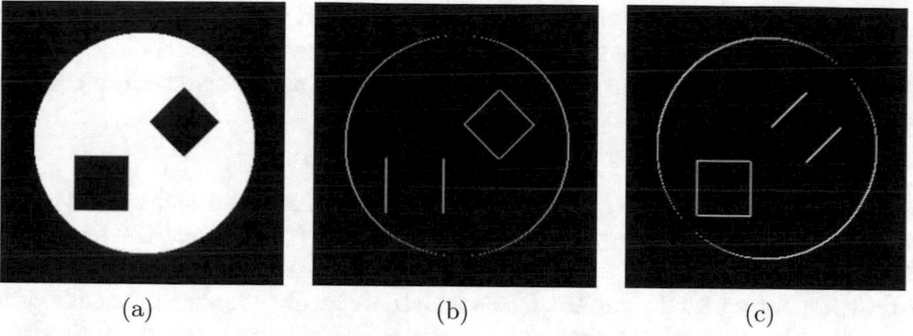

Fig. 1. An example of (a) image **I**, and the residual maps of the image (b) $\mathbf{R}^{\rightarrow}$ and (c) $\mathbf{R}^{\searrow}$. According to (b) and (c), the residual maps of different directions emphasize different edges.

2.2 Construction of LRP Statistic Models

In our proposed scheme, the local residual pattern (LRP) which describes the texture property of the local area on the residual map, is used to construct the statistic model of binary image. Once a pixel is flipped, the involved LRPs will be affected and transfer to other LRPs, and the probabilities of the original LRPs and new LRPs will change. Covers and stegos can be distinguished according to the change of probabilities of LRPs.

Different windows are designed to scan the residual maps of different directions to construct their corresponding LRP statistic models. e.g., the LRP statistic model $\mathbf{M}^{\rightarrow}$, $\mathbf{M}^{\leftarrow}$, $\mathbf{M}^{\downarrow}$, $\mathbf{M}^{\uparrow}$, $\mathbf{M}^{\searrow}$, $\mathbf{M}^{\nwarrow}$, $\mathbf{M}^{\swarrow}$, $\mathbf{M}^{\nearrow}$, which correspond to $\mathbf{R}^{\rightarrow}$, $\mathbf{R}^{\leftarrow}$, $\mathbf{R}^{\downarrow}$, $\mathbf{R}^{\uparrow}$, $\mathbf{R}^{\searrow}$, $\mathbf{R}^{\nwarrow}$, $\mathbf{R}^{\swarrow}$, $\mathbf{R}^{\nearrow}$, are respectively written as

$$M_{\mathbf{p}}^{\rightarrow} = Pr(\mathbf{P}^{\rightarrow}{}_{i,j} = \mathbf{p}) \tag{2}$$

in which

$$\mathbf{P}_{i,j}^{\rightarrow} = \begin{bmatrix} R^{\rightarrow}{}_{i-1,j-1} & R^{\rightarrow}{}_{i-1,j} & R^{\rightarrow}{}_{i-1,j+1} \\ R^{\rightarrow}{}_{i,j-1} & R^{\rightarrow}{}_{i,j} & R^{\rightarrow}{}_{i,j+1} \\ R^{\rightarrow}{}_{i+1,j-1} & R^{\rightarrow}{}_{i+1,j} & R^{\rightarrow}{}_{i+1,j+1} \end{bmatrix} \tag{3}$$

$\mathbf{P}_{i,j}^{\rightarrow}$ is the LRP centres at (i,j), which is determined by not only the point $R^{\rightarrow}{}_{i,j}$ but also the local area around it. As for $M_{\mathbf{p}}^{\searrow}$, $P_{\mathbf{i,j}}^{\searrow}$ is written as

$$\mathbf{P}_{i,j}^{\searrow} = \begin{bmatrix} R^{\searrow}{}_{i-2,j} & R^{\searrow}{}_{i-1,j+1} & R^{\searrow}{}_{i,j+2} \\ R^{\searrow}{}_{i-1,j-1} & R^{\searrow}{}_{i,j} & R^{\searrow}{}_{i+1,j+1} \\ R^{\searrow}{}_{i,j-2} & R^{\searrow}{}_{i+1,j-1} & R^{\searrow}{}_{i+2,j} \end{bmatrix} \tag{4}$$

Other models are designed similarly. Each window is specific to the direction of residual map.

All the models of different directions will be comprehensively considered and integrated as a single model, which is written as

$$\mathbf{M} = [\mathbf{M}^{\rightarrow} + \mathbf{M}^{\leftarrow} + \mathbf{M}^{\downarrow} + \mathbf{M}^{\uparrow}, \mathbf{M}^{\searrow} + \mathbf{M}^{\nwarrow} + \mathbf{M}^{\swarrow} + \mathbf{M}^{\nearrow}] \tag{5}$$

The summation of models of horizontal directions and vertical directions, and the summation of models of diagonal directions and opposite diagonal directions help to reduce the dimensionality of the statistic model, and make it more compact.

2.3 Selecting Features with Fisher Criterion

The amount of kinds of LRPs is huge, which means that the dimensionality of the model is high if we use the probabilities of all LRPs to construct the feature set. Furthermore, some LRPs are less helpful for the task of classification. e.g., the seldom appearing LRPs, and the LRPs which are rarely affected after embedding message. The *Fisher criterion* can be utilized to measure the dissimilarity of the features (probabilities of LRP) between covers and stegos, and select some suitable features for classification.

The d-th Fisher score, which corresponds to the d-th dimension of the sample is calculated as

$$f_d = \frac{(m_d^c - m_d^s)^2}{{s_d^c}^2 + {s_d^s}^2} \tag{6}$$

in which m_d is the mean of the d-th features of samples

$$m_d = \frac{1}{N} \sum_{n=1}^{N} X_{n,d} \tag{7}$$

and s_d is the variance of the d-th features of samples

$$s_d = \sum_{n=1}^{N} (X_{n,d} - m_d)^2 \tag{8}$$

The numerator and denominator of Eq. (6) correspond to the *between-class variance* and the *within-class variance* of the d-th features.

The Fisher scores of features are independent. We use this independent criterion based on the assume that the correlationship among most of the features of different dimensions are relatively weak. However, there exists some LRPs which can mutually transfer by flipping pixel. The disappearance of an LRP is accompanied with the appearance of another LRP. It is possible for some LRPs to participate in different transition pairs, so the increments or decrements of these LRPs are complexly relevant. The good news is that we can simply find out the LRPs which change the most, despite the intricate relationships among them. The method of evaluating LRPs separately with Fisher criterion may cause the redundancy of features (e.g., selecting both LRPs of a transition pair, which have equal changes), but it will not miss the LRPs with the potential to help classification.

3 Experimental Result

3.1 Experiment Setup

We use an open binary image dataset named BIVC [9] which contains 5000 binary images with size 256 × 256. Most of these images are from Google image.

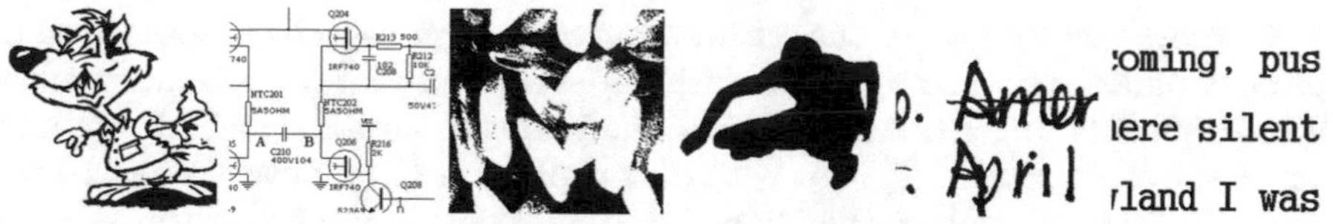

Fig. 2. Demonstrations of different types of binary images. The types of binary images are (from left to right) "cartoon", "CAD", "texture", "mask", "handwriting" and "document".

These images comprise CAD, texture, mask, handwriting, and document images. Figure 2 shows some images from BIVC. In [8,10], they also use this dataset for demonstrating the effectiveness of their steganalysis schemes. In our experiments, we randomly select 2500 images from the dataset as the training set, the rest 2500 images are the test set.

The error rate P_{E} is used for evaluating the performance of steganalysis schemes.

$$P_{\mathrm{E}} = \frac{1}{2}(P_{\mathrm{F_p}} + P_{\mathrm{F_n}}) \tag{9}$$

where $P_{\mathrm{F_p}}$ and $P_{\mathrm{F_n}}$ are the false-alarm and missed-detection rates, respectively.

3.2 Performance Comparison

In this section, we conduct many experiments to show our model performance. Our steganalysis scheme would be used for attacking several binary image steganography schemes or data hiding schemes such as PPCM [9], FusedDMS [13], SHUFFLE [16], GIM [12], ConnPre [17], EAG [1] and DPDC [18]. PPCM [9] exploits the weighted sum of crmiLTP changes presenting the distortion of flipping pixels. And in the embedding phase, syndrome-trellis code (STC) is used for minimizing the embedding distortion. The flipping distortion measurement and the edge adaptive grid method and the "Connectivity Preserving" criterion are integrated in FusedDMS, which would get a high statistical security and less distortion in human visual quality. But for SHUFFLE, the embedding pixels are restricted specific block-based relationship. Before embedding, shuffling will be applied. GIM with less distortion on human visual system aims at a larger embedding capacity. GIM uses matrix to embed the pixels. ConnPre is based on connectivity-preserving of pixels in a local neighborhood. ConnPre also with a matrix size 3×3 evaluates the pixels which whether are suitable flipping. EAG employs dense edge-adaptive grid to efficiently select good flipping pixel locations. DPDC flips the pixels by using the wavelet transform tracking the shifted edges.

For these steganography methods mentioned above, they have some parameters. In PPCM [9] and FusedDMS [13], θ_{c}, θ_{s} and θ_{m} stand for the number of elements in cover vectors, the number of superpixel and the number of message segment respectively. For the SHUFFLE [16], θ_w stands for the block size. θ_r stands for the cardinality of the complete set in GIM [12]. θ_{u} and θ_{o} in ConnPre [17] stand for the block size and the overlap mode.

To prove the validity of our scheme, for each steganography method we set different parameters and compare with different steganalysis schemes such as BISLTP [3], LP [10], RLCM [5], RLCL [6], PHD [7]. BISLTP and LP are mentioned above. In RLCM [5], Chiew et al. proposed a Gray Level Co-occurrence matrix(GLCM) to capture the inter-pixel relationship, and transform the two-dimensional GLCM to one-dimensional histogram, which would be added to the feature set. For RLCL [6], by extracting several sets of matrix including run length matrix, gap length matrix, and pixel difference, the characteristic function is applied to calculate the statistics. After that, the feature set would be set by the statistic. PHD [7] computes the histogram difference to detect the steganographic traces.

The experimental results are displayed on Tables 1, 2, 3, 4, 5, 6 and 7. Tables 1 and 2 compare different features to evaluate the detection performance of the proposed features when attacking PPCM and FusedDMS. These two binary image steganography methods achieve the great security and preserve the visual imperceptibly with using STC for minimizing additive distortion.

Table 1. Comparison of different features when attacking PPCM [9]

Parameter setting		$\theta_c = 8^2$	$\theta_c = 8^2$	$\theta_c = 8^2$	$\theta_c = 8^2$	$\theta_c = 8^2$	$\theta_c = 8^2$
		$\theta_{\mathfrak{Z}} = 5^2$	$\theta_{\mathfrak{Z}} = 4^2$	$\theta_{\mathfrak{Z}} = 5^2$	$\theta_{\mathfrak{Z}} = 3^2$	$\theta_{\mathfrak{Z}} = 4^2$	$\theta_{\mathfrak{Z}} = 3^2$
		$\theta_m = 8$	$\theta_m = 8$	$\theta_m = 16$	$\theta_m = 8$	$\theta_m = 16$	$\theta_m = 16$
Avg. payload		218.6	354.3	437.3	499.1	708.7	998.3
P_E	Proposed	**25.44**	**16.76**	**12.55**	9.55	**6.61**	**3.40**
	BISLTP [3]	25.70	17.46	13.79	**9.45**	7.62	4.98
	LP [10]	30.74	22.82	18.32	14.88	10.38	5.92
	RLCM [5]	46.50	41.67	36.55	34.41	27.24	16.44
	RLGL [6]	44.06	40.29	33.92	30.35	25.12	14.22
	PHD [7]	46.51	43.01	37.69	34.69	26.96	15.55

Table 2. Comparison of different features when attacking FusedDMS [13]

Parameter setting		$\theta_c = 8^2$	$\theta_c = 8^2$	$\theta_c = 8^2$	$\theta_c = 8^2$	$\theta_c = 8^2$	$\theta_c = 8^2$
		$\theta_{\mathfrak{Z}} = 5^2$	$\theta_{\mathfrak{Z}} = 4^2$	$\theta_{\mathfrak{Z}} = 5^2$	$\theta_{\mathfrak{Z}} = 3^2$	$\theta_{\mathfrak{Z}} = 4^2$	$\theta_{\mathfrak{Z}} = 3^2$
		$\theta_m = 8$	$\theta_m = 8$	$\theta_m = 16$	$\theta_m = 8$	$\theta_m = 16$	$\theta_m = 16$
Avg. payload		218.6	354.3	437.3	499.1	708.7	998.3
P_E	Proposed	**27.56**	**19.45**	**15.50**	**11.11**	**8.29**	**4.03**
	BISLTP [3]	44.03	29.73	24.27	17.21	12.75	5.88
	LP [10]	31.15	22.64	19.20	15.44	11.18	6.22
	RLCM [5]	48.80	42.67	38.45	37.87	29.36	18.68
	RLGL [6]	46.36	41.79	35.90	33.50	28.70	19.05
	PHD [7]	39.50	36.40	32.50	31.20	26.80	17.50

Table 3. Comparison of different features when attacking GIM [12]

Parameter setting		$\theta_r = 2$	$\theta_r = 3$	$\theta_r = 4$	$\theta_r = 5$	$\theta_r = 6$	$\theta_r = 7$
Avg. payload		259.4	389.8	520.0	650.3	780.6	910.7
P_E	Proposed	**13.60**	**9.24**	**5.87**	**4.41**	**3.78**	**3.19**
	BISLTP [3]	15.01	13.87	10.07	8.44	5.08	4.08
	LP [10]	17.92	12.88	9.56	7.51	5.97	4.79
	RLCM [5]	19.59	14.02	12.12	9.54	6.79	6.65
	RLGL [6]	18.87	12.85	11.14	8.91	5.95	5.55
	PHD [7]	21.60	13.72	11.45	9.59	7.38	6.86

Table 4. Comparison of different features when attacking ConnPre [17]

Parameter setting		$\theta_u = 3$ $\theta_o = 1$	$\theta_u = 4$ $\theta_o = 1$	$\theta_u = 5$ $\theta_o = 1$	$\theta_u = 5$ $\theta_o = 0$	$\theta_u = 4$ $\theta_o = 0$	$\theta_u = 3$ $\theta_o = 0$
Avg. payload		192.4	308.5	327.2	447.5	516.6	549.2
P_E	Proposed	17.50	11.89	11.80	8.27	6.65	6.33
	BISLTP [3]	**16.64**	14.55	12.64	9.55	8.00	6.50
	LP [10]	17.04	**10.25**	**9.52**	**6.04**	**4.54**	**4.00**
	RLCM [5]	17.94	16.02	15.13	11.95	10.97	10.05
	RLGL [6]	16.30	16.15	16.04	12.77	11.73	10.84
	PHD [7]	22.46	20.83	20.31	16.01	15.55	14.77

It can be observed that when attacking these two steganography, the proposed features outperform other compared steganalysis features, which indicates that the steganographic traces caused by STC encoder are effectively captured by the proposed features. The reason would be that our eight kinds of residual maps highlight the image content edge information which can well reflect the differences of content edge between cover and stego images. In addition, the proposed features also achieve the greatest performance compared with other features when attacking GIM, shown in Table 3.

Nevertheless, in the comparison of other binary image steganography, the proposed features behave less satisfactorily. Especially, Table 4 illustrates that the proposed features do not effectively reduce the detection error compared with LP which obtains the low detection error when attacking ConnPre. We consider that the "Flippability Criteria" set by ConnPre can avoid to reveal obvious embedding traces in the image content edges. Although the detection error of the proposed features is not lower than that of LP when attacking SHUFFLE, EAG, and DPDC, the proposed features gain the better detection performance, shown in Tables 5, 6 and 7.

Table 5. Comparison of different features when attacking SHUFFLE [16]

Parameter setting		$\theta_w = 13$	$\theta_w = 12$	$\theta_w = 11$	$\theta_w = 10$	$\theta_w = 9$	$\theta_w = 8$
Avg. payload		361	441	529	625	784	1024
P_E	Proposed	6.61	4.80	3.57	2.65	1.57	1.11
	BISLTP [3]	6.51	4.69	4.55	3.79	3.26	2.80
	LP [10]	**5.04**	**3.81**	**2.85**	**2.02**	**1.55**	**1.04**
	RLCM [5]	20.18	16.69	13.58	11.68	9.68	7.40
	RLGL [6]	15.26	12.51	9.83	8.09	5.89	3.69
	PHD [7]	15.73	12.58	9.74	7.81	6.21	4.06

Table 6. Comparison of different features when attacking EAG [1]

Scheme	Proposed	BISLTP [3]	LP [10]	RLCM [5]	RLGL [6]	PHD [7]
P_E	8.97	11.40	**7.11**	17.39	17.40	16.87

Table 7. Comparison of different features when attacking DPDC [18]

Scheme	Proposed	BISLTP [3]	LP [10]	RLCM [5]	RLGL [6]	PHD [7]
P_E	3.06	3.80	**2.33**	5.52	5.60	8.07

4 Conclusion

In this paper, we propose a new steganalysis scheme for binary image. This scheme contains three main processes: residual extraction, residual pattern counting and feature selection. The 8 directions residual maps of binary image are calculated with the XOR operation, which are sensitive to the difference between adjacent pixels. After that, the probabilities of local residual patterns are estimated based on 8 directions residual maps. Finally, the features would be selected according to fisher criterions which are relevant to the features independently. Other steganalysis schemes compared with our scheme, the experiment results show that our scheme has a better performance. In our future works, we may alter the generation form of the proposed LRPs which is the key point for reflecting the edge information to effectively improve the detection performance.

Acknowledgements. This work is supported by the National Natural Science Foundation of China (No. U1736118), the Natural Science Foundation of Guangdong (No. 2016A030313350), the Special Funds for Science and Technology Development of Guangdong (No. 2016KZ010103), the Key Project of Scientific Research Plan of Guangzhou (No. 201804020068), Shanghai Minsheng Science and Technology Support Program (17DZ1205500), Shanghai Sailing Program (17YF1420000), the Fundamental Research Funds for the Central Universities (No. 16lgjc83 and No. 17lgjc45).

References

1. Cao, H., Kot, A.: On establishing edge adaptive grid for bi-level image data hiding. IEEE Trans. Inf. Forensics Secur. **8**(9), 1508–1518 (2013)
2. Chang, C.C., Lin, C.J.: Libsvm: a library for support vector machines. ACM Trans. Intell. Syst. Technol. (TIST) **2**(3), 27 (2011)
3. Chen, J., Lu, W., Fang, Y., Liu, X., Yeung, Y., Xue, Y.: Binary image steganalysis based on local texture pattern. J. Vis. Commun. Image Represent. **55**, 149–156 (2018)
4. Chen, J., Lu, W., Yeung, Y., Xue, Y., Liu, X., Lin, C., Zhang, Y.: Binary image steganalysis based on distortion level co-occurrence matrix. Comput. Mater. Continua **55**(2), 201–211 (2018)
5. Chiew, K.L., Pieprzyk, J.: Binary image steganographic techniques classification based on multi-class steganalysis. In: Information Security, Practice and Experience, pp. 341–358. Springer (2010)
6. Chiew, K.L., Pieprzyk, J.: Blind steganalysis: a countermeasure for binary image steganography. In: International Conference on Availability, Reliability and Security. pp. 653–658. IEEE Computer Society, March 2010
7. Chiew, K.L., Pieprzyk, J.: Estimating hidden message length in binary image embedded by using boundary pixels steganography. In: International Conference on Availability, Reliability and Security, pp. 683–688. IEEE Computer Society, March 2010
8. Feng, B., Lu, W., Sun, W.: Binary image steganalysis based on pixel mesh markov transition matrix. J. Vis. Commun. Image Represent. **26**, 284–295 (2015)
9. Feng, B., Lu, W., Sun, W.: Secure binary image steganography based on minimizing the distortion on the texture. IEEE Trans. Inf. Forensics Secur. **10**(2), 243–255 (2015)
10. Feng, B., Weng, J., Lu, W., Pei, B.: Steganalysis of content-adaptive binary image data hiding. J. Vis. Commun. Image Represent. **46**, 119–127 (2017). https://doi.org/10.1016/j.jvcir.2017.01.008
11. Fridrich, J.J., Kodovsky, J.: Rich models for steganalysis of digital images. IEEE Trans. Inf. Forensics Secur. **7**(3), 868–882 (2012)
12. Guo, M., Zhang, H.: High capacity data hiding for binary image authentication. In: International Conference on Pattern Recognition, pp. 1441–1444. IEEE (2010)
13. Lu, W., He, L., Yeung, Y., Xue, Y., Liu, H., Feng, B.: Secure binary image steganography based on fused distortion measurement. IEEE Trans. Circuits Syst. Video Technol. (2018). https://doi.org/10.1109/TCSVT.2019.2903432
14. Pevny, T., Bas, P., Fridrich, J.: Steganalysis by subtractive pixel adjacency matrix. IEEE Trans. Inf. Forensics Secur. **5**(2), 215–224 (2010)
15. Theodoridis, S., Koutroumbas, K.: Pattern Recognition, 4th edn. Elsevier Pte Ltd., San Diego (2009)
16. Wu, M., Liu, B.: Data hiding in binary image for authentication and annotation. IEEE Trans. Multimedia **6**(4), 528–538 (2004)
17. Yang, H., Kot, A.C.: Pattern-based data hiding for binary image authentication by connectivity-preserving. IEEE Trans. Multimedia **9**(3), 475–486 (2007)
18. Yang, H., Kot, A.C., Rahardja, S.: Orthogonal data embedding for binary images in morphological transform domain-a high-capacity approach. IEEE Trans. Multimedia **10**(3), 339–351 (2008)

Robust Steganography Using Texture Synthesis Based on LBP

Weiyi Wei and Chengfeng A(✉)

College of Computer Science and Engineering, Northwest Normal University, Lanzhou 730070, Gansu, China
1640707657@qq.com

Abstract. In order to improve the embedded capacity and anti-interference capability of the coverless steganography algorithm, this paper proposes a texture synthesis information hiding method based on LBP texture analysis. Firstly, selects the original texture image and divides it into uniform pixel blocks, calculates the LBP value of each pixel in an image block and takes the LBP value with the largest LBP distribution as the information represented for the image block. Secondly, the pseudo-random sequence is generated with specified key to determine the position of the texture candidate block placed on the white paper, then select the candidate block based on the value of the secret message and place it on the designated position on the white paper, meanwhile, the remaining blank areas are filled with texture synthesis method. Inversely, during the procedure of extracting the secret message, the position of the steganography image block is obtained according to the pseudo-random sequence generated by the customary key, and then the LBP value of each image block with the largest distribution is calculated to obtain secret information. Experimental results show that the constructed stego-image has good visual effects, more embedded capacities and robustness to some interference.

Keywords: LBP · Texture synthesis · Steganography · Coverless information hiding

1 Introduction

Information hiding is hiding secret information in a host signal in an invisible way to achieve the purpose of covert communications and copyright protection [1]. Because digital images contain a lot of information and are used most widely, they are often regarded as ideal information hiding cover. The current information hiding technology mostly embeds the secret information in the cover by slightly modifying the cover data, that is, a ready-made digital cover as a shield. There are many classic image steganography schemes such as LSB, Jsteg algorithm [2, 3], ZZW [4, 5], and some transform domain method like DFT domain [6], DCT domain [7] and DWT domain steganography method [8], etc. Most of these methods are limited by certain distortions caused by modifying the cover pixels and the modification trace is inevitably left on the cover, and that the hidden information is also difficult to resist the detection of various steganalysis [9–13].

C.-N. Yang et al. (Eds.): SICBS 2018, AISC 895, pp. 914–924, 2020.
https://doi.org/10.1007/978-3-030-16946-6_75

In order to fundamentally resist the detection of various types of writing analysis, some scholars proposed the new concept of "coverless information hiding". Compared with the traditional information hiding, the coverless information hiding no longer embeds the secret information into the cover, but directly uses the secret information as the driving to "generate/acquire" the stego cover. Representative work is a steganography scheme for generating texture images. Texture synthesis techniques can generate large texture images from a small sample texture image [14–17]. Texture synthesis information hiding is to implement information hiding in the process of texture synthesis, so the resulting large-scale texture image is related to secret information. Pioneering works were done by Otori and Kuriyama [18, 19]. Secret messages are regularly arranged into colored dotted pattern using the colors picked from a texture sample with features corresponding to the embedded data. These dotted patterns are written onto a blank canvas, the blank regions of which are synthesized using the texture samples. Embedding capacity of the method is determined by the dotted patterns painted on the image. Cohen proposed a patch-based algorithm to improve texture images visual quality by means of the pixel-based composite [20, 21]. Wu and Wang proposed another synthesis based steganography method [22]. A texture synthesis process re-samples a smaller texture image to construct a new texture image. Qian proposed an information hiding method based on texture synthesis that not only has the robust performance against JPEG compression, but also overcomes the disadvantages of going to have to reconstruct the original texture before extracting the information [23]. Zhou proposed a coverless information hiding method based on the bag-of-words (BOW) model [24]. This method uses the BOW model to extract the visual words (VW) of the image to express the text information to be hidden so as to hide the text information in the image [24]. In order to further improve the embedding capacity and anti-interference capability of steganography schemes, this paper proposes a texture synthesis information hiding scheme based on LBP code. Firstly, selects the small-size texture image and divides it into uniform pixel blocks, and then calculates the LBP value of each image block as its information represented. During the procedure of extracting the secret message, the position of the steganography image block is obtained according to the pseudo-random sequence generated by the customary key, and then the LBP value of each image block with the largest distribution is calculated to obtain secret information.

The remainder of this paper is organized as follows:

In Sect. 2, we introduce original LBP feature description and calculation method. In Sect. 3, we detail our algorithm including embedding and extracting procedures. We describe experimental results and theoretical analysis in Sect. 4, followed by our conclusions presented in the concluding section.

2 Original LBP Feature Description

LBP (Local Binary Pattern) [25] is an operator used to describe the local texture features of an image, It has significant advantages such as rotation invariance and gray invariance for the extraction of texture features. The original LBP operator is defined in the neighborhood of pixel, the neighborhood center pixel is invoked as a threshold, the adjacent eight pixel gray values are compared with the pixel values in the neighborhood center, if the surrounding pixels are larger than the center pixel value, the position of the pixel is marked as 1, otherwise it is 0. In this way, 8 points in the neighborhood can be compared to generate 8-bit binary numbers, and the 8-bit binary numbers are arranged in sequence to form a binary number, this binary number is the LBP value of the center pixel. There are 2^8 possible LBP values for every center pixel. The LBP value of the center pixel reflects the texture information of the area around the pixel. The above process is represented graphically as:

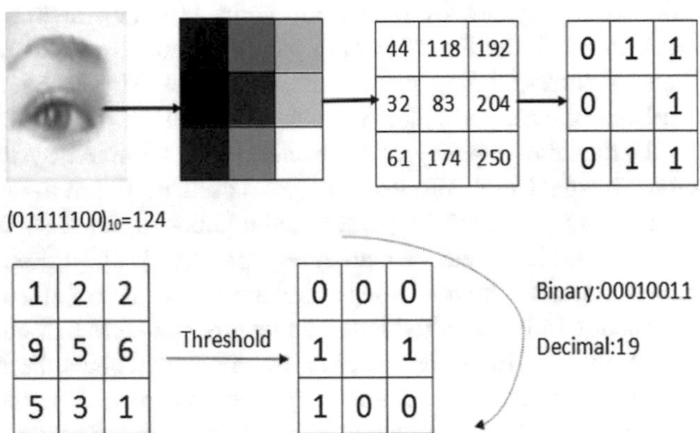

Fig. 1. Original LBP feature description and calculation method

Formulate the above process as:

$$LBP(x_c, y_c) = \sum_{p=0}^{p-1} 2^p s(i_p - i_c) \tag{1}$$

(x_c, y_c) is the coordinates of the center pixel, p is the pth pixel of the neighborhood, i_p is the gray value of the neighborhood pixel, i_c is the gray value of the center pixel, and $s(x)$ is a symbolic function defined as:

$$s(x)\begin{cases} 1 & if\ x \geq 0 \\ 0 & else \end{cases} \tag{2}$$

LBP can be used to represent texture features well. Therefore, this paper proposes a texture synthesis information hiding scheme based on LBP code, which further improves the algorithm's embedded capacity and anti-jamming capability. However, the LBP value of a single point is very vulnerable to interference, especially in the case of some small gradients with slight disturbances. Therefore, the LBP value of a single pixel is not used as the representative information value of this pixel block in this paper. Instead, take the LBP value with the largest LBP value distribution as the representative information for the image block. At the same time, in order to reduce the error during the extraction, this paper carries out some artificial noise interference on the original texture map.

3 Texture Synthesis Information Hiding Scheme Based on LBP Code

Framework of the proposed method is showed in Fig. 1. To hide a secret message, a data hider uses a source texture pattern and divides it into uniform pixel blocks, and then calculates LBP of each pixel in an image block and takes the LBP value with the largest LBP distribution as the information represented for the image block. The data hider arrange the selected candidate block to a blank canvas randomly with a secret key. Each candidate block is selected according to the value of the secret message. Other regions in blank are then synthesized by choosing the best candidate block from all candidates (Fig. 2).

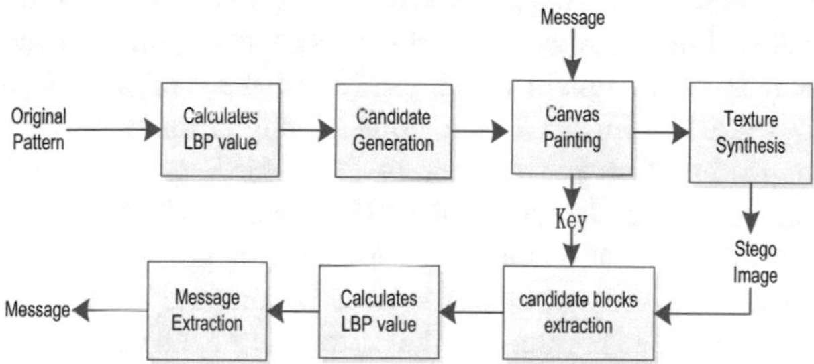

Fig. 2. Framework of the proposed method

The process of extracting secret information is the inverse process of information hiding. On the receiver side, the position of the steganography image block is obtained according to the pseudo-random sequence generated by the customary key, and then the LBP value of each image block with the largest distribution is calculated to obtain secret information.

3.1 Embedded Program

Assume that I_{in} is the input texture and I_{out} is the synthesized texture image. The integrated process for I_{out} synthesis is described as follows:

- **Collect candidate blocks:**

  Step1: Select an original texture I_{in}, divide and generate a uniform pixel block b, meanwhile, I_{in} is added some artificial noise and divide into block b' with same method.

  Step2: Use Eq. (1) to calculate the LBP representation of each image block b, b'.

  Step3: Selected candidate block. If the representative values of b, b' are consistent, the image block is added to the candidate block. Otherwise, discard it to ensure that the selected candidate block has a definite anti-jamming capability.

- **Embed information**

  Step1: Read secret information and convert it into a decimal number V.
  Step2: A pseudo-random sequence generated from the specified key is used to determine the position where the texture candidate block is placed on the white paper.
  Step3: randomly select a candidate block with representation value V from the candidate block, and place the candidate block on the designated position of white paper according to a pseudo-random sequence.
  Step4: Repeat the above process until all secret information is processed.
  Step5: The remaining blank areas are filled with a texture synthesis method.

We fill blank areas by the image quilting algorithm proposed by Efros and Freeman in [14]. Synthesis is realized by iteratively padding chosen identical sized candidates to a blank window. Due to the existence of overlapping areas, calculate the error between the selected block and the existing block at the overlapping area. Generally, a best tile that has the smallest mean square errors (MSE) of the overlapped parts is selected. A diagram is illustrated in Fig. 3. The regions in gray color stand for the synthesized contents. When synthesizing the content of "C", MSE of the overlapping regions between "C" and the upper tile "A", and MSE between "C" and the left tile "B" are calculated. One candidate that has the smallest MSE is chosen as the best. Then, the minimum cost path along the overlapped surface is computed to find the seam, see the red curves on the overlapped region, and the content is pasted onto the canvas along the seams. Details of the algorithm can be stated on [14].

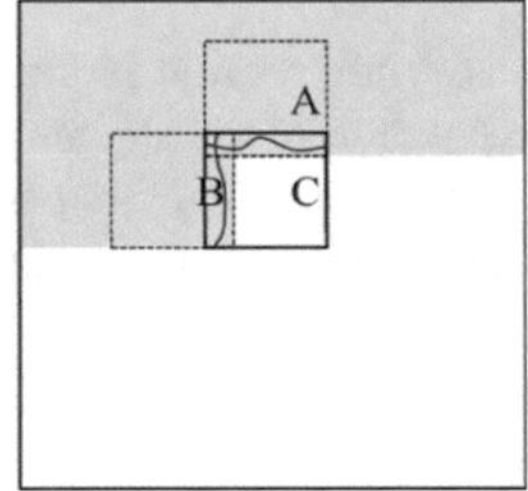

Fig. 3. Synthetic white space

3.2 Extraction Scheme

Step 1: the position of the steganography image block is obtained according to the pseudo-random sequence generated by the customary key and steganography image block is extracted.
Step2: Calculates the LBP value of each pixel in an image block and takes the LBP value with the largest LBP distribution as the information represented for the image block to obtain secret information.

4 Experimental Results and Analysis

4.1 Embedding Capacity

To verify the performance of the proposed algorithm, experiments loaded several texture images from Brodatz Textures [14] texture library. At present, LBP supports BPP (bits per pixel) = 6 maximum, because for thin texture maps, the distribution of candidate blocks of different LBP values is very uneven. When the BPP setting is relatively large, some candidate values are not found. The algorithm optimizes the parameters of LBP, the result is shown in the following figure:

The above three graphs is distribution histograms of the number of candidate blocks of LBP values. As can be observed in the figure, most of the candidate blocks are concentrated in a certain number of values, and the proportion of candidate blocks of other values is very small, and fewer candidate blocks can be robustly screened. Figure 4 shows that when the vertical axis is replaced with a logarithmic scale. It can be clearly seen that the ratio of most candidate blocks is less than 1‰. Therefore, when

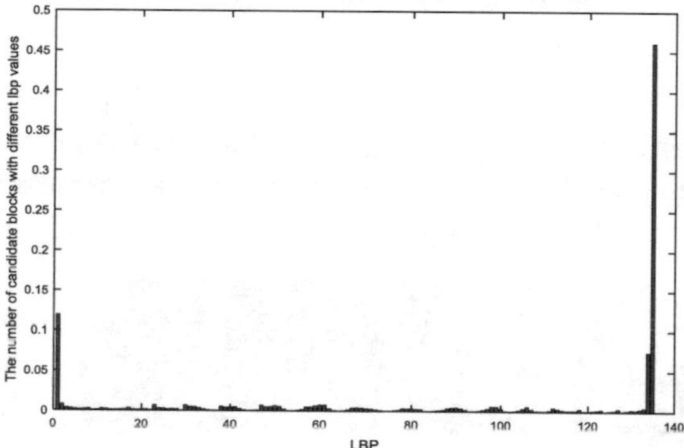

a) Candidate block distributions for different LBP values without robust filtering

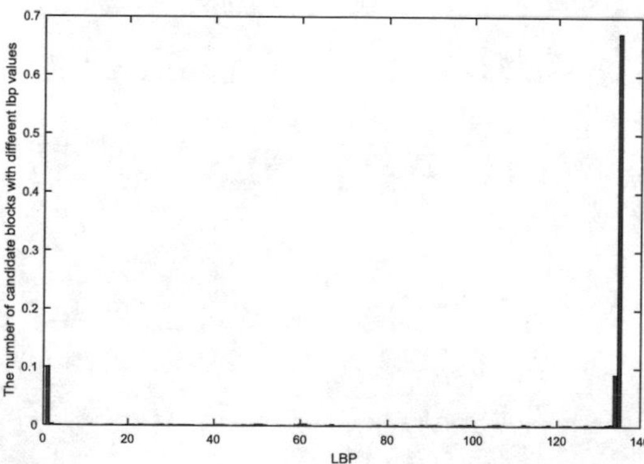

b) Candidate block distributions for different LBP values after robust screening

Fig. 4. (a) Candidate block distribution with different LBP values without robust filtering (b) Block candidate with different LBP values after filtering (c) Candidate blocks with different LBP values after robust filtering distributed

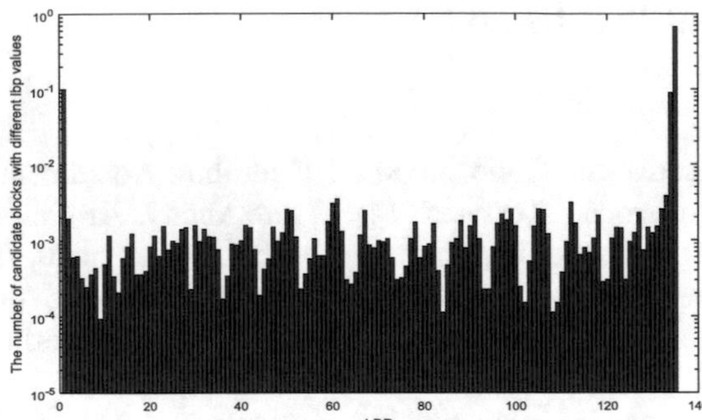

c) Candidate blocks with different LBP values after robust filtering distributed

Fig. 4. (*continued*)

BPP = 6, there are 64 types of values, which can be 16 cases compared to the original algorithm BPP = 4, which obviously increases the embedded capacity.

4.2 Visual Effects

In this paper, different texture images are selected from the Brodatz Textures [14] texture libraries. Figure 5(a)–(h) are for two types of source texture images. Their sizes

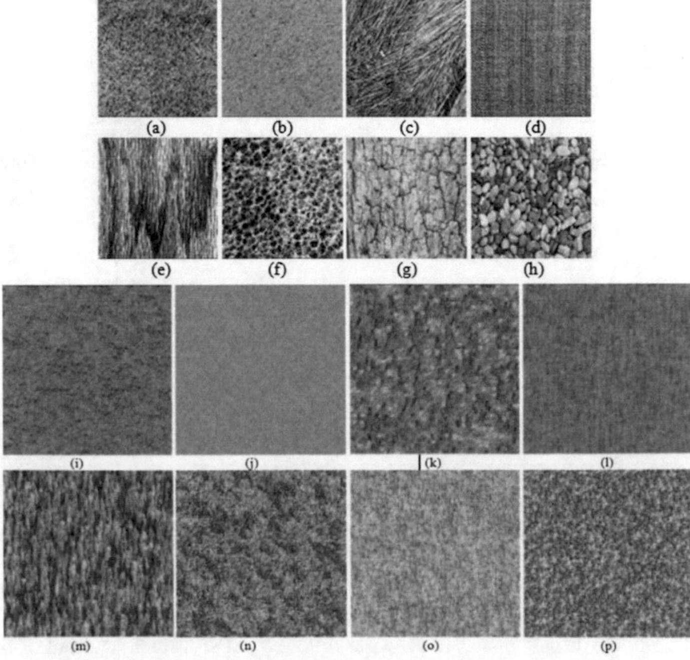

Fig. 5. Source texture and hidden texture. (a)–(h) is the source texture, and (i)–(p) is the synthetic texture containing the secret message

are 128 × 128, Figs. 5(i)–(p) are generated stego images of size are 653 × 653. The results show that the generated dense images have a real good visual effect.

4.3 Anti-interference Ability

To further verify the performance of our algorithm. This paper compares the anti-interference ability of the algorithm and the algorithm [23] when embedding different bits, as shown in the following figure (Fig. 6):

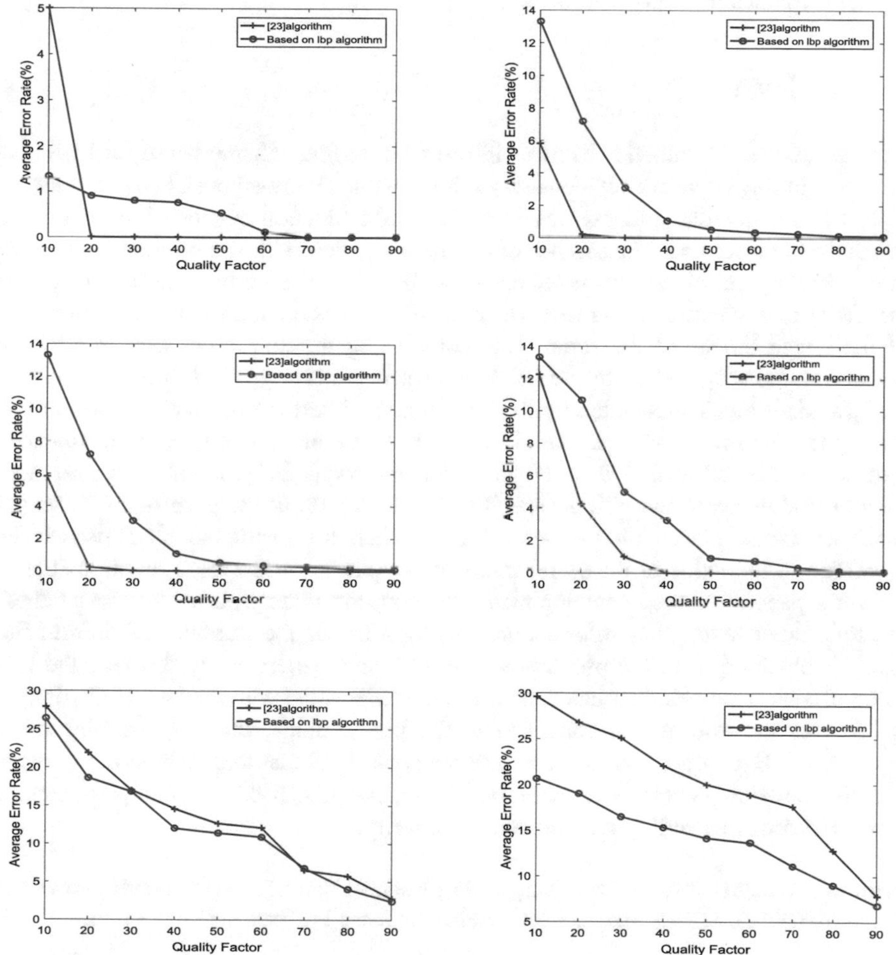

Fig. 6. Digit embedded median anti-interference test

Experiments carried out anti-interference tests with different embedded digits. In the fixed quality factor, the smaller the average error rates of the algorithm the better. Because the smaller the average error rate, it means that there is less error in the secret information that can be extracted. The QF value is inversely proportional to the anti-interference ability of the algorithm. From the six experimental maps, it is found that the anti-interference ability of the algorithm does not significantly decline with the increase of the number of embedded bits. However, the anti-interference ability of the algorithm [23] has been significantly reduced. Therefore, the advantage of the LBP based method is not obvious when the number of embedded bits is low, but with the gradual increase in the number of embedded bits until after 4bit, the performance of the algorithm in [23] has dropped significantly, the LBP algorithm has almost no decline. Especially in the case of 6 bit, the anti-interference ability of the algorithm of this paper clearly surpasses the algorithm of [23].

5 Conclusion

The texture-based synthetic coverless information hiding scheme based on LBP code proposed in this paper mainly consists of three parts. The selection of LBP values, the collection of candidate blocks, and the hiding and extraction of secret information are respectively performed. The essence of the method is to divide the original texture map into uniform pixel blocks and calculate the LBP value of each pixel in the block, take the LBP value with the largest LBP value distribution as the representative information of the image block. At the same time, considering the anti-interference ability and embedded capacity, calculate the LBP value using the original texture map and the uniform pixel blocks generated by the human noise interference graph. If they are the same ones, the image block is added to the candidate block queue, and the candidate blocks are selected in this way. When hiding the secret information, first generate a pseudo-random sequence with the specified key to determine the position of the texture candidate block placed on the white paper. Then the candidate block is selected according to the value of the secret information, placed on the designated position of the white paper, and the remaining blank positions are filled with the texture synthesis method. When extracting information, the location of the confidential information block is obtained from the pseudo-random sequence generated by the key, the LBP value of each image block is calculated, and the LBP value with the largest distribution is taken as the information contained in the image block, thereby obtaining secret information. Experimental and theoretical analysis show that this method not only can effectively realize coverless information hiding, but also has further improvement in anti-interference capability and embedded capacity.

Acknowledgement. This work was supported by National Natural Science Foundation of China (Grant 61762080), Science and Technology Plan of Gansu Province (17YF1FA119).

References

1. Shen, C.X., Zhang, H.G., Feng, D.G.: A survey of information security. Chinese Science. Sci. China **37**(2), 129–150 (2007)
2. Wang, H., Wang, S.: Cyber warfare: steganography vs. steganalysis. Commun. ACM **47** (10), 76–82 (2004)
3. Fridrich, J., Goljan, M.: Practical steganalysis of digital images: state of the art. In: Security and Watermarking of Multimedia Contents IV Security and Watermarking of Multimedia Contents IV, pp. 1–13 (2002)
4. Fridrich, J.: Asymptotic behavior of the ZZW embedding construction. IEEE Trans. Inf. Forensics Secur. **4**(1), 151–154 (2009)
5. Zhang, W., Wang, X.: Generalization of the ZZW embedding construction for steganography. IEEE Trans. Inf. Forensics Secur. **4**(3), 564–569 (2009)
6. Xu, D.H., Zhu, C.Q., Wang, Q.S.: A construction of digital watermarking model for the vector geospatial data based on magnitude and phase of DF. J. Beijing Univ. Posts Telecommun. **34**(5), 25–28 (2011)
7. Cox, I.J., Kilian, J., Leighton, F.T.: Secure spread spectrum watermarking for multimedia. Secur. Spread Spectr. Watermark. Images Audio Video **3**, 243–246 (1997)
8. Hsieh, M.S., Tseng, D.C., Huang, Y.H.: Hiding digital watermarks using multiresolution wavelet transform. IEEE Trans. Industr. Electron. **48**(5), 875–882 (2010)
9. Pevny, T., Fridrich, J.: Merging Markov and DCT features for multi-class JPEG steganalysis. In: Proceedings of SPIE - The International Society for Optical Engineering, pp. 650503-650503-13 (2007)
10. Shi, Y.Q., Chen, C., Chen, W.: A markov process based approach to effective attacking JPEG steganography. In: Information Hiding, International Workshop, IH 2006, Alexandria, VA, USA, 10–12 July 2006. Revised Selected Papers DBLP, pp. 249–264 (2006)
11. Lyu, S., Farid, H.: Steganalysis using higher-order image statistics. IEEE Trans. Inf. Forensics Secur. **1**(1), 111–119 (2006)
12. Pevny, T., Bas, P., Fridrich, J.: Steganalysis by subtractive pixel adjacency matrix. In: ACM Workshop on Multimedia and Security, pp. 75–84. ACM (2009)
13. Fridrich, J., Kodovsky, J.: Rich models for steganalysis of digital images. IEEE Trans. Inf. Forensics Secur. **7**(3), 868–882 (2012)
14. Efros, A.A., Freeman, W.T.: Image quilting for texture synthesis and transfer. In: Proceedings of SIGGRAPH, pp. 341–346 (2001)
15. Ashikhmin, M.: Synthesizing natural textures. In: Symposium on Interactive 3d Graphics, pp. 217–226. ACM (2001)
16. Kwatra, V.: Texture optimization for example-based synthesis. In: ACM SIGGRAPH, pp. 795–802. ACM (2005)
17. Dong, F., Ye, X.: Multiscaled texture synthesis using multisized pixel neighborhoods. IEEE Comput. Graph. Appl. **27**, 41–47 (2007)
18. Otori, H., Kuriyama, S.: Data-embeddable texture synthesis. In: Smart Graphics, International Symposium, SG 2007, Kyoto, Japan, 25–27 June 2007, Proceedings DBLP, pp. 146–157 (2007)
19. Otori, H., Kuriyama, S.: Texture synthesis for mobile data communications. IEEE Comput. Graph. Appl. **29**(6), 74–81 (2009)
20. Cohen, M.F., Shade, J., Hiller, S., Deussen, O.: Wang tiles for image and texture generation. ACM Trans. Graph. **22**(3), 287–294 (2003)
21. Xu, K.: Feature-aligned shape texturing. In: ACM SIGGRAPH Asia, p. 108. ACM (2009)

22. Wu, K.C., Wang, C.M.: Steganography using reversible texture synthesis. IEEE Trans. Image Process. **24**(1), 130–139 (2015)
23. Qian, Z., et al.: Robust steganography using texture synthesis. In: Advances in Intelligent Information Hiding and Multimedia Signal Processing. Springer, Cham (2017)
24. Zhou, Z., Yu, C., Sun, X.: The coverless information hiding based on image bag-of-words model. J. Appl. Sci. **34**(5), 527–536 (2016)
25. Yan, Y.K.: 2D Image Texture Synthesis Techniques Based on LBP. Hebei Normal University (2009)

Author Index

C.-N. Yang et al. (Eds.): SICBS 2018, AISC 895, pp. 925–928, 2020.
https://doi.org/10.1007/978-3-030-16946-6

Printed in the United States
By Bookmasters